Mass
Spectrometry

CLINICAL BIOCHEMISTRY

PRINCIPLES · METHODS APPLICATIONS

1

Series Editors
H.Ch.Curtius
M.Roth

Walter de Gruyter
Berlin · New York
1989

Mass Spectrometry

Editor
A. M. Lawson

W DE G

Walter de Gruyter
Berlin · New York
1989

Editor

Dr. Alexander M. Lawson
Head, Section of Clinical Mass Spectrometry
Clinical Research Centre
Watford Road
Harrow
GB-Middlesex HA1 3UJ
Great Britain

Series Editors

Dr. Hans-Christoph Curtius
Head, Division of Clinical Chemsitry
Department of Pediatrics
University of Zurich
Steinwiesstrasse 75
CH-8032 Zurich
Switzerland

Dr. Marc Roth
Director, Central Laboratory of
Clinical Chemistry
Cantonal Hospital
University of Geneva
CH-1211 Geneva 4
Switzerland

Publishers

Walter de Gruyter & Co.
Genthiner Straße 13
D-1000 Berlin 30
Fed. Republic of Germany
Telephone (030) 26005-0
Telefax (030) 26005-251 · Telex 184027

Walter de Gruyter, Inc.
200 Saw Mill River Road
Hawthorne, N.Y. 10532
U.S.A.
Telephone (914) 747-0110
Telefax (914) 747-1326 · Telex 646677

Library of Congress Cataloging in Publication Data

Mass spectrometry.

(Clinical biochemistry ; 1)
Includes bibliographies and index.
1. Mass spectrometry. 2. Clinical biochemistry—Methodology. I. Lawson, A. M. II. Series: Clinical biochemistry (Berlin, Germany) ; I. [DNLM: 1. Spectrum Analysis, Mass. QC 454.M3 M4141]
QP519.9.M3M35 1989 543'.0873 88-33414
ISBN 0-89925-488-8 (U.S.)

Deutsche Bibliothek Cataloging in Publication Data

Mass spectrometry / ed. A. M. Lawson. — Berlin ; New York : de Gruyter, 1989
(Clinical biochemistry ; 1)
Mit 186 ill. u. 55 tab.
ISBN 3-11-007751-5
NE: Lawson, Alexander M. [Hrsg.]; GT

Typesetting and printing: Arthur Collignon GmbH, Berlin. Binding: Dieter Mikolai, Berlin.
Cover design: Hansbernd Lindemann, Berlin. — Printed in Germany.

Preface

The first volume of this Series, published in 1974*, included a chapter on mass spectrometry by the late J. A. Völlmin. Attesting to the enormous expansion in the application of the technique in subsequent years is the need for the present volume which was written both to highlight applications in clinical biochemistry and to include a broader body of information on principles and methodological aspects from which will come the developments of the future. At no time, since the advent of combined gas chromatography-mass spectrometry, has a fresh wave of advance in biochemical application seemed more likely than at present with the introduction of new ionization methods, such as thermospray and fast atom bombardment, and new multiple sector techniques.

While the continuing hope is for low cost instruments to bring mass spectrometers within the realm of hospital clinical biochemistry departments, it is a technique confined largely to research laboratories specialising in topics such as endocrinology, paediatrics, psychiatry, nutrition and others. It was considered, however, that the book would have more general use if chapters were ordered under compound name headings despite differences in the extent to which individual compound classes are of current clinical interest. In some cases authors have had to stress the potential value of mass spectrometry rather than its implementation.

Three of the chapters are not classified by compound type. Chapter 1, for example, is included for those new to mass spectrometry as an introduction to the principles and techniques employed and, accordingly, these are discussed without resort to detailed physical and mathematical concepts. The two other subjects dealt with separately are Reference Methods and Stable Isotope Applications and are of general interest to quantification in clinical biochemistry.

In other volumes of the "Clinical Biochemistry" series it is the intention to incorporate practical descriptions of methodology and experimental procedure. With some exceptions this seemed less appropriate to the present volume where emphasis has been placed on demonstrating the scope of mass spectrometry to those with interests in the chemistry of diseased states. Existing mass spectrometry users are likely to prefer accessing selected experimental methods from the original literature.

The authors of chapters are from a range of backgrounds but each with extensive knowledge of mass spectrometric methods and their application. They have had the freedom of experts to include material they considered most relevant but were all charged with referencing aspects which could not be dealt with in detail. Indeed some chapters will serve as useful reviews. Limitations on the size of the book have meant that several topics could not be included directly and it is hoped that subjects such as pharmacology and toxicology may appear in a later volume.

---

* Clinical Biochemistry – Principles and Methods
Editors: H. Ch. Curtius/M. Roth
Walter de Gruyter; Berlin · New York 1974; Volume 1 and Volume 2

I would like to express my warm appreciation to contributors for their selfless effort in passing their hard fought experience to others. My thanks also go to the Series editors, Dr. H.-Ch. Curtius and Dr. M. Roth, and to those who helped in other ways in the preparation of the book, not least to my wife Alison and children Elaine and Neil, for their encouragement and indulgent forbearance during this time.

November, 1988

Alexander M. Lawson

Index of Contributors

Blau, K., Ph. D.
Queen Charlotte's Hospital for Women
Prenatal Biochemistry Unit
Department of Chemical Pathology
Queen Charlotte's Maternity Hospital
Goldhawk Road
London W6 OXG
Great Britain

Chalmers, R. A., Ph. D.
Section of Perinatal and Child Health
MRC Clinical Research Centre
Watford Road
Harrow Middlesex HA1 3UJ
Great Britain

Desiderio, D. M., Ph. D.
Stout Neuroscience
Mass Spectrometry Laboratory
University of Tennessee
800 Madison Avenue
Memphis, Tennessee 38163
U.S.A.

Ford, G. C., M. Sc.
Nutrition Research Group
Division of Clinical Sciences
MRC Clinical Research Centre
Watford Road
Harrow Middlesex HA1 3UJ
Great Britain

Gaskell, S. J., Ph. D.
Center for Experimental Therapeutics
Baylor College of Medicine
One Baylor Plaza
Houston, Texas 77030
U.S.A.

Halliday, D., Ph. D.
Nutrition Research Group
Division of Clinical Sciences
MRC Clinical Research Centre
Watford Road
Harrow Middlesex HA1 3UJ
Great Britain

Jackson, A. H., Ph. D.
Department of Chemistry
University College
P.O. Box 78
Cardiff CF1 1XL
Great Britain

Kamerling, J. P., Ph. D.
Department of Bio-Organic Chemistry
Transitorium III
Utrecht University
P.O. Box 80.075
NL-3508 TB Utrecht
The Netherlands

Kelly, R. W., Ph. D.
MRC Reproductive Biology Unit
Centre for Reproductive Biology
37 Chalmers Street
Edinburgh EH3 9EW
Great Britain

Kuksis, A., Ph. D.
Charles H. Best Institute
Banting and Best Department
of Medical Research
University of Toronto
112 College Street
Toronto, Ontario M5G 1L6
Canada

Lawson, A. M., Ph. D.
Section of Clinical Mass Spectrometry
MRC Clinical Research Centre
Watford Road
Harrow Middlesex HA1 3UJ
Great Britain

Myher, J. J., Ph. D.
Charles H. Best Institute
Banting and Best Department
of Medical Research
University of Toronto
112 College Street
Toronto, Ontario M5G 1L6
Canada

Ramsden, D., Ph. D.
Dept. of Medicine
University of Birmingham
Queen Elizabeth Hospital
Edgbaston
Birmingham B15 2TH
Great Britain

Schram, K. H., Ph. D.
College of Pharmacy
Department of Pharmaceutical Sciences
University of Arizona
Tucson, Arizona 85721
U.S.A.

Setchell, K. D. R., Ph. D.
Department of Pediatrics
Children's Hospital Medical Center
University of Cincinnati
Elland & Bethesda Avenues
Cincinnati, Ohio 045229
U.S.A.

Siekmann, L., Ph. D.
Institut für Klinische Biochemie
der Universität Bonn
Sigmund-Freud-Straße 25
D-5300 Bonn 1
Fed. Republic of Germany

Vliegenthart, J. F. G., Ph. D.
Department of Bio-Organic Chemistry
Transitorium III
Utrecht University
P.O. Box 80.075
NL-3508 TB Utrecht
The Netherlands

Abbreviations

ADC	analogue-to-digital converter
t-BDMS	t-butyldimethylsilyl ethers
B	magnetic field strength or magnetic sector
B/E	ratio of magnetic-to-electric fields
CAD	collision activated dissociation
CID	collision induced dissociation
CI	chemical ionization
CI-MS	chemical ionization-mass spectrometry
CI-SIM	chemical ionization-selected ion monitoring
CSF	cerebrospinal fluid
CNS	central nervous system
CV	coefficient of variation
DAC	digital-to-analogue converter
DADI	direct analysis of daughter ions
DCI	direct/desorption chemical ionization
DCI-MS	direct/desorption chemical ionization-mass spectrometry
DEI	direct probe electron ionization
DEI-MS	direct probe electron impact-mass spectrometry
DLI	direct liquid introduction
E	electric sector voltage or electric sector
EI	electron ionization or electron impact
FAB	fast atom bombardment
FAB-MS	fast atom bombardment ionization-mass spectrometry
FD	field desorption
FD-MS	field desorption-mass spectrometry
FFR	field-free region
GC	gas chromatography
GC-CI-MS	gas-liquid chromatography with chemical ionization-mass spectrometry
GC-EI-MS	gas chromatography with electron impact-mass spectrometry
GC-MS	gas liquid chromatography with mass spectrometry
HFB	heptafluorobutyryl
HPLC	high pressure or high performance liquid chromatography
ID-MS	isotope dilution-mass spectrometry
IKE(S)	ion kinetic energy (spectroscopy)
LC	liquid chromatography
LC-MS	liquid chromatography-mass spectrometry
LD	laser desorption
MIKES	mass-analysed ion kinetic energy spectroscopy
MF	mass fragmentography
MS	mass spectrometry
MS/MS	mass spectrometry combined with mass spectrometry
m/z	mass/charge ratio
NICI	negative ion chemical ionization

NICI-MS	negative ion chemical ionization-mass spectrometry
ODS	octadecylsilyl
PPINICI	pulsed positive-ion/negative-ion chemical ionization
Q	quadrupole
RIA	radioimmunoassay
RRA	radioreceptor assay
SEM	standard error of the mean
SFC	supercritical fluid chromatography
SIM	selected ion monitoring
SIMS	secondary ion mass spectrometry
TIC	total ion current
TFA	trifluoroacetyl
TLC	thin-layer chromatography
TOF	time-of-flight
TMS	trimethylsilyl
TSP	thermospray
TSQ	triple stage quadrupole
V	ion accelerating voltage

Contents

Chapter 2 Bile Acids (K. D. R. Setchell and A. M. Lawson)

Chapter 6 Organic Acids (R. A. Chalmers)

Chapter 7 Endogenous Brain Peptides (D. M. Desiderio)

Chapter 1
Mass Spectrometry
The Fundamental Principles

A. M. Lawson

1 Introduction

The success of mass spectrometry as an analytical technique, first in physics and chemistry and then in many of the biological sciences, is due to its potential to supply definitive qualitative and quantitative information on molecules based on their structural compositions. The technique has been extensively modified and developed since its introduction by J. J. Thompson and the application by F. W. Aston and A. J. Dempster, although the underlying concept remains unchanged. It is based on the separation, detection and recording of the ions formed by ionization of molecules. Depending on the nature of this ionization and the molecular structure, intact ionized molecules and/or fragments of these molecules are produced which provide the means to recognise or identify the original molecule.

Advances which have been important to the development of mass spectrometry for the investigation of molecules of relevance in biochemistry, include those associated with initial sample introduction such as gas chromatography and liquid chromatography, soft ionization methods compatible with the lability of biomolecules and the computerised techniques for processing mass spectral data.

While one might question the need for a chapter on the basic aspects of mass spectrometry when many excellent texts have published over the years (e. g. 1 – 8), the present chapter has been included to provide some of the fundamental information required by those with limited knowledge of mass spectrometry, leaving other authors in the book free to elaborate on the specific application and refinements of the technique relating to their topic. In this way readers embarking on the use of mass spectrometry for the first time should gain some understanding of their requirements although further advice will undoubtedly be needed for selection of an instrument from the ever-changing range available (Section 5). Developments in mass spectrometry are frequently being made and while the most recent of these are discussed briefly in this and other chapters, more detailed information should be sought in the general literature and in the proceedings of the Triennial International Mass Spectrometry Conference (e. g. 9), the Annual Conference of the American Society for Mass Spectrometry (e. g. 10) and other national and international meetings. A comprehensive biennial review of mass spectrometry is also available (11).

The analytical capability of a mass spectrometric system is dependent on the individual components of the system and these are selected and combined to provide the most appropriate information for each application. In the following sections the component parts of mass spectrometers are discussed in relation to basic function. The three primary elements of a mass spectrometer are the ionization region into which sample molecules are introduced and the ionized species extracted, a mass/charge separation region in which ionized species are separated according to mass and a detection system for detecting and recording ions.

2 Sample Introduction Systems and Ionization Methods

The selection of the best combination of sample introduction and ionization methods is made on several grounds. While an element of trial and error is often applied, data on how most compound types react under different ionization and introduction conditions are available in the literature. This information coupled with a knowledge of the sample type (gas, liquid, solid, single component, complex mixture etc.) will usually dictate the most appropriate method.

Subsequent chapters indicate the conditions most suited to the different compound classes and the purpose of this section will be to briefly explain the fundamental principles of the different ionization methods. These methods often require a specific mode of introduction (Table 1). The gas chromatographic, liquid chromatographic and supercritical fluid chromatographic inlets are described separately.

2.1 Electron impact ionization (EI)

As EI was one of the earliest methods of ionizing organic molecules, the vast majority of mass spectra in the literature were obtained in this way and hence the comparison of EI spectra for identification purposes and other reasons will continue to be important.

In the ion source of an EI mass spectrometer (Fig. 1), ionization is induced in volatilised sample molecules by collision with a beam of energetic electrons (approximately 20–70 eV) produced from a heated rhenium filament. Removal of an electron from the molecule in the collision process produces a radical molecular ion:

$$M + e \rightarrow M^{+\bullet} + 2e$$

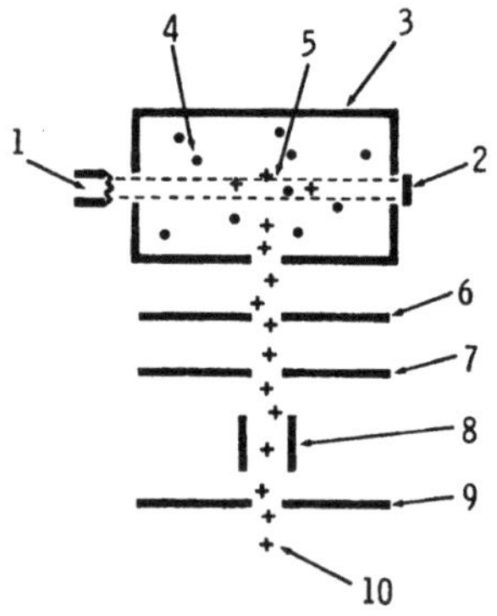

Figure 1. Schematic diagram of an electron impact ion source. The inlet line for the sample enters at right angles to the electron beam and is not shown. (1) Electron emitting filament; (2) electron trap; (3) ion source block held at high positive voltage for positive ion extraction (or negative voltage for extraction of negative ions); (4) sample molecules; (5) region of ionization; (6) extraction plates; (7) focusing slits; (8) beam deflection plates; (9) source slit and (10) ion beam.

The energy of the ionizing electrons and the stability of the chemical bonds in the molecule, determine whether the molecular ions subsequently undergo decomposition by internal bond cleavage and atomic rearrangements to give a plasma of ions, radicals and neutral species. Both positive and negative ions can be produced, but the former are usually orders of magnitude more numerous. Under normal EI conditions with electron energies above 10 eV, negative ions are formed by ion-pair production, i.e. $MN + e^- \rightarrow M^+ + N^- + e^-$, but the anions are generally of low mass and of little value to interpretation of structure. However, many organic compounds will give negative ions by electron capture when ionizing electrons of low energy (thermal energy) are used. This is best achieved at higher ion source pressure under chemical ionization conditions (Section 2.2.1).

The decomposition of the molecular ion by fragmentation and rearrangement is strongly dependent on its molecular structure and hence by understanding the structural basis for these processes it is often possible to predict the way in which similar or related molecules will react to EI. However, while the structure of a new compound may be suggested directly from its mass spectrum, definitive identity requires analysis of an authentic standard or comparison with an authentic spectrum. With a number of exceptions, paricularly for isomeric molecules, a compound's mass spectrum is unique to that compound. Also, the intensity of ions produced under fixed conditions is proportional to the absolute quantity of the compound within a fairly wide range.

Ion source conditions such as electron energy and temperature, and conditions which affect the residence time of ionic species in the ion source, influence the relative abundance of individual ions. For example it may be possible to reduce the electron energy to a level at which only the molecular ion is produced. Primary fragment and rearrangement ions will appear on increasing the energy and then secondary breakdown of these ions by increasing the energy still further. As the degree of ionization for many molecules is approximately constant with an electron voltage of between 60 and 100 eV, a value of about 70 eV is used to obtain reproducible spectra which can be compared usefully with data obtained by other laboratories. The ion source temperature is adjusted to prevent adsorption of sample molecules (typically 100–250 °C), but for thermally unstable components this may have to be reduced or at least held constant to give uniform spectra.

There are several drawbacks to the use of electron impact ionization which affect its applicability. A limited number of molecules are sufficiently thermally stable to be volatilised and thus suitable for EI. This number can be extended by conversion to stable and volatile derivatives which will survive introduction and ion source temperatures. The overall effect of a volatility requirement limits EI analysis to compounds of lower molecular weight, generally below 1000 daltons although there are noteable exceptions to this. The stability of molecular ions of many compounds is low and as a result they may be absent or of low relative abundance to fragment ions. As the molecular ion is potentially the most informative ion in a spectrum its absence reduces the utility of the spectrum. However, ions closely related to the molecular ion may help to negate this effect and much of the structural information is available from the fragment and other ions. In addition several of the ionization methods discussed later give predominant molecular or related ions and can be used to compliment EI spectra.

Figure 2 shows the distinctive EI mass spectrum of the N-heptafluorobutyryl-isobutyl ester derivative of phenylalanine from which it can be positively identified by

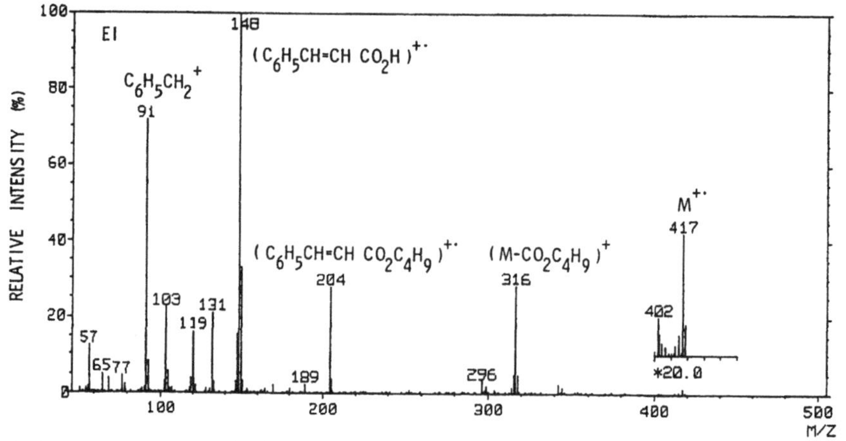

Figure 2. EI mass spectrum of *N*-heptafluorobutyryl-isobutyl ester of phenylalanine.

comparison with a reference spectrum. If the spectrum had been of an unknown compound then interpretation of the fragmentation pattern in terms of structure would have been possible in part, although the low intensity of the molecular ion at m/z 417 (3% of base peak m/z 148) would have introduced some doubt about the molecular mass.

Electron impact ionization can be used with a variety of inlet systems (Table 1). The direct insertion probe allows relatively involatile samples to be vapourised directly into the ion source from a heatable crucible held close to the electron beam by a probe which enters the mass spectrometer through a vacuum lock. It is often employed for handling isolated single component samples and, if necessary, can deliver sample molecules for ionization extremely rapidly or over an extended time period by controlling the volatilisation temperature. The direct insertion probe tip can be cooled to cope with unstable samples and the temperature computer controlled to maintain a constant sample flow.

Less commonly, reservoir inlets are used which expand a gas sample or a volatile solid or liquid into an evacuated chamber from which the sample passes steadily into the ion source through a molecular leak. This system requires comparatively large sample sizes, but is important when very precise measurements are being made (see isotope ratio instruments, (Chapter 14). Small reservoir inlets are suitable for introducing mass calibration standards.

The inlet system employed most frequently for biochemical investigations with EI is the gas chromatograph and this is described in Section 2.10. The liquid chromatographic inlet is discussed in Section 2.11.

2.2 Chemical ionization (CI)

The high energy transfer encountered during ionization of molecules by electron impact often produce an abundance of fragmentation and rearrangement ions. Chemical ionization, on the other hand was developed (10) to ionize molecules with a much reduced energy transfer and give a more stable molecular ion species. In this method

Table 1. Inlet systems and ionization methods used in mass spectrometry

Inlet System	Ionization Method*	Comment
Direct probe	EI CI	Useful for single component samples which lack volatility for GC introduction. Mixture analysis by temperature programmed probe. Mixture analysis by MS/MS.
Reservoir inlet	EI, CI and FI	For gases and volatile liquids and solids which are not thermally stable.
Gas chromatograph	EI, CI and FI	Separation and analysis of volatile mixtures – derivatisation usually required – wide application in clinical biochemistry.
Supercritical fluid chromatograph	EI, CI	Useful for separation of thermally labile compounds not amenable to GC.
HPLC inlet	EI, CI, TSP, FAB	Separation of mixture without prior derivatisation – great potential application in clinical biochemistry.
FD emitter probe	FD	Useful for polar or thermally labile compounds.
DCI probe	CI	For non-volatile samples
FAB target probe	FAB	Simpler and with wider applicability than FD for polar and thermally labile molecules and for large molecular weight compounds. Continuous flow system gives reduced background.
LD probe	LD	Useful for compounds of low volatility and thermal stability. Not widely available.
Aluminium foil or nitrocellulose on aluminized polyester foil	PD	Potentially suitable for large molecules. Not widely available.

* Electron impact (EI); chemical ionization (CI); field ionization (FI); field desorption (FD); thermospray (TSP); fast atom bombardment (FAB); desorption chemical ionization (DCI); laser desorption (LD); plasma desorption (PD).

a reagent gas is ionized at an ion source pressure of about 0.1 kPa and the reactive ionized plasma used to ionized sample molecules by ion molecule interations. Frequently the same ion source is used for CI and EI (Fig. 1), but in CI mode the ionization region is made sufficiently gas tight to maintain the required reagent gas pressure.

The ionization sequence can be illustrated for simple reagent gas, methane, which gives principally the EI products $CH_4^{+\bullet}$ and CH_3^+ that react with neutral CH_4 molecules, i.e.:

$$CH_4^{+\bullet} + CH_4 \rightarrow CH_5^+ + CH_3^\bullet$$
$$CH_3^+ + CH_4 \rightarrow C_2H_5^+ + H_2$$

These ionic species are strongly acidic and react with sample molecules chiefly by proton transfer, e. g. $M + CH_5^+$ (or $C_2H_5^+$) $\rightarrow MH^+ + CH_4$ (or C_2H_4), the protonated molecular ions being referred to as quasi-molecular ions.

Addition products may also arise: $M + CH_5^+ \rightarrow [M+CH_5]^+$ and hydride may be abstracted from some compounds to give $[M-H]^+$ quasi-molecular ions.

Other reagent gases provide a range of ionizing conditions and much of the power of CI mass spectrometry lies in the dependence of the CI spectrum on the reagent gas and varying the selection of reagent gas to provide a control on the information available. Increasing the exothermicity of the ion molecule reactions by changing the nature of the reactant ions increases the extent of fragmentation.

Figure 3 shows the CI spectra of the *N*-heptafluorobutyryl-isobutyl ester derivative of phenylalanine using three different reagent gases. The more exothermic protonation by methane in comparison to isobutane is evident from the degree of fragmentation in the methane spectrum which shows some similarity with the EI spectrum (Fig. 2). Reagent gas addition ions are present in the methane spectrum and help to confirm the quasi-molecular ion at m/z 418. It is common with isobutane to produce almost no fragmentation and similarly with ammonia although in this case NH_4^+ has added to the molecule. Proton transfer from NH_4^+ requires the compound to have a proton affinity greater that than of ammonia (e. g. amines and amides). Reagent gases that do not contain hydrogen to permit ionization by proton transfer produce molecular ions of the sample by charge exchange. Gases such as helium, argon and nitrogen cause extensive fragmentation in addition and are often used in combination with other reagents (e. g. H_2O, NO) to give the most informative spectra. The converse enhances the quasi-molecular ion abundances.

The subtleties of selection of reagent gases for particular applications is beyond the scope of this chapter and can be found elsewhere (e. g. 13). Commonly used reagent gases in positive ion formation are methane, isobutane, ammonia, nitric oxide/nitrogen, hydrogen and water ((7) and references therein).

The pressure of the reagent gas and the ion source temperature can influence CI spectra. Increasing pressure of reagent gas, for example, reduces the formation of fragment and addition ions but has little effect on proton transfer reactions. CI spectra should be acquired at the minimum ion source temperature consistent with preventing condensation of the sample. Figure 4 illustrates the result of increasing temperature from 100 to 160 °C on the methane and isobutane spectra of the phenylalanine derivative recorded in Figure 3. Fragmentation is seen in the isobutane spectrum and increased in the methane spectrum. This effect is usually greater for milder reagent gases such as isobutane and ammonia and not so pronounced for methane.

The stability of the quasi-molecular ion in CI often gives a gain in sensitivity over EI which, although having overall ion abundance of the same order, is reduced by the greater number of ions present. CI is used to provide molecular weight information and for the production of ions prior to collision induced dissociation experiments (Section 3.5). It is of particular value in quantitative measurements where the specificity of the molecular ion or related species compliments the increased sensitivity. The reduction in structurally informative fragment ions can be a disadvantage for identification of unknown molecules, particularly when dealing with compound classes with many isomeric structures. The complementary nature of information from EI and CI have made it common practice to include both ionization methods as standard features

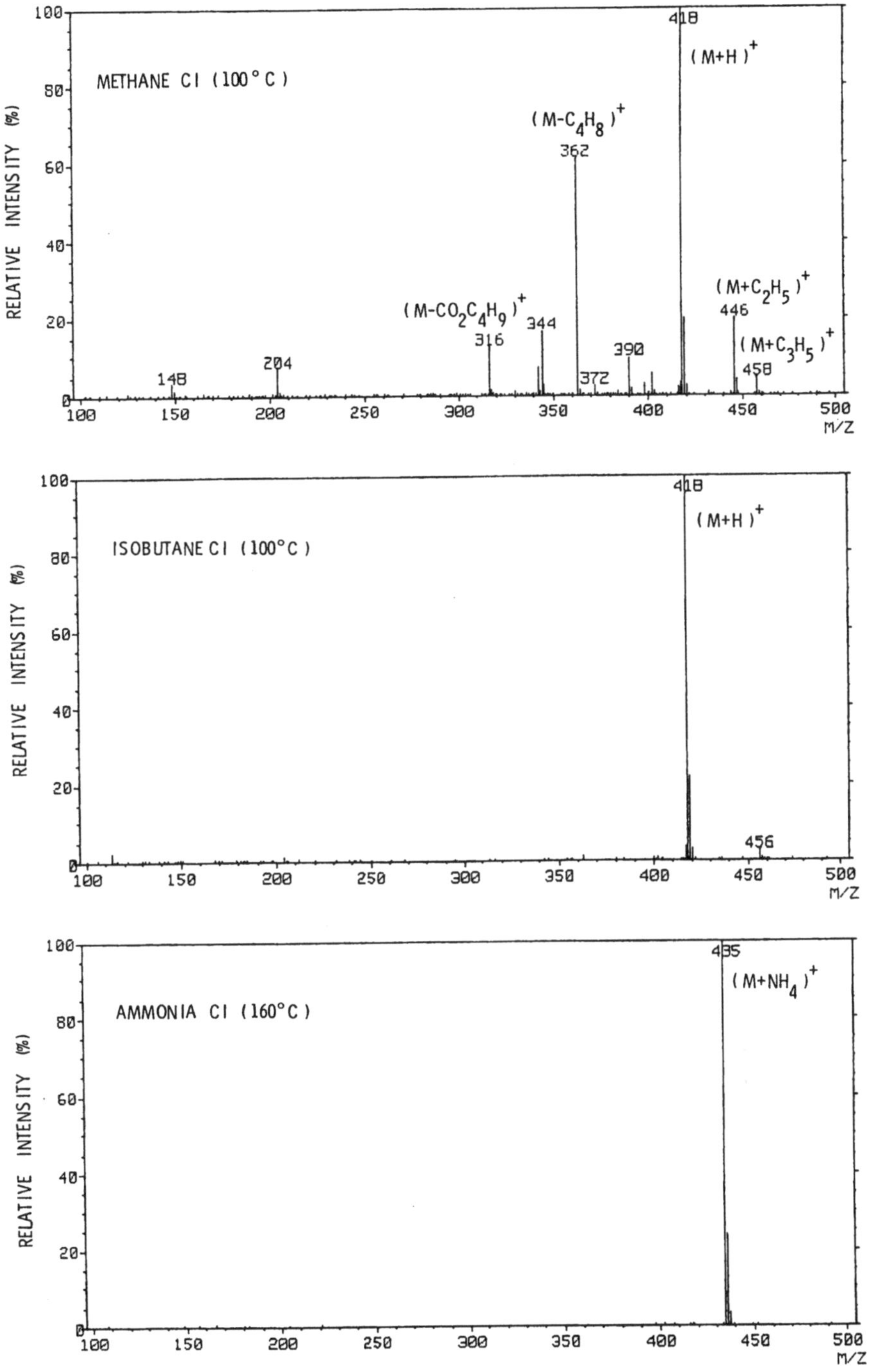

Figure 3. CI mass spectra of *N*-heptafluorobutyryl-isobutyl ester of phenylalanine using different reagent gases: methane at ion source temperature of 100 °C (top), isobutane at 100 °C (middle) and ammonia at 160 °C (bottom).

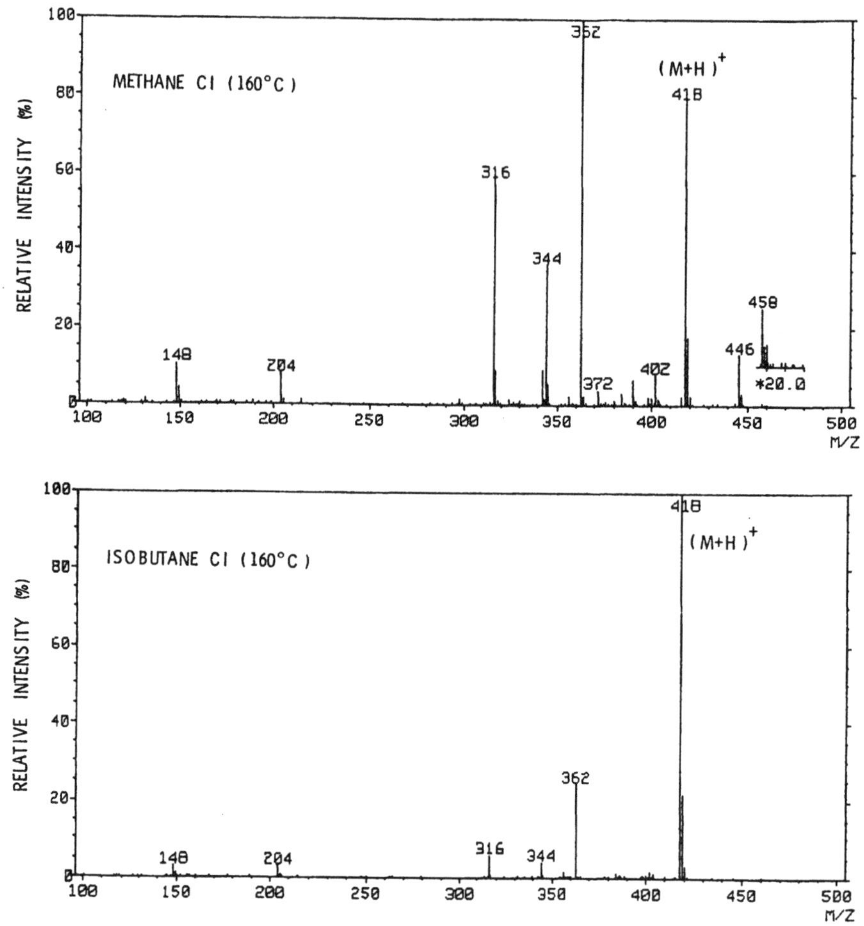

Figure 4. Methane and isobutane CI mass spectra of *N*-heptafluorobutyryl-isobutyl ester of phenylalanine recorded at an ion source temperature of 160 °C.

on most commercial mass spectrometers and some provide the facility to alternate continuously between EI and CI during a single analysis. It should not be overlooked, however, that by appropriate selection of reagent gases, fragment ions can be produced by CI and these will frequently afford structural and sterochemical information which is not available under EI conditions.

The inlet systems suitable for EI can all be used for CI, the gas chromatograph being especially important to give additional specificity to analysis.

2.2.1 **Negative ion chemical ionization (NICI)**

Renewed interest in negative ion production has been of importance to the development of CI application to compounds of biochemical and toxicological interest (14–17). There are several ways in which negative ions can be formed. Resonance electron

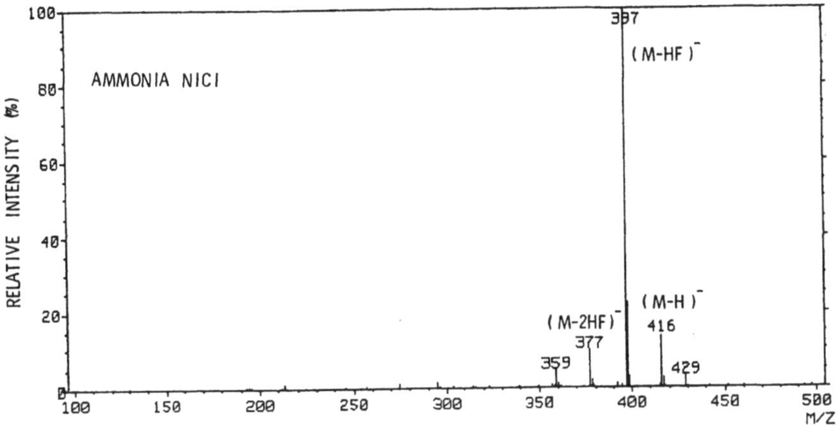

Figure 5. Ammonia NICI mass spectrum of *N*-heptafluorobutyryl-isobutyl ester of phenylalanine.

capture, for example, will produce negative ions when electrons of near thermal energy are available:

$$M + e^- \rightarrow M^{-\bullet}$$

This can be achieved in a CI source where the reagent gas serves to remove excess energy from the more energetic electrons and to collisionally stabilse the ions formed (16). In compounds without electron capturing ability, electron affinity can be induced by preparing derivatives similar to those used in GC for electron capture detection. Frequently organic samples will undergo dissociation following electron capture to give low mass fragments. Negative ions can also be produced by ion molecule reactions using anionic reagent gases under CI conditions. This gives flexibility by permitting selection of specific NICI mechanisms to provide the most suitable analytical information. Depending on the nature of the sample to be analysed proton abstraction, charge exchange, nucleophilic addition or nucleophilic substitution may be appropriate to ion formation.

The proton affinity of the reagent anion affects its ability to produce sample anions by proton abstraction. Reactive species from reagent gases which abstract a proton from the molecule to give abundant negative quasi-molecular ions with little fragmentation include NH_2^-, H^-, OH^-, $O^{-\bullet}$, MeO^-, Cl^- and others (listed in order of decreasing proton affinity) (see (11)). The spectrum of the *N*-heptafluorobutyryl-isobutyl ester derivative of phenylalanine obtained under ammonia NICI conditions is shown in Figure 5. The $[M - H]^-$ quasi-molecular at m/z 416 is present but is less intense than usual for NICI spectra due to the facile loss of HF from the heptafluorobutyryl group.

Charge exchange reactions require that an anionic reagent species is of lower electron affinity than the sample molecule. The $O_2^{-\bullet}$ anion has a low electron affinity and can be used for this purpose, e. g. $M + O_2^{-\bullet} \rightarrow M^- + O_2^{\bullet}$

Among anions which will form addition complexes with some compounds by nucleophilic attack are Cl^- and $O_2^{-\bullet}$ while OH^- and $O^{-\bullet}$ produce $[M - H]^-$ ions by proton abstraction or ionize some compounds by nucleophilic substitution.

NICI has extended the number of compounds amenable to MS analysis and frequently provides different and complementary structural information to either EI or positive CI. For this reason it can be usefully combined in the same source with these methods and spectra from the different modes acquired quasi-simultaneously during a GC-MS run. Pulsed positive ion negative ion chemical ionization (PPINICI) (15) and positive EI combined with NICI (17) have considerable value in structure elucidation although assays for specific compounds utilise the ionization mechanism which produces the most relevant specificity and sensitivity. Electron capture NICI, for example, will give very high sensitivity for compounds possessing strong electron affinity and would be the method of choice for quantifying many trace level components. In favourable cases the 10^{-18} mol range is accessible. The features of NICI and application to different compound classes (7) and to biomedical problems (18) have been summarised.

2.3 Desorption chemical ionization (DCI)

Following introduction of the DCI technique (19), a variety of methods for vapourising sample molecules in close proximity to the ionization region have been suggested (for review see 20). In general the sample is coated on a wire or similar support and inserted directly into the ionization region of a CI source, close to the electron beam. Neutral sample molecules are desorbed and ionized by the ion plasma before suffering collision with other surfaces. The process may be assisted by rapid heating of the probe tip, although the transfer of heat from the hot ion source and ion plasma to bring about vapourisation may often be adequate. Pyrolytic processes are diminished and it has been possible to obtain spectra of ionic salts and thermally labile organic molecules of relevance to biological systems.

2.4 Field ionization (FI)

Molecular ions are formed in FI by the removal of an electron from the molecule under the influence of a high electrostatic field. The electron transfer process takes place at energies below that of most chemical bonding energies and as a result molecular ions have high stability. The field is created by a metal electrode or pre-activated tungsten wire held at high voltage (≈ 10 kV) and the sample introduced by direct insertion probe or GC inlet. Fragmentation reactions take place by unimolecular decomposition in the gas phase, although field dissociation or surface reactions on the field emitter may also occur. The mass spectra produced are relatively simple and uncomplicated by rearrangement reactions and further fragmentation (21).

The FI technique has a lower sensitivity than EI or CI, in general 10^2 – 10^3 fold less sensitive, and is not widely used as it has not been as successful for biological application as the related method of field desorption described below.

2.5 Field desorption (FD)

The field desorption process of ionization was introduced (22) to facilitate the analysis of compounds susceptible to decomposition as a result of thermal volatilisation. It was found that sample molecules placed in a high electric field gradient (approximately $10^8 V \cdot cm^{-1}$) would either lose or gain an electron to produce ions which, by ion molecule reactions, produced $[M + H]^+$ species or under certain conditions [M + Na] and [M + K] anions and cations.

The high electric field gradient to produce ionization is developed at the end of needle like structures on a tungsten emitter wire. These structures, or dendrites, are created by pre-conditioning the wire in the presence of a suitable compound, such as benzonitrile, under controlled temperature and electric field conditions. The sample is placed on the emitter, any solvent removed and, after insertion into the ion source, a high voltage applied. Ionization may then take place although it is usually necessary to pass an electric current (several mA) through the emitter to melt the compound and encourage its movement to the tips of the dendrites where ionization and desorption occur. The basic FD ionization mechanisms have been studied, but some aspects remain a matter for discussion (22–25).

As the energy transferred to the molecule during ionization is small, FD spectra generally consist of the quasi-molecular ion and no fragment ions. The obvious value of this to molecular weight determination has been applied to many compound groups, but in particular to thermally labile or involatile molecules that are intractable to EI and CI or to molecules which give uninformative spectra in these modes. A noteable advantage of FD is that samples need not be derivatised to increase volatility and hence fractions isolated by TLC, LC or other chromatographic method can be analysed directly. While the lack of fragment ions is unhelpful for the deduction of structural features from a spectrum, the simplicity of the spectra allows mixtures of compounds to be assessed without prior separation. Care has to be taken, however, when isomeric compounds are present as these will often give identical spectra. Some FD spectra contain fragment ions and increased emitter currents may assist in increasing their intensities.

The ability of the FD process to ionize some molecules of several thousand molecular weight increased interest in mass spectrometers with much greater mass ranges, a development of importance to biological applications in the peptide/protein, carbohydrate and other fields where large molecular weight compounds are encountered.

There are disadvantages associated with the FD technique. The activation of the emitter wire is crucial to obtaining satisfactory spectra and their preparation is a matter of some experience. Also the emitters for FD are fragile and require care to limit breakages. These practical limitations coupled with the often variable and short lived production of ions in FD, have added to the rapid acceptance of fast atom bombardment ionization as an alternative procedure.

2.6 Fast atom bombardment (FAB)

Fast atom bombardment ionization developed from the technique of secondary ion mass spectrometry (SIMS) in which the sample molecules held on a surface are struck by ions of high translational energy (26). This has the effect of sputtering secondary ions from the surface which can then be analysed. Many organic molecules are

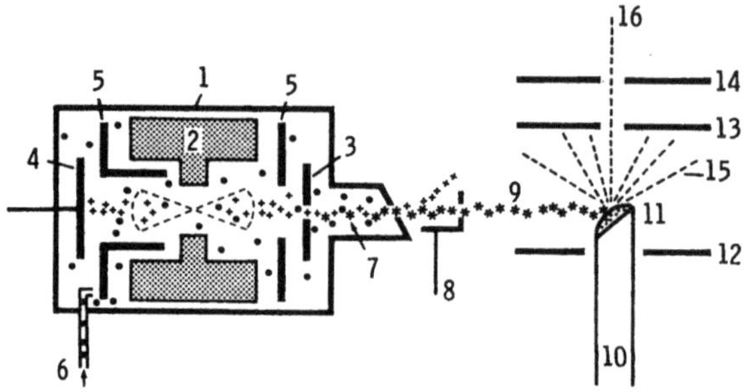

Figure 6. Schematic diagram of a FAB saddle-field atom gun and FAB ion source. (1) Outer casing of FAB gun; (2) circular anode showing region of electron density (dotted lines); (3) front cathode; (4) rear cathode; (5) screens; (6) gas inlet (•) (7) main region of resonant charge transfer (+ → *); (8) deflector electrode (+ve), (9) target probe; (10) sample dissolved in matrix; (11) source extractor plates; (12) focusing slits; (13) source slit; (14) sputtering sample and matrix ions and (15) ions passing out of the source.

destroyed during thie process, but it was found that by dissolving the sample in a suitable liquid matrix and utilising neutral bombarding atoms, much longer lived ion beams were produced (20). Continuous replenishment of the surface layer of molecules in direct contact with the atom beam is achieved in the liquid matrix and this greatly prolongs the time of ion production by reducing the degree of sample destruction and assisting in stabilising ion formation. This technique is known as FAB or liquid SIMS when an ion beam is employed.

A noble gas, initially argon, but now more commonly xenon, due to its greater mass and resulting greater sensitivity, is used as a source of bombarding atoms. A beam of xenon ions of high translational energy (2 – 10 keV), produced from electron collision in a saddle-field source (Fig. 6), is neutralised in a dense cloud of the xenon atoms by resonant charge exchange to produce a fast atom beam following the removal of the residual ionic component. The fast atom beam is passed into the ion source region to impinge on the sample placed in its matrix on a small metal target mounted on a removeable probe (28). While it was believed initially that the bombarding species should not be charged, subsequent studies have shown that equivalent spectra can be obtained with a mixture of ions and neutrals (e. g. Townsend discharge) or ions alone (Caesium ion gun (29)).

The angle of incidence of the beam with the sample surface and alignment of the target and the source exit slit are necessary to maximise sensitivity of the sample ions reaching the analyser. Both positive and negative ions can be produced by FAB. The energy imparted to quasi-molecular ions is often sufficient to create internal bond cleavage and the production of fragment ions, although the extent to which this takes place varies with compound type.

The nature of the matrix is of central importance to FAB ionization. Primarily the sample must be soluble in the matrix. Solubility in a matrix is often improved by first dissolving the compound in a suitable organic solvent and then entraining this in the matrix. Samples which are polar can be ionized effectively in glycerol, while fairly hydrophobic conmpounds will be dissolved more readily in a mixture of glycerol/thioglycerol. Other matrix materials have important advantages in some specific instances, but glycerol of thioglycerol are adequate for the vast majority of samples. The criteria for selection of a suitable matrix for a wide range of compound types have been considered (30). A highly volatile matrix will be used up rapidly and might not give sufficient time for data collection and a cooled sample introduction probe has been shown to increase sample duration and sensitivity in carbohydrate analysis (31).

The theory underlying ion production by FAB has not been fully elucidated and the protonation or hydrogen abstraction mechanisms which produce quasi-molecular ions may operate within the matrix or in a gaseous layer just above the surface. In practice several procedures for increasing the sensitivity of ion production are possible. Positive ions may be increased from compounds with basic character, for example, by pre-protonating the sample with acid (e. g. 0.1 $mol \cdot l^{-1}$ HCl). An increase in negative ions may be achieved by addition of ammonium or sodium hydroxide to the sample. When several anionic groups are present, it is advantageous to maintain a low pH and run in the negative mode.

The best conditions for analysing a particular type of compound by FAB is a matter of experimentation when standard procedures are ineffective. The presence of salts may suppress ionization and require chromotagraphic clean-up while in other cases they may assist in distinguishing protonated and cationized ions or improve sensitivity (32).

Mixed samples give rise to mixed FAB spectra although competitive effects on the matrix surface may lead to preferential ionization of individual compounds irrespective of concentration. Indeed complete ion suppression of some components in a mixture is often encountered. An additional disadvantage of FAB ionization is the relatively high background level of chemical noise giving ions at every mass in the spectrum in addition to intense cluster ions from the matrix. This leads to much lower sensitivity of FAB ionization in comparison with EI or CI.

Chromatographic separation of components by HPLC and the use of FAB to directly ionize components from a moving belt interface has been demonstrated (33, 34). Due to the motion of the belt and the continuous exposure of fresh sample the glycerol matrix is not essential and spectra with reduced background can be obtained. Of even greater promise is the use of a continuous-flow FAB probe (35,36) which allows the sample in solution to be deposited on the target probe. Only a relatively small amount of liquid matrix ($\approx$20%) is required in the solution thus reducing chemical background and matrix peaks. A 150 fold increase in sensitivity from this method over the standard FAB probe was obtained on the molecular ion of substance P at m/z 1348 (37).

Although FAB ionization is applied mainly to identification and structural problems its use in clinical biochemistry has great scope for the future. When coupled with suitable isolation methods its simplicity and speed make it ideally suited to screening biological fluid extracts for labile and polar molecules which are difficult or impossible by other procedures (38, 39).

2.7 Laser desorption (LD)

LD is still a developing, but promising, technique for biochemical compounds and will only be mentioned briefly (for review see (40)).

Interest in LD has been stimulated by the demonstration that quasi-molecular species could be desorbed and ionized from many organic molecules by short (<microsecond) laser pulses (41). Both bulk solids and thin films coated on a metal foil can be used for LD, although the mechanisms of ion formation may be different. The resulting spectra are similar with molecular ions produced by protonation or cationization or at times desorbed as preformed ions. Molecules of masses up to M_r 3600 and beyond have been ionized and, as with other desorption techniques, are particularly suitable for molecules of poor volatility or thermal stability.

A novel method of introducing a sample to the LD ion source is the moving belt interface developed for LC effluents (42) (Section 2.10).

Lasers are being used increasingly to induce secondary fragmentation of ions separated by MS/MS methods (Section 3.5), in FT-MS experiments (Section 3.4) and most recently in ion trap detectors.

2.8 Californium-252 plasma desorption (PD)

The nuclear fission products of ^{252}Cf can be used to induce desorption and ionization of sample molecules deposited as a thin film on a nickel foil (43). The energies of typical fission products (e.g. $^{106}Tc^{22+}$ and $^{142}Ba^{18+}$) are in the MeV range (104 MeV and 79 MeV respectively) and produce higher yields of quasi-molecular ions than FAB ionization, although this advantage is offset by the ease of producing high abundances of the keV primary ions used in FAB.

The ^{252}Cf source is aligned such that each time a ^{252}Cf nucleus undergoes fission, one fragment is detected and used to initiate acquisition of the spectrum of ions produced from the sample by the complementary fission fragment. The sample ions are separated in a time-of-flight analyser (Section 3.3) and summed over a period of time (0.24–10 h) to give a time-averaged spectrum.

Applications of PD have been reviewed (44) but as yet the technique has little to offer clinical biochemistry. Of most interest in the biological field is the potential of PD to analyse large molecules and this has been applied principally to peptides and proteins. Molecular weight information has been obtained on an increasing number of proteins above M_r 10000. The upper limit of detection is presently about M_r 30000 but should extend beyond this in future (45).

It is a technique with high sensitivity. Molecular weight determination, for example, has been obtained on as little as 1 picomole of porcine insulin at M_r 5777.6.

2.9 Thermospray (TSP)

Thermospray ionization is a recently developed method for the ionization of compounds contained in the effluent from an LC column. The liquid stream from the column is passed through a small diameter capillary and vapourised by heating. The jet of fine liquid particles produced may be passed directly into the ion source and molecules ionized by CI using a normal filament, excess vapour being pumped away. The thermospray process however, takes place without a filament (46,47), but requires an electrolytic solution. Within the supersonic jet of vapour containing liquid droplets and solid particles, primary ions are produced from the rapidly desolvating charged droplets by surface fields. These fields encourage the ejection of ions from the liquid phase. Ammonium acetate is the most widely used electrolyte due to its volatility and provision of a good source of gas phase proton donor (NH_4^+) and acceptor (CH_3COO^-) ions. In additionn to field effects the increased pressure within the thermospray jet brings about proton transfer reactions by gas-phase collisions. In the presence of ammonium acetate many compounds so far tested give $[M + H]^+$ and/or $[M + NH_4]^+$ quasi-molecular ions with no fragmentation. Molecules with more than one ionizeable site may give rise to multiply charged ions (48).

Despite the high temperatures used in the resistively heated capillary, compounds are ionized without apparent pyrolysis largely due to the protective effect of the solvent. Flow rates into the ion source are maintained about $1\ ml \cdot min^{-1}$. A possible limitation of TSP is a dependence on optimum temperature conditions of the vapouriser for different types of compounds. Changes in temperature may result in poor peak shape and a degree of pyrolytic fragmentation. TSP is potentially the ionization method of choice for obtaining molecular weight data on a wide variety of polar samples and best suited to cope with a broad range of flow rates and LC conditions. At first it could be used only with quadrupole instruments as the vapouriser is at ion source potential, however is it now possible to operate magnetic instruments with a source voltage of 3.5 kV (49) and above in the thermospray mode.

The lack of significant fragmentation in TSP spectra is a disadvantage for structure eludication although the incorporation of a discharge electrode close to the vapouriser and a repeller electrode have been shown to give fragmentation reactions (50). These arise from collision induced dissociation as a result of ion acceleration between the electrodes and can be reproducibly controlled by varying the repeller voltage. This and further development of TSP will undoubtedly take place although the limitation that both vapourisation and ionization is strongly affected by the LC mobile phase and operating temperature will remain. Nevertheless TSP will be of importance in future LC-MS analyses of biomedically relevant compounds.

2.10 Gas chromatographic (GC) inlet

The GC inlet is the single most important inlet for compounds of interest to the clinical chemist. The introduction of combined gas chromatography-mass spectrometry (GC-MS) made it possible to separate mixtures of compounds and obtain pure mass spectra of the individual components (e. g. 51,52). This greatly stimulated the application of mass spectrometry to the study of biologically important compounds.

2.10.1 **Capillary GC columns**

The early metal capillary columns had limited value for most of the chemically sensitive compounds encountered in biological systems and it was the much wider bore glass columns of 1 – 2 m ($\approx$5 mm internal diameter) packed with phase-coated support that were most suited to these compounds. The high carrier gas flow rates (30–60 ml · min^{-1}) of packed GC columns required an enrichment device to permit coupling of the column with the high vacuum region of the ion source (10^{-7} kPa) and led to the introduction of a variety of molecular separators to increase the ratio of sample molecules to carrier gas giving flow rates of about 1 ml · min^{-1} into the ion source (53). While packed columns are still retained for some well established procedures the availability of open-tubular glass capillary columns has made the packed columns largely redundant.

There are a variety of capillary column types which have been introduced during the technical development of the capillary column (see e. g. 54) but GC-MS users, in most cases, employ the current state-of-the-art columns. These are fused silica wall-coated open tubular columns with chemically bonded stationary phases. Fused silica is a very inert material and when drawn out as a thin capillary, although naturally straight, is sufficiently flexible to be bent by hand. On the outside the column is coated with polyimide to protect it from shattering. The liquid phase is formed as a thin film on the inner wall of the tubing to which it is chemically bonded, a feature which retains the phase in position, increases the inertness of the column, decreases bleeding and maintains the efficiency of the column for much longer periods than previously. Contaminants can be removed by washing with solvents.

A number of phases of varying polarity are available and as most manufacturers supply the same range of liquid phases, but under different names, their individual literature should be consulted. Information on the selection of a phase for separation of particular compound classes are referenced in their chapters. As a general rule, as the polarity of a compound increases, phases of increasing polarity are necessary for its separation from compounds of similar polarity. Low polarity compounds, such as aliphatic hydrocarbons, are best run on the least polar phases. However as the separation efficiency of present day columns is now so high many workers do not find the need for different phases and carry out much of their work on a single non-polar methyl silicone phase (e. g. OV-1).

The separation efficiency of a capillary column depends on the column length, inside diameter, and film thickness. An optimal combination of these parameters should be determined for individual applications but this is not generally necessary unless specific separation difficulties are encountered. Short columns reduce analysis time but will degrade resolution compared to the more usual 25 m length columns. Columns of 0.2–0.3 mm internal diameter are suited to general use and if greater efficiency is required then the diameter should be reduced. Depending on column bore, phase thickness and the test components used, the theoretical plate number of an average column is of the order of 5000 m^{-1}.

Thermal stability of the liquid phase is important for GC-MS operation to maintain low bleed and hence a low density of background ions which will not interfere with sample spectra. This stability depends on the nature of the phase with more polar phases having lower temperature ranges. Capillary columns with chemically bonded phases give much lower bleed rates than their packed column counterparts.

2.10.2 **Column coupling to MS**

The wide acceptance of capillary GC columns for GC-MS operation is due, not only to their improved resolution, but to the ease with which they can be interfaced with MS. Carrier gas flow rates of capillary columns are sufficiently low (0.5–2.0 ml · min^{-1})

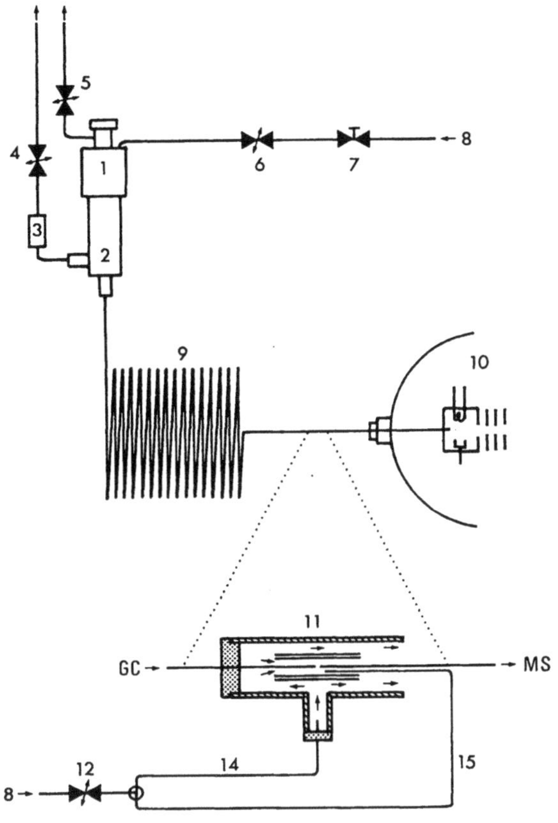

Figure 7. Schematic diagram of the direct and 'open-split' capillary GC-MS interfaces: (1) Injector block; (2) injection split zone; (3) filter; (4) split metering value; (5) septum flush metering value; (6) pressure controller; (7) shut-off valve; (8) helium carrier gas supply; (9) capillary column; (10) MS ion source; (11) open-split interface; (12) flow controller; (13) two way solenoid valve; (14) flushing gas line and (15) solvent diverter gas line.

that there is no requirement for an enrichment interface with its attendant disadvantages, in some cases, of blockage, thermal degradation and sample loss. The carrier gas can be handled directly by the MS pumping system although with some columns the gas flow may be greater and pumping must be adequate to maintain a pressure of $< 10^{-5} - 10^{-6}$ kPa in the source to avoid sensitivity loss.

In most instances the fused silica capillary column is passed through a simple seal on the ion source housing and terminated within the ion source close to the ionization region (Fig. 7). Thus eluting compounds pass without loss directly into the source and mixing in dead volumes or contact with active surfaces is avoided. The section of transferring capillary must be heated to eliminate adsorption. GC chromatographic resolution is slightly degraded because of the reduced pressure in the column but this is acceptable when set against the simplicity of the connection.

Earlier methods of column coupling, still used when high column flow rates are necessary, involved capillary restriction at the entrance to the transfer line into the source and some form of coupling device with the end of the column (Fig. 7). The interface shown is known as the open-split coupling and can cope with a range of carrier gas flow rates. When the column flow exceeds the sampling rate of the capillary restriction, excess flow vents to the atmosphere but when less than the transfer rate, it is supplemented by helium. The sample solvent front can be diverted from passing to the MS by applying a secondary flow of helium to the interface. Although a flexible and convenient interface for capillary columns (also packed columns with some modification), the open-split has the disadvantage of limiting sample transfer to the ion source by the flow through the capillary restriction. This means that in cases of higher flow rate only a proportion of eluting components are transferred to the mass spectrometer. Nevertheless, the pressure at the end of the GC column in this interface is at atmospheric pressure and maximum chromatogoraphic integrity is maintained.

2.10.3 Injectors

Table 2 lists the major types of injection system for glass and fused silica capillary columns. Each injection system has advantages and selection of the most appropriate injector will depend on the application. The most commonly employed procedure is liquid injection in the split mode due to ease of use and, as it restricts the sample size entering the column, column life is prolonged and reproducible retention times can be sustained. Retention times depend on the linear carrier gas velocity determined by the pressure drop through the column, column temperature, liquid phase thickness and column length, and all these parameters should be standardised for individual analysis. Some care is necessary however when the components of the sample have a range of boiling points, as the split injector may discriminate against either high or low boiling constituents if the split ratio (controlled mostly by the flow rate through the splitter) and the temperature are not optimised. Used in combination with MS a possible limitation of the injector is the large proportion passed to waste and hence not available for detection. Sufficient MS sensitivity may not be available for samples with low concentration components.

Much depends on the compound type requiring analysis and, in general, any of the injection systems are suitable assuming there is adequate sample and the injection conditions are compatible with volatilisation. On-column injection should be considered for biological samples which are prone to thermal degradation. The lower temperatures possible in the split injector may also have value for such samples. The solid sampling injector of the all-glass falling needle type (55) is particularly useful for high boiling point compounds and the removal of solvent before injection results in almost complete elimination of the solvent front. Volatile solvent in a sample is vented prior to sample volatilisation and hence the total sample can be loaded on to the needle which may be essential to detect minor or low concentration components. An important advantage of the dropping needle injector is the retention of non-volatile impurities on the needle after volatilisation of the sample which prevents contamination and increases the lifetime of the column.

2.10.4 Derivatives

The formation of volatile derivatives is crucial to the analysis of the large majority of biomedically important compounds. It is necessary to increase their volatility to permit passage and separation on the GC-column and in many cases to give appropriate

Table 2. GC Injection Systems

Injection method	Injection temperature	Initial Column temperature	Procedure	Comments and problems	Application
1. Cold trapping	Dependent on solutes (e. g. 150 – 300 °C)	ambient or at least 50 – 100 °C below boiling points of solutes	Sample vaporised in heated zone. Condensed in liquid phase at top of column. Chromatography begins with column-oven heating.	Almost complete sample enters column. Contaminants from carrier gas, pipework and septum bleed can also concentrate during trapping period and chromatograph with sample. Septum purge necessary.	Particularly useful for high boiling compounds and to cope with large injection volumes.
2. Splitless injection	Dependent on solutes (e. g. 150 – 250 °C)	≈20 °C below the boiling point of solvent	Sample vaporised and condensed in "solvent plug" at top of column. After delay (≃ 60 s) purge gas flow is applied to sweep out vaporisation zone and to provide carrier gas uncontaminated by residual sample components.	Sample degradation and septum bleed minimised by reduced injection temperature. Liquid phase may dissolve in solvent resulting in column deterioration – avoid methanol and acetone with non-chemically bonded silicon phases.	Important for trace analysis.
3. Split injection	Above temp. of highest boiling component	Column operating temperature or higher.	Sample is vaporised, mixed, expanded and split (typically 1:100) between column and atmosphere. Rapid flow through injector clears out residual sample quickly and minimises band broadening.	Decrease in sensitivity. Chance of non-linearity. Splitting behaviour affected by nature of solutes and their boiling points, size of injection, injection temperature, carrier gas and split ratio.	Best for major component analysis i. e. between 0.01 – 10% of injected sample. Unsuitable for minor components.
4. On-column	No heating	Ambient	Sample injected directly on to top of column and dissolved in liquid phase. Chromatography begins with column-oven heating.	No non-linearity. Sample volatilises at lowest temperature possible. Greater column contamination with this method. Difficult for use with narrow bore columns.	Best for avoiding discrimination also for samples containing unstable compounds.
5. Solid injection falling needle	Dependent on solutes (e. g. 150 – 300 °C)	Column operating temperature	Sample placed in solution on tip of needle and dried in carrier gas stream. Needle lowered into heated vaporisation region and sample swept on to column by carrier gas.	Removes solvent front from chromatogram. Involatile materials left behind on needle. Sample must be chemically stable when dried on needle. Solvent must be very volatile.	Useful for concentration of components in dilute samples or for injection of total sample volume.

ionization and fragmentation for MS analysis. These two aspects, volatility and MS properties, have to be considered together and details for each compound class are included where appropriate in the following chapters. Derivative formation is discussed in a comprehensive book by Blau and King (56).

2.10.5 **GC-MS operation**

Use of the capillary column has placed greater demands on the scan speed and sensitivity of the MS due to the short duration and rapidly changing concentration of a sample when delivered from a capillary column. Many peaks are only a few seconds wide and complete mass scans need to be made several times a second to maintain the integrity of the data. The increased scan speed reduces the time that individual mass values are sampled during a scan with the statistical chance of detection and recognition of an ion peak finally reaching a limit. The small elution volume of a capillary peak, however, results in a higher maximum concentration of the sample peak in the source and has a favourable effect on the sensitivity of detection of a given amount of compound. Nevertheless, the low loading capacity of capillary columns places an increased general demand on the instruments' sensitivity.

2.11 Liquid chromatographic (LC) inlet

The advantages of coupling LC to MS are of potentially greater importance than of GC to MS for application in the biochemical field. This stems from the rapid development in recent years of column packings and instrumentation which have given the technique an enormous utility for separating compounds of a wide range of size, polarity and stability without resort to chemical pretreatment (for a bibliography of LC-MS see 57 and reviews 58, 59 and 60). While other LC detectors have considerable scope, a further dimension is added by the specific detection and identification of eluting compounds by MS. The combination of LC-MS has proved a more difficult problem than GC-MS and, although commercial systems are available, several limitations still remain. This situation is rapidly improving but, in many cases, workers use LC in an off-line mode collecting fractions and submitting these to direct introduction for EI, CI, FD or FAB analysis or even to GC-MS following derivatisation.

A very large effort has been made since the first reports of LC-MS (e. g. 61–64) and two on-line methods of sample transfer are used.

2.11.1 **Mechanical transfer**

Following the earlier description of a moving wire for transporting solute from the column into the ion source (63), a belt system provided a much higher transfer rate (65). A more fully developed system is still used at present, with the eluant from the column being deposited on the belt and solvent removed as it passes through evacuated chambers on its way into the ion source. Flash evaporation of the sample is effected at this point and either EI or CI used for ionization. The sample needs to be sufficiently volatile and thermally stable to be introduced in this way, although ^{252}Cf desorption ionization (66), laser desorption (67) or a system in which FAB may be applied following application of a matrix to the sample on the belt have been demonstrated. FAB has proved successful in some cases without the use of a liquid matrix.

A conventional LC column operates at a flow rate of approximately 1 $ml \cdot min^{-1}$ and application of the total effluent to the belt is important to maintain maximum sensitivity. This can be achieved by spray deposition with suitable solvent systems. Total effluent from microbore columns (30–150 $\mu l \cdot min^{-1}$ flow) and microcapillary columns (5–10 $\mu l \cdot min^{-1}$) can be accommodated readily by the belt system. The peak volumes of microcapillary columns are in the range of 10–30 μl and hence provide a more concentrated solution and improve detection limits over conventional columns (68). The reduced-bore columns are restricted however, by lower sample loading and may not always have an advantage over conventional columns.

A mechanical transfer device suffers from a number of problems of varying seriousness. These include uneven spreading of the sample (particularly with aqueous solvent systems) giving ion beam fluctuations, inability to cope with non-volatile buffers, and difficulty in continuous laundry of the belt to eliminate memory effects and avoid a build up of involatile components. Hence the mechanical transfer system with thermal desorption has been effective with volatile solvents using EI and CI sources and with a wide range of low volatility thermally labile compounds (69). One solution to the problem of inducing rapid vaporisation of polar solvents from the belt has been spray deposition (70, 71). By this means almost complete evaporation of solvent can be achieved as the sample is being deposited on the surface. A thermospray vaporiser has proved successful in high sample transfer efficiency with complete solvent removal when the boiling points of solute and solvent differ by more than 50 °C (67). This permitted the use of aqueous buffers and almost any LC mobile phase. Desorption ionization spectra (LD-MS and FAB-MS) obtained using this deposition method give molecular weight information although further development is required to establish good sensitivity in all cases in relation to the matrix and other parameters (42).

2.11.2 **Direct liquid introduction (DLI)**

As the name suggests, DLI involves passing the LC effluent directly into the mass spectrometer via a capillary and some form of heated desolvation region before solvent and solute molecules enter the ion source (e.g. 62, 72, 73). In the source chemical ionization takes place in a plasma of solvent derived ions. There are limitations on the solvents and buffers which can be used, although aqueous solvents are suitable. In comparison to mechanical transfer, more labile molecules can be analysed successfully but restricted to more volatile compounds and to lower LC flow rates.

Micro-LC columns again give an increased detection sensitivity when the total effluent is transferred to the ion source (74) and several suggestions for sample enrichment prior to introduction have been made (e.g. 75, 76).

LC-MS is still a developing area which warrants the attention being given to it (60). The use of thermospray ionization of LC effluents is an exciting development (Section 2.9), but it seems likely that a combination of the present LC-MS interfacing techniques will be used for some time to come to derive the maximum analytical information.

2.12 Supercritical fluid chromatographic (SFC) inlet

As an inlet system for mass spectrometry, SFC is still at an early stage in development and its potential advantages for some applications over GC and LC remain to be exploited. Reviews of SFC (77) and SFC-MS (78) are available and only brief comment will be made here as use for compounds of clinical biochemistry interest have still to be demonstrated. SFC-MS should complement GC-MS and LC-MS, in particular by handling organic compounds with inadequate thermal stability for GC, while giving greater chromatographic efficiency and speed of analysis than LC. Supercritical fluids, consisting of highly compressed gas near or above the critical temperature and pressure points, have been used as chromatographic mobile phases since the early 1960's but experimental problems, availability of commercial instrumentation and competition from LC have inhibited their widespread introduction. They have densities similar to liquids and viscosities close to gases but diffusivities between these of liquids and gases and these features place SFC in an intermediate position between GC and LC.

Both packed and capillary columns are used in SFC (77) with each having their own predictable advantages based on flow rate, chromatographic resolution and sample capacity, availability of phases and other factors. With appropriate modification the direct liquid introduction, the moving belt with spray deposition and thermospray are effective MS interfaces with SFC (79–81).

3 Mass Analysis

Charged particles can be manipulated in a variety of ways, but for mass spectrometers involved in biochemical applications either magnetic deflection or electrostatic field filters are most commonly used. Other mass analysers such as time-of-flight mass spectrometers and Fourier transform instruments have advantages for specialised application. In this Section tandem mass analysers, linked scan modes and collision activation will also be outlined.

3.1 Magnetic field deflection

Positive or negative ions produced in the ion source are accelerated out of this region by a potential difference of between 1000 and 10 000 volts, depending on the instrument design, and into a magnetic field region created between the poles of a variable electromagnet. The deflection which ions of different charged masses, and hence

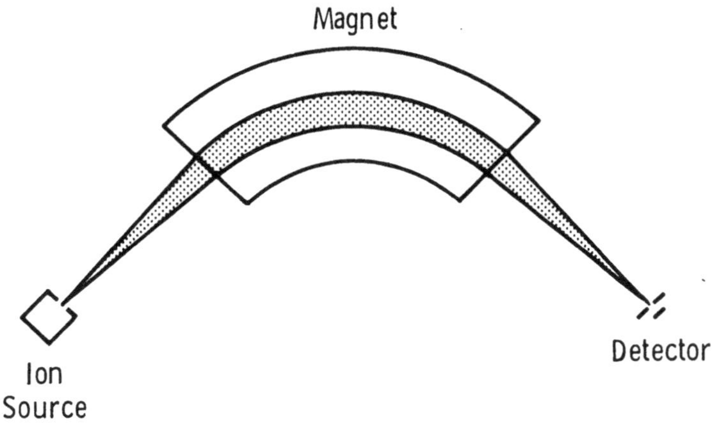

Figure 8. Schematic diagram of a 90° magnetic single sector instrument illustrating the direction focusing of the divergent ion beam.

different momenta, experience is related to the accelerating voltage (V), and the magnetic field strength (B). Ions of larger m/z are deflected to a lesser extent than those of small m/z with doubly charged ions being recorded at half their mass value. In order to focus an ion at the detector (Fig. 8) the general equation $m/z = B^2r^2/2V$, where r is the magnet radius, must be met. A spectrum of a specified mass range is recorded by keeping V constant and decreasing the magnetic field strength from a value capable of focusing the highest mass down to a level to detect the smallest mass. Upfield scanning is also used. Changing the accelerating voltage while maintaining the magnetic field constant will also produce a mass spectrum, but has limited value for acquiring complete spectra due to mass discrimination. However accelerating voltage switching is used extensively in the selected ion monitoring technique because of its faster response (Section 4.1.2).

The fast scanning speed necessary for capillary GC-MS operation has required improved magnet design and the ability to quickly scan the magnet through its field range and reset to the starting point. Scan speeds of 0.1 s · decade^{-1} (e. g., 0.1 s for mass range 50–500 daltons) are possible so that 4–5 complete scans a second can be made with a fully laminated magnet.

3.1.1 **Resolving power (RP)**

The mass resolution of a mass spectrometer is its ability to separate ions of different mass, the resolution to separate two masses, M and $M + \Delta M$, being defined by $RP = M/\Delta M$. The resolution necessary to separate the molecular ion of pyridine from the ^{13}C isotope of the molecular ion of benzene, for example, can be calculated from the mass difference (i. e. difference between 79.0422 and 79.0503) and is approximately 9750. The extent to which the ions are separated has also to be specified. The "10 per cent valley" definition is most commonly used in which ions of the same intensity are considered to be resolved when the overlap is no more than 10% of the maximum intensity of either peak.

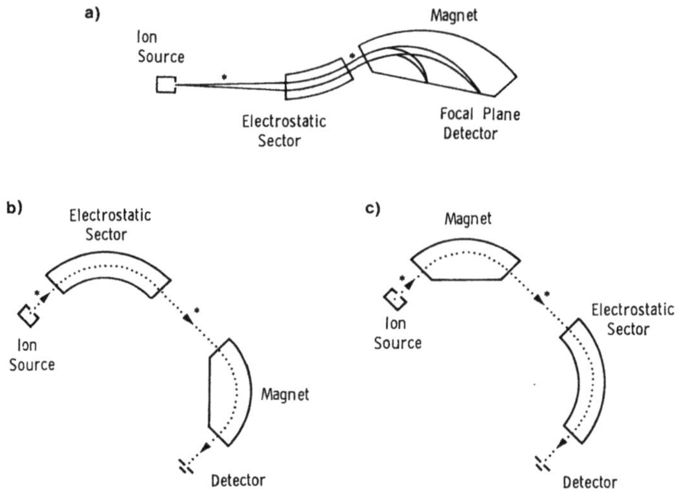

Figure 9. Schematics of double focusing magnetic sector instruments. a) Mattauch-Herzog geometry, b) Nier-Johnson geometry and c) reversed Nier-Johnson geometry. (* indicates field free regions).

Resolution in a magnetic instrument is limited by the angular dispersion of the ion beam and by the kinetic energy spread of monoisotopic ions. The magnetic sector focuses the angular beam divergence and on single sector instruments may give low *RP* (less than 1000). In order to counteract the kinetic energy spread or velocity dispersion of the beam a second sector consisting of a radial electrostatic field is used and dependent on design, double sector instruments have *RP*'s ranging up to 100 000 and beyond. Several different combinations of sectors are shown in Figure 9. The Mattauch-Herzog design focuses the ions in a plane to permit simultaneous detection and recording on a photoplate detector. While photoplate instruments are not now commonly applied to samples of biological interest, the potential increase in sensitivity from the integration of all ions reaching a focal plane detector is likely to be exploited in future. Mattauch-Herzog instruments are fitted also with 'single-point' detectors and can acquire data by field scanning. In the Nier-Johnson double focusing geometry the collector is placed at the point at which both mass and energy focusing occur. The ratio of the electrostatic sector voltage to the accelerating voltage is kept at a constant value for normal mass scanning operation. The reversed Nier-Johnson design where the magnet precedes the electrostatic sector makes mass-analysed ion kinetic energy spectrometry (MIKES) possible (82) (Section 3.5).

In the high resolution mode, the exact masses of ions are determined to within a few parts per million by recording the spectrum of a standard compound simultaneously with the sample spectrum and using the known masses of the standard for calibration. The possible elemental composition of each ion is calculated by computer programme which, by trial and error, fits all combinations of the given exact atomic masses (i.e, H = 1.00782, ^{12}C = 12.00000, ^{14}N = 14.00307, ^{16}O = 15.99491, ^{31}P = 30.97376, ^{32}S = 31.97207 etc.) to the ion's exact mass within a specified error limit. The approach is limited by the accuracy of mass determination, but for ions of less than a few hundred daltons it is a particularly powerful method to assist in the interpretion of spectra from unknown structures.

The increased mass range now possible by magnetic scanning machines has lead to a realisation that high mass ions (e.g. m/z 10000) cannot be determined accurately enough for the useful application of high resolution to the calculation of elemental composition. The large number of carbon, hydrogen and hetero-atoms in these molecules give rise to many more permutations of composition which could fit the measured mass. It may assist in excluding possible structures, but cannot be used to determine the composition of the ion. The smaller flux of ions at these high masses and the reduction of sensitivity in the high resolution mode now make low resolution acquisition and calculation of the average isotopic molecular mass of an unresolved isotope cluster a more realistic approach. Nevertheless the solution of many structural problems involving high mass compounds will require accurate mass determination and future design of instruments may be directed at this area.

3.1.2 **Metastable ions and collision induced dissociation (CID)**

A proportion of the ions that have sufficient stability to be accelerated from the ion source, subsequently decompose prior to detection and are called metastable ions. Their products appear in the spectrum usually as broad low intensity peaks of non-integral mass value. A metastable ion of mass m_1 dissociating to give mass m_2 appears in the spectrum at mass m^* where $m^* = m_2^2/m_1$. Metastable transition products in a spectrum can be used to establish product/precursor ion relationships and hence are extremely helpful in the interpretation of spectra for structure elucidation and understanding fragmentation. Ambiguity can arise when assigning the metastable ion products to particular decompositions from data obtained on a single sector machine but this difficulty does not arise when these transition ions are measured in a double sector magnetic instrument. In this case unimolecular metastable decompositions that take place in a field free region (FFR) can be recorded and uniquely assigned. This is achieved by a series of scanning methods in which the accelerating voltage (V), electrostatic analyser voltage (E) and the magnetic field (B) are controlled in concert with each other to detect metastable decompositions and provide information on parent-daughter ion relationships. The scan methods used to study metastable transitions are listed in Table 3.

The spectrum obtained when E is scanned while keeping V and B constant is known as an ion kinetic energy (IKE) spectrum. However in a reverse geometry instrument the main ion beam is first mass-analysed by the magnetic sector before the electrostatic sector and the resulting spectrum is called a mass-analysed ion kinetic energy spectrum (MIKES). The source focusing conditions remain constant throughout the scan, unlike the V-scan, but as the decompositions observed are in the 2nd FFR the smaller number of ions dissociating give lower sensitivities.

There are several problems associated with linked scanning, some of which are indicated in Table 3. A limitation of 1st FFR dissociations is the low resolution of the parent ion and linked scans can give interference peaks when ions of masses close to the selected parent ion of interest are present. Alternating current modulation of either

Table 3. Linked scan methods for double focusing mass spectrometers

Scan type*	Region of metastable transition (FFR)**	Information available	Comment
B scan	2	Normal mass spectrum containing all products ions from metastable transitions ($M_1^+ \rightarrow M_2^+$)	From single sector or EB geometry. Low intensity diffuse peaks not always uniquely assignable from $M^* = M_2^2/M_1$.
V scan	1	All M_1^+ from selected M_2^+ – translational energy released	Ion source focusing conditions change during scan, i. e. practical mass scan range limited.
E scan	2	All M_2^+ from selected M_1^+ – translational energy released	For *BE* geometry only, MIKES scan. Daughter ion resolution <150, sensitivity lower than *V* scan.
B/*E* linked scan	1	All M_2^+ from selected M_1^+	Source extraction efficiency constant and mass range limited.
$\sqrt{V}/E$ linked scan	1	All M_2^+ from selected M_1^+	Variation of source tuning during scan. Limited mass range.
B^2/E linked scan	1	All M_2^+ from selected M_1^+ – translational energy released	Fewer limiations than *V* scan, but poor resolution
$(B/E)\sqrt{1-E}$ scan	1	Loss of selected neutral from precursor ion	Constant neutral loss scan – useful for structural interpretation of unknown.

* For either *EB* or *BE* geometries unless otherwise stated.
** First FFR is between the ion source and first analyser sector and the second FFR is the region between the analyser sectors.

V or *E* with appropriate phase-shift detection methods can identify there artifacts and minimise their interference.

The introduction of a gas collison cell, containing a neutral gas, in one of the field free regions is used to enhance the number and intensity of ions produced by dissociation. This collision induced dissociation (CID), or alternatively collisionally-activated dissociation (CAD), is brought about by conversion of a proportion of the ion's kinetic energy to internal energy. The ion energy, the dimensions of the collision cell, the type of neutral gas and its pressure all influence this fragmentation and produce ions which are then analysed. The resulting spectra resemble those from EI and are particularly useful for obtaining structural information from compounds initially ionized by a soft ionization process (CI, FD, FAB) which give little fragmentation. The technique is rapidly gaining importance in biomedical applications particularly in conjunction with the development of multiple sector instruments (see MS/MS in Section 3.5) as it provides the potential to analyse individual components of a mixture present simultaneously in the ion source (i. e. without prior chromatographic separation). CID is employed in studies described in Chapter 7 (Section 6.4) and is discussed further there.

An example of linked scanning and CID is illustratzed in Figure 10 (83) which shows the daughter ion spectrum of the molecular ion of hexanoylcarnitine isobutyl ester (*m*/*z* 316). The spectrum was produced by maintaining *V* constant and scanning a fixed ratio of *B*/*E* selected to only transmit ions arising from *m*/*z* 316. A forward geometry (EB) mass spectrometer was used with a collision cell in the 1st FFR. The spectrum has several ions which are of value to interpretation of structure although in this case the conventional FAB mass spectrum contains useful fragment ions also. Nevertheless the *B*/*E* scan provides an individual and characteristic spectrum of the compound even if present in an unseparated mixture. The bottom spectrum in Figure 10 is a constant neutral loss spectrum from the same compound, in which ions arising from any other ion by loss of 115 daltons are recorded. The two major ions are seen at *m*/*z* 201 and 85 and come from the molecular ion and *m*/*z* 200 respectively by elimination of $CH_2CO_2C_4H_9$. The much lower level of chemical background ions compared to a normal FAB spectrum should be noted. A potentially useful application of the constant neutral loss scan to the detection of individual acylcarnitines present in mixtures has been demonstrated (83). FAB ionization of a sample containing a crude extract of a number of acylcarnitines (derivatised) followed by a 115 dalton constant neutral loss scan provides a clean spectrum containing only the [M−115] ions of each acylcarnitine present together with the common *m*/*z* 85 ion.

3.1.3 **Mass range**

Many biologically important molecules can be accommodated by instruments with mass ranges limited to less than 1000 daltons as will be seen in subsequent chapters. The development of FD ionization however, and more recently FAB ionization for the analysis of high molecular weight molecules, have increased interest in instruments of greater mass range.

Reducing the accelerating voltage will increase the mass range of an instrument, but often with an unacceptable decrease in sensitivity. The alternative of increasing

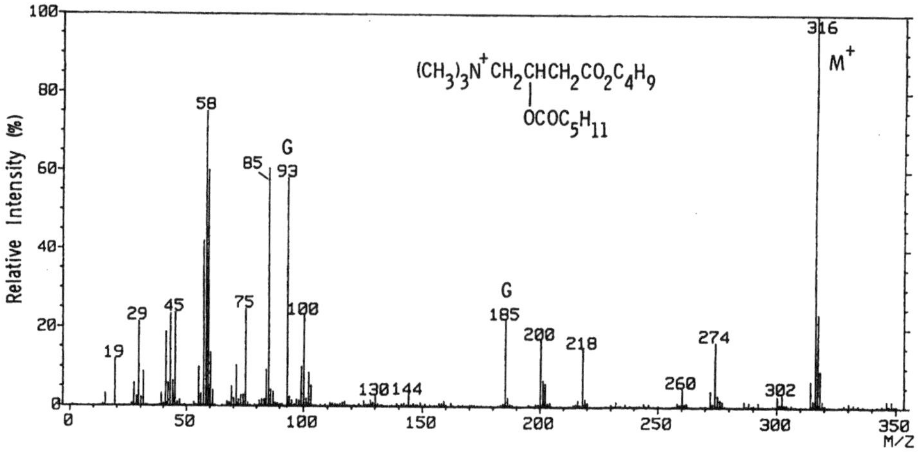

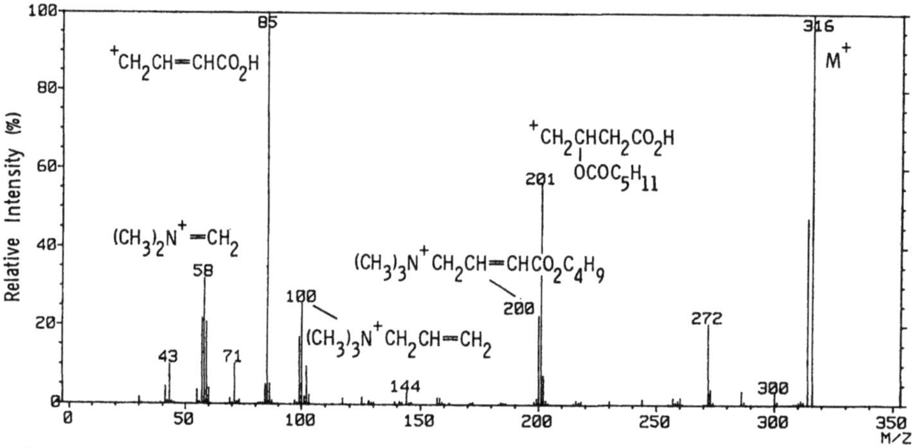

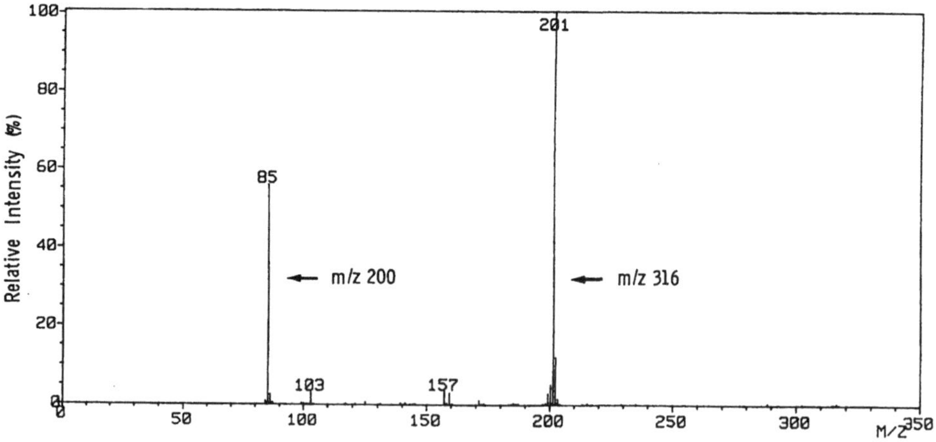

Figure 10. Mass spectra obtained from hexanoylcarnitine isobutyl ester. Conventional positive ion FAB spectrum (top); *B*/*E* linked scan of the daughter ions arising from 1st FFR collision of the molecular ion *m*/*z* 316 (middle) and the constant neutral loss (−115 daltons) spectrum (bottom).

field strength without saturation has been developed by improvements in magnet technology which include full lamination of the magnet and modified pole faces. Mass ranges of 15,000 daltons and beyond at full accelerating voltage are now possible and present limitations in recording spectra of high mass compounds are imposed by the small number of ions produced from such compounds by present ionization methods.

3.2 Electrostatic mass filters

The most usual form of this system is the quadrupole mass filter in which alternating electric fields are created in a square array of four electrodes by combinations of radio frequency (V) and dc (U) voltage applied between adjacent pairs of electrodes. Ions from the ion source are introduced into the filter along the axis of the rods by a small voltage and they undergo an oscillating trajectory of increasing amplitude and are finally collected on the electrodes. However, for selected rod voltages a particular mass is transmitted through the filter. The ratio of U to V is held constant, but increased in amplitude to allow increasingly higher mass ions to survive passage through the filter.

Their lower cost than magnetic machines has encouraged use in routine applications and further development to three dimensional systems (Ion Trap) has the potential to provide simplified mass spectrometers at moderate cost. The success of quadrupole instruments also stems from the ease with which the ion masses can be manipulated by computer control of the rf and dc power supplies. As a result they can be scanned very rapidly and are particularly suited to selected ion monitoring application (Section 4).

The low accelerating voltage necessary in the quadrupole filter (10 – 20 V) has disadvantages for the detection of higher mass ions due to lower translational energy and techniques such as FD and FAB are more suited to magnetic instruments. The low voltage source is compatible with thermospray ionization.

The mass range of a quadurpole filter is limited by the magnitude of rf voltages which can be applied to the rods but the most advanced systems can reach 2000 – 3000 daltons. While unit resolution is achieved in this mass range, the high resolution capability of magnetic instruments is not available.

3.3 Time-of-flight (TOF)

Time-of-flight analysis of ion masses is based on the different flight times of ions of the same kinetic energy, but different velocities, to pass down a drift tube. The ions are pulsed from the ion source through a control grid and accelerated by an electric field to give each a velocity dependent on its mass to charge ratio. During passage down the field-free drift tube, the ions separate and are recorded at the detector before the next pulse of ions is released from the source. The time-of-flight of each ion mass is used to establish the mass scale ($t \propto \sqrt{2m/z}$). Unlike the magnetic and quadrupole analysers, all ions from a sample are collected in a very short time (less than 0.1 ms) which makes TOF particularly suited to studying short lived species and to GC-MS and selected ion monitoring operations. However, it has not been widely adopted by laboratories interested in biological molecules due principally to lack of mass resolution which at normal accelerating voltage, is limited by the length of the flight tube.

As indicated in Section 2.8, the TOF principle has been revived for use with plasma desorption in which each fission in the ^{252}Cf source is used for ionization of the sample and to trigger the TOF detector system to record the spectrum. Flight tubes of various lengths are used and resolving powers of 1000–2000 are possible. LD is also used in conjunction with TOF analysis (84). As has been pointed out (44), this resolution is adequate to record the envelope of isotopic ions from large biomolecules and permit measurement of their average isotopic mass.

3.4 Fourier transform analysis

Fourier transform mass spectrometry (FT-MS) is not widely available and will only be mentioned briefly here (see 85 for review). Ions are formed and analysed in a small cell positioned in a strong constant magnetic field (several Tesla). They are formed in pulse mode by a normal electron filament and move in circular orbits within the cell with a frequency related to the field strength and the mass to charge ratio. An excitation rf signal is applied to the cell which, by promotion of ions to orbits of larger radius and their oscillation between two receiver plates in the cell, produce image currents which have frequency components characteristic of the ions present. Fourier transformation is used to establish the mass spectrum from the image currents.

The most important advantage of FT-MS is its high detection efficiency as no ions are 'lost' between ionization and detection. The sensitivity can be improved by increased measurement time as can resolution ($RP = 1500000$). These features, in combination with multiple sector instruments have exciting implications for the application of FT-MS (86).

3.5 Multiple sector analysers

In recent years a variety of multiple sector instruments have been devised to permit analysis of the product ions generated by dissociation of primary ions (87). Mass spectrometry/mass spectrometry (MS/MS) is the generalised name given to the overall technique in which several mass analysers are employed in sequence. The different combinations of magnet (B), electrostatic sector (E) and quadrupole filter (Q) are specified by the order of the ion optical arrangement. For example, the conventional Nier-Johnson is denoted by EB and the reversed Nier-Johnson geometry BE. New combinations include triple sector (e.g. EBE, BEB, QQQ, EBQ) and four sector instruments (e.g. BEEB, BEQQ and EBQQ). The merits of individual multi-analyser instruments require a more detailed treatment than is possible here particularly in relation to parent and daughter ion resolution, collision energy and the production of diagnostically useful fragments from molecules of interest (87,88). The limitations of linked scanning can be circumvented and unit or higher resolution achieved of both parent and daughter ions depending on which analyser is used. Observing CID processes in relation to the discrimination of ions formed from biomolecules offers another dimension in specificity and will be especially important when dealing with the simultaneous presence in the ion source of several compounds.

A triple stage quadrupole was one of the first commercially available MS/MS analysers (e. g. 87) and can be operated in several ways. When the first filter Q_1 is adjusted to transmit a selected ion mass, Q_2 operated as a collision cell and Q_3 scanned conventionally, a CID spectrum of the selected ion is obtained. This provides the opportunity to obtain spectra from individual ions present in the ion source and hence to investigate mixed samples without resort to chemical pretreatment or separation. NICI in which little secondary fragmentation takes place is useful for generating suitable primary ions.

There are several other scan modes of triple sector quadrupoles (cf. magnetic sector instruments) that provide additional information: parent ions from a particular daughter ion are obtained in a 'parent scan', the 'neutral loss scan' contains the daughter ions arising from the loss of a specific neutral fragment from the parent ions and in 'selected reaction monitoring' the specific daughter ions are monitored from selected parent ions. It will be clear that this MS/MS system is extremely powerful and provides wide scope to tackle problems of identification, quantification and screening of biological compounds. The application of the technique is still in its infancy but will unquestionably expand in future. Its potential in drug and drug metabolite analysis has already been demonstrated (e. g. 90) where rapid introduction of samples, with minimal sample preparation and simultaneous vaporisation, has proved a faster means of metabolite identification than hitherto possible.

4 Ion Detection and Recording

The current carried by the total number of ions of each mass is small and varies considerably depending on their flux and on mode of operation of the instrument. The secondary electron multiplier (SEM) is the most commonly used detector in mass spectrometers to detect these currents ($10^{-10}-10^{-18}$ A) and amplify them with a gain of approximately 10^6. Photoelectron multipliers are employed in some instrument designs and the Faraday cup detector is discussed in Chapter 14. The SEM consists of a series of dynodes electrically connected through a resistive network. Secondary electrons are produced when the ion beam strikes the first dynode and these electrons are accelerated towards the second dynode where further electrons are emitted. This process continues down each stage of the multiplier and the final output current is converted into a voltage suitable for further amplification and recording. Positive or negative ions are detected by maintaining a negative potential (approximately 2 kV) on the first dynode. An improvement in sensivity can be obtained by incorporating a conversion dynode at high potential (5 – 20 kV) just prior to the SEM to produce secondary electrons from positively charged species (or positive ions from negative ions) on collision with the dynode which then pass into the SEM to produce an electron current as described. The signal from the SEM, after pre-amplification and amplification, is recorded, the overall response time having to be very fast to enable rapid mass scanning.

The sensitivity of a mass spectrometer is a function of the number of ions striking the detector per molecule introduced into the ion source and the gain of the detection system. Only one ion reaches the detector on average from every 10^6 molecules introduced into the ion source at low resolution.

Methyl stearate is used to specify sensitivity in terms of ion current for a given sample consumption rate and, for low resolution MS, the ion current in EI mode for the molecular ion (*m/z* 298) at 1000 *RP* and electron voltage 70 eV would be expected to be in the range $10^{-8}-10^{-9}$ C · μg^{-1}. The efficiency of sample utilisation in GC-MS operation is of practical importance to obtaining maximum sensitivity. As indicated earlier, the capillary column delivers sample at a higher rate per unit time than packed columns, but requires a faster scanning rate. The ultimate detection limit of an ion current depends on a reasonable statistical sample of this ion current being recorded at the detector and simple calculation will give the theoretical lower limit of sample size which can be detected by the instrument in scanning mode under specified conditions. In the case of methyl stearate a sample in the low nanogram range would permit detection of the molecular ion with an acceptable signal to noise ratio. The low detection limits possible by selected ion monitoring (Section 4.1.2), are directly attributable to the longer integration time of individual ion currents. The gain factor achieved by the detection/amplification system is about 10^6, but increasing this factor does not necessarily increase the detection sensitivity as the signal to noise ratio remains unchanged.

In cases where a computer is not available for acquisition of data, an oscillographic recorder can provide the necessary response time and dynamic range. This recorder projects light from a hydrogen lamp via mirror galvanometers receiving the signal on to UV sensitive paper. The paper is motor driven and ion signals are recorded as peaks. A series of galvanometers of different gains allows the recording of peaks of intensity ratio of about 1 to 50000. UV oscillographic recorders are useful as a back up to computer acquisition, but are impractical when large numbers of mass spectra have to be processed.

The computer data system is integral to the present generation of mass spectrometers with control function in the operation of the instrument together with its role in collecting and processing mass spectral ion and intensity information. A dedicated on-line computer is essential to convenient operation and the acquisition of data to disc or tape for processing on a remote computer is no longer used. The instrument control aspect of computer operation has become increasingly important, especially for automated operation on GC-MS and the control of individual sectors in linked scanning modes or for MS/MS.

A number of general features of computer operation will be given in the following sections, but detailed discussion of computer systems should be sought in the literature (e.g. 91).

4.1 Acquisition of MS data

4.1.1 Mass scans

In current systems the computer is used to initiate and control the scanning of the MS with feedback control in some cases to optimise ion source conditions. The signals from the SEM amplifier are passed through an analogue filter before passing to the analogue-to-digital converter (ADC). The digitisation rate is selected to obtain a sufficient number of data points across each ion signal to permit definition of the peak and the accurate determination of its position and intensity. The scan speed and mass resolution of the MS dictate the digitisation rate required and for rapid scanning instruments ($0.1\ s \cdot decade^{-1}$) a conversion rate of up to 250 kHz may be necessary. The interface between the MS and computer, in addition to housing and A/D converter, will carry out several other operations depending on design. The raw digitised data for example may be reduced to a time and intensity value representing each peak or even converted to a mass value at this state. Acquisition of data from magnetic and quadrupole machines are handled in the same way although the low voltages involved in the operation of a quadrupole make it more amenable to computer control. The filter rod voltages, for example, may be stepped under digital-to-analogue converter (DAC) control and the ion signals digitised for each step. Depending on the DAC, the latter will represent a precise fraction of each atomic mass unit of the linear mass range.

The facility with which a computer handles large volumes of data has encouraged the use of repetitive scanning (92–94). Repeatedly scanning the mass range during the time that a sample is introduced into the instrument ensures that data collection is maximised. The mode is applicable in conjunction with most sample introduction systems and ionization methods. It has been particularly important for GC-MS operation where several thousand spectra may be acquired during a run.

4.1.2 Selected ion monitoring (SIM)

In the SIM mode selected individual ion masses are monitored continuously, or if several ions are to be measured, each is detected in sequence for a fixed period of time. Selected ions can be maintained in optimum focus under computer feedback control. The sensitivity achievable in this way is greatly increased over a mass scan where each ion in the spectrum is recorded for only a short time (0.05 – 1 ms) dictated by the scan speed and resolution used.

For magnetic instruments each ion is detected in turn by selection of the appropriate accelerating voltage at constant magnetic field strength (i. e. $m_1/z = B^2r^2/2V_1$; $m_2/z = B^2r^2/2V_2$ etc.). Switching between ions can be carried out rapidly, but the mass range of switching is limited due to mass discrimination. This is not a limitation in the field switching mode where at constant accelerating voltage the magnetic field is changed to focus the required ions and a linear response over the mass range is achieved. Quadrupole instruments are particularly suited to SIM as the rod voltages to transmit selected ions can be changed accurately and rapidly.

The important difference in acquiring SIM data in comparison with mass scan data is the much larger dynamic range of signals experienced in SIM. Various methods

Table 4. Mass calibration standards used in mass spectrometry

Ionization Mode	Reference Compound	Structure	Major ions* (*m/z*)	Comment
EI	Heptacosafluorotri-butylamine	$(C_4F_9)_3N$	502, 414, 264, 219, 131, 100, 69.	Limited to low mass range.
	Perfluorokerosine	$CF_3(CF_2)_nCF_3$	893, 881, 867, 855, 843, 831, 817, 805, 793, 781, ..., 181, 169, 131, 119, 69,	See Figure 11. Range 0–900 daltons.
	Fomblin®**-diffusion pump oil	$CF_3O[CF(CF_3)CF_2O]_m(CF_2O)_nCF_3$	1795, 1729, 1679, 1629, 1563, 1513, 1463, 1397, 1347, 1297, 1231, 1181, 1131, 1065, 1015, 999, 965, 949, 899, 849, 833, 799, 783, 733, 667, 633, 567, 501, 401, 335, 285, 235, 169, 150, 147, 119, 97, 69.	May contaminate the ion source. Useful for high masses (95) up to ≃ 5000 daltons (96)
	Ultramark® 1621***	Mixtures of perfluoroalkoxycyclo-triphosphazenes containing different fluoroalkoxy side chains.	3142, 3042, 2942, 2842, 2835, 2735, 2635, 2535, 2435, 2335, 2328, 2235, 2228, 2128, 2028, 1928, 1921, 1828, 1821, 1797, 1728, 1721, 1670, 1628, 1621, 1570, 1521, 1470, 1421, 1370, 1321, 1270, 1221, 1170, 1121, 1070, 1066, 1021, 990, 970.	Range 0–3000 daltons although larger gaps between ions at higher masses (96). Useful standard for FD (97) and FAB
	Tris(perfluoroalkyl)-S-triazine	Individual or mixed triazines with CF_3, C_2F_5, C_3F_7, C_7F_{15} or C_9F_{19} side chains	1485, 1185, 1166, 1066, 966, 866, 771, 616, 571, 566, 376, 281, 269, 281, 181, 169, 131, 119, 100, 76, 69.	Spectrum dependent on composition of mixture. See ref. 98 for exact masses of perfluorotri-heptyltriazine.
CI(+ve)	Perfluorokerosine	As above	As for EI but increased intensity of ions 643–681.	Reduced pressure of reagent gas (99). Use EI calibration for nominal mass.

NICI	Perfluorokerosine	As above	Similar to EI spectrum – some additional ions e. g. 674, 624, 562, 512, 474.	Range 0–800 daltons. Used with methane; few ions below *m/z* 231.
	Fomblin®	As above	965, 915, 849, 799, 749, 683, 633, 583, 567, 533, 517, 467, 417, 401, 367, 351, 319, 301, 251, 235, 185, 163, 147, 135, 128, 85.	Range 0–1000 daltons. Moderator gas N_2. See ref. 100.
FAB (+ve)	Glycerol	$CH_2OH.CHOH.CH_2OH$	... 737, 645, 553, 461, 369, 277, 185, 93.	Range 0–1000 daltons. Additional cluster peaks if sodium present.
	Polyethylene glycol	$H(OCH_2.CH_2)_nOH$	1031, 987, 943, 899, 855, 811, 767, 723, 679, 635, 591, 547, 503, 459, 415, 371, 327, 283, 239, 195, 177, 151, 133, 107, 89, 71.	Mixture of PEG 200:PEG 400: PEG 600 of 1:2:4 gives good distribution of ions 50–1000 daltons
	Caesium iodide	CsI	$[Cs(CsI)_n]^+$clusters	Range 20000 daltons and beyond. Closer spacing of reference peaks by addition of rubidium iodide (101) or glycerol (102).
FAB (–ve)	Glycerol	As above	... 735, 643, 551, 459, 367, 275, 183, 91.	Range 0–1000 daltons.
	Polyethylene glycol	As above	1029, 985, 941, 897, 853, 809, 765, 721, 677, 633, 589, 545, 501, 457, 413, 369, 325, 281, 237, 193, 175, 149, 131, 105, 87, 69.	Mixture of PEG 200:PEG 400: PEG 600 of 1:2:4 gives good distribution of ions 50–1000 daltons.
	Caesium iodide	As above	$[(CsI)_nI]^-$clusters	High mass range standard. Rubidium iodide or glycerol (103) increases number of ions.
	Ultramark® 1621	As above	1606, 1506, 1406, 1306, 1206, 1106, 1006, 906, 706, 692, 592, ...	Also for +ve FAB.

* Nominal mass values. For tables of accurate masses of some standards see e. g., 7, 97, 98, 102, 103

** Registered trade name of Montedison Co.

*** Registered trade name of PCR Research Chemicals Inc.

have been developed to cope with this problem. In some cases SIM operation is controlled by a hardware device in which the parameters for ion selection are preset and data acquired by sample and hold circuitry, the amplifier gain and monitoring time for each ion channel being selected to give sufficient dynamic range. With a computer system the limited range of the ADC is accommodated by reducing the dynamic range of the signal. The ion signal is integrated for a variable period dependent on ion intensity and then sampled by the ADC, the integrator being resent to zero prior to the next sampling period. Alternatively a very fast sampling of the signal directly at relatively high band width permits each ion pulse to be sampled several times. A dynamic range of $10^{-7}:1$ is possibly by this approach.

4.2 Processing of MS data

The primary stage in processing acquired scan data is to determine a mass calibration. This may be done in several ways, but in principle the computer develops a list of reference masses obtained by scanning a calibration compound, and uses this to assign mass values to sample ions. A time scale is usually employed or, as indicated above for the quadrupole instrument, the reference masses are related to a voltage scale derived from the quadrupole rod voltage. Similarly a Hall probe placed in the field of a magnetic machine can provide a voltage for correlation with reference peaks. If only nominal or integer mass assignment is needed the calibrant compound is recorded separately from the sample. The calibration reference standard can be any compound that gives a sufficient number of ions of high relative intensity regularly spaced across the mass range of interest. In practice a volatile liquid is usually selected that is chemically inert and can be easily introduced and pumped away from the ion source. Reference standards are available to cover most required mass ranges and for use with different ionization methods (Table 4).

Perfluorokerosene is the most widely used mass calibration standard in EI for masses up to $\approx$800 daltons and its spectrum is shown in Figure 11. Despite only having a few intense masses heptacosafluorotributylamine [$(C_4F_9)_3N$: M_r 671] is also a frequently used calibrant, particularly with quadrupole instruments where the linear mass scale simplifies mass assignment.

The reduction in fragmentation of many compounds under CI conditions reduces the number of reference standards which can be used in CI. This may be overcome in low resolution by calibrating in EI before changing to CI. Alternatively the pressure of reactant gas may be reduced below its operational level to allow fragmentation of perfluorokerosine or Fomblin to be observed. The situation is more straightforward in NICI where conditions for good references masses from perfluorinated reference standards can be readily found.

Several compounds are suitable for mass calibration in FAB mode in addition to those in Table 4 and include Ultramark®. Mass assignment at very high mass is possible using the cluster ions of caesium iodide for reference although the average protonated molecular mass of the isotopic cluster is usually determined due to difficulty in identifying the monoisotopic $[M + H]^+$. For bovine insulin for example, the values for $[M + H]^+$ of the protonated average and monoisotopic masses are 5734.5 and 5730.6 daltons respectively (101). Calibration of high resolution data requires the

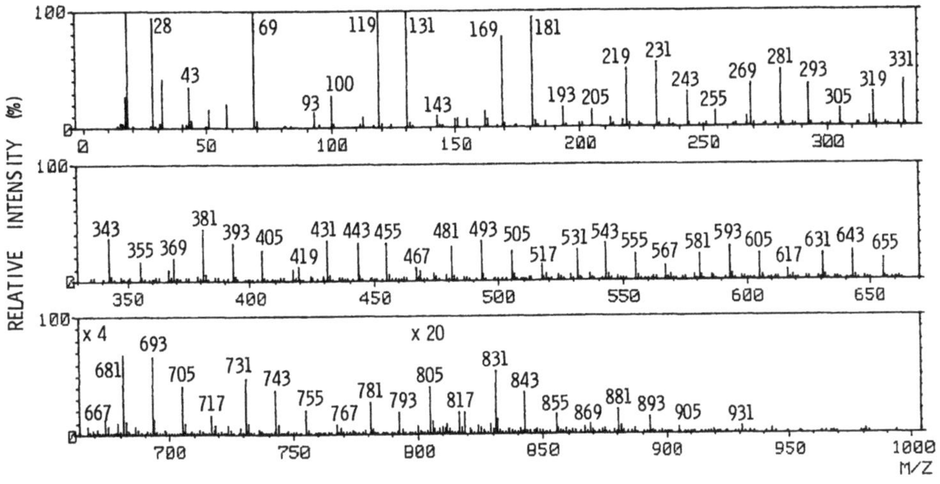

Figure 11. EI mass spectrum of the mass calibration standard perfluorokerosine.

simultaneous recording of both sample and standard to give the necessary accuracy for calculation of sample masses by interpolation between calibrant reference masses.

The reference compound must not give ions that interfere with sample ions if it is to be useful. Perfluorinated reference compounds, for example, due to the negative mass defect of fluorine (mass = 18.9984) and the resulting mass deficiency of fluorine containing ions normally allows their resolution from sample ions. In addition their simple composition of carbon and fluorine assists in the correct assignment of reference masses by minimising the presence of multiple peaks.

4.3 Reduction and presentation of MS data

The processing and manipulation of ion mass and intensity files for presentation in a form suitable for interpretation either by computer or the user is known as data reduction. A selection of routines are discussed below.

4.3.1 **Mass spectrum**

The graphical presentation of a mass spectrum as a histogram is shown in Figure 12 with the mass/charge ratio (m/z) plotted on the abscissa and the relative abundance of each ion to the most intense peak in the spectrum (base peak) recorded on the ordinate. The right hand axis records the % Σ value which indicates the ion intensity as a percentage of the total summed ion intensities. A tabulated list of ions and intensities is also available, but is less useful for a rapid visual assessment of the spectrum. The programme used to draw the spectrum in Figure 12 has automatically labelled the most intense ions within each mass range and, in addition to other features, will magnify intensities of specified mass regions.

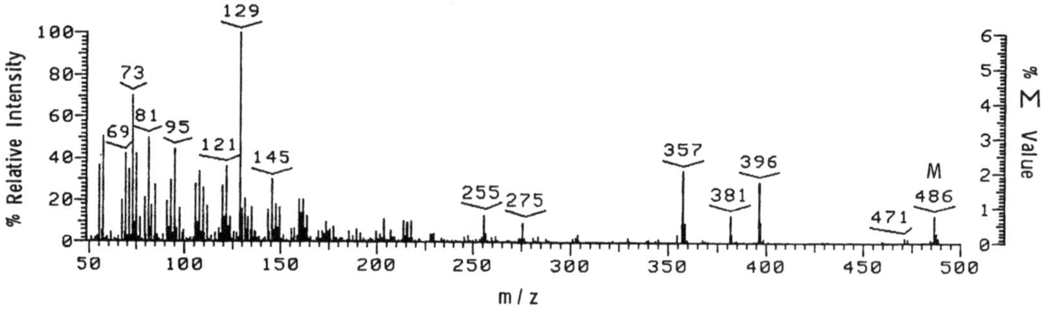

Figure 12. Graphical presentation of a mass spectrum acquired and plotted by computer. It is the spectrum of 3β-hydroxy-24β-ethyl-cholest-5-ene (β-sitosterol) as its trimethylsilyl ether derivative.

4.3.2 **Spectrum subtraction**

The presence in a spectrum of ions from another compound, from instrumental background and from GC column phase bleed, can be removed by substracting the interfering ions from the composite spectrum. The programme will take account of varying concentrations of contamination and is particularly useful for 'cleaning-up' mass spectra taken of partially resolved components from a GC column.

4.3.3 **Repetitively scanned data**

The way in which repetitive scan data are reduced depends on the objective of the analysis and two output formats are illustrated in Figures 13 and 14. Figure 13 is a total ion intensity plot (or total ion chromatogram — TIC) in which the summed ion intensity of each mass scan is plotted against scan number. Individual spectra can be selected and plotted (e. g., Figure 13, scan number 398) or compared with a reference file (see Section 4.3.5 — library search). A common approach to evaluating repetitively scanned data is to construct ion intensity chromatograms. Single ions or groups of ions can be selected and plotted for the detection of known compounds, homologous series, artefacts etc., for the comparison of selected components in multiple analyses, to search for structurally related analogues and suspected components, for the quantification of individual compounds and groups of compounds and for many other purposes. The ion profiles illustrated in Figure 14 show the specificity with which a component may be detected. Other presentation formats can be used for the normalised plots, alternative gain factors, summed plots of selected ions, summed plots of ions in selected mass ranges etc.

GC retention indices can be assigned automatically to each peak in the chromatogram following the computer recognition of co-injected hydrocarbon standards (104). Knowledge of a retention index greatly speeds up and assists in the automatic computer identification (and quantitation) of components in a GC-MS analysis of complex mixtures (e. g., 105).

Algorithms have been devised to improve the evaluation of repetitive scanned data from complex GC samples containing many partially resolved components (106). The

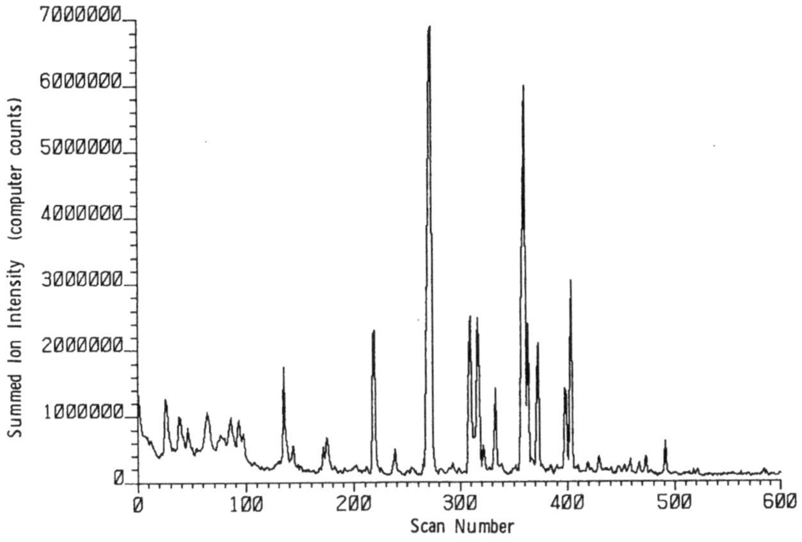

Figure 13. Summed ion intensity plot (or total ionization chromatogram – TIC) from a repetitively scanned sample containing a mixture of sterols as their trimethylsilyl ether derivatives introduced by capillary GC inlet. The spectrum in Figure 12 was acquired from this analysis – scan number 398.

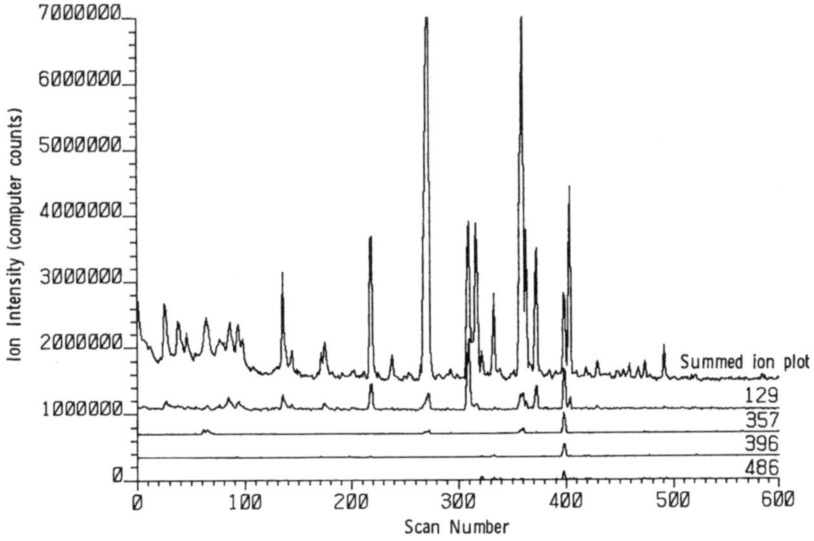

Figure 14. Selected ion plot to illustrate the detection of a compound from its characteristic ions. In this case the ions m/z 486, 396, 357 and 129 indicate the presence of β-sitosterol.

computer inspects blocks of consecutive spectra and determines the scans in which some of the ions present are reaching maximum intensity and hence indicate the centre of an eluting peak. A spectrum plot containing only the 'maximising ions' is free of interference from closely eluting components and this assists in interpretation or automatic identification by library searching. The total ion chromatogram from repetitive scan data produced in this way such that ions which are not maximising are eliminated shows a much improved apparent GC resolution. Rapid scanning can allow the differentiation of components with elution times which differ by only one or two seconds. An improved algorithm has been written to cope with the more difficult situations when components of very similar structure elute closely from one another and when background contributions from column bleed are not constant (107).

Many of the software programmes mentioned in this section are contained in a package for use in an interactive mode with the user selecting the routines appropriate to the analysis.

4.3.4 **Selected ion monitoring data**

SIM data, acquired for the sensitive quantitative detection of specific components, is often represented in graphic form as a chromatogram containing plots of each ion against time. In this respect it is very similar to the ion profiles shown in Figure 14, although the sensitivity of detection would be 10^2-10^3 times greater.

SIM is used principally for quantitation in which case the signal from each ion is measured and related to the signal from a calibration compound. While this may be done by monitoring the ions of interest directly on a pen recorder and manually evaluating the data, computer reduction of SIM information is convenient when large numbers of samples are involved. Interactive programmes permit the definition of the baseline and specification of the limits of the peak to be measured. These are then calculated for sample and calibration peaks and for the calibration standards. Depending on the particular calibration procedure adopted the computer will determine the sample concentration, more advanced programmes carrying out data reduction automatically.

4.4 Library searching

Library searching is the technique used to identify a compound by comparison of its mass spectrum with a data base of mass spectra of known compounds (108). Even when the spectrum is not in the library and a true fit is not obtained, the search algorithm will often pick out spectra of structurally similar compounds on the basis of a partial match, which may give a clue to the nature of the unknown.

The development of mini computers with extensive core memory and substantial random access disc storage, has meant that large library files can be searched by the MS user. Previously such large libraries could only be contained in a central time-shared computer. A variety of library files have been compiled (109), with the U.S. Environmental Protections Agency/National Institutes of Health, USA and the Wiley/National Bureau of Standards, USA (120,000 spectra) being among the largest and

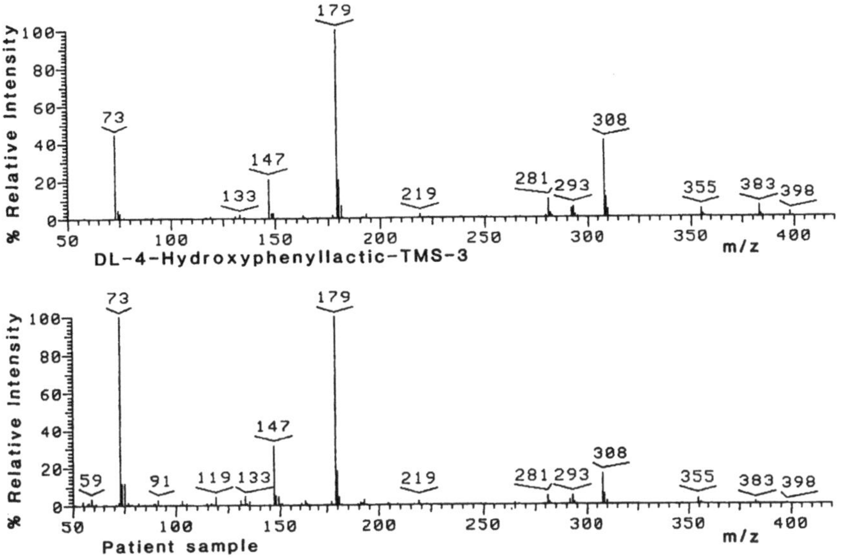

Figure 15. Comparison of the mass spectrum of a derivatised component from a patient's urine sample and the computer library spectrum of an authentic standard of 4-hydroxyphenyllactic acid as its trimethylsilyl ether-trimethylsilyl ester (M_r = 398).

being continuously expanded. The chance of a successful identification does not correlate necessarily with the size of the data base and compounds of biological interest, often in derivative form, have been under-represented in major collections in the past. It is likely to prove of more value for routine analysis of biological compounds, where many of the same compounds are found repeatedly, to prepare a library of mass spectra which are limited to particular compound classes of interest. Libraries constructed from user generated spectra avoid potential difficulties associated with the variability in comparing mass spectra recorded on different instrumentation and under different conditions. Nevertheless mass spectra obtained during GC-MS of complex biological extracts only infrequently give a perfect match with a library spectrum due to interference from co-eluting compounds and background. The computer output reproduced below from matching a patient sample spectrum, acquired from a derivatised urinary organic acid extract, against a small library ($\approx$400 spectra) correctly identified it as 4-hydroxyphenyllactic acid. However the matching "score" was reduced by the increased abundance of some ions from background e. g. *m/z* 73,147 – Fig. 15). The search also lists the next four most similar spectra in the library.

Spectrum	Score	M_r	Compound
307	0.385	398	DL-4-hydroxyphenyllactic-TMS-3
	0.237	296	mandelic-TMS-2
	0.143	439	N-acetyl-tyrosine-TMS-3 (L)
	0.130	310	3-(4-hydroxyphenyl)-propionic-TMS-2
	0.120	328	2-methoxyphenyl-acetic-TMS-1

This indicates the potential difficulty in library searching that can arise from spectra obtained from incompletely resolved components particularly when dealing with spectra which have strong similarities within a compound class.

Care has to be taken when compiling a library in order to eliminate as many errors in the data as possible and to avoid distorted (biased) spectra and those containing interfering ions from other sources. Some of the problems of biased and contaminated spectra can be overcome by the process of reverse searching (110). In this approach the normalised unknown spectrum is not compared with each normalised spectrum in the library (forward searching), but instead the ions for each library spectrum in turn are searched for in the unknown spectrum. In reverse searching the unknown spectrum is normalised by setting the ion corresponding to the base peak in the library spectrum equal to that peak and adjusting the remaining ions in relation.

Mass spectra are represented usually in encoded or abbreviated forms in libraries to reduce the storage capacity required and to speed up searching. Selection of a fixed number of peaks in a given masss range (usually 14 mass units), is a common method of encoding a spectrum, although this may retain many ions which are not useful for discrimination. Another method is to encode a specified number of the most intense peaks in the spectrum (111), but the similarity of spectra of similar compounds and the relatively common occurrence of many ions, can reduce the reliability of some matches. The most useful method of reducing the number of ions to be stored would be one in which the ions of greatest characteristic significance were kept. One suggestion to obviate the retention of homologous ions which can be found in spectra of similar compounds is to only encode ions whose intensity exceeds the arithmetic mean of its homologous neighbours at ± 14 mass units (112).

The search time required for a library search procedure is dependent not only on the size of the library, but on the way in which the spectra are stored and organised in the library and on the matching algorithm.

While many library search routines compare the unknown against all spectra in the data base, search time can be improved by organising the library in such a way that only limited sections need to be searched. Molecular weight data, retention index, ion series information, a mass spectrum dictionary (113) and other filters can be used for this purpose. Fast search times are possible by the use of inverted files in which against each mass is listed the spectra containing that mass. Searching is then a matter or selecting out spectra which contain a specified number of the ions and their intensity ranges in the unknown spectrum (113—116). This system requires more effort to update and is best used in interactive mode.

The encoding of spectra using structural feature information, such as characteristic ions, ion series and neutral losses, can be helpful for detecting component parts of the structure from the spectrum of an unknown compound when it is not contained in the library (e. g., 116).

4.5 Interpretation of MS data

The interpretation of MS data has been covered in many books and articles (e. g., 1—5, 117—119) and is discussed where appropriate in relation to compound classes in subsequent chapters in this book. Interpretation of the data from a qualitative mass spectral analysis depends on the information already available and the problem to be solved. These range from the determination of a molecular weight, the definitive comparison with a known compound, confirmation of the suspected identity of a component or a complete range of components to the more difficult problem of establishing the molecular structure of a previously unencountered compound. Indeed, the latter may not be possible without the help of other types of analytical data and in any event, would require a final comparison with a mass spectrum of established authenticity.

Various MS methods can be used to assist in understanding an unknown spectrum and some of these are mentioned in previous sections (e. g., high resolution measurement of ions, linked scanning techniques, the use of derivatives or stable isotope incorporation, alternative ionization methods, MS/MS, etc.).

The rules governing the fragmentation of molecules, under particular ionization conditions, in relation to structural elements have been formulated experimentally over many years. Several attempts have been made to encompass this knowledge within computer programmes in an effort to establish the class and finally the structure of a compound from its mass spectrum (see 91 and references therein). The greatest sustained effort to address this problem was undertaken by a group working at Stanford University who added structure generators to the generalised fragmentation rules. After generation of all possible structures from a given molecular composition and prediction of their fragmentation, the resulting spectra are compared with the unknown spectrum (120–122). The number of structures which are of value to test in this way can be reduced by interaction with the user (123) and the knowledge base extended by automating the acquisition of fragmentation rules (124). The final programmes are extensive and complex and best suited to running on large capactiy computer systems.

A further approach to spectrum interpretation is the use of pattern recognition techniques (see (91)).

5 Selection of Instrumentation

Selection of a mass spectrometer from the many commercial systems available can be a difficult task and made increasing difficult as new developments have taken place. Also the specifications and facilities of individual instruments are upgraded regularly and new instruments introduced frequently. Any detailed summary of available systems is likely to be outdated too quickly to be of value.

Manufacturers of mass spectrometers have been faced always with the dilemma of deciding whether to maximise the facilities available on an instrument such that its flexibility to tackle problems is greatest, or to simplify instrumentation for more specific application and hence bring down cost. In general they have taken the former option although a current trend towards automated and simplified instruments should help to foster application in clinical biochemistry by making more practical and affordable systems available for routine use in hospital pathology and other departments.

Those embarking on the application of mass spectrometry to compounds of biochemical interest will have to be advised from other chapters in this book of the appropriate MS facilities for their individual requirements. Several aspects should be given particular attention at the outset.

1. It is essential before purchasing an instrument to have a selection of representative samples of interest analysed by the manufacturer to test the instrument's suitability.
2. While a variety of MS techniques may be beneficial the greater the complexity of instrumentation, the more highly qualified and experienced personnel have to be to maintain and run such a system.
3. If many samples of a particular type need to be analysed then a simplified dedicated system for that purpose would favour high throughput and minimise downtime.

Two simple but powerful GC-MS systems, which hopefully represent the first in a generation of machines suited to "routine" operation are the Mass Selective Detector (MSD) made by Hewlett Packard (HP5970B), and the Ion Trap Detector (ITD) produced by Finnigan-MAT (Series 800).

The MSD was conceived as a stand alone capillary or packed column GC detector requiring no cooling water, compressed air or facilities other than a mains power supply. It is a quadrupole instrument limited to EI and is controlled exclusively by a computer operated system with application programmes for tuning, data acquisition, data editing and reports. Important features of the system are its reduced demand on specialised technical skill for operation and that instrumental conditions for individual GC-MS methods, once logged, can be used repeatedly and reproducibly.

The ITD was also designed as a GC detector for use with capillary GC columns to give highly specific MS data for detection, identification and quantification. The low cost of the system is due in part to the ionization and mass analysis regions being one and the same leading to reduced component requirements. It is constructed from a ring electrode and two end cap electrodes. Ions, formed and trapped in the central cavity following electron impact, become unstable as the RF voltage on the ring electrode is increased and exit axially through the end electrodes to be detected by a conventional electron multiplier. The theory of ion traps is described elsewhere (125). The ITD can be fitted with a CI source and the system is controlled and data acquired and processed by an IBM PC supporting extensive applications software.

Perhaps the most significant development in MS instrumentation in the last few years has been the introduction of multiple sector instruments and the technique of MS/MS. While the large capital cost of the experimental machines used in pioneering this advance are beyond the reach of working laboratories the concepts of MS/MS are of great potential value to studies of biochemical systems. It remains to be seen if this will be fully realised with intermediately priced research/routine instruments such as the triple sector quadrupoles or indeed by ion trap detectors that have the capability of a 'time-based' form of MS/MS (126, 127).

Acknowledgement

The author gratefully acknowledges Dr. C. N. Cheng's help and suggestions during preparation of the manuscript and for supplying the data for inclusion in Figures 2–5 and 10.

References

1. Waller, G. R. (1972) Biochemical Applications of Mass Spectrometry. Wiley-Interscience, New York and London.
2. Waller, G. R. & Dermer, O. C. (1980) Biochemical Applications of Mass Spectrometry: First Supplementary Volume. Wiley-Interscience, New York and London.
3. McLafferty, F. W. (1980) Interpretation of Mass Spectra. University Science Books, Mill Valley California.
4. Howe, I., Williams, D. H. & Bowen, R. D. (1981) Mass Spectrometry. Principles and Applications. McGraw-Hill Inc.
5. Rose, M. E. & Johnstone, R. A. W. (1982) Mass Spectrometry for Chemists and Biochemists. Cambridge University Press.
6. Chapman, J. R. (1985) Practical Organic Mass Spectrometry. John Wiley and Sons.
7. Rose, M. E. (1986) Senior Reporter, Mass Spectrometry, Specialist Periodical Report, Vol. 8, Royal Society of Chemistry, London. (see also earlier volumes).
8. Gaskell, S. J., ed. (1986) Mass Spectrometry in Biomedical Research, J. Wiley and Sons.
9. Todd, J. F. J. (1986) Advances in Mass Spectrometry 1985, J. Wiley and Sons, Chichester, England.
10. McEwen, C. N., ed. (1987) Thirty-fifth Annual Conference on Mass Spectrometry and Allied Topics. Denver (Colorado) ASMS, East Lansing (Michigan).
11. Burlingame, A. L., Baillie, T. A. & Derrick, P. J. (1986) Mass Spectrometry. Anal. Chem. 58, 165R–211R.
12. Field, F. H. (1968) Chemical ionization mass spectrometry. Accounts. Chem. Res. 1, 42–49.
13. Harrison, A. G. (1983) Chemical Ionization Mass Spectrometry. CRC Press, Inc. Boca Raton, Florida.
14. Hunt, D. F., McEwen, C. N. & Harvey, T. M. (1975) Positive and negative chemical ionization mass spectrometry using a Townsend discharge ion source. Anal. Chem. 47, 1730–1734.
15. Hunt, D. F., Stafford, G. C. Crow, F. W. & Russell, J. W. (1976) Pulsed positive negative ion chemical ionization mass spectrometry. Anal. Chem. 48, 2098–2105.
16. Hunt, D. F. & Crow, F. W. (1978) Electon capture negative ion chemical ionization mass spectrometry. Anal. Chem. 50, 1781–1784.
17. Brandenberger, H. & Ryhage, R. (1978) New developments in the field of mass spectrometry and GC-MS and their applications to toxicological analysis. In: Recent Developments in Mass Spectrometry in Biochemistry and Medicinc (A. Frigerio, ed.) pp. 327–341, Plenum Press, New York, London.
18. Faull, K. F. & Beck, O. (1986) Quantification of neurotransmitters and related compounds by electron capture chemical ionization GC-MS. In: Mass Spectrometry in Biomedical Research (S. J. Gaskell, ed.) pp. 403–421, John Wiley and Sons Ltd.
19. Baldwin, M. A. & McLafferty, F. W. (1973) Direct chemical ionization of relatively involatile samples. Application to underivatised oligopeptides. Org. Mass Spectrom. 7, 1353–1356.
20. Cotter, R. J. (1980) Mass spectrometry of non volatile compounds. Anal. Chem. 52, 1589A–1606A.
21. Beckey, H. D. (1977) Principles of Field Ionization and Field Desorption Mass Spectrometry. Pergamon, Oxford, England.
22. Beckey, H. D. (1969) Field desorption mass spectrometry: A technique for the study of thermally unstable substances of low volatility. Int. J. Mass Spectrom. Ion Phys. 2, 500–503.
23. Holland, J. F., Soltmann, B. & Sweeley, C. C. (1976) A model for ionization mechanisms in field desporption mass spectrometry. Biomed. Mass Spectrom. 3, 340–345.
24. Beckey, H. D. & Rollgen, F. W. (1979) Field desorption 'Without Fields'? Org. Mass Spectrom. 14, 188–190.
25. Giessmann, U. & Rollgen, F. W. (1981) Electrodynamic effects in field desorption mass spectrometry. Int. J. Mass Spectrom. Ion Phys. 38, 267–297.
26. Benninghoven, A., Jaspers, D. & Sichtermann, W. K. (1976) Secondary-ion emission of amino acids. Appl. Phys. 11, 35–39.
27. Barber, M., Bordoli, R. S., Garner, G. V., Gordon, D. B., Sedgwick. R. D. & Tetler, L. W. (1981) Fast atom bombardment mass spectra of enkephalins. Biochem. J. 197, 401–404.
28. Franks, J. (1983) F.A.B.: the fast atom beam source. Int. J. Mass spectrom. Ion Phys. 46, 343–346.
29. Arbeth, W., Straub, K. & Burlingame, A. L. (1982) Secondary ion mass spectrometry with cesium ion primary ion beam and liq-

uid target matrix for analysis of bio-organic compounds. Anal. Chem. 54, 2029–2034.
30. Gower, J. L. (1985) Matrix compounds for fast atom bombardment mass spectrometry. Biomed. Mass Spectrom. 12, 191–196.
31. Falick, A. M., Walls, F. C. & Laine, R. A. (1986) Cooled sample introduction probe for liquid secondary ionization mass spectrmetry. Anal. Biochem. 159, 132–137.
32. Costello, C. E. (1986) Sample preparation for FAB and FD mass spectrometry of polar analytes. In: Mass Spectrometry in Biomedical Research (S. J. Gaskell, ed.) pp. 31–40, J. Wiley and Sons Ltd.
33. Dobberstein, P., Karte, E., Meyerhoff, G. & Pesch, R. (1983) Investigation of an LC/MS interface for EI-, CI- and FAB-ionization. Int. J. Mass Spectrom. Ion Phys. 46, 185–188.
34. Stroh, J. G., Cook, J. C., Milberg, R. M., Brayton, L., Kihara, T., Huang, Z., Rinehart, K. L. & Lewis, I. A. S. (1985) On-line liquid chromatography/fast atom bombardment mass spectrometry. Anal. Chem. 57, 985–991.
35. Ito, Y., Takeichi, T., Ishii, D. & Goto, M. (1985) Direct coupling of micro high-performance liquid chromatography with fast atom bombardment mass spectrometry. J. Chromatogr. 346, 161–166.
36. Caprioli, R. M., Fan, T. & Cottrell, J. S. (1986) Continuous-flow sample probe for fast atom bombardment mass spectrometry. Anal. Chem. 58, 2949–2954.
37. Caprioli, B. M. & Fan, T. (1986) High sensitivity mass spectrometric determination of peptides: direct analysis of aqueous solutions. Biochem. Biophys. Res. Comm. 141, 1058–1065.
38. Egestad, B., Pettersson, P., Skrede, S. & Sjövall, J. (1985) Fast atom bombardment mass spectrometry in the diagnosis of cerebrotendinous xanthomatosis. Scand. J. Lab. Invest. 45, 443–336.
39. Lawson, A. M., Madigan, M. J. & Clayton, P. T. (1986) Rapid diagnosis of Zellweger syndrome and infantile Refsum's disease by fast atom bombardment-mass spectrometry of urine bile acids. Clin. Chim. Acta 161, 221–231.
40. Hillenkamp, F. (1983) Laser Induced Ion Formation from Organic Solids. In: Proc. 2nd Internat. Confer. Ion Formation from Organic Solids (Benninghoven, A., ed.). Springer-Verlag: New York, Springer Series in Chemical Physics 25, p. 190.
41. Posthumus, M. A., Kristemaker, P. G., Meuselaar, H. L. C. & DeBrauw, M. C. T. N. (1978) Laser desorption-mass spectrometry of polar non-volatile bio-organic molecules. Anal. Chem. 50, 985–991.
42. Fan, T. P., Hardin, E. D. & Vestal, M. L. (1984) Direct comparison of secondary ion and laser desorption mass spectrometry on bio-organic molecules in a moving belt liquid chromatogoraphy/mass spectrometry system. Anal. Chem. 56, 1870–1876.
43. Macfarlane, R. D. & Torgerson, D. F. (1976) Californium-252 plasma desorption mass spectrometry. Science 191, 920–925.
44. Sundqvist, B & Macfarlane, R. D. (1985) ^{252}Cf-Plasma desorption mass spectrometry. Mass Spectrometry Reviews 4, 421–460.
45. Roepstorff, P. & Sundqvist, B. (1986) Plasma desorption mass spectrometry of high molecular weight biomolecules. In: Mass Spectrometry in Biomedical Research (S. J. Gaskell, ed.) pp. 269–285, John Wiley and Sons Ltd.
46. Blakley, C. R., Carmody, J. J. & Vestal, M. L. (1980) A new soft ionization technique for mass spectrometry of complex molecules. J. Amer. Chem. Soc. 102, 5931–5933.
47. Blakley, C. R. & Vestal, M. L. (1983) Thermospray interface for liquid chromatography/mass spectrometry. Anal. Chem. 55, 750–754.
48. Pilosof, D., Kim, H. K., Dyckes, D. F. & Vestal, M. L. (1984) Determination of nonderivatized peptides by thermospray liquid chromatography/mass spectrometry. Anal. Chem. 56, 1236–1240.
49. Vestal, M. L., (1984) Thermospray liquid chromatographic interface for magnetic mass spectrometers. Anal. Chem. 56, 2590–2592.
50. Zakett, D., Kallos, G. J. & Savickas, P. J. (1986) Modification of a Finnigan thermospray interface to allow discharge chemical ionization of the thermospray vapor. 34th Ann. Conf. Mass Spectrom. and Allied Topics, June, Cincinnati, Ohio, pp. 3–4.
51. Ryhage, R. (1964) Use of mass spectrometer as a detector and analyser for effluents emerging from high temperature gas liquid chromatography columns. Anal. Chem. 36, 759–764.
52. Watson, J. T. & Biemann, K. (1965) Direct recording of high resolution mass spectra from gas chromatographic effluents. Anal. Chem. 37, 844–851.

53. Brooks, C. J. W. & Edmonds, C. G. (1979) Combined gas chromatography – mass spectrometry. In: Practical Mass Spectrometry (B. S. Middleditch, ed.) pp. 57–126, Plenum Press, New York and London.
54. Jennings, W. (1980) Gas chromatography with glass capillary columns. Academic Press, New York.
55. Van den Berg, P. M. J. & Cox, T. P. H. (1972) All-glass solid sampling device for open tubular columns in gas chromatography. Chromatographia 5, 301–305.
56. Blau, K. & King, G., eds. (1977, 1978) Handbook of Derivatives for Chromatography. Heyden and Son Ltd., London.
57. Edmonds, C. G., McCloskey, J. A. & Edmonds, V. A. (1983) A bibliography of combined liquid chromatography mass spectrometry. Biomed. Mass Spectrom 10, 237–252.
58. Games, D. E., Eckers, C., Gower, J. L., Hirter, P., Knight, M. E., Lewis, E., Rao, K. R. N. & Weerasinghe, N. C. (1980) Combined HPLC-MS and its potential in clinical studies. In: Current Developments in the Clinical Application of HPLC. GC and MS (A. M. Lawson, C. K. Lim & W. Richmond, eds.) pp. 97–118, Academic Press, London.
59. Nibberring, N. M. M. (1982) Survey of ionization methods with emphasis on liquid chromatography-mass spectrometry. J. Chromatogr. 251, 93–104.
60. Vestal, M. L. (1984) High-performance liquid chromatography-mass spectrometry. Science 226, 275–281.
61. Tal'roze, V. L., Grishin, V. D., Skurat, V. E. & Tantsyrev, G. D. (1970) In: Recent Developments in Mass Spectrometry (K. Ogata & T. Hayakawa, eds.) pp. 1218–1226, University Park Press, Baltimore.
62. Baldwin, M. A. & McLafferty, F. W. (1973) Liquid chromatography-mass spectrometry interface – I. The direct introduction of liquid solutions into a chemical ionization mass spectrometer. Org. Mass Spectrom. 7, 1111–1112.
63. Scott, R. P. W., Scott, C. G., Munroe, M. & Hess, J. (1974) Interface for on-line liquid chromatography-mass spectrometry analysis. J. Chromatogr. 99, 395–405.
64. Horning, E. C., Carroll, D. I., Dzidic, I., Haegele, K. D., Horning, M. G. & Stillwell, R. N. (1974) Atmospheric pressure ionization (API) mass spectrometry. Solvent-mediated ionization of samples introduced in solution and in a liquid chromatograph effluent stream. J. Chromatogr. 99, 13–21.
65. McFadden, W. H., Schwartz, H. L. & Evans, S. (1976) Direct analysis of liquid chromatographic effluents. J. Chromatogr. 122, 389–396.
66. Jungclas, H., Danigel, H. & Schmidt. L. (1982) Quantitative ^{252}Cf Plasma desorption mass spectrometry for pharmaceuticals. A new approach to coupling liquid chromatography with mass spectrometry. Org. Mass Spectrom. 17, 86–90.
67. Hardin, E. D., Fan, T. P. & Vestal, M. L. (1984) Laser desorption mass spectrometry with thermospray sample deposition for determination of non-volatile biomolecules. Anal. Chem. 56, 2–7.
68. Games, D. E., Lant, M. S., Westwood, S. A., Cocksedge, H. J., Evans, N., Williamson, J. & Woodhall, B. H. (1982) Microbore high-performance liquid chromatography mass spectreometry. Biomed. Mass Spectrom. 9, 215–224.
69. Games, D. E., McDavall, M. A., Larsen, K., Shafer, K. H., Dobberstein, P. & Gower, J. L. (1984) A comparison of moving belt interfaces for liquid chromatography mass spectrometry. Biomed. Mass Spectrom. 11, 87–95.
70. Smith, R. D. & Johnson, A. L. (1981) Deposition method for moving ribbon liquid chromatograph-mass spectrometer interfaces. Anal. Chem. 53, 739–742.
71. Hayes, M. J., Lankmayer, E. P., Vouros, P., Karger, B. L. & McGuire, J. M. (1983) Moving belt interface with spray deposition for liquid chromatography/mass spectrometry. Anal. Chem. 55, 1745–1752.
72. Apffel, J. A., Brinkman, U. A. I., Frei, E. W. & Evers, E. A. I. M. (1983) Gas-nebulized direct liquid introduction interface for liquid chromatography/mass spectrometry. Anal. Chem. 55, 2280–2284.
73. Dedieu, M., Juin, C., Arpino, P. J., Bounine, J. P. & Guiochon, G. (1982) Application of a combined liquid chromatographic mass spectrometric instrument using an interface for direct liquid introduction. J. Chromatogr. 251, 203–213.
74. Henion, J. D. & Wachs, T. (1981) Micro liquid chromatography/mass spectrometry diaphram probe interface. Anal. Chem. 53, 1963–1965.
75. Takeuchi, T., Hirata, Y. & Okumura, A. (1978) On-line coupling of a micro liquid chromatograph and mass spectrometer through a jet separator. Anal. Chem. 50, 659–660.

76. Christensen, R. G., Hertz, H. S., Melselman, S. & White, V. E. (1981) Liquid chromatograph/mass spectrometer interface with continuous sample preconcentration. Anal. Chem. 53, 171–174.
77. White, C. M. & Houck, R. K. (1986) Supercritical fluid chromatography and some of its applications: a review. J. High Resolut. Chromatogr. Chromatogr. Commun. 9, 4–17.
78. Games, D. E., Berry, A. J., Mylchreest, I. C., Perkins, J. R. & Pleasance, S. (1987) Packed and capillary column supercritical fluid chromatography/mass spectrometry. Europ. Chromatogr. News 1, 10–14.
79. Smith, R. D., Udseth, H. R. & Kalinoski, H. T. (1984) Capillary supercritical fluid chromatography/mass spectreometry with electron impact ionization. Anal. Chem. 56, 2971–2973.
80. Growther, J. B. & Henion, J. E. (1985) Supercritical fluid chromatography of polar drugs using small-particle packed columns with mass spectrometric detection. Anal. Chem. 57, 2711–2716.
81. Wright, B. W., Kalinoski, H. T., Udseth, H. R. & Smith, J. (1986) Capillary supercritical fluid chromatography-mass spectrometry. J. High Resolut. Chromatogr. Chromatogr. Commun. 9, 145–153.
82. Beynon, J. H., Cooks, R. G., Amy, J. W., Baitinger, W. E. & Ridley, T. (1973) Design and performance of a mass-analysed ion kinetic energy (MIKE) spectrometer. Anal. Chem. 45, 1023A–1031A.
83. Cheng, K. N., Tracey, B. M., Chalmers, R. A. & Rosankiewicz, H. (1987) Identification of acylcarnitines by fast atom bombardment using various linked scan techniques. Proc. 17th Meeting Brit. Mass Spectrom. Soc., York, England, September 1987.
84. Tabet, J.-C. & Cotter, R. J. (1984) Laser desorption time-of-flight mass spectrometry of high mass molecules. Anal. Chem. 56, 1662–1667.
85. Wilkins, C. L. & Gross, M. L. (1981) Fourier transform mass spectrometry for analysis. Anal. Chem. 53, 1661A–1676A.
86. Russel, D. H. & Castro, M. E. (1986) High mass fourier transform mass spectrometry: present capabilities and future potential. In: Spectrometry in Biomedical Research (S. J. Gaskell, ed.) pp. 313–338, John Wiley & Sons Ltd., New York.
87. McLafferty, F. W. (1983) Tandem Mass Spectrometry. John Wiley & Sons Ltd., New York.
88. Gaskell, S. J. (1986) Overview to Part II. In: Mass Spectrometry in Biomedical Research (S. J. Gaskell, ed.) pp. 137–147, John Wiley and Sons Ltd., New York.
89. Hunt, D. F., Shabanowitz, J. & Giordani, A. B. (1980) Collision activated decompositions of negative ions in mixture analysis with a triple quadrupole mass spectrometer. Anal. Chem. 52, 386–390.
90. Johnson, J. V., Lee, M. S., Lee, M. R., Brotherton, H. O. & Yost, R. A. (1986) Triple quadrupole MS/MS in biomedical research. In: Mass Spectrometry in Biomedical Research (S. J. Gaskell, ed.) pp. 459–475, John Wiley & Sons Ltd., New York.
91. Chapman, J. R. (1978) Computers in Mass Spectrometry. Academic Press, London, New York, San Francisco.
92. Hites, R. A. & Biemann, K. (1968) Mass spectrometer-computer system particularly suited for gas chromatography of complex mixtures. Anal. Chem. 40, 1217–1221.
93. Hites, R. A. & Biemann, K. (1970) Computer evaluation of continuously scanned mass spectra of gas chromatographic effluents. Anal. Chem. 42, 855–860.
94. Reimendal, R. & Sjövall, J. B. (1973) Computer evaluation of gas chromatographic-mass spectrometric analysis of steroids from biological materials. Anal. Chem. 45, 1083–1089.
95. Ligon, W. V. (1978) Poly(perfluoropro-pylene oxide) as a reference standard for mass measurement to mass 2000. Anal. Chem. 50, 1228–1229.
96. Morris, H. R., Dell, A. & McDowell, R. A. (1981) Extended performance using high field magnet mass spectrometer. Biomed. Mass Spectrom. 8, 463–473.
97. Olson, K. L., Rinehart, K. L. & Cook, J. C. (1977) Phosphazenes: High molecular weight reference compounds for field desorption mass spectrometry. Biomed. Mass Spectrom. 4, 284–290.
98. Aczel, T. (1968) Perfluorotriheptyltriazine: reference standard for precise mass measurement up to m/e 1600. Anal. Chem. 40, 1917–1918.
99. Dzidic, I., Desiderio, D. M., Wilson, M. S., Crain, P. F. & McCloskey, J. A. (1971) Mass standards for chemical ionization mass spectrometry. Anal. Chem. 43, 1877–1879.
100. Bruins, A. P. (1980) Calibration substances for negative ion chemical ionization mass spectrometry. Biomed. Mass Spectrom. 7, 454–456.

101. Green, B. N. & Bordoli, R. S. (1986) High mass capabilities of sector mass spectrometers. In: Mass Spectrometry in Biomedical Research (S. J. Gaskell, ed.) pp. 235–250, John Wiley and Sons Ltd., New York.
102. Buko, A. M., Phillips, L. R. & Fraser, B. A. (1983) Peptide studies using a fast atom bombardment high field mass spectrometer and data system. 1. Sample introduction, data acquisition and mass calibration. Biomed. Mass Spectrom. 10, 324–333.
103. Buko, A. M., Phillips, L. R. & Fraser, B. A. (1983) Peptide studies using fast atom bombardment high field mass spectrometer and data system. 3. Negative ionization: mass calibration, data acquisition and structural characterization. Biomed. Mass Spectrom. 10, 387–393.
104. Nau , H. & Biemann, K. (1974) Computer assisted assignment of retention indices in gas chromatography-mass spectrometry and its application to mixtures of biological origin. Anal. Chem. 46, 426–434.
105. Gates, S. C., Smisko, M. J., Ashendel, C. L., Young, N. D., Holland, J. F. & Sweeley, C. C. (1978) Automated simultaneous qualitative and quantitative analysis of complex organic mixtures with a gas chromatography-mass spectrometry computer system. Anal. Cham. 50, 433–441.
106. Biller, J. E. & Biemann, K. (1974) Reconstructed mass spectra, a novel approach for the utilisation of gas chromatograph-mass spectrometer data. Anal. Lett. 7, 515–528.
107. Dromey, R. G., Stefik, M. J., Rindfleisch, T. & Duffield, A. M. (1976) Extraction of mass spectra free of background and neighbouring component contributions from gas chromatography/mass spectrometry data. Anal. Chem. 48, 1368–1375.
108. Heller, S. R., McCormick, A. & Sargent, T. (1980) Compound identification by computer matching of mass spectra. In: Biochemical Applications of Mass Spectrometry. First Supplementary Volume. (G. R. Waller & O. C. Dermer, eds.) pp. 103–123, John Wiley & Sons, New York, Chichester, Brisbane, Toronto.
109. Middleditch, B. S. (1979) Collections of mass spectral data. In: Practical Mass Spectrometry – A Contemporary Introduction (B. S. Middleditch, ed.) pp. 179–187, Plenum Press, New York, London.
110. Abramson, F. P. (1975) Automated identification of mass spectra by the reverse search. Anal. Chem. 47, 45–49.
111. Crawford, L. R. & Morrison, J. D. (1968) Computer methods in analytical mass spectrometry. Anal. Chem. 40, 1464–1469.
112. Henneberg, D. (1980) Computerization and library search systems. In: Advances in Mass spectrometry (A. Quale, ed.) pp. 1511–1531, Heyden & Son Ltd., London.
113. Dromey, R. G. (1979) Mass spectrum dictionary for library searching. Anal. Chem. 51, 229–232.
114. Heller, S. R. (1972) Conversational mass spectral retrieval system and its use as an aid in structure determination. Anal. Chem. 44, 1951–1961.
115. Heller, S. R. (1979) The mass spectra search system. In: Practical Mass Spectrometry (B. S. Middleditch, ed.) pp. 209–217, Plenum Press, New York, London.
116. Kwok, K. S., Venkataragharan, K. & McLafferty, F. W. (1973) Computer-aided interpretation of mass spectra III. A self-training interpretive and retrieval system. J. Amer. Chem. Soc. 95, 4185–4194.
117. Budzikiewicz, H., Djerassi, C. & Williams, D. H. (1964) Interpretation of Mass Spectra of Organic Compounds. Holden-Day, San Francisco.
118. Budzukiewicz, H., Djerassi, C. & Williams, D. H. (1964) Structure Elucidation of Natural Products by Mass Spectrometry, Vols. 1, 2, Holden-Day, San Francisco.
119. Budzikiewicz, H., Djerassi, C. & Williams, D. H. (1967) Mass Spectrometry of Organic Compounds. Holden-Day, San Francisco.
120. Schroll, G., Duffield, A. M., Djerassi, C., Buchanan, B. G., Sutherland, G. L., Feigenbaum, E. A. & Lederberg, J. (1969) Applications of artificial intelligence for chemical inference III. Aliphatic ethers diagnosed by their low-resolution mass spectra and nuclear magnetic resonance data. J. Amer. Chem. Soc. 91, 7440–7445.
121. Buchs, A., Duffield, A. M., Schroll, G., Djerassi, C., Delfino, A. B., Buchanan, B. G., Sutherland, G. L., Feigenbaum, E. A. & Lederberg, J. (1970) Applications of artificial intelligence for chemical inference IV. Saturated amines diagnosed by their low resolution mass spectra and nuclear magnetic resonance spectra. J. Amer. Chem. Soc. 92, 6831–6838.
122. Masinter, L. M., Sridharam, N. S., Lederberg, J. & Smith, D. H. (1974) Application of artificial intelligence for chemical inference XII. Exhaustive generation of cyclic

and acyclic isomers. J. Amer. Chem. Soc. 96, 7702–7714.

123. Carhart, R. E., Smith, D. H., Brown, H. & Djerassi, C. (1974) Application of artificial intelligence for chemical inference XVII. An approach to computer-assisted elucidation of molecular structure. J. Amer. Chem. Soc. 96, 5755–5762.

124. Smith, D. H., Buchanan, B. G., White, W. C., Feigenbaum, E. A., Lederberg, J. & Djerassi, C. (1973) Application of artificial intelligence for chemical inference X. INSTUM, a data interpretation and summary programme applied to the collected mass spectra of estrogenic steroids. Tetrahedron 29, 3117–3134.

125. Stafford, G., Kelly, P., Syka, J., Reynold, W. & Todd, J. (1984) Recent improvements in and analytical applications of advanced ion trap technology. Int. J. Mass Spectrom. Ion Phys. 60, 85–98.

126. Louris, J. N., Amy, J., Ridley, T., Kascheres, C. & Cooks, R. G. (1987) The quadrupole ion trap as a tandem mass spectrometer-instrumentation and applications. Proc. 35th ASMS Conf. Mass Spectrom. Allied Topics, Denver, pp. 769–770.

127. Johnson, J. V. & Yost, R. A. (1987) Comparison of MS/MS with triple quadrupole and ion trap tandem mass spectrometers. Proc. 35th ASMS Conf. Mass Spectrom. Allied Topics, Denver, pp. 773–774.

Chapter 2
Bile Acids

K. D. R. Setchell and A. M. Lawson

1 Introduction

Bile acid methodology has advanced considerably over the last decade and many procedures, reviewed in Volume 1 of the excellent book series devoted to "The Bile Acids" (1), have been superseded or improved.

Mass spectrometry was first applied over 20 years ago as a tool for confirming the identity of several bile acids and bile alcohols (2–4), however it was only after the technical difficulties of combining the gas chromatograph with the mass spectrometer had been overcome that the use of this technique expanded and proved an invaluable tool for the qualitative and quantitative analysis of bile acids.

In this chapter the applications of mass spectrometry to bile acid analysis will be described together with the essential pre-instrumental procedures for bile acid analysis based on current state-of-the-art methodology.

1.1 Chemistry

Bile acids or bile salts are a group of compounds belonging to the steroid class. Structurally, they all possess a cyclopentanoperhydrophenanthrene nucleus, a side chain of variable length, most commonly 5 carbon atoms, attached at the C-17 position and a terminal carboxylic acid group. The configuration of the H atom at the C-5 position influences the stereochemistry of the A/B rings and hence the physicochemical properties of the molecule. When this proton is axial (β) to the plane of the nucleus, and *cis* to the C-19 methyl group, the A/B rings are *cis* to each other, however, when the H is in the equatorial (α) position and *trans* to the C-19 methyl group the A/B ring junction has a *trans* configuration.

Quantitatively, the major proportion of bile acids found in higher vertebrates are of the 5β configuration, however, in certain biological fluids, notably feces and urine, a variety of 5α (*allo*) bile acids are found.

A great variety of bile acids occur naturally (5), differing from each other by the number and orientation of substituent groups in the nucleus. Hydroxyl and oxo functions are most commonly found at C-3, C-7 and C-12 of the nucleus. Less commonly, and generally only in pathophysiological conditions or in early life, bile acids with substituents at the C-1, C-2, C-6 and side chain can be found.

In addition to the configuration of the A/B ring junction, the number, orientation and type of substituents are also key determinants of the physicochemical properties. In man, the common bile acids have hydroxyl groups in the α configuration which confer a hydrophobic and hydrophilic side to the molecule, thus giving bile acids their physiologically important amphiphilic nature.

The side chain of the common bile acids is 5 carbon atoms in length and has the 17β configuration. Several other important groups of acidic steroids with side chains longer and shorter in length also occur (Table 1). For example, C_{27} bile acids (cholestanoic acids) are found in cholestatic conditions and in some inborn errors of metabolism (6, 7) while etianic acids (C_{20} steroids) were recently shown to occur in the meconium of newborn infants (8). Furthermore, a C_{29} dicarboxylic acid has also been described (9). In all instances, these compounds have in common a terminal carboxylic

Table 1. Groups of acidic steroids referred to as bile acids

Number of carbon atoms	Side chain length	Trivial name*	Nomenclature
C_{20}	C_1	Etianic acids	5β-Androstan-17-oic
C_{21}	C_2	Cortolic acids	5β-Pregnan-21-oic
C_{22}	C_3	Bis-nor-cholanoic	24-Nor-5β-cholan-22-oic
C_{23}	C_4	Nor-cholanoic	24-Nor-5β-cholan-23-oic
C_{24}	C_5	Cholanoic	5β-Cholan-24-oic
C_{25}	C_6	Homo-cholanoic	25-Homo-5β-cholan-24-oic
C_{26}	C_7	Dihomo-cholanoic	25,26-Dihomo-5β-cholan-24-oic
C_{27}	C_8	Cholestanoic (Coprostanoic)	5β-Cholestanoic

* Bile acids having a 5α configuration are referred to by the previx *allo*.

group, which gives the molecule its weakly acidic nature and assists its water solubility; the pK_a for unconjugated bile acids is approximately 5.0 and consequently they have extremely low solubilities in aqueous solutions when protonated. It is outside the scope of this chapter to discuss the physicochemical characteristics of bile acids and for further information there are several excellent reviews and articles on this subject (e. g. (10–14)).

After synthesis by the liver and prior to secretion into bile, unconjugated bile acids are modified by conjugation. Conjugation can take several forms; the important pathways involve amidation of the carboxylic acid group with either glycine or taurine (15–17) although conjugates with ornithine, arginine and lysine have been described (18). Additionally, conjugation in the nucleus with sulphuric acid (19) and glucuronic acid (20, 21) also takes place yielding bile acid sulphates and glucuronides either alone or as a mixed conjugate with the amino acid. In all cases the effect of conjugation is to lower the pK_a, i. e. increase the acidity and polarity of the molecule thereby rendering the bile acid more water soluble. The complex array of structures differ widely in polarity, solubility and acidity, depending upon the number of oxo and/or hydroxyl groups and the type of conjugate moiety. While these differing physicochemical properties can present many analytical problems, they can be used to advantage for the selective isolation and separation of the individual bile acid species, as will be evident from the later discussion of the pre-instrumental methodological approaches to bile acid determination.

1.2 Biosynthesis

A wealth of literature exists on the biochemical pathways and physiology of bile acids in man and for this reason this section will be confined to a relatively concise description of the principal features of their biochemistry and physiology. For more information the reader is directed to the review of Björkhem (22).

In man, the two "primary" bile acids (Fig. 1) are cholic acid and chenodeoxycholic acid so called because they are synthesized from cholesterol by hepatic enzymes. The

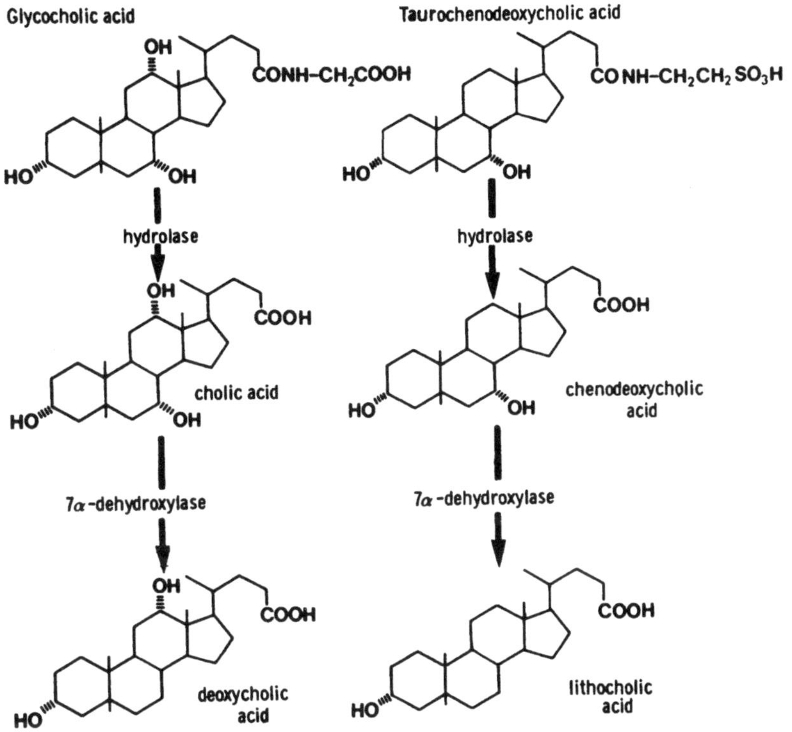

Figure 1. Structures of the principal primary bile acids in man and the reactions involved in their transformation to the principal secondary bile acids. Note that cholic and chenodeoxycholic may also be conjugated with glycine and taurine respectively and that other conjugates may occur, e. g. sulphates and glucuronides.

steps involved in bile acid synthesis (Fig. 2) have been elucidated in detail largely from experimental studies using animals, the use of sub-cellular fractions of liver preparations and from studies of patients with cerebrotendinous xanthomatosis (CTX) disease which results from a unique defect in bile acid metabolism.

The synthesis of bile acids from cholesterol was first established by Bloch et al. in 1943 from studies using deuterium labeled cholesterol (23) and later Bergström and co-workers and Chaikoff et al. in the 1950's, confirmed this pathway to be a major one in the metabolism of cholesterol (24–28).

Conversion of the C_{27} steroid, cholesterol, to the C_{24} bile acids requires a series of changes which include epimerization of the 3β-hydroxyl group, reduction of the 5-ene bond, introduction of hydroxyl groups at the positions C-7 (in chenodeoxycholic and cholic acids) and C-12 (in cholic acid), reduction in the length of the side chain by hydroxylation and oxidation to yield a 5-carbon chain with a terminal carboxylic acid group. Under normal conditions, the nuclear changes precede the side chain degradation.

The first step in the conversion involves a 7α-hydroxylase enzyme (29). This is the major rate limiting step in the formation of bile acids and leads to the formation of

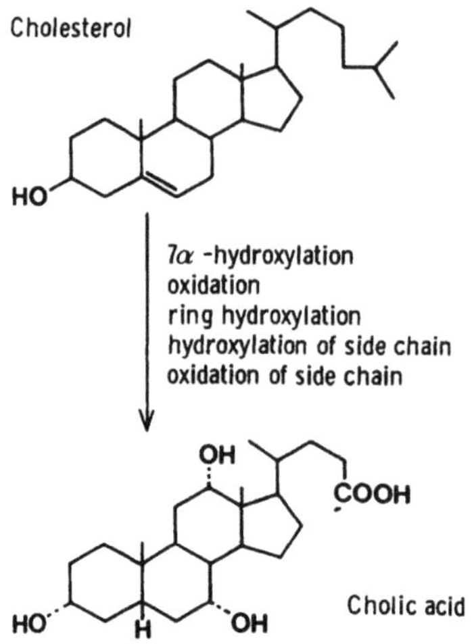

Figure 2. Biosynthetic reactions involved in the conversion of cholesterol to bile acids.

5-cholestene-3β,7α-diol (30, 31). Oxidation of the 3β-hydroxyl group and isomerization of the 5-ene to 4-ene structure (32–34), yields 7α-hydroxy-4-cholesten-3-one which acts as a substrate for 12α-hydroxylation (30, 35) in the pathway to cholic acid synthesis. By the action of the two enzymes, a 4-ene-3-oxosteroid 5β-reductase and a 3α-hydroxysteroid dehydrogenase (36), 5β-cholestane-3α,7α,12α-triol is formed and the required alterations to the nucleus are achieved.

Oxidation of the side chain starts with the introduction of a 26-hydroxyl group (37) although a 25-hydroxylation has been shown to be an alternative pathway in man (38–40). Conversion of 5β-cholestane-3α,7α,12α,26-tetrol to the respective 5β-cholestanoic acids is catalyzed by soluble ethanol and acetaldehyde dehydrogenase, and after formation of the coenzyme A ester, the product is hydroxylated at the C-24 position (41). β-Oxidation with release of propionyl coenzyme A yields the coenzyme A ester of cholic acid.

In the case of chenodeoxycholic acid biosynthesis, the structural changes are the same as those for cholic acid except that 12α-hydroxylation does not occur. An alternative pathway involving prior 26-hydroxylation of cholesterol followed by oxidation of the side chain to yield 3β-hydroxy-5-cholenoic acid has also been identified (42, 43). This bile acid is then converted to lithocholic acid which in turn is 7α-hydroxylated to yield chenodeoxycholic acid. Although in adults this is not an important pathway in bile acid synthesis, in early life and in cholestatic conditions this pathway may be significant in accounting for the increased presence of monohydroxy bile acids.

Conjugation of bile acids with glycine and taurine occurs after the formation of a coenzyme A derivative and is catalyzed by a bile acid coenzyme A: amino acid *N*-acyltransferase. In normal man, the ratio of glycine/taurine conjugates in bile is approximately 3:1, (15–17) however, this ratio is greatly influenced by diet, drugs and may be altered in disease.

1.3 Physiology

The most important feature of the physiology of bile acids is their unique enterohepatic circulation and description of this can be found in review articles and books (44–47).

In man, cholic and chenodeoxycholic are secreted into bile primarily as the glycine and taurine conjugates (15) with the rate of cholic acid synthesis being approximately twice that of chenodeoxycholic acid (48, 49). Bile is stored in the gallbladder and under normal physiological conditions in response to meal stimulus, is secreted into the upper small intestine to perform its major function of aiding the absorption of fat and fat soluble vitamins A, D, E, and K. During passage through the gastrointestinal tract, bile acids are efficiently reabsorbed and transported back to the liver via the portal venous circulation. Up to 95% of the bile acid pool is conserved in this way and the small percentage escaping this reabsorption process is excreted in the feces (50). To maintain a constant bile acid pool size, hepatic synthesis of new bile acids takes place to replace fecal loss (50–500 $mg \cdot day^{-1}$). On returning to the liver, bile acids are extracted by the hepatocyte, metabolized and resecreted into bile. This cyclical process is referred to as the enterohepatic circulation and since this takes place 6–10 times a day (51, 52), the small bile acid pool, 3–5 g, is efficiently utilized (53).

Intestinal reabsorption occurs by two mechanisms: 1. active transport, 2. passive diffusion. Active transport (54–56), so called because it takes place against a concentration gradient, is quantitatively the more important mechanism and its site is located in the ileum. Although both unconjugated and conjugated bile acids are absorbed in this way, there are kinetic differences between bile acid types. As a general rule the efficiency of transport is related to both polarity, trihydroxy > dihydroxy > monohydroxy, and conjugation state, taurine > glycine > unconjugated (57, 58).

While active transport of bile acids is confined to the ileum, passive diffusion can take place at any level of the gastrointestinal tract (56), including the stomach (59). Depending upon the bile acids species, passive absorption, which mainly occurs in the proximal small bowel, takes place by either a non-ionic or an ionic diffusion process. At the normal pH of the intestinal contents (pH 5–7), unconjugated bile acids (pK_a 5.0–6.0) will be protonated (11) and will therefore be absorbed by non-ionic diffusion. Conjugated bile acids with their lower pK_a's will be present in ionized form and will, if absorbed, do so by passive ionic diffusion. Non-ionic diffusion however is more efficient than ionic diffusion and thus ionized conjugated bile acids, particularly taurine conjugates will be favored by an active transport process and therefore be reabsorbed largely at the terminal ileum.

The two mechanisms of absorption complement each other and consequently lead to an efficient absorption of the pool by the time bile acids reach the terminal ileum and the loss of bile acids into the colon is consequently small.

During passage through the gastrointestinal tract, primary bile acids are metabolized by intestinal bacteria to "secondary" bile acids. The major bacterial reactions which occur are deconjugation and 7α-dehydroxylation, however, dehydroxylation at positions other than C-7, oxidation, reduction and isomerization reactions can all take place.

Under normal conditions, the number of anerobic bacteria in the upper small bowel are small (60, 61) but these increase towards the terminal ileum and cecum, while the colon is rich in bacteria. Deconjugation usually precedes other metabolic transfor-

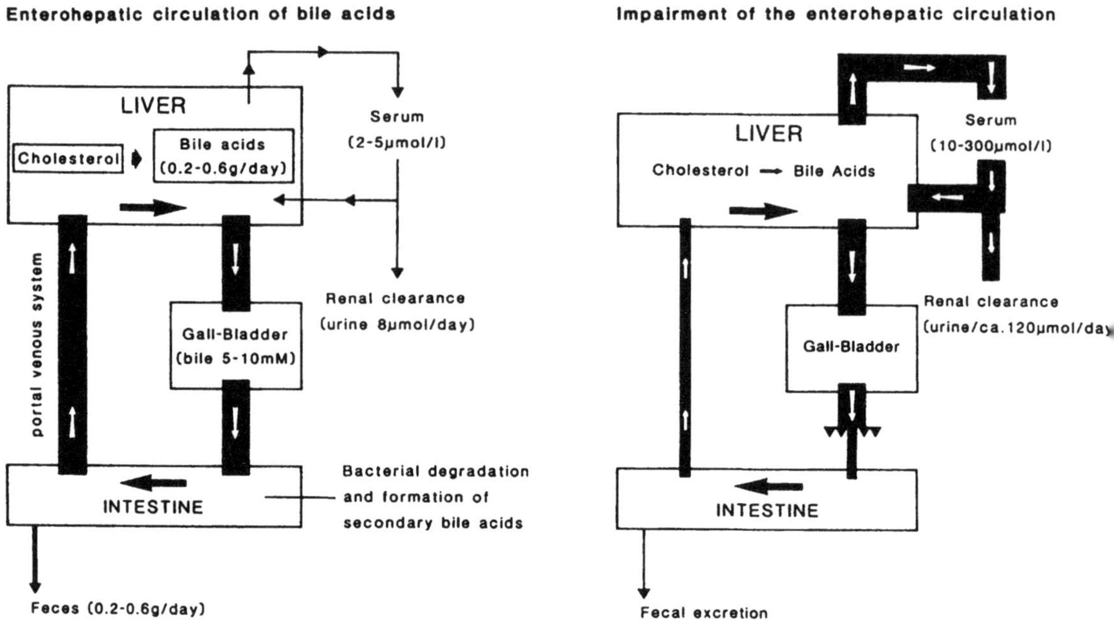

Figure 3. Simplified schematic of the enterohepatic circulation of bile acids in man in health (left) and in cholestatic liver disease (right). The boldness of the line indicates the relative quantitative importance of the various routes of bile acid excretion.

mations and the bacteria responsible for this reaction can be found in detectable numbers at the ileocecal sphincter and terminal ileum (62) resulting in significant concentrations of unconjugated bile acids in the lumen of the ileum and jejunum.

The principal secondary bile acids in man are lithocholic and deoxycholic acids (Fig. 1) formed from 7α-dehydroxylation of the primary bile acids chenodeoxycholic and cholic acids respectively (63). In addition, other bacterial reactions such as dehydroxylation at positions other than C-7, oxido-reduction and epimerization take place in the colon which account for the complex array of bile acids normally found in the feces (64).

In the normal healthy subject, bile acid physiology can be summarized schematically in Figure 3. Bile acid synthesis takes place in the hepatocyte and bile acids are secreted and concentrated in bile resulting in $mmol \cdot l^{-1}$ concentrations. During the enterohepatic cycling of the small bile acid pool (3 – 5 g), some loss into the colon and subsequent excretion in the feces occurs. Conservation of the pool size occurs by new synthesis so that daily synthesis rate approximates to fecal loss ($50-500\ mg \cdot day^{-1}$). The relatively low concentrations (less than $5\ \mu mol \cdot l^{-1}$) of bile acids in the peripheral circulation, reflects the efficiency of their hepatic extraction from the portal venous input. Urinary bile acid excretion, particularly of more hydrophilic bile acids represents a further means of eliminating bile acids, but this route is a relatively minor one in the healthy individual (65). Very small amounts of bile acids have been identified in sebaceous gland secretions (66) but this excretory route is of negligible importance when considering bile acid kinetics in healthy subjects.

1.4 Pathophysiology

Abnormalities which result in alterations in bile acid metabolism can be divided into several categories: (1) inborn errors affecting the conversion of cholesterol to bile acids, (2) altered hepatic function as in liver disease, (3) impairment in biliary secretion and gallbladder function, and (4) conditions affecting intestinal reabsorption. In each of these categories, profound and readily detectable changes in bile acid homeostasis results. Other conditions in which bile acids have either been implicated in the etiology of, or are considered to play a role, include colonic carcinogenesis, gastric ulceration, atherosclerosis, and hyperlipoproteinemia, however, the alterations in bile acid metabolism and physiology in these conditions may be more subtle and less readily detectable.

The following sections will briefly discuss the expected disturbances in bile acid synthesis and metabolism in these conditions and serve to illustrate the continuing importance of mass spectrometry in this area of clinical chemistry.

1.4.1 **Metabolic errors affecting bile acid synthesis**

One of the most extensively studied congenital conditions that affects bile acid metabolism is cerebrotendinous xanthomatosis (CTX) (67) and investigations of patients with this condition assisted greatly the understanding of pathways in bile acid synthesis in man. The disease is characterized by localized deposits of lipids, generally in the tendons and skin and high plasma cholesterol levels. There continues to be controversy as to the exact nature of the metabolic defect(s), however it is evident from analysis of biological fluids of these patients, that there is an impairment or absence of hepatic C-26 hydroxylase activity. This enzyme which is essential for normal synthesis of cholic acid initiates the events leading to the reduction in the length of the side chain. Biological fluids from these patients reveal excessive quantities of polyhydroxylated C_{27} bile alcohols (68 – 70).

Phytosterolemia or β-sitosterolemia is a rare condition, clinically similar to CTX in that tendinous xanthomas occur, however plasma cholesterol levels are normal and increased levels of plant sterols are found (71). Other inborn errors involving bile acid metabolism have been described which include a congenital defect in 12α-hydroxylase enzyme (72) and conditions resulting from peroxisomal dysfunction (73 – 76) and these have been recently reviewed (77).

1.4.2 **Liver disease**

The most marked influence on bile acid homeostasis is found in cholestasis since this produces significant alterations to the enterohepatic circulation of bile acids (Fig. 3). Cholestasis, whether complete or partial, leads to a reduction in biliary excretion, and decreased bile acid input to the liver. Bile acids continue to be synthesized that concentrate in bile, spill over into the peripheral circulation and result in high serum concentrations. The bile acid pool is redistributed and cholic and chenodeoxycholic acid synthesis rates are altered (78). In addition, alternative pathways in bile acid metabolism are "switched on", in particular hydroxylations, glucuronidation and sulphation pathways thereby increasing the water solubility of bile acids formed. Significant increase in renal and urinary excretion result and the bile acid composition of biological fluids becomes complex and, interestingly, resembles that found in the normal neonate. Mass spectrometry has become an essential tool for the analysis of these bile acid profiles.

In chronic diseases such as cirrhosis, bile acid metabolism also differs from that of the healthy subject. A reduced total bile acid pool size with a marked decrease in the cholic acid pool size and biliary concentration of deoxycholic acid occurs. Chenodeoxycholic acid synthesis is less affected, and measurement of trihydroxy:dihydroxy bile acid ratio has been proposed to be useful in the differential diagnosis of liver disease (79). The exact nature of the metabolic defect occurring in cirrhosis is unclear but since chenodeoxycholic synthesis is relatively unaffected, these changes may be due to an enzymatic defect in the conversion of the key intermediates in cholic acid synthesis. In this context only with more specific and sophisticated techniques would it be possible to examine this hypothesis.

1.4.3 **Cholelithiasis**

It has been reported that $\approx 10\%$ of the population of developed countries have gallstones and this incidence increases with increasing age (80). The occurrence of gallstones is more frequent in women and in patients with diseases of the terminal ileum or with ileal resection. This led to the suggestion that gallstone formation is related to a defective reabsorption of bile acids (81).

Cholesterol cholelithiasis results from the formation of bile super-saturated with cholesterol, which could occur by either an increased synthesis of cholesterol or a decreased synthesis of bile acids or phospholipids or both (81 – 87). Detectable changes in biliary bile acid and lipid composition (88) and alterations in bile acid kinetics occur. In a proportion of patients effective dissolution of cholesterol stones can be achieved by the oral administration of the unconjugated bile acids, chenodeoxycholic and ursodeoxycholic acids (82, 84). The factors involved in the formation of gallstones and their medical treatment involving bile acid administration have been reviewed in detail elsewhere (82 – 84, 87).

1.4.4 **Gastrointestinal diseases**

Two common gastrointestinal conditions affecting bile acid metabolism are bacterial overgrowth of the small intestine and ileal dysfunction. In both conditions, the integrity of the enterohepatic circulation is compromised. Bacterial overgrowth of the small intestine leads to an increased deconjugation and dehydroxylation of bile salts in the proximal small bowel and a reduction in intraluminal concentrations of conjugated bile salts, which if excessive (89, 90) results in malabsorption of fats and fat soluble vitamins. The unconjugated bile acids formed are reabsorbed by passive non-ionic diffusion in the jejunum. Increased bile acid deconjugation is most commonly detected using [^{14}C]cholylglycine breath test (91). In this test, [^{14}C] cholylglycine is administered orally and the $^{14}CO_2$ released from the glycine moiety, is measured. In small bowel bacterial overgrowth the specific activity of the exhaled $^{14}CO_2$ increases significantly. While the test generally used radioactively labeled bile acids for use in children and infants stable-labeled bile acids may be employed. Alternatively an increased serum concentration of unconjugated bile acids has been proposed as a useful diagnostic test for bacterial overgrowth and ileal dysfunction (92, 93).

2 Mass Spectrometry and Gas Chromatography-Mass Spectrometry in Bile Acid Analysis

2.1 Conditions for mass spectrometry

The diversity of chemical and physical characteristics of bile acids and their conjugates has lead to the use of several different sample introduction and ionization methods for their mass spectrometric analysis. The selection of these methods largely depends on whether information is required regarding molecular weight, diagnostic fragmentation for identification or characteristic fragments for quantitative measurement. In the following sections, unconjugated and conjugated bile acids and their derivatives are discussed in terms of available ionization methods and modes of sample introduction. It will be clear from this that many different mass spectrometric techniques have been applied to analyses. In a clinical biochemistry selecting simplified, reliable and easy-to-use instrumentation is recommended. The majority of relevant studies involving qualitative and quantitative analysis can be tackled satisfactorily by GC-MS with electron impact ionization.

2.1.1 **Underivatized unconjugated bile acids**

Bile acids are insufficiently volatile in the free form for introduction to the mass spectrometer via the gas chromatographic (GC) inlet, leaving direct insertion probe, liquid chromatographic (LC) interface or soft ionization target as the only suitable modes of sample insertion. This limits analysis, with the exception of LC-MS, to isolated and purified samples of bile acids.

Most unconjugated bile acids volatilize below 120 °C from the direct probe, but the ion source needs to be maintained 20 – 30 °C above this temperature to prevent adsorption effects. Electron impact (EI) spectra are useful for structural identification with ions arising from successive elminiation of the hydroxyl functions as water and loss of the side chain. Molecular ions are also observed when the volatilization temperature is properly controlled (94). A potential advantage of chemical ionization (CI) in this respect is the abundant quasimolecular ions obtained with ammonia as reagent gas, which can be used for sensitive detection of underivatized unconjugated bile acids even in incompletely purified samples (95). The lack of fragment ions in CI spectra, however, reduces their value for structure determination and discrimination of stereoisomers. Field desorption (FD) also has limitations particularly in distinguishing stereoisomers but does give molecular weight information on underivatized bile acids.

The quasimolecular ions resulting from fast atom bombardment (FAB) ionization or liquid secondary ion mass spectrometry (liquid SIMS) of the sodium salts of free bile acids are the $[M + H]^+$ and $[M + Na]^+$ in the positive ion mode and the $[M - Na]^-$ and $[M - H]^-$ in the negative ion mode. Sensitivity of detection is better in the latter mode, although the positive ion spectra may be more helpful due to the presence of ions arising from the elimination of all of the hydroxyl groups from the molecule as water; i.e. *m/z* 359, 357, and 355 for the monohydroxy, dihydroxy and trihydroxy bile acids respectively (96–99). In the present context little distinction need be drawn between FAB and liquid SIMS (99–100).

Despite the ability to analyse underivatized bile acids by mass spectrometry with a variety of ionization techniques, the complexity of bile acids in biological fluids limits this analytical approach. Derivatization followed by GC or GC-MS will generally give more complete information. Alternatively LC in combination with ammonia CI or the thermospray interface obviates the need for derivatization and could be applied to the routine detection or quantification of underivatized bile acids. At the present time, however, LC is more often used in a discontinuous mode with MS for bile acid analysis.

2.1.2 Derivatized unconjugated bile acids

The protection of carboxyl and hydroxyl functions in the bile acid molecule by derivative formation reduces polarity and increases greatly their volatility and thermal stability. The derivatized molecule is thus more suitable for direct probe introduction to the mass spectrometer under EI or CI conditions. However, the principal intention in developing derivatives was to facilitate their gas chromatographic analysis and permit the separation of mixtures of bile acids and at a later stage to direct fragmentation in order to aid their identification in EI or CI mass spectrometry. Several of these derivatives together with conditions for gas chromatography are discussed in Section 2.2.

2.1.3 Conjugated bile acids

2.1.3.1 *Amino acid conjugates*

In many studies the taurine and glycine conjugates are isolated by group separation (Section 3) and following hydrolysis, are identified as free or derivatized unconjugated bile acids by direct probe EI mass spectrometry or GC-MS respectively. The latter is likely to remain the method of choice for studying mixtures of amino acid conjugates.

Efforts to analyse directly derivatized taurine conjugates by GC have not been successful. A number of derivatized glycine conjugates can be separated by GC but they have long retention times and some isomers are incompletely resolved (101). More recent work has demonstrated a satisfactory separation of methyl ester dimethylethylsilyl (DMES) ether derivatives of glycine conjugates of the principal primary and secondary bile acids on high resolution fused silica capillary columns (Fig. 4). However at present this approach is limited to these conjugates and is unsuitable if the molecules are also conjugated to sulphuric acid.

The need to isolate individual conjugates and obtain their mass spectra can be met in several ways. The combined technique of LC-MS has the potential to do this in a single step but as LC is often employed as a stand alone method this is not being achieved routinely. Isolation of individual LC fractions permits the selection of the most suitable treatment (untreated, derivatized, hydrolyzed, etc.) and ionization (EI, CI, FD, FAB) of the conjugated bile acid component. Free glycine and taurine conjugated bile acids have been isolated by LC, introduced on the direct insertion probe under EI conditions and useful spectra obtained (103–106). The molecular

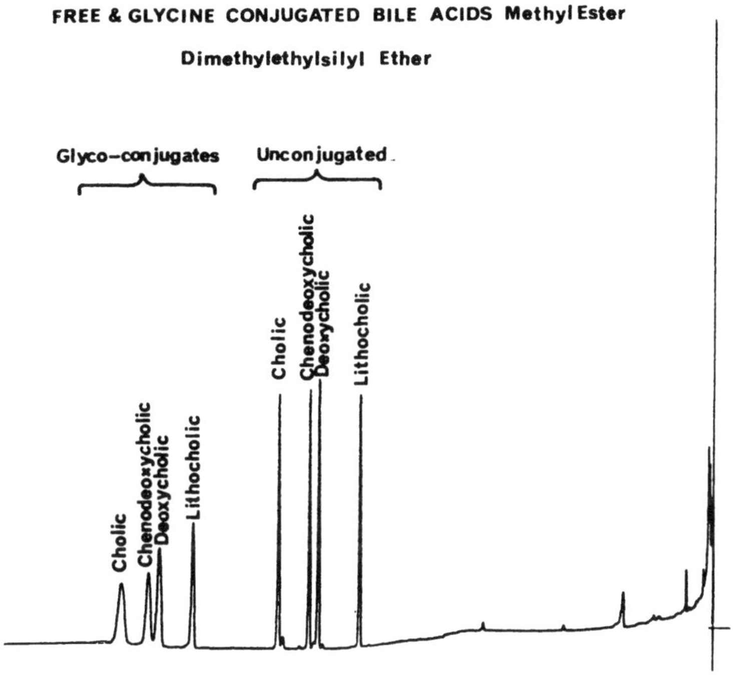

Figure 4. Capillary column GC separation of methyl ester-dimethylethylsilyl ether derivatives of unconjugated and glycine conjugated bile acids. GC was performed on a Carlo Erba 2920 housing a 30 metre CP-Sil 5 capillary column. Hydrogen was used as the carrier gas (flow rate approximately 2 ml $\cdot$ min^{-1}) and the analysis was carried out using temperature programmed operation from 150 °C – 280 °C with increments of 2 °C $\cdot$ min^{-1} (102).

ions of the glycine conjugates are small and the major ions result from loss of water and loss of the entire side chain or parts of the side chain. Most glycine conjugates volatilize in the ion source below 180 °C whereas temperatures in excess of 225 °C are required for the taurine conjugates. As a result, molecular ions of the latter are not observed (105, 106). The mass spectra of methyl esters of glycine conjugates have more intense molecular ions but give similar spectra to the free compounds. Methyl esters cannot be formed of the taurine conjugates. Further derivatization of the taurine conjugates and the methylated glycine conjugates by trifluoroacetylation has been suggested to give improved sensitivity for MS identification and quantification (107).

Soft ionization methods, including field desorption (FD), FAB, liquid SIMS and laser desorption (LD) have also been used to obtain spectra of intact glycine and taurine conjugates. The molecular weights are determined from the $[M + Na]^+$ quasi-molecular ions in FD spectra of the sodium salts (96, 103, 108) and from FAB spectra in both positive and negative ion mode (96, 98, 99, 109 – 111). It is not possible from FAB spectra to differentiate geometric or positional isomers, however their advantage over other ionization methods is in providing molecular weight information on the intact conjugated molecule and their use for rapidly screening samples to detect those requiring more detailed analysis (e. g. Sections 4.1 and 4.5). LD may prove to be of value in the same context but requires further development (112, 113).

Some caution is necessary when interpreting negative ion FAB spectra obtained on mixed samples as [M − Na − 2]-ions from hydrogen loss may be confused with [M − Na]-ions of oxo or unsaturated bile acids.

Little has been published on quantification of bile acids and their conjugates by FAB. It appears to be essential to use stable isotopically labeled internal standards for individual bile acids although competitive surface effects and differences in ionization of different conjugates in the matrix may still interfere with accurate measurement (98).

2.1.3.2 *Sulphates (with and without amidation)*

The involatility of sulphated bile acids has precluded the use of most ionization methods including FD, to obtain mass spectra. FAB, however, will effectively ionize bile acid sulphates giving good molecular weight information, particularly in the negative ion mode (98, 110, 111).

2.1.3.3 *Glucuronides*

Direct probe introduction and CI give satisfactory mass spectra of the methyl ester acetate derivatives of bile acid glucuronides, suitable for identification (114). Similarly the methyl ether trimethylsilyl ether derivatives give EI spectra which confirm the number of hydroxyl groups in the molecule and indicate the presence of the glucuronyl moiety (115, 116). Furthermore, these derivatives are suitable for GC introduction. Liquid SIMS spectra of glucuronides in negative ion mode contain $[M - H]^-$ ions and a fragment ion due to loss of the glucuronyl group at $[M - 176 - H]^-$ (98).

2.2 Conditions for gas chromatography-mass spectrometry

2.2.1 Types of derivatives and their preparation

As no single derivative is ideal for all applications and no one GC-phase can completely resolve the wide range of naturally occurring bile acids, many combinations of derivatives and GC columns have been advocated. Detailed reviews and discussion of these can be found in the literature (e. g. 104, 106, 110, 117−120) and a useful "Handbook of Derivatives for Chromatography" has been published (121).

Common to most derivatization procedures is the prior methylation of the carboxylic acid before derivatization of the hydroxyl or oxo groups. This may be achieved by use of one of several treatments, e. g. HCl or H_2SO_4/methanol, ethereal diazomethane, HCl/dimethoxypropane/methanol, acetylchloride/methanol and others). Despite the potential toxicity and hazards in preparing diazomethane, this is the most frequently used method and is described in more detail below. It is important that the diazomethane solution is freshly prepared and that the reaction time is short to prevent undesirable reactions or the formation of unwanted side products. Unfortunately, since several of the methylation reactions have yet to be fully evaluated for a wide spectrum of bile acids, it is not possible to draw definitive conclusions as to their relative merits for routine application.

Suggested Derivatization Procedures

1. Methyl esters. Diazomethane is prepared in cold (4 °C) ether solution by the reaction between sodium hydroxide and *N*-methyl-*N*-nitroso-*p*-toluene-sulphonamide (Diazald, Aldridge Chemical Co., Gillingham, UK) by the method described by Fales et al. (122). This solution is added to the sample of bile acid or bile acid extract dissolved in a small volume of methanol. The reaction mixture is left at room temperature for 15–20 min and should remain yellow in color indicating an excess of reagent after which it is evaporated to dryness under nitrogen and is then ready for analysis or further derivatization.

The following derivatives are usually prepared after methylation of the carboxylic acid:

2. Acetates. These derivatives are readily formed by treatment of dried bile acid extract or standard after methylation with 500 µl mixture of acetic anhydride/pyridine (5:1 v/v) at room temperature overnight. The acetate derivative can be obtained directly by blowing to dryness under a stream of nitrogen or by extraction with diethyl ether after dilution of the reaction mixture with water. Perchloric acid may be used as a catalyst and acetic acid substituted for pyridin (123).

3. Trifluoroacetates (TFA). The methylated bile acid extract or standard is dissolved in 250 µl trifluoroacetic anhydride containing a trace of pyridine and left overnight or heated for 30 min at 50 °C. The reagents are removed under a stream of nitrogen. It has been recommended that TFA derivatives are analyzed shortly after preparation to avoid decomposition (124).

4. Trimethylsilyl (TMS) ethers. A dried sample of methylated bile acid extract or standard is treated with an excess of a mixture of pyridine/hexamethyldisilazane/trimethylchlorosilane (3:2:1 by vol.) in a tightly stoppered tube for 30 min at 60 °C (125). Excess reagents are removed under nitrogen and the derivative can be purified if necessary by passage through a small column of Lipidex 5000 (126).

This method gives complete silylation of all hydroxyl groups, but the varying rates of silylation of specific configurations of hydroxyl groups with other methods may be of value in the MS analysis of unknown bile acids (117).

5. Dimethylethysilyl (DMES) ethers. After methylation, the dried bile acid extract or standard is reacted with 50 µl of dimethylethylsilylimidazole for 30 min at 60 °C in a glass stoppered vial (127). The derivatizing reagents are then removed by passage through a small column of Lipidex 5000 (126):

Esterification of the carboxyl group can be affected with other alkyl groups including the hexafluoroisopropyl (128, 129) and heptafluorobutyl groups (129, 130).

The hydroxyl functions in bile acids can be converted to acetate, TFA or TMS ether derivatives. The TFA and TMS ethers are the most frequently used although the former tend to lose trifluoroacetic acid on GC while TMS ether derivatives have limited storage stability. Alkyldimethylsilyl ether derivatives, despite their longer GC retention times give high mass ions suitable for selected ion monitoring (127, 131, 132).

The conversion of oxo-functions to *O*-methyloximes or oximes may be useful in the interpretation of mass spectra of oxo-bile acids (133, 134).

2.2.2 **Liquid phases for gas chromatography**

A variety of liquid phases have been used in bile acid analysis (for references see e. g. 104, 106, 110, 117–119) but in combination with mass spectrometry liquid phases should have low bleed characteristics within the temperature range required for chromatographic separation (104–106).

Open tubular capillary columns are steadily replacing conventional glass packed columns and while the latter may continue to be used in specific analyses, the advantages of capillary columns are too beneficial to ignore. This is evident when bile acid profiles for many biological fluids obtained using capillary columns are compared with respective analysis on packed columns (64, 135). The merits of capillary columns and their incorporation into a GC-MS system are discussed in Chapter 1 of this volume.

The stationary phases which are used for the capillary column GC separation of the most commonly used derivatives, the Me-TMS ethers, include the methyl- (SE-30, OV-1), phenyl-methyl- (OV-17) and cyanopropyl- (SP- 2340) silicones and polyethylene glycols (PEG 20,000, PEG-HT, Carbowax 20M) (120, 136–141). The non-selective phases such as OV-1, OV-101, and SE-30 require relatively high temperatures for the elution of bile acid Me-TMS ether derivatives but are more useful than other phases for the separation of complex mixtures especially where they include oxo bile acids. The latter have particularly long retention times on PEG 20,000, PEG-HT and Carbowax 20M and therefore phases of this type are more suitable for samples where oxo-acids are unlikely to be present in significant amounts. Hi-Eff 8BP has proved to be one of the best liquid phases for the separation of hydroxylated bile acid derivatives using packed columns but this liquid phase has yet to be successfully incorporated into capillary columns. It was often used in parallel with SE-30 to give complementary and confirmatory information on bile acid samples. At the present time, no single phase will separate all known bile acids and it may be necessary to use a combination of columns when making very detailed investigations of bile acids from biological sources. However, a non-selective phase of OV-1 type, or its equivalent, is likely to have most general application and its high temperature and chemical stability in a chemically-bonded capillary columm will give prolonged service in routine use.

2.3 Mass spectrometric fragmentation and identification of bile acids

The purpose of this section is to point out briefly some aspects of the electron impact fragmentation of derivatized bile acids which assist in their structural characterization.

The GC retention characteristics of derivatized bile acids are useful and in some cases essential when identifying individual bile acids with closely similar spectra. However the reader should always be cautious in relying upon published retention data because there are many discrepancies in the literature values. On this point, retention data should be established in each laboratory under strictly controlled experimental conditions. Obtaining GC-MS data before and after chemical modification of the sample may help in establishing structures for which no authentic standards are available (110).

The mass spectral fragmentation of bile acids has been studied extensively and if a detailed knowledge is required suitable texts (104, 106, 110, 142–144), and references therein should be consulted. EI has been the most widely applied method of obtaining bile acid spectra of derivatives, with the Me-TMS ether derivatives being favored. These derivatives often are more informative than others for structure identification (110, 144) and for this reason the discussion here is limited mainly to these spectra.

2.3.1 Molecular ions

The molecular ions of the Me-TMS ether derivatives are frequently absent and the molecular weight has to be determined from the [M−15] ion if present or from fragment ions. Other silyl ethers, the alkyldimethylsilyl ethers for example (127, 131, 132) have intense ions due to loss of the alkyl group from the molecular ion and readily confirm the molecular weight.

2.3.2 Loss of hydroxyl groups

The hydroxyl groups are eliminated under EI as consecutive losses of water in the free compounds, as acetic acid or trifluoroacetic acid in the acetates or trifluoroacetates respectively or as trimethylsilanol in the spectra of Me-TMS ether derivatives. These losses are also influenced by temperature and the ionizing voltage.

Depending on the configuration and position of the TMS ethers groups the trimethylsilanol losses give ions of varying intensities. The spectra of the Me-TMS ether derivatives of 3α,7α-dihydroxy-5β-cholan-24-oic (chenodeoxycholic), 3α,12α-dihydroxy-5β-cholan-24-oic (deoxycholic) and 3α,7β-dihydroxy-5β-cholan-24-oic (ursodeoxycholic) are compared in Figure 5 and demonstrate some of these effects. The *m/z* 370 ion represents the fragment remaining after loss of both silyl groups from the molecule and indicates the original number of these groups. Similarly, the ions *m/z* 372, 370 and 368 arise in spectra of bile acids containing one, two or three derivatized hydroxyl groups respectively. However, the presence of unsaturation in the molecule should not be overlooked and bile acids containing an oxo group or modified side chain would show appropriate shifts of these ions to other mass values.

In the spectra of bile acids containing viscinal hydroxyls, such as hyocholic (3α,6α,7α-trihydroxy), the last TMS group is eleminated without a hydrogen transfer (i.e. −89 daltons) to give a specific ion at *m/z* 369, $(M-[(2 \times 90) + 89])$.

2.3.3 Fragmentation of the side chain

A distinctive ion in most bile acid spectra arises from cleavage and removal of the complete side chain, following loss of the hydroxyl functions. Thus, in ring hydroxylated monohydroxy, dihydroxy, trihydroxy and tetrahydroxy bile acids, the resulting fragment ion due to the ABCD rings is at *m/z* 257, 255, 253 and 251 respectively. Once again it should be remembered that oxo substitution or unsaturation in the nucleus will alter these mass values. Depending on the configuration and position of the hydroxyl groups and the configuration of the hydrogen at C-5, side chain cleavage may also take place before all hydroxyl groups are eliminated. In the spectrum of the Me-TMS ether derivative of deoxycholic acid, for example, the side chain (115 daltons) is lost from the ion of *m/z* 345 (Fig. 5). The 5α isomer of deoxycholic acid gives a spectrum with an intense *m/z* 345 ion and small *m/z* 370 ion which indicates a more facile loss of the side chain than the final silylated hydroxyl (3α-hydroxyl). Similar intensity differences are observed in the spectra of other bile acids which help not only to differentiate isomeric compounds but assist in identifying unknown bile acids.

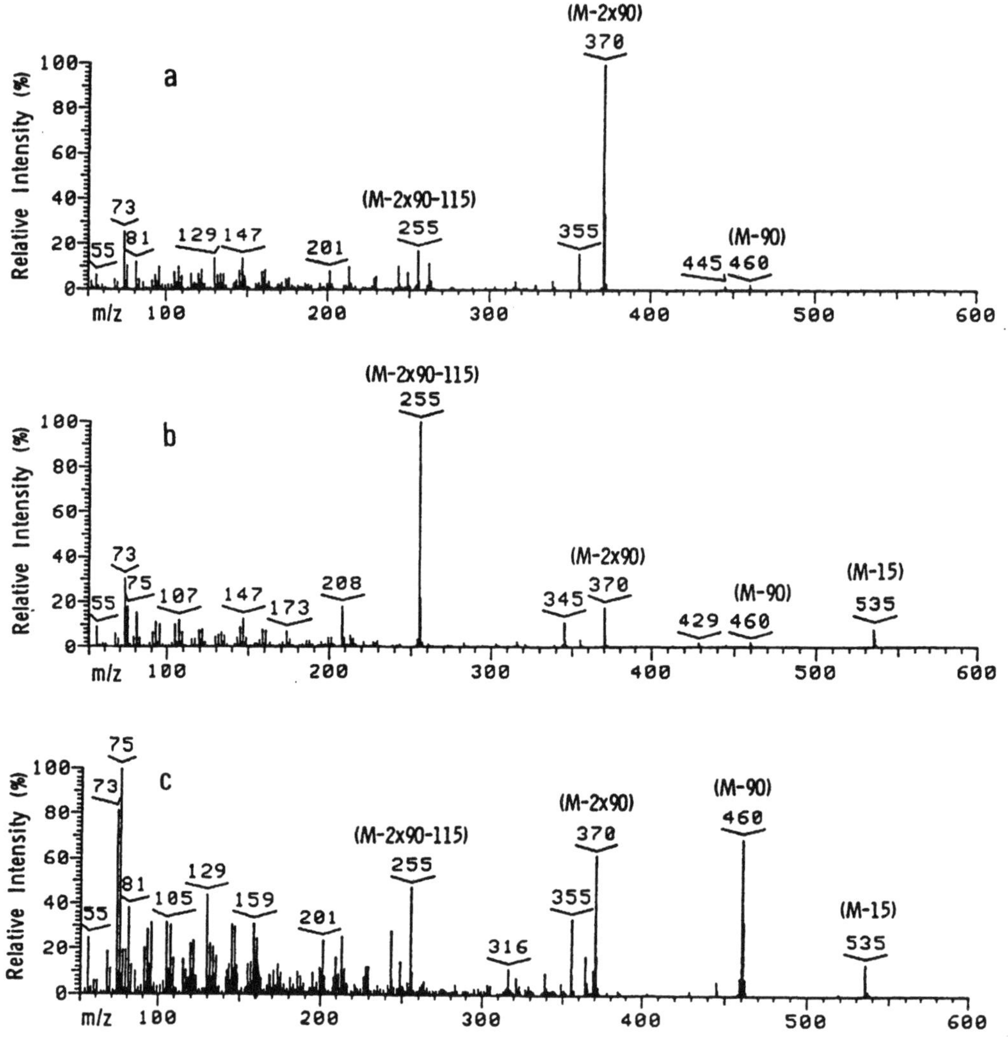

Figure 5. Mass spectra of the methyl ester-trimethylsilyl ether derivatives of three dihydroxy bile acids a) chenodeoxycholic, b) deoxycholic and c) ursodeoxycholic acid taken under EI (70 eV) conditions.

2.3.4 D-ring cleavage

Subsequent to hydroxy group and side chain elimination, cleavage through the D-ring often takes place such that carbon atoms C-16 and C-17 or C-15 to C-18 are lost to give ions at *m/z* 230, 228 and 226 or *m/z* 215, 213 and 211 respectively for the mono-, di-, and trihydroxy bile acids. These ions are only of minor intensity when other more facile nuclear cleavages are possible.

Table 2. A selection of characteristic ions in the Me-TMS ether derivative spectra of bile acids

Observed ion* (*m/z*)	Origin	Possible structure indicated
129	A-ring fragmentation	3-TMS-ether-5-ene
195	Contains A-ring and C-6 plus one OTMS group	3,6,7-Tri-TMS-ether 7β epimers have intense ion
208	Cleavage of C-11,12 and C-8,14 bonds with TMSOH loss	12-TMS-ether
217	A-ring fragment	1,3-Bis-TMS-ether
243	Partial AB-rings fragment	3,7-Bis-TMS-ether more intense if 2-OTMS present
261	B-ring fragmentation -contains C and D rings	7,12-Bis-TMS-ether (large ion in allocholic acid)
285	As for *m/z* 195 but containing two OTMS groups	3,6,7-Tri-TMS-ether 7β epimers have intense ions
314	A-ring retro-Diels Alder following TMSOH loss from A-ring	Tris-TMS-cholanoate
316	A-ring retro-Diels Alder following TMSOH loss from A-ring	Di-TMS-cholanoate
318	A-ring retro-Diels Alder following TMSOH loss from A-ring	Mono-TMS-cholanoate

* Unsaturation, additional functionality or side chain modification will alter mass of some ions.

2.3.5 **Miscellaneous diagnostic fragmentations**

As indicated earlier, full discussion of bile acid fragmentation is available elsewhere (104, 106, 110, 142–144). Listed in Table 2 are some selected ions which in addition to those ions described above are useful in the characterization of bile acid spectra.

2.4 Quantitative mass spectrometry of bile acids

Selected ion monitoring (SIM) and repetitive scanning methods have both been used in bile acid quantification and their application is discussed later. When low concentrations of bile acids are present, such as those found in tissue and serum then the high sensitivity of SIM may be essential (145, 146). The sensitivity can be maximized by choice of derivative or ionization method such as the use of alkyldimethylsilyl ether derivatives (127) or CI in conjunction with methyl ester acetate derivatives (147). More frequently, the Me-TMS ether derivatives have been used and general fragment ions common to the spectra of mono-, di- and trihydroxy bile acids, (e. g. *m/z* 372, 370, 368 respectively), employed to detect each bile acid, using also the GC retention data (145). The most suitable internal standards for accurate quantification are isotopically labeled analogues of the compounds to be measured, but availability of these standards has limited their application to the more commonly encountered bile acids.

The alternative technique of repetitive scanning is less often used but has been fully exploited by Sjövall and co-workers (148, 149). An important advantage of this approach is the ability to measure a wide range of bile acids without the requirement to pre-select ion masses as in SIM. Despite less favorable detection limits, repetitive scanning can be applied readily to extracts of urine and bile or concentrated preparations of other biological material.

2.5 Stable isotopes

In addition to the use of bile acids labeled with stable isotopes as internal standards for quantification, they have great potential for studying the metabolism, turnover rates, pool sizes and synthesis rates of bile acids and a combination of improved isolation procedures and instrumentation now allows these measurements to be made from serum samples (Section 4).

While the importance of radioisotopes in studying the metabolic rate of administered compounds cannot be disregarded, the use of stable isotopically labeled compounds and mass spectrometric detection of their metabolites has several advantages. Using only nanogram amounts of material, information can be obtained on the number and abundance of isotopes in the metabolite and frequently the position of these isotopic atoms in the molecule can be deduced from the mass spectrum.

3 Extraction and Isolation of Bile Acids from Biological Samples

At the present time, it is not possible to satisfactorily analyze bile acids directly in biological samples using mass spectrometry. The diversity of structures and the presence of vast quantities of interfering extraneous components necessitates the inclusion of extraction, isolation, hydrolytic and purification steps prior to derivatization and/or sample introduction to the mass spectrometer. Preinstrumental procedures have been reviewed in several useful articles (118, 120, 135, 150–152) and this section will discuss current techniques for the preparation of samples for bile acid analysis by GC-MS.

A general scheme was published (135) which affords a high degree of flexibility for analyses (Fig. 6). In later sections, similar schemes are discussed for fecal and urinary bile acid analysis. Routine measurements of total bile acids or total non-sulphated bile acids, specific measurements of unconjugated bile acids or a scheme, primarily for use in research whereby different bile acids are examined in detail providing information about types of conjugates present may be chosen. Individual steps essential in the analysis of serum bile acids are discussed in more detail below. Similar schemes have been reported for urinary bile acids and fecal bile acids (Section 4).

The primary purpose of the extraction step is to afford a means of removing proteins and/or inorganic ions, as well as other classes of compounds which might

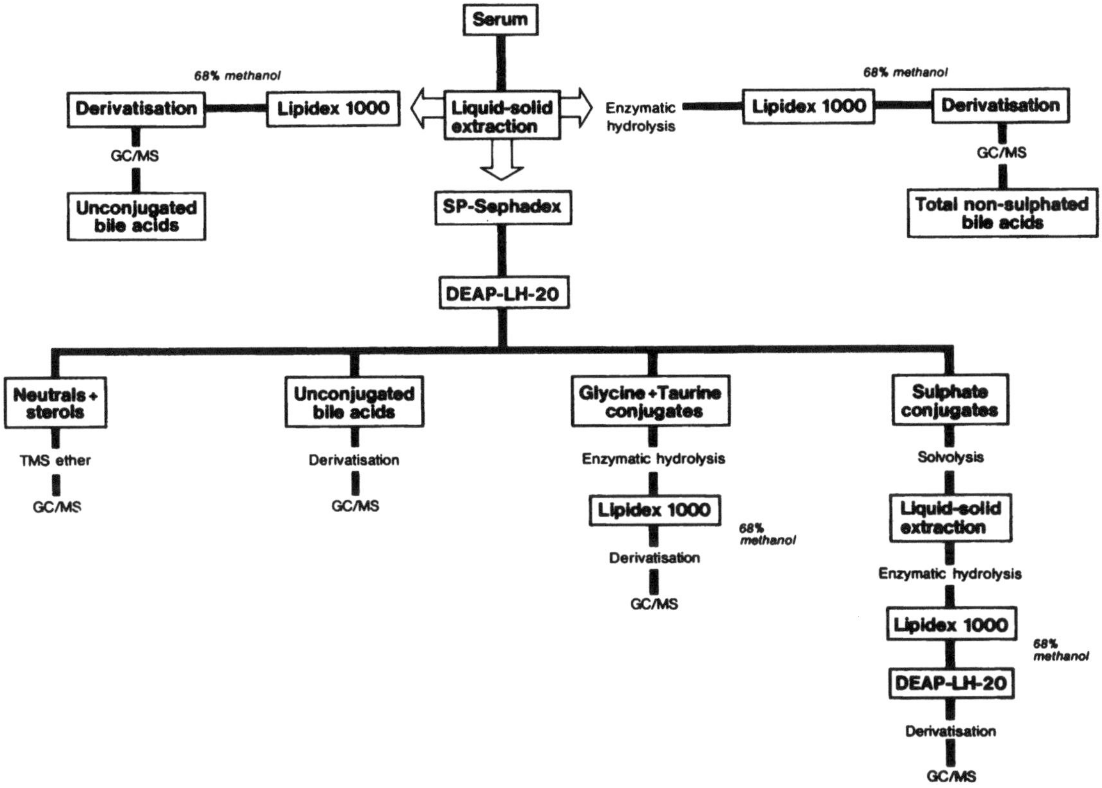

Figure 6. General analytical scheme for the analysis of bile acids in serum. (Taken from ref. 135 with permission.)

interfere in the subsequent analysis and to concentrate the sample. Many techniques have been adopted to extract bile acids from biological fluids and these have included solvent extraction, liquid-solid extraction, and liquid-gel extraction and have been reviewed in detail (152).

3.1 Liquid-liquid partitioning

These techniques although extensively employed are less desirable particularly where a quantitative extraction of a wide spectrum of bile acids is required. The extraction efficiency is dependent upon the mode of conjugation, the number of hydroxyl groups, the degree of ionization of the carboxylic acid group and the extent of protein binding and because bile acids exhibit a wide variation in these properties, polar solvents and the use of ion-pair reagents are necessary to ensure quantitative extraction of all classes of bile acids (153, 154). Care should be taken in the selection of solvent and on the extraction conditions because unwanted side reactions such as deconjugation, solvolysis, derivative formation or degradation may occur while the formation of emulsions is a frequent practical disadvantage.

3.2 Liquid-solid extraction

Liquid-solid extraction procedures eliminate the need for large volumes of toxic solvents, tedious shaking and separation steps and overcome emulsion formation. One of the earliest examples of a liquid-solid extraction procedure was the use of the anion exchange resin Amberlyst A-26 (155), however, the large volume of strongly acidic aqueous solvents and undesirable irreversible adsorption effects observed with polystyrene resins led to their replacement once the neutral cross-linked polystyrene polymer Amberlite XAD-2 was introduced (156).

Amberlite XAD-2 has beer shown to be suitable for the quantitative extraction of neutral steroids from urine (156, 157) and later Makino and Sjövall (158) described a column method for the extraction of bile acids from plasma. In a critical evaluation of Amberlite XAD-2, Almé et al. (65) demonstrated the importance of controlling the flow rate and pH to obtain maximum recoveries of bile acids; a poor recovery of polar sulphated bile acids occurs (65, 120, 159, 160) due to their irreversible binding because of the weak anion exchange properties of this neutral resin (161). The closely related Amberlite XAD-7 resin has also been used for bile acid extraction. A simplified batch method was described using 1 g of resin facilitating the extraction of large numbers of samples (162).

Over the last few years, with the advent of commercially available octadecylsilane bonded silica cartridges, liquid-solid extraction of bile acids from serum has become commonplace and offers the most rapid and simple method for bile acid extraction (120, 163–166). These cartridges, of which there are now many different types commercially available (152), were first applied to the extraction of steroids from urine (167, 168) and serum (161, 168) and later similar methods were developed for bile acid extractions (120, 163–166).

Reversed phase octadecylsilane bonded silica cartridges (Bond Elut C_{18} or Sep-Pak C_{18}) behave like polar absorbants in that they extract from aqueous solution medium and high polarity amphiphilic molecules and exclude non-polar lipids, salts and proteins (120). Cholesterol, which is present in serum at concentrations approximately 10^5 fold greater than bile acids, will cause interference with bile acids during their GC-MS analysis, because the TMS ether co-elutes with the Me-TMS ether derivatives of lithocholic acid on non-selective GC phases and cholic acid on selective GC phases. These cartridges are effective in removing the majority of the cholesterol in a single step (165).

To obtain a quantitative extraction of bile acids from serum the main problem to overcome is their strong binding to serum proteins (169). It is not sufficient to dilute the serum with water as has been advocated (163) because the recovery of unconjugated and sulphated bile acids which are tightly bound to proteins will be poor (165). The binding of bile acids to protein can be decreased by increasing the pH (169) and temperature, and a method has been developed in which serum is diluted with 0.1 $mol \cdot l^{-1}$ sodium hydroxide and heated to 64 °C before passage through the cartridge (120, 165).

Extraction of Bile Acids from Serum using Liquid-solid Extraction (165)

All solvents and reagents used should be analytical grade and redistilled. Reverse phase octadecylsilane bonded silica cartridges (Bond Elut from Analytichem Int., Harbor City, CA or Sep-Pak C_{18} obtained from Water Associates, Milford, MA, USA) are first washed and activated with 5 ml each of distilled water, methanol and distilled water. Serum (1 vol) is diluted with 0.1 mol $\cdot$ l^{-1} sodium hydroxide (4 vol) and placed in a water bath heated at 64 °C for 15 min. The diluted serum sample is passed through the cartridge of octadecylsilane bonded silica. The cartridge is then washed with distilled water and the sample and washings discarded. Bile acids absorbed to the silica are removed by elution with methanol (3 – 5 ml).

Variations of this method have also been reported (152), where the sample is diluted differently or bile acids recovered by elution with methanol:water (75:25 v/v) and cartridges washed with hexane prior to use (166). The recovery of a wide range of bile acids, including the very polar tri-sulphate conjugate of taurocholic acid has been shown to be quantitative (165), because these solid-phase cartridges do not have the weak anion exchange capacity inherent in Amberlite XAD resins. In practice, the extraction procedure is rapid, takes only a few minutes to perform and using a vacuum apparatus it is possible to batch extract up to 60 samples $\cdot$ h^{-1}. The large capacity of these cartridges (0.2 mmol of bile acid $\cdot$ 100 mg^{-1}) and the ease and speed of the technique make it superior to other extraction techniques currently available.

3.3 Liquid-gel extraction

A liquid-gel offers an alternative to a liquid-solid phase for the extraction of bile acids and is particularly useful for samples with a high lipid content. Lipidex 1000 is an alkylated cross-linked dextran gel which has a much higher capacity than octadecylsilane bonded silica and efficiently extracts non-polar steroids and lipids from aqueous acidified solutions (170 – 173).

In contrast to octadecylsilane bonded silica cartridges, Lipidex 1000 behaves as a non-polar adsorbent, in that it extracts from acidified aqueous solution, lipophilic molecules of low and medium polarity and excludes polar lipids, salts and proteins (120). Unconjugated bile acids are therefore quantitatively removed from acidic aqueous solutions and adsorbed to the gel while the more polar conjugated bile acids pass through the gel. Once adsorbed, the lipids, bile acids and related compounds can be recovered in a single fraction by elution with methanol or alternatively, by reverse phase chromatography using aqueous methanol mixtures as eluting solvents, a more selective and highly purified bile acid fraction can be obtained (64, 135).

One of the most useful practical applications of Lipidex 1000 is in the specific extraction of unconjugated bile acids from aqueous enzymic or alkaline hydrolysates (64, 135). The pH of the sample is adjusted to 3 – 4 by addition of acid and then passed through a bed of Lipidex 1000. Hydrolyzed (and unconjugated) bile acids are adsorbed on the gel and recovered by elution with 68% methanol; the method obviates the need for using flammable and hazardous solvents and results in a relatively pure sample free from interfering monohydroxy sterols.

Extraction of Unconjugated Bile Acids from Aqueous Hydrolysates

Lipidex 1000 gel is washed with methanol and slurried in a glass column (10 × 1 cm size) to give a bed size of 5 × 1 cm. The gel is then washed with 20 ml 0.01 mol $\cdot$ l^{-1} acetic (or hydrochloric) acid followed by 20 ml distilled water and is ready for use. The pH of the aqueous hydrolysate (either enzymic or alkaline) is adjusted to between pH 3–4 by dropwise addition of 0.1 mol·l^{-1} acetic (or hydrochloric) acid and rapidly passed through the bed of Lipidex 1000. The gel is washed with 20 ml of 0.01 mol·l^{-1} acetic acid and 20 ml distilled water. Bile acids are recovered by elution of the Lipidex 1000 gel bed with methanol (20 ml) or aqueous methanol.

The quantitative extraction of conjugated bile acids with Lipidex 1000 requires a suitable counter-ion. Following experiments with quaternary ammonium salts, which have been used in solvent extraction (153, 154), Dyfverman and Sjövall (174) developed a simple system for the extraction of conjugated and unconjugated bile acids from urine, bile and serum using dodecyltrimethylammonium bromide as a counter-ion. Extraction of bile acids from plasma requires dilution of the sample with an equal volume of phosphate buffer containing 0.06 mol·l^{-1} dodecyltrimethylammonium bromide and passage of this sample through a small bed of Lipidex 1000. After washing with aqueous buffer and distilled water, conjugated bile acids are recovered by elution with methanol. The recoveries of unconjugated, glycine, taurine, sulphate and glucuronide conjugates were found to be close to 100% (174). In contrast to liquid-solid techniques for extraction, Lipidex 1000 will also remove lipids from solution (172), which may be advantageous in certain instances. Where bile acid analysis is of primary interst however, lipids which interfere can be retained on the gel bed and bile acids recovered with a higher degree of purity of elution with solutions of aqueous methanol, e. g. 68% methanol (64, 135). The extraction technique is applicable to all types of biological fluids, however for serum samples, the flow rate of sample through the Lipidex 1000 column is more critical probably because of protein binding effects of bile acids.

3.4 Chromatographic separation and purification of bile acids

A battery of techniques are available for the separation of bile acids, either individually or by groups (152). Early methods used include, counter-current distribution (175), partition chromatography on cellulose (176) adsorption chromatography on silica, aluminium oxide or florisil of either the free or esterified bile acids (177–180), electrophoresis (181), and paper chromatography (15, 150, 182, 183), but these techniques are now less frequently employed. Thin layer chromatography (TLC) (150, 184, 185), lipophilic gel chromatography, high performance liquid chromatography and GC currently comprise the most commonly employed methods for bile acid separation, purification and isolation (152).

In an extensive review of TLC of steroids, Lisboa (185) described many solvent systems for steroids which are applicable to the separation of bile acids, while separations of isomeric bile acids (184), conjugated bile acids (186–189), unconjugated bile acids (184, 185, 190–197), and methyl esters of bile acids (198) have all been described. Although of limited use for direct quantitative determination of bile acids (196, 199–204), TLC continues to be extremely valuable for small scale preparative work for the rapid visualization of complex mixtures of bile acids prior to more detailed studies using more sophisticated techniques, and as a purification step prior to GC and MS determination.

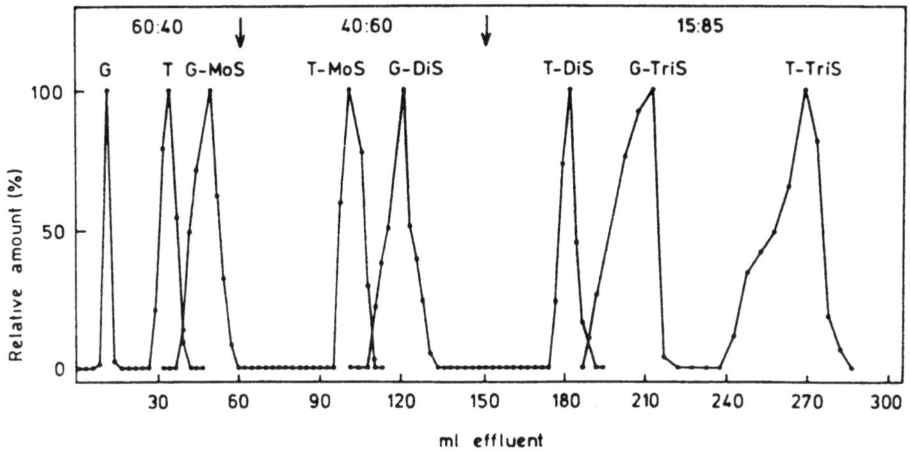

Figure 7. Separation of a synthetic mixture of sodium salts of glycine (G) and taurine (T) conjugated mono (MoS), di (DiS) and tri-sulfated (TriS) bile acids. *Column*: 4 g of Sephadex LH-20; *solvent*: stepwise elution with mixtures of chloroform and $0.02\,mol \cdot l^{-1}$ NaCl in methanol as follows: 60:40 (0 – 60 ml), 40:60 (60 – 150 ml), 15:85 (150 – 200 ml). (Reproduced from ref. 65 with permission.)

In most instances where mass spectrometric analysis of bile acids is performed, the gas chromatograph provides the means of sample introduction to the mass spectrometer and of resolving the components in the biological sample. With the advent of capillary column GC, the chromatographic resolution now possible is considerably greater than other chromatographic techniques, however, in many instances it is of help to be able to fractionate individual groups of bile acids. This simplifies the GC profile and may help in the interpretation of acquired mass spectra. In this respect, group fractionation based upon the type and number of functional groups in the molecule is most valuable and lipophilic gel chromatography affords a useful method for these separations.

A variety of lipophilic gels based upon the cross-linked polysaccharide gel Sephadex are now commercially available (120, 152) and these are useful for group separations of bile acids and steroids. Sephadex gels substituted with long alkyl chains (205 – 207) provide excellent stationary phases in alcohol/water mixtures (208) and have been used effectively for the purification of steroid sulphates, bile acids and lipids (208 – 215).

Sephadex LH-20 is a β-hydroxypropylated derivative of Sephadex prepared by reaction of Sephadex with 1,2-epoxypropane under alkaline conditions. This gel has been commercially available (Pharmacia Fine Chemicals, Uppsala, Sweden) for over a decade, but has proved to be more useful for the separation of individual and conjugated steroids (e. g. ref. 215 and references therein) than for bile acids. Its high polarity and poor swelling in non-polar solvents restricts its practical use to straight phase chromatography but solvent systems for the separation of conjugated bile salts have been described (216) and applied to plasma, feces and urine samples (114, 160, 217 – 220). The major disadvantages with this gel are the inability to satisfactorily separate the different conjugate classes of bile acids (Fig. 7), the high dependency on

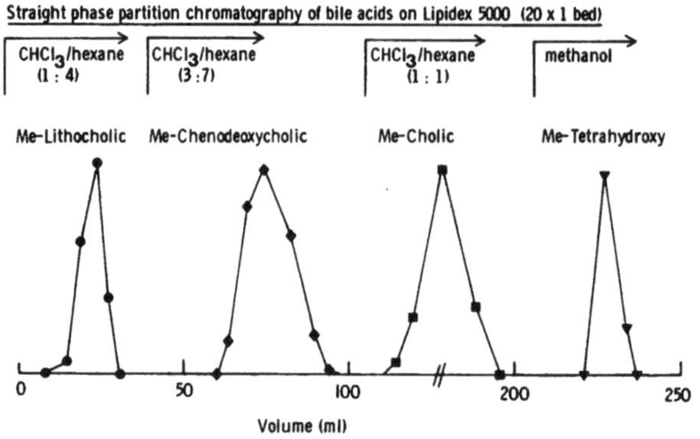

Figure 8. Separation of methyl esters of representative standards of mono-, di-, tri-, and tetra-hydroxy bile acids on columns of Lipidex 5000 gel. Stepwise elution of each fraction was achieved using the solvents indicated.

electrolyte type and overloading associated with crude extracts which create separation artefacts. Nevertheless it should be realized that where the isolation of a specific bile acid type is required, provided a radiolabeled marker is added to the biological extract and the column is calibrated, this gel is a useful complement to other chromatographic systems.

The separation of unconjugated bile acids into groups based upon the number of hydroxy groups (221) provides a method for simplifying a complex GC profile of unknown bile acid structures. Lipidex 5000, which is a hydroxyalkyl Sephadex, can be used for both straight phase and reverse phase chromatography depending upon the selection of the mobile phase (208). When polar solvent systems incorporating mixtures of a hydrocarbon with methanol or water, or both are employed, reversed phase chromatography is possible and this was illustrated in early applications by separations of chenodeoxycholic, lithocholic, *iso*-lithocholic and 3-oxo-5β-cholanoic acids.

A practical application of this gel utilizes a straight phase partition system in which monohydroxy, dihydroxy, trihydroxy and tetrahydroxy bile acids can be isolated as specific groups (Fig. 8) as their less polar methyl ester derivatives (221).

The value of this gel is evident from the following example (unpublished work). The chromatogram in Figure 9 shows the capillary column GC profile of the total bile acids in the feces from an adult female Sprague-Dawley rat. Peak 9 in the chromatogram was indicated from mass spectrometric analysis (of the Me-TMS ether derivative) to be a monohydroxy bile acid, yet its GC retention index was longer than previously published data for monohydroxy bile acids, eluting later than the derivative of deoxycholic acid (Fig. 9). High resolution GC-MS of the [M^+] ion revealed its structure to contain three oxygen atoms (2 belonging to the carboxylic acid) consistent with a monohydroxy structure. Confirmation that this compound was a monohydroxy bile acid was obtained from its chromatographic behavior on Lipidex 5000. Fractionation of the extract in the system described and GC analysis of the Me-TMS ether showed that this peak to be a monohydroxy bile acid and subsequent preparation of the authentic compound confirmed it to be 3β-hydroxy-5α-cholanoic acid. Consistent with TMS ether derivatives of $3\beta,5\alpha$ neutral steroids, this bile acid is retained on non-selective GC stationary phases, a feature not recognized in previously published retention data for this bile acid.

Separation of Bile Acids into Groups Based on Number of Hydroxyl Functions (221)

Lipidex 5000 is first washed with methanol followed by hexane in a glass scintered filter funnel. The gel is allowed to swell in the solvent mixture chloroform:hexane (1:4 v/v) before being slurried into a glass chromatography column (30 × 1 cm with teflon tap and scintered disc) to give a bed size of 20 × 1 cm. After washing the gel bed with 50 ml of this solvent, the column is ready for use.

The bile acid extract is first methylated using freshly prepared diazomethane and bile acids converted to methyl esters. After removal of the excess diazomethane under a stream of nitrogen, the dried extract is dissolved in 4 × 0.5 ml of the solvent chloroform:hexane (1:4 v/v) and applied to the column of Lipidex 5000. Monohydroxy bile acids are eluted with 40 ml of this solvent. Stepwise elution of dihydroxy, trihydroxy and tetrahydroxy bile acid methyl esters is achieved by elution with 70 ml of chloroform:hexane (3:7 v/v), 100 ml of chloroform:hexane (1:1 v/v) and 50 ml of methanol respectively.

Group separation of bile acids based upon their mode of conjugation is best achieved by ion exchange chromatography. The principles and early applications of polystyrene resins to the separation of bile acids have been reviewed elsewhere (222) but due to the large column sizes and elution volumes, and irreversible adsorption effects, these resins found only limited practical applications (155, 222–224) and were soon superseded by the introduction of lipophilic ion exchange gels synthesized from Sephadex LH-20 (206, 207). Although DEAE-Sephadex has been used for the separation of lipids and phenolic steroids (225–227) its hydrophilicity makes this gel less suitable for chromatography of lipophilic compounds in organic solvents. For this reason, less polar derivatives of Sephadex were prepared (206, 207) that gave good swelling characteristics in organic/aqueous solvent mixtures and permitted rapid ion exchange with high capacity. Diethylaminohydroxypropyl Sephadex LH-20 (DEAP-LH-20 as it was originally known, but now commercially available under the trade name Lipidex-DEAP) was first used to obtain separations of unconjugated steroids, monoglucuronides, monosulfates and disulphates of neutral steroids (228) and later of phenolic steroids (229) by employing aqueous ethanol solvent systems buffered with formic and acetic acids. The buffer system was later used to separate glucuronide conjugates of bile acids (230, 231) by employing slight variations in the eluting solvents. Alternatively the piperidinohydroxypropyl Sephadex LH-20 (PHP-LH-20) has been employed (232).

In a comprehensive publication Almé et al. (65) described the problems encountered with Lipidex-DEAP for the group separation of bile acid conjugates isolated from urine. With this gel, bile acids can be separated into neutral species, unconjugated, glycine, taurine, monosulphate and disulphate conjugate groups using small columns of the gel and aqueous ethanolic buffers.

Although the separation of individual bile acids within each conjugate group is not possible, this is not a serious disadvantage, because following deconjugation information of individual bile acids within the conjugate groups can be obtained by capillary column GC-MS analysis. The combined use of these group separation techniques and capillary column GC-MS affords the most appropriate technique for the detailed

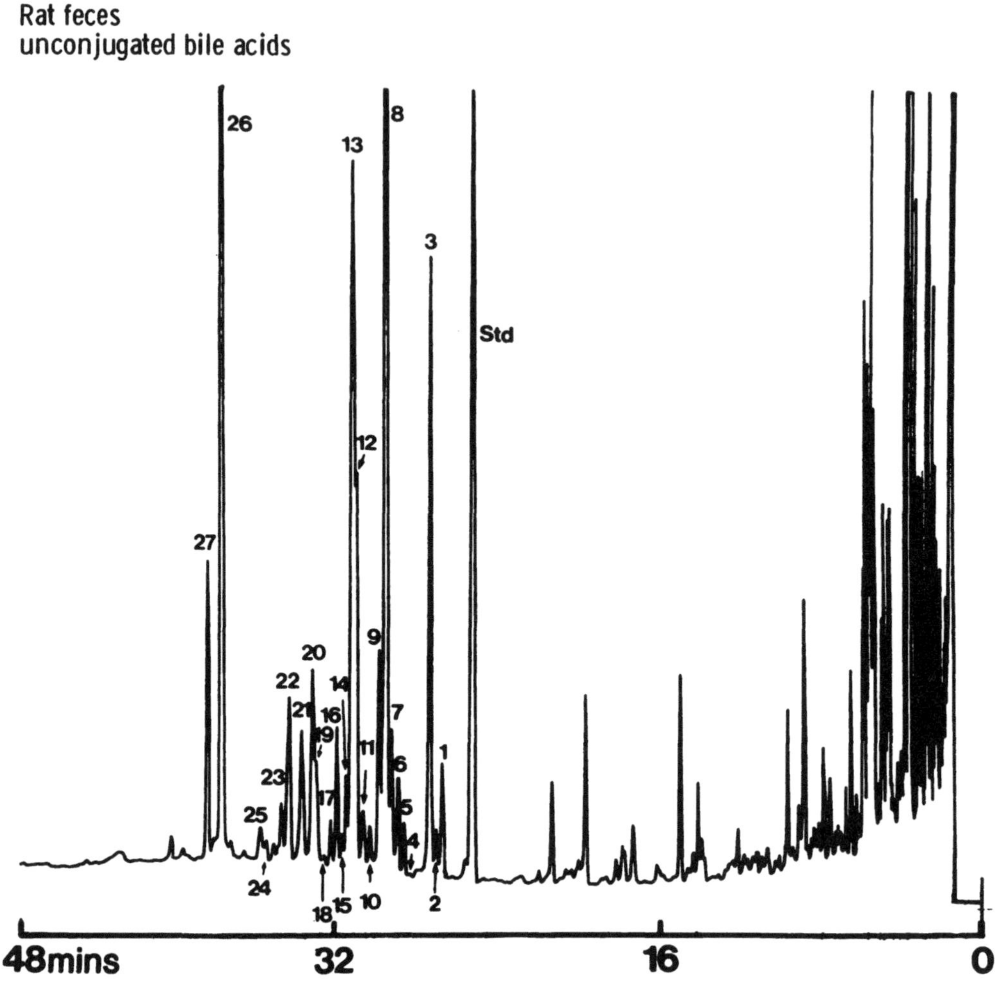

Figure 9. GC profile of the methyl ester TMS ether derivatives of unconjugated bile acids excreted in the feces of an adult female Sprague-Dawley rat. Chromatography was carried out on a 25 m × 0.4 i.d. capillary column coated with OV-1. Helium was used as carrier gas with a flow rate of 2.5 ml · min^{-1} and the column temperature programmed from 220 °C to 275 °C at 2 °C · min^{-1}. The following compounds are indicated: (1) 3β-hydroxy-5β-cholanoic; (2) 3α-hydroxy-5α-cholanoic; (3) 3α-hydroxy-5β-cholanoic (lithocholic); (4) unknown monohydroxy bile acid; (5) 3α,12α-dihydroxy-5α-cholanoic; (6) 3α,12β-dihydroxy-5β-cholanoic; (7) 3α,12β-dihydroxy-5α-cholanoic; (8) 3α,12α-dihydroxy-5β-cholanoic (deoxycholic); (9) 3β-hydroxy-5α-cholanoic; (10) 3α,6β-dihydroxy-5β-cholanoic and 12α-hydroxy-3-oxo-5β-cholanoic; (11) unknown dihydroxy + trihydroxy bile acids; (12) 3α,6β,7α-trihydroxy-5β-cholanoic (α-muricholic); (13) 3α,6α-dihydroxy-5β-cholanoic (hyodeoxycholic); (14) 3β,6α-dihydroxy-5β-cholanoic; (15) unknown tetrahydroxy bile acid; (16) 3β,12α-dihydroxy-5α-cholanoic; (17) 3α-hydroxy-12-oxo-5α-cholanoic; (18) unknown trihydroxy bile acid; (19) 3β,12β-dihydroxy-5α-cholanoic; (20) 3α-hydroxy-12-oxo-5β-cholanoic; (21) unknown mono-oxo-monohydroxy bile acid; (22) 3α,6β,7β-trihydroxy-5β-cholanoic (β-muricholic); (23) unknown compound; (24) mono-oxo-dihydroxy and mono-oxo-monohydroxy bile acids; (25) 3β-hydroxy-12-oxo-5α-cholanoic; (26) 3α,6α,7β-trihydroxy-5β-cholanoic (ω-muricholic; (27) 3α,6α,7β-trihydroxy-5α-cholanoic.

profiling of bile acid composition of biological fluids and this is evident from recent publications (64, 65, 233, 234).

Group Separation of Bile Acids Based on Mode of Conjugation

Lipdex-DEAP (10 g) is washed copiously with 72% ethanol (100 ml) in a glass scintered funnel and converted to the acetate form by washing with a solution of 0.1 mol · l^{-1} acetic acid in 72% ethanol followed by 72% ethanol until neutral pH. The gel is slurried into a glass column with Teflon tap to give a bed size of 13 cm × 0.4 cm (being equivalent to approximately 0.6 g dry weight of gel. The dried extract is suspended in 72% ethanol (3 × 5 ml), passed through the gel bed and collected. The bed is washed with 5 ml 72% ethanol and the combined effluents which contain neutral compounds collected. Unconjugated bile acids and glycine, taurine and sulphate conjugates are recovered respectively by the stepwise elution with solution of 0.1 mol · l^{-1} acetic acid in 72% ethanol; 0.3 mol · l^{-1} acetic acid in 72% ethanol adjusted to pH 5.0 by addition of ammonium hydroxide; 0.15 mol · l^{-1} acetic acid in 72% ethanol adjusted to pH 6.6 with ammonium hydroxide and 0.3 mol · l^{-1} acetic acid in 72% ethanol adjusted to pH 10. Between each change of solvent the column should be washed with several ml of 72% ethanol to minimize memory effects or overlapping of fractions.

For some types of samples the inclusion of a prior cation exchange chromatographic step may be necessary to remove interfering organic cations and convert the conjugates to their free acids. Amberlyst A-15 or SP-Sephadex may be used. Chromatography on SP-Sephadex, although more time consuming gives a greater degree of purification. It is important to apply samples in aqueous alcohol mixtures to avoid any unwanted side reactions occurring on column e. g. hydrolysis or methylation, and the effluent must not be too acidic when applied to the anion exchange gel otherwise uptake of unconjugated bile acids and separation of the conjugated groups will be affected.

3.5 High pressure liquid chromatography of bile acids

The application of HPLC to bile acid analysis is rapidly expanding (235) and the technique has a number of attractive advantages over other chromatographic procedures. It readily lends itself to the rapid separation of many individual bile acid conjugates in a single run. Group separations of glycine and taurine conjugates (236, 237), 5α- and 5β-cholanoates (105), unconjugated bile acids (238 – 240) and sulphates (241, 242) have been reported. Most of the applications of HPLC have been developed for the routine separation and measurement of bile acids in biological fluids and while this has worked well for samples such as bile, where there are high concentrations or serum from patients with liver disease, the detection limits for quantitative analysis are often insufficient to permit a satisfactory analysis of serum from healthy individuals or of samples where low concentrations are encountered. Attempts to improve sensitivity using detectors other than UV detectors or by post-column reactions or preparation of derivatives have met with mixed success. In many instances the extraction and group fractionation prior to HPLC has been recommended to overcome these problems and those of interference by other components in the samples (235).

It is evident that HPLC will play an increasingly important role in bile acid analysis in the future. With the improvement in design of LC-MS interfaces, particularly with the introduction of thermospray and plasmaspray ionization (243), the mass spectrometer will provide a more specific and sensitive detector. Because of the acidic nature of bile acids, ionization readily occurs in the thermospray process and mass spectra resembling those obtained by chemical ionization and FAB ionization are obtained (Setchell, unpublished data). HPLC-MS should provide an additional tool for the direct analysis of intact bile acid conjugates.

3.6 Hydrolysis of bile acid conjugates

Since volatility is a requirement for mass spectrometry and bile acid conjugates are relatively involatile, analysis of conjugated bile acids is generally carried out following hydrolysis and derivatization. A variety of techniques for hydrolysis have been employed, the method of choice depending upon the objectives of the assay. For example, where specific glycine and taurine conjugated bile acids have been isolated by chromatography or if a total non-sulphated bile acid determination on the crude extract is required, enzymatic hydrolysis with cholylglycine hydrolase may be adequate. The enzyme purified from *Clostridium perfringens* (244, 245) has an optimum pH at 5.5 and exhibits selectivity in rates of hydrolysis; the taurine conjugate being more efficiently hydrolyzed than glycine conjugates. Provided an excess of enzyme is added and the incubation period is prolonged this limitation is not a serious one. Enzymes have been prepared which show specificity towards glycine or taurine conjugates (246), however they have not been applied in bile acid analysis. Some limitations of cholylglycine hydrolase include the inability to cleave the glycine and taurine moiety from C_{27}-bile acids. Furthermore, it has been suggested that an inefficient hydrolysis of tetrahydroxycholanoic acids with hydroxy groups at positions C-1 and C-6 occurs (247) but our own experiences, however, indicate that tetrahydroxy cholanoic acids which are glycine/taurine conjugates hydrolyze adequately with this enzyme.

Chemical hydrolysis using aqueous alkali provides a rigorous method of deconjugation but, to minimize artefact formation and loss of bile acids, hydrolysis should be performed in Teflon lined tubes using aqueous alkali and heat.

Attempts to obtain a sulphatase enzyme active on bile acid sulphates have largely been unsuccessful. Although sulphatase activity has been found in species of intestinal bacteria, its specificity is confined to sulphates of C_{24} and C_{26} bile acids with equatorial groups only (248–251). At the present time the most appropriate method of hydrolyzing bile acid sulphates is chemical solvolysis followed by an appropriate hydrolytic step. No one solvolytic method merits specific recommmendation as the best and there are several available, each having drawbacks.

Solvolysis in methanol:acetone:HCl (252) may lead to transesterification, is time consuming, requiring 1–3 days, may lead to partial formation of acetonides of *cis*-glycol structures, as exist in hyocholic acid, and partial hydrolysis of glycine and taurine conjugates may occur. Dimethoxypropane/HCl has been used for the simulataneous solvolysis and methylation of bile acid sulphates (253, 254) but, depending upon the incubation conditions may also lead to acetonide formation. Ethyl acetate/H_2SO_4 (255) has been successfully employed for the hydrolysis of steroid sulphates and is suitable for bile acid sulphates (65). During solvolysis the partial formation of acetates and ethyl esters occurs, necessitating mild alkaline hydrolysis. An inefficient solvolysis of C-7 sulphates has been observed (256). Hydrolysis of bile acid sulphates at position C-3 can be effected by extraction into ether acidified with HCl (257), although the method was validated only for non-amidated bile acid sulphates. The technique is rapid and may therefore be suitable for routine use, however, it will not hydrolyze sulphates at positions C-7 and C-12 (160).

A solvolytic procedure for bile acid sulphates using tetrahydrofuran/1 $mol \cdot l^{-1}$ trifluoracetic acid in dioxan was recently described (258). The technique is rapid (2h/ 45 °C) and the reaction terminated by evaporation of the volatile reagents. Quantitative hydrolysis of glycolithocholic acid-3-sulphate and glycochenodeoxycholic acid-3-sulphate were demonstrated with no evidence for ester or artefact formation (258). This technique appears to be the most suitable general procedure at present, but further systematic studies are required to evaluate the efficiency of solvolysis for other bile acid conjugates.

After deconjugation the change in the physicochemical properties may lead to precipitation of unconjugated bile acids in samples with high concentrations of bile acids. For reasons discussed earlier, the extraction of the unconjugated bile acids into organic solvents, e. g. ether or ethyl acetate, after adjustment of the solution to pH 1 – 2 by addition of acid, is not recommended. Extraction is best performed by passage of the aqueous hydrolysate through either a cartridge of octadecylsilane bonded silica, or by the use of Lipidex 1000 after first adjusting the hydrolysate to pH 3 – 4 by addition of acid. The latter technique is preferred, because using Lipidex 1000 and a reverse phase solvent system (i. e. 68% methanol), any monohydroxy sterols remaining in the sample can be retained on the lipophilic gel (64, 135, 170, 171).

Hydrolysis of bile acid glucuronides is best achieved by mild enzymic hydrolysis using a β-glucuronidase enzyme, of which there are several types commercially available. When the bovine β-glucuronidase enzyme is used it should be realized that this preparation is contaminated with cholic acid and small amounts of deoxycholic, *allo*-cholic, oxo-bile acids, glyco- and taurocholic acids. The β-glucuronidase from the digestive juice of *Helix Pomatia* is free of bile acid contaminants, but this preparation also has sulphatase activity which is effective on neutral steroid sulphates. Removal of these unwanted components and particulate material may be effected by prior passage of the enzyme and buffer mixture through a cartridge of octadecylsilane bonded silica before incubation with the bile acid extract.

4 Detection and Measurement of Bile Acids

4.1 Serum

Due to the relatively low concentrations of bile acids in the peripheral circulation, mass spectrometric analysis of serum bile acids has largely been restricted to measurement of the principal bile acids utilizing the high sensitivity and selectivity of stable isotope dilution and SIM. Important contributions have been made using repetitive scanning techniques to characterize individual bile acids in a profile and more recently, direct inlet fast atom bombardment (FAB) mass spectrometry has proved is proving to be a viable technique for the analysis of intact bile acid conjugates (77, 96, 97, 99, 111). Some of the more exciting and clinically useful developments in the mass spectrometry of serum bile acids concern the determination of bile acid kinetics following administration of stable isotopically labeled bile acids to patients.

Single or repetitive scanning affords the most appropriate means of obtaining qualitative information about the composition of biological fluids. To a lesser extent scanning techniques have been employed for quantitative bile acid analysis (65, 233, 259, 260).

Rapid repetitive scanning is usually necessary during the emergence of the bile acid components in the GC effluent to enable the detection of minor or coeluting compounds in biological samples. Scanning over a wide mass range, e. g. 0 – 1000 daltons, is adequate for the detection of all of the commonly employed bile acid derivatives and will provide maximum information to allow structural elucidation. For example, the molecular ion of the Me-TMS ether of trihydroxy bile acids is at m/z 638, for tetrahydroxy bile acids the M^+ is at m/z 726, while the hexafluoroisopropyl derivative of a trihydroxy bile acid will have a M^+ of m/z 840. In some instances, particularly when a limited number of pre-selected bile acids are to be determined, it may be adequate to restrict the mass range which is scanned. In this way an improvement in sensitivity may be achieved.

The identification of a bile acid is made on the basis of a number of criteria: (1) inspection of the complete mass spectrum of the compound and where possible comparison with the spectrum of the authentic bile acid derivative, (2) when the gas chromatograph is used as the inlet, the retention time of the peak is often invaluable in differentiating the structures of stereoisomers which exhibit closely similar spectra and (3) from mass chromatograms of ions specific for the bile acid derivative. Such chromatograms should reveal coincidence in peaking of the selected m/z values at the correct retention time or scan number and the relative areas or height of each m/z value will provide useful information of the purity of the GC peak. In some instances this is the only way of elucidating the structures of compounds which co-elute from the GC column (65). The selection of ions is based upon general and specific bile acid fragmentation patterns (Section 2) and useful guidelines are reported in the literature (65, 110). A compilation of 111 reference mass spectra was recently published (144).

The repetitive scanning mode of GC-MS analysis has been reviewed in detail by many workers and the principles of qualitative and quantitative analysis of data acquired by this method have been described (110, 148, 261). Unlike the SIM mode repetitive scanning permits an unbiased approach to the analysis since all masses are recorded and stored on the computer, thereby allowing data retrieval at a later date. Quantification of compounds can be carried out by the inclusion of internal or external standards and measuring and comparing the areas or peak heights of one or a number of the ions specific for the compound of interest with those of an appropriate standard (125, 142, 143).

Although repetitive scanning is frequently used for qualitative analysis, applications to the quantitative determination of bile acids are limited. Since the lower limit of reliable quantification is 0.01 – 0.05 nmol (4 – 20 ng), subject to the ions chosen and the GC characteristics, this precludes its use for serum bile acid determinations unless relatively large volumes of serum are taken for analysis and the scanning range is restricted (259).

In a study of plasma bile acids in patients with intrahepatic cholestasis of pregnancy (259) the power of the repetitive scanning GC-MS mode of operation is apparent (Fig. 10). In this early study, 5 ml of plasma was taken and following extraction and fractionation of bile acids into non-sulphated and sulphated groups, bile acids were analyzed as their Me-TMS ethers using repetitive magnetic scanning GC-MS. In addition to cholic, chenodeoxycholic and deoxycholic acids previously found in similar samples, lithocholic, 3β-hydroxy-5-cholenoic and ursodeoxycholic acids were also identified and quantified. Of interest was the finding that these latter bile acids were predominantly in the sulphate fraction, an observation also made when SIM was recently employed for the determination of serum bile acids in patients with cholestasis, with and without pruritic liver disease (145). The increased concentrations of lithocholic, 3β-hydroxy-5-cholenoic and ursodeoxycholic acids, (also found in the meconium of newborn infants and in non-pregnant patients with cholestasis) poses the question of their origins, since these bile acids are usually considered to be 'secondary' bile acids. These data are indicative of alternative pathways in hepatic metabolism.

In a more recent similar study of patients with liver disease (233), bile acids were first fractionated into groups based upon their mode of conjugation using Lipidex-DEAP and measured using repetitive magnetic scanning. Although there were only a small number of samples analyzed, serum bile acid concentrations ranged from 14 – 252 $\mu mol \cdot l^{-1}$. Cholic acid comprised 44 – 89% of the total bile acids, exceeding the concentration of chenodeoxycholic acid in all patients with primary biliary cirrhosis. Several monohydroxy bile acids were found with 3β-hydroxy-5-cholenoic acid accounting for approximately 13% of the total serum bile acids. Lesser amounts of trihydroxy and tetrahydroxy bile acids several tentatively identified as 6-hydroxylated isomers, were also present. Only two healthy subjects were studied. Chenodeoxycholic acid was the predominant serum bile acid and unconjugated bile acids accounted for 30% and 53% respectively of the total serum bile acids. This latter observation is higher than that reported in an early study by Sandberg et al. (155) using GC alone but is in accord with observations of Makino et al. (262) and recent studies by Setchell et al. using GC-MS (263), and highlights the advantage of using group fractionation techniques in detailed metabolic studies.

In a study of patients with cholestasis, bile acids from plasma (300 ml) were extracted, hydrolyzed, methylated and fractionated into groups on Sephadex LH-20 and evidence for 6α-hydroxylation was obtained (264). Hyocholic (3α,6α,7α-trihydroxy-5β-cholanoic) was identified in the serum of these patients. The problems of interference from the TMS ethers of neutral sterols with the Me-TMS ethers of bile acids during GC was apparent in this study, and a chromatographic separation using Lipidex-DEAP would have assisted the identification of minor metabolites and allowed the separate determination of neutral sterols. In a later study using SIM, hyocholic acid was consistently present in the serum of patients with primary biliary cirrhosis occurring exclusively as a taurine conjugate (145).

The identification of the intact glucuronide conjugate of chenodeoxycholic in the plasma (8 ml) of a patient with chronic intrahepatic cholestasis was achieved using isobutane CI mass spectrometry (114). The conjugate was isolated by Sephadex LH-20 chromatography and TLC of the methyl ester polyacetate and the purified conjugate

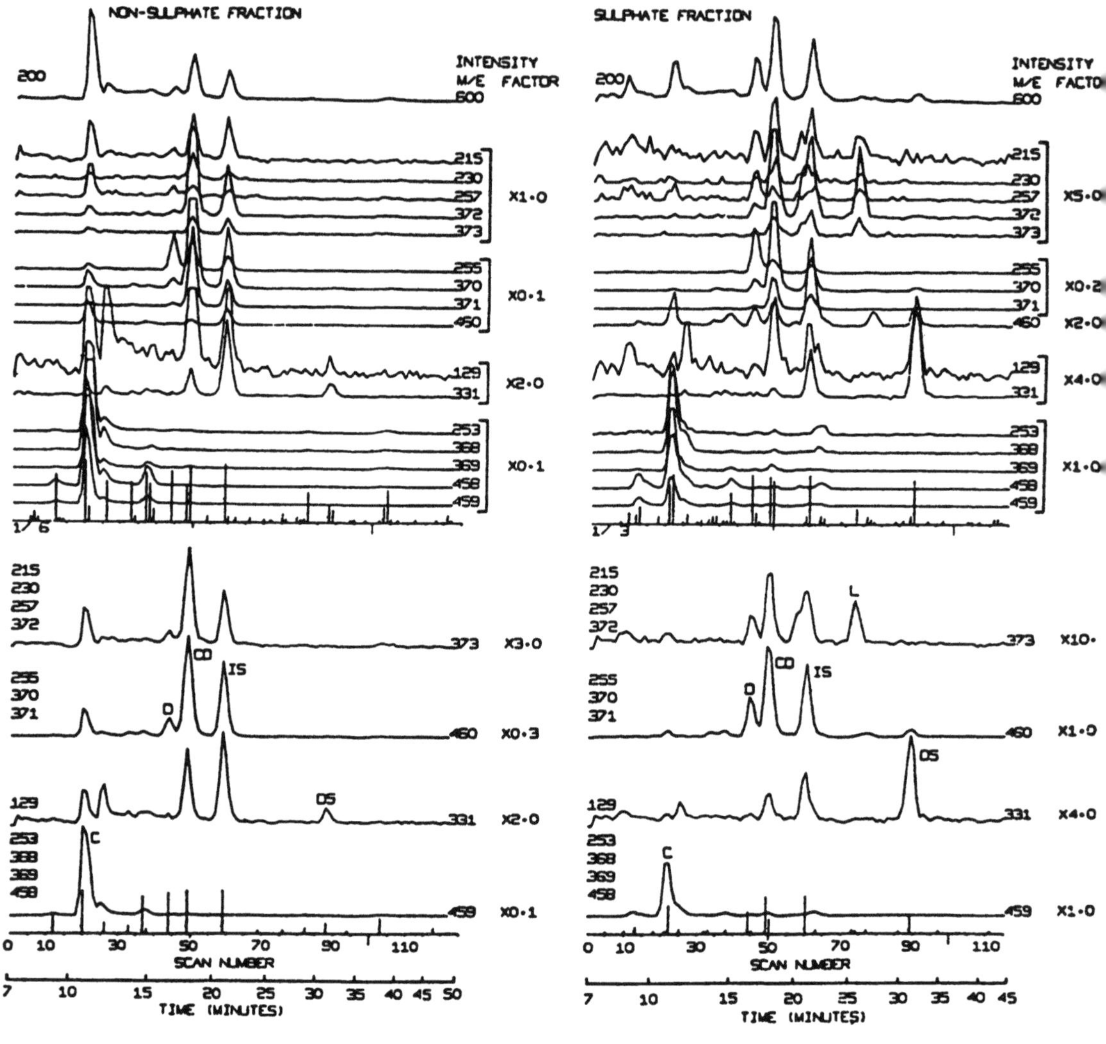

Figure 10. Computer evaluation of the GC-MS analysis of Me-TMS ether derivatives of bile acids in the sulphate and non-sulphate fractions of plasma from a patient with intrahepatic cholestasis of pregnancy. Shown are total ion currents (*m/z* 200 – 600) and fragment ion current chromatograms of series of ions selected for saturated monohydroxycholanoates (*m/z* 215, 230, 257, 372, 373), saturated dihydroxy- and monounsaturated mono-hydroxycholanoates (*m/z* 253, 368, 369, 458, 459) and a 5,6-double bond in 3-trimethylsiloxycholanoate (*m/z* 129, 331). Lower traces are the summed ion current chromatograms of each series of ions above. The internal standard used was hyodeoxycholic acid and hyocholic acid was an impurity in this standard. (Taken from ref. 259 with permission.)

inserted on a probe directly into the mass spectrometer. The mass spectrum revealed fragment ions typical of the 3α-*O*-β-D-glucuronide and was comparable to authentic chenodeoxycholic acid 3α-glucuronide. The concentration of chenodeoxycholicglucuronide and 3β-hydroxy-5-cholenoic acid glucuronide in the plasma were 0.4 and 0.8 $\mu g \cdot ml^{-1}$ respectively. These studies provided direct chemical evidence for the coupling of glucuronic acid to bile acids in man and although glucuronides constitute only a small percentage of the total serum bile acids their increased water solubility may be an important factor in the elimination of hepatotoxic monohydroxy bile acids. More accurate measurements of bile acid glucuronides were later made using SIM with stable isotopically labeled glucuronide conjugates as internal standards (265, 266).

In a qualitative study of serum bile acids employing methyl ester acetates (267), a number of unusual unidentified bile acids were found in the sera from 7 patients with myotonic muscular dystrophy. A significant increase in deoxycholic acid (2.1 $\mu mol \cdot l^{-1}$ compared with 0.7 $\mu mol \cdot l^{-1}$ for the control group) was also noted which led to speculation on the implications of these findings in relation to the disease with respect to the effects of bile acids on cellular membrane fluidity (267).

The inherent sensitivity of SIM has made this the technique of choice in the measurement of bile acids in serum (refs. 110, 136). The principles of the technique (268) have been described in Chapter 1 of this volume.

It has to be recognized that the prior selection of the masses of ions to be monitored introduces a significant bias to the analysis and disregards the presence of bile acids other than those selected for measurement. The SIM mode is primarily used to quantify bile acids and the specificity of the assay relies also upon the GC retention data. Quantification is performed by using an internal standard and measuring the ratio of the peak height or peak area response for the bile acid compared to the peak height or area of the ion monitored for the internal standard. Using a series of calibration standards the amount of the bile acid in the sample is determined from the derived calibration curve.

A variety of internal standards have been employed in SIM assays of bile acids (110, 136). In a comprehensive study of serum and interstitial fluid of patients with cholestatic liver disease, with and without associated pruritus, ions were chosen which were non-specific but characteristically found in the spectra of Me-TMS ether derivatives of all mono-, di- or trihydroxy bile acids. These ions, m/z 368, 370 and 372, and GC retention values were used for the identification of each bile acid. The serum samples were fractionated, following extraction, into groups based upon their conjugation state using Lipidex-DEAP and the bile acids in each of the unconjugated, glycine, taurine and sulphate fractions quantified against coprostanol as an internal standard. There was no correlation between serum or interstitial fluid total bile acid or individual bile acid concentrations and pruritus, and no correlation with the conjugation state of bile acids. Furthermore, the qualitative bile acid profiles of interstitial fluid and serum were similar and concentrations of total bile acids were linearly and positively correlated. These findings did not support a causative role for bile acids in the pruritic condition associated with cholestatic liver disease (145) and the etiology of this condition still remains unknown.

Diurnal changes in serum unconjugated bile acids in normal man have been investigated (263). Unconjugated bile acids were isolated by Lipidex 1000 chromatography (as described in Section 2) and analyzed by SIM as the Me-TMS ether derivatives

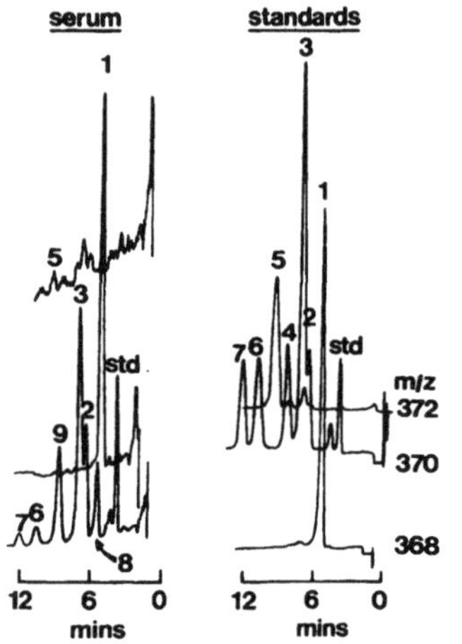

Figure 11
Qualitative GC-MS analysis with SIM of the methyl ester-trimethylsilyl ether derivatives of a mixture of authentic bile acids (50 pmol) and the unconjugated bile acid fraction from the serum (1 ml) of a normal adult. Chromatograms of the ions *m/z* 368, 370 and 372 are shown and the following bile acids are indicated: (1) cholic acid, (2) deoxycholic acid, (3) chenodeoxycholic acid, (4) hyodeoxycholic acid, (5) lithocholic acid, (6) ursodeoxycholic acid, (7) 3β-hydroxy-5-cholenoic acid, (8) 3β,7α-dihydroxy-5β-cholanoic acid, (9) 3β,7β-dihydroxy-5β-cholanoic acid. Coprostanol (10 nmol) was the internal standard used and gas chromatography was performed on a 1% Hi-eff 8BP packed column at 235 °C isothermal. (Taken from ref. 263 with permission.)

(Fig. 11). This study demonstrated that unconjugated bile acids constitute up to 30–40% of the total bile acids in serum, confirming the early reports of Makino et al. (262). The presence of unusual secondary bile acids in serum would appear to be consistent with colonic absorption and suggests that measurements of this fraction of bile acids may be of value in assessing gastrointestinal function (93, 136). The methodology is relatively simple and rapid and lends itself to studies of bile acid absorption where oral loads of unconjugated bile acids were administered.

The most appropriate internal standard for SIM to attain high accuracy is a stable labeled analogue of the compound to be measured since this can be added directly to the serum (or other fluid) and will then compensate for losses which may occur during the extraction, isolation, purification and derivatization steps, and in addition could be used in excess to act as a carrier to limit adsorption losses in the GC-MS. In the selection of an internal standard consideration should be given to (1) the chemical stability of the stable isotopic atoms incorporated, (2) the degree of isotopic enrichment and the number of atoms incorporated, which should be high to minimize interference from naturally occurring isotopes and vice versa, and to (3) isotope effects, which are greater for ^{2}H atoms than for ^{13}C or ^{18}O atoms.

Most of the early stable isotope dilution SIM assays for bile acids utilized deuterium labeled analogues of the principle primary and secondary bile acids (146, 269–272) which took no account of possible losses during hydrolysis steps, and in many instances bile acid sulphates were not determined by the procedure. More recently conjugated bile acids have been employed to overcome this problem (265, 266, 273).

In an attempt to measure mono-oxo bile acids in portal venous and peripheral blood Björkhem et al. (274) used sodium borodeuteride to chemically reduce oxo bile acids to the corresponding hydroxy analogues. These then differ from the natural hydroxylated bile acids by one mass unit due to incorporation of 1 deuterium atom following reduction. The ratio of the stable isotope labeled products to the endogenous hydroxy bile acids can be measured and if this is done after prior determination of the endogenous hydroxy bile acids, levels a semi-quantitative measurement of the original mono-oxo bile acids in serum can be deduced. This approach requires two separate analyses of the serum and provides no information of the structure of oxo-bile acids measured.

Table 3 lists some of the published methods for serum bile acids which employ SIM and scanning techniques including the types of internal standards, GC conditions and the reported ranges for healthy subjects.

Applying GC-MS and SIM, bile acids have been measured in 10 patients with cerebrotendinous xanthomatosis (278). In this lipid storage disease, the metabolic defect results in a diminished bile acid synthesis and a small bile acid pool size with a concomitant increase in bile alcohols (C_{27}-sterols). Circulating levels of bile acids are consequently low and under these conditions their measurement required the higher sensitivity of the SIM mode. Cholic acid (0.34 ± 0.29 $\mu g \cdot ml^{-1}$) was the major serum bile acid constituting 69.5% of the total bile acids. Although chenodeoxycholic acid was present, its concentration (0.11 ± 0.13 $\mu g \cdot ml^{-1}$) was markedly reduced compared with the levels of this bile acid in healthy control sera, and only traces of deoxycholic, lithocholic and ursodeoxycholic acid were reported (278).

In a series of articles by Björkhem and co-workers (269, 271, 276, 279–281), fasting and postprandial concentrations of bile acids in the serum of healthy subjects and patients with hyperlipoproteinemias in response to meal stimulus have been reported and these studies essentially illustrate how the mass spectrometer is utilized as a specific GC detector in more routine bile acid measurements. This approach has been made easier by the increased commercial availability of bile acids labeled with stable isotopic atoms, and has lead to a rapid expansion in publications where GC-MS and SIM have been used.

The combined application of the scanning mode and SIM allows for a better understanding of the qualitative and quantitative composition of serum bile acids than is possible by more routine techniques such as radioimmunoassay. Table 4 lists those bile acids which have been found in adult human serum.

Mention should be made of the potential of FAB ionization mass spectrometry for the direct determination of bile acid conjugates. Bile acid sulphates in particular are difficult to analyze unless they are first separated, solvolyzed and analyzed in their deconjugated state. It has been demonstrated that FAB ionization of bile acid sulphates leads to excellent mass spectra in both the positive and negative ion mode (110, 111) and spectra for other amidated conjugated bile acids have been described previously (96–99, 111, 282).

FAB is a useful rapid means of determining the mode of conjugation and the number of hydroxyl groups in the bile acid molecule. The technique has been successfully applied to the qualitative analysis of conjugated bile salts in duodenal bile from children with failure 'to thrive' and cystic fibrosis (96). More recently FAB-MS was shown to have adequate sensitivity to detect bile acid conjugates in serum of patients with cholestatic liver disease (111). Figure 12 shows the FAB ionization mass spectra obtained following a simple extraction of the serum from three patients with cholestatic liver disease. The bile acids were analyzed after conversion to triethylamine salts and mass spectra obtained in negative ion mode. The procedure takes only minutes to perform and simplified spectra are obtained also in positive FAB.

One of the more attractive and potentially most useful applications of GC-MS is studying bile acid kinetics. Pool sizes and turnover rates can be determined after the intravenous administration of radiolabeled tracers and collection of duodenal bile

Table 3. Methods for gas chromatography-mass spectrometry analysis of bile acids in serum

Method	Ref.	Bile acid determined	GLC phase	Derivative	Internal standard	Mode of operation	Normal range ($\mu mol \cdot l^{-1}$)
Angelin et al. (1977)	269	Cholic	1.5% SE-30	Me-TMS	$^2H_5$-CA	SIM	0.46 ± 0.06
		Chenodeoxycholic			$^2H_3$-DCA		1.06 ± 0.11
		Deoxycholic			$^2H_3$-DCA		1.22 ± 0.21
Bremmelgaard and Almé (1980)	233	Total, individual bile acids	SE-30 and Hi-eff 8 BP	Me-TMS	External standards	Repetitive scanning *m/z* 0-750 daltons	0.27 ± 1.32 (*n* = 2)
Beppu et al. (1981)	272	Cholic	3% QF-1	HFIP-TFA	$^2H_3$-DCA	SIM	0.22 – 1.88
		Chenodeoxycholic					0.64 – 3.34
		Deoxycholic					0.24 – 2.64
		Ursodeoxycholic					0.04 – 0.39
		Lithocholic					0.05 – 0.23
Barnes et al. (1981)	275	Cholic	25 m SE-30 capillary	Me-TMS	23-Nor-DCA	SIM	
		Chenodeoxycholic					22.3 ± 4.1
Björkhem et al. (1982)	274	Mono-oxo-bile acids	1.5% SE-30	Me-TMS	–	SIM	0.08 ± 0.02
Bartholomew et al. (1982)	145	Mono-,di-,tri-hydroxy bile acids	1% Hi-eff ABP	Me-TMS	Coprostanol	SIM	0.28 – 1.30
Takikawa et al. (1982)	265	Lithocholic	3% QF-1	HFIP-TFA	$^2H_3$-LCA-gluc	SIM	0.03 ± 0.03
		Deoxycholic			$^2H_3$-DCA-gluc		0.21 ± 0.31
		Chenodeoxycholic					0.21 ± 0.19
		Ursodeoxycholic					0.03 ± 0.02
		Cholic					0.02 ± 0.02
		Glucuronides					
Setchell et al. (1982)	263 135	Unconjugated mono-, di-,trihydroxy bile acids	1% Hi-eff 8BP	Me-TMS	Coprostanol	SIM	0.1 – 1.5
Björkhem and Falk (1983)	276	Cholic	1.5% SE-30	Me-TMS	$^2H_5$-CA	SIM	0.29 ± 0.81
		Chenodeoxycholic			$^2H_5$-CDCA		0.18 ± 0.32
		Deoxycholic			$^2H_5$-DCA		0.77 ± 0.38
Tohma et al. (1984)	277	3β-Hydroxy-5-cholenoic sulphate	1.5% Poly I.110	Me-TMS	[$^2H_5$]glyco-sulphate analogues	SIM	(0.48 $\mu g \cdot ml^{-1}$)

Table 4. List of bile acids reported to occur in human serum*

Systematic name	Trivial name
Monohydroxy	
3α-Hydroxy-5β-cholanoic	Lithocholic
3α-Hydroxy-5α-cholanoic	*allo*-Lithocholic
3β-Hydroxy-5-cholenoic	
3β-Hydroxy-5β-cholanoic	*iso*-Lithocholic
3β-Hydroxy-5α-androstane-17β-carboxylic	
3β-Hydroxy-5-androstene-17β-carboxyl	
3β-Hydroxybisnorchol-5-en-22-oic	
Dihydroxy	
3α,12α-Dihydroxy-5β-cholanoic	Deoxycholic
3α,12α-Dihydroxy-5α-cholanoic	
3β,12α-Dihydroxy-5β-cholanoic	
3β,12α-Dihydroxy-5-cholenoic	
3α,7α-Dihydroxy-5β-cholanoic	Chenodeoxycholic
3β-7α-Dihydroxy-5β-cholanoic	
3α,7β-Dihydroxy-5β-cholanoic	Ursodeoxycholic
3β,7β-Dihydroxy-5β-cholanoic	
3α,7α-Dihydroxy-5α-cholanoic	
3α,6α-Dihydroxy-5β-cholanoic	Hyodeoxycholic
3α,6β-Dihydroxy-5β-cholanoic	
Trihydroxy	
3α,6α,12α-Trihydroxy-5β-cholanoic	Cholic
3α,7α,12α-Trihydroxy-5α-cholanoic	
23-nor-3α,7α,12α-Trihydroxy-5β-cholanoic	Nor-cholic
3α,6α,7α-Trihydroxy-5β-cholanoic	Hyocholic
3α,6β,12α-Trihydroxy-5β-cholanoic	
3α,6α,7β-Trihydroxy-5β-cholanoic	ω-Muricholic
3α,7β,12α-Trihydroxy-5β-cholanoic	
3α,7α,12α-Trihydroxy-27a,27b-dihomo-5β-cholestane-26a,27b-dioic	
Tetrahydroxy	
3,6,7,12-Tetrahydroxy-5β-cholanoic	

* Compiled from the following references: 145, 165, 233, 259, 263, 264, 267, 275, 347, 362.

samples over a prolonged time (28, 53). This approach is unacceptable in pregnant women and newborn infants but identical information can be obtained using stable isotopically labeled bile acid tracers. Bile acids labeled with deuterium and ^{13}C have been synthesized (e.g. 283–286) and stable labeled analogues of the primary bile acids, cholic and chenodeoxycholic are now commercially available. Most of the early investigations were carried out on bile samples (287–291) because of the requirement to obtain sufficient concentrations of bile acids for analysis. With the recent improvements in methodology and the development of more sensitive and stable instrumentation, bile acid kinetics can be determined from serum samples, thus obviating the need for invasive duodenal intubation procedures (166, 292, 293). In practice, these studies require the collection of a blood sample before and on separate days after the

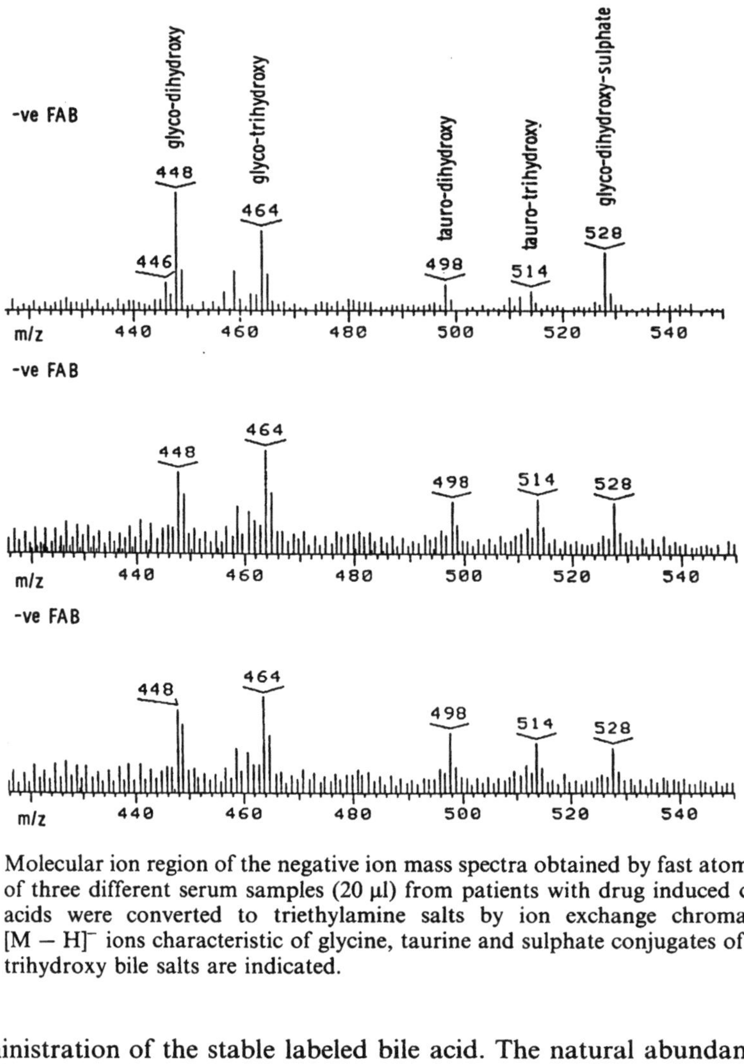

Figure 12. Molecular ion region of the negative ion mass spectra obtained by fast atom bombardment of three different serum samples (20 μl) from patients with drug induced cholestasis. Bile acids were converted to triethylamine salts by ion exchange chromatography. The $[M - H]^-$ ions characteristic of glycine, taurine and sulphate conjugates of dihydroxy and trihydroxy bile salts are indicated.

oral administration of the stable labeled bile acid. The natural abundance of $^{13}C/^{12}C$ is determined from the isolated serum bile acids and the atom% excess calculated and plotted versus time (294).

DeMark et al. (166) used ammonia CI to analyze the methyl ester acetate derivatives of chenodeoxycholic and cholic acids and carried out SIM of the predominant high mass ammonium adduct ion $[M + 18]^+$. The ratios of the ions *m/z* 509/508 (chenodeoxycholic) and 567/566 (cholic) were determined and the mean natural abundance of molecules with one ^{13}C atom was 33.24% and 35.25% respectively for the authentic [^{13}C] labeled bile acids. Similar values were found when the two bile acids were isolated from serum and the precision of these natural abundance measurements $\pm$ 0.6% (coefficient of variation). Similarly, [24-^{13}C]chenodeoxycholic acid kinetics were determined using the equivalent of 40 μl of serum (2% of a 2 ml serum sample injected) with EI and capillary column GC-MS of the Me-TMS ether derivatives by scanning over narrow mass ranges, *m/z* 368–373 and *m/z* 460–461 (292). Using a quadrupole mass spectrometer and computing the average of 10 values over each peak, natural abundance measurements were possible with a high precision, 0.1 – 0.9% CV. From the curves obtained it is possible to determine the pool size, fractional turnover and daily synthesis rates. In one healthy individual, measurements of these values taken on three occasions over a four month period were reproducible (CV $\pm$ 20%). Mean values for chenodeoxycholic acid pool size of 22.4 $\pm$ 4.6 $\mu mol \cdot kg^{-1} \cdot day^{-1}$ compared well with values obtained using radioactive tracers and biliary bile acid analysis. Satisfactory comparisons between serum and bile were also reported when [11,12-$^{2}H_{2}$]chenodeoxycholic acid was used (295). These studies were further extended to allow measurements of deoxycholic acid pool size and input rate to be determined from serum samples following the oral administration of [24-^{13}C]deoxycholic acid (296, 297). In all instances good correlations have been demonstrated between serum and biliary derived data and data obtained using radioactivity. This approach clearly offers a useful non-invasive and non-hazardous method of characterizing bile acid kinetics.

4.2 Bile

The major non-sulphated bile acids in bile of human adults with normal liver function are cholic, chenodeoxycholic and deoxycholic acids with small amounts of lithocholic, ursodeoxycholic, *allo*-cholic and trace amounts of other metabolites. Biliary bile acids are almost all conjugated to either glycine or taurine with the ratio of glycine to taurine conjugation being approximately 3:1 in healthy subjects (15, 16). As bile is a less available fluid than serum or urine it is less favored for use in assessment of patients with diseases affecting bile acid metabolism. Cholestatic conditions, in particular, provide difficulties for collection of fluid. Nevertheless bile has been extensively investigated by many workers using a variety of techniques. GC-MS has been useful for the identification of the less common bile acid metabolites in various disease states, in metabolic studies and also in the quantification of bile acids in relation to kinetic studies (287 – 289, 295, 298, 299). In a recent report, GC-MS was used to identify unconjugated bile acids in human bile (300) of six healthy adults and three patients with cholelithiasis. Unconjugated bile acids constituted 0.1 – 0.4% of the total biliary bile acids and this fraction contained in addition to the conventional bile acids, a number of nor-bile acids and cholestanoic acid isomers.

A GC-MS method for the micro determination of bile acids in liver tissue (146) is discussed in Section 4.7 and was applied to correlate bile acid composition in both liver tissue and bile (273). The method employed polydeuterated internal standards of glycine conjugates of lithocholic, deoxycholic, chenodeoxycholic, ursodeoxycholic and cholic acids, which were added to the samples. Following extraction and hydrolysis the ratio of individual labeled to unlabeled acids were determined by selectively monitoring characteristic ions in the mass spectra of their ethyl ester dimethylethylsilyl ether derivatives throughout their elution from the capillary GC system. The bile acid composition correlated well between liver tissue and bile but a higher ratio of cholic to chenodeoxycholic and primary to secondary bile acids was found for liver tissue.

A stable-isotope tracer technique using [$^{2}H_{2}$] labeled chenodeoxycholic acid and repetitive scanning GC-MS was developed to study the steady state kinetics of the enterohepatic circulation of bile acids in patients with Crohn's disease (291). Capsules

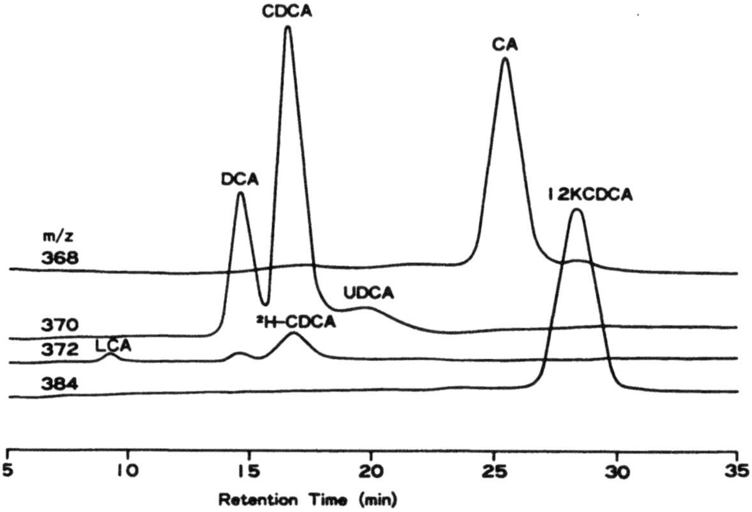

Figure 13. Mass fragmentograms of methyl ester propionate derivatives of bile acids in bile obtained from a normal subject. (Taken from ref. 291 with permission.)

(50 mg) of [11,12-$^2H_2$]chenodeoxycholic acid were administered to patients and controls and sequential samples of fasting bile obtained after 1, 2, 3 and 4 days. The methyl ester propionate derivaties of the extracted bile acids were submitted to GC-MS with repetitive scanning analysis. Figure 13 shows the trace obtained by plotting selected ions to determine the bile acids present in a normal bile and for 12-oxo-chenodeoxycholic acid used as the internal standard. The ion m/z 372 was selected to measure [$^2H_2$]chenodeoxycholic acid and account taken of the natural isotopic contribution at this mass from unlabeled chenodeoxycholic acid. The pool size of chenodeoxycholic acid, the total bile acid pool size and the biological half-life of [$^2H_2$] chenodeoxycholic were calculated and in Crohn's disease were all shown to be significantly reduced as compared with normal subjects. This was attributed to impaired bile acid absorption in these patients.

A similar methodological approach for the quantification of bile acids, and for ursodeoxycholic acid in particular, in bile samples has been reported (301). In this case ursodeoxycholic acid was analyzed in bile samples obtained from a patient with complete bile fistula following oral administration of [$^2H_2$]ursodeoxycholic acid. Within the first 2 hours 88% of the [$^2H_2$]ursodeoxycholic acid was recovered and after 6 hours the recovery was 95% and without significant metabolism to other bile acids. The method has obvious application to the pharmacokinetic study of administered ursodeoxycholic acid in normal and diseased states.

The hepatic and renal excretion of individual bile acids has been studied in patients with extrahepatic cholestasis (302). A careful investigation of individual bile acids as glycine, taurine and sulphate conjugates in bile, serum and urine was made during a period of percutaneous transhepatic drainage with bile refeeding. Bile acids were quantified by GC and identified by GC-MS. This provided an opportunity to observe the changing composition and concentrations of bile acids resulting from the restoration of an enterohepatic circulation. Concentrations of glycine and taurine conjugates of cholic and chenodeoxycholic in serum decreased rapidly but sulphated chenodeoxycholic acid concentrations decreased slowly. In concert with this there was a gradual increase in biliary sulphated bile acids. Bacterial production of deoxycholic and lithocholic acids was restored.

GC-MS is frequently used as a confirmatory technique to positively identify components in GC or LC chromatograms to enable the confident use of these methods for routine use. This may be of particular value for the LC analysis of intact conjugates for example (303).

Investigations of bile acid metabolism in newborns and children during health and disease have highlighted differences in synthesis and composition from that of the adult (304, 305). Bile acids have been reported in gallbladder bile as early as 14 to 16 weeks gestation (306) with chenodeoxycholic acid concentrations greater than cholic acid. In a GC-MS study of gallbladder bile derived from human fetuses during early gestation (week 14 – 20) bile acid synthesis by the fetal liver has shown to be reasonably well developed (307, 308). Chenodeoxycholic acid concentrations exceeded those of cholic acid indicating an immaturity in hepatic 12α-hydroxylase activity at this stage of development. This was also demonstrated from *in vitro* studies using human fetal liver (309). These biliary bile acid profiles also served to highlight an efficient C-6, C-4 and C-1 hydroxylation during gestation. The capacity of the fetal liver to 1-hydroxylate *in vitro* was confirmed by incubations of taurodeoxycholic acid with human fetal liver microsomes and GC-MS identification of the products formed (310) while both C-6 and C-1 hydroxylation of lithocholic acid was also demonstrated (311). Rationalization of the increased diversity of bile acids observed in fetal and early life requires further understanding of fetal synthesis, placental transport, maturation of liver enzymes and other factors (312) and it is evident from the above examples that GC-MS will continue to play a key role in such studies.

In an attempt to identify bile acids which might reflect primary products of neonatal bile acid synthesis, a GC-MS study was made of gastric fluid aspirated from infants with high intestinal obstruction prior to surgery (313, 314). These newborns had atresia of the duodenum and thus bile which is unable to pass down the intestine, refluxes into the stomach. The low pH inhibits bacterial metabolism and precipitates bile acids. Bile acids present are therefore most likely to reflect products of neonatal hepatic synthesis and not of bacterial metabolism. A list of some of the non-sulphated bile acids identified in these samples is given in Table 5 and is similar to metabolites found in fetal gallbladder bile (307, 308). Subsequently an investigation of the distribution of bile acids within the various conjugate groups and further identifications have been made with the aid of GC-MS (314).

Several unusual features of neonatal bile acid metabolism were evident. The presence of a 7α,12α-dihydroxy-3-oxo-chol-4-en-24-oic and 7α,12α-dihydroxy-3-oxo-5β-cholan-24-oic may represent intermediates in the production of cholic acid from cholesterol indicating that side chain oxidation can precede some of the nuclear changes. The 3-oxo-bile acids and 3β-epimers of cholate and chenodeoxycholate, however, may arise from the action of hepatic 3α-hydroxysteroid dehydrogenase enzymes. The ability of fetal liver microsomes to hydroxylate the steroid nucleus at a number of positions in addition to C-3, C-7 and C-12 (309, 310) was reflected by the presence in the gastric

Table 5. List of bile acids identified in gastric aspirate of newborn infants

7α-Hydroxy-3-oxo-5β-cholanoic
3α-Hydroxy-7-oxo-5β-cholanoic
3β,7α-Dihydroxy-5β-cholanoic
3α,7α-Dihydroxy-5β-cholanoic
7α,12α-Dihydroxy-3-oxo-5β-cholanoic
7α,12α-Dihydroxy-3-oxo-4-cholenoic
3β,7α,12α-Trihydroxy-5β-cholanoic
3α,12α-Dihydroxy-7-oxo-5β-cholanoic
3α,6α,7α-Trihydroxy-5β-cholanoic
3α,7α,12α-Trihydroxy-5β-cholanoic
3α,7α,12α-Trihydroxy-5α-cholanoic
3α,6α,7α12α-Tetrahydroxy-5β-cholanoic
2β,3α,7α,12α-Tetrahydroxy-5β-cholanoic
1β,3α,7α,12α-Tetrahydroxy-5β-cholanoic
24-nor-3α,7α,12α-Trihydroxy-5β-cholan-23-oic
3α,7α,12α,25-Tetrahydroxycholestan-26-oic (2 isomers)
(25ξ)-3α,7α,12α,25-Tetrahydroxy-5β-cholestan-26-oic
(25ξ)-3α,7α,12α,26-Tetrahydroxy-5β-cholestan-27-oic

aspirate of bile acids with hydroxyl groups at C-1, C-2 and C-6. A series of cholestanoic acids containing the 3α,7α,12α-trihydroxy-5β nucleus and hydroxyl groups at C-24, C-25 and C-26 were all indicative of fetal pathways of metabolism which are not evident in the normal adult.

4.3 Feces

There is an extensive literature on methods for the determination of fecal bile acids and equally on their measurement in various pathological states (315). From a quantitative standpoint, mass spectrometry to date has proved to be of little direct value in fecal bile acid analysis, but has nevertheless been important for confirming the identification of peaks in GC chromatograms of fecal bile acids. The methods most commonly used for the analysis of fecal bile acids are based upon the GC techniques developed in the 1960's (50, 316–318). In all of these methods rigorous extraction, saponification and/or chemical modification of the bile acids are carried out while liquid-liquid extraction and TLC steps are used to remove unwanted classes of interfering compounds from the extract prior to GC determination of bile acids. Unwanted side reactions and poor recoveries of bile acids often result by some of these treatments, and information on the exact nature of bile acids in feces is lost.

In some of the earliest examples of the use of GC-MS analysis, Eneroth et al. described the detailed composition of bile acids in feces (318–320). Extracts of feces were fractionated on silicic acid columns and bile acids isolated by preparative TLC and analyzed as their Me-TMS ether and Me-TFA derivatives. Tentative identification of bile acids by GC relied heavily on the retention data on different liquid phases and the shift in retention values of compounds in response to chemical reaction, e.g.

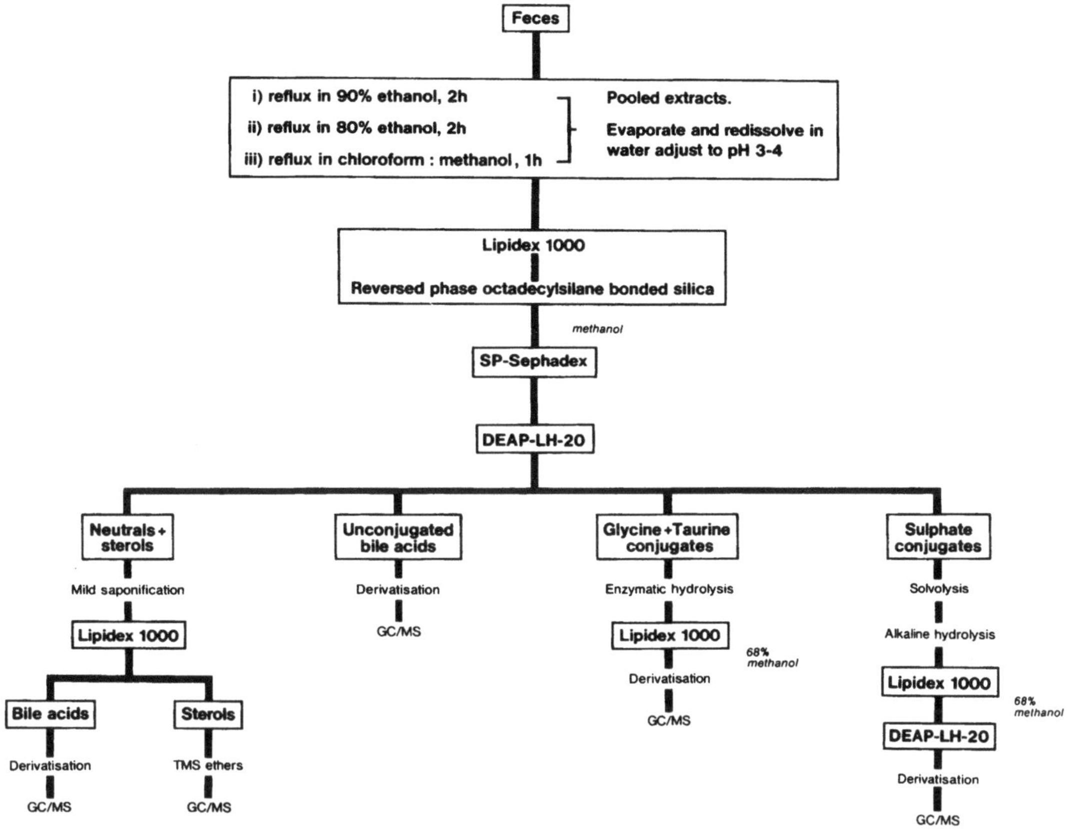

Figure 14. General scheme developed for the analysis of bile acids, steroids and related compounds in feces and described in detail in ref. 64. (Reproduced with permission.)

oxidation, silylation. Such retention data gives valuable information regarding the number, nature and configuration of substituents in the steroid nucleus necessary for the characterization of new or unknown metabolites.

Human feces from healthy subjects is characterized by the presence of the principal secondary bile acids, deoxycholic and lithocholic, formed from conjugated cholic and chenodeoxycholic acids after bacterial deconjugation and 7α-dehydroxylation (Fig. 1). Hydroxy bile acids (at positions 3, 7, 12) are quantitatively the most important group of metabolites excreted (319, 320), but significant amounts of oxo-bile acids are to be found; the 3-hydroxy-12-oxo isomers being the major ones excreted (318).

In these early studies, because of the inclusion of a hydrolysis step (also the case for other published methods at the time), the determination of conjugated bile acids and esters was not possible. In an attempt to gain more information on the composition of bile acids in feces and to minimize the possibility of artefact formation, a general method (Fig. 14) for the analysis of metabolic profiles of bile acids and related compounds was developed (64) which currently represents perhaps the most definitive approach to fecal bile acid analysis. The method involved an initial organic extraction by sequential solvent reflux followed by a selective extraction, isolation and partial

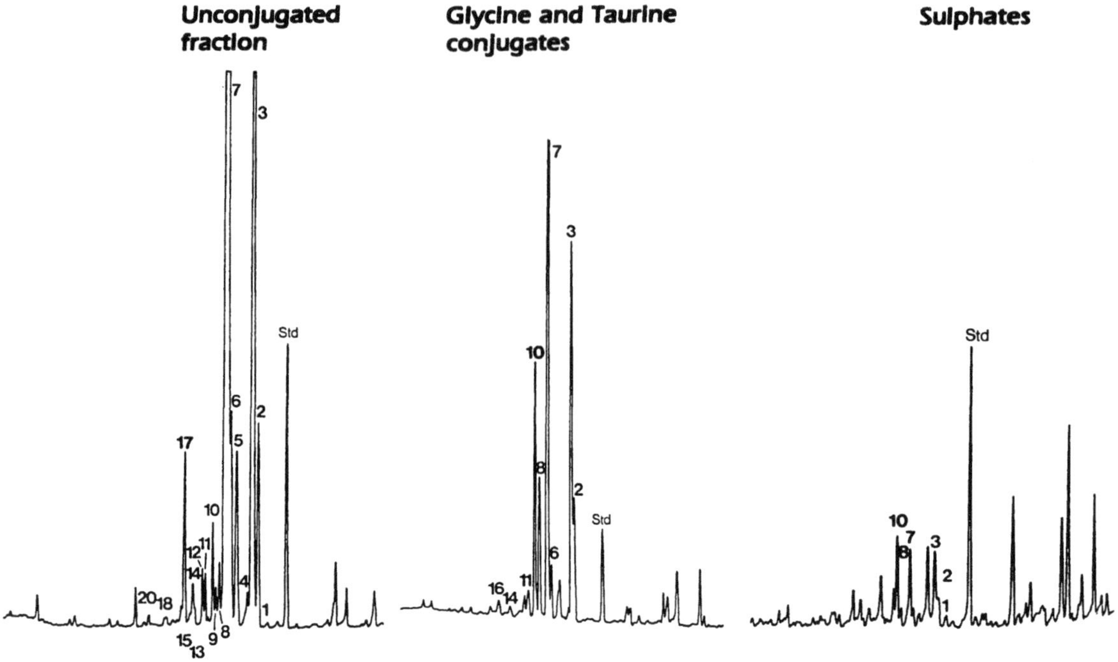

Figure 15. Gas chromatographic analysis of the bile acids excreted in the feces of a healthy adult. Bile acids were separated into a) unconjugated bile acids b) glycine and taurine conjugates combined and c) sulphates, using lipophilic gel chromatography on Lipidex-DEAP. After deconjugation and/or preparation of methyl ester-trimethylsilyl ether derivatives analysis was performed on a 30-m DB-1 chemically bonded fused silica column using temperature programmed operation 225 °C – 295 °C with increments of 2 °C/min. The internal standard (Std), coprostanol, was added and a list of those bile acids which were identified by mass spectrometry are as follows: (1) 3α-hydroxy-5α-cholanoic; (2) 3β-hydroxy-5β-cholanoic; (3) 3α-hydroxy-5β-cholanoic; (4) unknown monohydroxy bile acid; (5) 3α,12β-dihydroxy-5β-cholanoic; (6) 3β,12α-dihydroxy-5β-cholanoic; (7) 3α,12α-dihydroxy-5β-cholanoic; (8) 3α,7α-dihydroxy-5β-cholanoic; (9) 12α-hydroxy-3-oxo-5β-cholanoic; (10) 3α,7α,12α-trihydroxy-5β-cholanoic; (11) 3α,6α-dihydroxy-5β-cholanoic; (12) 3β,6α-dihydroxy-5β-cholanoic; (13) dihydroxy-monooxo bile acid; (14) 3β,12α-dihydroxy-5α-cholanoic; (15) 3α-hydroxy-12-oxo-5α-cholanoic; (16) 3β,12β-dihydroxy-5α-cholanoic; (17) 3α-hydroxy-12-oxo-5β-cholanoic; (18) unknown bile acid; (19) 7α,12α-dihydroxy-3-oxo-5β-cholanoic; (20) unknown bile acid; (21) dihydroxy-monooxo bile acid.

purification of the extract using a combination of liquid-gel (Lipidex 1000) and liquid-solid extraction steps (Bond Elut cartridges). Interfering cations are removed on a cation exchanger and bile acids fractionated according to their mode of conjugation using Lipidex-DEAP. Each conjugate fraction is hydrolyzed under mild conditions and analyzed by capillary column GC and mass spectrometry. Similar schemes using Amberlite XAD-2 for the desalting step after an organic extraction have also been reported (234, 321), but if neutral sterols are to be analyzed (which is often the case for studies of cholesterol metabolism) these methods are less suitable. The complexity in composition of bile acids excreted in feces is evident when typical chromatograms shown in Figure 15 are compared with those published earlier using conventional packed column GC. Using these procedures, studies of the bile acid composition of feces from rats (64), normal adults (64, 322–326) and patients with adenomatous polyps and colonic cancer (324, 325) were reported. For the first time an accurate assessment of the quantitative proportion of conjugated bile acids was possible. More than 90% of the bile acids excreted in feces were present in the unconjugated fraction, the remainder being distributed between the glycine, taurine and sulphate conjugate fractions. It should be realized that while this conjugation pattern is typical of healthy subjects, there may be conditions (e.g. diarrhea, antibiotic therapy) where marked changes in the conjugation state will occur, and in this respect this methodology unlike earlier techniques permits the study of these altered states.

Attention has been given to the role of bile acid sulphates in the intestine, following reports that up to one third of the total bile acids excreted in feces were sulphate conjugates (307, 328). These findings have yet to be confirmed by other groups. It is generally found that less than 10% and most usually 2–5% of the bile acids are present as sulphates and this discrepancy may result from the use of an indirect method (i.e. differential hydrolysis) for the determination of sulphate conjugates (328). Unfortunately at present the only satisfactory technique for the determination of fecal bile acid sulphates is their chromatographic isolation as a specific group prior to mild chemical and enzymatic hydrolysis.

Mass spectrometry was used to confirm the existence of a number of esterified bile acids (329, 330) first observed by Norman and Palmer (331) as saponifiable products of *iso*-lithocholic acid in human feces. The major ester was reported as 5β-cholan-24-oyl-3β-palmitate although evidence for palmitoleyl, stearyl and oleyl esters was presented (332). While the existence of ethyl esters of bile acids have also been reported (329), but these may result from transesterification during prolonged refluxing of fecal homogenates with alcoholic solvents.

Unsaturated bile acid metabolites are of particular interest in view of the suggested links between bile acid excretion, bacteria and colonic carcinogenesis (333, 334), however few studies have been able to find significant amounts of these metabolites. 5β-Chol-3-en-24-oic acid was identified from the mass spectrum of the methyl ester derivative and shown to be formed *in vitro* by gut microflora (330) supporting the early reports of Palmer (19). In patients with colon cancer and those at high risk for this disease, 5β-chol-3-en-24-oic was found at levels between traces and 11.2 $\mu mol \cdot day^{-1}$ and with increased frequency relative to healthy adults thus raising questions as to the possible significance of this bile acid in the pathogenesis of carcinogenesis of the large bowel (323, 324).

3-Methylcholanthrene, a highly unsaturated compound is an extremely potent carcinogen, which *in vitro* can be synthesized from deoxycholic acid (335). This metabolite may be formed from deoxycholic acid in the gut by bacteria and play a role in the etiology of colonic carcinogenesis (333, 334, 336–338). Using a GC-MS SIM technique, 3-methylcholanthrene could not be detected in the stools from five colon cancer patients and two healthy controls (339). Extraction from feces and isolation by HPLC followed by GC-MS analysis using a 3% OV-17 column at 290 °C and SIM of the molecular ion *m/z* 268 and the corresponding ion *m/z* 270 for [^{14}C] methylcholanthrene used as an internal standard, afforded a method with a detection limit of $<35\ ng \cdot g^{-1}$ of stool.

Important information regarding bile acid synthesis rates can be obtained from the determination of fecal bile acid excretion and estimates show a fair agreement with values obtained by the isotope dilution technique (340 and references therein). Using capillary column gas chromatography to resolve the complex mixture of bile acids in feces and mass spectrometry to enable each metabolite to be identified, accurate information regarding not only bile acid synthesis rates but also the relative rates of synthesis of cholic acid chenodeoxycholic is possible (324). The validity of this information is dependent upon accurate fecal collections (341). Investigations of the day-to-day variation in fecal bile acid excretion and synthesis rates of cholic and chenodeoxycholic acids in two healthy subjects and confirmed that values were remarkably consistent provided determinations were calculated from at least 3 day collections of stools. Considerable scatter in these data are found when only single day or single stools are collected (341). The degree of complexity in composition of fecal bile acids (Fig. 15) indicates a continuing need for mass spectrometry to characterize peaks in the GC profiles of fecal extracts.

4.4 Meconium

Since bile acid metabolism in the newborn infant differs from that of the adult, it is not unexpected that a variety of unusual bile acids should be present in meconium. Indeed, bile acid composition of meconium resembles that of fetal gallbladder bile (307, 308). In order to obtain an indication of fetal biosynthetic pathways, Back and Walter (342) investigated the bile acid content of meconium using GC-MS. In addition to cholic, chenodeoxycholic, deoxycholic and lithocholic acids, hyocholic, hyodeoxycholic, 1,3,7,12-tetrahydroxycholanoic, *allo*-cholic, norcholic, 3β-hydroxy-5-cholenoic and 3β,12α-dihydroxy-5-cholenoic acids were also identified. These findings have been confirmed by others (343 – 345). As meconium is sterile *in utero*, products of bacterial metabolism such as deoxycholic acid are presumed to be indicative of maternal placental transfer. Other components reflect the activation of hepatic enzymes which are not quantitatively important later in life. However, since patients with cholestatic liver disease excrete many of the same unusual bile acids this indicates a reversion to some aspects of a fetal metabolism in the disease state.

Sulphated bile acid conjugates predominate in the meconium of babies from uncomplicated pregnancies. In women with intrahepatic cholestasis of pregnancy, however, sulphated and non-sulphated bile acid concentrations are elevated in amniotic fluid and hence are found in increased amounts in meconium (346). The importance of sulphation in man to the detoxification of monohydroxy bile acids (e. g. lithocholic, 3β-hydroxy-5-cholenoic) which are hepatotoxic is still a subject of debate.

A series of saturated and unsaturated monohydroxy steroid acids containing 20, 21 and 22 carbon atoms have been isolated from meconium and identified by mass spectrometry (8, 343, 344). The C-20 acids (etianic acids) have been reported to be present in similar or larger concentrations than the two principal conventional monohydroxy bile acids in meconium, lithocholic and 3β-hydroxy-5-cholenoic acids (77, 344, 345). The accumulation of these short chain acids in meconium may arise from the high concentrations of steroidal hormones in the fetus acting as substrate for side chain oxidation, but as their biosynthesis, metabolism and function is still unclear, further work is necessary to determine their origin. It is interesting to note that a similar series of short-chain bile acids has been reported in sera from cholestatic patients (347).

4.5 Urine

In comparison with fecal excretion, elimination of bile acids in urine is a very minor pathway in the healthy adult. The daily excretion of bile acids in urine is less than 50 μmol (i.e. approximately 20 $mg \cdot day^{-1}$) (65) while fecal excretion is between 100 – 1000 $\mu mol \cdot day^{-1}$. In patients with liver disease, however, urinary bile acid excretion is often influenced as a result of changes in bile acid synthesis, conjugation and a reduced enterohepatic circulation. Their impaired hepatic clearance results in an increased urinary excretion in addition a spill-over into the peripheral circulation.

The low levels of bile acids in urine and the complexity of the conjugation state has made their comprehensive analysis difficult in the past. The introduction of the anion exchange gel, Lipidex-DEAP (Section 3) has greatly assisted the separation of unconjugated and conjugated forms and a comprehensive fractionation and analysis procedure was reported by Almé et al. (65). More recently the Amberlite XAD-2 step has been replaced by extraction with octadecylsilane bonded silica cartridges (Section 3). Quantitative GC-MS is the only method with sufficient specificity to make measurements on the wide range of bile acids present in urine, although other techniques may be adequate to determine specific components as long as conjugate separation is first carried out. The detection and identification of previously unsuspected bile acids has also relied heavily on mass spectrometry and GC-MS.

The most comprehensive GC-MS study of urinary bile acids in normal subjects was made by Sjövall and his co-workers (65). In this study the unconjugated, taurine and glycine conjugates and sulphated forms of 24 bile acids were quantified. The repetitive scanning GC-MS technique which was used has been discussed earlier. The findings of this study are of interest and useful to summarize as a background against which modified excretion profiles of bile acids arising from liver disease may be considered.

Table 6 lists a number of the bile acids present in normal urine and their predominant conjugation state. In general all monohydroxy and most dihydroxy bile acids were present as monosulphates, the sulphate ester being at position C-3. These monosulphated bile acids were all amidated. Trihydroxy bile acids were non-sulphated and the major unconjugated bile acids were cholic and nor-cholic acids. Bile acid glucuronides have been reported to be present only in trace amounts in the urine of normal healthy subjects (348).

The bile acid composition of the urine of normal infants is of interest as it consists of a number of bile acid metabolites which are not present in adults but some of which are found in patients with cholestasis. Tetrahydroxylated bile acids substituted at carbon positions 1,3,7,12 and 3,6,7,12 were shown to be present in equivalent concentrations to cholic acid, for example, and a further one, 2,3,6,7-tetrahydroxycholanoic, was tentatively identified (349). In addition 3β-hydroxy-5-cholenoic acid

Table 6. List of bile acids and their principal mode of conjugation identified in human urine (65)

Bile acid*	Unconjugated	Glycine	Taurine	Sulphate
Monosubstituted				
3α				+
3α (5α)				+
3β-5-ene				+
Disubstituted				
3α,12α				+
3α,12α (5α)				+
3α,12β		+		
3β,12α				+
3β,12α (5α)				+
3β,12α-5-ene				+
3α,7α (5α)				+
3α,7α				+
3α,7β				+
3α,6α				+
3β-12-oxo		+		
3α-12-oxo		+		+
Trisubstituted				
3α,7α,12α	+	+	+	+
3α,7α,12α (5α)		+		
3α,7α,12α(24-nor)	+			
3β,7β,12α (5α)	(+)		(+)	+
3β,7β,12α	+	+		
3α,7β,12α		+		
3α,6β,12α	+	+		
3α,6α,7α	(+)		+	+
1α,3α,12α	(+)	+		

* 5β-cholan-24-oic structure unless otherwise indicated. Numbers indicate positions of hydroxyl substitution.
\+ indicates presence and (+) low concentration

was found in all infant samples and this bile acid is frequently increased in association with liver disease.

The reason for the increased hydroxylation of bile acids in infants is thought to arise from differences from the adult enzyme activity in the neonatal liver (342) or to an increase in hydroxylation as an additional pathway for elimination of bile acids during a period of physiological cholestasis (349).

4.5.1 **Hepatobiliary Diseases**

Numerous studies have been undertaken to evaluate changes in the biosynthesis, circulation and excretion of bile acids as a result of liver disease. While many of these changes are secondary to disturbed liver function, changes in excretion patterns and conjugation state, the relative concentrations in bile, serum and urine and the presence of unusual bile acid metabolites, may have clinical value for diagnosis or assessment. As urinary bile acid composition is the net result of transport processes in the liver, kidney and intestine in addition to synthetic and metabolic alterations, interpretation of urinary bile acid excretion should be approached with caution. In the following sections several diseases of the hepatobiliary system will be briefly summarized in relation to modified urinary excretion of bile acids and the application of mass spectrometry.

4.5.1.1. *Cirrhosis*

Marked changes in bile acid metabolism are usually confined to patients with advanced cirrhosis and depend on the extent of liver cell disease and the presence of porta-systemic shunts. Reduced pools of cholic acid result in a reduction in intestinal metabolism and hence a reduced deoxycholic acid pool size. Chenodeoxycholic acid is the main urinary bile acid excreted in primary biliary cirrhosis (350).

GC-MS has been used in the quantitative determination of bile acids in the urine of cirrhotic patients (350) by the method discussed earlier (65). This has permitted an assessment of the bile acid composition and the distribution of conjugates in patients with portal and primary biliary cirrhosis. Several bile acids, not found in healthy subjects, were identified as hydroxylation products of common biliary bile acids. Both C-1 and C-6 hydroxylated bile acids were evident and also two tetrahydroxylated cholanoic acids substituted at the 1,3,7,12- and 3,6,7,12-carbon atom positions, and two trihydroxylated cholanoics the major one being 1,3,12-substituted and the other tentatively identified as a 1,3,7-trihydroxycholanoic. The mass spectra of the Me-TMS ether derivatives of these components are shown in Figure 16.

The presence of nor-cholic acid (24 nor-3α,7α,12α-trihydroxy-5β-cholan-23-oic) can be readily detected by GC-MS and this metabolite, although of unknown origin, is frequently found in the urine of patients in which the lobular structure of the liver is damaged due to portal and biliary cirrhosis (65, 350, 351). In addition to nor-cholic acid, other unusual trihydroxy bile acids have been identified in urine of patients with liver cirrhosis including hyocholic acid, 1β,3α,12α-trihydroxy-5β-cholanoic and *allo*-cholic (351).

GC-MS was also used in a study of the bile acid composition of serum and urine of two patients with primary biliary cirrhosis and severe cholestasis (264). In these cases 3β,7α- and 3α,6α-dihydroxycholanoics and 3α,6α,7α- and 3α,7β,12α-trihydroxy-cholanoates were identified as minor components which up to that time had not been detected in such samples.

3β-hydroxy-5-cholenoic acid is present in small amounts in normal urine but is frequently elevated in primary biliary cirrhosis. Whether this reflects reduced capacity for hepatic excretion or of increased synthesis is not known.

The detailed classification of conjugated bile acids (Table 6) is based on the initial chromatographic separation of individual classes of conjugates and subsequent enzymic or chemical hydrolysis, prior to identification by GC-MS. Modifications of the lipophilic gel chromatographic system (230) permitted analysis of bile acid glucuronides and demonstrated the presence of significant amounts of glucuronidated and sulphated bile acids in patients with alcoholic cirrhosis (247).

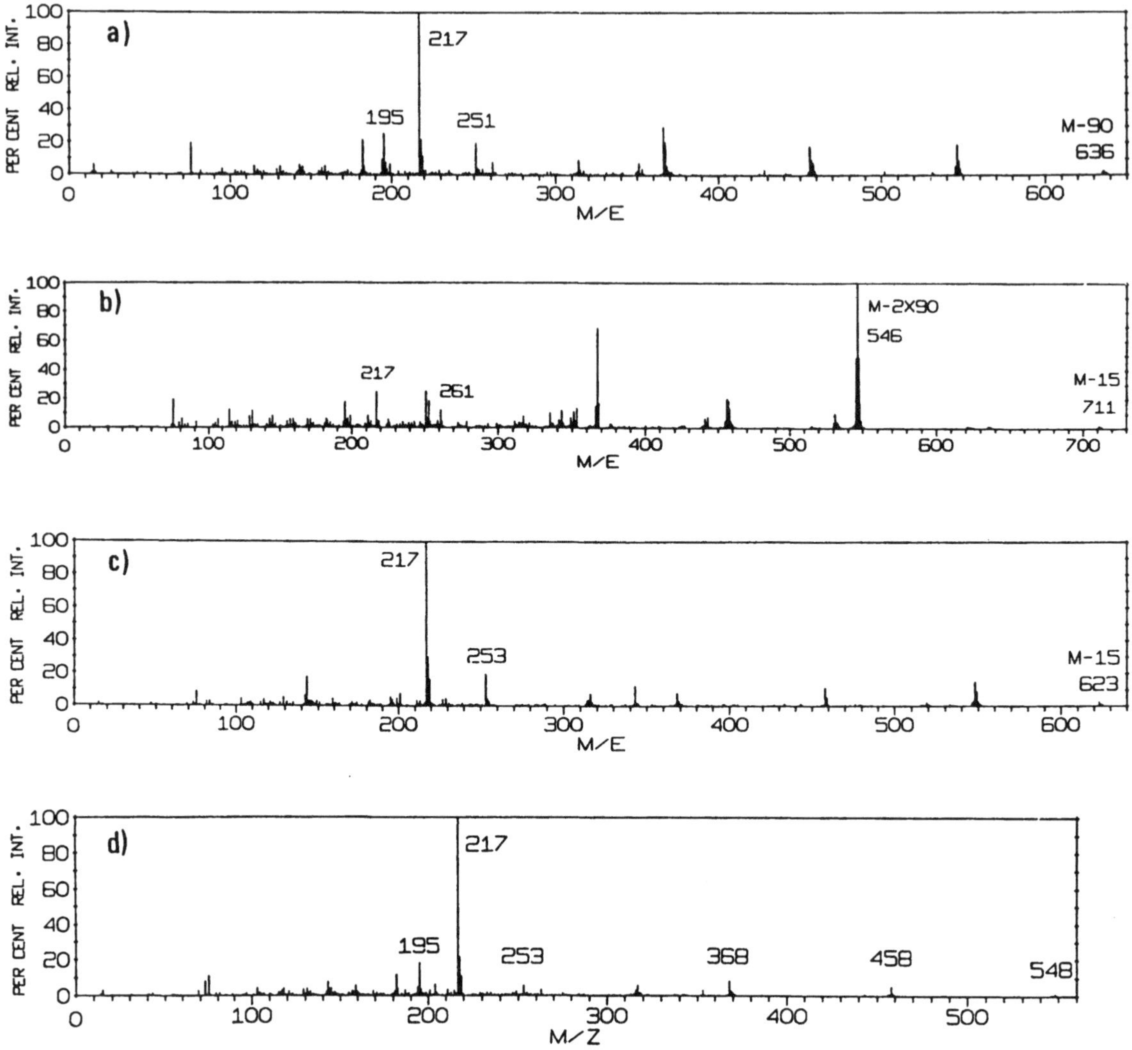

Figure 16. Mass spectra (a–d) of 1,3,7,12-tetrahydroxy-, 3,6,7,12-tetrahydroxy-, 1,3,12-trihydroxy- and 1,3,7-trihydroxy-5β-cholanoates as their methyl ester-trimethylsilyl ether derivatives, respectively. (Taken from refs. 65, 350 with permission.)

4.5.1.2 *Cholestasis*

In conditions of intra- and extrahepatic cholestasis urinary bile acid excretion is increased. A number of bile acids are found in addition to the usual components of cholic, chenodeoxycholic, deoxycholic and lithocholic and these include 3β-hydroxy-5-cholenoic, ursodeoxycholic, hyodeoxycholic, hyocholic, 3α,7α,12α-trihydroxy-5β-cholestanoic and others (231, 352–354). Most of these are likely to be formed in the liver as a result of reduced pool sizes, larger turnover rates and the increased peripheral circulation of bile acids.

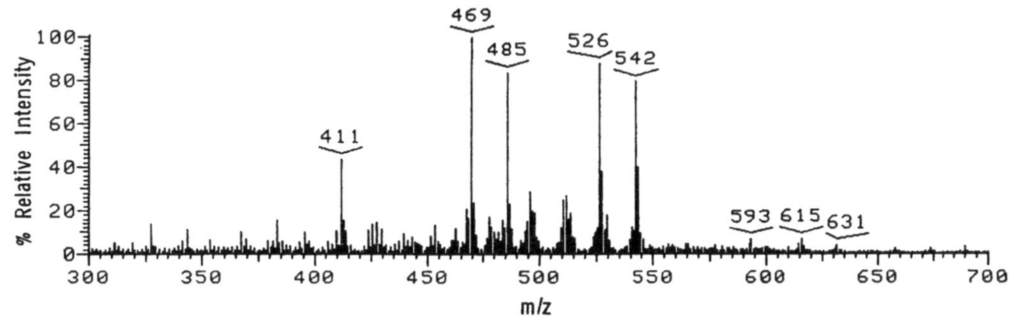

Figure 17. The partial negative ion FAB-MS spectrum, of a crude urine extract from a patient with a suspected defect in bile acid biosynthesis. The major ions, *m/z* 469, 485, 526 and 542 represent the molecular ion species, $[M-H]^-$, of 3β,7α-dihydroxy-5-cholenoic 3-sulphate, 3β,7α,12α-trihydroxy-5-cholenoic mono-sulphate and their glycine conjugates respectively (357).

Sulphation is the predominant conjugated form of bile acids in cholestasis but glucuronidation can also be extensive (20, 353). In future fast atom bombardment may be a useful technique for looking at the modified excretion pattern of conjugates associated with cholestasis.

A detailed GC-MS investigation has been made of urinary bile acids in patients with recurrent cholestasis of pregnancy (65, 221, 350, 355, 356). In this intermittent intrahepatic cholestasis of unknown etiology, urinary bile acid excretion increased during the last trimester of gestation from approximately 10 $\mu mol \cdot day^{-1}$ to 100 $\mu mol \cdot day^{-1}$. This increase is mainly in the sulphate fraction with cholic, chenodeoxycholic and deoxycholic acids being the predominant bile acids excreted. Taurine conjugates were quantitatively more important than glycine conjugates.

The comprehensiveness of a quantitative GC-MS method for the determination of a complex mixture of bile acids is illustrated well from these studies, making it possible to establish the concentration and conjugation state of the majority of components without prior knowledge of the constituents. This approach is important when unexpected or even unknown bile acid metabolites appear in samples and permits their quantitative relationships to other compounds to be established. Of interest were the presence of the 1- and 6-hydroxylation products of cholic, 1,3,7,12- and 3,6,7,12-tetrahydroxycholanoic acids, which as they increase in concentration change from being mainly glycine to mainly taurine conjugates. The equivalent hydroxylation products of deoxycholic acid, (e. g. 1β,3α,12α-, 3α,6β,12α- and 3α,6α,12α-trihydroxy-5β-cholanoic acids) were also observed.

The monosulphates of 3β,7α-dihydroxy-5-cholenoic and 3β,7α,12α-trihydroxy-5-cholenoic and their glycine conjugates were identified as the sole urinary bile acids in a 3 month old boy with cholestatic jaundice (357). The chemical lability of these compounds prevented their initial identification by the standard GC-MS procedures and FAB-MS was invaluable for preliminary characterization and for monitoring modified methods of isolation (Fig. 17). It was thought that this patient had a previously undescribed inborn error of bile acid biosynthesis, affecting the conversion of 7α-hydroxy-cholesterol to 7α-hydroxy-4-cholesten-3-one by the microsomal enzyme 3β-hydroxysteroid dehydrogenase/isomerase, although this remains to be proven conclusively.

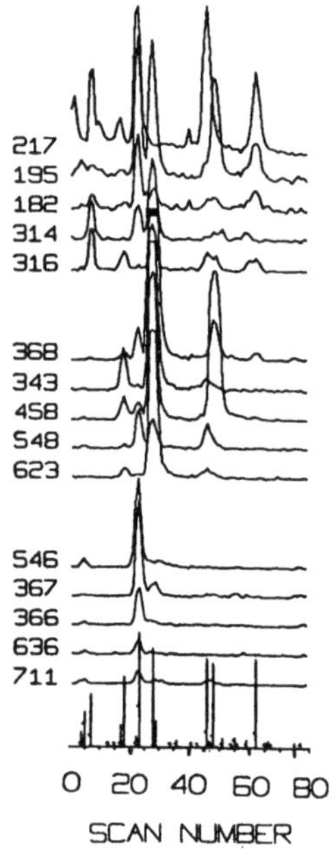

Figure 18
Fragment ion current chromatograms obtained from the analysis of methyl ester TMS ether derivatives of bile acids in the glycine conjugate fraction from urine of a patient with toxic hepatitis; *m/z* values were selected to indicate the presence of tri- (*m/z* 368, 343, 458, 548, 623) and tetrahydroxylated (*m/z* 217, 195, 182, 314, 316). Numbers to the right are values with which the ion intensities have been multiplied. Vertical lines at the bottom indicate computer localization of peaks of the derivatives of unknown tetrahydroxycholanoate (scan 5), internal standard (scan 7), hyocholate (scan 18), tetrahydroxycholanoates (scan 23), cholate (scan 28), 1,3,12-trihydroxycholanolate (scan 46), hyocholate (scan 47) and the tentative 1,3,7,12-tetrahydroxycholanoate (scan 62). (Taken from ref. 350 with permission.)

4.5.1.3 *Hepatitis*

In patients with hepatitis an increased urinary excretion of bile acids (mostly as sulphates) generally accompanies the increased levels in serum (353). Chenodeoxycholic acid is the predominant metabolite in both viral and toxic hepatitis and 3*β*-hydroxy-12-oxo-5*β*-cholanoic and 3*β*-hydroxy-5-cholenoic acids were also observed (350). Products of hydroxylation at C-1 and C-6 are found in these samples and an illustration of the detection of a compound tentatively identified as 1,3,7,-trihydroxy-cholanoic by the repetitive scanning technique is shown in Figure 18. The series of ions selected to be characteristic of the 1,3-bis-TMS structure were *m/z* 217, 195, 182, 314, and 316.

4.5.1.4 *Biliary atresia*

In extrahepatic bile duct atresia when there is no enterohepatic circulation of bile acids, cholic and chenodeoxycholic acids are excreted as conjugates in increased amounts in urine and *allo*-cholic and 3*β*-hydroxy-5-cholenoic acids have been identified by GC-MS in the urine of these patients (358).

4.5.2 **Defects in bile acid biosynthesis**

Mass spectrometry has been applied to the identification of urinary bile acids in patients suffering from disorders in which bile acid biosynthesis is affected (77). Studies of such conditions, while of interest in investigating the particular defect, may also be of value in establishing normal and minor pathways of bile acid metabolism. In patients with the cerebrohepato-renal syndrome of Zellweger, for example, a variety of precursors on the pathway from cholesterol to bile acids have been identified which reflect the inability to carry out peroxisomal oxidative cleavage of the steroid side chain (7, 75, 76). These intermediates to chenodeoxycholic and cholic acids included 3α,7α-dihydroxy-5β-cholestan-26-oic (DHCA), 3α,7α,12α-trihydroxy-5β-cholestan-26-oic (THCA) and 3α,7α,12α,24ξ-tetrahydroxy-5β-cholestan-26-oic (varanic acid). Further nuclear oxidation of DHCA and THCA can occur and this provided the basis for a rapid screening procedure for patients with this syndrome and related peroxisomal disorders (359, 360). This involved the extraction of bile acids from urine onto small cartridges of octadecylsilane-bonded silica, elution with methanol, evaporation to dryness and reconstitution in methanol for submission to negative ion FAB-MS. The presence of taurine conjugated tetrahydroxycholestanoic acids in the urine was a unique feature of these patients and was detected by an ion at m/z 572. Although this ion arose from several hydroxy-THCA's which FAB-MS could not directly distinguish, the value of the FAB approach as a screening technique to detect unusual urinary bile acid conjugates was clearly evident.

An unusual C_{29} dicarboxylic acid was identified in urine and also in serum from infants with coprostanic acidemia and Zellweger syndrome. After careful examination by mass spectrometry and nuclear magnetic resonance spectroscopy its structure was established as 3α,7α,12α-trihydroxy-27-carboxymethyl-5β-cholestan-26-oic acid (362).

Cerebrotendinous xanthomatosis (CTX) is another rare disorder in which the GC-MS study of bile acid and bile alcohol excretion has clarified aspects of bile acid biosynthesis. A primary defect in 26-hydroxylase activity leads to excessive production and excretion of 25-hydroxylated sterols (69) which are poor substrates for bile acid synthesis. The detection by FAB-MS of glucuronidated bile alcohols in urine with four to seven hydroxy groups has been suggested as a rapid means of diagnosing patients with CTX disease (77, 361).

4.6 Amniotic fluid

Amonitic fluid contains the major biliary bile acids of the adult: cholic, chenodeoxycholic, deoxycholic and lithocholic acids and in addition 3β-hydroxy-5-cholenoic acid (363). The transplacental transfer of primary and secondary bile acids complicates an interpretation of this profile of bile acids, however, the apparent absence of 3β-hydroxy-5-cholenoic in maternal serum (259) supports its fetal production. In premature pregnancies 3β-hydroxy-5-cholenoic is present in amniotic fluid in a higher proportion than in pregnancies at term and hence has been suggested as a measure of the maturity of the fetal liver.

4.7 Tissue

The early reports that bile acids, in particular monohydroxy bile acids, are hepatotoxic and induce cholestasis in animals (364–366) and that they may also act as carcinogens or co-carcinogens in inducing liver (367) or colonic tumors (368–370) has aroused interest in the determination of their tissue levels.

Nair et al. (371, 372) reported the existence of a tissue-bound fraction of bile acids and lithocholic-ξ-L-lysine was the predominant bile acid species identified. This tissue bound species of lithocholic acid was not extractable into a 0.1% solution of ammonia in 95% ethanol (373) and was only recovered after hydrolysis with the enzyme cholylglycine hydrolase. It was shown that trypsin digestion increased the yield of tissue-bound lithocholate. In twelve human liver biopsy samples the concentrations of tissue-bound lithocholic acid was reported to be in the range 4–116 $\mu g \cdot g^{-1}$ tissue and in half of the subjects studies represented >50% of the total lithocholic acid in tissue (371, 372).

In a study of tissue bile acids in colonic polyps and colonic carcinoma tissue using the methods described by Nair et al. (371, 372), chenodeoxycholic acid was the predominant bile acid detected and no lithocholic acid was observed in either the soluble or tissue bound fraction. Furthermore there was no significant difference in the tissue bile acid composition betweeen polyp or carcinoma tissue and normal bowel tissue resected adjacent to the site of the carcinoma (374).

The existence of tissue-bound lithocholic acid was recently questioned following a SIM GC-MS study of human liver tissue (375). Pure lithocholyl-ξ-L-lysine was chemically synthesized and its characteristics checked with TLC in two solvent systems, by its reaction with ninhydrin, and using field desorption (FD) mass spectrometry. The FD-mass spectrum showed the presence of a $[M + H]^+$ ion at *m/z* 505 which was the base peak accompanied by the Na^+ adduct ion at *m/z* 527 and an intense ion at *m/z* 461 considered to arise from the loss of the carboxylic acid. The corresponding EI spectrum gave a molecular ion at *m/z* 504 and the synthesized compound was identical in characteristics to the lithocholyl-ξ-L-lysine first described by Nair et al. (371). Alkaline hydrolysis of this bile acid resulted in the release of lithocholic acid and this was confirmed from the mass spectrum as its ethyl ester-DMES ether derivative.

Lithocholyl-ξ-L-lysine was subjected to a mild aqueous alkaline hydrolysis and enzymatic hydrolysis using three different cholylglycine hydrolase preparations and various incubation conditions but, contrary to the findings of Nair et al. (371), all of these conditions failed to liberate free lithocholate and the lysine conjugate was recovered unchanged (375). More vigorous and prolonged alkaline hydrolysis however resulted in complete hydrolysis and release of unconjugated lithocholic acid.

Using an isotope dilution technique and GC-MS with SIM soluble and tissue bound lithocholic acid was measured in 10 samples of human liver tissue. $[^2H_4]$lithocholic acid was used as the internal standard and the pairs of ions *m/z* 386/461 and *m/z* 390/465 were monitored for the methyl ester-DMES ether of lithocholic and $[^2H_4]$lithocholic respectively. Soluble lithocholic acid concentrations were reported in the range 0.28 to 12.1 $pmol \cdot mg^{-1}$ liver tissue (compared with 29–493 $pmol \cdot mg^{-1}$ liver reported earlier (371)) while bound lithocholic acid levels were <0.2 $pmol \cdot mg^{-1}$ liver and close to the limit of detection of the method. It was suggested that lithocholic acid found in the residue reflected traces that remained following organic extraction. While the results of Yanagisawa et al. (375) using reliable methodology did not support the presence of a tissue bound form of lithocholic acid (376), Nair et al. (377) subsequently indicated that this was the result of the instability of the enzyme preparation. The sub-species of cholylglycine hydrolase active on the *N*-ξ-lysllithocholate was shown to be unstable at temperatures above 0 °C during storage (377).

In an earlier report (146) SIM GC-MS was employed to determine tissue levels of the bile acids, deoxycholic, chenodeoxycholic, ursodeoxycholic and cholic acids. Total bile acid concentrations were 29.56 ± 8.62 $\mu g \cdot g^{-1}$ liver and these were made up of primarily chenodeoxycholic (38%) and cholic (41%) acids. Similar values were also reported in a study comparing the composition of liver tissue with bile when conjugated bile acids labeled with deuterium were employed as internal standards (273). Such standards are preferable to unconjugated bile acids because they take a more accurate account of procedural losses during the hydrolytic steps.

5 Conclusions

In the past 15 years GC-MS has proved to be an invaluable tool for the study of bile acid metabolism in health and disease. Unfortunately the high cost of sophisticated instrumentation has generally precluded its use in routine assays, however with the advent of lower cost instruments, the introduction of specific GC detectors, such as the ion trap detector, and general improvements in instrument design we can expect an increase in use of the technique in clinical studies. Considerable strides have been made in the application of soft ionization processes, opening up areas of analysis which were previously not possible, e. g. direct analysis of bile acid conjugates, and with the increased availability of stable isotopically labeled bile acids important contributions to the understanding of bile acid kinetics in disease is anticipated. With the introduction of improved LC-MS interfaces such as the thermospray technique and the excellent separations of bile acids which are now possible using HPLC, bile acid analysis may be considerable simplified in the future (378, 379). While it may never replace other routine techniques such as radioimmunoassay, mass spectrometry will continue to play an important complementary role in a wide range of bile acid investigations.

References

1. Nair, P. P. and Kritchevsky, D. (1971) The Bile Acids, Chemistry, Vol. 1, Plenum Publishers, New York.
2. Bergström, S., Ryhage, R. & Stenhagen, E. (1958) Mass spectrometric studies on bile acids and other steroid derivatives. Acta Chem. Scand. *12*, 1349–1358.
3. Ryhage, R. & Stenhagen, E. (1960) Mass spectrometry in lipid research. J. Lipid Res. *1*, 361–390.
4. Cross, A. D. (1964) Nuclear-magnetic-resonance and mass spectral analysis of ranol tetra-acetate. Biochem. J. *90*, 314–315.
5. Haslewood, G. A. D. (1978) The Biological Importance of Bile Salts, North Holland Publishing Co., Amsterdam.
6. Eyssen, H., Parmentier, G., Compernolle, F., Boon, J. & Eggermont, E. (1972) Trihydroxycoprostanic acid in the duodenal fluid of two children with intrahepatic bile duct anomalies. Biochem. Biophys. Acta *273*, 212–221.
7. Hanson, R. F., Isenberg, J. J., Williams, G. C., Hachey, D., Szczepanik, P., Klein, P. D. & Sharp, H. L., (1975) The metabolism of 3α,7α,12α-trihydroxy-5β-cholestan-26-oic acid in two siblings with cholestasis due to intrahepatic bile duct anomalies. An apparent inborn error of cholic acid biosynthesis. J. Clin. Invest. *56*, 577–587.
8. Lester, R., Pyrek, J. S. E., Adcock, E. W., Grimberg, A., Boros, J. & Sanghvi, A. (1980) Etianic acid, a C_{20} bile acid, identified in meconium (Abstract). Gastroenterology *79*, 1034.
9. Parmentier, G. G., Janssen, G. A., Eggermont, E. A. & Eyssen, H. J. (1979) C_{27} bile acids in infants with coprostanic acidemia

and occurrence of a 3α,7α,12α-trihydroxy-5β-C_{29}-dicarboxylic bile acid as a major component in their serum. Eur. J. Biochem. *102*, 173–183.
10. Hofmann, A. F. & Small, D. M. (1967) Detergent properties of bile salts: correlation with physiological function. Ann. Rev. Med. *18*, 333–376.
11. Small, D. M. (1971) The physical chemistry of cholanic acids. In: The Bile Acids Vol. 1 (Nair, P. P. and Kritchevsky, D., eds.) pp. 249–356, Plenum Publishers, New York.
12. Carey, M. C. (1982) Measurement of the physical-chemical properties of bile salt solutions. In: Bile Acids in Gastroenterology (Barbara, L., Dowling, R. H., Hofmann, A. F., Roda, E., eds.) pp. 19–57, MTP Press, Lancaster.
13. Roda, A., Hofmann, A. F. & Mysels, K. J. (1983) The influence of bile salt structure on self association in aqueous solutions. J. Biol. Chem. *258*, 6362–6370.
14. Carey, M. C. (1985) Physical-chemical properties of bile acids and their salts. In: Sterols and Bile Acids (Danielsson, H. and Sjövall, J., eds.) pp. 345–403, Elsevier, Amsterdam, New York, Oxford.
15. Sjövall, J. (1959) The determination of bile acids in bile and duodenal contents by quantitative paper chromatography. Clin. Chim. Acta *4*, 652–664.
16. Sjövall, J. (1960) Bile acids in man under normal and pathological conditions. Clin. Chim. Acta *5*, 33–41.
17. Sjövall, J. (1959) Dietary glycine and taurine on bile acid conjugation in man. Bile acids and steroids. Proc. Soc. Exptl. Biol. Med. *100*, 676–678.
18. PericGolia, L. & Jones, R. F. (1962) Ornilithocholanoic acid-abnormal conjugates of bile acid. Proc. Soc. Exptl. Biol. Med. *110*, 327–331.
19. Palmer, R. H. (1967) Formation of bile acid sulfates: a new pathway of bile acid metabolism in humans. Proc. Natl. Acad. Sci. (USA) *58*, 1047–1050.
20. Back, P., Spaczynski, K. & Gerok, W. (1974) Bile salt glucuronides in urine. Hoppe-Seyler's Z. Physiol. Chem. *355*, 749–752.
21. Back, P. (1974) Evidence for bile salt glucuronides in cholestasis. In: Advances in Bile Acid Research (Matern, S., Hackenschmidt, J., Back, P. & Gerok, W., eds.) pp. 148–152, Schattaner, Stuttgart, New York.
22. Björkhem, I. (1985) Mechanism of bile acid biosynthesis in mammalian liver. In: Sterols and Bile Acids (Danielsson, H. & Sjövall, J., eds.) pp. 231–287, Elsevier, Amsterdam, New York, Oxford.
23. Bloch, K., Berg, B. N. & Rittenberg, D. (1943) The biological conversion of cholesterol to cholic acid. J. Biol. Chem. *149*, 511–517.
24. Bergström, S. (1952) The formation of bile acids from cholesterol in the rat. Proc. Roy. Physiograph. Soc. Lund *22*, No. 16, 91–96.
25. Chaikoff, I. L., Siperstein, M. D., Dauben, W. G., Bradlow, H. L., Eastman, G. M., Tomkins, J. B., Meier, J. B., Chen, R. W., Hotta, S. & Srere, P. A. (1952) ^{14}C-cholesterol; oxidation of carbons 4 and 26 to carbon dioxide by intact rat. J. Biol. Chem. *194*, 413–416.
26. Siperstein, M. D. & Chaikoff, I. L. (1952) ^{14}C-cholesterol; excretion of carbons 4 and 26 in feces, urine and bile. J. Biol. Chem. *198*, 93–104.
27. Bergström, S. & Lindstedt, S. (1956) Formation of cholic acid from 3α,7α-dihydroxycoprostane in the rat. Biochim. Biophys. Acta *19*, 556–557.
28. Lindstedt, S. (1957) The formation of bile acids from 7α-hydroxy-cholesterol in the rat. Acta Scand. *11*, 417–420.
29. Myant, N. B. & Mitropoulos, K. A. (1977) Cholesterol 7α-hydroxylase. J. Lipid Res. *18*, 135–153.
30. Danielsson, H. & Einarsson, K. (1966) On the conversion of cholesterol to 7α,12α-dihydroxy cholest-4-en-3-one. J. Biol. Chem. *21*, 1449–1454.
31. Mendelsohn, D., Mendelsohn, L. & Staple, E. (1965) The catabolism in vitro of cholesterol: formation of the 7-epimeric hydroxycholesterols from cholesterol in rat liver. Biochim. Biophys. Acta *97*, 379–381.
32. Mendelsohn, D., Mendelsohn, L. & Staple, E. (1966) The in vitro catabolism of cholesterol: a comparison of the formation of cholest-4-en-7α-ol-3-one and 5β-cholestan-7α-ol-3-one from cholesterol in rat liver. Biochem. *5*, 1286–1290.
33. Björkhem, I. & Danielsson, H. (1967) Formation and metabolism of some Δ^4-cholestanols in rat. Eur. J. Biochem. *2*, 403–413.
34. Björkhem, I. (1969) On the mechanism of the enzymatic conversion of cholest-5-ene-3β-7α-diol into 7α-hydroxy-4-en-3-one. Eur. J. Biochem. *8*, 337–344.
35. Einarsson, K. (1968) On the properties of the 12α-hydroxylase in cholic acid biosynthesis: Bile acids in steroids 198. Eur. J. Biochem. *5*, 101–108.

36. Berséus, O. (1967) On the conversion of cholesterol to bile acids in rat: purification and properties of a Δ^4-3-ketosteroid 5β-reductase and a 3α-hydroxysteroid dehydrogenase. Eur. J. Biochem. *2*, 493–502.
37. Björkhem, I., Gustafsson, J., Johansson, G. & Persson, G. (1975) Biosynthesis of bile acids in man: Hydroxylation of the C_{27}-steroid side chain. J. Clin. Invest. *55*, 478–486.
38. Setoguchi, T., Salen, G., Tint, G. S. & Mosbach, E. H. (1974) A biochemical abnormality in cerebrotendinous xanthomatosis impairment of bile acid synthesis associated with incomplete degradation of the cholesterol side chain. J. Clin. Invest. *53*, 1393–1401.
39. Salen, G., Shefer, S., Setoguchi, T. & Mosbach, E. H. (1975) Bile alcohol metabolism in man. Conversion of 5β-cholestane-3α,7α,12α,25-tetrol to cholic acid. J. Clin. Invest. *56*, 226–231.
40. Shefer, S., Cheng, F. W., Dayal, B., Hanser, S., Tint, G. S., Salen, G. & Mosbach, E. H. (1976) A 25-hydroxylation pathway of cholic acid biosynthesis in man and rat. J. Clin. Invest. *57*, 897–903.
41. Björkhem, I. (1972) Reaction mechanisms in bile acid biosynthesis. In: Bile Acids in Human Diseases (Back, P., Gerok, W., eds.), pp. 9–16, Schattaner, Stuttgart, New York.
42. Mitropoulos, K. A. & Myant, N. B. (1967) The formation of lithocholic acid, chenodeoxycholic acid and α- and β-muricholic acids from cholesterol incubated with rat liver mitochondria. Biochem. J. *103*, 472–479.
43. Mitropoulos, K. A. (1972) The formation of cholest-5-ene-3α-26-diol as an intermediate in the conversion of cholesterol into bile acids by rat liver mitochondria. Biochem. J. *130*, 363–371.
44. Nair, P. P. & Kritchevsky, D. (1974) The Bile Acids, Physiology and Metabolism, Vol. 2, Plenum Publishers, New York.
45. Hofmann, A. F. (1977) The enterohepatic circulation of bile acids in man. In: Bile Acids. Clinics in Gastroenterology (Paumgartner, G., ed.), Vol. 6/1, pp. 3–24, Saunders, London, Philadelphia, Toronto.
46. Fromm, H. & Hofmann, A. F. (1975) The importance of bile salts in human diseases. In: Ergebnisse der Inneren Medizin und Kinderheilkunde (Frick. P., Harnack, G.-A. von, Marbini, G. A., Prader, A., Schoen, R., Wolff, H. P, eds.) Vol. 37, pp. 149–192, Springer, Berlin, Heidelberg, New York.
47. Matern, S. & Gerok, W. (1979) Pathophysiology of the enterohepatic circulation of bile acids. Rev. Physiol. Biochem. Pharmacol. *85*, 125–204.
48. Vlahcevic, Z. R., Miller, J. R., Farras, J. T. & Sewell, L. (1971) Kinetics and pool size of primary bile acids in man. Gastroenterology *61*, 85–90.
49. Danzinger, R. G., Hofmann, A. F., Thistle, J. L. & Schoenfield, L. J. (1973) Effect of oral chenodeoxycholic acid on bile acid kinetics and biliary lipid composition in women with cholelithiasis. J. Clin. Invest. *52*, 2809–2821.
50. Grundy. S. M., Ahrens, E. H., Jr. & Miettinen, T. A. (1965) Quantitative isolation and gas-liquid chromatographic analysis of total fecal bile acids. J. Lipid Res. *6*, 397–410.
51. Borgström, B., Dahlquist, A., Lundh, G. & Sjövall, J. (1957) Studies of intestinal digestion and absorption in the human. J. Clin. Invest. *36*, 1521–1536.
52. Borgström, B., Lundh, G. & Hofmann, A. F. (1963) The site of absorption of conjugated bile salts in man. Gastroenterology *45*, 229–238.
53. Lindstedt, S. (1957) The turnover of cholic acid in man. Acta Physiol. Scand. *40*, 1–9.
54. Lack, L. & Weiner, I. M. (1961) In vitro absorption of bile salts by small intestine of rats and guinea pigs. Am. J. Physiol. *200*, 313–317.
55. Holt, D. R. (1964) Intestinal absorption of bile salts in the rat. Am. J. Physiol. *207*, 1–7.
56. Dietschy, J. M., Salomon, H. S. & Siperstein, M. D. (1966) Bile acid metabolism. 1. Studies on the mechanisms of intestinal transport. J. Clin. Invest. *45*, 832–846.
57. Dietschy, J. M. (1974) Bile acids: their absorption from the gastrointestinal tract and role during fat absorption. Verh. Dtsch. Ges. Inn. Med. *80*, 399–407.
58. Krag, E. & Phillips, S. F. (1974) Active and passive bile acid absorption in man. Perfusion studies of the ileum and jejunum. J. Clin. Invest. *53*, 1681–1694.
59. Davenport, H. W. (1965) Damage to the gastric mucosa effects of salicylates and stimulation. Gastroenterology 49, 183–196.
60. Gorbach, S. L., Plaut, A. G., Nahas, L., Weinstein, L., Spanknebel, G. & Levitan, R. (1967) Microorganisms of the small intestine and their relations to oral and fecal flora. Gastroenterology *53*, 856–867.

61. Gorbach, S. L. & Tabaqchali, S. (1969) Bacteria, bile acids and the small bowel. Gut *10*, 963–972.
62. Northfield, T. C. & McColl, I. (1973) Postprandial concentrations of free and conjugated bile acids down the length of the normal human intestine. Gut *14*, 513–518.
63. Danielsson, H., Eneroth, P., Hellström, K., Lindstedt, S. & Sjövall, J. (1963) On the turnover and excretory products of cholic and chenodeoxycholic acid in man. Bile acids and steroids 134. J. Biol. Chem. *238*, 2299–2304.
64. Setchell, K. D. R., Lawson, A. M., Tanida, N. & Sjövall, J. (1983) General methods for the analysis of metabolic profiles of bile acids and related compounds in feces. J. Lipid Res. *24*, 1085–1100.
65. Almé, B., Bremmelgaard, A., Sjövall, J. & Thomassen, P. (1977) Analysis of metabolic profiles of bile acids in urine using a lipopholic anion exchanger and computerized gas-liquid chromatography-mass spectrometry. J. Lipid. Res. *18*, 339–362.
66. Schoenfield, L. J., Sjövall, J. & Perman, E. (1967) Bile acids on the skin of patients with pruritic hepatobiliary disease. Nature *213*, 93–94.
67. Van Bogaert, L., Scherer, H. J. & Epstein, E. (1937) Une forme cérébralé de la cholesterinase généralisée, Masson et cie, Paris.
68. Salen, G. & Mosbach, E. H. (1976) The metabolism of sterols and bile acids in cerebrotendinous xanthomatosis. In: The Bile Acids – Chemistry, Physiology and Metabolism (Nair, P. P. & Kritchevsky, D., eds.), Vol. 3, pp. 115–153, Plenum Press, New York.
69. Wolthers, B. G., Volmer, M., Van der Molen, J., Koopman, B. J., deJager, A. E. J. & Waterreus, R. J. (1983) Diagnosis of cerebrotendinous xanthomatosis (CTX) and effect of chenodeoxycholic acid therapy by analysis of urine capillary gas chromatography. Clin. Chim. Acta *131*, 53–65.
70. Oftebro, H., Björkhem, I., Skrede, S., Schreiner, A. & Pedersen, J. I. (1980) Cerebrotendinous xanthomatosis: a defect in mitochondrial 26-hydroxylation required for normal biosynthesis of cholic acid. J. Clin. Invest. *65*, 1448–1430.
71. Bhattacharyya, A. K. & Conners, W. E. (1974) β-Sitosterolemia and xanthomatosis: a newly described lipid storage disease in two sisters. J. Clin. Invest. *53*, 1033–1043.
72. Isler, J. H., Dowling, R. H., Murphy, G. M., Ponz de Leon, M. & Mitropoulous, K. A. (1977) Congenital bile salt deficiency associated with 28 years of intractable constipation. In: Bile Acid Metabolism in Health and Disease (Paumgartner, G., Stiehl, A., eds.) pp. 231–234, MTP Press, Lancaster.
73. Bowen, P., Lee, C. S., Zellweger, H. & Lindenberg, (1964) A familial syndrome of multiple congenital defects. Bull. Johns Hopkins Hosp. *114*, 402–414.
74. Collins, J., Javitt, N., Cohen, B. & Mosbach, E. H. (1978) Bile acid synthesis in man. III. Identification of 26-hydroxycholesterol in serum in Zellweger syndrome (Abstract). Gastroenterology *75*, 958.
75. Hanson, R. F., Szczepanik-VanLeeuwen, P., Williams, G. C., Yvabowski, G. & Sharp, H. L. (1979) Defects in bile synthesis in Zellweger's syndrome. Science *203*, 1107–1108.
76. Gustafsson, J., Gustavson, K.-H., Karlaganis, G. & Sjövall, J. (1983) Zellweger's cerebro-hepato-renal syndrome – variations in expressivity and in defects of bile acid synthesis. Clinical Genetics *24*, 313–319.
77. Setchell, K. D. R, & Street, J. M. (1987) Inborn errors of bile acid synthesis. Seminars in Liver Disease. Vol. *7.2*, 87–99.
78. Engelking, L. R., Barnes, S., Hirschowitz, B. I., Dasher, C. A., Spenney, J. G. & Naftel, D. (1980) Determination of the pool size and synthesis rate of bile acids by measurements in blood of patients with liver disease. Clin Sci. 58, 485–492.
80. Pearlman, B. J. & Schoenfield, L. J. (1978) Gallstones: The present and future of medical dissolution. Med. Clin. North Am. *62*, 87.
81. Dowling, R. H., Bell, H. D. & White, J. (1972) Lithogenic bile in patients with ileal dysformation. Gut *13*, 415–420.
82. Bacharach, W. H. & Hofmann, A. F. (1982) Ursodeoxycholic acid in the treatment of cholesterol cholelithiasis. Parts I and II. Dig. Dis. Sci. *27*, 737–856.
83. Dowling, R. H. (1983) Management of stones in the biliary tree. Gut *24*, 599–658.
84. Dowling, R. H., Hofmann, A. F. & Barbara, L. (eds.) (1979) Workshop on ursodeoxycholic acid, University Park Press, Baltimore.
85. Vlachevic, Z. R., Bell, C. C., Jr., Butiac, I., Farrar, J. T. & Swell, L. (1970) Diminished bile acid pool size in patients with gallstones. Gastroenterology *59*, 165–173.
86. Grundy, S. M., Metzger, A. L. & Adler, R. D. (1972) Mechanism of lithogenic bile formation in American Indian women with

cholesterol gallstones. J. Clin. Invest. *51*, 3026–3043.
87. Coyne, M. J., Marks, J. & Schoenfield, L. J. (1977) Mechanism of cholesterol gallstone information. In: Clinics in Gastroenterology (Paumgartner, G., ed.) pp. 129–130, Saunders, London, Philadelphia, Toronto.
88. Grundy, S. M., Duane, W. C., Adler, R. D., Avon, J. M. & Metzger, A. L. (1974) Biliary lipid output in young women with cholesterol gallstones. Metabolism *23*, 67–73.
89. Northfield, T. C., Drasar, B. S. & Wright, J. T. (1973) Value of small intestinal bile acid analysis in the diagnosis of the stagnant loop syndrome. Gut *14*, 341– 347.
90. Tabaqchali, S., Hatzioannou, J. & Booth, C. C. (1968) Bile salt deconjugation and steatorrhea in patients with the stagnant loop syndrome. Lancet *2*, 12–16.
91. Fromm, H. & Hofmann, A. F. (1971) Breath test for altered bile acid metabolism. Lancet *2*, 621–625.
92. Lewis, B., Panveliwalla, D., Tabaqchali, S. & Wootton, D. P, (1969) Serum bile acids in the stagnant loop syndrome. Lancet *1*, 219–220.
93. Setchell, K. D. R., Harrison, D. L., Gilbert, J. M. & Murphy, G. M. Serum unconjugated bile acids: Qualitative and quantitative profiles in ileal resection and bacterial overgrowth. Clin. Chim. Acta *152*, 297–306.
94. Dean, P. D. G. & Aplin, R. T. (1966) Mass spectrometric studies of bile acids: the differentiation between chenodeoxycholic acid and cholic acid and the identification of 3α,7α-dihydroxy-5β-cholestanoic acid in alligator bile. Steroids *8*, 565–579.
95. Shefer, S., Salen, G., Cheng. F. W., Dayal, B., Batta, A. K, Tint, G. S., Bose, A. K. & Pramanik, B. N. (1982) Chemical ionization – mass spectrometric approach to structure determination of an intermediate in bile acid biosynthesis. Anal. Biochem. *121*, 23–30.
96. Whitney, J., Lewis, S., Straub, K. M., Thaler, M. M. & Burlingame, A. L. (1981) Analysis of conjugated bile salts in human duodenal bile using fast atom bombardment and field desorption mass spectrometry (Proc. Jap. Soc. Med. Mass Spectrom.) Koeushu-Iyo Masu Kenkyukai *6*, 33–44.
97. Whitney, J. O. & Burlingame, A. L. (1982) Bile salt analysis by fast atom bombardment. Koeushu-Iyo Masu Kenkyukai *7*, 3–17.
98. Whitney, J. O. (1986) Novel MS approaches in the analysis of free and conjugated bile salts. In: Mass Spectrometry in Biomedical Research (Gaskell, S., ed.) pp. 61–74, John Wiley and Sons Ltd.
99. Whitney, J. O., Ling, V., Grunberger, D., Thaler, M. M. & Burlingame, A. L. (1983) Identification of hepatic bile acid metabolites in bile by liquid secondary ion mass spectrometry. Koeushu-Iyo Masu Kenkyukai *8*, 47–59.
100. Aberth, W., Straub, K. & Burlingame, A. L. (1982) Secondary ion mass spectrometry with cesium ion primary beam and liquid target matric for analysis of bio-organic compounds. Anal. Chem. *54*, 2029–2034.
101. Hanaineh, L. & Brooks, C. J. W. (1964) Gas-liquid chromatographic behavior of conjugated bile acids. Biochem. J. *92*, 9–10.
102. Street, J. M., Trafford, D. J. H. & Makin, H. L. J. (1986) Capillary gas-liquid chromatography of glycine-conjugated bile acids without prior hydrolysis. J. Lipid. Res. *27*, 208–224.
103. Shaw, R. & Elliott, W. H. (1978) Bile acids. LIV. Mass spectra of conjugated bile acids. Biomed. Mass Spectrom. *5*, 433–438.
104. Elliott, W. H. (1972) Mass spectra of bile acids. In: Biochemical applications of mass spectrometry (Waller, G. R., ed.), pp. 291–311, Wiley Interscience, New York, London.
105. Shaw, R. & Elliott, W. H. (1978) Bile acids. LVII. Analysis of bile acids by high-pressure liquid chromatography and mass spectrometry. Lipids *13*, 971–975.
106. Elliott, W. H. (1980) Mass spectra of bile acids. In: Biochemical applications of mass spectrometry, First Supplementary Volume (Waller, G. R. & Dermer, O. C., eds.), pp. 229–253, Wiley-Interscience, New York, London.
107. Mingrove, G., Greco, A. V., Bonifortia, L. & Passi, S. (1983) Analysis of conjugated bile acids by high performance liquid chromatography and mass spectrometry. Lipids *18*, 90–95.
108. Games, D. E., Games, M. P., Jackson, A. H., Olavesen, A. H., Rossiter, M. & Winterburn, P. J. (1974) Field desorption-mass spectrometry of sodium and potassium salts. Tetrahedron Letts. *27*, 2377–2380.
109. Ballatore, A. M., Beckner, C. F., Caprioli, R. M., Hoffman, N. E. & Liehr, J. G. (1983) Synthesis and spectroscopic analysis of modified bile salts. Steroids *41*, 198–206.
110. Sjövall, J., Lawson, A. M. & Setchell, K. D. R. (1984) Mass spectrometry of bile acids. In: Methods in Enzymology (Law, J.

& Rilling, H. C., eds.) Vol. 111, pp. 63–113, Academic Press, New York.
111. Lawson, A. M., Bayliss, C. & Setchell, K. D. R. (1984) The direct analysis of bile acid conjugates in biological samples by fast atom bombardment (Abstract). Falk Symposium No. 42, VIII International Bile Acid Meeting on Enterohepatic Circulation of Bile Acids and Sterol Metabolism, Bern, p. 260.
112. Cotter, R. J. (1981) Cationized species in laser desorption mass spectrometry. Anal. Chem. *53*, 719–720.
113. Day, R. J. Zimmermann, J. & Hercules, D. M. (1981) Laser desorption mass spectrometry of underivatized bile acids. Spectroscopy Letters *14*, 773–786.
114. Back, P. (1976) Bile acid glucuronides. II. Isolation of a chenodeoxycholic acid glucuronide from plasma in intrahepatic cholestasis. Hoppe-Seyler's Z. Physiol. Chem. *357*, 213–217.
115. Almé, B., Norden, A. & Sjövall, J. (1978) Glucuronides of unconjugated 6-hydroxylated bile acids in urine of a patient with malabsorption. Clin. Chim. Acta *86*, 251–259.
116. Almé, B. & Sjövall, J. (1980) Analysis of bile acid glucuronides in urine. Identification of 3α,6α,12α-trihydroxy-5β-cholanoic acid. J. Steroid Biochem. *13*, 907–916.
117. Sjövall, J. (1964) Gas chromatography of bile acids. In: Biomedical Applications of Gas Chromatography (Szymanski, H. A., ed.), pp. 151–167, Plenum, New York.
118. Eneroth, P. & Sjövall, J. (1971) Extraction, purification, and chromatographic analysis of bile acids in biological materials. In: The Bile Acids (Nair, P. P. and Kritchevsky, D., eds.) pp. 133–134, Plenum Press, New York.
119. Kuksis, A. (1976) Gas chromatography of bile acids. In: Lipid Chromatographic Analysis (Marinetti, G. V., ed.) Vol. 2, pp., 479–610, Dekker, New York.
120. Setchell, K. D. R. (1982) Liquid-solid extraction, liquid-gel chromatography, and capillary column gas chromatography in the analysis of bile acids from biological samples. In: Bile Acids in Gastroenterology. (Barbara, L., Dowling, R. H., Hofmann, A. F., & Roda, E., eds.) pp. 1–18, MTP Press Ldt. Lancaster, U.K.
121. Blau, K. & King, G. (1978) Handbook of Derivatives for Chromatography. Heyden, London, Philadelphia PA, Rheine.
122. Fales, H. M., Jaouni, J. M. & Babashak, J. F. (1973) Simple device for preparing ethereal diazomethane without resorting to co-distillation. Anal. Chem. *45*, 2302–2303.
123. Roovers, R. E., Evrard, E. & Vanderhaeghe, H. (1968) An improved method of measuring human blood bile acids. Clin. Chim. Acta *19*, 449–457.
124. Kuksis, A. & Gordon, B. A. (1963) Quantitative gas-liquid chromatography of selected bile acid esters on SE-30, QF-1, XE-60 silicone polymers. Can. J. Biochem. Physiol. *41*, 1355–1366.
125. Makita, M. & Wells, W. W. (1963) Quantitative analysis of fecal bile acids by gas-liquid chromatography. Anal. Biochem. *5*, 523–530.
126. Axelson, M. & Sjövall, J. (1974) Separation and computerized gas chromatography-mass spectrometry of unconjugated neutral steroids in plasma. J. Steroid Biochem. *5*, 733–738.
127. Miyazaki, H., Ishibashi, M. & Yamashita, K. (1978) Use of new silylating agent for separation of bile acids and cholesterol by selected ion monitoring with computer controlled intensity matching technique. Biomed. Mass Spectrom. *5*, 469–476.
128. Edenharder, R. & Slemr, J. (1981) Gas chromatographic and mass spectrometric analysis of bile acids as trifluoroacetylhexafluoroisopropyl and heptafluorobutyl derivatives. J. Chromatogr. *222*, 1–12.
129. Imai, K., Tamura, Z., Mashige, F. & Osuga, T. (1976) Gas chromatography of bile acids as their hexafluoroisopropyl ester-trifluoroacetyl derivatives. J. Chromatogr. *120*, 181–186.
130. Musial, B. C. & Williams, C. N. (1979) Quantitative assay of conjugated and free bile acids as heptafluorobutyrate derivatives by gas-liquid chromatography. J. Lipid Res. *20*, 78–85.
131. Fukunager, A., Hatta, Y., Ishibashi, M. & Miyazaki, H. (1980) Gas chromatography of dimethylalkylsilyl ether derivatives of bile acid methyl esters. J. Chromatogr. *190*, 339–345.
132. Andersson, S. H. G. & Sjövall, J. (1984) Preparation of enol-tert-butyldimethylsilyl and mixed tert-butyl dimethylsilyl ethers of steroids and bile acids. J. Chromatogr. *289*, 195–206.
133. Cronholm, T., Makino, I. & Sjövall, J. (1972) Steroid metabolism in rats given [1-$^{2}H_{2}$]-ethanol. Biosynthesis of bile acids and

reduction of 3-keto-5β-cholanoic acid. Eur. J. Biochem. *24*, 507–519.

134. Carlström, K., Kirk, D. N. & Sjövall, J. (1981) Microbial synthesis of 1β and 15β-hydroxylated bile acids. J. Lipid Res. *22*, 1225–1234.

135. Setchell, K. D. R. & Matsui, A. (1983) Serum bile acid analysis. The application of liquid-gel chromatographic techniques and capillary column gas chromatography and mass spectrometry. Clin. Chim. Acta. *127*, 1–17.

136. Setchell, K. D. R. (1984) Gas chromatography and mass spectrometry of serum bile acids. In: Proceedings of the First International Symposium on Bile Acids in Hepatobiliary and Gastrointestinal Diseases, pp. 63–78, IRL Press Ltd., Oxford, England.

137. Iida., T., Chang, F. C., Matsumoto, T. & Tamura, T. (1983) Separation of mono-, di- and trihydroxy stereoisomers of bile acids by capillary gas-liquid chromatography. J. Lipid Res. *24*, 211–215.

138. Laatikainen, T. & Hesso, A. (1975) Determination of serum bile acids by glass capillary gas-liquid chromatography. Clin. Chim. Acta *64*, 63–68.

139. Tanida, N., Hikasa, Y. & Shimoyama, T. (1982) Evaluation of polytheylene glycol-HT as a stationary phase for capillary column gas-liquid chromatography of trimethylsilyl ethers of bile acid methyl esters. J. Chromatogr. *240*, 75–79.

140. Karlaganis, G. & Paumgartner, G. (1978) Analysis of bile acids in serum and bile by capillary column gas-liquid chromatography. J. Lipid. Res. *19*, 71–774.

141. Karlaganis, G., Schwarzenbach, R. P. & Paumgartner, G. (1980) Analysis of serum bile acids by capillary gas-liquid chromatography-mass spectrometry. J. Lipid Res. *21*, 377–381.

142. Sjövall, J., Eneroth, P. & Ryhage, R. (1971) Mass spectra of bile acids. In: The Bile Acids – Chemistry, Physiology and Metabolism. Volume 1: Chemistry (Nair, P. P. & Kritchevsky, D., eds.) pp. 209–248, Plenum, New York.

143. Szczepanik, P. A., Hachey, D. L. & Klein, P. D. (1976) Characterization of bile acid methyl ester acetate derivatives using gas-liquid chromatography, electron impact, and chemical ionization mass spectrometry. J. Lipid Res. *17*, 314–334.

144. Lawson, A. M. & Setchell, K. D. R. (1988) Mass spectrometry of bile acids. In: The Bile Acids. Volume 4: Methods and Applications (Setchell, K. D. R., Kritchevsky, D. & Nair, P. P, eds.) pp. 167–267, Plenum, New York.

145. Bartholomew, T. C., Summerfield, J. A., Billing, B. H., Lawson, A. M. & Setchell, K. D. R. (1982) Bile acid profiles of human serum and skin interstitial fluid and their relationship to pruritus studied by gas chromatography-mass spectrometry. Clin. Sci. *63*, 65–73.

146. Yanagisawa, J., Itoh, M., Ishibashi, M., Miyazaki, H. & Nakayama, F. (1980) Microanalysis of bile acid in human liver tissue by selected ion monitoring. Anal. Biochem. *107*, 75–86.

147. DeMark, B. R. & Klein, P. D. (1981) Ammonia gas: an improved reagent for chemical ionization mass spectrometry of bile acid methyl ester acetates. J. Lipid Res. *22*, 166–171.

148. Reimendal, R. & Sjövall, J. (1972) Analysis of steroids by off-line computerized gas chromatography-mass spectrometry. Anal. Chem. *44*, 21–29.

149. Axelson, M., Cronholm, T., Curstedt, T., Reimendal, R. & Sjövall, J. (1974) Quantitative analysis of unlabelled and polydeuterated compounds by gas chromatography-mass spectrometry. Chromatographia 7, 502–509.

150. Eneroth, P. & Sjövall, J. (1969) Methods of analysis in the biochemistry of bile acids. Methods Enzymol. *15*, 237–280.

151. Street, J. M., Trafford, D. J. H. & Makin, H. L. J. (1983) The quantitative estimation of bile acids and their conjugates in human biological fluids. J. Lipid Res. *24*, 491–511.

152. Sjövall, J. & Setchell, K. D. R. (1988) Techniques for extraction and isolation of bile acids. In: The Bile Acids. Volume 4: Methods and Applications (Setchell, K. D. R., Kritchevsky, D. & Nair, P. P., eds.), pp. 1–42, Plenum, New York.

153. Fransson, B. & Schill, G. (1975) Isolation of acidic conjugates by ion pair extraction. II. Extraction of taurocholic acid derivatives. Acta Pharmacol. Suec. *12*, 417–424.

154. Hofmann, A. F. (1967) Efficient extraction of bile acid conjugates with tetraheptylammonium chloride, a liquid ion exchange. J. Lipid Res. *8*, 55–58.

155. Sandberg, D. H., Sjövall, J., Sjövall, K. & Turner, D.A. (1965) Measurement of human serum bile acids by gas-liquid chromatography. J. Lipid Red. *6*, 182–192.

156. Bradlow, H. L. (1968) Extraction of steroid conjugates with a neutral resin. Steroids *11*, 265–272.
157. Shackleton, C. H. L. Sjövall, J., & Wisen, O. (1970) A simple method for the extraction of steroids from urine. Clin. Chim. Acta *27*, 354–356.
158. Makino, I. & Sjövall, J. (1972) A versatile method for analysis of bile acids in plasma, Anal. Lett. *5*, 341–349.
159. Matsui, M., Hakozaki, M. & Kinuyama, Y. (1975) Extraction of steroid diconjugates using Amberlite XAD-2 resins. J. Chromatography *115*, 625–628.
160. Pageaux, J. F., Duperray, B., Anker, D. & Dubois, M. (1979) Bile acid sulphates in serum bile acid determination. Steroids *34*, 73–88.
161. Axelson, M. & Sahlberg, B.-L. (1981) Solid extraction of steroid conjugates from plasma and milk. Anal. Lett. *14*, 771–782.
162. Barnes, S. & Chitranukroh, A. (1977) A simplified procedure for the isolation of bile acids from serum based on a batch extraction with the non-ionic resin Amberlite XAD 7. Ann. Clin. Biochem. *14*, 235–239.
163. Whitney, J. O. & Thaler, M. D. (1980) A simple chromatographic method for quantitative extraction of hydrophobic compounds from aqueous solutions. J. Liq. Chromatogr. *3*, 545–556.
164. Goto, J., Kato, H., Saruta, Y. & Nambara, T. (1980) Separation and determination of bile acids in human bile by high pressure liquid chromatography. J. Liq. Chromatogr. *3*, 991–1003.
165. Setchell, K. D. R. & Worthington, J. (1982) A rapid method for the quantitative extraction of bile acids and their conjugates from serum using commercially available reverse phase octadecylsilane-bonded silica cartridges. Clin. Chim. Acta. *125*, 135–144.
166. DeMark, B. R., Everson, G. T., Klein, P. D., Showalter, R. B. & Kern, F., Jr. (1982) A method for the accurate measurement of isotope ratios of chenodeoxycholic and cholic acids in serum. J. Lipid Res. *23*, 204–210.
167. Shackleton, C. H. L. & Whitney, J. O. (1980) Use of Sep-Pak® cartridges for urinary steroid extraction: evaluation of methods for use prior to gas chromatographic analysis. Clin. Chim. Acta. *107*, 231–243.
168. Heikkinen, R., Fotsis, T. & Adlercreutz, H. (1981) Reversed-phase C_{18} cartridge for extraction of estrogens from urine and plasma. Clin. Chem. *27*, 1186–1189.
169. Rudman, D. & Kendall, F. E. (1957) Bile acid content of human serum. 1. Serum bile acids in patients with hepatic disease. J. Clin. Invest. *36*, 530–537.
170. Dyfverman, A. & Sjövall, J. (1978) Liquid-gel extraction of bile acids. In: Biological Effects of Bile Acids (Paumgartner, G., Stiehl, A. & Gerok, W., eds.) pp. 281–286, MTP Press Ltd., Lancaster, U.K.
171. Dyfverman, A. & Sjövall, J. (1978) A novel liquid-gel chromatographic method for extraction of unconjugated steroids from aqueous solutions. Anal. Lett. *6*, 485–499.
172. Egestad, B., Curstedt., T. & Sjövall, J. (1982) Simple procedures for the enrichment of chlorinated aromatic pollutants from fat, water, and milk for subsequent analysis by high resolution methods. Anal. Lett. *15*, 243–307.
173. Dahlberg, E., Snochowski, M. & Gustafsson, J.-A. (1980) Removal of hydrophobic compounds from biological fluids by a simple method. Anal. Biochem. *106*, 380–388.
174. Dyfverman, A. & Sjövall, J. (1983) Ion pair extraction of bile acids with Lipidex gel. Anal. Biochem. *134*, *2*, 303–308.
175. Ahrens, E. H., Jr. & Craig, L. C. (1952) Extraction and separation of bile acids. J. Biol. Chem. *195*, 763–778.
176. Gregg, J. A. (1968) Urinary excretion of bile acids in patients with obstructive jaundice and hepatocellular disease. Am. J. Clin. Path. *49*, 404–409.
177. Wooton, I. D. P. (1953) Studies in the bile acids. 1. A method of separation and identification. Biochem. J. *53*, 85–88.
178. Danielsson, H., Kallner, A. & Sjövall, J. (1963) On the composition of the bile acid fraction of rabbit feces and the isolation of a new bile acid: 3α,12α-dihydroxy-5α-cholanic acid. Bile acids and steroids 136. J. Biol. Chem. *283*, 3846–3852.
179. Karavolas, H. J., Elliott, W. H., Hsia, S. L., Doisy, E. A., Jr., Matshiner, J. T., Thayer, S. A. & Doisy, E. A. (1965) *Allo*-cholic, a metabolite of 5α-cholestan-3β-ol in the rat. J. Biol. Chem. *240*, 1568–1572.
180. Ikawa, S. (1978) Group separation of bile acids and salts by silicic acid column chromatography. Anal. Biochem. *85*, 197–208.
181. Kaplowitz, N. & Javitt, N. B. (1973) Quantitative analysis of unconjugated and conjugated bile acids in duodenal fluid by densitometry after paper electrophoresis. J. Lipid. Res. *14*, 224–228.
182. Sjövall, J. (1952) Separation of bile acids by paper chromatography. Bile acids and steroids 3. Acta Chem. Scand. *6*, 1552–1553.

183. Sjövall, J. (1964) Separation and determination of bile acids. Methods Biochem. Anal. *12*, 97–141.
184. Hofmann, A. F. (1964) Thin-layer chromatography of bile acids and their derivatives. In: New Biochemical Separations (James, A. J. & Morris, L. J., eds.) pp. 261–282, D. Van Nostrand Co. Ltd., London.
185. Lisboa, B. P, (1969) Thin layer chromatography of steroids and related compounds. In: Methods in Enzymology (Clayton, R. B., ed.) Vol XV, pp. 3–157, Academic Press, New York.
186. Hofmann, A. F. (1962) Thin-layer adsorption chromatography of free and conjugated bile acids on silicic acid. J. Lipid Res. *3*, 127–128.
187. Hofmann, A. F. & Mosbach, E. H. (1964) Identification of *allo*-deoxycholic acid as a major compound of gallstones induced in rabbit by 5α-cholestan-3β-ol. J. Biol. Chem. *239*, 2813–2821.
188. Hoshita, T., Nagayoshi, S., Kouchi, M. & Kazuno, T. (1964) Stereo-bile acids and bile sterols. 63. Studies in the bile of the family cyprinidae. J. Biochem. (Tokyo) *56*, 177–181.
189. Goswami, S. K. & Frey, C. F. (1978) Thin-layer chromatographic method for the separation of conjugated deoxycholic acid from conjugated chenodeoxycholic acid. J. Chromatogr. *145*, 147–150.
190. Eneroth, P. (1963) Thin-layer chromatography of bile acids. J. Lipid. Res. *4*, 11–16.
191. Hamilton, J. G. (1963) The effect of oral neomycin on the conversion of cholic acid to deoxycholic in man. Arch. Biochem. Biophys. *101*, 7–13.
192. Freimuth, U., Zawta, B. & Buchner, M. (1967) New chromatographic system for the separation of structurally isomeric free bile acids. J. Chromatogr. *30*, 607–610.
193. Bruusgaard, A. (1970) Quantitative determination of the major 3-hydroxy bile acids in biological material after thin-layer chromatographic separation. Clin. Chim. Acta. *28*, 495–504.
194. Cass, O. W., Cowen, A. E., Hofmann, A. F. & Coffin, S. B. (1975) Thin-layer chromatographic separation of sulfated and nonsulfated lithocholic acids and their glycine and taurine conjugates. J. Lipid. Res. *16*, 159–160.
195. Batta, A. K., Salen, G. & Shefer, S. (1980) Thin-layer chromatography of bile alcohols, bile acids and conjugated bile acids. J. Liq. Chromatogr. *3*, 1865–1879.
196. Touchstone, J. C., Levitt, R. E., Levin, S. S. & and Soloway, R. D. (1980) Separation of conjugated bile acids by reverse phase thin-layer chromatography. Lipids *15*, 386–387.
197. Kirkpatrick. R. B., Lack, I. & Killenberg, P. G. (1980) Identification of the 3-sulfate isomer as the major product of enzymatic sulfation of chenodeoxycholic acid conjugates. J. Biol. Chem. *255*, 10157–10159.
198. Bolt, M. J. G. (1987) Separation of methylated free bile acids from their taurine and methyl glycine conjugates by thin layer chromatography. J. Lipid Res. *28*, 1013–1015.
199. Norman, A. & Palmer, R. H. (1964) Metabolites of lithocholic acid-24-^{14}C in human bile and feces. J. Lab. Clin. Med. *63*, 986–1001.
200. Semenuk, G. & Beher, W. T. (1966) Quantitative determination of bile acids by direct densitometry thin-layer chromatograms. J. Chromatogr. *21*, 27–31.
201. Kottke, B. A., Wollenweber, J. & Owen, C. A., Jr., (1966) Quantitative thin layer chromatography of free and conjugated cholic acid in human bile and duodenal contents. J. Chromatogr. *21*, 439–447.
202. Panveliwalla, D., Lewis, B., Wootton, I. D. P. & Tabaqchali, S. (1970) Determination of individual bile acids in biological fluids by thin-layer chromatography and fluorimetry. J. Clin. Pathol. *23*, 309–314.
203. O'Moore, R. R. L. & Percy-Robb, I. W. (1973) Analysis of bile acids and their conjugates in jejunal juice by thin-layer chromatography and direct densitometry. Clin. Chim. Acta *43*, 39–47.
204. Boucard, M., Massa, V., Serrano, J. J., Sustuges, P. & Sucholla, D. (1974) Chromatography of bile I. Direct fluorimetry of biliary bile acid chromatograms. Trav. Soc. Pharm. Montpellier *34*, 19–25.
205. Nyström, E. (1968) Preparation of methylated Sephadex and lipophilic sephadex ion exchangers. Ark. Kemi. *29*, 99–106.
206. Almé, B. & Nyström, E. (1971) Preparation of lipophilic anion exchangers from chlorohydroxypropylated Sephadex and cellulose. J. Chromatogr. *59*, 45–52.
207. Ellingboe, J., Almé, B. & Sjövall, J. (1970) Introduction of specific groups into polysaccharide supports for liquid chromatography. Acta Chem. Scand. *24*, 463–467.
208. Ellingboe, J., Nyström, E. & Sjövall, J. (1970) Liquid-gel chromatography on lipophilic-hydrophobic sephadex derivatives. J. Lipid Res. *11*, 266–273.

209. Nyström, E. & Sjövall, J. (1965) Methylated Sephadex as support in reversed-phase partition chromatography. J. Chromatogr. *17*, 574–576.
210. Nyström, E. & Sjövall, J. (1965) Separation of lipids as methylated Sephadex. Anal. Biochem. *12*, 235.
211. Eneroth, P. & Nyström, E. (1967) A study on liquid-gel partition of steroids and steroid derivatives on lipophilic Sephadex gels. Biochim. Biophys. Acta *144*, 149–161.
212. Eneroth, P. & Nyström, E. (1968) Sterol composition of the steryl sulphate fraction in human feces. Steroids *11*, 417–427.
213. Eneroth, P. & Nyström, E. (1968) Quantification of cholesteryl sulfate and neutral sterol derivatives in human feces after purification on lipophilic Sephadex gels. Steroids *11*, 187–205.
214. Ellingboe, J., Nyström, E. & Sjövall, J. (1968) A versatile lipophilic Sephadex derivative for "reversed-phase" chromatography. Biochim. Biophys. Acta *152*, 803–805.
215. Sjövall, J. & Axelson, M. (1982) Newer approaches to the isolation, identification and quantitation of steroids in biological materials. Vitamins & Hormones *39*, 31–144.
216. Sjövall, J. & Vihko, R. (1966) Chromatography of conjugated steroids on lipophilic Sephadex. Acta Chem. Scand. *20*, 1419–1421.
217. Stiehl, A. (1974) Bile salt sulphates in cholestasis. Eur. J. Clin. Invest. *4*, 59–63.
218. Makino, I., Hashimoto, H., Shinozaki, K., Yoshino, K. & Nakagawa, S. (1972) Sulfated and nonsulfated bile acids in urine, serum and bile of patients with hepatobiliary diseases. Gastroenterology *68*, 545–553.
219. Norman, A., Strandvik, B., & Ojamäe, O. (1974) Urinary bile acid conjugates in extrahepatic biliary atresia. Acta Pediatr. Scand. *63*, 92–96.
220. Islam, M. A., Raicht, R. F. & Cohen, B. I. (1981) Isolation and quantitation of sulfated and unsulfated steroids in human feces. Anal. Biochem. *112*, 371–377.
221. Bremmelgaard, A. & Sjövall, J. (1980) Hydroxylation of cholic, chenodeoxycholic and deoxycholic acids in patients with intrahepatic cholestasis. J. Lipid Res. *21*, 1072–1081.
222. Kuksis, A. (1971) Ion-exchange chromatography of bile acids. In: The Bile Acids – Chemistry, Physiology and Metabolism. Vol. 1 (Nair, P. P. and Kritchevsky, D., eds.) pp. 173–208, Plenum Press, New York.
223. Lambiotte, M. (1957) Metabolisme des acides biliares. Revue Internationale d'Hepatologie *7*, 541–556.
224. Gordon, B. A., Kuksis, A. & Beveridge, J. M. R. (1963) Separation of bile acid conjugates by ion exchange chromatography. Can. J. Biochem. Physiol. *41*, 77–89.
225. Rouser, G., Kritchevsky, G., Yamamoto, A., Simon, G., Galli, C. & Bauman, A. J. (1969) Diethylaminoethyl and triethylaminoethyl cellulose column chromatographic procedures for phospholipids, glycolipids and pigments. In: Methods in Enzymology, Vol. XIV (Lowenstein, J. M., ed.) pp. 272–317, Academic Press, New York.
226. Hähnel, R. (1965) Use of Sephadex ion exchanger for the separation of conjugated urinary oestrogens. Anal. Biochem. *10*, 184–192.
227. Hähnel, R. (1962) Separation of oestrogen conjugates from urine. Clin. Chim. Acta *7*, 768–775.
228. Setchell, K. D. R., Almé, B., Axelson, M. & Sjövall, J. (1976) The multicomponent analysis of conjugates of neutral steroids in urine by lipophilic ion exchange chromatography and computerized gas chromatography-mass spectrometry. J. Steroid Biochem. *7*, 615–629.
229. Setchell, K. D. R., Taylor, N. S., Adlercreutz, H., Axelson, M. & Sjövall, J. (1979) The group separation of conjugates of oestrogens using lipophilic ion exchange chromatography. In: Research on Steroids (Klopper, A., Lerner, L., Van der Molen, H. J. & Sciarra, F., eds.) Vol. VIII, pp. 131–135, Academic Press, New York.
230. Stiehl, A., Raedsch, R., Rudolf, G., Czygan, P. & Walker, S. (1982) Analysis of bile acid glucuronides in urine: group separation on a lipophilic anion exchanger. Clin. Chim. Acta *123*, 275–285.
231. Hedenborg, G. & Norman, A. (1984) The nature of urinary bile acid conjugates in patients with extrahepatic cholestasis. Scand. J. Clin. Lab. Invest. *44*, 725–733.
232. Goto, J., Hasegawa, M., Kato, H. & Nambara, T. (1978) A new method for simultaneous determination of bile acids in human bile without hydrolysis. Clin. Chim. Acta *87*, 141–147.
233. Bremmelgaard, A. & Almé, B. (1980) Analysis of plasma bile acid profiles in patients with liver diseases associated with cholestasis. Scand. J. Gastroenterol. *15*, 593–600.
234. Tanida, N., Hikasa, Y., Hosomi, M., Satomi, M., Oohama, I. & Shimoyama, T.

(1981) Fecal bile acid analysis in healthy Japanese subjects using a lipophilic anion exchanger, capillary column gas chromatography, and mass spectrometry. Gastroenterol. Jpn. *16*, 363–371.

235. Nambara, T. & Goto, J. (1988) High performance liquid chromatography of bile acids. In: The Bile Acids: Methods and Applications (Setchell, K. D. R., Nair, P. P. & Kritchevsky, D., eds.) Vol. 4, pp. 43–64, Plenum Publishers, New York.

236. Miyazaki, S., Tanaka, H., Horikawa, R., Tsuchiya, H. & Imai, K. (1980) Rapid determination of glycine- and taurine-conjugated bile acids in human bile by high-performance liquid chromatography. J. Chromatogr. *181*, 177–185.

237. Rossi, S. S., Converse, J. L. & Hofmann, A. F. (1987) High pressure liquid chromatographic analysis of conjugated bile acids in human bile: simultaneous resolution of sulphated and unsulphated lithocholyl amidates and the common conjugated bile acids. J. Lipid Res. *28*, 589–595.

238. Okuyama, S., Uemura, D. & Hirata, Y. (1979) The improved method of high performance liquid chromatographic separation of individual bile acids: free and glycine conjugated bile acids. Chem. Lett. 461–462.

239. Stellaard, F., Hachey, D. L. & Klein, P. D. (1978) Separation of bile acids as their phenacyl esters by high-pressure liquid chromatography. Anal. Biochem. *87*, 359–366.

240. Parris, N. (1979) Determination of free and conjugated bile acids as ultraviolet-absorbing ion pairs by reverse-phase high-performance liquid chromatography. Anal. Biochem. *100*, 260–263.

241. Goto, J., Kato, H. & Nambara, T. (1978) Separation of sulfated bile acids by high-performance liquid chromatography. Lipids *13*, 908–909.

242. Goto, J., Kato, H. & Nambara, T. (1980) Separation of monosulphated bile acids by high-performance liquid chromatography. J. Liq. Chromatogr. *3*, 645–655.

243. Vestal, M. L. (1984) Higher performance liquid chromatography-mass spectrometry. Science *226*, 275–281.

244. Nair, P. P., Gordon, M. & Reback, J. (1967) The enzymatic cleavage of the carbon-nitrogen bond in 3β,7β,12β-trihydroxy-5β-cholan-24-oyl glycine. J. Biol. Chem. *242*, 7–11.

245. Nair, P. P. (1969) Enzymatic cleavage of bile acid conjugates. In: Bile Salt Metabolism (Schiff, L., Carey, J. B. & Dietschy, J., eds.) pp. 172–183, Charles C. Thomas, Springfield.

246. Kobashi, K., Nishizawa, I., Yamada, T. & Hase, J. (1978) A new hydrolase specific for taurine conjugates of bile acids. J. Biochem. (Tokyo) *84*, 495–497.

247. Stiehl, A., Raedsch, R., Rudolf, G., Gundert-Remy, U. & Senin, M. (1985) Biliary and urinary excretion of sulphated glucuronidated and tetrahydroxylated bile acids in cirrhotic patients. Hepatology *5*, 492–495.

248. Imperato, T. J., Wong, G. C., Chen, L. J. & Bolt, R. J. (1977) Hydrolysis of lithocholate sulphate by *Pseudomonas aeruginosa*. J. Bacteriol. *130*, 545–547.

249. Eyssen, H. J. & Huijghebaert, S. M. (1981) Metabolism of bile acid sulphates by strictly anaerobic intestinal microorganisms. In: Recent Advances in Germ Free Research (Sasaki, S. et al., eds.) pp. 231–236, Tokai University Press, Tokyo.

250. Huijghebaert, S. M. & Eyssen, H. J. (1982) Specificity of bile salt sulfatase activity from *Clostridium sp.* strain S-1. Appl. Environ. Microbiol. *44*, 1030–1034.

251. Huijghebaert, S. M., Mertens, J. A. & Eyssen, H. J. (1982) Isolation of bile salt sulfatase producing Clostridium strain from rat intestinal microflora. Appl. Environ. Microbiol. *43*, 185–192.

252. Palmer, R. H. & Bolt, M. G. (1971) Bile acid sulfates, I. Synthesis of lithocholic acid sulfates and their identification in human bile. J. Lipid Res. *12*, 671–679.

253. Galeazzi, R. & Javitt, N. B. (1977) Bile acid excretion: the alternative pathway in the hamster. J. Clin. Invest *60*, 693–701.

254. Shaw, R. & Elliott, W. H. (1978) Bile acids. LV. 2,2-Dimethoxypropane: an esterifying agent preferred to diazomethane for chenodeoxycholic acid. J. Lipid Res. *19*, 783–787.

255. Kornel, L. (1965) Studies on steroid conjugates. IV. Demonstration and identification of solvolyzable corticosteroids in human urine and plasma. Biochem. *4*, 444–452.

256. Cohen, B. I., Budai, K. & Javitt, N. B. (1981) Solvolysis of chenodeoxycholic acid sulfates. Steroids *37*, 621–626.

257. Van Berge Henegouwen, G. P, Allan, R. N., Hofmann, A. F. & Yu, P. Y. S. (1977) A facile hydrolysis-solvolysis procedure for conjugated bile acid sulfates. J. Lipid Res. *18*, 118–122.

258. Hirano, Y., Miyazaki, H., Higashidate, S. & Nakayama, F. (1987) Analysis of 3-sulphated and non-sulphated bile acids by one step solvolysis and high performance liquid chromatography. J. Lipid. Res. *28*, 1524–1529.
259. Back, P., Sjövall, J. & Sjövall, K. (1974) Monohydroxy bile acids in plasma in intrahepatic cholestasis of pregnancy. Identification by computerized gas chromatography-mass spectrometry. Med. Biol. *52*, 31–38.
260. Miyazaki, H., Ishibashi, M., Inoue, M., Itoh, M. & Kubodera, T. (1974) Simultaneous qualitative and quantitative analysis of bile acids in mass chromatography. J. Chromatogr. *99*, 553–565.
261. Vrbanac, J. J., Sweeley, C. C. & Pinkson, J. C. (1983) Automated metabolic profiling analysis of urinary steroids by gas chromatography mass spectrometry data system. Biomed. Mass Spectrom. *10*, 155–161.
262. Makino, I., Nakagawa, S. & Mashimo, K. (1969) Conjugated and unconjugated serum bile acid levels in patients with hepatobiliary diseases. Gastroenterology *56*, 1033–1039.
263. Setchell, K. D. R., Lawson, A. M., Blackstock, E. J. & Murphy, G. M. (1982) Diurnal changes in serum unconjugated bile acids in normal man. Gut *23*, 637–642.
264. Summerfield, J. A., Billing, B. H. & Shackleton, C. H. L. (1976) Identification of bile acids in serum and urine in cholestasis. Evidence for 6β-hydroxylation of bile acids in man. Biochem. J. *154*, 507–516.
265. Takikawa, H., Otsuka, H., Beppu, T., Seyama, S. & Yamakawa, T. (1982) Quantitative determination of bile acid glucuronides in serum by mass fragmentography. J. Biochem. *92*, 985–998.
266. Takikawa, H., Otsuka, H., Beppu, T., Seyama, S. & Yamakawa, T. (1983) Serum concentrations of bile acid glucuronides in hepatobiliary diseases. Digestion *27*, 189–195.
267. Tanaka, K., Takeshita, K. & Takita, M. (1982) Abnormalities of bile acids in serum and bile from patients with myotonic muscular dystrophy. Clin. Sci. *62*, 627–642.
268. Henneberg, D. (1961) Eine Kombination von Gaschromatograph and Massenspektrometer zur Analyse organischer Stoffgemische. Fresenius Z. Anal. Chem. *183*, 12–23.
269. Angelin, B., & Björkhem, I. (1977) Postprandial serum bile acids in healthy man. Gut *18*, 606–609.
270. Ahlberg, J., Angelin, B., Björkhem, I. & Einarsson, K. (1977) Individual bile acids in portal venous and systemic blood serum of fasting man. Gastroenterol. *73*, 1377–1382.
271. Angelin, B., Björkhem, I. & Einarsson, K. (1978) Individual bile acid concentrations in normo- and hyperlipoproteinemias as determined by mass fragmentography – relation to bile acid pool size. J. Lipid Res. *19*, 527–537.
272. Beppu, T., Seyama, Y., Kasama, T. & Yamakawa, T. (1981) Quantitative determination of non-sulphated bile acids and sulphated lithocholic acid in serum by mass fragmentography. J. Biochem. *89*, 1963–1973.
273. Akashi, Y., Miyazaki, H. & Nakayama, F., (1983) Correlation of bile acid composition between liver tissue and bile. Clin. Chim. Acta *133*, 125–132.
274. Björkhem, I., Angelin, B., Einarsson, K. & Ewerth, S. (1982) Fasting levels of monoketonic bile acids in human peripheral and portal circulation. J. Lipid Res. *23*, 1020–1025.
275. Barnes, S., Berkowitz, G., Hirschowitz, I & Wirtschafter, D. (1981) Postnatal physiologic hypercholemia in both premature and full-term infants. J. Clin. Invest. *68*, 775–782.
276. Björkhem, I. & Falk, O. (1983) Assay of the major bile acids in serum by isotope dilution-mass-spectrometry. Scand. J. Gastroenterol. *43*, 163–170.
277. Tohma, M., Wajima, H., Mahara, R., Kurosawa, T. & Makino, I. (1984) Determination of 3β-hydroxy-5-cholen-24-oic and its sulphate in human serum by gas chromatography-mass spectrometry. Steroids *44*, 47–55.
278. Beppu, T., Seyama, Y., Kasama, T., Serizawa, S. & Yamakawa, T. (1982) Serum bile acid profiles in cerebrotendinous xanthomatosis. Clin. Chim. Acta *118*, 167–175.
279. Ewerth, S. (1982) Postprandial serum concentrations of individual bile acids in man. Scand. J. Gastroenterol. *17*, 781–784.
280. Angelin, B., Björkhem, I., Einarsson, K. & Ewerth, S. (1982) Hepatic uptake of bile acids in man. Fasting and postprandial concentrations of individual bile acids in portal venous and systemic blood serum. J. Clin. Invest. *70*, 724–731.
281. Angelin, B., Björkhem, I., Einarsson, K. & Ewerth, S. (1985) Cholestyramine treatment reduces post-prandial but not fasting serum

bile acid levels in humans. Gastoenterol. *83*, 1097–1101.

282. Liehr, J. G., Beckner, C. F., Ballatore, A. M. & Caprioli, R. M. (1982) Proc. No. 30 Annual Conf. Mass Spectrom. Allied Topics, pp. 236.
283. Hachey, D. L., Szczepanik, P. A., Berngrubes, O. W. & Klein, P. D. (1973) Synthesis of stable isotopes: synthesis of deuterium and ^{13}C labeled bile acid. J. Labelled Compds. *9*, 703–718.
284. Cowen, A. E., Hofmann, A. F., Hachey, D. L., Thomas, P. J., Belobaba, D. T. E., Klein, P. D. & Tokes, L. (1976) Synthesis of 11,12-$^2H_2$- and 11,12-$^3H_2$-labelled chenodeoxycholic and lithocholic acids. J. Lipid Res. *17*, 231–238.
285. Tserng, K.-Y. & Klein, P. D. (1977) An improved synthesis of 24-^{13}C-labeled bile acids using formyl esters and a modified lead tetraacetate procedure. J. Lipid Res. *18*, 400–403.
286. Baillie, T. A., Karls, M. & Sjövall, J. (1978) Synthesis of deuterium-labelled lithocholic acid J. Labelled Compd. Radiopharm. *14*, 849–859.
287. Balistreri, W. F., Cowen, A. E., Hofmann, A. F., Szczepanik, P. A. & Klein, P. D. (1975) Validation of use of 11,12-$^2H_2$-labelled chenodeoxycholic acid in isotope dilution measurements of bile acid kinetics in man. Pediatr. Res. *9*, 757–760.
288. Watkins, J. B., Ingall, D., Szczepanik, P. A., Klein, P. D. & Lester, R. (1973) Bile salt metabolism in the newborn: measurement of pool size and synthesis of stable isotope technique. N. Engl. J. Med. *288*, 431–434.
289. Klein, P. D., Haumann, J. R. & Hachey, D. L. (1981) A stable isotope ratiometer-multiple ion detector (SIRMID) used for quantitative and qualitative stable isotope studies by electron impact and chemical ionization mass spectrometry. J. Lipid. Res. *22*, 166–171.
290. Kern, F., Jr., Everson, G. T., DeMark, B., McKinley, C., Showalter, R., Erfling, W., Braverman, D. Z., Szczepanik-Van Leeuwen, P. & Klein, P. D. (1981) Biliary lipids, bile acids, and gallbladder function in the human female. Effects of pregnancy and the ovulatory cycle. J. Clin. Invest. *68*, 1229–1242.
291. Nishida. T., Miwa, H., Yamamoto, M., Koga, T. & Yao, T. (1981) Bile acid absorption kinetics in Crohn's disease on elemental diet after oral administration of stable-isotope tracer with chenodeoxycholic-11,12-d_2 acid. Gut *23*, 751–757.
292. Stellaard, F., Schubert, R. & Paumgartner, G. (1983) Measurement of bile acid kinetics in human serum using stable isotope labelled chenodeoxycholic and capillary gas chromatography electron impact mass spectrometry. Biomed. Mass Spectrom. *10*, 187–191.
293. Stellaard, F., Sackman, M., Sauerbruch, T. & Paumgartner, G. (1984) Simultaneous determination of cholic acid and chenodeoxycholic acid pool sizes and fractional turnover rates in human serum using ^{13}C-labelled bile acids. J. Lipid. Res. *25*, 1313–1319.
294. Buckley, W. T., Huckin, S. N. & Eigendorf, G. K. (1985) Calculation of stable isotope enrichment for tracer kinetic procedures. Biomed. Mass Spectrom. *12*, 1–5.
295. Tateyama, T., Nezu, Y., Shino, M., Sakaguchi, K., Katayama, K. & Tsutsumi, J. (1978) Studies on metabolic fate of chenodeoxycholic acid in man. Koeushu-Iyo Masu Kenkyukai *3*, 207–214.
296. Stellaard, F., Paumgartner, G., van Berge Henegouwen, G. P. & van der Werf, S. D. J. (1986) Determination of deoxycholic acid pool size and input rate using [24-^{13}C] deoxycholic acid serum sampling. J. Lipid Res. *27*, 1222–1225.
297. Stellaard, F. & Paumgartner, G. (1987) A new model to assess deoxycholic acid metabolism in health using stable isotope dilution technique. Eur. J. Clin. Invest. *17*, 63–67.
298. Danzinger, R. G., Hofmann, A. F., Schoenfield, L. J., Berngruber, O. W., Szczepanik, P. A. & Klein, P. D. (1971) Measurement of the bile acid kinetics in man using stable isotopes: application to cholelithiasis. Gastroenterol. *60*, 192 (Abstract).
299. Norman, E. J., Heubi, J. E., Gelfand, M. J., Dan, P. & Matthews, D. W. (1984) Chenodeoxycholic acid pool size determination from children using isotope dilution mass spectrometry. Biomed. Mass Spectrom. *11*, 269–272.
300. Matoba, N., Une, M. & Hoshita, T. (1986) Identification of unconjugated bile acids in human bile. J. Lipid Res. *27*, 1154–1162.
301. Tohma, M., Nakata, Y., Yamada, H., Kurosawa, T., Makino, I. & Nakagawa, S. (1981) Quantitative determination of ursodeoxycholic acid and its deuterated derivatives in human bile by gas chromatography-

mass fragmentography. Chem. Pharm. Bull. *29*, 137–145.
302. Hedenborg, G., Norlander, A. & Norman, A. (1982) Bile acid excretion during drainage of extrahepatic cholestasis with refeeding of bile. Scand. J. Clin. Lab. Invest. *42*, 551–556.
303. Laatikainen, T., Lehtonen, P. & Hesso, A. (1978) Biliary bile acids in uncomplicated pregnancy and in cholestasis of pregnancy. Clin. Chim. Acta *85*, 145–150.
304. Murphy, G. M. & Singer, E. (1974) Bile acid metabolism in infants and children. Gut *15*, 151–163.
305. Lester, R., Pyrek, J. St., Little, J. M. & Adcock, E. W. (1983) Diversity of bile acids in the fetus and newborn infant. J. Pediatr. Gastroenterol. Nutr. *2*, 355–364.
306. Bongiovani, A. M. (1965) Bile acid content of gallbladder of infants, children and adults. J. Clin. Endocrinol. Metab. *25*, 678–685.
307. Colombo, C., Zuliani, G., Ronchi, M., Breidenstein, J. & Setchell, K. D. R. (1987) Biliary bile acid composition of the human fetus in early gestation. Ped. Res. *21*, 197–200.
308. Setchell, K. D. R., Dumaswala, R., Ronchi, G. & Colombo, C. (1988) Bile acid synthesis and metabolism during development determined from analysis of fetal gallbladder bile. J. Biol. Chem. (in press).
309. Gustafsson, J. (1985) Bile acid synthesis during development. Mitochondrial 12α-hydroxylation in human liver. J. Clin. Invest. *75*, 604–607.
310. Gustafsson, J., Andersson, S. & Sjövall, J. (1985) Bile acid metabolism during development: metabolism of taurodeoxycholic acid in human fetal liver. Biol. Neonate *47*, 26–31.
311. Gustafsson, J., Andersson, S. & Sjövall, J. (1987) Bile acid metabolism during development: metabolism of lithocholic acid in human fetal liver. Ped. Res. *21*, 99–103.
312. Watkins, J. B. (1983) Placental transport: Bile acid conjugation and sulphation in the fetus. J. Pediatr. Gastroenterol. Nutr. *2*, 363–373.
313. Clayton, P. T., Muller, D. P. R. & Lawson, A. M. (1982) The bile acid composition of gastric contents from neonates with high intestinal obstruction. Biochem. J. *206*, 489–498.
314. Lawson, A. M., Setchell, K. D. R., Clayton, P. T. & Muller, D. P. R. (1983) Neonatal bile acid biosynthesis. Proc. 31st. Ann. Confer. Mass Spectrom. Allied Topics, pp. 510–511, Boston.
315. Setchell, K. D. R., Street, J. M. & Sjövall, J. (1987) Fecal bile acids. In: The Bile Acids: Methods and Applications (Setchell, K. D. R., Kritchevsky, D. & Nair, P. P., eds.) Vol. 4, pp. 441–570, Plenum Press, New York.
315. Ali, S. S., Kuksis, A. & Beveridge, J. M. R. (1965) Excretion of bile acids by three men on a fat-free diet. Can. J. Biochem. *44*, 957–969.
317. Evrard, E. & Janssen, G. (1968) Gas-liquid chromatographic determination of human fecal bile acids J. Lipid Res. *9*, 226–236.
318. Eneroth, P., Hellström, K. & Sjövall, J. (1968) A method for quantitative determination of bile acids in human feces. Acta Chem. Scand. *22*, 1729–1744.
319. Eneroth, P., Gordon, B., Ryhage, R. & Sjövall, J. (1966) Identification of mono- and dihydroxy bile acids in human feces by gas-liquid chromatography and mass spectrometry. J. Lipid Res. *7*, 511–523.
320. Eneroth, P., Gordon, B. & Sjövall, J. (1966) Characterization of trisubstituted cholanoic acids in human feces. J. Lipid Res. *7*, 524–530.
321. DeWeerdt, G. A., Beke, R. & Barbier, F. (1980) Quantitative determination of fecal bile acids as their methyl ether methyl esters by repetitive scan technique. Biomed. Mass Spectrom *7*, 515–520.
322. Setchell, K. D. R., Gilbert, J. M. & Lawson, A. M. (1983) Fat and cancer. Brit. Med. J. *286*, 1750.
323. Hikasa, Y., Tanida, N., Ohno, T. & Shimoyama, T. (1984) Fecal bile acid profiles in patients with large bowel cancer in Japan. Gut *25*, 833–838,
324. Tanida, N., Hikasa, Y., Shimoyama, T. & Setchell, K. D. R. (1982) Comparison of the fecal bile and profiles between patients with adenomatous polyps of the large bowel and healthy subjects in Japan. Gut *25*, 824–832.
325. Tanida, N., Hikasa, Y., Shimoyama, T. & Setchell, K. D. R. (1985) Fecal bile acid profiles of Japanese patients with adenomatous polyps of the large bowel: special reference to distribution multiplicity, size and degree of dysplasia of the polyps. Japanese J. Cancer Res. (Gann) *76*, 104–112.
326. DeWeerdt, G. A., Verdievel, H., Beke, R. & Barbier, F. (1984) Fecal bile acid excretion in gastrointestinal diseases: mass spectrometric and enzymatic analysis. In: Proceedings of First International Symposium on

Bile Acids in Hepatobiliary and Gastrointestinal Diseases. pp. 185–201. IRL Press, Oxford, U.K.
327. Podesta, M. T., Murphy, G. M., Sladen, G. E., Breuer, N. F. & Dowling, R. H. (1978) Fecal extraction in diarrhea: effect of sulfated and nonsulfated bile acids on colonic structure and function. In: Biological Effects of Bile Acids (Paumgartner, G., Stiehl, A., & Gerok, W., eds.) pp. 245–256, MTP Press Ltd., Lancaster, U.K.
328. Podesta, M. T., Murphy, G. M. & Dowling, R. H. (1980) Measurement of fecal bile acid sulfates. Clin. Chim. Acta *182*, 243–300.
329. Kelsey, M. I. & Sexton, S. A. (1976) The biosynthesis of ethyl esters of lithocholic acid and isolithocholic acid by rat intestinal microflora. J. Steroid Biochem. *7*, 641–646.
330. Kelsey, M. I., Molina, J. E., Huang, S.-K. S. & Huang, K.-K. (1980) The identification of microbial metabolites of sulfolithocholic acid. J. Lipid Res. *21*, 751–759.
331. Norman, A. & Palmer, R. H. (1964) Metabolites of lithocholic acid-24-^{14}C in human bile and feces. J. Lab. Clin. Med. *63*, 986–1001.
332. Kelsey, M. I. & Sexton, S. A. (1977) Isolation and purification of lithocholic acid metabolites produced by the intestinal microflora. J. Chromatogr. *133*, 327–334.
333. Hill, M. J., Crowther, J. S., Drasar, B. S., Hawksworth, G., Aries, V. & Williams, R. E. O., (1971) Bacteria and aetiology of cancer of the large bowel. Lancet *1*, 95–100.
334. Hill, M. J. (1974) Bacteria and the etiology of colonic cancer. Cancer *34*, 815–818.
335. Cook, J. W. & Haslewood, G. A. D. (1933) The conversion of a bile acid into a hydrocarbon derived from 1,2-benzanthracene. Chem. Indust. *38*, 758–759.
336. Hill, M. J., Drasar, B. S., Williams, R. E. O., Meade, T. W., Cox, A. G., Simpson, J. E. P. & Morson, B. C. (1975) Fecal bile acids and clostridia in patients with cancer of the large bowel. Lancet *2*, 535–539.
337. Hill, M. J. (1980) Bacterial metabolism and human carcinogenesis. Br. Med. Bull. *36*, 89–94.
338. Wynder, E. I. & Reddy, B. S. (1974) Metabolic epidemiology of colorectal cancer. Cancer *34*, 801–806.
339. Duane, W. C., Behreas, J. C., Kelly, S. G. & Levine, A. S. (1984) A method for measurement of nanogram quantities of 3-methylcholanthrene in stool samples. J. Lipid. Res. *25*, 523–526.
340. Hofmann, A. F. & Cummings, S. A. (1982) Measurement of bile acid and cholesterol kinetics in man by isotope dilution: principles and applications. In: Bile Acids in Gastroenterology (Barbara, L., Dowling, R. H., Hofmann, A. F. & Roda, E., eds.) pp. 75–117, MTP Press, Lancaster, U.K.
341. Setchell, K. D. R., Ives, J. A., Cashmore, G. C. & Lawson, A. M. (1987) On the homogeniety of stools with respect to bile acid composition and the normal day-to-day variation in bile acid excretion: A detailed qualitative and quantitative study using capillary column gas chromatography-mass spectrometry. Clin. Chim. Acta *162*, 257–275.
342. Back., P. & Walter, K. (1980) Developmental pattern of bile acid metabolism as revealed by bile acid analysis of meconium. Gastroenterol. *78*, 671–676.
343. Pyrek, J. St., Sterzycki, R., Lester, R. & Adcock, E. W. (1982) Constituents of human meconium. II. Identification of steroidal acids with 21 and 22 carbon atoms. Lipids *17*, 241–249.
344. Pyrek, J. St., Lester, R., Adcock, E. W. & Sanghvi, A. (1983) Constituents of human meconium. I. Identification of 3-hydroxy-etianic acids. J. Steroid Biochem. *18*, 341–351.
345. Street, J. M., Balistreri, W. F. & Setchell, K. D. R. (1986) Bile acid metabolism in the perinatal period – excretion of conventional and atypical bile acids in meconium. Gastroenterol. *90*, 5 (Abstract).
346. Laatikainen, T., Lehtonen, P. & Hesso, A. (1978) Fetal sulphated and nonsulphated bile acids in intrahepatic cholestasis of pregnancy. J. Lab. Clin. Med. *91*, 185–193.
347. Pyrek, J. St., Little, J. M. & Lester, R. (1984) Detection of 3-hydroxy-etianic and 3-hydroxy-bisnorcholanoic acids in human serum. J. Lipid Res. *25*, 1324–1329.
348. Fröhling, W. & Stiehl, A. (1976) Bile salt glucuronides. Identification and quantitative analysis in the urine of patients with cholestasis. Eur. J. Clin. Invest. *6*, 67–74.
349. Standvik, B. & Wikström, S.-A. (1982) Tetrahydroxylated bile acids in healthy human newborns. Eur. J. Clin. Invest. *12*, 301–305.
350. Bremmelgaard, A. & Sjövall, J. (1979) Bile acid profiles in liver diseases. Eur. J. Clin. Invest. *9*, 341–348.
351. Amuno, Y., Hayashi, E., Endo, T., Higashino, K. & Kishimoto, S. (1983) Unusual

trihydroxylated bile acids in urine of patients with liver cirrhosis. Clin. Chim. Acta *127*, 61–67.

352. Back, P. (1973) Identification and quantitative determination of urinary bile acids excreted in cholestasis. Clin. Chim. Acta *44*, 199–207.
353. Stiehl, A. (1977) Disturbances of bile acid metabolism in cholestasis. In: Clinics in Gastroenterology (Paumgartner, G., ed.) Vol. 6, pp. 45–67, Saunders, London, Philadelphia, Toronto.
354. Hedenborg, G., Norlander, A. & Norman, A. (1985) Conjugation, metabolism and excretion of [24-^{14}C]chenodeoxycholic in patients with extrahepatic cholestasis before and after biliary drainage – analysis of conjugated bile acids by HPLC. Scand. J. Clin. Lab. Invest. *45*, 157–164.
355. Thomassen, P. A. (1979) Urinary bile acids in late pregnancy and in recurrent cholestasis of pregnancy. Eur. J. Clin. Invest. *9*, 425–432.
356. Thomassen, P. A. (1979) Urinary bile acid excretion in recurrent cholestasis of pregnancy. A study with computerized gas chromatography-mass spectrometry. Thesis, Karolinska Institute and Karolinska Sjukhuset, Stockholm, Sweden.
357. Clayton, P. T., Leonard, J. V., Lawson, A. M., Setchell, K. D. R., Andersson, S., Egestad, B. & Sjövall, J. (1987) Familial giant cell hepatitis associated with synthesis of 3β,7α-dihydroxy- and 3β,7α,12α-trihydroxy-5-cholenoic acids. J. Clin. Invest. *79*, 1031–1038.
358. Makino, I., Sjövall, J., Norman, A. & Strandvik, B. (1971) Excretion of 3β-hydroxy-5-cholenoic and 3α-hydroxy-5α-cholanoic acids in urine of infants with biliary atresia. FEBS Letts. *15*, 161–164.
359. Lawson, A. M., Madigan, M. J. & Clayton, P. T. (1986) Rapid diagnosis of Zellweger syndrome and infantile Refsum's disease by fast atom bombardment-mass spectrometry of urine bile salts. Clin. Chim. Acta *161*, 221–231.
360. Lawson, A. M., Madigan, M. J. & Clayton, P. T. (1987) Abnormal bile salts in patients with peroxisomal dysfunction – analysis by fast atom bombardment mass spectrometry and gas-liquid chromatography mass spectrometry. In: Bile Acids and Liver Disease (Stiehl, A. & Raedsch, R., eds.) pp. 269–276, MTP Press Ltd., Lancaster.
361. Egestad, B., Pettersson, P., Skrede, S. & Sjövall, J. (1985) Fast atom bombardment mass spectrometry in the diagnosis of cerebrotendinous xanthomatosis. Scand. J. Lab. Invest. *45*, 443–446.
362. Janssen, G., Toppet, S. & Parmentier, G. (1982) Structure of the side chain of the C_{29} dicarboxylic bile acid occurring in infants with coprostanic acidaemia. J. Lipid Res. *23*, 456–465.
363. Délèze, G., Paumgartner, G., Karlaganis, G., Giger, W., Reinhard, M. & Sidiropoulos, D. (1978) Bile acid pattern in human amniotic fluid. Eur. J. Clin. Invest. *8*, 41–45.
364. Javitt, N. B. & Emerman, S. (1968) Effect of sodium taurolithocholate on bile flow and bile acid excretion. J. Clin. Invest. *47*, 1002–1014.
365. Fisher, M. M., Magnusson, R. & Miyai, K. (1971) Bile acid metabolism in mammals. I. Bile acid-induced intrahepatic cholestasis. Lab. Invest. *25*, 88–91.
366. Jenner, R. E. & Howard, E. R. (1975) Unsaturated monohydroxy bile acids as a cause of idiopathic obstructive cholangiopathy. Lancet *2*, 1073–1075.
367. Hiasa, Y., Konishi, Y., Kamamoto, Y., Watanabe, T. & Ito, N. (1971) Effect of lithocholic acid on DL-ethionine carcinogenesis in rat liver. Japan. J. Cancer Res. (Gann) *62*, 239–245.
368. Reddy, B. S. (1975) Role of bile metabolites in colonic carcinogenesis. Cancer *36*, 2401–2406.
369. Reddy, B. S., Watanabe, K., Weisburger, J. H. & Wynder, E. L. (1977) Promoting effect of bile acids in colonic carcinogenesis in germ free and conventional F344 rats. Cancer Res. *37*, 3238–3242.
370. Cohen, B. I., Raicht, R. F., Deschner, E. E., Takahashi, M., Sarwal, A. N. & Fazzini, E. (1980) Effect of cholic acid feeding on N-methyl-N-nitrosourea induced colon tumors and cell kinetics in rat. J. Natl. Cancer. Inst. *64*, 573–578.
371. Nair, P. P., Mendeloff, A. I., Vocci, M., Bankoski, J., Gorelik, M., Heman, G. & Plapinger, R. (1977) Lithocholic acid in human liver: identification of ξ-lithocholyl lysine in tissue protein. Lipids *12*, 922–929.
372. Nair, P. P., Solomon, R., Bankoski, J. & Plapinger, R. (1978) Bile acids in tissues: binding of lithocholic acid to protein. Lipids *13*, 966–970.
373. Okishio, T., Nair, P. P. & Gordon, M. (1967) Studies on bile acids. The microquantitative separation of cellular bile acids by gas liquid

chromatography. Biochem. J. *102*, 654–659.

374. Gelb, A. M., McSherry, C. K., Sadowsky, J. R. & Mosbach, E. H. (1982) Tissue bile acids in patients with colon cancer and colonic polyps. American J. Gastroenterol. *77*, 314–317.

375. Yanagisawa, J., Akashi, Y., Miyazaki, H. & Nakayama, F. (1984) Critical evaluation of the existence of so-called tissue bound lithocholate in human liver tissue by selected ion monitoring. J. Lipid Res. *25*, 1263–1271.

376. Turjman, N. & Nair, P. P. (1981) Nature of tissue bound lithocholic acid and its implications in the role of bile acids in carcinogenesis. Cancer Res. *41*, 3761–3763.

377. Nair, P. P., Kessie, G. & Flanagan, V. P. (1986) Reaffirmation of the validity of enzymatic cleavage of lithocholic acid from N-ξ-lithocholyl-L-lysine and N-α-CBZ-N-ξ-lithocholyl-L-lysine. J. Lipid. Res. *27*, 905–909.

378. Street, J. M. & Setchell, K. D. R., (1988) Chromatographic methods for bile acid analysis. J. Chromatogr. (in press).

379. Vestal, C. H. & Setchell, K. D. R. (1988) Application of thermospray liquid chromatography-mass spectrometry to the analysis of bile acids. Proc. 36th Ann. Confec. Mass spectrom. Allied Topics San Francisco (in press).

Chapter 3
Biogenic Amines

K. Blau

1 Introduction

The biogenic amines (1) have for a long time provided chemists, pharmacologists, neurologists and other clinicians with a fruitful field of study, because many biogenic amines have powerful physiological or pharmacological activities. This is particularly true of the catecholamines, which are frequently analysed in the clinical setting, and whose physiological significance as neurotransmitters is quite well-established. On the other hand the normal physiological role of some of the other amines, such as the "trace" amines (2), is not so clearly understood, and their investigation is therefore of continuing interest to researchers trying to work out the significance of their presence in various tissues, the levels and metabolic patterns related to that presence, and the meaning of changes in those parameters that have been found to occur in certain circumstances. The biosynthesis and metabolism of the biogenic amines have also been actively investigated. For all of these reasons there has been a high level of interest and a very active research effort put into the analysis of these compounds. Generally the concentrations of biogenic amines in biological fluids and tissues are very low, because they are subject to oxidation catalysed by ubiquitous monoamine oxidases. Also, some amines such as the catecholamines and indoleamines are chemically not very stable. There are therefore difficulties in getting reliable analytical results, and the plethora of analytical methods in the literature is evidence that really satisfactory methods of analysis are still being sought (3). There are also problems in the isolation of biogenic amines from complex biological matrices, free of interfering contaminants and in good yield. What probably causes the most trouble in the isolation and analysis of biogenic amines is their polar character, and the fact that many of them have more than one polar grouping. For satisfactory analytical performance it has been found necessary to try to mask the polarity of the biogenic amines by making a variety of chemical derivatives (4). This approach also discriminates against interfering substances, and may improve specificity of detection and stabilize amines against decomposition.

It is not surprising that the demanding nature of the problems of analysing biogenic amines has led to the application of the most powerful analytical methods, and gas chromatography linked to mass spectrometry was enthusiastically put into service by workers in the biogenic amine field in the hope that this application would solve some of the major problems (5), particularly those of sensitivity and of interference (6). It was also quickly realised that improved specificity could be exploited to cut corners in the work-up, and that the resulting simplified isolation procedures would give quicker results and, it was hoped, less decomposition of sensitive compounds.

Many of the mass spectrometric methods for biogenic amines were developed for analysis of amines in animal tissues, and since many of the amines are associated with the function of neuronal structures, a great deal of the work was with brain tissue. Although not strictly clinical, it may still be of potential clinical value and will be included where appropriate.

1.1 General aspects and isolation procedures for biogenic amines

Biogenic amines are all to varying degrees basic, and so they can be isolated from their biochemical matrices (tissues, blood, urine, cerebrospinal fluid etc.) by modulation of their basic character. For example, at high pH values they exist in the free-base form, and amines without other polar groups can be extracted into immiscible organic solvents and back-extracted into aqueous acid: this is a convenient isolation procedure which effectively discriminates against acidic and neutral substances (7). However, catecholamines would be rapidly oxidised by atmospheric oxygen at high pH values, and isolation by absorption on activated alumina is used because of its specific affinity for the catechol grouping (8). Catecholamines and indoleamines are sensitive to oxidation, and are therefore protected by addition of acid (HCl or acetic acid, respectively) in which they are more stable, and by antioxidants such as ascorbic acid. The presence of a basic grouping in amines generally is of course a direct invitation to pull them out of whatever matrix they may inhibit by acidic ion-exchange resins (9), and this has been widely successful. However, it is important not to use strongly acidic resins because some amines, particularly those with more than one polar group, may be hard to recover completely, while amines with aromatic systems will tend to be non-specifically absorbed to resins with aromatic backbones, and these effects will reduce recovery of the amines. The application of liquid ion exchangers, (alias ion-pair extraction) is also useful for isolating amines (9, 10).

Difficulties may arise from the nature of the biological matrix: plasma proteins are an example. Some way of precipitating them is usually necessary, and the alcohol precipitation method of Bozzi et al. seems to work well (11). Albumin is well-known to be a carrier-protein, and tryptophan and other indoles are known to bind to it, and possibly some of the other biogenic amines, too. Denaturing procedures will release bound compounds, but care must be taken to rinse them clear of the precipitated protein. Alternatively it may be more useful to analyse only the unbound fraction, in which case some form of diafiltration is the best way of separating it from the bound fraction and of deproteinization at the same time.

After isolation comes derivatization, and acylation of some sort is the most favoured, because this most effectively eliminates the basic nature of the amino group and the polarity which is so troublesome a cause of non-specific absorption and consequent tailing attendant on attempts at the chromatographic separation of underivatized amines. A number of derivatization schemes have the first reaction step done in the original aqueous matrix, and this approach helps to minimize losses by making a non-polar derivative as soon as possible and only thereafter going on to the extraction step (12, 13).

1.2 Mass spectrometry and gas chromatography-mass spectrometry (GC-MS) in the measurement of biogenic amines

Relatively little has been done in the design of derivatives specifically for mass spectrometric analysis, nor have many studies tried to compare derivatives to select those that work best with these techniques, and so a variety of acyl derivatives is in general use (e. g. 7–9, 12, 14–16). Other workers have valued the greater convenience of silylation (17–19), while Narasimhachari and Vouros preferred isothiocyanate derivatives (20, 21). The general requirement of derivatization for GC-MS go beyond those of making biogenic amines amenable to good chromatography or to sensitive electron-capture detection: they are to produce characteristic ions, to increase the mass numbers of suitable fragments as an aid to eliminating interference, and to have the characteristic ions carry as high a proportion of the total ion-current as possible to increase the sensitivity of GC-MS analyses.

Anggard and Hankey compared a variety of different derivatives of sympathomimetic amines and their metabolites (trifluoroacetyl, pentafluoropropionyl, heptafluorobutyryl, perfluorooctanoyl, chloroacetyl, trichloroacetyl, isothiocyanate, *p*-trifluoromethylbenzoyl and pentafluorobenzoyl) not only for their suitability for GC and their sensitivity of electron-capture detection, but also for their relative usefulness in GC-MS (22, 23). There was a tendency for electron bombardment to produce *β*-cleavage, and derivatization greatly increased the proportion of diagnostically useful fragments and their relative abundance. In these aspects the acyl derivatives were generally better than the trimethylsilyl (TMS) derivatives, and most of the recent work on the GC-MS of biogenic amines has been done with pentafluoropropionyl (PFP) derivatives (e. g. 24). The cleavage of these under electron impact (EI) tends to occur between the alpha and beta positions, with charge-retention on the aromatic moiety, giving ring-containing fragments which account for a considerable proportion of the total ion current, for sensitive detection. Coutts et al. (25) have studied a range of perfluoroalkylated derivatives of biogenic amine metabolites in detail and were able to assign unequivocal structures to most of the fragments produced, and to elucidate the fragmentation mechanisms. A somewhat different range of derivatives was compared by a number of workers who used a functional group approach: rather than selecting fragment ions specific for the derivatives of individual amines, they monitored the common fragment ion of m/z 174 found in all TMS-amines (18, 19).

An interesting approach, perhaps originally making a virtue of necessity, is that of Boulton's group, who pioneered the use of the integrated ion current technique (26), in which the samples are not introduced via a GC inlet, but by direct probe insertion into a high-resolution mass spectrometer. The technique of accelerating voltage alternation (27) was adapted to focus very precisely on a fragment ion in the high-resolution mass spectrum of a compound to be analysed (originally the dimethylaminonaphthalenesulphonyl derivative of *p*-tyramine) in rapid alternation with the corresponding fragment in the spectrum of the deuterated analogue introduced simultaneously as an internal standard. A double-focussing instrument was used, and the ion-current of each fragment ion was recorded during the time the probe was heated to evaporate off the sample. By integrating the envelope of the curves for standard and sample derivatives, which resemble broad chromatographic peaks, excellent quantitation was achieved, while the specificity was very high, because of the precision with which one can focus on the two different masses (Fig. 1).

The derivatives selected were the *N,N*-dimethylaminonaphthalene-5-sulphonyl (dansyl) derivatives (28), which are highly fluorescent, and can be visualized and quantitatively recovered after TLC, which gives an extra purification step, and still further improves the analytical specificity. The mass spectra of the dansylated amines have been recorded in detail by several groups (29–33). Boulton's group have made a great deal of use of the integrated ion current technique to study various aspects of biogenic amine metabolism (34–36), particularly that of the tyramines. It is an interesting comment on the integrated ion current approach that the same group has recently described the simultaneous analysis of 12 biogenic amine metabolites in plasma, CSF

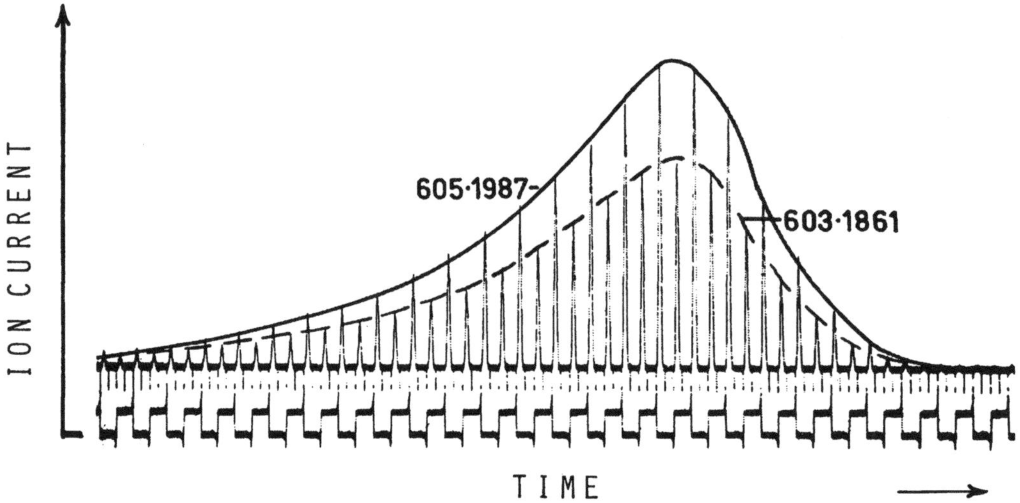

Figure 1. Integrated ion current recording of the *O,N*-bis-dansyl derivative of *p*-tyramine (*m/z* 603.1861) and of its deuterated analogue (*m/z* 605.1987).

and urine by capillary GC-high resolution mass spectrometry with selected ion monitoring (SIM) (37).

The potential of chemical ionization (CI) in the study of biogenic amines was recognized early (38), and with isobutane as reagent gas the majority gave mainly protonated molecular ions (together with the corresponding satellite ions). Underivatized amines with hydroxyls in the side-chains fragmented after protonation of the amino group and subsequent loss of ammonia from the protonated molecular ion. GC-CI-MS of derivatized amines has developed rather slowly, presumably because the instrumentation is not yet widely available, but the more uncluttered fragmentation patterns offer distinct advantages of specificity in the analysis of complex mixtures, as first realized in the biogenic amines by Miyazaki et al. (39) in their mass fragmentographic analysis of cerebrospinal fluid amines as their PFP-derivatives. Because the protonated molecular ions carried the major share of the total ion current, there was a useful 10-fold improvement in detection sensitivity. These authors attached a CI-source to an LKB-9000S GC-MS instrument and studied methane, isobutane and ammonia as reagent gases. Although ammonia was useful because it provided quasi-molecular ions as base peaks, its use failed to differentiate the base peaks of noradrenaline and 6-hydroxydopamine, and so they settled on isobutane. Edwards et al. (40) developed a procedure for using methane both as carrier gas for GC and as reagent gas for CI in a quadrupole GC-MS instrument, for the analysis of the urinary excretion of phenyl glycol- and phenyl ethanol-metabolites of catecholamines. These were analysed as PFP-derivatives. These all had base peaks corresponding to the loss of pentafluoropropionic acid from the protonated molecular ion, except for the dihydroxyphenylethanol derivative, where the base-peak resulted from the loss of a second PFP group. Edwards and his colleagues have used this technique to good effect in the investigation of the metabolism of dopamine. Mizuno and Ariga (41) measured brain catecholamines as PFP-derivatives in a conventional GC-MS fitted with a CI source. For adrenaline and noradrenaline the (M + 1 −PFPOH) ions (*m/z*

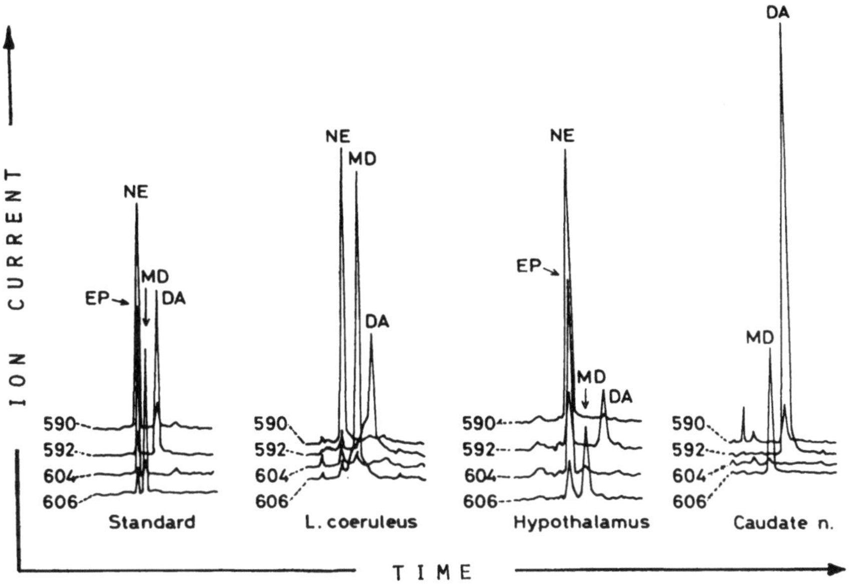

Figure 2. Selected ion recording of derivatized human brain extracts for the GC-MS determination of catecholamines.

590 and 604 respectively) were the base peaks, but for dopamine and α-methyldopamine they were the protonated molecular ions of *m/z* 592 and 606 respectively (Fig. 2).

Bergner and Dougherty have investigated the potential of negative ion chemical ionization (NICI) mass spectrometry for analysis of biogenic amines (42). They measure the mass spectra of dansyl, fluorescamine, 2-methoxy-2,4-diphenyl-3-(2H)-furanone (MDPF), 2,4-dinitrophenyl, 4-nitrophenyl, and 3,5-dinitrobenzoyl derivatives of *p*-tyramine, octopamine, dopamine, normetadrenaline and serotonin by probe injection into the heated ion-source with isobutane as reagent gas. Tyramine was investigated most, as a model substance. A problem with the nitro-substituted reagents was that mixtures of singly and multiply derivatized amines were obtained, and the latter had very low volatility. Molecules that fluoresce in the visible spectrum have positive electron affinities, because fluorescence is associated with a low-energy vacant orbital. It was found that MDPF and fluorescamine yielded single derivatives with tyramine that gave an intense $[M-H]^-$ ion in NICI-MS. However, while the MDPF-derivative had an equally abundant $[M-H-H_2O]^-$ ion, the $[M-H]^-$ ion of the fluorescamine derivative carried a much higher proportion of the total charge. The detection limit for this derivative was about 50 ng, about twenty times better than with CI-MS, and the derivative was much more volatile than the bis-dansyl derivative. Some of these comparisons are shown in Table 1.

The advantage of the NICI approach is that the isolation of the amines from their biological matrix is greatly simplified: in the case of the fluorescent fluorescamine derivatives a simple TLC step would probably suffice. It is not clear, however, whether the degree of sensitivity obtainable could match that of the integrated ion current method used with the dansyl derivatives, unless an integrated ion current method was applied to the NICI procedure.

Table 1. A comparison of mass spectral data for derivatives of *p*-tyramine. The percentages of the total ion current carried are only approximate and have been calculated from published spectra.

Derivative	Base peak *m/z*	% of total ion current
O,N-Di-PFP	266	28
O,N-Di-TMS	102	40
O,N-Bis-dansyl (EI)	170	15
O,N-Bis-dansyl (NICI)	234	62
MDPF	370	24
Fluorescamine	397	45

2 The Analysis of the Catecholamines and of their Metabolites

The catecholamines dihydroxyphenylalanine (dopamine), noradrenaline (norepinephrine) and adrenaline (epinephrine) are important as neurotransmitters in the normal function of the central and sympathetic nervous systems. They are synthesised there and in the adrenal medulla by the metabolic pathway shown in Figure 3.

Noradrenaline is a major neurotransmitter in the central nervous system and also in the sympathetic neurones. Dopamine too plays a neurotransmitter role in the central

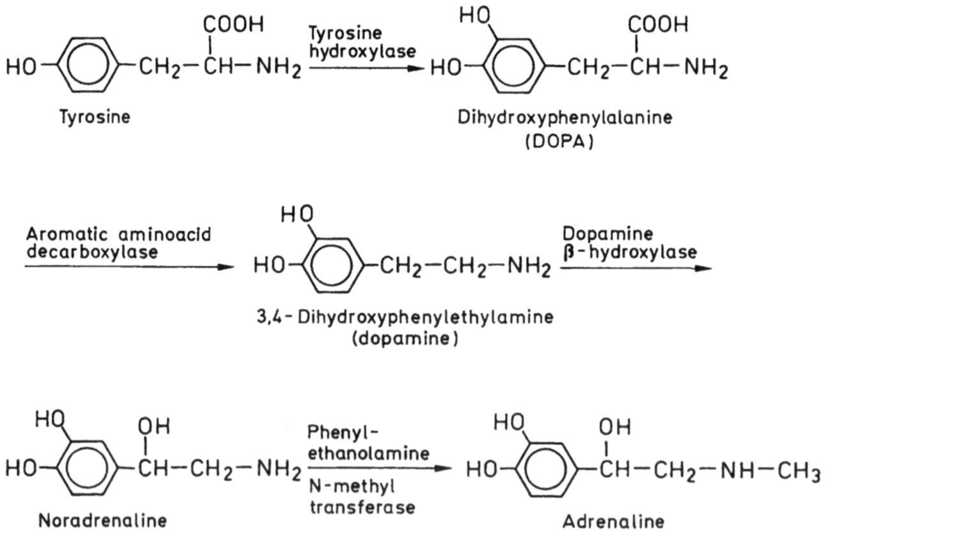

Figure 3. The biosynthesis of the catecholamines.

nervous system, while adrenaline is more prominent in peripheral neurones. The catecholamines are substrates of the type A monoamine oxidase (MAO-A) and the wide distribution of this type of enzyme activity results in very low concentrations of these amines being normally present in tissues and body fluids. The metabolic pathway for the breakdown of these amines is schematized in Figure 4.

The enzymes MAO, catechol *O*-methyltransferase (COMT), aldehyde reductase and aldehyde dehydrogenase give rise to a series of dihydroxyphenyl (catechol-) and the corresponding 3-*O*-methyl (vanil-) glycols and acids as the major catecholamine metabolites, and the concentrations of these in body fluids and tissues are to some

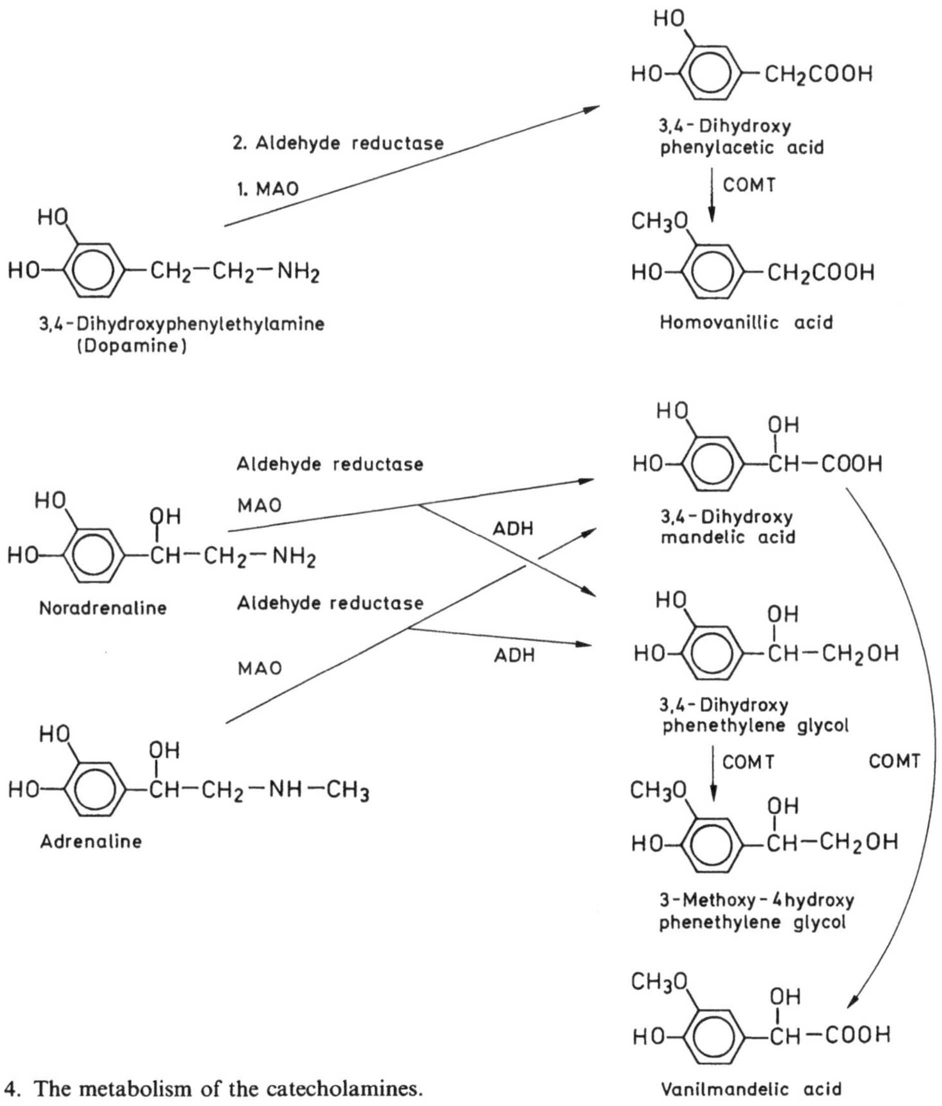

Figure 4. The metabolism of the catecholamines.

extent usable as indices of catecholamine turnover. These metabolites are generally much more abundant, more stable and easier to analyse than their parent catecholamines, and this is widely exploited in a variety of clinical conditions where catecholamine turnover is of interest.

2.1 The biochemical diagnosis of catecholamine secreting tumours

A number of catecholamine-secreting tumours have been described, and patients with these tumours have increasing concentrations of catecholamine in their circulation, and also excrete abnormally high amounts of catecholamines and particularly of their metabolites in the urine (43). The analysis of catecholamines in blood and urine, together with that of their metabolites, is therefore clinically useful, and permits a differential diagnosis to be made. The most important of the catecholamine-secreting tumours is phaeochromocytoma, a rare tumour which is most frequently located in an adrenal gland, although in a small proportion both glands may be affected, as may some other organs. The age of the patients at onset is variable, but the majority are affected between 20 and 50 years of age. Early diagnosis is important, because although the tumour is benign in 90% of cases and can be cured by prompt surgical excision, delay may eventually lead to premature death. The most prominent clinical sign is abnormally high blood-pressure, caused by the increased circulating catecholamines, whose output usually exceeds the normal output of the adrenal medulla, although individual variations among patients occur (44). Generally the most prominent of the catecholamines secreted by a phaeochromocytoma is noradrenaline, followed by adrenaline, with usually only a small amount of dopamine; there is a correspondingly increased urinary output of the appropriate metabolic end-products. Neuroblastomas are also rare, and as the name suggests, are tumours of neuronal tissue. They occur earlier in life, with the highest incidence in children of about two years of age. These tumours are very malignant, but in some cases they may regress spontaneously. The natural history of these tumours has been well reviewed (45). The tumour tissue is active both in the secretion and in the metabolism of catecholamines, so although there is an abnormally fast rate of biosynthesis of catecholamines, their tissue and body fluid concentrations are not nearly as prominent as with a phaeochromocytoma, and their pressor effects are consequently less marked. On the other hand the excretion of catecholamine metabolites is correspondingly greater. Of the amines in neuroblastoma, dopamine is characteristically the most prominent, and its increased rate of synthesis and metabolism is accompanied by increased excretion of dopamine metabolites (46). A number of very rare tumours of non-chromaffin paraganglionic cells are defined as paragangliomas, and also secrete catecholamines (43).

In addition to the clinical diagnosis of phaeochromocytoma and neuroblastoma, the biochemical confirmation by a demonstration of increased secretion of catecholamines is of course essential to the diagnosis and confirmation of these tumours and requires demonstration of increased circulating concentrations of the catecholamines, and what may be more prominent, demonstration of an increased urinary excretion of catecholamines and of their metabolites. A wide variety of analytical techniques have been applied to these diagnostic purposes, including all the various types of chromatography, and also specific colorimetric, fluorometric and radiometric assays.

Table 2. The gas chromatographic and mass spectral characteristics of the PFP-derivatives of noradrenaline and dopamine and of their α-methylated analogues (14)

PFP derivative	Retention time (min)	Relative intensity at m/z 176[a]	190[a]	428[b]	442[b]	590[b]	604[b]
Noradrenaline	2.83	100	–	–	–	40	–
α-Methyl noradrenaline	2.08	–	100	–	–	–	10
Dopamine	5.83	40	–	100	–	–	–
α-Methyl dopamine	4.36	–	100	–	33	–	–

[a] Cleavage between the α- and β-carbon atoms
[b] Cleavage between the nitrogen and the α-carbon atom

Clearly, however, mass spectrometry potentially offers the greatest specificity and the highest sensitivity, and the first demonstration of the feasibility of using selected ion monitoring was by Koslow et al., (14), who analysed catecholamines in rat brain as their pentafluoropropionyl (PFP) derivatives. Table 2 gives the GC-MS characteristics of the derivatives used in this pioneering venture, which demonstrated regional variations in the noradrenaline and dopamine concentrations in rat brain, measured on as little as 0.1 mg wet weight of tissue. This meant that even small areas of brain could be separately analysed.

A useful way of studying catecholamine metabolism, including both synthesis and turnover, was developed by Lhuguenot and Maume (47), who analysed dopamine and noradrenaline in extracts of the culture medium in which adrenal medullary cells had been grown. They used CI-MS with methane as the reagent gas, and the derivatives were made by silylating all the hydroxyls and making the Schiff-base derivatives of the amino groups with pentafluorobenzaldehyde. The internal standards were the ^{14}C analogues, and samples were introduced via a linked capillary GC. SIM was carried out at m/z 460 and 462 for dopamine and its internal standard and at m/z 548 and 550 for noradrenaline and its standard, corresponding to the $[M-15]^+$ ion produced by the loss of a methyl from the TMS groups. A number of workers compared the PFP derivatives with the TFA and TMS derivatives, and in general the PFP group has emerged as probably the most suitable for GC-MS of biogenic amines (48, 49). Application to the analysis of catecholamines in the blood of patients with catecholamine-secreting tumours was described by Wang et al., who use the derivatives for mass fragmentography, both of catecholamines (50) and of their 3-*O*-methyl derivatives (51). Plasma was deproteinized with perchloric acid, and the catecholamines were isolated with alumina and the 3-*O*-methylated catecholamines with Amberlite CG-50 and XAD-4. The eluates from the columns were evaporated to dryness and the residues derivatized with TFA-anhydride. Phenylephrine and synephrine were used as internal standards, and Table 3 gives the GC-MS characteristics of the ions used.

These were reliable and interference-free procedures, and gave good resolution of the derivatives and quantitative blood levels for the catecholamines studied. They demonstrated the increased levels of metadrenalines in a case of phaeochromocytoma and the increased adrenaline in essential hypertension. Of course few laboratories can apply mass spectrometric methods routinely for such diagnostic purposes, and therefore studies have been done to compare such methods with other techniques (50, 52):

Table 3. Mass spectral characteristics of catecholamines and related compounds (50, 51)

TFA-derivatives	Relative intensities (%) at *m/z*														
	126	140	233	245	246	328	342	345	358	359	427	440	454	471	485
Noradrenaline	100[a]	–	–	–	–	–	–	–	–	–	14.8[a]	22.2[b]	–	–	–
Adrenaline	–	100[a]	–	–	–	–	–	–	–	–	–	5.7[b]	8.0	–	–
Dopamine	61[a]	–	–	–	–	100[b]	–	–	–	–	–	–	–	–	–
Synephrine	–	100[a]	–	–	–	21.4[b]	–	–	–	–	–	–	–	–	–
Normetadrenaline	30.3	–	–	–	–	–	–	59.1[a]	100[b]	–	–	–	–	9[c]	–
Metadrenaline	–	23	–	–	–	–	–	19.0[a]	100[b]	–	–	–	–	–	–
3-Methoxytyramine	13.8	–	32.1[a]	–	100	–	–	–	–	10.9[c]	–	–	–	–	–
Phenylephrine	–	–	–	50	–	–	100	–	–	–	–	–	–	–	–

[a] Cleavage between the α- and β-carbon atoms
[b] Cleavage between the nitrogen and α-carbon atom
[c] Molecular ion

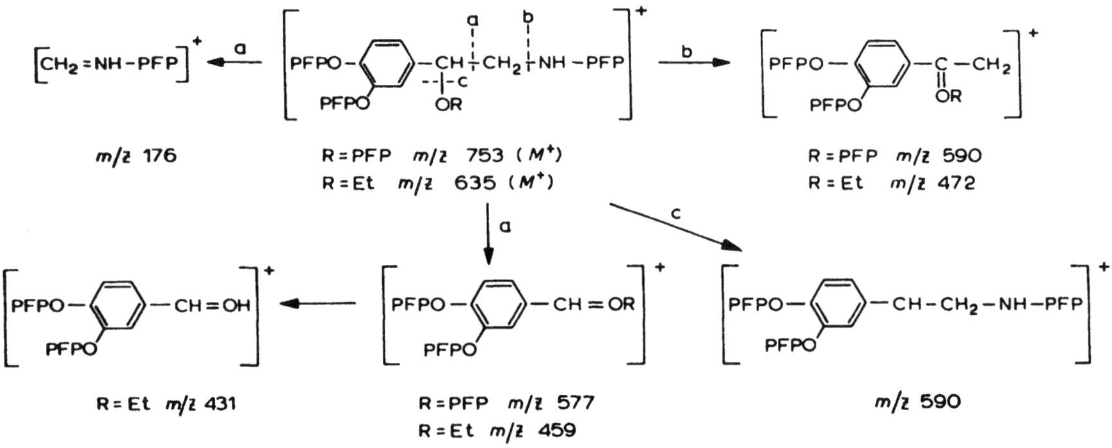

Figure 5. Fragmentation of the PFP-derivatives of noradrenaline.

mass spectrometry is generally considered to be the benchmark against which the other methods may be assessed.

A number of other reports have described the GC-MS of catecholamines in plasma (53, 54) and of catecholamines and metacatecholamines in urine (55, 56). Ehrhardt and Schwartz (53) in the main followed the procedure of Koslow et al. (14), except that they first converted the benzylic hydroxyls of the catecholamines to the ethyl ether with ethanolic HCl, and then made the PFP derivatives. This was a useful method to make the *m/z* values of the fragment ions measured in the mass spectra of dopamine, adrenaline and noradrenaline come close together, so that they could be measured by SIM in the LKB-9000S instrument, while the accelerating voltages for the fragments in the mass spectra of the fully PFP-derivatized compounds would have been too far apart. The fragmentation of the noradrenaline PFP-derivatives is schematized in Figure 5, and the fragments are listed in Table 4.

Table 4. Fragmentation of the catecholamine derivatives. Only peaks higher than 5% of the base peak are included (53)

Derivative	Major fragments – *m/z* (relative intensity)		
Dopamine $(PFP)_3$	176 (50%)	281 (26%)	428 (100%)
D_4 Dopamine $(PFP)_3$	178 (56%)	284 (30%)	431 (100%)
Noradrenaline $(PFP)_4$	176 (100%)	577 (13%)	590 (35%)
Noradrenaline ethyl ether $(PFP)_3$	176 (10%)	431 (100%)	459 (70%)
Adrenaline $(PFP)_4$	190 (100%)		
Adrenaline ethyl ether $(PFP)_3$	190 (30%)	431 (100%)	459 (80%)
N-n-Propyl-noradrenaline	176 (100%)	218 (63%)	
N-n-Propyl-noradrenaline	176 (32%)	218 (8%)	
ethyl ether $(PFP)_3$		431 (100%)	459 (85%)

Yoshida et al. (54) used GC-MS, and improved the procedure by the use of a column of boric acid gel (57) to isolate the catecholamines, a procedure they regarded as better than alumina. Muskiet et al. (55) made the TMS-ethers on the benzylic hydroxyls and then trifluoroacetylated all the other groups, but for the metadrenalines made the per-PFP-derivatives. Their procedures were applied to following the urinary excretion of the catecholamines over the whole human age-span from the newborn to over seventy years of age, and they also showed the expected increased catecholamine excretion in patients with catecholamine-secreting tumours and in patients with Parkinson's disease receiving L-dopa. Canfell et al. (56) showed increased urinary excretion of the metacatecholamines in patients with essential hypertension. Their work was notable for its use of reverse-phase C-18 separating column cartridges for the extraction of the metacatecholamines; otherwise they followed Muskiet's procedure (55).

The metabolic fate of the catecholamines has also received a great deal of attention, and several analyses have been reported using SIM for catecholamine metabolites (40, 58, 59). This, as already mentioned, is because these end-products of catecholamine metabolism are more stable and easier to isolate, derivatize and analyse than their parent amines. Two main types of derivative have been used: generally the acidic groups were esterified first, usually to the methyl or ethyl esters, and acylation to the PFP-derivatives then follows. However, the HFB- (58) and TFA- (61, 67) derivatives have also been tried. Alternatively TMS-derivatives were used, either the *O*-TMS ether-TMS ester type (25, 65) or methyl ester-*O*-TMS ethers (62). Deuterated analogues were used as internal standards. An interesting variation was the use of propionylation for initial derivatization in the aqueous phase, followed by solvent extraction of the catechol- and vanil-glycols and reaction with *t*-butyldimethylsilyl imidazole, which displaced the propionyl group and gave the *t*-butyldimethylsilyl (*t*-BDMS) derivatives. These were determined by SIM of the ions *m/z* 481 and 484 for the dihydroxyphenethylene glycol and its deuterated analogue, and of *m/z* 381 and 384 for the 3-methoxy-4-hydroxyphenethylene glycol (MHPG) and its analogue (69). The most extensive recent studies have come from Muskiet and his colleagues. They have used mass fragmentography and deuterium-labeled internal standards to determine urinary DOPA and dopamine metabolites (65), vanilpyruvic acid, 3,4-dihydroxyphenethylene glycol and dihydroxymandelic acid (66), and 4-*O*-methylated catecholamine derivatives

(67). They have also reported the simultaneous analysis of homovanillic acid, vanilmandelic acid and vanilglycol (MHPG) in serum (68). Vanil-lactic and dihydroxyphenylacetic acids, 3-methoxy-4-hydroxyphenylethanol and 3,4-dihydroxyyphenylethanol were trimethylsilylated, and run together with the deuterated analogues of *m/z* 209 and 212; 384 and 387; 312 and 319 and 370 and 375 respectively (65). A similar procedure was used for the catecholamine metabolites homovanillic and isohomovanillic acids, vanilmandelic and isovanilmandelic acid, MHPG and 4-methoxy-3-hydroxyphenethylene glycol, except that the trifluoroacetyl derivatives were used. For the homovanillic acids the ions *m/z* 196 and 201 were used, and for the vanilmandelic acids as well as for the glycols, *m/z* 345 and 348 (67). For the serum studies the authors returned to using PFP-derivatives: for vanilmandelic acid and MHPG they used the ions of *m/z* 445 and 448, and for homovanillic acid and isohomovanillic acid *m/z* 283 and 286 (68 – 70).

The metabolic fate of the catecholamines, and the rate of production and excretion of their metabolites is also important in other clinical situations. For example, patients with essential hypertension have high concentrations of circulating catecholamines (71, 72), whose pressor effects are the cause of the raised blood pressure, and these increased concentrations have been documented by mass fragmentography (50), as has the increased urinary excretion of catecholamine metabolites (65, 67). Another situation in which there is an increased level of circulating dopamine is in patients with Parkinson's disease who are being treated by administration of DOPA (73). GC-MS with SIM, among other methods, has been used to follow the increased urinary excretion of dopamine metabolites in these patients (65).

2.2 Neurochemical aspects of catecholamine analysis

Since the catecholamines function as neurotransmitters, a great deal of work has been done on establishing their distribution, not only in nerve tracts, but particularly in the central nervous system (74). The rat has been the most widely used experimental animal, but primate brains and human brains removed post mortem in various clinical contexts have also been investigated. Curtius and his group were the first to apply GC-MS to the analysis of dopamine in human brain using monitoring of selected ions in the mass spectrum of the N,O-trifluoroacetyl derivative: *m/z* 328 and 329 were used, and *m/z* 331 for the deuterated internal standard. They were able to demonstrate a significant decrease in brain dopamine in two phenylketonuric patients, which they attributed to inhibition of tyrosine hydroxylase by high phenylalanine levels. In the same paper a SIM assay for 6-hydroxydopamine was also described (75). Almost simultaneously Miyazaki et al. described their GC-CI-MS method for biogenic amines in cerebrospinal fluid (39). Brain catecholamine metabolites were recently investigated in a comparison of rat and primate brains (76): MHPG and the corresponding 3,4-dihydroxy compound, central nervous system metabolites of noradrenaline and dopamine respectively, were analysed by GC-MS with deuterium-labeled internal standards. The brains of rats, monkeys and humans were compared, and analyses were also done on cerebrospinal fluid. These compounds were liberated from their sulphate conjugates by the usual enzyme ("glusulase") procedure, simultaneously acetylated and extracted into an immiscible solvent, dried down and trifluoroacetylated. Deuterated standards were added at the outset of the analytical procedure to give reliable quantitation, and the ions monitored were *m/z* 362 and 365 for dihydroxyphenethylene glycol and *m/z* 376 and 379 for MHPG. While rat brain contained significant amounts of the dihydroxy compound, vervet monkey and man had very little, suggesting that in these species MHPG is a more suitable measure of cerebral noradrenaline metabolism (76). Much of the work on the distribution of catecholamines has been done by fluorescence microscopy, which gives a better overall view, but is not quantitative. Mass spectral analyses are quantitative, but have not been so widely used, and this aspect will be dealt with when discussing the regional distribution of amines in general. There has been a tendency for catecholamine turnover to be determined by analysis of metabolites, which are present in higher concentrations, are more stable and are easier to chromatograph than the amine precursors.

Studies of catecholamines in cerebrospinal fluid are complicated by the oxidative losses that accompany the rapid rise in pH that occurs after sampling, and ascorbic acid or even more powerful antioxidants have to be added to minimize these losses. The general neurochemical factors affecting the entry of metabolites from the central nervous system into the cerebrospinal fluid and via the blood-brain barrier have been reviewed by Wood (77). These factors include diurnal variations, posture, stress, age and possibly sex. The points of major interest as far as catecholamines are concerned are how the brain levels of noradrenaline and of dopamine may vary in different physiological and pathological states, and how those levels may be modulated by drugs.

One way of studying this is with the "probenecid test": the administration of the drug probenecid blocks transport of organic acids from the brain and cerebrospinal fluid into the bloodstream, and so enables a more reliable measure of their rate of production to be made. This was studied by Gordon et al. (78) with dihydroxyphenylacetic acid (DOPAC) using GC-MS of the PFP derivatives of the pentafluoropropyl esters, which were made simultaneously in a single derivatization step with pentafluoropropyl alcohol and PFP-anhydride, a useful technique originally developed for making strongly electron-capturing derivatives (79). The deuterated analogue was the internal standard, and quantitation was by SIM of the ions *m/z* 387 and 415 for the DOPAC derivative and *m/z* 392 and 420 for the internal standard. The effect of the probenecid was to increase the total DOPAC concentration about tenfold, mostly in the conjugated form. This work was repeated by Faull et al. (80) and extended to other metabolites, including indole derivatives and to the levels of probenecid itself (Table 5). The internal standard for probenecid was the *meta*-substituted di-isopropyl analogue with a slightly longer retention time: the fragments that were monitored were *m/z* 388 for probenecid and *m/z* 402 for the analogue.

As already noted in other connections, considerable more has been done in the determination of metabolites than in that of the parent amines, but a number of studies have been reported, such as the determination of dopamine and 3-methoxytyramine in the rat caudate nucleus by SIM using the PFP-derivatives (81). Edwards and Rizk have used a GC-CI-MS method (40) to follow brain catecholamine synthesis (82) and the metabolism of DOPA to alcoholic metabolites (83). GC-MS of the TFA-derivatives of dopamine was used to determine the proportion of conjugated dopamine in monkey cerebrospinal fluid (84), and the Finnish group has published an encyclo-

Table 5. Gas chromatographic retention times and mass spectral characteristics of the PFP derivatives of the monoamine metabolites

Compound	Structure of derivative and probable fragmentation pattern	Retention time (min)*	Major mass spectral ions (m/z) and their relative intensities (%)
DOPAC	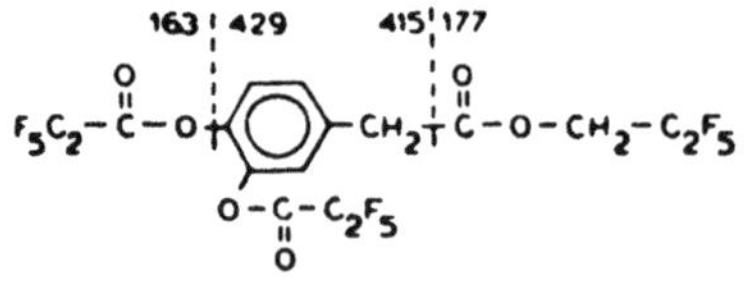	2.35	592 (44%,M^+), 429 (19%,$M^+ - 163$), 415 (100%,$M^+ - 177$), 387 (28%)
VMA	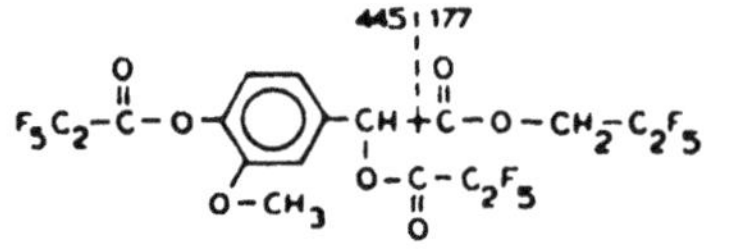	2.63	622 (33%,M^+), 472 (28%,$M^+ - 150$), 445 (100%,$M^+ - 177$), 417 (28%)
MHPG	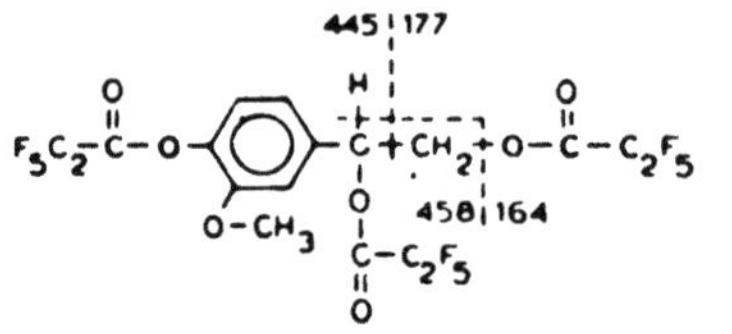	3.04	622 (43%,M^+), 458 (100%,$M^+ - 164$), 445 (52%,$M^+ - 177$), 417 (26%)
HVA	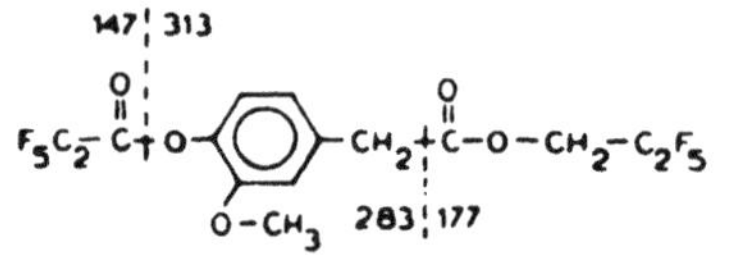	3.90	460 (100%,M^+), 313 (13%,$M^+ - 147$), 283 (6%,$M^+ - 177$)
5-HIAA	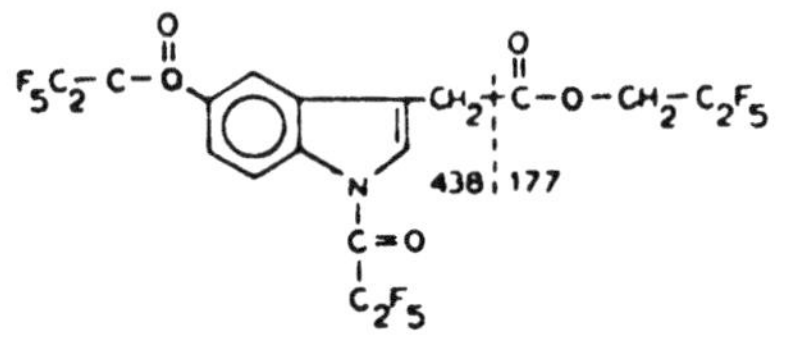	6.37	615 (39%,M^+), 438 (100%,$M^+ - 177$)
5-CH_3OIAA	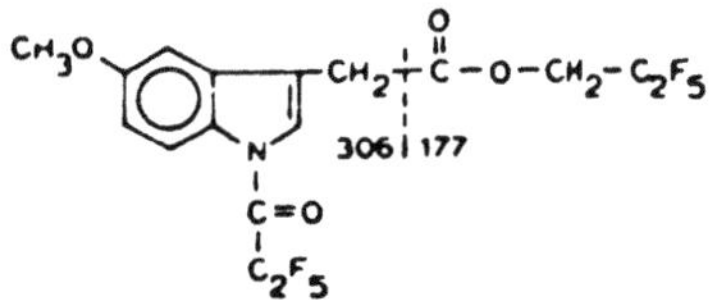	9.21	483 (100%,M^+), 306 (88%,$M^+ - 177$)

* Column: 3% OV-17 initially 115 °C and increased linearly at the time of sample injection at a rate of 6 °C · min^{-1}

paedic study of catecholamines, serotonin and of their metabolites in brain homogenates: essentially the procedure of Faull et al. (80) was used (84). GC-MS has also been used to verify the radioenzymatic assay for cerebrospinal fluid noradrenaline (85). The correlation between fluorimetry and SIM for homovanillic acid and MHPG was established by Muskiet and his colleagues (86), and these and other metabolites have been determined in cerebrospinal fluid by others (59, 87, 88). MHPG is generally regarded as the metabolite whose production best reflects the turnover of noradrenaline in the central nervous system (89), and the concentration of MHPG in cerebrospinal fluid is about 40-fold that of noradrenaline (77). GC with NICI-MS has recently been applied to the determination of MHPG in plasma (90).

2.3 The analysis of catecholamines and their metabolites in amniotic fluid

Although with amniotic fluid there is an increase in pH on storage (91), the increasing accessibility of this material recently has led to a considerable amount of work to establish the levels of the catecholamines and of their metabolites. The underlying rationale is that the development of the fetal adrenergic systems might be paralleled by detectable changes in amniotic fluid catecholamine levels, because in middle and late pregnancy there is an increasing contribution to amniotic fluid from fetal sources. It thus seemed possible that problems of the fetus in pregnancy, such as intrauterine growth retardation, and problems of the mother, such as toxaemia of pregnancy, might put stress on the fetus which could be reflected by changes in amniotic fluid catecholamine content. This was put on a factual basis by the identification of adrenaline and noradrenaline in amniotic fluid, as measured by the trihydroxyindole fluorescence method (91), and by evidence that amniotic fluid catecholamines are principally derived from fetal sources (92). Concentrations of catecholamines and of their metabolites in amniotic fluid increase with gestational age (93, 94), but these results were all obtained by radioenzymatic methods. The effect of intrauterine growth retardation was to increase the fetal output of catecholamines and their metabolites, particularly of 3,4-dihydroxyphenylethylene glycol, probably as a response to chronic stress, by activation of adrenergic systems (95, 96). The first of these was done by a radio-enzymatic assay, but the second (96) was by SIM of fragments from PFP-derivatives. For MHPG m/z 445 and 448 were used for the glycol and the deuterated analogue, and for the PFP-vanilmandelic acid methyl ester m/z 504 and 507 for the ester and deuterated analogue respectively. GC-MS has not been widely applied to these problems, and then only to catecholamine metabolites, as seems to be the usual practice. There has been confirmation of the increased fetal output of catecholamine metabolites into amniotic fluid with increasing gestational age by GC-MS of PFP-methyl ester derivatives (97), and of TFA-ethyl ester derivatives (98). These assays were done by SIM. Although 3-methoxy-4-hydroxyphenylethanol, 3,4-dihydroxymandelic acid and 3,4-dihydroxyphenylacetic acid were specifically looked for, they did not appear to be present. Increased ratios of MHPG to vanilmandelic acid were found in amniotic fluid from pregnancies involving fetuses with the neural tube defects spina bifida and anencephaly (99). (These amniotic fluids all had raised α-fetoprotein concentrations). These assays were carried out by the previously described procedures (96). Increased catecholamine levels were suspected to occur in amniotic fluids where the fetus had a congenital neuroblastoma: the catecholamines elicited pressor effects via entry into the maternal compartment (100), and the presence of catecholamine metabolites in the amniotic fluid in such cases has been confirmed by direct analysis (101).

3 Analysis of Indole Amines and of their Metabolites

A number of indole amines are derived from the amino acid tryptophan (Figure 6). These amines have an indole ring system, and are intriguing because their pharmacological activities may have considerable clinical significance, and because they may be plausibly connected, via metabolic relationships, with certain hallucinogenic substances that are the basis of hypotheses about the aetiology of various affective disorders such as schizophrenia (102). There has been considerable activity involving the testing of such hypotheses, but this has not led to as firm conclusions as with the catecholamines, because the precise functions of the indoleamines are less well established. However, the extensive analytical effort has produced many methods for the sensitive analysis of indoleamines and related compounds, and mass spectrometry has been involved whenever positive chemical identity became an essential feature of the investigation.

3.1 Analysis of tryptamine

Much of the attention paid to tryptamine rests on the possibility that it may function as a neurotransmitter or neuromodulator in the central nervous system: a useful review of this aspect of tryptamine metabolism has been written by Jones (103). Difficulties in this work are first, the low concentrations of the amine, and second, its rapid inactivation by the enzyme monoamine oxidase (MAO), which is largely responsible, and requires that an appropriate MAO-inhibitor be administered to experimental animals before the analytical studies or that the inhibitor should at least be included

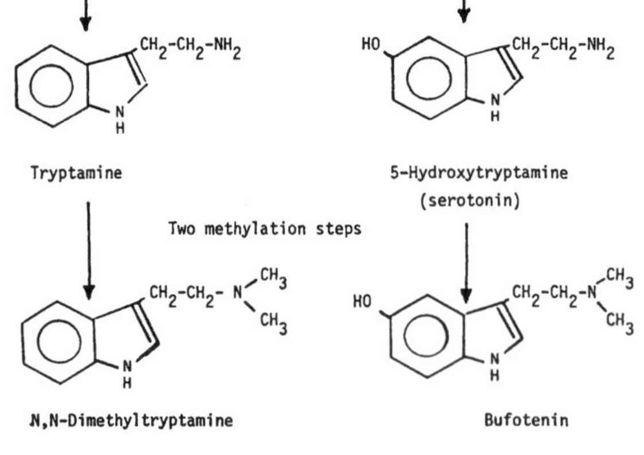

Figure 6
The biosynthesis of the indoleamines from tryptophan.

during isolation procedures. Since tryptamine is a characteristic substrate for the B-form of this enzyme, changes in tryptamine excretion are sometimes taken to reflect changes in endogenous MAO-activity, such that increases in its urinary excretion are regarded as a measure of the *in vivo* inhibition of endogenous MAO-B (104, 105). The physiological role of tryptamine is not known, but it has a very rapid turnover, and this may be because it does not appear to enter any storage sites. The urinary excretion of the amine has been of interest from earliest times (106, 107), and was found to be decreased in depression (108), but elevated in some patients with Parkinson's disease (109, 110).

Tryptamine is a relatively non-polar amine, and so it is thought to be readily able to cross biological membranes, even in the absence of active transport, because it can dissolve in lipid membrane constituents unlike more polar amines such as e. g. the catecholamines. Another consequence is that urinary excretion is pH-dependent (111): when an alkaline urine is secreted, there is a proportionally greater amount of tryptamine in the free-base form, in which it is presumably preferentially retained dissolved in lipid depots, while in "acidified" subjects the ionized proportion increases, which favours excretion in the aqueous phase (112, 113).

Early studies on the fragmentation patterns of tryptamines and their acetylated derivatives under electron impact were reported by Couch and Williams (114). The main tendency was for the acetyl group to come off first, followed by progressive cleavage of the side-chain: there was then the possibility of the methylene-indole fragment to rearrange to a quinolinium ion before the pyrrole ring opened up. But acetyl derivatives have not been applied analytically, and analytical applications of mass spectrometry to biological systems were first done on rat brain by Boulton's group, with the dansyl-derivatives which were analysed by their high-resolution integrated ion current procedure (115). The dansylated deuterated analogue was added as internal standard, and the instrument was alternatively focussed on ions of m/z 393.1511 and 395.1637. Gelpi and his group, on the other hand, used PFP-derivatives to determine tryptophan, tryptamine, serotonin and their metabolites in rat brain (116), while Warsh's team did an extractive acetylation of the rat brain homogenate first, and then made the TFA derivative of the dried residues (117). Gelpi's group has also recently used capillary GC-MS (118) and has critically reviewed the practical aspects of the isolation and trace analysis of these amines and related metabolites (119). Aside from the analysis of rat brain there have been relatively few applications of the mass spectrometric analysis of tryptamine to actual clinical problems. However, Huebert and Boulton have investigated the urinary excretion of tryptamine among other trace amines in a volunteer over a 28-day period. The results could not easily be interpreted: there were fairly wide fluctuations about a mean daily urinary output of about 80 micrograms, but it was not clear whether the amine originated in the diet or from endogenous metabolism or both (120). Gelpi and his colleagues have developed a useful isolation procedure for tryptophan metabolites by adsorption to the nonionic macroreticular resin XAD-2, and have applied this procedure to the simultaneous analysis in physiological fluids by single-ion monitoring in the same GC-MS analysis (116, 121). This resin acts in the manner of a "solid solvent extraction", and has affinity for non-polar groupings. Since it has a polystyrene-based structure, aromatic ring-systems too, in this case the indole ring, are adsorbed preferentially. XAD-2 needs to be pre-treated with solvents before it can be used, and some care is needed to optimize the amount of resin for a given application, to minimize interference by non-polar impurities that may be present in the biological matrix. The resin can be washed free of most impurities before the adsorbed material is eluted from the resin with methanol. The addition of 5-methoxytryptamine as a carrier was found to improve the recoveries.

3.2 Serotonin (5-hydroxytryptamine)

Serotonin (5-HT) is one of the few unequivocally established neurotransmitters in the central nervous system, and this is of itself sufficient to lend importance to studies involving the determination of changes in its concentration there, and indeed elsewhere in the body. However, its central role is not its only one, and so there are many other reasons for further investigation of its biological functions (122, 123). Clinically, however, 5-HT is most frequently analysed, together with its major metabolite 5-hydroxyindole-3-acetic acid (5-HIAA), for the diagnosis of the carcinoid syndrome (124). This is a combination of symptoms which include flushing of the skin, diarrhoea, intestinal hyperperistalsis, and cardiac insufficiency, all to some extent (though not entirely) reflections of increased blood 5-HT concentrations caused by the presence of 5-HT-secreting carcinoid tumours. They are commonly found in the gastrointestinal tract, particularly the appendix, while the oat-cell carcinomas of the lung may represent a more malignant variety of the carcinoid tumour (125). In addition to the patients' raised blood 5-HT levels, there is increased urinary 5-HIAA excretion.

Apart from these 5-HT-secreting tumours, the most significant clinical finding involving altered 5-HT blood levels is the significant decrease found in patients with Down's syndrome, mongolism (126), and in patients with uncontrolled phenylketonuria (127). Serotonin in blood is predominantly contained within the blood platelets, which have an active-uptake system for concentrating it. For this reason many studies on blood 5-HT concentrations involve determination of the amine in previously isolated platelets. The results of such studies have to be interpreted with caution, however, because it is possible for changes in the total number of platelets to occur, and this must be taken into consideration in evaluating the result of 5-HT determinations. These factors have considerably hampered investigations into the possible significance of changes in the serum-level of 5-HT in patients with infantile autism (128).

Although platelets are widely used as a model system for looking at the serotoninergic function of the central nervous system (129), the development of mass spectrometric methods has permitted more direct studies on the concentrations of 5-HT in the brain itself. One of the earliest reports on the mass spectrometric identification of serotonin metabolites used thin-layer chromatography for their preliminary separation, and detected the indole compounds by their formation of coloured spots with the electron-acceptor tetranitrofluorenone prior to their insertion into the MS via the direct probe inlet (130). Here the complexes dissociated, and the underivatized indoles sublimed into the source, underwent fragmentation by electron bombardment, and yielded their own characteristic mass spectra. Later workers published descriptions of the mass spectra of 5-HT and of its metabolites (15, 50, 114), and the first studies on biological material, rat brain in this instance (131). Studies on rat brain (52, 116) and primate brain and cerebrospinal fluid (131) have continued to refine the methodology. In a recent example, plasma 5-hydroxytryptamine was spiked with the ^{14}C-analogue and three different isolation procedures were followed by a radio-HPLC system, while a radio-GC system was used to follow the derivatization with PFP-anhydride. This enabled the authors to choose the best isolation and derivatization procedure for GC-MS, using a deuterated internal standard and SIM of the ions at m/z 451 and 454 for the amine and its internal standard. The results obtained by SIM were about 50% higher than with the fluorimetric ninhydrin method, where the extraction efficiency was probably lower. The revised method was therefore considered better in accuracy and precision. The use of XAD-2 for isolation of the indoles from physiological fluids, followed by single ion monitoring (116, 121) has already been mentioned. Clinical studies have been facilitated by the development of these more efficient procedures for the isolation and quantitation in biological fluids of serotonin by SIM (132) and of tryptamine and 5-HT in urine (133) as well as of 5-HIAA in cerebrospinal fluid (80).

It is important, when analysing brain tissue for 5-HT, to realize that brain 5-HT concentrations are governed by the availability of the precursor amino acid tryptophan, and by the rate of tryptophan transport into the brain (134, 135). GC-MS was used to show such a relationship for tryptamine also (136).

3.3 Methylated indole-ethylamines

The discovery that some hallucinogenic drugs were methylated derivatives of tryptamine, and could on administration elicit schizophrenia-like symptoms, led to a search for such compounds in the urine and other body fluids of schizophrenics. The rationale for this search was the hypothesis that schizophrenia might be caused by changes in the normal metabolism of indole ethylamines leading to the abnormal endogenous synthesis of psychotogens. Alternatively it was hoped that even if such a causal relationship could not be established, nevertheless the abnormal presence of such compounds in patients might give a clue to the aetiology of the disease or might at least provide useful dignostic criteria (137, 138). The most frequently sought compound was *N,N*-dimethyltryptamine, followed by the 5-hydroxylated analogue bufotenin.

These are thought to derive from tryptamine and 5-HT respectively, and are known to be hallucinogenic because they have been isolated as the active principles of natural products valued by some primitive societies for their psychoactive properties (139). This idea gained support from the demonstration of the appropriate enzyme activity to catalyse these methylations in human brain (140). Not surprisingly, among other methods GC-MS was used to confirm the chemical nature of the products (141), and to measure them in human plasma (142), and in the urine of schizophrenics (143). GC-MS has also been used to assay the activity of the enzyme (144).

In spite of all these efforts, however, there is no very good evidence that dimethyltryptamine is overproduced in schizophrenia (145), and this probably applies equally to bufotenin, which in any case is a less potent hallucinogen. Doubt has been cast on whether the enzymatic methylation processes can actually occur in brain (146). The controversy appears to have been settled by the recent study of Walker et al. (147). They used capillary GC-CIMS, with isobutane as reagent gas, to measure doubly

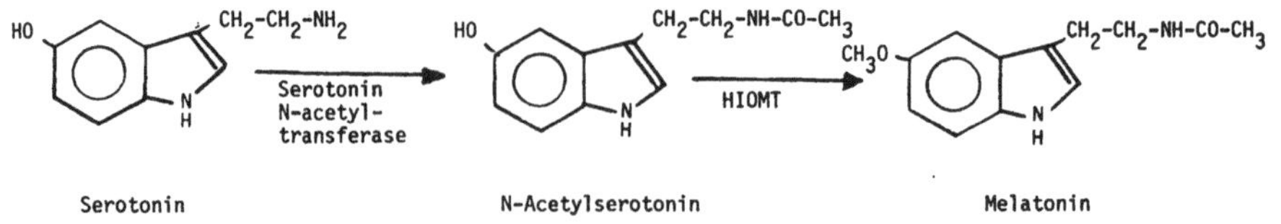

Figure 7. The biosynthesis of melatonin in the pineal gland.

labeled [^{13}C,^{15}N]-*N*-methyltryptamine in human urine to establish the extent of the conversion of administered [^{13}C,^{15}N]tryptamine. As internal standard they used *N*-methyltryptamine labeled with three deuterium atoms in the methyl group. The amines were run as the di-TMS-derivatives, prepared with BSTFA. The ions monitored for the derivative were m/z 320 and 321, and for the tri-deuterated analogue, m/z 324. The limit of detection, determined by spiking the urine with the doubly-labeled *N*-methyltryptamine, was ca. 0.00005%. Only one out of seven schizophrenic patients showed any degree of conversion greater than this limit: to the extent of twice the limit or 0.0001%, and this may reasonably be regarded as negligible. It has in general been difficult to obtain clear-cut evidence supporting biochemically-oriented hypotheses about the causation of schizophrenia, partly because of the low concentrations of the substances being looked for, partly because of the difficulty of unequivocally classifying the patients on the basis of their symptomatology, and partly because apparently normal healthy subjects cannot always be reliably distinguished and may also show evidence of overproduction of the compounds under investigation. Nevertheless, because there is a familial component to schizophrenia, and because genetic effects are manifested via biochemical mechanisms, the search will probably continue, and mass spectrometry will certainly also continue to play its part.

3.4 Melatonin and related compounds

Descartes considered that the pineal gland was the seat of the soul, because no other function of the pineal was known in his day, but nowadays it is known to be concerned with the control and development of sexual activity and maturation. Nevertheless, many studies were needed to establish the role of the pineal gland, and the greatest fillip to this effort came when Lerner and his colleagues discovered that the pineal contained melatonin (148). Melatonin, *N*-acetyl-5-methoxytryptamine, is biosynthesised in the pineal gland from it precursor serotonin in a two stage pathway (Fig. 7).

There is wide diurnal variation in melatonin production, mediated via the variation in the activity of the first enzyme, serotonin *N*-acetyltransferase, which is most active at night (149): 5-hydroxyindole *O*-methyltransferase activity also follows this pattern. These 24-hour rhythms are endogenous, but are "entrained" by the 24-hour day-night cue, and lead to corresponding rhythms in the blood levels of melatonin (150) and in its urinary excretion (151). There are also changes in serum melatonin in the menstrual cycle, with peak values during menstruation (152), and also seasonal and age-related changes (153). As early as 1898 Heubner suggested that the pineal produced an antigonadotrophic substance (154), a role for the most part filled by melatonin, whose night-time peak values have now been found to decline at puberty (155, 156).

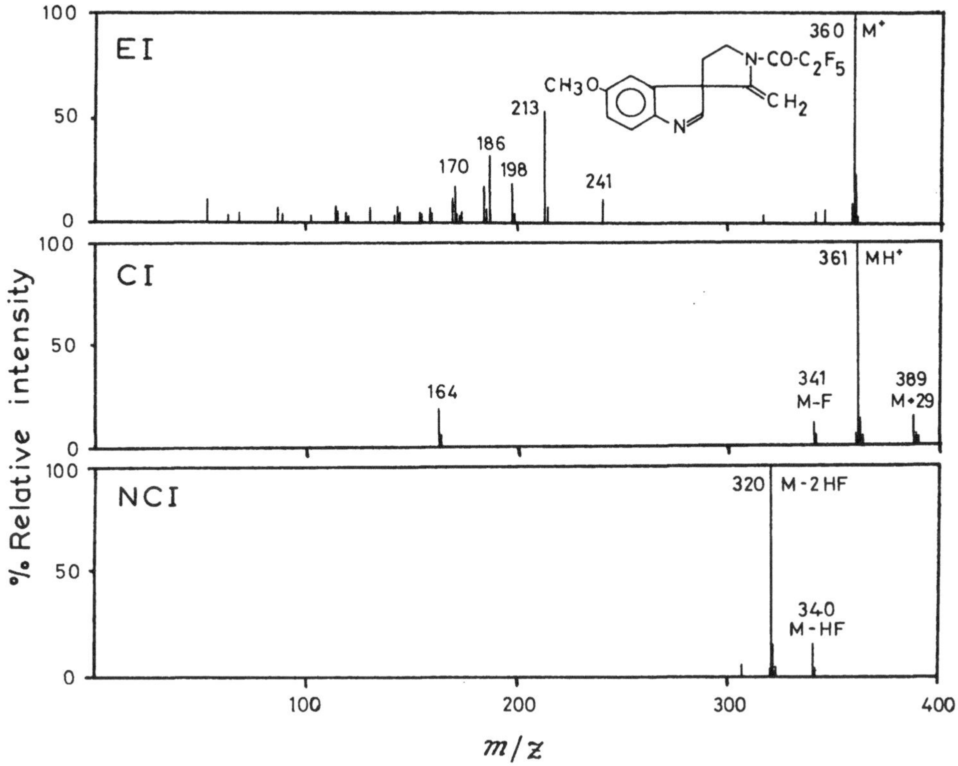

Figure 8. A comparison of the EI, CI and NICI mass spectra of the PFP-derivative of melatonin.

Although the necessary sensitivity and specificity of present-day melatonin assays is achieved by radioimmunoassay, mass spectrometry has been used at times to provide absolute confirmation of chemical identity of the isolated materials (157–159), while GC-MS (160) and GC-NICI-MS (161) have been used for melatonin assay in plasma following solvent extraction, usually with chloroform. The first absolute identification of melatonin in human plasma and cerebrospinal fluid was by mass spectrometry of the isolated melatonin as the TMS derivative (159), and the same authors used a GC-MS assay with *ω-N-n*-hexanoyl-5-methoxytryptamine as internal standard and using the fragment of *m/z* 232 for quantitation. A fragment of this mass was also contributed by the mass spectrum of cholesterol, which was extracted at the same time, but gave a peak with a different retention time. This could have been used as an endogenous internal standard (160). The serum values for melatonin were given as up to 80 $ng \cdot l^{-1}$ in the light and 120–224 $ng \cdot l^{-1}$ in the dark. The NICI method (161) used a side-chain deuterated melatonin as internal standard. The EI, positive CI and NICI spectra of the PFP-derivative are shown for comparison in Figure 8.

The sensitivity with the NICI-MS method using ions of *m/z* 320 $(M-2HF)^-$ and 340 $(M-HF)^-$ was about 150 times greater than with the positive ion CI-MS procedure using the ion of *m/z* 361 $[M+H]^+$, but remarkably, gave values for blood melatonin levels which were about one tenth of those quoted above (160): 1.5–4.9 $ng \cdot l^{-1}$ during

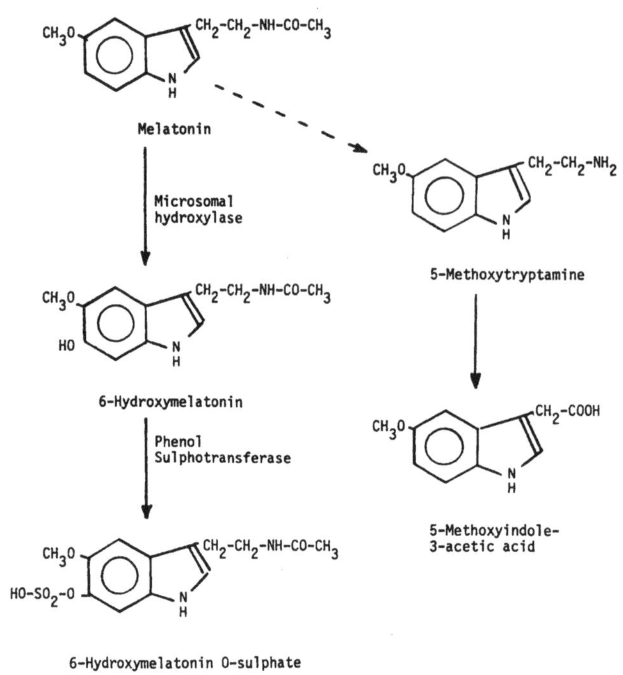

Figure 9
The catabolic pathways of melatonin.

daytime and 19.4–42.6 ng · l^{-1} at night. Finally, radioimmunoassay has been directly compared with analysis by GC-MS for plasma melatonin determination (162).

The pineal produces other antigonadotrophic substances besides melatonin, some of them peptides with which we are not concerned here, but there has been interest in the melatonin-related methoxyindoles 5-methoxytryptamine (5-MT) and 5-methoxytryptophol. Melatonin is probably de-acetylated to 5-MT (163). GC-MS with SIM has been used to quantify 5-MT in the pineal (157, 164), in sheep brain (165) and in human plasma (17), but much of the 5-MT that was found could be an artefact of the conditions used during the derivatization (166). The PFP-derivative of 5-MT was quantitated by monitoring the ions at m/z 306 and 319 (157, 164, 165, 167), together with the corresponding ions of the side-chain deuterated analogue of m/z 308 and 322, also the molecular ions of m/z 482 and 486 respectively (164). The tri-TMS-derivative has also been used (17), but one of the ions used, of m/z 174, is commonly found in all sorts of TMS-derivatives, and can only be recommended for SIM by the "functional group" approach. In this connection, however, the other ion monitored, m/z 232, is common to 5-methoxyindoles, and thus leads itself to the study of these compounds as a group, provided that they are chromatographically resolved, as for example by capillary column GC (17).

Melatonin is metabolized (Fig. 9) almost entirely to 6-hydroxymelatonin-*O*-sulphate (168) and only about 2% goes to 5-MT. The urinary excretion of the former has been measured using GC-MS (169). The urine was spiked with deuterated 6-hydroxymelatonin sulphate (ring-labeled at the 2,4 and 7 positions) and the metabolites were isolated by adsorption first on XAD-2 and then on Florisil. Derivatization with PFP-anhydride displaced the sulphate group, and gave the spirocyclic derivative (170). The

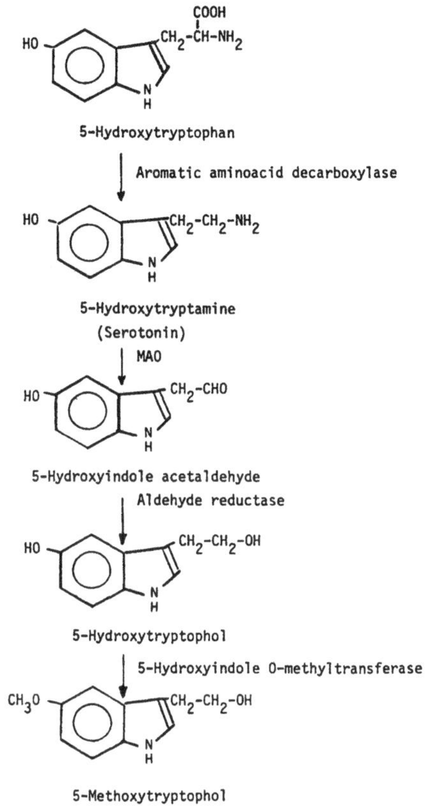

Figure 10
Biosynthesis of 5-methoxytryptophol.

ions monitored had *m/z* 522 and 525 respectively for the authentic compound and its deuterated analogue. The end-product of 5-MT metabolism is 5-methoxyindole-3-acetic acid, whose urinary excretion has likewise been studied by GC-MS with selected ion monitoring (171).

3.5 5-Methoxytryptophol

Although at first sight related to melatonin, the biosynthesis of this substance (Fig. 10) is probably via 5-hydroxytryptophol and derives directly from serotonin (172). It has been proposed that these alcohol metabolites are decreased in depression, and a sensitive GC-MS method for their analysis in CSF was therefore developed by Curtius's group (49) using SIM of the PFP-derivatives. An internal standard, 5-fluoro-α-methyltryptamine, was added and the ions monitored were *m/z* 321 for the internal standard, *m/z* 319 for 5-methoxytryptophol and *m/z 305* for 5-hydroxytryptophol. Other workers have used mass spectrometric methods for analysing these compounds in cerebrospinal fluid (173) and plasma (172), while the need to analyse these substances in plasma and tissues was the occasion for detailed analytical studies including mass spectral data on many different derivatives (174). However, in spite of these efforts, it has not been possible to get very clear-cut results, and there appears to be wide overlapping between control subjects and depressed patients as far as 5-methoxytryptophol output is concerned (173). Changes in the production of these compounds probably reflect changes in the brain's oxidation-reduction state or energy charge (175).

4 The Trace Amines

As analytical methods became more sensitive and more specific, various substituted phenylethylamines were looked for and found in the central nervous system and in physiological fluids. Like the catecholamines and indoleamines, these various amines are for the most part directly derived from aromatic amino acids, in this case phenylalanine and tyrosine, by decarboxylation (Fig. 11).

4.1 Clinical significance of the trace amines

The term "trace amines", although not ideal, does reflect the fact that they are present in the brain at only trace levels, their concentration there being almost two orders of magnitude less than those of the catecholamines or serotonin (176). This raised the question of whether the trace amines were there merely because the precursor amino acids happened to interact to some extent with aromatic amino acid decarboxylases which normally work on different substrates, or whether the trace amines were produced in the brain for the purpose of exerting their own specific actions (177). It is fair to conclude that the balance is tilting in the direction of the latter view, and this is mainly based on what happens when patients or volunteers are given drugs which inhibit MAO or other enzymes involved in the metabolism of the trace amines.

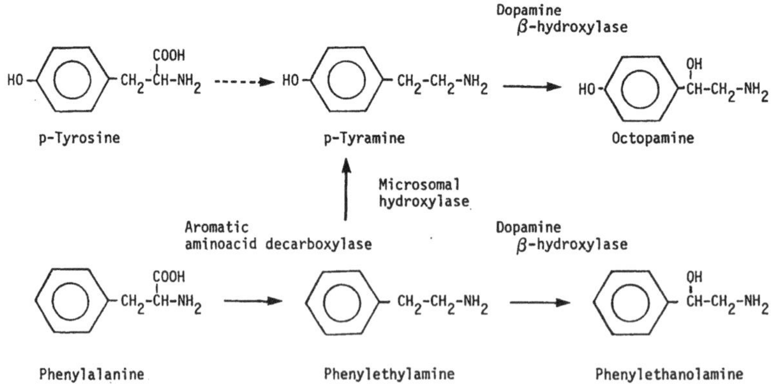

Figure 11. The trace amines: metabolic interrelationships.

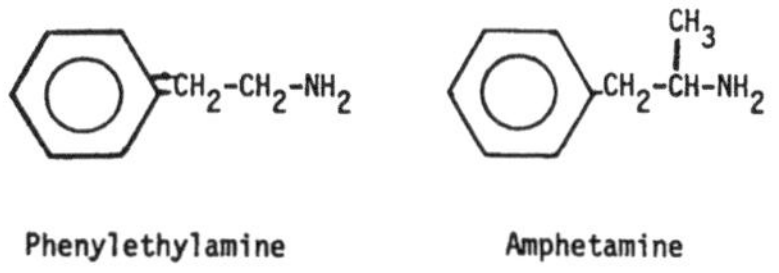

Figure 12
Phenylethylamine and amphetamine.

This has been summarised in the historical section of a recent symposium (178, 179). Such experiments indicate for example that their great susceptibility to MAO is responsible for the low residual concentrations of the trace amines everywhere in the body, and that after inhibition of MAO the normal turnover of some of these amines in the brain is found to be just as high as that of the catecholamines or serotonin. Boulton in particular has put forward hypotheses about possible modes of action (176, 179). He envisages a trace amine as a "synaptic activator", acting either directly or indirectly during the process of neurotransmission. It could produce its effect directly, by continuously creating miniature synaptic potentials, or perhaps more plausibly, indirectly by maintaining the synaptic region in a "state of readiness". One can readily appreciate how in some such manner the presence of one (or more) trace amines could modulate the threshold for firing, or the rate and intensity of firing of a "classical" neurotransmitter amine.

We can distinguish three main sub-groups among the trace amines: phenylethylamine (PE) and phenylethanolamine; the tyramines; and the octopamines. Phenylethylamine has probably been studied most. Because of its structure and pharmacological properties (180) it has been called "nature's amphetamine", the only difference between them being an α-methyl group which renders the drug resistant to the action of MAO (Fig. 12).

Phenylethylamine is therefore generally regarded as an excitatory neuromodulator. From the clinical viewpoint the interest focusses, as usual, on abnormal situations involving either overproduction or underproduction. The biochemical correlates of these effects are reflected in increased or decreased concentrations at various key locations within the body, and increased or decreased urinary excretion of the major metabolic end-products, which are phenylacetic acid and phenacetylglutamine. The activity of the B-form of MAO, which is the main metabolizing enzyme, also plays a decisive role.

The most important overproduction syndrome involving PE is phenylketonuria (PKU). In this inherited metabolic disease the liver enzyme phenylalanine hydroxylase is not present. Its normal function is to convert phenylalanine to tyrosine, and in its absence there is a massive accumulation of phenylalanine, which permeates all tissues and physiological fluids (181). Conversion to tyrosine is the major pathway for all phenylalanine surplus to the body's needs, and treatment of PKU is designed to restrict dietary phenylalanine intake and to minimize any surplus. The activity of MAO-B prevents a correspondingly great accumulation of PE, although there are measurable increases in blood and tissue concentrations and in urinary excretion, and there is a very great increase in the production and excretion of the metabolic end-products which is evidence for a greatly increased PE-turnover in PKU (182). There are also abnormal metabolites, not only the well-known aromatic acids (183), but also for example the abnormally produced condensation product of PE and vitamin B_6 (184). It is difficult to ascribe very much significance to the overproduction of PE in PKU as a contributory cause to the pathology of the disease, except to say that excess PE probably makes the patients unsettled and harder to manage, and even in treated patients the increased turnover of PE probably contributes to learning difficulties and shortness of attention span. However, the excessive forced synthesis of phenacetylglutamine is the basic feature of Perry's glutamine-depletion hypothesis for the pathogenetic mechanism in PKU (185).

Deficiency of PE was first postulated as a concomitant of endogenous clinical depression by Fischer et al. (186) and these workers found a greatly decreased excretion of PE compared to normal even after MAO inhibition in depressed patients. They proposed that amphetamine works by depleting PE (187) an that the anti-depressive effect of some MAO-inhibitors was a PE-sparing one (188). As an elaboration of these ideas Sabelli and Mosnaim proposed a PE-hypothesis of affective behaviour in which PE and its congener phenylethanolamine are cast in the role of "neuromodulators responsible for triggering or sustaining wakefulness, alertness and excitement," and their deficiency has a corrsponding depressive effect (189). Apart from this there is little information about penylethanolamine. It increases in PKU also, and this is attested by the increased excretion of its metabolic end-product mandelic acid (190). It is a substrate of MAO-B: the products of this reaction are phenylethylene glycol (PEG) in the central nervous system, and PEG together with mandelic acid peripherally (191). At present phenylethanolamine may be regarded as a PE-like neuromodulator contributing to all the PE-effects, but it may eventually turn out to have specific functions of its own. Its role in hepatic coma will be mentioned in connection with that of octopamine.

4.2 Analysis of phenylethylamines

The early methods for analysis of PE were not sufficiently specific and gave results which are now known to be much too high. This situation was improved by the application of GC (192) and GC-MS (193) and it is now clear that prior separation is essential to free PE from interfering substances. Mass spectrometric methods have been developed by a number of groups. Boulton and his colleagues have extensively applied their time-integrated high resolution mass spectrometric technique to phenylethylamine among the other trace amines and for this the amines are again first dansylated (194). Other workers have used PFP derivatives to measure PE and phenylethanolamine as well as other trace amines (195, 196). The mass spectra of the HFB- and PFP-derivatives of PE are not very informative, because the abundance of the molecular ions is negligible and most of the ion-current is shared between ions of m/z 91 and 104. The former ion, tropylium, is present in the mass spectra of many aromatic compounds. These ions were used for measurement of PE (193, 195, 196), with the ring-deuterated analogue being measured at m/z 93 and 107. PE, phenylethanolamine and the tyramines have also been determined in rat brain by SIM of the *N*-dinitrophenyl,*O*-trimethylsilyl derivatives (197), which are made by a procedure in which isolation of the amines follows rather than precedes the first derivatization step. Here again, GC was used with CI-MS and methane was both carrier and reagent gas. Table 6 compares the corresponding EI and CI mass spectra.

Table 6. EI and CI mass spectra of some representative N-DNP, O-TMS phenylethylamines. Isotope peaks and minor peaks are not included (197)

Amine	Fragment m/z	% Abundance EI	% Abundance CI	Structure
Phenylethylamine	77	29		$C_6H_5^+$
	91	68		$C_7H_7^+$
	104	19		196-$(NO_2)_2$
	150	15		196-NO_2
	196	100		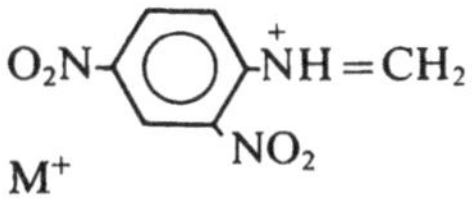
	287	9		M^+
	288		100	MH^+
	289		18	
Phenylethanolamine	73	60		TMS^+
	77	12		$C_6H_5^+$
	91	9		$C_7H_7^+$
	179	100	42	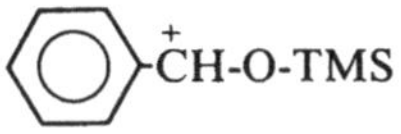
	286		29	MH^+-TMS-OH
	376		100	MH^+
	377		53	
p-Tyramine	73	39		TMS^+
	179	100		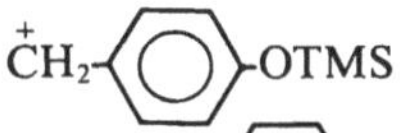
	193		100	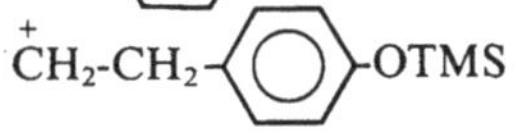
	360		18	M-CH_3^+
	375	8		M^+
	376		71	MH^+
	377		22	

The ion of m/z 288 together with that of m/z 292 for the ring-deuterated standard was used for quantitation. The isothiocyanate derivative has also been used for the GC-MS of PE (198): the molecular ion, m/z 163, carried about one third of the total ion current, and so this ion, together with that for the deuterated internal standard, m/z 167, was used for mass fragmentography.

4.3 The analysis and brain distribution of the tyramines

The distribution of the tyramines, and indeed of several other trace amines, in rat and human brain is peculiarly the province of Boulton's team, who have very successfully exploited their time-integrated-high-resolution-mass spectrometry technique together with the use of the dansyl derivatives (35, 36) to trace out the levels of the amines in different brain regions and also to determine their concentrations in physiological fluids. The first result of these studies was to revise downwards the values of these concentrations compared to earlier values obtained with evidently less specific techniques (35). A paradoxical consequence was that these more realistic estimates, far from affecting the credibility of such studies, put them on a sounder basis than hitherto, and not only created confidence in the results, but also made the point that trace amines were likely to have a role in the central nervous system and were not there by chance (177). The work done by Boulton's group (but by other workers also) has been well reviewed by Davis & Durden (36), quoting among others what is a by no means exhaustive list of their own reports: a brief quotation best indicates their scope. "These studies have included the identification and distribution of the trace amines phenylethylamine, tryptamine and *m*- and *p*-tyramine in rat organs and brain regions (115, 194, 199–201), in rabbit brain (202), in human brain (203), their subcellular distribution in rat brain (204), and their urinary excretion in the rat and human (120, 205–207). Other amines measured by this method include phenylethanolamine (208), *m*- and *p*-synephrine (209, 210) and *m*- and *p*-octopamine (210) ..."

Reversing the normal sequence, it must now be considered what these results indicate, particularly as far as the possible clinical significance of the tyramines is concerned. In addition to the general models such as those propounded by Boulton earlier on, there is also the concept of the "false" or perhaps more appropriately "alternative" neurotransmitter. "Alternative neurotransmitter" is a more non-committal way of describing a transmitter molecule which can interact with a receptor or enter a storage site in much the same way as the preferred transmitter, but possibly with a somewhat different effect. These notions are now being explored in more detail (211, 212). At the present time it is not at all clear how the results obtained for the distribution of trace amines in the central nervous system fit these and other ideas about their function or possible significance in brain metabolism: it is still at the information-gathering stage, at which mass spectrometry is playing an unusually prominent part. It is reassuring that the results obtained by this technique and those obtained by other sensitive and specific techniques are generally in reasonable agreement.

There are some more clear-cut clinical correlates of *p*-tyramine overproduction. It has long been known that this amine is excreted in increased amounts in PKU, and this was always a paradoxical finding in view of the block in tyrosine synthesis. It is now known that *p*-tyramine can be directly made from phenylethylamine (113, 206, 213), probably by the action of a microsomal hydroxylase (214), and this biosynthetic route is apparently much more prominent than production from tyrosine by decarboxylation. Tyramine is also excreted in increased amounts in the hereditary metabolic disease tyrosinaemia, where the basic biochemical defect is the deficiency of the liver enzyme fumarylacetoacetase at the end of the tyrosine catabolic pathway (215). A further cause of increased tyramine production is acute liver failure, which is accompanied by a generalised tyrosinosis-tyrosyluria caused by the deleterious effects of cell necrosis on the activity of *p*-hydroxyphenylpyruvic acid oxidase, an enzyme lying earlier on the same pathway (216).

An interesting aspect of *p*-tyramine metabolism is its metabolic conjugation via *O*-sulphation catalysed by phenolsulphotransferase. Evidence is accumulating for a deficiency in the activity of this enzyme in patients with depressive illnesses (217), and there are indications that this may be a predisposing factor detectable even in the absence of symptoms, but that some external stimulus, whether physical, environmental or psychological, may trigger actual depressive episodes. Studies of this kind on the significance of trace amines in affective disorders are notoriously difficult, because groups of patients are heterogeneous and their symptomatology may be irreproducible, because the numbers are limited and statistically insignificant, because appropriate controls may be difficult to arrange, and particularly because symptoms such as small increases or decreases in the excretion of some amine or its metabolic end-product are very hard to interpret, and may be subject to influences that may be unsuspected, ill-understood or poorly controlled.

4.4 Octopamine and hepatic encephalopathy

Patients in acute liver failure may go into a coma, and may suffer brain damage or encephalopathy. Fischer and Baldessarini were the first to discover that there were significantly increased concentrations of phenylethanolamine and octopamine in the serum and urine in such patients (218). They suggested that these amines were acting as false neurotransmitters, and since the effects of e. g. octopamine at nerve endings are much less than those of noradrenaline, the increase in the local concentration of the false neurotransmitter would have the effect of competitively diluting out the action of the true one. A number of other workers have confirmed these findings (219–211, 11). In general, the blood-levels of the false neurotransmitters broadly reflected the severity of the patients' condition. There is still no consensus on the significance of octopamine in acute liver failure, nor is it settled whether the false neurotransmitter is cause or symptom, although on balance the latter seems more likely. Most of this work was done by radio-enzymatic procedure (222), but mass spectrometric methods for these amines were also developed. Early methods (12, 18) employed existing approaches, but assays for these amines in brain and urine have now been developed using CI-MS, one positive (223) and one negative (224). In the first of these the PFP-derivatives of brain amines were made, and tri-deuterated octopamine was added before the beginning of the isolation procedure. Methane was the reagent gas, and the ions monitored were m/z 428 and 431 for octopamine and its deuterated internal standard, corresponding to the loss of PFP-group from the molecular ion. The GC-MS method of Ibrahim et al. (224) involved a capillary column to resolve the three isometric octopamines, and NICI for greatest sensitivity. Deuterated analogues were used as internal standards, and derivatization was again to the PFP-derivatives. For the undeuterated derivatives the molecular ion of m/z 591 and the $[M-HF]^-$ ion of m/z 571 was used; for the deuterated analogues the molecular ion of m/z 594 was used. Of the three octopamines studied in 10 normal subjects, mean excretions were: *ortho*-octopamine, 0.3 $ng \cdot mg^{-1}$ creatinine (and about an equal amount in conjugated form); the *meta*-isomer, 1.0 $ng \cdot mg^{-1}$ creatinine, and again about an equal amount conjugated; and the *para*-isomer 0.9 $ng \cdot mg^{-1}$ creatinine, but over 20 times this amount in conjugated forms.

5 The Di- and Polyamines

The di- and polyamines putrescine, cadaverine, spermidine and spermine are aliphatic amines whose biosynthesis is controlled by the activity of the enzyme ornithine decarboxylase (ODC) (Fig. 13). Polyamines (and their metabolic enzymes) are widely distributed in the body, and have various roles, not all of which are clearly understood at the moment. One function is to organize and stabilize the structure of transfer-RNA (225), and they are also associated with growth. Growth stimuli increase the activitiy of ODC, perhaps in order to keep pace with increases in turnover of t-RNA. However, from a clinical viewpoint these amines have received most attention as diagnostic markers for cancer, and their increased excretion by cancer patients was first reported by Russell (226). Apart from the association with the growth involved in cell proliferation, it has also been suggested that tumour cell death may be a cause of increased polyamine levels in physiological fluids (227). Polyamine excretion has been measured as a correlate of the response of cancer patients to chemotherapy (228, 229). A great deal of polyamine research has been done by dansylation and measurement of the fluorescent derivatives. However, GC-MS has also been applied successfully, for example using the TFA derivatives (230, 231). Although the base peaks varied from *m/z* 167 for putrescine, *m/z* 336 for spermidine to *m/z* 517 for spermine, all had a common ion of *m/z* 154, which could be used for monitoring, to produce a functional group profile of these amines (230). Deuterated analogues were used by Smith and Daves, and good quantitation at the picomole level was obtained with e. g. the ions of *m/z* 517 for spermine and *m/z* 525 for its internal standard (231).

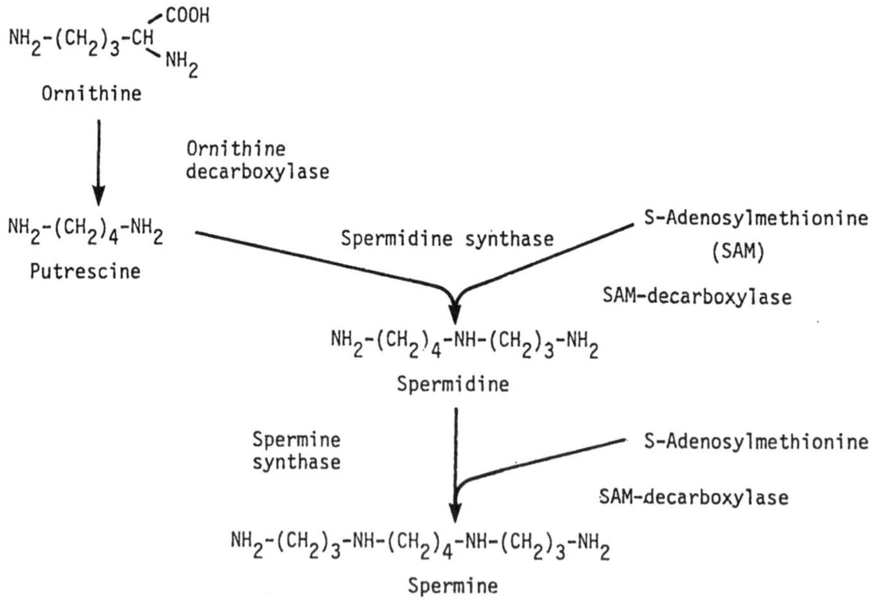

Figure 13. Biosynthesis of the di- and polyamines.

Giumanini et al. devised a reductive methylation and extraction procedure which simultaneously affected derivatization and isolation. The permethylated derivatives have some advantages for mass spectrometry, but do not chromatograph quite as well as acyl derivatives (232). The rationale for permethylation was to minimize the increase in molecular weight attendant on the addition of more than one derivatizing group to polyfunctional molecules, which leads to problems in the chromatographic separations because of very considerable increased retention times. The procedure had the advantage that derivatization in aqueous medium (by reaction with formaldehyde with sodium borohydride reduction) was followed by solvent extraction, a convenient combination of derivatization and isolation. The mass spectral details were tabulated but the base peak for all the derivatives was *m/z* 58, [trimethylamine − H]$^+$, which clearly again lends itself to functional group profiling. No applications of this rather original approach have so far been reported. An interesting finding is the discovery and mass spectrometric identification of isoputreanine, a metabolite of spermidine, in human urine (233).

6 The Analysis of Histamine and of its Metabolites

Histamine is important as a mediator of gastric secretion and is of course involved in allergic syndromes and anaphylactic shock. Antihistamine drugs are widely used to combat allergic reactions and to alleviate hay-fever. For these reasons considerable effort has been applied to analytical studies to follow individual patients' histamine status. However, it was found that the radioenzymatic and fluorimetric methods which were applied originally lacked specificity (234), so that lately a number of mass spectrometric methods have been published for histamine using NICI-MS (235, 236). Figure 14 shows the mass spectrum of the tri(pentafluorobenzyl) derivative of histamine. Figure 15 shows the selected ion recording of the peak of *m/z* 430 for this

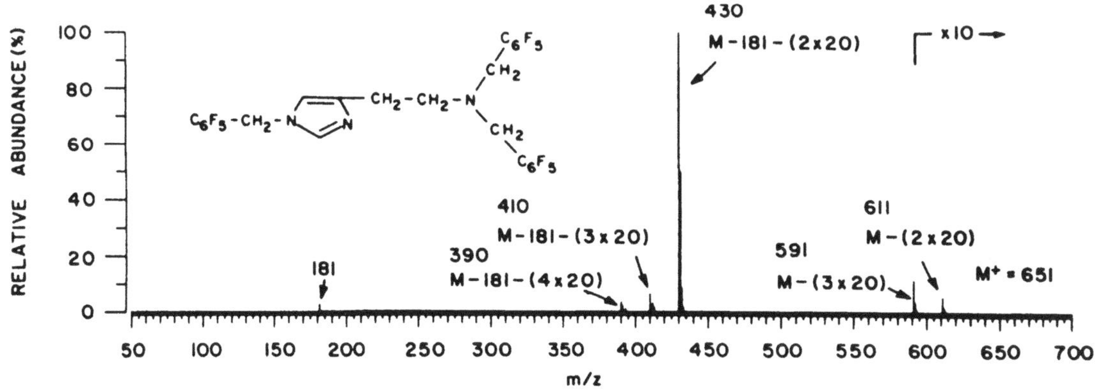

Figure 14. The negative ion CI mass spectrum of tri(pentafluorobenzyl)histamine.

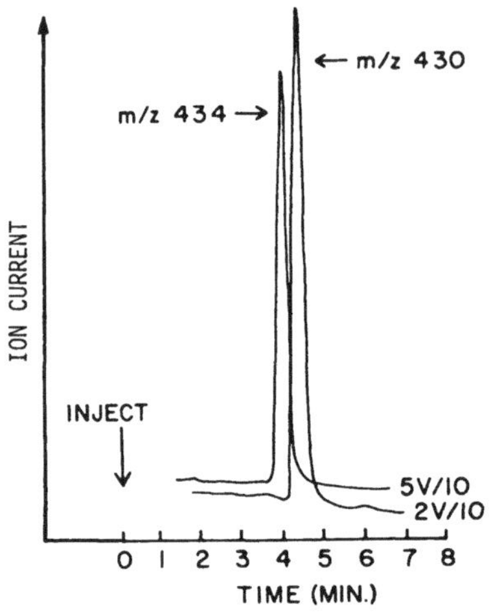

Figure 15
Selected ion recording applied to the analysis of histamine in normal urine.

derivative, together with that for the deuterated internal standard (*m/z* 434), applied to the analysis of histamine in normal human urine. There are also papers describing more conventional MS methods for histamine (237), and for *N*-methylhistamine in human plasma, urine and cerebrospinal fluid (238, 239) and in rat brain (240), and for the metabolite *N*-methylimidazole acetic acid in urine (241).

7 The Analysis of Other Biogenic Amines

The work described so far covers clinical and mass spectrometric investigations on those amines that are generally regarded as biogenic, although what constitutes a biogenic amine is not universally agreed, and is to some extent governed by individual preference. For the sake of completeness we will interpret it generously and include a number of other amines which have been investigated as brain constituents or urinary excretion products. Of some interest is piperidine, which is produced in the human body from lysine, by one pathway in the brain, and by a different one in the intestine through the action of the bacterial microflora (Fig. 16).

The only significant amounts of piperidine to be excreted in the urine are those produced in the intestine, and if the precursor, lysine, is eliminated from the diet, piperidine is no longer excreted in the urine (242). A reduction in piperidine excretion can also be achieved by sterilizing the gut with e. g. neomycin. Two mass spectrometric methods for the analysis of piperidine have been developed (243, 244). The first method is interesting in its use of the di-*n*-butylaminonaphthalenesulphonyl (BNS) derivative, specially developed for the purpose. This derivative was less polar than the dansyl one, and therefore easier to extract into organic solvents. Better fluorescence quantum yield was also found, but the main advantage was that the mass spectrum showed a base peak of $[M-43]^+$, which still contained most of the information about

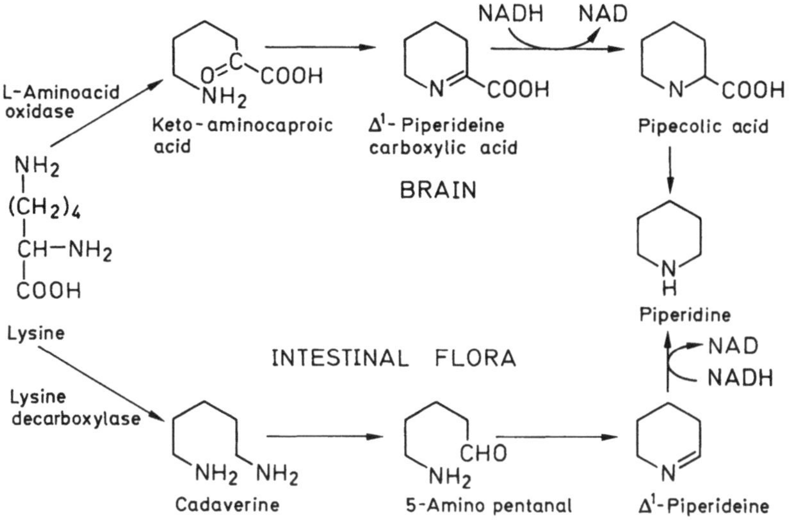

Figure 16. The biosynthesis of piperidine.

the main molecule, while in dansyl derivatives the S—C cleavage is very prominent, yielding *m/z* 170 (dimethylaminonaphthalene) as the main fragment. The di-BNS-piperidine was measured by the integrated ion current procedure, using the lower homologue di-BNS-pyrrolidine as internal standard, and measuring the ions of *m/z* 345 and 359 for internal standard and di-BNS-piperidine respectively (243). Regional distribution of piperidine showed kidney to have the highest content, and brain the lowest. Levels in the brains of various animal species were determined, and frog, rabbit and snail brains were found to have the highest piperidine levels. No evidence for the existence of conjugated piperidine was found. The second method used the classical integrated ion current procedure with the dansyl derivative (fragment of *m/z* 318); the pyrrolidine analogue was again used as internal standard (*m/z* 304). The rationale for the investigation of brain piperidine levels was the idea that piperidine in the brain increases during sleep (243, 244) and that it can induce sleep (245). The brain concentrations were all in the region of 50 pmol $\cdot$ g^{-1} fresh tissue. To settle the question one would of course have to get values for waking and sleeping subjects: this would clearly be hard to set up.

Other amines found in the central nervous system include ethanolamine (246—248), pyrrolidine and dimethylaminoethanol (249). Mass spectrometry has been used to determine the brain pyrrolidine concentration (250). The free base was steam distilled from basified brain homogenates into excess dilute acid, and the amine was derivatized to the dinitrophenyl derivative *in situ*, followed by solvent extraction of the derivative. The perdeuterated analogue of pyrrolidine was added at the outset and carried through the whole procedure as internal standard. The ions monitored were *m/z* 220 and 227 for amine and standard respectively. As has generally been found, the improved specificity gave values (95 pmol $\cdot$ g^{-1} wet weight) that were lower than those determined by less specific methods. Several other small aliphatic amines, some of them such as *iso*-amylamine, methylamine and dimethylamine produced by the decarboxylation of aliphatic amino acids, have received relatively little attention and have not been analysed by mass spectrometry, with one exception: trimethylamine. This fishy smelling amine is excreted in increased amounts (and indeed may pervade the tissues and immediate vicinity) in patients with the rare hereditary metabolic disease trimethylaminuria (251), and an ingenious mass spectrometric method of headspace analysis by direct insertion has been developed for it (252). Small urine samples were spiked with perdeuterated trimethylamine as internal standard and sealed into 2 ml vials, with teflon-lined screw caps, under helium. After heating to 60 °C for ten minutes and cooling, the headspace vapour was taken up into a disposable syringe, and allowed passively to drift into the MS ion sourse through the GC inlet via a long glass tube, ahead of a slow trickle of helium. The mass spectrometer was alternatively focussed onto the $[M-H]^+$ ion of the trimethylamine (*m/z* 58.065, see also ref. 184) and the $[M-D]^+$ ion of the deuterated analogue at *m/z* 66.116. A background ion, *m/z* 57.070, $C_4H_9^+$, was used as a lock-mass to stabilize the instrument. Because of the non-polar nature of the amine and its tendency to cling to surfaces, it was found essential to use the syringes once only.

8 Conclusion

The biogenic amines are still being very actively investigated, particularly as far as their significance in neurochemistry and psychopharmacology is concerned, and mass spectrometric methods find good use in this effort, because they guarantee chemical identity and possess the necessary sensitivity. In routine clinical applications mass spectrometry has played a part in the past, particularly in putting analytical studies on an unequivocal basis. However, high technology instrumentation of this kind is not sufficiently widely available yet, nor can it be tied up for continuous periods in routine assays. There are indications that this is likely to change: clinical chemistry apparatus is becoming more complex and sophisticated, while GC-MS equipment is becoming more compact and "user-friendly". As these trends converge, purpose designed GC-MS equipment is becoming a more familiar sight in clinical chemistry laboratories. Demanding assays such as those for some of the biogenic amines and their metabolites are now being routinely performed on GC-MS machines in research institutes and even large hospitals. The unrivalled properties of mass spectrometric methods will continue to be creatively applied not only to research on biogenic amines, but more and more to the investigation of the clinical aspects of their significance.

References

1. Guggenheim, M. (1951) Die Biogenen Amine, 4th Ed., Simon Karger, Basel.
2. Usdin, E. & Sandler, M. (1976) Trace Amines and the Brain, Marcel Dekker, New York.
3. Baker, G. B. & Coutts, T. (1982) in: Analysis of Biogenic Amines (Baker, G. B. & Coutts, T., eds.) pp. 1–13, Elsevier, Amsterdam, Oxford, New York.
4. Blau, K. & King, G. S. (1977) in: Handbook of Derivatives for Chromatography (Blau, K. & King, G. S., eds.) pp. 104–151, Heyden, London.
5. Gelpi, E. (1982) in: Analysis of Biogenic Amines (Baker, G. B. & Coutts, T., eds.) pp. 151–181, Elsevier, Amsterdam, Oxford, New York.
6. Caprioli, R. M., Liehr, J. G. & Seifert, W. S. Jr. (1980) in: Biomedical Applications of Mass Spectrometry (Waller, G. R. & Derner, O. C., eds.) pp. 1039–1050, Wiley Interscience, New York.
7. Blau, K., Claxton, I. M., Ismahan, G. & Sandler, M. (1979) Urinary phenylethylamine excretion: gas chromatographic assay with electron capture detection of the pentafluorobenzoyl derivative. Clin. Chim. Acta *163*, 135–142.
8. Wong, K. P., Ruthven, C. R. J. & Sandler, M. (1973) Gas chromatographic measurement of urinary catecholamines by an electron capture procedure. Clin. Chim. Acta *47*, 215–222.
9. Nelson, L. M., Bubb, F. A., Lax, P. M., Weg, M. W. & Sandler, M. (1979) An improved method for the differential assay of 3-O-methylated catecholamines in human urine using ion-pair extraction and gas chromatography electron capture detection. Clin. Chim. Acta *92*, 235–240.
10. Schill, G., Modin, R., Borg, K. O. & Persson, B.-A. (1977) in: Handbook of Derivatives for Chromatography (Blau, K. & King, G. S., eds.) pp. 500–529, Heyden, London.
11. Bozzi, A., Cardelli-Cangiano, P., Strom, R., Riggio, O., Rossi-Fanelli, F., Cangiano, C., Gentile, S. & Capocaccia, L. (1981) Simultaneous radio-enzymatic determination of phenylethylamine, phenylethanolamine, tyramine and octopamine in plasma of patients with hepatic encephalopathy. Clin. Biochem. *14*, 187–190.
12. Le Gatt, D. F., Baker, G. B. & Coutts, R. T. (1981) Simultaneous extraction and separation of trace amines of biological interest. J. Chromatogr. *225*, 301–308.
13. Buu, N. T., Angers, M., Chavalier, D. & Kuchel, O. (1984) A new method for the simultaneous analysis of free and sulfoconjugated normetanephrine, metanephrine, and 3-methoxytyramine in human urine by HPLC with electrochemical detection. J. Lab. Clin. Med. *104*, 425–432.
14. Koslow, S. H., Cattabeni, F. & Costa, E. (1972) Norepinephrine and dopamine: assay by mass fragmentography in the picomole range. Science *176*, 177–180.
15. Creveling, C. R., Kondo, K. & Daly, J. W. (1968) Use of dansyl derivatives and mass spectrometry for identification of biogenic amines. Clin. Chem. *14*, 302–309.
16. Andersen, H. & Braestrup, C. (1977) Mass fragmentographic demonstration of low amounts of β-phenylethylamine in human urine. Scand. J. Lab. Invest. *37*, 33–37.
17. Wilson, B. W. & Snedden, W. (1979) Capillary column GC-MS detection of 5-methoxytryptamine in human plasma using selected ion monitoring. J. Neurochem. *33*, 939–941.
18. Abramson, F. P., McCaman, M. W. & McCaman, R. E. (1974) Femtomole level of analysis of biogenic amines and amino acids using functional group mass spectrometry. Anal. Biochem. *51*, 482–499.
19. Iwase, H. & Murai, A. (1977) On the derivatives for the ultramicro-determination of amines by mass fragmentography. Chem. Pharm. Bull. *25*, 3129–3136.
20. Narasimhachari, N. & Vouros, P. (1972) Gas-liquid chromatography and mass spectrometry of biogenic amines and amphetamines as their isothiocyanate derivatives. Anal. Biochem. *45*, 154–163.
21. Narasimhachari, N. & Vouros, P. (1972) 3- and 4-O-methyl ethers of dopamine: gas-liquid chromatography and mass spectrometric studies of their isothiocyanate derivatives. J. Chromatogr. *70*, 135–140.
22. Anggard, E. & Hankey, A. (1969) Gas chromatography of catecholamine metabolites using electron capture detection and mass spectrometry. Anal. Chem. *41*, 1250–1256.
23. Anggard, E. & Hankey, A. (1969) Derivatives of sympathomimetic amines for gas

chromatography with electron capture detection and mass spectrometry. Acta Chem. Scand. *23*, 3110–3119.

24. Tas, A. C., Odink, J., Ten Noever de Brauw, M. C., Schrijver, J. & Jonk, R. J. G. (1984) Derivatization and mass spectrometric behaviour of catecholamines and their 3-O-methylated metabolites. J. Chromatogr. *310*, 243–250.
25. Coutts, R. T., Baker, G. B., Pasutto, F. M., Lin, S.-F., LeGatt, F. & Prelusky, D. B. (1984) Mass spectral analysis of perfluoroalkylated derivatives of some arylalkylamines of biological interest. Biomed. Mass Spectrom. *11*, 441–449.
26. Majer, J. R. & Boulton, A. A. (1970) Absolute, unambiguous ultramicroanalysis of metabolites present in complex biological extracts. Nature *225*, 658–660.
27. Sweeley, C. C., Elliott, W. H., Fries, I. & Ryhage, R. (1966) Mass spectrometric determination of unresolved components in gas chromatographic effluents. Anal. Chem. *38*, 1549–1553.
28. Seiler, N. & Wiechmann, M. (1965) Zum Nachweis von Aminen im 10^{-10} Mol. Maßstab. Experientia *21*, 203–207.
29. Reisch, J., Pagnocco, R., Alfes, H., Jantos, N. & Möllmann, H. (1968) Mass spectra of derivatives of phenylalkylamines. J. Pharm. Pharmacol. *20*, 81–86.
30. Reisch, J., Alfes, H., Jantos, N. & Möllmann, H. (1968) Mass spectra of 1-dimethylaminonaphthalene-5-sulphonyl derivatives of biogenic amines. Acta Pharm. Suec. *5*, 393–397.
31. Creveling, C. R., Kondo, K. & Daly, J. W. (1968) Use of dansyl derivatives and mass spectrometry for identification of biogenic amines. Clin. Chem. *14*, 302–309.
32. Seiler, N., Schneider, H. & Sonnenberg, K.-D. (1970) Massenspektrometrische Identifizierung von biogenen Aminen in Form ihrer 1-dimethylaminonaphthalin-5-sulfonsäureamide auf Dünnschichtchromatogrammen. Z. analyt. Chem. *220*, 109–127.
33. Chapman, D. I., Chapman, J. R. & Clark, J. (1972) Synthesis and mass spectra of dansyl derivative of some physiologically important B-phenylaklylamine derivatives. Int. J. Biochem. *3*, 66–72.
34. Durden, D. A., Davis, B. A. & Boulton, A. A. (1974) Qualitative and quantitative mass spectrometry of some noncatecholic biogenic amines and related compounds as their dansyl derivatives. Biomed. Mass Spectrom. *1*, 83–95.
35. Durden, D. A. & Boulton, A. A. (1979) in: Techniques in Metabolic Research (Kornberg, H. L., Metcalfe, J. C., Northcote, D. H., Pogson, C. J. & Tipton, K. F., eds.) pp. 1–25, Elsevier/North Holland, Amsterdam.
36. Davis, B. A. & Durden, D. A. (1982) in: Analysis of Biogenic Amines (Baker, G. B. & Coutts, T., eds.) pp. 129–149, Elsevier, Amsterdam, Oxford, New York.
37. Davis, B. A., Durden, D. A. & Boulton, A. A. (1986) Simultaneous analysis of 12 biogenic amine metabolites in plasma, cerebrospinal fluid and urine by capillary gas chromatography-high resolution mass spectrometry with selected ion monitoring. J. Chromatogr. *374*, 227–238.
38. Milne, G. W. A., Fales, H. M. & Colburn, R. W. (1973) Chemical ionization mass spectrometry of complex molecules: biogenic amines. Anal. Chem. *45*, 1952–1954.
39. Miyazaki, H., Hashimoto, M., Iwanaga, M. & Kubodera, T. (1974) Analysis of biogenic amines and their metabolites by gas chromatography-chemical ionization mass spectrometry. J. Chromatogr. *99*, 575–586.
40. Edwards, D. J., Rizk, M. & Neil, J. (1979) Simultaneous analysis of phenylglycols and phenylethanols in human urine by gas chromatography-mass spectrometry. J. Chromatogr. *164*, 407–416.
41. Mizuno, Y. & Ariga, T. (1979) Gas chromatographic chemical ionization-mass spectrometric assay of catecholamines in the brain. Clin. Chim. Acta *98*, 217–224.
42. Bergner, E. A. & Doughy, R. C. (1981) Primary amine derivatives for negative chemical ionization mass spectrometry. Biomed. Mass Spectrom. *8*, 204–207.
43. Ruthven, C. R. J. & Sandler, M. (1979) in: Chemical Diagnosis of Disease (Brown, S. S., Mitchell, F. L. & Young, D. S., eds.) pp. 1217–1290, Elsevier/North Holland, Amsterdam.
44. Hermann, H. & Mornex, R. (1964) Human Tumours Secreting Catecholamines, Pergamon, Oxford.
45. Kinnear Wilson, L. M. & Draper, G. J. (1974) Neuroblastoma, its natural history and prognosis: a study of 487 cases. Brit. Med. J. *3*, 301–307.
46. Page, L. B. & Jacoby, G. A. (1964) Catecholamine metabolism and storage granules in pheochromocytoma and neuroblastoma. Medicine (Baltimore) *43*, 379–386.
47. Lhuguenot, J. C. & Maume, B. F. (1980) A method for the analysis of catecholamines

by selective ion monitoring and ^{14}C isotope dilution in adrenal medullary cell culture. Biomed. Mass Spectrom. *7*, 529–532.

48. Gelpi, E., Peralta, E. & Segura, J. (1974) Gas chromatography-mass spectrometry of catecholamines and tryptamines: determination of gas chromatographic profiles of the amines, their precursors and their metabolites. J. Chromatogr. Sci. *12*, 701–709.
49. Curtius, H. Ch., Wolfensberger, M., Redweik, U., Leimbacher, W., Maibach, R. A. & Isler, W. (1975) Mass fragmentography of 5-hydroxytryptophol and 5-methoxytryptophol in human cerebrospinal fluid. J. Chromatogr. *112*, 523–533.
50. Wang, M.-T., Imai, K., Yoshioka, M. & Tamura, Z. (1975) Gas-liquid chromatographic and mass fragmentographic determination of catecholamines in human plasma. Clin. Chim. Acta *63*, 13–19.
51. Wang, M.-T., Yoshioka, M., Imai, K. & Tamura, Z. (1975) Gas-liquid chromatographic and mass fragmentographic determination of 3-O-methylated catecholamines in human plasma. Clin. Chim. Acta *63*, 21–27.
52. Warsh, J. J., Chiu, A., Li, P. P. & Godse, D. (1980) Comparison of liquid chromatography-electrochemical and gas chromatography-mass spectrometric methods for brain dopamine and serotonin. J. Chromatogr. *183*, 483–486.
53. Ehrhardt, J.-D. & Schwartz, J. (1978) A gas chromatography-mass spectrometry assay of human plasma catecholamines. Clin. Chim. Acta *88*, 71–79.
54. Yoshida, J.-I., Yoshino, K., Matsunaga, T., Higa, S., Suzuki, T., Hayashi, A. & Yamamura, Y. (1980) An improved method for determination of plasma norepinephrine: isolation by boric acid gel and assay by selected ion monitoring. Biomed. Mass Spectrom. *7*, 396–398.
55. Muskiet, F. A. J., Thomasson, C. G., Gerding, A. M., Fremouw-Ottevangers, D. C., Nagel, G. T. & Wolthers, B. G. (1979) Determination of catecholamines and their 3-O-methylated-metabolites in urine by mass fragmentography with use of deuterium internal standards. Clin. Chem. *25*, 453–560.
56. Canfell, C., Binder, S. R. & Khayam-Bashi, H. (1982) Quantitation of urinary normetanephrine and metanephrine by reversed-phase extraction and mass fragmentographic analysis. Clin. Chem. *28*, 25–28.
57. Higa, S., Suzuki, T., Hayashi, A., Tsuge, I. & Yamamura, Y. (1977) Isolation of catecholamines in biological fluids by boric acid gel. Anal. Biochem. *77*, 18–24.
58. Sjoquist, B. & Anggard, E. (1972) Gas chromatographic determination of homovanillic acid in human cerebrospinal fluid by electron capture detection and by mass fragmentography with a deuterated internal standard. Anal. Chem. *44*, 2297–2301.
59. Gordon, E. K., Oliver, J., Black, K. & Kopin, I. J. (1974) Simultaneous assay by mass fragmentography of vanillylmandelic acid, homovanillic acid, and 3-methoxy-4-hydroxyphenethylene glycol in cerebrospinal fluid and urine. Biochem. Med. *11*, 32–40.
60. Narasimhachari, N., Leiner, K., Plaut, J. M. & Lin, R.-L. (1974) Selective and specific gas-liquid chromatographic and gas-liquid chromatographic-mass spectrometric methods for the separation and quantitative determination of homovanillic and iso-homovanilllic acids in urine samples. Clin. Chim. Acta *50*, 337–344.
61. Sjoquist, B., Lindstrom, B. & Anggard, E. (1975) Mass fragmentographic determination of 4-hydroxy-3-methoxyphenylglycol (HMPG) in urine, cerebrospinal fluid, plasma and tissues using a deuterium-labelled internal standard. J. Chromatogr. *105*, 309–316.
62. Narasimhachari, N., Leiner, K. & Brown, C. (1975) The simultaneous determination by selective ion monitoring of the levels of homovanillic, iso-homovanillic, 3,4-dihydroxyphenylacetic and 3-methoxy-4-hydroxymandelic acids in single biological samples. Clin. Chim. Acta *62*, 245–253.
63. Takahashi, S., Yoshioka, M., Yoshino, S. & Tamura, Z. (1978) Mass fragmentographic determination of vanilmandelic acid and isohomovanillic acid in human body fluids. J. Chromatogr. *145*, 1–9.
64. Murray, S., Baillie, T. A. & Davies, D. S. (1977) A non-enzymic procedure for the quantitative analysis of (3-methoxy-4-sulphophenyl)-ethylene glycol (MHPG sulphate) in human urine using stable isotope dilution and gas chromatography-mass spectrometry. J. Chromatogr. *148*, 541–551.
65. Muskiet, F. A. J., Fremouw-Ottevangers, D. C., van der Meulen, J., Wolthers, B. G. & de Vries, J. (1978) Determination of some 3,4-dihydroxyphenylalanine and dopamine metabolites in urine by means of mass fragmentography. Clin. Chem. *24*, 122–127.
66. Muskiet, F. A. J., Fremouw-Ottevangers, D. C., Nagel, G. T., Wolthers, B. G. &

de Vries, J. (1978) Determination of 3-methoxy-4-hydroxyphenylpyruvic acid, 3,4-dihydroxyphenylethyleneglycol and 3,4-dihydroxyphenylmandelic acid urine by mass fragmentography with use of deuterium-labelled internal standard. Clin. Chem. *24*, 2001–2008.

67. Muskiet, F. A. J., Fremouw-Ottevangers, C. D., Nagel, G. T. & Wolthers, B. G. (1979) Determination of 4-O-methylated catecholamine metabolites in urine my mass fragmentography. Clin. Chem. *25*, 1708–1713.
68. Muskiet, F. A. J., Nagel, G. T. & Wolthers, B. G. (1980) Simultaneous determination of unconjugated homovanillic acid, vanilmandelic acid and 3-methoxy-4-hydroxyphenylethylene glycol in serum by mass fragmentography with deuterated internal standards. Anal. Biochem. *109*, 130–136.
69. Jimerson, D. C., Markey, S. P., Oliver, J. A. & Kopin, I. J. (1981) Simultaneous measurement of plasma 4-hydroxy-3-methoxyphenylethylene glycol and 3,4-dihydroxyphenylethylene glycol by gas chromatography mass spectrometry. Biomed. Mass Spectrom. *8*, 256–259.
70. Jakob, K., Vogt, W. & Schwerdtfeger, G. (1984) Sensitive mass fragmentographic determination of acid catecholamine metabolites in human body fluids. J. Chromatogr. *290*, 331–339.
71. Christensen, M. S. & Christensen, N. J. (1972) Plasma catecholamines in hypertension. Scand. J. Clin. Lab. Invest., *30*, 169–173.
72. Louis, W. J., Doyle, A. E. & Anavekar, S. (1973) Plasma norepinephrine levels in essential hypertension. New Engl. J. Med. *288*, 599–601.
73. Calne, D. B., Karoum, F., Ruthven, C. R. J. & Sandler, M. (1969) The metabolism of orally administered L-dopa in Parkinsonism. Br. J. Pharmacol. *37*, 57–68.
74. Sandler, M. & Halket, J. M. (1986) Mass Spectrometry and the brain — the monoamine oxidase connection. Proc. Jap. Soc. Med. Mass Spectrom. *11*, 7–19.
75. Curtius, H. Ch., Wolfensberger, B., Steinmann, B., Redweik, U. & Siegfried, J. (1974) Mass fragmentography of dopamine and 6-hydroxydopamine. Application to the determination of dopamine in human brain biopsies from the caudate nucleus. J. Chromatogr. *99*, 529–540.
76. Elsworth, J. D., Roth, R. H. & Redmond, D. E. Jr. (1983) Relative importance of 3-methoxy-4-hydroxyphenylglycol and 3,4-dihydroxyphenylglycol as norepinephrine metabolites in rat, monkey, and humans. J. Neurochem. *41*, 786–793.
77. Wood, J. H. (1982) in: Handbook of Neurochemistry Vol. 1, 2nd. Ed. (Lajtha, A., ed.) pp. 415–487, Plenum, New York.
78. Gordon, E. K., Markey, S. P., Sherman, R. L. & Kopin, I. J. (1976) Conjugated 3,4-dihydroxyphenylacetic acid (dopac) in human and monkey cerebrospinal fluid and rat brain, and the effects of probenecid treatment. Life Sci. *18*, 1285, 1292.
79. Watson, E., Wilk, S. & Roboz, J. (1974) Derivatization and gas chromatographic determination of some biologically important acids in cerebrospinal fluid. Anal. Biochem. *59*, 441–451.
80. Faull, K. F., Anderson, P. J., Barchas, J. D. & Berger, P. A. (1979) Selected ion monitoring assay for biogenic amine metabolites and probenecid in human lumbar cerebrospinal fluid. J. Chromatogr. *163*, 337–349.
81. Kilts, C. D., Vrbanac, J. J., Rickert, D. E. & Reich, R. H. (1977) Mass fragmentographic determination of 3,4-dihydroxyphenylethylamine and 4-hydroxy-3-methoxythenylethylamine in the caudate nucleus. J. Neurochem. *28*, 465–467.
82. Edwards, D. J. & Rizk, M. (1981) Effects of amino acid precursors on catecholamine synthesis in the brain. Prog. Neuro-Psychopharmacol. *5*, 569–572.
83. Edwards, D. J. & Rizk, M. (1981) Conversion of 3,4-dihydroxyphenylalanine and deuterated 3,4-dihydroxyphenylalanine to alcoholic metabolites of catecholamines in rat brain. J. Neurochem. *36*, 1641–1647.
84. Herranen, J., Huhtikangas, A., Tirronen, H., Halonen, T., Huuskonen, M., Rainikainen, K. & Riekkinen, P. (1984) Quantification of a number of biogenic amines and their metabolites in brain homogenates by gas chromatography-mass spectrometry. J. Chromatogr. *307*, 241–249.
85. Ziegler, M. G., Lake, C. R., Foppen, F. J., Shoulson, I. & Kopin, I. J. (1976) Norepinephrine in cerebrospinal fluid. Brain Res. *108*, 436–440.
86. Muskiet, F. A. J., Jeuring, H. J., Korf, J., Sedvall, G., Westerink, B. H. C., Teelken, A. W. & Wolthers, B. G. (1979) Correlations between a fluorimetric and mass fragmentographic method for the determination of 3-methoxy-4-hydroxyphenylacetic acid and two mass fragmentographic methods for the determination of 3-methoxy-4-hydroxy-

phenylethylene glycol in cerebrospinal fluid. J. Neurochem. *32*, 191 – 194.

87. Karoum, F., Gillin, J. C., Wyatt, R. J. & Costa, E. (1975) Mass fragmentography of nanogram quantities of biogenic amine metabolites in human cerebrospinal fluid and whole rat brain. Biomed. Mass Spectrom. *2*, 183 – 189.
88. Sjöquist, B. (1975) Mass fragmentographic determination of 4-hydroxy-3-methoxymandelic acid in human urine, cerebrospinal fluid, brain and serum using a deuterium-labelled internal standard. J. Neurochem. *24*, 199 – 201.
89. Schanberg, S. M., Schildkraut, J. J., Breese, G. R. & Kopin, I. J. (1968) Metabolism of normetanephrine-^{3}H in rat brain — identification of conjugated 3-methoxy-4-hydroxyphenylglycol as the major metabolite. Biochem. Pharmacol. *17*, 247 – 254.
90. Vajta, S., Le Moing, J. P., Rovei, V. & Strolin Benedetti, M. (1985) Determination of low plasma concentrations of 3-methoxy-4-hydroxyphenylethylene glycol using gas chromatography — negative chemical ionization mass spectrometry. J. Chromatogr. *343*, 239 – 245.
91. Zuspan, F. P. & Abbott, M. (1970) Identification of a pressor substance in amniotic fluid 1. Role of epinephrine and norepinephrine. Amer. J. Obstet. Gynecol. *107*, 664 – 672.
92. Zuspan, F. (1974) in: Amniotic Fluid (Natelson, S., Scommegna, A. & Epstein, M. B. eds.) pp. 101 – 108, Wiley, New York.
93. Phillippe, M. & Ryan, K. J. (1981) Catecholamines in human amniotic fluid. Amer. J. Obstet. Gynecol. *139*, 204 – 208.
94. Divers, W. A. Jr., Wilkes, M., Babaknia, A. & Yen, S. S. C. (1981) An increase in catecholamines and metabolites in the amniotic fluid compartment from middle to late gestation. Amer. J. Obstet. Gynecol. *139*, 483 – 486.
95. Divers, W. A., Wilkes, M. M., Babaknia, A., Hill, L. M., Quilligan, E. J. & Yen, S. S. C. (1981) Amniotic fluid catecholamines and metabolites in intrauterine growth retardation. Amer. J. Obstet. Gynecol. *141*, 608 – 610.
96. Lagercrantz, H., Sjoquist, B., Bremme, K., Lunell, N.-O. & Somell, C. (1980) Catecholamine metabolites in amniotic fluid as indicators of intrauterine stress. Amer. J. Obstet. Gynecol. *136*, 1067 – 1070.
97. Zambotti, F., Blau, K., King, G. S., Campbell, S. & Sandler, M. (1975) Monoamine metabolites and related compounds in human amniotic fluid: assay by gas chromatography and gas chromatography-mass spectrometry. Clin. Chim. Acta *61*, 247 – 256.
98. Muskiet, F. A. J., Jeuring, H. J., Nagel, G. T., de Bruyn, H. W. A. & Wolthers, B. G. (1978) Mass fragmentographic determination of catecholamine metabolites in amniotic fluid and its possible clinical usefulness. Clin. Chem. *24*, 1899 – 1904.
99. Lagercrantz, H. & Sjöquist, B. (1981) Catecholamine metabolites in amniotic fluid from fetuses with neural tube defects. Prenatal Diagn. *1*, 157 – 159.
100. Voute, P. A., Jr., Wadman, S. K. & van Putten, W. J. (1970) Congenital neuroblastoma. Symptoms in the mother during pregnancy. J. Clin. Endocrinol. Metab. *22*, 575 – 580.
101. Lagercrantz, H., Sjöquist, B., Lindsten, J., Lunell, N.-O., Bremme, K. & Hagenfeldt, L. (1977) Biokemisk undersökning av fostrets nervsystem; catecholaminmetaboliter i fostretsvattenprov. Acta Soc. Med. Suec. *86*, 503 – 506.
102. Martin, W. R. & Sloan, J. W. (1974) The possible role of tryptamine in brain function and its relationship to the action of LSD-like hallucinogens. Mt. Sinai J. Med. *41*, 276 – 282.
103. Jones, R. S. G. (1982) Tryptamine: a neuromodulator or neurotransmitter in the mammalian brain? Progr. Neurobiol. *19*, 117 – 139.
104. Tabakoff, B., Moses, F., Phillipps, S. R. & Boulton, A. A. (1977) Effects of tranylcypromine and pargyline on brain tryptamine. Experientia *33*, 380.
105. Bieck, P. R., Nilsson, E., Schick, C., Waldmeier, P. C. & Lauber, J. (1983) in: Neurobiology of the Trace Amines (Boulton, A. A., Baker, G. B., Dewhurst, W. G. & Sandler, M., eds.) pp. 525—539, Humana Press, Clifton N. J.
106. Sullivan, M. X. (1922) Indolylethylamine in the urine of pellagrins. J. Biol. Chem. *50*, xxxix.
107. Rodnight, R. (1956) Separation and characterization of urinary indoles resembling 5-hydroxytryptamine and tryptamine. Biochem. J. *64*, 621 – 626.
108. Coppen, A., Shaw, D. M., Malleson, A., Eccleston, E. & Gundy, G. (1965) Tryptamine metabolism in depression. Brit. J. Psychiat. *111*, 993 – 998.
109. Kuehl, F. A., VandenHeuvel, W. J. A. & Ormond, R. E. (1968) Urinary metabolites in Parkinson's disease. Nature *217*, 136 – 138.

110. Smith, I. & Kellow, A. H. (1969) Aromatic amines and Parkinson's disease. Nature *221*, 1261.
111. Price, J. (1975) Decreased tryptaminuria after an alkali load. Clin. Chim. Acta *65*, 339–342.
112. Price, J. (1976) Increased tryptaminuria after an acid load. Clin. Chim. Acta *72*, 269–272.
113. Reynolds, G. P., Ceasar, P. M., Ruthven, C. R. J. & Sandler, M. (1978) The effect of urinary pH and flow rate on monoamine output. Clin. Chim. Acta *84*, 225–231.
114. Couch, M.W. & Williams, C. M. (1972) Mass spectrometry of tryptamines and acetylated derivatives. Anal. Biochem. *50*, 612–622.
115. Philips, S. R., Durden, D. A. & Boulton, A. A. (1974) Identification and distribution of tryptamine in the rat. Canad. J. Biochem. *52*, 447–451.
116. Artigas, F. & Gelpi, E. (1979) A new mass fragmentographic method for the simultaneous analysis of tryptophan, tryptamine, indole-3-acetic acid, serotonin, and 5-hydroxyindole-3-acetic acid in the same sample of rat brain. Anal. Biochem. *92*, 233–242.
117. Warsh, J. J., Godse, D. D., Stancer, H. C., Chan, P.W. & Coscina, D.V. (1977) Brain tryptamine in rats by a new GC-MS method. Biochem. Med. *18*, 10–20.
118. Artigas, F., Sunol, C., Tusell, J. M., Martinez, E. & Gelpi, E. (1984) Determination of tryptamine in brain tissue by capillary gas chromatography-mass-spectrometry (selective ion monitoring). Biomed. Mass Spectrom. *11*, 142–144.
119. Artigas, F., Martinez, E. & Gelpi, E. (1982) Neglected practical aspects in the trace analysis of indoleamines and related metabolites by GC and GC/MS (selected ion monitoring). J. Chromatogr. Sci. *75*, 75–82.
120. Huebert, N. D. & Boulton, A. A. (1979) Longitudinal urinary trace amine excretion in a human male. J. Chromatogr. *162*, 169–176.
121. Segura, J., Artigas, F., Martinez, E. & Gelpi, E. (1976) Adsorption of tryptophan metabolites from physiological fluids on XAD-2 and determination by single ion monitoring. Biomed. Mass Spectrom. *3*, 91–96.
122. Costa, E., Gessa, G. L. & Sandler, M., eds. (1974) Serotonin – New Vistas. Advances in Biochemical Psychopharmacology Vol. 11, Raven Pres, New York.
123. Osborne, N. N., ed. (1982) Biology of Serotonergic Transmission, Wiley, Chichester, New York, Brisbane, Toronto, Singapore.
124. Brown, H. (1977) in: Serotonin in Health and Disease. Vol. IV: Clinical Correlates (Essman, W. B., ed.) pp. 393–423, Spectrum Publications, New York.
125. Bensch, K. G., Corrin, B., Pariente, R. & Spencer, H. (1968) Oat-cell carcinoma of the lung. Cancer *22*, 1163–1172.
126. Kamoun, P. & Jerome, H. (1978) in: Serotonin in Mental Abnormalities (Boullin, J., ed.) pp. 119–154, Wiley, Chichester, New York, Brisbane, Toronto.
127. Pare, C. M. B., Sandler, M. & Stacey, R. S. (1957) 5-Hydroxytryptamine deficiency in phenylketonuria. Lancet *i*, 551–553.
128. Boullin, J. (1978) in: Serotonin in Mental Abnormalities (Boullin, J., ed.) pp. 1–28, Wiley, Chichester, New York, Brisbane.
129. Paasonen, M. K. (1968) Platelet 5-hydroxytryptamine as a model in pharmacology. Ann. Med. exp. Fenn. *46*, 416–422.
130. Jamieson, W. D. & Hutzinger, O. (1970) Mass spectral identification of serotonin metabolites. Anal. Biochem. *37*, 182–187.
131. Bacopoulos, N. G., Redmond, D. E. & Roth, R. H. (1979) Serotonin and dopamine metabolites in brain regions and cerebrospinal fluid of a primate species: effects of ketamine and fluphenazine. J. Neurochem. *32*, 1215–1218.
132. Markey, S. P., Colburn, R. W. & Johannesen, J. N. (1981) Efficient extraction and mass spectrometric assay of serotonin in biological fluids. Biomed. Mass Spectrom. *8*, 301–304.
133. Curtius, H.-Ch., Farner, H. & Rey, F. (1980) Quantification of serotonin and tryptamine in urine by gas chromatography-mass fragmentography. J. Chromatogr. *199*, 171–179.
134. Fernstrom, J. D. & Wurtman, R. J. (1971) Brain serotonin content: physiological dependence on plasma tryptophan levels. Science *173*, 149–150.
135. Fernstrom, J. D. & Wurtman, R. J. (1972) Brain serotonin content: physiological regulation by plasma neutral amino acids. Science *178*, 414–416.
136. Warsh, J. J., Coscina, D. V., Godse, D. D. & Chan, P. W. (1979) Dependence of brain tryptamine formation on tryptophan availability. J. Neurochem. *32*, 1191–1196.
137. Woolley, D. W. (1962) The biochemical basis of the serotonin hypothesis about mental diseases. Wiley, New York.

138. Tanimukai, H., Ginther, R., Spaide, J. & Himwich, H. E. (1967) Psychotomimetic indole compounds in the urine of schizophrenics and mentally defective patients. Nature *216*, 490–491.
139. Airaksinen, M. M. & McIsaac, W. M. (1968) Indoleakylamines and behavior. Ann. Med. exp. Fenn. *46*, 367–381.
140. Mandell, A. J. & Morgan, M. (1971) Indole(ethyl)amine N-methyltransferase in human brain. Nature New Biol. *230*, 85–87.
141. Walker, R. W., Ahn, H. S., Mandel, L. R. & VandenHeuvel, W. J. A. (1972) Identification of N,N-dimethyltryptamine as the product of an in vitro enzymic methylation. Anal. Biochem. *47*, 228–234.
142. Walker, R. W., Ahn, H. S., Albers-Schonberg, G., Mandel, L. R. & VandenHeuvel, W. J.A. (1973) Gas chromatographic-mass spectrometric isotope dilution assay for N,N-dimethyltryptamine in human plasma. Biochem. Med. *8*, 105–113.
143. Narasimhachari, N. & Himwich, H. E. (1973) GC-MS-identification of bufotenin in urine samples form patients with schizophrenia or infantile autism. Life Sci. *12*, 475–478.
144. Narasimhachari, N., Lin, R.-L., Plaut, J. & Leimer, K. (1973) Selective and specific TLC, GC and GC-MS methods for the assay of indole ethylamine N-ethyltransferase activity. J. Chromatogr. *86*, 123–128.
145. Wyatt, R. J., Gillin, J. C., Kaplan, J., Stillman, R., Mandel, L., Ahn, H. S., VandenHeuvel, W. J. A. & Walker, R. W. (1974) N,N-dimethyltryptamine – a possible relationship to schizophrenia. Advanc. Biochem. Psychopharmacol. *11*, 299–313.
146. Gomes, U. R., Neethling, A. A. & Shanley, B. C. (1976) Enzymatic N-methylation of indoleamines in mammalian brain: fact or artefact? J. Neurochem. *27*, 701–705.
147. Walker, R. W., Mandel, L. R., DeLisi, L., Wyatt, R. J. & VandenHeuvel, W. J. A. (1984) Capillary gas-liquid chromatography selected ion monitoring assay for [^{13}C,^{15}N]N-methyltryptamine in human urine: failure to detect conversion of [^{13}C,^{15}N]tryptamine in schizophrenia patients. J. Chromatogr. *289*, 223 –229.
148. Lerner, A. B., Case, J. D., Takahashi, Y., Lee, T. H. & Mori, W. (1958) Isolation of melatonin, the pineal gland factor that lightens melanocytes. J. Am. Chem. Soc. *80*, 2587.
149. Klein, D. C. & Weller, J. L. (1970) Indole metabolism in the pineal gland: a circadian rhythm in N-acetyltransferase. Science *169*, 1093–1094.
150. Pelham, R. W., Vaughan, G. M., Sandock. K. L. & Vaughan, M. K. (1973) Twenty-four-hour cycle of a melatonin-like substance in the plasma of human males. J. Clin. Endocrinol. Metab. *37*, 341–344.
151. Lynch, H. J., Wurtman, R. J., Moskowitz, M. A., Archer, M. C. & Ho, M. H. (1975) Daily rhythm in human urinary melatonin. Science *187*, 169–171.
152. Wetterberg, L., Arendt, J., Paunier, L., Sizonenko, P. C., van Donselaar, W. & Heyden. T. (1976) Human serum melatonin during the menstrual cycle. J. Clin. Endocrinol. Metab. *42*, 185–188.
153. Arendt, J., Wirz-Justice, A., Bradtke, J. & Kornemark, M. (1979) Long term studies on immunoreactive human melatonin. Ann. Clin. Biochem. *16*, 307–312.
154. Heubner, O. (1898) Tumor der glandula pinealis. Dtsch. Med. Wochenschr. *24*, 214–215.
155. Silman, R. E., Leone, R. M., Hooper, R. J. L. & Preece, M. A. (1979) Melatonin, the pineal gland and human puberty. Nature *282*, 301–303.
156. Waldhauser, F., Weiszenbacher, G., Frisch, H., Zeitlhuber, U. & Wurtman, R. J. (1984) Fall in nocturnal serum melatonin during prepuberty and pubescence. Lancet *i*, 362–365.
157. Cattabeni, F., Koslow, S. H. & Costa, E. (1972) Gas chromatogoraphic-mass spectrometric assay of four indole alkylamines in rat pineal. Science *178*, 166–168.
158. Pelham, R. W., Ralph, C. L. & Campbell, I. M. (1972) Mass spectral identification of melatonin in blood. Biochem. Biophys. Res. Commun. *46*, 1236–1241.
159. Smith, I., Mullen, P. E., Silman, R. E., Snedden, W. & Wilson, B. W. (1976) Absolute identification of melatonin in human plasma. Nature *260*, 718–719.
160. Wilson, B. W., Snedden, W., Silman, R. E., Smith, I. & Mullen, P. (1977) A gas chromatography-mass spectrometry method for the quantitative analysis of melatonin in plasma and cerebrospinal fluid. Anal. Biochem. *81*, 283–291.
161. Lewy, A. J. & Markey, S. P. (1978) Analysis of melatonin in human plasma by gas chromatography-negative chemical ionization mass spectrometry. Science *201*, 741–743.

162. Fraser, S., Cowen, P., Franklin, M. & Lewy, A. J. (1983) Direct radioimmunoassay and gas chromatography-mass spectrometry compared for determination of melatonin in plasma. Clin. Chem. *29*, 1703–1704.
163. Rogawski, M. A., Roth, R. H. & Aghajanian, G. K. (1979) Melatonin: deacetylation to 5-methoxytryptamine by liver but not brain aryl acylamidase. J. Neurochem. *32*, 1219–1226.
164. Bosin, T. R. & Beck, O. (1979) 5-Methoxytryptamine in the human pineal gland: identification and quantification by mass fragmentography. J. Neurochem. *32*, 1853–1855.
165. Beck, O. & Bosin, T. R. (1979) Analysis of 5-methoxytryptamine in brain by gas chromatography-mass spectrometry. Biomed. Mass Spectrom. *6*, 19–22.
166. Narasimhachari, N., Kempster, E. & Anbar, M. (1980) 5-Methoxytryptamine in rat hypothalamus and human cerebrospinal fluid. A fact or artifact? Biomed. Mass Spectrom. *7*, 231–235.
167. Bosin, T. R., Jonsson, G. & Beck, O. (1979) On the occurrence of 5-methoxytryptamine in brain. Brain Res. *173*, 79–88.
168. Kveder, S. & McIsaac, W. M. (1961) The metabolism of melatonin (N-acetyl-5-methoxytryptamine) and 5-methoxytryptamine. J. Biol. Chem. *236*, 3214–3220.
169. Fellenberg, A. J., Phillipou, G. & Seamark, R. F. (1980) Specific quantitation of urinary 6-hydroxymelatonin sulfate by gas chromatography-mass spectrometry. Biomed. Mass Spectrom. *7*, 84–87.
170. Blau, K., King, G. S. & Sandler, M. (1977) Mass spectrometric and nuclear magnetic resonance confirmation of a 3,3-spirocyclic indole derivative formed from melatonin and related acyl-tryptamines. Biomed. Mass Spectrom. *7*, 232–236.
171. Hoskins, J. A. & Pollitt, R. J. (1978) The determination of 5-methoxyindole-3-acetic acid in human urine by mass fragmentography. J. Chromatogr. *145*, 285–289.
172. Leone, R. M., Silman, R. E & Hooper, R. J. L. (1979) A sensitive and specific assay for 5-methoxytryptophol in plasma. J. Endocrinol. *82*, 243–251.
173. Takahashi, S., Godse, D. D., Naqvi, A., Warsh, J. J. & Stancer, H. C. (1978) 5-Hydroxytryptophol in human cerebrospinal fluid: quantitative determination by gas chromatography-mass spectrometry using a deuterated internal standard. Clin. Chim. Acta *84*, 55–62.
174. Francis, P., Smith, I. & Finnie, M. D. A. (1980) Mass spectrometric and chromatographic characteristics of neutral and acidic 5-methoxyindoles and some related compounds. Biomed. Mass Spectrom. *7*, 294–300.
175. Atkinson, D. E. (1968) The energy charge of the adenylate pool as a regulatory parameter. Interaction with feedback modifiers. Biochemistry *7*, 4030–4034.
176. Boulton, A. A. (1976) in: Trace Amines and the Brain (Usdin, E. & Sandler, M., eds.) pp. 21–39, Marcel Dekker, New York, Basel.
177. Boulton, A. A. (1978) The tyramines: functionally significant biogenic amines or metabolic accidents? Life Sci. *23*, 659–672.
178. Dewhurst, W. G. (1984) in: Neurobiology of the Trace Amines (Boulton, A. A., Baker, G. B., Dewhurst, W. G., & Sandler, M., eds.) pp. 3–12, Humana Press, Clifton N. J.
179. Boulton, A. A. (1984) in: Neurobiology of the Trace Amines (Boulton, A. A., Baker, G. B., Dewhurst, W. G. & Sandler, M., eds.) pp. 13–24, Humana Press, Clifton N. J.
180. Mantegazza, P. & Riva, M. (1963) Amphetamine-like activity of β-phenylethylamine after monoamine oxidase inhibitor in vivo. J. Pharm. Pharmacol. *15*, 472–478.
181. Tourian, A. & Sidbury, J. B. (1983) in: The Metabolic Basis of Inherited Disease, 5th ed. (Stanbury, J. B., Wyngaarden, J. B., Fredrickson, D. S., Goldstein, J. L. & Brown, M. S., eds.) pp. 270–286, McGraw-Hill, New York.
182. Blau, K. (1978) in: Noncatecholic Phenylethylamines, Part 1 (Mosnaim, A. D. & Wolf, M. E., eds.) pp. 385–396, Marcel Dekker, New York, Basel.
183. Blau, K. (1979) in: Aromatic Amino Acid Hydroxylases and Mental Disease (Youdim, M. B. H., ed.) pp. 77–139, John Wiley, Chichester, New York, Brisbane, Toronto.
184. Loo, Y. H. (1967) Characterization of a new phenylalanine metabolite in phenylketonuria. J. Neurochem. *14*, 813–821.
185. Perry, T. L., Hansen, S., Tischler, B., Bunting, R. & Diamond, S. (1970) Glutamine depletion in phenylketonuria. A possible cause of the mental defect. New Engl. J. Med. *282*, 761–766.
186. Fischer, E., Heller, B. & Miro, A. H. (1968) β-Phenylethylamine in human urine. Arzneimittel-Forschung *18*, 1486–1487.
187. Fischer, E. & Heller, B. (1972) Phenethylamine as a neurohumoral agent in brain. Behav. Neuropsychiatry *4*, 8–11.

188. Fischer, E. (1965) Monoamine-oxidase inhibitors. Lancet *ii*, 245–246.
189. Sabelli, H. C. & Mosnaim, A. D. (1974) Phenylethylamine hypothesis of affective behavior. Am. J. Psychiat. *131*, 695–699.
190. Blau, K. (1970) Aromatic acid excretion in phenylketonuria. Analysis of the unconjugated aromatic acids derived from phenylalanine. Clin. Chim. Acta *27*, 5–18.
191. Edwards, D. J. & Antelman, S. (1980) in: Noncatecholic Phenylethylamines, Part 2 (Mosnaim, A. D. & Wolf, M. E., eds.) pp. 53–80, Marcel Dekker, New York, Basel.
192. Schweitzer, J. W., Friedhoff, A. J. & Schwartz, R. (1975) Phenethylamine in human urine: failure to verify high values. Biol. Psychiat. *10*, 277–285.
193. Andersen, H. & Braestrup, C. (1977) Mass fragmentographic demonstration of low amounts of β-phenylethylamine in human urine. Scand. J. Clin. Lab. Invest. *37*, 33–37.
194. Durden, D. A., Davis, B. A. & Boulton, A. A. (1973) Identification and distribution of β-phenylethylamine in the rat. Can. J. Biochem. *51*, 995–1002.
195. Willner, J., LeFevre, H. F. & Costa, E. (1974) Assay by multiple ion detection of phenylethylamine and phenylethanolamine in rat brain. J. Neurochem. *23*, 857–859.
196. Karoum, F., Nasrallah, H., Potkin, S., Chuang, L., Moyer-Schwing, J., Phillips, I. & Wyatt, R. J. (1979) Mass fragmentography of phenylethylamine, m- and p-tyramine and related amines in plasma, cerebrospinal fluid, urine and brain. J. Neurochem. *33*, 201–212.
197. Edwards, D. J., Doshi, P. S. & Hanin, I. (1979) Analysis of phenylethylamines by gas chromatography-chemical ionization mass spectrometry. Anal. Biochem. *96*, 308–316.
198. Suzuki, O. & Hattori, H. (1983) Determination of β-phenylethylamine as its thiocyanate derivative in biological samples by gas chromatography-mass spectrometry. Biomed. Mass Spectrom. *10*, 430–433.
199. Boulton, A. A., Philips,. S. R. & Durden, D. A. (1973) The analysis of certain amines in tissues and body fluids as their dansyl derivatives. J. Chromatogr. *82*, 137–142.
200. Philips, S. R., Durden, D. A. & Boulton, A. A. (1974) Identification and distribution of p-tyramine in the rat. Can. J. Biochem. *52*, 366–373.
201. Philips, S. R., Davis, B. A., Durden, D. A. & Boulton, A. A. (1975) Identification and distribution of m-tyramine in the rat. Can. J. Biochem. *53*, 65–69.
202. Boulton, A. A., Juorio, A. V., Philips, S. R. & Wu, P. H. (1975) Some arylalkylamines in rabbit brain. Brain Res. *96*, 212–216.
203. Philips, S. R., Rozdilsky, B. & Boulton, A. A. (1978) Evidence for the presence of m-tyramine, p-tyramine, tryptamine, and phenylethylamine in the rat brain and several areas of the human brain. Biol. Psychiat. *13*, 51–57.
204. Boulton, A. A. & Baker, G. B. (1975) The subcellular distribution of B-phenylethylamine, p-tyramine and tryptamine in rat brain. J. Neurochem. *25*, 477–481.
205. Slingsby, J. M. & Boulton, A. A. (1976) Separation and quantitation of some urinary arylalkylamines. J. Chromatogr. *123*, 51–56.
206. Boulton, A. A., Dyck, L. E. & Durden, D. A. (1974) Hydroxylation of β-phenylethylamine in the rat. Life Sci. *15*, 1673–1683.
207. Boulton, A. A. & Dyck, L. E. (1974) Biosynthesis and excretion of meta and para tyramine in the rat. Life Sci. *14*, 2497–2506.
208. Durden, D. A. (1978) in: Research Methods in Neurochemistry, Vol. 4 (Marks, N. & Rodnight, R., eds.) pp. 205–250, Plenum Press, New York.
209. Durden, D. A., Juorio, A. V. & Davis, B. A. (1980) Thin layer chromatographic and high-resolution mass spectrometric determination of β-hydroxyphenylethylamines in tissues as their dansyl-acetyl derivatives. Anal. Chem. *52*, 1815–1820.
210. Durden, D. A., Juorio, A. V. & Davis, B. A. (1978) in: Quantitative Mass Spectrometry in the Life Sciences, Vol. 2 (de Leenheer, A. P., Roncucci, R. R. & van Peteghem, C., eds.) pp. 389–397, Elsevier, Amsterdam.
211. Baldessarini, R. (1980) in: Noncatecholic Phenylethylamines Part 2 (Mosnaim, A. D. & Wolf, M. E., eds.) pp. 1–19, Marcel Dekker, New York, Basel.
212. Jones, R. S. G. (1984) in: Neurobiology of the Trace Amines (Boulton, A. A., Baker, G. B., Dewhurst,. W. G. & Sandler, M., eds.) pp. 205–223, Humana Press, Clifton, N. J.
213. Silkaitis, R. P. & Mosnaim, A. D. (1976) Pathways linking l-phenylalanine and 2-phenylethylamine with p-tyramine in rabbit brain. Brain Res. *114*, 105–115.
214. Jonsson, J., Lindeke, B. & Cho, A. K. (1975) Oxidation of phenylethylamine yielding tyramine by rat liver microsomes. Acta Pharmacol. Toxicol. Kbh. *37*, 352–360.

215. Lundblad, B., Lindstedt, S. & Steen, G. (1977) On the enzymic defects in hereditary tyrosinemia. Proc. Natl. Acad. Sci. U.S.A. *74*, 4641–4645.

216. Faraj, B. A., Bethel, R., Ali, F. M., Rudman, D. & Galambos, J. (1980) in: Non-catecholic Phenylethylamines Part 2 (Mosnaim, A. D. & Wolf, M. E., eds.) pp. 81–95, Marcel Dekker, New York, Basel.

217. Sandler, M., Bonham-Carter, S. M. & Walker, P. L. (1984) in: Neurobiology of the Trace Amines (Boulton, A. A., Baker, G. B., Dewhurst, W. G. & Sandler, M., eds.) pp. 487–498, Humana Press, Clifton, N. J.

218. Fischer, J. E. & Baldessarini, R. (1971) False neurotransmitters and hepatic failure. Lancet *ii*, 75–79.

219. Lam, K. C., Tall, A. R., Goldstein, G. B. & Mistilis, S. P. (1973) Role of a false neurotransmitter, octopamine, in the pathogenesis of hepatic and renal failure. Scand. J. Gastroenterol. *8*, 465–472.

220. Manghani, K. K., Lunzer, M. R, Billing, B. H. & Sherlock, S. (1975) Urinary and serum octopamine in patients with portal-systemic encephalopathy. Lancet *i*, 943–946.

221. Capocaccia, L., Cangiano, C., Attili, A. F., Angelico, M., Cascino, A. & Rossi-Fanelli, F. (1977) Octopamine and ammonia plasma levels in hepatic encephalopathy. Clin. Chim. Acta *75*, 99–105.

222. Molinoff, P. B., Landsberg, L. & Axelrod, J. (1969) An enzymatic assay for octopamine and other β-hydroxylated phenylethylamines. J. Pharmacol. Exp. Ther. *170*, 253–261.

223. Duffield, P. H., Dougan, D. F. H., Wade, D. N. & Duffield, A. M. (1981) A chemical ionization gas chromatographic mass spectrometric assay for octopamine and tyramine in rat brain. Biomed. Mass Spectrom. *8*, 170–173.

224. Ibrahim, K. E., Couch, M. W., Williams, C. M., Budd, M. B., Yost, R. A. & Midgley, J. M. (1984) Quantitative measurement of octopamines and synephrines in urine using capillary negative chemical ionization mass spectrometry. Anal. Chem. *56*, 1695–1699.

225. Cohen, S. S. (1978) What do the polyamines do? Nature *274*, 209–210.

226. Russell, D. H. (1971) Increased polyamine concentrations in the urine of human cancer patients. Nature New Biol. *233*, 144–145.

227. Heby, O. & Andersson, G. (1978) Tumor cell death: the probable cause of increased polyamine levels in physiological fluids. Acta Pathol. Microbiol. Scand. Sect. A *86*, 17–20.

228. Durie, B. G. M., Salmon, S. E. & Russell, D. H. (1977) Polyamines as markers of response to disease activity in cancer chemotherapy. Cancer Res. *37*, 214–221.

229. Bakowski, M. T., Toseland, P. A., Wicks, J. F. C. & Trounce, J. R. (1981) A rapid gas chromatographic method for the determination of plasma polyamine and its application to the prediction of tumour response to chemotherapy. Clin. Chim. Acta *110*, 273–286.

330. Walle, T. (1973) in: Polyamines in Normal and Neoplastic Growth (Russell, D. H., ed.) pp. 355–365, Raven Press, New York.

231. Smith, R. G. & Daves, G. D., Jr. (1977) Gas chromatography-mass spectrometry analysis of polyamines using deuterated analogs as internal standards. Biomed. Mass Spectrom. *4*, 146–151.

232. Guimanini, A. G., Chiavari, G. & Scarponi, F. L. (1976) N-Permethylation of polyamines for gas chromatographic and mass spectrometric analysis. Anal. Chem. *48*, 484–488.

233. Muskiet, F. A., Stratingh, C. M. & Fremouw-Ottevangers, C. M. (1982) Mass spectrometric identification of isoputreanine, a metabolite of spermidine and/or spermine in human urine. J. Chromatogr. *230*, 142–147,

234. Gleich, G. J. & Hull, W. M. (1980) Measurement of histamine: a quality control study. J. Allergy Clin. Immunol. *66*, 295–298.

235. Roberts, L. J. II (1984) Pentafluorobenzyl derivative of histamine for determination by gas chromatography negative ion chemical ionization mass spectrometry. J. Chromatogr. *287*, 155–160.

236. Roberts, L. J. II & Oates, J. A. (1984) Accurate and efficient method for quantitation of urinary histamine by gas chromatography negative ion chemical ionization mass spectrometry. Anal. Biochem. *136*, 258–263.

237. Mita, H., Yasueda, H. & Shida, T. (1980) Quantitative analysis of histamine in biological samples by gas chromatography-mass spectrometry. J. Chromatogr. *181*, 153–159.

238. Mita, H., Yasueda, H. & Shida, T. (1980) Simultaneous determination of histamine and N-methylhistamine in human plasma urine by gas chromatography-mass spectrometry. J. Chromatogr. *221*, 1–7.

239. Keyzer, J. J., Wolthers, B. G., Muskiet, F. A. J., Kauffman, H. F. & Groen, A. (1981) Determination of N-methylhistamine in plasma and urine by isotope dilution mass fragmentography. Clin. Chim. Acta *113*, 165–173.

240. Hough, L. B. & Domino, E. F. (1979) *Tele*-methylhistamine in rat brain. J. Neurochem. *32*, 1865–1866.

241. Duggan, D. E., Hooke, K. F. & Ramjit, H. G. (1984) Sensitive gas-liquid chromatography procedure for urinary N-methylimidazole acetic acid, an index of histamine turnover. J. Chromatogr. *306*, 69–74.

242. Blau, K. (1961) Chromatographic methods for the study of amines from biological material. Biochem. J. *80*, 193–200.

243. Seiler, N. & Schneider, H. H. (1974) Estimation of picomole quantities of piperidine in tissues. Biomed. Mass Spectrom. *1*, 381–385.

244. Stepita-Klauco, M., Dalezalova, H. & Fairweather, R. (1974) Piperidine increase in the brain of dormant mice. Science *183*, 536–537.

245. Drucker-Colin, R. R. & Giacobini, E. (1975) Sleep-inducing effects of piperidine. Brain Res. *88*, 186–189.

246. Perry, T. L., Hansen, S., Foulks, J. G. & Ling, G. M. (1965) Aliphatic and aromatic amines of cat brain. J. Neurochem. *12*, 397–405.

247. Spanner, S. & Ansell, G. B. (1978) The determination of free ethanolamine in brain tissue and its release on incubation. J. Neurochem. *30*, 497–498.

248. Baba, S., Watanabe, Y., Gejyo, F. & Arakawa, M. (1984) High-performance liquid chromatographic determination of serum aliphatic amines in chronic renal failure. Clin. Chim. Acta *136*, 49–56.

249. Honegger, C. R. & Honegger, R. (1960) Volatile amines in brain. Nature *185*, 530–532.

250. Okano, Y., Miyata, T., Iwasaki, K., Takahama, K., Hitoshi, T., Kase, Y., Matsumoto, I. & Shinka, T. (1981) Application of selected ion monitoring of picomole levels of pyrrolidine in the brain. Anal. Biochem. *115*, 254–259.

251. Spellacy. E., Watts, R. W. E. & Goolamali, S. K. (1980) Trimethylaminuria. J. Inher. Metab. Dis. *2*, 83–88.

252. King, G. S., Pettit, B. R. & Leonard, J. V. (1984) The analysis of trimethylamine in body fluids by a direct medium resolution mass spectrometric method. Biomed. Mass Spectrom. *11*, 1–3.

Chapter 4
Carbohydrates

J. P. Kamerling and J. F. G. Vliegenthart

1 Introduction

Structural analysis of carbohydrates is highly complicated. In nature a series of monosaccharides occur, both in free form and as part of linear or branched (poly)saccharides. They are also constituents of e. g. lipopolysaccharides, proteoglycans, glycoproteins and glycolipids.

The primary structure of carbohydrate chains is defined by several parameters: 1. nature and number of the constituent monosaccharides; 2. sequence and ring size of the monosaccharides and 3. type and anomeric configuration of the glycosidic linkages (Figure 1). In the case of proteoglycans and glycoproteins two additional parameters have to be determined, namely the type of the carbohydrate-peptide linkage and the nature and position in the polypeptide backbone of the amino acids involved (Figure 2). For the determination of these parameters mass spectrometry (MS) frequently combined with gas-liquid chromatography (GC) plays a key role. Several comprehensive reviews on MS and GC of carbohydrates have already appeared (1 – 8). For this reason the present chapter is focused on those aspects of the use of GC-MS which are of relevance in (clinical) biochemistry, biomedicine and biology.

Carbohydrates are of importance in several contexts e. g. in hereditary metabolic diseases, recognition phenomena on the cellular level, cancer and immunological features. Saccharides isolated from physiological fluids, tissues, bacteria, etc., are subjected to structural analysis in several ways. Most of the oligo- and polysaccharides and the glycolipids can be analysed directly. However, in general it is not possible to study the structures of high molecular weight carbohydrate chains directly in intact lipopoly-saccharides, proteoglycans and glycoproteins. In lipopolysaccharides the lipid A part has to split off; if present the *O*-antigen and core oligosaccharide are also split (9). Proteoglycans and glycoproteins often contain more than one carbohydrate chain attached to separate amino acids in the polypeptide backbone. Frequently the chains show (micro-)heterogeneity. The presence of the polypeptide backbone as such can interfere with the analysis of carbohydrates. To obtain suitable products for the analysis of the carbohydrate chains of these glycoconjugates a number of degradation methods

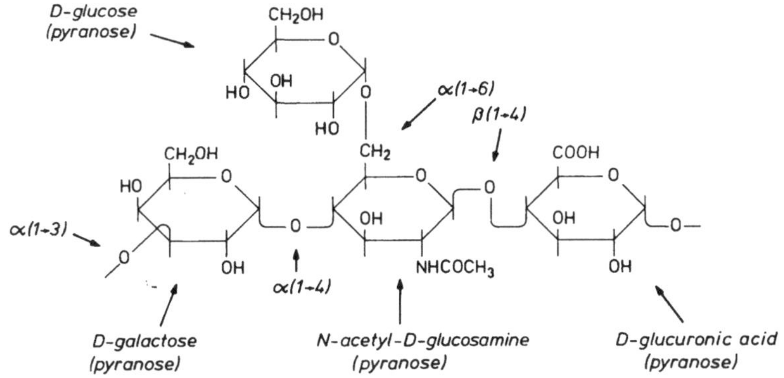

Figure 1. Schematic presentation of the parameters, which define the primary structure of a carbohydrate chain.

are commonly applied: 1. degradation of glycoproteins by exhaustive protease (pronase) digestion to glycopeptides having only a few amino acids; 2. cleavage of the *N*-acetylgalactosamine-serine/threonine linkage and formation of oligosaccharide-alditols using alkaline borohydride (*β*-elimination reaction); 3. cleavage of the *N*-acetylglucosamine-asparagine linkage by hydrazinolysis followed by *N*-acetylation and reduction, yielding oligosaccharide-alditols; 4. cleavage by endoglycosidases like endo-*β*-*N*-acetylglucosaminidases and endo-*α*-*N*-acetylgalactosaminidase or by peptide-N[4]-(*N*-acetyl-*β*-glucosaminyl)-asparagine amidases, releasing oligosaccharides. For the preparation of carbohydrate oligomers, polymeric structures are frequently subjected to further chemical degradations such as periodate oxidation, partial acetolysis, partial hydrolysis or hydrazinolysis/nitrous acid deamination, and to enzymic degradations with exo- or endo-glycosidases (10,11).

In the following, experimental procedures and mass spectrometric data in relation to sugar analysis and methylation analysis, and MS of oligosaccharides, oligosaccharide-alditols, glycopeptides and glycolipids will be presented.

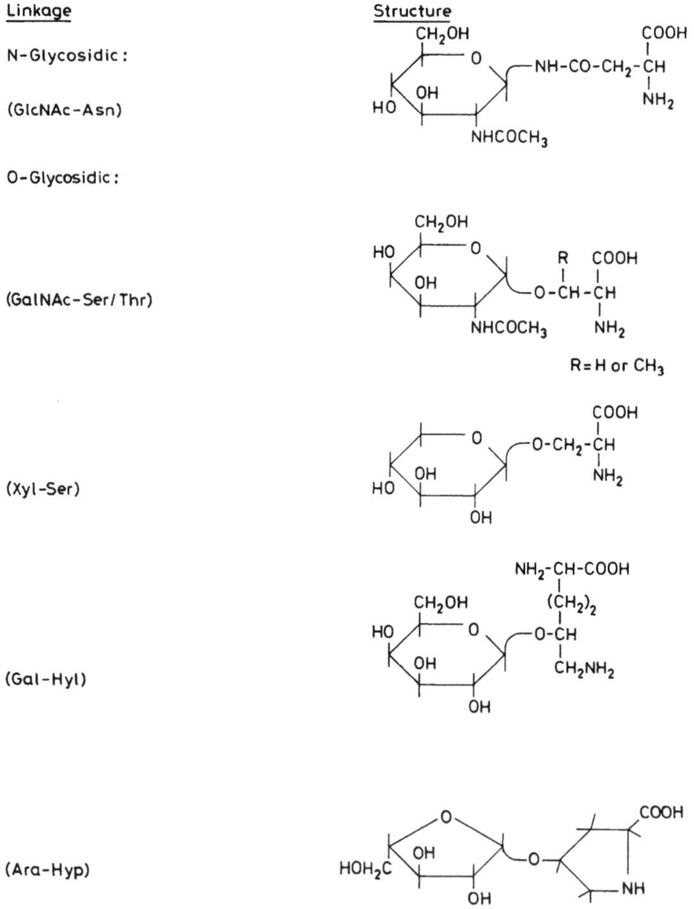

Figure 2. Major types of carbohydrate-peptide linkages in glycoproteins.

2 Sugar Analysis

The approaches to the determination of the sugar composition of complex carbohydrates are mainly based on GC-(MS) procedures. Sugar analysis of oligosaccharides, oligosaccharide-alditols, polysaccharides, glycoproteins, glycopeptides, glycolipids and proteoglycans is currently established after hydrolysis or methanolysis. Monosaccharides, occurring as such in biological sources, can be analysed directly or after treatment with methanolic HCl. To obtain volatile derivatives, many derivatisation procedures have been reported (4).

The hydrolysis procedure involves cleavage of the glycosidic linkages by mineral acid and, after conversion into suitable derivatives, analysis of the resulting monosaccharide mixture by GC-(MS). The conditions for complete hydrolysis of all glycosidic bonds with minimal destruction of the constituents have to be determined experimentally in view of the different stabilities of the monomers. For many polysaccharides the use of 0.25 $mol \cdot l^{-1}$ H_2SO_4 for 17 h at 100 °C seems to be a good compromise (12). Glycoproteins have been hydrolysed using 4 $mol \cdot l^{-1}$ HCl for 4–6 h at 100 °C (13). However, for quantification of sialic acid, very mild acidic conditions have to be applied (14). The hydrolysis of glycosidic bonds of 2-acetamido sugars and uronic acids can give rise to severe problems due to incomplete splitting. Acetamido sugar-containing polysaccharides can be hydrolysed in very good yields with anhydrous HF (2 h, room temperature) (15). Also 0.5 $mol \cdot l^{-1}$ trifluoroacetic acid (19 h, 100 °C) and 4 $mol \cdot l^{-1}$ trifluoroacetic acid (1 h, 125 °C /4 h, 100 °C) have been shown to give good results (16, 17). Ketose-containing carbohydrates have to be hydrolysed under extremely mild conditions (18).

Before analysis by GC-(MS) neutral monosaccharides and amino sugars are most commonly converted into the corresponding alditol acetates (4, 6, 13, 19–23). Other applied derivatives are i. a. trimethylsilylated alditols (4, 6, 24), trifluoroacetylated alditols (4, 6, 25), aldononitrile acetates (4, 6, 7, 26–29) and oxime derivatives (4, 6, 7, 17, 28, 30–35). Also direct derivatisations such as pertrimethylsilylation, peracetylation and pertrifluoroacetylation are sometimes used (3, 4, 6). *N,O*-acylneuraminic acids can be analysed by GC(-MS) after trimethylsilylation of the free acids or methyl esters (36).

Methanolysis, usually carried out with methanolic 1 $mol \cdot l^{-1}$ HCl for 24 h at 85 °C, is applied to the simultaneous analysis of neutral monosaccharides, acetamido sugars, uronic acids, octulosonic acids and sialic acids (37). Carboxyl groups of the acidic sugars are blocked by methyl ester formation. Owing to anomerisation, each monosaccharide gives rise to a characteristic GC peak pattern of methyl glycosides, facilitating the identification of a sugar. This procedure causes less destruction of the monosaccharides than hydrolysis, but it is very effective in cleaving glycosidic linkages. The mixture of methyl glycosides of the different types of sugars can be determined by GC(-MS) after *N*-(re)acetylation/trimethylsilylation (4, 36–49) or trifluoroacetylation (4, 50–54). It has to be noted that in glycolipid analysis the ceramide residues are degraded to sphingosine and fatty acid methyl esters. Similar procedures based on the glycosidation with (–)-2-butanol or (+)-2-octanol are in use for the GC(-MS) determination of the absolute configuration of monosaccharides (55–57).

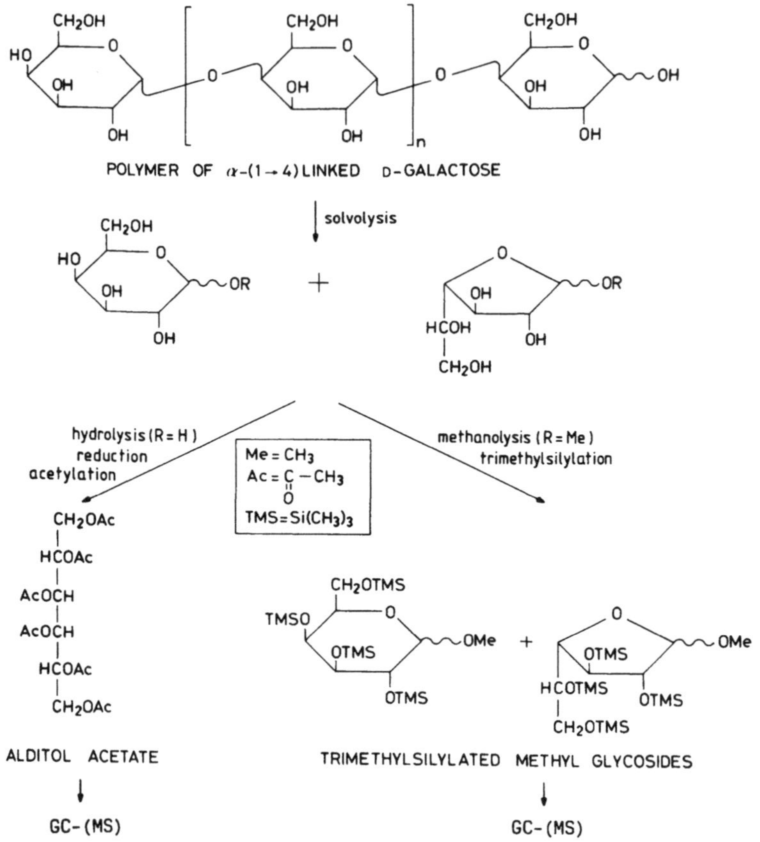

Figure 3. Schematic hydrolysis and methanolysis procedures for the sugar analysis.

In Figure 3 the sugar analysis procedures based on the preparation of alditol acetates and of trimethylsilylated methyl glycosides (methyl esters) are presented schematically. These procedures will be discussed in more detail, including the mass spectrometric characteristics of both types of derivatives. The same holds for the already mentioned analysis of *N,O*-acylneuraminic acids and the absolute configuration determinations.

2.1 Hydrolysis procedure/alditol acetates

In an ampoule the carbohydrate material (0.5 mg) is dissolved in 0.2 ml 4 $mol \cdot l^{-1}$ trifluoroacetic acid and heated for 4 h at 100 °C. Any alditol which does not interfere with the analysis of the constituting monosaccharides can be used as internal standard. The solution is evaporated with a stream of nitrogen and the residue washed twice with methanol followed by evaporation. The residue is taken up in 0.2 ml water and treated with 5 mg $NaBH_4$ for 2 h at room temperature. After decomposition of the excess of $NaBH_4$ with a few drops of 20% acetic acid (to pH 4), the solution is

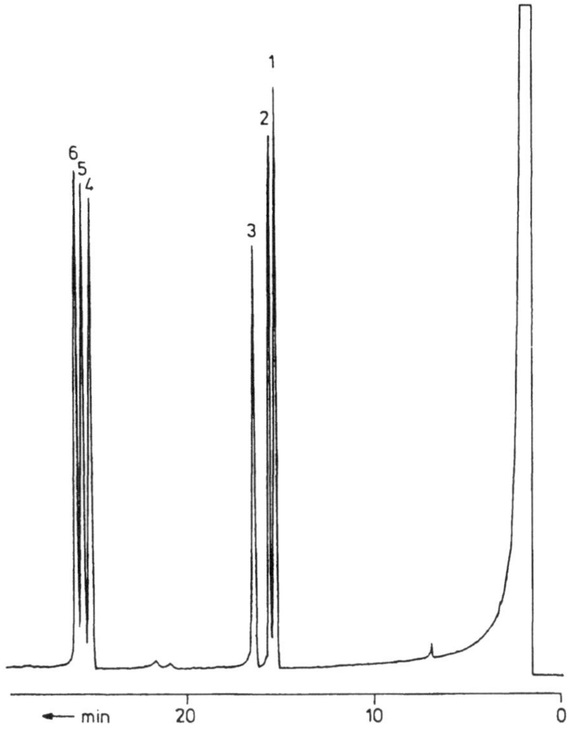

Figure 4. Gas chromatogram of alditol acetates on a CP-Sil 5 WCOT fused-silica capillary column (25 m × 0.32 mm). Oven temperature program: 130 °C to 220 °C at 2 °C · min^{-1}; 1 min at 220 °C. The peaks are numbered in their order of elution and are assigned as follows: (1) rhamnose; (2) fucose; (3) xylose; (4) mannose; (5) glucose and (6) galactose.

evaporated *in vacuo*. Boric acid is removed as trimethylborate by co-evaporation with 0.2 ml 1% methanolic acetic acid (five times). The mixture of alditols is acetylated with 0.2 ml acetic anhydride/pyridine (1:1) for 1 h at 100 °C. Finally the solution is evaporated with a stream of nitrogen in the presence of toluene. For analysis by GC (-MS) the residue is dissolved in 0.1 ml methylene chloride.

The sugar analysis can be performed on a CP-Sil 5 WCOT fused-silica capillary column (25 m × 0.32 mm) using flame-ionisation detection (58) and the following conditions: carrier gas nitrogen flow-rate 1.5 ml · min^{-1} and the make-up gas nitrogen flow-rate 35 ml · min^{-1}; split-ratio of 1:10; injection-port temperature 210 °C, and the detector temperature 230 °C; oven temperature programmed from 130 °C to 220 °C at 2 °C · min^{-1} and kept isothermally at 220 °C for 1 min. A typical gas chromatogram of a mixture of neutral monosaccharides is presented in Figure 4.

For the alditol acetate analysis several other types of stationary phases have been recommended, e. g. SP-1000 (12), Silar 10C (13), OV-1 (21), SE-54 (21), Carbowax 20M (21), SP-2330 (23), Dexsil 410 (59), SE-30 (59), OV-101 (59), and OV-275 (60). The more polar column materials are not suitable for the analysis of the amino sugar alditols.

2.2 Mass spectrometry of alditol acetates

The EI fragmentation patterns of alditol acetates have been extensively described in the literature (1–3, 6, 61–64). Some basic information is presented in Scheme 1. In Figures 5–8 the EI mass spectra of a series of acetylated alditols obtained from different classes of monosaccharides are depicted: xylitol-[1-^{2}H], glucitol, fucitol-[1-^{2}H] and 2-acetamido-2-deoxy-glucitol-[1-^{2}H]. Under the usual EI mass spectrometric conditions, it is almost impossible to discriminate between stereochemical differences within one class of alditol acetates like pentitols, hexitols, heptitols, etc. Combination of the mass spectrometric data with the GC retention times enhances the reliability of the assignments. However, it has to be noted that owing to symmetry properties two different aldoses can yield the same alditol upon reduction (e. g. glucose and gulose or arabinose and lyxose), whereas ketoses are converted into two epimeric alditol structures. To differentiate between alditols obtained from aldoses and ketoses, the reduction of these monosaccharides can be carried out with NaB^2H_4 leading to deuterium-labeled C1 or C2 atoms, respectively (see also Figs. 5, 7 and 8).

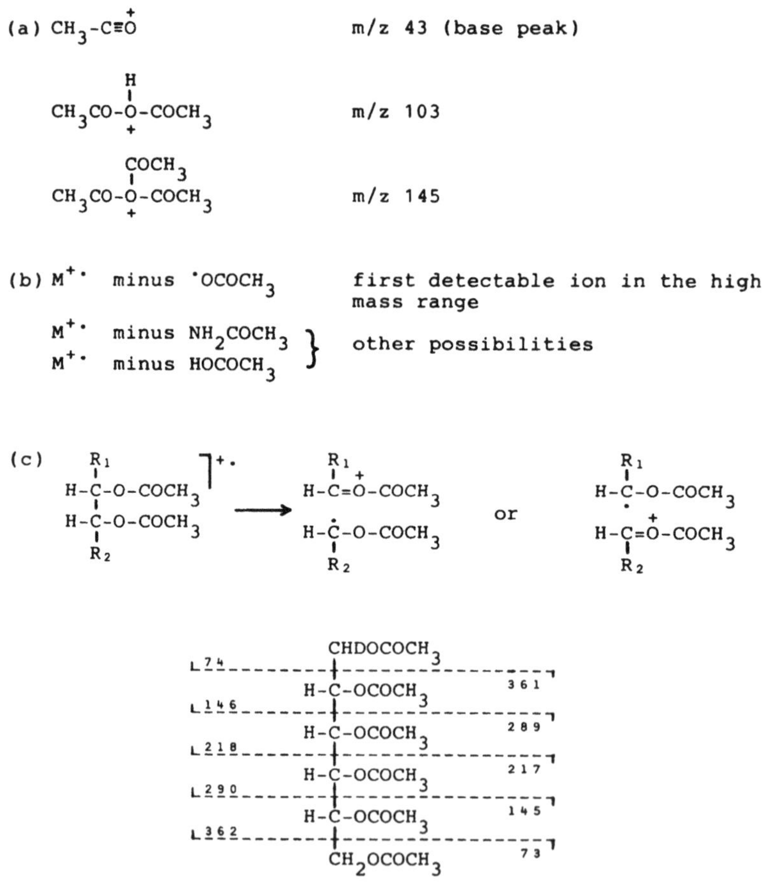

Scheme 1. General information on alditol acetate EI fragmentation (a–f).

(d)

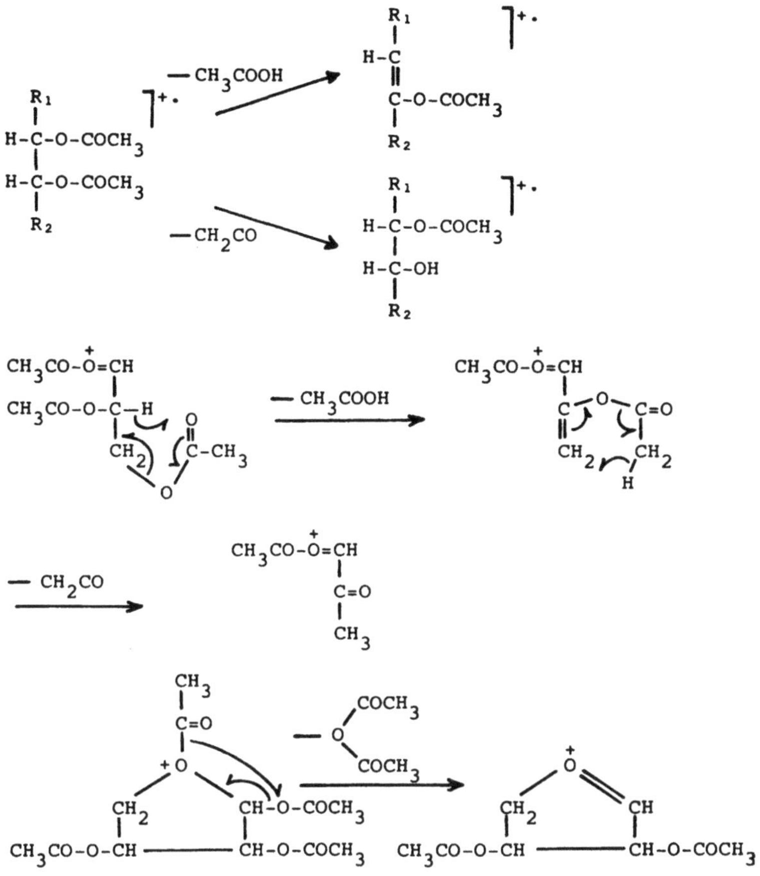

(cyclic form of a primary fragment)

(e)

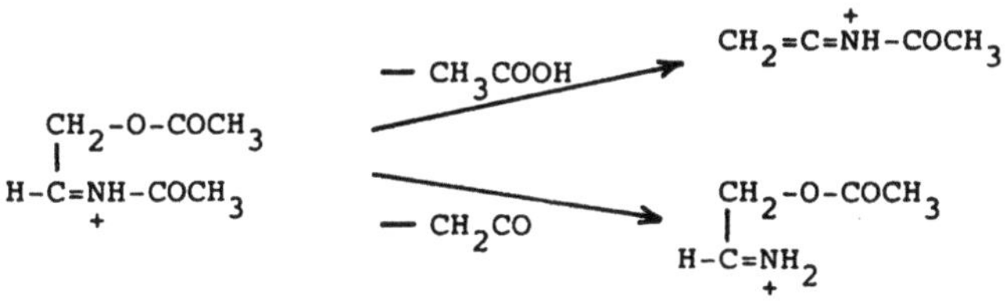

(f)

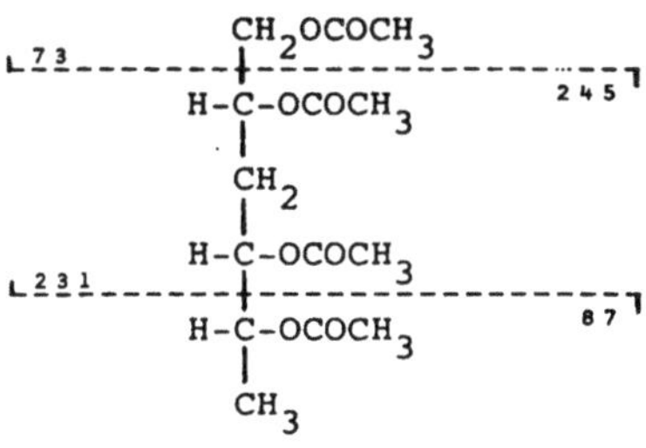

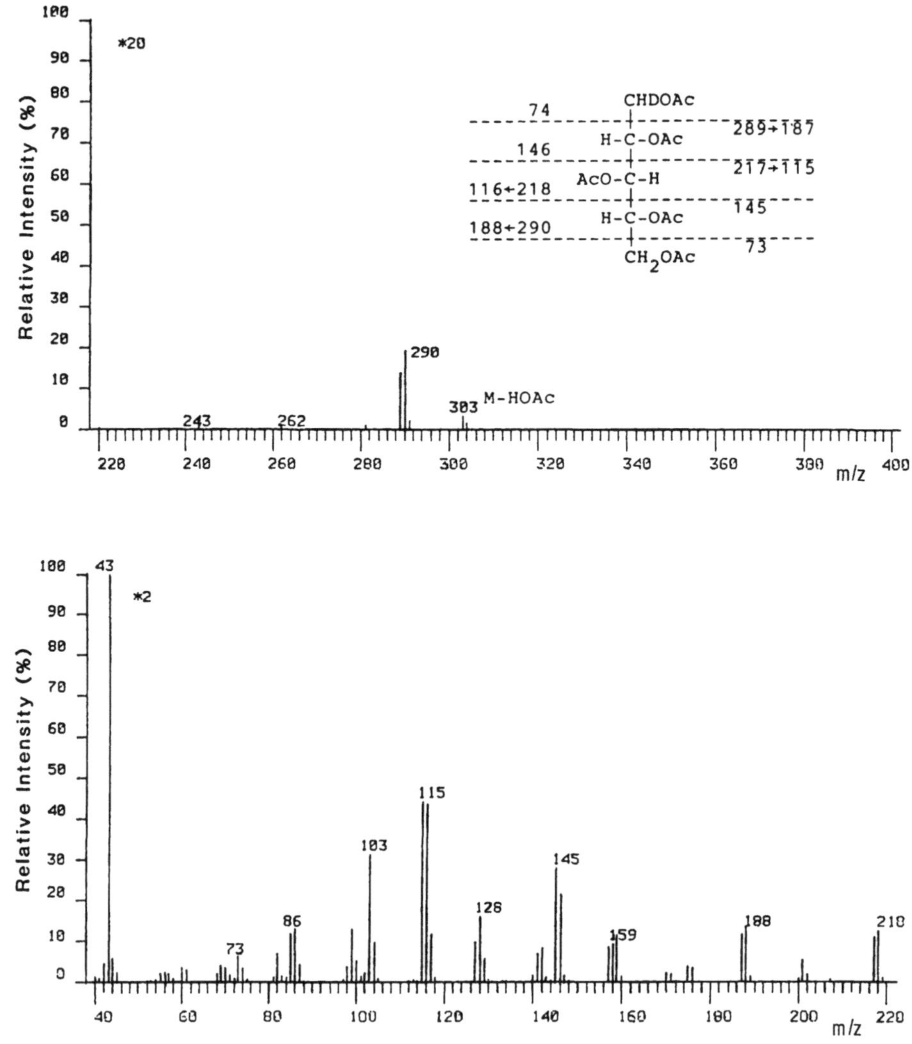

Figure 5. 70 eV EI mass spectrum of peracetylated xylitol-[1-^{2}H], as an example of a pentitol acetate.

Molecular ions are rarely observed in EI-MS. The first detectable fragment ion in the high-mass-region corresponds to the elimination of an acetoxyl radical or acetic acid from the molecular ion. For acetamido-alditols the elimination of acetamide cannot be excluded without high-resolution exact mass measurements (Fig. 8). As indicated in Scheme 1 and Figures 5–8, several primary fragmentations do occur, either by elimination of an acetoxyl group, e. g. as acetic acid, or by cleavage of the alditol chain. The primary fragments are degraded by elimination(s) of acetic acid, ketene, acetic anhydride or acetamide. The presence of methylene groups (deoxy functions) in the alditol chain as well as acetamido and methoxyl substituents have a

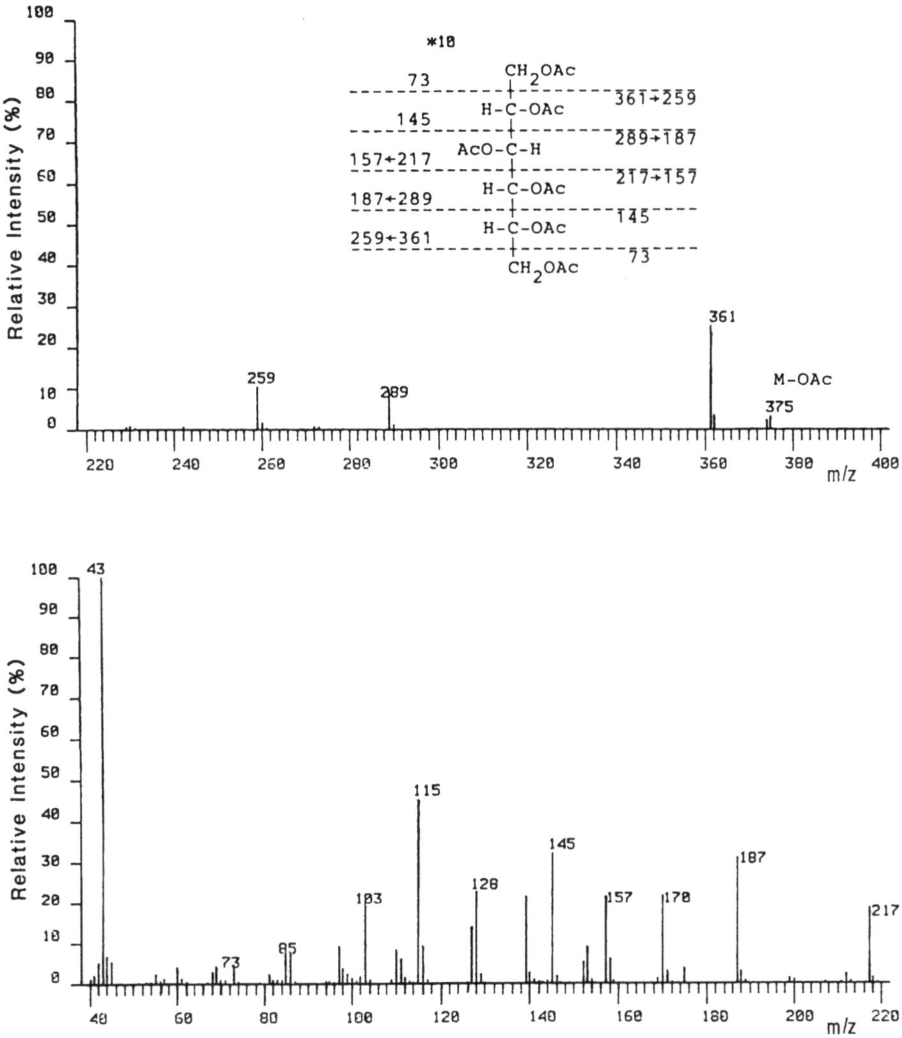

Figure 6. 70 eV EI mass spectrum of peracetylated glucitol, as an example of a hexitol acetate.

significant influence on the peak patterns of the mass spectra. The alditol chain of an acetylated deoxyalditol is little cleaved next to the methylene group. The occurrence of acetamido or methoxyl functions give rise to a high preference of cleavage next to the carbon atoms at which the specific functions are bound (see also methylation analysis section).

The use of NaB^2H_4 has already been mentioned above for discrimination purposes. When the carbohydrate chain contains *O*-methylated monosaccharides (12), it is also advisable to carry out the reduction with NaB^2H_4, making assignment of the C1 atom possible. Reduction of uronic acids, carried out with NaB^2H_4 in the presence of a

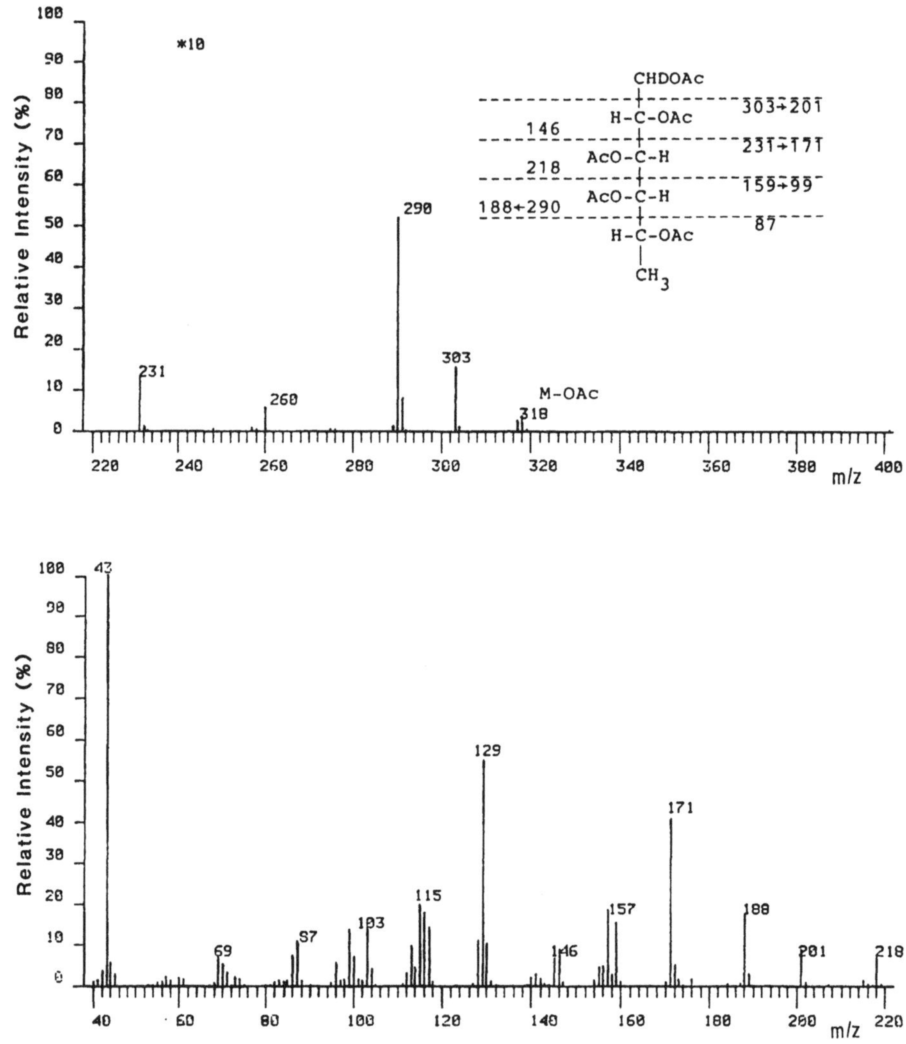

Figure 7. 70 eV EI mass spectrum of peracetylated fucitol-[1-^{2}H], as an example of a 6-deoxyhexitol acetate.

carbodiimide before hydrolysis (65, 66), yields an the alditol acetate analysis products with two deuterium atoms at C6. For the identification of *N*-acetyl groups of amino sugars and the determination of the degree of *N*-deacetylation in these sugars during acid hydrolysis, the acetylation should be carried out with trideuteroacetic anhydride (67).

Biological fluids contain several free alditols. To analyse these alditols by GC(-MS) use has been made of the corresponding peracetylated derivatives. A selected ion monitoring assay has been reported for the analysis of peracetylated mannitol, gal-

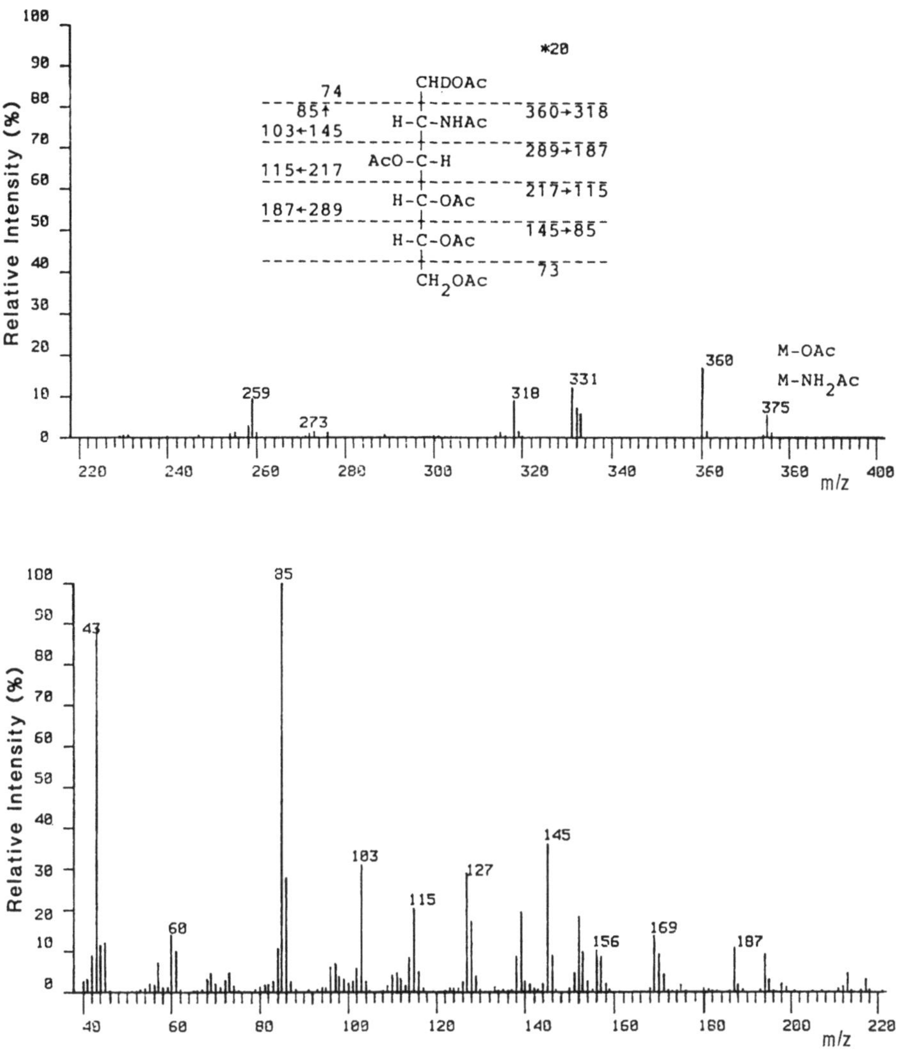

Figure 8. 70 eV EI mass spectrum of peracetylated 2-acetamido-2-deoxy-glucitol-[1-^{2}H], as an example of a *N*-acetyl-aminohexitol acetate.

actitol and glucitol (*m/z* 217 and *m/z* 259) as well as inositol (*m/z* 210) in serum, in urine and in amniotic fluid from pregnant women, applying iditol as an internal standard (68). A similar approach has been described for the determination of glucitol (*m/z* 361) in erythrocytes of diabetic and healthy subjects (58). For the determination of galactitol in amniotic fluid a selected ion monitoring method has been developed, utilizing GC-CI(NH_3)MS with galactitol-[6,6'-$^2H_2$] as internal standard (*m/z* 452 / *m/z* 454 ≡ $[M + NH_4]^+$) (69).

2.3 Methanolysis procedure/trimethylsilylated methyl glycosides

In an ampoule the carbohydrate material (0.1 – 2 mg) is mixed with a mannitol solution (internal standard; 10 – 100 nmol). After lyophilisation and drying over P_2O_5 in a vacuum desiccator for 18 h, the residue is dissolved in 1.0 $mol \cdot l^{-1}$ methanolic HCl (0.5 ml). Nitrogen is bubbled through the solution for 30 s, and then the ampoule is sealed. The solution is heated for 24 h at 85 °C; subsequently, neutralisation is carried out by addition of solid silver carbonate (pH-paper). *N*-(Re)acetylation is performed by the addition of acetic anhydride (10 – 50 µl). After mixing, the resulting suspension is kept at room temperature for 24 h in the dark. The precipitate is then triturated thoroughly and after centrifugation, the supernatant is collected. The residue of silver salts is washed twice with 0.5 ml dry methanol.

The pooled supernatants are evaporated under reduced pressure at 35 °C. In the case of glycolipids, fatty acid methyl esters can be removed by hexane extraction of the residue. The final residue is dried for 12 h in a vacuum desiccator over P_2O_5. Before GC-analysis, the sample is trimethylsilylated with a mixture of pyridine-hexamethyldisilazane-chlorotrimethylsilane (5:1:1; 100 µl) for 30 min at room temperature.

The quantitative sugar analysis is carried out on a CP-Sil 5 WCOT fused-silica capillary column (25 m × 0.32 mm) using flame-ionization detection with carrier gas nitrogen flow-rate 1.5 $ml \cdot min^{-1}$, make-up gas nitrogen flow-rate 35 $ml \cdot min^{-1}$, split-ratio of 1:10, injection-port temperature 210 °C, and the detector temperature 230 °C. The oven temperature is programmed from 130 °C to 220 °C at 2 $°C \cdot min^{-1}$ and kept isothermally at 220 °C for 1 min. A typical gas chromatogram of a standard mixture is presented in Figure 9.

Notes:

1. Methanolic HCl is prepared by bubbling HCl gas (obtained by mixing conc. HCl with conc. H_2SO_4) through cooled dry methanol. After determination of the substance concentration by titration, the solution is diluted with dry methanol until 1.0 $mol \cdot l^{-1}$ HCl, and stored in a desiccator at – 18 °C.
2. The amount of added internal standard depends on the carbohydrate-content of the material analysed.
3. Under the applied conditions of methanolysis, the *N*-acyl group in amino sugars is cleaved nearly completely, giving rise to additional peaks in the neutral hexose region of the gas chromatogram. Therefore a *N*-(re)acetylation step is incorporated in the procedure. However, when too much acetic anhydride is added, one of the primary hydroxyl functions of mannitol is *O*-acetylated giving rise to an additional small peak in the gas chromatogram. The same holds for other alditols and for the primary hydroxyl function of the methyl ester methyl glycoside of *N*-acetylneuraminic acid.
4. The internal standard used should not occur naturally in the material being investigated. In this connexion it has to be noted that human urine and other physiological fluids contain mannitol (68 – 70).
5. Molar adjustment factors of monosaccharides, except *N*-acetylneuraminic acid, are determined by application of the methanolysis procedure on standard mixtures of free sugars and internal standard. For *N*-acetylneuraminic acid the molar adjustment factor is determined by subjecting a known sialo-oligosaccharide to methanolysis.

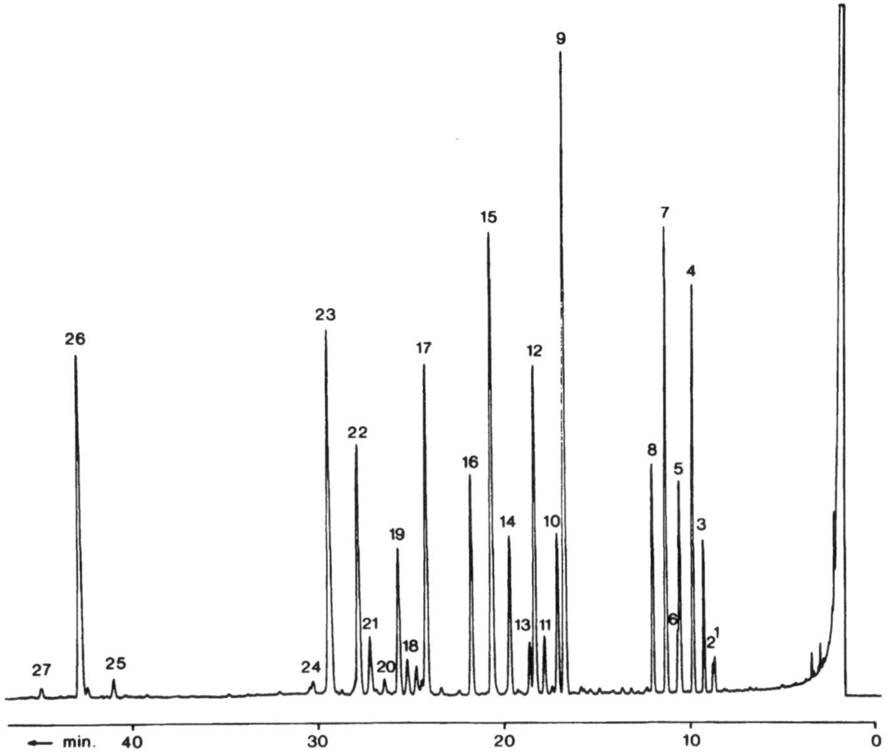

Figure 9. Gas chromatogram of trimethylsilylated (methyl ester) methyl glycosides on a CP-Sil 5 WCOT fused-silica capillary column (25 m × 0.32 mm). Oven temperature program: 130 °C to 220 °C at 2 °C · min^{-1}; 1 min at 220 °C. The peaks are numbered in their order of elution and are assigned as follows: (1) xylose (*β-f*); (2) xylose (*α-f*); (3) fucose (*β-f*); (4) fucose (*α-p*); (5) fucose (*β-p*); (6) fucose (*α-f*); (7) xylose (*α-p*); (8) xylose (*β-p*); (9) mannose (*α-p*); (10) galactose (*β-f*); (11) mannose (*β-p*); (12) galactose (*α-p*); (13) galactose (*α-f*); (14) galactose (*β-p*); (15) glucose (*α-p*); (16) glucose (*β-p*); (17) mannitol (internal standard); (18) *N*-acetylglucosamine (*α-f*); (19) *N*-acetylgalactosamine (*α,β-f*); (20) mono-*O*-acetyl-mannitol; (21) *N*-acetylglucosamine (*β-p*); (22) *N*-acetylgalactosamine (*α,β-p*); (23) *N*-acetylglucosamine (*α-p*); (24) *N*-acetylglucosamine (*α,β-p*; no methyl glycoside); (25) *N*-acetylneuraminic acid (*α*); (26) *N*-acetylneuraminic acid (*β*); (27) 9-*O*-acetyl-*N*-acetylneuraminic acid. *f*, furanoside; *p*, pyranoside.

6. The sugar analysis of *N*-glycosidically bound carbohydrate chains of glycoproteins and glycopeptides (GlcNAc-Asn type) has shown that, under the usual conditions, the linkage between *N*-acetylglucosamine (GlcNAc) and asparagine (Asn) is split only to a very limited extent and mainly the free monosaccharide is liberated instead of its methyl glycoside. This has to be taken into account when calculating molar ratios.
7. Under the applied conditions of methanolysis several alditols give rise to anhydro derivatives, e. g. xylitol (23%), arabinitol (5%), fucitol (10%), glucitol (20%), galactitol (14%), 2-acetamido-2-deoxy-galactitol (35%) and the epimeric alditols

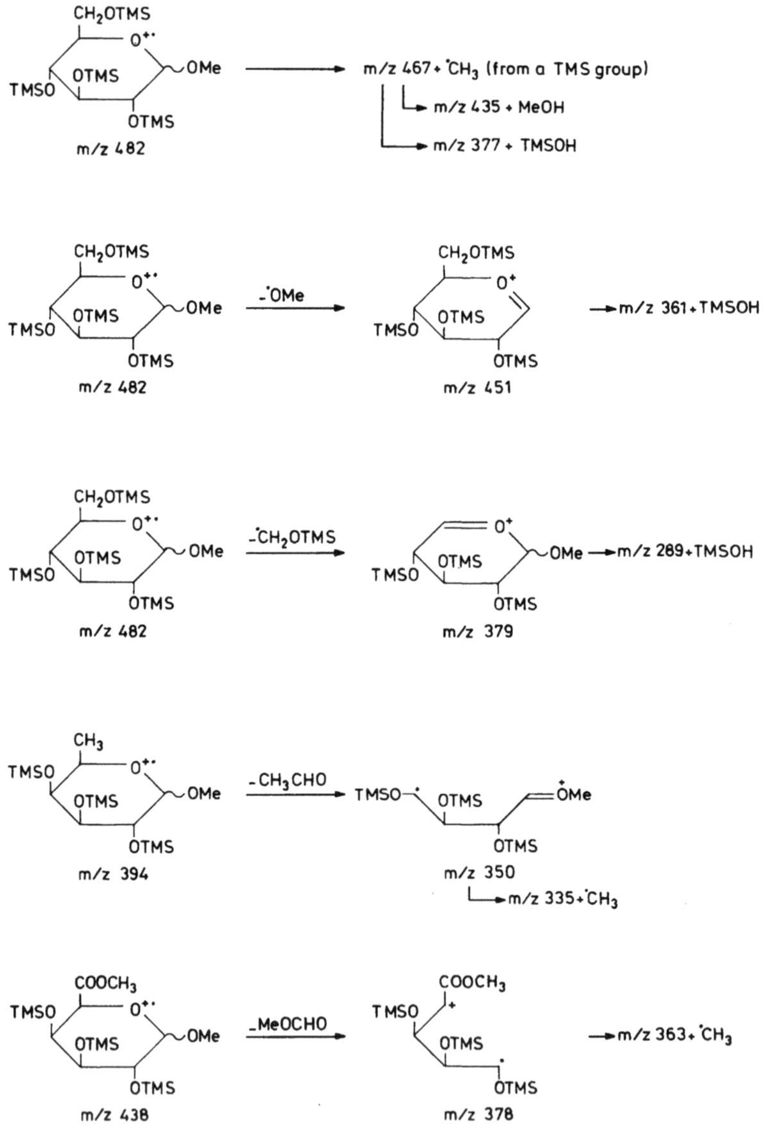

Scheme 2. General information on the fragmentation of trimethylsilylated methyl glycosides.

$CH_2O\overset{+}{Si}(CH_3)_2$
|
$CH{=}O$

m/z 117

$CH{=}\overset{+}{O}TMS$
|
CH_3

m/z 117

$TMSO{-}\overset{+}{Si}(CH_3)_2$

m/z 147

$CH_2{=}\overset{+}{O}TMS$

m/z 103
(is not specific for the presence of a primary hydroxyl function)

$CH_3CO\overset{+}{N}H{=}CH{-}\dot{C}HOTMS$

m/z 173

$CH_3CO\overset{+}{N}H{=}CH{-}CH{=}CHOTMS$

m/z 186

$TMSO\dot{C}H{-}CH{=}\overset{+}{O}TMS$

m/z 204
(major C2-C3, minor C3-C4 in pyranose ringforms)

$TMSOCH_2{-}CH{=}\overset{+}{O}TMS$

m/z 205
(C5-C6 fragment in furanose ringforms)

$TMSOCH{=}CH{-}CH{=}\overset{+}{O}TMS$

m/z 217
(mainly C2-C3-C4 in pyranose ringforms)

$TMSOCH{=}C(OTMS){-}CH{=}\overset{+}{O}TMS$

m/z 305
(mainly C2-C3-C4 in pyranose ringforms)

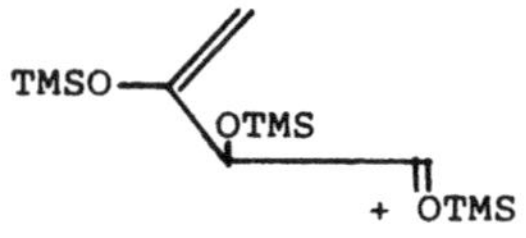

m/z 319
(major contribution in pyranose ringforms)

$TMSOCH{=}CH{-}CH(OTMS){-}CH{=}\overset{+}{O}TMS$

m/z 319
(furanose ringforms)

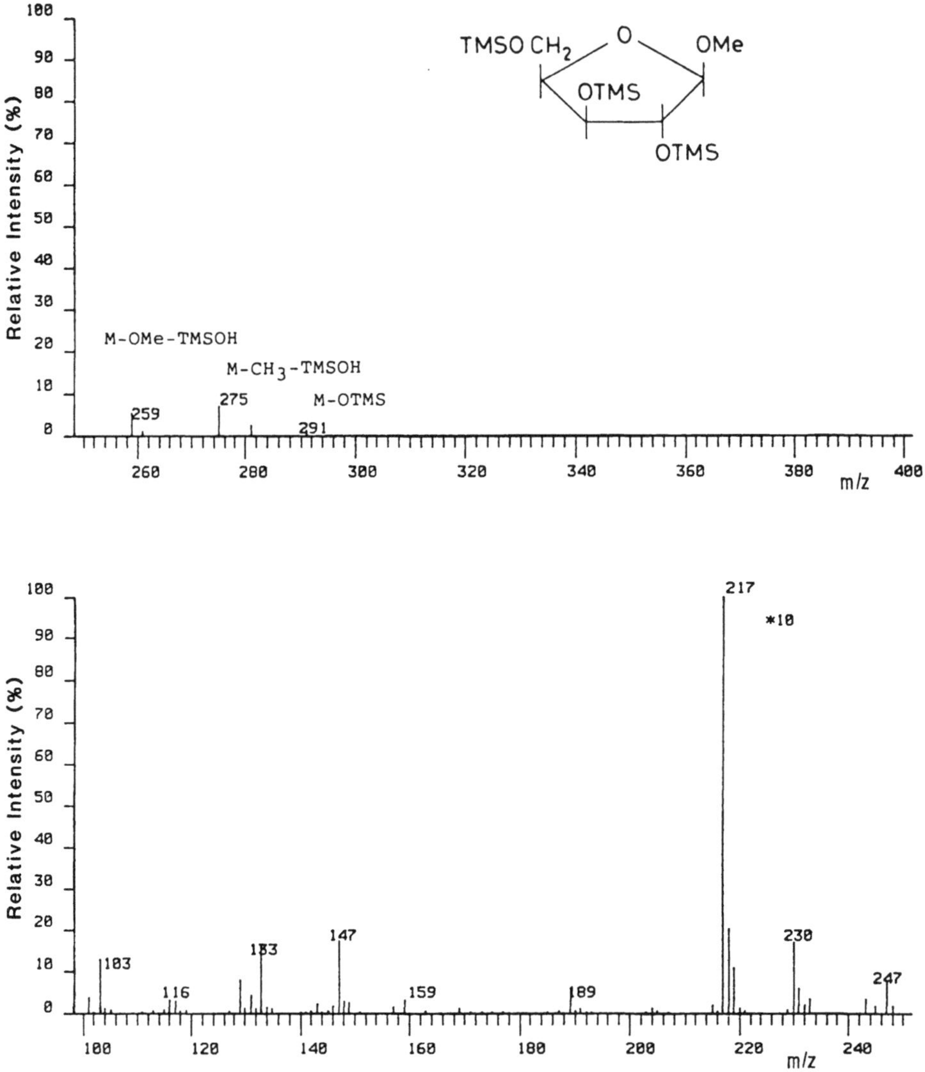

Figure 10. 70 eV EI mass spectrum of trimethylsilylated methyl β-D-xylofuranoside.

of *N*-acetylneuraminic acid (43%). Mannitol and 2-acetamido-2-deoxy-glucitol form little anhydride (<1%) (49). These findings are of importance for the structural analysis of *O*-glycosidic carbohydrate chains, which are linked to serine (Ser) or threonine (Thr) through *N*-acetylgalactosamine (GalNAc), and for the analysis of alditols together with aldoses from for instance physiological fluids. Mass spectra of trimethylsilylated 1,4-anhydro-D-xylitol-[1-^{2}H], 2,5-anhydro-D-xylitol-[1-^{2}H], 1,4-anhydro-L-fucitol-[1-^{2}H], 1,4-anhydro-D-galactitol-[1-^{2}H], 3,6-anhydro-D-galactitol-[1-^{2}H], 2-acetamido-3,6-anhydro-2-deoxy-D-galactitol-[1-^{2}H] and 2-acetamido-1,4-anhydro-2-deoxy-D-galactitol-[1-^{2}H] are included in (49).

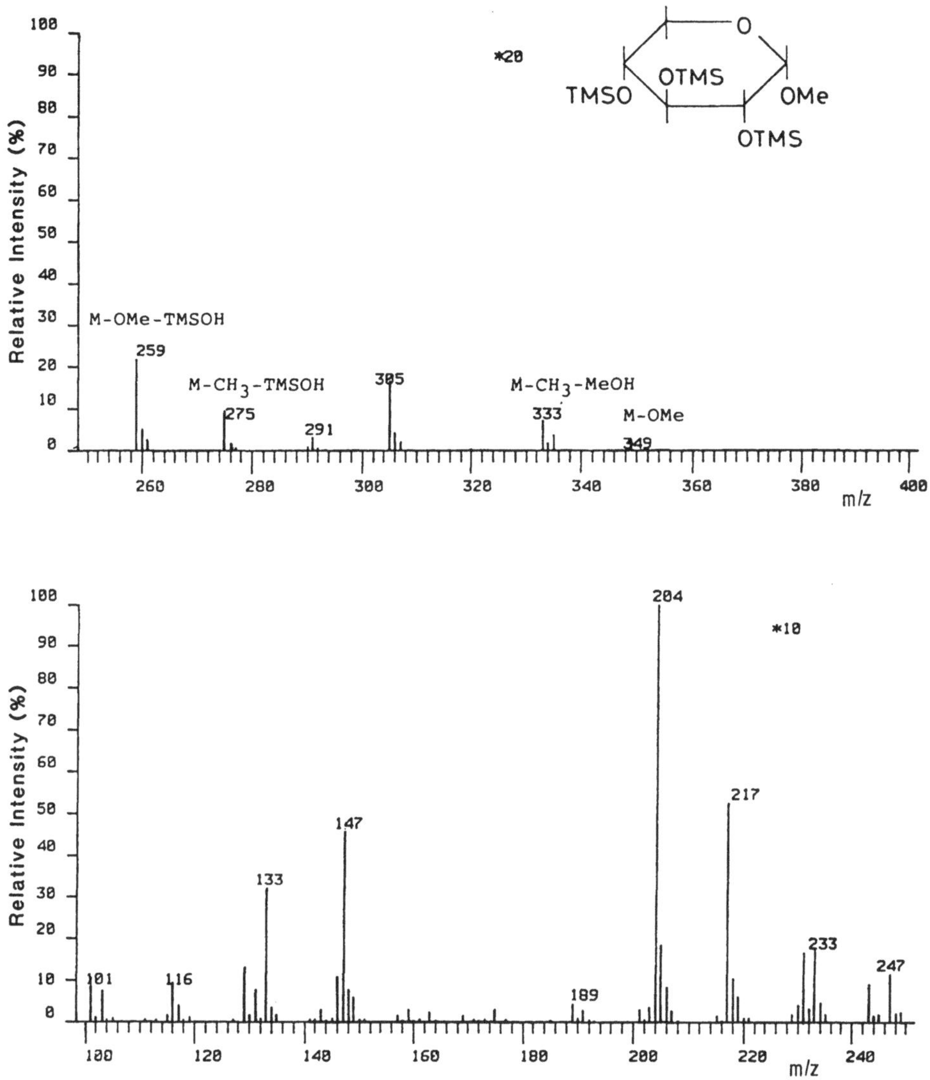

Figure 11. 70 eV EI mass spectrum of trimethylsilylated methyl α-D-xylopyranoside.

8. The applied methanolysis conditions lead to total degradation of fructose (2-ketohexose). Several uronic acids give rise to 3,6-lactone formation. Gulose yields also the 1,6-anhydride (30%) (71), whereas *N*-acetylneuraminic acid gives a little 2,7-anhydride (3%). Anhydro-formation has also been observed for heptoses (72). The cleavage of the rather stable glycosidic linkages of uronic acid residues may be incomplete, as in the hydrolysis procedure.
9. For data on 3-deoxy-D-*manno*-2-octulosonic acid (KDO), see (54, 73, 74).

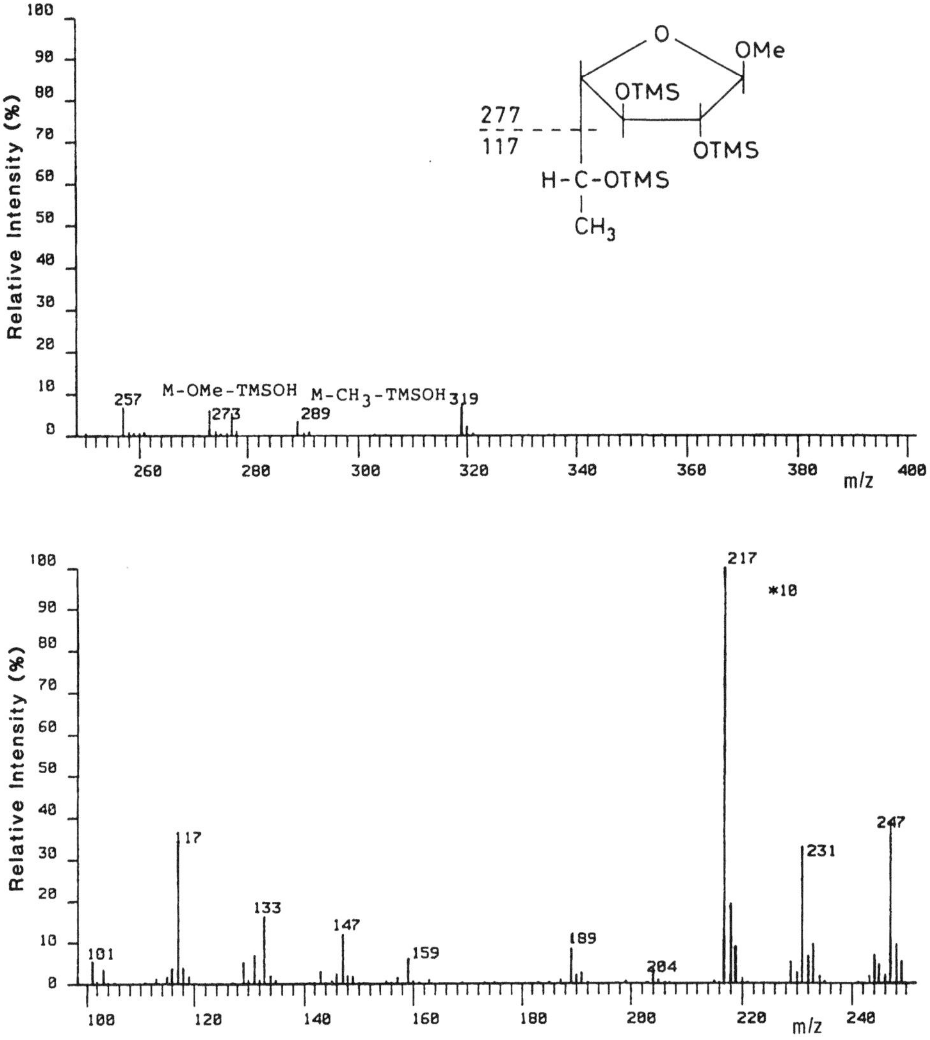

Figure 12. 70 eV EI mass spectrum of trimethylsilylated methyl β-D-fucofuranoside.

2.4 Mass spectrometry of trimethylsilylated methyl glycosides

The EI mass spectra of trimethylsilylated methyl glycosides have been studied in great detail (2, 3, 6, 7, 40, 75–82). Some basic fragmentation patterns are presented in Scheme 2. A series of examples of mass spectra obtained from different types of monosaccharides in furanose, pyranose and lactone forms is depicted in Figures 10–21. In the high-mass-region the spectra are usually characterised by [M – CH_3] fragment ions. Sometimes, the molecular ion itself is observed in low intensity. Just as for the alditol acetates, differences in stereochemistry within one class of monosaccharides are not reflected in the mass spectra.

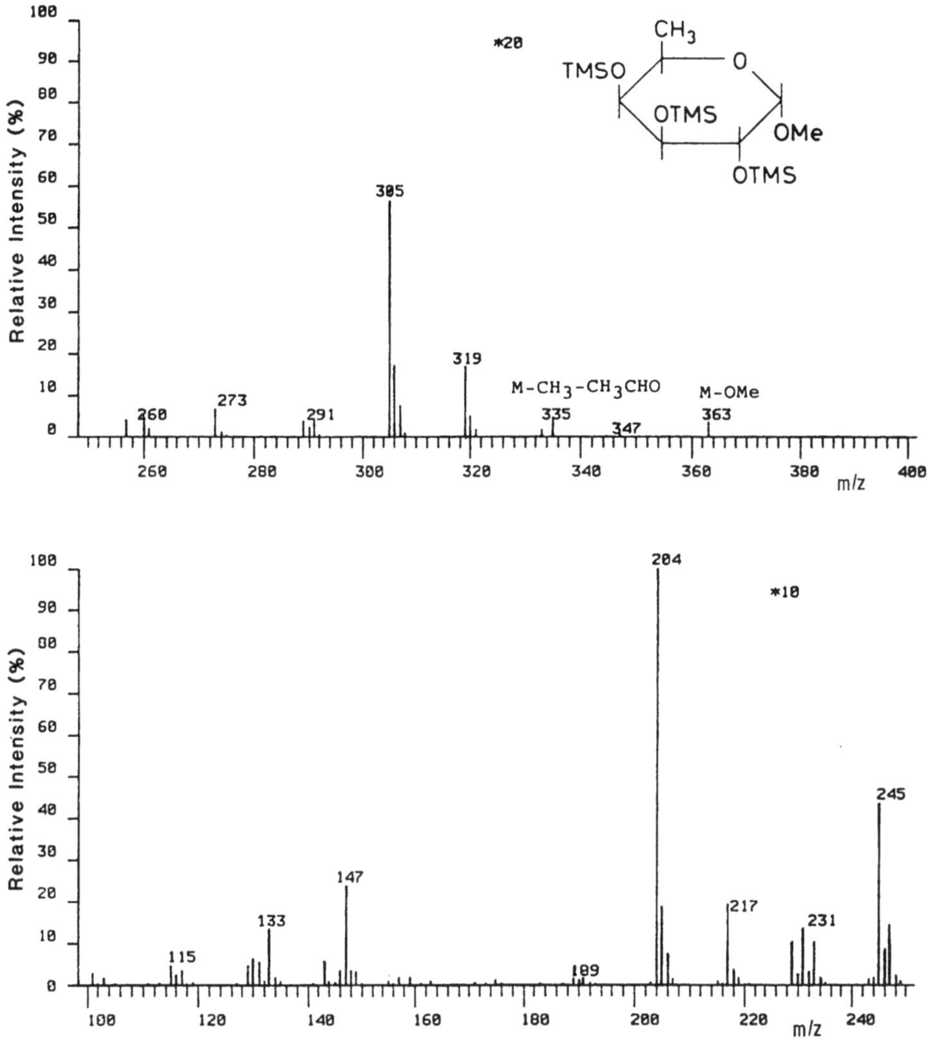

Figure 13. 70 eV EI mass spectrum of trimethylsilylated methyl α-D-fucopyranoside.

In general it is possible to discriminate between pyranose and furanose ring forms in one class of monosaccharides. For trimethylsilylated methyl aldohexosides it has been found that the ratio of the intensities of the peaks at *m*/*z* 204 and *m*/*z* 217 is useful (40, 76, 77). In pyranose ring forms the ion *m*/*z* 204 corresponds with the fragments TMSO$\overset{2}{C}$H$-$$\overset{3}{C}$HOTMS (major) and TMSO$\overset{3}{C}H-$$\overset{4}{C}$HOTMS (minor); *m*/*z* 217 corresponds mainly with TMSO$\overset{2}{C}$H$-$$\overset{3}{C}H-$$\overset{4}{C}$HOTMS. In furanose ring forms *m*/*z* 204 can be explained as TMSO$\overset{2}{C}$H$-$$\overset{3}{C}$HOTMS and *m*/*z* 217 as TMSO$\overset{2}{C}$H$-$$\overset{3}{C}$HOTMS$-$$\overset{4}{C}$H (83). The studied pyranose rings give rise to a ratio of $I_{204}/I_{217} > 1$,

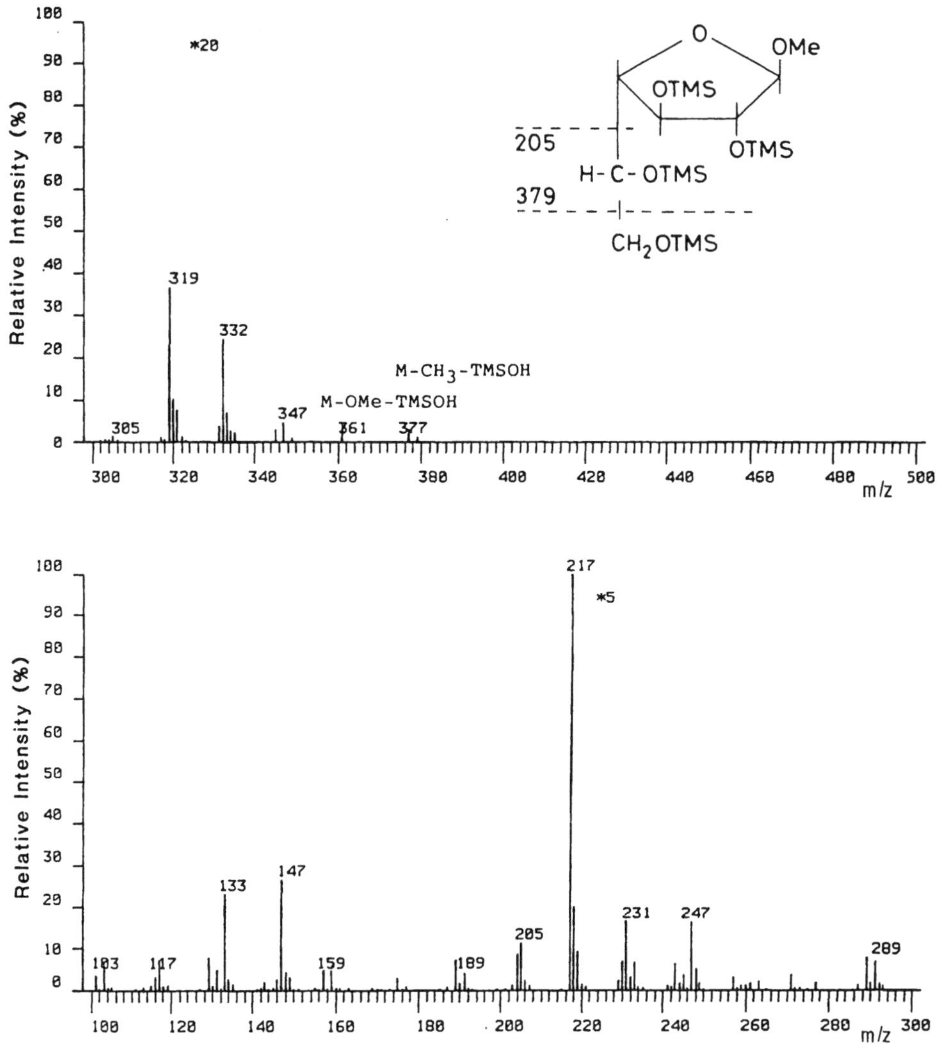

Figure 14. 70 eV EI mass spectrum of trimethylsilylated methyl β-D-galactofuranoside.

whereas the furanose rings show a ratio $\ll 1$. In addition the presence of a relatively intense peak at *m/z* 205 ($TMSO\overset{6}{C}H_2-\overset{5}{C}HOTMS$) is characteristic of a furanose ring. The concept of discrimination between pyranose and furanose ring forms on the basis of the intensities of *m/z* 204 and *m/z* 217 is also applicable to trimethylsilylated methyl glycosides of uronic acid methyl esters (78) and to most of the aldopentosides (40, 75, 77). In the same way for trimethylsilylated methyl 2-acetamido-2-deoxy-aldohexosides use can be made of the intensities of the fragment ions at *m/z* 173 ($CH_3CONH\overset{2}{C}H-\overset{3}{C}HOTMS$) and *m/z* 186 ($CH_3CONH\overset{2}{C}H-\overset{3}{C}H-\overset{4}{C}HOTMS$; $CH_3CONH\overset{2}{C}H-\overset{3}{C}HOTMS-\overset{4}{C}H$) (40, 77, 79, 81). A ratio of $I_{173}/I_{186} > 1$ points to

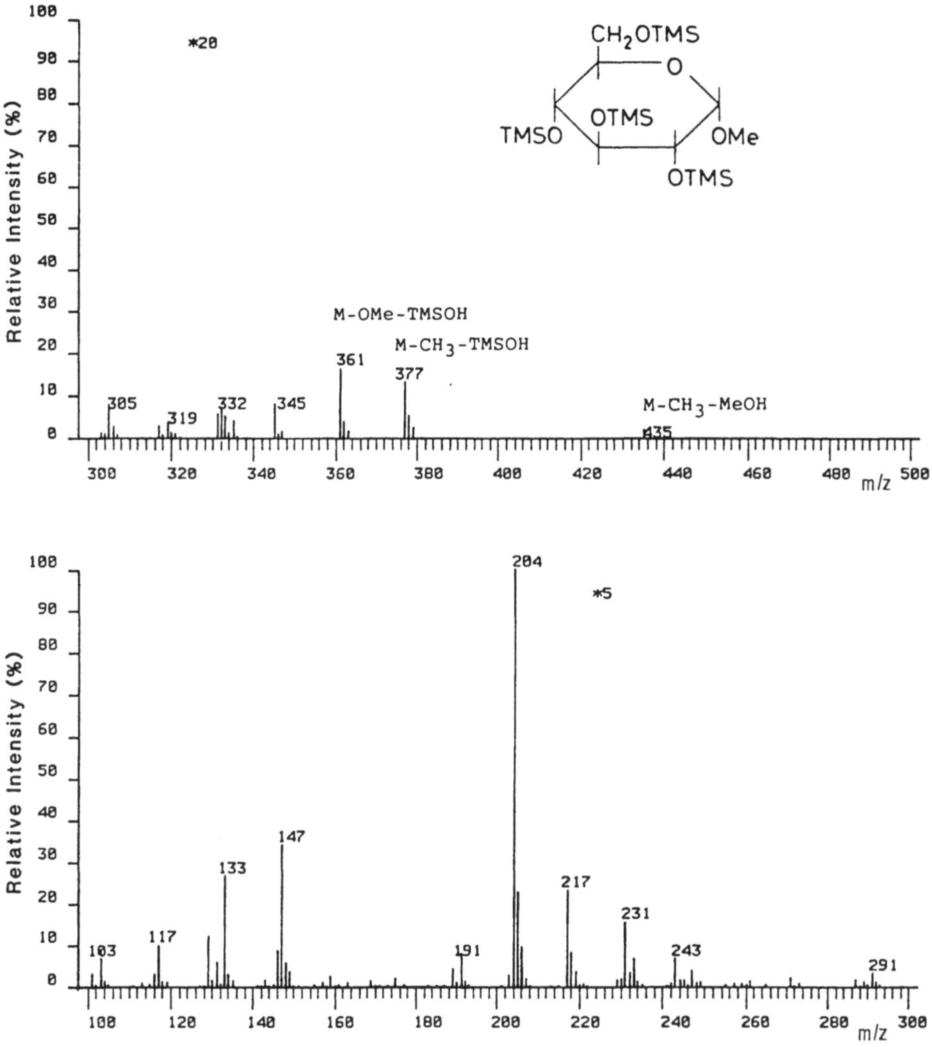

Figure 15. 70 eV EI mass spectrum of trimethylsilylated methyl α-D-glucopyranoside.

a pyranose ring form and a ratio of $I_{173}/I_{186} \ll 1$ to a furanose ring form. In the latter case also an intense peak at *m/z* 205 is present.

Naturally occurring *O*-methylated monosaccharides have also been analysed after applying the methanolysis procedure. The presence of *O*-methyl groups at specific positions have a distinct influence on the EI mass spectrometric fragmentation patterns (12, 71, 75–77, 79, 84). Especially the fragment ions mentioned above together with those in the high-mass-region are very diagnostic. Frequently, the introduction of deuterated trimethylsilyl, *N*-acetyl and/or methyl groups in the monosaccharide is very helpful for the interpretation of mass spectra of unknown monosaccharide residues.

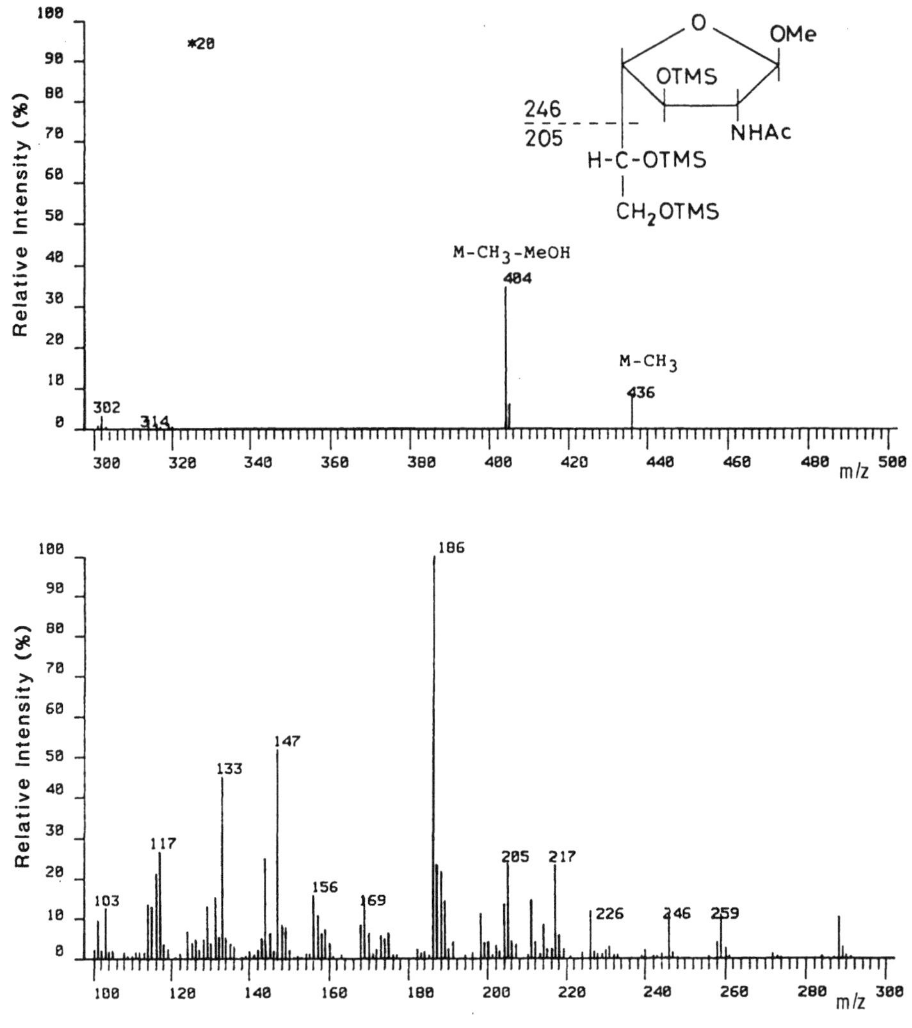

Figure 16. 70 eV EI mass spectrum of trimethylsilylated methyl *β*-*N*-acetyl-D-galactofuranoside.

2.5 Mass spectrometry of trimethylsilylated alditols and anhydro-alditols

The EI mass spectra of a trimethylsilylated hexitol and a trimethylsilylated 2-acetamido-2-deoxy-hexitol are presented in Figures 22 and 23, respectively. The amino hexitol derivative is frequently observed when analysing carbohydrate chains derived from GlcNAc-Asn type glycoproteins subjected to the hydrazinolysis procedure and from GalNAc-Ser/Thr type glycoproteins subjected to the *β*-elimination procedure, respectively. Some specific framentations have been reported (85). As indicated already above, a series of EI mass spectra of anhydro-alditols obtained from different alditols have been discussed (49).

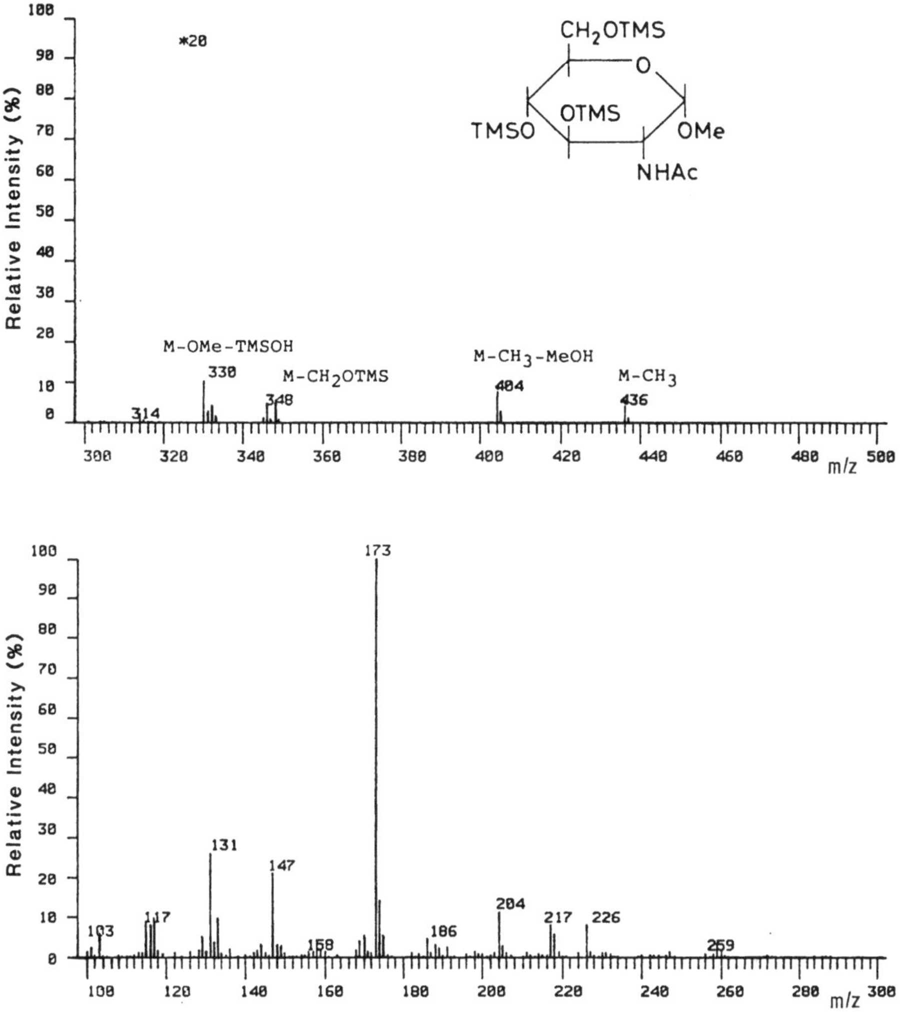

Figure 17. 70 eV EI mass spectrum of trimethylsilylated methyl α-*N*-acetyl-D-glucopyranoside.

In connexion with the analysis of alditols in uremic serum by GC-MS, attention has been given to the use of CI-MS with isobutane as reagent gas (86). The latter study also describes the analysis of aldoses and aldonic acids as their trimethylsilylated derivatives. For additional CI studies of trimethylsilylated aldohexoses (NH_3 as reagent gas) and amino sugars (CH_4 as reagent gas), see (87) and (88), respectively. Another GC-MS study of the polyols present in urine and serum of healthy subjects and uremic patients has led to the finding of several new deoxyalditols and inositol isomers. Both EI- and CI-MS were used for the analysis of the trimethylsilylated derivatives (89). For additional MS data of trimethylsilylated (and acetylated) inositols and trimethylsilylated naturally occurring *O*-methylinositols, see (90) and (91), respectively.

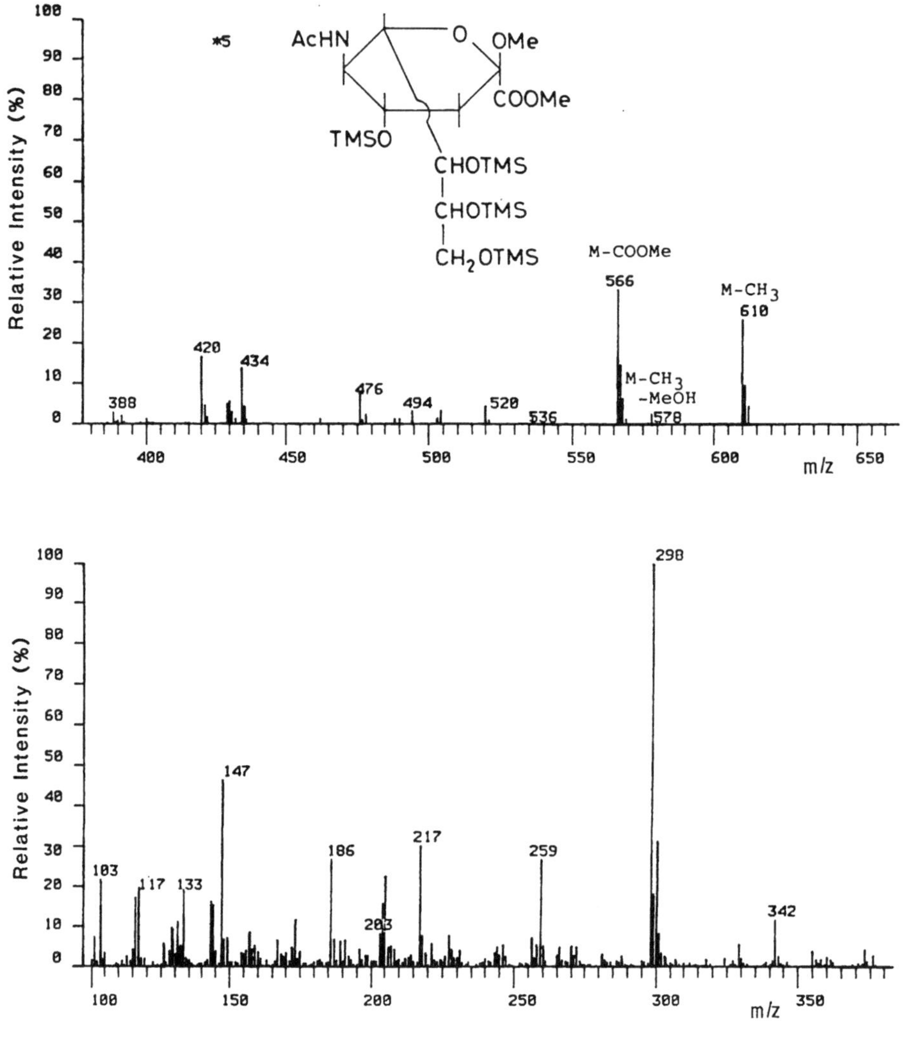

Figure 18. 70 eV EI mass spectrum of trimethylsilylated methyl ester *β*-methyl glycoside of *N*-acetylneuraminic acid.

2.6 GC-MS of *N,O*-acylneuraminic acids

The analysis of *N,O*-acylneuraminic acids (sialic acids) requires special attention to retain the *N*- and *O*-acyl substituents during isolation and analysis procedures. Routine sugar analysis methods as described above lead to a complete *N*- and *O*-deacylation of sialic acids. For this reason several attempts have been made to develop analytical methods keeping the specific sialic acids unimpaired (36, 92). Sialic acids are released in good yield under mild acidic conditions like formic acid, pH 2 (60 min, 70 °C). These conditions do not cause *N*-deacylation, whereas only 20–30% of the *O*-acyl groups are lost. Also the possibility of *O*-acyl migration has to be taken into account.

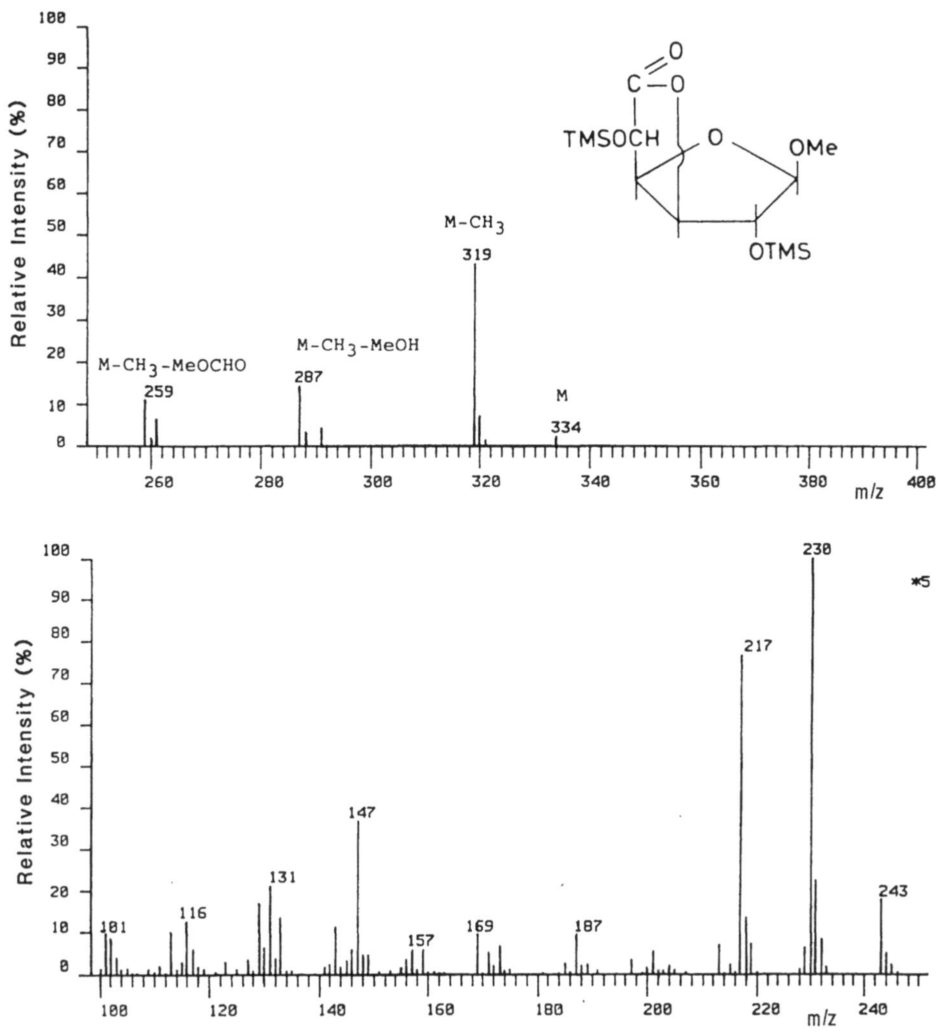

Figure 19. 70 eV EI mass spectrum of trimethylsilylated methyl β-D-glucofuranurono-6,3-lactone.

In several cases enzymic cleavages by neuraminidase have been applied (92). The released sialic acids are isolated by ultrafiltration or dialysis at 2–4 °C. Purification and separation of the isolated sialic acid fraction is based on various chromatographic techniques (14, 92).

Sialic acids are currently analysed by GC-MS after esterification with diazomethane followed by trimethylsilylation of the free hydroxyl groups, or after trimethylsilylation of carboxyl and free hydroxyl groups (36, 93–95). Depending on the applied silylation reagent, also the secondary amide group (*N*-acyl group) can be trimethylsilylated (96). Because of partial *O*-deacylation of the *N,O*-acylneuraminic acids and/or migration of the *O*-acyl groups in the glycerol side chain (7-*O*-acetyl → 9-*O*-acetyl; 7,9-di-*O*-acetyl → 8,9-di-*O*-acetyl) (96a) which can occur during isolation procedures, quanti-

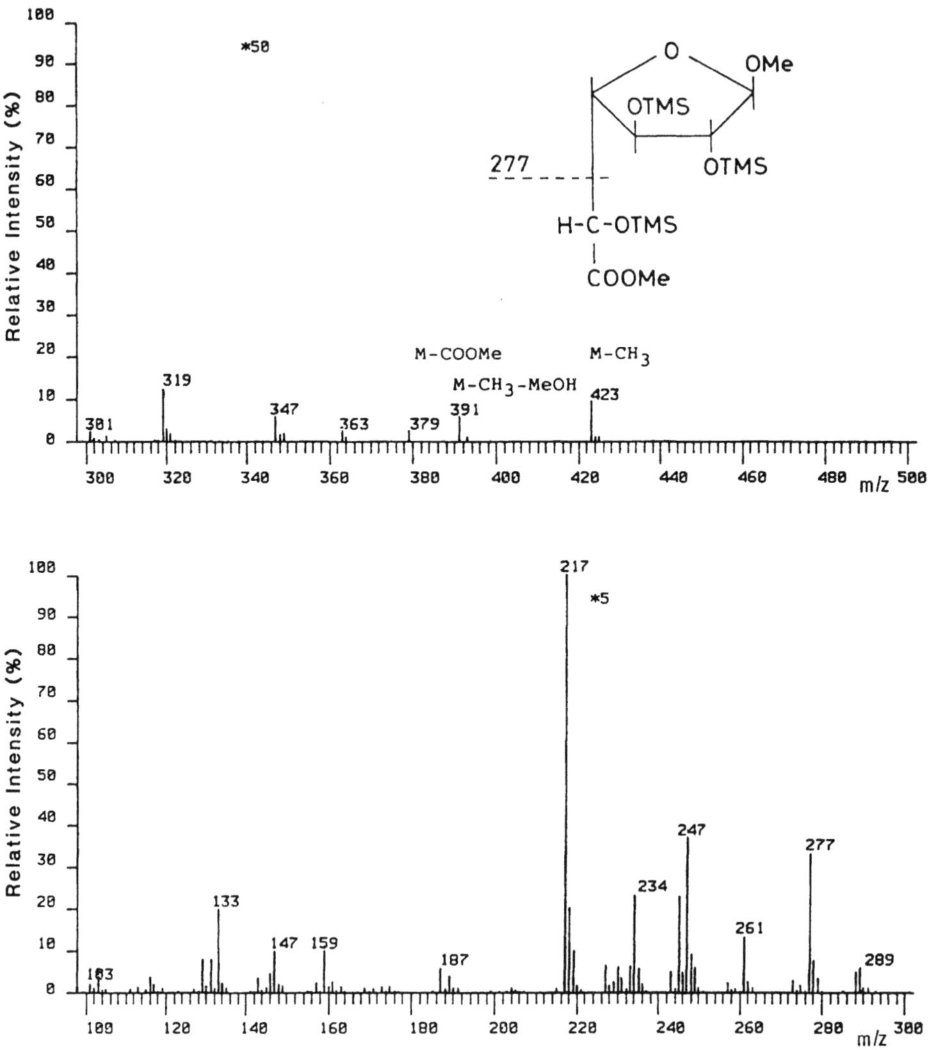

Figure 20. 70 eV EI mass spectrum of trimethylsilylated methyl ester methyl β-D-galactofuranosiduronic acid.

tative analysis of the various *N,O*-acylneuraminic acids present in native biological material is still a serious problem. For quantitative determinations based on GC-MS, see (36, 97).

The determination of the number, type and position of *O*-acyl groups (and also of *O*-alkyl groups) as well as of the type of the *N*-acyl group in sialic acids is facilitated by the highly specific EI mass spectra of these derivatised compounds. Based on the spectra of the trimethylsilylated methyl esters a micromethod using seven selected fragment ions A–G has been reported, which furnish the information (abundances and *m*/*z* values of the ions) necessary to deduce the complete structure of the sialic

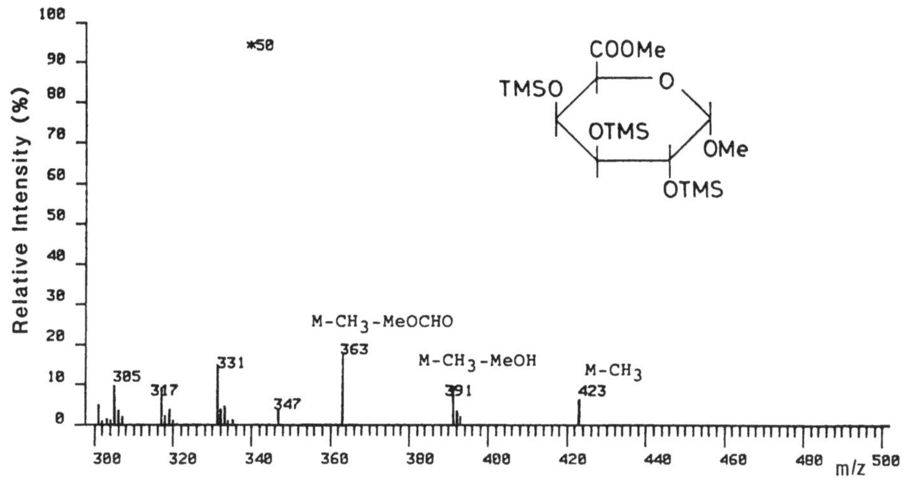

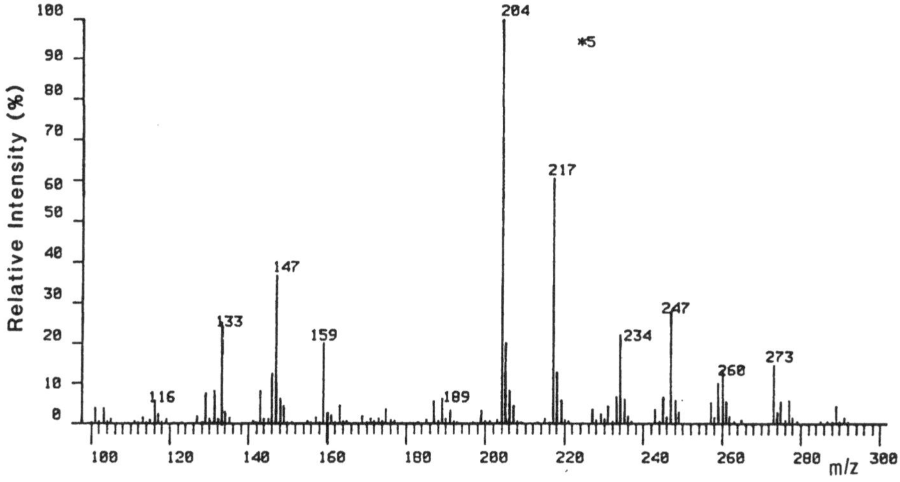

Figure 21. 70 eV EI mass spectrum of trimethylsilylated methyl ester methyl α-D-galactopyranosiduronic acid.

acids (Scheme 3). In Figure 24 the mass spectrum of one example, namely the trimethylsilylated methyl ester of *β-N*-acetylneuraminic acid, is depicted. Although it has not been checked in detail, it is highly probable that the use of trimethylsilyl esters instead of methyl esters do not change the identification procedure. As will be discussed in the methylation analysis part of this chapter, the scheme is also of value in determining substitution patterns of sialic acids in carbohydrate chains. In the latter case an additional fragment H is included (36).

Fragments A and B indicate the molecular weight of the sialic acid derivatives and thereby the number and type of substituents. Fragments C-H can be used for determination of the positions of the different substituents (36).

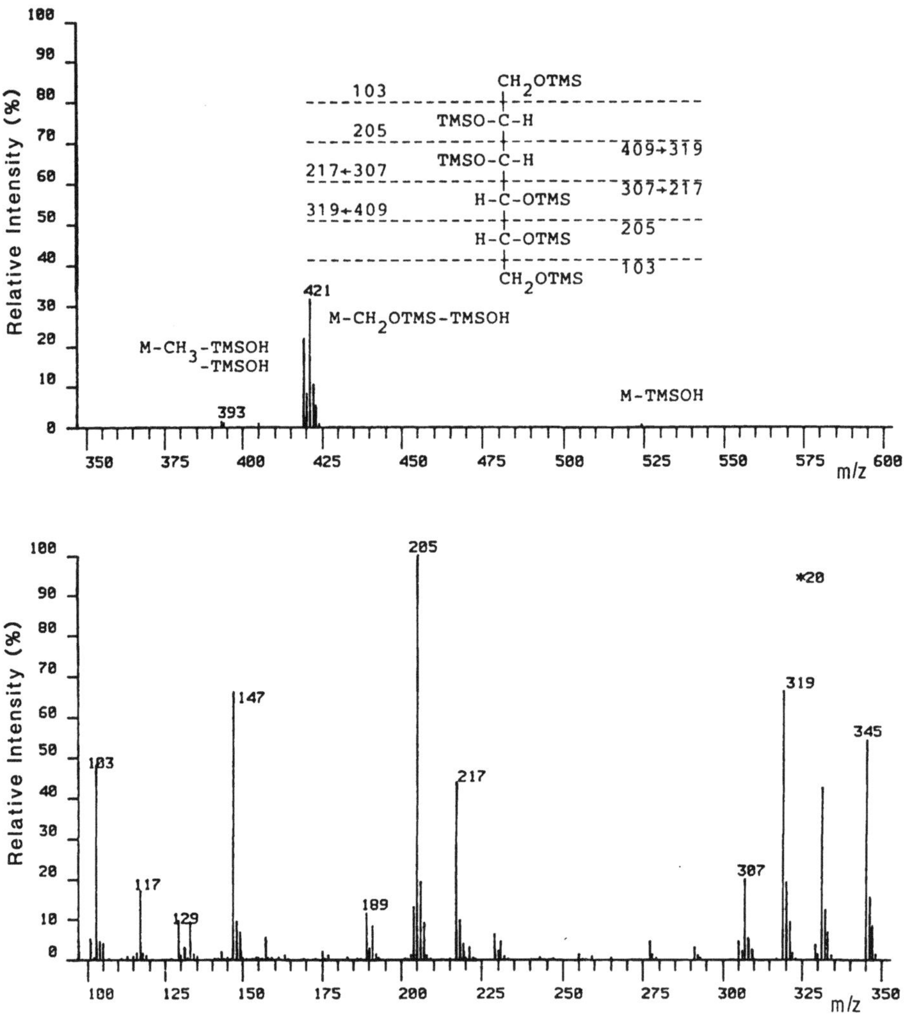

Figure 22. 70 eV EI mass spectrum of trimethylsilylated mannitol.

Fragment A is formed from the molecular ion by the elimination of a methyl group originating from a trimethylsilyl substituent in trimethylsilylated (*O*-acylated) *N*-acylneuraminic acid derivatives. When $R_{5'}$ = CH_3 (methylation analysis procedure), the eliminated methyl group can also originate from the *N*,*N*-acyl,methyl group.

Fragment B is formed by elimination of the C1 part of the molecule. Eliminations of $OCOCH_3$ in *O*-acetylated sialic acid derivatives and of NH_2COCH_3 in *N*-acetylneuraminic acid derivatives, which in principle give rise to the same *m*/*z* value as fragment B in the case of R_1 = CH_3, can be neglected.

Fragment C is formed by elimination of C8-C9, with localisation of the charge on position 7. In general, cleavage occurs between two alkoxylated carbon atoms, or between an acetoxylated and an alkoxylated carbon atom, rather than between two

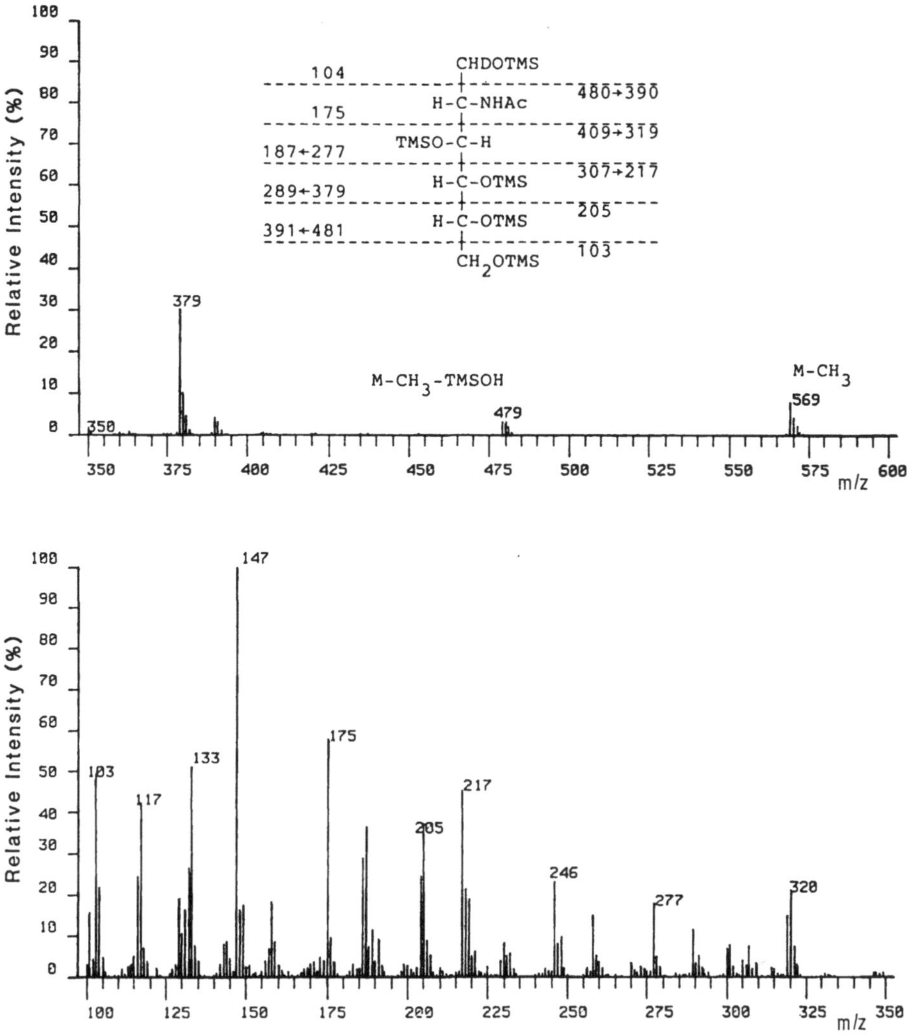

Figure 23. 70 eV EI mass spectrum of trimethylsilylated 2-acetamido-2-deoxy-glucitol-[1-^{2}H].

acetoxylated carbon atoms. It has been found that the charge is preferentially located on an ether oxygen instead of on an ester oxygen (6). Therefore fragment C has only significant abundance if C7 bears an ether group. When an ester group is present at C7, this fragment ion is absent or hardly observable.

Fragment D is formed from fragment C by consecutive eliminations of R_2OH and R_4OH. It is evident that the occurrence of this fragment ion is dependent on the presence of fragment C.

Fragment E is formed by elimination of the whole side-chain C7-C8-C9 and the substituent at C5. This fragment ion is not observed if an *O*-acyl group is attached to C4, illustrating that the transition state in the McLafferty rearrangement is more favoured when the substituent at C4 is an ether group rather than an ester group.

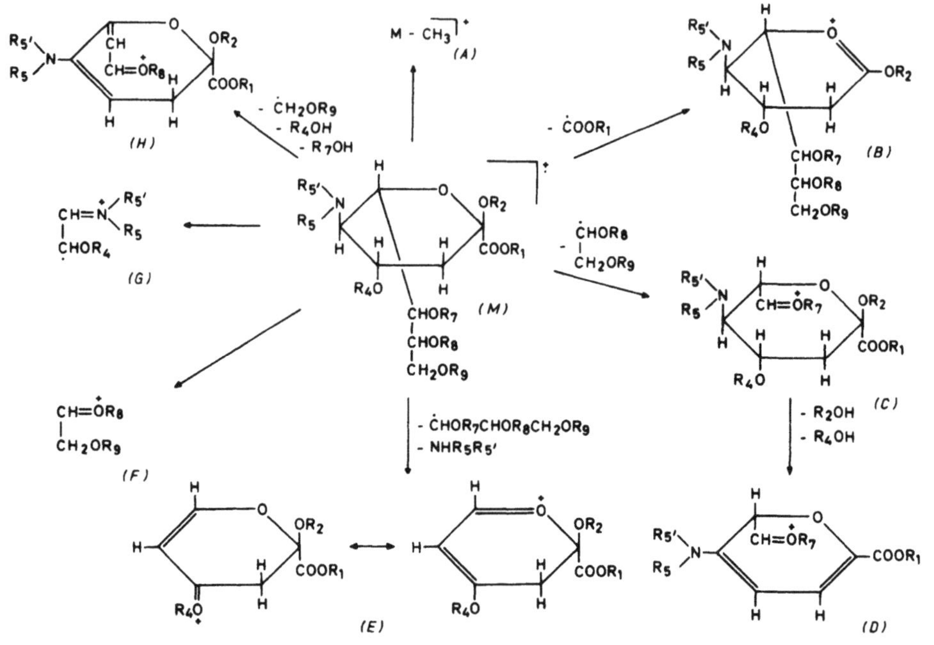

$R_1 = CH_3$

$R_2 = CH_3$ or $Si(CH_3)_3$

$R_4, R_7, R_8, R_9 = CH_3$, $Si(CH_3)_3$, $COCH_3$ and / or $COCH[OSi(CH_3)_3]CH_3$

$R_5 = COCH_3$ or $COCH_2OSi(CH_3)_3$

$R_{5'} = H$ or CH_3

Scheme 3. Survey of the selected fragment ions A-H determined for trimethylsilylated (*O*-acylated) *N*-acylneuraminic acid methyl ester(s) (methyl glycosides) and for (partially) *O*-methylated *N,N*-acetyl,methyl-neuraminic acid methyl ester methyl glycosides. Partially *O*-methylated compounds are trimethylsilylated or acetylated.

Fragment F contains C8-C9. Based on the same fragmentation rules as mentioned above for fragments C and D, this ion can only readily be formed if an ether group is attached to C8.

Fragment G consists of the C4-C5 part of the molecule.

Fragment H, only necessary to use in methylation analysis studies, is formed by elimination of the C9 part, followed by elimination of R_4OH and R_7OH. The fragment is useful to discriminate between an OTMS group at C8 or C9 in trimethylsilylated partially methylated *N*-acylneuraminic acids.

A series of naturally occurring *N,O*-acylneuraminic acids analysed by GC-MS as their trimethylsilylated methyl ester derivatives is summarized in Table 1. The Table includes retention times on 3.8% SE-30 and the *m/z* values of the diagnostic fragments A-G. For more detailed information including high-resolution exact mass measurements and the EI mass spectra of the various sialic acid derivatives, see (36).

Chemical ionization mass spectra of derivatised and underivatised sialic acids have also been reported (97a, 97b).

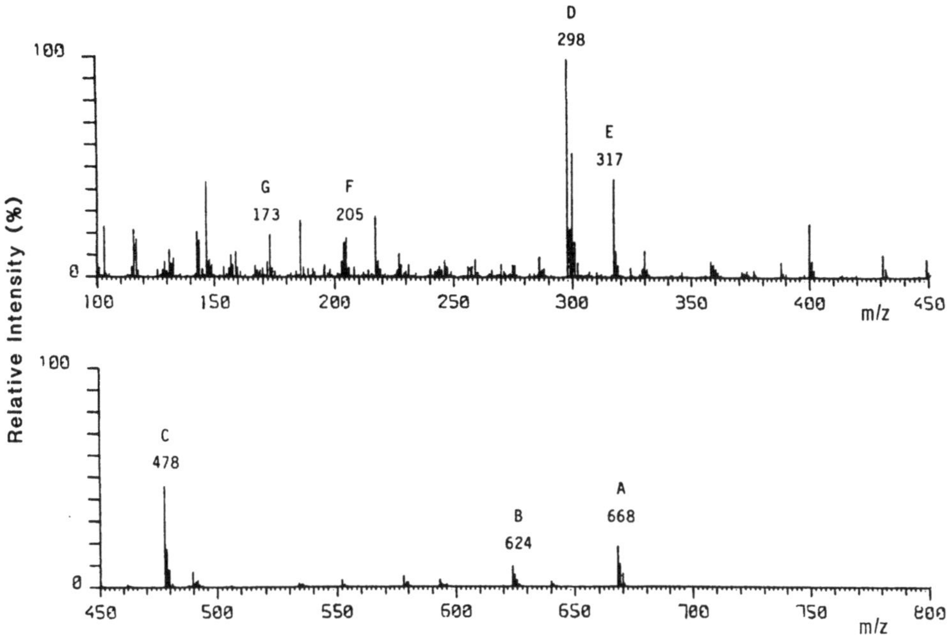

Figure 24. 70 eV EI mass spectrum of the trimethylsilylated methyl ester of *β*-*N*-acetylneuraminic acid.

2.7 Absolute configuration determination of monosaccharides

Monosaccharides can occur both as D- and L-enantiomers. It is even possible that the D- and L-enantiomers of one sugar form part of one carbohydrate chain (12). The resolution of racemic mixtures on GC can be achieved on a non-chiral phase after conversion of the enantiomers into diastereomers by glycosidation with a chiral alcohol, and subsequent protection of additional polar functions. Two approaches have been reported for the resolution of D- and L-monosaccharides: 1. GC(-MS) of the trimethylsilylated (−)-2-butyl glycosides (55, 57), and 2. GC of the acetylated (+)-2-octyl glycosides (56). In the case of the trimethylsilylated (−)-2-butyl glycosides attention has been given to the EI mass spectral data in relation to pyranose and furanose ring forms (55, 57). These GC(-MS) approaches can be applied directly to mixtures of monosaccharides, and can be incorporated easily in the sugar analysis procedure using methanolysis (see above).

2.8 Miscellaneous data

By using the trimethylsilylated methyloxime derivatives of glucose and glucose-[$^2H_7$], a mass fragmentographic reference method has been developed for the determination of serum glucose (98). *n*-Butyldiboronate derivatives of mannitol and glucitol have been applied for the simultaneous determination of these alditols in plasma with GC-

Table 1. GLC and MS data of trimethylsilylated (*O*-acylated) *N*-acylneuraminic acid methyl esters. The R_{Neu5Ac}-values on 3.8% SE-30 at 215 °C are given relative to the trimethylsilylated methyl ester of Neu5Ac (β-anomeric form). For an explanation of the '-' signs, see text. EI-mass spectra are depicted in (36).

Sialic acid*	R_{Neu5Ac} β-anomer	R_{Neu5Ac} α-anomer	*m/z* values						
			A	B	C	D	E	F	G
Neu5Ac	1.00	1.05	668	624	478	298	317	205	173
Neu4,5Ac_2	1.18	1.21	638	594	448	298	–	205	143
Neu5,7Ac_2	1.04	1.00	638	594	–	–	317	205	173
Neu5,8Ac_2	1.05		638	594	478	298	317	–	173
Neu5,9Ac_2	1.13		638	594	478	298	317	175	173
Neu4,5,9Ac_3	1.31		608	564	448	298	–	175	143
Neu5,7,9Ac_3	1.14	1.07	608	564	–	–	317	175	173
Neu5,8,9Ac_3	1.19		608	564	478	298	317	– #	173
Neu5,7,8,9Ac_4	1.15		578	534	–	–	317	–	173
Neu5Ac9Lt	2.55		740	696	478	298	317	277	173
Neu4,5$Ac_2$9Lt	3.01		710	666	448	298	–	277	143
Neu5Gc	1.81	1.90	756	712	566	386	317	205	261
Neu4Ac5Gc	2.02		726	682	536	386	–	205	231
Neu7Ac5Gc	1.83		726	682	–	–	317	205	261
Neu9Ac5Gc	2.04		726	682	566	386	317	175	261
Neu7,9$Ac_2$5Gc	2.01		696	652	–	–	317	175	261
Neu8,9$Ac_2$5Gc	1.99		696	652	566	386	317	– #	261
Neu7,8,9$Ac_3$5Gc	1.93		666	622	–	–	317	–	261

* Neu5Ac = *N*-acetylneuraminic acid; Neu5Gc = *N*-glycolylneuraminic acid; Neu4,5Ac_2 = 4-*O*-acetyl-*N*-acetylneuraminic acid; Neu5Ac9Lt = 9-*O*-lactyl-*N*-acetylneuraminic acid; etc.

The small peak at *m/z* 145, also present in the mass spectra of other sialic acids, has not been checked by exact mass measurements (93).

mass fragmentography (99). The isotopic dilution technique was used for determining the content of myo-inositol (as hexa-acetate) in human urine, plasma, and haemolysed erythrocyte samples using myoinositol-[$^2H_6$] as standard (100). Application of EI- and CI-MS for the GC-analysis of sugar components in human eye lenses has led to the identification of a heptitol and an octitol besides several other alditols (101).

3 Methylation Analysis

To gain insight into the substitution pattern (positions of glycosidic linkages, terminal units, branching points) and into the ring size of the monomeric units of a carbohydrate chain, methylation analysis including GC-MS has proved to be of great importance. A frequently applied procedure for neutral and acetamido sugars is presented in Figure 25. The first step involves the methylation of the free hydroxyl groups and of

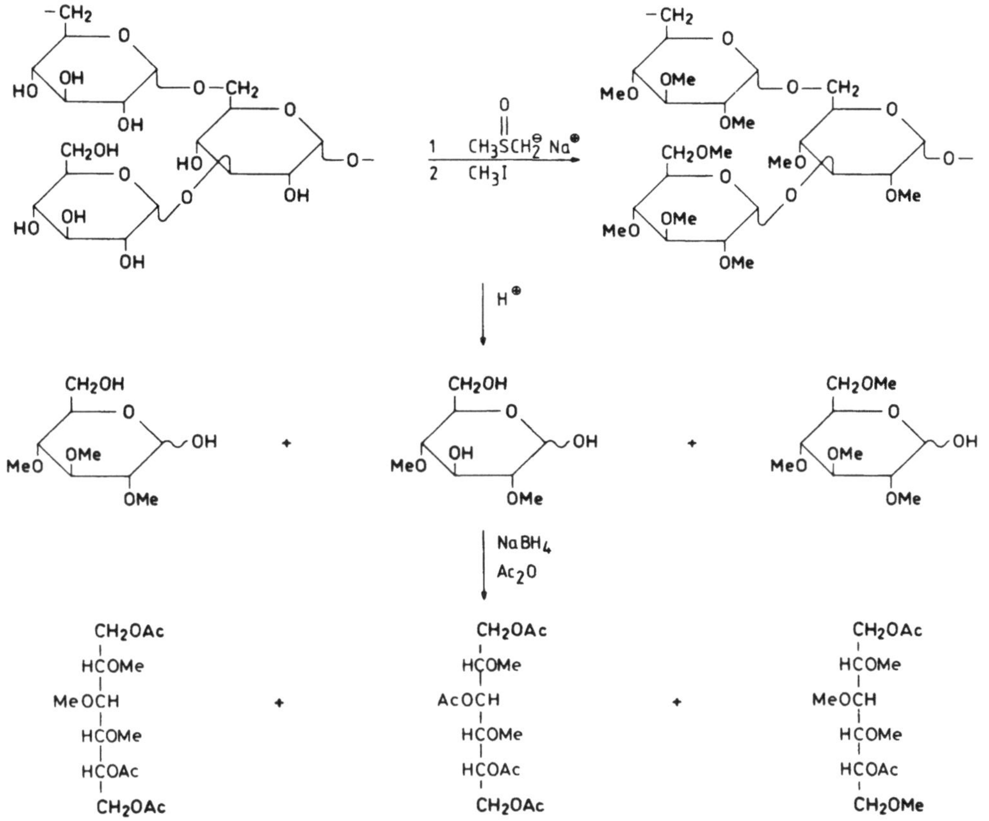

Figure 25. Schematic presentation of the methylation analysis of an artificial polysaccharide.

the acetamido groups. Carboxyl functions of uronic acids, sialic acids, etc. are esterified. For a review of the various developed methylation methods, see (5). This reaction is carried out most often with methyl iodide using sodium methylsulfinylmethanide in dimethylsulfoxide as base (102–106). Recently other bases have been applied to this methylation, e. g. potassium *tert*-butoxide/potassium methylsulfinylmethanide in dimethylsulfoxide (107), and potassium (108, 109) and lithium (110, 110a, 110b) methylsulfinylmethanide in dimethylsulfoxide, mainly because the latter reagents seem to suffer less from impurities when compared with the sodium salt preparation. In addition, the use of methyl trifluoromethanesulfonate in trimethylphosphate with 2,6-di-(*tert*-butyl)pyridine as proton scavenger has been mentioned (111). Although a complete methylation in a single operation is essential for reliable results, it is sometimes impossible to overcome a certain degree of undermethylation, even after several successive methylation steps. When uronic acid-containing carbohydrate chains are investigated, repeated treatment with base leads to β-elimination reactions in the carbohydrate chain (112). Such alkaline degradations can also be expected if reducing carbohydrates are methylated in the presence of strong bases as mentioned above.

After permethylation the material is hydrolysed, yielding a mixture of partially methylated monosaccharides. Routine solvolysis conditions are e. g. 90% formic acid/

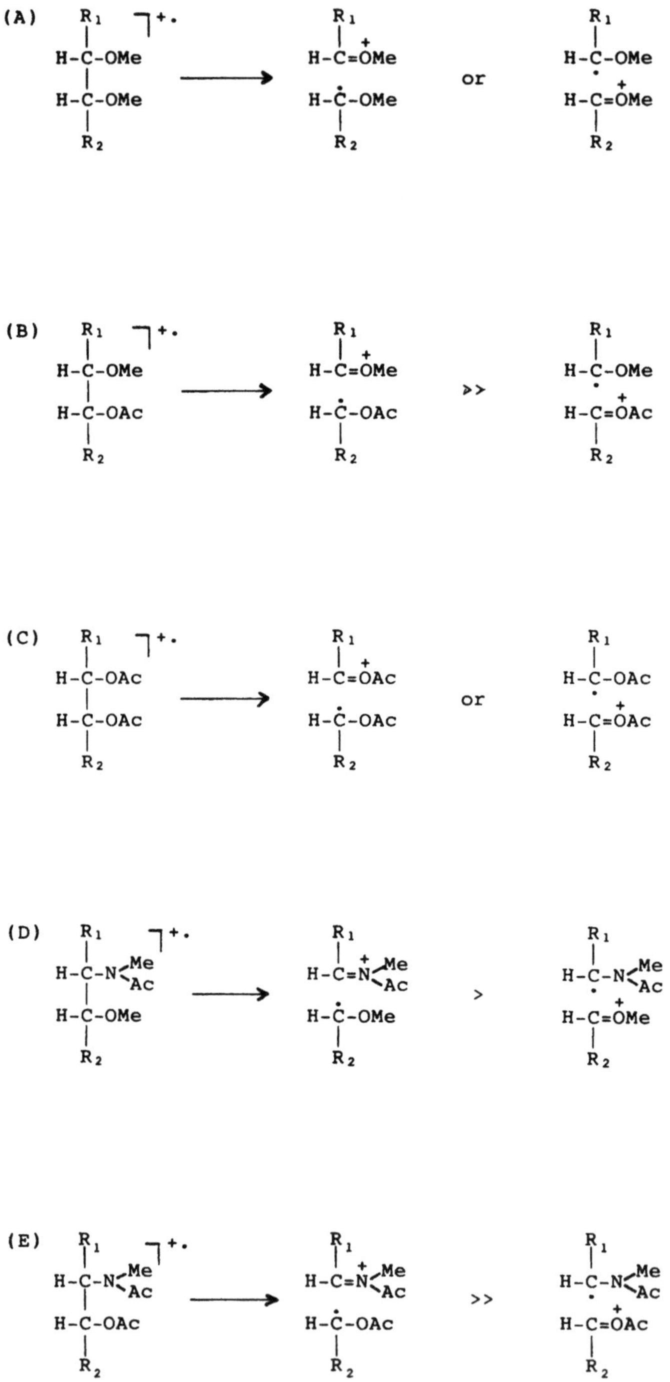

Scheme 4. Genral information on the fragmentation of partially methylated alditol acetates.

(F) $CH_3\text{-}C{\equiv}\overset{+}{O}$ (m/z 43)
Generally the most intense ion, derived from the acetyl function by α-cleavage

(G) $CH_2{=}\overset{+}{O}Me$ (m/z 45)
Not characteristic for a primary methoxyl group, but can also be formed by migration of a hydrogen atom to any charged one-carbon fragment bearing an OMe group

(H)

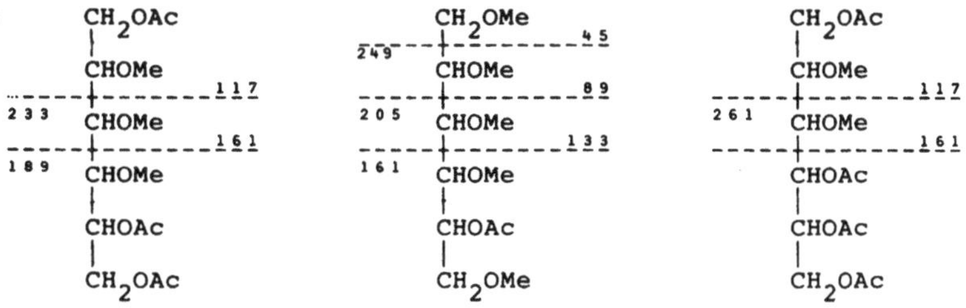

(I)

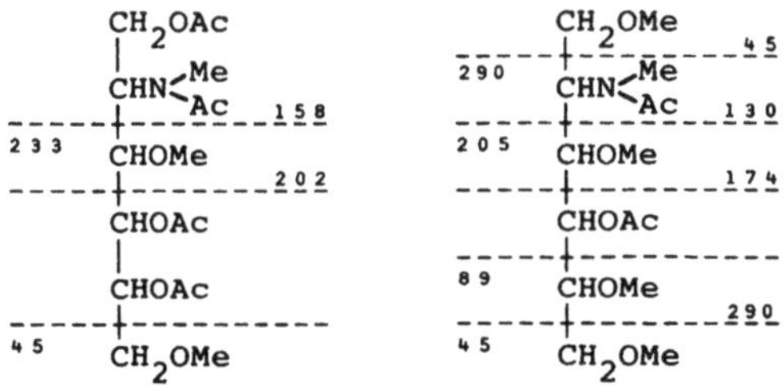

(J)

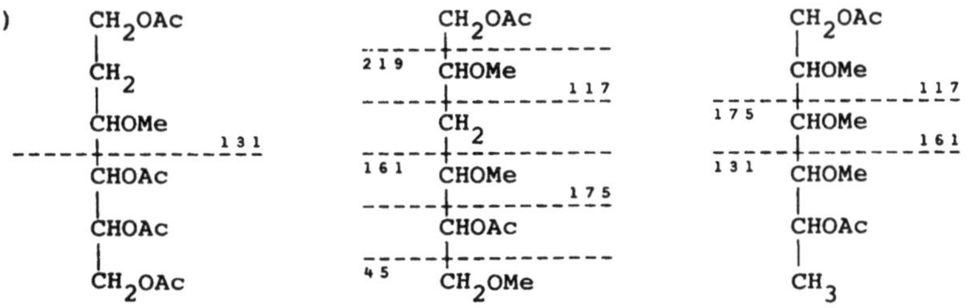

(K)

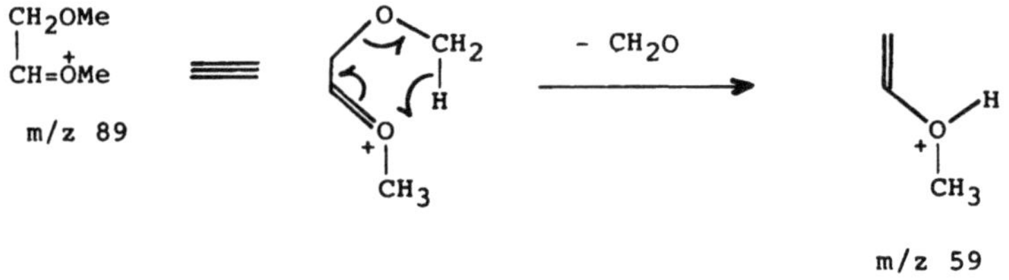

(L)

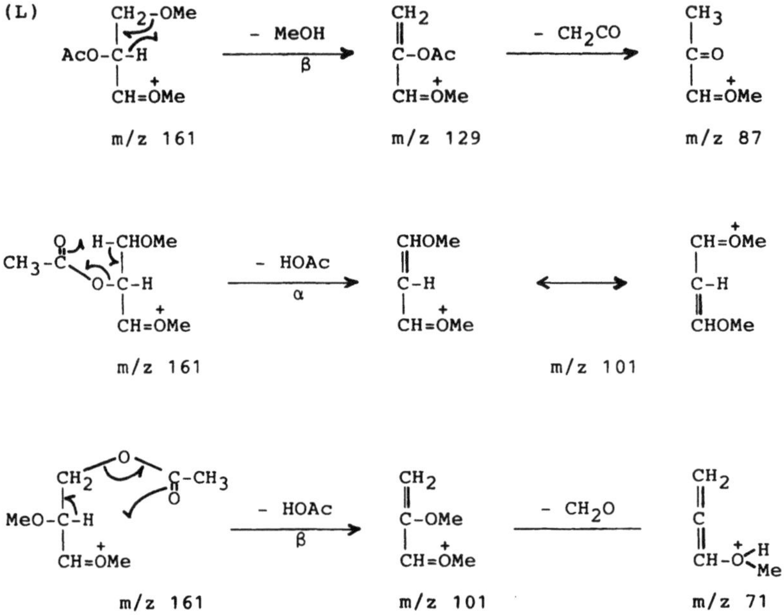

(M)

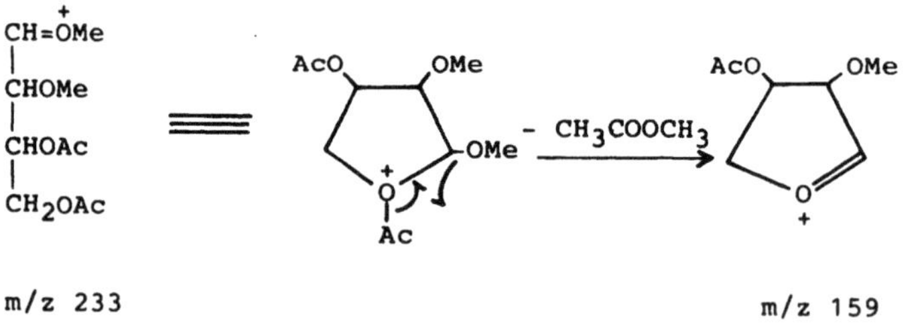

Elimination of methanol is only observed when the methoxyl group is situated at the β-position to the carbon atom having the formal charge. Elimination of acetic acid is possible from β- and α-positions, in the latter case provided that the resulting ion is stabilized by resonance. When an acetoxyl group is linked to an unsaturated carbon atom, ketene is eliminated with the formation of a carbonyl group.

(N) 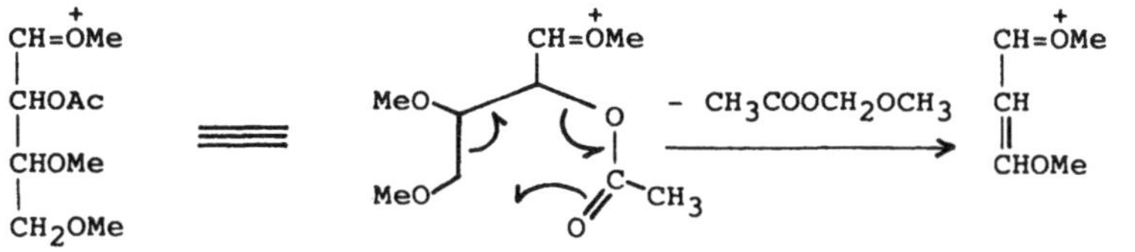

m/z 205 m/z 101

(O) 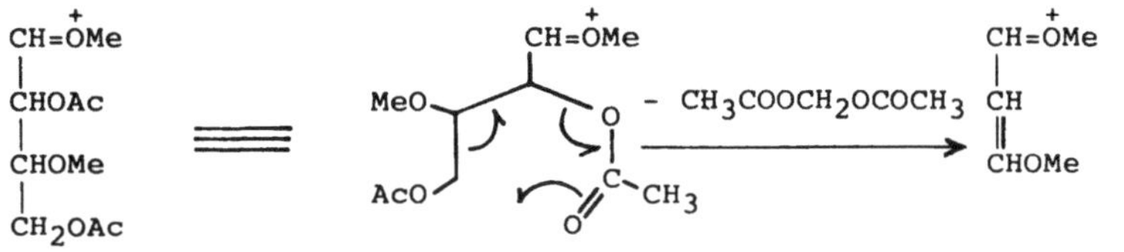

m/z 233 m/z 101

(P) 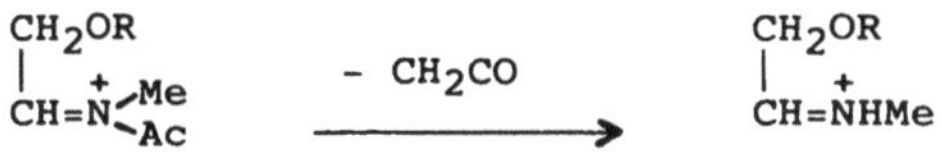

R=Ac : m/z 158 → m/z 116
R=Me : m/z 130 → m/z 88
Intense and highly characteristic peaks for assignments of partially methylated N-methyl-acetamido alditol acetates

(Q) 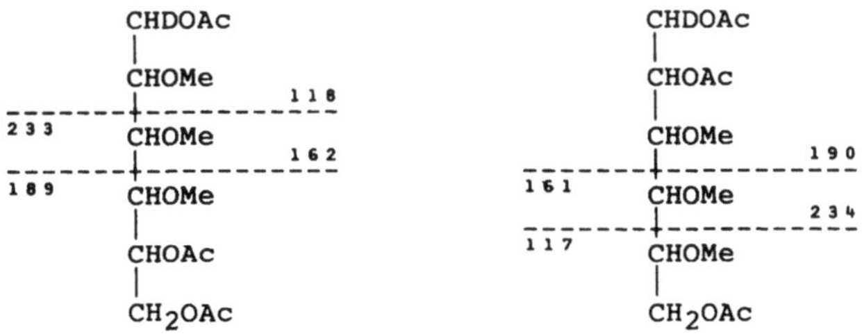

0.13 mol · l^{-1} sulfuric acid (103), 2–4 mol · l^{-1} trifluoroacetic acid (106), and 0.25 mol · l^{-1} sulfuric acid in 90% acetic acid (113–115). The mixture is reduced with $NaBH_4$ or NaB^2H_4, followed by acetylation of the liberated hydroxyl groups with acetic anhydride. The use of NaB^2H_4 facilitates the discrimination between primary hydroxyl groups. The partially methylated alditol acetates derived from neutral and amino sugars are analysed by GC-MS (4, 6, 8, 18, 59, 62, 103–106, 113, 116–122). Uronic acids, sialic acids and KDO can not be analysed in this way because of hydrolysis of their methyl ester groups.

3.1 EI-MS of partially methylated alditol acetates

EI-MS of partially methylated alditol acetates gives rise to very characteristic mass spectra, mostly without molecular ion peaks. The peak patterns obtained from the various derivatives yield information on the positions of the *O*-methyl and *O*-acetyl groups in the alditol chains. The *O*-methyl groups reflect the free hydroxyl groups in the corresponding monosaccharides of the native material. The mass spectra of stereoisomeric, partially methylated alditol acetates show only minor differences making an assignment of the sugar configuration (*gluco, galacto*, etc.) impossible. However, the nature of the parent monosaccharides can be derived from the retention times of the alditol derivatives on GC.

Comprehensive data concerning the mass spectra of partially methylated alditol acetates are available in the literature (6, 62, 103, 113, 116–121, 123–129). The relevant possibilities for the primary fragmentation of these derivatives have been summarized in Scheme 4 (A-J). Primary fragments are formed by α-cleavage, resulting in fission between the carbon atoms in the alditol chain. In principle, either of the two fragments formed can carry the positive charge. In the case of alditols derived from neutral sugars the charge is preferentially located on an ether oxygen instead of on an ester oxygen. The following rules can be formulated: 1. formation of ions of lower mass is preferred; 2. formation of ions from cleavage between two methoxylated carbon atoms is predominant, with no marked preference for one of the two possible cations (Scheme 4A); 3. in the formation of ions from cleavage between a methoxylated and an acetoxylated carbon atom, there is a high preference for the methoxyl-bearing cations (Scheme 4B), (in general the fission as shown in Scheme 4A is preferred over that in Scheme 4B, because the methoxyl radical formed seems to be better stabilised than the acetoxyl radical) and 4. ions formed by cleavage between two acetoxylated carbon atoms are generally of low abundance (Scheme 4C).

For alditols derived from *N,N*-methyl,acetyl amino sugars the same rules hold as mentioned above. However, the most preferable α-cleavage in partially methylated alditol acetates derived from 2-acetamido-2-deoxy sugars stems from the scission between C2, bearing the *N,N*-methyl,acetyl amino group, and C3, bearing the methoxyl or acetoxyl function, with the predominant localisation of the charge at the amino fragment (Scheme 4D–4E). α-Cleavages adjacent to deoxygenated carbon atoms are only significant when the neighbouring carbon atom bears a methoxy group (Scheme 4J).

The primary fragments give rise to secondary fragments, generally by single or successive eliminations of formaldehyde, methanol, ketene, acetic acid, methyl acetate, methoxymethyl acetate or acetoxymethyl acetate (Scheme 4K-4P). It is advantageous to use NaB^2H_4 for the preparation of the alditols from the mass spectrometric point of view, because it enables the differentiation between primary hydroxyl groups. The primary fragmentation patterns of 1,5,6-tri-*O*-acetyl-2,3,4-tri-*O*-methyl-hexitol-[1 – ^{2}H] and 1,2,6-tri-*O*-acetyl-3,4,5-tri-*O*-methyl-hexitol-[1-^{2}H] are given in Scheme 4Q to illustrate this feature.

To determine the substitution patterns and ring sizes of uronic acid moieties it is necessary to reduce these residues to the corresponding aldoses. The reduction can be carried out with $NaBH_4$ after complexation of the carboxyl group(s) with a carbodiimide in water (65). It is also possible to reduce the methyl ester groups in the permethylated carbohydrate material with $LiAlH_4$ in diethyl ether or tetrahydrofuran (130), or with $NaBH_4$ in 95% ethanol : oxolane (27:73, v/v) (106). For microgram scale procedures it has been advised to reduce the methyl esters of uronic acid residues in permethylated polysaccharides instead of to carboxyl-reduce the native polysaccharide followed by permethylation (106). The application of NaB^2H_4 and $LiAl^2H_4$ as reducing agents lead to the incorporation of two deuterium atoms at the carboxyl-group derived primary hydroxyl functions. In this way alditols derived from uronic acids can be distinguished by MS from the corresponding native aldoses.

If the carbohydrate chain contains methoxyl groups initially, the use of trideuteromethyl iodide is important to discriminate between originally present and chemically introduced methyl groups (131). For several specific degradation procedures in which methylation analysis plays a role, the use of trideuteromethyl iodide and/or ethyl iodide have been reported (132–135). The location of a methoxyl function at C4 or C5 in aldoses defines the ring size of the sugar residue. When acetoxyl groups are present at C4 and C5, 4-linked aldopyranosyl or 5-linked aldofuranosyl residues cannot be distinguished from each other. To this end specific degradation methods in combination with alkylation analysis have been developed (136).

Since the use of methylsulfinylmethanide causes only marginal desulfation in the methylation procedure, the location of ester sulfate groups can be established by application of the methylation technique before and after desulfation. The same holds for acetals. For phosphate substituents it has to be taken into account that they may migrate under alkaline conditions via intramolecular cyclic esters. Because *O*-acyl groups are cleaved during base treatment, other procedures have to be followed e. g. as described in (137), applying methylvinyl ether.

As was already indicated, a large series of EI mass spectra of partially methylated alditol acetates have been depicted in several references, e. g. in (103, 113, 116–121, 123–127, 120a). To give an impression of the highly characteristic peak patterns, the mass spectra of 1,5-di-*O*-acetyl-2,3,4,6-tetra-*O*-methyl-mannitol-[1-^{2}H], 1,3,5-tri-*O*-acetyl-2,4,6-tri-*O*-methyl-mannitol-[1-^{2}H] and 4-mono-*O*-acetyl-1,3,5,6-tetra-*O*-methyl-2-*N*-methyl-acetamido-2-deoxy-glucitol-[1-^{2}H] are shown in Figures 26–28. In Tables 2 and 3 fragment ions of several partially methylated alditol acetates of neutral and amino sugars, respectively, are summarized. Mass fragmentography has been shown to be of value for the analysis of complex mixtures.

Mass spectra of partially methylated alditol acetates of neutral sugars are not greatly influenced by the mass spectrometer used. However, for amino sugar derivatives the spectra can be affected by the type of instrument. The intensities of the diagnostic fragments of larger mass tend to be higher in the mass spectra of magnetically scanning

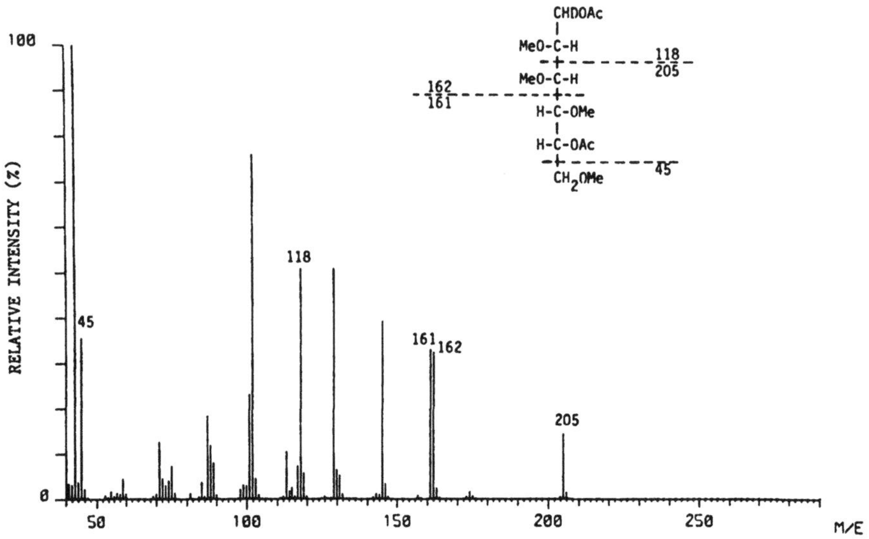

Figure 26. 70 eV EI mass spectrum of 1,5-di-*O*-acetyl-2,3,4,6-tetra-*O*-methyl-mannitol-[1-^{2}H].

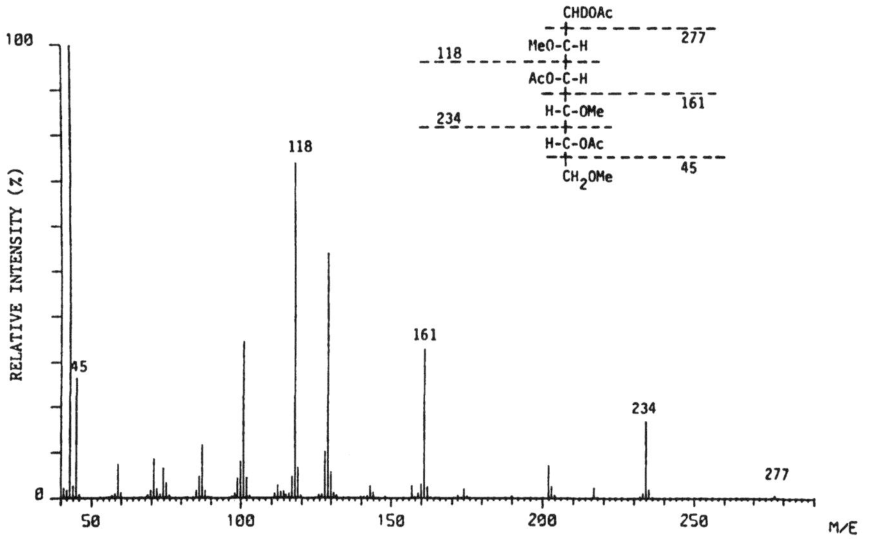

Figure 27. 70 eV EI mass spectrum of 1,3,5-tri-*O*-acetyl-2,4,6-tri-*O*-methyl-mannitol-[1-^{2}H].

instruments than in the mass spectra obtained with quadrupole instruments (8, 119). The temperature of the separator can also influence the peak pattern, probably because of pyrolysis (138). Furthermore it has been reported that the GC column phase can have an influence on the mass spectrometric result (121). Therefore, the availability of reference spectra of authentic amino sugar derivatives recorded with the same GC-MS combination is useful.

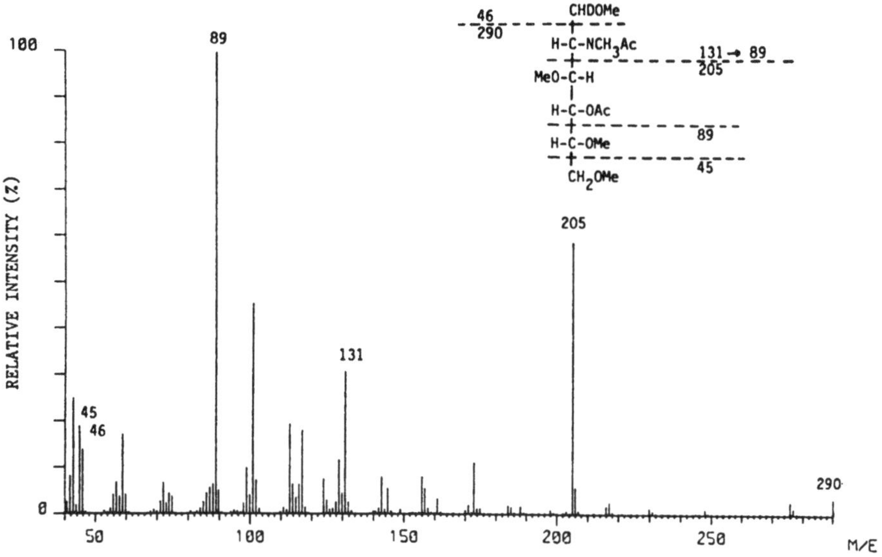

Figure 28. 70 eV EI mass spectrum of 4-mono-*O*-acetyl-1,3,5,6-tetra-*O*-methyl-2-*N*-methyl-acetamido-2-deoxy-glucitol-[1-^{2}H].

3.2 CI-MS of partially methylated alditol acetates

Using isobutane as the ionizing gas, a series of partially methylated alditol acetates of glucose, galactose and mannose has been investigated (139). The various derivatives give rise to abundant MH^+, $[MH-MeOH]^+$ and $[MH-HOAc]^+$ ions. The spectra yield direct information on the molecular weight of the substances and therefore also on the number of *O*-methyl and *O*-acetyl substituents in the alditol chain. Fragmentations of the alditol chain are very limited. The relative intensities of the ions are found to be sensitive to the source temperature. At higher temperatures there is a preference for the formation of $[MH-MeOH]^+$ and $[MH-HOAc]^+$ ions whereas the MH^+ ions are enhanced at lower temperatures. In contrast to the mentioned insensitivity of EI to stereochemical differences, CI(isobutane) has shown that it is possible to differentiate between diastereomers having identical locations of *O*-methyl and *O*-acetyl substituents in the alditol chain (139). Additional data on CI(isobutane)-MS of deoxyhexitol, hexitol and pentitol derivatives have been reported in (60). For a study of alditols derived from fructose, see (140). It is evident that CI-MS is highly suited for the recording of mass chromatograms based on quasi-molecular ions.

Partially methylated alditol acetates of neutral and amino sugars have also been investigated by CI(methane)-MS (141). The mass spectra show intense $[MH-HOAc]^+$ ions for neutral sugar derivatives, and MH^+ ions for amino sugar derivatives. In addition some fragmentation is observed.

Finally, data on the use of CI(ammonia)-MS have been reported (105). Ammonia may be considered to be a more suitable reagent gas than isobutane or methane. At an ion-source temperature of about 120 °C partially methylated alditol acetates of neutral sugars form mainly $[M + NH_4]^+$ ions. In the case of amino sugar derivatives there is a preference for the formation of MH^+ ions.

Table 2. Primary fragment ions (EI-MS) characteristic for the substitution pattern of partially methylated alditol acetates

Position of OMe groups	*m/z* values	Position of OMe groups	*m/z* values
Pentitol		*6-Deoxyhexitol*	
2 (4)	261,117	2	275,117
3	189	3	203,189
5	45	4	261,131
2,3 (3,4)	189,161,117	2,3	203,161,117
2,4	233,117	2,4	247,233,131,117
2,5	233,117,45	3,4	189,131
3,5	189,161,45	2,3,4	175,161,131,117
2,3,4	161,117	2,3,5	175,161,117,59
2,3,5	161,117,45	*Heptitol*	
Hexitol		3	333,189
2 (5)	333,117	2,6	117
3 (4)	261,189	2,7	117,45
6	45	3,6	189,117
2,3	261,161,117	4,6	261,233,117
2,4 (3,5)	305,233,189,117	6,7	89,45
2,5	305,117	2,3,6	305,161,117
2,6	305,117,45	2,3,7	305,161,117,45
3,4	233,189	2,4,6	233,117
3,6	233,189,45	2,4,7	233,117,45
4,6	261,161,45	2,6,7	377,349,117,89,45
5,6	333,89,45	4,6,7	261,205,89,45
2,3,4	233,189,161,117	2,3,4,6	321,277,233,161,117
2,3,5	233,161,117	2,3,4,7	277,233,205,161,117,45
2,3,6	277,233,161,117,45	2,3,6,7	277,161,117,89,45
2,4,6	277,233,161,117,45	2,4,6,7	349,321,233,205,117,89,45
2,5,6	117,89,45	3,4,6,7	205,189,89,45
3,4,6	233,205,189,161,45	2,3,4,6,7	249,205,161,117,89,45
3,5,6	305,205,189,89,45		
1,3,4,6	205,161,45		
2,3,4,6	205,161,117,45		
2,3,5,6	277,205,161,117,89,45		
1,2,3,4,5	177,161,133,117,89,45		
1,2,3,5,6	249,205,133,89,45		
1,2,4,5,6	249,205,133,89,45		
1,3,4,5,6	249,205,161,133,89,45		

3.3 Methylation analysis procedures based on partially methylated alditol acetates

Several detailed practical procedures for the methylation analysis of oligosaccharide-alditols, polysaccharides, glycoproteins, glycopeptides and glycolipids have been reported (103–106, 109, 142). Some of them can be used for quantities below 1 µg. Here, procedures will be presented essentially as described in (105) and (106).

Table 3. Primary fragment ions (EI-MS) characteristic for the substitution pattern of partially methylated 2-deoxy-2-(*N*-methyl)acetamidohexitol acetates

Position of OMe groups	*m/z* values
2-deoxy-2-(N-methyl)acetamidohexitol	
3	261,202,158
4	274,189,158
6	158,45
3,4	246,233,202,189,158
3,6	233,202,158,45
4,6	274,161,158,45
1,3,5	318,290,233,174,130,117,45
1,4,5	318,290,246,161,130,117,45
3,4,6	246,205,202,161,158,45
1,3,5,6	290,205,174,130,89,45
1,3,4,5	290,218,174,161,130,117,45
1,4,5,6	290,246,133,130,89,45

An aqueous solution of 50 µg glycan is transferred to a 1 ml reacti-vial. After lyophilisation the residue is dried for 18 h *in vacuo* over P_2O_5. The screw cap, septum and reacti-vial magnetic stirrer are also kept in the vacuum desiccator. The vial is flushed with dry argon or nitrogen and subsequently sealed. Dry dimethylsulfoxide (250 µl) is added with a syringe through the septum of the vial. The mixture is stirred magnetically (or sonicated) for 2 h at room temperature. For polysaccharides a much longer time may be required to dissolve the material. Subsequently, 60 µl 2 $mol \cdot l^{-1}$ sodium methylsulfinylmethanide in dimethylsulfoxide is added, followed by 35 µl methyl iodide after 1 h. Before adding the freshly distilled methyl iodide, the solution is frozen. During the whole procedure the mixture is continuously stirred. After 2 h a clear solution is obtained.

For purification of the permethylated glycan use is made of reversed-phase chromatography on Sep-Pak C_{18} cartridges. Before use, the cartridge is washed with 40 ml 100% ethanol. Subsequently, 2 ml 100% acetonitrile (HPLC-grade) followed by 4 ml water (HPLC-grade) are passed through the column. The methylation-reaction mixture is diluted with water, resulting in a dimethylsulfoxide/water (1:1, v/v) solution. The latter solution is slowly pushed (1 – 2 drops/s) with a syringe plunger, through the cartridge bed until the syringe is just empty. Take care to avoid air bubbles! The reacti-vial is washed with 500 µl dimethylsulfoxide/water (1:1, v/v), which is then passed through the cartridge in the same way. To remove more-polar contaminants from the cartridge, four 2 ml portions of water are added. The first three portions are pushed through the cartridge with the syringe plunger until the syringe is just empty; the fourth one is pushed completely through the cartridge. Subsequently, less-polar contaminants can be eluted with acetonitrile/water (3:17, v/v, solvent A; 1:4, v/v, solvent B). In the case of disaccharide-alditols four 2 ml portions of solvent A are added. For intermediate-sized glycans (d. p. 3 – 10) the cartridge is washed with three 2 ml portions of solvent A followed by one 2 ml portion of solvent B. Larger glycans need washing with two 2 ml portions of solvent A followed by two 2 ml portions of solvent B. Then, permethylated glycans are eluted from the cartridge with 2 ml 100% acetonitrile, followed by 4 ml 100% ethanol. The collected solution containing the permethylated material is evaporated to dryness with a stream of nitrogen. To transfer the glycan material to the bottom of the tube, the residue is dissolved in 1 ml methylene chloride and evaporated again. The product is transferred with methylene chloride (five times 0.2 ml portions) to a 0.3 ml reacti-vial (procedure 1) or to a small Pyrex glass tube (procedure 2). Methylene chloride is evaporated with a stream of nitrogen.

Procedure 1:

To the residue 100 µl 4 mol · l^{-1} trifluoroacetic acid is added. After thorough mixing, the solution is heated in an aluminum block for 4 h at 100 °C. Then the cooled trifluoroacetic acid solution is evaporated with a stream of nitrogen at room temperature until the sample is just dry. In order to concentrate the mixture of partially methylated monosaccharides at the bottom of the vial cone, methanol is added to the reacti-vial and evaporated with a stream of nitrogen. Also traces of residual trifluoroacetic acid are removed in this way.

The residue is mixed with 50 µl aqueous $NaBH_4$ or NaB^2H_4 (10 µg reducing agent/µl water) and kept for 2 h at room temperature. To convert the excess of reducing agent into boric acid, 5 µl acetic acid is added. After the addition of 50 µl methanol the solvent is evaporated with nitrogen.

Boric acid is removed as trimethylborate by co-evaporation with 50 µl 10% methanolic acetic acid (four times) using a stream of nitrogen. Finally, one evaporation to dryness with 75 µl methanol is carried out.

For the acetylation 75 µl acetic anhydride is added to the reacti-vial. After mixing, the vial is heated for 3 h at 120 °C in an aluminum block. The vial is then allowed to cool down to room temperature, and 50 µl toluene is added. The mixed solution is evaporated with a stream of nitrogen just to dryness. One additional evaporation just to dryness with 50 µl toluene is then performed. The residue is mixed with 25 µl methylene chloride and equilibrated for at least 3 h before GC-MS analysis.

Procedure 2:

To the residue 500 µl 0.25 mol · l^{-1} H_2SO_4 in 90% aqueous acetic acid is added. The tube is flushed with argon and heated in an aluminum block for 4–6 h at 80 °C. After cooling and neutralisation with 550 µl 0.5 mol · l^{-1} NaOH, the mixture is evaporated under reduced pressure.

Two additional evaporations are carried out with 200 µl water.

The residue is dissolved in 300 µl water containing 3 mg $NaBH_4$ or NaB^2H_4 and kept for 3 h at room temperature. The solution is mixed with 2 mol · l^{-1} acetic acid until pH $\approx$ 5 is reached and then evaporated under reduced pressure. Boric acid is removed as trimethylborate by co-evaporation with 1.5 ml 1% methanolic acetic acid (four times); the solvent is blown dry with nitrogen each time. The residue is dried *in vacuo* over P_2O_5 for 1 h.

For the acetylation 400 µl acetic anhydride is added. The mixture is heated for 4 h at 105 °C under argon in an aluminum block and excess reagent is removed by blowing with nitrogen just to dryness. The residue is taken up in 4 ml methylene chloride. This solution is extracted three times with 2 ml water. Finally, the methylene chloride phase is transferred to a Pyrex glass tube with a conical bottom and evaporated with nitrogen just to dryness. For GC-MS analysis the residue is taken up in 50 µl methylene chloride.

Several stationary phases for GC analysis of partially methylated alditol acetates have been recommended in the literature, e. g. Silar 9CP (59, 105), Dexsil 410 (59), SE-30 (59), OV-101 (59), OV-275 (60), OV-225 (103, 143, 144), ECNSS-M (103), SP-1000 (103, 145), OV-17 (104, 143, 144), DB-1 (105, 106), OV-1 (115, 146), CP-Sil 5 (120), Silar 10C (147). In several investigations it has been advised to use the retention data from more than one column. It has to be noted that the more polar columns as SP-1000 and Silar 9CP are not suitable for the analysis of amino sugar derivatives.

Notes:

1. For the preparation of sodium methylsulfinylmethanide, see (103).
2. The excess of methylsulfinylmethanide in the permethylation reaction can be checked with triphenylmethane. The latter reagent gives a red colour in the presence of the carbanion (148).
3. The Sep-Pak C_{18} purification procedure for permethylated glycans has been reported recently (106). Other procedures include dialysis (only applicable for permethylated polymers) (103), filtration of the methylation-reaction mixture over Sephadex LH-20 (105), chloroform or methylene chloride extraction of the permethylated material (103), eventually followed by filtration over silica gel (104) or Sephadex LH-20.
4. The hydrolysis of amino sugar-containing permethylated glycan chains needs special attention. When *N*-deacetylation of the methylated 2-acetamido-2-deoxyhexose units occurs as the first step, the adjacent glycosidic linkages at C1 become resistant to acid hydrolysis. For this reason several hydrolysis conditions have been studied (113, 114). The frequently used formolysis followed by hydrolysis with H_2SO_4 (103) results in good yields of partially methylated alditol acetates of neutral sugars, but give low and more variable yields of the amino sugar derivatives (115). The acetic acid/H_2SO_4 procedure described above gives good recoveries of amino sugar derivatives and of most of the neutral sugar derivatives (113, 114). Low molar ratios have been reported for trisubstituted *β*-D-mannose residues (2-*O*-methyl-mannitol) in glycoprotein glycans (115). Furthermore, the hydrolysis with 4 $mol \cdot l^{-1}$ trifluoroacetic acid as presented above gives good results (see also (116)).
5. In the case of hydrolysis with acetic acid/H_2SO_4, NaOH has been used for neutralisation. It is also possible to neutralise via anion-exchange chromatography on Bio-Rad AG-3 (acetate form) (104, 113).
6. Hydrolysis of permethylated carbohydrate chains containing a 2-acetamido-2-deoxyhexitol unit can give rise to some demethylation of the alditol residue. Some authors state that an *O*-demethylation at C1 occurs (121), whereas others conclude to a *N*-demethylation at C2 (118, 119). The extent of demethylation is also influenced by the substitution pattern of the alditols. Recently, *O*-demethylation at C1 and C3 has been discussed (121a).
7. Generally, the reduction with $NaBH_4$ or NaB^2H_4 is carried out in water. However, procedures have been published wherein the reduction is performed in alkaline medium with or without ethanol (see for instance (106).

8. For the analysis of partially methylated alditol acetates obtained from permethylated glycoproteins a final purification step on a silica gel G column is necessary (142). After washing the column with petroleum ether/ethyl acetate (2:1, v/v), neutral sugar derivatives are eluted with petroleum ether/ethyl acetate (1:1, v/v), and, subsequently, amino sugar derivatives with methanol.
9. The quantitative aspects of methylation analysis procedures are poorly understood. Several factors influence the recovery of partially methylated alditol acetates, e. g., undermethylation, incomplete hydrolysis, degradation and demethylation during hydrolysis, incomplete reduction, incomplete acetylation and contaminants. Also the degradation on the GC-column phase is important, especially for amino sugar derivatives (122). In general molar ratios of the different sugar derivatives are calculated on the basis of FID responses, taking into account similar molar adjustment factors for the various partially methylated alditol acetates. A better approach is the use of molar response-factors based on the effective-carbon-response theory (134). It has to be noted that in several examples the FID responses of certain amino sugar derivatives have shown to be lower than those of the neutral ones. To overcome these problems it is suggested to estimate the ratio of the total amount of amino sugar residues to that of the neutral sugar residues by sugar analysis. Other quantification procedures are based on selected ion-monitoring, especially in CI-approaches, which seem to give very acceptable results (105).

3.4 Other methylation analysis procedures

Instead of preparing alditol acetate derivatives after the hydrolysis step, other working-up procedures can be followed, e. g. leading to trimethylsilylated *O*-methyl oximes (149), acetylated oxime derivatives (30), and acetylated aldononitriles (26, 30, 150–154). The acetylated aldononitriles have been studied extensively by EI- and CI-MS. They give rise to EI fragmentation patterns comparable to those presented for the partially methylated alditol acetates.

In some approaches hydrolysis has been replaced by methanolysis, resulting in the formation of partially methylated methyl glycosides. The monosaccharide derivatives can be analysed in their underivatised forms, or as the corresponding trimethylsilylated or acetylated compounds (1–7, 75, 76, 155–162). A detailed procedure has been worked out for the methylation analysis of glycoprotein glycans using acetylated, methylated methyl glycosides (160). The report includes GC-retention times of various α and/or β glycosides and a detailed EI-fragmentation scheme for the determination of the positions of the methyl and acetyl substituents in galactose, glucose, mannose and *N*-acetylglucosamine derivatives. Moreover, a series of mass spectra have been depicted. It has to be noted that methanolysis (0.5 $mol \cdot l^{-1}$ methanolic HCl, 24 h, 80 °C) does not split quantitatively the linkage between *N*-acetylglucosamine and asparagine.

For the linkage analysis of sialic acids in non-reducing or internal positions, methanolysis is the appropriate method of solvolysis. The resulting (partially) methylated methyl ester β-methyl glycosides show very characteristic EI mass spectra (163, 164), after treatment with reagents for acetylation or *N*-acetylation/trimethylsilylation,

Table 4. GC and MS data of trimethylsilylated/methylated *N,N*-acyl,methyl-neuraminic acid methyl ester β-methyl glycosides. The R_{Neu5Ac}-values on a packed column (2 m × 4 mm, i.d.) of 3.8% SE-30 at 220 °C (163) and on a capillary column (80 m × 0.35 mm, i.d.) wall-coated with OV-101 at 215 °C (164) are given relative to Neu5Ac4,5,7,8,9Me$_5$ methyl ester β-methyl glycoside.

Sialic acid* (as methyl ester β-methyl glycoside)	R_{Neu5Ac} 3.8% SE-30	R_{Neu5Ac} OV-101	*m/z* values							
			A	B	C	D	E	F	G	H
Neu5Ac4,5,7,8,9Me$_5$	1.00	1.00	392	348	318	254	201	89	129	298
Neu5Ac4,5,7,8Me$_4$	1.30	1.31	450	406	318	254	201	147	129	298
Neu5Ac4,5,7,9Me$_4$	1.14	1.14	450	406	318	254	201	147	129	356
Neu5Ac4,5,8,9Me$_4$	1.07	1.06	450	406	376	312	201	89	129	298
Neu5Ac5,7,8,9Me$_4$		1.20	450	406	376	254	259	89	187	298
Neu5Ac4,5,7Me$_3$	1.55	1.52	508	464	318	254	201	205	129	356
Neu5Ac4,5,9Me$_3$	1.27	1.27	508	464	376	312	201	147	129	356
Neu5Ac5,7,8Me$_3$		1.60	508	464	376	254	259	147	187	298
Neu5Ac5,7,9Me$_3$		1.39	508	464	376	254	259	147	187	356
Neu5Ac5,8,9Me$_3$		1.23	508	464	434	312	259	89	187	298
Neu5Ac4,5Me$_2$	1.70	1.78	566	522	376	312	201	205	129	356
Neu5Ac5,7Me$_2$		1.91	566	522	376	254	259	205	187	356
Neu5Ac5,9Me$_2$	1.43	1.50	566	522	434	312	259	147	187	356
Neu5Ac5Me	1.89	2.05	624	580	434	312	259	205	187	356
Neu5MeGc4,5,7,8,9Me$_5$#			422	378	348	284	201	89	159	328
Neu5MeGc4,5,7,9Me$_4$			480	436	348	284	201	147	159	386

* Neu5Ac4,5,7,8,9Me$_5$ = 4,7,8,9-tetra-*O*-methyl-*N,N*-acetyl,methyl-neuraminic acid; etc.
# MeGc = methylated glycolyl group.

(for a comprehensive review, see (36)). Tables 4 and 5 summarize the GC and MS data of a series of trimethylsilylated/methylated *N,N*-acyl,methyl-neuraminic acid methyl ester β-methyl glycosides and of a series of acetylated/methylated *N,N*-acyl,methyl-neuraminic acid methyl ester β-methyl glycosides, respectively. The presented fragment ions A-H have already been discussed in relation to Scheme 3. Typical mass spectra of some sialic acid derivatives are depicted in Figures 29–33. The choice of the applied methanolysis conditions in relation to the possible release of the *N*-acyl group seems to be very important. It has been shown that the *N*-acyl (*N*-acetyl or methylated *N*-glycolyl) groups of sialic acids in terminal positions of the permethylated carbohydrate chain are resistant to cleavage in 0.5 mol·l^{-1} methanolic HCl (18 h, 80 °C). However, internal sialic acids are *N*-deacylated to a large extent. Therefore, after methanolysis *N*-(re)acetylation is necessary in the working-up procedure for trimethylsilylated derivatives. The use of 0.05 mol·l^{-1} methanolic HCl (1 h, 80 °C) does not give *N*-deacylation, but these milder conditions do not liberate sialic acid quantitatively from the sialo-carbohydrate chain (36).

Table 5. GC and MS data of acetylated/methylated *N,N*.acyl,methyl-neuraminic acid methyl ester *β*-methyl glycosides. The R_{Neu5Ac}-values on a packed column (2 m × 4 mm, i.d.) of 3.8% SE-30 at 220 °C (163) are given relative to Neu5Ac4,5,7,8,9Me_5 methyl ester *β*-methyl glycoside. For an explanation of the '-' signs, see the discussion in relation to Scheme 3. For high-resolution exact mass measurements, see (36) and (163).

Sialic acid* (as methyl ester *β*-methyl glycoside)	R_{Neu5Ac}	*m/z* values						
		A	B	C	D	E	F	G
Neu5Ac4,5,7,8Me_4	1.47	420	376	318	254	201	117	129
Neu5Ac4,5,7,9Me_4	1.25	420	376	318	254	201	–	129
Neu5Ac4,5,8,9Me_4	1.08	420	376	–	–	201	89	129
Neu5Ac5,7,8,9Me_4		420	376	346	254	–	89	157
Neu5Ac4,5,7Me_3	1.75	448	404	318	254	201	–	129
Neu5Ac4,5,9Me_3	1.26	448	404	–	–	201	–	129
Neu5Ac4,5Me_2	1.70	476	432	–	–	201	–	129
Neu5Ac5,9Me_2	1.63	476	432	–	–	–	–	157
Neu5Ac5Me	2.17	504	460	–	–	–	–	157
Neu5MeGc4,5,7,9Me_4		450	406	348	284	201	–	159

* For an explanation of the abbreviations, see Table 4.

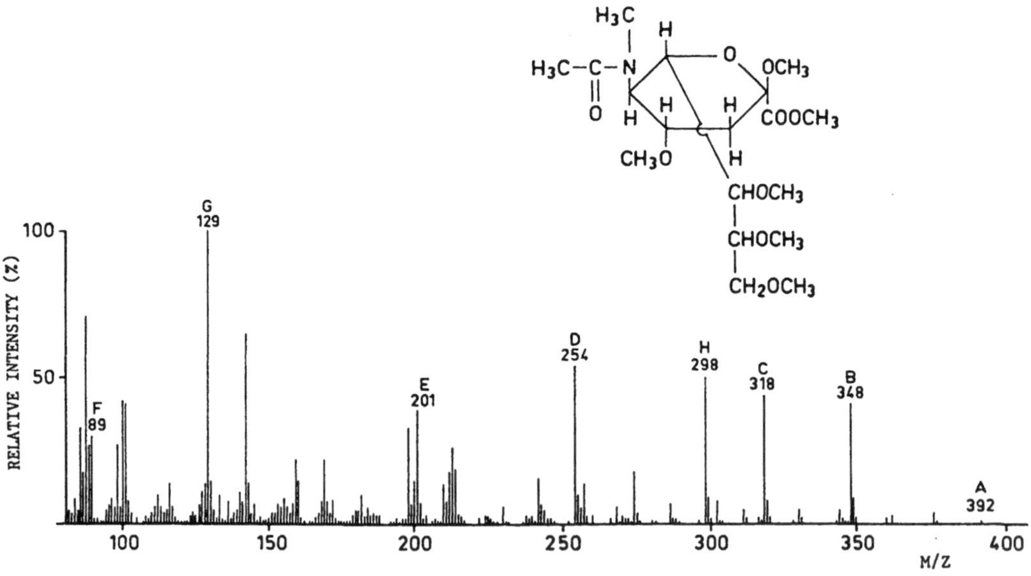

Figure 29. 70 eV EI mass spectrum of 4,7,8,9-tetra-*O*-methyl-*N,N*-acetyl,methyl-neuraminic acid methyl ester *β*-methyl glycoside.

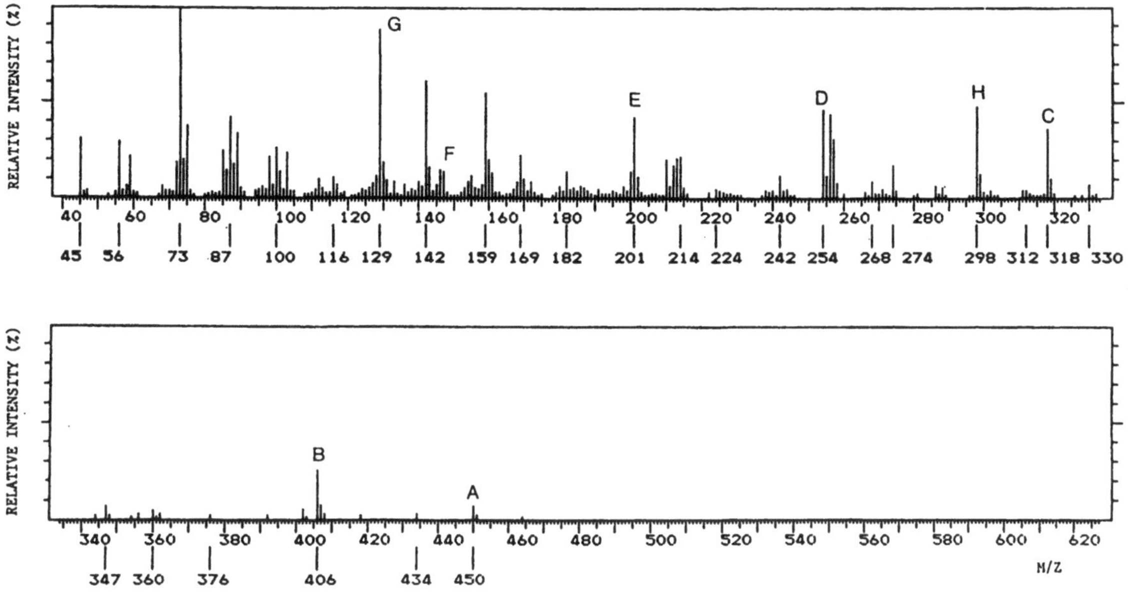

Figure 30. 70 eV EI mass spectrum of 4,7,8-tri-*O*-methyl-9-mono-*O*-trimethylsilyl-*N*,*N*-acetyl,methyl-neuraminic acid methyl ester *β*-methyl glycoside.

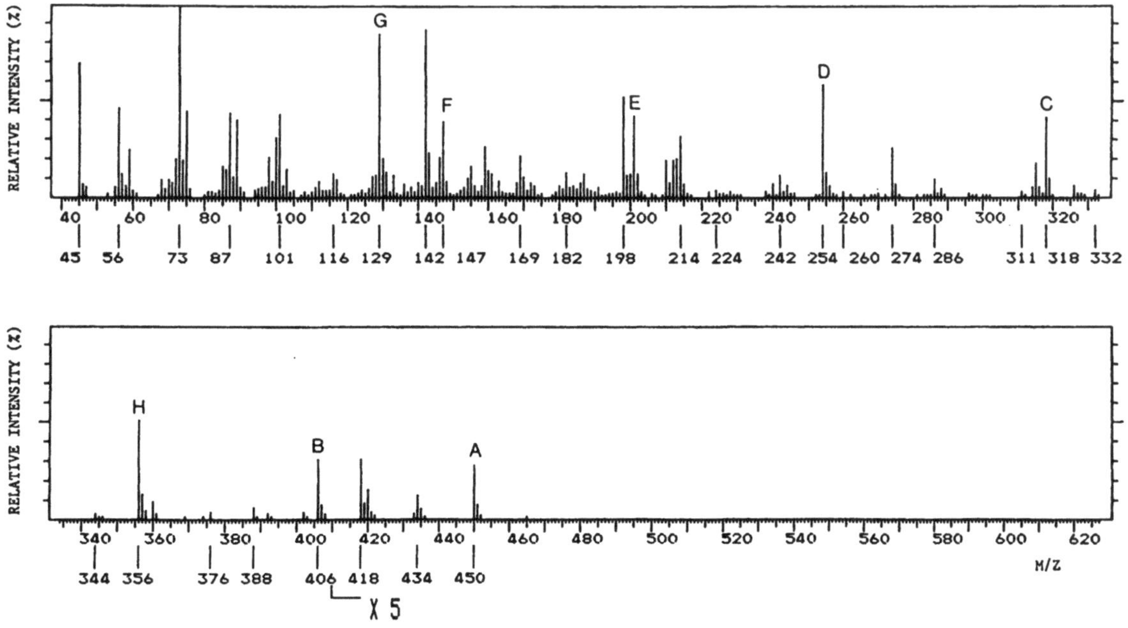

Figure 31. 70 eV EI mass spectrum of 4,7,9-tri-*O*-methyl-8-mono-*O*-trimethylsilyl-*N*,*N*-acetyl, methyl-neuraminic acid methyl ester *β*-methyl glycoside.

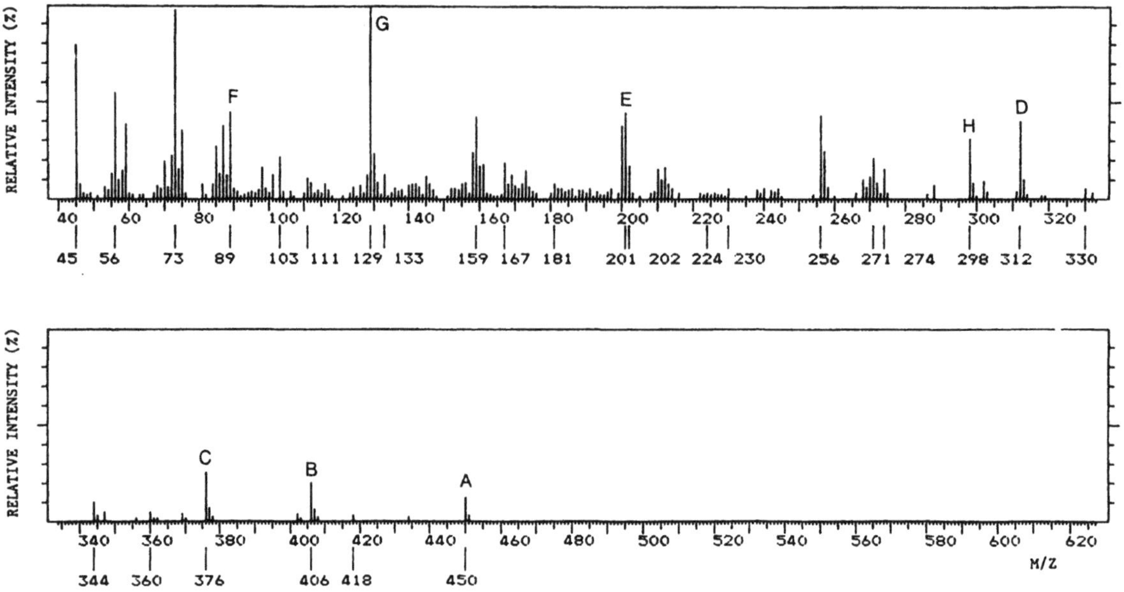

Figure 32. 70 eV EI mass spectrum of 4,8,9-tri-*O*-methyl-7-mono-*O*-trimethylsilyl-*N*,*N*-acetyl,methyl-neuraminic acid methyl ester *β*-methyl glycoside.

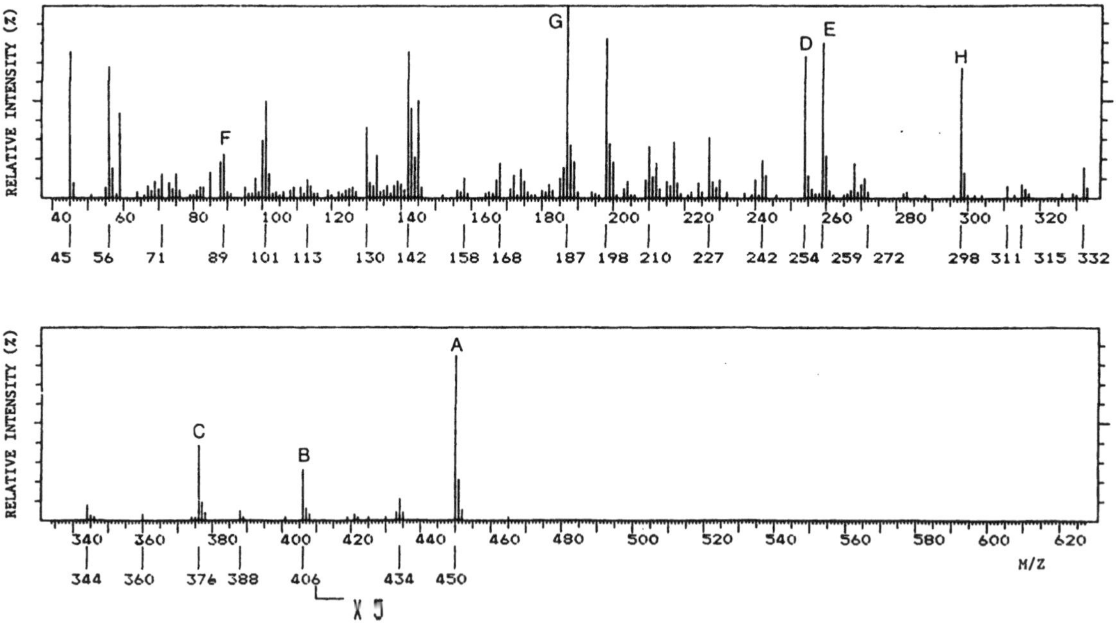

Figure 33. 70 eV EI mass spectrum of 7,8,9-tri-*O*-methyl-4-mono-*O*-trimethylsilyl-*N*,*N*-acetyl,methyl-neuraminic acid methyl ester *β*-methyl glycoside.

3.5 Methylation analysis of oligosaccharides isolated from physiological fluids

Methylation analysis is one of the methods incorporated in studies concerning the structure elucidation of oligosaccharides isolated from physiological fluids like milk and urine. In the case of patients with inborn errors of metabolism related to the carbohydrate catabolism of polysaccharides, glycoproteins and glycolipids relatively large amounts of carbohydrate material is available from the urine. For applications in the field of clinical biochemistry, a series of methylation analysis examples can be found in (165–199, 199a, 199b, 199c).

4 Sequence Analysis

MS of oligosaccharides, oligosaccharide-alditols, glycopeptides and glycolipids have been reviewed regularly (1–3, 6–8, 200–202). Permethylated derivatives are among the most commonly used derivatives. However, trimethylsilylated and acetylated derivatives have also shown their suitability for structural studies. Most of the published MS data entail EI-MS, but CI-MS, DCI-MS, FI-MS, FD-MS, SI-MS and FAB-MS have been employed. For EI, CI and FI conversion of polar saccharides into thermally more stable, volatile derivatives is important (203–276) and when possible GC is used. DCI, FD, SI and FAB are ionization methods which are applicable for the analysis of derivatised and underivatised saccharides (277–310).

Fundamental studies have been carried out with the aim of developing general MS methods for the determination of the sequence of monosaccharide units in terms of hexoses, pentoses, heptoses, deoxyhexoses, uronic acids, aminohexoses, sialic acids, etc. Furthermore for disaccharide derivatives discrimination rules to establish the type of glycosidic linkage have been developed (see below). Higher saccharides were sometimes incorporated in the latter studies.

4.1 Characterisation of permethylated oligosaccharide-alditols

Analysis of reducing oligosaccharides is most conveniently made after conversion of these compounds into the corresponding oligosaccharide-alditols. The reduction can be carried out with $NaBH_4$ or NaB^2H_4. Then the oligosaccharide-alditols are permethylated as already described in the section on methylation analysis.

Permethylated oligosaccharide-alditols rarely give rise to molecular ions in EI-MS. Fortunately, molecular weights can often be deduced from specific fragment ions. EI mass spectra contain a great deal of structural information due to extensive fragmentation and, for the interpretation of these spectra, the general principles developed for methylated ring forms and alditol chains are applicable (1, 6). In Scheme 5 a number of mass spectral fragmentation pathways in permethylated methyl glucopyranoside

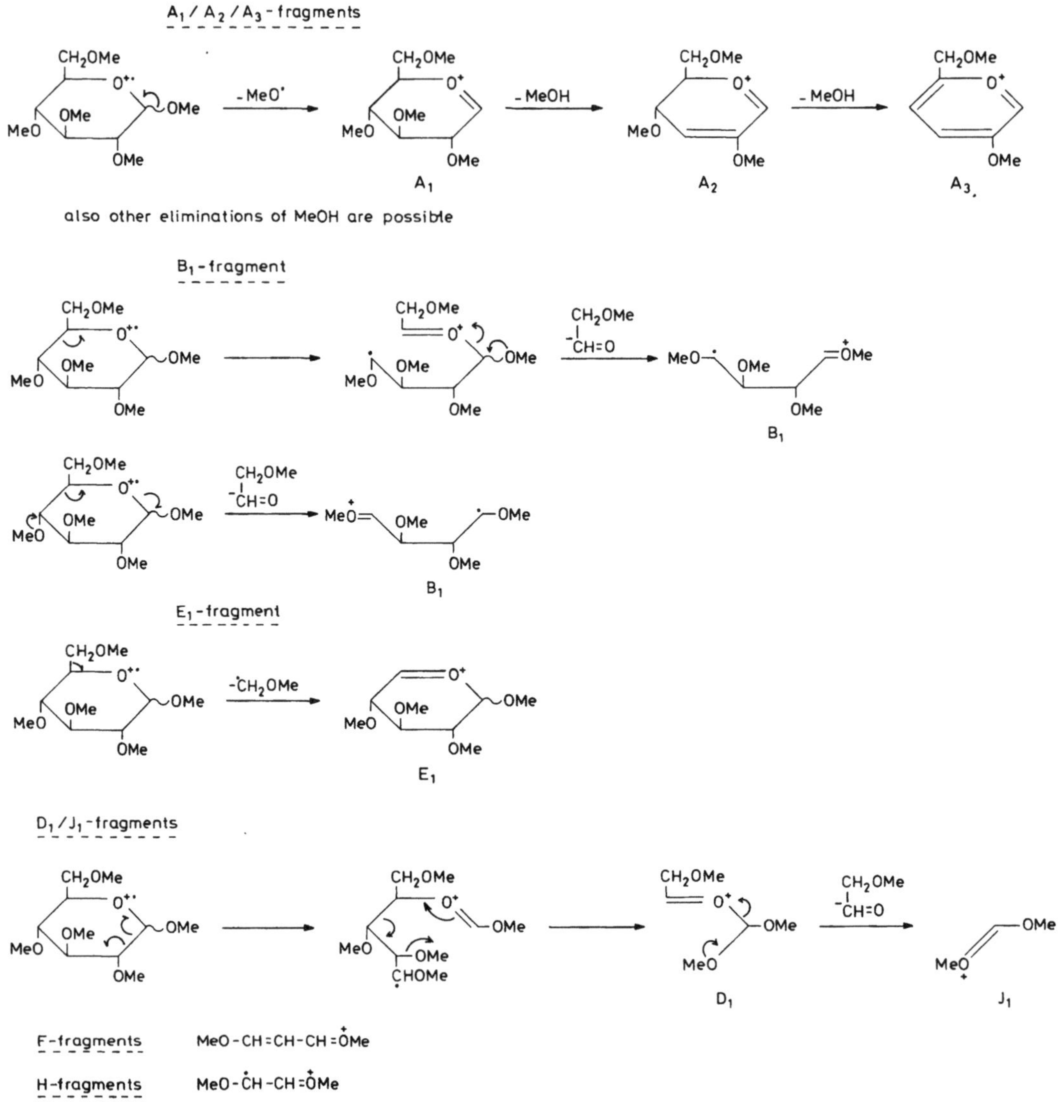

Scheme 5. Some important examples of fragmentation in permethylated methyl glucopyranoside. For detailed information, see (1).

are shown. Comparable breakdown pathways are followed in deoxy sugars, amino sugars and uronic acids. In furanose ring forms similar fragmentations are observed. The introduced nomenclature system with upper-case letters A-K to describe the various fragment ions (1), has also been used for the development of a fragmentation scheme for unbranched oligosaccharides (6, 203, 204) and oligosaccharide-alditols (6, 18). As has been depicted in Scheme 6 for the sequence fragment ions, the upper-case letters are supplied with lower-case letters a, b, c, etc., to denote the various sugar

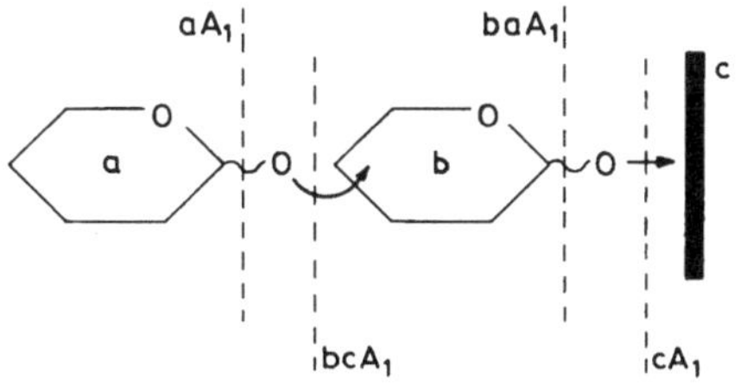

Scheme 6
Sequence fragment ions of the A-series in a linear trisaccharide-alditol a-b-c.

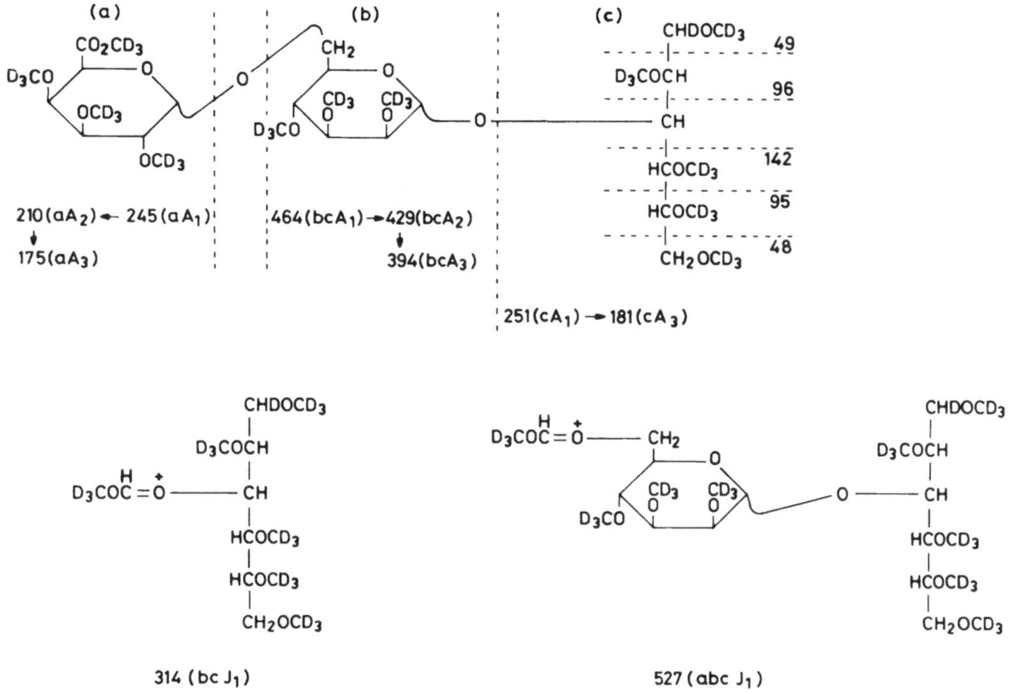

Figure 34. Important fragmentation pathways in the pertrideuteromethylated trisaccharide-alditol-[1-^{2}H] prepared from α-D-Gal*p*A-(1 → 6)-α-D-Man*p*-(1 → 3)-D-Man.

units starting from the non-reducing monosaccharide. When more than one lower-case letter has to be used, the first lower-case letter characterises the part of the molecule from which the fragment originates, whereas the second designates the unchanged part of the molecule, etc. For instance, baA$_1$ denotes an ion that has arisen from cleavage of ring b following pathway A and is substituted by ring a (Scheme 6). It may be evident that the A-series is highly valuable for the sequence analysis of oligosaccharide-alditols. In Figures 34 and 35 two examples are shown which include this type of fragmentation: the pertrideuteromethylated trisaccharide-alditol-[1-^{2}H] prepared from α-D-Gal*p*A-(1 → 6)-α-D-Man*p*-(1 → 3)-D-Man, which has been isolated from a partial hydrolysate of the polysaccharide associated with the coccoliths of the alga *Emiliania huxleyi* (Lohmann) Kamptner (205), and the permethylated trisaccharide-alditol-[1-^{2}H] prepared from α-D-Man*p*-(1 → 3)-β-D-Man*p*-(1 → 4)-D-GlcNAc,

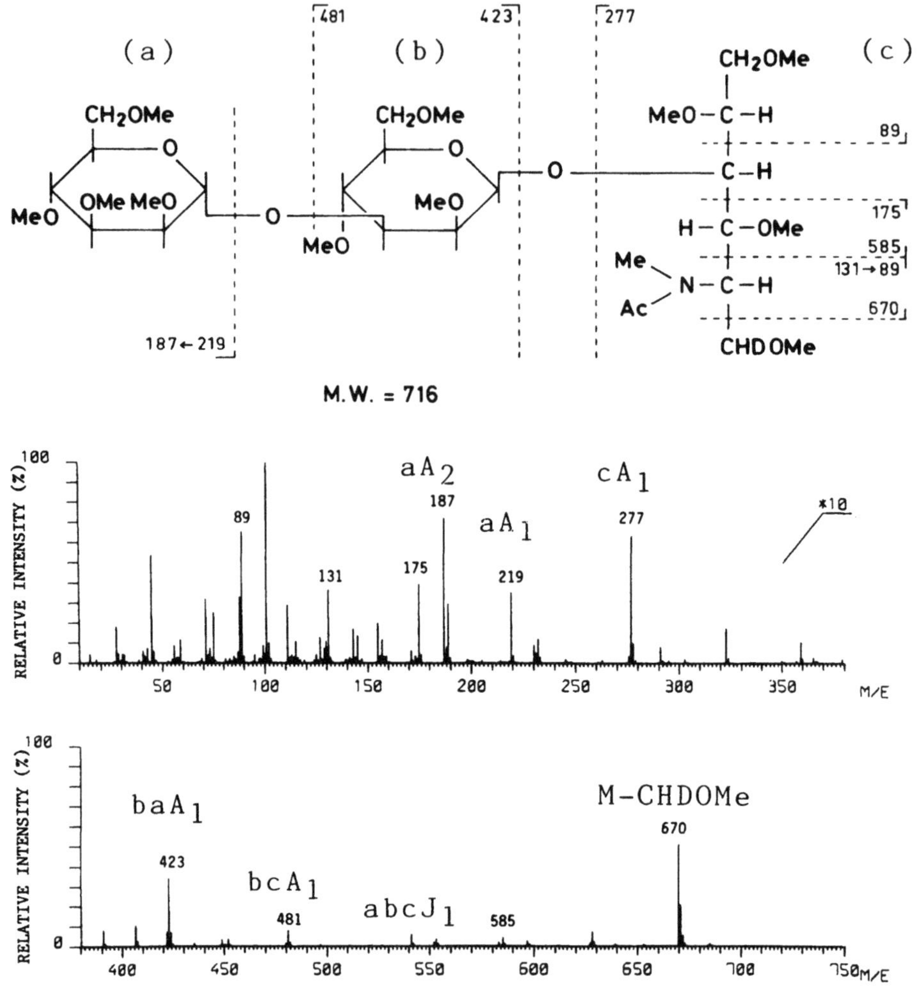

Figure 35. 70 eV EI mass spectrum of the permethylated trisaccharide-alditol-[1-^{2}H] prepared from urinary α-D-Man*p*-(1 → 3)-*β*-D-Man*p*-(1 → 4)-D-GlcNAc.

which has been isolated from the urine of a a-mannosidosis patient (206). Table 6 presents the *m/z* values of the A_1 fragments and the mass increments for internal sugar residues of different monosaccharide classes. It has to be noted that the expected theoretical A_1 fragments are not always observed.

Although model studies of di- and trisaccharides have shown that the EI mass spectra do contain information with regard to the type of glycosidic linkages, in most cases only the J_1 fragment is useful for this purpose in unknown saccharides. The reason is that discrimination rules are mainly based on intensity differences between selected peaks in the spectrum. The J_1 fragment ion is indicative for the presence or absence of 1 → 3 linkages (Scheme 5). For instance the absence of *m/z* 337 (bcJ_1 fragment ≡ CH_3OCHO-HexNAc-ol-[1-^{2}H] in the mass spectrum of Figure 35 suggests the occurrence of a → 3)-Hex-HexNAc-ol-[1-^{2}H] unit.

Table 6. Survey of m/z-values of A_1 fragments and mass increments, corresponding to internal sugar residues, of a series of different monosaccharide classes

Non-reducing units m/z A_1 fragments		Internal units mass increments		Alditol-[1-^{2}H] units m/z A_1 fragments	
Hex →	219	→ Hex →	204	→ Hexol	236
Deoxyhex →	189	→ Deoxyhex →	174	→ Deoxyhexol	206
Pent →	175	→ Pent →	160	→ Pentol	192
HexA →	233	→ HexA →	218	→ HexAol	250
HexNAc →	260	→ HexNAc →	245	→ HexNAcol	277

Hex = hexose; Deoxyhex = deoxyhexose; Pent = pentose; HexA = hexuronic acid; HexNAc = *N*-acetylhexosamine.

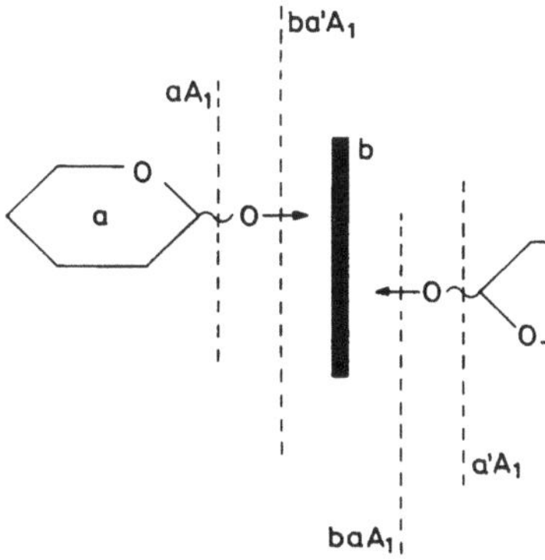

Scheme 7
Sequence fragment ions of the A-series in a branched trisaccharide-alditol a-(a'-)b.

The primary cleavages of the C-C bonds in the alditol unit give information about the type of glycosidic linkage next to this residue (207–214), although sometimes the intensity of characteristic fragments is not sufficiently high to serve as reliable criteria for the identification of the type of linkage. Comparison of the MS data of non-labeled and ^{2}H-labeled hexitols is useful. Discrimination rules have been reported for a series of (1→2)-, (1→3)-, (1→4)- and (1→6)-linked hexosyl-hexitols-[1-^{2}H] (207, 208), a series of (1→3)-, (1→4)- and (1→6)-linked hexosyl-2-acetamido-2-deoxy-hexitols (211), and a series of (1→3)-, (1→4)- and (1→6)-linked *N*-acetylglucosaminyl-*N*-acetylglucosaminitols (212). The presence of *N,N*-methyl,acetyl functions in the alditol chains strongly influences the peak patterns. See also Figures 34 and 35 for primary cleavages of the alditol C-C bonds.

The sequence analysis of branched oligosaccharide-alditols with a branching point at the alditol unit takes place in a similar way (Scheme 7). The m/z values of the A_1 fragments being indicative for the alditol units in linear oligosaccharide-alditols are absent (see cA_1 in Scheme 6 and Table 6). Instead, new fragment ions at higher m/z values corresponding with a substituted alditol unit appear in the mass spectra (e.g. baA_1 and $ba'A_1$ in Scheme 7). Furthermore, an additional fragmentation pathway has been observed, as indicated in Scheme 8. For a hexitol-[1-^{2}H] this fragmentation leads to the formation of m/z 250 (214). In the case of 2-acetamido-2-deoxy-hexitols this peak is observed at m/z 291 (215).

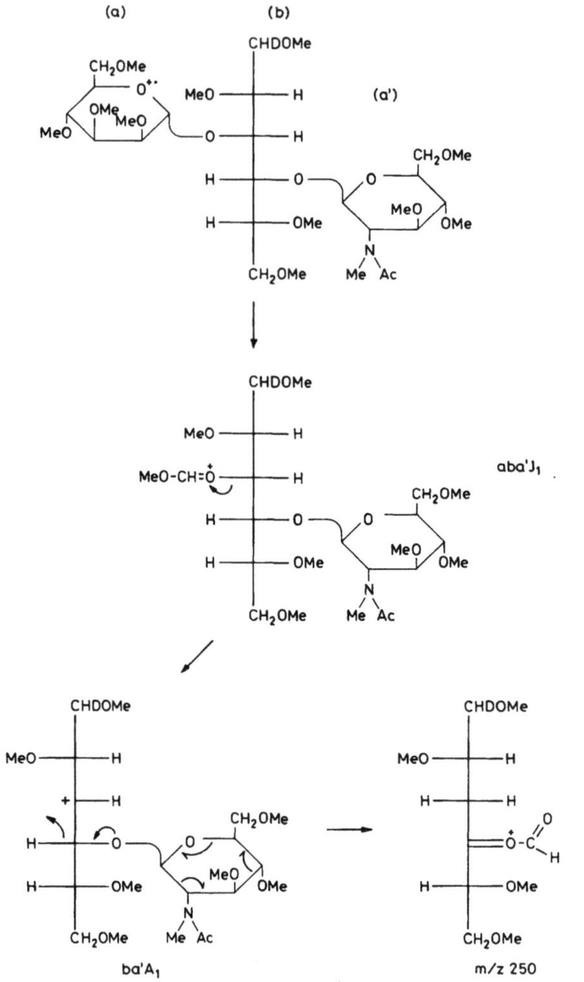

Scheme 8. Formation of the characteristic peak at *m/z* 250 in a branched oligosaccharide-alditol-[1-^{2}H].

Also CI-MS is currently used for the characterisation of permethylated oligosaccharide-alditols (214, 216–211). CI(isobutane)-MS of (1→2)-, (1→3)-, (1→4)- and (1→6)-linked aldosyl-alditols give rise to intense quasi-molecular MH^+ ions and characteristic sequence ions (216). Close inspection of the spectra of the various disaccharide-alditols showed that the intensity ratios of the alditol A_1^+ ions and the alditolhydrate $A_1OH_2^+$ ions could be correlated with the linkage positions between glucose and glucitol. Furthermore, the values found for the (1→4)-linked Glc*p*-Glc-ol and the trisaccharide-alditol Glc*p*(1→4)Glc*p*(1→4)Glc-ol were in the same range. It is not evident if these rules are generally applicable. Figure 36 shows the fragmentation in the mentioned trisaccharide-alditol. The formation of the cA_1^+ ion can be explained by protonation of the glycosidic oxygen followed by assisted cleavage of the bond at the alditol site of the molecule. The $cA_1OH_2^+$ ions probably arise by

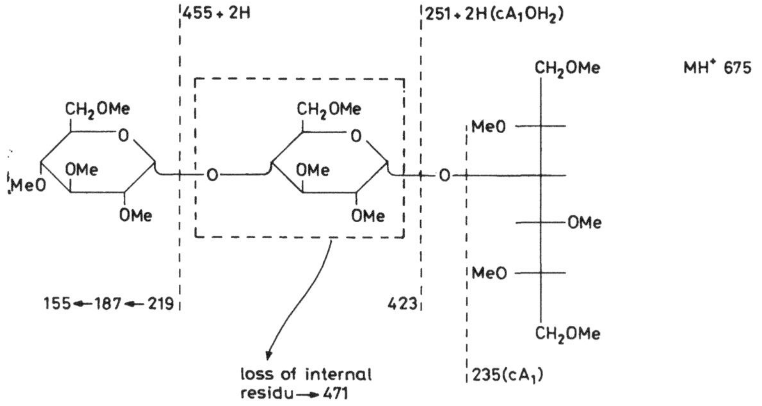

Figure 36. CI(isobutane) mass fragmentation pathways in a trisaccharide-alditol.

protonation at a site remote from the glycosidic oxygen followed by elimination of an unsaturated sugar molecule (H-rearrangement), but direct formation of cA_1^+ from $cA_1OH_2^+$ can not be excluded. The interpretation of the CI(isobutane) mass spectra of the tri- and higher oligosaccharide-alditols is complicated by the occurrence of an unusual elimination reaction (Fig. 36). The MH^+ ion can eliminate the glycosyl residue directly attached to the alditol. This results in a glycosyl-residue-elimination fragment ion corresponding with the MH^+ ion of an oligosaccharide-alditol with one sugar residue less (220). In this way a tetrasaccharide-alditol can eliminate successively two internal glycose residues. The elimination reaction is influenced by the substitution pattern of the sugar unit linked to the alditol. In the case of a $(1 \rightarrow 3)$-linked residue the relative abundance of the glycosyl-residue-elimination fragment ion decreases dramatically. The loss of an internal residue has not been detected in the ammonia/isobutane mass spectra (216). A detailed study on the CI(ammonia)- and CI(isobutane)-MS of sialo- and corresponding asialo-oligosaccharide-alditols derived from gangliosides has been published (221). The CI(ammonia) mass spectra show both protonated and ammonia adduct molecular ions. In addition several major fragment (sequence) ions are observed, formed by direct cleavage of the glycosidic bond as well as by cleavage of this bond involving H-rearrangement. They can occur as such in the protonated form, or as ammonia adduct ion. Also other more complicated pathways have been detected and detailed information can be obtained from the spectra with regard to the branching structures, CI(ammonia) and CI(isobutane) being complementary in the structural eludication of these oligosaccharides.

Hydrazinolysis-nitrous acid deamination of oligosaccharide chains containing 2-acetamido-2-deoxy-glucose residues leads to specific cleavage at C1 of the latter residues with formation of oligosaccharides ending on 2,5-anhydromannose. The permethylated oligosaccharide-2,5-anhydromannitols have been extensively studied by EI-MS, giving structural information comparable to that obtained from normal oligosaccharide-alditols (222–224).

An enormous number of papers in the literature report on the structural analysis of oligosaccharides and their corresponding alditols using MS of the permethylated derivatives. In the field of clinical biochemistry structural studies have been carried out on a great variety of carbohydrate chains. Physiological fluids like milk and urine (especially in the case of lysosomal storage diseases) are main sources of oligosaccharide material. A series of interesting examples containing many reports of mass spectra can be found in (215, 222, 225–253, 199a, 253a, 253b).

4.2 Characterisation of permethylated glycopeptides

Only a few studies on the mass spectrometric analysis of glycopeptides are known. Permethylated sialo- and asialo-biantennary glycopeptides derived from *N*-glycosidic glycoproteins give rise to complicated EI-mass spectra with several sequence ions (254). Also the methylated-reduced and methylated-reduced-trimethylsilylated derivatives are incorporated in this study (see below, characterisation of permethylated glycosphingolipids). Furthermore, MS data of smaller glycopeptides are available (255, 256). *O*-Glycosidic glycopeptides will degrade via *β*-elimination under the conditions of permethylation, leading to peptides with one or more unsaturated amino acids and one or more oligosaccharide chains (257), (for an EI- and CI(methane)-MS study of permethylated glycopeptides containing *N*-acetylmuramyl residues, see (258)). Finally, one of the main urinary compounds excreted by patients with aspartylglycosaminuria is the glycopeptide 4-*N*-acetamido-2-deoxy-*β*-D-glucopyranosyl-L-asparagine. GC-MS data of the permethylated compound are available from (255, 259).

4.3 Characterisation of permethylated glycosphingolipids

The structural analysis of glycosphingolipids can conveniently be performed by EI-MS on permethylated derivatives. Besides the use of these derivatives permethylated-reduced and permethylated-reduced-trimethylsilylated compounds have been incorporated in several studies. The reduction of permethylated glycosphingolipids is carried out with $LiAlH_4$. In this way amide groups of ceramide and amino sugars are converted into amines, whereas esterified carboxyl groups of sialic acids are transformed into primary hydroxyl functions. The latter groups can be trimethylsilylated. In Figure 37 the different derivatives have been depicted for GM_1 ganglioside (260) containing C_{20}-sphingosine and stearic acid.

Detailed fragmentations of the glycosphingolipid derivatives have been discussed extensively (200–202, 260–265). In principle four series of characteristic primary fragments can be distinguished: 1. fragment ions derived from the molecular ion; 2. fragment ions related to the ceramide moiety; 3. fragment ions originating from the carbohydrate chain and 4. fragment ions consisting of parts from both units.

The molecular ion of derivatised glycosphingolipids is rarely observed in EI-mass spectra. However, in several cases, depending on the size of the structure and of the type of derivative, $[M-H]^+$ ions can be detected. Furthermore, loss of MeOH, $COOCH_3$ (in permethylated gangliosides) or CH_2OTMS (in permethylated-reduced-trimethylsilylated gangliosides) can take place. Several series of primary and rearrangement ions allow a conclusive interpretation of the monosaccharide sequence, and of the branching points in the carbohydrate chain. These fragments originate predominantly from cleavages around the glycosidic bonds but they can also contain parts of the ceramide residue (Fig. 37). Although the expected sequence ions are not always detectable for one of the derivatives, the application of three types of derivatives gives rise to reliable overlapping sequence data. Sometimes, additional information can be obtained about certain types of glycosidic linkages in individual glycosphingolipids.

Because of the frequently occurring heterogeneity of the constituent fatty acids and long-chain-bases, the EI mass spectra of glycosphingolipids can be very complicated. Fragments characteristic of the ceramide components, elaborated for sphingosine and 4D-hydroxysphinganine, are presented schematically in Figures 38 and 39. Type A fragments are generally observed for permethylated glycosphingolipids and represent the entire ceramide moiety. In the permethylated-reduced derivatives fragment A' prevails. The type of sphingosine base is characterised by a fragment ion formed from A after cleavage between C3 and C4 of the fatty acid chain in the permethylated glycosphingolipid with rearrangement of one hydrogen atom, yielding m/z 364 for sphingosine and m/z 396 for 4D-hydroxysphinganine. Fragments of the B-type (minus 1 H) or B'-type (plus 1 H) (Fig. 38) furnish information concerning the fatty acid composition of the glycolipids with sphingosine as long-chain-base. Because of its low intensity, fragment C is not very useful for identification purposes. In the case of glycolipids with 4D-hydroxysphinganine (Fig. 39) the fragments D and D' are significant. In the non-reduced form there is a preferential fission leading to D' fragments, whereas the reduced form gives predominantly rise to D fragments. In the case of 2-hydroxy fatty acids as *N*-acyl substituents more complicated peak patterns have been observed.

Mixtures of permethylated glycosphingolipids and corresponding derivatives have been analysed by selected ion monitoring in the EI-mode. Samples were evaporated slowly from the direct inlet probe using a temperature program. A successive partial separation was obtained for glycosphingolipids containing one up to seven sugars (264, 265).

The various glycosphingolipid derivatives have also been studied with CI(ammonia)-MS. The neutral permethylated glycosphingolipids show intense $MH^+ \cdot MeOH$ ions. The quasi-molecular MH^+ and $(M + NH_4)^+$ ions have a low abundance. Fragment ions, produced by the successive loss of sugar moieties from the non-reducing end are prominent (266). Mass spectra of permethylated gangliosides furnish also peak patterns with information about molecular weight, ceramide structure and sugar sequences (267, 268). In Figure 40 the peak patterns of the permethylated, the permethylated-reduced and the permethylated-reduced-trimethylsilylated GM_1 ganglioside (stearic acid; C_{20}-sphingosine) are compared. It may be evident that overlapping sequence data are obtained for the three types of derivatives. An impression of the differences between the EI and CI(ammonia) mass spectra of the GM_1 ganglioside derivatives is evident from a comparison of Figures 37 and 40.

The mass spectrometric data obtained in the field of glycolipid research is vast and has to remain outwith the scope of this chapter. Some typical recent data can be found in (269–276). For results on oligosaccharides released from glycolipids via ozonolysis or trifluoroacetolysis, see (217, 221, 247, 252).

(a)

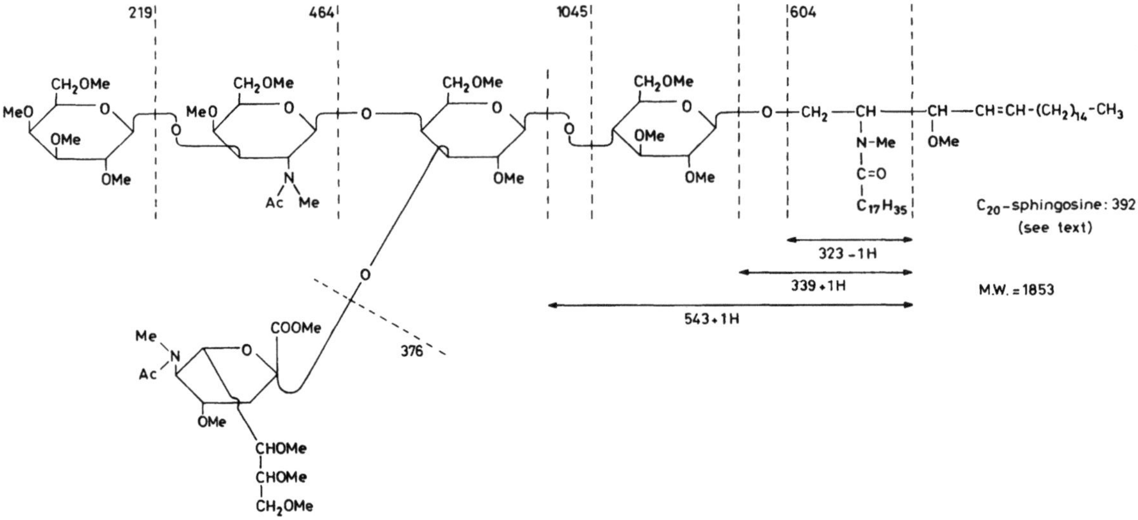

(b)

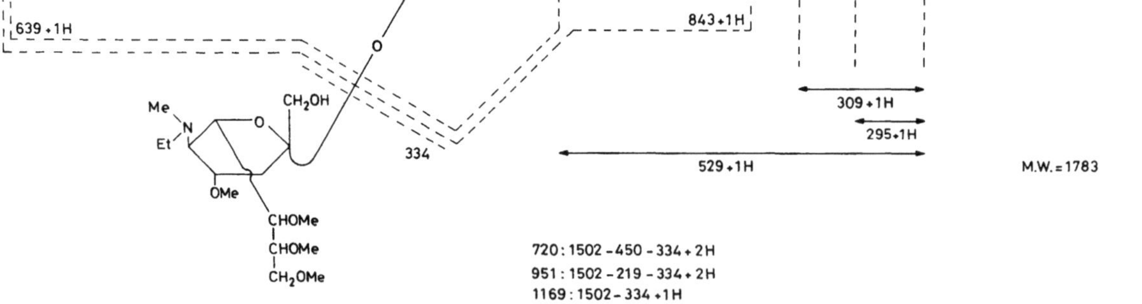

Figure 37. Fragmentation diagrams of GM_1 ganglioside (stearic acid; C_{20}-sphingosine): (a) permethylated; (b) permethylated-reduced; (c) permethylated-reduced-trimethylsilylated. The mass spectra were recorded in the EI-mode.

(c)

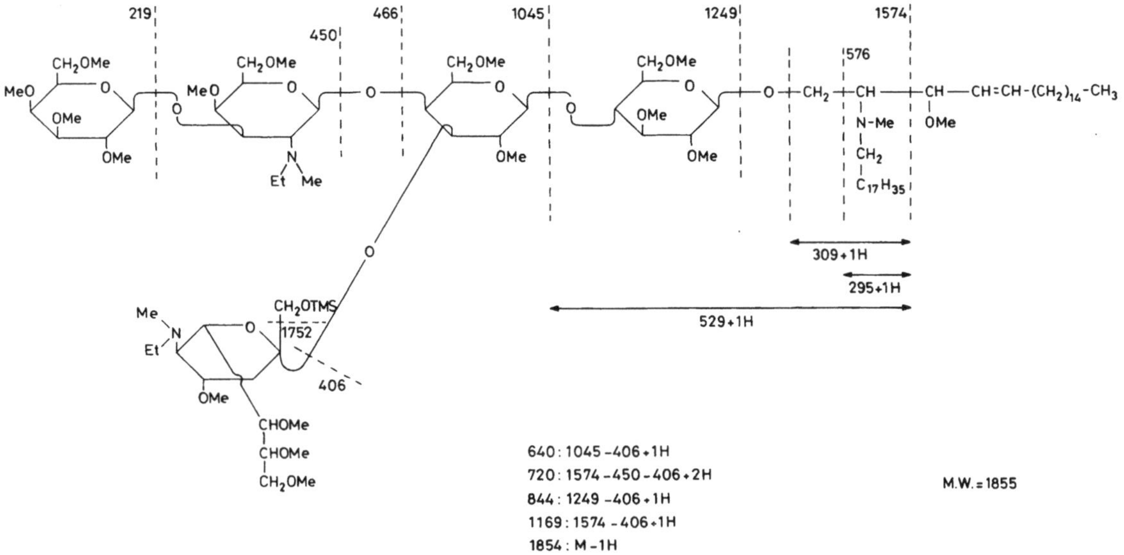

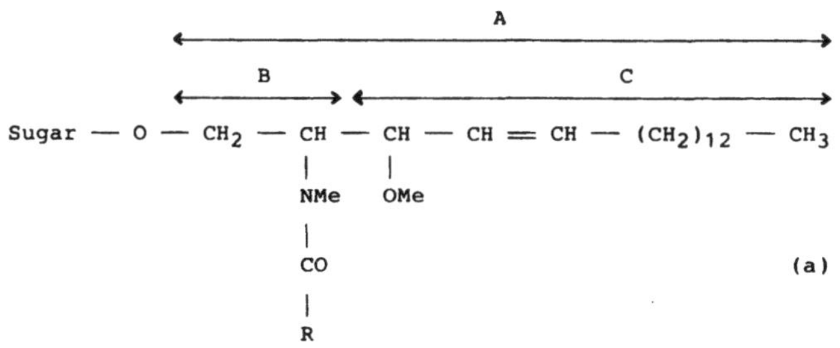

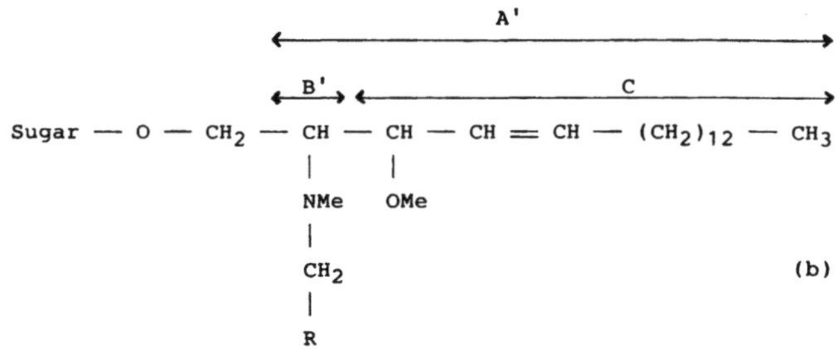

Figure 38. Schematic fragmentation patterns of permethylated (a) and permethylated and reduced (b) glycosphingolipids whose ceramide residue is composed of sphingosine and *n*-fatty acids (R = acyl group).

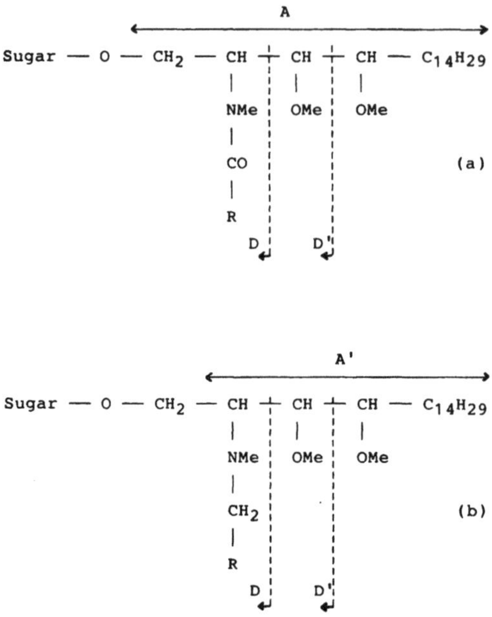

Figure 39. Schematic fragmentation patterns of permethylated (a) and permethylated and reduced (b) glycosphingolipids whose ceramide residue is composed of 4D-hydroxysphinganine and *n*-fatty acids (R = acyl group).

4.4 Fast atom bombardment mass spectrometry

The recent introduction of FAB-MS in carbohydrate chemistry has proven to be a valuable additional technique for the structural analysis of sugar chains, especially with respect to molecular weight determination and sequence analysis. Its combination with high field and high resolution magnetic sector mass spectrometers makes it possible to investigate polar non-volatile compounds or derivatives with relatively high molecular weights. Positive as well as negative ion FAB-MS data have been published for all categories of carbohydrate chains discussed in this review: oligosaccharides, oligosaccharide-alditols, glycopeptides and glycosphingolipids (277–304). Mainly underivatised, permethylated and peracetylated derivatives have been studied.

In the positive ion FAB mass spectra the molecular weight can be deduced from the relatively abundant protonated molecular ion $[M + H]^+$, and from the frequently present cationized molecular ions $[M + Na]^+$, $[M + K]^+$. The amounts of the latter species can be enhanced by the addition of suitable salts to the matrix. The negative ion FAB mass spectra are generally characterized by relatively abundant $[M - H]^-$ ions. The intense quasi-molecular ions in FAB-MS mass spectra may be used with great advantage to ascertain the homogeneity of the compounds under investigation (282), or just to study mixtures of oligosaccharide chains (281). Several solvents have been employed as matrix i. a. glycerol, 1-mercapto-2,3-propanediol (thioglycerol), polyethylene glycol and triethanolamine. Frequently use is made of additives like sodium acetate in methanol, acetic acid and 1,1,3,3,-tetramethylurea.

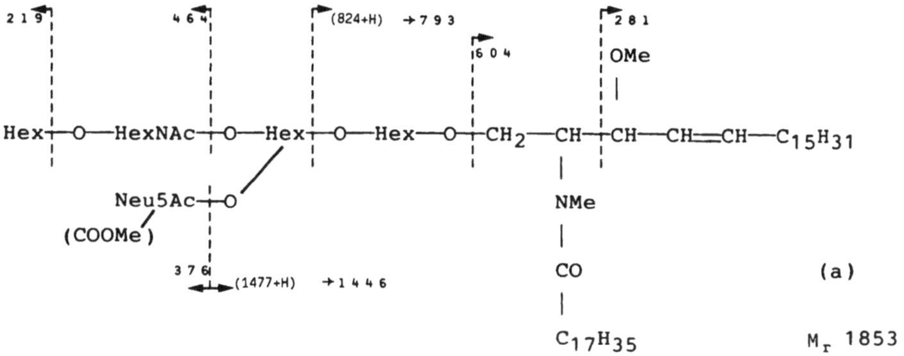

982 : 1478 - 464 - MeOH
1227 : 1478 - 219 - MeOH
1822 : MH - MeOH
1854 : MH

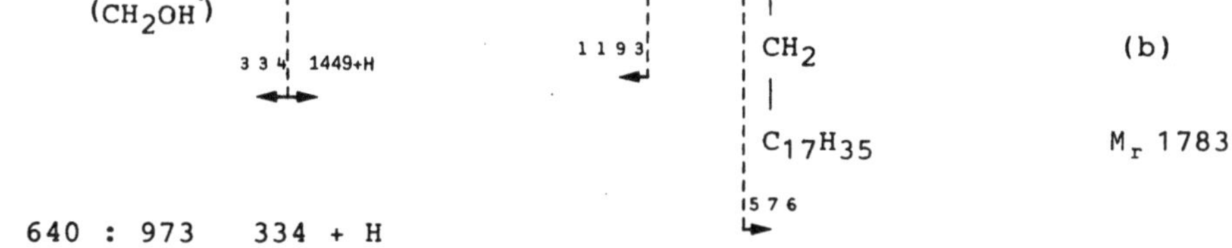

640 : 973 334 + H
844 : 1177 - 334 + H
1000 : 1450 - 450
1784 : MH

Figure 40. Fragmentation diagrams of GM_1 ganglioside (stearic acid; C_{20}-sphingosine); (a) permethylated; (b) permethylated-reduced: (c) permethylated-reduced-trimethylsilylated. The mass spectra were recorded in the CI(ammonia)-mode. For complete structures, see Figure 37.

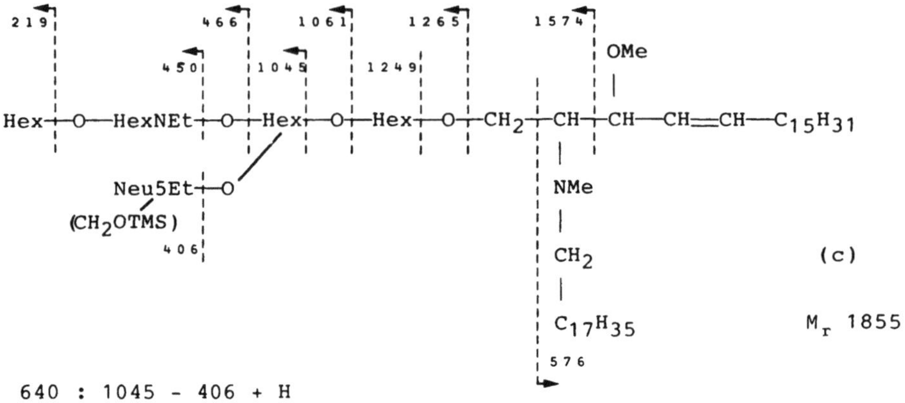

640 : 1045 - 406 + H
844 : 1249 - 406 + H
1856 : MH
1896 : M + TMS - MeOH
1928 : M + TMS

Figure 40 (Continued)

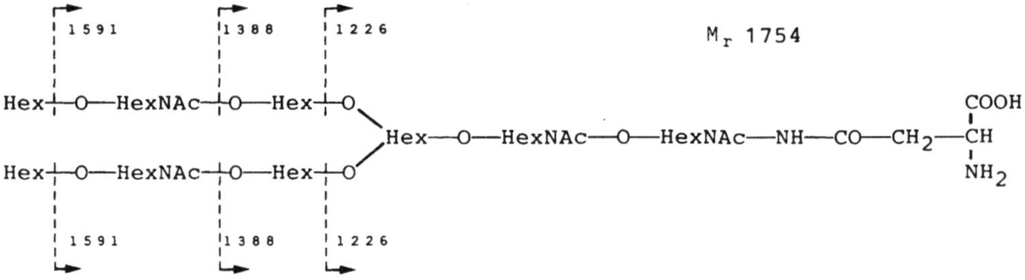

1753 :	M - H	1416 :	(1388 + HCO) - H
1735 :	(M - H_2O) - H	1388 :	(1388 + H) - H
1619 :	(1591 + HCO) - H	1370 :	(1388 + H - H_2O) - H
1591 :	(1591 + H) - H	1254 :	(1226 + HCO) - H
1573 :	(1591 + H - H_2O) - H	1226 :	(1226 + H) - H

Figure 41. Molecular weight and sequence information from the negative ion FAB mass spectrum of Gal*p*-β(1$\rightarrow$4)-GlcNAc*p*-β-(1$\rightarrow$2)-Man*p*-α(1$\rightarrow$3)-[Gal*p*-β(1$\rightarrow$4)-GlcNAc*p*-β(1$\rightarrow$2)-Man*p*-α(1$\rightarrow$6)]Man*p*-β(1$\rightarrow$4)-GlcNAc*p*-β(1$\rightarrow$4)-GlcNAc*p*-β(1$\rightarrow$N)Asn (glycerol as matrix).

In Figure 41 the schematic presentation of a biantennary asialoglycopeptide together with the molecular weight and sequence information deduced from the negative ion FAB mass spectrum is shown (280). The most important fragmentation reaction involves cleavage of the glycosidic bond between the anomeric carbon atom and the interglycosidic oxygen atom. This reaction is accompanied by a hydrogen migration to the interglycosidic oxygen, resulting in negative fragment ions ($[S_n + H] - H)^-$; see Scheme 9. In the case of analysis of underivatized glycosphingolipids, the matrix

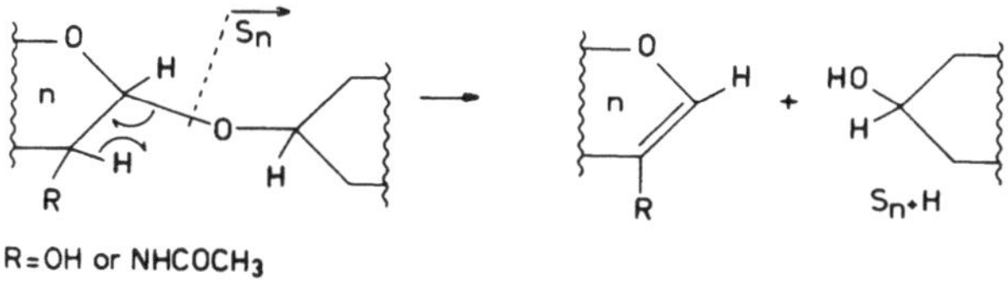

Scheme 9. Sequence ion formation by glycosidic bond cleavage. The index *n* of S*n* refers to the cleaved glycosidic bond at the anomeric carbon atom of monosaccharide *n*.

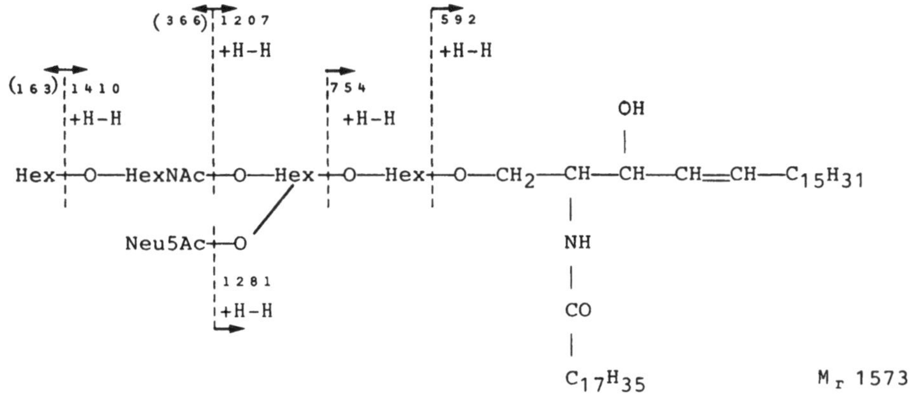

1572 : M - H
1119 : (1281 + H) - 163
916 : (1281 + H) - 366

Figure 42. Fragmentation diagram of underivatized GM_1 ganglioside (stearic acid; C_{20}-sphingosine) using negative ion FAB-MS with triethanolamine/1,1,3,3-tetramethylurea as matrix.

triethanolamine/1,1,3,3-tetramethylurea has shown to give excellent negative ion FAB mass spectra (283, 284). In Figure 42 the fragmentation for GM_1 ganglioside (with stearic acid and C_{20}-sphingosine) is presented (284). Structural isomers having different binding positions of sialic acid can be discriminated (293). Natural and synthetic neutral glycosphingolipids are discussed in (304). It has to be noted that examples have been reported showing no sequence information (292).

Positive ion FAB-MS has been successfully used for the analysis of permethylated and peracetylated derivatives. Besides the occurrence of quasi-molecular ions sequence ions are observed. Cleavages have been detected on either site of the glycosidic oxygen. Charged fragments can result from the non-reducing end of the chain as well as from the reducing end of the molecule. In the case of hexosamine-containing compounds, it has been found that fragment ions are predominantly formed by cleavage of the glycosidic bonds at C1 of the amino sugar residues, starting from the non-reducing terminal of the carbohydrate chain (278, 279, 294–299). This makes the technique highly effective for sequence analysis of permethylated polylactosamine carbohydrate

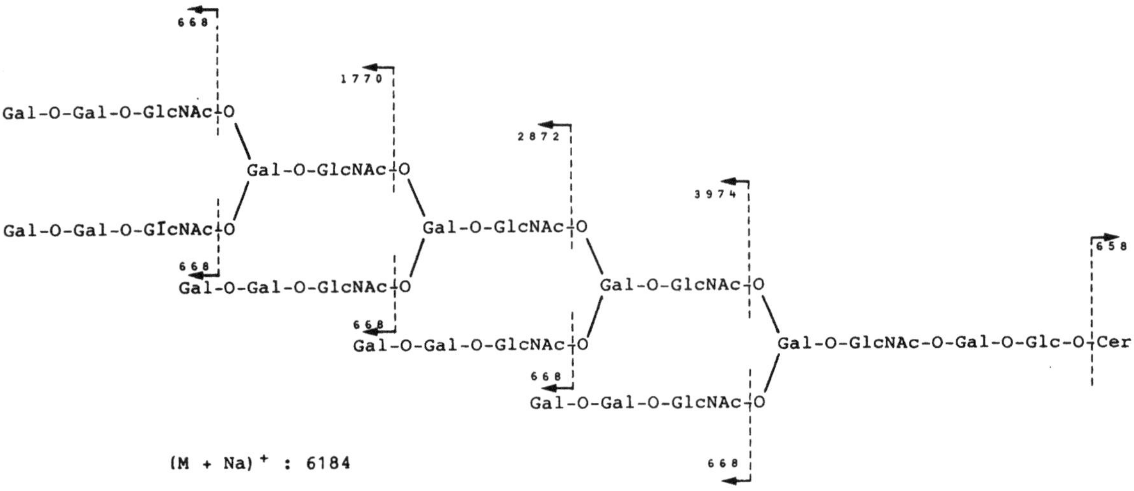

Figure 43. Fragmentation diagram of a permethylated glycolipid with molecular weight 6161 using positive ion FAB-MS with thioglycerol as matrix.

chains (296–299). An example of a permethylated glycolipid with a molecular weight of 6161 is given in Figure 43 (294, 295). Besides the cleavages in the carbohydrate chain, a specific cleavage occurs at the glycosidic linkage between the chain and the ceramide moiety. Intense ions are observed representing the whole ceramide residue. Analysis of peracetylated oligosaccharides using positive ion FAB-MS gives better results with thioglycerol as matrix than with glycerol (278, 279, 296).

FAB-MS has become a well established method for the structural analysis of carbohydrate chains and many papers have been published on its application. Space does not permit a detailed discussion of these, but data are available on glycoproteins (313–336), proteoglycans (337, 338), polysaccharides (339) and glycolipids (340–356). Recently, a review has apeared (357).

5 A General Strategy for Oligosaccharide-Analysis

To give an impression of the general strategy followed in the structural analysis of oligosaccharides isolated from biological fluids, tissues and cell systems, the characterisation of the simple mannosidosis trisaccharide α-D-Man*p*-(1→3)-β-D-Man*p*-(1→4)-D-GlcNAc will be discussed in more detail.

Sugar analysis (Section 2). The mannosidosis-saccharide was subjected to the methanolysis procedure, leading to a mixture of trimethylsilylated methyl glycosides. Quantitative GC(-MS) analysis demonstrated the occurrence of Man and GlcNAc residues in a molar ratio of 2:1. In the case where sugar analysis was carried out on the reduced oligosaccharide, Man and GlcNAc-ol were also found in the molar ratio 2:1. These data indicate the oligosaccharide to be a trisaccharide having GlcNAc as reducing unit. Determination of the absolute configuration by GC-MS of the trimethylsilylated (−)-2-butyl glycosides (Section 2.7) showed D-configuration for both monosaccharides.

Permethylation analysis (Section 3). In order to get information about the substitution patterns of the three monosaccharides, the trisaccharide-alditol-[1-^{2}H] was permethylated with CH_3I in the presence of sodium methylsulfinylmethanide. GC-MS analysis of the partially methylated alditol-[1-^{2}H] acetates showed the presence of 1,5-di-*O*-acetyl-2,3,4,6-tetra-*O*-methyl-mannitol-[1-^{2}H], 1,3,5-tri-*O*-acetyl-2,4,6-tri-*O*-methyl-mannitol-[1-^{2}H] and 4-mono-*O*-acetyl-1,3,5,6-tetra-*O*-methyl-2-*N*-methyl-acetamido-2-deoxy-glucitol-[1-^{2}H] (for EI mass spectra Figs. 26−28). The finding of these alditols points to the occurrence of a terminal Man*p*-residue, a 3-substituted Man*p*-residue and a 4-substituted GlcNAc-ol residue. Therefore the structure of the trisaccharide-alditol is: Man*p*-(1 → 3)-Man*p*-(1 → 4)-GlcNAc-ol[1-^{2}H].

Sequence analysis (Section 4). The permethylated trisaccharide-alditol-[1-^{2}H] was also studied directly by EI-MS (Figure 35). Although the spectrum does not show a molecular ion peak, the molecular weight can be deduced easily from the abundant primary fragment ion *m*/*z* 670 (M minus CHDOMe). The monosaccharide sequence Hex → Hex → HexNAc-ol-[1-^{2}H] is proved by the presence of the sequence fragment ions *m*/*z* 481, 423, 277, 219 and 187. Information concerning the (1 → 4)-linkage between Hex and HexNAc-ol-[1-^{2}H] is indicated by the *m*/*z* values 89 and 175. The absence of *m*/*z* 337 (bcJ_1) suggests the occurrence of a 3-linked Hex → HexNAc-ol-[1-^{2}H] unit. Although not discussed here, in addition other ionization techniques such as CI and FAB can be applied. For FAB-studies also the underivatized trisaccharide can be used.

Finally, the determination of the configuration of the glycosidic linkages has to be performed along other routes. Nowadays, mainly NMR techniques and enzymic approaches are incorporated for this purpose. It should be stressed that besides MS these methods are also of more general importance in carbohydrate analysis. For ^{1}H-NMR and ^{13}C-NMR data of the trisaccharide, see (311, 312). From these data it can be concluded that the terminal Man*p*-residue is coupled α-glycosidically to the internal Man*p*-residue, which in turn is coupled *β*-glycosidically to the reducing GlcNAc unit.

In conclusion, the oligosaccharide isolated from urine of a patient with α-mannosidosis has the structure α-D-Man*p*-(1 → 3)-*β*-D-Man*p*-(1 → 4)-D-GlcNAc.

Acknowledgements

We thank Mr G. J. Gerwig, Mr A. A. M. Maas, Mrs I. Rijkse and Mrs A. van der Kerk-van Hoof for technical assistance. The investigations carried out in the authors' laboratory were supported by the Netherlands Foundation for Chemical Research (SON) with financial aid from the Netherlands Organization for the Advancement of Pure Research (ZWO), and by the Netherlands Foundation for Cancer Research (KWF, Grant UUKC 83-13).

References

1. Kochetkov, N. K. & Chizhov, O. S. (1966) Mass spectrometry of carbohydrate derivatives. Adv. Carbohydr. Chem. *21*, 39–94.
2. Hanessian, S. (1971) Mass spectrometry in the determination of structure of certain natural products containing sugars. Meth. Biochem. Anal. *19*, 105–228.
3. Radford, T. & DeJongh, D. C. (1972) Carbohydrates, in: Biochemical Applications of Mass Spectrometry (Waller, G., ed.) pp. 313–333, Wiley-Interscience, New York.
4. Dutton, G. G. S. (1973) Applications of gas-liquid chromatography to carbohydrates: Part I. Adv. Carbohydr. Chem. Biochem. *28*, 11–160.
5. Dutton, G. G. S. (1974) Applications of gas-liquid chromatography to carbohydrates: Part II. Adv. Carbohydr. Chem. Biochem. *30*, 9–110.
6. Lönngren, J. & Svensson, S. (1974) Mass spectrometry in structural analysis of natural carbohydrates. Adv. Carbohydr. Chem. Biochem. *29*, 41–106.
7. Radford, T. & DeJongh, D. C. (1980) Carbohydrates, in: Biochemical Applications of Mass Spectrometry, suppl. vol. 1 (Waller, G. R. & Dermer, O. C., eds.) pp. 255–310, Wiley-Interscience, New York.
8. Rauvala, H., Finne, J., Krusius, T., Kärkkäinen, J. & Järnefelt, J. (1981) Methylation techniques in the structural analysis of glycoproteins and glycolipids. Adv. Carbohydr. Chem. Biochem. *38*, 389–416.
9. Kenne, L. & Lindberg, B. (1983) Bacterial polysaccharides, in: The Polysaccharides, vol. 2 (Aspinall, G. O., ed.) pp. 287–363, Academic Press, Inc., New York.
10. Berger, E. G., Buddecke, E., Kamerling, J. P., Kobata, A., Paulson, J. C. & Vliegenthart, J. F. G. (1982) Structure, biosynthesis and functions of glycoprotein glycans. Experientia *38*, 1129–1162.
11. Baumann, H. & Doyle, D. (1984) Determination of carbohydrate structures in glycoproteins and glycolipids, in: Molecular and Chemical Characterization of Membrane Receptors pp. 125–160, Alan R. Liss, Inc., New York.
12. Fichtinger-Schepman, A. M. J., Kamerling, J. P., Vliegenthart, J. F. G., de Jong, E. W., Bosch, L. & Westbroek, P (1979) Composition of a methylated, acidic polysaccharide associated with coccoliths of *Emiliania huxleyi* (Lohmann) Kamptner. Carbohydr. Res. *69*, 181–189.
13. Henry, R. J., Blakeney, A. B., Harris, P. J. & Stone, B. A. (1983) Detection of neutral and aminosugars from glycoproteins and polysaccharides as their alditol acetates. J. Chromatogr. *256*, 419–427.
14. Schauer, R. & Corfield, A. P. (1982) Isolation and purification of sialic acids. Cell Biol. Monogr. *10*, 51–57.
15. Jansson, P.-E., Lindberg, B., Lönngren, J., Ortega, C. & Svenson, S. B. (1984) Structural studies of the *Escherichia coli O*-antigen 6. Carbohydr. Res. *131*, 277–283.
16. Kenne, L., Lindberg, B., Madden, J. K., Lindberg, A. A. & Gemski, P. (1983) Structural studies of the *Escherichia coli O*-antigen 25. Carbohydr. Res. *122*, 249–256.
17. Neeser, J. R. & Schweizer, T. F. (1984) A quantitative determination by capillary gas-liquid chromatography of neutral and amino sugars (as *O*-methyloxime acetates), and a study on hydrolytic conditions for glycoproteins and polysaccharides in order to increase sugar recoveries. Anal. Biochem. *142*, 58–67.
18. Aspinall, G. O. (1982) Chemical characterization and structure determination of polysaccharides, in: The Polysaccharides, vol. 1 (Aspinall, G. O., ed.) pp. 35–131, Academic Press, New York.
19. Sawardeker, J. S., Sloneker, J. H. & Jeanes, A. (1965) Quantitative determination of monosaccharides as their alditol acetates by gas liquid chromatography. Anal. Chem. *37*, 1602–1604.
20. Torello, L. A., Yates, A. J. & Thompson, D. K. (1980) Critical study of the alditol acetate method for quantitating small quantities of hexoses and hexosamines in gangliosides. J. Chromatogr. *202*, 195–209.
21. Oshima, R., Kumanotani, J. & Watanabe, C. (1982) Fused-silica capillary gas chromatographic separation of alditol acetates of neutral and amino sugars. J. Chromatogr. *250*, 90–95.
22. Henry, R. J., Harris, P. J., Blakeney, A. B. & Stone, B. A. (1983) Separation of alditol acetates from plasticizers and other contaminants by capillary gas chromatography. J. Chromatogr. *262*, 249–256.
23. Fox, A., Morgan, S. L., Hudson, J. R., Zhong Tao Zho & Lau, P. Y. (1983) Capillary gas chromatographic analysis of alditol acetates of neutral and amino sugars in bacterial cell walls. J. Chromatogr. *256*, 429–438.

24. Bradbury, A. G. W., Halliday, D. J. & Medcalf, D. G. (1981) Separation of monosaccharides as TMS alditols on fused-silica capillary columns. J. Chromatogr. *213*, 146–150.
25. Arakawa, Y., Imanari, T. & Tamura, Z. (1976) Determination of neutral and amino sugars in glycoproteins by GLC. Chem. Pharm. Bull. *24*, 2032–2037.
26. Li, B. W., Cochran, T. W. & Vercellotti, J. R. (1977) Chemical-ionization mass spectra of per-*O*-acetylaldononitriles and methylated aldononitrile acetates. Carbohydr. Res. *59*, 567–570.
27. Seymour, F. R., Chen, E. C. M. & Bishop, S. H. (1979) Identification of aldoses by use of their peracetylated aldononitrile derivatives: A g.l.c.-m.s. approach. Carbohydr. Res. *73*, 19–45.
28. Mawhinney, T. P., Feather, M. S., Barbero, G. J. & Martinez, J. R. (1980) The rapid, quantitative determination of neutral sugars (as aldononitrile acetates) and amino sugars (as *O*-methyloxime acetates) in glycoproteins by gas-liquid chromatography. Anal. Biochem. *101*, 112–117.
29. McGinnis, G. D. (1982) Preparation of aldononitrile acetates using *N*-methylimidazole as catalyst and solvent. Carbohydr. Res. *108*, 284–292.
30. Seymour, F. R., Chen, E. C. M. & Stouffer, J. E. (1980) Identification of ketoses by use of their peracetylated oxime derivatives: A g.l.c.-m.s. approach. Carbohydr. Res. *83*, 201–242.
31. Pelletier, O. & Cadieux, S. (1982) Glass capillary or fused-silica gas chromatography-mass spectrometry of several monosaccharides and related sugars: Improved resolution. J. Chromatogr. *231*, 225–235.
32. Mawhinney, T. P., Madson, M. A., Rice, R. H., Feather, M. S. & Barbero, G. J. (1982) Gas-liquid chromatography and mass-spectral analysis of per-*O*-trimethylsilyl acyclic ketoxime derivatives of neuraminic acid. Carbohydr. Res. *104*, 169–181.
33. Schweer, H. (1982) Gas chromatography-mass spectrometry of aldoses as *O*-methoxime, *O*-2-methyl-2-propoxime and *O*-*n*-butoxime pertrifluoroacetyl derivatives on OV-225 with methylpropane as ionization agent. I. Pentoses. J. Chromatogr. *236*, 355–360.
34. Schweer, H. (1982) Gas chromatography-mass spectrometry of aldoses as *O*-methoxime, *O*-2-methyl-2-propoxime and *O*-*n*-butoxime pertrifluoroacetyl derivatives on OV-225 with methylpropane as ionization agent. II. Hexoses. J. Chromatogr. *236*, 361–367.
35. Willis, D. E. (1983) GC analysis of C_2-C_7 carbohydrates as the trimethylsilyl oxime derivatives on packed and capillary columns. J. Chromatogr. Sc. *21*, 132–138.
36. Kamerling, J. P. & Vliegenthart, J. F. G. (1982) Gas-liquid chromatography and mass spectrometry of sialic acids. Cell Biol. Monogr. *10*, 95–125.
37. Chambers, R. E. & Clamp, J. R. (1971) An assessment of methanolysis and other factors used in the analysis of carbohydrate-containing materials. Biochem. J. *125*, 1009–1018.
38. Clamp, J. R., Bhatti, T. & Chambers, R. E. (1971) The determination of carbohydrates in biological materials by gas-liquid chromatography. Meth. Biochem. Anal. *19*, 229–344.
39. Jamieson, G. R. & Reid, E. H. (1974) Use of mannitol as an internal standard in the gas-liquid chromatography of methyl glycosides. J. Chromatogr. *101*, 185–188.
40. Kamerling, J. P., Gerwig, G. J., Vliegenthart, J. F. G. & Clamp, J. R. (1975) Characterization by gas-liquid chromatography-mass spectrometry and proton-magnetic-resonance spectroscopy of pertrimethylsilyl methyl glycosides obtained in the methanolysis of glycoproteins and glycopeptides. Biochem. J. *151*, 491–495.
41. Pritchard, D. G. & Todd, C. W. (1977) Gas chromatography of methyl glycosides as their trimethylsilyl ethers. The methanolysis and re-*N*-acetylation steps. J. Chromatogr. *133*, 133–139.
42. Rickert, S. J. & Sweeley, C. C. (1978) Quantitative analysis of carbohydrate residues of glycoproteins and glycolipids by gas-liquid chromatography. An appraisal of experimental details. J. Chromatogr. *147*, 317–326.
43. Akhrem, A. A., Avvakumov, G. V., Sviridov, O. V. & Strel'chyonok, O. A. (1978) Analysis of methyl glycosides as their trimethylsilyl ethers: On-column re-*N*-acetylation and improved gas-liquid chromatographic separation. J. Chromatogr. *166*, 123–131.
44. Kozulić, B., Ries, B. & Mildner, P. (1979) *N*-Acetylation of amino sugar methyl glycosides for gas-liquid chromatographic analysis. Anal. Biochem. *94*, 36–39.
45. Mononen, I. (1981) Quantitative analysis, by gas-liquid chromatography and mass

fragmentography, of monosaccharides after methanolysis and deamination. Carbohydr. Res. *88*, 39–50.
46. Mega, T. & Ikenaka, T. (1982) Methanolysis products of asparagine-linked-*N*-acetylglucosamine and a new method for determination of *N*- and *O*-glycosidic *N*-acetylglucosamine in glycoproteins that contain asparagine-linked carbohydrates. Anal. Biochem. *119*, 17–24.
47. Lifely, M. R. & Cottee, F. H. (1982) Formation and identification of two novel anhydro compounds obtained by methanolysis of *N*-acetylneuraminic acid and carboxyl-reduced, meningococcal B polysaccharide. Carbohydr. Res. *107*, 187–197.
48. Dierckxsens, G. C., de Meyer, L. & Tonino, G. J. (1983) Simultaneous determination of uronic acids, hexosamines, and galactose of glycosaminoglycans by gas-liquid chromatography. Anal. Biochem. *130*, 120–127.
49. Gerwig, G. J., Kamerling, J. P. & Vliegenthart, J. F. G. (1984) Anhydroalditols in the sugar analysis of methanolysates of alditols and oligosaccharide-alditols. Carbohydr. Res. *129*, 149–157.
50. Zanetta, J. P., Breckenridge, W. C. & Vincendon, G. (1972) Analysis of monosaccharides by gas-liquid chromatography of the *O*-methyl glycosides as trifluoroacetate derivatives. Application to glycoproteins and glycolipids. J. Chromatogr. *69*, 291–304.
51. Ando, S., Ariga, T. & Yamakawa, T. (1976) Identification of stereoisomers of methyl hexopyranosides by mass spectrometry of their trifluoroacetates. Bull. Chem. Soc. Japan *49*, 1335–1338.
52. Wrann, M. M. & Todd, C. W. (1978) Sensitive determination of sugars utilizing packed capillary columns and electron capture detection. J. Chromatogr. *147*, 309–316.
53. Pritchard, D. G. & Niedermeier, W. (1978) Sensitive gas chromatographic determination of the monosaccharide composition of glycoproteins using electron capture detection. J. Chromatogr. *152*, 487–494.
54. Bryn, K. & Jantzen, E. (1982) Analysis of lipopolysaccharides by methanolysis, trifluoroacetylation, and gas chromatography on a fused-silica capillary column. J. Chromatogr. *240*, 405–413.
55. Gerwig, G. J., Kamerling, J. P. & Vliegenthart, J. F. G. (1978) Determination of the D and L configuration of neutral monosaccharides by high-resolution capillary g.l.c.. Carbohydr. Res. *62*, 349–357.
56. Leontein, K., Lindberg, B. & Lönngren, J. (1978) Assignment of absolute configuration of sugars by g.l.c. of their acetylated glycosides formed from chiral alcohols. Carbohydr. Res. *62*, 359–362.
57. Gerwig, G. J., Kamerling, J. P. & Vliegenthart, J. F. G. (1979) Determination of the absolute configuration of monosaccharides in complex carbohydrates by capillary g.l.c.. Carbohydr. Res. *77*, 1–7.
58. Popp-Snijders, C., Lomecky, M. Z. & de Jong, A. P. (1983) Determination of sorbitol in erythrocytes of diabetic and healthy subjects by capillary gas chromatography. Clin. Chim. Acta *132*, 83–89.
59. Geyer, R., Geyer, H., Kühnhardt, S., Mink, W. & Stirm, S. (1982) Capillary gas chromatography of methylhexitol acetates obtained upon methylation of *N*-glycosidically linked glycoprotein oligosaccharides. Anal. Biochem. *121*, 263–274.
60. Klok, J., Cox, H. C., de Leeuw, J. W. & Schenck, P. A. (1982) Analysis of synthetic mixtures of partially methylated alditol acetates by capillary gas chromatography, gas chromatography-electron impact mass spectrometry and gas chromatography-chemical ionization mass spectrometry. J. Chromatogr. *253*, 55–64.
61. Golovkina, L. S., Chizhov, O. S. & Vul'fson, N. S. (1966) Mass spectrometer investigation of carbohydrates. Comm. 9. Acetates of polyols. Ivz. Akad. Nauk SSSR, Ser. Khim. 1915–1926.
62. Björndal, H., Hellerqvist, C. G., Lindberg, B. & Svensson, S. (1970) Gas-Flüssigkeits-Chromatographie und Massenspektrometrie bei der Methylierungsanalyse von Polysacchariden. Angew. Chem. *82*, 643–652.
63. Jansson, P.-E. & Lindberg, B. (1980) Mass spectrometry of alditol acetates: origin of the fragments having even mass numbers. Carbohydr. Res. *86*, 287–292.
64. Weber, P. L. & Carlson, D. M. (1982) A simultaneous assay system for *N*-acetylgalactosamine and *N*-acetylgalactosaminitol using gas chromatography/mass spectrometry. Anal. Biochem. *121*, 140–145.
65. Taylor, R. L., Shively, J. E. & Conrad, H. E. (1976) Stoichiometric reduction of uronic acid carboxyl groups in polysaccharides. Meth. Carbohydr. Chem. *7*, 149–151.

66. McNeil, M., Szalecki, W. & Albersheim, P. (1984) Quantitative methods for determining the points of *O*-(carboxymethyl) substitution in *O*-(carboxymethyl)-guar. Carbohydr. Res. *131*, 139–148.
67. Banoub, J. H. & Michon, F. (1982) A rapid, g.l.c.-m.s. method for identification of the *N*-acetyl group of amino sugars in complex carbohydrates. Carbohydr. Res. *100*, C24–C26.
68. Pettit, B. R., King, G. S. & Blau, K. (1980) The analysis of hexitols in biological fluid by selected ion monitoring. Biomed. Mass Spectrom. *7*, 309–313.
69. Jakobs, C., Warner, T. G., Sweetman, L. & Nyhan, W. L. (1984) Stable isotope dilution analysis of galactitol in amniotic fluid: an accurate approach to the prenatal diagnosis of galactosemia. Pediatr. Res. *18*, 714–718.
70. Laker, M. F. & Gunn, W. G. (1979) Natural occurrence disqualifies mannitol as an internal standard when urinary monosaccharides are determined by gas-liquid chromatography. Clin. Chim. Acta *96*, 265–267,
71. Gerwig, G. J., Kamerling, J. P., Vliegenthart, J. F. G., Homan, W. L., van Egmond, P. & van den Ende, H. (1984) Characteristic differences in monosaccharide composition of glycoconjugates from opposite mating types of *Chlamydomonas eugametos*. Carbohydr. Res. *127*, 245–251.
72. Chaby, R. & Szabó, P. (1976) Formation of 1,6-anhydro-L-*glycero*-D-*manno*-heptopyranose during acidic hydrolysis and methanolysis of a polysaccharide from *Bordetella pertussis*. Carbohydr. Res. *49*, 489–493.
73. Jantzen, E., Bryn, K. & Boure, K. (1976) Cellular monosaccharide patterns of *Neisseriaceae* Acta Pathol. Microbiol. Scand. *84B*, 177–188.
74. Banoub, J. H., Shaw, D. H. & Michon, F. (1983) Hydrolytic release, and identification by g.l.c.-m.s., of 3-deoxy-D-*manno*-2-octulosonic acid in the lipopolysaccharides isolated from bacteria of the Vibrionaceae. Carbohydr. Res. *123*, 117–122.
75. Petersson, G. & Samuelson, O. (1968) Determination of the number and position of methoxyl groups in methylated aldopentoses by mass spectrometry of their trimethylsilyl derivatives. Svensk Papperstidn. *71*, 77–84.
76. Petersson, G. & Samuelson, O. (1968) Determination of the number and position of methoxyl groups in methylated aldohexoses by mass spectrometry of their trimethylsilyl derivatives. Svensk Papperstidn. *71*, 731–738.
77. DeJongh, D. C., Radford, T., Hribar, J. D., Hanessian, S., Bieber, M., Dawson, G. & Sweeley, C. C. (1969) Analysis of trimethylsilyl derivatives of carbohydrates by gas chromatography and mass spectrometry. J. Amer. Chem. Soc. *91*, 1728–1740.
78. Larsson, K. & Petersson, G. (1974) Methanolysis of D-galacturonic acid: dimethyl acetals. Carbohydr. Res. *34*, 323–329.
79. Coduti, P. L. & Bush, C. A. (1977) Structure determination of *N*-acetylamino sugar derivatives and disaccharides by gas chromatography and mass spectrometry. Anal. Biochem. *78*, 21–38.
80. Kennedy, J. F. & Robertson, S. M. (1978) Mass spectrometric fragmentation pathways of the *O*-trimethylsilyl derivatives of hexuronic acids and their lactones. Carbohydr. Res. *67*, 1–15.
81. Kärkkäinen, J. & Vihko, R. (1969) Characterisation of 2-amino-2-deoxy-D-glucose, 2-amino-2-deoxy-D-galactose, and related compounds, as their trimethylsilyl derivatives by gas-liquid chromatography-mass spectrometry. Carbohydr. Res. *10*, 113–120.
82. Vink, J. (1974) Mass spectra of monosaccharides. Pattern recognition as an identification method. Thesis State University of Utrecht (The Netherlands).
83. De Jong, E. G., Heerma, W., Sicherer, C. A. X. G. F. & Dijkstra, G. (1980) The structure of the *m*/*z* 101 ion in the spectra of permethylated saccharides. Biomed. Mass Spectrom. *7*, 122–126.
84. Hall, R. L., Wood, E. J., Kamerling, J. P., Gerwig, G. J. & Vliegenthart, J. F. G. (1977) 3-*O*-Methyl sugars as constituents of glycoproteins. Identification of 3-*O*-methylgalactose and 3-*O*-methylmannose in pulmonate gastropod haemocyanins. Biochem. J. *165*, 173–176.
85. Petersson, G. (1969) Mass spectrometry of alditols as trimethylsilyl derivatives. Tetrahedron *25*, 4437–4443.

86. Schoots, A. C. & Leclercq, P. A. (1979) Chemical ionization mass spectrometry of trimethylsilylated carbohydrates and organic acids retained in uremic serum. Biomed. Mass Spectrom. *6*, 502–507.
87. Murata, T. & Takahashi, S. (1978) Characterization of *O*-trimethylsilyl derivatives of D-glucose, D-galactose and D-mannose by gas-liquid chromatography-chemical ionization mass spectrometry with ammonia as reagent gas. Carbohydr. Res. *62*, 1–9.
88. Bowser, D. V., Teece, R. G. & Somani, S. M. (1978) Identification of amino sugars from bacterial lipopolysaccharides by gas chromatography electron impact and chemical ionization mass spectrometry. Biomed. Mass Spectrom. *5*, 627–633.
89. Niwa, T., Yamamoto, N., Maeda, K., Yamada, K., Ohki, T. & Mori, M. (1983) Gas chromatographic-mass spectrometric analysis of polyols in urine and serum of uremic patients. Identification of new deoxyalditols and inositol isomers. J. Chromatogr. *277*, 25–39.
90. Sherman, W. R., Eilers, N. C. & Goodwin, S. L. (1970) Combined gas chromatography-mass spectrometry of the inositol trimethylsilyl ethers and acetate esters. Org. Mass Spectrom. *3*, 829–840.
91. Binder, R. G. & Haddon, W. F. (1984) Analysis of *O*-methylinositols by gas-liquid chromatography-mass spectrometry. Carbohydr. Res. *129*, 21–32.
92. Schauer, R. (1978) Characterization of sialic acids. Meth. Enzymol. *50*, 64–216.
93. Kamerling, J. P., Vliegenthart, J. F. G., Versluis, C. & Schauer, R. (1975) Identification of *O*-acetylated *N*-acylneuraminic acids by mass spectrometry. Carbohydr. Res. *41*, 7–17.
94. Schauer, R., Haverkamp, J., Wember, M., Vliegenthart, J. F. G. & Kamerling, J. P. (1976) *N*-Acetyl-9-*O*-L-lactylneuraminic acid, a new acylneuraminic acid from bovine submandibular gland. Eur. J. Biochem. *62*, 237–242.
95. Casals-Stenzel, J., Buscher, H.-P. & Schauer, R. (1975) Gas-liquid chromatography of *N*- and *O*-acylated neuraminic acids. Anal. Biochem. *65*, 507–524.
96. Schröder, C. & Schauer, R. (1982) Qualitative and quantitative Bestimmung von Acylneuraminsäuren mit Hilfe der Capillarsäulen-Gas-Chromatographie. Fresenius Z. Anal. Chem. *311*, 385–386.
96a. Kamerling, J. P., Schauer, R., Shukla, A. K., Stoll, S., Van Halbeek, H. & Vliegenthart, J. F. G. (1987) Migration of *O*-acetyl groups in *N,O*-acetylneuraminic acids. Eur. J. Biochem. *162*, 601–607.
97. Sugawara, Y., Iwamori, M., Portoukalian, J. & Nagai, Y. (1983) Determination of *N*-acetylneuraminic acid by gas chromatography-mass spectrometry with a stable isotope as internal standard. Anal. Biochem. *132*, 147–151.
97a. Reuter, G. & Schauer, R. (1986) Comparison of electron and chemical ionization mass spectrometry of sialic acids. Anal. Biochem. *157*, 39–46.
97b. Shukla, A. K., Schauer, R., Schade, U., Moll, M. & Rietschel, E. Th. (1985) Structural analysis of underivatized sialic acids by combined high-performance liquid chromatography-mass spectrometry. J. Chromatogr. *337*, 231–238.
98. Björkhem, I., Blomstrand, R., Falk, O. & Öhman, G. (1976) The use of mass fragmentography in the evaluation of routine methods for glucose determination. Clin. Chim. Acta *72*, 353–362.
99. Marunaka, T., Matsushima, E., Umeno, Y. & Minami, Y. (1983) GLC-mass fragmentographic determination of mannitol and sorbitol in plasma. J. Pharm. Sc. *72*, 87–89.
100. Andersen, J. R., Larsen, E., Harbo, H., Bertelsen, B., Christensen, J. E. J. & Gregersen, G. (1982) Gas chromatographic mass spectrometric determination of myo-inositol in humans utilizing a deuterated internal standard. Biomed. Mass Spectrom. *9*, 135–140.
101. Barker, S. A., Fish, F. P., Tomana, M., Garner, L. C., Settine, R. L. & Prchal, J. (1983) Gas chromatographic/mass spectrometric evidence for the identification of a heptitol and an octitol in human and *Octodon degu* eye lenses. Biochem. Biophys. Res. Commun. *116*, 988–993.
102. Hakomori, S.-I. (1964) A rapid permethylation of glycolipid, and polysaccharide catalyzed by methylsulfinyl carbanion in dimethyl sulfoxide. J. Biochem. *55*, 205–208.

103. Jansson, P.-E., Kenne, L., Liedgren, H., Lindberg, B. & Lönngren, J. (1976) A practical guide to the methylation analysis of carbohydrates. Chem. Commun. Univ. Stockholm 8, 1–75.
104. Endo, Y., Yamashita, K., Tachibana, Y., Tojo, S. & Kobata, A. (1979) Structures of the asparagine-linked sugar chains of human chorionic gonadotropin. J. Biochem. *85*, 669–679.
105. Geyer, R., Geyer, H., Kühnhardt, S., Mink, W. & Stirm, S. (1983) Methylation analysis of complex carbohydrates in small amounts: Capillary gas chromatography-mass fragmentography of methylalditol acetates obtained from *N*-glycosidically linked glycoprotein oligosaccharides. Anal. Biochem. *133*, 197–207.
106. Waeghe, T. J., Darvill, A. G., McNeil, M. & Albersheim, P. (1983) Determination, by methylation analysis, of the glycosyl-linkage compositions of microgram quantitites of complex carbohydrates. Carbohydr. Res. *123*, 281–304.
107. Finne, J., Krusius, T. & Rauvala, H. (1980) Use of potassium *tert*-butoxide in the methylation of carbohydrates. Carbohydr. Res. *80*, 336–339.
108. Phillips, L. R. & Fraser, B. A. (1981) Methylation of carbohydrates with dimsyl potassium in dimethyl sulfoxide. Carbohydr. Res. *90*, 149–152.
109. Harris, P. J., Henry, R. J., Blakeney, A. B. & Stone, B. A. (1984) An improved procedure for the methylation analysis of oligosaccharides and polysaccharides. Carbohydr. Res. *127*, 59–73.
110. Paz Parente, J., Cardon, P., Leroy, Y., Montreuil, J., Fournet, B. & Ricart, G. (1985) A convenient method for methylation of glycoprotein glycans in small amounts by using lithium methylsulfinyl carbanion. Carbohydr. Res. *141*, 41–47.
110a. Blakeney, A. B. & Stone, B. A. (1985) Methylation of carbohydrates with lithium methylsulphinyl carbanion. Carbohydr. Res. *140*, 319–324.
110b. Kvernheim, A. L. (1987) Methylation analysis of polysaccharides with butyllithium in dimethylsulfoxide. Acta Chem. Scand. *B41*, 150–152.
111. Prehm. P. (1980) Methylation of carbohydrates by methyl trifluoromethanesulfonate in trimethyl phosphate. Carbohydr. Res. *78*, 372–374.
112. Lindberg, B., Lönngren, J. & Svensson, S. (1975) Specific degradation of polysaccharides. Adv. Carbohydr. Chem. Biochem. *31*, 185–240.
113. Stellner, K., Saito, H. & Hakomori, S.-I. (1973) Determination of aminosugar linkages in glycolipids by methylation. Aminosugar linkages of ceramide pentasaccharides of rabbit erythrocytes and of Forssman antigen. Arch. Biochem. Biophys. *155*, 464–472.
114. Fukuda, M. & Hakomori, S.-I. (1979) Carbohydrate structure of a galactoprotein a, a major transformation-sensitive glycoprotein released from hamster embryo fibroblasts. J. Biol. Chem. *254*, 5451–5457.
115. Conchie, J., Hay, A. J. & Lomax, J. A. (1982) A comparison of some hydrolytic and gas chromatographic procedures used in methylation analysis of the carbohydrate units of glycopeptides. Carbohydr. Res. *103*, 129–132.
116. Schwarzman, G. O. H. & Jeanloz, R. W. (1974) Separation by gas-liquid chromatography, and identification by mass spectrometry, of the methyl ethers of 2-deoxy-2-(*N*-methylacetamido)-D-glucose. Carbohydr. Res. *34*, 161–168.
117. Tai, T., Yamashita, K. & Kobata, A. (1975) Synthesis and mass fragmentographic analysis of partially *O*-methylated 2-*N*-methylglucosamines. J. Biochem. *78*, 679–686.
118. Hase, S. & Rietschel, E. Th. (1976) Methylation analysis of glucosaminitol and glucosaminyl-glucosaminitol disaccharides. Formation of 2-deoxy-2-(*N*-acetylacetamido)-glucitol derivatives. Eur. J. Biochem. *63*, 93–99.
119. Finne, J. & Rauvala, H. (1977) Determination (by methylation analysis) of the substitution pattern of 2-amino-2-deoxyhexitols obtained from *O*-glycosylic carbohydrate units of glycoproteins. Carbohydr. Res. *58*, 57–64.
120. Radziejewska-Lebrecht, J., Shaw, D. H., Borowiak, D., Fromme, I. & Mayer, H. (1979) Gas chromatographic-mass spectro-

metric identification of a range of methyl ethers from L-glycero-D-mannoheptose and D-glycero-D-mannoheptose. J. Chromatogr. *179*, 113–122.

120a. Tacken, A., Brade, H., Unger, F. M. & Charon, D. (1986) GLC-MS of partially methylated and acetylated derivatives of 3-deoxy-octitols. Carbohydr. Res. *149*, 263–277.

121. Caroff, M. & Szabo, L. (1980) *O*-Demethylation of per-*O*-methyl derivatives of 2-amino-2-deoxyhexitols during acid hydrolysis and acetolysis. Carbohydr. Res. *84*, 43–52.

121a. Nilsson, B. (1985) *O*-Demethylation of methylated 2-acetamido-2-deoxy-D-hexitols during acid hydrolysis. Glycoconj. J. *2*, 335–341.

122. Funakoshi, I. & Yamashina, I. (1980) quantitative determination of partially methylated alditol acetate of amino sugar by gas chromatography-mass spectrometry. Anal. Biochem. *107*, 265–270.

123. Björndal, H., Lindberg, B. & Svensson, S. (1967) Mass spectrometry of partially methylated alditol acetates. Carbohydr. Res. *5*, 433–440.

124. Björndal, H., Lindberg, B., Pilotti, A. & Svensson, S. (1970) Mass spectra of partially methylated alditol acetates. Part II. Deuterium labelling experiments. Carbohydr. Res. *15*, 339–349.

125. Borén, H. B., Garegg, P. J., Lindberg, B. & Svensson, S. (1971) Mass spectra of partially methylated alditol acetates. Part III. Labelling experiments in the mass spectrometry of partially methylated deoxyalditol acetates. Acta Chem. Scand. *25*, 3299–3308.

126. Axberg, K., Björndal, H., Pilotti, A. & Svensson, S. (1972) Mass spectra of partially methylated alditol acetates. Part IV. Deuterium labelling experiments on some higher fragments. Acta Chem. Scand. *26*, 1319–1325.

127. Fournet, B., Leroy, Y., Montreuil, J. & Mayer, H. (1974) Analytical and preparative gas-liquid chromatography of methyl α-D-mannoside monomethyl ethers. J. Chromatogr. *92*, 185–190.

128. Schwarzman, G., Reinhold, V. & Jeanloz, R. W. (1974) Identification of methylated sugar derivatives by gas-liquid chromatography and mass spectrometry. Coll. Int. CNRS *211*, 85–94.

129. Wong, C. G., Sung, S. J. & Sweeley, C. C. (1980) Analysis and structural characterization of aminosugars by gas-liquid chromatography and mass spectrometry. Meth. Carbohydr. Chem. *8*, 55–65.

130. Lindberg, B., Lönngren, J., Thompson, J. L. & Nimmich, W. (1972) Structural studies of the *Klebsiella* type 9 capsular polysaccharide. Carbohydr. Res. *25*, 49–57.

131. Fichtinger-Schepman, A. M. J., Kamerling, J. P., Versluis, C. & Vliegenthart, J. F. G. (1981) Structural studies of the methylated, acidic polysaccharide associated with coccoliths of *Emiliania huxleyi* (Lohmann) Kamptner. Carbohydr. Res. *93*, 105–123.

132. Sweet, D. P., Shapiro, R. H. & Albersheim, P. (1974) The mass spectral fragmentation of partially ethylated alditol acetates, a derivative used in determining the glycosyl linkage composition of polysaccharides. Biomed. Mass Spectrom. *1*, 263–268,

133. Sweet, D. P., Albersheim, R. & Shapiro, R. H. (1975) Partially ethylated alditol acetates as derivatives for elucidation of the glycosyl linkage-composition of polysaccharides. Carbohydr. Res. *40*, 199–216.

134. Sweet, D. P., Shapiro, R. H. & Albersheim, P. (1975) Quantitative analysis by various g.l.c. response-factor theories for partially methylated and partially ethylated alditol acetates. Carbohydr. Res. *40*, 217–225.

135. Valent, B. S., Darvill, A. G., McNeil, M., Robertsen, B. K. & Albersheim, P. (1980) A general and sensitive chemical method for sequencing the glycosyl residues of complex carbohydrates. Carbohydr. Res. *79*, 165–192.

136. Darvill, A. G., McNeil, M. & Albersheim, P. (1980) General and facile method for distinguishing 4-linked aldopyranosyl residues from 5-linked aldofuranosyl residues. Carbohydr. Res. *86*, 309–315.

137. De Belder, A. N. & Norrman, B. (1968) The distribution of substituents in partially acetylated dextran. Carbohydr. Res. *8*, 1–6.

138. Lindberg, B. & Lönngren, J. (1978) Methylation analysis of complex carbohydrates: General procedure and application for sequence analysis. Meth. Enzymol. *50*, 3–33.

139. McNeil, M. & Albersheim, P. (1977) Chemical-ionization mass spectrometry of methylated hexitol acetates. Carbohydr. Res. *56*, 239–248.
140. Hancock, R. A., Marshall, K. & Weigel, H. (1976) Structure of the levan elaborated by *Streptococcus salivarius* strain 51: An application of chemical-ionisation mass-spectrometry. Carbohydr. Res. *49*, 351–360.
141. Laine, R. A. (1981) Enhancement of detection of partially methylated alditol acetates by chemical ionization mass spectrometry. Anal. Biochem. *116*, 383–388,
142. Lowe, M. E. & Nilsson, B. (1984) A method of purification of partially methylated alditol acetates in the methylation analysis of glycoproteins and glycopeptides. Anal. Biochem. *136*, 187–191.
143. Barreto-Bergter, E., Hogge, L. & Gorin, P. A. J. (1981) Gas-liquid chromatography of partially methylated alditol acetates on capillary columns of OV-17 and OV-225. Carbohydr. Res. *97*, 147–150.
144. Gorin, P. A. J., Giblin, E. M., Slater, G. P. & Hogge, L. (1982) Improved resolution of *O*-methyl alditol acetates by gas-liquid chromatography on capillary columns with blended coatings of OV-17 and OV-225. Carbohydr. Res. *106*, 235–238.
145. Lomax, J. A. & Conchie, J. (1982) Separation of methylated alditol acetates by glass capillary gas chromatography and their identification by computer. J. Chromatogr. *236*, 385–394.
146. Lawson, A. M., Hounsell, E. F. & Feizi, T. (1983) Structural analysis of mixed carbohydrate chains released from glycoproteins by base-borohydride degradation: Capillary GC-MS analysis of partially methylated alditol acetate derivatives. Int. J. Mass Spectrom. Ion Physics *48*, 149–152.
147. Shibuya, N. (1981) Gas-liquid chromatographic analysis of partially methylated alditol acetates on a glass capillary column. J. Chromatogr. *208*, 96–99.
148. Rauvala, H. (1979) Use of triphenylmethane as an indicator of complete methylation of glycolipids and glycopeptides. Carbohydr. Res. *72*, 257–260.
149. Laine, R. A. & Sweeley, C. C. (1973) *O*-Methyl oximes of sugars. Analysis as *0*-trimethylsilyl derivatives by gas-liquid chromatography and mass spectrometry. Carbohydr. Res. *27*, 199–213.
150. Dmitriev, B. A., Backinowsky, L. V., Chizhov, O. S., Zolotarev, B. M. & Kochetkov, N. K. (1971) Gas-liquid chromatography and mass spectrometry of aldononitrile acetates and partially methylated aldononitrile acetates. Carbohydr. Res. *19*. 432–435.
151. Seymour, F. R., Plattner, R. D. & Slodki, M. E. (1975) Gas-liquid chromatography-mass spectrometry of methylated and deuteriomethylated per-*O*-acetylaldononitriles from D-mannose. Carbohydr. Res. *44*, 181–198.
152. Seymour. F. R., Slodki, M. E., Plattner, R. D. & Jeanes, A. (1977) Six unusual dextrans: Methylation structural analysis by combined g.l.c.-m.s. of per-*O*-acetyl-aldononitriles. Carbohydr. Res. *53*, 153–166.
153. Stortz, C. A., Matulewicz, M. C. & Cerezo, A. S. (1982) Separation and identification of *O*-acetyl-*O*-methyl-galactononitriles by gas-liquid chromatography and mass spectrometry. Carbohydr. Res. *111*, 31–39.
154. Tanner, G. R. & Morrison, I. M. (1984) Gas chromatography-mass spectrometry of partially methylated glycoses as their aldononitrile peracetates. J. Chromatogr. *299*, 252–258.
155. Matsubara, T. & Hayashi, A. (1974) Determination of the structure of partially methylated sugars as *O*-trimethylsilyl ethers by gas chromatography-mass spectrometry. Biomed. Mass Spectrom. *1*, 62–65.
156. Kundu, S. K., Ledeen, R. W. & Gorin, P. A. J. (1975) Determination of position of substitution on 2-acetamido-2-deoxy-D-galactosyl residues in glycolipids. Carbohydr. Res. *39*, 179–191.
157. Kundu, S. K., Ledeen, R. W. & Gorin, P. A. J. (1975) Determination of position of substitution on 2-acetamido-2-deoxy-D-glucosyl residues in glycolipids. Carbohydr. Res. *39*, 329–334.
158. Elkin, Yu. N. (1979) Gas chromatographic separation of the methyl ether methylglycopyranoside series hexose, 6-deoxyhexose and pentose acetates. J. Chromatogr. *180*, 163–169.
159. Elkin, Yu. N., Tomshich, S. V. & Zurabyan, S. E. (1981) GC-MS study on methyl ethers

of methyl 2-deoxy-2-(*N*-methylacetamido)-glucopyranoside acetates. Bioorg. Khim. *7*, 1859.

160. Fournet, B., Strecker, G., Leroy, Y. & Montreuil, J. (1981) Gas-liquid chromatography and mass spectrometry of methylated and acetylated methyl glycosides. Application to the structural analysis of glycoprotein glycans. Anal. Biochem. *116*, 489–502.
161. Kadentsev, V. I., Kaymarazov, A. G., Chizhov, O. S., Derevitskaya, V. A., Evstigneev, A. Yu. & Botnikov, M. Ya. (1981) Chemical ionization mass spectra of 2-acetamido-2-deoxy-D-glucopyranoside methyl ethers. Bioorg. Khim. *7*, 422.
162. Grishkovets, V. I., Zemlyakov, A. E. & Chirna, V. Ya. (1983) Mass spectrometric identification of methyl (methyl-*O*-methyl-α-D-mannopyranoside)-uronates. Bioorg. Khim. *7*, 966.
163. Van Halbeek, H., Haverkamp, J., Kamerling, J. P., Vliegenthart, J. F. G., Versluis, C. & Schauer, R. (1978) Sialic acids in permethylation analysis: Preparation and identification of partially *O*-methylated derivatives of methyl *N*-acetyl-*N*-methyl-β-D-neuraminate methyl glycoside. Carbohydr. Res. *60*, 51–62.
164. Bruvier, C., Leroy, Y., Montreuil, J., Fournet, B. & Kamerling, J. P. (1981) Gas-liquid chromatography and mass spectrometry of methyl ethers of methyl *N*-acetyl-*N*-methyl-β-D-neuraminate methyl glycoside. J. Chromatogr. *210*, 487–504.
165. Lundblad, A., Hallgren, P., Rudmark, A. & Svensson, S. (1973) Structures and serological activities of three oligosaccharides isolated from urines of nonstarved secretors and from secretors on lactose diet. Biochemistry *12*, 3341–3345.
166. Lundblad, A., Masson, P. K., Nordén, N. E., Svensson, S., Öckerman, P.-A. & Palo, J. (1976) Structural determination of three glycoasparagines isolated from the urine of a patient with aspartylglycosaminuria. Eur. J. Biochem. *67*, 209–214.
167. Strecker, G., Fournet, B., Bouquelet, S., Montreuil, J., Dhondt, J. L. & Farriaux, J.-P. (1976) Etude chimique des mannosides urinaires excrétés au cours de la mannosidose. Biochimie *58*, 579–586.
168. Strecker, G., Riazi-Farzad, T., Fournet, B., Bouquelet, S. & Montreuil, J. (1976) Structure et propriétés immunochimiques des oligosaccharides urinaires excrétés au cours de la galactosurie provoquée. Biochimie *58*, 815–825.
169. Strecker, G., Hondi-Assah, T., Fournet, B., Spik, G., Montreuil, J., Maroteaux, P., Durand, P. & Farriaux, J.-P. (1976) Structure of the three major sialyl-oligosaccharides excreted in the urine of five patients with three distinct inborn diseases: "I-cell disease" and two new types of mucolipidosis. Biochim. Biophys. Acta *444*, 349–358.
170. Lemonnier, M. & Bourrillon, R. (1976) Characterization and structure of a sialic acid-containing hexasaccharide isolated from human pregnancy urine. Carbohydr. Res. *51*, 99–106.
171. Tsay, G. C., Dawson, G. & Sung, S.-S. J. (1976) Structure of the accumulating oligosaccharide in fucosidosis. J. Biol. Chem. *251*, 5852–5859.
172. Lemonnier, M., Bourrillon, R. & Fournet, B. (1977) Isolation and characterization of monofucosyl monosialyl derivative of lacto-*N*-neohexaose obtained from human pregnancy urine. Biochem. Biophys. Res. Commun. *77*, 767–774.
173. Michalski, J. C., Strecker, G., Fournet, B., Cantz, M. & Spranger, J. (1977) Structures of sialyl-oligosaccharides excreted in the urine of a patient with mucolipidosis I. FEBS Lett. *79*, 101–104.
174. Yamashita, K., Tachibana, Y. & Kobata, A. (1977) Oligosaccharides of human milk. Structural studies of two new octasaccharides, difucosyl derivatives of para-lacto-*N*-hexaose and para-lacto-*N*-neohexaose. J. Biol. Chem. *252*, 5408–5411.
175. Yamashita, K., Tachibana, Y. & Kobata, A. (1977) Oligosaccharides of human milk: Structures of three lacto-*N*-hexaose derivatives with H-haptenic structure. Arch. Biochem. Biophys. *182*, 546–555.
176. Sugahara, K., Akasaki, M., Funakoshi, I., Aula, P. & Yamashina, I. (1977) Structural studies of glycoasparagines from urine of a patient with aspartylglycosylaminuria (AGU). FEBS Lett. *78*, 284–286.
177. Sugahara, K., Akasaki, M., Funakoshi, I., Aula, P. & Yamashina, I. (1977) Structure

of two glycoasparagines isolated from the urine of patients with aspartylglycosylaminuria (AGU). J. Biochem. *82*, 1499–1501.
178. Strecker, G., Peers, M.-C., Michalski, J.-C., Hondi-Assah, T., Fournet, B., Spik, G., Montreuil, J., Farriaux, J.-P., Maroteaux, P. & Durand, P. (1977) Structure of nine sialyl-oligosaccharides accumulated in urine of eleven patients with three different types of sialidosis. Eur. J. Biochem. *75*, 391–403.
179. Strecker, G., Herlant-Peers, M.-C., Fournet, B., Montreuil, J., Dorland, L., Haverkamp, J., Vliegenthart, J. F. G. & Farriaux, J.-P. (1977) Structure of seven oligosaccharides excreted in the urine of a patient with Sandhoff's disease (GM_2-gangliosidosis-variant O). Eur. J. Biochem. *81*, 165–171.
180. Koseki, M. & Tsurumi, K. (1978) Structure of a novel sialooligosaccharide from the urine of a patient with mucolipidosis. Tohoku J. Exp. Med. *124*, 361–366.
181. Lundblad, A., Sjöblad, S. & Svensson, S. (1978) Characterization of a penta- and an octasaccharide from urine of a patient with juvenile GM_1-gangliosidosis. Arch. Biochem. Biophys. *188*, 130–136.
182. Ng Ying Kin, N. M. K. & Wolfe, L. S. (1978) Structure of the minor oligosaccharides in the liver of a patient with GM_2-gangliosidosis variant O (Sandhoff-Jatzkewitz disease). Carbohydr. Res. *67*, 522–526.
183. Nishigaki, M., Yamashita, K., Matsuda, I., Arashima, S. & Kobata, A. (1978) Urinary oligosaccharides of fucosidosis. Evidence of the occurrence of X-antigenic determinant in serum-type sugar chains of glycoproteins. J. Biochem. *84*, 823–834.
184. Tachibana, Y., Yamashita, K. & Kobata, A. (1978) Oligosaccharides of human milk: Structural studies of di- and trifucosyl derivatives of lacto-*N*-octaose and lacto-*N*-neooctaose. Arch. Biochem. Biophys. *188*, 83–89.
185. Yamashita, K., Tachibana, Y., Takada, S., Matsuda, I., Arashima, S. & Kobata, A. (1979) Urinary glycopeptides of fucosidosis. J. Biol. Chen. *254*, 4820–4827.
186. Ng Ying Kin, N. M. K. & Wolfe, L. S. (1979) Urinary excretion of a novel hexasaccharide and a glycopeptide analogue in fucosidosis. Biochem. Biophys. Res. Commun. *88*, 696–705.
187. Maury, P. (1979) Accumulation of two glycoasparagines in the liver in aspartylglycosaminuria. J. Biol. Chem. *254*, 1513–1515.
188. Koseki, M. & Tsurumi, K. (1979) The structure of a new sialic acid-containing decasaccharide from the urine of a patient with mucolipidosis. Tohoku J. Exp. Med. *128*, 39–49.
189. Yamashita, K., Tachibana, Y., Mihara, K., Okada, S., Yabuuchi, H. & Kobata, A. (1980) Urinary oligosaccharides of mannosidosis. J. Biol. Chem. *255*, 5126–5133.
190. Svennerholm, L., Vanier, M.-Th. & Månsson, J.-E (1980) Krabbe disease: A galactosylsphingosine (psychosine) lipidosis. J. Lip. Res. *21*, 53–64.
191. Matsuura, F., Nunez, H. A., Grabowski, G. A. & Sweeley, C. C. (1981) Structural studies of urinary oligosaccharides from patients with mannosidosis. Arch. Biochem. Biophys. *207*, 337–352.
192. Jones, M. Z. & Laine, R. A. (1981) Caprine oligosaccharide storage disease. Accumulation of β-mannosyl(1 → 4)β-*N*-acetylglucosaminyl(1 → 4)β-*N*-acetylglucosamine in brain. J. Biol. Chem. *256*, 5181–5184.
193. Yamashita, K., Ohkura, T., Okada, S., Yabuuchi, H. & Kobata, A. (1981) Urinary oligosaccharides of GM_1-gangliosidosis. Different excretion patterns of oligosaccharides in the urine of type 1 and type 2 subgroups. J. Biol. Chem. *256*, 4789–4798.
194. Ohkura, T., Yamashita, K. & Kobata, A. (1981) Urinary oligosaccharides of GM_1-gangliosidosis. Structures of oligosaccharides excreted in the urine of type 1 but not in the urine of type 2 patients. J. Biol. Chem. *256*, 8485–8490.
195. Michalski, J.-C., Strecker, G., van Halbeek, H., Dorland, L. & Vliegenthart, J. F. G (1982) The structure of six urinary oligosaccharides that are characteristic for a patient with Morquio syndrome type B. Carbohydr. Res. *100*, 351–363.
196. Warner, T. G. & O'Brien, J. S. (1982) Structure analysis of the major oligosaccharide accumulating in canine GM_1 gangliosidosis liver. J. Biol. Chem. *257*, 224–232.
197. Michalski, J.-C., Montreuil, J., Strecker, G., van Halbeek, H., Dorland, L., Vliegenthart, J. F. G., Cartigny, B. & Farriaux, J.-P.

(1983) Primary structure of three mannosyl-glycoasparagines and nine sialyl-oligosaccharides isolated from the urine of two patients with Gaucher's disease (infantile form). Eur. J. Biochem. *132*, 375–381.

198. Parkkinen, J. & Finne, J. (1983) Isolation and structural characterization of five major sialyloligosaccharides and a sialylglycopeptide from normal human urine. Eur. J. Biochem. *136*, 355–361.

199. Parkkinnen, J. & Finne, J. (1984) Isolation and characterization of novel phosphate-containing sialyloligosaccharides from normal human urine. Eur. J. Biochem. *140*, 427–431.

199a. Matsuura, F. & Jones, M. Z. (1985) Structural characterization of novel complex oligosaccharides accumulated in the caprine β-mannosidosis kidney. Occurrence of tetra- and pentasaccharides containing a β-linked mannose residue at the non-reducing terminus. J. Biol. Chem. *260*, 15239–15245.

199b. Wenger, D. A., Sujansky, E., Fennessey, P. V., Thompson, J. N. (1986) Human β-mannosidase deficiency. New Engl. J. Med. *315*, 1201–1205.

199c. Herkt, F., Paz Parente, J., Leroy, Y., Fournet, B., Blanchard, D., Cartron, J.-P., Van Halbeek, H. & Vliegenthart, J. F. G. (1985) Structure determination of oligosaccharides isolated from Cad erythrocyte membranes by permethylation analysis and 500-MHz ^{1}H-NMR spectroscopy. Eur. J. Biochem. *146*, 125–129.

200. Karlsson, K.-A. (1976) Microscale fingerprinting of blood-group fucolipids by mass spectrometry, in: Glycolipid Methodology (Witting, L., ed.) pp. 97–122, Am. Oil Chem. Soc. Press, Champaign.

201. Karlsson, K.-A. (1978) Mass-spectrometric sequence studies of lipid-linked oligosaccharides, blood-group fucolipids, gangliosides and related cell-surface receptors. Prog. Chem. Fats other Lipids *16*, 207–230.

202. Egge, H. (1978) The application of mass spectrometry in the structural elucidation of glycosphingolipids. Chem. Phys. Lipids *21*, 349–360.

203. Kováčik, V., Bauer, Š., Rosík, J. & Kováč, P. (1968) Mass spectrometry of uronic acid derivatives. Part III. The fragmentation of methyl ester methyl glycosides of methylated uronic and aldobiouronic acids. Carbohydr. Res. *8*, 282–290.

204. Kováčik, V., Bauer, Š. & Rosík, J. (1968) Mass spectrometry of uronic acid derivatives. Part IV. The fragmentation of methyl ester methyl glycosides of methylated aldotriouronic acids. Carbohydr. Res. *8*, 291–294.

205. Fichtinger-Schepman, A. M. J., Kamerling, J. P., Versluis, C. & Vliegenthart, J. F. G. (1980) Structural analysis of acidic oligosaccharides derived from the methylated, acidic polysaccharide associated with coccoliths of *Emiliania huxleyi* (Lohmann) Kamptner. Carbohydr. Res. *86*, 215–225.

206. Van Halbeek, H., Dorland, L., Veldink, G. A., Vliegenthart, J. F. G., Strecker, G., Michalski, J.-C., Montreuil, J. & Hull, W. E. (1980) A 500-MHz ^{1}H-NMR study of urinary oligosaccharides from patients with mannosidosis. FEBS Lett. *121*, 71–77.

207. Chizhov, O. S., Malysheva, N. N., Kadentsev, V. I. & Kochetkov, N. K. (1970) Mass spectra of derivatives of glycosylpolyols. Dokl. Akad. Nauk SSSR *194*, 836–839.

208. Kärkkäinen, J. (1970) Analysis of disaccharides as permethylated disaccharide alditols by gas-liquid chromatography-mass spectrometry. Carbohydr. Res. *14*, 27–33.

209. Kärkkäinen, J. (1971) Structural analysis of trisaccharides as permethylated trisaccharide alditols by gas-liquid chromatography-mass spectrometry. Carbohydr. Res. *17*, 11–18.

210. Whyte, J. N. C. (1973) Gas-liquid chromatography-mass spectrometry of permethylated aldosyl aldonates. Can. J. Chem. *51*, 3197–3202.

211. Mononen, J., Finne, J. & Kärkkäinen, J. (1978) Analysis of permethylated hexopyranosyl-2-acetamido-2-deoxyhexitols by g.l.c.-m.s. Carbohydr. Res. *60*, 371–375.

212. Jensen, M., Borowiak, D., Paulsen, H. & Rietschel, E. Th. (1979) Analysis of permethylated glucosaminyl-glucosaminitol disaccharides by combined gas-liquid chromatography mass spectrometry. Biomed. Mass Spectrom. *6*, 559–565.

213. Fraser, B. A., Tsui, F.-P. & Egan, W. (1979) Mass-spectral studies of isomeric D-ribo-

furanosylribitol disaccharides from the capsular polysaccharides of *Haemophilus influenzae* type b and *Escherichia coli* K100. Carbohydr. Res. *73*, 59–65.

214. Fournet, B., Dhalluin, J.-M., Strecker, G., Montreuil, J., Bosso, C. & Defaye, J. (1980) Gas-liquid chromatography and mass spectrometry of oligosaccharides obtained by partial acetolysis of glycans of glycoproteins. Anal. Biochem. *108*, 35–56.
215. Herlant-Peers, M.-C., Montreuil, J., Strecker, G., Dorland, L., van Halbeek, H., Veldink, G. A. & Vliegenthart, J. F. G. (1981) Structures of fifteen oligosaccharides isolated from new-born meconium. Eur. J. Biochem. *117*, 291–300.
216. Chizhov, O. S., Kadentsev, V. I., Solv'yov, A. A., Levonowich, P. F. & Dougherty, R. C. (1976) Polysaccharide sequencing by mass spectrometry: Chemical ionization spectra of permethyl glycosylalditols. J. Org. Chem. *41*, 3425–3428.
217. Ando, S., Kon, K., Nagai, Y. & Murata, T. (1977) Chemical ionization and electron impact mass spectra of oligosaccharides derived from sphingoglycolipids. J. Biochem. *82*, 1623–1631.
218. Åman, P., McNeil, M., Franzén, L.-E., Darvill, A. G. & Albersheim, P. (1981) Structural elucidation, using H.P.L.C.-M.S. and G.L.C.-M.S., of the acidic polysaccharide secreted by *Rhizobium meliloti* strain 1021. Carbohydr. Res. *95*, 263–282.
219. McNeil, M., Darvill, A. G., Åman, P., Franzén, L.-E. & Albersheim, P. (1982) Structural analysis of complex carbohydrates using high-performance liquid chromatography, gas chromatography; and mass spectrometry. Meth. Enzymol. *83*, 3–45.
220. McNeil, M. (1983) Elimination of internal glycosyl residues during chemical ionization-mass spectrometry of per-*O*-alkylated oligosaccharide-alditols. Carbohydr. Res. *123*, 31–40.
221. Tanaka, Y., Yu, R. K., Ando, S., Ariga, T. & Itoh, T. (1984) Chemical-ionization mass spectra of the permethylated sialo-oligosaccharides liberated from gangliosides. Carbohydr. Res. *126*, 1–14.
222. Strecker, G., Fournet, B., Montreuil, J., Dorland, L., Haverkamp, J., Vliegenthart, J. F. G. & Dubesset, D. (1978) Structure of the three major fucosyl-glycoasparagines accumulating in the urine of a patient with fucosidosis. Biochimie *60*, 725–734.
223. Strecker, G., Pierce-Cretel, A., Fournet, B., Spik, G. & Montreuil, J. (1981) Characterization by gas-liquid chromatography-mass spectrometry of oligosaccharides resulting from the hydrazinolysis-nitrous acid deamination reaction of glycopeptides. Anal. Biochem. *111*, 17–26.
224. Mononen, I. & Krusius, T. (1983) Isolation and g.l.c.-m.s. analysis of the terminal, sialic acid-containing oligosaccharides obtained from intact glycoproteins by nitrous acid deamination. Carbohydr. Res. *112*, 165–170.
225. Hallgren, P., Lundblad, A. & Svensson, S. (1975) A new type of carbohydrate-protein linkage in a glycopeptide from normal human urine. J. Biol. Chem. *250*, 5312–5314.
226. Lundblad, A., Masson, P. K., Nordén, N. E., Svensson, S. & Öckerman, P.-A. (1975) Determination of a mannose-containing trisaccharide in the urine of patients and heterozygotes of mannosidosis. Biomed. Mass Spectrom. *2*, 285–287.
227. Lennartson, G., Lundblad, A., Sjöblad, S., Svensson, S. & Öckerman, P.-A. (1976) Quantitation of a urinary tetrasaccharide by gas chromatography and mass spectrometry. Biomed. Mass Spectrom. *3*, 51–54.
228. Hallgren, P. & Lundblad, A. (1977) Structural analysis of nine oligosaccharides isolated from the urine of a blood group O, nonsecretor, woman during pregnancy and lactation. J. Biol. Chem. *252*, 1014–1022.
229. Hallgren, P. & Lundblad, A. (1977) Structural analysis of oligosaccharides isolated from the urine of a blood group A, secretor, woman during pregnancy and lactation. J. Biol. Chem. *252*, 1023–1033.
230. Hallgren, P., Lindberg, B. S. & Lundblad, A. (1977) Quantitation of some urinary oligosaccharides during pregnancy and lactation. J. Biol. Chem. *252*, 1034–1040.
231. Lennartson, G., Lundblad, A., Lundsten, J., Svensson, S. & Häger, A. (1978) Glucose-containing oligosaccharides in the urine of patients with glycogen storage disease type II and type III. Eur. J. Biochem. *83*, 325–334.

232. Lundblad, A., Lundsten, J., Nordén, N. E., Sjöblad, S., Svensson, S., Öckerman, P.-A. & Gehloff, M. (1978) Urinary abnormalities in fucosidosis. Characterization of a disaccharide and two glycoasparagines. Eur. J. Biochem. *83*, 513–521.
233. Chester, M. A., Hallgren, P., Lundblad, A. & Messeter, L. (1979) Urinary excretion of oligosaccharides induced by galactose given orally or intravenously. Eur. J. Biochem. *100*, 385–392.
234. Lundblad, A., Svensson, S., Yamashina, I. & Ohta, M. (1979) Increased urinary excretion of a glucose-containing tetrasaccharide in patients with Duchenne muscular dystrophy. FEBS Lett. *97*, 249–252.
235. Lundblad, A., Svensson, S., Löw, B., Messeter, L. & Cedergren, B. (1980) Release of oligosaccharides from human erythrocyte membranes of different blood-group-P phenotypes by trifluoroacetolysis. Eur. J. Biochem. *104*, 323–330.
236. Derappe, C., Lundblad, A., Messeter, L. & Svensson, S. (1980) Isolation and characterization of a blood group A-specific urinary tetrasaccharide. FEBS Lett. *119*, 177–180.
237. Kuriyama, M., Ariga, T., Ando, S., Suzuki, M., Yamada, T. & Miyatake, T. (1981) Four positional isomers of sialyloligosaccharides isolated from the urine of a patient with sialidosis. J. Biol. Chem. *256*, 12316–12321.
238. Van Halbeek, H., Dorland, L., Haverkamp, J., Veldink, G. A., Vliegenthart, J. F. G., Fournet, B., Ricart, G., Montreuil, J., Gathmann, W. D. & Aminoff, D. (1981) Structure determination of oligosaccharides isolated from A^+, H^+ and A^- H^- hog-submaxillary-gland mucin glycoproteins, by 360-MHz ^{1}-H-NMR spectroscopy, permethylation analysis and mass spectrometry. Eur. J. Biochem. *118*, 487–495.
239. Matsuura, F., Laine, R. A. & Jones, M. Z. (1981) Oligosaccharides accumulated in the kidney of a goat with β-mannosidosis: Mass spectrometry of intact permethylated derivatives. Arch. Biochem. Biophys. *211*, 485–493.
240. Pierce-Cretel, A., Pamblanco, M., Strecker, G., Montreuil, J. & Spik, G. (1981) Heterogeneity of the glycans *O*-glycosidically linked to the hinge region of secretory immunoglobulins from human milk. Eur. J. Biochem. *114*, 169–178.
241. Hounsell, E. F., Wood, E., Feizi, T., Fukuda, M., Powell, M. E. & Hakomori, S.-I. (1981) Structural analysis of hexa- to octasaccharide fractions isolated from sheep gastric-glycoproteins having bloodgroup I and i activities. Carbohydr. Res. *90*, 283–307.
242. Lemonnier, M., Derappe, C., Poenaru, L., Chester, M. A., Lundblad, A., Svensson, S. & Öckerman, P.-A. (1982) Isolation and characterization of 4-*O*-β-D-mannopyranosyl-2-acetamido-2-deoxy-D-glucose from the urine of a patient with mucolipidosis II and its occurrence in normal and pathological urine. FEBS Lett. *141*, 263–266.
243. Egge, H., Michalski, J.-C. & Strecker, G. (1982) Heterogeneity of urinary oligosaccharides from mannosidosis: Mass spectrometric analysis of permethylated Man_9, Man_8, and Man_7 derivatives. Arch. Biochem. Biophys. *213*, 318–326.
244. Derappe, C., Bauvy, C., Meyer, D. M., Lemonnier, M., Lhermitte, M. & Platzer, N. (1983) Characterization of a new oligosaccharide containing myo-inositol found in pregnancy urine. Carbohydr. Res. *115*, 221–229.
245. Parkkinen, J. (1983) Characterization of a *scyllo*-inositol-containing sialyloligosaccharide from normal human urine. FEBS Lett. *163*, 10–13.
246. Sadeh, S., Warren, C. D., Daniel, P. F., Bugge, B., James, L. F. & Jeanloz, R. W. (1983) Characterization of oligosaccharides from the urine of locointoxicated sheep. FEBS Lett. *163*, 104–109.
247. Nilsson, B. & Zopf, D. (1983) Oligosaccharides released from glycolipids by trifluoroacetolysis can be analyzed by gas chromatography-mass spectrometry. Arch. Biochem. Biophys. *222*, 628–648.
248. Matsuura, F., Jones, M. Z. & Frazier, S. E. (1983) Structural analysis of the major caprine β-mannosidosis urinary oligosaccharides. Biochim. Biophys. Acta *759*, 67–73.
249. Meyer, D. M., Lemonnier, M., Derappe, C., Sellier, N. & Platzer, N. (1984) Isolation and characterization of 1-*O*-α-2-acetamido-

2-deoxy-D-galactopyranosyl-myo-inositol from pregnancy urine. FEBS Lett. *172*, 99–102.

250. Michalski, J.-C., Bouquelet, S., Montreuil, J., Strecker, G., Dulac, O. & Munnich, A. (1984) Abnormal galactoside excretion in urine of a patient with early myoclonic epileptic encephalopathy. Clin. Chim. Acta *137*, 43–51.
251. Lecat, D., Lemonnier, M., Derappe, C., Lhermitte, M., van Halbeek, H., Dorland, L. & Vliegenthart, J. F. G. (1984) The structure of sialyl-glycopeptides of the *O*-glycosidic type, isolated from sialidosis (mucolipidosis I) urine. Eur. J. Biochem. *140*, 415–420.
252. Ugorski, M., Nilsson, B., Schroer, K., Cashel, J. A. & Zopf, D. (1984) Gas chromatography/mass spectrometry analysis of oligosaccharides from neutral glycosphingolipids of murine B cell hybridomas. J. Biol. Chem. *259*, 481–486.
253. Wang, W. T., Matsuura, F., Nunez, H., Ledonne, N., Baltzer, B. & Sweeley, C. C. (1984) Glycerol-, inositol-, and reducing end hexose-containing oligosaccharides in human urine. Glycoconj. J. *1*, 17–35.
253a. Jardine, I., Matsuura, F. & Sweeley, C. C. (1984) Electron ionization mass spectra of reduced and permethylated urinary oligosaccharides from patients with mannosidosis. Biomed. Mass Spectrom. *11*, 562–568.
253b. Kuriyama, M., Ariga, T., Ando, S., Suzuki, M., Yamado, T., Miyatake, T. & Igata, A. (1985) Two positional isomers of sialylheptasaccharides isolated from the urine of a patient with sialidosis. J. Biochem. *98*, 1049–1054.
254. Karlsson, K.-A., Pascher, I., Samuelsson, B. E., Finne, J., Krusius, T. & Rauvala, H. (1978) Mass spectrometric sequence study of the oligosaccharide of human transferrin. FEBS Lett. *94*, 413–417.
255. Maury, C. P. J. & Kärkkäinen, J. (1979) Identification of 4-*N*-2-acetamido-2-deoxy-*β*-D-glucopyranosyl-L-asparagine in biological materials by gas chromatography-mass spectrometry. Clin. Chim. Acta *91*, 75–79.
256. Fukuda, M. N., Papermaster, D. S., Hargrave, P. A. (1979) Rhodopsin carbohydrate. Structure of small oligosaccharides attached at two sites near the NH_2 terminus. J. Biol. Chem. *254*, 8201–8207.
257. Morris, H. R., Thompson, M. R., Osuga, D. T., Ahmed, A. I., Chan, S. M., Vandenheede, J. R. & Feeney, R. E. (1978) Antifreeze glycoproteins from the blood of an antarctic fish. The structure of the proline-containing glycopeptides. J. Biol. Chem. *253*, 5155–5162.
258. Nebelin, E. & Das, B. C. (1979) Mass spectrometry of synthetic glycopeptide analogs of bacterial peptidoglycans (MDP and derivatives). FEBS Lett. *107*, 254–258.
259. Maury, C. P. J. (1979) Quantitative determination of 4-*N*-2-acetamido-2-deoxy-*β*-D-glucopyranosyl-L-asparagine in the urine of patients with aspartylglycosaminuria by gas-liquid chromatography. J. Lab. Clin. Med. *93*, 718–723.
260. Karlsson, K.-A., Pascher, I. & Samuelsson, B. E. (1974) Analysis of intact gangliosides by mass spectrometry. Comparison of different derivatives of a hematoside of a tumour and the major monosialoganglioside of brain. Chem. Phys. Lipids *12*, 271–286.
261. Karlsson, K.-A. (1974) Carbohydrate composition and sequence analysis of a derivative of brain disialoganglioside by mass spectrometry, with molecular weight ion at *m/z* 2245. Potential use in the specific microanalysis of cell surface components. Biochemistry *13*, 3643-3647.
262. Karlsson, K.-A. (1974) Use of mass spectrometry for the carbohydrate composition and sequence analysis of glycosphingolipids. Biomed. Mass Spectrom. *1*, 49–56.
263. Ledeen, R. W., Kundu, S. K., Price, H. C. & Fong, J. W. (1974) Mass spectra of permethyl derivatives of glycosphingolipids. Chem. Phys. Lipids *13*, 429–446.
264. Breimer, M. E., Hansson, G. C., Karlsson, K.-A., Leffler, H., Pimlott, W. & Samuelsson, B. E. (1978) Structure determination of blood group type glycolipids of cat small intestine by mass fragmentography. FEBS Lett. *89*, 42–46.
265. Breimer, M. E., Hansson, G. C., Karlsson, K.-A., Leffler, H., Pimlott, W. & Samuelsson, B. E. (1979) Selected ion monitoring of glycosphingolipid mixtures. Identification

of several blood group type glycolipids in the small intestine of an individual rabbit. Biomed. Mass Spectrom. *6*, 231–241.

266. Ariga, T., Murata, T., Oshima, M., Maezawa, M. & Miyatake, T. (1980) Characterization of glycosphingolipids by direct inlet chemical ionization mass spectrometry. J. Lip. Res. *21*, 879–887.
267. Ariga, T., Yu, R. K., Suzuki, M., Ando, S. & Miyatake, T. (1982) Characterization of GM_1 ganglioside by direct inlet chemical ionization mass spectrometry. J. Lip. Res. *23*, 437–442.
268. Ariga, T., Yu, R. K. & Miyatake, T. (1984) Characterization of gangliosides by direct inlet chemical ionization mass spectrometry. J. Lip. Res. *25*, 1096–1101.
269. Fredman, P., Månsson, J.-E., Svennerholm, L., Karlsson, K.-A., Pascher, I. & Samuelsson, B. E. (1980) The structure of the tetrasialoganglioside from human brain. FEBS. Lett. *110*, 80–84.
270. Egge, H. & Hanfland, P. (1981) Immunochemistry of the Lewis-blood-group-system. Mass spectrometric analysis of permethylated Le^a-, Le^b-, and H-type 1 (Le^{dH}) bloodgroup active and related glycosphingolipids from human plasma. Arch. Biochem. Biophys. *210*, 396–404.
271. Hanfland, P., Egge, H., Dabrowski, U., Kuhn, S., Roelcke, D. & Dabrowski, J. (1981) Isolation and characterization of an I-active ceramide decasaccharide from rabbit erythrocyte membranes. Biochemistry *20*, 5310–5319.
272. Ishizuka, I. & Tadano-Aritomi, K. (1984) Characterization of sulfoglycosphingolipids by direct probe electron impact mass spectrometry. J. Biochem. *96*, 829–839.
273. Kannagi, R., Levery, S. B., Ishigami, F., Hakomori, S.-I., Shevinsky, L. H., Knowles, B. B. & Solter, D. (1983) New globoseries glycosphingolipids in human teratocarcinoma reactive with the monoclonal antibody directed to a developmentally regulated antigen, stage-specific embryonic antigen 3. J. Biol. Chem. *258*, 8934–8942.
274. Kundu, S. K., Samuelsson, B. E., Pascher, I. & Marcus, D. M. (1983) New gangliosides from human erythrocytes. J. Biol. Chem. *258*, 13857–13866.
275. Matsumoto, M., Taki, T., Samuelsson, B., Pascher, I., Hirabayashi, Y., Li, S.-C. & Li, Y.-T. (1981) Further characterization of the structure of GM_{1b} ganglioside from rat ascites hepatoma. J. Biol. Chem. *256*, 9737–9741.
276. Breimer, M. E. (1984) Tissue specificity of glycosphingolipids as expressed in pancreas and small intestine of blood group A and B human individuals. Arch. Biochem. Biophys. *228*, 71–85.
277. Forsberg, L. S., Dell, A., Walton, D. J. & Ballou, C. E. (1982) Revised structure for the 6-*O*-methylglucose polysaccharide of *Mycobacterium smegmatis*. J. Biol. Chem. *257*, 3555–3563.
278. Dell, A., Morris, H. R., Egge, H., Von Nicolai, H. & Strecker, G. (1983) Fast-atom-bombardment mass-spectrometry for carbohydrate-structure determination. Carbohydr. Res. *115*, 41–52.
279. Egge, H., Dell, A. & Von Nicolai, H. (1983) Fucose containing oligosaccharides from human milk. I. Separation and identification of new constituents. Arch. Biochem. Biophys. *224*, 235–253.
280. Kamerling, J. P., Heerma, W., Vliegenthart, J. F. G., Green, B. N., Lewis, I. A. S., Strecker, G. & Spik, G. (1983) Fast atom bombardment mass spectrometry of carbohydrate chains derived from glycoproteins. Biomed. Mass Spectrom. *10*, 420–425.
281. Abraham, D. J., Sidebothom, R., Winchester, B. G., Dorling, P. R. & Dell, A. (1983) Swainsonine affects the processing of glycoproteins in vivo. FEBS Lett. *163*, 110–113.
282. Egge, H., Peter-Katalinić, J., Paz-Parente, J., Strecker, G., Montreuil, J. & Fournet, B. (1983) Carbohydrate structures of hen ovomucoid. A mass spectrometric analysis. FEBS Lett. *156*, 357–362.
283. Arita, M., Iwamori, M., Higuchi, T. & Nagai, Y. (1983) 1,1,3,3,-Tetramethylurea and triethanolamine as a new useful matrix for fast atom bombardment mass spectrometry of gangliosides and neutral glycosphingolipids. J. Biochem. *93*, 319–322.
284. Arita, M., Iwamori, M., Higuchi, T. & Nagai, Y. (1983) Negative ion fast atom bom-

bardment mass spectrometry of gangliosides and asialo gangliosides: A useful method for the structural elucidation of gangliosides and related neutral glycosphingolipids. J. Biochem. *94*, 249–256.

285. Dell, A., Oates, J. E., Morris, H. R. & Egge, H. (1983) Structure determination of carbohydrates and glycosphingolipids by fast atom bombardment mass spectrometry. Int. J. Mass Spectrom. & Ion Phys. *46*, 415–418.
286. Dell, A. & Ballou, C. E. (1983) Fast atom bombardment mass spectrometry of a 6-*O*-methylglucose polysaccharide. Biomed. Mass Spectrom. *10*, 50–56.
287. Dell, A. & Ballou, C. E. (1983) FAB, negative-ion MS of the mycobacterial *O*-methyl-D-glucose polysaccharide and lipopolysaccharides. Carbohydr. Res. *120*, 95–111.
288. Dell, A., York, W. S., McNeil, M., Darvill, A. G. & Albersheim, P. (1983) The cyclic structure of β-D-(1 $\rightarrow$ 2)-linked D-glucans secreted by *Rhizobia* and *Agrobacteria*. Carbohydr. Res. *117*, 185–200.
289. Phillips, L. R., Nishimura, O. & Fraser, B. A. (1983) The structure of the repeating oligosaccharide unit of the pneumococcal capsular polysaccharide type 18C. Carbohydr. Res. *121*, 243–255.
290. Yamashita, K., Ueda, I. & Kobata, A. (1983) Sulfated asparagine-linked sugar chains of hen egg albumin. J. Biol. Chem. *258*, 14144–14147.
291. Hounsell, E. F., Madigan, M. J. & Lawson, A. M. (1984) Fast-atom-bombardment mass spectrometry and electron-impact direct-probe mass spectrometry in the structural analysis of oligosaccharides with special reference to the poly(*N*-acetyl-lactosamine) series. Biochem. J. *219*, 947–952.
292. Bosso, C., Defaye, J., Heyraud, A. & Ulrich, J. (1984) Fast-atom bombardment mass spectrometry for the characterization of cello- and malto-oligosaccharides. Carbohydr. Res. *125*, 309–317.
293. Arita, M., Iwamori, M., Higuchi, T. & Nagai, Y. (1984) Positive and negative ion fast atom bombardment mass spectrometry of glycosphingolipids. Discrimination of the positional isomers of gangliosides with sialic acids. J. Biochem. *95*, 971–981.
294. Egge, H., Dabrowski, J. & Hanfland, P. (1984) Structural analysis of blood-group ABH, I, i, Lewis and related glycosphingolipids. Application of FAB mass spectrometry and high resolution proton NMR. Pure Appl. Chem. *56*, 807–819.
295. Egge, H., Peter-Katalinić, J. & Hanfland, P. (1984) Structure analysis of glycosphingolipids using fast atom bombardment (FAB) techniques. Adv. Exp. Med. Biol. *174*, 55–63.
296. Fukuda, M., Dell, A. & Fukuda, M. N. (1984) Structure of fetal lactosaminoglycan. The carbohydrate moiety of band 3 isolated from human umbilical cord erythrocytes. J. Biol. Chem. *259*, 4782–4791.
297. Spooncer, E., Fukuda, M., Klock, J. C., Oates, J. E. & Dell, A. (1984) Isolation and characterization of polyfucosylated lactosaminoglycan from human granulocytes. J. Biol. Chem. *259*, 4792–4801.
298. Fukuda, M., Dell, A., Oates, J. E. & Fukuda, M. N. (1984) Structure of branched lactosaminoglycan, the carbohydrate moiety of band 3 isolated from adult human erythrocytes. J. Biol. Chem. *259*, 8260–8273.
299. Fukuda, M., Spooncer, E., Oates, J. E., Dell, A. & Klock, J. C. (1984) Structure of sialylated fucosyl lactosaminoglycan isolated from human granulocytes. J. Biol. Chem. *259*, 10925–10935.
300. Dell, A., Oates, J., Lugowski, C., Romanowska, E., Kenne, L. & Lindberg, B. (1984) The enterobacterial common-antigen, a cyclic polysaccharide. Carbohydr. Res. *133*, 95–104.
301. Geyer, R., Geyer, H., Egge, H., Schott, H. H. & Stirm, S. (1984) Structure of the oligosaccharides sensitive to endo-β-*N*-acetylglucosaminidase H in the glycoprotein of Friend murine leukemia virus. Eur. J. Biochem. *143*, 531–539.
302. Gabel, C. A., Costello, C. E., Reinhold, V. N., Kurz, L. & Kornfeld, S. (1984) Identification of methylphosphomannosyl residues as components of the high mannose oligosaccharides of *Dictyostelium discoideum* glycoproteins. J. Biol. Chem. *259*, 13762–13769.
303. Derappe, C., Bauvy, C., Lemonnier, M., Lhermitte, M., Platzer, N., Egge, H., Peter-

Katalinić, J., van Halbeek, H., Kamerling, J. P. & Vliegenthart, J. F. G. (1986) Structure determination of a phosphate-containing fucosyl-trisaccharide isolated from human urine. Carbohydr. Res. *150*, 273–284.

304. Hemling, M. E., Yu, R. K., Sedgwick, R. D. & Rinehart Jr., K. L. (1984) Fast atom bombardment mass spectrometry of glycosphingolipids. Glycosphingolipids containing neutral sugars. Biochemistry *23*, 5706–5713.

305. Linscheid, M., D'Angona, J., Burlingame, A. L., Dell, A. & Ballou, C. E. (1981) Field desorption mass spectrometry of oligosaccharides. Proc. Natl. Acad. Sci. USA *78*, 1471–1475.

306. Kushi, Y. & Handa, S. (1982) Application of field desorption mass spectrometry for the analysis of sphingoglycolipids. J. Biochem. *91*, 923–931.

307. Handa, S., Kushi, Y., Kambara, H. & Shizukuishi, K. (1983) Secondary ion mass spectra of neutral sphingoglycolipids. J. Biochem. *93*, 315–318.

308. Kushi, Y., Handa, S., Kambara, H. & Shizukuishi, K. (1983) Comparative study of acidic glycosphingolipids by field desorption and secondary ion mass spectrometry. J. Biochem. *94*, 1841–1850.

309. Shimamura, M., Endo, T., Inoue, Y. & Inoue, S. (1983) A novel neutral oligosaccharide chain found in polysialoglycoproteins isolated from pacific salmon eggs. Structural studies by secondary ion mass spectrometry, proton nuclear magnetic resonance spectroscopy, and chemical methods. Biochemistry *22*, 959–963.

310. Shimamura, M., Endo, T., Inoue, Y., Inoue, S. & Kambara, H. (1984) Fish egg polysialoglycoproteins: Structures of new sialooligosaccharide chains isolated from the eggs of *Oncorhynchus keta* (Walbaum). Fucose-containing units with oligosialyl groups. Biochemistry *23*, 317–322.

311. Van Halbeek, H., Dorland, L., Veldink, G. A., Vliegenthart, J. F. G., Strecker, G., Michalski, J.-C., Montreuil, J. & Hull, W. E. (1980) A 500 MHz ^{1}H-NMR study of urinary oligosaccharides from patients with mannosidosis. FEBS Lett. *121*, 71–77.

312. Nunez, H. A., Matsuura, F. & Sweeley, C. C. (1981) Complete structure elucidation of a trisaccharide from a patient with mannosidosis by carbon-13 nuclear magnetic resonance spectrometry. Arch. Biochem. Biophys. *212*, 638–643.

313. Townsend, R. R., Heller, D. N., Fenselau, C. C. & Lee, Y. C. (1984) Determination of the sialylation pattern of human fibrinogen glycopeptides with fast atom bombardment mass spectrometry. Biochemistry *23*, 6389–6392.

314. Abraham, D., Blakemoore, W. F., Dell, A., Herrtage, M. E., Jones, J., Littlewood, J. T., Oates, J., Palmer, A. C., Sidebotham, R. & Winchester, B. (1984) The enzymic defect and storage products in canine fucosidosis. Biochem. J. *221*, 25–33.

315. Sturman, J. A., Lin, Y. Y., Higuchi, T. & Fellman, J. H. (1985) *N*-Acetylneuramin lactose sulfate: A newly identified nutrient in milk. Ped. Res. *19*, 216–219.

316. Wieruszeski, J.-M., Chekkor, A., Bouquelet, S., Montreuil, J., Strecker, G., Peter-Katalinić, J. & Egge, H. (1985) Structure of two new oligosaccharides isolated from human milk: Sialylated lacto-*N*-fucopentaoses I and II. Carbohydr. Res. *137*, 127–138.

317. Mutsaers, J. H. G. M., van Halbeek, H., Kamerling, J. P. & Vliegenthart, J. F. G. (1985) Determination of the structure of the carbohydrate chains of prostaglandin endoperoxide synthase from sheep. Eur. J. Biochem. *147*, 569–574.

318. Oates, J. E., Dell, A., Fukuda, M. & Fukuda, M. N. (1985) A rapid mass-spectrometric procedure for probing the non-reducing structures of lactosaminoglycan-containing glycoconjugates. Carbohydr. Res. *141*, 149–152.

319. Naik, S., Oates, J. E., Dell, A., Taylor, G. W., Dey, P. M. & Pridham, J. B. (1985) A novel mass spectrometric procedure for the rapid determination of the types of carbohydrate chains present in glycoproteins: Application to α-galactosidase I from *Vicia faba* seeds. Biochem. Biophys. Res. Commun. *132*, 1–7.

320. Hanisch, F.-G., Egge, H., Peter-Katalinic, J., Uhlenbruck, G., Dienst, C. & Fangmann, R. (1985) Primary structures and Lewis blood-group-dependent expression of major sialylated saccharides from mucus

glycoproteins of human seminal plasma. Eur. J. Biochem. *152*, 343–351.

321. Krusius, T., Fukuda, M., Dell, A. & Ruoslahti, E. (1985) Structure of the carbohydrate units of human amniotic fluid fibronectin. J. Biol. Chem. *260*, 4110–4116.
322. Kuriyama, M., Hiwatari, R., Ariga, T., Sakano, Y., Abe, J., Osame, M. & Igata, A. (1985) Neutral oligosaccharides in the urine of a patient with glycogen storage disease type II. J. Biochem. *98*, 1041–1047.
323. Fukuda, M., Bothner, B., Ramsamooj, P., Dell, A., Tiller, P. R., Varki, A. & Klock, J. C. (1985) Structures of sialylated fucosyl polylactosaminoglycans isolated from chronic myelogenous leukemia cells. J. Biol. Chem. *260*, 12957–12967.
324. Fukuda, M. N., Dell, A., Oates, J. E. & Fukuda, M. (1985) Embryonal lactosaminoglycan. The structure of branched lactosaminoglycans with novel disialosyl (sialylα2→9sialyl) terminals isolated from PA1 human embryonal carcinoma cells. J. Biol. Chem. *260*, 6623–6631.
325. Fukuda, M., Carlsson, S. R., Klock, J. C. & Dell, A. (1986) Structures of *O*-linked oligosaccharides isolated from normal granulocytes, chronic myelogenous leukemia cells, and acute myelogenous leukemia cells. J. Biol. Chem. *261*, 12796–12806.
326. Hanisch, F.-G., Egge, H., Peter-Katalinić, J. & Uhlenbruck, G. (1986) Primary structure of a major sialyl-saccharide alditol from human amniotic mucins expressing the tumor-associated sialyl-X antigenic determinant. FEBS Lett. *200*, 42–46.
327. Hanish, F.-G., Egge, H., Peter-Katalinić, J. & Uhlenbruck, G. (1986) Structure of neutral oligosaccharides derived from mucus glycoproteins of human seminal plasma. Eur. J. Biochem. *155*, 239–247.
328. Trimble, R. B. & Atkinson, P. H. (1986) Structures of yeast external invertase $Man_{8-14}GlcNAc$ processing intermediates by 500-MHz ^1H-NMR spectroscopy. J. Biol. Chem. *261*, 9815–9824.
329. Dennis, J. W., Laferté, S., Fukuda, M., Dell, A. & Carver, J. P. (1986) Asn-linked oligosaccharides in lectin-resistant tumor-cell mutants with varying metastatic potential. Eur. J. Biochem. *161*, 359–373.
330. Carr, S. A. & Roberts, G. D. (1986) Carbohydrate mapping my mass spectrometry: A novel method for identifying attachment sites of Asn-linked sugars in glycoproteins. Anal. Biochem. *157*, 396–406.
331. Trimble, R. B., Atkinson, P. H., Tarentino, A. L., Plummer Jr., T. H., Maley, F. & Tomer, K. B. (1986) Transfer of glycerol by endo-β-*N*-acetylglucosaminidase F to oligosaccharides during chitobiose core cleavage. J. Biol. Chem. *261*, 12000–12005.
332. Tsai, P.-K., Dell, A. & Ballou, C. E. (1986) Characterization of acetylated and acetolyzed glycoprotein high-mannose core oligosaccharides by fast-atom-bombardment mass spectrometry. Proc. Natl. Acad. Sci. USA *82*, 4119–4123.
333. Dell, A. & Tiller, P. R. (1986) A novel mass spectrometric procedure to rapidly determine the position of *O*-acylated residues in the sequence of naturally occurring oligosaccharides. Biochem. Biophys. Res. Commun. *135*, 1126–1134.
334. Abraham, D., Daniel, P., Dell, A., Oates, J., Sidebotham, R. & Winchester, B. (1986) Structural analysis of the major urinary oligosaccharides in feline α-mannosidosis. Biochem. J. *233*, 899–904.
335. Barker, C. G., Blakemore, W. F., Dell, A., Palmer, A. C., Tiller, P. R. & Winchester, B. G. (1986) G_{M1} Gangliosidosis (type 1) in a cat. Biochem. J. *235*, 151–158.
336. Palcic, M. M., Srivastava, O. P. & Hindsgaul, O. (1987) Transfer of D-galactosyl groups to 6-*O*-substituted 2-acetamido-2-deoxy-D-glucose residues by use of bovine D-galactosyltransferase. Carbohydr. Res. *159*, 315–324.
337. Scudder, P., Tang, P. W., Hounsell, E. F., Lawson, A. M., Mehmet, H. & Feizi, T. (1986) Isolation and characterization of sulphated oligosaccharides released from bovine corneal keratan sulphate by the action of endo-β-galactosidase. Eur. J. Biochem. *157*, 365–373.
338. Reinhold, V. N., Carr, S. A., Green, B. N., Petitou, M., Choay, J. & Sinaÿ, P. (1987) Structural characterization of sulfated glycosaminoglycans by fast atom bombardment mass spectrometry: Application to heparin fragments prepared by chemical synthesis. Carbohydr. Res. *161*, 305–313.

339. Ballou, C. E. & Dell, A. (1985) Polysaccharide-lipid interaction analyzed by fast-atom bombardment mass spectrometry. Carbohydr. Res. *140*, 139–143.
340. Egge, H., Peter-Katalinić, J., Reuter, G., Schauer, R., Ghidoni, R., Sonnino, S. & Tettamanti, G. (1985) Analysis of gangliosides using fast atom bombardment mass spectrometry. Chem. Phys. Lipids *37*, 127–141.
341. Rösner, H., Rahmann, H., Reuter, G., Schauer, R., Peter-Katalinić, J. & Egge, H. (1985) Mass spectrometric identification of the pentasialoganglioside G_{P1c} of embryonic chicken brain. Biol. Chem. Hoppe-Seyler *366*, 1177–1181.
342. Thurin, J., Herlyn, M., Hindsgaul, O., Strömberg, N., Karlsson, K.-A., Elder, D., Steplewski, Z. & Koprowski, H. (1985) Proton NMR and fast-atom bombardment mass spectrometry analysis of the melanoma-associated ganglioside 9-*O*-acetyl-G_{D3}. J. Biol. Chem. *260*, 14556–14563.
343. Dennis, R. D., Geyer, R., Egge, H., Peter-Katalinić, J., Li, S.-C., Stirm, S. & Wiegandt, H. (1985) Glycosphingolipids in insects. Chemical structures of ceramide tetra-, penta-, hexa-, and heptasaccharides from *Calliphora vicina* pupae (insecta: diptera). J. Biol. Chem. *260*, 5370–5375.
344. Egge, H., Kordowicz, M., Peter-Katalinić, J. & Hanfland, P. (1985) Immunochemistry of I/i-active oligo- and polyglycosylceramides from rabbit erythrocyte membranes. Characterization of linear, di- and tri-antennary neolactoglycosphingolipids. J. Biol. Chem. *260*, 4927–4935.
345. Kamisango, K., Saadat, S., Dell, A. & Ballou, C. E. (1985) Pyruvylated glycolipids from *Mycobacterium smegmatis*. Nature and location of the lipid components. J. Biol. Chem. *260*, 4117–4121.
346. Neuenhofer, S., Schwarzmann, G., Egge, H. & Sandhoff, K. (1985) Synthesis of lysogangliosides. Biochemistry 24, 525–532.
347. Fukuda, M. N., Dell, A., Oates, J. E., Wu, P., Klock, J. C. & Fukuda, M. (1985) Structures of glycosphingolipids isolated from human granulocytes. The presence of a series of linear poly-*N*-acetyllactosaminylceramide and its significance in glycolipids of whole blood cells. J. Biol. Chem. *260*, 1067–1082.
348. Iwamori. M., Kiguchi, K., Kanno, J., Kitagawa, M. & Nagai, Y. (1986) Gangliosides as markers of cortisone-sensitive and cortisone-resistant rabbit thymocytes: Characterization of thymus-specific gangliosides and preferential changes of particular gangliosides in the thymus of cortisone-treated rabbits. Biochemistry *25*, 889–896.
349. Fukuda, M. N., Bothner, B., Lloyd, K. O., Rettig, W. J., Tiller, P. R. & Dell, A. (1986) Structures of glycosphingolipids isolated from human embryonal carcinoma cells. The presence of mono- and disialosyl glycolipids with blood group type 1 sequence. J. Biol. Chem. *261*, 5145–5153.
350. Hanfland, P., Kordowicz, M., Peter-Katalinić, J., Pfannschmidt, G., Crawford, R. J., Graham, H. A. & Egge, H. (1986) Immunochemistry of the Lewis blood-group system: Isolation and structures of Lewis-c active and related glycosphingolipids from the plasma of blood-group O Le(a^-b^-) nonsecretors. Arch. Biochem. Biophys. *246*, 655–672.
351. Leffler, H., Hansson, G. C. & Strömberg, N. (1986) A novel sulfoglycosphingolipid of mouse small intestine, IV^3-sulfogangliotetraosylceramide, demonstrated by negative ion fast atom bombardment mass spectrometry. J. Biol. Chem. *261*, 1440–1444.
352. Fukuda, M. N., Dell, A., Tiller, P. R., Varki, A., Klock, J. C. & Fukuda, M. (1986) Structure of a novel sialylated fucosyl lacto-*N*-norhexaosylceramide isolated from chronic myelogenous leukemia cells. J. Biol. Chem. *261*, 2376–2383.
353. Månsson, J.-E., Mo, H., Egge, H. & Svennerholm, L. (1986) Trisialosyllactosylceramide (G_{T3}) is a ganglioside of human lung. FEBS Lett. *196*, 259–262.
354. Araki, S., Satake, M., Ando, S., Hayashi, A. & Fujii, N. (1986) Characterization of a diphosphonopentaosylceramide containing 3-*O*-methylgalactose from the skin of *Aplysia kurodai* (sea hare). J. Biol. Chem. *261*, 5138–5144.
355. Ariga, T., Kohriyama, T., Freddo, L., Latov, N., Saito, M., Kon, K., Ando, S., Suzuki, M., Hemling, M. E., Rinehardt Jr., K. L., Kusunoki, S. & Yu, R. K. (1987) Charac-

terization of sulfated glucuronic acid containing glycolipids reacting with IgM M-proteins in patients with neuropathy. J. Biol. Chem. *262*, 848–853.

356. Chatterjee, D., Aspinall, G. O. & Brennan, P. J. (1987) The presence of novel glucuronic acid-containing, type-specific glycolipid antigens within *Mycobacterium* spp.. Revision of earlier structures. J. Biol. Chem. *262*, 3528–3533.

357. Egge, H. & Peter-Katalinić (1987) Fast atom bombardment mass spectrometry for structural elucidation of glycoconjugates. Mass Spectrom. Rev. *6*, 331–393.

Chapter 5
Lipids

A. Kuksis and J. J. Myher

1 Introduction

MS is eminently suited for the analysis of lipids, and extensive applications have been made in structural and quantitative analyses, but its utilization in clinical biochemistry has been limited. This has been due to a lack of disorders of lipid metabolism that can be readily diagnosed by generally recognized lipid markers, although the relatively high expense of the instrumentation and the need for highly skilled personnel has also been a problem. Exceptions are the MS and GC/MS routines developed for the chemical assay of bacterial infection based on characteristic fatty acid components, and the quantitation of abnormal levels of normal lipid metabolites by means of definitive methods of analysis employing stable isotope labeled reference compounds.

The principles of MS and GC-MS are outlined in Chapter 1. The developments in GC-MS have been stimulated by the need to analyze samples of biological origin, which frequently contain complex mixtures of components in a wide range of concentrations. The place of GC-MS (1, 2)and LC-MS (3) in clinical chemistry, as well as the scope of MS in clinical chemistry in general (4) also have been reviewed elsewhere. Likewise, the quantitative aspects of MS as they apply to clinical analysis have been detailed previously (5, 6), including stable isotope dilution MS (7).

In view of the limited number of specific applications in clinical biochemistry, the present chapter deals largely with research routines, which have been selected for their use of clinical material as well as for their potential for a future utilization in the clinical laboratory. In several instances it has been possible to provide specific suggestions for improved sensitivity based on progress made in related areas of research. However, the coverage is not uniform even in the selected areas for potential clinical application because of the large variety and number of lipid molecules involved. Since MS in combination with chromatographic techniques is the most sensitive and specific methodology available for the determination of the structure of fatty acids and fatty acid esters, it will obviously remain indispensable for their identification and measurement in clinical samples. As the relevance of such measurements is more fully recognized MS will be utilized routinely in diagnosis and therapy.

2 Strategy of Analysis

MS is an analytical technique that is especially well suited for the identification of unknown lipids and for a sensitive and highly specific detection and quantitation of known lipids. For clinical biochemistry GC-MS or LC-MS constitute an ideal analytical system. The chromatographic step provides an effective separation of the components of interest from related lipids, thus avoiding the need for extensive preliminary purification and isolation of pure substances. Depending on the chemical properties of the lipid component, 10 – 100 ng of injected material may be sufficient for analysis. Specific qualitative detection may require no more than 1 – 100 pg, when the instrument is operated in the selected ion monitoring (SIM) mode.

2.1 Preparation of sample

There is a great need for simple and efficient methods of pretreatment of samples for MS analysis. The requirement for a hydrophobic medium and a non-polar nature of the analyte in GC-MS may be circumvented by LC-MS, which is compatible with aqueous solutions of polar substances. Most classes of lipids, however, require derivatization to stabilize them and to make them more amenable to a meaningful interpretation of the mass spectra.

Prior to analyis, the lipids are usually resolved into classes of neutral and polar lipids, which may be further segregated into individual neutral and polar lipid classes by column chromatography, TLC or HPLC. A preliminary separation into classes helps to identify the origin of the analyte as well as simplifies the method of analysis and interpretation of the MS data. Separation into chemical classes is also necessary for the purpose of subsequent chemical treatment. Fatty acids, free of their ester, are readily characterized after methylation and silylation. Transesterification in alkaline alpha-medium can produce artefacts of epimerization when the mixture contains α-substituted fatty acids (8) while transesterification with boron trihalides destroys cyclopropane fatty acids (8, 9). Furthermore, α,β-unsaturated fatty acids can add diazomethane across the double bond to give pyrazolines (10). It may be necessary to first saponify the lipid esters and then prepare the methyl esters or TMS ethers (esters). Saponification is rapid and selective, but any carbohydrate present in the sample is degraded and the yields of amide-bound hydroxy fatty acids are low (11, 12). Depolymerization of complex lipids by acidic methanolysis, e. g. treatment with anhydrous methanolic HCl ($2\ mol \cdot l^{-1}$, 85 °C) seems to be a reasonably mild method of bond splitting (13). This reagent liberates the fatty acids directly as methyl esters and the yields of amide-linked constituents are about quantitative. However, cyclopropane substituted fatty acids are degraded. Enzymatic dephosphorylation of glycerophospholipids and of sphingomyelin with phospholipase C yields diacylglycerol and ceramide moieties, which can be identified and quantitated by GC-MS (14, 15) and LC-MS (16, 17). Dephosphorylation of other sphingophospholipids also gives ceramides. After derivatization, most complex lipids can be directly analyzed. These include (mono-, di- and triacylglycerols (18, 19), ceramides (20) and glycosphingolipids (cerebrosides) (21 – 23). The sphingosine moiety isolated from the lipid after hydrolysis gives an interpretable mass spectrum (24).

If possible, however, it is preferable to record the mass spectra on the intact molecules. Glycerophospholipids eliminate the phosphate group under GC conditions (25) and have to be chemically (25, 26) or enzymatically (14 – 17) dephosphorylated before analysis. The structure of the phosphate ester can be established in the 1,2-diacyl-*sn*-glycerol-3-phosphate ester by initial deacylation and mass spectrometry of the resulting *sn*-glycerol-3-phosphate (27, 28). Glycosphingolipids are best handled by direct insertion probe, and polyglycosyl forms with up to 12-sugar units have been studied without decomposition as the permethylated derivatives (29).

In GC analyses of hydoxylated fatty acids, which are present in most gram-negative bacteria, a derivatization step prior to GC is required to obtain a reproducible detector signal from the constituents (13). TMS and TFA derivatives are commonly used (13). However, the preparation of the *t*-BDMS derivatives (30) and pentafluorobenzoates (31) and p-bromophenacetates (32) may have special advantages for improved sensitivity of detection. Likewise, mono- and di-acyl glycerols and ceramides require conversion into either TMS or *t*-BDMS ethers for effective GC-MS (14, 15) or LC-MS (16, 17). The MS determination of the positional location of the methyl branching, cyclopropane ring and/or a double bond requires the preparation of pyrrolidides (33 – 35). Alternatively, double bonds may be determined following oxidation to the corresponding polyhydroxy compounds and permethylation (36) or trimethylsilylation (37).

2.2 Structural analysis

The technique of MS surpasses all others in the structural identification of lipids. In simple compounds the molecular ion and the fragmentation pattern enable the analyst to identify known substances by matching against reference spectra. Prior GC or HPLC of the compounds provide retention times which may be used as a further indication of the identity of the solute. Packed glass and metal conventional columns originally employed have now been almost completely replaced by glass capillary columns, which are much more efficient and flexible quartz capillary columns with chemically bonded liquid phases have eliminated much of the breakage hazard of working with capillary columns. These methods have been effective in the identification of unknown fatty acid components in tissue and cell membrane extracts, including phytanic (38, 39), tuberculostearic (40), various hydroxy fatty acids (41) and extra long normal chain fatty acids (42). In other instances, abnormal quantities of the normal fatty acids have been recognized in specific disease conditions (43). In a few cases the characterization of the abnormal fatty acids has been extended to the glycerol (44) and steryl esters (45) of these acids. In the analysis of very complex lipid molecules, such as the free ceramides of human platelets (46), ceramides from plasma sphingomyelin (20) and glycosphingolipids with blood group specificity (47, 48) MS has been indispensable. Both sequence and bond position of oligosaccharides and glycosphingolipids have been established.

2.3 Lipid profiling

GC-MS and LC-MS are the most powerful methods for the determination of lipid profiles. GC-MS has been used intensively for recognition of normal and abnormal patterns of fatty acids (41), methyl branched fatty acids (40), long chain fatty acids (42, 43), hydroxy fatty acids (41), as well as molecular species of the diacylglycerol (14, 15) and ceramide (15, 49), moieties of glycerophospholipids and sphingomyelins, respectively. More recently, LC-MS has been used for the analysis diacylglycerols (16, 17). Likewise, profiling of the molecular species of triacylglycerols is possible either by GC-MS (50) or LC-MS (51, 52). Whereas conventional clinical chemistry tests determine individual lipid species or classes, which are known or expected to be present, profile analyses provide an opportunity to detect and characterize a large number of substances simultaneously, find new and unexpected substances, determine changes in the ratios of different constituents, as well as give much more precise information about the nature of the lipid class than a simple group test (e. g. triacylglycerols) (53, 54).

Furthermore, profile analyses add to the understanding of metabolic abnormalities and diseases. Analyses of metabolic profiles of molecular species of lipids have been most frequently performed by a combination of TLC and GC methods, but GC-MS and LC-MS provide an opportunity to work with smaller samples and to complete the work more rapidly. More recently, polar capillary GC of diacylglycerols as the TMS ethers has allowed rapid and essentially complete profiling of the molecular species of all glycerophospholipid classes of tissues and cell membranes and lipoproteins.

Although no significant differences have been discovered between the triacylglycerol profiles of normal and hyperlipoproteinemic plasma (53, 54), the analyses have provided substantial new information about the triacylglycerol structure of plasma lipoproteins in general.

2.4 Mass chromatography

SIM has been extensively utilized for the highly sensitive and specific detection and quantitation of lipids. The detection limits are in the low picogram range (e. g. tuberculostearic acid (40)) and low femtogram range (e. g. halogenated derivatives of hydroxyfatty acids) when combined with NICI (31). The ultimate response is observed when only one ion is studied. The retrieval of individual ions from the total mass spectra recorded by repetitive scanning, sometimes known as MF has a sensitivity of detection for individual lipid components very similar to that achieved by the mass spectrometer in recording the full spectrum (generally 100 ng or greater). This technique is valuable for scanning unknown lipid mixtures (14, 15). It permits identification and quantitation of essentially all components of a lipid mixture, provided that suitable calibration is performed (14, 52). Appropriate illustrations of the application of this method may be found in publications dealing with the identification of molecular species of diacylglycerols and triacylglycerols within the partition numbers of acylglycerols eluted from HPLC columns and within carbon numbers of acylglycerols recovered from GC columns.

In quantitative analyses of lipids the internal standards used are isomers, homologues or lipids labeled with stable isotopes (^{2}H or ^{13}C) (4, 55–57). Using SIM together with isotope dilution, reference and definitive methods have been established for clinical chemistry of various lipids. Thus, fatty acids (58, 59), triacylglycerols (4) and acyl carnitines (56, 60) have been quantitated using deuterium labeled internal standards. Variation coefficients have ranged from 1.3 to 4% for the various parameters, with the pipetting errors contributing substantially to the variation.

In both detection and quantitation of lipids by SIM, special attention needs to be directed towards the type of derivative employed, because it affects the sensitivity of detection and the precision of identification of the solute. Also, a milder method of ionization may be an advantage under certain conditions, as may be switching from positive to negative ion detection. NICIMS has been found to yield especially high sensitivity for the heptafluorobutyrates and pentafluorobenzoates (31) of the hydroxy fatty acid methyl esters.

Correlations between SIM and radioimmunological or chemical methods have shown either good agreement or have given lower values for SIM, probably as a result of its higher specificity (57).

2.5 Clinical studies with stable isotope labeled lipids

Another application of MS in clinical chemistry arises from the use of stable isotope-labeled lipids in clinical investigation. Because of possible radiation hazard, the use of radioactive tracers in human experimentation has been curtailed, and stable isotope labeling has been substituted. Lipid molecules labeled with stable isotopes (^{2}H or ^{13}C) can be easily detected and quantitated in blood, urine and tissues by MS and MF, especially if more than one deuterium or ^{13}C-atom occurs per molecule (57, 58). Deuterium-labeled fatty acids have been administered to normal human subjects (59) and to patients with cystic fibrosis (61) by intestinal perfusion and the fatty acids recovered in the various plasma lipid classes have been determined at various times as an indication of the relative rates of absorption and clearance of the different fatty acids. In other studies, human subjects have been fed triacylglycerols containing deuterium labeled *cis*- and *trans*-fatty acids or fatty acids having double bonds at different locations, and the plasma appearance and disappearance of the fatty acids have been determined (62, 63).

In animal model experiments, deuterated ethanol has been extensively used for the labeling of biliary (64), hepatic (65) and plasma (66) glycerophospholipids in rats, while in other studies isolated liver has been perfused with deuterium oxide (67) or the animals have been given D_2O in drinking water (68). Similar experiments may be applicable to human experimentation. The use of stable isotope labeled lipids for clinical diagnosis is in an early stage of development, although the future would appear to be promising. Potentially, stable isotope-labeled lipids could be administered to patients for the measurement of labeled metabolites to assess a defective enzyme system or metabolic parameter, such as pool size or turnover rate (69). Furthermore, with an appropriate route of administration, it might be possible to investigate or assess the function of a specific organ with respect to any or all of the parameters.

3 Neutral Lipids and Fatty Acids

3.1 Hydrocarbons

Normal and branched-chain hydrocarbons and cycloparaffins have been identified in abnormal amounts in human pathological conditions such as malignancy, (70, 71), follicular lipidosis (72–74,), paraffin oil pneumonia (75) and atherosclerotic plaques (76, 77).

The hydocarbons are usually isolated from total lipid extracts by chromatography either on columns of alumina (78) or silicic acid (70, 71, 74), or by TLC (72, 75, 77).

Prior to chromatography it may be convenient to saponify the lipid extract (70, 78). The resolution of the hydrocarbons on the basis of molecular weight is performed on conventional (71, 75, 77) or capillary (79) GC columns. The elution order of the isomeric hydrocarbons from the OV-17 or Apiezon L stationary phases is as follows (80): *iso-iso, anteiso-iso, anteiso-anteiso, iso-normal* and *anteiso-normal, normal-normal.* The identification of the peaks is confirmed by EI-MS. The chromatographic resolution and MS identification of complex mixtures of hydrocarbons is difficult (81) and requires both GC and MS of high resolution (82).

The straight chain paraffins yield characteristic EI spectra, which have been extensively studied. A number of empirical rules have been developed based on the spectra of 18 isomeric octanes (83), which have been found to be applicable to describe the spectra of other paraffinic hydrocarbons. The relative height of the parent peak is greatest for the straight chain compounds and decreases as the degree of branching increases. Fragmentation takes place most readily at highly branched carbon atoms. Ions of odd mass tend to be more abundant than those of even mass number, and secondary fragmentation involving loss of hydrogens or larger fragments also tends to give odd mass ions, especially for straight-chain compounds. Prominent peaks at even mass number are indicative of fragmentation of two separate side chains and hence a higher degree of branching. The peaks corresponding to C_3 and C_4 ions are always large in the spectra of paraffins. In the mass spectra of the longer chain hydrocarbons, the most prominent peaks of ionized fragments are the highly charged alkyl-type ions of empirical formula $[C_nH_{2n+1}]^+$. These ions have odd mass number and contain an even number of electrons. The peak heights of the alkyl-type ions decrease in a regular manner as the number of carbon atoms in the ionizing fragment increases. The peaks of formula $[C_nH_{2n+1}]^+$ have been shown to arise by simple cleavage of the chain and partly from $[(CH_2)_n + H]$, a process of double cleavage and capture of one hydrogen atom from one of the fragments lost. Other prominent peaks occur at C_nH_{2n-1}. The correlations between structure and spectra are less satisfactory for olefins and acetylenes. This is because rearrangement peaks, which usually make identification more difficult, are more abundant and the spectra of many isomers are so similar that they are difficult to distinguish (84, 85). The position of a double bond in or near the center of the chain is determined by osmium tetroxide dihydroxylation followed by mass spectrometry of the bis-trimethylsilyl derivatives (80).

Gazzarini and Nagy (76) have reported the characterization of octadecane from femoral arterial plaques. Brooks et al. (77) identified a series of *n*-alkanes ($C_{17}-C_{31}$) in the atheromatous aortic plaques, with no particular preference for odd or even carbon numbers. There occurred a prominent base line hump in the GC elution pattern, which was thought to be due to a complex mixture of unresolved isomeric hydrocarbons, some of which form a homologous series as can be seen from an examination of the small peaks which appear between successive *n*-alkanes of the larger series. The latter compounds were not further identified. Alkanes were found to be present in plaques in quantities of the order of 700–800 µg/g of total lipid extract. No explanation was advanced for the finding.

Primarily a homologous series of saturated paraffins, *n*-alkanes ($C_{15}-C_{34}$) along with two other much smaller series of alkanes, probably of an *iso-* and *anteiso*series, have been identified by Cain et al. (71) in normal brain meninges (*dura mater*) and neoplastic tissue (meningiomas) from human brain. The hydrocarbons were analyzed by TLC and GC.

The hydrocarbon mixture isolated by Cain et al. (71) from human brain exhibits some similarity to the hydrocarbons studied by Rose and Liber (74) in human spleen. The latter workers reported the isolation of a complex mixture of long-chain saturated hydrocarbons from human spleens after death in which lipid vacuoles were seen during histochemical examination. The splenic hydrocarbon mixture contained few components smaller than C_{21}. Beginning with C_{21} and extending to C_{28} there was a complex mixture of largely unresolved hydrocarbon peaks reaching a broad maximum between C_{24} and C_{26}. Mineral oil, in contrast, showed distinct peaks from C_{15} up to C_{21}. It was concluded that mineral oil as such could not account for the paraffins isolated from these human spleens. The hydrocarbons isolated by Riley et al. (70) from cancerous human liver were found to be essentially identical to those with mineral oil.

The hydrocarbons present in such preparations of medical mineral oil as Nujol are extremely difficult to separate and to identify. Usually only a type analysis is obtained. A procedure for the type analysis by MS has been discussed in a review (81). The method employs characteristic fragment masses of paraffins, non-condensed naphthenes and condensed naphthenes containing two to six condensed rings per molecule as a basis for their identification. The naphthene masses are chosen so that the lowest mass in each group represents the minimum molecular mass, assuming all C_5 rings. The series advances by 14 mass units (CH_2 groups) to the molecular mass corresponding to all C_6 rings, then four additional peaks in the same C_nH_{2n+x} series are added in order that alkyl substitution in the nucleus might contribute to the summation. Characteristic mass groupings for the non-condensed naphthenes were 69, 83, 97, 111, 125, 139. For the condensed naphthenes there were: with two rings 109, 123, 137 etc.; with three rings 149, 163, 177 etc.; with four rings 189, 203, 217 etc.; with five rings 229, 243, 257 etc.; and with six rings 269, 283, 297 etc.

Likewise, Boitnott and Margolis (72, 73) have published studies reporting the isolation of saturated hydrocarbons from human tissues in which lipid vacuoles were identified histologically. MS showed a close similarity to mineral oil. Cycloalkanes with one to six rings were identified, as were significant amounts of normal-chain alkanes. It was suggested that the saturated hydrocarbons represented absorbed mineral oil. It was speculated that the high incidence of mineral oil droplets found in the lymph nodes are out of proportion to laxative and other medicinal uses of mineral oil, although pertinent quantitative data on this matter were not available. The authors point out that there are rich sources of non-medicinal origin. The extensive use of mineral oil in the food processing industry was estimated to result in the ingestion of an average of 47.5 g of mineral oil per capita in 1964 (86).

Somewhat less detailed MS studies have confirmed the presence of mineral oil in paraffin oil pneumonia studied by Heckers et al. (75). The mass spectra (highest observed mole peak at 436) demonstrated the characteristic fragmentation pattern: a stepwise split-off of CH_2-groups of linear (C_nH_{2n+2}) or cyclic (C_nH_{2n}) hydrocarbons (87). Chromatograms recorded from sputum extracts showed the same pattern as liquid paraffin Persido. The demonstration of complex paraffin in the sputum permits the rapid and definite diagnosis of paraffin oil pneumonia or pulmonary paraffinoma without resorting to lung biopsy. The authors (75) also identified ingested paraffin oil in the lymph following administration of liquid paraffin to a patient with chyle fistula.

TLC and capillary GC of human epidermal lipids, isolated from human skin, have been supplemented by EIMS and FAB-MS of *n*-alkanes and squalene by Lampe et al. (88, 89) and by Williams and Elias (90). Cerbulis et al. (91) have identified the major components of the hydrocarbon fraction of goat milk as 3,7,11,15-tetramethylhexadec-2-ene (phytene-2), squalene and n-$C_{29}H_{60}$. In addition, a series of odd and even carbon number *n*-alkanes (C_{15}-C_{33}), a series of alkenes (C_{16}-C_{23}) and a series of branched-chain hydrocarbons were found. Nachman (92) has reported an unusual predominance of even-carbon hydrocarbons in an Antarctic food chain by GC-MS. The bulk of the hydrocarbons were due to the saturated 16:0, 18:0 and 20:0, and the unsaturated 18:1 and 21:1 species. Pristane was also found in significant amounts. It was suggested that a different biochemical pathway to hydrocarbons may be operative in members of this food chain that is well adopted to low temperatures.

3.2 Alcohols

The saturated normal-chain aliphatic alcohols occur in wax esters, which have been shown to be secreted in abnormal amounts and in abnormal composition in psoriasis (93). The unsaturated normal chain aliphatic alcohols, their acetates and aldehydes are physiologically active chemicals in insects (94) and much effort has been expanded in determining their structural characteristics (95). Although unsaturated alcohols have not been implicated in the physiological or disease process in man, the progress made in the mass spectrometric elucidation of their structures is pertinent to the discussion of the MS of lipids and may be of interest in the identification of unsaturated fatty acids in complex lipids.

The presence of a hydroxyl group in an alkane lowers the ionization potential, but the molecular ion abundance decreases and often vanishes altogether (96). The alcohols undergo, however, ion-molecule interactions to give a pressure dependent $[M+H]^+$ peak, which is useful for molecular mass determination. Thermal and catalytic dehydrogenation and dehydration of the alcohol can give spurious peaks, such as $[M-2]^+$, $[M-18]^+$ and $[M-20]^+$, although these can also arise through EIMS. The rest of the spectrum closely resembles that of the corresponding alkene. The prominent C_nH_{2n-1}, C_nH_{2n} and C_nH_{2n+1} ion series of 1-alkanols can give evidence of branching or other structural features of the carbon skeleton. This fragmentation mechanism generally follows that outlined for the coresponding hydrocarbons.

Usually the heaviest ion in all spectra is $[M-18]^+$, due to loss of water. The mechanism for this elminination is a ring closure with extraction of one hydrogen mainly from carbon atom 4 (97). For normal 18 : 0 alcohol the base peak is at m/z 43, but this is not the rule for other homologues (98). In the high mass region $[M-18-28]^+$ is of about equal abundance to $[M-18]^+$. The $[M-18-28]^+$ ion is probably formed by a mechanism analogous to those postulated for the fragmentation of methyl esters (98). Of the branched-chain isomers of the 18:0 alcohols, the (*n*-2) isomer shows two characteristic fragments at m/z 237 and m/z 196 (99). The first one is probably formed from the molecular ion by a successive loss of water and one of the terminal methyl groups, that is $[M-18-15]^+$. The mechanism responsible for the formation of the m/z 196 ion has been attributed to a successive loss of CH_2OH

and the terminal isopropyl group, that is $[M-31-43]^+$. The base peak in the spectrum of the branched (n − 2) alcohol is at m/z 57. The peak at m/z 56 increases its intensity from about 50% for the normal and branched (n − 3) isomer to about 85% for the branched (n − 2) isomer. The mass spectrum of the branched (n − 3) alcohol also shows some characteristic fragments, such as m/z $[M-18-29]^+$ and also one corresponding to $[M-18-28-29]^+$ or m/z 195. The base peak at m/z 70 probably contains the radical ion $[CH_3CH_2CH(CH_3)CH]^+$ as in the methyl ether (see below).

Conversion of the aliphatic alcohols into the corresponding ethers tends to yield more of the M^+ peak, but the intensity may still be low (96, 98). Again, ion-molecule interaction gives the $[M+1]^+$ peak, which can be useful for molecular mass determination. The fragmentation of ethers resembles that of amines, but with a smaller propensity for α-cleavage and a much larger tendency for charge site cleavage. The methyl ether gives a base peak at m/z 45, which is the ion $[CH_3OCH_2]^+$, due to a β-cleavage. Molecular ions are of very low intensity. The heaviest fragment is found at m/z $[M-32]^+$, due to a loss of methanol.

Karlsson et al. (99) have proposed the transformation of the methyl esters of fatty acids into the methyl ethers of the corresponding fatty alcohols for the analysis of the *iso* and *anteiso* esters. The transformation is accomplished by reduction with $LiAlH_4$, followed by methylation with CH_3I and Ag_2O. Intense peaks at m/z 56 $[CH_3CH(CH_3)CH_2-1H]^+$ and m/z 70 $[CH_3CH_2CH(CH_3)CH_2-1H]^+$ are observed in the mass spectra, as well as peaks at $[M-(32+28)]$ and $[M-(32+56)]^+$ for *iso* esters, and $[M-(32+29)]^+$ and $[M-32+57)]^+$ for *anteiso* esters. These ions are best seen on mass spectra recorded at low voltage: 25 eV.

Trimethylsilylation of an alcohol also has a beneficial effect on its fragmentation. The large $[M-15]^+$ peak is indicative of the molecular mass (96). The TMS ethers, however, are unstable to moisture and cannot be purified before analysis. Corey and Venkanteswarlu (100) and Kelly and Taylor (30) have demonstrated that the *t*-BDMS ethers of simple alcohols yield prominent $[M-57]^+$ fragment ions, which can be utilized for accurate measurement of molecular mass. The *t*-BDMS ethers are stable to moisture and can be isolated from the reaction mixture and purified prior to analysis. The latter derivatives have been shown (101) to possess special advantages for the analysis of the mycocerosic alcohols derived from the phthiocerol dimycocerosates found in the cell envelopes of *Mycobacteria*. These derivatives chromatograph well on both packed and capillary GC columns and are much less susceptible to hydrolysis than the corresponding TMS ethers. The mass spectra were entirely in agreement with those expected for simple *t*-BDMS ethers, showing a prominent ion at high mass, which was produced by loss of a *t*-butyl group from the molecular ion. At lower mass a series of hydrocarbon fragment ions were seen with a few low abundance ions being produced at the branching points on the molecule. The $[M-57]^+$ ion is particularly well suited for SIM. The reductive fission of the parent phthiocerol dimycocerosates also leads to the formation of di-*t*-BDMS ethers of phthiocerols A and B and to a tri-*t*-BDMS ether of the reduced phthiodiolone A (phthiotriol A). In every case a small $[M-57]^+$ ion is observed together with low abundance ions formed by subsequent loss of either 32 or 132 mass units (CH_3OH or *t*-BDMS-OH) from these ions. At lower m/z values ions produced by fissions α to the *t*-BDMS ether groups were found but the most intense peak in this region was derived via a complex six-membered ring rearrangement of the $[M-57]^+$ ion, which had not been previously reported for the *t*-BDMS ethers. The novel rearrangement of di-*t*-BDMS ethers of *β*-diols provides very characteristic ions for the identification of members of the phthiocerol family. Mallet et al. (101) have shown the potential of SIM of the *t*-BDMS ethers of the phthiocerols and mycocerosic alcohols derived from the mycobacterial phthiocerol dimycocerosates. The present procedure offers an alternative method providing more detail concerning profiles of both mycocerosate and phthiocerol components of mycobacterial waxes, previously determined by SIM of the C_{32} of mycocerosic acid following methanolysis (102–104). In recent studies (105) phthiocerol dimycocerosates were characterized from *M. leprae*, the causative agent of leprosy. The phthiocerol A fraction from *M. leprae* was found to be significantly different from that found in *M. tuberculosis*, the former having molecules of 30 and 32 carbon atoms and the latter having molecules of 34 and 36 carbon atoms.

Horiike and Hirano (106) have reported the identification of double bond position in dodecenyl acetate without prior chemical modification by conventional single-focusing EIMS. They have tested the generality of their previous findings by reporting the spectra of seven positional isomers of dodecenol with the double bond at C_5-C_{11}, which constitute the sex attractants of many moths (106). In the spectra of all isomers, the molecular ion at *m/z* 184 was not detectable, but the ion of the highest *m/z* value was observed at *m/z* 166, which is probably a fragment ion derived from the loss of water from the molecular ion. The most striking feature of the seven spectra was the base peak of *m/z* 67 in 5-ene- and 6-ene-isomers and *m/z* 55 in 7-ene- to 11-ene-isomers. The intensity ratio of the fragment peaks at *m/z* 55 and 67 in each spectrum was always unchanged. Each isomer could therefore be clearly characterized by the intensity ratio of these two fragment peaks i. e. *m/z* 55 to 67, depending upon whether or not the value of the ratio is over unity. On this basis the authors have proposed a systematic procedure for the elucidation of positional isomerism in these alcohols.

Nicolaides et al. (107) have used GC-MS to confirm the structure assignments of hydroxy fatty acids and alcohols found in steer and human meibum based on previous GC analyses. The meibum lipid components involved are unsubstituted fatty acids, *α*-hydroxy, *ω*-hydroxy fatty acids, fatty alcohols and some other incompletely characterized lipids.

During the last few years, several improved methods have been developed for the structural determination of fatty alcohols by MS (108). Especially convenient were found to be the nicotinates, which are capable of directing the fragmentation of the fatty chain (109). These esters fragment by random intramolecular hydrogen abstraction from the aliphatic chain following initial charge localization on the pyridine ring or the nicotinyl group. The masses and relative abundances of the fragment ions give information on chain branching and double bond position. Recently, Harvey (110) has combined the advantages of this nicotinate ester with the ease of preparation of silyl ethers by linking the 3-pyridyl carbinyl group to alcohols via a dimethylsilyl bridge. The mass spectra of the resulting derivatives were characterized by very abundant (base peak) molecular ions and a series of abundant radical-induced cleavage ions of masses and abundances, which were characteristic of the position of double bonds in both mono-and poly-unsaturated alcohols.

3.3 Fatty acids

Frequently unusual fatty acids are detected in bacterial lipids and these may serve as a basis for recognition of bacterial infection (111). In other instances abnormalities in the tissue metabolism of fatty acids may lead to the accumulation of characteristic fatty acids or fatty acid patterns, which may be valuable in clinical diagnosis (103, 112–115).

Due to easy volatilization of their methyl esters, fatty acids are conveniently analyzed by GC, but precise assignment of structure to a fatty acid detected as a peak in GC requires MS analysis. Similar problems are encountered in the identification of fatty acids by HPLC, which is further complicated by difficulties in quantitation, unless ultraviolet absorbing or fluorescent derivatives are employed.

Much work has been performed on the analysis of fatty acids by GC-MS on bacterial lipids and their identification has become a relatively easy task, provided sufficient material is available. The MS identification of bacterial fatty acids has been recently reviewed by Asselineau and Asselineau (116). GC-MS in SIM mode is capable of detecting bacterial constituents in amounts down to about 20 pg. This sensitivity is sufficient to detect unique bacterial marker molecules in body fluids and specimens (102, 117–119). SIM-GC-MS methods of identification and quantitation of the common fatty acids in plasma (120, 121), tissues (122) and sputum (123) have been developed and discussed by other laboratories.

3.3.1 **Saturated normal chain fatty acids**

Methyl esters are the derivatives of fatty acids most often used for MS (124), but silyl esters (125, 126), pyrrolidides (33) and picolinyl derivatives (127) are of special interest for the identification of the hydoxyl and methyl substituted and unsaturated fatty acids. Mass spectra of normal long-chain saturated methyl esters are characterized (124) by peaks due to ions containing oxygen, corresponding to $(CH_2)_nCOOCH_3^+$ (*m/z* 87, 101, 115, etc.). In the high mass region, the molecular peak, usually well defined, is accompanied by a peak at M−31, arising from the loss of a methoxy group and formation of an acylium ion. A peak at M−43 is formed by a complex intramolecular rearrangement followed by the loss of a propyl radical. The base peak of the methyl ester is observed at *m/z* 74. If a branch R occurs at position 2, this peak is observed at *m/z* 73 + R; in case of a methyl branch, the peak is at *m/z* (73 + 15). When one or several branches of the molecular ion next to one of the tertiary carbon atoms are observed, the positive charge is kept by the fragment with the highest relative stability. A competition between such fragmentation and the McLafferty rearrangement may occur, so that the peak at *m/z* 74 may not be the base peak of the spectrum (124).

The major mass fragments of the TMS esters of the saturated fatty acids have been reported by Markey (128). All the fatty acids yield characteristic mass spectra including a significant proportion of molecular ion as well as an especially prominent peak for M−15. The TMS esters give a base peak at *m/z* 73 $[(CH_3)_3Si^+]$ and a large peak (approx. 80%) at *m/z* 75 $[(CH_3)_2Si = OH^+]$. Also a common feature of the TMS esters are prominent ions having masses 117, 129, 132 and 145. The latter two ions are analogous to the ions found at *m/z* 74 and 87 in the spectra of the methyl esters, but shifted 58 mass units higher by the larger mass of the TMS group.

The *t*-BDMS esters of fatty acids lose both a *t*-butyl and methyl radical, the loss of the bulkier group being favoured. Other prominent peaks were seen at *m/z* 75 [$(CH_3)_2Si = OH^+$] and *m/z* 73 (126). In addition, analysis of fatty acids as their *t*-BDMS esters has been shown to be potentially useful in the quantitation by GC-MS (126).

CIMS has been effectively used for the determination of small amounts of unusual fatty acids in blood plasma and other tissues of patients with a variety of diseases. Adrenoleukodystrophy (ALD) is a hereditary demylinating disease that is transmitted as an X-linked recessive trait (129). Tsuji et al. (113) have devised an analytical method for improved determination of the fatty acid composition of the ALD patients. The method involves the additon of trideuteriomethyl docosanoate and trideuteriomethyl hexacosanoate as internal standards to the fatty acid methyl esters prepared from erythrocyte membrane sphingomyelin. MS is then performed by monitoring $[M + 18]^+$ ions, *m/z* 372 and 375 (for methyl docosanoate and trideuteriomethyl docosanoate), and *m/z* 428 and 431 (for methyl hexacosanoate and trideuteriomethyl hexacosanoate), respectively. The diagnosis of ALD is based on the finding that the ratio of $C_{26:0}/C_{22:0}$ in erythrocyte membrane sphingomyelin is 2.6-fold higher in ALD patients than in control subjects. The analysis requires only 1 ml of whole blood.

GC-MS has also been utilized to establish the biosynthetic origin of the long chain fatty acids in the ALD patients. Thus, Tsuji et al. (115) demonstrated that ALD or AMN fibroblasts synthesize $C_{26:0}$ fatty acid from stearic acid, while Brown et al. (130) have cited experiments in which an ALD patient incorporated deuterated water into plasma and brain $C_{26:0}$ fatty acid *in vivo*. Dietary restriction of C_{26} acid intake had no apparent effect on the tissue or plasma levels of the long chain fatty acids.

The cerebro-hepato-renal syndrome of Zellweger is characterized by the absence of peroxisomes in liver tissue and in proximal tubules of the kidney (131). Brown et al. (132) have reported increased levels of C_{24}-C_{26} fatty acids in brain white matter, adrenal gland and cultured fibroblasts from some Zellweger patients. Bakkeren et al. (133) have reported a clear increase in the very long chain fatty acid levels of the serum of six Zellweger patients compared with age-matched controls. In the latter instances the fatty acids were analyzed by capillary GC without MS. Brown et al. (132) point out that the long chain fatty acid analyses may be meaningfully performed on cultured amniocytes to permit a prenatal diagnosis of neonatal ALD and the Zellweger syndrome, using the same techniques that have proven successful for the prenatal diagnosis of childhood ALD (132).

Bougneres and Bier (121) have developed an SIM-GC-MS micromethod, which allows measurement of plasma enrichment of tracer ^{13}C-palmitate as well as total palmitate content through the concurrent use of deuterated palmitate as internal standard. Ions at *m/z* 270, 271 and 274, representing the molecular ions of unlabeled, [1-^{13}C], and [$^2H_4$]palmitic acid methyl esters, respectively, were selectively monitored. The use of the method was illustrated by measuring palmitate turnover in a 12 h old human infant.

Parsons et al. (59) and Parsons and Emken (61) have reported a sensitive and accurate method for detection and quantitation of deuterated saturated and unsaturated fatty acids as *t*-BDMS esters. Mass chromatograms were obtained by scanning search in the range *m/z* 200–400 for the $[M + 57]^+$ ions. Thus, stearate, for example, was measured at *m/z* 341 (d_0) and at *m/z* 347 (d_6).

3.3.2 Unsaturated normal chain fatty acids

Mass spectra of unsaturated fatty esters differ from those of saturated esters by the fact that the peak at *m/z* 74 is not the base peak any longer and that the peak at M − 31 is accompanied by a peak at M − 32, representing a loss of methanol (134). The mass spectra of methyl *cis* and *trans*-2-octadecenoate show a characteristic peak at *m/z* 113 (with a high intensity in the case of the *cis* isomer) (134). Figure 1 illustrates some of the steps involved in the GC-MS identification of 2-hexadecenoic acid as a minor component in amniotic fluid phospholipids (135). The identification of the unknown peak as the conjugated 2-hexadecenoic acid was consistent with its characteristic absorption in the ultraviolet in contrast to the 3-hexadecenoic.

Location of a double bond on the carbon chain requires its chemical modification under non-isomerizing conditions, by the addition of groups able to induce specific fragmentation. A detailed review of the problem has been published (136). The simplest and most reliable method of double bond location is based on the successive transformation of the monoenoic ester into a dihydroxy ester and a TMS derivative. The mass spectrum of the TMS derivative exhibits intense peaks due to cleavage between the TMS ether functions. For example, in the case of methyl octadec-9-enoate, peaks at *m/z* 215 and 259 are formed. (Parasitic peaks at *m/z* 332 and 155 are also observed due to rearrangement by cyclization of the molecule). These derivatives of unsaturated fatty esters have been analyzed by GC-CI-MS using isobutane as the reagent gas (137). In the case of methyl oleate, the base peak at *m/z* 385, corresponding to elimination of one trimethylsilanol; the peaks at *m/z* 215 and 259 allow the location of the double bond at position 9, 10. Small differences of peak intensities might be used to determine the double bond geometry, *cis* or *trans*.

Dihydroxylation by osmium tetroxide and formation of the bis-TMS derivatives has been used for the location by MS of the double bond of tetradec-7-enoic acid, a fatty acid isolated from phosphatidyglycerol of *E. coli* (138). Osmium tetroxide is also a convenient and an efficient reagent for the oxidation of mixtures of polyunsaturated fatty acids methyl esters. The resulting polyhydroxy compounds are easily converted to apolar derivatives for GC. However, only the TMS ethers have proved to yield all the information necessary for determination of chain length as well as number and position of double bonds of polyunsaturated fatty acids (37). More recently, Janssen et al. (139) have used GC-MS to identify the TMS ether derivatives of polyols derived from the methyl octadecadienoates and octadecatrienoates by osmium tetroxide oxidation as a means of locating double bonds in microbial isomerization and hydrogenation products of α- and γ-linolenic acids.

The methoxymercuration-demercuration sequence is a facile reaction of olefinic linkages that represents another effective approach in derivatization of double bonds for MS. Plattner et al. (140) have modified the original procedure of Minnikin (135) to produce partial methoxymercuration. When a mole of a methylene interrupted polyunsaturated fatty acid is treated with one mole of methanolic mercuric acetate reagent, the various double bond positions are randomly attacked. After removal of the acetoxymercuri substituents with sodium borohydride, remaining double bonds are reduced by catalytic hydrogenation to provide a mixture of saturated monomethoxy esters in which each original double bond carbon is labeled in a separate isomer. MS of these monomethoxy derivatives is simpler and more definitive than those from

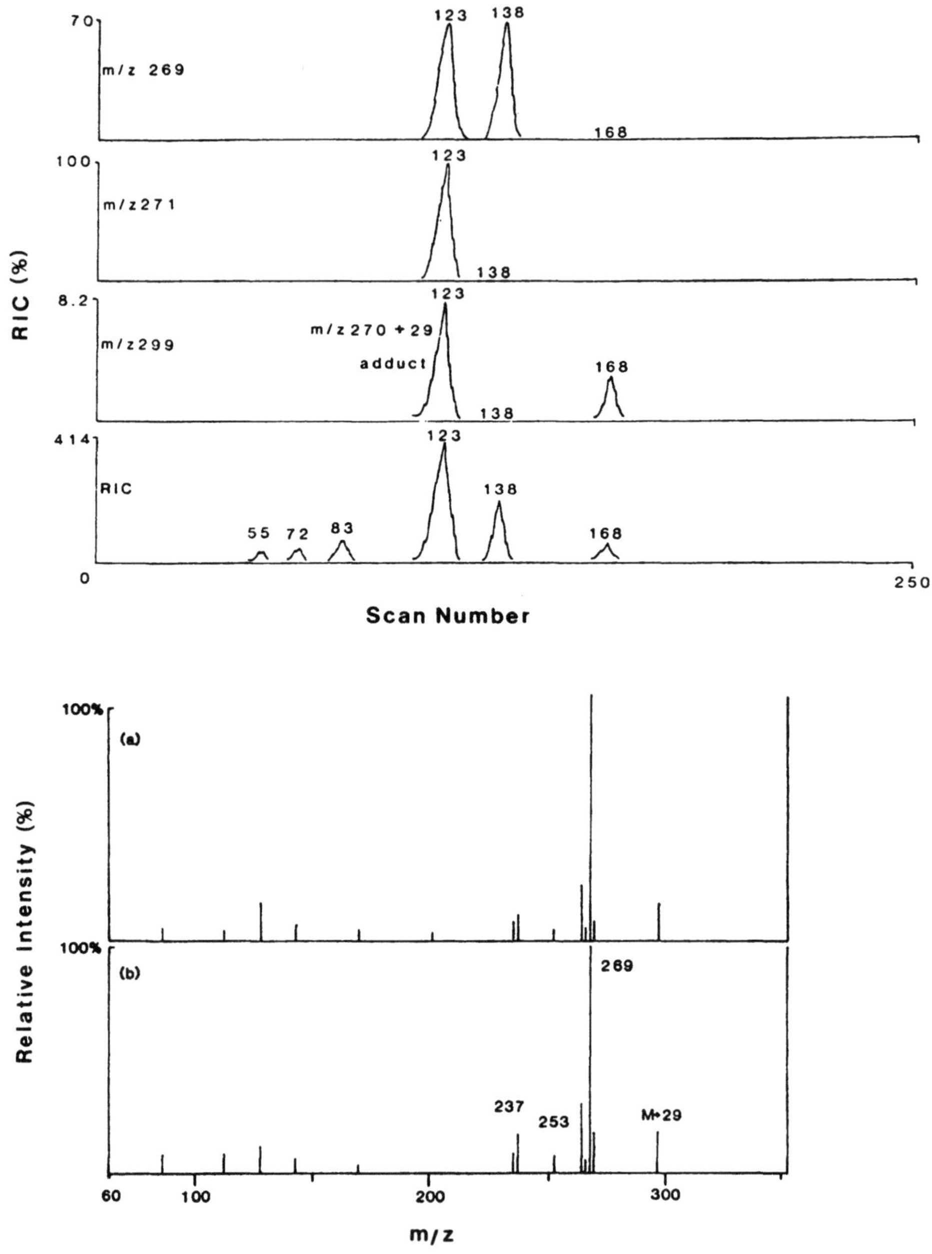

Figure 1. Limited chemical ionization mass scan chromatograms. Upper: parent ions of methyl esters of 2- and 3-hexadecenoic acids (*m*/*z* 269), hexadecanoic acid (*m*/*z* 271), and octadecanoic acid (*m*/*z* 299) and Lower: chemical ionization mass spectra of an amniotic fluid peak (a) and of a synthetic 2-hexadecenoic acid (b) as the methyl esters. GC-MS conditions: Finnigan 3200 quadrupole mass spectrometer-gas chromatograph system; methane chemical ionization (135 eV) Reproduced with permission from: Coombe, R. G., Vine, J. H. and Yip, H. (Ref. 135).

completely methoxylated polyunsaturates. Methoxymercuration also can be applied to MS of acetylenic polyunsaturates, and can be used to differentiate between double and triple bonds in the same molecule (141).

Another approach for double bond location has been developed by Anderson and Holman (34) and Anderson (33) using the pyrrolidides. This group is ionized in preference to the double bond, resulting in a fragmentation in the α-position to the double bond. The base peak is due to the usual McLafferty rearrangement. Between the base peak and the peak for the molecular ion are cluster peaks, with the main masses spaced 14 mass units apart, except in the vicinity of the double bond, where the interval is 12 mass units. The double bond occurs between carbons n and $n+1$ in the molecule (33). Thus, in *N*-octadec-9-enoyl-pyrrolidide, the ions m/z 196 and 208 correspond to fragments containing the 8th and 9th carbon atoms of the acyl chain, respectively, while in *N*-octadec-6-enoyl- and *N*-octadec-11-enoyl-pyrrolidide the ions m/z 154 and 166, and m/z 244 and 266, correspond to the 5th and 6th, and the 10th and 11th carbon atoms of the acyl chain (33).

The pyrrolidide method can be conveniently applied to unsaturated fatty acids of medium chain length, but the results obtained are more difficult to interpret in the case of the very long chain fatty acids. Thus, Takayama et al. (142) used the pyrrolidide method to identify $C_{15}-C_{21}$ monounsaturated esters, but preferred to study the ozonolysis products of $C_{22}-C_{34}$ unsaturated esters. Usually bacterial fatty acids do not contain polyunsaturated homologues. Klein and Schmitz (35) have reported a critical analysis of the feasibility of using specifically deuterated pyrrolidides for mass spectrometric localization of double bonds. It was concluded that this method may give correct results for up to six double bonds, but that the type of instrument used should be considered among the variables.

Location of double bond in olefinic compounds can be directly performed by CIMS, without using chemically modified derivatives. The use of the reagent gas system: 75% N_2/20% CS_2/5% vinyl methyl ether gives spectra containing easily identified ions characteristic of the location of the double bond, as well as abundant molecular ions (143). Myher et al. (144) have noted that in the case of methylene interrupted polyenoic acids it is possible to identify the $(n-3)$, $(n-6)$ and $(n-9)$ fatty acid types by the presence of ions at m/z 108, 150 and 192, respectively, using EIMS of the methyl esters. Fellenberg et al. (145) have applied this method to designating the double bond positional series of the polyenoic acid methyl esters obtained from the brains of Zellweger syndrome patients. These differences are also seen in the mass spectra of the very long chain polyunsaturated fatty acids isolated from the phosphatidylcholines of vertebrate retina and analyzed by Aveldano and Sprecher (42). It is apparent upon inspection of their data that the mass spectrum of 32:4 $(n-6)$ shows an increased abundance of m/z 150, whereas the mass spectra of 32:5 $(n-3)$ and 32:6 $(n-3)$ show relative increases in the abundances of m/z 108. Still other methods of double bond location in methylene interrupted and conjugated polyunsaturated fatty acids have been referred to by Jenssen et al. (139).

Fatty acids in blood lipids from subjects fed triacylglycerols containing deuterium-labeled fatty acids have been analyzed by capillary SIM-GC-MS using isobutane as the reagent gas (120). A special feature was the development of a computer system to correct intensities resulting from what appeared to be hydrogenation of unsaturated fatty acids during CIMS. Hunneman and Schweickhardt (122) have developed an

SIM-GC-MS method of quantitation of myocardial free fatty acids using methyl nonadecanoate as internal standard. Linear standard curves were obtained from 62 pg to 5 ng for oleic, linoleic and linolenic acids. Arachidonic acid could not be detected below 250 pg.

3.3.3 Hydroxy fatty acids

There are several clinical conditions where determination of hydroxy fatty acids is desirable. Patients with the common skin disease psoriasis, show significantly increased concentration of certain monohydroxy fatty acids in their lesional as opposed to non-involved skin. Woollard and Mallet (146) have presented a novel MS method for the profiling of monohydroxy fatty acids and have applied it to the analysis of lesional skin of patients with psoriasis. By a combination of vapour phase hydrogenation and GC-MS with SIM the position of the substituted hydroxyl group of each fatty acid was determined in their methyl TMS ether derivative. Reduction was instantaneous and quantitative allowing detection of less than 100 picograms of each compound. Biological extracts, derivatized as methyl ester *t*-BDMS ether, were prepared by reversed phase HPLC. Separation was dependent on carbon chain length and degree of unsaturation, but not on the position of the silyl ether group. Subsequent conversion of the *t*-BDMS ether to TMS ether facilitated detection of each of the positional hydroxyl isomers as described. Distinction between double bond isomers was possible when they were separated on GC prior to reduction.

The EI mass spectra of the TMS and *t*-BDMS ether derivatives of hydroxy fatty acids are strongly influenced by the presence or absence of unsaturation within the molecule (126). Both derivatives of the unsaturated methyl esters predominantly yield a single intense ion due to α-cleavage at the ether group. However, only the α-cleavage fragment is usually produced by this fission, which probably reflects the greater stability of the ion in conjugation with the diene in the hydrocarbon chain. On hydrogenation of the molecule, however, the methyl ester TMS ether derivatives yield both α-cleavage fragment ions, which are widely used determinants for the position of the hydroxyl substitution in the parent molecule. Thus the hydrogenated methyl ester of ricinoleic acid as the TMS ether (147) gave major ions with masses 301 and 187 in agreement with fragments expected to arise from a 12-TMS group; namely, $CH_3O_2(CH_2)_{10}CHOSi(CH_3)_3^+$ and $CH_3(CH_2)_5CHOSi(CH_3)_3^+$. A prominent peak at *m/z* 371 represented M(386) − 15, which is usual for the TMS derivatives. In contrast, the *t*-BDMS ether derivatives of the saturated compounds exhibit characteristically intense high mass ions where the $[M-(C_4H_9+CH_3OH)]^+$ ion is the base peak (126).

Abe and Tamai (148) have described the determination of all classes of fatty acids as the methyl esters, including the α-hydroxy derivatives from nerve tissue in a single injection by the use of capillary fused silica a column without previous separation or further derivatization.

Nicolaides et al. (149) have presented EI mass spectra of all 15 positional isomers of the acetoxy and the TMS derivatives of the methyl hydroxypalmitates. The authors have published tables of comparative abundances of selected *m/z* values from the mass spectra of 2-hydroxy to 16-hydroxy isomers. The *m/z* values selected for inclusion in these tables are those having a significant difference in abundance for one or more particular positions of the hydroxyl group as compared to the abundances of that *m/z* value for most of all of the remaining positions. The mass spectra of the entire series of positional isomers of the methyl hydroxypalmitates as the acetoxy and TSM derivatives have been deposited with the FPA/NIH Spectra Data Base and the UCLA Geochemical Data Base.

Whereas prominent ions in the mass spectra of TMS derivatives of hydroxy fatty acids are due to α-cleavage at the ether group, the major directing influences in the expression of the *t*-BDMS group in its derivatives arise from both the functional group derivatized and the presence or absence of unsaturation within the molecule (146). Although characteristically the $[M-C_4H_9]^+$ ion was present in all derivatives containing the *t*-BDMS group further fragmentation initiated by this group was prevalent. Thus the methyl ester *t*-BDMS ether derivatives led to the formation of the $[M-(C_4H_9+CH_3OH)]^+$ ion as the base peak in the saturated compound, 12-hydroxystearic acid, while the corresponding 12-hydroxy-5,8,10,14-eicosatetraenoic acid led to the formation of the $[M-132]^+$ ion by the loss of *t*-BDMSOH. The *t*-BDMS derivatives lost both a *t*-butyl and methyl radical, the loss of the bulkier group being favoured. In general the shorter the chain length of the fatty acid and the longer the alkyl substituents attached to silicon, the greater the proportion of the total ion current carried by the $[M-\text{alkyl}]^+$ ion. Increasing unsaturation of the molecule caused greater hydrocarbon fragmentation and a loss of the ability of the silyl group to suppress this. The position of the double bond had little or no influence on either of these parameters (126).

Sud and Feingold (118) suggested the use electron capture detection of 3-hydroxydecanoic acid (heptafluorobutyrated methyl ester) as a screening method for detection of gonococcal infection. The lower detection limits of the technique are about 10^5 gonococci, but interfering substances make it problematic to work at maximum sensitivity. However, the method is sensitive enough for a direct detection of gonorrhoea in vaginal specimens. It may be possible to combine the use of the GC of this derivative with NICI-MS as already done for the pentafluorobenzoates (31) which has provided a pg range of sensitivity in the detection of other hydroxy fatty acid metabolites, when combined with SIM.

In gram negative bacteria, the endotoxin called lipid A has a C_{14} fatty acid with an OH group on its third carbon atom (β-hydroxymyristic acid). Maitra et al. (150) identified this acid by GC-MS of its TMS derivative using SIM at mass 315.4 (M − 15) in endotoxic human serum. Quantitative estimates were obtained by means of the TMS pentadeuteroderivative of methyl β-hydroxymyristate added to the sample (*m/z* 320) as internal standard.

Although problems of high lipid background were experienced and a TLC step was required, the sensitivity was determined to be about 60 pg, which is claimed to be sufficient for detection of gram-negative bacteria in serum. Marom et al. (151) have used EI-SIM at *m/z* 229 vs 225 11- and/or 15-HETE, *m/z* 301 vs 295 for 12-HETE and *m/z* 313 vs 305 *for* 5-HETE generated from D_8 and unlabeled HETE in cultured human airways. Low et al. (152) have studied the instrumental parameters and possible correction procedures for hydrogen transfer phenomena during SIM and have made specific recommendations to avoid isotope effects.

Mycolic acids are a series of high molecular mass 3-hydroxy fatty acids with a long chain alkyl branch at the 2-position. These fatty acids are now known to be widely distributed in *Mycobacterium, Nocardia, Corynebacterium* and related taxa (114, 153). The general structure of these acids apparently produced by all different strains of *Mycobacteria* has been shown (for a review see ref. 114) to be

R'CH(OH)CH(R")COOH

where R" is usually a normal C_{22} or C_{24} chain and R' is approximately 40–60 carbon atoms long. These acids decompose thermally above 250 °C into the meromycolal R'CHO and R"CH_2COOH. Much useful information is available from the spectra of both methyl mycolates and methyl meromycolates (i. e. methyl ester derived from R'CHO). The TLC, GLC and GC/MS of mycolic acids has been recently reviewed (156). Yano et al. (154) have reported on the GC-MS identification of mycolic acids rearranging from C_{28} to C_{68} in *Nocardia* and related acid-fast groups of bacteria. The molecular species of mycolic acids were clearly separated as the TMS derivatives of the methyl esters, according to their carbon number. From the $[M)^+$, $[M-15]^+$ and $[M-90]^+$ ions produced by GC-MS, the total carbon number and double bond number were determined, while the straight and branched chain structures were identified by the mass fragment ions due to C_2-C_3 cleavage $[RCHOSi(CH_3)_3]^+$, and C_3-C_4 cleavage $[(CH_3)_3SiOCHCH(R')COOCH_3]^+$, respectively.

The cell wall of leprosy-derived *Corynebacteria* (isolated from human leprosy lesions and patients' blood) has been shown (155) to contain mycolic acids. These α-branched β-hydroxy fatty acids have been also fractionated by TLC and GC-MS has been used to establish their structures. In all leprosy-derived *Corynebacteria*, some 20 components containing 24–36 carbon atoms and 0–4 double bonds were obtained. The three major groups had 32, 34 and 36 carbons, and the frequency of unsaturated vs saturated chains increased proportionally to the molecular mass. The fractionation of intact corynemycolic acid methyl ester with free or silylated β-hydroxy group on capillary GC proved to be the best analytical procedure at present available for resolving this complex mixture of mycolic acid isomers. Structural determination of silylated samples by MS is preferred because they give more diagnostic fragments.

Steck et al. (32) have fractionated the mycolic acids derived from the cell walls of various strains of *Mycobacteria* by HPLC following their conversion into the *p*-bromophenacyl esters. Using a C_{18}-reversed phase column, they isolated some 24 mycolic acids from each bacterium instead of the two to four that had been described previously. When subjected to EIMS each component gave the expected ester and aldehyde pyrolysis products, the molecular compositions of R', R" and the intact mycolic acid were derived.

As a first step in elucidation of the pathway of mycolic acid biosynthesis, HPLC methods have been developed to purify the $C_{30}-C_{56}$ fatty acids from *Mycobacterium* (157). A combination of NMR, MS and chemical analysis enabled the characterization of several fatty acids including *cis,cis*-3-4- and 15,16-dimethylenetetratriacontanoic acid.

Recent interest in mycolic acids has focused on their biological activity, as they are constituents of the glycolipid "cord factor" (trehalose 6,6'-dimycolate), which has been found to been an essential ingredient of various preparations that, in an experimental animal model, are effective as immunotherapeutic agents in the treatment of cancer (158).

3.3.4 Methyl-branched fatty acids

Since many bacterial and microbial products are unique for a particular genus or species, SIM constitutes a diagnostic tool of great potential and wide scope for use in the field of clinical microbiology. Thus, bacterial fatty acids often contain high proportions of branched-chain analogues, with the branching in the *iso* and *anteiso* position. Mass spectra of *anteiso* methyl esters exhibit a peak at M − 57 [elimination of $CH(CH_3)CH_2CH_3$], but a similar peak is also observed in the spectrum of *n*-5-methyl-branched esters (due to the elimination of $CH_2CH_2CH_2CH_3$), and at M − 29 (elimination of CH_2CH_3). The *anteiso* structure is suggested by the unusual relative intensities of peaks M − 29 greater than M − 31 (for a review see ref. 116). Karlsson et al. (159) have suggested the transformation of the methyl esters into the coresponding alcohols for the *anteiso*-derivative detection.

Spectra for iso esters exhibit a peak at M − 43 [elimination of $CH(CH_3)CH_3$], but such a peak is found also in the spectrum of any kind of fatty acid methyl ester, sometimes with a higher intensity (160). A peak of very low intensity at M − 65 [due to M − (CH_3 + CH_3OH + H_2O)] seems to be more specific (161). The use of pyrrolidide derivatives of fatty acids have been found to be beneficial also for the location of methyl branching in the iso and at the anteiso positions (33). Jensen and Gross (162) have observed that *iso*- and *anteiso*-branched fatty acids can be distinguished from isomeric straight chain fatty acids by collisionally activating the $[M-H]^-$ ions obtained by using FAB-MS. In contrast to the *iso*-acids, the *anteiso*-acids undergo loss of the elements of CH_4 and C_2H_6 but not C_3H_8. Mixtures of homologs and isomers could be investigated by using the combination of FAB and MS-MS.

Harvey (163, 164) has used GC-MS to establish branching points and centers of unsaturation in 3-picolinyl esters of *normal*, *iso*-, *anteiso*- and unsaturated long chain fatty acids. These spectra contained diagnostic ions defining the position of both methyl branching and double bonds, which resulted from the extraction of hydrogen from the chain by the nitrogen of the pyridine ring and cleavage at the resulting radical site. The spectra of the 3-picolinyl esters were much more informative than those of the other picolinyl isomers as well as the mass spectra of pyrrolidides used previously (33). The ion at *m*/*z* 108 has the base peak, representing the pyridione ring plus part of the side chain. Ryhage and Stenhagen (165) have shown that the mass spectra of methyl 2-methylalkanoates exhibit an intense peak at *m*/*z* 88, due to $CH_3CH = C(OH)OCH_3$ [instead of *m*/*z* 74; $CH_2 = C(OH)OCH_3$] through a McLafferty rearrangement. When a second methyl branch is present in position 4 (2,4-dimethylalkanoates), the peak at *m*/*z* 74 is still strong, but two important peaks at *m*/*z* 101 and 129 are also observed arising from cleavage of the bonds on both sides of the tertiary carbon atoms at C_4; these latter peaks are characteristic of the presence of a methyl branch in position 4. By using the characteristic peaks, Julak et al. (166) have identified 2-methyltetradecanoic acid in the fatty acids of *M. gordonae*, and 2,4-dimethyltetradecanoic acid in those of *M. kansasii*. When the methyl branch is in position 3, the usual peak at *m*/*z* 74 constitutes the base peak of the spectrum, but a fragment at *m*/*z* 101 [$CH(CH_3(CH_2COOCH_3]$ is observed, with a lower intensity.

Refsum's disease is an inherited lipid storage disease in which phytanic acid (3,7,11,15-tetramethylhexadecanoic acid) accumulates in various organs and fluids (167). The accumulation is due to the failure of the α-oxidation system, which allows the conversion of phytanic acid to the α-hydroxy-derivative. Stenhagen (168) has analyzed the mass spectra of methyl phytanate and has shown that it has a base peak at *m/z* 101, with a strong peak at *m/z* 74 and other major ions at 143, 171, 213, 241 and 283, consistent with cleavages of the hydrocarbon chain on both sides of the 3,7,11 and 15 methyl branching.

Pileire et al. (169) have developed a rapid GC-CIMS method for the detection of phytanic acid in plasma of Refsum's disease. After a brief alkaline hydrolysis of lipids, the biological sample is placed directly onto a pre-column. An acid carrier gas (formic acid in nitrogen) is used to displace the long chain fatty acids from the sodium salts and from their binding to proteins. The formic acid introduced into the GC column also serves as a reagent gas for chemical ionization in combined GC-CIMS. Fatty acids (14:0 to 18:2 and phytanic acid) are easily identified by the M + 1 (base peak) and M − 17 peaks. The presence of phytanic acid is monitored by the *m/z* 313 (M + 1) base peak. The fatty acid composition of plasma lipids of one patient (untreated Refsum's disease) was compared with two other procedures, which gave similar results and confirmed the high percentage of phytanic acid already noted by Skrbic and Cummings (170) and Steinberg (167). This procedure, which bypasses the tedious extraction and derivatization steps, should be much more rapid and simpler than previously described methods for the identification and estimation of phytanic acid in the diagnosis and care of Refsum's disease. Greter et al. (171) have observed that the urine of patients with Refsum's disease contains 2,6-dimethyloctanedioic acid, along with increased concentration of 3-methylhexanedioic acid. These acids were identified by GC-MS as the methyl and TMS esters.

Mee et al. (58) have employed CI-MS for the quantitation of phytanic acid by means of the corresponding deuterium (D_2)-labeled analogue. By using fixed amounts of standards deuterated in position 2, calibration curves were constructed for the saturated C_{12}, C_{14}, C_{16} and C_{18} acids. Linear relationship was found between the [M + 1] peak height ratio (D_0/D_2) of the unlabeled and labeled acid and the amounts of fatty acid present. Phytanic acid was determined in microgram quantities of dried blood spots from patients with Refsum's disease. Yao et al. (172) have used GC-MS to identify phytanic and pristanic acids in the triacylglycerols from plasma of patients with multineuronal degeneration, hepatosplenomegaly, and adrenocortical deficiency. Elevated levels of these acids were thought to represent possibly an undetermined error in metabolism. Recently, Yao and Dyck (39) have used HPTLC, GC and MS to identify three of six kin as having phytanic acid storage disease. Phytanic acid was found in the triacylglycerol and to a lesser degree in phosphatidylcholine and free fatty acid fractions. An unsaturated analogue of phytanic acid was identified in plasma and erythrocyte triacylglycerols. In plasma the branched-chain acids were primarily localized in LDL. In contrast to plasma only small amounts of phytanic acid were found in cerebrospinal fluid and biopsied sural nerve. After conversion to their methyl esters, oxidation of polyhydroxy fatty acids, and trimethylsilylation, a monounsaturated and triunsaturated form of phytanic acid has been identified in plasma of a patient with Refsum's disease by GC/MS using both EI and Ci methods (173). The CI spectrum provided a number of ions that indicated a molecular mass of 502. These included ions at *m/z* 531 [M + C_2H_5], *m/z* 501 [M − H], *m/z* 487 [M − CH_3], *m/z* 471 [M − OCH_3] and *m/z* 413 [M − OTMS]. The CI spectrum also contained ions that indicated the position of the vicinal OTMS groups in the molecule. The most indicative of them were at m/z 399, which results from loss of the terminal CH_2OTMS, and at m/z 219, which contains the terminal vicinal OTMS formed by cleavage between carbons 14 and 15. These data established that the monounsaturated phytanic acid was 3,7,11,15-tetramethylhexadeca-15-monoenoic acid. The isobutane CI spectrum of the triunsaturated phytanic acid derivative did not show a pseudomolecular ion, or any ions resulting from a loss of less than 89 daltons (OTMS) from the molecular ion. The EI spectrum of the triunsaturated phytanic acid derivative provided the clearest evidence for location of its vicinal OTMS groups, which supported the conclusion that the triunsaturated phytanic acid is 3,7,11,15-tetramethyl hexadeca-6,10,14-trienoic acid.

The α,β-unsaturated fatty acids having their methyl branches at positions 2, 4 and 6 are found in the lipids of the virulent strains of *M. tuberculosis* (174). The main representatives of this group of acids are the C_{25} and C_{27} phthienoic acids. The methyl esters of these acids give rise to highly characteristic peaks at m/z 169 $[CH(CH_3)CH_2CH(CH_3)CH = (CH_3)COOCH_3]$ in the EI spectrum. In addition, the EI spectra of these acids possess the high intensity peaks at m/z 88 and 101, showing that a double bond shift occurs before cleavage of the C_2-C_3 and C_3-C_4 bonds.

Several bacterial species belonging to the *Actinomycetes* synthesize fatty acids having one methyl branch in the middle of the aliphatic chain. The best known of these is tuberculostearic acid (10D-methyloctadecanoic acid) (116, 174). The EI spectrum of this acid shows characteristic ion fragments arising from cleavage of the carbon-carbon bonds on both sides of the methyl-bearing methine group, giving rise to a strong peak at m/z 199 $[CH(CH_3)(CH_2)_8COOCH_3]$ and a smaller one at m/z 171 $[(CH_2)_8COOCH_3]$ (169). This latter peak is accompanied by peaks at m/z 172 and 173 arising from rearrangement of one or two hydrogen atoms, respectively (175, 176).

Tuberculostearic acid has been identified in cells of several species of *Mycobacteria* by capillary GC-MS using SIM. Monitoring at m/z 312 (molecular ion) gives a single peak in the chromatograms. It is possible to detect tuberculostearate by this technique in as little as 2μg lyophilized cells (containing about 20 pg of tuberculostearate) (40).

Larsson et al. (40) examined a series of strains of *Mycobacteria*, pneumocci, staphylococci, streptococci, various anaerobic bacteria and *E. coli* and have shown that monitoring m/z 167 and m/z 312 gave only one peak for *Mycobacteria* and *Nocardia* species, while all other bacteria studied were completely free of these fragments. Odham et al. (117) have used this method to detect tuberculostearic acid in lyophilized samples of autoclaved sputum (2–4 ml) of patients with pulmonary tuberculosis.

Larsson et al. (123) reported a potential method of increasing the sensitivity in the detection of tuberculostearic acid by using EI-MS with SIM of the *t*-BDMS esters, as compared to the corresponding methyl ester, and by CIMS of methyl tuberculostearate using methane and isobutane as reactant gases. Furthermore, the capillary GC system with OV-17 as stationary phase allowed the optimization of the separation of various methyl branched C_{18} positional isomers possibly present in clinical specimens. It was found that in the EI-SIM technique, the detection limit of the methyl ester of tuberculostearic acid (monitored at m/z 312) was 5 to 10 times higher than that of the *t*-BDMS ester (monitored at m/z 335); roughly the same increase in sensitivity was gained when focusing at m/z 313 $[M + 1]^+$, employing isobutane as reactant gas in the CI-SIM mode. The detection limit was estimated to be approximately 1 pg for the methyl ester as detected by CI-SIM (isobutane) or for the *t*-BDMS ester using EI.

However, preparation of the *t*-BDMS ester is more laborious than methylation, and also seems to give lower yields. The almost exclusive formation of the quasi-molecular ion in the CI mode gives fragmentograms not complicated by additional fragments from cleavage of other ionized large molecules present in the sample. An EI-SIM method for the detection of tuberculostearic acid (*m/z* 312 and 167) in cerebrospinal fluid has been applied to the diagnosis of tuberculous meningitis (119).

3.3.5 **Cyclopropane fatty acids**

Although proper food processing tends to destroy them, the cyclopropenoid fatty acids present in some plant oils cause adverse physiological effects when ingested (177).

The location of a cyclopropane ring cannot be directly deduced from the mass spectrum of a fatty acid ester, because the ring is unstable under electron impact, and the EI spectrum of a cyclopropane fatty acid methyl ester is similar to that of the corresponding monoenoic ester with the same molecular weight. Conventional GC or capillary GC allow a distinction between cyclopropane fatty acid esters and isomeric unsaturated fatty esters (178). The EI-MS of the pyrrolidides of cyclopropane fatty acids has been found to be helpful. According to Anderson (33) a 12 dalton interval, instead of the usual 14, is observed between the most intense peaks or clusters of fragments containing $n-2$ and $n-1$ and n, or and $n+1$ carbon atoms of the acid moiety, when a cyclopropane ring occurs between carbon n and $n+1$ in the molecule.

3.4 Dimethylacetals

The ether bond in plasmalogens is cleaved by acids to yield, in the presence of methanol, dimethylacetal drivatives of aldehydes (179). Christiansen et al. (180) reported EI mass spectra of the dimethylacetals of myrist- and palmit-aldehydes. The dimethylacetals yield very small molecular ions, but the $[M-31]^+$ and $[M-64]^+$ ions can be readily seen. The ion at *m/z* 71 $[CH_2CHCHOCH_3]^+$ provided the base peak for the two compounds with the ion *m/z* 75 $[CH_3OCHOCH_3]^+$ having relative intensities of 15% and 3.5%, respectively. Using GC-EI-MS, Myher et al. (144), however, obtained nearly equal intensities for the *m/z* 71 and *m/z* 75 ions. Since the ion at *m/z* 75 is absent from the spectra of the corresponding alkyl ether, it alone may be taken as a measure of undegraded dimethylacetal. Hence the presence of *m/z* 71 in great abundance may indicate extensive degradation of the dimethylacetal to the alkenyl ether. For the diethylacetals of myristaldehyde the two corresponding major ions had *m/z* values of 85 and 103 providing additional evidence for the reported structure of these ions. Bjorkhem et al. (181) have proposed a simple diagnosis of the Zellweger syndrome by a GC determination of the ratio of dimethylacetals to the methyl esters of the corresponding saturated fatty acids. In this study the identity of the dimethylacetal of palmitaldehyde was established by GC-MS. The mass spectrum of the unknown was identical to that of the authentic standard and showed characteristic peaks at *m/z* 286 $[M]^+$, 254 $[M-32]^+$, 222 $[M-64]^+$ and *m/z* 124, 110, 96, 82, and 68 (alkadiene fragments of various chain lengths), as well as *m/z* 71 and 75. The natural stearaldehyde dimethylacetal also gave the expected mass spectrum with peaks at *m/z* 282 $[M-32]^+$, 250 $[M-64]^+$, as well as the fragment ions at *m/z* 124, 110, 96, 75, 71 and 68.

3.5 Wax esters

Wax esters, composed of long chain acids (RCOOH) and long chain alcohols (R'OH), occur principally in a variety of plant and marine species. Wax esters in mammalian systems are produced mainly by the sebacious glands, whose activity appears to change with age, hormone levels and disease (88–90, 93). The EI spectra of wax esters (160, 182, 183) are composed of a number of diagnostic ions primarily $[M]^+$, $[RCO_2]^+$ and $[R'-1]^+$. Octadecyl myristate, palmitate and stearate have been identified by GC-MS in the nail of subjects having psoriasis but not in the nail of normal subjects (93). GC-MS analysis has also been used to characterize the way ester of Jojoba oil (184) and sperm whale oil (185, 186), which are components of various skin care preparations. Methane CIMS of both saturated and unsaturated wax esters (187, 188) characteristically yield $[M+H]^+$, $[M+C_2H_5]^+$, $[M+C_3H_5]^+$ and $[M-H]^+$ type ions. Identification is aided by the occurrence of $[RCO_2H_2]^+$ and $[R'-2H]^+$ ions. Isobutane and ammonia CIMS of wax esters yields predominantly $[M+H]^+$ and $[M+NH_4]^+$ ions, respectively, with very little fragmentation.

3.6 Glyceryl esters and ethers

The neutral glyceryl esters represent largely an energy storage form in the body, and normally they are found in the fatty tissues. The neutral glyceryl ethers occur naturally at low concentrations and their physiological role and clinical importance is unknown. In a specific instance, however, the mutagenic properties of the polyunsaturated monoalkylglyceryl ethers have been described (189). Although characteristic diradylglycerol structures have been identified in different tissues (14, 16), there are no known pathological conditions associated with the presence of specific diacylglycerol species in a tissue. The main interest in the MS, GC-MS and LC-MS of these compounds stems from the need for detailed structural analyses of the diradylglycerol moieties of glycerophospholipids. In certain hyperlipoproteinemias abnormally high amounts of triacylglycerols accumulate in plasma lipoproteins, and attempts have been made to associate this with abnormalities in their composition and positional distribution of fatty acids (53, 54).

3.6.1 **Monoradylglycerols**

Monoradylglycerols are a class of compounds consisting of monoacyl-, monoalkyl- and monoalk-l-enyl glycerols. Although they may occur naturally at low concentration, these lipids are generally derived from more complex glycerolipids by chemical or enzymatic means.

Direct MS of monoacylglycerols, monoacylglycerol diacetates and monoacylglycerol di-TMS ethers was first conducted by Johnson and Holman (190). Curstedt used deuterium labeled monoacylglycerols to clarify the structure of a number of fragment ions in the mass spectra of the TMS ethers of monoacylglycerols (191). Myher et al. (192) reported a GC-MS study comparing the spectra of the TMS ethers of monoacyl-, monoalkyl- and monoalk-l-enyl-glycerols. Gaskell and Brooks (193) prepared methaneboronate derivatives for GC-MS investigation of monoacylglycerols derived from egg lysophosphatidyl choline.

A variety of derivatives have been employed for the GC-MS analysis of monoalkyl- and monoalk-1-enylglycerols. Dimethoxy derivatives have been used for monoalkylglycerols (194) and mono(methoxyalkyl) glycerols (195). Isopropylidene derivatives have been used for the analyses of monoalkylglycerols (195–198), mono(methoxyalkyl) glycerols (195), mono(hydroxyalkyl)glycerols (199, 200), mono(ketotoalkyl)glycerols (199) and monophytanylglycerols (201). Monoalkylglycerol diacetates have also been analyzed by MS (202–204). Trimethylsilyl derivatives are suitable for GC-MS analysis of monoalkylglycerols (192, 203) and monoalk-1-enylglycerols (192). The MS of monoradylglycerols has been reviewed a number of times (203, 205–207). Spener (208) has recently reviewed the clinical and medical aspects of ether lipids.

A search for chemicals that may be responsible for colon cancer has led to the discovery of a strongly mutagenic fraction that can be isolated from the feces of 40% of the Caucasian population (209). Two unusual glyceryl ethers, 1-(1-glycero) tetradeca-1,3,5,7,9-pentaene and 1-(1-glycero)dodeca-1,3,5,7,9-pentaene, have been identified in the mutagenic fraction by chemical and chromatographic procedures and instrumental techniques, including NMR and LC-MS (189). Under LC-MS conditions using a ODS reversed phase column and an elution solvent consisting of acetonitrile/water/methanol/tetrahydrofuran (36.2:32:25.4:6.4) a number of characteristic ion fragments were observed. The spectra of the $R-OCH_2CH(OH)CH_2OH$ type compounds showed ions corresponding to MH^+, ROH^+, and R^+. NMR also showed the presence of various geometrical isomers of the double bonds in at least one of the compounds.

In support of the enediol concept of *O*-alkyl ether bond synthesis (210) it has been shown that the loss of the fatty acid of acyldihydroxyacetone phosphate occurs via an unusual cleavage whereby the C-1 oxygen of dihydroxyacetone phosphate remains with the fatty acid (211). The investigation included the synthesis of MS of normal hexadecanoyldihydroxyacetone phosphate and the compound having ^{18}O exclusively at the C-1 oxygen. MS of the methoxime-TMS derivative of hexadecanoyldihydroxyacetone exhibited ions corresponding to $[M-CH_3]^+$ (17%), $[M-CH_2OSi(CH_3)_3]^+$ (5%), $[RCO + 74]^+$ (8%), $[M-(RCO-1)]^+$ (6%), $[M-(RCO + (CH_3)_3SiO)]^+$ (19%) all of which increased in mass by two *m/z* units for the 1-^{18}O labeled compound, and $[RCO]^+$ (11%), $[M-RCOOH]^+$ (100%) and $[CH_3)_3SiOCH_2]^+$ (26%) which are invariant under transformation to the labeled compound.

3.6.2 **Diradylglycerols**

Diradylglycerols (diacyl-, alkylacyl-, alkenylacylglycerols, etc.) may be isolated in free form from natural sources or they may be derived from more complex glycerolipids by chemical or enzymatic means. Diradylglycerols derived from phospholipids by phospholipase C are more amenable to analysis by chromatography and MS. Diacylglycerols may also be derived from triacylglycerols by pancreatic lipase or Grignard reagents in order to simplify the analyses. MS of various types of diradylglycerols has been reviewed previously (203, 205, 206, 207, 212).

Underivatized diacylglycerols may be used for direct injection-MS (DIMS) (213) but preparation of acetates (214, 215), TMS (25, 216), or *t*-BDMS (14, 217) ethers is required for GC-MS. ^{252}Cf plasma desorption MS has recently been used to confirm the identity of diarachidonyl glycerol, an unusual product which is synthesized by rat neutrophils from exogenously added arachidonic acid (218). MS of the underivatized sample exhibited an ion at *m*/*z* 691 which corresponds to the sodium adduct $[M + Na]^+$ and a less abundant M^+ ion at *m*/*z* 668 (the molecule contains one ^{14}C atom per arachidonyl group).

The EI spectra of diacylglycerol acetates exhibit mainly ions retaining a single acyl group, namely $[M - RCOO]^+$, $[M - ROOH]^+$, $[RCO]^+$ and $[RCO - 1]^+$ (214, 215). The fragment resulting from loss of acetic acid may be used to determine the molecular mass although its relative abundance is generally less than 10% of the base peak. EI MS of diacylglycerol acetates has been used to study complex mixtures (214, 215) and also to study the stereochemical course of hydrolysis by lipoprotein lipase using deuterium labeled substrates (217). Yergey et al. (26) have measured the levels of disaturated phosphatidylcholines in human plasma as the diacylglycerol acetates by ammonia CI-MS.

TMS ethers of diacylglycerols are easily made, have good GC properties and generate characteristic fragments upon EI-MS. The EI-MS of the TMS ethers of 1,2-diacylglycerols show peaks corresponding to $[M - CH_3]^+$, $[M - Si(CH_3)_3OH]^+$, $[M - RCOO]^+$, $[RCO + 74]^+$, $[RCO + 90]^+$, $[RCO]^+$ and $[RCO - 1]$, as well as ions at *m*/*z* 129 and 145. When one of the fatty acids is unsaturated, the other fatty acid may be lost primarily as RCOOH rather than RCOO. MS of the TMS ethers of 1,3-diacylglycerols show a major ion, $[M - RCOOCH_2]^+$, due to loss of an acyloxymethylene group (25, 216) which is not found in the spectra of the 1,2-isomer. Deuterium labeled substrates have also been utilized to investigate the structure of the ions in the spectra of the TMS ethers (191). By analogy with the fragmentation pattern of triacylglycerols (see later) the $[RCO + 74]^+$ ion might be expected to retain the glycerol backbone and have a structure corresponding to $[RCOOCH_2CH(OH)CH_{2]}^+$. However, for the TMS ether of 1,2-dipalmitoylglycerol this structure accounts for only 10% of the $[RCO + 74]^+$ peak. The remaining 90% of the peak is due to a rearrangement ion having the structure $[RCOOSi(CH_3)_2]^+$. The $[RCO + 90]^+$ ions were shown to have the structure $[RC(OH)OSi(CH_3)_3]^+$. The ion at *m*/*z* 129, corresponding to the structure $[CH_2CHCHOSi(CH_3)_3]^+$ is apparently of mixed origin since a portion retains four hydrogens from carbons one and three and another portion retains the hydrogen from carbon two, and three hydrogens from carbons one and three. The ion at *m*/*z* 145 has the structure $[CH_2(O)CHCH_2OSi(CH_3)_3]^+$ and retains the hydrogen at carbon two and three hydrogens from carbons one and three.

Curstedt and Sjövall (219) developed a detailed method of determining the deuterium content of all three carbons of glycerol in *sn*-1,2-diacylglycerols derived from phospholipids. Selected *sn*-1,2-diacylglycerols were converted into perdeuterosilyl derivatives, hydrolyzed by pancreatic lipase and finally converted into 2-acyl-*sn*-glycerol-1-TMS-3-perdeutero-TMS ethers. Analyses of the $[CH_2OSi(CH_3)_3]^+$ type of ions at *m/z* 103 and *m/z* 112 were used to determine the deuterium content at carbons one and three of *sn*-glycerol, respectively. The labeling at carbon two was determined by subtracting the deuterium at these positions from the total. The deuterium content of saturated acyl residues can be determined via the RCO^+ ions in the spectra of the diacylglycerols.

The distribution of deuterium in the molecular species of the phosphatidylcholines of bile and liver in bile fistula rats given [1,1-^{2}H]ethanol (64, 66, 220) and [2,2,2-$^2H_3$]ethanol (221) has been reported in a series of papers. These studies have recently been extended to the determination of deuterium incorporation into hepatic phosphatidylethanolamines, phosphatidylserines and phosphatidylinositols (65) as well as into the ether analogues of the hepatic phosphatidylcholines and phosphatidylethanolamines (222) during metabolism of [1,1-^{2}H] ethanol in rats. The results indicated that the same pool of dihydroxyacetone phosphate and pyridine nucleotide was used for the biosynthesis of the various phospholipids. The results also confirm the earlier conclusion that the major part of the *sn*-glycerol 3-phosphate used in the biosynthesis of the different phospholipids appears to have a gluconeogenic origin. The incorporation of deuterium into the glycerol moiety of the phosphatidylcholines was much smaller in the heart than in the liver. The labeling may result via labeled glucose or lactate transported from liver or from the incorporation of labeled phosphatidylcholines from plasma.

The molecular species of the hepatic and biliary phosphatidylcholines of gallstone patients and gallstone-free controls have been determined by GC-MS (223). The composition ratio for 16:0 18:1/16:0 18:2 was found to be higher (0.7) in gallstone subjects than in the controls (0.45). In gallstone subjects treated with either cholic or chenodeoxycholic acid the ratio was similar to the controls. In a study of the lung phosphatidylcholines from rabbit and rat fetuses at different times of gestation it was found that dipalmitoylglycerophosphorylcholine increased at the later times (224).

Alkylacylglycerols may be analyzed by GC-MS as the TMS ethers (224–226). Ions corresponding to $[M-15]^+$ and $[M-90]^+$ indicate the molecular mass, whereas $[M-R'O]^+$, $[RCO+74]^+$ and $[RCO]^+$ type ions determine the acyl chain. The identity of the alkyl chain can be determined from the $[R'O+72]^+$ ion. It should be noted that there is a coincidence in mass for $[R'O+72]^+$ and $[RCO+74]^+$ fragments when alkyl and acyl chains have the same carbon number and unsaturation.

Alk-1-enylacylglycerols can also be analyzed by GC-MS as the TMS derivatives (225, 227). The spectra have a base peak at *m/z* 129 but only small peaks corresponding to $[M-15]^+$ and $[M-90]^+$. The acyl chain can be established by an intense ion, $[M-R''O]^+$, as well as less abundant $[RCO+74]^+$ and $[RCO]^+$ ions. The alk-1-enyl chains are characterized by the low intensity ions, $[R''O+72]$ and $[R''-1]$.

The EI-MS of the TMS ethers of dialkylglycerols (228) exhibit a base peak at *m/z* 133 and diagnostic ions corresponding to losses of CH_3, $CH_3Si(CH_3)_3$, R'OH and $R'OCH_3$ from the molecular ion. There are also ions corresponding to $[R']^+$ and $[R'O+72]^+$.

The EI-MS of the *t*-BDMS ethers of diacylglycerols have been reported by Myher et al. (14) and subsequently by Satouchi and Saito (229). One of the important characteristics of the spectra is the abundant ion $[M-57]^+$ resulting from loss of the *t*-butyl radical from the *t*-BDMS group. Even though the intensity of the $[M-57]^+$ ion diminishes as unsaturation increases it remains at a practical level even for species containing docosahexaenoic acid. The acyl groups are indicated by the very abundant $[RCO + 74]^+$ ions and the less abundant $[M-RCOO]^+$ ions. The acyl groups are also indicated by the $[M-RCOO]^+$ ions resulting from loss of an acyloxy radical. For at least one set of reverse isomers namely 1-palmitoyl-2-stearoyl-rac-glycerol and 1-stearoyl-2-palmitoyl-rac-glycerol the intensity ratios of the two $[M-RCOO]^+$ ions, *m/z* 427 (loss of stearate) to *m/z* 455 (loss of palmitate), are 2.0 and 0.6, respectively (14). Therefore the $[M-RCOO]^+$ fragment bearing the acyl group at position 1 is more intense than the corresponding fragment bearing the acyl group at position 2. In common with the MS of fatty acid methyl esters (144) the MS of diacylglycerols (14) having polyunsaturated fatty acids shows an enhanced intensity at *m/z* 108 for $(n-3)$ series acids and at *m/z* 150 for $(n-6)$ series acids. The spectra of *t*-BDMS ethers of 1,3-diacylglycerols are distinguished by the presence of $[M-RCOOCH_2]^+$ fragments.

Since the intensity of the $[M-57]^+$ ions various from species to species it is necessary to calibrate these responses. By comparing the $[M-57]^+$ intensities obtained for a series of mixtures with predetermined compositions it was possible to establish a series of relative correction factors for species ranging from saturates (correction factor = 1) to hexaenoates containing docosahexaenoic acid (correction factor = 6.1) (230). Unknown diacylglycerol mixtures are then quantitated first according to carbon number by conventional GC and then for each carbon number according to unsaturation by GC-MS. The $[M-57]^+$ intensities are first corrected for contribution due to naturally occurring heavy isotopes and then by the appropriate predetermined correction factors. By these methods it has been possible to show that 70% of the endogenous phospholipids of all rat plasma lipoproteins was replaced by egg yolk phospholipids after a 24 h infusion with Intralipid.

It was also possible to demonstrate the similarity of the molecular species of phosphatidylcholine in the major human lipoprotein fractions (15). The effect of diet on the molecular species was also determined. The relative changes in the molecular species are not quantitatively predictable from the changes in total fatty acid composition. Although oleic acid decreased only 10% in going from a saturated to an unsaturated diet the monoenoic species 16:0 18:1 and 18:0 18:1 showed a drop of 50%. The remainder of the oleic acid was found mainly in the 18:1 18:2 species.

A variety of techniques have also been used to determine the molecular species of triacylglycerols (231). Most methods however resolve a complex mixture into a very limited number of fractions. The fatty acid composition of all three positions can be determined by the methods of stereospecific analysis (232–235) and the molecular association of fatty acids can be measured in pairs by generating representative diacylglycerols from the triacylglycerols. The *sn*-1,2- and *sn*-2,3-diacylglycerols can be resolved after they have been chemically converted into phosphatidylcholines by stepwise hydrolysis with phospholipase C (235). Thus MS can be used to probe triacylglycerol structures via the derived diacylglycerols. Studies have been conducted on dietary fats of varying degrees of atherogenic potential (236), on cardiac triacylglycerols in dietary lipidosis (237) and on the triacylglycerols from lipoproteins of normolipemic subjects and patients with Type II, Type III and Type IV hyperlipoproteinemia (53, 54). GC-MS combined with stereospecific analysis has also been used to examine the course of acylation of 2-monoacylglycerols with deuterium labeled palmitic acid (55).

It is also possible to monitor relative changes in molecular species by GC-MS without using response factors. Thus the phosphatidylcholine composition of rat brain (238) and rat liver (239) have been determined at various stages of development and the molecular species of the choline and ethanolamine phosphatides of dog liver have been determined before and after the induction of ischemic hypoxic conditions (240).

It has been suggested that one way of eliminating the need for GC-MS response factors would be to catalytically reduce the samples with deuterium (206, 241). All the molecules would then be fully saturated and the original number of double bonds could still be determined from the deuterium content. Such a procedure using a Wilkinson homogeneous catalyst [tris(triphenylphosphine) rhodium(I) chloride], has been developed for the quantitation of the *t*-BDMS ethers of diacylglycerols (242) and applied to the phosphilipid species of *Tetrahymena* pyriformis.

Unter hydrogen CI conditions the *t*-BDMS ethers of diacylclycerols (243) also yield practical mass spectra. The base peaks in the CI spectra are the $[MH-RCOOH]^+$ ions formed from the more saturated of the two fatty acids in the diacylglycerol species. In contrast to EI conditions, only low abundances of $[RCO + 74]^+$ ions are generated in the hydrogen CI mode. $[RCO]^+$ ions increase in abundance from 2% for palmitic acid to 13% for docosahexaenoic acid. The spectra also contain ions representing $[MH]^+$, $[M-57]^+$ and $[MH-132]^+$ type ions.

The *t*-BDMS of diacylglycerols have recently been analyzed by LC-MS using a direct liquid inlet interface and C−18 reversed phase column. The MS is carried out in the CI mode using the HPLC elution solvent (acetonitrile/propionitrile) as reagent gas (16). MS indicated the presence of $[MH]^+$ ions which increased in abundance with the degree of unsaturation. $[MH-132]^+$ was quite strong but decreased with unsaturation and there were only trace amounts of the $[M-57]^+$ ions. The $[MH-RCOOH]^+$ ions resulting from loss of the fatty acid from the *sn*-2 position dominated the spectra with the other $[MH-RCOOH]^+$ ions having a 3 to 4 fold lower abundance. The total ion response was found to be approximately the same for all species but the use of specific ions requires calibration. Species such as 18:1 18:1 and 18:0 18:2 that are not separated by HPLC can be effectively resolved using the $[MH-RCOOH]^+$ ions. Analysis of several test mixtures by LC-MS (16) gave results in good agreement with analysis by GC using polar capillary columns (244). Since polyunsaturated species are eluted early when chromatographed on reverse phase columns, the LC-MS method is well suited to the unequivocal identification and measurement of small quantities of species having more than 6 double bonds which are eluted at the end of a GC run. The LC-MS method is also well suited for studies involving stable isotopes. It has recently been used to measure the relative rates of deacylation of the various molecular species of endogenously labeled rat liver microsomal phospholipids by endogenous phospholipases (17).

Alkyl- and alk-1-enylacylglycerols may also by analyzed be EI mass spectrometry as the *t*-BDMS ethers (229, 241, 245). The molecular masses are determined from the masses of the abundant $[M-57]^+$ ions. The structure of the acyl chains is indicated by very intense ions corresponding to $[RCO + 74]^+$ and by the less abundant $[RCO]^+$, $[M-R'O]^+$ or $[M-R''O]^+$ ions. The alkyl or alk-1-enyl chains can be determined by monitoring the [R'O + 114] or [R″ + 114] ions resulting from the consecutive losses of the *t*-butyl group and the fatty acid (RCOOH). There is also an ion corresponding to $[R'O + 40]^+$ or $[R''O + 40]^+$ in the spectra of the alkylacyl- and alk-1-enylacylglycerols respectively but the latter ion has the higher relative intensity. The spectra of *t*-BDMS ethers of an isomeric pair of alkylacylglycerols is shown in Figure 2 (241). In addition to the ions contained in the spectrum of the 1′,2-isomer, the spectrum of the 1′,3-isomer includes a prominent ion at *m/z* 439 corresponding to $[M-R'OCH_2]^+$. There is also an enhancement of the $[RCO + (2 \times 74)]^+$ ion at *m/z* 413. In the case of the *t*-BDMS of a dialkylglycerol (229, 241) the abundance of the $[M-57]^+$ ion is very low. The alkyl chains however are still indicated by the presence of $[R']^+$, $[R'O + 40]^+$, $[R'O + 114]^+$ and $[R'O + 132]^+$ ions. In a detailed study, including both diacyl- and alk-1-enylacylglycerol classes, Kino et al. (246) have compared the molecular species of both choline and ethanolamine phosphatides from rat brain myelin and synaptosomes. In order to minimize the elution temperatures and times during GC-MS short (2 m) capillary columns can be used with hydrogen as carrier gas (247). The hydrogen CI spectra consisted of ions such as $[M-57]^+$, $[R''O + 40]^+$ and $[M-131]^+$ which are also present in the EI spectra. The $[M-R''O]^+$ ion is base peak under hydrogen CI conditions for all the species studied including 18:1′ 22:6 and 18:2″ 22:6.

FD-MS of either the TMS or *t*-BDMS ethers of alk-1-enylacylglycerols is characterized by intense M^+ ions with little fragmentation except for $[M-15]^+$ and $[M-57]^+$ ions (248). The spectrum of the *t*-BDMS ether also contains a small amount of the $[M-R''O]^+$ ion.

Chloropropanediol, dichloropropanol and their fatty acid esters have been identified in a number of food related products (249–251). Esters of 1,3-dichloro-2-propanol and diesters of 3-chloro-1,2-propanediol have been identified in samples of total protein hydrolysates but were apparently absent from the final commercial product that is used to enhance the flavor of various foods (249). Adulterated rapeseed oil has been suspected of causing 20000 illnesses and 300 deaths in Spain since 1981. Although aniline contamination is thought to be responsible for the toxic effects, samples also contain fatty acid monoesters and diesters of chloropropanediol (250). Diesters of 3-chloro-1,2-propanediol have also been identified in goat milk (251).

The EI-MS of 3-chloro-1,2-propanediol diesters shows diagnostic fragments corresponding to $[M-RCOO]^+$, $[RCO]^+$ and $[RCO-1]^+$ (249). $[RCO]^+$ fragments predominate for saturated fatty acids whereas $[RCO-1]^+$ is more abundant for unsaturated fatty acids. High resolution EI-MS was used to confirm the presence of chlorine in the $[M-RCOO]^+$ fragments in the diester fraction from the adulterated Spanish cooking oil (250). The methane CI spectrum for 3-chloro-1,2-propane diol contains ions corresponding to $[MH]^+$, $[MH-H_2O]^+$ and $[MH-HCl]^+$ (251). Figure 3 shows the spectrum of one peak in the LC-MS profile of the 1,2-diacyl-3-chloro propanediols from goat's milk fat (252). The spectrum shows a prominent $[MH]^+$ ion at *m/z* 587 as well as $[MH-RCOOH]^+$ ions at *m/z* 303, 331 and 359.

3.6.3 **Triradylglycerols**

Triacylglycerols from natural sources are extremely complex mixtures. Since the number of possible molecular species is equal to the number of fatty acids cubed, a mixture composed of only ten different fatty acids would have a thousand possible components including positional and stereochemical isomers. The problem is therefore

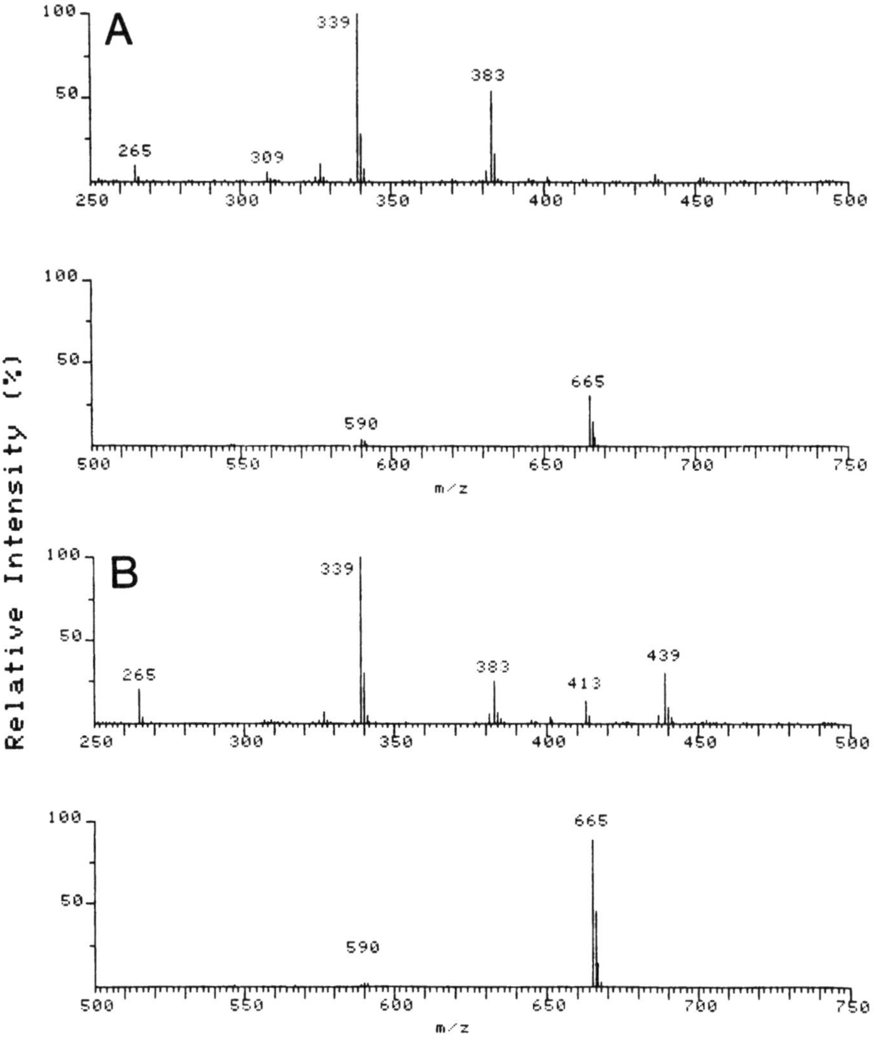

Figure 2. Mass spectra of *t*-BDMS ethers of alkylacylglycerols: (A) 1-stearyl-2-oleoyl-3-*t*-BDMS-*sn*-glycerol; (B) 1-stearyl-2-*t*-BDMS-3-oleoyl-*sn*-glycerol. GC-MS conditions: Hewlett-Packard 5985B quadrupole mass spectrometer-gas chromatograph system; electron impact ionization (70 eV) (241).

extremely difficult and analyses of intact triacylglycerols generally yield partial answers. If more detail is required it is necessary to generate and resolve representative *sn*-1,2-, *sn*-2,3- and *sn*-1,3-diacylglycerols (235). This latter approach has been discussed in the section dealing with diacylglycerols.

Direct probe insertion CI-MS (253, 254) is a rapid method of determining a molecular mass profile of a triacylglycerol mixture. Such a molecular weight profile can also be determined by FD-MS and has been applied to the study of plasma triacylglycerols (255). In this case a sample of the total lipid extract was applied to

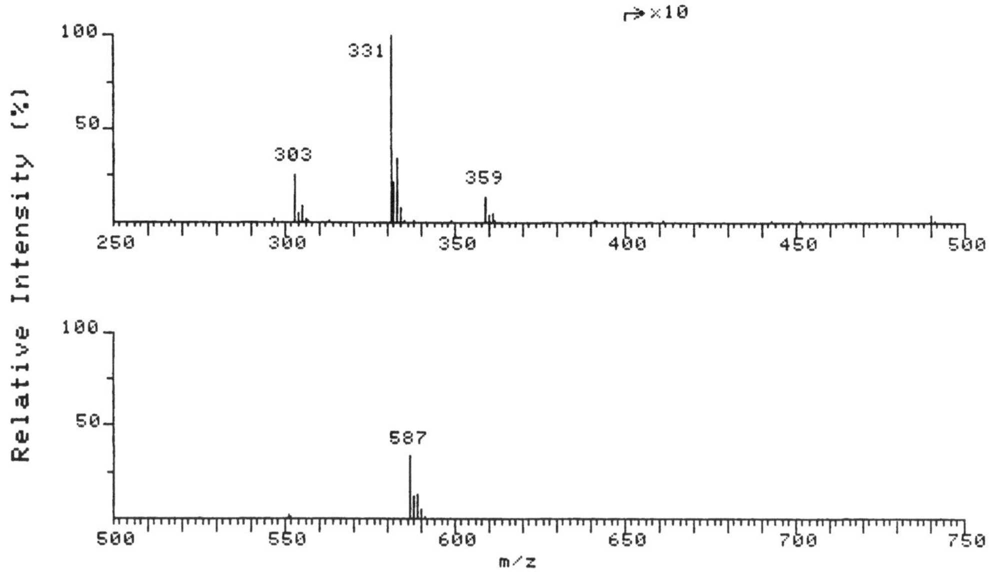

Figure 3. Mass spectrum of one HPLC peak in the LC-MS profile of 1,2-diacyl-3-chloropropanediols from goat milk fat. Principal components: 16:0 16:0 and 14:0 18:0 species. LC-MS conditions: Hewlett-Packard 5985B quadrupole mass spectrometer-liquid chromatograph system with direct liquid inlet interface: acetonitrile/propionitrile chemical ionization (70 eV) (252).

the activated carbon emitters. Triacylglycerol molecular ions were observed at heating currents between 18 and 24 mA. With the addition of a known amount of ($^{13}C_3$)trioleoylglycerol as internal standard it was also possible to determine the concentration of triacylglycerols in the plasma. In a comparison of the plasma triacylglycerols from two subjects receiving animal fat in one case and polyunsaturated fat in the other, it was possible to demonstrate that the polyunsaturated diet led to plasma triacylglycerols having a higher proportion of unsaturated species.

The combination of GC with ammonia CI is a powerful technique (50). The spectra of triacylglycerols obtained using ammonia as reagent gas include abundant ions corresponding to $[M + NH_4]^+$ from which the molecular mass can be obtained. In contrast, EI and CI using other reagent gases such as methane and isobutane generate only ions of low abundance in the high mass region. The ammonia CI spectra also include important diagnostic ions corresponding to $[MH - RCOOH]^+$ from which fatty acid composition can be determined. Since the ion intensities are sensitive to the structure of the triacylglycerols, especially the degree of unsaturation, accurate quantitation requires calibration. The direct probe insertion CI technique leads only to pseudo molecular ions having intensities that are not sensitive to unsaturation although there is a fall off in response as the molecular weight increases.

LC-MS is an even more powerful technique (51, 52). Unlike GC, which brings about resolution according to total acyl carbon number HPLC on reversed phase (octadecyl) columns also brings about resolution according to unsaturation and, since more species are chromatographically resolved, the MS data can be used to even greater effect. Because a portion (1%) of the total HPLC effluent goes directly to the

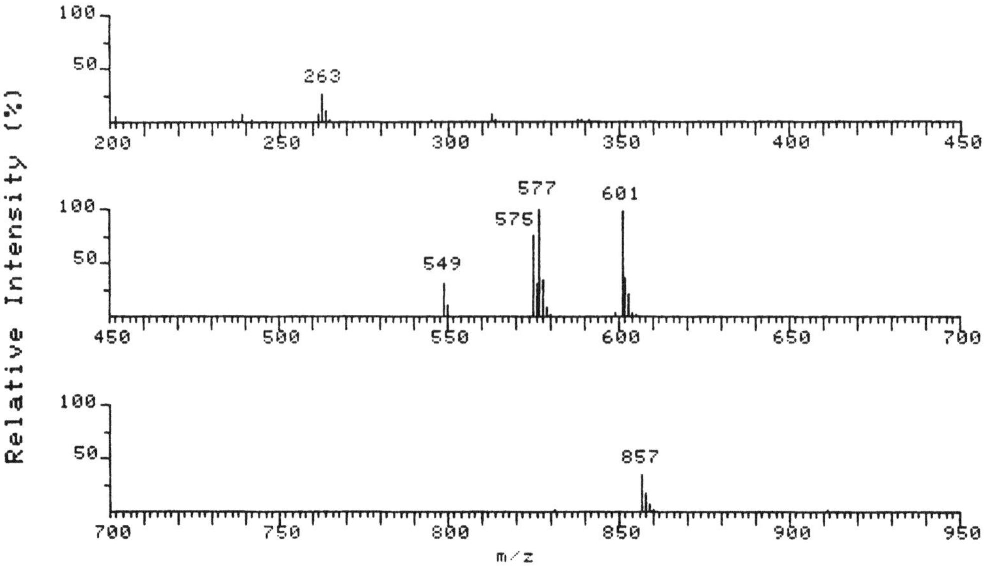

Figure 4. Mass spectrum of one HPLC peak (Peak 5, ref. 231) in the LC-MS profile of human plasma triacylglycerols. Principal components: 16:0 18:1 18:2 and 16:0 16:1 18:1 species. LC-MS conditions as in Figure 3.

ion source, it is necessary to operate in the CI mode using the HPLC elution solvent as reagent gas. When acetonitrile plus propionitrile are used as the mobile phase two prominent ion types are produced in the ion source: the protonated molecular ion $[MH]^+$, and a set of "diacylglycerol" ions, $[MH-RCOOH]^+$ resulting from the loss of a fatty acid from each of the 3 positions on the glycerol backbone. Thus the spectrum of a triacylglycerol made up of 3 different fatty acids would have 3 different $[MH-RCOOH]^+$ ions whereas the spectra of triacylglycerols made up of 1 or 2 fatty acids would have 1 and 2 such ions respectively. The mass spectrum of one HPLC peak in the LC-MS profile of human plasma triacylglycerols is shown in Figure 4 (51, 231). The abundance ratio of the $[MH]^+$ ions relative to the sum of the $[MH-RCOOOH]^+$ ions is very sensitive to unsaturation varying from 0.87 for trilinoleoylglycerol to an undetectable level for tripalmitoylglycerol. The relative abundances of the $[MH-RCOOH]^+$ ions vary over a much smaller range, being dependent on the positional distribution of the fatty acids and on the total unsaturation. Thus for positionally randomized palmitoyloleoyllinoleoylglycerol, ion fragments resulting from the loss of palmitic acids, oleic acid and linoleic acids have abundances in the proportion of 1.0, 0.84 and 0.88 respectively. Ions resulting from the loss of fatty acid from either the *sn*-1- or *sn*-3-position have abundances that are about four times higher than those resulting from the loss of a fatty acid from the *sn*-2-position. Therefore positional effects dominate the abundances of the $[MH-RCOOH]^+$ ions.

The triacylglycerols from a number of dietary oils including corn oil, soybean oil and peanut oil have been analyzed by LC-MS and approximately 35 molecular species have been identified in peanut oil. Quantitation was also possible after calibration with triacylglycerol mixtures of known composition (52).

Instead of chromatographic techniques it is possible to use the so called MS/MS group of procedures. One such method requires a high resolution MS having reversed geometry; the identification of ions separated in the first stage magnetic sector is carried out using metastable ions analyzed by the direct analysis of daughter ion (DADI) technique. This method was applied to the triacylglycerols isolated from the bacterial strains *Mycobacterium paraffinicum, Actinomyces aureoverticilus* 1306 and *Actinomyces streptomycini* (255). In order to illustrate the procedures involved, consider one species in the triacylglycerol fraction from *M. paraffinicum*. The $[M-RCOO]^+$ fragment at *m/z* 523 in the EIMS was found to break down further into $[RCO]^+$ ions at *m/z* 211 and 239 corresponding to myristic (14:0) and palmitic (16:0) acids respectively. The ion at *m/z* 523 was also shown to arise from the molecular ion at *m/z* 776 through loss of the acyloxy radical from palmitoleic acid (16:1). These fragments then established the presence of a triacylglycerol species containing 14:0, 16:0 and 16:1. In addition the presence of an $[M-RCOOCH_2]^+$ ion at *m/z* 509 made up of fatty acids 14:0 and 16:0 suggests that 16:1 must be located at one of the primary positions of the glycerol backbone. Therefore, distinct species were identified in the complex mixture but quantitative results were not reported for these triacylglycerols. Since this report has not been comprehensive the reader may wish to consult a number of reviews which cover the earlier work in this area (256–258).

The spectra of 1,3-diacyl-2-*O*-alkylglycerols, 1,3-di-*O*-alkyl-2-acylglycerols and 1,2,3 tri-*O*-alkylglycerols have also been reported (203).

3.7 Ceramides

Free ceramides normally occur in plasma and tissues in small amounts and differ in their composition from those of the ceramide moieties of sphingomyelins and other sphingolipids (15). In certain pathological conditions, however, the free ceramide content may increase significantly and there may be alterations in the composition of the component molecular species. Since the ceramides are made up of two chemical entities, which may vary independently, they are notably difficult to analyze by chromatographic techniques alone and MS, GC-MS and LC-MS are required for their accurate identification and quantitation.

The ceramides represent a combination of a long chain base (LCB) and a fatty acid (FA) and must be studied intact in order to determine the pairing of these components. The first detailed GC-MS analyses of ceramides containing normal fatty acids (20, 259, 260) or 2-hydroxy fatty acids (260, 261) were made using the di-TMS ethers. Subsequently the experience gained from the study of synthetic standards allowed the characterization and quantitation of ceramides from natural sources (262–265). Thus, in human plasma the free ceramides (263, 264) were not quantitatively the same as those derived from sphingomyelin (262). The free ceramides had considerably less palmitic acid than did the ceramides from sphingomyelin. Ceramides from plasma and various tissues in a patient with lipogranulomatosis (Farber's disease) were also studied by GC-MS. The concentration was found to be seven times higher in the kidney compared to the control and the increase was due primarily to ceramides containing hydroxy fatty acids. There was no ceramide accumulation in plasma, spleen, lung, liver or brain. In the granuloma taken from the eyelid of the patient the ceramide content was twenty times that found in any normal tissue. The accumulated ceramides were mainly sphingosine combined with palmitic, behenic and nervonic acids and none containing hydroxy acids were detected. The fact that there are compositional differences between the two pools of accumulated ceramides indicates that they are formed locally.

The EI spectra of the di-*0*-TMS ethers of ceramides (20, 260) contain considerable structural information. Ions at $[M-15]^+$, $[M-90]^+$ and $[M-103]^+$ due to losses of a methyl group, trimethylsilanol and the terminal $CH_2OSi(CH_3)_3$ group respectively can be used to determine the molecular mass of the ceramides. A cleavage between carbons 2 and 3 of the sphingoid base leads to the formation of major ion fragments corresponding to $[CH_3(CH_2)_n(CHCH)_m(CHOSi(CH_3)_3]^+$ and $[RCONHCHCHCH_2CH_2OSi(CH_3)_3]^+$ which are indicative of the long chain base and fatty acyl (RCO) chain. For *N*-palmitoyl sphingosine these ions occur at *m/z* 311 and *m/z* 370 respectively. There are also fragments corresponding to $[M-RCONH_2]^+$ and $[RCONHCHCHOH]^+$.

In the case of ceramides containing 2-hydroxy fatty acids (260, 261) many of the fragments in mass spectra are analogous to those observed for normal fatty acids, except for the pattern of relative abundances. In the principle cleavage between the C−2 and C−3 carbons of the LCB there is a 75% chance that the charge is retained by the fragment containing the fatty acid. For ceramides with normal fatty acids both ion fragments occur with approximately equal probability. The features of the MS are described in much greater detail elsewhere (260).

Markey and Wenger (266) demonstrated the use of direct CI (methane as reagent gas) for the analysis of underivatized ceramides. The spectra of ceramides containing hydroxy fatty acids exhibited a weak MH^+ ion and a strong ion resulting from the loss of water, $[MH-18]^+$. Cleavage between carbons 2 and 3 of the sphingosine base resulted in a strong ion for the LCB-containing fragment and a weaker ion for the FA-containing fragment. The major FA-containing fragment resulted from a cleavage between the C−1 and C−2 carbons of the acyl chain. The spectra for permethylated and acetylated ceramides were also reported.

The structures of free ceramides in human erythrocytes and the identification of minor bases were confirmed by EI and ammonia CI (267). Much of the fragmentation pattern was very similar in CI compared to EI with some major differences in the high mass portion of the spectra. Most significant were the presence of ions approaching 100% of the base peak in intensity corresponding to $[M+1]^+$ and $[M-90+1]^+$. The di-*t*-BDMS ethers of ceramides are also good derivatives for GC-MS (15, 49). The greater stability of this derivative also allows prior separation by TLC if required.

The molecular species of sphingomyelins in the various normal human lipoprotein fraction (15, 49) and the effect of diet (15) has been studied by GC-MS of the *t*-BDMS ethers of the derived ceramides. It was demonstrated that the HDL fractions contain a significantly higher proportion of the longer chain sphingomyelins than do the VLDL and LDL fractions. It was also found that HDL_3 contained a significantly higher proportion of the long chain species compared to the HDL_2 fraction. In the case of hyperlipoproteinemia types II, III and IV the sphingomyelin composition in HDL was very similar to that in the normal subjects. In type II subjects the sphingomyelin composition of LDL was very similar to that in normals, in type IV subjects it was very similar to that in HDL and in type III subjects the results were variable.

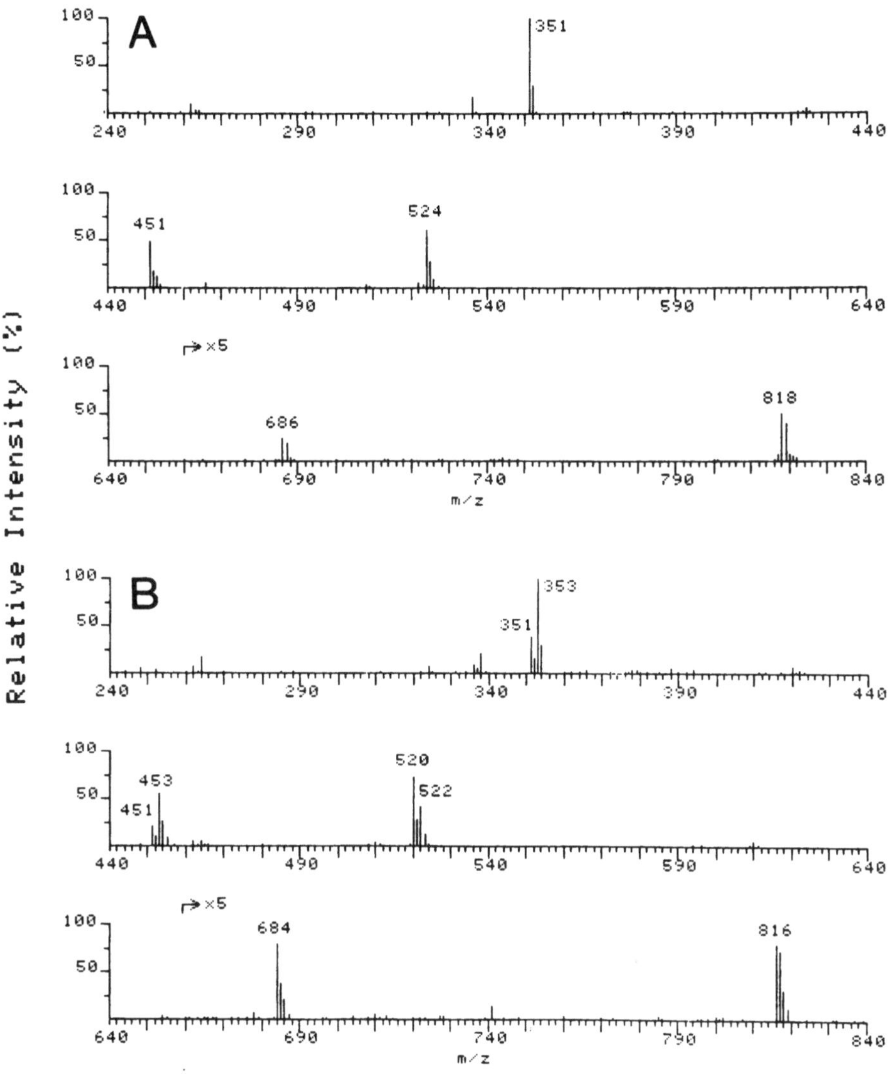

Figure 5. Mass spectra of the di-*t*-BDMS ethers of ceramides derived from the sphingomyelins of human low density lipoproteins: (A) d18:2 24:0; (B) d18:1 24:2 plus d18:2 24:1. GC-MS conditions: Hewlett-Packard 5985B quadrupole mass spectrometer-gas chromatograph system; electron impact ionization (70 eV) (15, 49).

The response to diet of the spingomyelin species was very small compared to that of the phosphatidylcholine species. However there were detectable increases in the relative amounts of the C_{42} species along with a decrease of the C_{41} species in all the lipoprotein classes in response to a polyunsaturated diet as compared to a saturated diet. Figure 5 shows the spectra of several species of ceramides present in the sphingomyelins of human plasma (15, 49).

The fragmentation pattern is very similar to that described for the TMS ethers. There are major fragments resulting from cleavage between C−2 and C−3 of the long chain base. Depending on which fragment retains the positive charge this cleavage gives rise to two major fragment ions of similar intensity having formulas $[CH_3(CH_2)_n(CHCH)_mCHO\text{-}t\text{-}BDMS]^+$ and $[CH_3(CH_2)_nCONHCHCH_2O\text{-}t\text{-}BDMS]$. The former fragment is indicative of the long chain base and occurs at *m/z* 453 for sphingosine whereas the latter contains the fatty acyl chain and occurs at *m/z* 412 for palmitic acid. Also indicative of the long chain base is an abundant ion, $[CH_3(CH_2)_n(CHCH)_mCHO(t\text{-}BDMS)CHCHOSi(CH_3)_2]^+$, found 100 daltons higher than the other long chain base fragment. This ion results from the losses of a *t*-butyl group and an acyl amide molecule. The molecular weight is derived from the mass of the $[M-57]^+$ ion and the relative abundance of this ion is higher than the $[M-15]^+$, $[M-90]^+$ or $[M-103]^+$ ions found in the EI-MS of the TMS ethers of ceramides. Relative specific ion response factors were determined by analyzing mixtures made up of ceramide fractions that had been resolved by argentation TLC. It was found that for the species within any single GC peak the ion responses were within 5% of each other. In the case of ceramides the molecular species are more uniquely determined by GC-MS than are those of diacylglycerols where there is positional ambiguity. Even though there may be as many as 9 molecular species in a given GC peak, the identification and quantitation of all these species can be accomplished by solving a set of simultaneous equations containing the relevant intensity data. In this way it has been possible to identify and quantitate 65 molecular species of sphingomyelins in human plasma lipoproteins (15).

It is also possible to study ceramides by GC-MS as the methaneboronate derivates (268) or by direct probe as the benzoyl and *p*-nitrobenzoyl derivatives (269).

4 Polar Lipids

Glycerophospholipids are the major building blocks of all biological membranes. It is well documented that membrane integrity and proper cellular function, including enzyme activation and the normal conductivity of electrical signals across the membrane, is dependent on the exact content and composition of the phospholipids. Plasmalogens may also act as membrane stabilizers and enzyme activators. Higher than normal values of plasmalogens have been associated with neoplastic tissues (208). Brains of multiple sclerosis patients show lower levels of ethanolamine plasmalogens. In addition, the free or acetylated lyso (alkyl)glycerophospholipids have been recognized to possess physiological properties generally not associated with the bulk properties of membrane components. Like the glycerophospholipids, the sphingomyelins and glycolipids are integral membrane components and their composition is believed to affect the properties of cell membranes and the antigenic receptors located there. MS and LC-MS provide effective means for the analyses of these lipids in the intact form.

The molecular species of phospholipids are usually analyzed after appropriate hydrolytic and derivation procedures. As described earlier diradylglycerols or ceramides, obtained from phospholipids by hydrolysis with phospholipase C, can be converted into TMS, *t*-BDMS or acetyl derivatives. The characteristic bases of the glycerophospholipids have been identified by GC-MS after the phospholipid is partially hydrolyzed and converted into TMS derivatives (27, 270–272). The long chain sphingoid base of sphingolipids can be analyzed after appropriate hydrolysis and derivatization (24, 273–278). Intact phospholipids are not compatible with GC and must be introduced into a MS via direct probe or HPLC. When identification is critical it is desirable to analyze the intact molecule in order to avoid any artifact production by chemical degradation reactions. Since the original work of Klein (279–281) on the EI-MS of intact phospholipids much progress has been made in the development of soft ionization techniques (282) and the linking of HPLC with a MS through direct liquid injection (DLI) interface (283).

4.1 Glycerophospholipids

The EI spectra of glycerophospholipids obtained by direct injection (DEI) have been reported by Klein (279–281). The principal high mass ions $[M-OP(O_2)OX]^+$ result from the loss of the phosphorylated base. Batrakov et al. (256) have used an MS-MS (DADI) procedure for determining the molecular species of phospholipids. By this method it is possible to determine the fatty acid composition corresponding to each $[M-OP(O)_2)X]^+$ ion.

Desorption chemical ionization (DCI) studies of intact phospholipids have been carried out in a number of laboratories in the last few years (283–291). The samples are applied on inert probe tips which can be inserted directly into the ion source. The samples can be volatilized within 10 s by rapidly heating the probe. Sugnaux and Djerassi (283) have compared the spectra of dimyristoylglycerophosphorylcholine obtained by DEI (ion beam, 70 eV), methane DCI, ammonia DCI and on-line LC-MS using a DLI interface. As expected from previous results (279–281), the EI-MS gave a base peak, $[M-183]^+$, at *m/z* 494 due to loss of the phosphorylcholine group. The molecular ion occurred with an abundance 0.6% of the base peak. Whereas methane CI also leads to $[M+H]^+$ ions of low abundance the most intense higher mass pseudomolecular ion $[MH-183]^+$ is missing the phosphorylcholine group. Ammonia CI, however, leads to a very intense ion corresponding to $[MH-42]^+$. When $^{15}NH_3$ or ND_3 is used as reagent gas the *m/z* value of this fragment is increased by one (285) or 3 (286). This direct evidence and that cited in references (283, 286) support the suggestion that the $[MH-42]^+$ ion is the result of replacement of the $(CH_3)_3N^+$ groups of choline by NH_4^+ from the reagent gas. Since a minor ion corresponding to $[MH-42]^+$ occurs in the methane CI spectra of phosphatidylcholine, Jungalwala et al. (287) dispute this substitution mechanism. However, it is possible that the fragmentation mechanism responsible for the $[MH-42]^+$ during methane CI contributes very little to the $[MH-42]^+$ abundance in the ammonia CI spectra. The fragmentation under ammonia DCI conditions has been studied in detail for a number of molecular species of phosphatidylcholine and a number of model synthetic phosphatidylethanolamine, *N*-methylphosphatidylethanolamine, *N,N*-dimethylphosphatidylethanolamine, phosphatidylserine and phosphatidylglycerol (285, 286). All the phospholipids excepting phosphatidylserine exhibited $[M+H]^+$ ions. Sugnaux and

Djerassi (283) have also demonstrated a time dependence for the relative abundance of the protonated molecular ion. Apparently desorption of intact neutral molecules takes place for a short period followed immediately by desorption of decomposition products formed on the surface. These authors favor the more reproducible conditions of on-line DLI-LC-MS. Using ethanol as solvent and ammonia as make-up reagent gas they obtained a spectrum that had characteristics of both methane and ammonia DCI. The $[MH-183]^+$ ion was the most abundant high mass ion followed by $[MH-42]^+$ and $[M + H]^+$. Bisseret et al. (288) have used a gold wire positioned about 2−4 mm from the electron beam instead of the usual ceramic or silica sample holder. They obtained ammonia DCI spectra for a number of phosphatidylcholines including a pair of reverse isomers, 1-palmitoyl-2-oleoyl-*sn*-glycerol-3-phosphorylcholine and 1-oleoyl-2-palmitoyl-*sn*-glycerol-3-phosphorylcholine. The spectra were very similar for the two isomers except for an inversion of intensities for the $[RCO + 128]^+$ type ions at *m/z* 367 and 393. This feature in which the ion corresponding to the acyl group from the *sn*-1-position of phosphatidylcholine is more abundant than that from the *sn*-2-position has also been reported for the MS of diacylglycerols (14). This would not be applicable to a complex mixture. Ammonia DCI of phosphatidylsulfocholines yielded $[M + NH_4]^+$ instead of the $[M + H]^+$ observed for phosphatidylcholine (289). Other fragments such as $[MH-42]^+$ and $[MH-183]^+$ were the same for both types of compounds. Dessort et al. (290) have shown the feasibility of using DCI for studying several natural mixtures of phosphatidylcholine and phosphatidylethanolamine. Whereas ammonia DCI was found to be preferable for phosphatidiylcholines, methane DCI was the preferred reagent for the analysis of phosphatidylethanolamines. Crawford and Plattner (291) have reported a detailed method for quantitating the molecular species of phospholipids. With ammonia DCI they found that the specific response of the $[M-183]^+$ ion declined with increasing molecular weight. This fact dictated the use of a complex external standard mixture consisting of all the disaturated phosphatidylcholines containing the even and odd carbon fatty acids from pentadecanoic (15:0) to docosanoic (22:0) acids. However when ammonia was used there were some interferences due to overlapping ions which made precise quantitation difficult. The interference was eliminated by switching to isobutane as reagent gas. Although the $[MH]^+$ ions have reduced intensity there are no ions between the $[MH]^+$ ion and the $[MH-183]^+$ ions to cause interference between different molecular species. The peak area of the $[MH-183]^+$ ion for any species relative to the area of the $[MH-183]^+$ ion for dipentadecanoylglycerophosphorylcholine (internal standard) decreases exponentially with molecular mass. However, with an external standard for calibrating the ion response for different molecular weights and an internal standard it is possible to quantitate the molecular species in a complex mixture. The authors have also solved another major problem associated with direct exposure techniques, which arises when two different molecular species have the same molecular weight. Even though there are fatty acid containing fragments it is not normally possible to associated these with a particular molecular species. The solution reported by Crawford and Plattner (291) involved MS-MS whereby a particular $[MH-183]^+$ ion selected by the first mass filter was subjected to collision-induced dissociation (CID) and the acyl fragments separated in the second analyser. In this way the fatty acid composition of any particular $[MH-183]^+$ ion can be determined. Using these techniques the molecular species of the phosphatidylcholines from egg and soybean and the phosphatidylethanolamines from soybean were determined.

Jungalwala et al. (287) analyzed phospholipids by means of LC-MS with a moving-belt interface. Positive and negative ion mass spectra were obtained in the CI mode with methane or ammonia as the reagent gas. Extensive information on the molecular species of phospholipids could be obtained and the phospholipids clearly identified using specific ions for the phospholipid bases. Kim and Salem (292) have recently reported the analysis of natural mixtures of phosphatidylcholines and phosphatidylethanolamines by thermospray LC-MS. Since phospholipids were resolved on line by a C_{18} reversed phase HPLC column using a highly organic solvent mixture of hexane/methanol/0.1 $mol \cdot l^{-1}$ ammonium acetate, it was necessary to operate the MS ion source in the filament-on mode. The spectra were similar to those obtained using ammonia CI showing $[MH]^+$ ions, "diglyceride" type ions and $[RCO + 74]^+$ ions.

Field desorption (FD) was the first of the soft ionization techniques to be applied to the analysis of intact phospholipids. In a series of papers Wood and coworkers (293–295) outlined the conditions necessary to produce useable spectra and characterized the major fragmentation pathways. One of the principal problems with this technique has been the presence of ionic contaminants, such as sodium chloride which may lead to low ion intensities, increased fragmentation and increased ion cluster formation. Ions corresponding to $[M + H]^+$, $[M + CH_3]^+$, $[M + Na]^+$, $[M + \text{choline}]^+$ and $[M + \text{choline} - \text{water}]^+$ are the main high mass ions found in the FD spectra of phosphatidylcholines. Wood and Perkins (296) reported an adaptation of the Bligh-Dyer extraction technique which removes metal ions from samples of phospholipids and leads to a great improvement in the quality of their FDMS. The spectra of treated phospholipids generally exhibited $[M + H]^+$ ions as base peaks. Even phosphatidic acid, which previously proved troublesome, behaved well after ion removal.

Lehman and Kessler (297) chose to add choline chloride to samples in order to promote the formation of $[M + \text{choline}]^+$ ions at m/z (M + 104) as base peaks. They demonstrated the possibility of obtaining molecular weight profiles of the phosphatidylcholines from human plasma after placing crude lipid extracts on the FD probe. Lehman and Kessler (255) used an acylcholine ion (m/z 342 for palmitate) in the FDMS of phosphatidylcholine to determine a fatty acid profile of the phosphatidylcholines in human plasma. Dialkyl- (298) and alkenylacyl- (248) glycerophosphorylcholines have also been investigated by FD-MS.

It has recently been shown that FAB-MS can also be used for the analysis of intact phospholipid (255, 298–300). Fenwick et al. (298) studied various diacyl homologs of phosphatidylcholines, lysophosphatidylcholines and phosphatidylethanolamines, and phosphatidyl-*N*-methylethanolamines, phosphatidyl-*N*,*N*-dimethylethanolamines, phosphatidylserines and phosphatidylglycerol. The quality of the spectra for the phosphatidylcholines was very good but decreased in the order phosphatidylcholines > phosphatidyl-*N*,*N*-dimethylethanolamines, phosphatidyl-*N*-methylethanolamine > phosphatidylethanolamine ≫ phosphatidylglycerol. The addition of HCl to the samples improved the spectra of the non-choline phosphatides. The spectra of the phosphatidylcholines exhibited peaks corresponding to $[MH]]^+$, $[M - H]^+$, $[2M + H]^+$, $[MH + H - RCO]^+$, $[MH + H - RCOO]^+$ and a strong peak for choline phosphate ion at m/z 184.

The analysis of molecular species was made more complicated by the fact that the $[M-H]^+/[M+H]^+$ abundance ratio appeared to increase as the acyl chains were made more unsaturated. The FAB spectra appeared to be stable over a considerable time period but the background tended to be higher than for FD-MS. Traitler and Nikiforov (301) compared FD and positive FAB techniques in the analysis of complex mixtures of phosphatidylcholines and concluded that FD was not suitable for quantitative work due to poor reproducibility of the spectra. However, FAB spectra of phosphatidylcholines in a 1,2,4-butanetriol/85% phosphoric acid (50:50, v/v) matrix were stable over longer periods of time and showed strong $[MH]^+$ ion intensities. Secondary ion mass spectrometry (SIMS) is similar to FAB except that an incident beam of ions replaces the beam of atoms. Aberth et al. (300) compared the positive ion FAB-MS and SIMS spectra of dipalmitoylglycerophosphorylcholine supported in liquid matrix. They showed that the two techniques generate identical spectra provided that the bombarding neutral and charged particles, such as Xe versus Cs^+, have similar kinetic energy and mass. Ohashi (302) using matrix assisted SIMS has shown some differences in spectra of phosphatidylcholine reverse isomers such as *sn*-1-palmitoyl-2-linoleoyl- and *sn*-1-linoleoyl-2-palmitoyl-glycerophosphorylcholines. Several ion fragments resulting from loss of acyl or acyloxy groups from the *sn*-2-position had higher abundances than those resulting from the loss at the *sn*-1-position. Recently the analyses of a number of phospholipids by negative FAB and SIMS have been explored (303–306). Good $[M-H]^+$ intensities were observed for all phospholipids including phosphatidylserine and with proper sample preparation the spectra are stable for extended periods of time. MS-MS techniques have also been used along with negative FAB to determine isobaric molecular species (303, 306).

The prediction of fetal lung maturation is made on the basis of the phosphatidylcholine/sphingomyelin (L/S) ratio in amniotic fluid. However, it has been shown (307) that the concentration of dipalmitoylglycerophosphorylcholine (DPPC), which is a major constituent of lung surfactant, is a better indicator of respiratory distress syndrome. Ho et al. (308) have applied positive FAB to the measurement of DPPC in amniotic fluid using d_9-DPPC as internal standard. Using the intensities of the corresponding $[MH]^+$ ions they were able to measure DPPC with a CV of about one percent. If the total DPPC ($d_0 + d_9$) was less than 0.2 μg or the ratio of d_0- to d_9-DPPC exceeded 100, the precision deteriorated.

The cerebro-hepato-renal syndrome (Zellweger syndrome) is a lethal autosomal recessive disorder characterized by the absence of peroxisomes in liver and kidney (309). It has been shown that there is a severe deficiency of acyl-CoA dihydroxyacetone phosphate acyltransferase in Zellweger patients which results in an almost complete absence of plasmalogenic phospholipids (310). Although much simpler methods (311) have been developed, it is possible to measure plasmalogen content in phospholipids by means of positive FAB-MS (312). Plasmalogenic phosphatidylcholines of canine myocardial sarcolemma could be detected by the presence of $[MH]^+$ ion at *m/z* 742, 744, 766, 770, 772, 792 and 794 as well as by the presence of lysoplasmalogen type fragments resulting from the loss of long chain ketenes from position 2. Thus an ion at *m/z* 480 is representative of species containing a 16:0 alkenyl group, whereas ions at *m/z* 508 and 506 indicate the presence of 18:0 and 18:1 alkenyl chains respectively.

Both platelet-activating factor (PAF), which causes platelets to change shape, aggregate and release their granule contents (313, 314), and antihypertensive polar

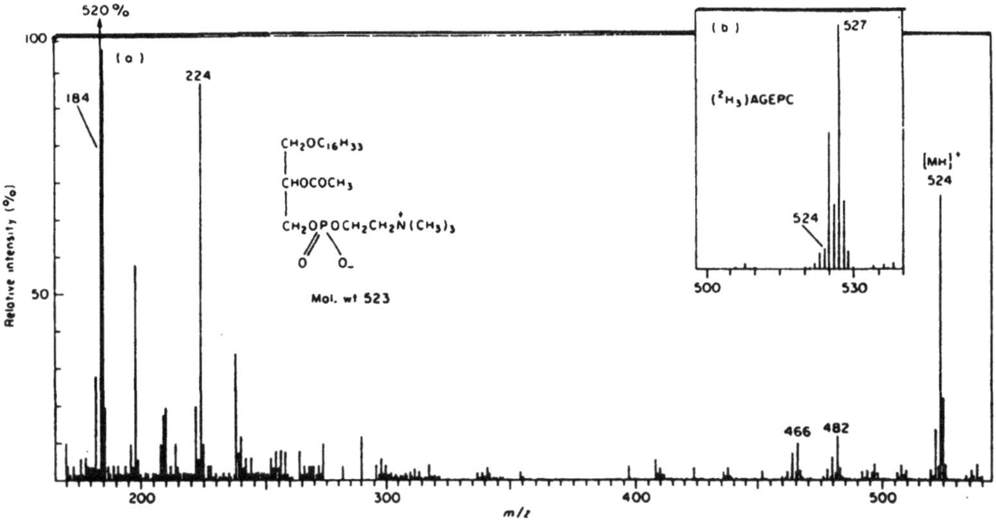

Figure 6. FAB mass spectrum of platelet activating factor (a) and the molecular ion region of [$^2H_3$]platelet activating factor (b). Conditions: VG Micromass 7070H mass spectrometer in FAB mode. Glycerol ions were not subtracted but suppressed by the amount of platelet activating factor on the target (323).

renomedullary lipid (APRL) from kidney, which lowers the blood pressure of artificially hypertensive animals (315), have the same structure, namely, 1-*O*-alkyl-2-acetyl-*sn*-glycerol-3-phosphorylcholine (316). Using methods including CIMS it has been shown that 1-*O*-alkyl-*sn*-glycerol-3-phosphorylcholine (lyso PAF) is released simultaneously with PAF from hog leukocytes (317). The isobutane CI spectrum of the natural hexadecyl homologue contained ions corresponding to MH^+ at *m/z* 482, $[MH-CH_3OH]^+$ at *m/z* 450, $[M-PO_3CH_2CH_2N(CH_3)_3+2]^+$ at *m/z* 317 and others. PAF has been measured in human neutrophils (318) and in a number of rat tissues (319, 320) by GC-MS. PAF, isolated by TLC, is hydrolyzed by phospholipase C, converted to the *t*-BDMS ethers and analyzed by SIM GC-MS. Quantitation is achieved by comparing the appropriate ion signals obtained with those arising from a deuterated internal standard. It is possible also to monitor PAF by GC-SI-MS following propionolysis (321) or by GC-MS-NICI-MS following hydrolysis with phospholipase C and derivatization with pentafluorobenzoyl chloride (322).

A method for quantitating PAF using $^2H_3$-labeled PAF as internal standard and FAB-MS has also been reported (323). When 60 ng of the deuterated PAF was used a standard plot for the intensity ratio of the $[MH]^+$ ions for the natural and the deuterated PAF against the quantity of PAF was linear from 10 to 200 ng. Quantitation was not possible without an internal standard due to the signal saturation effect of FAB. The FAB spectrum of the platelet activating factor and of $^2H_3$-labeled platelet activating factor is shown in Figure 6 (323).

By means of chromatographic techniques combined with both MS and biological assays, Clay et al. (324) have demonstrated that PAF in activated human neutrophils consists primarily of 1-*0*-hexadecyl-2-acetyl-*sn*-glycerol-3-phosphocholine. Other active molecular species were absent. Lyso PAF can be assayed by FAB-MS but analysis by

GC-MS following hydrolysis and derivatization is a more sensitive method (325). A recent study of PAF and related phospholipids by CI (isobutane) and FAB has also been reported (326). Finally, a vasoactive lipid produced by the incubation of plasma has been identified as lysophosphatidic acid by GC-MS (327). Tokumura et al. (328, 329) found that it was possible to analyze the TMS derivatives of lysophosphatidic acid and its analogues directly by GC-MS without any prior hydrolytic pretreatment.

4.2 Sphingomyelins

The MS analysis of intact spingomyelin is very similar to that of phosphatidylcholine. Batrakov et al. (256) have shown that the major high mass ion in the DEIMS of sphingomyelins, $[M-184]^+$, is due to the loss of $HO_2P(O)OCH_2CH_2NCH_3)_3$. They were also able to determine the fatty acid composition corresponding to each $[M-184]^+$ species by means of the direct analysis of daughter ions (DADI) technique. Jungalwala et al. (287) reported that the ammonia CI spectra of sphingomyelins exhibit abundant $[M-182]^+$ ions and less intense $[M+18-182]^+$ ions. It has been demonstrated by Lehman and Kessler (255, 297) that the FD spectra of bovine brain sphingomyelins in the presence of choline chloride exhibit a series of $[M+104]^+$ ions resulting from the addition of choline to each sphingomyelin species. They found that it was necessary to isolate sphingomyelin prior to MS analysis. The FAB spectra of sphingomyelins exhibit $[MH]^+$ ions (255).

4.3 Glycolipids

Glycolipids are an important class of compounds having antigenic and receptor properties (330) and interest has been stimulated by the observation that the glycosphingolipid content is dramatically altered in malignant cells (331). Sphingolipidoses, a class of inherited disorders, results in the pathological accumulation of certain sphingolipids (see reference 332 for a review). The accumulation of glycolipids in various tissues and organs occurs in Gaucher's disease, Krabbe's disease (Globoid leukodystrophy), metachromatic leukodystrophy, ceramide lactoside lipidosis, Fabry's disease, Tay-Sachs disease, Tay-Sachs variant, generalized gangliosidosis and fucosidosis.

One class of glycolipid is glycerol based consisting of one or more sugar residues bound to a diradylglycerol. This class is generally a minor component in most mammalian tissues except for testes. In adult human testes 1-*O*-alkyl-2-*O*-acyl-3-[β-3'-sulfogalactosyl]-glycerol and its non sulfated species are principal glycolipids (333, 334). The other class of glycolipids consists of molecules with one or more sugar residues bound to a ceramide.

Although a number of derivatization procedures have been used (335), permethylated (336–338) and permethylated and reduced (338, 339) glycolipids are now the favored derivatives for MS. All hydroxyls and amide nitrogens are first methylated and then reduction with $LiAlH_4$ converts all amide groups into the corresponding amines. Methyl ethers neither produce many rearrangement ions upon electron impact nor increase the molecular mass as much as other derivatives. Glycolipids with up to 12 sugar residues require MS with a 3000 dalton mass range (29) and the spectra of permethylated and reduced-permethylated samples give information on the size, sugar sequence and ceramide species (29, 340). Reduced samples contain more high mass information.

The MS of the permethylated or reduced-permethylated glycosphingolipid is relatively straightforward (355, 340–343). The number of sugars and fatty acids is obtained from the F fragments that result from cleavage between the C–2 and C–3 carbons of the long chain base of the reduced-permethylated derivative. The carbohydrate sequence is derived from a series of S_n fragments that result from cleavage of each of the glycosidic linkages. Thus, starting from the terminal sugar residue there are fragments containing 1,2,...*n* sugar residues. Branch points will lead to gaps in the spectrum equivalent to 2 or more sugar residues and each sugar can be identified by its molecular mass. The presence of hexosamines can ben confirmed by noting a loss of 14 daltons following reduction. There are also daughter ions, $[S_n - 32]^+$, due to the loss of methanol from each of the S_n fragments. The alcohol formed from reduction of the methyl ester of sialic acid residues should be further derivatized for optimum fragmentation (335). Therefore, it is preferable to prepare the TMS ethers of the gangliosides after reduction with $LiAlH_4$.

A large number of glycosphingolipids have now been characterized by combinations of chemical, MS and NMR techniques. The latter technique can be used to determine the anomeric structure (344–346). Blood group structures can be analyzed for single human individuals (347). The mass spectrum of a permethylated-reduced glycosphingolipid mixture from human plasma is shown in Figure 7 (347).

Blood group activities have long been associated with lipid extracts of erythrocytes but specific identification of glycolipid antigens came much later (for reviews see ref. 348, 349). The findings of Hakomori and Strycharz (350) and of Marcus and Cass (351) have shown that the Lewis antigens in plasma and erythrocytes are glycosphingolipids. The antigens are apparently acquired from plasma and are not synthesized *in situ*. Many glycolipids with blood group activity have now been characterized with the aid of MS. The glycolipid antigens are fucolipids based on two types of lactosyl ceramides (349); type I, a *lacto-N*-tetraosylceramide, Gal(β1 → 3)GlcNAc(β1 → 3)Gal(β1 → 4)Glc(β1 → 1)cer and type II, a *lacto-N*-neotetrosylceramide, Gal(β1 → 4)GlcNAc(β1 → 3)Gal(β1 → 4)Glc(β1 → 1)cer. The ABO and Lewis blood group antigens are derived from these precursors by addition of fucose, galactose and *N*-acetylgalactosamine residues. Blood type A glycolipids have from 6 to 12 sugar residues and blood type B glycolipids from 6 to 10 sugar residues.

Approximately 70% of the total glycolipids of human erythrocytes are derivatives of globoside (GalNAc(β1 → 3)Gal(α1 → 4)Gal(β1 →4)Glc(β1→1′)cer). Globoside, the major glycolipid component, represents the blood group P antigen, and is absent in rare individuals with blood group p phenotype. The major ganglioside of human erythrocyte has *N*-acetylneuraminic acid attached to the terminal β-galactosyl residue of the type II lactosylceramide.

Prompted by the known serological differences between human cord blood and human adult blood, Larson and Samuelsson (352) have analyzed the blood group type glycosphingolipids of human cord blood erythrocytes. There were no detectable glycolipids with carbohydrate chains longer than 8 sugar residues even though for

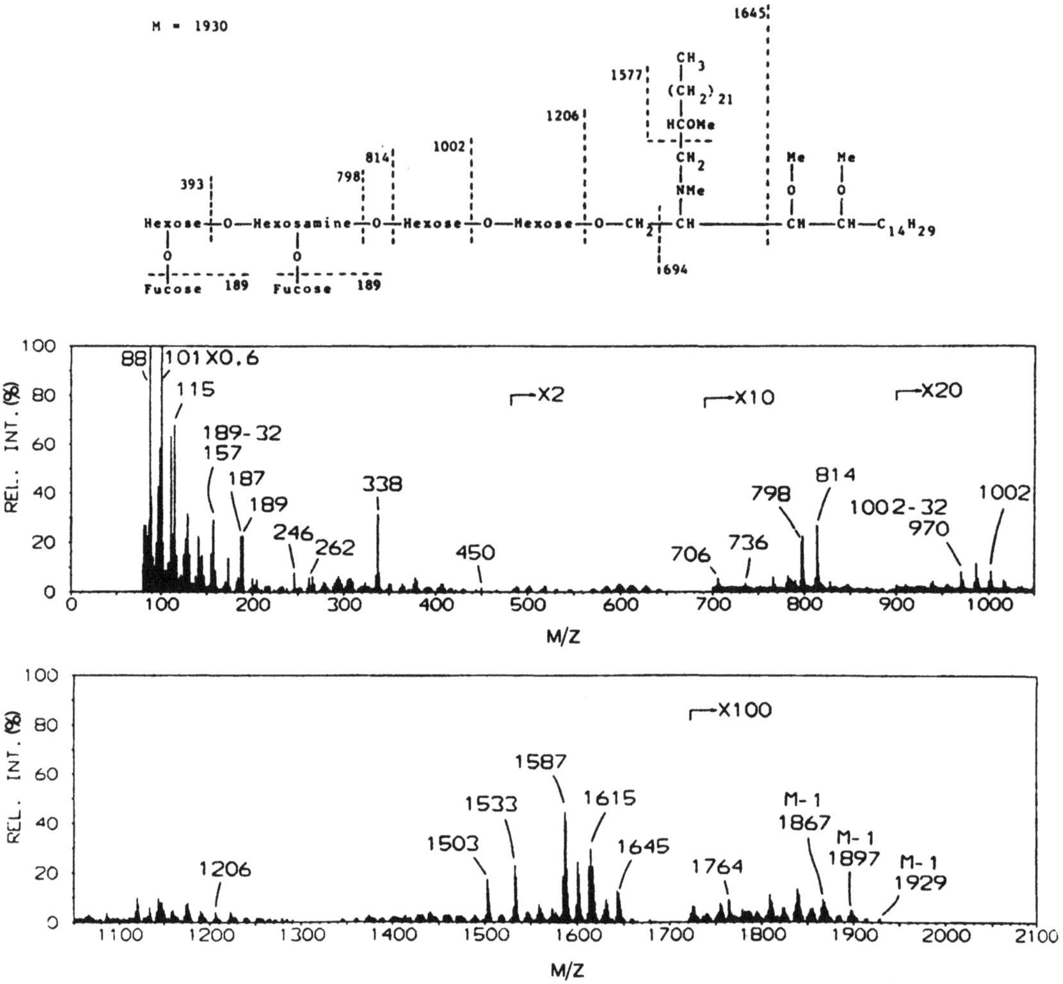

Figure 7. Mass spectrum of permethylated-reduced human plasma glycophingolipid mixture. Conditions: AEI MS902 mass spectrometer equipped with direct inlet probe on a separate heater. The sample (80 μg) was introduced into the ion source of the mass spectrometer and the probe temperature was stepwise increased from 100 – 280 °C. The spectrum was recorded at 275 °C using electron impact ionization (40 eV) Reproduced with permission from: Falk, K. E., Karlssan, K. E. and Samuelsson, B. E. (Ref. 347).

ABH and Lewis type glycosphingolipids in human adult erythrocytes no branched carbohydrate structure having less than 10 sugar residues is known (353). Larson and Samuelsson (352) have considered that the cord blood structures may be non-fucosylated branched precursors of the native structures in adult blood.

Malignant transformations of cells lead to alterations in glycolipid composition. These changes generally result in a simplification due to metabolic blocks (354, 355). However the differences in the neutral glycosphingolipid content of normal and leukemic cells are generally of a quantitative rather than a qualitative nature (356,

357). It is possible, however, that the structures of the neutral glycosphingolipids can be used to identify the origin of leukemic cells. Thus myeloid cells (neutrophils, chronic and acute myeloid leukemia cells) have gala- and lactoneo-type glycolipids. Lee et al. (358) have used mass spectrometry to investigate the structure of the neutral glycosphingolipids in hairy cell leukemia. They found compounds with one to four sugar residues belonging to the globo series of neutral glycosphingolipids. These structures are identical to those found in human lymphoid cells but unlike those of myeloid cells which have compounds of the lactoneo series (357).

Although direct probe EI-MS has been successfully applied to a large number of glycosphingolipids several laboratories have been investigating soft ionization methods in order to increase the abundance of high mass pseudomolecular ions. Markey and Wenger (266) have used methane CI to investigate peracetylated ceramide mono- and di-hexosides but the abundance of high mass ions was very low. Murata et al. (359) using DCI with isobutane as reagent gas and Oshima et al. (360) using GC combined with isobutane CI have analyzed the TMS derivatives of cerebrosides from the spleen of a patient with Gaucher's disease. Both $[MH-90]^+$ and $[M+73]^+$ type ions were found to have appreciable intensities. Ariga et al. (361) used ammonia DCI for the analysis of permethylated derivatives of cerebrosides and ceramide di-, tri-, tetra-, and penta-hexosides. The $[M-32]^+$ ion is the most prominent high mass ion. This ion is the base peak in the spectrum of a permethylated cerebroside but its relative abundance decreases as the number of carbohydrate residues increase. Fragments yielding sequence information are also present. Similarly ammonia DCI of permethylated and reduced-permethylated GM_1 ganglioside and the TMS derivative of the latter yielded spectra that were useful for structural studies (362). The method was extended to the study of a number of other permethylated gangliosides (363). Carr and Reinhold (364, 365) have applied a modified NH_3-DCI-MS technique using desorption from heated polyimide-coated nichrome wire, to study a number of underivatized, permethylated and permethylated-reduced glycosphingolipids. The spectra show abundant $[M+H]^+$ and $[M+NH_4]^+$ ions as well as sequence-specific oligosaccharide fragments. The method is complementary to EI-MS and may be preferable to FAB-MS.

FD-MS of underivatized ceramide mono-, di-, tri-, tetra- and penta-saccharides exhibited base peaks corresponding to $[M+H]^+$, $[M+Na]^+$ or $[M+H_2O]^+$ (366, 367). Fragments due to sequential cleavage of the glycosidic linkages were also observed.

The advantages of FAB-MS for the analysis of high molecular mass glycosphingolipids have been demonstrated (368–372). In the positive ion mode, in the presence of Na^+, only weak ions corresponding to $[M+Na]^+$ and $[M+2Na-H]^+$ are observed for native glycosphingolipids. Negative ion FAB gives intense $[M-1]^-$ ions and is preferred for the analysis of native gangliosides and neutral glycosphingolipids. Fragment ions, which provide sequencing information, are also present in negative FAB-MS. Intense $[M+Na]^+$ ions are observed in the positive FAB-MS of permethylated gangliosides and also for permethylated glycosphingolipids having up to 25 sugar residues and molecular masses in excess of 6000 daltons (372). The method may be useful for molecules with molecular masses up to 15000 daltons.

Glycosphingolipids have also been analzyed by SIMS (373). $[M+H]^+$ and $[M+H-H_2O]^+$ ions predominated in the high mass region spectra of mono- and di-glycosylceramides whereas $[M+Na]^+$ ions predominate in the high mass region of the spectra of the higher glycolipids. It has even been possible to obtain SIMS spectra by a combination of TLC and MS (374). Instead of eluting the sample, a strip of the TLC plate impregnated with a matrix liquid, such as glycerol or triethylamine, is attached to the probe tip. The spectra obtained in this way were comparable to those obtained with the conventional method. Some aspects of the MS of glycoconjugates have recently been reviewed (375).

4.4 Other polar lipids

Recently Millington et al. (56) described a mass spectrometric method for the determination of acylcarnitines in urine of children with propionic acidemia, methyl malonic aciduria and Reye's syndrome. The method was based on FAB-MS with SIM detection and daughter ion analysis. Deuterated acetyl- and propionylcarnitines were used as internal standards. The spectra of synthetic acylcarnitines exhibit very strong $[M + H]^+$ and fragment ions. The parent ion for propionylcarnitine has an m/z value of 218 and that for acetylcarnitine 204. Roe et al. (60) have identified excreted short-chain acylcarnitines in the urine of patients with propionic acidemia and in normal subjects and patients after administration of L-carnitine. The urine of patients showed a prominent ion at mass 218, while the normal subjects gave a prominent ion of mass 204. The presence of the ions corresponding to the propionyl and acetyl carnitine was confirmed by linked scanning at constant electric and magnetic field scanning ratio, to provide an independent fragmentation pattern for each major ion in the mass spectrum (376).

Am interesting recent application of MS in biochemical research is the identification of lipophilic peptides in the hydrolysis products of certain proteins with blocked *N*-terminals. There are now several well-documented reports indicating the covalent bonding of myristic acid to the NH_2-terminus of these proteins (377–380). Both DCI and FAB spectra were found to be useful for the identification of the lipophilic oligopeptides (di-, tri- and tetra-peptides). In most instances the FAB spectra were the most informative. They were obtained following the dissolution of the underivatized oligopeptides in appropriate glycerol and thioglycerol matrices. In one instance the lipophilic dipeptide was identified by GC-EI-MS following trimethylsilylation (379).

5 Fat-Soluble Vitamins and Analogues

The fat-soluble vitamins include: Vitamin A, Vitamin E, Vitamin K and Vitamin D. The metabolic and clinical importance of these fat-soluble vitamins is gradually becoming understood and there is an increasing interest in their identification and quantitation in tissues and in body fluids. These vitamins are bound to chylomicrons for transport from mucosal cells (381).

5.1 Vitamin A

Vitamin A is also known as retinol and it is the precursor of retinal, the light absorbing group in visual pigments. A deficiency of this vitamin results in night blindness. Furthermore, young animals require Vitamin A for growth. The photosensitive molecule in rods is rhodopsin, which consists of opsin, a protein, and 11-*cis*-retinal, a prosthetic group. It cannot be synthesized *de novo* by mammals and is derived from all-*trans*-retinol (Vitamin A) via retinol hydrogenase, and retinal isomerase (382).

The measurement of serum Vitamin A is now very appropriately made using HPLC (383, 384). According to Driskell et al. (383), who developed an improved method for this purpose, as little as 100 μl serum may be employed, and is compatible with assays of Vitamin A in infants. For serum, the lower detection limit is estimated to be 100 $\mu g \cdot l^{-1}$ for retinol. Although there appears to be little interference with the ultraviolet absorption determination of Vitamin A by HPLC, it is of interest to verify or to confirm the indirect estimates by means of MS (385).

Vecchi et al. (386) have reported the EIMS of all-*trans*-retinol and *retro*-retinol and of the TMS ether derivatives, and Lin et al. (387) have reported on a series of isomers related to all-*trans*-retinol. There was extensive fragmentation and the spectra contained ions at almost every mass value between 50 and 200. The molecular ion was the base peak for retinol (*m/z* 284), retinoic acid (*m/z* 300) and methyl retinoate (*m/z* 314). Retinol lost water (M − 18) more readily than its *retro*-isomer and retinyl acetate suffered deacylation (M − 60) less easily than is dehydration. The ions from retinal, *m/z* 255 (M − 29; CH_3OCO), which result from the loss of functional groups, can be of additional help in identification. Retinol, retinoic acid and methyl retinoate lost a methyl group (M − 15) before the loss of their functional groups, giving peaks at *m/z* 269, 285 and 299, respectively, whereas retinol and retinyl acetate did not. Other fragments arose from a combination of losses. In addition, all the Vitamin A derivatives and analogs showed hydrocarbon fragments.

Preparation of the TMS ether of the all-*trans*-retinol and *retro*-retinol resulted in a differential effect in the formation of the ion at *m/z* 255 from the parent compounds (more intense in the all-*trans*-retinol). Other notable differences were the intensities of peaks at *m/z* 168 and 169, the *retro*-isomer having the fragment *m/z* 221, and the *trans*-isomer showing about 20 times the intensity of the *retro*-isomer (385). Vecchi et al. (386) reported that differences between the mass spectra of the TMS derivatives of the 13-*cis*- and all *trans*-isomer of retinol were not significant.

Retinoic acid is an active component of various medicinal preparations for use in clinical oncology and dermatology and several laboratories have described rapid and sensitive methods of analysis for retinoic acid in blood of treated and untreated patients using GC-EI-MS (388, 389) and GC-CI-MS (390). Using CD_3-retinoate as internal standard, De Ruyter et al. (388) estimated that human plasma contained retinoic acid at about $2 \mu g \cdot ml^{-1}$, but Chiang (389) could not detect retinoic acid in a 10 ml sample, although the GC-MS method was validated for retinoic acid down to $1 \mu g \cdot ml^{-1}$ plasma. Chiang (389) compared the intensities of *m/z* 314 (molecular ion of retinoate) and *m/z* 316 (molecular ion of ^{14}C-labeled retinoate). For samples which also contained 13-*cis*-retinoic acid, the methyl 13-*cis*-retinoate was clearly separated from the methyl all-*trans*-retinoate. More recently, using GC-NICI-MS, Napoli et al. (390) reported $2.8 - 6.6 \mu g \cdot ml^{-1}$ of retinoic acid in healthy male volunteers taking no medication. The assay was based on SIM of the negative ions produced from methyl retinoate and methyl 4,4,18,18,18-pentadeuterated retinoate at *m/z* 314 and 319, respectively. Methane was the reagent gas. The lower limit of detection was 75 pg. A comparable quantitative determination of physiological levels of all-*trans*-retinoic acid in human serum has been obtained by De Leenheer et al. (391) using HPLC alone. A double phase extraction of 3.5 ml of serum followed by a specific and sensitive HPLC allowed detection of retinoic acid down to 1 $ng \cdot ml^{-1}$. They confirmed the presence of the methyl retinoate in the diazomethylation products of a silica gel trapping of the HPLC effluent. It gave a molecular ion at *m/z* 314 and fragment ions at *m/z* 299 and 255, $[M-CH_3]^+$ and $[M-COOCH_3]^+$ respectively. De Leenheer et al. (391) point out that the GC-EIMS method of Chiang (389) may have failed to detect retinoic acid under physiological conditions probably due to incomplete derivatization of the serum extract. It should be noted, however, that Napoli (392) claimed that the use of 0.2 $mol \cdot l^{-1}$ base in sample preparation, as reported by De Leenheer et al. (391), results in a partial breakdown or retinoids.

Zile et al. (393) have used EI-MS of the TMS and acetyl derivatives of a minor biliary metabolite of retinoic acid to establish its structure as a glucuronide. MS of all-*trans*- and 13-*cis*-retinoylglucuronide as the TMS derivative has been used to establish the identity of retinoyl glucuronides in rat bile (394), but the method could not distinguish between the two isomers. MS data has also been reported for the 5,6-epoxyretinol and 5,6-epoxyretinyl phosphate (395).

It has become increasingly important in clinical studies to be able to identify and follow the metabolism of substances used in anticancer therapy, many of which are lipophilic in nature. Sorg et al. (396) have employed EI-MS to identify synthetic intermediates and to confirm the synthesis of 12-*O*-retinoylphorbol-13-acetate. Retinoic acid is a strong inhibitor of skin tumour promotion, and was expected to weaken the tumour promoting efficacy of phorbol esters without impairing their mitogenic properties.

5.2 Vitamin E

Vitamin E is a generic name for the biologically active members of the tocopherol and tocotrienol families of compounds, of which α-tocopherol is the most abundant. It occurs in vegetable oils along with β- and γ-analogs. These compounds contain a hydroxyl-bearing aromatic ring system and an isoprenoid side-chain. A nutritional deficiency of Vitamin E in rodents leads to sterility and muscular weakness and atrophy. The specific molecular function of Vitamin E compounds is not known with certainty, but there is good evidence that they prevent the destructive non-enzymatic attack of molecular oxygen on the double bonds of polyunsaturated fatty acid components of tissue lipids (397). In man, Vitamin E deficiency has been linked to a variety of pathological conditions including broncho-pulmonary dysplasia, retrolental fibroplasia and hemolytic anemia (398). High doses of Vitamin E have been used therapeutically in the treatment of non-congenital spherocytic, hemolytic anemia secondary to glucose-6-phosphate dehydrogenase deficiency (399), while others have investigated the protective effect of Vitamin E in relation to andriamycin induced cardiomyopathy (400) and carcinogenicity (401).

In the past colorimetric, fluorometric and TLC and GC methods (402) have been employed for the determination of Vitamin E, while more recently HPLC methods have been shown to separate the naturally occurring Vitamin E analogs, α-, β- and γ-tocopherols, as well as other derivatives (403, 404). In serum, the lower limit for α-tocopherol by reversed phase HPLC was 1.9 $\mu mol \cdot .1^{-1}$ (383). For some applications, however, the chromatographic methods lack the specificity and sensitivity needed, and MS or GC-MS analysis is required.

A major difficulty with all analyses of the tocopherols is their instability. This can be significantly increased by chemically blocking the free hydroxyl group. The TMS, TFA and acetyl derivatives are easily prepared and have excellent GC and MS properties (385).

Mass spectra of the TFA of α- and γ-tocopherols were reported by Nair and Luna (405). The ion at *m/z* 165 is small in the TFA derivative, but the TFA derivative has the base peak at *m/z* 261. In the acetate, ketene is lost more readily and the ion at *m/z* 165 is very intense. α-Tocopherol fragments to give the base peak *m/z* 165, an ion formed by cleavage and hydrogen migration at the heterocyclic chroman ring. The ion at *m/z* 205 results from a loss of the side chain. The molecular ion varies in intensity with these derivatives and is the base peak for the β-isomer (385). The β- and γ-tocopherols exhibit molecular ions at *m/z* 416, a mass fragment ion at *m/z* 151, less intense ions at *m/z* 191, and very low intensity ions at *m/z* 401 (M − 15). The distinguishing features in the spectra are the base peak shifts between *m/z* 151 (γ) and the molecular ion (β) in the two tocopherol derivatives (385). The β-isomer has an ion at *m/z* 192 and the γ-isomer an ion at *m/z* 189.

Thomas et al. (406) described a definitive method of determination of α-tocopherol in rat lung as the TMS ether using a stable isotope-labeled α-tocotrienolpherol as internal standard. The other tocopherols were measured simultaneously by monitoring the molecular ions. The deuterated α-tocopherol standard was prepared by reducing α-tocopherol with deuterium at room temperature and pressure in ethanol-d_1 using Pd/C catalyst. The product, a mixture of deuterated α-tocopherols resulting from hydrogenation and exchange, was purified by TLC. The d_{13} species was the most abundant of the many labeled variants in this mixture. Masses *m/z* 502 (molecular ion of tocopherol TMS ether) and *m/z* 515 (molecular ion of tocopherol-d_{13} TMS ether) were monitored at 20 eV. The method gave excellent sensitivity in the picogram range (170 pg).

Kaseki et al. (407) reported a highly sensitive method for α-tocopherol (*m/z* 430) determination in biological tissues by mass chromatography using a scanning range of *m/z* 300 to 500. A hexane-refined ethanolic extract was directly injected into a MS source using a pipette. The detection limit of the method in EI-MS was less than 50 pg and that the vitamin could be detected in 1 μl plasma and gave an average value of 6.92 $\mu g \cdot ml^{-1}$. About 100 μl of the same plasma was required to obtain a comparable result by HPLC. Kaseki et al. (408) utilized this method to measure α-tocopherol in unfertilized eggs from mice and obtained levels of approximately 230 pg in an individual egg. A GC-MS procedure for quantitation of Vitamin E in lung tissue sections and in microdissected structures from tissue sections required only 3 μg dried tissue for analysis (409). Vitamin E was measured as the TMS ether in the concentration range 2 − 230 $\mu g \cdot g^{-1}$ tissue.

Interest in Vitamin E has been stimulated by clinical trials which suggested that it might be effective in reducing thrombolic disorders. Vitamin E quinone is more active than the parent compound and this substance has been synthesized in large amounts for clinical trials. The synthetic material was characterized by MS (410). Of interest also was the MS identification of novel urinary metabolite of d-δ-tocopherol in rats. On the basis of the appearance of major mass fragments at *m/z* 306, 264 and 137, the metabolite was identified as 2,8-dimethyl-2(2'-methoxycarbonylethyl)-6-chromanyl acetate (411). The proposed structure was confirmed by comparison with results obtained with a synthetic standard and by infrared and NMR.

5.3 Vitamin K

Vitamin K, which is required for normal clotting, participates in the carboxylation of glutamate residues to γ-carboxyglutamate. Vitamin K is found in two forms: Vitamin K_1 (phylloquinone) which is synthesized by plants, and Vitamin K_2 (menaquinone) synthesized by bacteria. The Vitamins K_1 and K_2 are naphthoquinones possessing long isoprenoid side chains of different length at carbon 3. Usually Vitamin K_1 has a saturated 20 carbon side chain (phytyl) derived from four isoprenoid units, while Vitamin K_2 shows greater variation, with an all-*trans*-farnesylgeranylgeranyl 35 carbon chain of 7 isoprenoid units being most common. The synthetic analogues are derivatives of menadione (Vitamin K_3), which lacks a side-chain substituent at position 3 (412).

The K vitamins have proved very difficult to measure in biological samples by physico-chemical methods due to their low concentration and the existence of several molecular species with biological activity. Their sensitivity to alkali precludes the use of saponification to remove other fats during isolation. Owing to the lack of sensitivity and/or selectivity of chemical assays, Vitamin K deficiency has been usually diagnosed indirectly by measuring its effect on blood coagulation (413), and metabolic studies have usually been performed with radioactive tracer techniques (414). Bechtold and Jahnchen (415) described a GC method, which permitted studies of the disposition of Vitamin K_1 and K_1 2,3-epoxide following therapeutic doses. For this purpose plasma samples were supplemented with Vitamin K_2 (415) internal standard and solvent extract analyzed by GC with an electron capture detector. The lower limit of detection was 3 $ng \cdot ml^{-1}$ plasma for both substances. Substitution of a capillary for the packed GC column did not decrease the lower limit of detection significantly, although the capillary column gave complete baseline separation between Vitamin K_1 and the isomers of Vitamin K_1 2,3-epoxide (416).

The introduction of HPLC has greatly facilitated the assay of the K vitamins and the subject has been reviewed (417, 418). Lefevere et al. (419, 420) described the application of HPLC to the determination of Vitamin K_1 (*trans*-phylloquinone) and its quantitation down to a level of 0.5 $ng \cdot ml^{-1}$ of serum. In healthy adults, a mean serum concentration of 2.6 $ng \cdot ml^{-1}$ was found with normal range of 0.9 to 7.8 $ng \cdot ml^{-1}$ (420). The selectivity and specificity of the assay were confirmed by GC with electron capture detection and LC-MS. Langenberg and Tjaden (421) desribed a more sensitive HPLC method based on coulometric reduction combined with fluorometric detection of Vitamin K_1 in human plasma. This method has been further improved (422) by allowing a detection of Vitamin K_1 and Vitamin K_2 down to 5 pg/injection for standards and 50 pg . ml^{-1} for each as components of human plasma.

The most sensitive and precise method for determining naphthoquinone structures is mass spectrometry. The isoprenoid quinones produce characteristic fragmentation patterns that provide structural information on the nature of the ring system and on the length and degree of saturation of the multiprenyl side chain. A detailed study of the EI-MS spectra of phylloquinone (Vitamin $K_{1(20)}$ (2-methyl-3-phytyl-1,4-naphthoquinone) and a number of deuterated analogs has been made by DiMari et al. (423). Major fragments included ions m/z 435 (M − 15), m/z 225 (M − 225), m/z 198 (M − 252) and m/z 186 (M − 264). Two processes, loss of either ring or side chain methyls, contributed to the formation of fragments at m/z 435. A major fragment (base peak in some spectra) at m/z 225 was attributed to a stable oxonium ion which increased in intensity with the size of the residual side chain. A fragment at m/z 186 was believed to be formed by cleavage of the α,β-bond of the phytyl side chain with a transfer of a hydrogen atom. m/z 198 was probably formed from the molecular ion or a fragment of high molecular weight in which the allylic double bond of the side chain isomerized to the vinylic position (423). The EI spectrum of the bis-TMS ether of dihydrophylloquinone showed the molecular ion as base peak (m/z 596) and ions of small intensity at m/z 581 (M − 15), m/z 524 (M − 72), m/z 331 $[M-(\text{side chain-}CH_2)]^+$ and the m/z 73 due to free TMS (368). The 2,3-epoxide metabolite of phylloquinone (Vitamin K_1 2,3-epoxide) gives a strong molecular ion (m/z 466, base peak) along with characteristic ion at m/z 423 andm/z 159, which are useful for identification, in addition to m/z 241 (the 2,3-epoxide analogue of m/z 225 in the spectrum of phylloquinone) (424).

The EI spectrum of Vitamin $K_{2(35)}$ (menaquinone-7) gave a prominent molecular ion (m/z 648) along with a small (M − 15) ion. The base peak occurred at m/z 225, with a second intense ion at m/z 187 derived from the napthoquinone nucleus. Fragmentation of the side chain under EI involves diallylic bonds, following the rule $(M-69)-(68 \times N)$, where N is the number of isoprene units in the side chain minus one (425). Thus an effective assessment of the prenyl groups (m/z 68) is afforded. The spectrum of the bis-TMS ether of dihydro MK-5 exhibited a large molecular ion and the ion at m/z 331, similar to phylloquinone (426, 427). Howarth et al. (428) have reported the structure of a new naturally occurring cyclic Vitamin K, which on the basis of MS and NMR corresponded to (2E,14E,18E,22E)-2-[3,7,11,15,19,23-hexamethyl-25-(2,6,6-trimethylcyclohex-2-enyl)pentacosa-2,14,18,22-tetraenyl]-3-methyl-1,4-naphthoquinone. In addition to the characteristic ions of the normal naphthoquinone ring system, the MS gave an unusual side-chain fragmentation pattern characterized by prominent peaks at m/z 584 and 596.

5.4 Vitamin D

Vitamin D is a hormone that regulates the metabolism of calcium and phosphorus. A deficiency in Vitamin D impairs bone formation in growing animals. 7-dehydrocholesterol (provitamin D_3) is photolyzed by UV light to previtamin D_3 by cleavage of ring B between carbons 9 and 10 and leads to a spontaneous isomerization to vitamin D_3 (cholecalciferol). *cis*-Vitamin D_3 formed in the skin is specifically complexed by the Vitamin D-binding protein and then transported in the blood stream to the liver microsomes, where it is converted to 25-hydroxyvitamin D_3 (diol), and then to the kidney tubule mitochondria, where α,25-dihydroxyvitamin D_3 (triol) is finally produced (see ref. 429 for a brief review). Zagalak et al. (430) demonstrated that approximately 50% of the lipophilic conjugates of Vitamin D_3 exist as acetate and the rest as long chain fatty acid esters. These authors have suggested that the lipophilic conjugates may represent a stored form of Vitamin D_3 and may be involved in an enzymatic regulation of Vitamin D_3 levels in the organism. Vitamin D_2 (ergocalciferol) is formed in a similar manner starting with ergosterol. Vitamin D_2 differs from Vitamin D_3 in having $C_{22}-C_{23}$ double bond and a methyl group at C_{24}. The occurrence of a large number of vitamin D metabolites, many of which are isomers (e. g. 23S,25-, 24R,25-, 25,26S and 1-α,25-dihydroxyvitamin D) complicates the measurement of vitamin D compounds, because of the need for a large number of specific assays and the risk of interference (431). Of the derivatives of vitamin D circulating in plasma, 25-hydroxyvitamin D is most widely used for the assessment of vitamin D status in a variety of clinical conditions (432). The majority of these assays have been derived from the competitive protein-binding assay of Haddad and Chyu (433) and do not distinguish between 25-hydroxyvitamin D_2 and 25-hydroxyvitamin D_3. The major difficulty in the assay of the hydroxylated vitamin D derivatives in biological fluids is related to the small amounts of the compounds and the lack of specificity in the methods. To a considerable extent these problems have been overcome by HPLC and GC-MS, especially GC-MS with SIM.

Using isomerization to the isotachysterols to increase the sensitivity of detection on HPLC, Trafford et al. (434) were able to analyze down to about 0.5 $mg \cdot ml^{-1}$ using 2 ml of plasma, which was of the order obtained by SIM. Thermal rearrangement, however, may give rise to pyrolysis. Using the SIM method an excellent correlation was obtained between HPLC and the GC-MS method (435). The HPLC separations were monitored at 301 nm. Only vitamin D derivatives were believed to be affected by the HCl isomerization to the isotachysterols to increase absorbance in this range. The sensitivity of the HPLC method for both secosteroids, 25-hydroxyvitamin D_3 and 25-hydroxyvitamin D_2, were estimated to be around 0.5 $\mu \cdot l^{-1}$, so that practical measurements could be made on 0.5 ml of plasma, more accurately on 2 ml of plasma. Trafford and Makin (436) have applied this HPLC method to the examination of human saliva for the presence of 25-hydroxyvitamin D_2 and 25-hydroxyvitamin D_3, which have been reported to occur there in high concentrations (437). Using 20 ml of saliva the HPLC method provided a minimum detectable limit for 25-hydroxyvitamin D of 0.05 $\mu g \cdot l^{-1}$ (130 $pmol \cdot l^{-1}$ for 25-hydroxyvitamin D_3). No 25-hydroxyvitamin D could be detected in any of the samples. Examination of larger quantities of saliva (100 ml) by a combination of HPLC and GC-MS showed the presence of a peak presumably derived from 25-hydroxyvitamin D_3 at a concentration not in excess of 65 $pmol \cdot l^{-1}$. Simple calculations, based on known binding characteristics of 25-hydroxyvitamin D and plasma vitamin D binding protein, and comparing them with similar characteristics for cortisol and plasma cortisol binding globulin and the known salivary concentration of cortisol, indicated that a likely concentration of 25-hydroxyvitamin D in saliva would be around 8 $pmol \cdot l^{-1}$. Previously reported values ranging from 130–4160 $pmol \cdot l^{-1}$ (50–1600 $\mu g \cdot l^{-1}$) using simple competitive protein binding assay could not be confirmed.

Bouillon (431) has recently described an HPLC procedure which allowed a separation of a total of three dihydroxylated vitamin D isomers (24,25-, 25,26-dihydroxyvitamin D) although the 24R,25-dihydroxyvitamin D_2 and the 25-hydroxyvitamin D-25,23-lactone were coeluted.

The Vitamin D derivatives yield characteristic MS as the free compounds and as the TMS ethers. The EI spectra of the D vitamins have been studied extensively (438, 439). The principal fragmentation pathways gave ions corresponding to loss of the side-chain *m/z* 271 [M − 113] or [M − 125]. A loss of water from this fragment produced *m/z* 253. The ring structure may be broken to yield *m/z* 135 although the detailed mechanism for its formation is not clear (385). The *trans*-Vitamin D_3 gave nearly the same spectrum (1). The mass spectrum of 25-hydroxyergocalciferol, a metabolite of Vitamin D_2, indicated the molecular mass of the metabolite to be 16 daltons higher than its precursor and, with an exact mass corresponding to $C_{28}H_{44}O_2$, represents an increase of one oxygen atom. The intense ions at *m/z* 136 and 118 were derived from ring A of the molecule and were the same as those formed in the precursor D_2 molecule. Both showed ions at *m/z* 271 resulting from loss of the entire side chain, indicating clearly that any changes during metabolism had occurred exclusively in the side chain. The position of the hydroxyl group on the side chain was demonstrated by means of the TMS ether of Vitamin D_2 and of the isolated metabolite. The base peak of the metabolite was an ion *m/z* 131 [$(CH_3)_2C = O^+Si(CH_3)_3$], which was not observed for Vitamin D_2 (438, 439). The same ion was present in the EI spectrum of the TMS ether of the metabolite of the Vitamin D_3, but not in the spectrum of the TMS ether of the parent compound. The ion at *m/z* 136 shifts to *m/z* 208 in the TMS ether as expected.

Bjorkhem and Holmberg (440 – 442), Zagalak et al. (443 – 445) and Seamark et al. (435) have overcome the problems associated with the lack of specificity by the use of SIM. In a standard SIM procedure, Bjorkhem and Holmberg (442) added 145 ng of 25-hydroxy[26-^{2}H]Vitamin D_2 dissolved in 50 µl of ethanol to 2 ml of serum. After a solvent extraction the hydroxy derivative of Vitamin D was converted into the 3-*t*-BDMS ether (the 25-hydroxyl group was slow to react) and purified by TLC. The recovered 3-*t*-BDMS ether derivatives were then converted into the 25-TMS ether derivatives, which were analyzed by GC-EI-MS using a SE-30 column. The spectrometer was focused on the ions *m/z* 586 and *m/z* 589, corresponding to the molecular peaks in the mass spectra of the derivatives of unlabeled and $^2H_3$-labeled 25-hydroxyVitamin D_3. The specificity of the method was checked by focusing on the ions at *m/z* 131 and at *m/z* 134. The amount of unlabeled 25-hydroxyvitamin D_3 was calculated from the ratio between the height of the peak recorded at *m/z* 586 and the height of the peak at *m/z* 589 by means of a standard curve. The mean values for concentration of 25-hydroxyvitamin D_3 in serum from healthy men and women were 21 $ng \cdot ml^{-1}$. The assay can determine 25-hydroxyvitamin D_3 down to a level of a few nanograms. It was suggested that the sensitivity of the method could be increased if an internal standard was used with a mass more than three mass units higher than that of the unlabeled 25-hydroxyvitamin D_3 (442). The mass spectrum of 3-*t*-BDMS, 25-TMS ether of 25-hydroxyvitamin D_3 gave prominent peaks for the molecular ion *m/z* 586, also for the ions *m/z* 439 [M − (90 − 57)], and *m/z* 131. The 26-$^2H_3$-derivative contained the corresponding *m/z* 134, 442 and 589 ions as prominent peaks.

Zagalak et al. (443) employed SIM to determine cholecalciferol and 25-hydroxycholecalciferol in human serum using deuterated cholecalciferol and deuterated 25-

hydroxycholecalciferol as internal standards. GC-SIM assay of the vitamins in serum was possible only after careful purification by column, TLC and HPLC methods. Filtration on LH-20 Sephadex of crude serum extracts was not adequate because of high excess of cholesterol. The method was applied to studies of the biosynthesis and excretion of Vitamin D in man (444). The overall yield of the purification procedure ranged from 76–86% when determined with [^{14}C]vitamin D_3. Use of deuterated internal standards allowed the simultaneous determination of natural and exogenous cholecalciferol. More recently, Zagalak et al. (445) have employed MF for the determination of Vitamin D in human blood and urine. After extensive purification of the extracts using alumina and digitonin-celite columns and HPLC, the samples were silylated and the molecular ion or the $[M-90]^+$ ion of the pyroform of ergocholecalciferol TMS peak was monitored. The deuterated internal standard was synthesized from *cis*-Vitamin D_2.

Berlin et al. (446) have used the method of Bjorkhem and Holmberg (442) in the assay of 25-hydroxyvitamin D_3 in patients with urolithiasis. The group of patients with a high urinary excretion of calcium (over 6.0 $mmol \cdot 24\ h^{-1}$) had a significantly higher level of 25-hydroxyvitamin D_3 ($26.2 + 1.5$ $ng \cdot ml^{-1}$) than those patients ($17.6 + 0.9$ $ng \cdot ml^{-1}$) with a normal urinary excretion of calcium (lower than 6.1 $mmol \cdot 24\ h^{-1}$). There was no overall correlation between level of 25-hydroxyvitamin D_3 and serum level of calcium.

Bjorkhem et al. (447) described a specific assay for 1,25-dihydroxyvitamin D_3 in human serum which employed [26-$^2H_3$]-1,25-dihydroxyvitamin D_3 as internal standard, HPLC extraction and TMS derivatisation. The TMS ether of 1,25-hydroxyvitamin D_3 gave small peaks at *m/z* 632 $[M]^+$ and at *m/z* 617 $[M-15]^+$ and more prominent peaks at *m/z* 542 $[M-90]$, 452 $[M-2 \times 90]$ and 362 $[M-3 \times 90]$. The base peak in the mass spectrum was at *m/z* 131, corresponding to cleavage between C_{24} and C_{25} (437). In the SIM, unlabeled 1,25-dihydroxyvitamin D_3 was quantitated from the ratio of *m/z* 452 and 455, $[M-2 \times 90]$. Essentially the same results were obtained when they used the more intense, but less specific ions, *m/z* 131 and 134 from cleavage between C_{24} and C_{25}. The estimated concentration of 1,25-dihydroxyvitamin D_3 in sera from healthy subjects was 55 ± 10 $ng \cdot l^{-1}$($\pm$ SD).

Seamark et al. (435) described the SIM measurement of Vitamin D_3, 25-hydroxyvitamin D_2, 25-hydroxyvitamin D_3 and of 24,25-dihydroxyvitamin D_3. The quantitation was facilitated following extraction in the presence of Vitamin D_2 and formation of the isotachysterol TMS ethers, which yield higher response in the MS. The $[M-57]^+$ ion of cholesterol *t*-BDMS (*m/z* 443) was used as a reference ion when monitoring molecular ions, and *m/z* 355 of perfluorokerosene, when monitoring $[M-90]^+$ ions. Analyte peaks were quantitated against the internal standard (isotachysterol$_2$-TMS) and the resulting peak ratios compared with standard curves run on the same day. The spectrum of isotachysterol$_3$-TMS ether (derived from Vitamin D_3) contained ions *m/z* 456 (M), *m/z* 366 $[M-90]$ and a base peak at *m/z* 252 $[M-(114 + 90)]$. Its molecular ion (*m/z* 456) was twice as intense as that of the TMS ether of either *pyro*- or *isopyro*-Vitamin D_3. Similar improvements in the intensity of the molecular ion were found when the spectra of 25-hydroxyisotachysterol$_3$-TMS and 24,25-dihydroxyisotachysterol$_3$-TMS were compared with the spectra of the *pyro*- and *isopyro*-forms. It was necessary to form derivatives of all three hydroxyl groups (at C_3, C_{24} and C_{25}), particularly the 25-hydroxyl group, which is difficult to derivatize and, if free, will dehydrate to varying degrees during GC. Using 5 ml of plasma the sensitivity was estimated for 24,25-dihydroxyvitamin D_3 to be 0.14 $\mu g \cdot l^{-1}$ and for Vitamin D_3, 25-hydroxyvitamin D_2 and 25-hydroxyvitamin D_3 0.1 $\mu g \cdot l^{-1}$, corresponding to approximately 0.1 ng isotachysterol-TMS injected into the GC-MS system. However, these estimates have been further qualified (448). The values for plasma 25-hydroxyvitamin D_3 were similar to those obtained by Bjorkhem and Holmberg (441) using SIM. Figure 8 illustrates the use of SIM in the detection and identification of 25,26$(OH)_2D_3$ using a pentadeuterated homologue as a carrier and internal standard (449).

Whitney et al. (450) developed a SIM-GC-MS method for the measurement of 25-hydroxy vitamin D_3, which uses the deuterated analogue as internal standard and a combination of Sep-Pak C_{18} and straight phase silica Sep-Pak for purification. 25-hydroxy vitamin D_3 was determined in the serum of children with hypophosphatemic vitamin D resistant rickets.

In addition to serum assays for Vitamin D, MS and GC-MS has been extensively used for the identification and quantitation of various metabolites of this vitamin. Napoli et al. (451) reported that 24-keto-1,25-$(OH)_2$vitamin D_3 is a rat intestinal metabolite of 1,25-$(OH)_2$vitamin D_3 under physiological conditions, while another metabolite, 24-keto-1,23,25-$(OH)_3$vitamin D_3 (452), was shown to be formed from 1,25-$(OH)_2$vitamin D_3 via 1,24,25-$(OH)_3$vitamin D_3 and 24-keto-1,25-$(OH)_2$vitamin D_3. (23S)-1,23,25-trihyroxyvitamin D_3 was isolated and identified from bovine kidney homogenates incubated with 1,25-dihydroxyvitamin D_3 by sequential chromatography and EI-CI-GC-MS (453), and 24,25,26,27-tetra-*nor*-23-hydroxyvitamin D_3 identified from rat kidney perfusate incubated with 24(R),25-dihydroxyvitamin D_3 (454). Other new metabolites of 1,25-dihydroxyvitamin D_2 produced by rat kidney include 1,24,25-trihydroxyvitamin D_2, 1,24,25,28-tetrahydroxyvitamin D_2 and 1,24,25,26-tetrahydroxyvitamin D_2 (455). The mass spectra have been obtained of 5(E)-19-*nor*-10-ketovitamin D_3, 5(E)-19-*nor*-ketovitamin D_2 and 5(E)-19-*nor*-10-keto-25hydroxyvitamin D_3, which are unique metabolites of Vitamin D_3, Vitamin D_2 and 25-hydroxyvitamin D_3, respectively, and represent a detoxification mechanism (456).

Of great interest in the vitamin D field has been the direct analysis and identification of intact sulfate and glucuronide conjugates of vitamin D_3. Litwiller et al. (457) have used MS to confirm the presence of a monoglucuronide of 1-α,25-hydroxyvitamin D_3 in rat bile. FAB-MS produced an intense ion at *m/z* 637 compatible with the cationized sodium salt of a 1-α,25-dihydroxyvitamin D_3 glucuronide. EI-MS analysis of the methyl ester/penta TMS derivative of the intact material also gave a molecular ion and fragmentation pattern consistent with this identification. Positive FAB spectra of standard vitamin D_3 glucuronide and 1-α,25-dihydroxyvitamin D_3 glucuronide, isolated from rat bile, were obtained by Jardine et al. (458). The spectra were dominated by $[M-H+2Na]^+$ ions but no clearly recognizable aglycone ions. The same authors have published positive and negative FAB spectra of the ammonium salts of vitamin D_2 recorded after addition of sodium acetate to the sample. The major ion in the positive spectrum was *m/z* 521, $[M-NH_4+2Na]^+$, while $[M-NH_4]^-$ dominated in negative ion mode. Vitamin D_3 sulfate was sysnthesised and a negative ion spectrum obtained by cesium ion liquid SIMS (459). The molecular ion at *m/z* 463 was the base peak. Selva et al. (460) reported the behaviour of 3-phosphoryl derivatives of vitamins D_2, D_3 and 7-dehydrocholesterol under different mass spectrometric conditions.

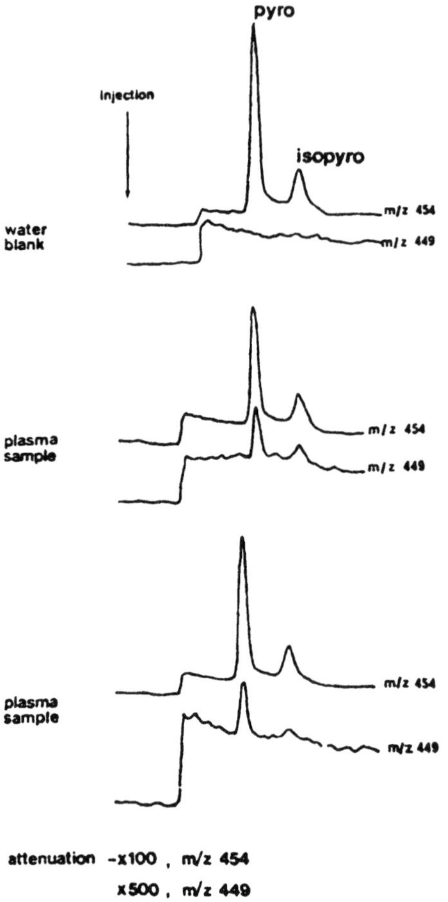

Figure 8. Representative mass fragmentograms from a nondeuterated (*m*/*z* 449) and a pentadeuterated (*m*/*z* 454) homologue of the *pyro*- and *isopyro*-products of 25,26(OH)$_2$D$_3$-*n*-butylboronate-3-trimethylsilyl ethers. The plasma samples had 25,26(OH)$_2$D$_3$ concentrations of 0.81 $\mu g \cdot l^{-1}$ (middle trace) and 0.78 $\mu \cdot l^{-1}$ (bottom trace). GC-MS conditions: LKB 2091 GC-MS system; electron impact ionization (24 eV) Reproduced with permission from: Coldwell, R. D., Trafford, D. J. H., Malcin, H. L. J., Verley, M. J. and Kirk, D. N. (Ref. 449).

Zagalak et al. (430, 444) and Takada (461) have demonstrated the formation of fatty acid esters of vitamin D$_3$ in rat skin exposed to artificial UV rays. The results indicated that the fatty acid esters of 7-dehydrocholesterol, which normally account for about 80% of total 7-dehydrocholesterol in rat skin, are also isomerized to D$_3$ esters. The MS of the free vitamin D$_3$ thermoisomers showed ion intensities for the molecular ion *m*/*z* 384 and fragment ions (*m*/*z* 366, 351, 325, 271, 145 and 143) corresponding to the GC peaks of pyrovitamin D$_3$ and isopyrovitamin D$_3$, which are known thermoisomers of vitamin D$_3$ (461).

The GC-MS properties of boronate esters of 24R,25-dihydroxycholecalciferol have been investigated to determine whether this derivative would be useful for quantitative purposes (462). The authors concluded that boronate esters were preferable to TMS ethers because of the increased abundance of high mass ions.

Recent reviews on the methods available for quantitating vitamin D and its metabolites in human plasma and tissues have been published by Makin and Trafford (448), Jones et al. (463) and Napoli et al. (464).

6 Fat-soluble Xenobiotics

The application of MS with SIM has been essential to the detection and identification in human tissues of chemicals that are foreign to the human body yet biologically active. The suitability of MS to the study of these xenobiotic compounds stems again from the high sensitivity of detection and the specificity of identification of small amounts of the material.

Dougherty (465) has reviewed the application of MS and GC-MS to the detection and identification of toxic residues and pollutants and has listed the toxic residues found by these techniques in humans and in their food chain. In the following sections an attempt has been made to update this review in regards to selected fat-soluble xenobiotics. The discussion has been restricted to those compounds of the greatest clinical interest.

6.1 Polycyclics

The aflatoxins are a group of mycotoxins, which have been shown to be potent carcinogens in certain laboratory animals (466). Rigorous proof that aflatoxins are human carcinogens is not available. However, epidemiologic reports about the incidence of hepatocellular carcinoma in man from various regions of the Asian (467, 468), African (469, 470) and Western (471, 472) countries suggest aflatoxin involvement. Ryan et al. (473) reported the presence of aflatoxin B_1 in seven Reye's syndrome patients.

EI-MS has been the most widely used tool in the analyses of samples for mycotoxins (474). Kingston (475) reported a systematic approach to the potential application of MS to mycotoxin structure determination, while a computer-based compilation of EI-MS data on mycotoxins has been published (476). EI-MS also has been the most frequently used technique for the structural characterization of mycotoxins suspected to be present on the basis of assays of relatively low specificity (e. g. GC, TLC and HPLC). FD-MS has also been applied (477).

Aflatoxins B_1 (M_r 312), B_2 (M_r 314), G_1 (M_r 328), G_2 (M_r 330) and M_1 (M_r 328) cannot be easily derivatized for GC or GC-MS. They can be separated, however, by TLC and the purified fractions analyzed by direct probe EI-MS (478). Peaks below m/z 200 comprised less than 2% of the total ionization while molecular ions for all the compounds carried up to 47% of total ion current greatly facilitating the sensitivity and specificity of detection. Sequential losses of CHO and CO, and loss of CH_3O accounted for most of the significant fragment peaks in all the compounds. Aflatoxins G_1 and M_1, although coincident in molecular weight and elemental composition, differ dramatically in the relative abundance of the peaks resulting from loss of CHO and CHO + CO from the molecular ion and can be differentiated easily from MS. Using this method it was possible to demonstrate the presence of alfatoxins B_1 and B_2 in cottonseed meal at 30 $\mu g \cdot l^{-1}$ range, with an estimated minimum level of detection below 1 μg. Subsequently, Haddon et al. (479) employed SIM as an improved method of characterization of suspected aflatoxins. To avoid the need for extensive purification of the samples, the resolution of the MS was increased from the normal values of 500−1000 to above 5000 and the scans were made over a series of narrowly defined mass ranges (0.3 dalton). The method was applied to analysis of aflatoxin B_1 added to a 24 hour sample of human urine.

Brumley et al. (480) described the use of NICI-MS of aflatoxins and related mycotoxins as a complementary method to EI-MS. With methane as the ionizing gas, the negative ion spectra of aflatoxins consisted of three major ions: $[M]^-$, $[M-H]^-$ and $[M-CH_3]^-$. Other mycotoxins related to aflatoxins also exhibited relatively simple spectra. The aflatoxins were detected at levels as low as 15 $\mu \cdot kg^{-1}$ in food matrices. Proctor (481) has reviewed the new developments in NICI-MS of aflatoxins.

The presence of aflatoxin B_1 in human tissues was demonstrated using HPLC in combination with DEI-MS (482). A prominent peak with *m/z* 312 appearing at a probe temperature of 165 °C was considered positive confirmation of the presence of aflatoxin B_1. Better than 79% recovery was obtained from human blood samples spiked with 1−50 $ng \cdot ml^{-1}$ of aflatoxin B_1. Using this procedure, aflatoxin B_1 was detected in 4 of the 6 liver samples from primary hepatocellular carcinoma patients, and in 11 of 14 liver samples, 4 of 6 blood samples and 4 of 5 urine samples from patients with Reye's syndrome.

In common with many toxic xenobiotics, the hepatocarcinogenic aflatoxin B_1 is believed to require metabolic activation to bind covalently to macromolecules in the target organ (483). Moss et al. (484) have used FD-MS for the study of the structure of aflatoxin B_1 glutathione conjugate formed *in vitro*. The spectrum was averaged by accumulation of 20 scans (16−21 mA) and showed an abundant $[MH]^+$ ion at *m/z* 636 associated with a smaller molecular ion at *m/z* 658 due to Na^+ attachment, $[MNa]^+$. The FD-MS fragment ions at *m/z* 308, 130 and 84 were characteristic of glutathione conjugates. The most intensive ion at *m/z* 228 was not definitely assigned but was thought to be a fragment of the aflatoxin B_1 skeleton, since a similar fragment had been observed in the MS of aflatoxin B_1 (485). The molecular ion obtained was consistent with a glutathione conjugate of aflatoxin B_1 formed via an epoxide (484).

Aflatoxin BS and GS were shown to be the respective sodium sulfonates of aflatoxin B and G by FAB analysis (486).

6.2 Polychlorinated biphenyls (PCBS)

GC-MS provides the most effective means of detecting PCBS in human tissues (487, 488) and is widely used to monitor plant and animal products for the presence of the ubiquitous PCBS, polycyclic aromatic hydrocarbons and polychlorinated dibenzo-*p*-dioxins, all by-products of modern chemical processing (489, 490). These compounds are lipophilic, highly resistant to biological breakdown and subject to bioconcentration and biomagnification in species at higher trophic levels of the food chain and can have mutagenic and carcinogenic effects (491). The qualitative MS of the PCBS and related chlorinated polycyclics has been extensively reviewed in the literature on pesticides and pollutants (489, 490, 492) and useful collection of EI spectra of chlorinated aromatic pesticides and pollutants has been published (493). The MS behaviour of standard PCBS (Aroclor 1254 and 1260) and of PCBS isolated from human adipose tissue utilizing standard analytical methodology has been reported (492, 493). Most components of the Aroclor standard produce mass spectra with molecular ion groups of high intensity, characteristic isotopic distribution patterns corresponding to the number of chlorine atoms in parent and fragment ions, relatively intense ion fragments from consecutive loss of chlorine atoms from the parent ion and intense doubly charged molecular ions. An important development, which has permitted the assessment of the potential health impact of these compounds, has been the identification of the major congeneric constituents by a combination of GC, GC-MS and comparative GC analyses using synthetic standards. On this basis the standard Aroclor 1260 was shown to consist of at least two pentachloro-, four hexachloro-, six heptachloro-, two octachloro-, two nonachloro- and a decachlorbiphenyl. A human adipose tissue extract was found to contain traces of PCB compounds, whose mass spectra were consistent with many of those found for Aroclor 1260. Watanabe et al. (494) identified a large number of PCB congeners in human tissue and reported average concentrations in the breast milk of women. The PCBS identified in human tissues and their relative concentrations have been listed (465, 487).

A GC-MS method has been described for the detection and quantitation of any of the 209 possible PCBS (495), as has been a GC-CIMS method for the characterization of the five FDA reference Aroclor standards (496).

NICI-MS is suited to screening microquantities of environmental and tissue samples for trace polychlorinated chemicals because of high sensitivity and a virtual transparency to potentially interfering biomolecules. Ions from PCBS are formed by electron capture, disassociative capture, chloride attachment, deprotonation and oxygen exchange (497). Kuehl et al. (498) described methods for the analysis of chlorinated chemical residues, including PCBS, found in milligram quantities of biological samples. Sample preparation involved liquid-liquid extraction or microgel permeation and chlorophenols and PCBS were confirmed in small human adipose samples (12.5 mg). These methods make it practical to screen milligram samples of biological substrates for contamination with persistent toxic substances at $ng \cdot g^{-1}$ levels. The predominant contaminants identified in human seminal plasma were PCB isomers and pentachlorophenol; compound identifications were based on a comparison with NICI spectra of standards. Dougherty et al. (499) discussed the utilization of NICI-MS for detection of xenobiotic chemical contamination, in particular the poly(chloro-naphthalenes), poly(chlorostyrenes), PCBS and related polychlorinated cyclic aromatics, in humans and in the food chain.

6.3 Plasticizers

Plastic medical devices are possible sources of plasticizers and several groups of investigators have measured the most commonly used plasticizers, di(2-ethylhexyl) phthalate (DEHP) and di-*n*-butyl phthalate (DBP) in patients. Measureable amounts of DEHP were detected in peripheral blood samples of hemodialyzed patients immediately after dialysis (500) and in autopsied heart and intestine of infants who had umbilical catheters inserted and received varying amounts of blood products (501). Plasticizer contamination has been reported in plasma (502) and during lipid analyses (503). Furthermore, several groups of workers (504—507) have noted plasticizer contamination of blood stored in plastic bags. The levels of extracted plasticizers increased with duration of storage and could be found in both the red cell and plasma fractions. In several instances the identification of the plasticizers in human plasma and tissues have been confirmed by GC-MS (501—508).

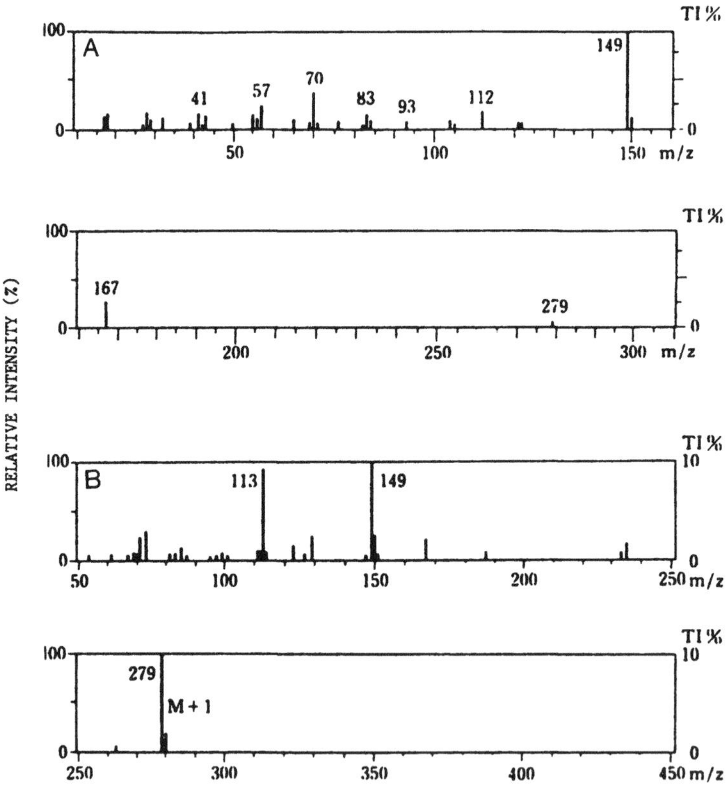

Figure 9. Electron impact (A) and chemical ionization (B) mass spectra of mono-(2-ethylhexyl)phthalate. Conditions: JMA-D 300 mass spectrometer with 70 eV (electron impact) and 300 eV (isobutane chemical ionization) ionization voltages Reproduced with permission from: Shintani, H. (Ref. 511).

The mass spectra of the dialkyl phthalates have been described (509, 510). The spectrum of dimethylphthalate was typical of most aromatic esters, with major fission products due to loss of OCH_3, $COOCH_3$, and $(COOCH_3)_2$ to give the *m/z* 163, 135 and 76 ions, respectively. The general fragmentation pathway for the dialkyl phthalates (except dimethylphthalate) are similar (509). The *m/z* 149 species was the most abundant ion in the mass spectra of many alkyl phthalates and is often accompanied by *m/z* 167. Most of the compounds give weak molecular ions and the main diagnostic high mass ions are $[M-(R-2H)]^+$ peaks. In the study of Ching et al. (508) the DBP and DEHP were identified by their GC retention times and mass spectra in which prominent ions were observed at *m/z* 149 and *m/z* 169 for DEHP and at *m/z* 149 for DBP. The background level in ethanol extracted samples was too high to be of value, and only the Dole's extract proved suitable for evaluation. The distilled water blank contained DEHP but not DBP and similar DEHP contamination of water has been reported (503). The circulating levels of DEHP were highest in patients immediately after receiving blood transfusion ($6-14\ \mu g \cdot ml^{-1}$) and when there was associated renal dysfunction. The DBP levels were relatively low in hospitalized patients (0.2 to 0.4 $\mu g \cdot ml^{-1}$) but higher in subjects 37 $\mu g \cdot ml^{-1}$) not hospitalized, suggesting contamination from environmental sources. A monomethylhexylphthalate ester was identified in human plasma that apparently arose by enzymatic hydrolysis of the diethylhexylphthalate which had migrated into blood from a flexible bag (511). Figure 9 shows the EI and CI mass spectra of the mono-(2-ethylhexyl)phthalate (511).

The plasticizer DBP is recognized (512) as being more toxic than DEHP but reviews of plasticizer toxicity studies concluded that only a low order of acute systemic toxicity has been demonstrated in animal and human experiments (513, 514). On the basis of animal studies, however, phthalate esters have been shown to significantly perturb lipid metabolism in liver, testes, heart, brain and fatty tissue, as well as alter the pattern of plasma lipoproteins and exert an effect on the developing fetus *in utero*.

6.4 Other fat-solubles

Many of the foreign compounds entering the human body are subject to enzymatic modification, including lipophilic conjugation, while certain drugs may be deliberately administered in the form of lipophilic conjugates in order to increase their efficacy (516). These conjugates may be identified and quantitated by many of the mass spectrometric methods described under the appropriate groups of compounds discussed above. FAB with or without collisional activation-tandem mass spectrometry appears particularly promising for work with these molecules. Fatty acid anions, alkylsulfonates, and long chain ammonium and phosphonium ions decompose by collisional activation to give structurally informative remote charge site fragmentations, which are useful for characterizing compounds containing long alkyl chains that are difficult to determine by other methods. The use of metal ions as reagents for CIMS is useful for obtaining molecular mass information and for structural analyses of lipophilic compounds of various degrees of complexity. Recently, Adams and Gross (517) have described an application of this method for determining double bond location in a large variety of fatty acids using FAB-MS-MS.

Other fat-soluble organic molecules have been identified and quantitated by GC-MS in the extracts of inorganic water-treatment chemicals (518).

7 Summary and Conclusions

MS alone or in combination with GC has provided the most powerful technique available for the identification and quantitation of lipid components in complex mixtures of biological origin. The recent combination of MS with HPLC expands the range of application to lipid components that are not volatile or that cannot be made volatile by derivatization. The ionization of lipid molecules by EI is often complemented by the softer methods of ionization which have special advantages for quantitative work because of the higher intensity of the parent ion. The major areas of application of MS in the clinical chemistry of lipids are concerned with the identification and quantitation of known lipids in biological fluids as a technique for the diagnosis of a variety of pathological conditions, and the determination of structure of unknown complex lipids. The quantitation of known lipids has been accomplished by means of stable isotope labeled derivatives, a technique which provides a definitive measurement of the lipid component in a sample and may be used for the standardization of other less direct methods. Identification of unknown lipids by MS has been less important in the practice of clinical chemistry, although several noteworthy applications have been made including fatty acids and xenobiotics along with their metabolites. The use of MS and GC-MS in the structure determination of complex lipids has been largely confined to research laboratories employing clinical material. Thus, the extensive analytical MS carried out on the glycosphingolipids in normal and pathological conditions has yet to be applied in routine clinical chemistry. In a more limited number of instances, GC-MS and LC-MS techniques have been employed in the identification and quantitation of the molecular species of the glycerolipids from normal subjects and patients, but the clinical significance of the differences demonstrated has yet to be established.

Despite the high accuracy and general applicability of the MS, GC-MS and LC-MS techniques, most of the activity has been confined to specialized laboratories with appropriate instrumentation and highly trained personnel. Initial instrument costs and maintenance expenses appear to be limiting the application of these new methods in clinical chemistry. Therefore, the promise of metabolic profiling of plasma and membrane lipids has not been realized and the use of the definitive methods of highly specific and sensitive measurement of individual lipids has remained at the demonstration-in-principle stage.

The comprehensive range of lipids that can be analyzed by MS, GC-MS and LC-MS ensures that this methodology will have a firm place in future work in clinical lipid chemistry and in the study of diseases as the biochemistry of lipids and their involvement in disease processes continues to be better understood.

Acknowledgment

The studies by the author and their collaborators were supported by the Heart and Stroke Foundation of Ontario, Toronto, Ontario and the Medical Research Council of Canada, Ottawa, Canada.

References

1. Liebich, H. M. (1978) The place of gas chromatography-mass spectrometry in clinical chemistry. J. Chromatogr. Biomed. Applic. *146*, 185–196.
2. Hill, R. E. & Whelan, D. T. (1984) Mass spectrometry and clinical chemistry. Clin. Chim. Acta *139*, 231–294.
3. Erni, F. (1982) Liquid chromatography-mass spectrometry in the pharmaceutical industry: objectives and needs. J. Chromatogr. Biomed. Applic. *251*, 141–151.
4. Lawson, A. M. (1975) The scope of mass spectrometry in clinical chemistry. Clinical Chem. *21*, 803–824.
5. Bjorkhem, I., Blomstrand, R. & Svensson, L. (1976) Determination of serum triglyceride by mass fragmentography. Clin. Chim. Acta *71*, 191–198.
6. Odham, G., Larsson, L. & Mardh, P.-A. (1984) Quantitative mass spectrometry and its application in microbiology, in: Gas Chromatography/Mass Spectrometry. Applications in Microbiology (Odham, R., Larsson, L. & Hardh, P. A., eds.), pp. 303–333 Plenum Press, New York, N. Y.
7. De Leenheer, A. P., Lefevere, M. F., Lambert, W. E. & Colinet, E. S. (1985) Isotope-dilution mass spectrometry in clinical chemistry. Adv. Clin. Chem. *24*, 111–161.
8. Moss, C. W. & Dees, S. B. (1975) Identification of fatty acids by gas chromatographic-mass spectrometric analysis of cellular fatty acids. J. Chromatogr. *12*, 595–604.
9. Marshall, J. L., Erickson, K. C. & Folsom, T. K. (1970) The esterification of carboxylic acids using a boron trifluoride-etherate-alcohol reagent. Tetrahedron Lett. 4011–4012.
10. Eistert, B. (1948) Syntheses with diazomethane, in: Newer Methods of Preparative Organic Chemistry, pp. 513–570, Wiley-Interscience, New York, N.Y.
11. Rietschel, E. T., Gottert, H., Luederitz, O. & Westphal, O. (1972) Nature and linkages of the fatty acids present in lipid A. component of Salmonella lipopolysaccharides. Eur. J. Biochem. *28*, 166–173.
12. Fautz, E., Rosenfelder, G. & Grotjahn, L. (1979) Isobranched 2- and 3-hydroxy fatty acids as characteristic lipid constituents of some gliding bacteria. J. Bact. *140*, 852–858.
13. Jantzen, E., Brynk, Hagen. N., Bergan, T. & Bovre, K. (1978) Fatty acids and monosaccharides of Neisseriaceae in relation to established taxonomy. NIPH Annals (Oslo) *1*, 59–71.
14. Myher, J. J., Kuksis, A., Marai, L. & Yeung, S. K. F. (1978) Microdetermination of molecular species of oligo- and polyunsaturated diacylglycerols by gas chromatography-mass spectrometry of their tert-butyl dimethylsilyl ethers. Anal. Chem. *50*, 557–561.
15. Myher, J. J., Kuksis, A., Shepherd, J., Packard, C. J., Morrisett, J. D., Taunton, O. D. & Gotto, A. M., Jr. (1981) Effect of saturated and unsaturated fat diets on molecular species of phosphatidylcholine and sphingomyelin of human plasma lipoproteins. Biochim. Biophys. Acta *666*, 110–119.
16. Pind, S., Kuksis, A., Myher, J. J. & Marai, L. (1984) Resolution and quantitation of diacylglycerol moieties of natural glycerophospholipids by reversed phase liquid chromatography with direct liquid inlet mass spectrometry. Can. J. Biochem. Cell Biol. *62*, 301–309.
17. Pind, S., Kuksis, A., Myher, J. J. & Marai, L. (1985) Deacylation of endogenously deuterated rat liver microsomal phospholipids by endogenous phospholipases. Can. J. Biochem. Cell Biol. *63*, 137–144.
18. Lauer, W. M., Aasen, A. J., Graff, G. & Holman, R. T. (1970) Mass spectrometry of triglycerides. I. Structural effects. Lipids *5*, 861–868.
19. Aasen, A. J., Lauer, W. M. & Holman, R. T. (1970) Mass spectrometry of triglycerides. II. Specifically deuterated triglycerides and elucidation of fragmentation mechanism. Lipids *5*, 869–877.
20. Samuelsson, B. & Samuelsson, K. (1969) Gas-liquid chromatography-mass spectrometry of synthetic ceramides. J. Lipid Res. *10*, 41–46.
21. Anderson, B. A., Karlsson, K. A., Pascher, I. (1972) Mass spectra of acetyl derivatives of homogeneous cerebrosides (monoglycosylceramides). Chem. Phys. Lipids *9*, 89–111.
22. Karlsson, K. A., Pascher, I., Samuelsson, B. E. & Steen, G. O. (1972) Mass spectra of trimethylsilyl derivatives of homogeneous cerebrosides (monoglycosylceramides). Chem. Phys. Lipids *9*, 230–246.
23. Karlsson, K. A., Samuelsson, B. E. & Steen, G. O. (1973) Separation of monoglycosylceramides (cerebrosides) of bovine kidney

into subgroups and characterization by mass spectrometry. Biochim. Biophys. Acta *306*, 317–328.
24. Polito, A. J., Akita, T. & Sweeley, C. C. (1968) Gas chromatography and mass spectrometry of sphingolipid bases. Characterization of sphinga-4,14-dienine from plasma sphingomyelin. Biochemistry *7*, 2609–2614.
25. Horning, M. G., Casparrini, G. & Horning, E. C. (1969) The use of gas phase analytical methods for the analysis of phospholipids. J. Chromatogr. Sci. *7*, 267–275.
26. Yergey, A. D., Kody, M. H. & Gershfeld, N. L. (1981) Quantitative analysis of disaturated lecithins in human plasma by ammonia chemical ionization mass spectrometry. Biomed. Mass Spectrom. *8*, 503–505.
27. Duncan, J. H., Lennarz, W. J. & Fenselau, C. C. (1971) Mass spectral analysis of glycerophospholipids. Biochemistry *10*, 927–932.
28. Cicero, T. J. & Sherman, W. R. (1971) Combined gas chromatography-mass spectrometry of brain polyphosphoinositides. Biochem. Biophys. Res. Commun. *42*, 428–433.
29. Breimer, M. E., Hansson, G. C., Karlsson, K. A., Leffler, H. (1981) Sequencing of oligosaccharides by mass spectrometry applied on a 12-sugar glycolipid. FEBS Lett. *124*, 299–303.
30. Kelly, R. W. & Taylor, P. L. (1976) Tert-Butyl dimethylsilyl ethers as derivatives for quantitative analysis of steroids and prostaglandins by gas phase methods. Anal. Chem. *48*, 465–467.
31. Barrow, S. E., Waddel, K. A., Ennis, M., Dollery, C. T. & Blair, I. A. (1982) Analysis of picomolar concentrations of 6-oxo-prostaglandin in biological fluids. J. Chromatogr. *239*, 71–80.
32. Steck, P. A., Schwartz, B. A., Rosendahl, M. S. & Gray, G. R. (1978) Mycolic acids. A reinvestigation. J. Biol. Chem. *253*, 5625–5629.
33. Anderson, B. A. (1978) Mass spectrometry of fatty pyrrolidides. Progr. Chem. Fats Other Lipids *16*, 279–308.
34. Anderson, B. A. & Holman. R. T. (1974) Pyrrolidides for mass spectrometric determination of the position of the double bond in monounsaturated fatty acids. Lipids *9*, 185–190.
35. Klein, R. A. & Schmitz, G. (1986) Double bond location in fatty acids. A critical analysis of the feasibility of using specifically deuterated pyrrolidides for mass spectral analysis. Biomed. Environm. Mass Spectrom. *13*, 429–437.
36. Suzuki, M., Ariga, T., Sekine, M., Araki, E. & Miyatake, T. (1981) Identification of double bond positions in polyunsaturated fatty acids by chemical ionization mass spectrometry. Anal. Chem. *53*, 985–988.
37. Dommes, V., Wirtz-Peitz, F. & Kunau, W. H. (1976) Structure determination of polyunsaturated fatty acids by gas chromatography-mass spectrometry. A comparison of fragmentation patterns of various derivatives. J. Chromatogr. Sci. *14*, 360–366.
38. Pileire, B., Beaune, Ph., Laudat, M. H. & Cartier, P. (1980) The use of formic acid in carrier gas. Rapid method for identification and determination of phytanic acid by gas chromatography-mass spectrometry-chemical ionization. J. Chromatogr. Biomed. Applic. *182*, 269–276.
39. Yao, J. K. & Dyck, P. J. (1987) Tissue distribution of phytanic acid and its analogues in a kinship with Refsum's disease. Lipids *22*, 69–75.
40. Larsson, K., Mardh, P. A. & Odham. G. (1979) Detection of tuberculostearic acid in Mycobacteria and Nocardiae by gas chromatogoraphy and mass spectrometry using selected ion monitoring. J. Chromatogr. Biomed. Applic. *163*, 221–224.
41. Abe, K. & Tamai, Y. (1982) Simultaneous determination of methyl esters of α-hydroxy and non-hydroxy fatty acids from brain cerebroside by fused-silica capillary gas chromatography. J. Chromatogr. Biomed. Applic. *232*, 400–405.
42. Alvedano, M. I. & Sprecher, H. (1987) Very long chain (C_{24} to C_{36}) polyenoic fatty acids of the n-3 and n-6 series in dipolyunsaturated phosphatidylcholines from bovine retina. J. Biol. Chem. *262*, 1180–1186.
43. Igarashi, M., Schaumburg, H. H., Powers, J., Kishimoto, Y., Kolodny, E. & Suzuki, K. (1976) Fatty acid abnormality in adrenoleukodystrophy. J. Neurochem. *26*, 851–860.
44. Kawamura, N., Mosere, A. B., Moser, H. W., Ogino, T., Suzuki, K., Schaumburg, H., Milunsky, A., Murphy, J. & Kishimoto, Y. (1978) High concentration of hexacosanoate in cultured skin fibroblast lipids from adrenoleukodystrophy patients. Biochem. Biophys. Res. Commun. *82*, 114–120.
45. Kishimoto, Y., Moser, H. W., Kawamura, N., Platt, M., Palante, S. L. & Fenselau, C.

(1980) Adrenoleukodystrophy: evidence that abnormal very long chain fatty acids of brain cholesteryl esters are of exogenous origin. Biochem. Biophys. Res. Commun. *96*, 69–76.

46. Krivit, W. & Hammarstrom, S. (1972) Identification and quantitation of free ceramides in human platelets. J. Lipid Res. *13*, 525–530.
47. Wherrett, J. R. & Hakamori, S. I. (1973) Characterization of a blood group B glycolipid, accumulating in the pancreas of a patient with Fabry's disease. J. Biol. Chem. *248*, 3046–3051.
48. Egge, H., Hanfland, P. & Furmuli, A. (1976) Sequenzanalyse biologisch relevanter Oligosacharide and Glycosphingolipide. Z. Anal. Chem. *279*, 165–166.
49. Myher, J., Kuksis, A., Breckenridge, W. C. & Little, J. A. (1981) Differential distribution of sphingomyelins among plasma lipoprotein classes. Can. J. Biochem. *59*, 626–636.
50. Murata, T. (1977) Analysis of triglycerides by gas chromatography chemical ionization mass spectrometry. Anal. Chem. *49*, 2209–2213.
51. Marai, L., Myher, J. J. & Kuksis, A. (1983) Analysis of triacylglycerols by reversed-phase high pressure liquid chromatography with direct liquid inlet mass spectrometry. Can. J. Biochem. Cell Biol. *61*, 840–849.
52. Myher, J. J., Kuksis, A., Marai, L. & Manganaro, F. (1984) Quantitation of natural triacylglycerols by reversed phase liquid chromatography with direct liquid inlet mass spectrometry. J. Chromatogr. *283*, 289–301.
53. Myher, J. J., Kuksis, A., Breckenridge, W. C. & Little, J. A. (1984) Studies of triacylglycerol structure of very low density lipoproteins of normal subjects and patients with Type III and Type IV hyperlipoproteinemia. Lipids *19*, 683–691.
54. Myher, J. J., Kuksis, A., Breckenridge, W. C., McGuire, V. & Little, J. A. (1985) Comparative studies of triacylglycerol structure of very low density lipoproteins and chylomicrons of normolipemic subjects and patients with Type II hyperlipoproteinemia. Lipids *20*, 90–101.
55. Bugaut, M., Myher, J. J., Kuksis, A. & Hoffman, A. G. D. (1984) An examination of the stereochemical course of acylation of 2-monoacylglycerols by rat intestinal villus cells using [$^2H_3$]palmitic acid. Biochim. Biophys. Acta *792*, 254–269.
56. Millington, D. S., Roe, C. R. & Maltby, D. A. (1984) Application of high resolution fast atom bombardment and constant B/E ratio linked scanning to the identification and analysis of acylcarnitines in metabolic disease. Biomed. Mass Spectrom. *11*, 236–241.
57. Bjorkhem, I., Blomstrand, R., Lantto, O., Svensson, L. & Ohman, G. (1976) Toward absolute methods in clinical chemistry: application of mass fragmentography to high accuracy analysis. Clin. Chem. *22*, 1789–1801.
58. Mee, J. M. L., Korth, J. & Halpern, N. (1976) Rapid quantitative blood analysis for free fatty acids by chemical ionization mass spectrometry. Anal. Lett. *9*, 1075–1083.
59. Parsons, H., Emken, E. A., Marai, L. & Kuksis, A. (1986) Simultaneous analysis of low plasma levels of deuterium-labeled saturated and unsaturated fatty acids as t-butyldimethylsilyl esters. Lipids *21*, 247–251.
60. Roe, C. R., Millington, D. S., Maltby, D. A., Bohan, T. P. & Hoppel, C. L. (1984) L-Carnitine enhances excretion of propionyl coenzyme A as propionylcarnitine in propionic acidemia. J. Clin. Invest. *73*, 1785–1788.
61. Parsons, H. G. & Emken, E. A. (1984) Simultaneous measurement of the assimilation of four-long-chain fatty acids in human chylomicron triglycerides and cholesteryl esters. Fed. Proc. *43*, 621. Abs. No. 1965.
62. Emken, E. A., Adlof, R. O., Rohwedder, W. K. and Gulley, R. M. (1983) Incorporation of deuterium-labeled trans- and cis-13-octadecanoic acids in human plasma lipids. J. Lipid Res. *24*, 34–46.
63. Emken, E. A., Rohwedder, W. K., Adlof, R. O., DeJarlais, W. J. & Gulley, R. M. (1986) Absorption and distribution of deuterium-labeled trans- and cis-11-octadecenoic acid in human plasma and lipoprotein lipids. Lipids *21*, 589–595.
64. Curstedt, T. & Sjovall, J. (1974) Biosynthetic pathways and turnover of individual biliary phosphatidylcholines during metabolism of [1,1-$^2H_2$]ethanol in rat. Biochim. Biophys. Acta *396*, 173–195.
65. Curstedt, T. (1982) Biosynthesis of acyl groups in molecular species of biliary phosphatidylcholines during metabolism of [2,2,2-$^2H_3$]ethanol in rats. Biochim. Biophys. Acta *713*, 589–601.
66. Curstedt, T. & Sjovall, J. (1974) Biosynthesis of molecular species of phosphatidylcholines in bile, liver and plasma of rats given

[1,1-$^2H_2$]ethanol. Biochim. Biophys. Acta *369*, 196–208.
67. Kuksis, A., Myher, J. J., Marai, L., Yeung, S. K. F., Steiman, I. & Mookerjea, S. (1975) Distribution of newly formed palmitate and stearate among molecular species of choline and ethanolamine phosphatides. Can. J. Biochem. *53*, 519–526.
68. Tanaka, Y. & Ando, S. (1981) Increased stearate formation by chain elongation in streptozotocin diabetic mice: compensation for decreased de novo fatty acid synthesis. Biomed. Res. *2*, 404–412.
69. Bougneres, P. I. & Bier, D. M. (1982) Stable isotope dilution method for measurement of palmitate content and labeled palmitate tracer enrichment in microliter plasma samples. J. Lipid Res. *23*, 502–507.
70. Riley, R. F., Hokama, Y. & Kratz, P. (1958) Nonsteroidal components of the unsaponifiable fraction of human liver. Cancer Res. *18*, 825–832.
71. Cain, C. E., Bell, O. E., Jr., White, H. B., Jr., Sulya, L. L. & Smith, R. R. (1967) Hydrocarbons from human meninges and meningiomas. Biochim. Biophys. Acta *144*, 493–500.
72. Boitnott, J. K. & Margolis, S. (1966) Mineral oil in human tissues. I. Detection of saturated hydrocarbons using thin layer chromatography. Bull. Johns Hopkins Hosp. *118*, 402–413.
73. Boitnott, J. K. & Margolis, S. (1966) Mineral oil in human tissues. II. Oil droplets in lymph nodes of the porta hepatis. Bull. Johns Hopkins Hosp. *118*, 414–422.
74. Rose, H. B. & Liber, A. F. (1966) Accumulation of saturated hydrocarbons in human spleens. J. Lab. Clin. Med. *68*, 475–483.
75. Heckers, H., Melcher, F. W., Dittmar, K. & Kalinowski, H. O. (1978) Paraffin oil pneumonia. Analysis of saturated hydrocarbons in different human tissues. J. Chromatogr. Biomed. Applic. *146*, 91–102.
76. Gazzarrini, F. & Nagy, B. (1966) Saturated hydrocarbons in human femoral arterial tissues and plaques. Arch. Biochem. Biophys. *113*, 245–247.
77. Brooks, C. J. W., Steel, G. & Harland, W. A. (1970) Lipids of human atheroma. VI. Hydrocarbons of the atheromatous plaque. Lipids *5*, 818–824.
78. Ferreti, A. & Flanagan, V. P. (1977) Tissue variation in hydrocarbon composition in the rabbit. Lipids *12*, 198–203.
79. Lintas, C., Balduzzi, A. M., Bernardini, M. P. & Di Muccio, A. (1979) Distribution of hydrocarbons in bovine tissues. Lipids *14*, 298–303.
80. Tornabene, T. G. & Markey, S. (1971) Characterization of monounsaturated hydrocarbons of Sarcina lutea and Sarcina flava. Lipids *6*, 190–195.
81. Schnoes, H. K. & Burlingame, A. L. (1970) Applications of mass spectrometry to organic geochemistry, in: Topics in Organic Mass Spectrometry (Burlingame, A. L., ed.) pp. 369–439. Wiley-Interscience, New York.
82. Meili, J., Wells, F. C., McPherron, R. & Burlingame, A. L. (1979) Design, implementation, and performance of a high resolution GC- high resolution MS-real-time computer system for the analysis of complex organic mixtures. J. Chromatogr. Sci. *17*, 29–42.
83. Henneberg, D. & Schomburg, G. (1968) Mass spectra of isoparaffins, obtained from a capillary GC/MS (gas chromatograph/mass spectrometer)-combination applied to isomeric mixtures made by insertion of methylene and methylene-d_2 into paraffins. Adv. Mass Spectrom. *4*, 333–343.
84. McCarthy, E. D., Han, J. & Calvin, M. (1968) Hydrogen atom transfer in mass spectrometric fragmentation patterns of saturated aliphatic hydrocarbons. Anal. Chem. *40*, 1475–1480.
85. McLafferty, F. W. (1973) Interpretation of mass spectra. Second Edition, W. A. Benjamin, Inc., Reading, MA.
86. Anon. (1968) Mineral oil in human tissues. Nutrition Revs. *25*, 46–49.
87. Williams, D. H. & Fleming, J. (1971) Spektrochemische Methoden in der Organischen Chemie. Georg Thieme Verlag, Berlin.
88. Lampe, M. A., Williams, M. L. & Elias, P. M. (1983) Human epidermal lipids: characterization and modulations during differentiation. J. Lipid Res. *24*, 131–140.
89. Lampe, M. A., Burlingame. A. L., Whitney, J., Williams, M. L., Brown, B. E., Roitman, E. & Elias, P. M. (1983) Human stratum corneum lipids: characterization and regional variations. J. Lipid Res. *24*, 120–130.
90. Williams, M. L. & Elias, P. M. (1982) N-alkanes in normal and pathological scalp. Biochem. Biophys. Res. Commun. *107*, 322–328.
91. Cerbulis, J., Flanagan, V. P. & Farrell, H. M., Jr. (1985) Composition of the hydrocabon fraction of goat's milk. J. Lipid Res. *26*, 1438–1443.

92. Nachman, R. J. (1985) Unusual predominance of even-carbon hydrocarbons in an antarctic food chain. Lipids *20*, 629–633.
93. Maeda, K., Kawaguchi, S., Niwa, T., Ohki, T. & Kobayashi, K. (1980) Identification of some abnormal metabolites in psoriatic nail using gas chromatography mass spectrometry. J. Chromatogr. *221*, 199–204.
94. Jacobson, M. (1972) Insect Sex Pheromones, pp. 170–193, Academic Press, New York, N.Y.
95. Horiike, M. & Hirano, C. (1980) A facile synthesis of the sex pheromones (Z)-7-dodecen-1-yl acetate and its homologs. Agric. Biol. Chem. *44*, 2229–2230.
96. Horiike, M., Norima, M. & Hirano, C. (1981) An empirical approach to the location of double bonds in tetradecen-1-yl acetates by mass spectrometry. Biomed. Mass Spectrom. 8, 41–42.
97. Benz, W. & Biemann, K. (1964) The specificity of the electron impact induced elimination of water from alchols and related reactions in the mass spectrometer. J. Am. Chem. Soc. *86*, 2375–2378.
98. Budzikiewicz, H., Djerassi, C. & Williams, D. M. (1967) Interpretation of mass spectra of organic compounds, pp. 94–128, Holden Day, Inc., San Francisco, CA.
99. Karlsson, K. A., Samuelsson, B. E. & Steen, G. O. (1973) Improved identification of monomethyl paraffin chain branching (close to the methyl end) of long-chain compounds by gas chromatography mass spectrometry. Chem. Phys. Lipids *1*, 17–38.
100. Corey, E. J. & Venkateswarlu, A. (1972) Protection of hydroxyl groups as tert-butyldimethylsilyl derivatives. J. Am. Chem. Soc. *94*, 6190–6191.
101. Mallet, A. I., Minnikin, D. E. & Dobson, G. (1984) Gas chromatography mass spectrometry of t-butyldimethylsilyl ethers of phthiocerols and mycocerosic alcohols from Mycobacterium tuberculosis. Biomed. Mass Spectrom. *11*, 79–86.
102. Larsson, L., Mardh, P. A., Odham, G. & Westerdahl, G. (1981) Use of selected ion monitoring for detection of tuberculostearic and C_{32} mycocerosic acid in Mycobacteria and in five day old cultures of sputum specimens from patients with pulmonary tuberculosis. Acta Path. Microbiol. Scand. Sect. *B89*, 245–251.
103. Mardh, P. A., Larsson, L., Odham, G. & Engbaek, H. C. (1983) Tuberculostearic acid as a diagnostic marker in tuberculous meningitis. Lancet *1*, 367.
104. Bergman, R., Larsson, L., Odham, G. & Westerdahl, G. (1983) Selected ion monitoring for rapid differentiation of mycobacteria isolated from domestic animals. J. Microbiol. Methods *1*, 19–22.
105. Draper, P., Payne, S. N., Dobson, G. & Minnikin, D. E. (1983) Isolation of a characteristic phthiocerol dimycocerosate from Mycobacterium leprae. J. Gen. Microbiol. *129*, 859–863.
106. Horiike, M. & Hirano, C. (1984) Characterization of double bond positions in dodecenols by electron impact mass spectrometry. Biomed. Mass Spectrom. *11*, 145–148.
107. Nicolaides, N., Santos, E. C. & Papadakis, K. (1984) Double bond patterns of fatty acids and alcohols in steer and human meibomian gland lipids. Lipids *19*, 264–277.
108. Harvey, D. J. & Tiffany, J. M. (1984) Comparison of derivatives for the characterization of branched long-chain alcohols and 1,2-diols by mass spectrometry. Biomed. Mass Spectrom. *11*, 353–359.
109. Vetter, W., Meister, W. & Oesterhelt, G. (1978) Alkylaminopyridines and some analogues as derivatives for the structure elucidation of long aliphatic chains by mass spectrometry. Helv. Chim. Acta *61*, 1287–1295.
110. Harvey, D. J. (1986) Mass spectometry of the picolinyldimethylsilyl (PICSI) ethers: New derivatives for the structural determination of unsaturated fatty alcohols by gas chromatography/mass spectrometry. Biomed. Environm. Mass Spectrom. *14*, 103–109.
111. Jantzen, E. (1984) Analysis of cellular components in bacterial classification and diagnosis, in: Gas chromatography/Mass Spectrometry. Applications in Microbiology (Odham, G., Larsson, L. & Mardh, P.-A., eds.), pp. 257–302, Plenum Press, New York, N.Y.
112. Igarashi, M., Schaumburg, H. H., Powers, J., Kishimoto, Y., Kolodny, E. & Suzuki, K. (1976) Fatty acid abnormalities in adrenoleukodystrophy. J. Neurochem. *26*, 851–860.
113. Tsuji, S., Suzuki, M., Ariga, T., Sekine, M., Kuriyama, M. & Miyatake, T. (1981) Abnormality of long-chain fatty acids in erythrocyte membrane sphingomyelin from patients with adrenoleukodystrophy. J. Neurochem. *36*, 1046–1049.
114. Boksdale, L. & Kim, K.-S. (1977) Mycobacterium. Bacteriol. Revs. 41, 217–372.

115. Tsuji, S. (1981) Increased synthesis of hexacosanoic acid ($C_{26:0}$) by cultured skin fibroblasts from patients with adrenoleukodystrophy (ALD) and adrenomyeloneuropathy (AMN). J. Biochem. *90*, 1233–1236.
116. Asselineau, C. & Asselineau, J. (1984) Fatty acids and complex lipids, in: Gas Chromatography/Mass spectrometry. Applications in microbiology (Odham, G., Larsson, L. & Mardh, P.-A., eds.) pp. 57–102, Plenum Press, New York, N.Y.
117. Odham, G., Larsson, L. & Mardh, P.-A. (1979) Determination of tuberculostearic acid in sputum from patients with pulmonary tuberculosis by selected ion monitoring, J. Clin Invest. *63*, 813–819.
118. Sud, I. J. & Feingold, D. S. (1979) Detection of 3-hydroxy fatty acids at picogram levels in biologic specimens. A chemical method for the detection of Neisseria gonorrhoeae? J. Investing. Dermat. *73*, 521–525.
119. French, G. L., Chan, C. Y., Cheung, S. W., Teoh, R., Humphries, M. J. & O'Mahony, G. (1987) Diagnosis of tuberculous meningitis by detection of tuberculostearic acid in cerebrospinal fluid. Lancet *ii*, 117–119.
120. Rohwedder, W. K., Emken, E. A. & Wolff, D. J. (1985) Analysis of deuterium labeled blood lipids by chemical ionization mass spectrometry. Lipids *20*, 303–311.
121. Bougneres, P. F. & Bier, D. M. (1982) Stable isotope dilution method for measurement of palmitate content and labeled palmitate tracer enrichment in microliter plasma samples. J. Lipid Res. *23*, 502–507.
122. Hunneman, D. H. & Schweickhardt, C. (1982) Mass fragmentographic determination of myocardial free fatty acids. J. Mol. Cellular Cardiol. *14*, 339–351.
123. Larsson, L., Mardh, P.-A., Odham, G. & Westerdahl, G. (1980) Detection of tuberculostearic acid in biological specimens by means of glass capillary gas chromatography-electron and chemical ionization mass spectrometry, utilizing selected ion monitoring. J. Chromatogr. Biomed. Applic. *182*, 402–408.
124. Odham. G. (1980) Fatty Acids, in: Biochemical Applications of Mass Spectrometry (Waller, G. R., Dermer, O. C., eds.) pp. 153–171, Wiley-Interscience, New York, N.Y.
125. Kuksis, A., Myher, J. J., Marai, L. & Geher, K. (1976) Estimation of plasma free fatty acids as trimethylsilyl (TMS) esters. Anal. Biochem. *70*, 302–312.
126. Woollard, P. M. (1983) Selective silylation using tert butyldimethylsilyl reagents: their use in the quantification of fatty acids. Biomed. Mass Spectrom. *10*, 143–154.
127. Harvey, D. J. (1984) Picolinyl derivatives for the structural determination of fatty acids by mass spectrometry: applications to polyenoic acids, hydroxy acids, di-acids and related compounds. Biomed. Mass Spectrom. *11*, 340–347.
128. Markey, S. P., Keyser, A. J. & Levine, S. P. (1974) in: Application of gas chromatography-mass spectrometry to the investigation of human disease (Mamer, O. A., Mitchell, W. J., Scriver, C. R., eds.) Appendix, p. 239, McGill University-Montreal Children's Hospital Research Institute, Montreal, Quebec.
129. Schaumburg, H. H., Powers, J. M., Raine, C. S., Suzuki, K. & Richardson, E. P. (1975) Adrenoleukodystrophy. A clinical and pathological study of 17 cases. Arch. Neurol. *32*, 577–591.
130. Brown, III F. R., Van Duyn, M. A. S., Moser, A. B., Schulman, J. D., Rizzo, W. B., Snyder, R. D., Murphy, J. V., Kamoshita, S., Migeon, C. J. & Moser, H. W. (1982) Adrenoleukodystrophy: Effects of dietary restriction of very long chain fatty acids and of administration of carnitine and clofibrate on clinical status and plasma fatty acids. Johns Hopkins Med. J. *151*, 164–172.
131. Goldfischer, S. (1982) Peroxisomes and human metabolic diseases: The cerebrohepatorenal syndrome (CHRS), cerebrotendinous xanthomatosis, and Schelder's disease (adrenoleukodystrophy). Ann. NY Acad. Sci. *386*, 526–529.
132. Brown, III F. R., McAdams, A. J., Cummins, J. W., Konkol, R., Singh, I., Moser, A. B. & Moser, H. W. (1982) Cerebro-hepato-renal (Zellweger) syndrome and neonatal adrenoleukodystrophy: similarities in phenotype and accumulation of very long chain fatty acids. Johns Hopkins Med. J. *151*, 344–361.
133. Bakkeren, J. A. J. M., Monnens, L. A. H., Trijbels, J. M. F. & Maas, J. M. (1984) Serum very long chain fatty acid pattern in Zellweger syndrome. Clin. Chim. Acta *138*, 325–331.
134. Ryhage, R., Stallberg-Stenhagen, R. & Stenhagen, E. (1961) Mass spectrometric studies. VII. Methyl esters of α, β-unsaturated long chain acids. On the structure of

C_{27}-phthienoic acid. Arkiv Kemi *18*, 179–187.

135. Coombe, R. G., Vine, J. H. & Yip. H. (1985) Unusual fatty acids from amiotic fluid phospholipids. J. Chromatogr. Biomed. Applic. *341*, 146–153.
136. Minnikin, D. E. (1978) Location of double bonds and cyclopropane rings in fatty acids by mass spectrometry. Chem. Phys. Lipids *21*, 313–347.
137. Murata, T., Ariga, T. & Araki, E. (1978) Determination of double bond positions of unsaturated fatty acids by a chemical ionization mass spectrometry computer system. J. Lipid Res. *19*, 172–176.
138. Kobayashi, T., Nishijima, M., Tamori, Y., Nojima, S., Sayama, S. & Yamakawa, T. (1980) Acylphosphatidylglycerol of *Escherichia coli*. Biochim. Biophys. Acta *620*, 356–363.
139. Janssen, G., Parmentier, G., Verhulst, A. & Eyssen, H. (1985) Location of the double bond positions in microbial isomerization and hydrogenation products of α- and γ-linolenic acids. Biomed. Mass Spectrom. *12*, 134–138.
140. Plattner, R. D., Spencer, G. F. & Kleiman, R. (1976) Double bond location in polyenoic fatty esters through partial oxymercuration. Lipids *11*, 222–227.
141. Smith, C. R. (1977) Structural analysis of polyunsaturated fatty acids, in: Polyunsaturated Fatty Acids (Kunau, W. H. & Holman, R. T., eds.). pp. 81–103 American Oil Chemists' Society, Champaign, IL.
142. Takayama, K., Qureshi, N. & Schnoes, H. K. (1978) Isolation and characterization of the monounsaturated long chain fatty acids of *Mycobacterium tuberculosis*. Lipids *13*, 575–579.
143. Chai, R. & Harrison, A. G. (1981) Location of double bonds by chemical ionization mass spectrometry. Anal. Chem. *53*, 34–37.
144. Myher, J. J., Marai, L. & Kuksis, A. (1974) Identification of fatty acids by GC-MS using polar siloxane liquid phases. Anal. Biochem. *62*, 188–203.
145. Fellenberg, A. J., Johnson, D. W., Poulos, A. & Sharp, P. (1986) Simple mass spectrometric differentiation of the n-3, n-6 and n-9 series of methylene interrupted polyenoic acids. Biomed. Environm. Mass Spectrom. *14*, 127–129.
146. Woollard, P. M. & Mallet, A. I. (1984) Lipoxygenase products. A novel gas chromatographic-mass spectrometric assay for monohydroxy fatty acids. J. Chromatogr. *306*, 1–21.
147. Kleiman, R. & Spencer, G. F. (1971) Ricinoleic acid in *Linum mucronatum* seed oil. Lipids *6*, 962–963.
148. Abe, K. & Tamai, Y. (1982) Simultaneous determination of methyl esters of α-hydroxy and non-hydroxy fatty acids from brain cerebroside by fused-silica capillary gas chromatography. J. Chromatogr. Biomed. Applic. *232*, 400–405.
149. Nicolaides, N., Soukup, V. G. & Ruth, E. C. (1983) Mass spectrometric fragmentation patterns of the acetoxy and trimethylsilyl derivatives of all the positional isomers of the methyl hydroxypalmitates. Biomed. Mass Spectrom. *10*, 441–449.
150. Maitra, S. K., Schotz, M. C., Yoshikawa, T. T. & Guze, L. B. (1978) Determination of lipid A and endotoxin in serum by mass spectroscopy. Proc. Natl. Acad. Sci. U.S.A. *72*, 3993–3997.
151. Marom, Z., Shelheimer, Sun, F. & Kaliner, M. (1983) Human airway monohydroxyeicosatrienoic acid generation and mucous release. J. Clin. Invest. 72, 122–127.
152. Low, I. A., Lui, R. H., Barker, S. A., Fish, F., Settine, R. L., Piotrowski, E. G., Damert, W. C. & Liu, J. Y. (1985) Selected ion monitoring mass spectrometry: parameters affecting quantitative determination. Biomed. Mass Spectrom. *12*, 633–637.
153. Lederer, E., Adam, A., Ciorbaru, R., Petit, J. F. & Wietzerbin, J. (1975) Cell walls of *Mycobacteria* and related organisms. Mol. Cell Biochem. *7*, 87.
154. Yano, I., Kageyama, K., Ohno, Y., Masui, M., Kusunose, E., Kusunose, M. & Akimori, N. (1978) Separation and analysis of molecular species of mycolic acids in *Nocardia* and related taxa by gas chromatography-mass spectrometry. Biomed. Mass Spectrom. *5*, 14–24.
155. Gailly, G., Sandra, P., Verzele, M. & Cocito, C. (1982) Analysis of mycolic acids from a group of *Corynebacteria* by capillary gas chromatography and mass spectrometry. Eur. J. Biochem. *125*, 83–94.
156. Kamagata, K. & Suzuki, K. I. (1987) Lipid and Cell wall analysis in bacterial systematics. Methods Microbiol. *19*, 161–207.
157. Qureshi, N., Takayama, K. & Schnoes, H. K. (1980) Purification of C_{30-56} fatty acids from *Mycobacterium tuberculosis* H37Ra. J. Biol. Chem. *255*, 182–189.

158. Ribi, E., Takayma, K., Milner, K., Gray, G. R., Goren, M., Parker, R., McLaughlin, C. & Kelly, M. (1976) Regression of tumours by an endotoxin combined with trihalose mycolates of differing structure. Cancer Immunol. Immunother. *1*, 265–270.
159. Karlsson, K. A., Samuelsson, B. E. & Steen, G. O. (1973) Improved identification of monomethyl paraffinic chain branching (close to the methyl end) of ion chain compounds. Chem. Phys. Lipids *11*, 17–38.
160. Ryhage, R. & Stenhagen, E. (1960) Mass spectrometry in lipid research. J. Lipid Res. *1*, 361–390.
161. McCloskey, J. A. (1970) Mass spectrometry of fatty acid derivatives, in: Topics in Lipid Chemistry Volume 1 (Gunstone, F. D., ed.). pp. 369–440 Logos Press, London, England.
162. Jensen, N. J. & Gross, M. L. (1986) Fast atom bombardment and tandem mass spectrometry for determining iso- and anteiso-fatty acids. Lipids *21*, 362–365.
163. Harvey, D. J. (1982) Picolinyl esters as derivatives for the structural determination of long chain branched and unsaturated fatty acids. Biomed. Mass Spectrom. *9*, 33–38.
164. Harvey, D. J. (1984) Picolinyl derivatives for the structural determination of fatty acids by mass spectrometry: Applications to polyenoic acids, hydroxy acids, dia-acids and related compounds. Biomed. Mass Spectrom. *11*, 340–347.
165. Ryhage, R. & Stenhagen, E. (1960) Mass spectrometric studies. IV. Esters of monomethyl-substituted long chain carboxylic acids. Arkiv Kemi *15*, 291–315.
166. Julak, J., Turecek, F. & Mikova, Z. (1980) Identification of characteristic branched chain fatty acids of *Mycobacterium kansasii* and *Mycobacterium gordonae* by gas chromatography mass spectrometry. J. Chromatogr. *190*, 183–187.
167. Steinberg, D. (1978) Phytanic acid storage disease: Refsum's syndrome, in: The Metabolic Basis of Inherited Disease (Stanbury, J. B., Wyngaarden, J. B., Fredrickson, D. S., eds.) pp. 688–706. McGraw-Hill Book Co., New York, New York.
168. Stenhagen, E. (1961) Mass spectrometry in determination of structure of organic compounds especially lipids and peptides. Z. Anal. Chem. *81*, 462–480.
169. Pileire, B., Beaune, Ph., Laudat, M. H. & Cartier, P. (1980) The use of formic acid in carrier gas. Rapid method for identification and determination of phytanic acid by gas chromatography-mass spectrometry-chemical ionization. J. Chromatogr. Biomed. Applic. *182*, 269–276.
170. Skrbic, T. R. &Cumings, N. (1969) Phytanic acid in tissue lipids in Refsum's disease. Clin. Chim. Acta *23*, 17–21.
171. Greter, J., Lindstedt, S. & Steen, G. (1983) 2,6-Dimethyloctanedioic acid – a metabolite of phytanic acid in Refsum's disease. Clin. Chem. *29*, 434–437.
172. Yao, J. K., Jardine, I., Dyck, P. J. (1982) Presence of plasma branched-chain fatty acids in multineuronal degeneration, hepatosplenomegaly and adrenocortical insufficiency. J. Neurol. Sci. *55*, 185–195.
173. Evans, J. E. & Dulaney, J. T. (1983) Location of double bonds in two unsaturated forms of phytanic acid from Refsum disease as determined by mass spectrometry. Biochem. Biophys. Acta *754*, 346–352.
174. Asselineau, J. (1966) in: The Bacterial Lipids (Asselineau, J., ed.) pp. 67–151. Hermann, Paris and Holden-Day, San Francisco.
175. Ryhage, R. (1962) Mass spectrometric analyses of the methyl stearic acid from *Mycobacterium phlei*. J. Biol. Chem. *237*, 670–671.
176. Odham, G. & Stenhagen, E. (1972) Fatty acids, in: Biochemical Applications of Mass Spectrometry (Waller, G. R., ed.), Wiley-Interscience, New York pp. 211–228.
177. Fisher, G. S. & Cherry, J. P. (1983) Variation of cyclopropanoid fatty acids in cottonseed lipids. Lipids *18*, 589–594.
178. Panos, C. & Henrickson, C. V. (1968) Resolution of positional isomers of bacterial long chain cyclopropane ring containing fatty acids by capillary column chromatography. J. Gas Chromatogr. *6*, 551–554.
179. Klenk, E. (1944) Eine Methode zur Darstellung der Acetal-phosphatide des Gehirn und ihrer Aldehydkomponent in Form der Dimethylacetale. Hoppe-Seyler's Z. Physiol. Chem. *281*, 25–28.
180. Christiansen, K., Mahadevan, V., Viswanathan, C. V. & Holman. R. T. (1969) Mass spectrometry of long-chain aliphatic aldehydes, dimethyl acetals and alk-1-enyl ethers. Lipids *4*, 421–427.
181. Bjorkhem, I., Sisfontes, L., Bostrom, B., Kase, B. F. & Blomstrand, R. (1986) Simple diagnosis of the Zellweger syndrome by gas-liquid chromatography of dimethylacetals. J. Lipid Res. *27*, 786–791.
182. Aasen, A. J., Hofstetter, H. H., Iyengar, B. T. R. & Holman, R. T. (1971) Identifica-

tion and analysis of wax esters by mass spectrometry. Lipids *6*, 502–507.
183. Vajdi, M., Nawar, W. W. & Merritt, Jr. C. (1981) GC/MS analysis of some long chain esters, ketones and propanediol diesters. J. Am. Oil Chemists Soc. *58*, 106–110.
184. Spencer, G. F., Plattner, R. D. & Miwa, T. (1977) Jojoba oil analysis by high pressure liquid chromatography and gas chromatography/mass spectometry. J. Am. Oil Chemists Soc. *54*, 187–189.
185. Spencer, G. F. & Tallent, W. H. (1973) Sperm whale oil analysis by gas chromatography and mass spectrometry. J. Am. Oil Chemists Soc. *50*, 202–206.
186. Spencer, G. F. (1979) Alkoxy-acyl combinations in the wax esters from winterized sperm whale oil by gas chromatography-mass spectrometry. J. Am. Oil Chemists Soc. *56*, 642–646.
187. Wakeham, S. G. & Frew, N. M. (1982) Glass capillary gas chromatography-mass spectrometry of wax esters, steryl esters and triacylglycerols. Lipids *17*, 831–843.
188. Plattner, R. D. & Spencer, G. F. (1983) Chemical ionization mass spectrometry of wax esters. Lipids *18*, 68–73.
189. Gupta, I., Baptista, J., Bruce, W. R., Che, C. T., Furrer, R., Gingerich, J. S., Grey, A. A., Marai, L., Yates, P. & Krepinsky, J. J. (1983) Structure of fecapentaenes, the mutagens of bacterial origin isolated from human feces. Biochemistry *22*, 241–245.
190. Johnson, C. B. & Holman. R. T. (1966) Mass spectrometry of lipids II. Monoglycerides, their diacetyl derivatives and their trimethylsilyl ethers. Lipids *1*, 371–380.
191. Curstedt, T. (1974) Mass spectra of trimethylsilyl ethers of ^{2}H-labeled mono- and diglycerides Biochim. Biophys. Acta *360*, 12–23.
192. Myher, J. J., Marai, L. & Kuksis, A. (1974) Identification of monoacyl- and monoalkylglycerols by gas-liquid chromatography-mass spectrometry using polar siloxane liquid phases. J. Lipid Res. *15*, 586–592.
193. Gaskell, S. J. & Brooks, C. J. W. (1977) Gas-liquid chromatography-mass spectrometry of phospholipid mixtures after enzymic hydrolysis. J. Chromatogr. *142*, 469–480.
194. Hallgren, B. & Larsson, S. (1962) The glyceryl ethers in the liver oils of elasmobranch fish. J. Lipid Res. *3*, 31–38.
195. Hallgren, B. & Stallberg, G. (1967) Methoxy-substituted glycerol ethers isolated from Greenland shark liver oil. Acta Chem. Scand. *21*,1519–1529.
196. Snyder, F., Rainey, W. T., Blank, M. L., Jr. & Christie, W. W. (1970) The source of oxygen in the ether bond of glycerolipids. J. Biol. Chem. *245*, 5853–5856.
197. Saito, K. & Gamo, M. (1973) The distribution of diol waxes in preen glands of some birds III. The occurrence of 1,2-diols. Comp. Biochem. Physiol. *45B*, 603–611.
198. Downing, D. T. & Sharaf, D. M. (1976) Skin surface lipids of guinea pig. Biochim. Biophys. Acta *431*, 378–389.
199. Muramatsu, T. & Schmid, H. H. (1972) 1-0-2^1-hydroxyalkyl and 1-0-2^1-ketoalkyl glycerols. Chem Phys. Lipids *9*, 123–132.
200. Kasama, K., Rainey, W. T. & Snyder, F. (1973) Chemical identification and enzymic synthesis of a newly discovered lipid class – hydroxyalkylglycerols. Arch. Biochem. Biophys. *154*, 648–658.
201. Hallgren, B., Niklasson, A., Stallberg, G. & Thorin, H. (1974) On the occurrence of 1-0-(2-methoxy-alkyl) glycerols and 1-0-phytanylglycerol in marine animals. Acta Chem. Scand. *28B*, 1035–1040.
202. Totani, N. & Muramatsu, T. (1981) A new procedure for the acetylation of lipids. Chem. Phys. Lipids *29*, 375–377.
203. Egge, H. (1983) Ether glycolipids, in: Ether Lipids: Biochemical and Biomedical Aspects (Mangold, H. K. & Paltauf, F., eds.) pp. 17–47, Academic Press, New York.
204. Kumar, K., Weintraub, S. T., McManus, L. M., Pinckard, R. V. & Hanahan, D. J. (1984) A facile route to semi-synthesis of acetylglycerylether phosphoethanolamine and its choline analogue. J. Lipids Res. *25*, 198–208.
205. Odham, G. & Stenhagen, E. (1972) Complex lipids, in: Biochemical Applications of Mass Spectrometry (Waller, G. R., ed.) pp. 229–249, J. Wiley and Sons, New York.
206. Myher, J. J. (1978) Separation and determination of the structure of acylglycerols and their ether analogues, in: Fatty Acids and Glycerides (Kuksis, A., ed.) pp. 123–196, Plenum Press, New York.
207. Wood, G. W. (1980) Complex lipids, in: Biochemical Applications of Mass Spectrometry. First Supplementary Volume (Waller, G. R. & Dermer, O. C., eds.) pp. 173–209, John Wiley and Sons, New York.
208. Spener, F. (1983) Ether lipids in clinical diagnosis and medical research, in: Ether Lipids: Biochemical and Biomedical Aspects (Mangold, H. K. & Paltauf, F., eds.) pp. 239–259, Academic Press, New York.

209. Bruce, W. R., Varghese, A. J., Furrer, R. & Land, P. C. (1977) A mutagen in the feces of normal humans, in : Origins of Human Cancer (Hiatt, H. H., Watson, J. D. & Winsten, J. S., eds.) pp. 1641–1646, Cold Spring Harbour Laboratory, Cold Spring Harbour, New York.
210. Friedberg, S. J., Heifetz, A. & Greene, R. C. (1971) Loss of hydrogen from dihydroxyacetone phosphate during glyceryl ether synthesis. J. Biol. Chem. *246*, 5822–5827.
211. Friedberg, S. J., Weintraub, S. T., Singer, M. R. & Greene, R. C. (1982) The mechanism of ether bond formation in O-alkyl lipid synthesis in Ehrlich Ascites tumor. J. Biol. Chem. *258*, 136–142.
212. Curstedt, T. (1983) Chromatographic analysis, isolation and characterization of ether lipids, in: Ether Lipids: Biochemical and Biomedical Aspects (Mangold, H. K. & Paltauf, F., eds.), pp. 1–15, Academic Press, New York.
213. Morrison, A., Barratt, M. D. & Aneja, R. (1970) Mass spectrometry of some deuterated 1,3-distearins. Chem. Phys. Lipids *4*, 47–59.
214. Hasegawa, K. & Suzuki, T. (1973) Determination of molecular species of ovolecithin using gas chromatography-mass spectrometry. Lipids *8*, 631–634.
215. Hasegawa, K. & Suzuki, T. (1975) Examination of acetolysis products of phosphatidylcholine by gas chromatography-mass spectrometry. Lipids *10*, 667–672.
216. Barber, M., Chapman, J. R. & Wolstenholme, W. A. (1968) Lipid analysis by coupled mass spectrometry-gas chromatography (MS-GLC) I. Diglycerides. J. Spectrom. Ion Phys. *1*, 98–101.
217. Morley, N. H., Kuksis, A., Buchnea, D. & Myher, J. J. (1975) Hydrolysis of diacylglycerols by lipoprotein lipase. J. Biol. Chem. *250*, 3414–3418.
218. Siegel, M. I., McConnell, R. T., Brent, D. A., Chandrasurin, P., Macfarlane, R. D. & McNeal, C. J. (1982) The formation of diarachidonyl diglyceride by rat neutrophils. Mol. Pharmacol. *21*, 688–693.
219. Curstedt, T. & Sjövall, J. (1974) Analysis of molecular species of ^{2}H-labeled phosphatidylcholines by liquid-gel chromatography and gas chromatography-mass spectrometry. Biochim. Biophys. Acta *360*, 24–37.
220. Cronholm, T., Curstedt, T. & Sjövall, J. (1980) Origin of biliary phosphatidylcholines studied by coenzyme labeling with [1,1-$^2H_2$] ethanol. Biochim. Biophys. Acta *753*, 276–279.
221. Curstedt, T. (1975) Biosynthesis of acyl groups in molecular species of biliary phosphatidylcholines during metabolism of [2,2,2$^2H_3$] ethanol. Biochim. Biophys Acta *398*, 265–274.
222. Curstedt, T. (1982) Transfer of ^{2}H atoms to molecular species of their analogues of hepatic phosphatidylcholines and phosphatidylethanolamines during metabolism of [1,1$^2H_2$] ethanol in rats. Biochim. Biophys. Acta *713*, 602–608.
223. Ahlberg, J., Curstedt, T., Einarson, K. & Sjövall, J. (1981) Molecular species of biliary phosphatidylcholines in gallstone patients. The influence of treatment with cholic acid and chenodeoxycholic acid. J. Lipid Res. *22*, 404–409.
224. Ojino, H., Matsamura, T., Satouchi, K. & Saito, K. (1977) Studies on molecular species of fetal lung lecithins by gas-liquid chromatography mass spectrometry. Biomed. Mass Spectrom. *4*, 326–329.
225. Curstedt, T. (1977) Analysis of molecular species of ether analogues of phosphatidylcholines from biological samples. Biochim. Biophys. Acta *489*, 79–88.
226. Satouchi, K. & Saito, K. (1976) Studies on the trimethylsilyl derivatives of 1-alkyl-2-acylglycerols by gas-liquid chromatography mass spectrometry. Biomed. Mass Spectrom. *3*, 122–126.
227. Satouchi, K. & Saito, K. (1977) Studies on trimethylsilyl derivatives of 1-alk-1-enyl-2-acyl glycerols by gas-liquid chromatography mass spectrometry. Biomed. Mass Spectrom. *4*, 107–112.
228. Satouchi, K., Saito, K. & Kates, M. (1978) Studies on trimethylsilyl derivatives of 1,2-dialkylglycerols by gas-liquid chromatography mass spectrometry. Biomed. Mass Spectrom. *5*, 87–88.
229. Satouchi, K. & Saito, K. (1979) Use of t-butyldimethylchlorosilane/imidazole reagent for identification of molecular species of phospholipids by gas-liquid chromatography mass spectrometry. Biomed. Mass Spectrom. *6*, 396–402.
230. Kuksis, A., Breckenridge, W. C., Myher, J. J. & Kakis, G. (1978) Replacement of endogenous phospholipids in rat plasma lipoproteins during intravenous infusion of an artificial lipid emulsion. Can. J. Biochem. *56*, 630–639.
231. Kuksis, A., Marai, L. & Myher, J. J. (1983) Strategy of glycerolipid separation and

quantitation by complementary analytical techniques J. Chromatogr. *273*, 43–66.
232. Brockerhoff, H. (1965) A stereospecific analysis of triglycerides. J. Lipid Res. *6*, 10–15.
233. Brockerhoff, H. (1967) A stereospecific analysis of triglycerides: An alternative method. J. Lipid Res. *8*, 167–169.
234. Breckenridge, W. C. (1979) Stereospecific analysis of triacylglycerols, in: Fatty acids and glycerides (Kuksis, A., ed.) pp. 197–231, Plenum Press, New York.
235. Myher, J. J. & Kuksis, A. (1979) Stereospecific analysis of triacylglycerols via racemic phosphatidylcholines and phospholipase C. Can. J. Biochem. *57*, 117–124.
236. Manganaro, F., Myher, J. J., Kuksis, A. & Kritchevsky, D. (1981) Acylglycerol structure of genetic varieties of peanut oils of varying atherogenic potential. Lipids *16*, 508–517.
237. Myher, J. J., Kuksis, A., Vasdev, S. C. & Kako, K. J. (1979) Acylglycerol structure of mustard seed oil and cardiac lipids of rats during dietary lipidosis. Can. J. Biochem. *57*, 1315–1327.
238. Ogino, H., Matsumura, T., Satouchi, K. & Saito, K. (1979) Studies on molecular species of choline glycerophospholipids of developing rat brain. Biochim. Biophys. Acta *574*, 57–63.
239. Ogino, H., Matsumura, T., Satouchi, K. & Saito, K. (1980) Changes in molecular species of rat liver choline glycerophospholipids with development. Biochim. Biophys. Acta *618*, 431–438.
240. Matsumoto, J., Tanaka, T., Gamo, M., Saito, K. & Honjo, I. (1981) Phospholipid metabolism of dog liver under hypoxic conditions induced by ligation of the hepatic artery. Biochim. Biophys. Acta *664*, 527–537.
241. Kuksis, A. & Myher, J. J. (1980) New approaches to lipid analysis of lipoproteins and cell membranes, in: Membrane Fluidity. Biophysical Techniques and Cellular Regulation. (Kates, M. & Kuksis, A., eds.) pp. 3–32, Humana Press, Clifton, N.J.
242. Dickens, B. F., Ramesha, C. S. & Thompson, G. A. (1982) Quantification of phospholipid molecular species by coupled gas chromatography-mass spectrometry of deuterated samples. Anal. Biochem. *127*, 37–48.
243. Kuksis, A., Myher, J. J. & Marai, L. (1984) Lipid methodology – Chromatography and beyond. Part 1. GC/MS and LC/MS of glycerolipids. J. Am. Oil Chemists Soc. *61*, 1582–1589.
244. Myher, J. J. & Kuksis, A. (1982) Resolution of diacylglycerol moieties of natural glycerophospholipids by gas-liquid chromatography on polar capillary columns. Can. J. Biochem. *60*, 638–650.
245. Saito, K., Ogino, H. & Satouchi, K. (1980) Determination of molecular species of glycerophospholipids by a GC-MS selected ion monitoring technique, in: Membrane fluidity. Biophysical techniques and cellular regulation. (Kates, M. & Kuksis, A., eds.) pp. 33–41, Humana Press, Clifton, N.J.
246. Kino, M., Matsumura, T., Gamo, H. & Saito, K. (1982) Studies on molecular species of choline and ethanolamine glycerophospholipids obtained from rat brain myelin and synaptosomes by gas-liquid chromatography mass spectrometry. Biomed. Mass Spectrom. *9*, 363–369.
247. Myher, J. J. & Kuksis, A. (1984) Resolution of alkenylacylglycerol moieties of natural glycerophospholipids by gas-liquid chromatography on polar capillary columns. Can. J. Biochem. Cell Biol. *62*, 352–362.
248. Sugatani, J., Kino, M., Saito, K., Matsuo, T., Matsuda, H. & Katakuse, I. (1982) Analysis of molecular species of phospholipids by field desorption mass spectrometry. Biomed. Mass Spectrom. *9*, 293–301.
249. Velisek, J., Davidek, J., Kukbelka, V., Janicek, G. & Svobodova, Z. (1980) New chlorine-containing organic compounds in protein hydrolysates. J. Agric. Food Chem. *28*, 1142–1144.
250. Gardner, A. M., Yurawecz, M. P., Cunningham, W. C., Diachenko, G. W., Mazzola, E. P. & Brumley, W. C. (1983) Isolation and identification of C16 and C18 fatty acid esters of chloropropanediol in adulterated Spanish cooking oils. Bull. Environ. Contam. Toxicol. *31*, 625–630.
251. Cerbulis, J., Parks, D. W., Liu, R. H., Piotrowski, E. G. & Farrell, Jr., H. M. (1984) Occurrence of diesters of 3-chloro-1,2-propanediol in the neutral lipid fraction of goats milk. J. Agric. Food Chem. *32*, 474–476.
252. Kuksis, A., Marai, L., Myher, J. J., Cerbulis, J. & Farrell, H. M., Jr. (1986) Comparative study of the molecular species of chloropropanediol diesters and triacylglycerols in milk fat. Lipids *21*, 183–190.
253. Murata, T. (1977) Qualitative and quantitative chemical ionization mass spectrome-

try of triglycerides. Anal. Chem. *49*, 728–733.

254. Schulte, E., Hohn, M. & Rapp, U. (1981) Mass spectrometic determination of triglyceride patterns of fats by the direct chemical ionization technique (DCI) Fresenius Z. Anal. Chem. *307*, 115–119.
255. Lehman, W. D. & Kessler, M. (1983) Fatty acid profiling of phospholipids by field desorption and fast atom bombardment mass spectrometry. Chem. Phys. Lipids *32*, 123–135.
256. Batrakov, S. G., Sadovskaya, V. L., Rozynov, B. V. & Bergelson, L. D. (1983) Molecular species analysis of lipids by metastable ion mass spectrometry. Chem. Phys. Lipids *33*, 331–353.
257. Hites, R. A. (1975) Mass spectrometry of triglycerides. Methods Enzymol. *35*, 348–359.
258. Wood, G. W. (1980) Complex lipids, in: Biochemical Applications of Mass Spectrometry, first supplementary volume (Waller, G. R. & Dermer, O. C., eds.) pp. 173–209, John Wiley and Sons, New York.
259. Casparrini, G., Horning, E. C. & Horning, M. G. (1969) Gas phase analytical separation and structural study of ceramides. Chem Phys. Lipids *3*, 1–10.
260. Samuelsson, K. & Samuelsson, B. (1970) Gas chromatographic and mass spectrometric studies of synthetic and natural occurring ceramides. Chem Phys. Lipids *5*, 44–79.
261. Hammarstrom, S., Samuelsson, B. & Samuelsson, K. (1970) Gas-liquid chromatography-mass spectrometry of synthetic ceramides containing 2-hydroxy acids. J. Lipid Res. *11*, 150–157.
262. Samuelsson, B. & Samuelsson, K. (1969) Separation and identification of ceramides derived from human plasma sphingomyelins. J. Lipid Res. *10*, 47–55.
263. Samuelsson, K. (1969) On the occurrence and nature of free ceramides in human plasma. Biochim. Biophys. Acta *176*, 211–213.
264. Samuelsson, K. (1971) Identification and quantitative determination of ceramides in human plasma. Scand. J. Clin. Lab. Invest. *27*, 371–380.
265. Samuelsson, K. & Zetterstrom, R. (1971) Ceramides in a patient with Lipogranulomatosis (Farber's disease) with chronic course. Scand. J. Clin. Lab. Invest. *27*, 393–405.
266. Markey, S. P. & Wenger, D. A. (1974) Mass spectra of complex molecules I Chemical ionization mass spectrometry of sphingolipids. Chem. Phys. Lipids *12*, 182–200.
267. Bonhours, J. & Bonhours, D. (1984) Identification of free ceramide in human erythrocyte membrane. J. Lipid Res. *25*, 613–619.
268. Gaskell, S. J., Edmonds, C. G. & Brooks, C. J. W (1976) Applications of boronate derivatives in the study of ceramides by gas-liquid chromatography-mass spectrometry. J. Chromatogr. *126*, 591–599.
269. Do, Un Hoi & Pei P. T. (1981) Separation of molecular species of ceramides as benzoyl and p-nitrobenzoyl derivatives by high performance liquid chromatography. Lipids *16*, 855–862.
270. Einolf, W. N. & Fenselau, C. (1974) Phospholipids in pulmonary alveolar proteinosis. Biomed. Mass Spectrom. *1*, 195–198.
271. Brown, A. E. & Elovson, J. (1974) Isolation and characterization of a novel lipid, 1(3),2-Diacylglycerol-(3)-0-4′-(N,N,N.-trimethyl) homoserine, from *Ochromonas danica*. Biochemistry *13*, 3476–3482.
272. Epps, D. E., Natarajan, V., Schmid, P. D. & Schmid, H. H. O. (1980) Accumulation of N-acylethanolamine glycerophospholipids in infarcted myocardium. Biochim. Biophys. Acta *618*, 420–430.
273. Karlsson, K. A. (1965) Studies on sphingosine. 10. Use of trimethylsilylethers for the gas chromatography-mass spectrometry of sphingosines. Acta Chem. Scand. *19*, 2425–2427.
274. Karlsson, K. A. (1970) On the chemistry and occurrence of sphingolipid long-chain bases. Chem. Phys. Lipids *5*, 6–43.
275. Gaskell, S. J. & Brooks, C. J. W. (1977) Electron impact and chemical ionization mass spectrometry of cyclic boronate derivatives of sphingosines and related compounds. Org. Mass Spectrom. *12*, 615–658.
276. Jungalwala, F. B., Evans, J. F., Bremer, E. & McCluer, R. H. (1983) Analysis of sphingoid bases by reversed-phase high performance liquid chromatography. J. Lipid Res. *24*, 1380–1388.
277. Kadawaki, H., Bremer, E. G., Evans, J. E., Jungalwala, F. B. & McCluer, R. H. (1983) Acetonitrile-hydrochloric acid hydrolysis of gangliosides for high performance liquid chromatographic analysis of their long chain bases. J. Lipid Res. *24*, 1389–1397.
278. Jungalwala, F. B., Evans, J. E., Kakowaki, H. & McCluer, R. H. (1984) High per-

formance liquid chromatography-chemical ionization mass spectrometry of sphingoid bases using moving-belt interface. J. Lipid Res. *25*, 209–216.

279. Klein, R. A. (1971) Mass spectrometry of the phosphatidylcholines: dipalmitoyl, dioleoyl, and stearoyl-oleoyl glycerolphosphorylcholines. J. Lipid Res. *12*, 123–131.
280. Klein, R. A. (1971) Mass spectrometry of the phosphatidylcholines: fragmentation processes for dioleoyl and stearoyl-oleoyl glycerylphosphorylcholine. J. Lipid Res. *12*, 628–634.
281. Klein, R. A. (1972) Mass spectrometry of the phosphatidyl amino alcohols: detection of molecular species and use of low voltage spectra and metastable scanning in the elucidation of structure. J. Lipid Res. *13*, 672–679.
282. Burlingame, A. L., Dell, A. & Russell, D. H. (1982) Mass spectrometry. Anal. Chem. *54*, 363R–409R.
283. Sugnaux, F. & Djerassi, C. (1982) Liquid chromatography-mass spectrometry of polar lipids. Comparison of on-line method using direct liquid introduction interface with off-line method using desorption chemical ionization. J. Chromatogr. *251*, 189–201.
284. Foltz, R. L. (1972) Chemical ionization mass spectrometry in structural analysis. Lloydia *35*, 344–353.
285. Crawford, C. G. & Plattner, R. D. (1983) Ammonia chemical ionization mass spectrometry of intact diacyl phosphatidylcholine. J. Lipid Res. *24*, 456–460.
286. Ayanoglu, E., Wegmann, A., Pilet, O., Marbury, G. D., Hass, J. R. & Djerassi, C. (1984) Mass spectrometry of phospholipids. Some applications of desorption chemical ionization and fast atom bombardment. J. Am. Chem. Soc. *106*, 5246–5251.
287. Jungalwala, F. B., Evans, J. E. & McCluer, R. H. (1984) Composition and molecular species analysis of phospholipids by high performence liquid chromatography coupled with chemical ionization mass spectrometry. J. Lipid Res. *25*, 738–749.
288. Bisseret, P., Nakatani, Y., Ourisson, G., Hueber, R. & Teller, G. (1983) Ammonia chemical ionization mass spectrometry of lecithins on a gold support. Chem. Phys. Lipids *33*, 383–392.
289. Dessort, D., Bisseret, P., Nakatani, Y., Ourisson, G. & Kates, M. (1983) Ammonia desorption chemical ionization mass spectrometry of phospholipids. Chem. Phys. Lipids *33*, 323–330.
290. Dessort, D., Mersel, M., Lepage, P. & Van Dorsselaer, A. (1984) Fast-heating mass spectrometry of phosphatidylcholine, lysophosphatidylcholine, phosphatidylethanolamine, and sphingomyelin. Anal. Biochem. *142*, 43-52.
291. Crawford, C. G. & Plattner, R. D. (1984) Phospholipid molecular species of underivatized lipids. J. Lipid Res. *25*, 518–522.
292. Kim, H.-Y. & Salem, N., Jr. (1985) Phospholipid molecular species analysis by thermospray liquid chromatography/mass spectrometry. Anal. Chem. *57*, 9–14.
293. Wood, G. W. & Lau, P. Y. (1974) Analysis of intact phospholipids by field desorption mass spectrometry. Biomed. Mass Spectrom. *1*, 154–155.
294. Wood, G. W., Law, P. Y. & Rao, G. N. S. (1976) Field desorption mass spectrometry of phospholipids. II. Fragmentation of dipalmitoylphosphatidylcholine form. Comparison of do, d_4 and d_9 species. Biomed. Mass Spectrom. *3*, 172–176.
295. Wood, G. W., Lou, P. Y., Morrow, G., Rao, G. N. S., Schmidt, D. E. & Tuebner, J. (1977) Field desorption mass spectrometry of phospholipids III. Survey of structural types. Chem. Phys. Lipids *18*, 316–333.
296. Wood, G. W. & Perkins, S. E. (1982) Purification technique for enhancement of field desorption mass spectra of phospholipids. Anal. Biochem. *122*, 368–371.
297. Lehman, W. D. & Kessler, M. (1983) Characterization and quantification of human plasma lipids from crude lipid extracts by field desorption mass spectrometry. Biomed. Mass Spectrom. *10*, 220–226.
298. Fenwick, G. R., Eagles, J. & Self, R. (1983) Fast atom bombardment mass spectrometry of intact phospholipids and related compounds. Biomed. Mass Spectrom. *10*, 382–386.
299. May, H.-E. & Desidero, D. M. (1983) Fast atom bombardment mass spectrometry of underivatized phosphatidylcholines, lysophosphatidylcholines, and diglycerides. J. Chem. Soc. Commun., 72–73.
300. Aberth, W., Straub, K. M. & Burlingame, A. L. (1982) Secondary ion mass spectrometry with cesium ion primary beam and liquid target matrix for analysis of bioorganic compounds. Anal. Chem. *54*, 2029–2034.
301. Traitler, H. & Nikiforov, A. (1984) Comparative determination of phosphatidylcholine

(lecithin) species in product by fast atom bombardment-computer averaging integration mass spectrometry (FAB-CAI), in: Chromatography and Mass Spectrometry in Nutrition Science and Food Safety (Frigerio, A. & Milon, H., eds.) pp. 299–304, Elsevier Science Publishers, Amsterdam.

302. Ohashi, Y. (1984) Structure determination of phospholipids by secondary ion mass spectrometric techniques: differentiation of isomeric esters. Biomed. Mass Spectrom. *11*, 383–385.

303. Sherman, W. R., Ackermann, K. E., Bateman, R. H., Green, B. N. & Lewis, I. (1985) Mass-analysed ion kinetic energy spectra and B_1E-B_2 triple sector mass spectrometric analysis of phosphoinositides by fast atom bombardment. Biomed. Mass Spectrom. *12*, 409–413.

304. Benfenati, E. & Reginato, R. (1985) A comparison of three methods of soft ionization mass spectrometry of crude phospholipid extracts. Biomed. Mass Spectrom. *12*, 643–651.

305. Munster, H., Stein, J. & Budzikiewicz, H. (1986) Structure analysis of underivatized phospholipids by negative ion fast atom bombardment mass spectrometry. Biomed. Environm. Mass Spectrom. *13*, 423–427.

306. Jensen, N. J., Tomer, K. B. & Gross, M. L. (1986) Fast atom bombardment and tandem mass spectrometry of phosphatidylserine and phosphatidylcholine. Lipids *21*, 580–588.

307. Torday, J., Carson, L. & Lawson, E. E. (1979) Saturated phosphatidylcholine in amniotic fluid and prediction of the respiratory distress syndrome. N. Engl. J. Med. *30*, 1013–1018.

308. Ho, B. C., Fenselau, C., Hansen, G., Larsen, J. & Daniel, A. (1983) Dipalmitoylphosphatidylcholine in amniotic fluid quantified by fast-atom-bombardment mass spectrometry. Clin. Chem. *29*, 1349–1353.

309. Heymans, H. S. A., Schutgens, R. B. H., Tan, R., Van den Bosch, H., & Borst, P. (1983) Severe plasmalogen deficiency in tissues of infants without peroxisomes (Zellweger syndrome). Nature *306*, 69–70.

310. Schutgens, R. B. H., Romeyn, G. J., Wanders, R. J. A., van den Bosch, H., Schrakomp, G. & Heymans, H. S. (1984) Deficiency of acyl-CoA: dihydroxyacetone phosphate acyltransferase in patients with Zellweger (cerebro-hepatorenal) syndrome. Biochem. Biophys. Res. Commun. *120*, 179–189.

311. Hanahan, D. J. (1972) Ether-linked lipids: chemistry and methods of measurement, in: Ether Lipids (Snyder, F., ed.) pp. 25–50. Academic Press, New York.

312. Gross, R. W. (1984) High plasmalogen and arachidonic acid content of canine myocardial sarcolemma: A fast atomic bombardment mass spectroscopic and gas chromatography-mass spectroscopic characterization. Biochem. *23*, 158–165.

313. Demopoulos, C., Pincard, R. N. & Hanahan, D. J. (1979) Platelet-activating-factor. Evidence for 1-0-alkyl-2-acetyl-sn-glyceryl-3-phosphorylcholine as the active component (A new class of lipid chemical mediators). J. Biol. Chem. *254*, 9355–9358.

314. Benveniste, J., Tence, M., Bidault, J., Boullet, C., Varenne, P. & Polonsky, J. (1979) Semi-synthese et structure proposee du facteur activant les plaquettes (P.A.F.): PAF-acetether, un alkylether analogue de la lysophosphatidylcholine. C.R. Acad. Sci. Ser. D. (Paris) *289*, 1037–1040.

315. Blank, M. L., Snyder, F., Byers, L. W., Brooks, B. & Muirhead, E. E. (1979) Antihypertensive activity of an alkyl ether analog of phosphatidylcholine. Biochem. Biophys. Res. Commun. *90*, 1194–1200.

316. Hanahan, D. J. (1986) Platelet activating factor: a biologically active phosphoglyceride. Ann. Rev. Biochem. *55*, 483–509.

317. Polonsky, J., Tence, M., Vareene, P., Das, B. C., Lunel, J. & Benveniste, J. (1980) Release of 1-0-alkylglyceryl 3-phosphorylcholine, 0-deacetyl platelet-activating factor, from leukocytes: Chemical ionization mass spectrometry of phospholipids. Proc. Natl. Acad. Sci. USA *77*, 7019–7023.

318. Satouchi, K., Oda, M., Yasunaga, K. & Saito, K. (1983) Application of selected ion monitoring to determination of platelet-activating factor. J. Biochem. *94*, 2067–2070.

319. Nakayama, R., Oda, M., Satouchi, K. & Saito, K. (1985) Generation of acetyl glyceryl ether phosphorylcholine from the rat skin and muscle tissues stimulated by moxibustion. Biochem. Biophys. Res. Commun. *127*, 629–634.

320. Yasuda, K., Satouchi, K. & Saito, K. (1986) Platelet-activating factor in normal rat uterus. Biochem. Biophys. Res. Commun. *138*, 1231–1236.

321. Tokumura, A., Suzuki, K., Takauchi, K. & Tsukatani, H. (1985) Gas chromatographic-mass spectrometric analysis of biologically active phospholipids having an sn-2-acetyl

group. J. Chromatogr. Biomed. Applic. *343*, 138–142.
322. Ramesha, C. S. & Pickett, W. C., (1986) Measurement of sub-picogram quantities of platelet activating factor (AGEPC) by gas chromatography/negative ion chemical ionization mass spectrometry. Biomed. Environm. Mass Spectrom. *13*, 107–111.
323. Clay, K. L., Stene, D. O. & Murphy, R. C. (1984) Quantitative analysis of platelet activating factor (AGEPC) by fast atom bombardment mass spectrometry. Biomed. Mass Spectrom. *11*, 47–49.
324. Clay, K. L., Murphy, R. C., Andres, J. L., Lynch, J. & Henson, P. M. (1984) Structure elucidation of platelet activating factor derived from human neutrophils. Biochem. Biophys. Res. Commun. *121*, 815–825.
325. Haroldsen, P. E., Clay, K. L. & Murphy, R. C. (1987) Quantitation of lyso-platelet activating factor molecular species from human neutrophils by mass spectrometry. J. Lipid Res. *28*, 42–49.
326. Varenne, P., Das, B. C., Polonsky, J. & Tence, M. (1985) Chemical ionization and fast atom bombardment mass spectrometry of platelet activating factor (PAF-acether) and related phospholipids. Biomed. Mass Spectrom. *12*, 6–10.
327. Tokumura, A., Harada, K., Fukuzawa, K. & Tsukatani, H. (1986) Involvement of lysophospholipase D in the production of lysophosphatidic acid in rat plasma. Biochim. Biophys. Acta *875*, 31–38.
328. Tokumura, A., Harada, K., Yoshioka, Y., Tsukatani, H. & Handa, Y. (1984) Analyses of lysophosphatidic acids by gas chromatography mass spectrometry without hydrolytic treatment. Biomed. Mass Spectrom. *11*, 167–171.
329. Tokumura, A., Yoshioka, Y., Tsukatani, H. & Handa, Y. (1986) Gas chromatography/mass spectrometric analyses of ether-linked lysophosphatidic acid and its analogues. Biomed. Environm. Mass Spectrom. *13*, 175–180.
330. Marcus, D. M. (1976) Applications of immunological techniques to the study of glycosphingolipids, in: Glycolipid Methodology (Witting, L. A., ed.) pp. 233–245, Am. Oil Chemists Soc., Champaign, Illinois.
331. Hakomori, S. (1973) Glycolipids of tumor cell membrane. Adv. Cancer Res. *18*, 265–313.
332. Brady, R. O. (1978) Sphingolipidoses. Ann. Rev. Biochem. *47*, 687–713.
333. Murray, R. K., Levine, M. & Kornblatt, M. J. (1976) Sulfatides: Principal glycolipids of the testes and spermatozoa of chordates, in: Glycolipid methodology (Witting, L. A., ed.) pp. 305–327 Am. Oil Chemists Soc., Champaign, Illinois.
334. Ishizuka, I. & Yamakawa, T. (1974) Absence of seminolipid in seminoma tissue with concomitant increase of sphingoglycolipids. J. Biochem. (Tokyo) *76*, 221–223.
335. Karlsson, K. A., Pascher, I., Pimlott, W. & Samuelsson, B. E. (1974) Use of mass spectrometry for the carbohydrate composition and sequence analysis of glycosphingolipids. Biomed. Mass Spectrom. *1*, 49–56.
336. Ledeen, R. W., Kundu, S. K., Price, H. C. & Fong, J. W. (1974) Mass spectra of permethyl derivatives of glycosphingolipids. Chem. Phys. Lipids *13*, 429–446.
337. Hanfland, P. & Egge, H. (1975) Mass spectrometric analysis of permethylated glycosphingolipids I. Sequence analysis of two blood-group B active glycosphingolipids from human B erythrocyte membranes. Chem. Phys. Lipids *15*, 243–247.
338. Karlsson, K. A. (1974) Carbohydrate composition and sequence analysis of a derivative of brain disialoganglioside by mass spectrometry with molecular weight ions at m/e 2245. Potential use in the specific microanalysis of cell surface components. Biochemistry *13*, 3643–3647.
339. Karlsson, K. A. (1973) Carbohydrate composition and sequence analysis of cell surface components by mass spectrometry. Characterization of the major monosialoganglioside of brain. FEBS Lett. *32*, 317–320.
340. Breimer, M. E., Hansson, G. C., Karlsson, K. A., Larson, G., Leffler, H., Pimlott, W., Samuelsson, B. E., Stromberg, N., Teneberg, S. & Thurin, J. (1983) Sequencing of large oligosaccharides by direct inlet mass spectrometry. Application to cell surface glycolipids. Int. J. Mass Spectrom. and Ion Phys. *48*, 113–116.
341. Hanfland, P. & Egge, H. (1976) Mass spectrometric analysis of permethylated glycosphingolipids II. Comparative studies on different blood-group active and related erythrocyte membrane glycosphingolipids. Chem. Phys. Lipids *16*, 201–214.
342. Egge, H. (1978) The application of mass spectrometry in the structural elucidation of glycosphingolipids. Chem. Phys. Lipids *21*, 349–360.

343. Breimer, M. E., Hansson, G. C., Karlsson, K. A., Leffler, H., Pimlott, W. & Samuelsson, B. E. (1979) Selected ion monitoring of glycosphingolipid mixtures. Identification of several blood group type glycolipids in the small intestine of an individual rabbit. Biomed. Mass Spectrom. *6*, 231–242.
344. Falk, K. E., Karlsson, K. A. & Samuelsson, B. E. (1979) Proton nuclear magnetic resonance analysis of anomeric structure of glycosphingolipids. The globo-series (one to five sugars). Arch. Biochem. Biophys. *192*, 164–176.
345. Falk, K. E., Karlsson, K. A. & Samuelsson, B. E. (1979) Proton nuclear magnetic resonance analysis of anomeric structure of glycosphingolipids. Blood group ABH-active substances. Arch. Biochem. Biophys. *192*, 177–190.
346. Falk, K. E., Karlsson, K. A. & Samuelsson, B. E. (1979) Proton nuclear magnetic resonance analysis of anomeric structure of glycosphingolipids. Lewis-active and Lewis-like substances. Arch. Biochem. Biophys. *192*, 191–202.
347. Falk, K. E., Karlsson, K. A. & Samuelsson, B. E. (1981) A new way of chemical bloodtyping of humam single individuals. FEBS Lett. *124*, 173–177.
348. Hakomori, S. & Watanabe, K. (1976) Blood group glycolipids of human erythrocytes, in: Glycolipid Methodology (Witting, L. A., ed.) pp. 13–47, Am. Oil Chemists Soc., Champaign, Illinois.
349. McKibbin, J. M. (1978) Fucolipids. J. Lipid Res. *19*, 131–147.
350. Hakomori, S. & Strycharz, G. D. (1968) Investigations on cellular blood group substances. I. Isolation and chemical composition of blood-group ABH and Le^b isoantigens of sphingoglycolipid nature. Biochem. *7*, 1279–1286.
351. Marcus, D. M. & Cass, L. E. (1969) Glycolipids with Lewis blood group activity: uptake by human erythrocyte. Science *164*, 553–555.
352. Larson, G. & Samuelsson, B. E. (1980) Blood group type glycosphingolipids of human cord blood erythrocytes. J. Biochem. *88*, 647–657.
353. Feizi, T., Childs, R. A., Watanabe, K. & Hakomori, S. I. (1979) Three types of blood group I specificity among monoclonal anti I antibodies revealed by analogues of a branched erythrocyte glycolipid. J. Exp. Med. *149*, 975–980.
354. Hakomori, S. I. (1975) Structures and organization of cell surface glycolipids dependency on cell growth and malignant transformation. Biochim. Biophys. Acta *417*, 55–89.
355. Brady, R. O. & Fishman, P. H. (1974) Biosynthesis of glycolipids in virus-transformed cells. Biochim. Biophys. Acta *355*, 121–148.
356. Lee, W. M. F., Klock, J. C. & Macher, B. A. (1981) Isolation and structural characterization of human lymphocyte neutral glycosphingolipids. Biochemistry *20*, 3810–3814.
357. Klock, J. C., D'Angona, J. L. & Macher, B. A. (1981) Chemical characterization of neutral glycolipids in the human myeloid leukemias. J. Lipid Research *22*, 1079–1083.
358. Lee, W. M. F., Klock, J. C. & Macher, B. A. (1981) Neutral glycosphingolipids in hairy cell leukemia. Biochemistry *20*, 6505–6508.
359. Murata, T., Ariga, T., Oshima, M. & Miyatake, T. (1978) Characterization of trimethylsilylderivatives of cerebrosides by direct inlet-chemical ionization mass spectometry. J. Lipid Research *19*, 370–374.
360. Oshima, M., Ariga, T. & Murata, T. (1977) Combined gas chromatography-chemical ionization mass spectrometry of sphingolipids I Glucosyl sphingosine, ceramides and cerebrosides of the spleen in Gaucher's disease. Chem. Phys. Lipids *19*, 289–299.
361. Ariga, T., Murata, T., Oshima, M., Maezawa, M. & Miyatake, T. (1980) Characterization of glycosphingolipids by direct inlet chemical ionization mass spectrometry. J. Lipid Res. *21*, 879–887.
362. Ariga, T., Yu, R. K., Suzuki, M., Ando, S. & Miyatake, T. (1982) Characterization of GMI ganglioside by direct inlet chemical ionization mass spectrometry. J. Lipid Res. *23*, 437–442.
363. Ariga, T., Yu, R. K., Miyatake, T. (1984) Characterization of gangliosides by direct inlet chemical ionization mass spectrometry. J. Lipid Res. *25*, 1096–1101.
364. Carr, S. A. & Reinhold, V. N. (1984) Structural characterization of glycosphingolipids by direct chemical ionization mass spectrometry. Biomed. Mass Spectrom. *11*, 633–642.
365. Reinhold, V. N. & Carr, S. A. (1982) Direct chemical ionization mass spectrometry with polyimide-coated wires. Anal. Chem. *54*, 499–503.

366. Kushi, Y. & Handa, S. (1982) Application of field desorption mass spectrometry for the analysis of sphingoglycolipids. J. Biochem. *91*, 923–931.
367. Handa, S. & Kushi, Y. (1982) High performance liquid chromatography and structural analysis by field desorption mass spectrometry of underivatized glycolipid. Adv. Exp. Med. Biol. *152*, 23–31.
368. Egge, H., Dabrowski, J., Hanfland, P., Dell, A. & Dabrowski, U. (1982) High resolution H nuclear magnetic resonance spectroscopy and "soft ionization" mass spectrometry of glycosphingolipids. Adv. Exp. Med. Biol. *152*, 33–40.
369. Arita, M., Iwamori, M., Higuchi, T. & Nagai, Y. (1983) Negative ion fast atom bombardment mass spectrometry of gangliosides and asialo gangliosides. A useful method for the structural elucidation of gangliosides and related neutral glycosphingolipids. J. Biochem. *94*, 249–256.
370. Arita, M., Iwamori, M., Higuchi, T. & Nagai, Y. (1983) 1,1,3,3-tetramethylurea and triethanolamine as a new useful matrix for fast atom bombardment mass spectrometry of gangliosides and neutral glycosphingolipids. J. Biochem. *93*, 319–322.
371. Handa, S., Kushi, Y., Kambara, H. & Shizukuishi, K. (1983) Secondary ion mass spectra of neutral sphingoglycolipids. J. Biochem. *93*, 315–318.
372. Egge, H., Peter-Katalini, J. & Hanfland, P. (1984) Structure analysis of glycosphingolipids using fast atom bombardment (FAB) techniques. Adv. Expt. Med. Biol. *174*, 55–63.
373. Handa, S., Kushi, Y., Kambara, H. & Shizukuishi, K. (1983) Secondary ion mass spectra of neutral sphingoglycolipids. J. Biochem. *93*, 315–318.
374. Kushi, Y. & Handa, S. (1985) Direct analysis of lipids on thin layer plates by matrix-assisted secondary ion mass spectrometry. J. Biochem. *98*, 265–268.
375. Samuelsson, B. E. (1986) Investigation of the structures of large molecules with special reference to glycoconjugates, in: Advances in Mass Spectrometry 1985 (Todd, J. F. J., ed.) pp. 225–241, John Wiley and Sons, New York.
376. Millington, D. S. & Smith, J. A. (1977) Fragmentation patterns by fast linked electric and magnetic field scanning. Organic Mass Spectrom. *12*, 264–265.
377. Carr, S. A., Biemann, K., Shoji, S., Parmelee, D. C. & Titani, K. (1982) n-Tetradecanoyl is the NH_2-terminal blocking group of the catalytic subunit of cyclic AMP-dependent protein kinase from bovine cardiac muscle. Proc. Natl. Acad. Sci. USA *79*, 6128–6131.
378. Aiken, A., Cohen, P., Santikarn, S., Williams, D. H., Calder, A. G., Smith, A. & Klee, C. B. (1982) Identification of the NH_2-terminal blocking group of calcineurin B as myristic acid. FEBS Lett. *150*, ^{2}314–318.
379. Ozols, J., Carr, S. A. & Strittmatter, P. (1984) Identification of thc NH_2-terminal blocking group of NADH-cytochrome b_5 reductase as myristic acid and the complete amino acid sequence of the membrane-binding domain. J. Biol. Chem. *259*, 13349–13354.
380. Henderson, L. E., Krutzsch, H. C. & Oroszlan, S. (1983) Myristyl amino-terminal acylation of murine retrovirus proteins: An unusual post-translational protein modification. Proc. Natl. Acad. Sci. USA *80*, 339–343.
381. Kuksis A. (1987) Absorption of Fat-Soluble Vitamins, in: Fat Absorption (Kuksis, A., ed.) pp. 65–86, CRC Press, Boca Raton, Florida.
382. DeLuca, L. M. (1978) Vitamin A, in: The Fat-Soluble Vitamins. Handbook of Lipid Research, Volume 2 (DeLuca, H. F., ed.) pp. 1–67, Plenum Press, New York, New York.
383. Driskell, W. J., Neese, J. W., Bryant, C. C. & Basher, M. M. (1982) Measurement of Vitamin A and Vitamin E in human serum by high pressure liquid chromatography. J. Chromatogr. Biomed. Applic. *231*, 439–444.
384. Catignani, G. L. (1986) An HPLC method for the simultaneous determination of retinol and α-tocopherol in plasma or serum. Methods Enzymol. *123*, 215–219.
385. Elliott, W. H. & Waller, G. R. (1972) Vitamins and Cofactors, in: Biochemical Applications of Mass Spectrometry (Waller, G. R., ed.) pp. 499–536, Wiley-Interscience, New York, NY.
386. Vecchi, M., Vetter, W., Waltner, W., Jermstad, S. F. & Schutt, G. W. (1967) Gaschromatographische und massenspektrometrische Untersuchungen ueber trimethylsilylaether von Vitamin A und einigen seiner Isomeren. Helv. Chim. Acta *50*, 1243–1248.
387. Lin, R. L., Waller, G. R., Mitchell, E. D., Yang, K. S. & Nelson, E. C. (1970) Mass spectra of retinol and related compounds. Anal. Biochem. *35*, 435–441.

388. De Ruyter, M. G., Lambert, W. E. & De Leenheer, A. P. (1979) Retinoic acid: an endogenous compound of human blood. Unequivocal demonstration of endogenous retinoic acid in normal physiological condition. Anal. Biochem. *98*, 402–409.
389. Chiang, T.-C. (1980) Gas chromatographic-mass spectrometric assay for low levels of retinoic acid in human blood. J. Chromatogr. Biomed. Applic. *182*, 335–340.
390. Napoli, J. L., Pramanik, B. C., Williams, J. B., Dawson, M. I. & Hobbs, P. D. (1985) Quantification of retinoic acid by gas-liquid chromatography-mass spectrometry: total versus all-*trans*-retinoic acid in human plasma. J. Lipid Res. *26*, 387–392.
391. De Leenheer, A. P., Lambert, W. E. & Claeys, I. (1982) All-*trans*-retinoic acid: measurement of reference values in human serum by high performance liquid chromatography. J. Lipid Res. *23*, 1362–1367.
392. Napoli, J. L. (1986) Quantification of physiological levels of retinoic acid. Methods Enzymol. *123*, 112–124.
393. Zile, M., Schnoes, H. K., DeLuca, H. F. (1980) Characterization of retinyl *β*-glucuronide as a minor metabolite of retinoic acid in bile. Proc. Natl. Acad. Sci. *77*, 3230–3233.
394. Zile, M. H., Inhorn, R. C. & De Luca, H. F. (1982) Metabolites of all-*trans*-retinoic acid in bile: identification of all-*trans*- and 13-*cis*-retinoyl glucuronides. J. Biol. Chem. *257*, 3537–3543.
395. Bhat, P. V., Roller, P. P. & DeLuca, L. M. (1981) Chemical and biological studies on 5,6-epoxyretinol, retinol and their phosphoryl esters. J. Lipid Res. *22*, 1069–1078.
396. Sorg, B., Furstenberger, G., Berry, D. L., Hecker, E. & Marks, F. (1982) Preparation of retinoic acid esters of phorbol derivatives. J. Lipid Res. *23*, 443–447.
397. Dillard, C. J., Litov, R. E. & Tappel, A. L. (1978) Effect of dietary Vitamin E, selenium and polyunsaturated fats on in vivo lipid peroxidation in the rat as measured by pentane production. Lipids *13*, 396–402.
398. Ehrenkrantz, R. A. (1980) Vitamin E and the neonate. Am. J. Dis. Child *134*, 1157–1166.
399. Spielberg, S. P., Boxer, L. A., Corash, L. M. & Schulman, J. D. (1979) Improved erythrocyte survival with high dose Vitamin E in chronic hemolyzing G6PD and glutathione synthetase deficiencies. Ann. Int. Med. *90*, 53–54.
400. Wang, Y.-M., Madanat, F. F., Kimball, J. C., Gleiser, C. A., Ali, M. K., Kaufman, M. W. & Van Eys, J. (1980) Effect of Vitamin E against adriamycin-induced toxicity in rabbits. Cancer Res. *40*, 1022–1027.
401. Wang, Y. M., Howell, S. K., Kimball, J. C., Tsai, C. C., Sato, J. & Gleiser, C. A. (1982) α-Tocopherol as a potential modifier of Daunomycin carcinogenicity in Sprague-Dawley rats, in: Molecular Interrelation of Nutrition and Cancer (Arnott, M., Van Eys, J. & Wang, Y. M., eds.) pp. 369–379, Raven Press, New York, NY.
402. Korman, A. W. (1980) Routine assay for determination of α-tocopherol in liver. J. Lipid Res. *21*, 780–783.
403. Howeel, S. K. & Wang, Y. M. (1982) Quantitation of physiological α-tocopherol metabolites and related compounds by reserved phase high performance liquid chromatography. J. Chromatogr. Biomed. Applic. *227*, 174–180.
404. Handelman, G. J., Machlin, L. J., Fitch, K., Weiter, J. J. & Dratz, E. A. (1985) Oral α-tocopherol supplements decrease plasma γ-tocopherol levels in humans. J. Nutr. *115*, 807–813.
405. Nair, P. P. & Luna, Z. (1968) Identification of α-tocopherol from tissues by combined gas-liquid chromatography, mass spectrometry and infrared spectroscopy. Arch. Biochem. Biophys. *127*, 413–418.
406. Thomas, D. W., Parkhurst, R. M., Negi, D. S., Lunan, K. D., Wen, A. C., Brandt, A. E. & Stephens, R. J. (1981) Improved assay of α-tocopherol in the picogram range using gas chromatography-mass spectrometry. J. Chromatogr. Biomed. Applic. *225*, 433–439.
407. Kaseki, H., Sato, K., Yoshino, Y., Tanaka, M., Sugawara, K. & Takeshima, T. (1983) Mass-chromatographic determination of vitamin E in biological tissus. Koenshu – Iyo Masu Kenkyukai *8*, 151–154. Chem. Abstr. *100*, 82274n. 1984.
408. Kaseki, H., Yoshino, Y., Takeshima, T., Tanaka, M., Sugahara, K. & Nakagata, N. (1984) Measurement of vitamin E in mammalian eggs – mass-chromatographic determination with simple clean-up. Nippon Funin Gakkai Zasshi *29*, 202–208. Chem. Abstr. *101*, 68721v. 1985.
409. Buntman, D. J., Thomas, D. W., Parkhurst, R. M., Negi, D. S. & Stephens, R. J. (1984) A practical microbiochemical procedure for analysis of vitamin E in either solid or fluid tissues. Microchem. J. *29*, 66–73,

410. Rao, G. H. R., Krick, T. P. & Whets, J. L. (1981) Preparation, separation and characterization of vitamin E quinone. J. Chromatogr. *196*, 506–511.
411. Chiku, S., Hamamura, K. & Nakamura, T. (1984) Novel urinary metabolite of d-δ-tocopherol in rats. J. Lipid Res. *25*, 40–48.
412. Elsas, J. J. & McCormick, D. B. (1986) Genetic defects in vitamin utilization. Part I: General aspects and fat-soluble vitamins. Vitamins and Hormones *43*, 103–144.
413. Olson, R. E. & Suttie, J. W. (1977) Vitamin K and γ-carboxyglutamate biosynthesis. Vitamins and Hormones *35*, 59–108.
414. Bjornsson, T., Swezey, S. E., Meffin, P. J. & Blaschke, T. F. (1978) Quantitation of radiolabeled Vitamin K1 and Vitamin K1 epoxide in plasma by high pressure liquid chromatography. Thromb. Haemostas. *39*, 466–473.
415. Bechtold, H. & Jahnchen, E. (1979) Quantitative analysis of Vitamin K_1 and Vitamin K_1 2,3-epoxide in plasma by electron capture gas-liquid chromatography. J. Chromatogr. Biomed. Applic. *164*, 85–90.
416. Bechtold, H., Klein, F., Trenk, D. & Jahnchen, E. (1984) Improved method for quantitative analysis of Vitamin K_1 and Vitamin K_1 2,3-epoxide in human plasma by electron capture gas-liquid capillary chromatography. J. Chromatogr. Biomed. Applic. *306*, 333–337.
417. Shearer, M. J. (1986) Assay of K vitamins in tissues by high-performance liquid chromatography with special reference to ultraviolet detection. Methods Enzymol. *123*, 235–251.
418. Tamaoka, J. (1986) Analysis of bacterial menaquinone mixtures by reverse-phase high-performance liquid chromatography. Methods Enzymol. *123*, 251–256.
419. Lefevere, M. F., De Leenheer, A. P. & Claeys, A. E. (1979) High performance liquid chromatographic assay of vitamin K in human serum. J. Chromatogr. *186*, 749–762.
420. Lefevere, M. F., De Leenheer, A. P., Claeys, A. E., Claeys, I. V. & Steyart, H. (1982) Multidimensional liquid chromatography: a breakthrough in the assessment of physiological vitamin K levels. J. Lipid Res. *23*, 1068–1072.
421. Langenberg, J. P. & Tjaden, U. R. (1984) Determination of (endogenous) vitamin K_1 in human plasma by reversed-phase high performance liquid chromatography using fluorometric detection after post-column electrochemical reduction. Comparison with ultraviolet, single and dual electrochemical detection. J. Chromatogr. *305*, 61–72.
422. Hirauchi, K., Sakano, T. & Morimoto, A. (1986) Measurement of K vitamins in human and animal plasma by high-performance liquid chromatography with fluorometric detection. Chem. Pharm. Bull. *34*, 845–849.
423. DiMari, S. J., Supple, J. H. & Rapoport, H. (1966) Mass spectra of naphthoquinones. Vitamin K1(20). J. Amer. Chem. Soc. *88*, 1226–1232.
424. Collins, M. D. (1985) Isoprenoid quinone analyses in bacterial classification and identification, in: Chemical Methods in Bacterial Systematics (Goodfellow, M. & Minnikin, D. E., eds.) pp. 267–287, Academic Press, London.
425. Matschiner, J. T., Bell, R. G., Amelotti, J. M. & Knauer, T. E. (1970) Isolation and characterization of a new metabolite of phylloquinone in the rat. Biochim. Biophys. Acta *201*, 309–315.
426. Vetter, W., Vecchi, M., Gutmann, H., Ruegg, R., Walther, W. & Myer, P. (1967) Gaschromatographische und massenspektrometrische Untersuchung von phytylubichinon, Vitamin K_1 and Vitamin K_2. Helv. Chim. Acta *50*, 1866–1879.
427. Dialameh, G. H. & Olson, R. E. (1969) Gas-liquid chromatography of phytyl ubiquinone, vitamin E, vitamin K_1, and homologs of vitamin K_2. Anal. Biochem. *32*, 263–272.
428. Howarth, O. W., Grund, E., Kroppenstedt, R. M. & Collins, M. D. (1986) Structural determination of a new naturally occurring cyclic vitamin K. Biochem. Biophys. Res. Commun. *140*, 916–923.
429. Norman, A. W. (1987) Studies on the vitamin D endocrine system in the avian. J. Nutr. *117*, 797–807.
430. Zagalak, B., Neuheiser, F. & Curtius, H.-Ch. (1984) Lipophilic conjugates of vitamin D_3 in humans. Naturwissenschaften *71*, 321–322.
431. Bouillon, R. (1983) Radiochemical assays for Vitamin D metabolites: technical possibilities and clinical applications. J. Steroid Biochem. *19*, 921–927.
432. Mawer, E. B. (1980) Clinical implications of measurements of circulating vitamin D metabolites. Clin. Endocrinol. Metab. *9*, 63–79.

433. Haddad, J. G. & Chyu, K. J. (1971) Competitive protein-binding radioassay for 25-hydroxycholecalciferol. J. Clin. Endocrinol. *33*, 922–995.

434. Trafford, D. J. H., Seamark, D. A., Turnbull, H. & Makin, H. L. J. (1981) High performance liquid chromatography of 25-hydroxyvitamin D_2 and 25-hydroxyvitamin D_3 in human plasma. Use of isotachysterols and a comparison with gas-chromatography-mass spectrometry. J. Chromatogr. Biomed. Applic. *226*, 351–360.

435. Seamark, D. A., Trafford, D. J. H. & Makin, H. L. J. (1980) The estimation of vitamin D and some metabolites in human plasma by mass fragmentography. Clin. Chim. Acta *106*, 51–62.

436. Trafford, D. J. H. & Makin, H. L. J. (1983) Studies on the presence of 25-hydroxyvitamin D in human saliva. Clin. Chim. Acta *129*, 19–25.

437. Fairney, A. & Saphier, P. W. (1981) The measurement of 25-hydroxyvitamin D in saliva. J. Clin. Chem. Clin. Biochem. *19*, 622–663.

438. Blunt, J. W., DeLuca, H. F. & Schnoes, H. K. (1968) 25-hydroxycholecalciferol. A biologically active metabolite of vitamin D_3. Biochemistry *7*, 3317–3322.

439. Suda, T., DeLuca, H. F., Schnoes, H. K. & Blunt, J. W. (1969) The isolation and identification of 25-hydroxyergocalciferol. Biochemistry *8*, 3515–3520.

440. Bjorkhem, I. & Holmberg, I. (1976) A novel specific assay of 25-hydroxy vitamin D_3. Clin. Chim. Acta *68*, 215–221.

441. Bjorkhem, I. & Holmberg, I. (1978) Mass fragmentography of 25-hydroxyvitamin D_3, in: Quantitative Mass Spectrometry. Life Sciences II, pp. 167–174. Elsevier, Amsterdam.

442. Bjorkhem, I. & Holmberg, I. (1980) Mass fragmentographic assay of 25-hydroxyvitamin D_3. Methods Enzymol. *67*, 385–393.

443. Zagalak, B., Curtius, H.-Ch., Foschi, R., Wipf, G., Redweik, U. & Zagalak, M. J. (1978) Mass fragmentographic determination of cholecalciferol and 25-hydroxycholecalciferol in human serum. Experientia *34*, 1537–1539.

444. Zagalak, B., Neuheiser, F., Zagalak, M. J., Kuster, T., Curtius, H. Ch., Exner, G. U., Fanconi, S. & Prader, A. (1983) Application of mass fragmentography in the study of biosynthesis and excretion of vitamin D in humans. Anal. Chem. Symp. Ser. *14* (Chromatogr. Mass Spectrom. Biomed. Sci., 2), 347–360.

445. Zagalak, B., Neuheiser, F., Curtius, H. Ch. & Borschberg, H. J. (1984) Mass fragmentographic determination of vitamin D_2 in blood and urine in humans. Spectroscop.: Int. J. *3(4–5)*, 210–225.

446. Berlin, T., Bjorkhem, I., Collste, L., Holmberg, I. & Wijkstrom, H. (1982) Relation between hypercalciuria and vitamin D_3 status in patients with urolithiasis. Scand. J. Urol. Nephrol. *16*, 269–273.

447. Bjorkhem, I., Holmberg, I., Kristiansen, T. & Pedersen, J. I. (1979) Assay of 1,25-dihydroxyvitamin D_3 by isotope dilution-mass fragmentography. Clin. Chem. *25*, 584–588.

448. Makin, H. L. & Trafford, D. J. H. (1984) Measurement of vitamin D and its metabolites by gas chromatography-mass spectrometry, in: Vitamin D: Basic and Clinical Aspects (Kumar, R., ed.) 497–521. Martinus Nijhoff Publishers, Boston, MA.

449. Coldwell, R. D., Trafford, D. J. H., Makin, H., L. J., Verley, M. J. & Kirk, D. N. (1985) Specific mass fragmentographic assay for 25,26-dihydroxyvitamin D in human plasma using a deuterated internal standard. J. Chromatogr. Biomed. Applic. *338*, 289–302.

450. Whitney, J. O., Lewis, S., Piel, C. F. & Burlingame, A. L. Presented at 31st Annual Conference on Mass Spectrometry and Allied Topics, Boston, MA, 1983, pp. 763–764.

451. Napoli, J. L., Pramanik, B. C., Royal, P. M., Reinhardt, T. A. & Horst, R. L. (1983) Intestinal synthesis of 24-keto-1,25-dihydroxyvitamin D_3. A metabolite formed in vivo with high affinity for the vitamin cytosolic receptor. J. Biol. Chem. *258*, 9100–9107.

452. Napoli, J. L. & Horst, R. L. (1983) C_{24}- and C_{23}-Oxidation, converging pathways of intestinal 1,25-dihydroxyvitamin D_3 metabolism: identification of 24-keto-1,23,25-trihydroxyvitamin D_3. Biochemistry *22*, 5848–5853.

453. Horst, R. L., Wovkulich, P. M., Baggiolini, E. G., Uskokovic, M. R., Engstrom, G. W. & Napoli, J. L. (1984) (23S)-1,23,25-trihydroxyvitamin D_3: Its biologic activity and role in α,25-dihydroxyvitamin D_3 26,23-lactone biosynthesis. Biochemistry *23*, 3973–3979.

454. Jones, G., Kano, K., Yamada, S., Furusawa, T., Takayama, H. & Suda, T. (1984)

Identification of 24,25,26,27-tetranor-23-hydroxyvitamin D_3 as a product of the renal metabolism of 24,25-dihydroxyvitamin D_3. Biochemistry *23*, 3749–3754.

455. Reddy, G. S. & Tserng, K.-Y. (1986) Isolation and identification of 1,24,25-trihydroxyvitamin D_2, 1,24,25,28-tetrahydroxyvitamin D_2 and 1,24,25,26-tetrahydroxyvitamin D_2: New Metabolites of 1,25-dihydroxyvitamin D_2 produced in rat kidney. Biochemistry *25*, 5328–5336.
456. Napoli, J. L., Sommerfeld, J. L., Pramanik, B. C., Gardner, R., Sherry, A. D., Partridge, J. J., Uskokovic, M. R. & Horst, R. L. (1983) 19-Nor-10-ketovitamin D derivatives: Unique metabolites of vitamin D_3, vitamin D_2 and 25-hydroxyvitamin D_3. Biochemistry *22*, 3636–3640.
457. Litwiller, R. D., Mattox, V. R., Jardine, I. & Kumar, R. (1982) Evidence for a monoglucuronide of 1,25-dihydroxyvitamin D_3 in rat bile. J. Biol. Chem. *257*, 7491–7494.
458. Jardine, I., Scanlan, G. F., Mattox, V. R. & Kumar, R. (1984) Analysis of steroid and Vitamin glucuronides and sulfates by fast atom bombardment mass-spectrometry. Biomed. Mass Spectrom. *11*, 4–9.
459. Epstein, E. H., Jr., Han, A. & Shackleton, C. H. L. (1982) Failure of steroid sulfatase to desulfate Vitamin D3 sulfate. J. Invest. Dermatol. *80*, 514–516.
460. Selva, A., Traldi, P., Rapi, G., Ginanneschi, M. & Chelli, M. (1983) Mass spectrometry of 3-phosphoryl derivatives of vitamins D_2, D_3 and 7-dehydrocholesterol. Biomed. Mass Spectrom. *10*, 74–77.
461. Takada, K. (1983) Formation of fatty acid esterified vitamin D_3 in rat skin by exposure to ultraviolet radiation. J. Lipid Res. *24*, 441–448.
462. Halket, J. M., Ganschow, I. & Lisban, B. P. (1980) Gas chromatographic-mass spectrometric properties of boronate esters of 24R, 25-dihydroxycholecalciferol. J. Chromatogr. *192*, 434–440.
463. Jones, G., Seamark, D. A., Trafford, D. J. H. & Makin, H. L. J. (1985) Vitamin D: cholecalciferol, ergocalciferol, and hydroxylated metabolites, in: Modern Chromatographic Analysis of the Vitamins (De Leenheer, A. P., Lambert, W. E. & De Ruyter, M. G. M., eds.) pp. 73–128, Marcel Dekker, New York.
464. Napoli, J. L., Koszewski, N. J. & Horst, R. L. (1986) Isolation and identification of vitamin D metabolites. Methods Enzymol. *123*, 127–139.
465. Dougherty, R. C. (1980) Toxic residues and pollutants, in: Biochemical applications of mass spectrometry (Waller, G. R. & Dermer, O., eds.) pp. 951–968, Wiley Interscience, New York, NY.
466. Rodricks, J. V., Hesseltine, C. W. & Mehlman, M. A. (1977) Mycotoxins in human and animal health, pp. 7–28, 597–603, Pathotox, Park Forest, IL.
467. Shank, R. C., Wogan, G. N., Gibson, J. B. & Nondasuta, A. (1972) Dietary aflatoxins and human liver cancer. II. Aflatoxins in market food stuff of Thailand and Hong Kong. Food Cosmet. Toxicol. *10*, 61–69.
468. Shank, R. C., Bhamapravati, N., Gordon, J. E. & Wogan, G. N. (1972) Dietary aflatoxins and human liver cancer. IV. Incidence of primary liver cancer in two municipal populations of Thailand. Food Cosmet. Toxicol. *10*, 171–179.
469. Peers, F. G., Gilman, G. A. & Linsell, C. A. (1976) Dietary aflatoxins and human liver cancer. A study in Swaziland. Int. J. Cancer *17*, 167–176.
470. Peers, F. G. & Linsell, C. A. (1977) Dietary aflatoxins and human primary liver cancer. Ann. Nutr. Alim. *31*, 1005–1018.
471. Phillips, D. L., Yourtee, D. M. & Serales, S. (1976) Presence of aflatoxin B_1 in human liver in the United States. Toxicol. Appl. Pharmacol. *36*, 403–406.
472. Sora, C., Dvorackova, I. & Ayraud, N. (1981) Characterization of aflatoxin B_1 in human liver cancer. Res. Commun. Chem. Pathol. Pharmacol. *31*, 77–84.
473. Ryan, N. J., Hogan, G. R., Hayes, A. W., Unger, P. D. & Siraji, M. Y. (1979) Aflatoxin B: Its role in the etiology of Reye's syndrome. Pediatrics *64*, 71–75.
474. Self, R. (1979) Recent developments in the estimation of trace toxic substances in food. Biomedical Mass Spectrom. *6*, 361–373.
475. Kingston, D. G. I. (1976) Applications of mass spectrometry to the analysis of mycotoxins. J. Assoc. Off. Anal. Chem. *59*, 101–1022.
476. Dusold, L. R., Pohland, A. E., Dreifuss, P. A. & Sphon, J. A. (1978) Mycotoxins Mass Spectra Data Bank, Association of Official Analytical Chemists, Washington, DC.
477. Sphon, J. A., Dreifuss, P. A. & Schulten, H. R. (1977) Field desorption mass spectrometry of mycotoxins and mycotoxin mixtures, and its application as a screening technique for foods. J. Assoc. Off. Anal. Chem. *60*, 73–82.

478. Haddon, W. F., Wiley, M. & Waiss, A. C., Jr. (1971) Aflatoxin detection by thin-layer chromatography-mass spectrometry. Anal. Chem. *43*, 268–270.
479. Haddon, W. F., Masri, M. S., Randall, V. G., Elsken, R. H. & Meneghelli, B. J. (1977) Mass spectral confirmation of aflatoxins. J. Assoc. Off. Anal. Chem. *60*, 107–113.
480. Brumley, W. C., Nesheim, S., Trucksess, M. W., Trucksess, E. W., Dreifuss, P. A., Roach, J. A. G., Andrzejewski, D., Eppley, R. M., Pohland, A: E., Thorpe, C. W. & Sphon, J. A. (1981) Negative ion chemical reaction mass spectrometry of aflatoxins and related mycotoxins. Anal. Chem. *53*, 2003–2006.
481. Proctor, B. L. (1983) Recent developments in mass spectrometry for analysis of complex mixtures. Ann. Rev. Pharmacol. Toxicol. *23*, 171–192.
482. Siraj, M. Y., Hayes, A. W., Unger, P. D., Hogan, G. R., Ryan, N. J. & Wray, B. B. (1981) Analysis of aflatoxin B_1 in human tissues with high-pressure liquid chromatography. Toxicol. Appl. Pharmacol. *58*, 422–430.
483. Garner, R. C., Martin, E. C. & Miller, J. A. (1972) Liver microsomal metabolism of aflatoxin B_1 to a reactive derivative toxic to Salmonella typhirmurium TA 1530. Cancer Res. *32*, 2058–2066.
484. Moss, E. J., Judah, D. J., Przbylski, M. & Neal, G. E. (1983) Some mass-spectral and n.m.r. analytical studies of a glutathione conjugate of aflatoxin B_1. Biochem. J. *210*, 227–233.
485. Lee, L. S., Stanley, J. B., Cucullu, A. F., Pons, W. A., Jr. & Goldbatt, L. A. (1974) Annoniation of aflatoxin B_1: isolation and identification of the major reaction product. J. Assoc. Off. Anal. Chem. *57*, 626–631.
486. Walther, H. J., Parker, C. E., Harvan, D. J., Voyksner, R. D., Hernandez, O., Haler, W., Hamilton, P. B. & Hass, J. R. (1984) Fast atom bombardment mass spectrometry of aflatoxins and reaction products of sodium bisulfite with aflatoxins. J. Agr. Food Chem. *31*, 168–171.
487. Safe, S. (1982) Halogenated hydrocarbons and aryl hydrocarbons identified in human tissues. Toxicol. Environmental Chem. *5*, 153–165.
488. Cairns, T., Siegmund, E. G., Jacobson, R. A., Barry, T., Petzinger, G., Morris, W. & Heikes, D. (1983) Application of mass spectrometry in the regulatory analysis of pesticides and industrial chemicals in food and feed commodities. Biomed. Mass Spectrom. *10*, 301–315.
489. Parkinson, A., Robertson, L. & Safe, S. (1980) Reconstituted human breast milk PCBs as potent inducers of aryl hydrocarbon hydroxylase. Biochem. Biophys. Res. Commun. *96*, 882–889.
490. Safe, S. & Hutzinger, O. (1973) Mass Spectrometry of pesticides and pollutants. CRC Press, Cleveland, Ohio.
491. Self, R. (1979) Recent developments in the estimation of trace toxic substances in food. Biomed. Mass Spectrom. *6*, 361–373.
492. Biros, F. J. & Walker, A. C. (1970) Pesticide residue analysis in human tissue by combined gas chromatography-mass spectrometry. J. Agr. Food Chem. *18*, 425–429.
493. Sissons, E. & Welti, D. (1971) Structural identification of polychlorinated biphenyls in commercial mixtures by gas-liquid chromatography, nuclear magnetic resonance and mass spectrometry. J. Chromatogr. *60*, 15–32.
494. Watanabe, I., Yakushiji, T., Kuwabara, K., Yoshida, S., Kashimoto, T., Koyama, K. & Dunita, N. (1979) Surveillance of the daily PCB intake in the diet of Japanese women for 1972–1976. Arch. Environ. Contam. Toxicol. *8*, 67–75.
495. Tindall, G. W. & Winninger, P. E. (1980) Gas chromatography mass spectrometry method for identifying and determining polychlorinated biphenyls. J. Chromatogr. *196*, 948–949.
496. Cairns, T. & Siegmund, E. G. (1981) Determination of polychlorinated biphenyls by chemical ionization mass spectrometry. Anal. Chem. *53*, 1599–1603.
497. Dougherty, R. C. & Petrowska, K. (1976) Screening by negative chemical ionization mass spectrometry for environmental contamination with toxic residues: Application to human urines. Proc. Natl. Acad. Sci. USA *73*, 1777–1781.
498. Kuehl, D. W., Whitaker, M. J. & Dougherty, R. C. (1980) Micromethods for toxic residue screening by negative chemical ionization mass spectrometry. Anal. Chem. *52*, 935–940.
499. Dougherty, R. C., Whitaker, M. J., Smith, L. M., Stalling, D. L. & Kuehl, D. W. (1980) Negative chemical ionization studies of humans and food chain contamination with xenobiotic chemicals. Environ. Health Perspec. *36*, 103–118.
500. Ono, L., Taksukuma, R. & Wakimoto, T. (1975) Migration of plasticizer from hemo-

dialysis blood tubing. Preliminary report. J. Amer. Med. Assoc. *234*, 948–949.
501. Hillamn, S., Goodwin, S. L. & Sherman, W. R. (1975) Identification and measurement of plasticizers in neonatal tissues after umbilical catheters and blood products. N. Engl. J. Med. *292*, 381–386.
502. Roll, D. B., Doublas, J. D., Peterson, P. V. (1974) GLC analysis of bis-(2-ethylhexyl) phthalate plasticizer in tissues and plasma. J. Pharm. Sci. *63*, 1628–1629.
503. Ishida, M., Suyama, K. & Adachi, S. (1980) Background concentrations of phthalates commonly encountered in the chromatographic analysis of lipid samples. J. Chromatogr. *189*, 421–424.
504. Marcel, Y. L. & Noel, S. P. (1970) Contamination of blood stored in plastic packs. Lancet i, 35–36.
505. Jaeger, R. J. & Rubin, R. J. (1970) Contamination of blood stored in plastic packs. Lancet ii, 151.
506. Jaeger, R. J. & Rubin, R. J. (1972) Migration of phthalate ester plasticizer from polyvinyl chloride blood bags into stored human blood and its localization in human tissues. N, Engl. J. Med. *287*, 1114–1118.
507. Ching, N. P. H., Jhma, G. N., Subbarayan, C., Grossi, C., Hicks, R. & Nealon, T. F., Jr. (1981) Gas chromatographic quantitation of two plasticizers contaminating intravenous fluids stored in plastic containers. J. Chromatogr. Biomed. Applic. *255*, 196–201.
508. Ching, N. P. N., Jham, G. N. & Subbarayan, C. (1981) Gas chromatographic-mass spectrometric detection of circulating plasticizers in surgical patients. J. Chromatogr. Biomed. Applic. *222*, 171–177.
509. McLafferty, F. W. & Gohlke, R. S. (1959) Mass spectrometric analysis of aromatic acids and esters. Anal. Chem. *31*, 2076–2082.
510. Djerassi, C. & Fenselau, C. (1965) Mass spectrometry in structural and stereo-chemical problems. LXXXVI. The hydrogen-transfer reaction in butyl propionate, benzoate and phthalate. J. Am. Chem. Soc. *87*, 5756–5762.
511. Shintani, H. (1985) Determination of phthalic acid, mono-(2-ethylhexyl)phthalate and di-(2-ethylhexyl)phthalate in human plasma and in blood products. J. Chromatogr. Biomed. Applic. *337*, 279–290.
512. Calley, D., Autian, J. & Guess, W. I. (1966) Toxicology of a series of phthalate esters. J. Pharm. Sci. *55*, 158–162.
513. Stetson, J. B. & Autian, J. (1976) Necrotizing enterocolitis and plastic catheters, in: Intensive Care in the Newborn (Stern, L., ed.) pp. 79–98, Mason Publishing Co., New York.
514. Reddy, J. K. & Lalwai, N. D. (1983) Carcinogenesis by hepatic peroxisome proliferators: evaluation of the risk of hypolipidemic drugs and industrial plasticizers to humans. Crit. Rev. Toxicol. *21*, 1–58.
515. Bell, F. P., Makowske, M., Schneider, D. & Patt, C. S. (1979) Inhibition of sterologenesis in brain and liver of fetal and suckling rats from dams fed di-2-ethylhexyl phthalate. Lipids *14*, 372–377.
516. Fears, R. (1985) Lipophilic xenobiotic conjugates: the pharmacological and toxicological consequences of the participation of drugs and other foreign compounds as substrates in lipid metabolism. Progr. Lipid Res. *24*, 177–195.
517. Adams, J. & Gross, M. L. (1987) Tandem mass spectrometry for collisional activation of alkali-metal-cationized fatty acids: a method for determining double bond location. Anal. Chem. *59*, 1576–1582.
518. Thompson, T. S. & Karasek, F. W. (1987) Identification of trace organic contaminants in inorganic water treatment chemicals by gas chromatography-mass spectrometry. J. Chromatogr. *388*, 351–363.

Chapter 6
Organic Acids

R. A. Chalmers

1 Introduction

The association of organic acids and clinical biochemistry is primarily represented by those inherited disorders of intermediary metabolism in which defects in the metabolism of acyl CoA esters and other acyl moieties result in the accumulation of organic acids in body tissues and fluids and lead to characteristic and diagnostic organic acidurias and organic acidaemias. Thus much of the development of interest in, and knowledge of, organic acids and their diagnostic utility and value in clinical biochemistry have resulted from the study of these relatively rare inborn errors of metabolism, although the application of the knowledge gained is of value in a wide variety of other clinical conditions. Study of organic acids and the organic acidurias is closely associated with gas chromatography and mass spectrometry (GC-MS) and the growth of knowledge in the former has paralleled the development in clinical biochemistry of the latter despite earlier advances using more simple chromatographic and identification procedures (1). While much of this development has been with use of packed column GC-MS, more recently the application of capillary gas chromatography, especially with fused silica columns, has revolutionised the analytical capabilities and hence ultility of this area of clinical biochemistry. The renaissance in liquid chromatography with the development of high performance column systems (HPLC) together with new methods for ionization and for interfacing liquid chromatography with mass spectrometry including thermospray and spray-deposition belt interfaces have also made possible the direct study of less volatile conjugated acyl moieties and hence the extension of the study of organic acids in human metabolism.

The complexity of organic acid patterns in body fluids, particularly urine, together with the wide range of substituted and unsubstituted acidic compounds present both in healthy individuals and in those with disease states has made the identification of individual organic acids extremely difficult and in many cases only soluble with the use of mass-spectrometry. Some 40 to 50 individual chromatographic peaks are commonly observed in acidic extracts of urine when packed column gas chromatography is used: with many of these peaks containing several co-eluting components, mass-spectrometry has become essential in the unambiguous identification of all of these organic acids. The complexity of the patterns observed has increased considerably with the use of the great separation power of capillary columns, with one hundred to several hundred individual acids now being observable in a single urinary organic acid analysis. While chromatographic retention data obtained with such columns is much more reliable and consistent than with packed columns, the use of mass spectrometry is essential for absolute identification of known components and the identification of previously-unrecorded acids. The use of capillary gas chromatography has in turn placed much greater demands upon the mass spectrometers used, especially in terms of scan speed for the fast eluting peaks and sensitivity for the small amounts of compounds concerned. Even with capillary GC-MS, individual components may remain unresolved from one another and the use of specific techniques, for example selected ion monitoring (SIM) and extracted ion profiling, is of value in the analysis of complex mixtures of organic acids. Similarly, for accurate and selective quantification, particularly of components at low concentrations, in such mixtures, the use of mass spectrometry with SIM is an essential component in the analysis of body fluids such as cerebrospinal fluid, amniotic fluid and blood plasma.

The first chapter of this book has detailed the general principles of mass spectrometry, the ionization procedures and mass separation methods available and aspects of the manipulation of the large amounts of data produced. Most work with organic acids in clinical biochemistry has been carried out using electron impact ionization (EI) and much of the data produced has been essentially qualitative for diagnostic purposes. Repetitive scanning with the mass spectrometer throughout the chromatographic analysis followed by library searching of significant spectra is commonly employed for such qualitative studies. Quantitative analysis in the SIM mode either with the use of chemically-similar internal standards or with stable isotope dilution analysis using analogues of the acid of interest labeled with deuterium is increasingly employed for studies on acids although the potential advantage of the increased sensitivity offered by chemical ionization (CI) has not been greatly utilised. However, prior to a more detailed discussion of these areas in the analysis of organic acids, it is necessary to define further the nature of the organic acids concerned and the methods necessary for their extraction and preparation for GC-MS analysis.

2 Organic Acids and Acyl Compounds in Body Fluids

2.1 Definitions

Organic acids are carboxylic acids that may contain as additional functional groups, hydroxyl and oxo (keto and aldo) and other non-amino groups. They share the common features of high water solubility, acidity and ninhydrin-negativity. Thus (α-)amino acids and other amino acids are excluded although nitrogen-containing acids, for example pyrrolidone carboxylic acid (pyroglutamic acid) and pyrrole and pyrrolidine carboxylic acids are included. Many of the organic acids are also excreted into urine as conjugates with other compounds, for example glycine, glutamine, glutamate, aspartate and glucuronic acid (glucuronides). These compounds, particularly the acylglycines, are also considered as organic acids since they share the common characteristics given above, are extracted concomitantly with organic acids in most systems and are important diagnostically and biochemically in the study of organic acid and acyl group metabolism. Organic acids arise from the degradation of amino acids in particular since all intermediates in the catabolic pathways of amino acids after removal of the α-amino nitrogen are organic acids or acyl CoA esters. They also arise from the metabolism of carbohydrates and lipids and from biogenic amines as well as directly from the tricarboxylic acid cycle. Organic acids are thus among the most important intermediary metabolites and their study is of particular importance in clinical biochemistry. Many of the intermediates in the pathways concerned are acyl CoA esters: these compounds contain a high energy bond and the CoA moiety is readily interchangeable with L-carnitine, with acylcarnitines occurring in increased amounts in a number of disorders of organic acid metabolism associated with acyl

CoA accumulation. While acylcarnitines are basic compounds rather than acids, they have also been considered briefly in the present chapter since they represent a class of acyl compounds the importance of which is increasingly becoming recognised in the study of metabolic disorders and of intermediary metabolism. Special methods are required for their analysis however and discussion of them has been confined to the final section of this chapter.

While short-chain (C_1-C_6) fatty acids are sufficiently volatile and thermally stable to be analysed directly by gas chromatography, the majority of the highly polar and thermally labile organic acids, particularly those with multiple functional groups, are not amenable to direct analysis and must be extracted from biological fluids and converted into suitable derivatives for gas chromatography and mass spectrometry. The methods used for extraction of acids from biological fluids and the choice of derivatives are fundamental to studies of organic acids using GC-MS and are described in more detail below.

2.2 Extraction and derivative preparation of organic acids

2.2.1 **Extraction**

Organic acids are usually extracted as a group from physiological fluids prior to further separation and analysis and the procedures will also extract conjugated acids, particularly the acylglycines. Urine specimens used range from random 'mid-stream' urine collections to timed 6 hour to 24 hour collections and the volumes selected for extraction are usually based upon the urinary creatinine concentration. Creatinine (2-imino-1-methyl-4-imidazolidinone) is the most universally accepted and employed endogenous reference metabolite (2) and results on urine specimens are frequently related to this parameter to enable comparison to be made between different specimens and patients (with results being given in terms of $mg \cdot g^{-1}$ creatinine or $mmol \cdot mol^{-1}$ creatinine for example). Blood specimens are usually collected into heparin or heparinised tubes as anticoagulant, preferably on ice, the erythrocytes and other cells removed by centrifugation within 30 minutes of collection, ideally at 4 °C, and the plasma is deproteinised prior to further extraction and analysis. The preferred method for deproteinisation of such samples is ultrafiltration at pH 8 using membrane cones: the slightly alkaline pH prevents loss of acids due to protein and membrane binding but still avoids hydrolysis of alkali-sensitive acyl conjugates such as the acylcarnitines.

Organic solvents are widely employed for the extraction of organic acids from urine and other protein-free physiological fluids, the general approach involving acidification of the fluid to pH 1, ideally with sulphuric acid rather than hydrochloric acid to avoid the formation of artefacts such as lactyllactate, followed by saturation with salt (sodium sulphate or sodium chloride, depending on the acid employed) to produce a slight excess of solid, and then extraction with diethylether and ethylacetate. The quality of the extraction achieved is determined by the limitations imposed by the relatively unfavourable partition coefficients between aqueous solution and organic solvents, particularly diethylether, that are characteristic of many of these highly hydrophilic acids. Ethylacetate provides much better efficiency of extraction than diethylether but still fails to extract acids with multiple functional groups such as the hydoxy and polyhydroxy acids [for example glyceric acid, erythronic acid (trihydroxybutanoic acid) and tricarboxylic acid cycle acids such as citrate]. In addition, direct solvent extraction of acidified urine will also result in extraction of considerable quantities of neutral components that may complicate the GC-MS analysis still further. Anion exchange extraction processes are increasingly favoured, with DEAE-Sephadex being perhaps most widely employed. In this procedure, the neutral physiological fluid is applied to a column of the DEAE-Sephadex, previously washed with aqueous pyridinium acetate buffer, the neutral and basic components are removed by water washing and the acidic components are eluted with an aqueous pyridinium acetate buffer. This procedure will quantitatively extract all acidic species present but suffers from the potential disadvantage that inorganic anions, for example sulphate and phosphate, are also extracted and other organic compounds such as urate and sugar phosphates are included in the acidic fraction, although the latter may also yield valuable metabolic information not otherwise readily available. The potential interference produced by these compounds on chromatography is minimised by the use of capillary GC columns and in practice the advantages of the total quantitative extraction achieved far outweigh any disadvantages.

Following extraction, these extracts must be reduced to dryness prior to derivative preparation and this is particularly important with aqueous extracts when subsequent trimethylsilylation is intended since the trimethylsilyl (TMS) organic acids are extremely sensitive to moisture. Solvent extracts are simply dried under a stream of nitrogen: prior drying of the solvent extracts with anhydrous sodium or magnesium sulphate is to be avoided to prevent further losses of hydrophilic acids by sequestration with adsorbed water. Aqueous extracts must be dried by lyophilisation (freeze drying): special conditions of $-10\,°C$ and 66.6 Pa (0.5 torr) have been recommended in order to retain quantitatively acids of low relative molecular mass (3) but careful attention during freeze drying to ensure the process is stopped as soon as the sample is just dry has also been used successfully. The dry extract free from all but minimal traces of moisture (and, in the case of pyridinium acetate buffers, also free from pyridine and acetic acid) is ready for derivative preparation.

Oxo acids present particular problems because of lability and the formation of multiple derivatives due to keto-enol tautomerism (see below) and are generally stabilised prior to extraction in the case of solvent based systems or on elution from anion exchange columns in the latter procedures. Volatile acids are also extracted and analysed using more specific methods and gas chromatography is suitable for determination of the short chain C_2-C_6 acids and other volatile constituents. These procedures and further details of the extraction of the less volatile acids are given elsewhere (2, 4).

2.2.2 **Derivative preparation for gas chromatography and mass spectrometry**

The high polarity, low volatility and thermal instability of most organic acids, particularly those with multiple functional groups, makes it essential to convert them into stable volatile derivatives for GC. The nature of the derivatives selected obviously affects the mass spectra that are obtained for the organic acids, different derivatives presenting different properties of fragmentation and stability. Oxo acids are also selectively derivatised to avoid keto-enol tautomerism with its consequent problems and other groups, for example hydroxyl and imino, are also derivatised to reduce the residual polarity of the molecule to a minimum for optimisation of GC properties.

The most universally used derivatives of organic acids are the esters, alkyl esters and trimethylsilyl esters being the most favoured. Alkyl esters, particularly methyl esters but also others including n-butyl esters (5) have been used, the former being introduced originally for fatty acid analysis and perhaps still being ideally suited to study of these monofunctional carboxylic acids. Esterification or methylation is achieved with use of diazomethane:

$$RCOOH + CH_2 = \overset{+}{N} = \overset{-}{N} \longrightarrow RCOOCH_3 + N_2$$

with methanol and a mineral acid or Lewis acid (boron trifluoride):

$$RCOOH + CH_3OH \xrightarrow{\text{acid}} RCOOCH_3 + H_2O$$

or with use of dimethylformamide dialkylacetals. However, in the presence of other functional groups, further partial derivatisation may occur, for example methyl ether formation with hydroxy acids and free oxo acids (6) and *N*-methyl derivatives of amino acids co-extracted on anion exchange columns (7). Other artefacts may also occur, for example methylglycidates with free oxo acids and diazomethane (8) and pyrrazolines with unsaturated acids and diazomethane (9). Despite these problems, some workers still prefer methyl esters, overcoming some problems with oxo acids by prior formation of methoximes (see below), but most workers today utilise the trimethylsilyl derivatives. Methyl esters are still favoured by others for specific determination of acylglycines because of dual derivative formation on trimethylsilylation (see below) and other esters, for example the *l*-menthyl esters are used for specific application with a variety of other derivatives, such as dicyclohexyl esters (10) and *t*-butyldimethylsilyl esters (11) being employed by individual workers.

Trimethylsilylation of carboxylic acids, and of hydroxyl groups in hydroxy acids, proceeds rapidly and quantitatively at ambient temperature with modern silylating reagents to form trimethylsilyl (TMS) esters and TMS ethers. Nitrogen containing compounds may also be further derivatised by formation of N-TMS derivatives with, for example, the acylglycines: with such compounds the disadvantageous formation of a mixture of the N-TMS and free NH compounds usually results, with associated problems for quantification. The trimethylsilylating reagents most generally employed are *N,O*-bis-trimethylsilylacetamide (BSA) and its derivative, *N,O*-bis-trimethylsilyl-trifluoroacetamide (BSTFA), often in the presence of small amounts of pyridine as catalyst and occasionally with addition of 1 – 10% of trimethylchlorosilane (TMCS) to enhance reactivity:

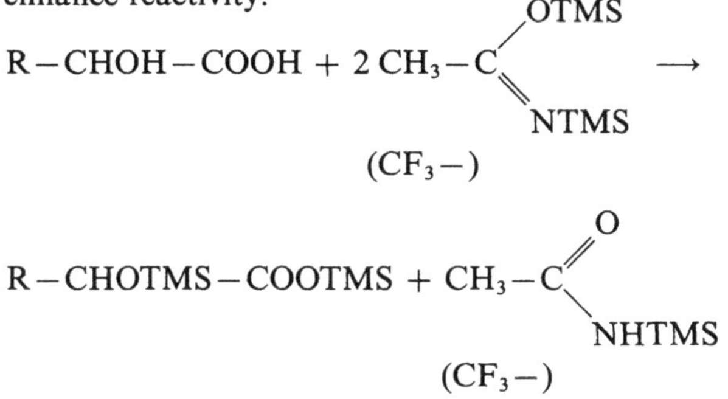

The volatile byproducts are N-TMS-acetamide and N-TMS-trifluoroacetamide from BSA and BSTFA respectively, with the trifluoro compounds showing greater reactivity and volatility. Free acids are preferred for trimethylsilylation although alkali metal salts and ammonium salts have been used. Dry conditions are essential since both reagents and the derivatives are highly sensitive to moisture, TMS carboxylic acids for example showing reactivity with water that is comparable to that of acid chlorides. Inorganic anions are also trimethylsilylated under the conditions used and triTMS-phosphate and diTMS-sulphate are observed in extracts of organic acids, particularly when anion exchange extraction methods are used.

Oxo acids exhibit keto-enol tautomerism and will form both oxo esters and enol-ether esters on derivatisation thereby complicating the chromatograms obtained and making quantification impossible. For this reason, and because of their inherent chemical reactivity and lability, these acids are stabilised prior to, or immediately after, extraction, generally by oximation with hydroxylamine or with *O*-alkylhydroxylamines to give the oxime or *O*-alkyloxime derivatives. The alkyloximes commonly used are the methyl and ethyl oximes although the *n*-butyloximes and benzyloximes have also been used for specific applications:

$$R-\underset{\underset{O}{\|}}{C}-COOH + NH_2OR' \longrightarrow R-\underset{\underset{N\diagdown OR'}{\|}}{C}-COOH \quad + H_2O$$

where $R' = H, CH_3, C_2H_5$ ....

Simple oxime formation from hydroxylamine has the disadvantage that heating (60 °C) may be required for their formation. The oximes formed will also react further with trimethylsilylating agents to give the O−TMS oximes as the final derivatives. These are relatively involatile, being comparable to *t*-butyloximes but have been used very successfully (12−15). They also have particular application to the determination of dioxoacids for example succinylacetone (dioxoheptanoic acid) (16). Most workers however favour the methoximes (17) and ethoximes (18, 19), the latter being generally more favoured because of the better separations achieved between oxo acids and their hydroxy acid analogues. The use of such derivatives introduces specific mass spectral characteristics to oxo acids that are of particular value in their structural identification although quantification is still not optimised because alkyl and TMS oximes form *syn* and *anti* isomers and two peaks are observed on GC for most branched-chain and dicarboxylic oxo acids. However one isomer generally predominates and effects on quantification are minimised.

Several workers have also utilised quinoxalinols of 2-oxo acids for the selective analysis of these compounds: ortho-phenylenediamine (1,2-diaminobenzene) reacts with dicarbonyl compounds to form quinoxalines and hence with 2-oxo carboxylic acids to form quinoxalinols:

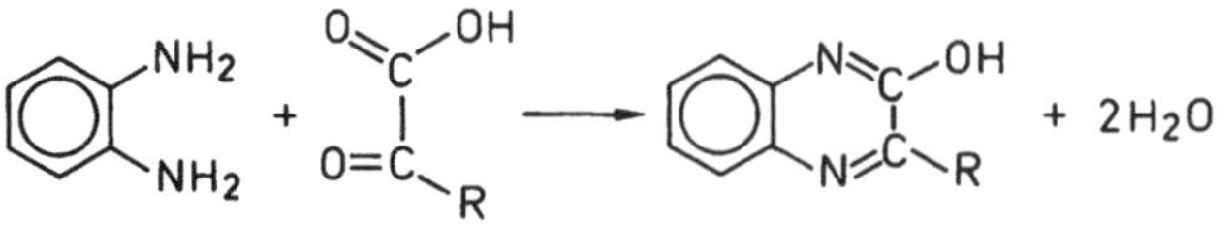

The resulting compounds react further with trimethylsilylating reagents to give TMS-quinoxalinols (20). These derivatives again have specific chromatographic and mass spectrometric characteristics that have been of value in the accurate determination of 2-oxo acids by GC-MS with SIM.

2.3 Chromatography of organic acids and their derivatives

2.3.1 **Gas chromatography of short-chain fatty acids**

Short chain fatty acids may be chromatographed as the free acids in aqueous solution by temperature-programmed GC on columns packed ester-coated (e. g. neopentyladipate) or silicone oil-coated materials treated with 1 – 2% phosphoric acid (21 – 23) or the equivalent phosphoric acid treated capillary columns. They may also be separated using gas-solid chromatography (24). Care must be taken with such methods to avoid artefact formation caused by dehydration of hydroxy acids and decarboxylation of dicarboxylic acids (25, 26) and prior esterification may be preferred. Short-chain carboxylic acids have been analysed most successfully as their methyl, n-butyl and trimethylsilyl esters (27) and a wide variety of liquid phases have been employed, with the use of silicone-based materials being generally favoured.

2.3.2 **Gas chromatography of organic acids**

Methyl esters and other alkyl esters are commonly separated using programmed-temperature GC on columns packed with polyester-phases (for example, polyethylene glycoladipate [PEGA], diethyleneglycoladipate [DEGA] and neopentylglycoladipate [NPGA], Carbowax 20M and silicone phases (for example, SE30, OV1, OV101) with the latter also being particularly suitable for capillary GC. Trimethylsilyl esters necessitate the use of silicone oil and gum based stationary phases (for example SE30, SE52, OV1, OV101, SP2100, OV17, OV22) usually in a 1 – 10% loading on specially acid-washed and silanised solid supports. These phases are also particularly useful for capillary GC of these derivatives and most workers today utilise fused silica capillary columns with cross-linked or chemically-bonded phases based on these silicone materials. The use of capillary columns has greatly improved separations and hence both quantification and identification of individual acids. The flexible, polyimide-coated fused silica columns offer particular advantages in GC-MS, allowing direct introduction of the GC column into the ion source and obviating the need for molecular separators. The much faster eluting chromatographic peaks observed with capillary columns have introduced the need for faster mass spectrometer scan speeds and the much greater number of spectra produced has in turn introduced the need for high capacity data systems. However, the great improvements in mass spectral quality and reduction in the number of mixed spectra in organic acid analysis has improved the quality of mass spectral library searching and the reliability of library identifications and most organic acid GC-MS analyses are performed today on fused silica capillary columns.

2.4 Mass spectra of organic acids and their derivatives

An understanding of the nature of the mass spectra of organic acids and particularly of their derivatives is essential to appreciate the value and applications of mass spectrometry to the study of organic acids in clinical biochemistry and some of the fundamental features of their EI-spectra are outlined below.

2.4.1 **Free acids**

As indicated in previous sections, acids in their free form are generally too polar and unstable for GC-MS analysis and discussion in this section is limited to consideration of the short-chain monocarboxylic ("fatty") acids. In general in their spectra the molecular ion is present but decreases in intensity with molecular mass to a minimum around C_5-C_6, before increasing again. Useful diagnostic ions are obtained by loss of 45 daltons due to the carboxylic acid group and loss of 17 daltons resulting from loss of carboxyl OH. Acetic and propionic acid also show an [M − 1] ion due to loss of a proton from the molecular ion. Rearrangement ions occur in longer chain length acids [> C_3], resulting from a McLafferty rearrangement of hydrogen from the carbon γ to the carboxyl group onto the carboxyl oxygen with cleavage of the β carbon-carbon bond (2, 28) to give a common rearrangement ion at m/z 60. Thus

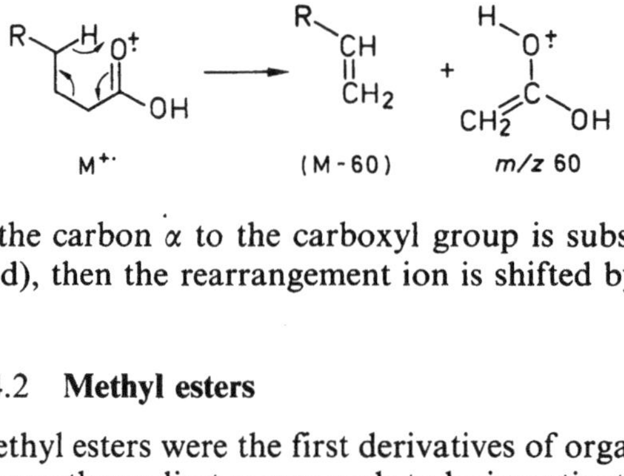

If the carbon α to the carboxyl group is substituted (for example in 2-methylbutyric acid), then the rearrangement ion is shifted by the mass of the substituent.

2.4.2 **Methyl esters**

Methyl esters were the first derivatives of organic acids studied by GC and hence were among the earliest compounds to be investigated by mass spectrometry. Several studies have been made of the spectra of aliphatic acid methyl esters (29 − 31) including high resolution spectra (32) and various aspects of ester spectra have been reviewed (33 − 35). With aliphatic methyl esters as with the free acids, the molecular ion is of significant intensity in the methyl esters of C_1-C_3 acids but diminishes sharply with increasing molecular weight to a minimum at C_5-C_6 thereafter increasing in intensity again with increase in molecular weight. Among the shorter chain length acids (C_1-C_5), cleavage α to the carboxylic group occurs:

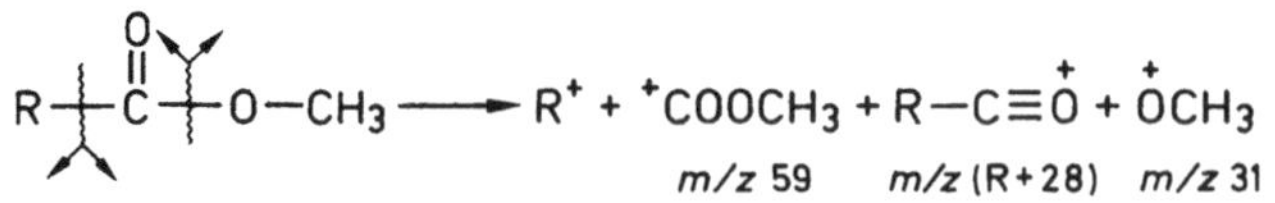

with the R^+ ions being fairly intense.

Rearrangement ions again occur in all the aliphatic methyl esters with the predominant *m/z* 74 being analogous to the *m/z* 60 in the free acid spectra and formed by the same γ hydrogen rearrangement:

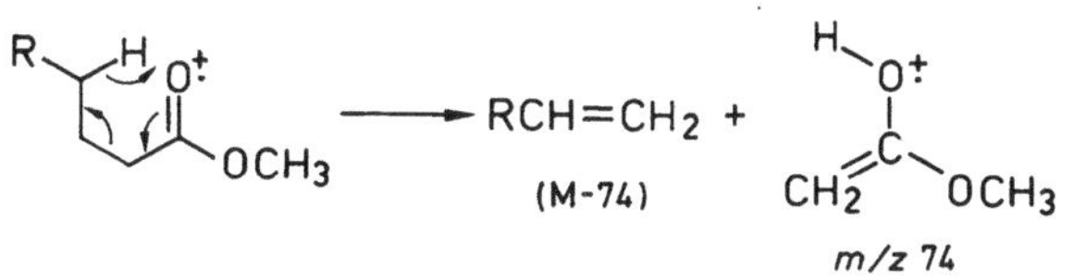

and as before, where the α-carbon carries a substituent, the mass of the rearrangement ion shifts by the mass of the substituent. Thus the isomeric pentanoates, isovaleric acid and 2-methylbutyric acid are readily distinguished as their methyl esters by the formation of rearrangement ions at *m/z* 74 and *m/z* 88 ($74 + CH_2$) respectively (36). Among the longer chain aliphatic acids, the molecular ions are well defined while the M−31 ion decreases in intensity. Oxygen-containing fragment ions are favoured over hydrocarbon ions and in addition to the *m/z* 74 base peak, a dominant ion occurs at *m/z* 87 as one of a series of ions corresponding to *m/z* (59 + 14n). In this series, the abundance of *m/z* 87 is high being favoured by resonance stabilisation with a periodic enhancement of the stabilisation occurring every four methylene groups at *m/z* 87, 143, 199, 255, 311 etc. (2).

The spectra of dicarboxylic acid methyl esters are more complex and differ considerably from those of the monocarboxylic acids, showing large M−31 ions due to loss of a methoxy group from one of the ester groups and low abundance molecular ions. A series of ions due to *m/z* (27 + 14n) occurs and above C_9 a further series due to *m/z* (84 + 14n) is also apparent. Interaction of the two carboxylic ester functions occurs with ions at M−64 ($-2CH_3OH$), M−92 ($-CH_3OOC-CH_3O-2H$) and M−105 ($-CH_3OOCCH_2-CH_3O-H$). Aromatic acids give prominent molecular ions and dominance of the aromatic ring in fragmentation and show losses of methoxyl (M−31) and carbomethoxyl (M−59). Benzoic acid analogues with hydrogen-containing ortho substituents (e. g. 2-methoxy benzoic acid) show an 'ortho effect', with a rearrangement loss of 32 ($-OCH_3-H$) and 60 ($-COOCH_3-H$) from the molecular ion. In aromatic acids with longer side chains, benzylic cleavage occurs with ions at *m/z* (M−77) predominating, but those with unsaturated side chains show similar spectra to the benzoates with M−31 (methoxyl loss) and M−59 (carbomethoxyl loss) occurring, presumably due to resonance stabilisation through the aromatic ring structure and unsaturated side chain.

Methyl esters are particularly useful for the study of acylglycines since single derivatives are obtained (see above) and molecular ions occur in almost all spectra. In contrast to the short-chain aliphatic acid methyl esters where a methoxyl radical is lost, N-acylglycines eliminate methanol to give an M−32 ion. Again carbomethoxyl loss occurs to give M−59 ions and a major ion occurs at M−88 due to cleavage of the amide linkage (37). Cleavage of the R-CO bond also occurs to yield M−116 and formation of *m/z* 30 ($CH_2 = \overset{+}{N}H_2$) with M−116 being a significant ion in most spectra:

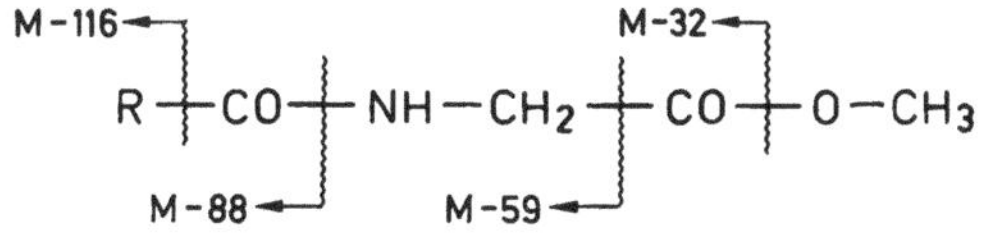

2.4.3 Trimethylsilyl derivatives

The trimethylsilyl (TMS) esters of carboxylic acids and the corresponding TMS-ether-TMS-ester derivatives of hydroxy acids are the most widely used in organic acid analysis, having low polarity, high volatility and being easy to prepare. The TMS esters do show very high reactivity to water however and dry extracts and conditions are required for their preparation and stable storage. Similarly their reactivity dictates stringent care in sample handling and particularly injection into the gas chromatograph since many of the problems encountered and reported with use of these derivatives can be ascribed to poor injection port configurations and design and to exposed metal surfaces.

In the TMS spectra, the base peak is almost invariably *m/z* 73, usually accompanied by an intense *m/z* 75 due to $(CH_3)_3Si^+$ and $(CH_3)_2Si = \overset{+}{O}H$ respectively. A ubiquitous ion in compounds with two or more TMS groups occurs at *m/z* 147 and is formed by rearrangement to give TMS-$\overset{+}{O} = Si(CH_3)_2$ and this is seen in hydroxy acid and dicarboxylic acids. The molecular ion is invariably small, with an ion at M − 15 (loss of CH_3) being the main indication of the molecular weight of the derivatised acid (see, for example, the spectrum given in Fig. 3).

Among the short-chain aliphatic monocarboxylic acids where *γ* hydroxy rearrangement is possible as in the corresponding methyl esters, an ion at *m/z* 132 occurs:

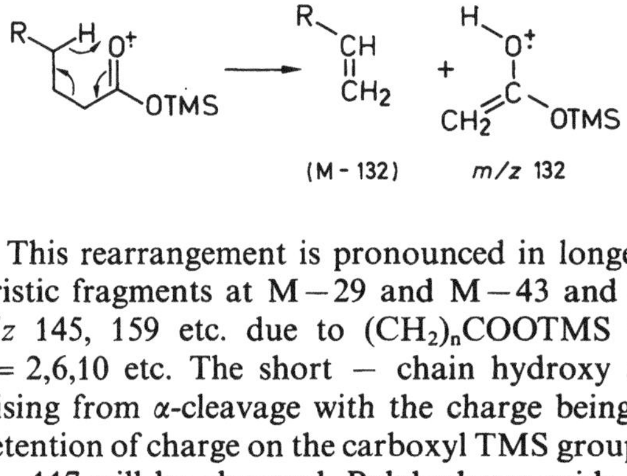

This rearrangement is pronounced in longer-chain acids which also show characteristic fragments at M − 29 and M − 43 and a series of carboxy TMS ester ions at *m/z* 145, 159 etc. due to $(CH_2)_nCOOTMS$ with ion intensity being enhanced at $n = 2,6,10$ etc. The short – chain hydroxy acids show a prominent *m/z* 131, this arising from *α*-cleavage with the charge being retained on the TMS ether fragment. Retention of charge on the carboxyl TMS group results in *m/z* 117 but more commonly, M − 117 will be observed. Polyhydroxy acids (aldonic and deoxyaldonic acids) show a number of fragmentations due to carbon chain cleavages *α* to the TMS ether groups, giving rise to a series of ions at *m/z* [103 + (*n* × 102)]. The ion at *m/z* 103 may indicate a terminal hydroxyl on the chain and *m/z* 205 indicates two terminal hydroxyl groups. A series of carboxyl containing fragments (M − 103) also occurs usually at low intensity with *m/z* 423 generally being the largest ion in the higher molecular weight acids. A ubiquitous ion at *m/z* 292 is present in spectra of the longer chain polyhydroxy acids, occurring due to a *γ* rearrangement process:

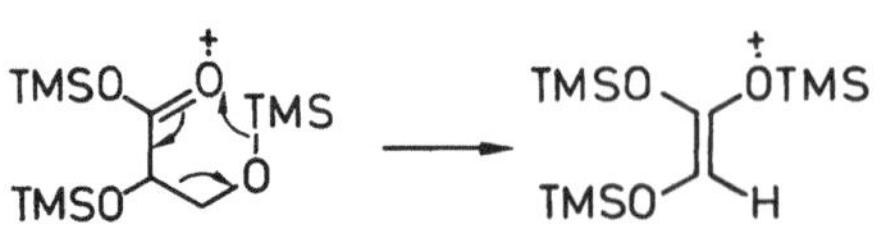

m/z 292

Among the deoxyaldonic acids, multiple fragmentation gives abundant ester-containing ions, for example in the 2-deoxyaldonic acids:

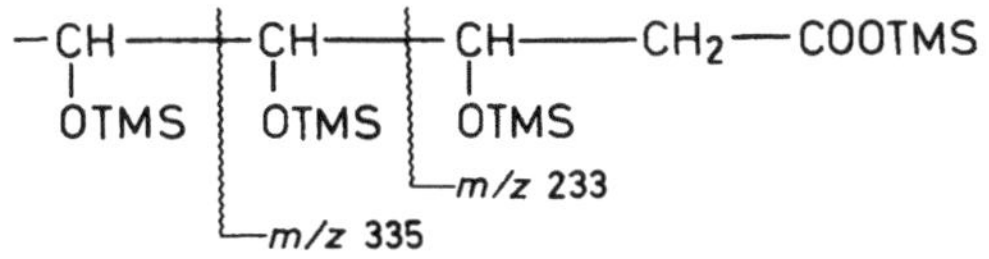

The ion at m/z 335 will also occur in 3-deoxypentonic acids and eliminates in turn trimethylsilanol to give m/z 245 and carbon monoxide to give m/z 217. 3-deoxyaldonic acids may also lose the ester function to give $M-117$ and related ions $[M-117-90]$. Spectra of polyhydroxyacids with two carboxylic acid groups (aldaric acids) are dominated by TMS rearrangement processes with m/z 292 occurring in 2,3-dihydroxy acids (and in hydroxy dicarboxylic acids) and ions arising from $M-15-CO$, $M-15-CO-CO_2$ etc. Deoxyaldaric acids show chain cleavage as with the deoxyaldonic acids with 2-deoxyaldaric acids giving ions at m/z 233 and 3-deoxyaldaric acids ions at m/z 335 (and at m/z 245 due to further loss of 90 daltons).

Simple dicarboxylic acids are also characterised, in addition to m/z 147, by ions arising from decarboxylation of the $M-15$ ion and by multiple ions, some caused by TMS group migration and rearrangement, with m/z 204 and m/z 217 being prominent in longer chain dicarboxylic acids due to

$$CH_2=\underset{\underset{+\cdot}{OTMS}}{\underset{|}{C}}-OTMS \quad \text{and} \quad CH_2=CH-\underset{+\,OTMS}{\underset{\|}{C}}-OTMS \qquad \text{respectively.}$$

Aromatic acids show much more prominent molecular ions together with more intense $M-15$ ions with fragmentation dictated by the stabilisation introduced by the aromatic ring. Thus benzoates all have a $M-OTMS$ $(M-89)$ ion, others being decarboxylated to give $M-117$ by cleavage of the benzylic bond. Hydroxyphenylacetic acids tend to show a prominent m/z 179 due to $\overset{+}{C}H_2-C_6H_4-OTMS$ or m/z $(179+R)$ where R is a ring substituent.

Phosphate esters, for example phosphoglyceric acids, sugar phosphates and TMS phosphate itself, show ions at m/z 315, 299 and 243 due to

$$(TMSO)_3-P=\overset{+\cdot}{O}H, \quad (TMSO)_2-P\begin{smallmatrix}\nearrow\!\!\!\!\!\nearrow O \\ \searrow \underset{+}{O}=Si(CH_3)_2\end{smallmatrix} \quad \text{and} \quad (TMSO)_2-P\begin{smallmatrix}\nearrow OH \\ \searrow\!\!\!\!\!\searrow \underset{+\cdot}{OH}\end{smallmatrix}$$

respectively, with that at m/z 299 being particularly prominent and 'diagnostic'. Similarly TMS sulphate shows a prominent and characteristic ion at m/z 227 due to

$$TMSO-\underset{\underset{O}{\|}}{\overset{\overset{O}{\|}}{S}}-\overset{+\cdot}{O}-Si(CH_3)_3$$

Glycine conjugates form a mixture of monoTMS and diTMS derivatives, the relative proportions being dependent upon the derivatisation conditions used. These derivatives are separated on GC and show different mass spectral characteristics. MonoTMS derivatives arise from trimethylsilylation of the carboxylic acid group and have prominent M−15 and M−44 ions with M−59 (M−44−15) and M−72 (M−TMS + H) also occurring. Information on the identity of the conjugated acid is given by cleavage of the amide linkage at M−146. This cleavage also occurs in diTMS glycine conjugates to give M−218. These diTMS derivatives show a prominent *m/z* 102 ion due to $CH_2 = \overset{+}{N}$ H-TMS arising from the M−117 ion with M−15 and M−43 ($M - CH_3 - CO$) also occurring (38). Aromatic glycine conjugates are again stabilised by the presence of the aromatic ring with hippuric acid (benzoylglycine) giving a prominent ion at *m/z* 206 due to $C_6H_5CO \cdot N(TMS) \cdot CH_2^+$.

2.4.4 **Oxo acids**

In most analyses, oxo acids are stabilised by conversion into their oxime and *O*-alkyloxime derivatives prior to trimethylsilylation, simple oximes becoming silylated themselves to give O-TMS oximes. The spectra of the O-TMS oxime-TMS esters of oxo acids show molecular ions, the intensities of which decrease as the chain lengths increase, and also show M−15, M−43 ($M - CH_3 - CO$) and M−117 ions, all the ions being distinctive because of their even masses arising from the odd valency of the nitrogen in the molecule. *O*-alkyloximes are again characterised by fragment ions with even masses, showing characteristic ions at M−15, M−31 (methoximes), M−45 (ethoximes) and M−59 (*n*-butyl-oximes) and M−117 (M−COOTMS). The latter oximes (ethoximes and *n*-butyloximes) show γ hydrogen rearrangement ions involving the ethyl and *n*-butyl groups with loss of C_2H_4 or C_3H_6 respectively to give M−28 and M−42 ions. Similar losses may also occur from the M−15 ion and rearrangement ions may occur due to $(CH_3)_2Si = \overset{+}{O}R$ at *m/z* 89 (methoximes), *m/z* 103 (ethoximes) and *m/z* 117 (*n*-butyloximes). Dicarboxylic keto acids show many of the fragmentation processes, 2-oxoglutaric acid for example having prominent ions at 156 and 198 due to M−15−(31 or 45 or 59)−117 and M−(31 or 45 or 59)−90 respectively, although the spectra are more complex with the relationship of the oxime γ to the TMS ester function promoting the production of several ions including M−90, M−117 and M−118 (39). Among the aromatic oxo acids, the molecular and M−15 ions are prominent with a series of ions at low intensity occurring due to loss of an alkoxyl radical ($M - OR - CO_2$, M−OR−45, $M - OR - CO_2TMS$, $M - OR - HCO_2TMS$).

Quinoxalinol derivatives of α-keto acids (2-oxo acids) trimethylsilylate to give OTMS quinoxalinols:

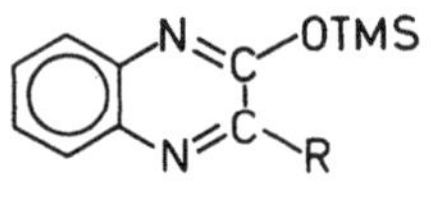

The spectra of these derivatives of both mono-carboxylic and dicarboxylic keto acids are characterised by several common ions among which *m/z* 217, *m/z* 232 and *m/z* 245 are prominent due to

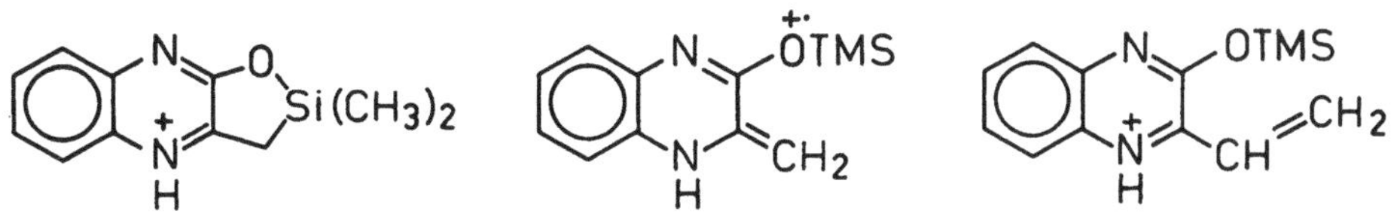

m/z 217 m/z 232 m/z 245

respectively, making these derivatives very suitable for quantitative SIM analysis of 2-oxo acids, although of less value in qualitative and profiling studies. The molecular ion is usually apparent and an ion at *m/z* 201 also occurs at low intensity due to ring stabilisation with retention of only one of the silylmethyl groups.

2.4.5 **CI spectra of TMS organic acids**

All of the above discussion relates to the EI spectra of the organic acids and their derivatives since EI is almost universally employed in studies of organic acids in biochemistry and metabolism. The CI spectra of TMS organic acids using ammonia as reagent gas are characterised by prominent $M + 1$ ($M + H$) ions with $M + 18$ ($MH + NH_3$) ions also occurring. Very little fragmentation occurs in these spectra although an ion at *m/z* 90 ($TMS + NH_3$) is generally the base peak due to inevitable loss of a TMS group from the molecular ion (see spectrum in Fig. 7). The lack of fragmentation makes these CI spectra of less value than the EI spectra for qualitative and profiling work but potentially they offer advantages of increased sensitivity and specificity for quantitative SIM analysis from monitoring the ($M + 1$) ion. Much greater use of this ionization mode may be expected in future and it has already been introduced in quantitative studies of TMS-quinoxalinols and of amniotic fluid organic acids for prenatal diagnosis (Section 5).

3 Mass Spectrometry in Profile Analysis of Organic Acid Mixtures

3.1 Methods

Early applications of chromatographic procedures to the study of organic acids were reviewed by Nordmann and Nordmann (1) who used the approach to study the acids in biological fluids from healthy subjects and those with various diseases states. The first use of mass spectrometry in the study of a disorder of organic acid metabolism was by Klenk and Kahlka in 1963 (40) who used GC and MS to identify phytanic acid in the serum of a patient who had died with Refsum's disease. However, this acid is not particularly water-soluble and the identification and study of organic acidurias was really initiated when Tanaka and his colleagues used GC-MS in 1966 to identify 3-hydroxy-isovaleric acid in the urine of a patient with the previously unidentified disease of isovaleric acidaemia (41) and opened up this area to systematic and more

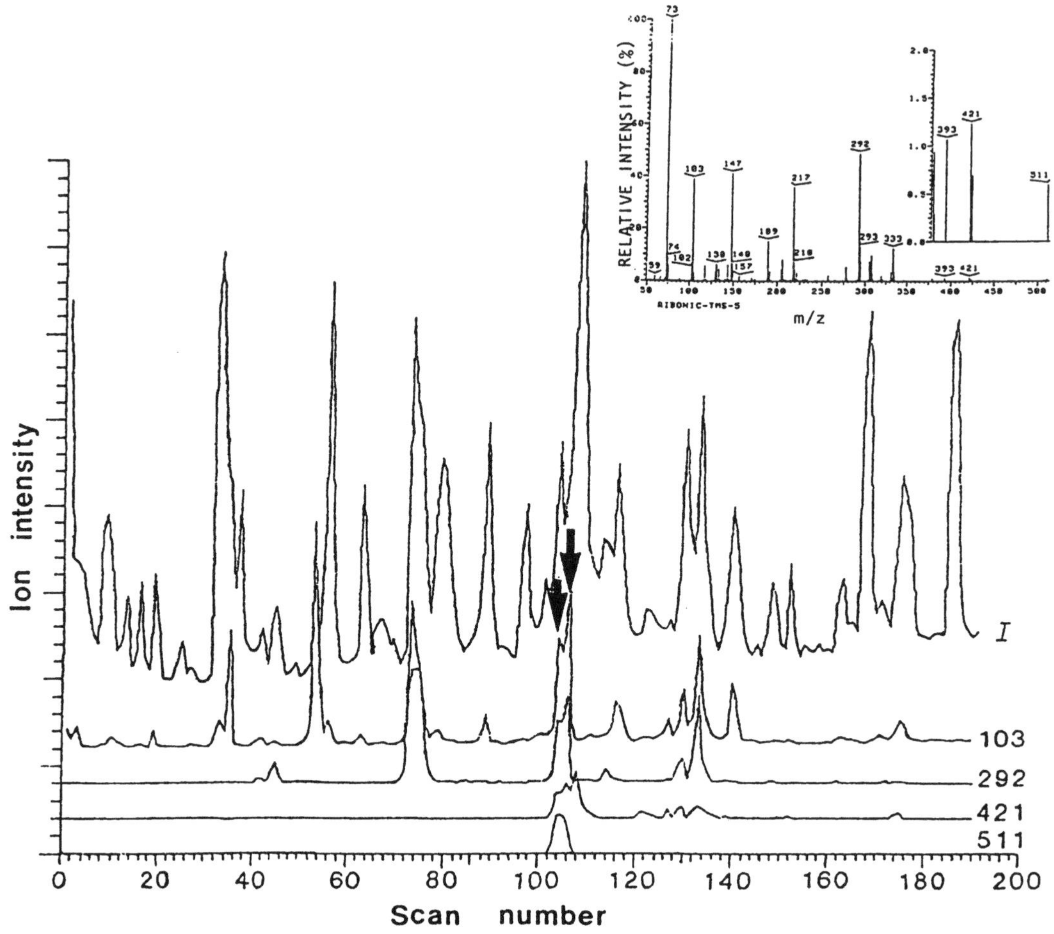

Figure 1. Use of extracted ion profiles to locate the positions of isomeric pentonic acids in a packed-column GC-MS analysis.
Showing the computer reconstruction of ion intensity plots for ions at *m/z* 103, 292, 421 and 511 and for total ion intensity (I) and showing clearly the presence at the positions indicated of two isomeric pentonic acids (TMS) for which these ions are characteristic. The inset mass spectrum is typical for such pentonic acids (ribonic acid penta TMS). (Reproduced in part from Chalmers and Lawson (2).)

detailed study. At the same time the seminal paper appeared by Dalgliesh and his colleagues, working in E. C. Horning's laboratory, on the use of GC to separate and identify in a single analysis a wide range of (acidic) substances in urine and other extracts of biological fluids and hence the study of metabolic patterns in such fluids (42). This paper together with succeeding papers (43, 44) led to the concepts of metabolic profiling ("multicomponent GC analysis that defines or describes metabolic patterns for a group of metabolically-related compounds") of physiological fluids using GC-MS by Horning and Horning (6, 45) and of screening for metabolic diseases using these techniques (5, 46). Early work utilised GC-MS instruments without on-

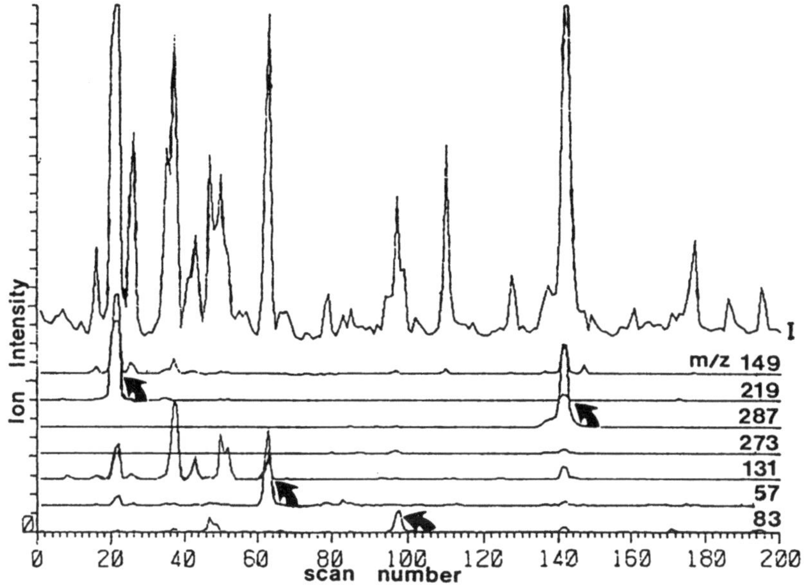

Figure 2. Extracted ion profiles as an aid to diagnosis of organic acidurias.
Showing the computer drawn ion intensity plots extracted from packed-column GC-MS data from analysis of urine from a patient with previously-undiagnosed propionic aciduria (total ion intensity chromatogram I): ions at *m/z* 287, 319, 83 and 57 are characteristic for TMS-methylcitrate, TMS-3-hydroxypropionate, TMS-tiglylglycine and TMS-propionylglycine respectively. Other ions shown are used to aid the differential diagnosis, for example *m/z* 273, characteristic of TMS-citrate. The combined extracted ion profile showing the presence of large amounts of these characteristic metabolites is diagnostic for propionic acidaemia.

line computerised data systems with only selected peaks in the profile being scanned with the mass spectrometer by manual activation. However, the value of metabolic profiling was only fully realized with the introduction of on-line computers (GC-MS-DS) to control and maintain the mass spectrometer in a repetitive scanning mode where spectra are continuously recorded over a defined mass range at a pre-selected scan rate, with the large volume of data created also being acquired and stored by the same computer system (47, 48). Such methods were pioneered particularly by Jellum and his co-workers (46, 49–51), by Hutterer et al. (52) and Mamer et al. (53), the latter authors providing one of the first compilations of mass spectra of TMS organic acids, and were extended to the study of profiles of organic acids in both healthy subjects and those with metabolic diseases by several groups in particular those of Chalmers and Lawson (54–58), Markey (59, 60) and Jellum (46, 49–51, 61). The majority of this metabolic profiling work was carried out using packed GC columns, although some early trials with capillary columns were reported (14, 61). Analyses with packed columns yield some 50–60 peaks from a single 50 minute chromatogram: with a scan rate of 1 spectrum per 5 seconds across the mass range required for most organic acids (*m/z* 50–500) (i. e. $5\,s \cdot decade^{-1}$) some 500–600 spectra, often of mixed components, may be obtained, requiring not only the use of

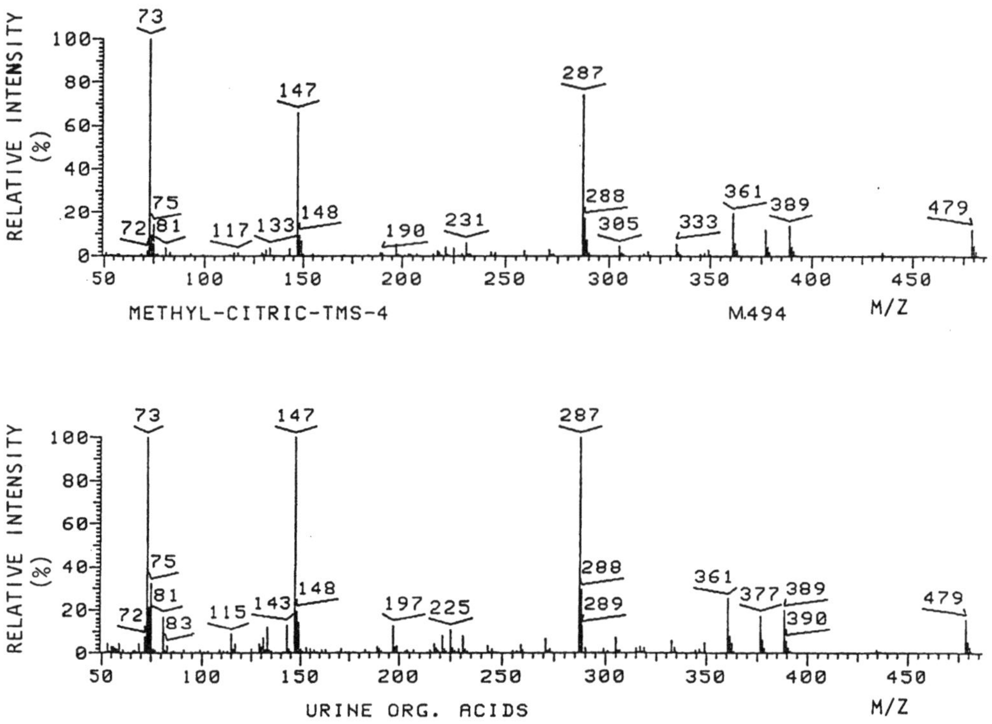

Figure 3. Library search and comparison of spectra of TMS-methylcitrate.
Spectra selected from extracted ion plot shown in Figure 2 are recalled from computer stored data files and library searched to give absolute identifications (upper: library spectrum; lower: spectrum from urine analysis).

automated peak identification, background subtraction and library search routines for identifications to be made but also considerable skill and experience from the mass spectroscopist in sorting out the (mixed) spectral data. Storage of the complete data allows retrospective processing of spectra for the latter purpose and the use of extracted ion profiles to identify individual components obscured by overlapping peaks (Fig. 1). Similarly, extracted ion profiles may be used to accentuate the position of a number of different but metabolically-related components in a complex mixture that is poorly resolved on packed column GC, as an aid both to identification and to diagnosis. Thus in Figure 2, the ion profile at m/z 287 has been used to detect TMS methylcitrate for which it is a characteristic ion (cf. m/z 273 in TMS citrate), at m/z 219 to pinpoint TMS 3-hydroxypropionate, while those at the low masses of m/z 57 and m/z 83, which are however characteristic of TMS acylglycines, have been used to determine propionylglycine and tiglylglycine respectively. The combined extracted ion profile is characteristic and diagnostic of propionic acidaemia. Specific spectra may be recalled from the stored data and their identity confirmed by comparison to library spectra for absolute confirmation (for example, methylcitrate in Fig. 3). Similarly the use of the Biller-Biemann algorithm (62) to detect ions which 'maximise' in each repetitive scan followed by reconstruction of the chromatograms from these maximising ions to improve the apparent resolution of the GC, may be used to obtain 'purer' spectra of

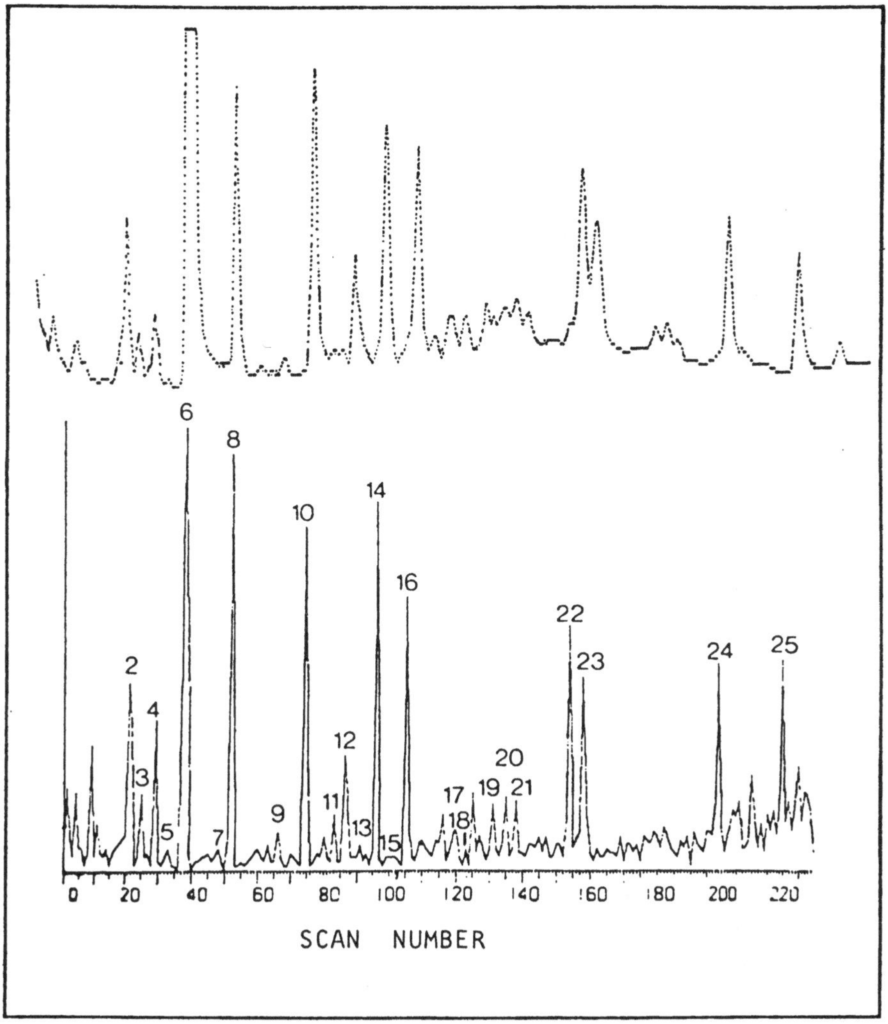

Figure 4. Use of the Biller-Biemann algorithm (61) to improve apparent resolution of a packed-column GC-MS analysis of urinary organic acids.
Obtained by inspecting the intensity of each ion in successive spectra in analysis with continuous repetitive scanning and retaining only those ions that maximise in each spectrum: summing the ion intensities of the reconstructed spectra and plotting against scan number gives a 'maximising ion chromatogram' in which an apparent improvement in GC resolution is obtained and 'purer' mass spectra may be obtained from recalled reconstructed data.
The upper plot shows the total ion chromatogram versus scan number and the lower part, the summed maximising ion intensity plot versus scan number.
(Reproduced from Chalmers et al. (148) with permission.)

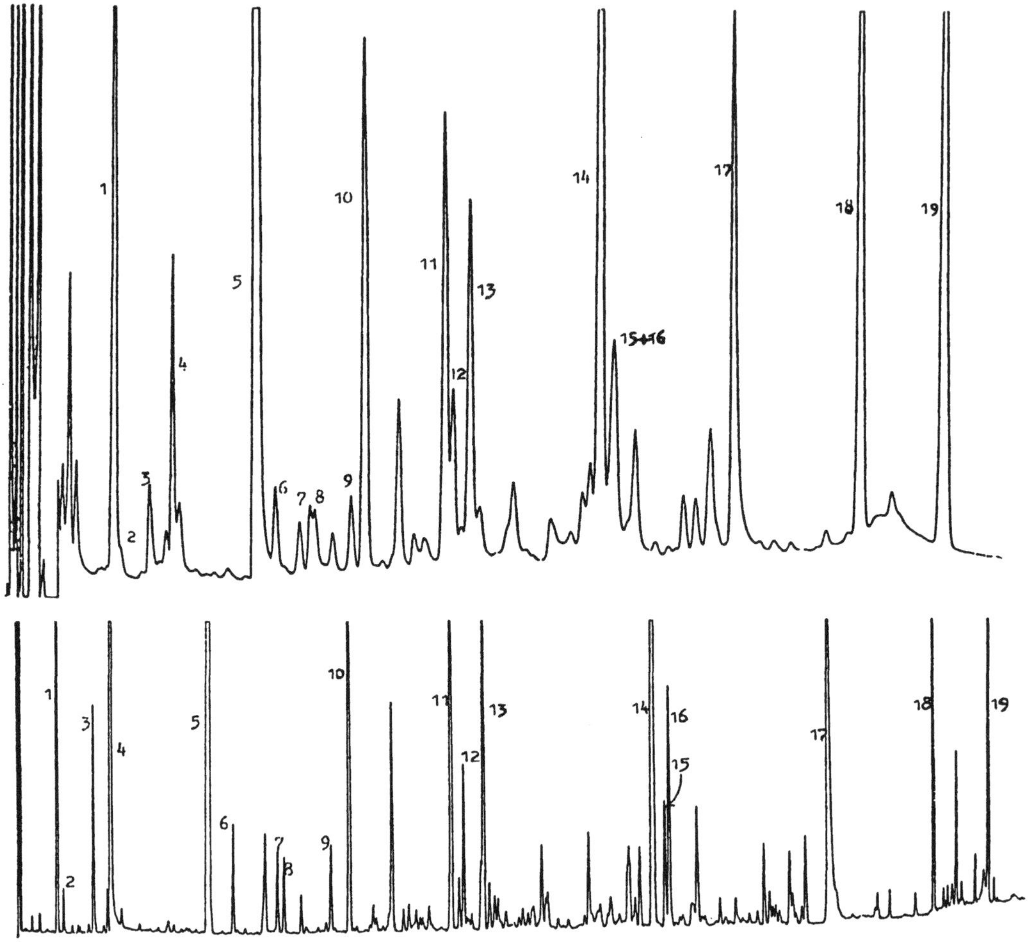

Figure 5. Improvements in GC resolution achieved by use of open-tubular capillary columns. Showing chromatograms of extracted urinary acidic metabolites separated as their trimethylsilyl (TMS) and TMS-ethoxime derivatives by temperature programming on a packed column (upper) [10% OV101 on HP Chromosorb W, 80 – 100 mesh] and a wall coated open tubular capillary column (OV101). Selected peaks are numbered for comparative purposes (Tracey and Chalmers, unpublished).

individual components and hence assist identification, either manual or computer aided (Fig. 4). Automated metabolic profiling methods have been introduced with production of both qualitative and quantitative data (63 – 65) but such processes have not found general acceptance or been introduced in many laboratories, perhaps because of the over-reliance placed upon the computer identifications made. These problems associated with lack of GC resolution and the methods used to partially overcome them have in turn been reduced considerably by the introduction of capillary GC and particularly by fused silica columns with chemically-bonded stationary phases that have revolutionised metabolic profiling and other studies with GC-MS. This can be

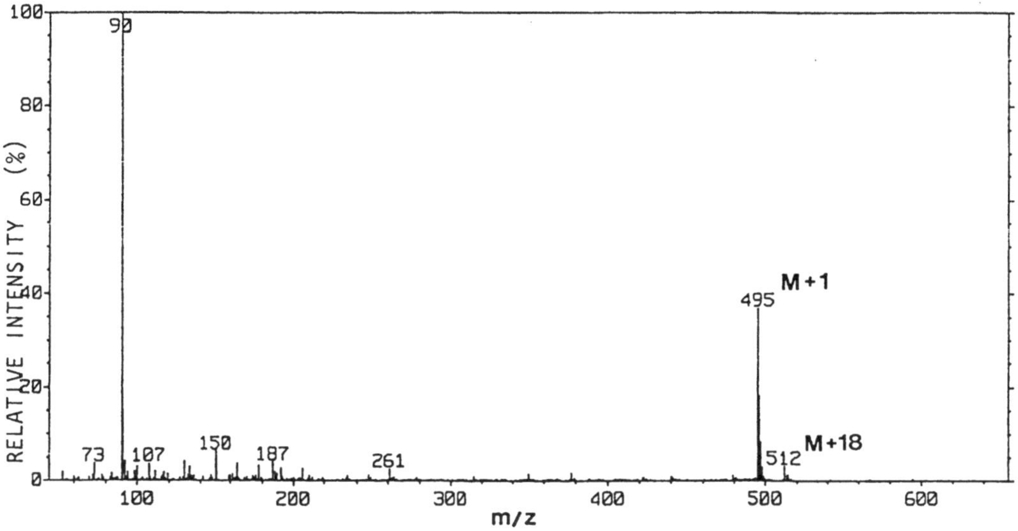

Figure 6. Ammonia CI spectrum of trimethylsilyl (TMS) methylcitrate, showing the characteristic [M + 1] ion obtained for TMS organic acids with a small [M + 18] and relatively little fragmentation. The base peak at *m/z* 90 is due to a rearrangement ion of trimethylsilanol.

illustrated by comparison of the chromatograms of the same sample analysed both using packed column and capillary column (Fig. 5) where the greatly improved separations given by the latter analysis are apparent. Such use of capillary columns allows considerable augmentation of the value of metabolic profile analysis of the organic acidurias and a further example of this is given in Section 3.3 below (Figs. 7 and 8). The use of capillary columns has in turn introduced new mass spectrometric requirements into GC-MS of organic acids, with much faster scan rates and cycle times being necessary to cope with the fast eluting components, a peak eluting in some 5 – 20 s and scan rates of 0.5 – 1 s · decade^{-1} ideally being required. Although analysis times are shorter on capillary GC (30 min compared to 60 min on a packed column for an equivalent analysis), with these scan rates some 1800 – 3600 spectra will be obtained in a single analysis. Even greater quantities of data may be generated by use of alternate EI/CI scanning which is possible with some instruments, the study of both spectra of components yielding useful information for identification both from the fragmentation (EI) pattern and the molecular mass given by CI [Fig. 6, showing the CI spectrum of TMS-methylcitrate (compare Fig. 3)]. Thus large data storage systems are required, especially when several analyses are made sequentially, and extensive use of automated library search routines and the use of selected libraries of organic acids (for examples, that of Chalmers and Lawson (2)) becomes necessary to increase processing speed. The use of retention indices to reduce still further library searching to only those spectra in restricted retention index windows (63) should increase the speed of library searching, with relatively few library spectra being searched for each component in the sample. Similarly the use of 'pre-search' algorithms that utilise only a '8-peak' index to reduce the number of spectra on which a full search is then carried out also increases the rate of data processing.

The large majority of studies on metabolic profiling have been carried out using low or medium resolution magnetic and double sector instruments or with quadrupole mass spectrometers, operating with unit nominal mass resolution but some limited applications with high resolution mass spectrometry have been reported (52, 66). Similarly most work has utilised EI because of the greater structural and identification information provided but some metabolic profiling applications of CI-MS have been reported (67) on unseparated mixtures and extracts of organic acids. MS-MS techniques have also received limited use with triple quadrupole mass-spectrometry (68) and collision activated dissociation-mass analysed ion kinetic energy (CAD-MIKES) spectrometry approaches (67) being reported for unresolved mixtures: however, for general metabolic profiling studies, work with unresolved mixtures is tedious and unnecessary, and capillary GC coupled with EI mass spectrometry or possibly rapid alternate EI/CI obviates any need for such expensive and complex instrumentation and the methods are primarily of novelty interest in this area.

3.2 Metabolic profiling of normal healthy subjects

Metabolic profiling has been used extensively to study the organic acids in physiological fluids, especially urine, and much information obtained on these compounds in normal healthy adults and children defined and selected using a variety of criteria. It is of interest however that many organic acids occurring in fluids from normal subjects have proved difficult to identify until observed and identified in a specific disease state in which they occur in increased concentrations; subsequent re-examination of normal fluids for the particular acid has often then revealed its presence at low concentrations. Metabolic profiling using GC-MS techniques was applied to the study of urine from normal subjects by Horning and Horning (6, 45) who reported on the occurrence of a variety of aromatic acids and from the same group Jaakonmaki et al. (43) reported the previously-unidentified urinary acids erythronic and threonic acids [*erythro* and *threo* trihydroxybutanoic acids (tetronic acids)]. Lawson et al. (53) confirmed the presence of these acids in normal urine and reported on a previously unrecorded series of urinary aldonic and deoxyaldonic acids including acids with four, five and six carbon atoms (the tetronic, pentonic and hexonic acids and their deoxy derivatives) which have been described in detail elsewhere (2, 15, 56, 59). Jellum et al. (46) identified a series of additional acids in normal urine including 3-hydroxy-isovaleric acid, oxalic, benzoic, furan-2,5-dicarboxylic, and indoleacetic acid and also the glycine conjugate, furoylglycine. Furoic acids were also observed by other workers (59, 70), and have been shown to be of exogenous origin from 5-hydroxymethylfuraldehyde and furaldehyde produced by heat-treatment of sugar-containing foodstuffs (2, 71, 72). Many other acids have now been reported from these profiling studies to occur in normal urine including acids of the tricarboxylic acid cycle (59), dicarboxylic acids (59, 73–75) and additional aromatic acids (76). The most complete metabolic profiling studies of normal urine were reported by Chalmers, Lawson and their colleagues (54, 56, 57) and by Thompson and Markey and their colleagues (15, 77) and some 250 organic acids are now known to occur in normal urine (2, 61, 78). Lists of the more important acids observed are given in the latter references above and particularly by Chalmers and Lawson (2). Many of these references also give quantitative data on the acids

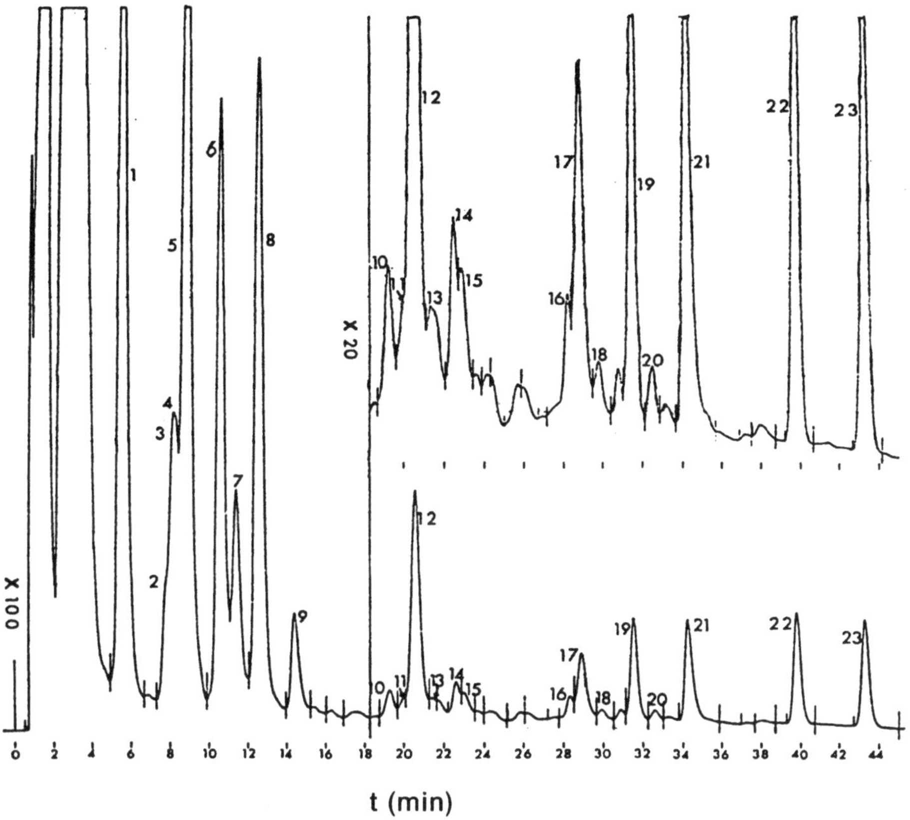

Figure 7. Chromatogram of urinary organic acids from a patient with multicarboxylase deficiency caused by holocarboxylase synthetase deficiency.
Extracted using DEAE-Sephadex and separated using temperature programming as their trimethylsilyl (TMS) and TMS-ethoxime derivatives on packed columns (10% OV101 on HP Chromosorb W, 80–100 mesh); temperature programme from 110 °C to 285 °C at 4 °C · min^{-1} with a 5 minute initial isothermal delay. The horizontal axis shows the time in minutes.

Peaks identified were:

(1) Lactate
(2) 2-hydroxy-*n*-butyrate
(3) 3-hydroxypropionate
(4) sulphate
(5) 3-hydroxy-*n*-butyrate
(6) 3-hydroxyisovalerate
(7) 3-hydroxy-*n*-valerate
(8) phosphate
(9) fumarate *plus* glycerate
(10) adipate *plus* malate
(12) 3-methylcrotonylglycine
(13) tetronates (erythronate *plus* threonate)
(14) 3-hydroxy-3-methylglutarate
(15) 4-hydroxyphenylacetate
(16) citrate
(17) methylcitrate
(18) 4-hydroxyphenyl-lactate
(19) undecandioate (internal standard)
(20) glucuronate
(21) urate
(22) *n*-tetracosane (standard)
(23) *n*-hexacosane (standard)

(Reproduced from Chalmers et al. (96) with permission.)

observed and on their known or putative metabolic origins. Fewer studies have been made on other fluids but metabolic profiling using GC-MS has been applied particularly to amniotic fluid (79–85) providing valuable information to form the basis for prenatal diagnosis of organic acidurias by quantitative GC-MS (Section 5.2 below).

Metabolic profiling has also been used extensively in the diagnosis and study of the organic acidurias and this is discussed in detail in Section 4 below.

4 Metabolic Profiling and Mass Spectrometry in the Diagnosis and Study of the Organic Acidurias

Metabolic profiling has been used extensively in the diagnosis and research for metabolic diseases, especially in infants and children, and also in the comprehensive study of the major and minor organic acids excreted by patients with known organic acidurias. In the role of diagnosis and search for (new) metabolic diseases, organic acid profiling has been shown to be more efficient than equivalent amino acid profiling (86). Unselective screening for such diseases is however of greatly limited value both in such paediatric populations (86) and in mentally-subnormal patients (46, 87) and diagnostic investigations on more carefully preselected paediatric patients following application of a preliminary screening protocol has been recommended (86) and is now generally accepted.

Metabolic profile analysis using GC and MS has essentially contributed most to our knowledge of the organic acids excreted by patients with a wide variety of metabolic diseases and an enormous literature has evolved reporting the results of this work (see for example Chalmers and Lawson (2)). As referred to earlier, the first studies using GC-MS were on isovaleric acidaemia (4) resulting in the identification of 3-hydroxyisovaleric acid and isovalerylglycine in the patient's urine (88, 89). Within a few years, other 'new' or previously-unrecorded organic acidurias had been identified using GC-MS including methylmalonic aciduria (90), propionic acidaemia (91), pyroglutamic aciduria (92) and 3-methylcrotonylglycinuria (93). More detailed GC-MS studies on patients with these diseases led to the identification of previously unrecorded organic acid metabolites including 3-hydroxypropionate and methylcitrate in the urine of patients with propionic acidaemia (94, 95) and in methylmalonic aciduria (96) and glycine conjugates of propionic acid and of tiglic acid (2-methylcrotonic acid) in these diseases (97, 98). The occurrence of a disorder affecting all three mitochondrial carboxylase enzymes (propionyl CoA, 3-methylcrotonyl CoA and pyruvate carboxylases) was indicated from the GC-MS profiling studies of Chalmers et al. (96) studying a patient originally diagnosed as having 3-methylcrotonylglycinuria (99) and subsequently confirmed as holocarboxylase synthetase deficiency by detailed enzyme studies on cultured skin fibroblasts from the patient (100). Further studies with GC-MS have revealed the complexity of the metabolite patterns associated with these inter-related disorders of branched-chain amino acid metabolism with numerous 'new' metabolites being identified including 3-hydroxy-*n*-valeric acid and 3-oxo-*n*-valeric acid that provided further evidence of the underlying abnormal biochemical process in these diseases in which propionyl CoA accumulates (101 – 106). As with many of the studies described below, these investigations were made using GC-MS profiling and with use of packed GC columns, with their relatively poor separation power in comparison to capillary columns: this can be illustrated by reference to the patient described above who is now known to have multicarboxylase deficiency and in whom earlier studies (96) had shown the presence of metabolites of pyruvate, propionyl CoA and 3-methylcrotonyl CoA (Fig. 7). Re-analysis of this urine using fused silica capillary columns coated with a non-polar chemically-bonded silica phase (Fig. 8) demonstrates the greatly improved separations achieved and hence the clearer spectra obtained by mass spectrometry and more complete identifications made. Thus for example, separation is achieved of the isomeric 3-methylcrotonylglycine and tiglylglycine, both being present in the sample, propionylglycine is clearly identified as are both methylcitrate isomers (2RS,3RS and 2RS,3SR) as well as many more previously unrecorded metabolites in this sample. Such analyses can considerably simplify library search routines and complimentary use of extracted ion profiles improves still further the quality of the separation achieved on GC-MS (Fig. 9). Similar examples could be provided of most of the known organic acidurias and such detailed studies of both urine and plasma have continued to be of extreme value in the study of individual patients as well as of the diseases collectively, with the resulting data being used in regular monitoring and assessment of the treatment provided (107 – 110).

In the same manner, exhaustive analysis of urine from patients with isovaleric acidaemia have permitted the identification of many previously unrecorded metabolites in that disorder including 4-hydroxyisovaleric acid (111, 112), 3-hydroxyisoheptanoic acid (113) and *N*-isovalerylglutamate (114) which occur during periods of metabolic decompensation but are absent in periods of remission (109). Metabolic profiling has also continued to lead to the recognition of previously unknown organic acidurias and has contributed much to their detailed study including evaluation of the effects of therapeutic manoeuvres: these include 3-hydroxy-3-methylglutaric aciduria (115 – 121), 2-methylacetoacetic aciduria (β-ketothiolase deficiency) (122 – 123), 2-oxoadipic aciduria (124, 125), glutaric aciduria type I (126, 127), D-glyceric aciduria (128, 129), and 4-hydroxybutyric aciduria (130, 131). GC-MS has also contributed to the understanding of the metabolic pathways in other disorders including formiminoglutamic aciduria (132), phenylketonuria (133 – 138) and hyperprolinaemia type II (139). The identification of 4,6-dioxoheptanoic acid (succinylacetone) and 3,5-dioxo-octanoic acid (succinylacetoacetate) in the urine of patients with hepatorenal tyrosinaemia ("type I") (140) has provided a biochemical basis for the multiple biochemical and clinical manifestations of the disease in many of the cases diagnosed, has aided considerably the characterisation and differentiation of new patients and provided the means for prenatal diagnosis of this condition (Section 5 below). The relative paucity of very recent data and reports of some of these conditions reflects not the reduction in identification and study of new patients and the disorders but more of the increasing difficulty in publishing such studies after the initial interest in the scientific press has diminished.

Mass spectrometry coupled with GC has been of particular value in the study and continuing investigation of patients with abnormal dicarboxylic acidurias. Dicarboxylic acids occur at low concentrations in the urine of normal adults and particularly of

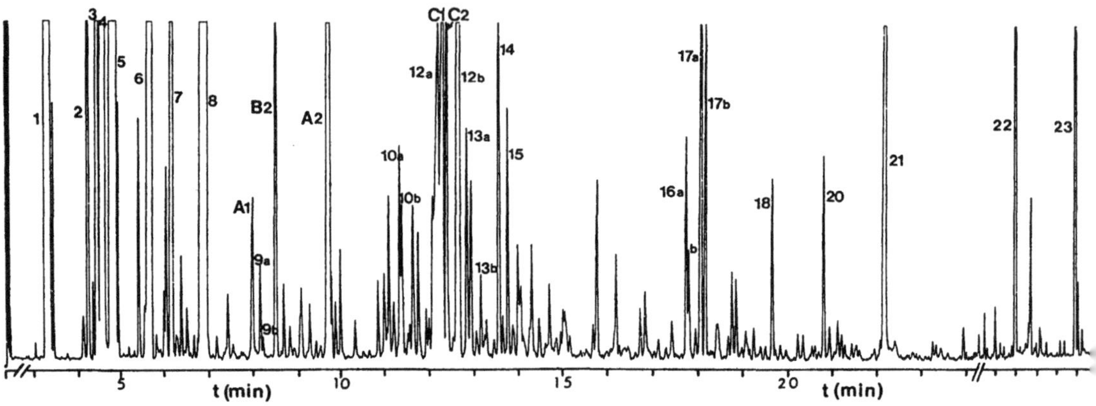

Figure 8. Capillary GC-MS analysis of urinary organic acids from a patient with multicarboxyl deficiency caused by holocarboxylase synthetase deficiency.
A similar urine extract to that shown in Figure 7 obtained using the same urine sam with the organic acids separated as their TMS and TMS-ethoxime derivatives on a CP 5 fused silica capillary column using temperature programming from 70 °C to 250 °C 6 °C · min^{-1} with a 70:1 split injection and helium carrier gas, 1.4 ml · min^{-1}
Peaks are numbered to correspond to those in Figure 7 with additional identifications given below. Peak identifications are:

(1) Lactate
(2) 2-hydroxy-*n*-butyrate
(3) 3-hydroxypropionate
(4) Sulphate
(5) 3-hydroxy-*n*-butyrate
(6) 3-hydroxy-isovalerate
(7) 3-hydroxy-*n*-valerate
(8) Phosphate
(A1) propionylglycine TMS-1
(9a) glycerate
(9b) fumarate
(B2) acetylglycine TMS-2
(A2) propionylglycine TMS-2
(10a) adipate
(10b) malate
(12a) 3-methylcrotonylglycine TMS-1
(C1) tiglylglycine TMS-1
(C2) tiglylglycine TMS-2
(12b) 3-methylcrotonylglycine TMS-2
(13a) erythronate
(13b) threonate
(14) 3-hydroxy-3-methylglutarate
(15) 3-hydroxyphenylacetate
(16a) citrate
(16b) isocitrate
(17a) 2RS,3RS methylcitrate
(17b) 2RS,3SR methylcitrate
(18) 4-hydroxyphenyl-lactate
(19) glucuronate
(20) urate
(22) *n*-tetracosane (standard)
(23) *n*-hexacosane (standard)

Most other peaks shown were also identified: this analysis demonstrates the greatly impro separations achieved using fused silica capillary columns and hence the greatly impro quality of identification including isomeric compounds and other closely-eluting com nents.

Figure 9. (opposite page) Extracted ion profile of organic acids in multicarboxylase deficiency cau by holocarboxylase synthetase deficiency.
From the same analysis as that shown in Figure 8. The upper part of this figure shows summed total ion chromatogram obtained on capillary GC-MS analysis with a relati slow scan speed of 2 s · $decade^{-1}$, corresponding to the capillary chromatogram show Figure 8. The lower part shows the computer drawn ion intensity plots of ions at m/z 83, 200, 219, 247, 287 and at m/z 73 to indicate TMS compounds in the TIC trace. Sh increases in ion intensity are obtained at specific elution times for m/z 287 characteristi the methylcitrates, m/z 247 indicated 3-hydroxyisovalerate, and 3-hydroxy-3-methylgl rate, m/z 219 characteristic for lactate and 3-hydroxypropionate, m/z 83 characteristic 3-methylcrotonylglycine and tiglylglycine and m/z 57 characteristic of propionylglycine latter two ions also indicate the hydrocarbon standards).

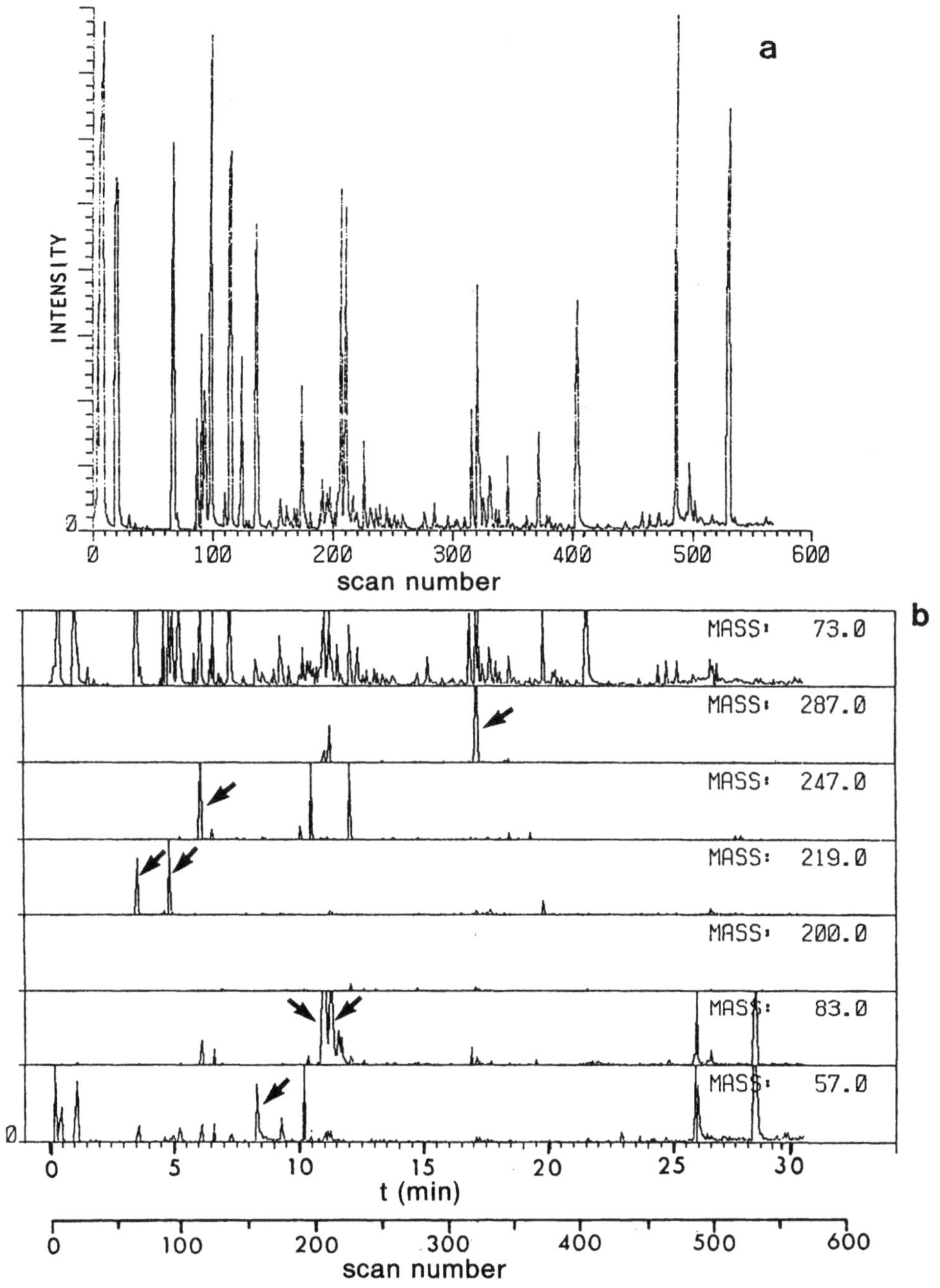

(Figure 9, continued) The complete extracted ion plot with indications of increased amounts of methylcitrates, 3-hydroxypropionate, propionylglycine and tiglylglycine (propionic acidaemia), of 3-hydroxyisovalerate and 3-methylcrotonylglycine (3-methylcrotonyl CoA carboxylase deficiency) and of lactate and (not shown) small increases in fumarate and malate and normal pyruvate (pyruvate carboxylase deficiency) is characteristic and diagnostic of multicarboxylase deficiency caused by holocarboxylase synthetase deficiency.

children as indicated earlier (Section 4.2) and may be increased as a secondary phenomenon in ketosis (141) and in the glycogen storage diseases (142) but of particular and continuing interest has been the application of detailed GC-MS profiling to the study of patients with non-ketotic hypoglycaemia asssociated with pathological concentrations of these acids. These studies were greatly stimulated by the detailed biochemical investigation by Tanaka and his colleagues (143) of two patients with 'Jamaican vomiting sickness' (ackee or hypoglycin poisoning) in whom a dicarboxylic aciduria together with metabolites associated with isovaleric acidaemia and glutaric aciduria (as well as methylene cyclopropylacetylglycine derived from hypoglycin itself) was observed. All of these metabolites were caused by inhibition by hypoglycin and its acyl CoA metabolites of a variety of acyl CoA dehydrogenases (2, 143, 144). At the same time, the description by Przyrembel and colleagues (145) of a patient in Europe with 'glutaric aciduria type II' with a very similar abnormal organic aciduria in whom an inherited metabolic disorder affecting simultaneously several acyl CoA dehydrogenases was postulated, together with reports of another patient with non-ketotic hypoglycaemia, dicarboxylic aciduria and suberylglycinuria (146) stimulated much interest in this area. Subsequent detailed GC-MS studies on other similar patients (147–151), together with supportive biochemistry (152), have shown that "glutaric aciduria type II" or better, multiple acyl CoA dehydrogenase deficiency, is caused by a primary deficiency of the electron transfer flavoprotein (ETF) dehydrogenase responsible for transfer of electrons from the acyl CoA dehydrogenases into the respiratory chain. Similarly patients with a dicarboxylic aciduria associated with suberylglycinuria, in whom GC-MS profiling has an important role in the identification and study of new patients and their differentiation from other causes of dicarboxylic aciduria including dietary sources such as medium chain triglycerides (153, 154), have been shown to have deficiencies of medium chain acyl CoA dehydrogenase activity (155–164). Mass spectrometric studies leading to the identification of 5-hydroxyhexanoic acid in urine from patients with dicarboxylic aciduria (165) lead to further studies on ω-1 oxidation of fatty acids (154, 166) with the identification of other newly described organic acids and elucidation of new metabolic pathways for these compounds (162).

The marked similarities between the clinical presentation of these non-ketotic hypoglycaemic syndromes (including 3-hydroxy-3-methylglutaric aciduria) to Reye's syndrome (encephalopathy with fatty degeneration of the viscera) stimulated GC-MS profiling studies on patients with Reye's syndrome (143, 167) with demonstration of a dicarboxylic aciduria without suberylglycinuria or other associated metabolites (e. g. octanoylcarnitine, Section 7) in samples taken from patients immediately on presentation (164). Again, the similarities of these conditions and Reye's syndrome to a disorder described as systemic carnitine deficiency in which a moderate dicarboxylic aciduria was observed (168) stimulated the growing interest in the role of L-carnitine in the organic acidurias and mass spectrometry has played an important part in much of the resultant research into acylcarnitine metabolism described in Section 7.

A further group of organic acidurias, the primary causes of many of which are still unclear, in which GC-MS has been of value are the congenital lactic acidoses. Lactic acidosis and lactic acidaemia occur as secondary phenomena in many other organic acidurias and other conditions (2, 110, 169) but are also characteristic of a group of disorders associated with the defects of pyruvate metabolism, the tricarboxylic acid cycle and in the respiratory chain. Detailed GC-MS studies of a patient with proven pyruvate decarboxylase deficiency demonstrated a possibly characteristic organic aciduria involving lactate, pyruvate and acids of the tricarboxylic acid cycle and led to the suggestion that detailed profiling studies in such disorders could permit identification of the areas of the underlying enzymatic deficiency and reduce considerably the amount of investigative enzymology required in their study (170). This concept was further supported by similar observations on patients with pyruvate carboxylase deficiency (171), with glycogen storage diseases (172) and with dihydrolipoyl dehydrogenase (E3) deficiency (173), with the results of further detailed GC-MS profiling in the lactic acidoses being summarised by Chalmers (169) and Matsumoto (110).

In many of these studies the various mass spectrometric techniques, for example extracted ion profiles, accurate mass measurements at high resolution and SIM (Section 5), have been extensively used in addition to metabolic profiling by GC-MS and have provided aids to the identification of new metabolites and their quantification. GC and MS have also been used in metabolic pathway mapping in both man and in subsidiary experiments in animals, the latter often stimulated because of discoveries made in patients with organic acidurias. For example, studies initiated because of the identification of 2-ethyl-hydracrylic acid (2-ethyl-3-hydroxypropionic acid) in the urine of patients with β-ketothiolase deficiency (174) led to further GC-MS and other studies which resulted in the elucidation of a new pathway of alloisoleucine catabolism in man yielding an R-series of metabolites as opposed to the S-series of metabolites from the normal pathway of L-isoleucine catabolism (175). Similar studies in the areas of ω-oxidation, ω-1 oxidation and in hypoglycin metabolism have been referred to above.

The diagnosis of new patients with abnormal organic acidurias and identification of new metabolites, exploration of new metabolic pathways and monitoring and evaluation of treatment regimens is a continually evolving process and it is apparent from the studies reported over the past ten years that mass spectrometry and particularly GC-MS will continue to play a central and vital role in this research.

5 Quantitative Mass Spectrometry of Organic Acids in Biological Fluids

5.1 Methods

Conventional and capillary GC and GC-MS analyses, normally with the use of appropriate internal standards, provide a substantial amount of quantitative data, particularly for urine samples, and more recently such quantitative data has also been produced for blood plasma samples, especially with capillary GC. Employing the mass spectrometer as the detector system for quantitative studies with such samples is unnecessary for metabolites excreted in large amounts and those unambiguously separated from other co-eluting components. However, for metabolites at low concentrations, for those obscured or overlapped by other components on chromatography, and for some biological fluids, the use of the GC-MS for quantitative analysis provides the invaluable advantages both of sensitivity and, particularly, selectivity. Quantitative mass spectroscopy and mass fragmentography (176, 177) have been applied previously to drug and drug metabolism studies (177–179) but have received particular application in the field of organic acid analysis in the study of amniotic fluid for prenatal diagnostic purposes. Both EI and ammonia CI have been used although the majority of applications have been by EI: both methods offer different advantages, EI providing a high total ion current (TIC) and characteristic and specific fragment ions while CI with a lower TIC offers higher sensitivity and specificity when the molecular ion is utilised. EI at lower ionization energies (e.g. 20–40 eV) rather than 70 eV commonly employed may also be advantageous resulting in reduced fragmentation and presenting larger characteristic ions for analysis. Quantitation may be achieved by repetitive scanning techniques, with quantification being made from the extracted ion profiles of selected ions recalled or selected from the mass of data obtained. However such procedures are relatively insensitive (nanogram quantities) because of the time wasted on measuring ions of no value and for most quantitative studies at high sensitivity (sub-nanogram quantities), SIM is employed. In this mode it becomes essential to select carefully the ions used such that they are specific for the component to be measured and do not occur in the background: thus ions of relatively high mass in a spectrum are preferred to avoid such interference. This, coupled with the high degree of component separation offered by capillary GC columns, will provide high specificity to the assay. Specificity could be improved still further in double focussing instruments by SIM at high resolution, with monitoring of ions at particular selected elemental compositions being possible. In practice, however, most work is carried out at unit nominal mass separation since as resolution is increased, sensitivity decreases and may become a limiting factor in high resolution SIM work. In addition, in the majority of analyses, more than one ion is monitored including those from appropriate internal standards and peak switching is employed in which the mass spectrometer is focussed on the selected masses in a rapid sequential fashion. In the quadrupole mass spectrometer this is relatively easy to achieve with the mass being selected by rapid switching of the rod voltages to allow selected mass (ion) measurements at any part of the mass range of the instrument. Settling time is virtually instantaneous hence switching can be very fast with a quadrupole instrument. With magnetic instruments, SIM is generally performed by switching the ion accelerating voltage at constant magnetic field strength: the smallest ion of interest is focussed by adjustment of the magnetic field at the optimum accelerating voltage setting and then focussing on the higher masses by reduction of the accelerating voltage to the appropriate values (179). However defocusing of the ion source can often limit the useable mass difference between ions of interest in a single analysis to 20–25% (179). In modern instruments with laminated core magnets (180), switching between different magnetic field strengths can be achieved rapidly, mass drift due to magnet instability is minimised and magnetic switching may be employed. Similarly, in recent developments with an "air cored" electromagnet, hysteresis-free instantaneous switching of the field strength is possible by switching the magnet power (voltage) supply and no mass drift occurs (181). Thus the differential between quadrupole and magnetic instruments for SIM work has been virtually eliminated, with the latter offering the added advantage of high resolution measurements. All of these processes are usually computer-controlled with the data system acquiring and storing the data and undertaking post analysis calculations.

The use of an internal standard is considered essential for quantitative SIM analysis as with quantitative GC work. Problems caused by sample preparation procedures are eliminated when the internal standard is chemically similar to the compound of interest, with similar extraction and the derivatisation properties, and is added to the biological fluid at the start of the analysis. Ideally, the selected standard should also elute from the chromatographic column relatively close to the compound of interest, to reduce analysis time, and should have characteristic ions that are relatively close in mass to those of the analyte that are to be utilised. A typical example is the use of 3-methylglutarate as an internal standard for the quantitation of glutaric acid in amniotic fluid (Fig. 10) where the chemically-similar glutarate and 3-methylglutarate extract similarly, elute as their TMS esters closely together on capillary GC, and have significant ions at m/z 261 and m/z 275 respectively. Alternatively, compounds that are structural analogues and thus also present identical ions may also be preferred in some applications, for example use of homocitrate as standard when methylcitrate is to be analysed: both have the characteristic ions at m/z 479 and m/z 287 in their trimethylsilyl derivatives (Fig. 3) and the derivatives elute relatively closely on gas chromatography. The optimum internal standards may be considered to be the same compound as the analyte but labeled with a stable isotope to produce a mass difference that can be uniquely monitored by SIM. Thus deuterated internal standards are frequently employed for quantitative SIM work (182–184) providing identical chemical properties for extraction and derivatisation. However, isotope effects may be observed by gas chromatography with the standard compound eluting from the GC column with a slightly different retention time particularly when capillary columns are used. The use of a stable labeled internal standard (isotope dilution analysis) offers increased accuracy over the use of other chemically similar standards but for best results the labeled internal standard should be of high isotopic purity for each compound of interest. A mass difference of at least three daltons between analyte and standard (185) is required to negate corrections necessary due to naturally-occurring isotopes and to obtain linear calibration curves. Sensitivity may be increased at very low concentrations by the 'carrier effect' exerted by the standard compound although this has been questioned (185). In the author's experience and that of others published in the literature, the use of chemically-similar internal standards often provides, for all practical purposes, satisfactory data for most applications, avoiding the costs and problems of synthesis of labeled standards and providing the advantage in some circumstances of quantification of more than one chemically similar analyte against a single internal standard (187, 188).

Quantification is carried out via preparation of a standard calibration curve obtained by analysis of samples containing known amounts of the analyte, with the relative peak area of the selected ion plot being plotted against the amount of compound introduced (Fig. 10). The calibration curves should ideally be prepared by addition of the analyte to a sample of the biological fluid in which the concentration of the analyte is known: the standard curve is thus unique to the particular analysis and a new calibration should be carried out for each new batch of samples. The concentration of the analyte in the samples is read from the calibration curve, SIM GC-MS analysis providing high accuracy and precision (i. e. reproducibility) of measurement.

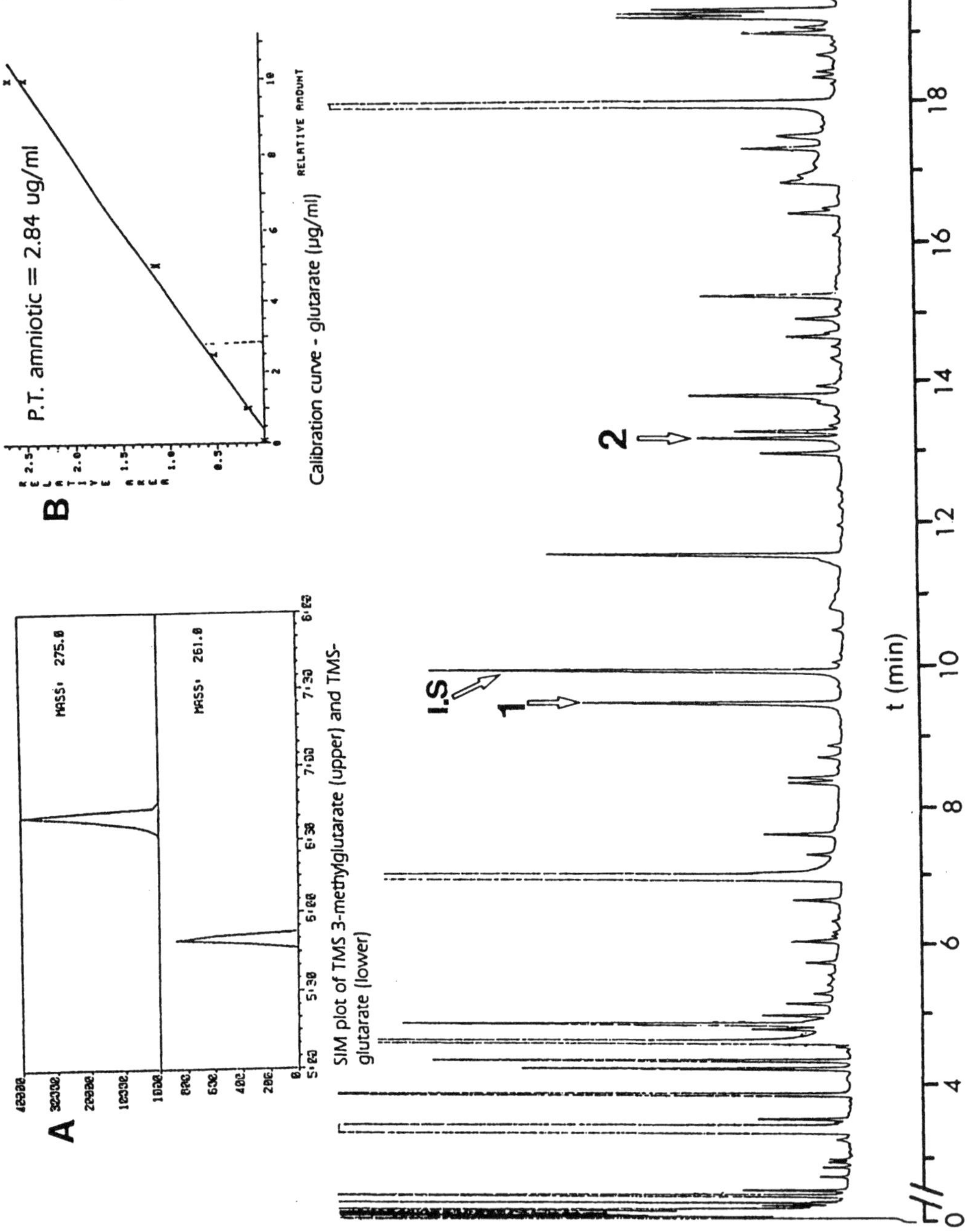

Figure 10.
Use of quantitative selected ion monitoring for organic acid analysis in amniotic fluid.
The figure shows the first part of the chromatogram of amniotic fluid organic acids extracted from cell-free fluid using DEAE-Sephadex and separated as their trimethylsilyl (TMS) and TMS-ethoxime derivatives on a CP Sil 5 fused silica capillary column by temperature programming from 70 °C to 250 °C at 6 °C $\cdot$ min^{-1} with a 70:1 split injection and helium carrier gas at 1.4 ml $\cdot$ min^{-1}.
Peak 1 is due to glutarate, peak 2 is 2-hydroxyglutarate and I.S is the 3-methylglutarate internal standard.

5.2 Applications

Quantitative GC-MS studies using SIM techniques have been used to study 2-oxo acids as their TMS-quinoxalinols (188, 189) in urine and other biological fluids (190, 191), the latter using ammonia CI with detection limits of $1-50\,ng \cdot ml^{-1}$. It has also been used to study the acidic metabolites of biogenic amines at low concentrations in cerebrospinal fluid, with detection limits of $2-5\ ng \cdot ml^{-1}$ (192), and in plasma (193), by isotope dilution analysis with deuterated internal standards and high resolution mass spectrometry. Urinary aliphatic dicarboxylic acids have also been determined as their dicyclohexyl derivatives in a variety of disease states (10). However the most important application of quantitative GC-MS to organic acid analysis has undoubtedly been in the field of prenatal diagnosis of the organic acidurias. This approach is based upon the composition of amniotic fluid being derived in part from fetal urine with the fetus producing some $7-17$ ml of urine per 24 hours by 18 weeks' gestation. Since many organic acidurias are characterised by greatly increased concentrations of organic acids in urine, it was suggested that amniotic fluid at some $16-17$ weeks' gestation could be useful in the prenatal diagnosis of such conditions in mothers who have previously produced an affected index case with a proven diagnosis (see Chalmers and Lawson (2)). This approach has attracted much attention since the first proposal in 1974 and considerable development has occurred in the quantitative determination of organic acids at low concentrations in amniotic fluid. Although the concentrations of some acids, for example lactate, are high in amniotic fluid (79, 83, 194), prenatal diagnosis by measurement of most acids characteristic of the organic acidurias in this field require highly sensitive, specific and reliable methods. GC-MS with SIM provides the ideal approach, either with stable isotope dilution analysis (195) or by chemically-related internal standards (187). Prenatal diagnosis has been carried out using such procedures for propionic acidaemia and methylmalonic aciduria based upon isotope dilution analysis of methylcitrate and methylmalonate with deuterated internal standards and CI (196–198), or by use of homocitrate internal standard (Chalmers et al., unpublished) or against prepared calibration curves (199). Similarly the diagnosis of glutaric aciduria type II (multiple acyl CoA dehydrogenase deficiency) has been based upon determination of glutarate and other dicarboxylic acids against a 3-methylglutarate internal standard (187) or against a 2-hydroxy-*n*-caproate internal standard using ammonia CI (186). Ammonia CI methods have been developed for hydroxy acids in amniotic fluid (200) and ammonia CI with stable isotope dilution analysis of 3-hydroxyisovaleric acid (201), for the prenatal diagnosis of multicarboxylase deficiency and other disorders in which this metabolite accumulates. Prenatal diagnosis using GC-MS has also recently been reported briefly for 3-hydroxy-3-methyl-glutaric aciduria (202) and for isovaleric acidaemia with stable isotope dilution analysis for isovalerylglycine (183). Finally prenatal diagnosis of tyrosinaemia type I has been reported using GC-MS with SIM for succinylacetone against a glutarylacetone internal standard, both as their pentafluorobenyl-isoxazole derivatives (203).

Inset A shows the selected ion plot for ions at *m/z* 261 (TMS-glutarate) and *m/z* 275 (TMS-3-methylglutarate) and inset B shows the calibration curve for known amounts of glutarate ($\mu g \cdot ml^{-1}$) analysed against the same amount of 3-methylglutarate (duplicate points), from which the amount of glutarate in the sample can be calculated. 2-hydroxyglutarate could be quantified similarly using the same internal standard.

6 *In vivo* Studies of Organic Acid Metabolism Using Stable Isotopes

Relatively few studies have been made of organic acid metabolism using precursors and substrates labeled with stable isotopes although a number of studies have been reported in the mid-1970s. This may perhaps be ascribed to the lack of close links and mutual interest between mass spectroscopists and the more clinical departments that would undertake such work and the lack of understanding of the mass spectroscopist of the needs of the metabolist and of the metabolist of the powerful research tool offered by use of stable-isotopes and mass spectrometry. However, clinical studies on humans, especially children, dictate the use of stable isotopes as opposed to radioactive tracers and one may expect a considerable increase in the number of reported applications in the future. Deuterium can be a poor choice of label for *in vivo* experiments because of the large isotope effects associated with it (18) and because of the relative ease of exchange with protons in living systems. In the future, use of ^{13}C as the isotope of choice for organic acid studies is to be expected. Deuterium has the advantages of cheapness and of availability of substrates with high isotopic purity but more substrates labeled with ^{13}C in high purity are becoming available and costs are becoming lower. The toxic effects of stable isotopes are negligible in the amounts used in metabolic studies, especially with ^{13}C, and compounds labeled with stable isotopes are no more toxic than their natural ^{12}C analogues.

Some of the earliest studies on organic acid metabolism using substrates labeled with stable isotopes were concerned with oxalate metabolism using ^{13}C-oxalate, ^{13}C-glycine and ^{13}C-ascorbate but utilised isotope ratio mass spectroscopy after conversion of the isolated compounds into $^{13}CO_2$ (204–206). One of the earliest GC-MS studies on organic acids reported with use of stable isotopes *in vivo* was by Curtius et al. (207) in 1972 who used 27% deuterated D,L-phenylalanine to study the metabolism of this compound in a normal child and patients with phenylketonuria and with hyperphenylalaninaemia and to measure the incorporation of the isotope into aromatic acid metabolites. The results were somewhat compromised from the metabolic viewpoint because D-phenylalanine is metabolised in a different manner to L-phenylalanine in man but the studies served to demonstrate the value of the use of stable isotopes for such metabolic studies in humans. These studies with phenylalanine were followed by a series of papers on phenylalanine and tyrosine metabolism using both deuterated and ^{13}C labeled substrates (207–214), with use of dideuterotyrosine and dideuterobenzoate to study intestinal tyrosine metabolism in man (208, 209), pentadeutero(aryl)-L-phenylalanine and L-tyrosine to study the metabolism of these compounds to aromatic acids in normal subjects (210) and dideutero-L-tyrosine in metabolic studies on two patients with "non-hepatic" tyrosinaemia (211), although the latter study provoked adverse comment and caution on the use of dideutero-L-tyrosine and the interpretation of the results (212). More recently, studies on seven patients with proven tyrosinaemia type I using 80% dideutero-L-tyrosine showed evidence for incorporation of the label into succinylacetone and succinylacetoacetate as well as into 4-hydroxyphenyl acids although substantial and unexplained losses of the deuterium label were observed (213).

The use of stable isotopes in metabolic studies of organic acid metabolism have also been pioneered by Tanaka and his colleagues and by Mamer and his colleagues. In a seminal paper in 1975 on ^{13}C-valine metabolism in patients with methylmalonic aciduria, albeit using NMR as the analytical tool, Tanaka et al. (215) indicated the advantage of ^{13}C-labeled substrates for *in vivo* metabolic studies and followed this with detailed studies by GC-MS on the metabolism in rats of ^{13}C-labeled isobutyrates in connection with methylmalonic aciduria (216, 217). Mamer et al. (175) studied alloisoleucine and isoleucine catabolism in rats, rabbits and dogs with pentadeutero butyric, methylbutyric and ethylmalonic acids to demonstrate an R-series of metabolites formed from allosioleucine and provide some evidence for 2-methylacetoacetyl CoA as an intermediate in isoleucine catabolism in connection with β-ketothiolase deficiency in man. GC-MS has been used in the study of ^{13}C-lactate metabolism and turnover in patients with coronary disease (218) and in normal subjects during rest and exercise (219). Deuterated 3-hydroxybutyrate has also been effective in ketone body kinetics (220).

Mass spectrometry has proved of value in other metabolic studies using stable isotopes when the incorporation of a ^{13}C-label into $^{13}CO_2$ has been measured. Fatty acid turnover and oxidation has been studied in dogs with [1-^{13}C]palmitate, and measurement of the ^{13}C-palmitate enrichment in plasma by GC-MS with SIM and measurement of the $^{13}CO_2$ in expired air using stable isotope ratio mass spectrometry (221). This technique has also been applied to human studies in a patient with methylmalonic aciduria when the effects of L-carnitine administration on long-chain fatty acid oxidation were measured using [1-^{13}C]trioleylglyceride and determining the concentration of $^{13}CO_2$ in expired air by isotope ratio mass spectrometry (222).

7 Mass Spectrometry in the Study of Basic Metabolites of Acyl Groups: the Acylcarnitines

L-carnitine [3-hydroxybutyrobetaine, *N*-trimethyl-3-hydroxy-butyrate, or R-(−)-3-carboxy-3-hydroxypropyl)-trimethylammonium (chloride)] is essential for the metabolism of long-chain fatty acids, mediating their transport across the inner mitochondrial membrane as acylcarnitines for β-oxidation. Observation of secondary carnitine deficiency in patients with organic acidurias (223, 224) and increased acylcarnitine excretion in methylmalonic aciduria (225, 226) led to the discovery of increased urinary acylcarnitine excretion as a general phenomenon in disorders of organic acid metabolism (227). The nature of acyl moiety in the excreted acylcarnitines is of great interest and Roe et al. used fast atom bombardment (FAB) mass spectrometry to show that a patient with methylmalonic aciduria excreted primarily propionylcarnitine, consistent with accumulation of propionyl groups *in vivo*, and that this excretion could be enhanced by administration of exogenous L-carnitine (226). Acylcarnitines are positively charged by virtue of their trimethylamino group:

$$(CH_3)_3\overset{+}{N}-CH_2-\underset{\displaystyle\underset{OR}{|}}{CH}-CH_2-COOH$$

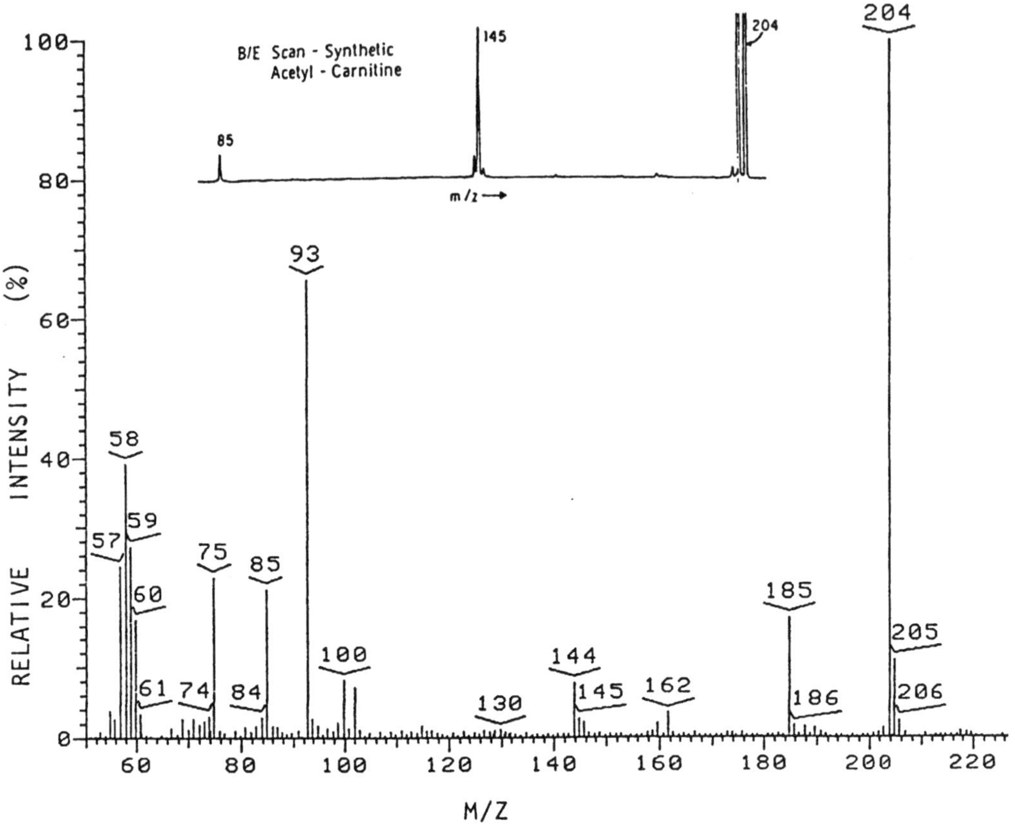

Figure 11. Positive ion fast atom bombardment (FAB) mass spectrum of acetylcarnitine. Obtained on a Varian MAT 731 mass spectrometer with a glycerol matrix on a stainless steel target and a beam of xenon atoms at 6 kV, using 1 μg D,L-acetylcarnitine in water. The ion at *m/z* 204 is the characteristic $[MH]^+$ ion for acetylcarnitine; ions at *m/z* 58, 85, 100 and 144 are common to acylcarnitine spectra; that at *m/z* 145 is the characteristic $[MH-59]^+$ daughter ion of the acetylcarnitine $[MH]^+$. Ions at *m/z* 93 ($[MH]^+$) and 185 ($[MH + M]^+$) are due to the glycerol matrix. The inset shows the constant *B/E* ratio linked scan spectrum for the parent $[MH]^+$ ion of acetylcarnitine (after Roe et al. (226)).

and are most suitable compounds for study by positive ion FAB-MS. This technique coupled with accurate mass measurements at high resolution to determine the elemental composition and linked scanning at constant *B/E* (magnetic/electrostatic) ratio to examine the spectra for characteristic daughter ions (228) has proved valuable for both qualitative and quantitative studies of acylcarnitines extracted from urine samples (109, 226, 229–232). Aqueous solutions of extracts containing 1–5 μg acylcarnitine are applied to the stainless steel target coated with glycerol matrix and exposed to a fast stream of xenon atoms at 6 kV translational energy. Accurate mass measurements are made by peak matching either to a glycerol matrix ion or to an added internal standard and quantitation is carried out by stable isotope dilution analysis and SIM using appropriate deuterated acylcarnitines as internal standards. The spectra show characteristic $[MH]^+$ parent ions due to $(CH_3)_3\overset{+}{N}-CH_2-CHOR-CH_2COOH$, where

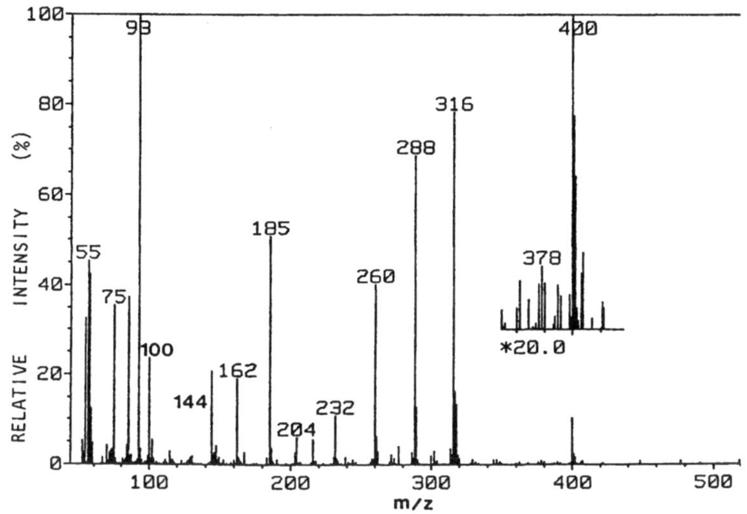

Figure 12. Positive ion fast atom bombardment (FAB) mass spectrum of a equimolar mixture of acylcarnitines.
Obtained on a JEOL DX 303 mass spectrometer with a glycerol matrix and xenon beam, using an equimolar mixture of carnitine, acetylcarnitine (C_2), butyrylcarnitine (C_4), hexanoylcarnitine (C_6), octanoylcarnitine (C_8), decanoylcarnitine (C_{10}) and palmitoylcarnitine (C_{16}) giving characteristic $[MH]^+$ ions at *m/z* 162, 204, 232, 260, 288, 316 and 400 respectively. Individual acylcarnitines, studied in equimolar amounts, show increasing intensities of their $[MH]^+$ ions as the side-chain length increases from C_2 to C_{10} after which intensity diminishes as the chain length increases further (to C_{16}) and this also occurs with the mixture shown here.

R is the acyl moiety, and major fragment ions at *m/z* 58, 85, 100 and 144 that are common to all acylcarnitines, together with less intense signals at $[MH-59]^+$ in which the acyl moiety is retained:

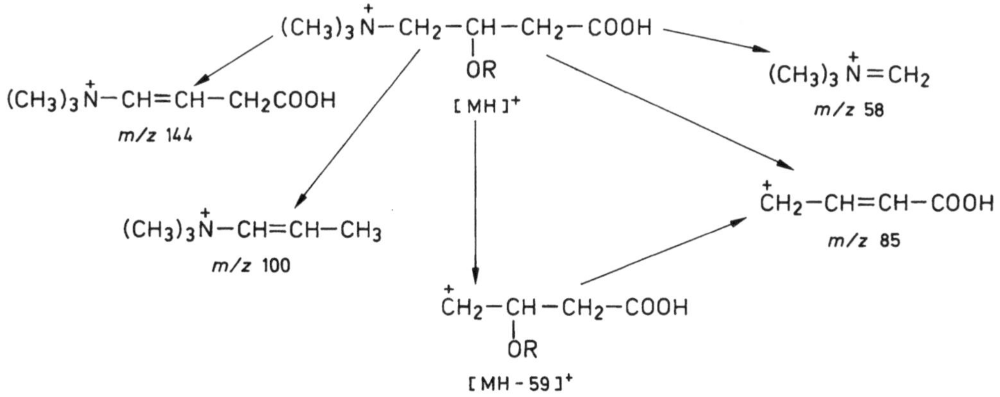

The spectrum of acetyl carnitine is typical of these spectra (Fig. 11): the $[MH]^+$ ions and characteristic $[MH-59]^+$ ions (at *m/z* 204 and 145 respectively in this example) are enhanced using linked scanning at constant *B/E* ratio which enhances the signal to noise ratio for the metabolite daughter ions in these spectra. Mixtures of acylcarnitines may be analysed (Fig. 12) both qualitatively and quantitatively (with isotope dilution analysis).

Studies have now been made of the acylcarnitines in urine from patients with a variety of organic acidurias both qualitatively and quantitatively. Patients with methylmalonic aciduria and propionic acidaemia have been shown to excrete predominantly propionylcarnitine (226, 227, 230, 233), those with isovaleric acidaemia excrete isovalerylcarnitine (109, 227, 234) and those with medium-chain acyl CoA dehydrogenase deficiency excrete octanoylcarnitine and hexanoylcarnitine (164, 227, 232), all consistent with the intramitochondrial accumulation of the corresponding acyl CoA esters in these conditions. Normal subjects and those with ketosis excrete primarily acetylcarnitine (227).

Desorption chemical ionisation (DCI) with ammonia as reagent gas has also been used to characterise urinary acylcarnitines: spectra show a $[M + 1]^+$ ion as the parent ion with ions at $[MH-14]^+$ and $[MH-42]^+$ and at *m/z* 144, 130, 102, 85, 60 with the latter being the base peak and representing the protonated trimethylammonium ion (235). These species probably arise from the fragmentation pattern:

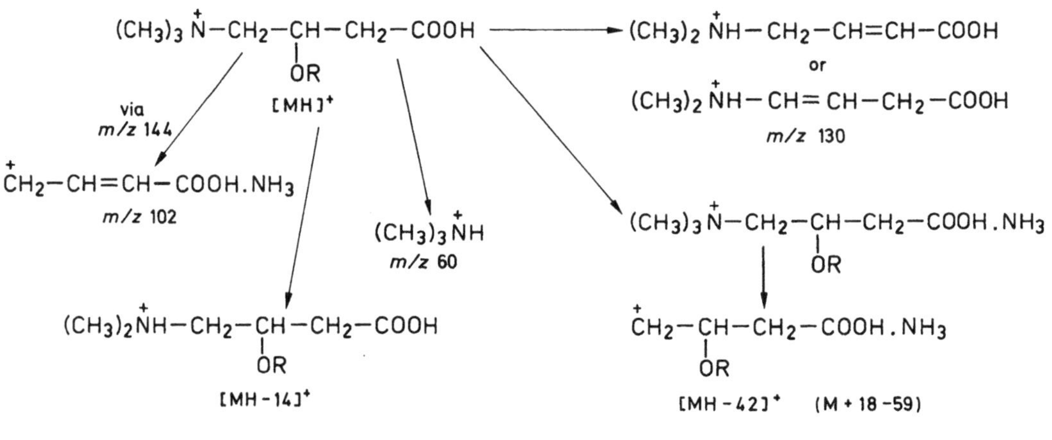

with the ions at *m/z* 144 and *m/z* 85 probably arising from the same process as in the FAB spectra. This procedure although requiring highly purified extracts of the individual acylcarnitines compared to the mixtures that can be analysed by FAB-MS (Fig. 12), has been employed to characterise octanoylcarnitine in patients with medium-chain acyl CoA dehydrogenase deficiency (236) and propionylcarnitine in patients with propionic acidaemia (237).

High performance liquid chromatography (HPLC)-mass spectrometry (LC-MS) has also been used to analyse urinary acylcarnitines with the thermospray interface (238): in this process the column effluent from the reverse-phase HPLC column is pumped at a flow of 1 $ml \cdot min^{-1}$ through a heated vapouriser block to produce a supersonic jet containing a fine droplet mist, in which a small percentage of the droplets are charged. The droplets disappear in crossing the the source region due to a combination of solvent evaporation and charge density-induced droplet disruption and ions are formed in the resulting highly directed gas stream both by direct (field) desorption and ion molecule reactions (resembling CI) (238). With ammonium ions in the buffer system, the spectra obtained resemble those produced by DCI described above but with more abundant $[M + H]^+$ ions and characteristic ions at $[MH-14]^+$ and $[MH-42]^+$, m/z 102 and m/z 130 respectively. Sodium and potassium adduct ions may also occur from other species in the buffers used, to produce $[M + 23]^+$ and $[M + 39]^+$ ions respectively (238), and such ions may also be observed in FAB spectra. Yergey et al. (238) showed the separation of carnitine from acetylcarnitine and propionylcarnitine and demonstrated the possible quantification of acylcarnitines using this technique together with isotope dilution analysis and SIM with deuterated acylcarnitine internal standards. This approach has been put to practical use in the identification of vaproylcarnitine in the urine of patients treated with valproate (dipropylpentanoate) with HPLC being used to separate the valproylcarnitine from the isomeric octanoylcarnitine prior to thermospray MS characterisation (239). It is apparent from these studies that this important class of acyl group metabolites will continue to be of much interest and that mass spectrometry, particularly the newer techniques, for example tandem mass spectrometry (240), and LC-MS will play an important role in their characterisation and accurate quantification.

8 Conclusions

This chapter has provided an overview of the major role mass spectrometry, particularly GC-MS, has played in opening the area of organic acids and their metabolism to systematic study. Metabolic profiling studies will continue to have a major part in the diagnosis of new patients, identification of new diseases and elucidation of metabolic pathways, the latter being further augmented by the use of substrates labeled with stable isotopes, the study of which makes mass spectrometry an essential analytical and investigational tool. Accurate and specific quantification with high precision also makes mass spectrometry essential in this area, particularly where stable isotope dilution analysis is employed. The newer techniques available in mass spectrometry, for example FAB-MS and particularly thermospray-MS and other LC-MS systems have opened new areas to investigation and study. There is no doubt that mass spectrometry which permitted the development of much research into the area of organic acids and the organic acidurias, will continue for the forseeable future to have a major part to play in the future development of this area of clinical biochemistry.

References

1. Nordmann, J. & Nordmann, R. (1961) Organic acids in blood and urine. Adv. Clin. Chem. *4*, 53–120.
2. Chalmers, R. A. & Lawson, A. M. (1982) Organic Acids in Man: The Analytical Chemistry, Biochemistry and Diagnosis of the Organic Acidurias, Chapman & Hall, London, New York.
3. Chalmers, R. A. & Watts, R. W. E. (1972) Studies on the quantitative freeze drying of aqueous solutions of some metabolically-important aliphatic acids prior to gas-liquid chromatographic analysis. Analyst *97*, 224–232.
4. Goodman, S. I. & Markey, S. P. (1981) Diagnosis of Organic Acidemias by Gas Chro-

matography-Mass Spectrometry, Alan. R. Liss, New York.
5. Jellum, E., Stokke, O. & Eldjarn, L. (1971) Screening for Metabolic disorders using gas-liquid chromatography-mass spectrometry, and computer technique. Scand. J. Clin. Lab. Invest. *27*, 273–285.
6. Horning, E. C. & Horning, M. G. (1971) Metabolic profiles: gas phase methods for analysis of metabolites. Clin. Chem. *17*, 802–809.
7. Liebich, H. M. & Förster, C. (1985) N-methylation and N,N-dimethylation of amino acids. An artefact production in the analysis of organic acids using diazomethane as derivatizing agent. J. Chromatogr. *228*, 33–40.
8. Simmonds, P. G. & Pettitt, B. C. & Zlatkis, A. (1967) Esterification, identification and gas chromatographic analysis of Krebs cycle keto acids. Anal. Chem. *39*, 163–167.
9. McKeown, G. G. & Read, S. I. (1965) Esterification and gas chromatography of some acids of the tricarboxylic acid cycle. Anal. Chem. *37*, 1780–1781.
10. Norman, E. J., Berry, H. K. & Denton, M. D. (1979) Identification and quantitation of urinary dicarboxylic acids as their dicyclohexyl esters in disease states by gas chromatography-mass spectrometry. Biomed. Mass Spectrom. *6*, 546–553.
11. de Jong, A. P. J. M., Elema, J. & van de Berg, B. J. J. (1980) Gas chromatography mass spectrometry of t-butyldimethylsilyl derivatives of organic acids. Biomed. Mass Spectrom. *7*, 359–364.
12. Sternowsky, H. J., Roboz, J., Hutterer, F. & Gaull, G. (1973) Determination of α-keto acids as silylated oximes in urine and serum by combined gas chromatography-mass spectrometry. Clin. Chim. Acta *47*, 371–379.
13. Lancaster, G., Lamm, P., Scriver, C. R., Tjoa, S. S. & Mamer, O. A. (1973) Quantitative analysis of branched-chain α-ketoacids as their trimethylsilylated oximes. Clin. Chim. Acta *48*, 279–285.
14. Jakobs, C., Solem, E., Ek, J., Halvorsen, K. & Jellum, E. (1977) Investigation of the metabolic pattern in maple syrup urine disease by means of glass capillary gas chromatography and mass spectrometry. J. Chromatogr. *143*, 31–38.
15. Thompson, J. A. & Markey, S. P. (1975) Quantitative metabolic profiling of urinary organic acids by gas chromatography-mass spectrometry: Comparison of isolation methods. Anal. Chem. *47*, 1313–1321.
16. Tuchman, M., Whitley, C. B., Ramnaraine, M.-L., Bowers, L. D., Fregian, K. D. & Krivit, W. (1984) Determination of urinary succinylacetone by capillary gas chromatography. J. Chromatogr. Sci. *22*, 211–215.
17. Horning, M. G., Boucher, E. A., Moss, A. M. & Horning, E. C. (1968) Gas chromatographic study of derivatives of acids of the Krebs cycle and related compounds. Anal. Lett. *1*, 713–723.
18. Chalmers, R. A. & Watts, R. W. E. (1972) Derivatives for the identification and quantitative determination of some keto and aldo-carboxylic acids by gas-liquid chromatography. Analyst *97*, 951–957.
19. Sweetman, L. (1984) Qualitative and quantitative analysis of organic acids in physiologic fluids for diagnosis of the organic acidurias. In: Abnormalities in Amino Acid Metabolism in Clinical Medicine, (Nyhan, W. L.), pp. 419–445, Appleton-Century-Crofts, Norwalk, Connecticut.
20. Langenbeck, U, Möhring, H.-U., Hinney, B. & Spiteller, M. (1977) Quinoxalinol derivatives of aliphatic 2-oxocarboxylic acids. Infrared and Mass spectra of the O-trimethylsilylated compounds. Biomed. Mass Spectrom. *4*, 197–202.
21. Perry, T. L., Hansen, S., Diamond, S., Bullis, B., Mok, C. & Melançon, S. B. (1970) Volatile fatty acids in normal human physiological fluids. Clin. Chim. Acta *29*, 369–374.
22. Ottenstein, D. M. & Bartley, D. S. (1971) Separation of free acids C_2-C_5 in dilute aqueous solutions. Column technology. J. Chromatogr. Sci. *9*, 673–676.
23. Van den Berg, H. & Hommes, F. A. (1974) A rapid and sensitive method for the determination of short chain fatty acids in serum. Clin. Chim. Acta 51, 225–232.
24. Chalmers, R. A., Bickle, S. & Watts, R. W. E. (1974) A method for the determination of volatile organic acids in aqueous solutions and urine, and the results obtained in propionic acidaemia, β-methylcrotonylglycinuria and methylmalonic aciduria. Clin. Chim. Acta *52*, 31–41.
25. Frenkel, E. P. & Kitchens, R. L. (1975) A simplified and rapid quantitative assay for propionic and methylmalonic acids. J. Lab. Clin. Med. *85*, 487–496.
26. Jellum, E. (1979) Profiling of human body fluids in healthy and diseased states using gas chromatography and mass spectrome-

try, with special reference to organic acids. J. Chromatogr. *143*, 427–462.
27. Mamer. O. A. & Gibbs, B. F. (1973) Simplified gas chromatography of trimethylsilyl esters of C_1 through C_5 fatty acids in serum and urine. Clin. Chem. *19*, 1006–1009.
28. Happ, G. P. & Stewart, D. W. (1952) Rearrangement peaks in the mass spectra of certain aliphatic acids. J. Am. Chem. Soc. *74*, 4404–4406.
29. Sharkey, A. G., Schultz, J. L. & Friedel, R. A. (1959) Mass spectra of esters. Formation of rearrangement ions. Anal. Chem. *31*, 87–94.
30. Ryhage, R. & Stenhagen, E. (1959) Mass spectrometric studies. I. Methyl esters of saturated normal chain carboxylic acids. Ark. Kem. *13*, 523–528.
31. Ryhage, R. & Stenhagen, E. (1963) Mass spectrometry of long chain esters. In: Mass Spectrometry of Organic Ions (McLafferty, F. W., ed.), pp. 399–452, Academic Press, New York.
32. Beynon, J. H., Saunders, R. A. & Willliams, A. E. (1961) The high resolution mass spectra of aliphatic esters. Anal. Chem. *33*, 221–225.
33. Budzikiewicz, H., Djerassi, C. & Williams, D. H. (1967) Mass Spectrometry of Organic Compounds, Holden-Day, San Francisco.
34. Beynon, J. H., Saunders, R. A. & Williams, A. E. (1968) The Mass Spectra of Organic Molecules, Elsevier, Amsterdam.
35. Keen, G. W. & Hamming, M. C. (1971) Mass spectrometric analysis of aliphatic esters, Mass Spectrosc. *19*, 308–315.
36. Tanaka, K. & Yu, G. M. (1973) A method for the separate determination of isovalerate and α-methylbutyrate by use of GLC-mass spectrometer. Clin. Chim. Acta *43*, 151–154.
37. Ramsdell, H. S., Baretz, B. H. & Tanaka, K. (1977) Mass spectrometric studies of twenty-one metabolically important acylglycines. Biomed. Mass Spectrom. *47*, 220–228.
38. Tjoa, S. S. & Fennessey, P. V. (1979) Acylglycines. The gas chromatograph/mass spectrometric identification and interpretation of the spectra. Clin. Chim. Acta 95, 35–45.
39. Lawson, A. M., Chalmers, R. A. & Watts, R. W. E. (1974) Studies of O-substituted oxime-trimethylsilyl ester derivatives of some metabolically important oxocarboxylic acids. Biomed. Mass Spectrom. *1*, 199–205.
40. Klenk, E. & Kahlke, W. (1963) Über das Vorkommen der 3,7,11,15-tetramethylhexadecansäure (Phytansäure) in den Cholesterinestern und anderen Lipoidfraktionen der Organe bei einem Krankheitsfall unbekannter Genese (Verdacht auf Heredopathia atactica polyneuritiformis, Refsum's syndrome). Hoppe Seyler's Z. Physiol. Chem. *333*, 133–143.
41. Tanaka, K., Budd, M. A., Efron, M. L. & Isselbacher, K. J. (1966) Isovaleric acidemia. A new genetic defect of leucine metabolism. Proc. Natl. Acad. Sci. USA *56*, 236–242.
42. Dalgliesh, C. E., Horning, E. C., Horning, M. G., Knox, K. L. & Yarger, K. (1966) A gas-liquid-chromatographic procedure for separating a wide range of metabolites occurring in urine or tissue extracts. Biochem. J. *101*, 792–810.
43. Jaakonmaki, P. I., Knox, K. L., Horning, E. C. & Horning, M. G. (1967) The characterisation by gas-liquid chromatography of ethyl β-D-glucosiduronic acid as a metabolite of ethanol in rat and man. Eur. J. Pharmacol. *1*, 63–67.
44. Zaura, D. S. & Metcoff, J. (1969) Quantification of seven tricarboxylic acid cycle and related acids in human urine by gas-liquid chromatography. Anal. Chem. *41*, 1781–1787.
45. Horning, E. C. & Horning, M. G. (1971) Human metabolic profiles obtained by GC and GC/MS. J. Chromatogr. Sci. *9*, 129–140.
46. Jellum, E., Stokke, O. & Eldjarn, L. (1972) Combined use of gas chromatography, mass spectrometry and computer in diagnosis and studies of metabolic disorders. Clin. Chem. *18*, 800–809.
47. Hites, R. A., & Biemann, K. (1968) Mass spectrometer-computer system particularly suited for gas chromatography of complex mixtures. Anal. Chem. *40*, 127–132.
48. Hites, R. A. & Biemann, K. (1970) Computer evaluation of continuously scanned mass spectra of gas chromatographic effluents. Anal. Chem. *42*, 855–860.
49. Jellum, E., Stokke, O. & Eldjarn, L. (1973) Application of gas chromatography-mass spectrometry, and computer methods in clinical biochemistry. Anal. Chem. *45*, 1099–1106.
50. Jellum, E., Stokke, O., Eldjarn, L. (1974) In: Applications of Gas Chromatography-Mass Spectrometry to the Investigation of Human Disease (Mamer, O. A., Mitchell, W. J. & Scriver, C. R., eds.), pp. 67–83, McGill

University – Montreal Children's Hospital Research Institute, Montreal.

51. Eldjarn, L., Jellum, E. & Stokke, O. (1974) Application of gas chromatography-mass spectrometry in routine and research in clinical chemistry. J. Chromatogr. *91*, 353–366.
52. Hutterer, F., Roboz, J., Sarkozi, L., Ruhig, A. & Bacchin, P. (1971) Gas chromatograph-mass spectrometer computer system for detection and identification of abnormal metabolic products in physiological fluids. Clin. Chem. *17*, 789–794.
53. Mamer, O. A., Crawhall, J. C. & Tjoa, S. S. (1971) The identification or urinary acids by coupled gas chromatography-mass spectrometry. Clin. Chim. Acta *32*, 171–184.
54. Lawson, A. M., Chalmers, R. A., Purkiss, P., Mitchell, F. L. & Watts, R. W. E. (1974) The application of GC-MS to the study of urinary acidic and neutral metabolites in normal subjects and in patients with some metabolic disorders. Adv. Mass Spectrom. *6*, 235–243.
55. Chalmers, R. A. & Lawson, A. M. (1975) Human metabolic diseases. Chem. Br. *11*, 290–295.
56. Lawson, A. M., Chalmers, R. A. & Watts, R. W. E. (1976) Urinary organic acids in man. I. Normal patterns. Clin. Chem. *22*, 1283–1287.
57. Chalmers, R. A., Healy, M. J. R., Lawson, A. M. & Watts, R. W. E. (1976) Urinary organic acids in man II. Effects of individual variation and diet on the urinary excretion of acidic metabolites. Clin. Chem. *22*, 1288–1291.
58. Chalmers, R. A. (1980) Organic Acidurias: Diagnosis and Investigation using GC-MS. In: Current Developments in the Clinical Applications of HPLC, GC and MS (Lawson, A. M., Lim, C. K. & Richmond, W., eds.), pp. 233–246, Academic Press, London, New York.
59. Witten, T. A., Levine, S. P., King, J. O. & Markey, S. P. (1973) Gas chromatographic mass spectrometric determination of urinary acid profiles of normal young adults on a controlled diet. Clin. Chem. *19*, 586–589.
60. Markey, S. P., Keyser, J. A. & Levine, S. P. (1974) Normal metabolic profiles in body fluids as determined by gas chromatography mass spectrometry. In: Application of Gas Chromatography-Mass Spectrometry to the Investigation of Human Disease (Mamer, O. A., Mitchell, W. J. & Scriver, C. R., eds.), pp. 39–66, McGill University-Montreal Children's Hospital Research Institute, Montreal.
61. Jellum, E., Størseth, P., Alexander, J., Helland, P., Stokke, O. & Teig, E. (1976) Application of glass capillary-column gas chromatography-mass spectrometry to studies of human disease. J. Chromatogr. *126*, 487–493.
62. Biller, J. E. & Biemann, K. (1974) Reconstructed mass spectra, a novel approach for the utilisation of gas chromatography-mass spectrometer. Anal. Lett. *7*, 515–520.
63. Gates, S. C., Smisko, M. J., Ashendal, C. L., Young, N. D., Holland, J. F. & Sweeley, C. (1978) Automated simultaneous qualitative and quantitative analysis of complex organic mixtures with a gas chromatography-mass spectrometry-computer system. Anal. Chem. *50*, 433–441.
64. Fitch, W. L., Anderson, P. J. & Smith, D. H. (1979) Isolation, identification and quantitation of urinary organic acids. J. Chromatogr. *162*, 249–259.
65. Muzino, T., Abe, N., Teshima, H., Yamauchi, E., Itagaki, Y., Matsumoto, I., Kuhara, T. & Shinka, T. (1981) Application of a gas chromatography-mass spectrometry computer system for clinical analysis. Biomed. Mass Spectrom. *8*, 593–597.
66. Lewis, S., Kenyon, C. H., Meili, J. & Burlinghame, A. L. (1979) High resolution gas chromatographic/real time high resolution mass spectrometric identification of organic acids in human urine. Anal. Chem. *51*, 1275–1285.
67. Issacher, D. & Yinon, J. (1979) Screening of organic acids in urine by chemical ionization mass spectrometry. Biomed. Mass Spectrom. *6*, 47–56.
68. Hunt, D. F., Giodani, A. B., Rhodes, G. & Herold, D. A. (1982) Mixture analysis by triple-quadrupole mass spectrometry: Metabolic profiling of urinary carboxylic acids. Clin. Chem. *28*, 2387–2392.
69. Rinaldo, P., Chiandetti, L., Zachello, F., Daolio, S. & Traldi, P. (1984) CAD MIKES: A new method for a rapid and unequivocal structural identification of organic acids in biological fluids. A first application to a case of methylmalonic aciduria. Biomed. Mass Spectrom. *11*, 643–646.
70. Mrochek, J. E. & Rainey, W. T. (1972) Identification and biochemical significance of substituted furans in human urine. Clin. Chem. *18*, 821–828.
71. Pettersen, J. E. & Jellum, E. (1972) The identification and metabolic origin of 2-fu-

roylglycine and 2,5-furandicarboxylic acid in human urine. Clin. Chim. Acta *41*, 199–207.
72. Jellum, E., Børresen, H. C. & Eldjarn, L. (1973) The presence of furan derivatives in patients receiving fructose-containing solutions intravenously. Clin. Chim. Acta *47*, 191–201.
73. Pettersen, J. E. & Stokke, O. (1973) Branched short-chain dicarboxylic acids in human urine. Biochim. Biophys. Acta *304*, 316–325.
74. Lindstedt, S. & Steen, G. (1975) 5-decynedioic acid, an acetylenic compound in human urine. Clin. Chem. *21*, 1964–1969.
75. Lindstedt, S., Norberg, K., Steen, G. & Wahl, E. (1976) Structure of some aliphatic dicarboxylic acids found in urine of an infant with congenital lactic acidosis. Clin. Chem. *22*, 1330–1338.
76. Heindl, A., Rau, O. & Spiteller, G. (1985) Identification of aromatic dihydroxy acids in biological fluids. Biomed. Mass Spectrom. *12*, 59–66.
77. Thompson, J. A., Miles, B. S. & Fennessey, P. V. (1977) Urinary organic acids quantitated by age groups in a healthy pediatric population. Clin. Chem. *23*, 1734–1738.
78. Knights, B. A., Legendre, M., Laseter, J. L. & Storer, J. S. (1975) Use of high resolution open tubular glass capillary columns to separate acidic metabolites in urine. Clin. Chem. *21*, 888–891.
79. Hagenfeldt, L. & Hagenfeldt, K. (1972) Gas chromatograpahic-mass spectrometric analysis of organic acids in amniotic fluid. Clin. Chim. Acta *42*, 219–220.
80. Nicholls, T., Hähnel, R., Wilkinson, S. & Wysocki, S. (1976) Identification of 2-hydroxybutyric acid in human amniotic fluid. Clin. Chem. Acta *69*, 127–130.
81. Nicholls, T. M., Hähnel, R. & Wilkinson, S. P. (1978) Organic acids in amniotic fluid. Clin. Chim. Acta *84*, 11–17.
82. Wilkinson, S. P., Niven, S. E. & Hähnel, R. (1979) Quantitation of organic acids in amniotic fluid by gas chromatography. Clin. Chim. Acta *99*, 241–245.
83. Chalmers, R. A., Lawson, A. M. & Raymond, F. D. (1978) Organic acids in human amniotic fluid and the prenatal diagnosis of the organic acidurias. Clin. Sci. *35*, 50.
84. Williams, V. P., Ching, D. K. & Cederbaum, S. D. (1979) Absorption of organic acids from amniotic fluid and urine onto silica gel before analysis by gas chromatography and combined gas chromatography-mass spectrometry. Clin. Chem. *24*, 1814–1820.
85. Ng, K. J., Andresen, B. D., Bianchine, J. R., Iams, J. D., O'Shaughessey, W., Stempel, L. E. & Zuspan, F. P. (1982) Capillary gas chromatographic-mass spectrometric profiles of trimethylsilyl derivatives of organic acids from amniotic fluids of different gestational age. J. Chromatogr. *228*, 43–50.
86. Chalmers, R A., Purkiss, P., Watts, R. W. E. & Lawson, A. M. (1980) Screening for organic acidurias and amino acidopathies in newborns and children. J. Inher. Metab. Dis. *3*, 27–43.
87. Watts, R. W. E., Baraitser, M., Chalmers, R. A. & Purkiss, P. (1980) Organic acidurias and amino acidurias in the aetiology of long-term mental handicap. J. Ment. Defic. Res. *24*, 257–270.
88. Tanaka, K., Orr, J. C. & Isselbacher, K. J. (1968) Identification of β-hydroxyisovaleric acid in the urine of a patient with isovaleric acidemia. Biochim. Biophys. Acta *152*, 638–641.
89. Tanaka, K. & Isselbacher, K. J. (1967) The isolation and identification of N-isovalerylglycine from urine of patients with isovaleric aciduria. J. Biol. Chem. *242*, 2966–2972.
90. Stokke, O., Eldjarn, L., Norum, K. R., Steen-Johnsen, J. & Halvorsen, S. (1967) Methylmalonic acidemia. A new inborn error of metabolism which may cause fatal acidosis in the neonatal period. Scand. J. Clin. Lab. Invest. *20*, 313–328.
91. Ando, T., Rasmussen, K., Nyhan, W. L., Donnell, G. N. & Barnes, N. D. (1971) Propionic acidemia in patients with ketotic hyperglycinemia. J. Pediatr. *78*, 827–832.
92. Jellum, E., Kluge, T., Børresen, H. C., Stokke, O. & Eldjarn, L. (1970) Pyroglutamic aciduria – a new inborn error of metabolism. Scand. J. Clin. Lab. Invest. *26*, 327–335.
93. Stokke, O., Eldjarn, L., Jellum, E., Pande, H. & Waaler, P. E. (1972) β-methylcrotonyl CoA carboxylase deficiency: A new metabolic error in leucine degradation. Pediatrics *49*, 726–735.
94. Ando, T., Rasmussen, K., Nyhan, W. L., Hull, D. (1972) 3-hydroxypropionate: significance of β-oxidation of propionate in patients with propionic acidemia and methylmalonic acidemia. Proc. Natl. Acad. Sci. USA *69*, 2807–2811.
95. Ando, T., Rasmussen, K., Wright, J. M. & Nyhan, W. L. (1972) Isolation and identifi-

cation of methylcitrate, a major metabolic product of propionate in patients with propionic acidemia. J. Biol. Chem. *247*, 2200–2204.
96. Chalmers, R. A., Lawson, A. M., Watts, R. W. E. (1974) Studies on the urinary organic acidic metabolites excreted by patients with β-methylcrotonylglycinuria, propionic acidemia and methylmalonic acidaemia, using gas-liquid chromatography and mass spectrometry. Clin. Chim. Acta *52*, 43–51.
97. Rasmussen, K., Ando, T., Nyhan, W. L., Hull, D., Cottom, D., Donnell, G., Wadlington, W. & Kilroy, A. W. (1972) Excretion of propionylglycine in propionic acidemia. Clin. Sci. *42*, 665–671.
98. Rasmussen, K., Ando, T., Nyhan, W. L., Hull, D., Cottom, D., Kilroy, A. W. & Wadlington, W. (1972) Excretion of tiglylglycine in propionic acidemia. J. Pediatr. *81*, 970–975.
99. Gompertz, D., Draffan, G. H., Watts, J. L. & Hull, D. (1971) Biotin-responsive β-methylcrotonylglycinuria. Lancet *ii*, 22–24.
100. Saunders, M., Sweetman, L., Robinson, B., Roth, K., Cohn, R. & Gravel, R. A. (1979) Biotin responsive organic aciduria. Multiple carboxylase defects and complementation studies with propionic acidemia in cultured fibroblasts. J. Clin. Invest. *64*, 1695–1702.
101. Lehnert, W., Schuchmann, L., Urbânek, R., Niederhoff, H. & Böhm, N. (1978) Excretion of 2-methyl-3-oxovaleric acid in propionic acidaemia. Eur. J. Pediatr. *128*, 197–205.
102. Duran, M., Gompertz, D., Bruinvis, L., Ketting, D. & Wadman, S. K. (1978) The variability of metabolite excretion in propionic acidemia. Clin. Chim. Acta *82*, 93–99.
103. Sweetman, L., Weyler, W., Nyhan, W. L., Cespedes, C. de, Loria, A. R. & Estrada, Y. (1978) Abnormal metabolites of isoleucine in a patient with propionyl-CoA carboxylase deficiency. Biomed. Mass Spectrom. *5*, 198–207.
104. Truscott, R. J. W., Pullin, C. J., Halpern, B., Hammond, J., Haan, E. & Danks, D. M. (1979) The identification of 3-keto-2-methylvaleric acid and 3-hydroxy-2-methylvaleric acid in a patient with propionic acidemia. Biomed. Mass Spectrom. *6*, 294–300.
105. Przyrembel, H., Bremer, H. J., Duran, M., Bruinvis, L., Ketting, D., Wadman, S. K., Baumgartner, R., Irle, U. & Bachmann, C. (1979) Propionyl-CoA carboxylase deficiency with overflow of metabolites of isoleucine catabolism at all levels. Eur. J. Pediatr. *130*, 1–14.
106. Kuhara, T., Inoue, Y., Shinka, T., Matsumoto, I. & Matsuo, M. (1983) Identification of 3-hydroxy-3-ethylglutaric acid in urine of patients with propionic acidaemia. Biomed. Mass Spectrom. *10*, 629–632.
107. Kuhara, T., Shinka, S., Matsuo, M. & Matsomoto, I. (1982) Increased excretion of lactate, glutarate, 3-hydroxyisovalerate and 3-methylglutaconate during clinical episodes of propionic acidaemia. Clin. Chim. Acta *123*, 101–109.
108. Shigematsu, Y., Sudo, M., Momoi, T., Inoue, Y., Suzuki, Y. & Kameyama, J. (1982) Changing plasma and urinary organic acid levels in a patient with isovaleric acidaemia during an attack. Pediatr. Res. *16*, 771–775.
109. de Sousa, C., Chalmers, R. A., Stacey, T. E., Tracey, B. M., Weaver, C. M. & Bradley, D. (1985) The response to L-carnitine and glycine therapy in isovaleric acidaemia. Eur. J. Pediatr. *144*, 451–456.
110. Matsumoto, I. & Kuhara, T. (1987) Gas chromatography-mass spectrometry for the chemical diagnosis of the inherited metabolic diseases – Differential chemical diagnosis of lactic acidosis. Mass Spectrom. Rev. *6*, 77–134.
111. Lehnert, W. & Niederhoff, H. (1981) 4-hydroxyisovaleric acid: A new metabolite in isovaleric acidemia. Eur. J. Pediatr. *136*, 281–283.
112. Truscott, R. J. W., Malegon, D., McCairns, E., Burke, D., Hick, L., Sims, P., Halpern, B., Tanaka, K., Sweetman, L., Nyhan, W. L., Hammond, J., Bumack, C., Haan, E. A. & Danks, D. M. (1981) New metabolites in isovaleric acidemia. Clin. Chim. Acta *110*, 187–203.
113. Lehnert, W. (1981) 3-hydroxyisoheptanoic acid: a new metabolite in isovaleric acidemia. Clin. Chim. Acta *113*, 101–103.
114. Lehner, W. (1981) Excretion of N-isovalerylglutamic acid in isovaleric acidemia. Clin. Chim. Acta *116*, 249–252.
115. Faull, K. F., Bolton, P. D., Halpern, B., Hammond, J. & Danks, D. M. (1976) The urinary organic acid profile associated with 3-hydroxy-3-methylglutaric aciduria. Clin. Chim. Acta *73*, 553–559.
116. Wysocki, S. J., Wilkinson, S. P., Hähnel, R., Wong, C. Y. B. & Panegyres, P. K. (1976) 3-hydroxy-3-methylglutaric aciduria, combined with 3-methylglutaconic aciduria. Clin. Chim. Acta *70*, 399–406.

117. Duran, M., Ketting, D., Wadman, S. K., Jakobs, C., Schutgens, R. B. H. & Veder, H. A. (1978) Organic acid excretion in a patient with 3-hydroxy-3-methylglutaryl CoA lyase deficiency: facts and artefacts. Clin. Chim. Acta *90*, 187–193.
118. Schutgens, R. B. H., Heymans, H., Ketel, A., Veder, H.A., Duran, M., Ketting, D. & Wadman, S. K. (1979) Lethal hypoglycaemia in a child with a deficiency of 3-hydroxy-3-methylglutarylcoenzyme A lyase. J. Pediatr. *94*, 89–91.
119. Norman, E. J., Denton, M. D. & Berry, H. K. (1982) Gas-chromatographic/mass spectrometric detection of 3-hydroxy-3-methylglutaryl CoA lyase deficiency in double first cousins. Clin. Chem. *28*, 137–140.
120. Tracey, B. M., Stacey, T. E., Chalmers, R. A. (1983) Urinary and plasma organic acids in dizygotic twin siblings with 3-hydroxy-3-methylglutaric aciduria, studied by gas chromatography and mass spectrometry using fused silica capillary columns. J. Inher. Metab. Dis. *6*, Suppl. 2, 125–126.
121. Stacey, T. E., de Sousa, C., Tracey, B. M., Whitelaw, A., Mistry, J., Timbrell, P. & Chalmers, R. A. (1985) Dizygotic twins with 3-hydroxy-3-methylglutaric aciduria: unusual presentation, family studies and dietary management. Eur. J. Pediatr. *144*, 177–181.
122. Daum, R. S., Scriver, C. R., Mamer, O. A., Delvin, E., Lamm, P. & Goldman, H. (1973) An inherited disorder of isoleucine catabolism causing accumulation of α-methylacetoacetate and α-methyl-β-hydroxybutyrate, and intermittent metabolic acidosis. Pediatr. Res. *7*, 149–160.
123. Gompertz, D., Saudubray, J. M., Charpentier, C., Bartlett, K., Goodey, P. A. & Draffan, G. H. (1974) A defect in L-isoleucine metabolism associated with α-methyl-β-hydroxybutyric and α-methylacetoacetic aciduria: quantitative in vivo amd in vitro studies. Clin. Chim. Acta *57*, 269–281.
124. Wilson, R. W., Wilson, C. M., Gates, S. C. & Higgins, J. V. (1975) α-Ketoadipic aciduria: A description of a new metabolic error in lysine-tryptophan degradation. Pediatr. Res. *9*, 522–526.
125. Przyrembel, H., Bachmann, D., Lombeck, I., Becker, K., Wendel, U., Wadman, S. K. & Bremer, H. J. (1975) Alpha-ketoadipic aciduria, a new inborn error of lysine metabolism; Biochemical studies. Clin. Chim. Acta *58*, 257–269.
126. Goodman, S. I., Markey, S. P., Moe, P. G., Miles, B. S. & Teng, C. C. (1975) Glutaric aciduria; A "new" disorder of amino acid metabolism. Biochem. Med. *12*, 12–21.
127. Gregersen, N., Brandt, N. J., Christensen, E., Grøn, I., Rasmussen, K. & Brandt, S. (1977) Glutaric aciduria: Clinical laboratory findings in two brothers. J. Pediatr. *90*, 740–745.
128. Brandt, N. J., Rasmussen, K., Brandt, S., Kølvraa, S. & Schønheyder, F. (1976) D-glyceric acidaemia and non-ketotic hyperglycinaemia. Acta Paediatr. Scand. *65*, 17–22.
129. Wadman, S. K., Duran, M., Ketting, D., Bruinvis, L., de Bree, P. K., Kamerling, J. P., Gerwig, G. J., Vliegenhardt, J. F. G., Przyrembel, H., Becker, K. & Bremer, H. J. (1976) D-glyceric acidemia in a patient with chronic metabolic acidosis. Clin. Chim. Acta *71*, 477–479.
130. Jakobs, C., Bojash, M., Mönch, E., Rating, D., Siemes, H. & Hanefeld, F. (1981) Urinary excretion of gamma-hydroxybutyric acid in a patient with neurological abnormalities. The possibility of a new inborn error of metabolism. Clin. Chim. Acta *111*, 169–178.
131. Jakobs, C., Kneer, J., Rating, D., Hanefeld, H., Divry, P. & Hermier, M. (1984) A new disorder: 4-hydroxybutyric aciduria. Biochemical findings. J. Inher. Metab. Dis. *7*, Suppl. *1*, 92–94.
132. Niederwieser, A., Matasović, A., Steinmann, B., Baerlocher, K. & Kempken, B. (1976) Hydantoin-5-propionic aciduria in folic acid non-dependant formiminoglutamic aciduria observed in two siblings. Pediatr. Res. *10*, 215–219.
133. Blau, K. (1970) Aromatic acid excretion in phenylketonuria. Analysis of the unconjugated aromatic acids derived from phenylalanine. Clin. Chim. Acta *27*, 5–18.
134. Wadman, S. K., van der Heiden, C., Ketting, D. & Van Sprang, F. J. (1971) Abnormal tyrosine and phenylalanine metabolism in patients with tyrosyluria and phenylketonuria: gas-liquid chromatographic analysis of urinary metabolites. Clin. Chim. Acta *34*, 277–287.
135. Koepp., P. & Hoffmann, B. (1974) Aromatic acid excretion in classical phenylketonuria and hyperphenylalaninemic variants. Helv. paediatr. Acta *29*, 489–498.
136. Rampini, S., Völlmin, J. A., Bosshard, H. R., Müller, M. & Curtius, H, C. (1974) Aromatic acids in urine of healthy infants, persistent hyperphenylalaninemia,

and phenylketonuria, before and after phenylalanine load. Pediatr. Res. *8*, 704–709.

137. Chalmers, R. A. & Watts, R. W. E. (1974) Quantitative studies on the urinary excretion of unconjugated aromatic acids in phenyletonuria. Clin. Chim. Acta *55*, 281–294.

138. Olek, K., Oyanagi, K. & Wardenbach, P. (1974) Quantitative analysis of phenylalanine metabolites in urine to detect heterozygotes of phenylketonuria. Humangenetik *22*, 85–88.

139. Applegarth, D. A., Goodman, S. I., Irvine, D. G. & Jellum, E. (1977) Hyperprolinemia type II: Identification of the glycine conjugate of pyrrole-2-carboxylic acid in urine. Clin. Biochem. *10*, 20–23.

140. Lindblad, B. & Steen, G. (1982) Identification of 4,6-dioxoheptanoic acid (succinylacetone), 3,5-dioxooctanedioic acid (succinylacetoacetate) and 4-oxo-6-hydroxyheptanoic acid in the urine from patients with hereditary tyrosinaemia. Biomed. Mass Spectrom. *9*, 419–424.

141. Pettersen, J. E., Jellum, E. & Eldjarn, L. (1972) The occurence of adipic and suberic acid in urine from ketotic patients. Clin. Chim. Acta *38*, 17–24.

142. Dosman, J., Crawhall, J. C., Klassen, G. S., Mamer, O. A. & Neumann, P. (1974) Urinary excretion of C_6-C_{10} dicarboxylic acids in glycogen storage diseases types I and III. Clin. Chim. Acta *51*, 93–101.

143. Tanaka, K., Kean, E. A., Johnson, B. (1976) Jamaican vomiting sickness. Biochemical investigation of two cases. New Engl. J. Med. *295*, 461–467.

144. Tanaka, K. (1975) Branched pentanoic aciduria and medium-chain dicarboxylic aciduria induced by hypoglycin A: Inhibition of several short-chain acyl CoA dehydrogenases. In: Hypoglycin (Kean, E. A., ed.), pp. 67–92, Academic Press, New York.

145. Przyrembel, H., Wendel, U., Becker, K., Bremer, H. J., Bruinvis, L., Ketting, D. & Wadman, S K. (1976) Glutaric aciduria type II: Report on a previously undescribed metabolic disorder. Clin. Chim. Acta *66*, 277–289.

146. Gregersen, N., Lauritzen, R. & Rasmussen, K. (1976) Suberylglycine excretion in the urine from a patient with dicarboxylic aciduria. Clin. Chim. Acta *70*, 417–425.

147. Chalmers, R. A., Lawson, A. M., Whitelaw, A. & Purkiss, P. (1977) Organic acids in Reye's-like syndrome: similarities with Jamaican vomiting sickness. Lancet *i*, 1156 (C).

148. Chalmers, R. A., Lawson, A. M., Whitelaw, A. & Purkiss, P. (1980) Twin siblings with a Reye's-like syndrome associated with an abnormal organic aciduria, hypoglycaemia, diarrhea, and vomiting with close similarities to Jamaican Vomiting Sickness. Pediatr. Res. *14*, 1097–1103.

149. Truscott, R. J., Hick, L., Pullin, C., Halpern, B., Wilcken, B., Griffiths, H., Silink, M., Kilham, H. & Grunseit, F. (1979) Dicarboxylic aciduria: the response to fasting. Clin. Chim. Acta *94*, 31–39.

150. Goodman, S. I., McCabe, E. R. B., Fennessey, P. V. & Mace, J. W. (1980) Multiple acyl CoA dehydrogenase deficiency (glutaric aciduria type II) with transient hypersarcosinemia and sarcosinuria: possible inherited deficiency of an electron transfer flavoprotein. Pediatr. Res. *14*, 12–17.

151. Niederwieser, A., Steinmann, B., Exner, V., Neuheiser, F., Redweik, U., Wang, M., Rampini, S. & Wendel, U. (1983) Multiple acyl-CoA dehydrogenation deficiency (MADD) in a boy with non-ketotic hypoglycemia, hepatomegaly, muscle hypotonia and cardiomyopathy. Helv. Paediatr. Acta *38*, 9–26.

152. Goodman, S. I. & Frerman, F. (1984) Glutaric acidaemia type II (Multiple acyl-CoA dehydrogenation deficiency). J. Inher. Metab. Dis. *7*, Suppl. 1, 33–37.

153. Mortensen, P. B. & Gregersen, N. (1980) Medium-chain triglyceride medication as a pitfall in the diagnosis of non-ketotic C_6-C_{10} dicarboxylic acidurias. Clin. Chim. Acta *103*, 33–37.

154. Shigematsu, Y., Momoi, T., Sudo, M. & Suzuki, Y. (1981) (ω-1)-hydroxymonocarboxylic acids in urine of infants fed medium-chain triglycerides. Clin. Chem. *27*, 1661–1664.

155. Saudubray, J. M., Coudé, F. X., Demaugre, F., Johnson, C., Gibson, K. M. & Nyhan, W. L. (1982) Oxidation of fatty acids in cultured fibroblasts: a model system for the detection and study of defects in oxidation. Pediatr. Res. *16*, 877–881.

156. Stanley, C. A., Hale, D. E., Coates, P. M., Hall, C. L., Corkey, B. E., Yang, W., Kelley, R. I., Gonzales, E. L., Williamson, J. R. & Baker, L. (1983) Medium chain acyl-CoA dehydrogenase deficiency in children with non-ketotic hypoglycaemia and low carnitine levels. Pediatr. Res. *17*, 877–884.

157. Divry, P., David, M., Gregersen, W., Kølvraa, S., Christensen, E., Collet, J. P., Dellamonica, C. & Cotte, J. (1983) Dicarboxylic aciduria due to medium-chain acyl-CoA dehydrogenase defect: A cause of hypoglycaemia in childhood. Acta Paediatr. Scand. *72*, 943–949.
158. Gregersen, N., Kølvraa, S., Rasmussen, K., Mortensen, P. B., Divry, P., David, M. & Hobolth, N. (1983) General (medium-chain) acyl-CoA dehydrogenase deficiency (non-ketotic dicarboxylic aciduria): quantitative urinary excretion pattern of 23 biologically significant organic acids in three cases. Clin. Chim. Acta *132*, 181–191.
159. Divry, P., Vianey-Liaud, C. & Cotte, J. (1984) Gas chromatography-mass spectrometry (GC-MS) diagnosis of two cases of medium chain acyl-CoA dehydrogenase deficiency. J. Inher. Metab. Dis. *7*, Supp. 1, 44–47.
160. del Valle, J. A., Garcia, M. J., Merinero, B., Pérez-Cerda C., Roman, E., Jimenez, A., Ugarte, M., Martinez-Pardo, M., Ludera, C., Camarero, C., del Olmo, R., Duran, M. & Wadman, S. K. (1984) A new patient with dicarboxylic aciduria suggestive of medium-chain acyl-CoA dehydrogenase deficiency presenting as Reye's syndrome. J. Inher. Metab. Dis. *7*, 62–64.
161. Duran, M., De Klerk, J. B. C., Wadman, S. K., Bruinvis, L. & Ketting, D. (1984) The differential diagnosis of dicarboxylic aciduria. J. Inher. Metab. Dis. *7*, Suppl. 1, 48–51.
162. Gregersen, G. (1985) The acyl CoA dehydrogenation deficiencies. Scand. J. Clin. Lab. Invest. *45*, Suppl. 174, 1–60.
163. Coates, P. M., Hale, D. E., Stanley, C. A., Corkey, B. E. & Cortner, J. A. (1985) Genetic deficiency of medium-chain acyl Coenzyme A dehydrogenase: Studies in cultured skin fibroblasts and peripheral mononuclear leukocytes. Pediatr. Res. *19*, 671–676.
164. Roe, C. R., Millington, D. S., Maltby, D. S., Bohun, T. P., Kahler, S. G. & Chalmers, R. A. (1985) Diagnostic and therapeutic implications of medium-chain acylcarnitines in the medium-chain acyl-CoA dehydrogenase deficiency. Pediatr. Res. *19*, 459–466.
165. Chalmers, R. A. & Lawson, A. M. (1979) Identification of 5-hydroxyhexanoic acid in the urine of twin siblings with a Reye's-like syndrome associated with dicarboxylic aciduria and hypoglycaemia and with similarities to Jamaican vomiting sickness. Biomed. Mass Spectrom. *6*, 444–446.
166. Mamer, O. A., Montgomery, J. A. & Colle, E. (1980) Profiles in altered metabolism III. Ω-1-hydroxyacid excretion in a case of episodic hypoglycaemia. Biomed. Mass Spectrom. *7*, 53–57.
167. Harrington, W., Lui, A. S., Lonsdale, D. & Igou, D. (1977) Urinary organic acid profiles of Reye's syndrome patients. Clin. Chim. Acta *74*, 247–253.
168. Karpati, G., Carpenter, S., Engel, A. G., Watters, G., Allen, J., Rothman, S., Klassen, G. & Mamer, O. A. (1975) The syndrome of systemic carnitine deficiency. Neurology *25*, 16–24.
169. Chalmers, R. A. (1984) Organic acids in urine of patients with congenital lactic acidoses: An aid to differential diagnosis. J. Inher. Metab. Dis. *7*, Suppl. 1, 79–89.
170. Chalmers, R. A., Lawson, A. M. & Borud, O. (1977) Gas chromatographic and mass spectrometric studies on urinary organic acids in a patient with congenital lactic acidosis due to pyruvate decarboxylase deficiency. Clin. Chim. Acta *77*, 117–124.
171. Van Biervliet, J. P. G. M., Bruinvis, L., Van der Heiden, C., Ketting, D., Wadman, S. K., Willems, J. L. & Monnens, L. (1977) Report of a patient with severe chronic lactic acidaemia and pyruvate carboxylase deficiency. Dev. Med. Child. Neurol. *19*, 392–401.
172. Chalmers, R. A., Ryman, B. E. & Watts, R. W. E. (1978) Studies on a patient with in vivo evidence of type I glycogenosis and normal enzyme activities in vitro. Acta Paediatr. Scand. *67*, 201–207.
173. Kuhara, T., Shinka, T., Inoue, Y., Matsumoto, M., Yoshino, M., Sakaguchi, Y. & Matsumoto, I. (1983) Studies of urinary organic acid profiles of a patient with dihydrolipoyl dehydrogenase deficiency. Clin. Chim. Acta *133*, 133–140.
174. Mamer, O. A. & Tjoa, S. S. (1974) 2-ethylhydracrylic acid: a newly described urinary organic acid. Clin. Chim. Acta *55*, 199–204.
175. Mamer, O. A., Tjoa, S. S., Scriver, C. R. & Klassen, G. A. (1976) Demonstration of a new mammalian isoleucine catabolic pathway yielding an R series of metabolites. Biochem. J. *160*, 417–426.
176. Sweeley, C. C., Elliot, W. A., Frits, I. & Ryhage, R. (1966) Mass spectrometric determination of unresolved components in chro-

matographic effluents. Anal. Chem. *38*, 1549–1553.
177. Hammer, C. G., Holmstedt, B. & Ryhage, R. (1968) Mass fragmentography. Identification of chlorpromazine and its metabolites in human blood by a new method Anal. Biochem. *25*, 532–535.
178. Millard, B. J. (1978) Quantitative mass spectrometry, Heyden, London.
179. Hill, R. E. & Whelan, D. T. (1984) Mass spectrometry and clinical chemistry. Clin. Chim. Acta *139*, 231–294.
180. Craig, R. D., Bateman, R. H., Green, B. N. & Millington, D. S. (1979) Mass spectrometry instrumentation for chemists and biologists. Philos. Trans. R. Soc. London *A293*, 135–155.
181. Bateman, R. H., Burns, P., Owen, R. & Parr, V. C. (1985) A novel general purpose double focussing mass spectrometer. Proceedings of the 10th International Mass Spectrometry Conference, Swansea, September 9–13, 1985.
182. Van den Berg, G. A., Breukelman, H., Elzinga, H., Trijbels, J. M. F., Monnens, L. A. H. & Muskiet, F. A. J. (1986) Determination of pipecolic acid in urine and plasma by isotope dilution mass fragmentography. Clin. Chem. Acta *159*, 229–237.
183. Hine, D. G., Hack, A. M., Goodman, S. I. & Tanaka, K. (1986) Stable isotope dilution analysis of isovalerylglycine in amniotic fluid and urine and its application for the prenatal diagnosis of isovaleric acidemia. Pediatr. Res. *20*, 222–226.
184. Mamer, O. A., Laschic, N. S. & Scriver, C. R. (1986) Stable isotope dilution assay for branched-chain alpha-hydroxy and alpha-ketoacids: serum concentrations for normal children. Biomed. Environ. Mass Spectrom. *13*, 553–558.
185. Haskins, N. J. (1982) The application of stable isotopes in biomedical research. Biomed. Mass Spectrom. *9*, 269–277.
186. Jakobs, C., Sweetman, L., Wadman, S. K., Duran, M., Saudubray, J.-M. & Nyhan, W. L. (1984) Prenatal diagnosis of glutaric aciduria type II by direct chemical analysis of dicarboxylic acids in amniotic fluid. Eur. J. Pediatr. *141*, 153–157.
187. Chalmers, R. A., Tracey, B. M., King, G. S., Pettit, B., Rocchiccioli, F., Saudubray, J.-M., Gray, R. G. F., Boué, J., Keeling, J. W. & Lindenbaum, R. H. (1985) The prenatal diagnosis of glutaric aciduria type II using quantitative GC-MS. J. Inher. Metab. Dis. *8*, Suppl. 2, 145–146.
188. Langenbeck. U., Hoinowski, A., Mantel, K. & Möhring, H. U. (1977) Quantitative gas chromatography and single ion detection of aliphatic α-keto acids from urine as their O-trimethylsilylquinoxalinol derivatives. J. Chromatogr. *143*, 39–50.
189. Langenbeck. U., Mench-Hoinowski, A., Dieckmann, K.-P., Möhring, H.-U. & Petersen, M. (1978) O-Trimethylsilylquinoxalinol derivatives of aromatic α-keto acids. Mass spectra and quantitative gas chromatography. J. Chromatogr. *145*, 185–193.
190. Rocchiccioli, F., Leroux, J. P. & Cartier, P. (1981) Quantitation of 2-ketoacids in biological fluids by gas chromatography-chemical ionization mass spectrometry of O-trimethylsilylquinoxalinol derivatives. Biomed. Mass Spectrom. *8*, 160–164.
191. Rocchiccioli, F., Leroux, J. P. & Cartier, P. (1981) Quantitative gas chromatography-chemical ionisation mass-spectrometry of 2-ketoglutarate from urine as its O-trimethylsilylquinoxalinol derivative. J. Chromatogr. *226*, 325–332.
192. Vogt, W., Jacob, K., Ohnesorge, A.-B. & Schwertfeger, G. (1980) Highly sensitive method for the quantitation of homovanillic acid in cerebrospinal fluid. J. Chromatogr. *199*, 191–197.
193. Davis, B. A., Durden, D. A. & Boulton, A. A. (1982) Plasma concentrations of p- and m-hydroxyphenylacetic acid and phenylacetic acid in humans. Gas chromatographic-high resolution mass spectrometric analysis. J. Chromatogr. *230*, 219–230.
194. Lind, T. (1973) The biochemistry of amniotic fluid. In: Amniotic Fluid, Research and Clinical Applications (Fairweather, D. V.I. & Eskes, T. K. A. B., eds.), pp. 60–81, Excerpta Medica, Amsterdam.
195. Sweetman, L. (1984) Prenatal diagnosis of the organic acidurias. J. Inher. Metab. Dis. *7*, Suppl. 1, 18–22.
196. Naylor, G., Sweetman, L., Nyhan, W. L., Hornbeck, C., Griffiths, J., Mörch, L. & Brandänge, S. (1980) Isotope dilution analysis of methylcitric acid in amniotic fluid for the prenatal diagnosis of propionic and methylmalonic acidemia. Clin. Chim. Acta *107*, 175–183.
197. Sweetman, L., Naylor, G., Ladner, T., Holm, J., Nyhan, W. L., Hornbeck, C., Griffiths, J., Mörch, L., Brandänge, S., Gruenke, L. & Craig, J. C. (1982) Prenatal diagnosis of propionic and methylmalonic acidemia by stable isotope dilution analysis of methylcitric and methylmalonic acids in

amniotic fluids. In: Stable Isotopes (Schmidt, H.-L., Förstel, H. & Heinzinger, J., eds.), pp. 287–293, Elsevier, Amsterdam.

198. Zinn, A. B., Hine, D. G., Mahoney, M. J. & Tanaka, K. (1982) The stable isotope dilution method for measurement of methylmalonic acid: a highly accurate approach to the prenatal diagnosis of methylmalonic aciduria. Pediatr. Res. *16*, 740–745.

199. Fensom, A. H., Benson, P. F., Chalmers, R. A., Tracey, B. M., Watson, D., King, G. S., Pettit, B. R., & Rodeck, C. H. (1984) Experience with prenatal diagnosis of propionic acidaemia and methylmalonic aciduria. J. Inher. Metab. Dis. *7*, Suppl. 2, 127–128.

200. Jakobs, C., Sweetman, L. & Nyhan, W. L. (1984) Hydroxyacid metabolites of branched-chain amino acids in amniotic fluid. Clin. Chim. Acta *140*, 157–166.

201. Jakobs, C., Sweetman, L., Nyhan, W. L. & Packman, S. (1984) Stable isotope dilution analysis of 3-hydroxyisovaleric acid in amniotic fluid: Contribution to the prenatal diagnosis of inherited disorders of leucine catabolism. J. Inher. Metab. Dis. *7*, 15–20.

202. Chalmers, R. A., Tracey, B. M., Mistry, J., Stacey, T. E., McFadyen, I. R. & Madigan, M. J. (1985) The prenatal diagnosis of 3-hydroxy-3-methylglutaric aciduria by GC-MS and enzymology on cultured amniocytes and on chorionic villi. Pediatr. Res. *19*, 1075 (A).

203. Pettit, B. R., MacKenzie, F., King, G. S. & Leonard, J. V. (1984) The antenatal diagnosis and aid to the management of hereditary tyrosinaemia by use of a specific and sensitive GC-MS assay for succinylacetone. J. Inher. Metab. Dis. *7*, Suppl. 2, 135–136.

204. Atkins, G. L., Dean, B. M., Griffin, W. J. & Watts, R. W. E. (1964) Quantitative aspects of ascorbic acid metabolism in man. J. Biol. Chem. *239*, 2975–2980.

205. Atkins, G. L., Dean, B. M., Griffin, W. J., Scowen, E. F. & Watts, R. W. E. (1965) Quantitative aspects of ascorbic acid metabolism in patients with primary hyperoxaluria. Clin. Sci. *29*, 305–314.

206. Dean, B. M., Watts, R. W. E. & Westwick. W. J. (1968) The conversion of [1-^{13}C] glycine and [2-^{13}C] glycine to [^{13}C] oxalate in primary hyperoxaluria: Evidence for the existence of more than one metabolic pathway from glycine to oxalate in man. Clin. Sci. *35*, 325–335.

207. Curtius, H. C., Völlmin, J. A. & Baerlocher, K. (1972) The use of deuterated phenylalanine for the elucidation of the phenylalanine-tyrosine metabolism: In: Organic Acidurias (Stern, J. & Toothill, C., eds.), pp. 146–158, Livingstone, Edinburgh.

208. Curtius, H. C., Völlmin, J. A. & Baerlocher, K. (1973) Study of metabolic pathways in vivo using stable isotopes. Anal. Chem. *45*, 1107–1110.

209. Curtius, H. C., Mettler, M. & Ettlinger, L. (1976) Study of the intestinal tyrosine metabolism using stable isotopes and gas chromatography-mass spectrometry. J. Chromatogr. *126*, 569–580.

210. Fell, V., Hoskins, J. A. & Pollitt, R. J. (1978) The labeling of urinary acids after oral doses of deuterated L-phenylalanine and L-tyrosine in normal subjects. Quantitative studies with implications for the deuterated phenylalanine load test in phenylketonuria. Clin. Chim. Acta *83*, 259–269.

211. Faull, K. F., Gan, I., Halpern, B., Hammond, J., Im, S., Cotton, R. G. H., Danks, D. M. & Freeman, R. (1977) Metabolic studies on two patients with nonhepatic tyrosinaemia using deuterated tyrosine loads. Pediatr. Res. *11*, 631–637.

212. Buist, N. R. M., Fellman, J. H. & Kennaway, N. (1978) Letter to the Editor: Metabolic Studies in Tyrosinemia. Pediatr. Res. *12*, 56–57.

213. Wadman, S. K., Duran, M., Ketting, D., Bruinvis, L., van Sprang, F. J., Berger, R., Smit, G. P. A., Steinmann, B., Leonard, J. V., Divry, P., Fariaux, J. P. & Cartigny, B. (1983) Urinary excretion of deuterated metabolites in patients with tyrosinemia type I after oral loading with deuterated L-tyrosine. Clin. Chim. Acta *130*, 231–238.

214. Shimamura, M., Kamada, S., Hayashi, T., Naruse, H. & Iida, Y. (1986) Sensitive determination of tyrosine metabolites, p-hydroxyphenylacetic acid, 4-hydroxy-3-methoxyphenylacetic and 4-hydroxy-3-methoxymandelic acid by gas chromatography-negative-ion chemical ionization-mass spectrometry. Application to stable isotopes labeled tracer experiments to investigate their metabolism in man. J. Chromatogr. *374*, 17–26.

215. Tanaka, K., Armitage, I. M., Ramsdell, H. S., Hsia, Y. E., Lipsky, S. R. & Rosenberg, L. E. (1975) [^{13}C]valine metabolism in methylmalanicacidemia using nuclear magnetic resonance: Propionate as an obligate

intermediate. Proc. Natl. Acad. Sci. USA *72*, 3692–3696.

216. Baretz, B. H. & Tanaka, K. (1978) Metabolism in rats in vivo of isobutyrates labeled with stable isotopes at various positions. Identification of propionate as an obligate intermediate. J. Biol. Chem. *253*, 4203–4213.

217. Amster, J. & Tanaka, K. (1980) Metabolism in rats in vivo of (2S) [3,3,3-$^3H_3$]-isobutyrate. J. Biol. Chem. *255*, 119–120.

218. Neese, R. A., Gertz, E. W., Wisneski, J. A., Gruenke, L. D. & Craig, J. C. (1983) A stable isotope technique for investigating lactate metabolism in humans. Biomed. Mass Spectrom. *10*, 458–462.

219. Mazzeo, R. S.,. Brooks, G. A., Schoeller, D. A. & Budinger, T. F. (1986) Disposal of blood (1-13C) lactate in humans during rest and exercise. J. Appl. Physiol. *60*, 232–241.

220. Bourgnères, P. F., Balasse, E. O., Ferne, P. & Bier, D. M. (1986) Determination of ketone body kinetics using a D-(–)-3-hydroxy (4,4,4-2H3) butyrate tracer. J. Lipid Res. *27*, 215–220.

221. Wolfe, R. R., Evans, J. E., Mullany, C. J. & Burke, J. F. (1980) Measurement of plasma free fatty acid turnover and oxidation using [1-^{13}C] palmitic acid. Biomed. Mass Spectrom. *7*, 168–171.

222. Kneer, T., Brockstedt, M., Paust, H., Jakobs, C., Penn, D., Mönch, E. & Helge, H. (1985) Abstracts of the 22nd Meeting of the Society for the Study of Inborn Errors of Metabolism, Newcastle, 1984; Abstract P39; J. Inher. Metab. Dis. *8*, Suppl. 2, 94. (Title)

223. Roe, C. R. & Bohun, T. P. (1982) L-carnitine therapy in propionic acidaemia. Lancet *1*, 1411 (C).

224. Allan, R. J., Hansch, D. B. & Wu, H. L. C. (1982) Hypocarnitinaemia in disorders of organic acid metabolism. Lancet *2*, 500 (C).

225. Seccombe, D. W., Snyder, F. & Parsons, H. G. (1982) L-carnitine for methylmalonic aciduria. Lancet *2*, 1401 (C).

226. Roe, C. R., Hoppel, C. L., Stacey, T. E., Chalmers, R. A., Tracey, B. M. & Millington, D. S. (1983) Metabolic response to carnitine in methylmalonic aciduria: an effective strategy for elimination of propionyl groups. Arch. Dis. Child. *58*, 916–920.

227. Chalmers, R. A., Roe, C. R., Stacey, T. E. & Hoppel, C. L. (1984) Urinary excretion of *l*-carnitine and acylcarnitines by patients with disorders of organic acid metabolism: Evidence for secondary insufficiency of *l*-carnitine. Pediatr. Res. *18*, 1325–1328.

228. Millington, D. S. & Smith, J. A. (1977) Fragmentation patterns of fast linked electric and magnetic field scanning. Org. Mass Spectrom. *12*, 264–265.

229. Millington, D. S., Roe, C. R. & Maltby, D. A. (1984) Applications of high resolution fast atom bombardment and constant B/E ratio linked scanning to the identification and analysis of acylcarnitines in metabolic disease. Biomed. Mass Spectrom. *11*, 236–241.

230. Chalmers, R. A., Stacey, T. E., Tracey, B. M., de Sousa, C., Roe, C. R., Millington, D. S. & Hoppel, C. L. (1984) L-carnitine insufficiency in disorders of organic acid metabolism: response to L-carnitine by patients with methylmalonic aciduria and 3-hydroxy-3-methylglutaric aciduria. J. Inher. Metab. Dis. *7*, Suppl. 2, 109–110.

231. Roe, C. R., Millington, D. S. & Maltby, D. A. (1986) Identification of 3-methylglutarylcarnitine. A new diagnostic metabolite of 3-hydroxy-3-methylglutaryl-Coenzyme A lyase deficiency. J. Clin. Invest. *77*, 1391–1394.

232. Chalmers, R. A., Tracey, B. M., Stacey, T. E., Cheng, K. N., Madigan, M. J. & Lawson, A. M. (1986) The use of fast atom bombardment mass spectrometry to identify and study urinary acylcarnitines in disorders of organic acid metabolism. Biochem. Soc. Trans. *14*, 967–969.

233. Roe, C. R., Millington, D. S., Maltby, D. A., Bohun, T. P. & Hoppel, C. L. (1984) L-carnitine enhances excretion of propionyl Coenzyme A as propionylcarnitine in propionic acidemia. J. Clin. Invest. *73*, 1785–1788.

234. Roe, C. R., Millington, D. S., Maltby, D. S., Kahler, S. G. & Bohun, T. P. (1984) L-carnitine therapy in isovaleric acidemia. J. Clin. Invest. *74*, 2290–2295.

235. Duran, M., Ketting, D., Dorland, L. & Wadman, S. K. (1985) The identification of acylcarnitines by desorption chemical ionisation mass spectrometry. J. Inher. Metab. Dis. *8*, Suppl. 2, 143–144.

236. Duran, M., Mitchell, G., de Klerk, J. B., de Jager, J. P., Hofkamp, M., Bruinvis, L., Ketting, D., Saudubray, J. M. & Wadman, S. K. (1985) Octanoic acidemia and octanoylcarnitine excretion with dicarboxylic aciduria due to defective oxidation of medium-chain fatty acids. J. Pediatr. *107*, 397–404.

237. Duran, M., Ketting, D., Beckeringh, T. E., Leupold, D. & Wadman, S. K. (1986) Direct identification of propionylcarnitine in pro-

pionic acidaemia: Biochemical and clinical results of oral carnitine supplementation. J. Inher. Metab. Dis. *9*, 202–207.
238. Yergey, A. L., Liberato, D. J. & Millington, D. S. (1984) Thermospray liquid chromatography/mass spectrometry for the analysis of L-carnitine and its short-chain acyl derivatives. Anal. Biochem. *139*, 278–283.
239. Millington, D. S., Bohun, T. P., Roe, C. R., Yergey, A. L. & Liberato, D. J. (1985) Valproylcarnitine: a novel drug metabolite identified by fast atom bombardment and thermospray liquid chromatography-mass spectrometry. Clin. Chim. Acta *145*, 69–76.
240. Gaskell, S. J., Guenat, C., Millington, D. S., Maltby, D. A. & Roe, C. R. (1986) Differentiation of isomeric acylcarnitines using tandem mass spectrometry. Anal. Chem. *58*, 2801–2805.

Chapter 7
Endogenous Brain Peptides

D. M. Desiderio

1 Introduction

The objective of this chapter is to describe state-of-the-art analytical instrumental methodology used to measure endogenous brain peptides with high specificity (1). The measurement process is based on an off-line combination of high performance liquid chromatography (HPLC) and mass spectrometry (MS) and was devised to investigate the molecular basis of stress and pain, especially in relation to neuropeptides. The methodology involves several aspects including the rapid acquisition of tissue, the careful biochemical extraction of a peptide-rich fraction, reversed phase (RP) HPLC to separate peptides in either the gradient or isocratic elution mode, the production of a protonated molecular ion $[M + H]^+$ of the peptide of interest by means of fast atom bombardment (FAB) MS, unimolecular or collision activated dissocation (CAD) processes to produce an amino acid sequence-determining ion that is unique to the peptide being measured, synthetic stable isotope-incorporated peptide internal standards for quantification, a linked-field (B/E) scan in the selected ion monitoring (SIM) mode to monitor that amino acid sequence-determining ion, and a minicomputer to collect and analyze the resulting mass spectrometric analog data. Currently, the detection level of this overall analytical procedure is at the picomolar level (ng peptide $\cdot$ g^{-1} tissue) for biological tissue extracts and the femtomolar level for solutions of synthetic peptides. Every effort is made to maximise the molecular specificity of measurement and a comparison is drawn with other commonly-used analysis methods including HPLC, radioimmunoassay (RIA), and radioreceptor assay (RRA). Commercial antibodies were used for RIA of methionine enkephalin (ME = YGGFM) and leucine enkephalin (LE = YGGFL), and a novel canine limbic synaptosomal preparation used for RRA, in a semi-quantitative mode or as a "screen" for the detection of opioid peptides that elute from an HPLC column.

2 Peptides

2.1 Neuroregulatory peptides

Peptides are well-suited to act as an interface that efficiently communicates between an external environment and a cell or its constituents. One reason for such an effective role in communication stems from the efficient storage of information in a peptide molecule. It is instructive to calculate the information density that exists in a peptide compared to other media. For example, a six-bit computer word can carry 2^6 or 64 different pieces of information, a hexanucleotide 4^6 or 4096 pieces. In contrast, a hexapeptide can carry 20^6 or 64 000 000 different pieces of information – a much greater information density for a comparable molecular size. This section will focus on only a few aspects of biologically important peptides, namely the opioid peptides.

Neuroregulators play a key role in communication between nerve cells and as a class of compounds can be subdivided into those that convey information between

Table 1. Amino acid sequences of selected biologically important neuropeptides

ACTH 1–39 (human)	SYSMEHFRWGKPVGKKRRPVKVYPN
α-Neoendorphin	YGGFLRKYPK
Angiotensin II	DRVYIHPF
α-MSH	SYSMEHFRWGKPV
β-Lipotropin (human)	ELAGAPPEPARDPEAPAEGAAARAELEYGLVAEAQA AEKKDEGPYKMEHFRWGSPPKDKRYGGFMTSEKS QTPLVTLFKNAIVKNAHKKGO
Bombesin	ZQRLGNQWAVGHLM
Bradykinin	RPPGFSPFR
Corticotropin releasing factor (ovine)	SQEPPISLDLTFHLLREVLEMTKADQLAQQAH SNRKLLDIA–NH_2
Dynorphin 1–8	YGGFLRRI
Dynorphin 1–13	YGGFLRRIRPKLK
Dynorphin 1–17	YGGFLRRIRPKLKWDNQ
Dynorphin 1–32	YGGFLRRIRPKLKWDNQKRYGGFLRRQFKVVT
Dynorphin A	YGGFLRRIRPKLKWDNQ
Dynorphin B	YGGFLRRQFKVVT
α-Endorphin	YGGFMTSEKSQTPLVT
β-Endorphin (1–31)	YGGMTSEKSQTPLVTLFKNAIIKNAYKKGQ
Leucine enkephalin	YGGFL
Methionine enkephalin	YGGFM
Met-Enkephalin sulfoxide	YGGFM (O)
Gastrin I (human)	ZGPWLEEEEEAYGWMDF
Insulin A (bovine)	GIVEQCCASVCSLYQLENYCN
Insulin B (bovine)	FVNQHLCGSHLVEALYLVCGERGFFYTPKA
Substance P	RPKPQQFFGLM
Arg Vasopressin	CYFQNCPRG–NH_2
Neurotensin	ZLYENKPRRPYIL
Oxytocin	CYIQNCPLG
Somatostatin	AGCKNFFWKTFTSC

The single letter amino acid code used for the amino acids is: A = ala, C = cys, D = asp, E = glu, F = phe, G = gly, H = his, I = ile, K = lys, L = leu, M = met, N = asn, P = pro, Q = gln, R = arg, S = ser, T = thr, V = val, W = trp, and Y = tyr.

adjacent nerve cells (neurotransmitters) and those that either amplify or decrease neuronal firing activity (neuromodulators) (2). In a sense, the information that relates to neurotransmitters derives from previous experience with classical neurotransmitters, whereas more recent data concerning neuropeptides from the basis for the neuromodulators.

Table 1 lists the amino acid sequences of some of the biologically important peptides that are discussed in this chapter.

Classically, it was thought that only one neurotransmitter was released from a presynaptic neuron to act with its target receptor and then be rapidly inactivated by metabolic degradation. It was found more recently, however, that the opiomelanotropinergic neuron system secretes at least seven peptides, which in turn may be readily biotransformed into as many as five bioactive peptides (3). These endorphin and

MARFLGLCTWLLALGPGLLATVRAECSQDC-
ATCSYRLARPTDLNPLACTLESEGKLPSLK-
TWETCKELLQLTKLELPPDATSALSKQEES-
HLLAKKYGGFMKRYGGFMKKMDELYPLEVE-
EEANGGEVLGKRYGGFMKKDAEEDDGLGNS-
SNLLKELLGALDQREGSLHQEGSDAEDVSK-
RYGGFMRGLKRSPHLEDETKELQKRYGGFM-
RRVGRPEWWMDYQKRYGGFLKRFAEPLPSE-
EEGESYSKEVPEMEKRYGGFMRF

Bovine Adrenal preproenkephalin

Figure 1. Amino acid sequence of bovine adrenal preproenkephalin. The amino acid sequences of four ME (YGGFM), one LE (YGGFL), and two C-terminally extended ME peptides (YGGFMRGL, YGGFMFR) are underlined. These seven enkephalin peptides are each flanked by pairs of basic amino acids, except the C-terminus YGGFMRF. See footnote to Table 1 for amino acid code. (Reproduced from ref. 1 with permission.)

melanocyte stimulating hormone (MSH) peptides have several biologically important activities. Theoretical and functional analyses are needed to understand why one cell would secrete multiple transmitters. A variety of post-translational processing events of propiomelanocortin (POMC) occurs that includes glycosylation, *O*-phosphorylation, acetylation, amidation, and perhaps others. This multiplicity of post-translational peptide structure modifications is one of the reasons that clearly demonstrates the need for unambiguous molecular specificity during an analytical measurement of peptides. Several different structural modifications to a peptide can provide compounds of different structures, but which may elicit equivalent responses from bioassay (BA), RRA, or RIA.

A current hypothesis in peptide neurochemistry is that the central nervous system (CNS) contains four distinct opioid peptidergic systems: enkephalinergic, endorphinergic, tachykininergic, and dynorphinergic, and that these independent systems are available to a cell to deal with stress, noxious and/or pleasurable stimuli. Figure 1 contains the sequence of the precursor proenkephalin A from bovine adrenal.

2.2 Peptide degradation

When naturally-occuring enkephalins are injected into a cell, they may be degraded rapidly by endogenous proteolytic enzymes. Replacing various amino acid residues of the enkephalin molecule by their stereoisomers or other amino acids greatly inhibits that enzymatic activity. For example, placing D-alanine at position 2 and *N*-methyl phenylalanine at position 4 and modifying the structure of the methionine at position 5 by oxidation produces an enkephalin analog that is 30 000 times more potent than ME. Peptide analog studies are very important because of the information that is gleaned from their properties in relation to tolerance and physical dependence. Tolerance is defined as a decreased responsiveness to a fixed dose of a drug after continuously repeating that administration, whereas physical dependence causes withdrawal symptoms once the drug is no longer taken.

Peptidases located in the brain are mainly found in lysosomes that are normally separate from the synaptic vessels where neuroactive peptides are presumably stored (4). Although the method of animal sacrifice should not alter that separation aspect, homogenization of tissues and/or cell disruption will bring into contact the components of these two structures. Thus it is important to recognise the experimental importance of irreversible inactivation of peptidases, prior to tissue homogenization and analysis. Heat is a common method for enzyme inactivation. Microwave irradition (1.33 kW, 2.45 GHz) proved to be a less satisfactory method than decapitation and boiling of the intact tissue. The latter method indicated that no α-endorphin was present, and that neither a higher morphine treatment nor morphine withdrawal induced significant change of this concentration (5).

2.3 Dynorphin

Dynorphin$_{1-13}$ (Table 1) is a peptide that displays typical opioid activity in several different assay systems and is extremely potent; 700 times more potent than LE, 200 times more than morphine, and 50 times more than β_C-endorphin. Naloxone has less potency ($^1/_{13}$) in blocking this peptide compared to either normorphine or LE. Dynorphin$_{1-13}$ competes variably with opiate alkaloid and peptide radioligands for stereospecific binding to rat brain membranes and, whereas it is not as highly potent, that reduced activity may be due to rapid degradation of the dynorphin by a membrane-bound enzyme. It is significant to realize that dynorphin$_{1-13}$ is a C-terminal octapeptide extension of LE, where that octapeptide extension (-arg-arg-ile-arg-pro-lys-leu-lys) contains five basic amino acid residues. The N-terminus of dynorphin$_{1-13}$ is necessary for biological potency because N-acetylation abolishes bioactivity. By deleting the ^{6}arg residue, the remaining carboxy terminus is shifted towards the N-terminus by 500 picometers (0.5 nm), whereas inserting a ^{6}gly shifts the remaining amino acid sequence towards the C-terminus by that same distance. The loss of potency in both of these dynorphin analogs demonstrates that the unique amino acid sequence in the octapeptide extension and the linear distance of the octapeptide sequence from the LE N-terminal pentapeptide are critically important structural parameters for optimal potency, and also confirm that the spatial relationships within the opiate receptor must be rigidly maintained within 0.5 nm (6).

The amino acid sequence of dynorphin$_{1-17}$ extracted from porcine pituitary is tyr-gly-gly-phe-leu-arg-arg-ile-arg-pro-lys-leu-lys-trp-asp-asn-gln. The full biological activity of this heptadecapeptide resides in the N-terminal tridecapeptide sequence. The usually susceptible ^{13}lys-^{14}trp bond is resistant to digestion with trypsin. Although trypsin-resistant lys bonds are rare, nevertheless, they have been reported (6). The resistance of this specific peptide bond to trypsinolysis may reflect some unique microenvironmental aspect of the secondary structure of this heptadecapeptide; one possibility is the hydrophobicity of the trp residue. Improvements in isolation methodology that can be made include the avoidance of lyophilization, where large losses of peptide had been noted, use of detergents to minimize absorption losses, monitoring peak effluents by RIA instead of BA, and use of an immunoaffinity chromatography column.

The existence of a highly specific dynorphin receptor in the guinea pig myenteric plexus and brain is known, where that receptor is presumably the same as the κ opioid receptor with which certain benzomorphans preferentially bind. Data were presented to support the hypothesis that dynorphin is a specific ligand for the κ subclass of opioid receptors (7).

During the isolation of the dynorphin heptadecapeptide, neutralization of the acetic acid extract precipitated most (60 – 70%) of the immunoreactive dynorphin, indicating that a larger dynorphin molecule was also present in that preparation (8). HPLC purification of the larger dynorphin precursor, followed by automatic Edman degradation, yielded the following amino acid sequence information that corresponded to a peptide with a molecular mass of 3986: tyr-gly-gly-phe-leu-arg-arg-ile-arg-pro-lys-leu-lys-trp-asp-asn-gln-lys-arg-tyr-gly-gly-phe-leu-arg-arg-gln-phe-lys-val-val-thr. This dynorphin$_{1-32}$ is one of the more interesting opioid peptide precursors that has been found to date and it is a precursor for several related peptides that are discussed below.

The peptide bond between ^{18}lys and ^{18}arg in this large dynorphin precursor is significant because trypsin-like cleavage of that bond generates an amino terminal heptadecapeptide (dynorphin A) and a carboxy terminal tridecapeptide (dynorphin B). The latter peptide, dynorphin B, represents the third "big" LE besides dynorphin A and α-neoendorphin (tyr-gly-gly-phe-leu-arg-lys-tyr-pro-lys). The dynorphin B molecule surprisingly displays the same high κ receptor activity as dynorphin A, a fact that indicates that the "address" portion of a peptide that is needed for receptor recognition and binding most probably resides in the *N*-termini of the two peptides. The three fragments have been mapped in rat brain with immunocytochemical techniques (9).

As stated above, recent biochemical and anatomical studies provide evidence that three of the opioid peptidergic systems in the brain are comprised of the enkephalin, dynorphin, and β-endorphin systems (10). The enkephalin and dynorphin peptidergic systems occur in many of the same areas, the enkephalin system but not the dynorphin system occurs in a large number of areas, and a few areas of nuclei contain the dynorphin but not the enkephalin system (11).

2.4 Concentrations of endogenous peptides

In general, peptide hormones are present in the brain at the level of pmol peptide · mg^{-1} protein compared to classical biogenenic amine neurotransmitters (nmol · mg^{-1}) and amino acid neurotransmitters (nmol · mg^{-1}) (12, 13). It is important to realize that the CNS concentration of peptides that is measured in homogenized tissues may not reliably indicate either the microenvironmental functional importance or the turnover characteristics of those peptides because regional, cellular, and subcellular distributions of peptides are probably of more *in vivo* biological relevance. Although some researchers believe that disparate tissues such as brain and gut derive from the neural crest and that this common source is the developmental reason why similar peptides are found at these two distant locations, other workers have shown that this rationalization may not be the case. There are several modes that are possible for peptide secretion in the gastrointestinal (GI) tract, pancreas, pituitary, and neural tissue. As mentioned above, neurotransmitters or neuromodulators may occur in the neuronal system. However, if the neuronal product is released into the blood stream, then it is a neurohormone.

A study involving regional concentrations of selected peptides indicated noticeable and, in some cases, marked concentration differences. For example, somatostatin (in pmol peptide $\cdot$ g^{-1} wet weight tissue) ranged from 1000 in the hypothalamus, 23 in the substantia nigra, and 1 in the cerebral cortex, whereas substance P had a maximum concentration of 922 in the substantia nigra, 518 in the globus pallidus, and 0.3 in the cerebellar cortex. These values differ from those of neurotensin, which were at a maximum in the hypothalamus median eminence (43 pmol), less in the substantia nigra (23 pmol) and least in the cerebella cortex (0.8 pmol) (12, 13). One can infer from this that regional distribution of these peptides is required to provide a particular microenvironment to act in an appropriate neurochemical fashion (with a receptor, precursor formation or metabolism, packaging for transport, etc.).

3 Techniques

This section describes several experimental factors that must be controlled during biochemical studies of endogenous peptides.

3.1 Time factors

The extremely rapid (ms) time-course of the biochemical events involved in pain transmission make it necessary to consider carefully the potential and observed time-dependent factors that can occur in any of the steps between the intact animal model, tissue procurement, compound purification, and analytical measurement. The most significant temporal factors include enzymolysis/hydrolysis, thioether oxidation/reduction, and other post-translational structural modifications (amidation, acetylation, phosphorylation, sulfate ester formation, etc.) of endogenous peptides.

After exsanguination, decapitation, or microwave treatment of the experimental animal, it is important to extract rapidly and cool (liquid nitrogen, dry ice acetone) the target tissue or fluid to minimize any biochemical interconversions. Experience with endogenous prostaglandins (14–16) indicates that rapid temperature lowering is required to ensure that only endogenous compounds remain present in this critical first step and that artefact formation is avoided. Enzymatic interconversion of peptides can also occur unintentionally, or even unknowingly by mobilizing the nociceptive process itself, an event that precipitates a cascade of biochemical events. The first step in the production of opioid peptides involves, in one particular pathway, a preproenkephalin molecule containing 264 amino acids that includes six copies of ME and one copy of LE. These enkephalin pentapeptides are bracketed in the precursor molecule by pairs of the basic amino acids, lys (K) and arg (R), that effectively provide target sites for trypsin-like activity. Subsequent carboxypeptidase B-like enzymatic activity then trims the peptide. Once enzymic excision of the selected peptide occurs, the enkephalin pentapeptides become sensitive to the further action of enkephalinases, aminopeptidases, dipeptidyl-aminopeptidases (cathepsins), and perhaps other, yet-to-be-described, peptidases. Exsanguination removes the blood enzymes, and if the tissue is not damaged, tissue proteases will not be released to initiate proteolysis.

It cannot be overemphasized that great care must be exercised in deriving the biological tissue that will be used for time-consuming and expensive analytical measurements upon which a number of physiological and clinical conclusions will be based, and which will be the source for scientific verification in many other laboratories around the world for years to come.

3.2 Post-translational modifications of peptides

Another set of biological events that can alter the structure of a target peptide includes chemical, biochemical, or enzymatic post-translational modifications to specific amino acid residues. These structural modifications of biological peptides demonstrate the presence of an elegant series of effective metabolic control points that are available to an animal to "fine-tune" its opioid peptide metabolic system to maintain efficient homeostatic control, or to rapidly mobilize a potential reservoir of peptide. In addition to the enzymolysis that is mentioned above for enkephalins, *N*-terminal tyrosine residues may be sulfated; amidation of the *C*-terminus, aspartic (D), or glutamic (E) residues may occur; and the amino terminus may be acetylated. For example, the modified *N*-acetylated β-endorphin$_{1-27}$ is the predominant immunoreactive species in brain tissue, even though it has no opioid reactivity and does not interact with the appropriate opiate receptor (17). This example of the endorphin molecule illustrates the type of severe experimental limination that may be experienced when only an immunoassay is used to measure a biologically important molecule. What is needed is an assay to monitor the entire structure of a molecule, and not just a selected interaction (with an antibody, or a receptor) of only part of the molecule. Lastly, the methylthioether group of a methionine amino acid residue is sensitive to oxidation, and may convert readily to a methionine sulfoxide group during extraction procedures if appropriate precautions are not taken. The more hydrophilic sulfoxide compound elutes at a shorter retention time under reverse phase (RP)-HPLC conditions. A known methionine sulfoxide reductase system may also exert metabolic control (18).

This brief discussion of selected examples illustrates the presence of an elegant and sensitive fine-tuning metabolic apparatus that is available in a tissue for rapid packaging, synthesis, and cleavage to produce the "working opioid peptide", which is then rapidly metabolized to inactive metabolites. This latter point is important to the neurotransmitter roles that the endogenous opioid peptides may play.

3.3 Species differences

Whether choosing an animal model for development of an analytical measurement method, or measuring a particular target compound in a tissue extract, it should be remembered that species differences may exist in peptide structure (19). Depending on the conservation of amino acid sequences across different species, certain amino acid residues may be replaced by others and these may be significant in relation to particular assay methods. For example, it may be possible that a RIA antibody may be sensitive to that portion of the molecule that underwent an amino acid replacement. The BA and RRA may or may not be more tolerant of such amino acid replacements, but again, the neurochemical differences must be kept in mind. The MS methodology for analysis discussed below is one of only two analytical measurement procedures that is highly sensitive to such structural differences – the other is amino acid analysis (Edman, or gas phase).

Species differences are also manifested in neuroanatomical relationships. For example, the pituitary is different in the human (no intermediate) versus other animals and important neurochemical differences also exist.

3.4 Method of sacrifice

The method chosen to sacrifice the animal also plays a significant role in peptide analysis, especially as it pertains to the overall metabolic cascade of various peptides that are produced under specific metabolic demands and by individual proteolytic enzymes. The microwave process for sacrifice is certainly rapid if it is appropriately focused and of sufficiently high energy. This method is restricted to small animals, in particular to the rodent. Furthermore, intense temperature in the internal brain structure may liberate unwanted lysosomal enzymes that may destroy the target compound. Nevertheless, for a small experimental animal, this methodology is extremely rapid.

Decapitation, especially of the rodent family, is perhaps the most satisfactory method for acquiring brain tissue. Although some cell damage may occur in the plane of decapitation, the other internal brain structures are relatively well-preserved and enzymatic degradation of the peptides of interest is minimal.

Sacrifice by exsanguination is suitable for studying the effect of blood levels and blood enzymes and for larger animals such as the dog, it is a very efficient means of obtaining tissue.

The "emotional state" of the animals at the moment of sacrifice must also be considered (20). Although this experimental factor is difficult to rationalize or measure and may not appear to be an important parameter when working with the lower order animals, it may be a very significant factor in an animal such as the dog. The biochemical and enzymatic disposition of metabolic precursors may be affected by emotional state, especially the larger opioid peptide precursors and the variety of proteolytic enzymes that are metabolically active during stress.

3.5 Internal brain structure

For neurochemical studies of brain peptides, it is important that a variety of internal brain structures (21) be neuroanatomically identified, neurosurgically removed, and rapidly frozen, a three-step process that is carried out quickly to minimize any potential metabolic conversions. Depending on the particular physiological or neurological parameter(s) to be investigated, any number of the internal brain structures may be excised. For example, if movement disorders (Parkinsonism, Huntington chorea) are of interest, then the basal ganglia are used, for elucidation of nociceptive processes, the spinal cord and thalamus, and for neuro-endocrinological interelationships, the hypothalamus and pituitary.

3.6 Neuropeptidase activity

Much work has been done recently on studying the enzymes that act upon neuropeptides and their metabolic precursors. The precursors generally contain pairs of lys and arg, which are sensitive to trypsin and/or trypsin-like enzymatic activity. Once this particular enzyme activity hydrolyzes a specific peptide bond in the precursor, carboxypeptidase B-like activity removes the remaining second basic *C*-terminal amino acid residue. Bioactive peptides such as β-endorphin, dynorphin, and the pentapeptide enkephalins are produced by these enzymatic processes, and the peptides interact with their respective receptors to presumably release a second messenger and to evoke a biological response. Following this receptor interaction, a variety of peptidases then acts upon the neuropeptide to rapidly and efficiently remove the active peptide from circulation. These peptidases include aminopeptidases, carboxypeptidases, and specific enkephalinases. Several inhibitors of peptidases may be employed for particular clinical or pharmacological studies. For example, thiorphan is known to be a specific inhibitor of enkephalinase (22); *N*-ethyl-maleimide, a sulfhydryl reagent, is an irreversible chemical method to inhibit particular enzymes; and bestatin and bacitracin are protease inhibitors. Alteration of the temperature and pH conditions may also suffice for enzyme inactivation.

Enzymatic activity has to be considered when providing biological tissue for particular studies. For example, if the objective of the experiment is to measure the amount of endogenous peptide by HPLC, MS, BA, or RIA, then it is important that the peptidase activity be minimized to provide an accurate measurement of the peptide *in vivo* at the moment of sacrifice. On the other hand, if RRA is the objective, then it is necessary to provide a population of biologically active receptors to interact with their appropriate endogenous ligands or with those ligands that were extracted and then purified by HPLC. In the case of RRA, enzymatic inhibitors may be utilized to retain the peptides to be measured but protease inactivators such as *N*-ethyl-maleimide would not be suitable because they irreversibly bind also to the receptor. Lastly, if the study of precursors is the objective of the investigation, then tissue homogenization must be performed in a neutral aqueous medium and not in an acidic organic medium that may precipitate larger peptides.

4 Reversed Phase Chromatography of Peptides

This section discusses some of the basic principles of RP-HPLC applied to peptides and provides several examples of peptide separation.

4.1 Peptide separations

Brain peptides have different relative hydrophobicities and consequently, different retention times on a RP-HPLC column when perfluorinated alkanoic acids (for example, trifluoroacetic and heptafluorobutyric acids) are utilized as the buffer. These differential hydrophobic effects can be used to advantage in the isolation and purification of endogenous peptides from biological sources (23). Two major forms of adrenocorticotropic hormone (ACTH) from the rat anterior pituitary were isolated using only RP chromatographic methodology and an extraction medium with an acidic pH (1 $mol \cdot l^{-1}$ HOAc at 4°C) to inhibit peptidase activity. The HPLC-grade water used in this study was prepared from de-ionized, glass-distilled water that was subsequently passed at a slow rate of elution under gravity over a small bed of octadecylsilyl (ODS) silica that was taken from one C_{18} Sep-Pak cartridge. Pituitaries were removed following decapitation, dissected into anterior and neurointermediate lobes, and rapidly (within one minute) homogenized (4°C). Supernatants were combined and passed five times through C_{18} Sep-Paks. Plasma was obtained, and 3 ml portions were extracted by passing them ten times through cartridges. These "multi-passages" through Sep-Paks were used to provide sufficient capacity of the ODS RP packing. Recoveries of the various forms of peptide were very high.

Seven synthetic peptides were separated by HPLC acetonitrile gradients in trifluoracetic (0.01 $mol \cdot l^{-1}$) or heptafluorobutyric acid (0.01 $mol \cdot l^{-1}$). Several differences were apparent in the two HPLC chromatograms, the most important being the order of elution of the peptide standards between the two chromatograms. The difference in chromatographic elution behavior was due to the hydrophobicity of the counter-ion, the perfluorinated carboxylic acid. The retention times of all peptides were greater in the heptafluorobutyric acid chromatogram compared to the smaller carboxylic acid. Peptides with a greater number of basic groups at pH 2 show large increases in retention times when comparing the heptafluorobutyric to the trifluoracetic acid system whereas peptides that contain relatively few basic groups show smaller increases.

A variety of workers has recently demonstrated that peptides can be readily separated by RP-HPLC. The most successful chromatographic separation methods depend on the use of an ionic-modifier or ion-pairing reagent in the solvent system to provide good separation. These modifiers can be divided conveniently into two groups; acids at pH 2–3 (such as trifluoracetic, phosphoric, and hydrochloric acids) and buffered salts (such as triethylammonium phosphate, ammonium acetate, or triethylamine formic acid).

Two major forms of the immunoreactive neuropeptide ACTH found in the rat neurointermediate pituitary are underivatized corticotropin-like intermediate peptide (CLIP) and *O*-phospho-^{31}seryl-CLIP, with the phosphorylated form predominating by a ratio of 2:1 (24). The main forms of immunoreactive α-MSH found were *N*-acetyl-

^{1}ser-α-MSH and *N,O*-diacetyl-^{1}ser-α-MSH, with the diacetyl form being the major component (9 : 1). All four of these neuropeptides were purified by only two RP-HPLC steps. Phosphorylation of the ^{31}ser-CLIP molecule is a post-translational modification that (most probably) serves as a mechanism to mask the recognition site for glycosylation of the ^{29}asn residue.

RP-HPLC techniques were utilized in studies of the *in vitro* incorporation of ^{32}P into ACTH-related peptides by rat pituitary tissue (23) and of α-*N*-acetyl-β-endorphin$_{1-26}$ from the rat pituitary neurointermediate lobe. Phosphorylation of proadrenocorticotropin and endorphin-derived peptides was studied (25).

4.2 High resolution reversed phase high performance liquid chromatography (RP-HPLC) columns

The last step recommended before any analytical measurement (MS, RIA, RRA, BA) of peptides is HPLC in the RP mode. RP-HPLC provides high sensitivity, resolution and speed of peptide separation with excellent recoveries; it is estimated that 80% of all separations being currently performed are in the RP mode. Solutions of the synthetic tetradecapeptide somatostatin have been quantified down to 500 fmol (0.9 ng) with good straight-line statistics in a study to demonstrate high sensitivity and resolution (26). The RP-HPLC resolution of individual peptides in a biological mixture is high. Many longer peptides are differentiated from their homologs, where one residue is replaced with another that differs only by one methylene unit. For example, two pentapeptides differing by only one methyl group are readily separated (27). LE is separated by seven minutes from its homolog (tyr-ala-gly-phe-leu).

RP-HPLC effectively separates enkephalin peptides (28). For example, the chromatogram in Figure 2 illustrates the isocratic RP-HPLC of enkephalins from an extract of canine hypothalamus. This RP-HPLC chromatogram demonstrates that the extract is relatively clean in the isocratic mode when monitored with UV detection. However, RRA and RIA of these extracts indicates the presence of pmol amounts of LE and ME. The "cleanliness" of a HPLC chromatogram may be misleading because only UV chromphores are detected. It is possible that non-absorbing material also elutes from the HPLC column and that these compounds may interfere with several types of subsequent assay methodologies (RIA, RRA, MS).

Although the basic molecular processes involved in RP-HPLC relating to the interaction of the ion-pair formed between peptide and buffer with the hydrocarbonaceous column material are not yet completely described, detailed mathematical and experimental data have been analyzed (29). The theory of ion-pair chromatography was investigated, including factors that participate in controlling the capacity factor, k'.

Comparing phosphoric acid ($1\ g \cdot l^{-1}$, pH 2.5) and acetic acid ($1\ g \cdot l^{-1}$, pH 4) on the retention times of various peptides demonstrated a significant reduction in retention times (30). A series of different ion-pairing reagents was used to show changes in selectivity of a RP system for peptides ranging in size from tripeptides to heptapeptides. These ion-pairing reagents included phosphoric acid, sodium hexanesulfonate, and sodium dodecylsulfate (31).

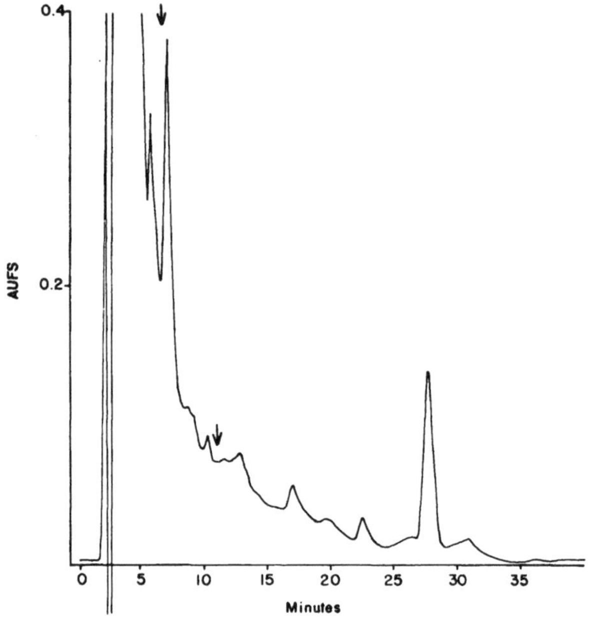

Figure 2. RP-HPLC of the peptide-rich fraction extracted from a canine hypothalamus. Arrows indicate the retention times of synthetic ME and LE, respectively. TEAF buffer; acetonitrile organic modifier; 1.5 ml · min^{-1} flow rate; 200 nm UV detection. (Reproduced from ref. 1 with permission).

An amphoteric molecule such as a peptide may participate readily in either hydrophilic or hydrophobic ion-pairing processes, and poor chromatographic resolution may be observed if there is insufficient ion-pairing. Such complications are not surprising in the light of the number of complex ionic equilibria possible with these amphoteric compounds.

One problem encountered occasionally is that smaller peptides are sufficiently polar so that they will not be retained on a hydrocarbonaceous column. In such cases, a series of perfluoroalkanoic acids may be utilized to extend the retention time of those small peptides (32). For certain tripeptide solutes, the retention time is increased when the mobile phase is changed from trifluoracetic acid to pentafluoropropionic acid, then increased further by heptafluorobutyric acid. This effect of increasing retention time may be rationalized by considering the lipophilicity of the perfluoroalkyl groups that results in a decreased polarity of the ion pair formed between the peptide solute and the perfluoroalkanoic acid present in the mobile phase.

RP-HPLC conditions were optimized for the cyclic tetradecapeptide somatostatin (33). A variety of phosphate, acetate, and sulfonic acids were utilized. Organic modifiers were tested and included acetonitrile and various alcohols. Temperature effects and buffer concentrations were changed to optimize the retention time of somatostatin. Sensitivity of detection permitted determinations down to 10 – 20 ng of somatostatin.

Two complementary chromatographic techniques, ion-exchange chromatography and RP-HPLC, were used in conjunction with RIA and BA for analysis of known peptides (34). HPLC was utilized on acid-boiled crude rat brain extract and, in conjunction with RIA, demonstrated that the enkephalins were present in the eluate, even though a high UV background was noted. Although these authors employed 280 nm for determination of peptides, other workers have shown that the 200 nm range is more sensitive towards peptides (26). Yields of peptide eluted from the HPLC column were excellent and, for ^{125}I-labeled VIP, gave recoveries in excess of 85%.

A simpler method has been developed for purification of gastrointestinal (GI) neuropeptides from relatively small amounts of GI tissue (35). The ion-exchange chromatography and HPLC without lyophilization produced good yields for somatostatin, bombesin, and VIP immunoreactivity. The authors noted that the rigorous maintenance of acidic conditions was required to achieve optimal recoveries of these basic peptides. Acidity ensured that the carboxylate, guanidinium, and amino groups (lys and *N*-terminus) remain protonated. VIP was first purified from hog intestine, somatostatin from ovine hypothalamus and bombesin from frog skin. Recent data indicate that these peptides may function as neurotransmitters.

The observation that trifluoroacetic acid enhanced the partition of opioid peptides into butanol during aqueous-organic phase extractions led to the use of trifluoroacetic acid as a counter-ion for ion-pair partition RP-HPLC. Trifluoroacetic acid is an effective counter-ion because it provides decreased nonspecific absorption of opioids and opioid peptides onto column matrices while decreasing elution time without loss of resolution (36).

The profiling of neuropeptides using RP-HPLC has been reviewed (37).

5 Assay of Peptides

5.1 Radioimmunoassay (RIA)

The basic principles of RIA have been described in Volume I of this series (38) and numerous other publications (see for example refs. 39 – 42).

Despite the RIA system used, it is important to evaluate fully the molecular specificity and detection sensitivity of that system to demonstrate the level of confidence to the placed on the analytical data. Rigorous experimental controls are needed for RIA because it cannot be assumed that immunoassays, which yield sensitive and reproducible results in a pure control buffer, will necessarily give valid results for measurement of the target endogenous molecule in the biological matrix or an appropriate extract. Although the antigen and/or the antibody may be heterogeneous, some proteins, polypeptide hormones, and enzymes are known to be secreted in multiple molecular precursor forms and may exhibit partial structural homology with other hormones and enzymes, undergo partial proteolysis, and become attached to naturally-occurring inhibitors or carriers that may or may not affect immunologic reactivity. For all of these reasons, procedures have been developed to validate immunoassay results.

5.2 Neuropeptides studied by combining RP-HPLC separation and RIA measurements

Electrostimulation (15 Hz, 5 ms, 10 V, 15 min) of the mandibular nerve influences the concentration of substance P-like immunoreactivity in the superfusate of dental pulp (43). The undecapeptide substance P may play a role in the development of the inflammatory response that is associated with injury and pain in peripheral tissues.

The optimal molecular specificity of the analytical measurement of a biological compound is not limited to peptides. Other workers have noted that the sensitivity and molecular specificity of RIA measurement of prostaglandins and other hormones cannot lead, but rather must follow in a parallel fashion, those developments that occur in the field of combined GS-MS, an instrumental method that offers optimal molecular specificity (44, 45).

5.3 Radioreceptor analysis (RRA)

5.3.1 **Background**

A receptor is defined as the macromolecular site on the surface of a cell that specifically interacts with and binds to a selected molecule. Putatively, this interaction is structurally unique. The macromolecular receptor site may be comprised of protein, lipid, and saccharide portions. The three-dimensional aspects of the receptor are crucial to interaction with a particular ligand molecular conformation. Various ligands have different experimental binding characteristics, which may be represented by their binding avidity and concentration range.

An assay method with broad molecular specificity is necessary to screen effectively for the presence of all opioid receptor-binding compounds in either a biological mixture or the fractions separated by RP-HPLC. An RRA method is the clear method of choice, with its speed, efficiency, sensitivity and range of response, to assay chromatographic fractions for opioid receptor activity.

The RRA is based on the experimental observation that opiate-receptor interactions occur with specific endogenous brain receptors, a feature of the *in vitro* studies that showed the presence of brain enkephalins. The polypeptide macromolecular opioid receptor complex is thought to be located at a synapse. Preliminary receptor studies demonstrated that some binding did occur to brain membranes, even though that binding was non-specific and only 2% was stereospecific. Later, opiate ligands with a high specific activity of radioactivity made it possible to examine opioid receptors that had a high affinity for the opiates, and also permitted the use of low ligand concentrations, an experimental feature that significantly decreased the amount of non-specific binding (46). Rapid separation and washing of the membrane reduced non-specific entrapment of radioactive ligands and up to 90% stereospecific binding was demonstrated.

Pharmacological data indicated that the binding of the opioid receptor population could be saturated (a phenomenon that demonstrated the presence of a finite number of receptors) and reversed, with association and dissociation constants that were independent of the ligand being used. Bacitracin was used to inhibit enkephalin destruction (47).

Initial studies indicated that interaction between an endogenous brain fraction and the opioid receptor in tissue was reversible and competitive (48). Human CSF and brain extracts were tested with a receptor (49, 50), and different receptor selectivities were found with various opioid enkephalins (51). Pain perception and endorphin levels in CSF were measured (52, 53) and electrophoretically-separable endorphins in human CSF were characterized (54). Inhibitors of the narcotic receptor binding in brain extracts and CSF were studied (48) as were brain-pituitary opioid mechanisms and specific influences and interactions of the opioid receptor binding (47).

Metabolic profiles have been obtained using the combination of RP-HPLC and RRA of the opioid peptides with tritiated etorphine as the displacing ligand (55). This study included opioid peptides in human tooth pulp (56); a human pituitary ACTH-secreting tumour (57); canine pituitary and selected brain regions (58); CSF from lower-back pain patients (59); and in 26 human post-mortem pituitary tissue extracts (60).

It has been shown by using a variety of immunohistochemical techniques that structures within the limbic system (61) of the brain contained some of the highest concentrations of opioid receptors (62). Following an assay of opioid peptides by inhibiting tritiated naloxone and tritiated LE binding in brain homogenates, it was concluded that the opioid peptidergic system has several agonists possessing different characteristics that interact with more than one type of receptor (63). The hydrophobic moiety of the benzene of the enkephalin molecule may be one of the major structural factors that is operative in differentiating binding sites (64). These multiple opiate receptors were measured in different brain regions, and the distribution and differential bindings of opiates and opioid peptides were studied (65). The binding, molecular and structural parameters have been reviewed (66).

Based on the assumption that the mu receptor is the mediator of analgesia, two distinct mu and delta receptors were located in an "opioid receptor complex" in which the two receptors interacted (67). The occupation of one receptor by its specific ligand was hypothesized to induce a conformational change in the second receptor, which in turn allowed the coupling of that receptor to its effector. Ethanol inhibited opiate receptor binding (68). Endogenous ligands for the κ opiate receptors have been studied (69).

The distribution and physiological significance of opiate receptors in the brain have been discussed (62) and the different types of opioid receptors classified (70).

5.3.2 **Measurement of endogenous enkephalins in human tooth pulp extracts**

In a study of pain (nociceptive) processes, tooth pulp was chosen as the model tissue based on the suggestion (71) that the only sensory output from this tissue was pain. The object of the study was to detect changes in LE and ME with time in tooth pulp as a result of applied physical stress and to compare RIA and RRA for the measurement of these peptides. The treated patients had a strong orthodontic force applied by springs to two maxillary and two mandibular premolar teeth for 30–45 min prior to extraction. A similar experiment was performed recently, and immunoreactive substance P (72) was shown to decrease with increasing orthodontic stress.

Endogenous LE and ME were measured in tooth pulp tissue extracts by a RRA method (28, 56), which involved measurement of ^{3}H-dihydromorphine used in the competitive displacement of the peptides from a synaptosomal receptor preparation following initial binding. A subcellular fractionation of tissue from the limbic system of dog brain was used to preferentially extract synaptosomes (64, 73). Following neurosurgical removal of the brain the opioid receptor-rich limbic system was neuroanatomically identified, excised rapidly in the cold, and the synaptosomal fraction isolated.

The RP-HPLC fraction eluting at the retention time of synthetic ME for one patient's (DD) first extracted tooth was also studied by FAB-MS to confirm the presence of the pentapeptide. An appropriate $[M + H]^+$ was observed at the nominal mass of 574 daltons and supported the conclusion that ME was collected from the HPLC column at the appropriate retention time and increased confidence that the target peptide was indeed the one being quantified by RRA.

Table 2 contains the results of the study (74). The seven patients differed in the amount of receptor binding ME equivalents measured. Patient TH (a) had the lowest amounts, whereas patient JG had the highest enkephalin-equivalent values for all four teeth and patient JW the highest value of enkephalin-equivalents for the first tooth extracted. An average of 3668 ng of RRA enkephalin $\cdot$ g^{-1} wet weight of tissue was found in the first tooth extracted from the four control patients (DD, JG, JW, and CB), whereas the three treated patients (TH, SG, and SS) had an average of 53 ng of RRA enkephalin $\cdot$ g^{-1} weight weight tooth pulp tissue. This difference represents a decrease of 69-fold between the control patients and the treated patients. This observed decrease in the amount of one selected opioid pentapeptide, ME, is not inconsistent with the working hypothesis that a homeostatic relationship exists among the three opioid peptidergic families (endorphinergic, enkephalinergic, dynorphinergic), each of which is composed of larger precursors that lead to the "working" peptide that is rapidly metabolized. The second part of the hypothesis states that a noxious stimulus will activate a multi-step peptidergic process, and that decreases in the relative concentrations of the enkephalins will occur. The enkephalins are thought to interact with and decrease the firing rate of substance P-containing neurons.

One of the experimental variables over which there was little control was the wet weight of the tooth pulp tissue derived from human teeth. The average weight was 0.067 g $\pm$ 0.031 ($n = 42$) and the coefficient of variation (49%) for this parameter was considered acceptable for the source of this tissue.

Clearly the RIA and RRA data that are given in Table 2 do not either agree with each other or indeed with data derived by HPLC analysis. However, on closer examination, the data do show parallel behavior. Although it might be expected that the three analytical modes should agree numerically with each other, further reflection on the basic principles of each procedure indicates that this is not necessarily the case. Conventional RP-HPLC measures only UV absorbance at one selected wave length and cannot observe non-absorbing but yet, coeluting material that is usually present; RIA measures an antibody interaction with an oligopeptide, where that antibody is raised against a peptide-protein conjugate and not the peptide; and in RRA, a

Table 2. Measurement of enkephalins in human tooth pulp extracts with RIA and RRA

Patient	Tooth no.	Wet weight	RIA	RRA
A. Controls				
DD	5	0.1181	68	**
	12	0.0576	4.9	521
	21	---	---	---
	28	0.1181	5	47
JG	1	0.0394	15.9	6726
	16*	0.0252	500	3155
	17	0.0543	7.2	5608
	32	0.0649	16.6	1495
JW	– –*	0.0123	1138	6098
	– –	0.0273	7.1	421
	– –	0.0478	1.3	1339
CB (a)	5	0.0710	176	292
	12*	0.0843	148	1751
	21	0.0496	51	107
	28	0.0556	155	79
CB (b)	5	0.0710	38	1215
	12*	0.0843	14	867
	21	0.0496	5	968
	28	0.0556	12	809
B. Treated				
TH (a)	5*	0.0448	19	91
	12	0.0456	4	52
	21	0.0396	n.d.	67
	28	0.0402	15	74
TH (b)	5*	0.0448	n.d.	81
	12	0.0456	4	86
	21	0.0396	n.d.	83
	28	0.0402	40	114
SG (a)	5*	0.0486	12	n.d.
	12	0.0489	33	22
	21	0.1107	17	29
	28	0.0925	1	n.d.
SG (b)	5*	0.0486	4	66
	12	0.0489	16	21
	21	0.1107	6	25
	28	0.0925	1	n.d.
SS (a)	5*	0.1303	32	68
	12	0.1056	2	28
	21	0.0773	5	29
	28	0.0732	22	80
SS (b)	5*	0.1303	2	39
	12	0.1056	2	19
	21	0.0773	3	28
	28	0.0732	n.d.	23

(a) HPLC fraction collected at synthetic ME (YGGFM) retention time.
(b) HPLC fraction collected at synthetic LE (YGGFL) retention time.
* first tooth extracted n.d. = not detected
** sample used for MS corroboration Patients TH, SG, and SS were fitted with springs.

lipoprotein macromolecular complex on a synaptosomal surface is carefully isolated and used in a competitive binding assay versus a radiolabeled ligand. It is crucial to realize that no rigid molecular proof is provided by, nor can ever be inferred from, any one of the three analytical methods HPLC, RRA, or RIA.

5.3.3 Radioreceptor assay screening

RRA was used for effective screening of chromatographically-separated opiates in human CSF (49–51, 54). For example, opiate peptides were purified from pooled human CSF (2 liters), and purification was effected by Sephadex G-10 and electrophoresis. HPLC and RIA were utilized for further characterization, and all experimental procedures were guided elegantly by a RRA (54). Binding properties of opioid receptors and whole rat cerebrum were studied with tritium-labeled dihydromorphine, naltrexone, and LE as radioindicators (51). Three different kinds of opiate binding sites were observed.

High concentrations of opiate receptors were found in the anterior amygdala, the hypothalamus, and the medial thalamus (62). The latter location is important because of its participation in behavioral functions and processing of pain sensations. Moderate levels of receptors were found in the basal ganglia, mid-brain (especially periaquaductal gray), cerebral cortex and the frontal, temporal, and hippocampal cortices. The basal ganglia location may be important in providing an understanding of movement disorders such as Parkinsonism and Huntington chorea. Low levels of receptors were found in the lower brain stem and spinal cord, and no receptors were found in the cerebellum. It is interesting to note that no correlation with opioid receptors was found with the distribution of other known neurotransmitters (acetylcholine, serotonin, norepinephrine, domapine, histamine, or GABA). This lack of correlation probably indicates that a separate "opiatergic" system exists. Two tritiated opiate ligands, diphrenorphine and etorphine, are expecially useful for autoradiographic localization of opiate receptors in rat brain because of their high affinity and their low dissociation rates. These two properties minimize diffusion of these ligands away from the receptor site during histochemical processing of the tissue.

A canine-limbic system substance P-receptor preparation has been developed to analyze for tachykinin receptor activity to investigate potential interactions among the opioid peptidergic systems (75).

5.3.4 Screening for opioids with a combination of gradient RP-HPLC, RRA and FAB-MS

A combination of gradient RP-HPLC, RRA, and FAB-MS can be used effectively to assign an opioid activity profile to a biological tissue extract (76). A peptide-rich fraction from a tissue extract was eluted from a C-18 Sep-Pak, subjected to a gradient RP-HPLC chromatographic separation, and the individual fractions were then incubated with a limbic system synaptosomal receptor preparation, where constituents compete with the radioactive ligand for binding to the receptors. Figure 3 shows the opioid receptor activity profile of the individual HPLC fractions, which were measured as picomoles of ME equivalents. Those RRA data are superimposed upon the corre-

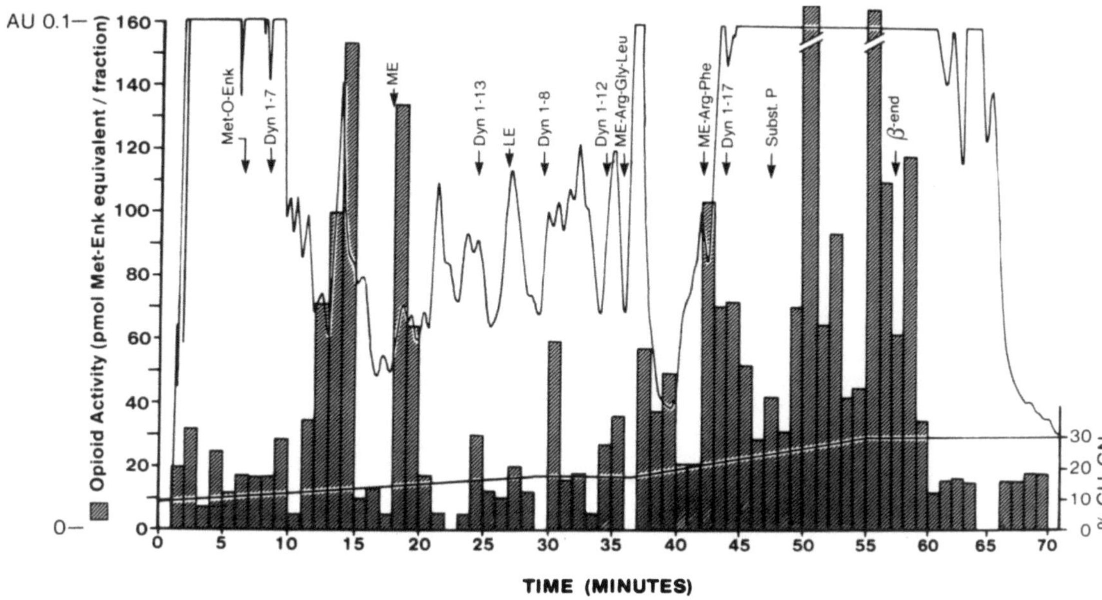

Figure 3. Gradient RP-HPLC and RRA of the peptide-rich fraction from canine hypothalamus extract. A 90-min gradient was employed as shown by the solid line. One-minute fractions were collected, and the opioid receptoractivity content in each fraction was measured with a canine limbic system synaptosomal receptor-rich preparation; ^{3}H-etorphine was the radioligand. The arrows indicate retention times of synthetic peptides determined in a separate chromatogram. (Reproduced from ref. 1 with permission.)

sponding RP-HPLC UV chromatogram of the peptide-rich fraction of a canine hypothalamic tissue extract. The gradient profile is shown, and the elution times for several selected synthetic peptide standards, determined in a separate chromatogram, are indicated by the arrows along the top of the chromatogram. In another experiment, very similar HPLC data were obtained for canine pituitary tissue (76).

Three areas of the gradient chromatogram in Figure 3 were then subjected to analysis by FAB-MS production of $[M + H]^+$ ions. In fraction 17, a compound, measured by FAB-MS, was noted at *m/z* 1259. α-neoendorphin has a molecular mass of 2229 daltons. By the use of MS peptide standards, it was estimated that the peak at *m/z* 1259 daltons corresponded to 3–4 mg of material. Fractions 21 and 22 showed the presence of ME (600 ng), a large peak at *m/z* 453, and a smaller peak at *m/z* 485, where the latter two were presumed to be $[M + H]^+$ ions. The mass at *m/z* 453 was found throughout the chromatogram, and apparently was a continuously-eluting HPLC component. This particular set of fractions indicated the need to provide a separate detection system with a high level of molecular specificity following HPLC separation. In other studies of RRA measurements in human teeth (77), it was noted that RRA analysis of this particular HPLC fraction gave analytical data with highly variable results. Fractions 36 and 37 indicated $[M + H]^+$ ion at *m/z* 900, corresponding to a level of approximately 50 ng of the pro-enkephalin molecule ME-arg-phe. No dynorphin$_{1-12}$ was observed at *m/z* 1571 mass units, but a very small and real peak at *m/z* 1755.

A combination of analytical methodologies is absolutely essential when analysing extracts of biologically complex materials such as brain peptides. Although many workers presume that RP-HPLC provides a pure fraction, this cannot be relied upon because of the great number of peptides possible and the limited detection system. Only by using a multiple detection ion approach, for example RRA, RIA, and FAB-MS, can any level of confidence in molecular structure be acquired (78).

6 Fast Atom Bombardment Mass Spectrometric (FAB-MS) Studies of Peptides

6.1 Experimental aspects of FAB-MS

Whereas FD-MS represented a major step forward in MS studies of biologically important peptides, a more significant advance soon followed with the introduction of FAB-MS. FAB-MS developed from secondary ion mass spectrometry (SIMS) in which an ion of high translational energy was directed at a sample target and the secondary ions sputtered from the target material were mass-analyzed (79 – 81). Amino acids, steroids, and folic acid derivatives were studied.

Although some organic and bioorganic compounds could be investigated by SIMS, the use of liquid matrix was found to minimize sample destruction and stabilize ion production. The liquid matrix effectively and continuously regenerated a fresh surface of molecules for ion/atom bombardment, while also reducing the degree of sample burning (82 – 88). This method is now known as FAB-MS and is discussed in Chapter 1. A few experimental points will be made here.

In FAB, a noble gas (commonly Ar^0 or Xe^0) is ionized, accelerated to high translational energy (8 – 10 keV), neutralized, and exits the FAB gun to impact onto a probe tip containing the sample dissolved in a non-volatile viscous matrix. The momentum of the impacting fast atom is transferred to the molecule and a $[M+H]^+$ ion results by a process involving this energy transfer.

The modification of a conventional Finnigan MAT 731 MS ion source housing (89) to accept a commercially available neutral ion gun (90, 91) has been described. The standard procedure used by those authors (89) for FAB analysis is to coat a stainless steel target tip with 50% (v/v) nitric acid for a few seconds, rinse with distilled water and nanograde methanol and allow to dry. Samples (usually at a concentration of $\mu g \cdot \mu l^{-1}$) are applied to the probe tip and dried. Glycerol (300 nl) is spread on the surface of the probe so that a thin, uniform film is visible to the unaided eye. When a larger volume of glycerol is applied a hemispherical shape is produced, which alters the angle-of-incidence of the impacting fast atoms and causes a decrease in sensitivity.

A rotatable probe was developed for accurate mass measurement ($\pm$ 0.001 dalton) during FAB-MS (92) and ionization efficiency of reagent gases has been studied (88, 89).

Sensitivity in FAB is not defined in the same way as for EI (i. e. coulombs of ion current produced per microgram of sample), but rather as a minimum amount of sample that is required to provide a signal for the $[M + H]^+$ ion of the compound that is clearly distinguishable from the background signal. The ratio of the signal versus background noise, for example, in the positive ion mode is greater than 40:1, 17:1, and 25:1 for the peptides LE, [Sar^1,Ala^8]angiotensin II, and bradykinin, respectively, at a sample concentration of 1.0 mg · ml^{-1}.

Two major problems were found in sample preparation for FAB: inefficient sample dissolution in a matrix-compatible solvent and the presence of salts (89). Solubility problems can be partially overcome by using acids for basic samples and bases for acidic samples to facilitate their dissolution prior to adding the matrix. Often, either sonication of the solution for a short time or allowing a sample to stand overnight will greatly increase the ion abundance, due presumably to an increase in sample concentration. The suppression of the FAB ionization by salts can be eliminated readily by utilizing either a Sep-Pak or an HPLC purification step prior to FAB-MS. Matrix-associated ions observed in FAB include the components glycerol, sodium, potassium, chloride, ammonium, formic acid, guanidine hydrochloride, phosphate, diaminobutane, oxalic acid, acetic acid, hydrochloric acid, heptafluorobutyric acid, and ammonium trifluoroacetate.

The abundant total ion current produced by FAB of a peptide is comprised of multiple ionic species, including protonated and cationized molecular ions, due to peptide and matrix. Fragment ions are also produced but are generally of lower intensity. A continuum of peaks at virtually every mass is observed in the FAB-MS of glycerol, with major peaks occurring at masses $(92n + H)$ daltons where $n = 1, 2, 3 \ldots$. If a glycerol cluster ion persistently interferes with a critical sample ion, then d_5-glycerol can be used as an alternate matrix to shift the matrix ion away from the sample ion. Pure d_5-glycerol $HO-CD_2-CD(OH)-CD_2OH$ can also be used as the matrix (1). The first glycerol monomeric ion, mass 93, shifts to m/z 98 in the mass spectrum of d_5-glycerol. Each succeeding multimer shifts by 10, 15, 20, 25 etc. daltons corresponding to m/z 195, 292, 389, 486 etc., respectively.

FAB mass spectra should be obtained as soon as possible after probe insertion to minimize the amount of fast atom irradiation (and potential damage) to the glycerol matrix (93) and solute molecules.

6.2 FAB-MS of proteins and peptides

The study of proteins and peptides by FAB-MS, and other MS techniques, has expanded dramatically in recent years but a review of this field is beyond the scope of the present chapter and should be sought elsewhere (e. g. 94–96). However a selection of examples of FAB-MS application is discussed below.

Proinsulin has an elemental composition of $C_{410}H_{638}N_{114}O_{127}S_6$ and a molecular mass of 9382.529 daltons and has been ionised by FAB (88). The initial FAB mass spectrum of this protein was obtained at reduced ion accelerating voltage with an atom beam of 1–2 mA using a matrix of monothioglycerol and glacial acetic acid. An unresolved cluster of ions in the $[M + H]^+$ ion region was observed but no amino acid sequence-determining ions could be seen due to high background. The monoisotopic $[M + H]^+$ ion at this high mass is only a minor component in the isotopic distribution (97). In the years since this spectrum was obtained improvements in sensitivity and mass range characteristics have greatly increased the quality of data available from high mass compounds. The experimental limitation to FAB-MS studies of high mass bioorganic molecules is due to the limited number of ions produced in the ionization process and to the sensitivity of detection. Ion transmission through the analyzer(s) of a mass spectrometer is independent of polarity, but when using an electron multiplier as an ion detector, negative ions give a lower response due to the reduced energy with which they impact the negatively-charged first dynode of the detector. This difference in detector response leads to a reduction in the available dynamic range when compared to positive ions. The solution to this problem has been the incorporation of a conversion dynode (+ 15 – 20 kV) placed before the detector. This dynode accelerates negative ions and the positive ions produced on impact consequently give increased numbers of secondary electrons in the multiplier. Several laboratories have reported the FAB spectrum of the intact insulin molecule containing both the A and B chains with M_r 5627 (98 – 100).

Monellin contains two peptide subunits of molecular mass 5248 and 5831 respectively. One subunit was shown by FAB-MS to be heterogeneous, with approximately 10% containing an additional phe residue at the N-terminus (98). Insulin was utilized as a test compound to optimize resolution and scan rate conditions of the instrument before studying monellin. The proteins were dissolved in a mixture (3 : 1, v/v) of dimethylsulfoxide and hydrochloric acid, and caesium iodide cluster ions were used for mass calibration at masses 5069.3 ($Cs_{20}I_{19}$), 5329.1 ($Cs_{21}I_{20}$), 5589.9 ($Cs_{22}I_{21}$), 5847.7 ($Cs_{23}I_{22}$), and 6108.5 ($Cs_{24}I_{23}$).

Other peptides studied include mellitin (M_r 2845), a constituent of bee venom, glucagon (3841), ACTH (4538) and the B-chain of bovine insulin (3494) (86). Using a VG ZAB 1-F reverse geometry mass spectrometer at full accelerating potential, an unresolved complex of ions was obtained of the molecular ion region of mellitin with an $[M + H]^+$ ion of 2845.8. A similar instrument but incorporating a high-field magnet with a mass range of 3000 daltons at full accelerating potential demonstrated baseline resolution of the $[M + H]^+$ ion region. The positive ion mass spectrum of the B-chain of bovine insulin (disulfonic acid derivative) showed ions in the $[M + H]^+$ ion region as did the spectrum of glucagon. Masses were counted manually by comparison with the FAB-MS of potassium iodide cluster ions $[K(KI)_n]^+$ and agreed to within one mass unit of the theoretical value. However improved results are possible with current instrumentation.

Useful analytical data have been obtained by FAB on a series of "difficult" peptides including cytochrome C-550 of the denitrifying aerobic bacterium *Paracoccus*, the *C*-terminus of that protein cytochrome C-550 and the *N*-terminus portion of a dolphin cytochrome C (102).

A detailed positive and negative FAB-MS study was reported for peptides that contained up to 21 amino acids. Mass spectra were obtained with sample size ranging from 2 – 15 nmol of peptide and in favorable cases molecular mass information was derived from 100 pmol of sample. Mass spectra were produced for substance P, human fibrinopeptide peptide A, gastrin-13, and various peptide enzymatic digests (103).

An adrenal zona glomerulosa-stimulating component of a posterior pituitary extract was characterized by FAB-MS as bis-acetyl-^{1}ser-α-melanocyte stimulating hormone (MSH) (104). HPLC fractionation yielded two major peaks of aldosterone-stimulating activity, the second component being characterized as the bis-acetyl hormone. FAB-

MS of the biologically active material gave an $[M + H]^+$ ion at mass 1706 while chymotrypsin treatment, followed by EI-MS of the *N*-terminal dipeptide, demonstrated the site of *O*-acetylation.

The analysis of eleven zervamicin and two emerimicin peptide antibiotics has been achieved by FAB-MS (105, 106).

A combination of protease, carboxypeptidase digestion, and subtractive Edman degradation was coupled with FAB-MS to obtain amino acid sequence-determining information from substance P and an eledoisin-related peptide (107).

Generally, 100 picomoles was sufficient for determining the molecular ion of the peptides in either the positive or negative ion mode and a larger sample amount (1 – 5 nmol) generated amino acid sequence-determining ions.

Mass spectra with abundant $[M + H]^+$ ions were found for somatostatin, gastrin, and substance P (108). Tetragol $[HO(CH_2CH_2O)_4H]$ and teracol $[HO(CH_2CH_2CH_2CH_2O)_nH]$ can be substituted for glycerol when increased polarity is required and an advantage was noted in higher sensitivity of compounds that have a net positive charge at pH 6.5 in the positive ion mode, and those that have a net negative charge at pH 6.5 in the negative ion mode. Substance P yielded a $[M + H]^+$ ion at 1347 for 1 µg (740 pmol) with a signal-to-noise ratio of 40 : 1.

Positive and negative ion FAB have been used in a study of the pentapeptides, ME and LE, and the negative ion mass spectra were noticeably less complex (102). Molecular ions and amino acid sequence-determining ions were found in both positive and negative spectra of bradykinin, lysyl-bradykinin, and methionyl-lysyl-bradykinin (83). No loss of biological activity of ^{5}Ile-angiotensin I and II was found following fast atom irradiation (84). Positive and negative ion mass spectra and, in one case a mass-analyzed ion kinetic energy spectrum of the $[M + H]^+$ ion, were obtained. In general, fragmentation was reduced in the negative ion spectra due to the differences in ion stabilities and energetics between positive and negative ions. Smaller amounts of the two angiotensins were derivatized with a 1 ; 1 mixture of d_3- and d_0-acetic acid anhydride to produce an ion doublet that indicated all of the amino terminus-containing ions.

6.3 Collision activated dissociation (CAD) mass spectrometry

6.3.1 **Unimolecular metastable decompositions**

A unimolecular metastable decomposition occurs whenever an ionized molecule fragments during the transit time between acceleration from the ion source and before the multiplier detector (110, 111). Metastable decompositions may occur in forward (EB) or reverse (BE) geometry double focusing mass spectrometers in either of the field-free regions (FFR). The study of these unimolecular decompositions is necessary to establish the relationship between the product and the precursor ions. Metastables are discussed in Chapter 1, Section 3.1.

6.3.2 **Collision activated dissociation processes**

In CAD spectroscopy, ions having a high translational energy (1 – 100 keV) collide inelastically with neutral atoms. Part of their translational energy is converted into excitation energy, a process that can lead to fragmentation of the excited ions (112).

During unimolecular dissociation of the excited ion, part of the excess energy is released as translational energy that results in a broadening of the signals in the CAD spectrum. Collision broadening reduces the effective resolution in CAD processes but this reduction can be offset to some extent by using a higher ion accelerating voltage.

Experimentally in a double focusing magnetic sector instrument, collisions are induced in a collision chamber placed in one of the FFRs between ions accelerated from the ion source and neutral atoms in the chamber. In a reverse geometry instrument (*BE*) the magnetic field performs mass analysis and, following collision in the second FFR, the electric sector is scanned to produce a MIKES spectrum (see Section 6.5). In a forward geometry instrument, linked scan methods (Section 6.5.2) are used to obtain the CAD spectrum.

CAD can be carried out also in an instrument constructed from three sets of quadrupoles (Q_1, Q_2 and Q_3). Q_1 is operated as a mass selector to transmit a selected ion mass to Q_2. The second quadrupole is pressurized and operates as a collision cell. CAD occurs at low energy in Q_2 and the products are mass analyzed in Q_3 (113). A collision gas with a higher atomic mass is more advantageous for use with the low energy ions produced in quadrupoles.

Most conventional collision cells for high energy MS/MS are a few centimeters in length and have ion entrance and exit slits. Collision gas can effuse out of the slits along the ion axis and increase the effective length of the cell. Ideally, an infinitely thin collision region at the object focus point of the mass spectrometer is needed. An efficient collision region using a hypodermic needle has been devised to address this problem (114). MS/MS can also be affected by utilizing combinations of high resolution mass spectrometers (115, 116). A double-focusing MS/MS instrument was described that had a unique molecular beam collision region in which the helium beam profile gave 95% of the ion collisions within a distance of less than one μm. The first mass spectrometer in this combination instrument had a resolution of 50 000 and ions up to 3600 daltons could be separated and accelerated to a collision energy of greater than 10 keV.

The major types of reactions that can occur when a singly-charged positive ion in a keV energy range collides with a neutral target include: collisional excitation followed by dissociation, scattering, charge exchange (neutralization), charge stripping and charge inversion. These reactions are in competition with each other and their occurrence will lead to an attenuation of the primary ion beam.

The following discussion is included to provide the experimentalist with an appreciation of the wide range of analytical applicability that is available for peptide measurements using these CAD processes.

6.3.2.1 **Scattering**

Scattering will lead to a decrease in the intensity of the primary ion beam if the ions are scattered beyond the acceptance angle of the mass spectrometer. Theoretical considerations predict that the cross-section for elastic scattering decreases with an increase in the ion velocity and a decrease in the size of the target atom. Therefore, for example, instruments operating at higher ion acceleration voltages (8 – 10 keV) are preferable, and helium is a better target molecule for CAD processes involving keV collisions. On the other hand, for collisions in the eV range (quadrupoles), large target masses are preferable. Furthermore, the amount of scattering increases with collision gas pressure.

6.3.2.2 **Charge exchange (neutralization)**

Neutralization of the primary ion beam by charge exchange is an important ion removal mechanism, and the cross-section for this collision process may be of the same order as, or even higher than, that for the CAD process itself. Thus it is possible that, after collision, up to 40% of the initial ion current is due to fast neutrals. Neutralization of the primary ion beam is minimized, again, if the helium is used as target gas.

The target gas pressure has a dramatic influence on the overall CAD efficiency. Depending on the system that is being studied and the instrument that is employed, the CAD efficiency optimizes at a target gas pressure that corresponds to an ion transmittance value of 20 – 60% of the original primary ion beam. This transmission value corresponds to those experimental conditions wherein an ion in the primary ion beam undergoes multiple collisions (on the average, 1.5 – 2.5 collisions).

CAD efficiency is affected by beam divergence of the fragment ions resulting from kinetic energy release and the increased scattering efficiency of the fragments of smaller mass. Proper instrumental design may enhance this ion transmission efficiency, in particular by confining the collision gas to a very small region near the focal point of the mass spectrometer, and/or by employing mass analyzers with a large acceptance angle for CAD fragments. Also, increasing the translational energy of the ion will decrease those losses that are due to the angular distribution and increase the mass resolution.

6.3.2.3 **Internal energy effects**

Determining ion structures using CAD assumes that the relative abundance of fragments that are generated in CAD processes reflects exclusively the structure of the precursor ion, and is not influenced by its internal energy prior to collision (however, see ref. 112).

6.3.2.4 **Related methods: angle-resolved mass spectrometry**

A range of energies is conferred on a polyatomic ion during a high energy collision. An approximate selection of the internal energy can be achieved by studying the CAD of a given species under different and well-defined scattering angles. This process is known as "angle-resolved MS" (117). Angular resolution of CAD products is made by a system of slits that can define the scattering angle (0° – 2°) and angular resolution (typically 0.1°). Increased angular resolution is achieved at the expense of sensitivity. The angular distribution of CAD products is due to scattering and to the release of translational energy that occurs during fragmentation.

6.3.3 **Analytical applications of CAD-MS**

The three types of analytical infomation available from a CAD mass spectrum are (1) the relative intensities of the secondary ions; (2) the translational energy which is released upon dissociation; and (3) the excitation energy which is transferred. The identification of organic compounds can be made by matching the CAD mass spectrum with the corresponding spectrum of a known reference compound (118).

CAD-MS can unambiguously differentiate between leucine and isoleucine residues in peptides, regardless of their relative position in the peptide, by studying an immonium ion found in the mass spectra of all oligopeptides that contain these amino acids (119). Amino acid sequence-determining ions in the EI mass spectra of peptides are created by cleavages on either side of the carbonyl group of the peptide bond, with charge retention by the *N*-terminal-containing portion. However the CAD mass spectra have peaks of higher abundance, more amino acid sequence-determining peaks, and fewer rearrangement peaks.

Studies have been made of the CAD mass spectra of field-ionized molecules including the acetylated, permethylated tripeptide leu-gly-gly (120), and also nine dipeptides and two tripeptides (121). In the latter report it was shown that whereas the FD mass spectra of the two isobaric and isomeric dipeptides glycylalanine and alanylglycine were similar, the CAD mass spectra of the two protonated isomers differed significantly. Several aromatic compounds and two mixtures (estrone and progesterone; phenacetine and methylsalicylate) were studied (120). These authors stated that the direct identification of the compounds in the mixture may be possible without any prior chromatographic separation. This concept of greatly increased molecular specificity in the analysis of an unresolved mixture is elaborated more fully later.

A triple quadrupole spectrometer (TSQ) was used in the direct amino acid sequence determination of complex mixtures of oligopeptides (122). *N*-acetylated and *N,O*-permethylated oligopeptides were analysed under CI conditions using isobutane as the reagent gas and the $[M + H]^+$ ions collisionally dissociated to produce amino acid sequence-determining fragment ions. A series of 26 oligopeptides was studied in the normal and CAD modes. Three limitations to previous work were noted: low abundance of ions in the high mass region due to the molecular ions of derivatized oligopeptides; the lower efficiency of production and collection of CAD ions in a double-focusing mass spectrometer; and the lower mass resolution available for measuring CAD products. The TSQ configuration overcame these experimental limitations by using CI to generate abundant $[M + H]^+$ ion currents, by maximizing the ion dissociation and fragment ion collection efficiency using the low energy collisions that occur inside a quadrupole MS, and by the use of a quadrupole instrument to effect unit resolution in the CAD mode. A mixture of d_0- and d_3-*N*-acetylated peptides produced doublets that were observed for all *N*-terminus-containing fragment ions. The amino acid sequences for 24 oligopeptides, produced from glucagon by elastase digestion, were obtained despite the mass range limitation of the instrument (0 – 1000 daltons). The scheme (122) to use amino acid sequence-determination data for measurement of neuropeptides, utilizing electron capture negative ion mass spectrometry and CAD of the dipeptide carboxylate anions at the 100 pmol (5.5 ng) level, was comparable in sensitivity to FAB-CAD-*B*/*E* methods (see below).

A combination of SIMS and CAD-MS, to determine amino acid sequence information on oligopeptides, utilized a triple quadrupole mass spectrometer fitted with an ion and a saddle-field neutral beam gun (123). Samples were introduced by a modified LC-MS moving belt interface (for review see 124). A tetrapeptide (5 nmol of met-arg-phe-ala) in glycerol matrix, containing oxalic acid to pre-protonate the peptide, produced an abundant quasimolecular ion by SIMS but few amino acid sequence-determining ions. The latter were produced by the CAD processes. In a second

Table 3. Mass spectrometric instrumental scanning methods that are avilable to observe metastable transitions deriving from the equation $M_1^+ \rightarrow M_2^+ + (M_1 - M_2)$

	Scanned fields	Fixed fields	Information obtained
1.	B	V,E	$(M_2)^2/M_1$
2.	V	E,B	All M_1^+ from selected M_2^+
3.	E	V,B	For (E,B), all M_1^+/M_2^+. For (B,E), all M_2^+ from selected M_1^+
4.	$V^{1/2}/E$	B	All M_2^+ from selected M_1^+
5.	B/E	B	All M_2^+ from selected M_1^+
6.	B^2/E	V	All M_1^+ from selected M_2^+
7.	$(B/E)(1 - E)^{1/2}$	V	Constant neutral fragment loss

experiment, d_0- and d_3-*N*-acetylated met-arg-phe-ala were studied in a mixture of four different peptides by analyzing the $[M + H]^+$ ions. Analysis of the CAD products showed that the mass spectra were free of ions from the other three peptides in the mixture.

6.4 Scanning methods to obtain metastable transitions

The three fields, accelerating voltage (V), the electric sector field (E), and the magnetic sector field (B) in a double-focusing mass spectrometer can be varied independently. These fields may be scanned either separately or in several combined or linked-field methods to collect the product ions from metastable transitions in the FFRs. These scans are listed in Table 3 and described in Chapter 1, Section 3.1. Of particular interest to the studies outlined in the following Section is the B/E scan. In the B/E linked-field scan (method 5), the acceleration voltage V is held constant while the (B) and (E) fields are scanned, maintaining the ratio B/E constant throughout the scan. This linked-field scan mode permits all product ions from a selected precursor ion to be determined. It has the advantages that the source extraction efficiency is constant (V is constant) and the mass range scanned is limited only by that range over which the Hall probe magnetic field relationship is operable. The sensitivity of this linked-field scan method is higher than the other scan methods due to the greater intensity of metastable ion products from decompositions in the first FFR. However, as in the E-scan (method 3) with a reverse geometry instrument, the response of the electron multiplier decreases as E decreases because the relative intensity of peaks within a given linked-field scan will depend on the energy released from each decomposition.

Further details of metastable decompositions and the linked scan modes for their detection can be obtained elsewhere (110, 111, 125–128).

An elegant set of experiments was performed in an effort to differentiate decompositions that occurred in the first FFR region of a double-focusing mass spectrometer from those in the second FFR (129). The genesis of the amino acid sequence-determining ions in tripeptides that contain an *N*-terminal trifunctional amino acid was analyzed by direct analysis of daughter ions/mass analyzed ion kinetic energy (DADI/MIKES) and linked-field scan (B/E) mass spectrometry. CAD in the first and second FFRs of a reversed geometry instrument were compared and the mass spectrum of the tripeptide *N*-acetyl glu-ala-leu-Me was investigated in detail. The DADI/MIKE mass spectra showed very few differences when they were compared to the *B*/*E* mass spectra, contrary to initial expectations. CAD processes enhanced the number and intensity of fragment ions. In this study, DADI/MIKE spectra were obtained at a magnetic field set to transmit one selected ion to the second FFR, where CAD processes occurred. The voltage on the electric sector was scanned and a normal MIKE mass spectrum obtained with the normally observed broad peaks. On the other hand, "younger" ion transitions were obtained with a linked-field (*B*/*E*) scan where a precursor ion was selected and the first FFR decompositions studied with and without CAD processes. A *B*/*E* scan was taken to detect all the product ions arising from that precursor ion. The peaks were comparatively narrow in the *B*/*E* scan mode compared to a MIKE scan.

7 Analytical Measurement of Peptides by Mass Spectrometry

7.1 Basic considerations

A rapid and direct method for the measurement of endogenous peptides with maximum molecular specificity is essential to the development of current research involving these compounds. Mass spectrometric techniques can provide such a method based upon the specific determination of the $[M + H]^+$ ion current of an individual peptide. This should be calibrated against an identical measurement on a known amount of the isotopically labeled form of the peptide as internal standard.

MS measurement of peptides can be performed routinely at the level of ng peptide $\cdot$ g^{-1} tissue (1, 27, 28, 130–145). This sensitivity corresponds to 1 pmol $\cdot$ g^{-1} for peptides with a molecular mass of 1000, but recent MS instrumental developments indicate that fmol sensitivity should be attained without loss of molecular specificity.

The molecular ion of a peptide alone conveys no amino acid sequence (structural) information. For example, 555 is the nominal mass, $C_{28}H_{37}N_5O_7$ the elemental composition, and 555.2693 the accurate mass of the molecular mass of LE (tyr-gly-gly-phe-leu). For any 5 given amino acids 120 different sequence permutations are possible. (This number is reduced in the case of LE as it has two glycine residues). Thus a significant increase in molecular specificity is required for an objective and structurally unambiguous analytical measurement of LE in a biological matrix. That increase in molecular specificity can be provided by selecting an amino acid sequence-determining ion in its mass spectrum as the basis for the analytical measurement.

An amino acid sequence-determining fragment ion can usually be produced from the $[M + H]^+$ ion from FAB of an endogenous neuropeptide by CAD or, in some cases, directly by unimolecular dissociation. The CAD mass spectrum of the $[M + H]^+$ ion can be acquired by linked-field scanning (*B*/*E*) – see Section 6.5. A unique amino acid sequence-determining ion monitored at a specific *B*/*E* setting is then used for quantitative measurement of the endogenous peptide.

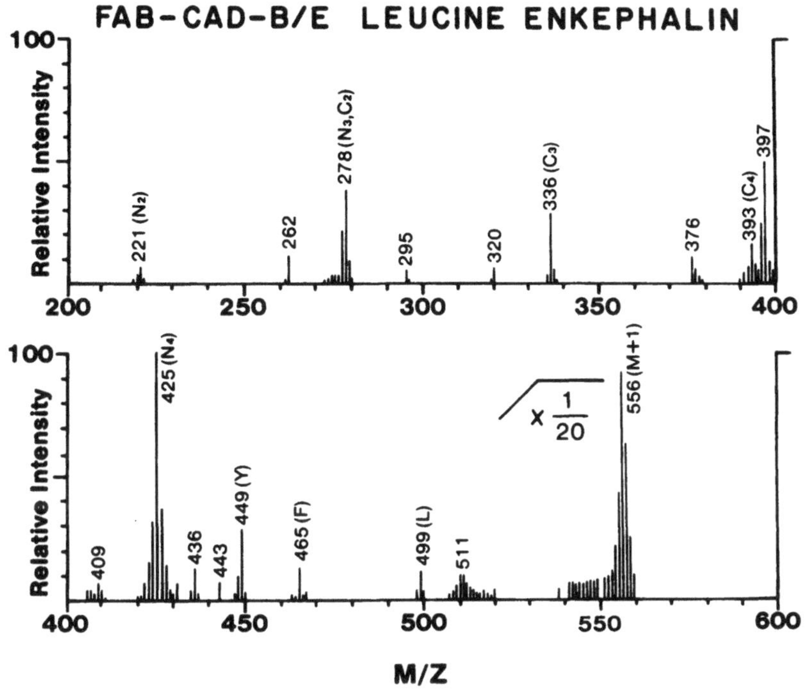

Figure 4. FAB-CAD-*B*/*E* mass spectrum of LE. One µg of synthetic LE in glycerol used for FAB-*B*/*E* linked-field scan of CAD product ions (1st FFR) of $[M + H]^+$ of LE at *m*/*z* 556. (Reproduced from ref. 1 with permission.)

7.2 Measurement of endogenous peptides

Figure 4 shows the FAB-CAD-*B*/*E* mass spectrum of the $[M + H]^+$ ion produced from LE (136), and Figure 5 the corresponding spectrum of LE containing two $[^{18}O]$ atoms located specifically in the carboxy terminus (1). It can be seen by comparing the two mass spectra that the *N*-terminus-containing ions, *m*/*z* 221, 278 and 425, remain unchanged in Figure 4 whereas all of the *C*-terminus-containing ions, *m*/*z* 262, part of 278, 336, 449, 465, 499 and 556, shift the appropriate four mass units. Clearly, the *C*-terminus-containing ions (Fig. 6) can be used to differentiate the two opioid pentapeptides ME and LE. The *C*-terminal tripeptide fragment corresponding to -gly-phe-leu at *m*/*z* 336 is particularly suitable for quantification as it occurs in a noise-free region with no neighboring peaks and shifts cleanly and completely in the FAB-CAD-*B*/*E* mass spectrum of the stable isotope-incorporated enkephalin to *m*/*z* 340.

7.3 Stable isotope-incorporated peptide internal standards

In principle, several types of compound may serve as internal standards for MS measurement of endogenous peptides such as any well-characterized organic compound. However the accuracy and precision achievable by these standards is not always adequate (146).

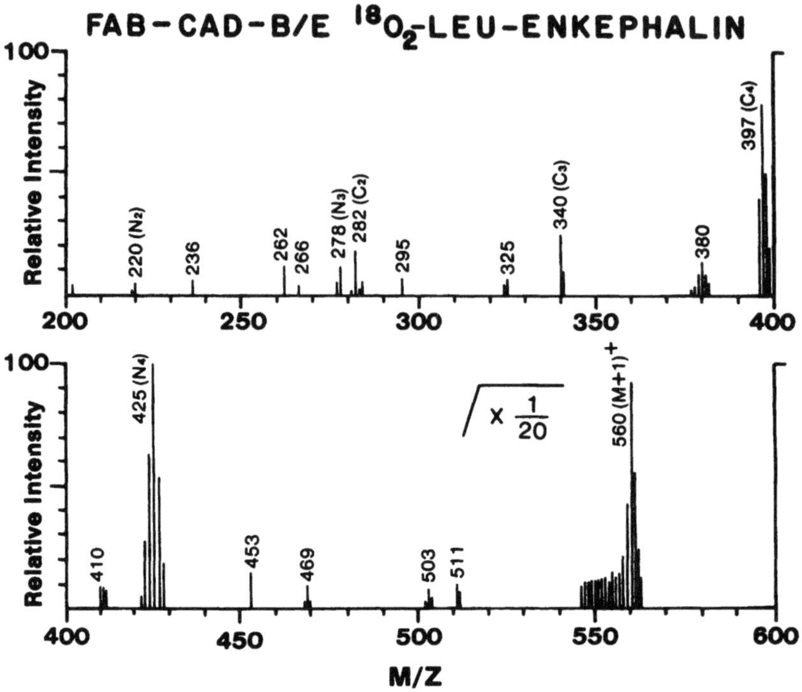

Figure 5. FAB-CAD-*B*/*E* mass spectrum of $^{18}O_2$-LE. One µg of synthetic $^{18}O_2$-LE in glycerol used FAB-*B*/*E* linked-field scan of CAD product ions (1st FFR) of $[M + H]^+$ of $^{18}O_2$ LE at *m*/*z* 560. (Reproduced from ref. 1 with permission.)

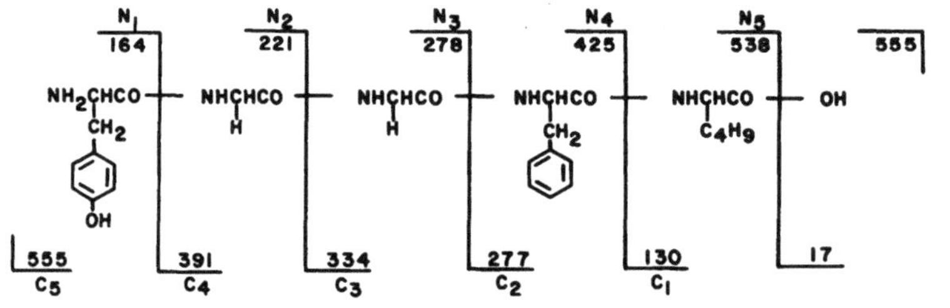

Figure 6. Fragmentation pattern of LE. This simplified fragmentation pattern indicates the preferentially cleaved amide bonds. C-terminal and N-terminal containing fragment masses are indicated. The C_3 fragment at 334, that adds two hydrogens to produce the ion found at 336, is the ion used for quantification. (Reproduced from ref. 1 with permission).

Based on a number of studies (e. g. 147, 148), the most appropriate internal standard for quantitative measurement of an endogenous peptide is a stable isotope-incorporated peptide. Such a compound has similar physico-chemical characteristics in terms of extraction efficiency, RP-HPLC retention time, and MS ionization efficiency and fragmentation to the unlabeled peptide. An increase of at least four daltons by isotope incorporation avoids interference between the molecular ion peaks on MS analysis.

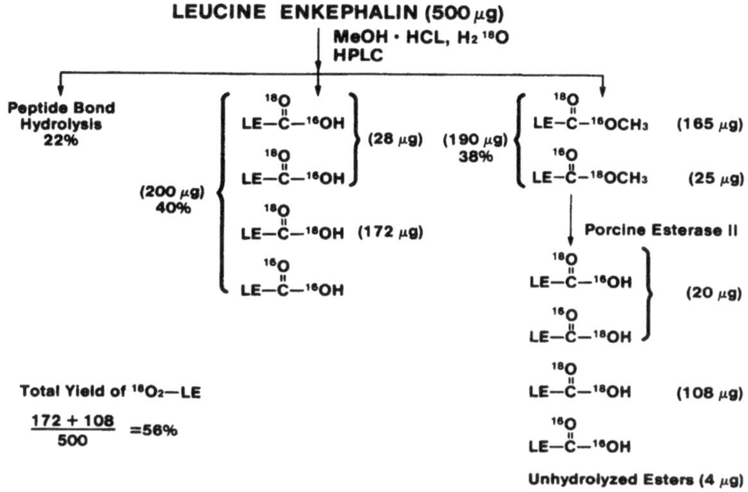

Figure 7. Chemical incorporation of ^{18}O into LE. Synthetic LE is treated with ^{18}O-enriched water and methanolic HCl to produce three sets of HPLC fractions: hydrolytic products, direct incorporation of 0, 1, or 2 ^{18}O atoms, and the methyl ester with 0 or 1 ^{18}O atom, which is subsequently treated with esterase. (Reproduced from ref. 136 with permission).

The preparation of individual isotope-incorporated standards from labeled amino acid precursors can be circumvented by incorporation of ^{18}O into the free carboxyl group of any internal amino acid residue (149). This method provides synthetic flexibility for the preparation of peptide internal standards. If no free carboxylic group is present, a deamidation-[^{18}O]-incorporation reamidation sequence may be appropriate. [^{18}O]-peptides possess virtually equivalent RP-HPLC and MS properties although stable isotope-labeled compounds can have slightly different hydrophobicity values (148).

Either acid-catalyzed or porcine esterase II-catalyzed (150) exchange reactions can be used to incorporate [^{18}O] from $H_2{}^{18}O$ (136). Figure 7 outlines the reactions to introduce [^{18}O] atoms into LE by Fischer esterification, to separate the products by HPLC, and to recover the incorporated [^{18}O] atoms by porcine esterase (136). The products of Fischer esterification yielded three peaks on HPLC due to: (1) peptide bond hydrolysis, (2) direct incorporation of none, one, or two oxygen atoms into the peptide carboxyl group, and (3) methyl esterification, with incorporation of none or one [^{18}O] atoms. In the latter case, exchange was with either oxygen atom. The expensive [^{18}O] stable isotope was recovered conveniently by digestion with porcine esterase II. The total yield of [^{18}O] LE was 56%.

Figure 8 shows the reaction kinetics (136) of [^{18}O]-incorporation into the C-terminus of LE (upper) and the LE methyl ester (lower). In the upper panel, the reaction kinetics demonstrate that the [$^{16}O_2$] species decreases rapidly, and is practically gone within three days. The mixed isotope species [$^{16}O^{18}O$] maximises within one day, and then decreases smoothly. The [$^{18}O_2$] species increases steadily and, after four days, levels out to approximately 90%. In the lower panel containing the methyl ester data, two complementary curves show incorporation of the [^{18}O] species and corresponding decrease of the [^{16}O] species.

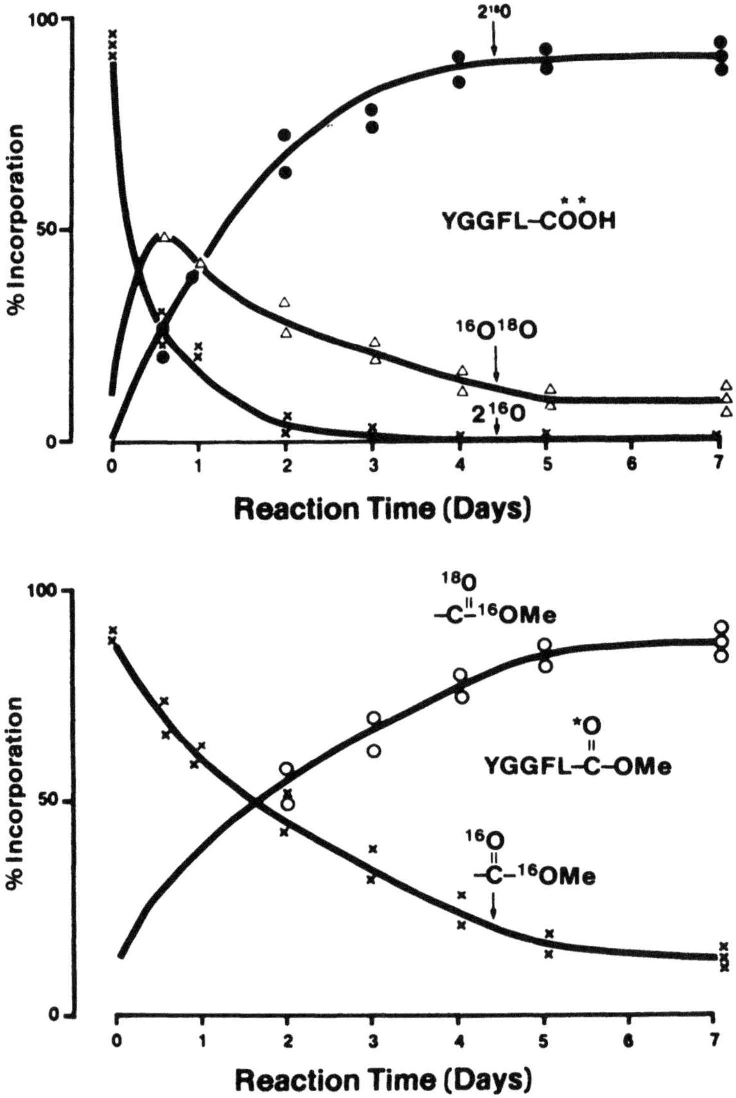

Figure 8. Kinetics of the incorporation of ^{18}O atoms into LE. Lower panel-incorporation of one ^{18}O atom into LE methyl ester. Samples are taken from the reaction mixture each day for seven days, and analyzed by MS for stable isotope content. (Reproduced from ref. 136 with permission).

The FAB-CAD-*B*/*E* scan mass spectra (Figs. 4 and 5) confirm unambiguously that the two [^{18}O] atoms are incorporated solely into the carboxy terminal group of LE, and not into any peptide amide bond or tyrosine hydroxy sidechain. Only the C-containing amino acid sequence-determining ions shift in the latter mass spectrum.

[^{18}O]-incorporated peptide internal standards were used first for the measurement of endogenous ME and LE in brain and tooth pulp tissues and CSF extracts. Several

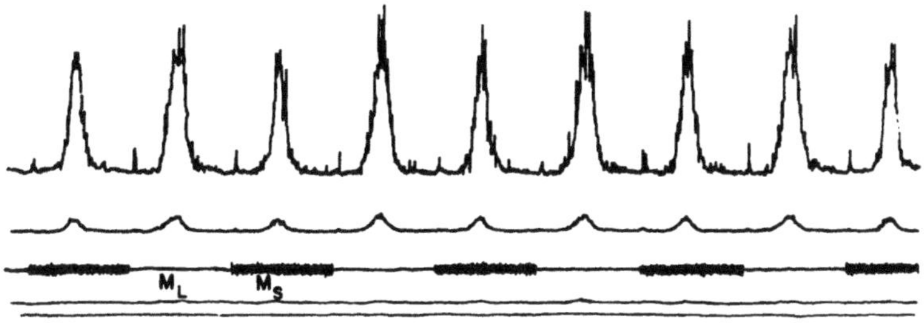

Figure 9. Trace of analog data that is obtained during the MS measurement LE (M_L, 336) and $^{18}O_2$-LE (M_S, 340). The top two analog traces differ only in sensitivity (top: x 10; second from top: x 1). (Reproduced from ref. 150 with permission).

tissues were studied in an effort to elucidate basic molecular mechanisms involved in pain. Caudate nucleus, hypothalamus, pituitary, tooth pulp, spinal cord, thalamus, and CSF extracts have been analyzed with this method (28). Currently, ^{13}C-containing residues are being incorporated into ME for use as internal standards.

7.4 Examples of analytical measurements of endogenous enkephalin peptides

7.4.1 Construction of calibration curve

The method of analysis employing the FAB-CAD-*B*/*E*′-SIM measurement approach must give a linear response over the range of concentrations corresponding to endogenous peptides. Raw data (150) from an accelerating voltage alternation experiment are shown in Figure 9 for LE (2 µg) versus ^{18}O-LE (2 µg). The monitored ions were *m*/*z* 336 and 340 respectively and correspond to the tripeptide sequence -gly-phe-leu (-GFL) derived from the CAD-*B*/*E* scan mode. As indicated earlier these amino acid sequence-determining ions were the product ions arising only from the $[M + H]^+$ ions *m*/*z* 556 of LE and *m*/*z* 560 of ^{18}O-LE. In the oscillographic recording (Fig. 9) the $[^{18}O]$-species is clearly indicated when the higher mass (M_S) was being monitored by a 60 Hz signal superimposed on the lower mass trace. The *m*/*z* 336 from endogenous LE is denoted by M_L. The microcomputer accepted a large number of signals to optimize ion statistics. Integrated areas were calculated, and the ratio of the known amounts of the $[^{16}O]$ and the $[^{18}O]$ species was plotted for a series of concentrations. Figure 10 contains the calibration lines for the ME (upper) and LE (lower) experiments (150). Both lines intercept at, or very near, the origin and have correlation coefficients near unity. The statistical parameters for the two best-fit straight-lines were, for ME: $y = 1.09\,x + 0.06, r^2 = 0.999$ and for LE: $y = 0.69\,x, r^2 = 0.995$. Other studies have demonstrated practical sensitivity down to 72 fmol (40 pg) for LE by FAB-MS/MS (151) and amol for negative ion FAB-MS (152).

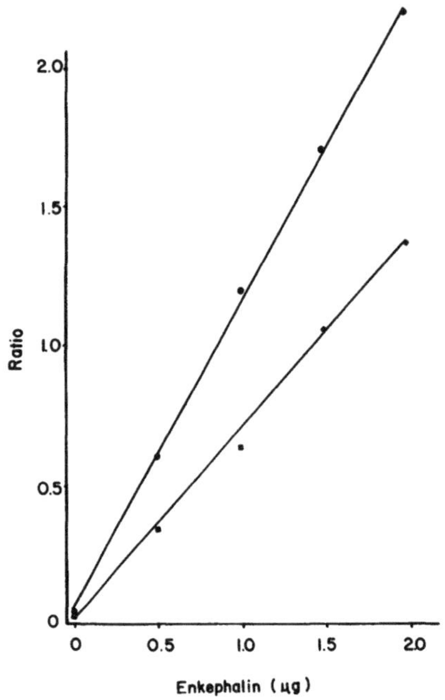

Figure 10
Calibration curves that are constructed from the integrated MS peaks areas vs. amount of ME (upper line) and LE (lower line) (150). These data cover the analytical range of zero to 2.0 µg of ME and LE. (Reproduced from ref. 150 with permission).

7.4.2 **Hypothalamus**

The RP-HPLC for canine hypothalamus tissue (2 g) is shown in Figure 2 (28) and the chromatogram of the sample spiked with the two enkephalin peptide internal standards $^{18}O_2$-ME and $^{18}O_2$-LE is shown in Figure 11. Arrows indicate the known retention times of synthetic ME and LE, respectively. It should be noted that, at the UV wavelength monitored (200 nm), the RP-HPLC chromatogram of a hypothalamic peptide-rich fraction was relatively clear of UV-absorbing material. However a wealth of other RRA, BA, or RIA responding material may, and generally does, coelute on this chromatogram (74). Experimentally with UV spectrophotometry microgram amounts of peptides are required for detection, whereas RRA, RIA, and BA are capable of detecting ng and pg amounts.

In the MS analysis virtually all chromatographic and chemical background noise disappeared and the resultant chromatographic resolution increased significantly as only one compound was monitored.

Table 4 lists the amounts of endogenous enkephalin that are determined in the peptide-rich fraction of a variety of canine tissue extracts.

7.4.3 **Pituitary**

Ten canine pituitaries were pooled and neuroanatomically separated into the anterior (1.9 g total wet weight) and posterior (0.44 g total wet weight) portions. The tissue was homogenized in acetic acid ($1 \text{ mol} \cdot l^{-1}$) and divided into three equal samples. Table 4 records the amounts of the enkephalins found in the anterior (LE = 70; ME = $2950 \text{ ng} \cdot g^{-1}$) and posterior (LE = 2; ME = $3760 \text{ ng} \cdot g^{-1}$) pituitary extracts.

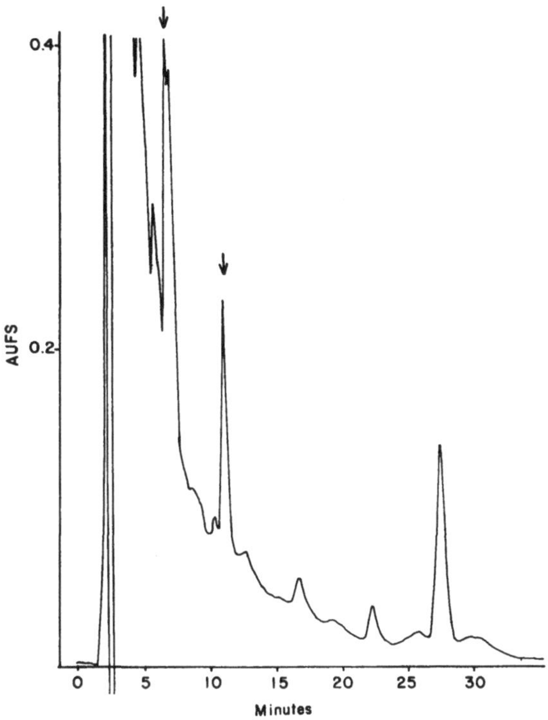

Figure 11
RP-HPLC of the peptide-rich fraction extracted from canine hypothalamus, which is spiked with ^{18}O-ME and ^{18}O-LE (denoted by arrows, respectively) (28). The two indicated fractions were collected, analyzed by MS, and the ratio of endogenous ^{16}O peptide to exogenous ^{18}O-internal standard measured. (Reproduced from ref. 28 with permission.)

7.4.4 **Caudate nucleus**

FD-MS methods have been used to measure the endogenous enkephalin in canine nucleus tissue extracts (144, 145). The amount of endogenous enkephalin in the canine caudate nucleus tissue extract was 190 ng of ME and 1500 ng LE $\cdot$ g^{-1} tissue. This measurement represented one of the few times where the concentration of LE exceeded that of ME.

7.4.5 **Tooth pulp**

Tooth pulp tissue was collected from four animals and pooled (four teeth from each animal; total = 16). Canine tooth pulp RP-HPLC chromatograms have been published (143). The endogenous amount of enkephalin for pooled tooth pulp tissue is 179 ng ME and 30 ng LE $\cdot$ g^{-1} tooth pulp tissue respectively.

7.4.6 **Electrostimulated tooth pulp**

It is thought that four peptidergic pathways (Section 2.1) maintain a dynamic homeostatic relationship and deal with noxious stimuli. These pathways involve large and intermediate precursor peptides, the opioid oligopeptide, and metabolites. Noxious

Table 4. Mass spectrometric measurement of enkephalins in canine tissue extracts ($ng \cdot ml^{-1}$) and CSF ($ng \cdot ml^{-1}$) (28)

Source	Enkephalin	Amount	Method*
Hypothalamus	LE	170	1
Cerebrospinal fluid	LE	44	1
Pituitary			
anterior	LE	70	2
	ME	2950	2
posterior	LE	2	2
	ME	3760	2
Caudate nucleus	LE	1500	2
Tooth pulp**	LE	30	2
	ME	179	2
Tooth pulp			
control	LE	20	2
	ME	487	2
electrostimulated	LE	45	2
	ME	390	2

* Method 1. FDMS-SIM-Microcomputer
2. FAB-CAD-*B*/*E*-*B'*/*E'*-SIM-Microcomputer

** Unstimulated pooled tissue from five dogs

stimuli and stress may activate the peptidergic pathways, and individual opioid peptides may have decreased concentrations following stimulation.

In a study to determine whether electrical stimulation affected endogenous enkephalin concentrations, three animals served as control and three for electrostimulation experiments. The results indicated that the opioid pentapeptides ME and LE were altered by electrostimulation (153, 154). The preliminary data suggested that the four pathways may be mobilized in the following sequence: large precursor → intermediate precursor(s) → "working" peptide → inactive metabolites. On one hand, there may be a naturally occurring pool of "working" peptide(s) that is electrostimulated towards metabolism or, on the other hand, the entire metabolic scheme noted above may be stimulated to produce a lowered endogenous amount of each constituent peptide. Other human and *in vivo* dynamic studies are needed to resolve that question.

7.5 Significance

Several conclusions can be derived from the studies described in relation to tissue peptide measurement. The method of tissue sample acquisition and procurement was important and should involve rapid freezing to avoid, or at least minimize, metabolic and chemical interconversions and to increase the probability of measuring only target compounds and not artefacts or chemical/enzyme products. An internal standard was essential for MS measurement to overcome the biological matrix effects (147, 148). The appropriate stable isotope-incorporated peptides standards should be added as soon as possible after tissue acquisition and given sufficient time for equilibration (147).

The most fundamental parameter of any assay of a biological compound is the molecular specificity of the measurement and this parameter is experimentally difficult to prove definitely (155). Many methods are employed because of the relative ease of use, low cost, high speed, high sensitivity, and (putatively) high molecular specificity. These methods include chromatography, colour reactions, enzymatic reactions, HPLC, BA, RRA and RIA. However the molecular specificity of all of these techniques is insufficient for the unambiguous structural proof during an analytical measurement. The uniqueness of the $[M + H]^+$ ion in the mass spectrum of an endogenous peptide is not in itself sufficient to confer unambiguous molecular specificity to quantitative measurement of the peptide. Presently, maximum specificity is derived by determining an amino acid sequence-determining fragment arising by unimolecular or CAD decomposition of the $[M + H]^+$ ion. Furthermore, the use of a stable-isotope-incorporated internal standard of the peptide is necessary to substantiate this molecular specificity.

The current instrumental limit of sensitivity corresponds to approximately 30 μg peptide $\cdot$ kg^{-1} tissue. This limit copes adequately with enkephalin peptides, for example, that are present in canine caudate nucleus tissue in the region of 200 – 400 μg $\cdot$ kg^{-1}. Significant instrumental increase (10 – 100 x) in the detection limits is likely to be realized in future, and current probe chemistry improvements indicate amol sensitivity (152).

It is important to undertake the analytical and physiological studies within one laboratory to ensure quality control over experimental manipulations that range from the live animal model through exsanguination, tissue procurement, homogenization, chromatography, MS, and data analysis.

7.6 Metabolic profiles of peptides in tooth pulp

It is necessary to be able to discern what effect, if any, a physiological experiment will have on the peptidergic pathways. Towards that end, metabolic profiles were obtained for peptides contained in canine tooth pulp extracts (156). Several immobilized proteolytic enzymes (α-chymotrypsin, trypsin, and carboxypeptidase A) were utilized to determine which HPLC peaks derive specifically from peptides. Results from the treatment of tooth pulp extracts with these enzymes indicated that almost all peaks in this HPLC system were due to peptides with many of the smaller peptides being produced in the digestions from larger peptides.

Figure 12 shows the RP-HPLC gradient chromatogram of an extract of control tooth pulp tissue, using a microBondpak C_{18} column (200 nm, 1.5 ml $\cdot$ min^{-1}) flow rate. The gradient profile is denoted by the solid line on the chromatogram and the shaded peaks indicate those peaks that disappeared following trypsin treatment. The digestion was carried out on lyophilized tooth pulp peptide extract (100 ml sample, 250 μg dry solid equivalent to 70 mg sample tissue) in TEAP buffer (0.06 mol $\cdot$ l^{-1}, pH 2.12) adjusted to pH 8.5 with concentrated ammonium hydroxide (1 – 2 μg) and stirred for one hour with TPCK treated trypsin (1.5 units). Figure 13 contains the RP-HPLC gradient chromatogram of the tooth pulp extract following treatment with

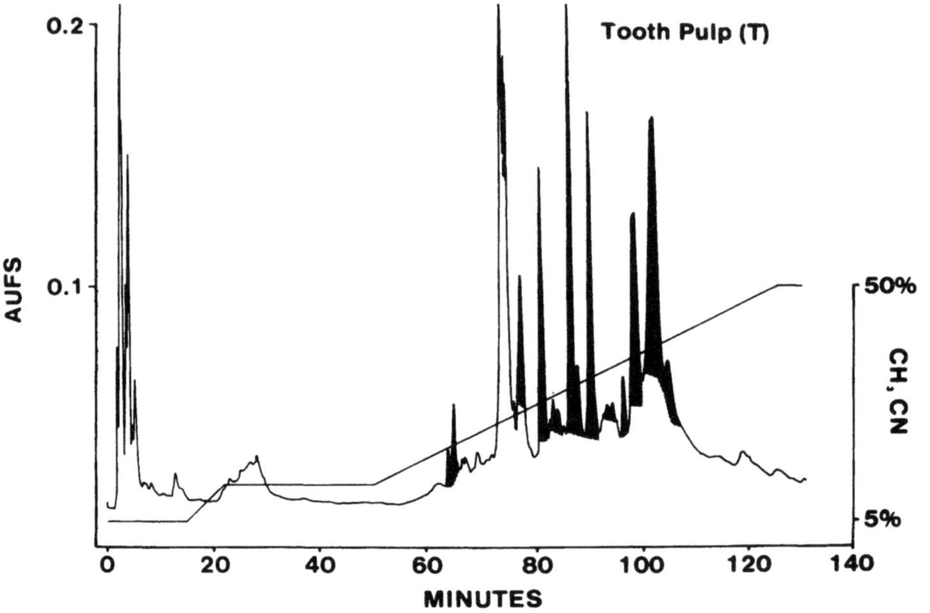

Figure 12. RP-HPLC of the peptide-rich fraction extracted from canine tooth pulp before trypsinolysis. The darkened peaks disappear after trypsinolysis, and shift to a more hydrophobic region. (Reproduced from ref. 156 with permission.)

trypsin and can be referred to as the peptide metabolic profile. It indicates new peaks deriving from larger peptides that contained trypsin-sensitive bonds.

This experiment was only one of a number of studies directed at monitoring and quantifying the enkephalin-, dynorphin-, and endorphin-related peptides in extracts of tooth pulp tissue and CSF, and utilized a variety of enzyme preparations: α-chymotrypsin on cellulose, carboxypeptidase A on agarose, and trypsin on 4% agarose. Following enzyme treatment of extracts nearly all of the higher molecular weight species shifted to a lower retention time, reflecting peptide bond cleavage and production of smaller peptides with decreased hydrophobicity. An HPLC gradient (5 – 50%) of organic modifier (acetonitrile) was utilized to elute larger peptides. Peptides with too large a molecular mass (e. g. β-endorphin = 3642), for the MS analysis on limited range mass spectrometers, can be analyzed, following enzyme treatment and isolation by gradient HPLC elution or isocratic purification (157). Newer high mass MS instrumentation and FAB will decrease the need for extensive enzymolysis studies.

An important strategy in the characterization of peptides from neuronal tissue is screening HPLC fractions for immunoreactive or opiate receptor activity prior to more detailed molecular analysis. Changes in concentration of ME, LE, β-endorphin, and other neuropeptides following physiological manipulation can be identified by RRA or RIA assays. RRA is more efficient for this purpose due to its wider range of "screening" reactivity. Experiments based on RRA employing a canine limbic system synaptosome fraction and gradient RP-HPLC separation have been described for opioid peptide analysis (37).

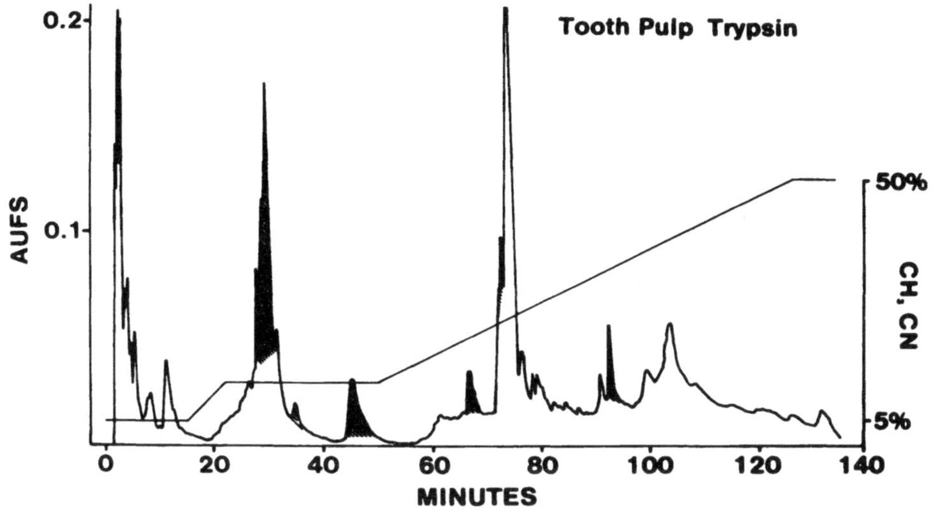

Figure 13. RP-HPLC of the peptide-rich fraction extracted from canine tooth pulp following trypsinolysis. The hatched peaks have appeared after trypsinolysis. (Reproduced from ref. 156 with permission.)

7.7 Comparison of FD and FAB measurement methods

Before FAB was introduced, FD was the only ionization method available for studies of peptides and many analytical measurements of endogenous peptides were made in this way of the $[M + H]^+$ ion in the SIM mode with a higher homolog of the peptide serving as internal standard.

However, the delicate FD emitters were easily broken and although not a serious problem with peptide standard solutions, the matrix in real biological samples increased the frequency of emitter breakage. FAB-MS circumvented these problems and provided a simpler and more convenient technique. High molecular specificity was obtained, as described in this chapter, by combining FAB ionization with CAD followed by monitoring selected product ions in the linked-field (B/E) scan mode. Amino acid sequence-determining information, for example, was obtained from nmol amounts of underivatized substance P in this way (158).

A comparison of FD-MS and FAB-MS for the measurement of LE showed a difference in the concentration values obtained. The FAB-CAD-B/E-B'/E'-SIM, method described in section 7.2, yielded a value for the endogenous amount of LE in canine caudate nucleus of 250 ng LE $\cdot$ g^{-1} wet weight tissue corresponding to 451 pmol (140). In a parallel experiment using the other half of the same tissue extract, FD-MS in the SIM mode of the $[M + H]^+$ ion gave a value 160 ng $\cdot$ g^{-1} of LE. The difference between these measurements reflected in part the difference in signal-to-noise ratios of the two ionization and instrumental analysis modes, where FD was found to have a higher signal-to-noise ratio compared to the linked-field procedure. The amount of LE determined in canine caudate tissue compared favorably to previously reported MS (423 and 351 ng $\cdot$ g^{-1} of tissue) (145) and RIA (210 ng $\cdot$ g^{-1}) data (159).

FAB-MS produced a steady and abundant $[M + H]^+$ ion current that lasted for minutes for endogenous amounts of peptides. FD, on the other hand, produced a spurious, rapidly fluctuating, and smaller ion current. However the high glycerol matrix background observed with FAB made it essential to use CAD and linked-field scanning to effectively increase the molecular specificity of the measurement (1, 141, 143).

8 Conclusions

The purpose of this chapter was to describe a recently developed MS methodology pertinent to the field of peptide chemistry and prompted by the molecular specificity demanded by current research. Although HPLC, RIA, and RRA are methods that may be relatively cheap, fast, and facile to use, there is no significant level of molecular specificity that can be attached to those measurement modes. It should be pointed out that even mass spectrometry, which can produce an $[M + H]^+$ ion of a peptide, still cannot confer an unambiguous level of molecular specificity to an analytical measurement, even in the high resolution mode, because several peptides that have different amino acid sequences may still produce the same molecular ion (nominal mass and accurate mass). Only when an amino acid sequence determining fragment ion can be obtained from a selected $[M + H]^+$ ion, which in turn is produced from a selected liquid chromatographic fraction, can there be a sufficient level of confidence in the peptide measurement. This higher level of molecular specificity is also needed in the study of other biological polymers, such as oligosaccharides, oligonucleotides, and oligopeptides.

Acknowledgements

The author gratefully acknowledges the financial assistance of NIH (GM 28 611, 26 666), typing assistance of D. Cubbins, editorial assistance of G. Fridland, and the scientific collaboration of colleagues.

References

1. Desiderio, D. M. (1984) Analysis of Neuropeptides with Liquid Chromatography Mass Spectrometry, Elsevier, Amsterdam.
2. Barchas, J. D., Akil, H., Elliott, G. R., Holman, R. B. & Watson, S. J. (1978) Behavorial neurochemistry: neuroregulators and behavioral states. Sience *200*, 964–973.
3. O'Donohue, T. L. & Dorsa, D. M. (1982) The opiomelanotropinergic neuronal and endocrine systems. Peptides *3*, 250–252.
4. Rossier, J., Bayon, A., Vargo, T. M., Ling, N., Guillemin, R. & Bloom, F. (1977) RIA of brain peptides: evaluation of a methodology for the assay of beta-endorphin and enkephalin. Life Sci. *21*, 847–852.
5. Wesche, D., Hollt, V. & Herz, A. (1977) RIA of enkephalins. Arch. Pharmacol. *201*, 79–82.
6. Goldstein, A., Fischli, W., Lowney, L. I., Hunkapiller, M. & Hood, L. (1981) Porcine

pituitary dynorphin: complete amino acid sequence of the biologically active heptadecapeptide. Proc. Natl. Acad. Sci. USA *78*, 7219 – 7223.
7. Chavin, C., James, I. F. & Goldstein, A. (1982) Dynorphin is a specific endogenous ligand of the kappa opioid receptor. Science *215*, 413 – 415.
8. Fischli, W., Goldstein, A., Hunkapiller, M. W. & Hood, L. E. (1982) Isolation and amino acid sequence analysis of a 4,000 dalton dynorphin from porcine pituitary. Proc. Natl. Acad. Sci. USA 79, 5435 – 5437.
9. Watson, S. J., Khachaturian, H., Taylor, L., Fischli, W. & Akil, H. (1983) Pro-dynorphin peptides are found in the same neurons throughout rat brain: immunocytochemical study. Proc. Natl. Acad. Sci. *80*, 891 – 894.
10. Watson, S. J., Khachaturian, H., Akil, H., Coy, D. H. & Goldstein, A. (1982) Comparison of the distribution of dynorphin systems and enkephalin systems in the brain. Science *218*, 1134 – 1136.
11. Vincent, S. R., Hokfelt, T., Christensson, I. & Terenius, L. (1982) Dyn-Immunoreactive neurons in rat CNS. Neurosci. Lett. *33*, 185 – 190.
12. Krieger, D. T. & Martin, J. B. (1981) Brain peptides. New Eng. J. Med. *304*, 876 – 885.
13. Krieger, D. T. & Martin, J. B. (1981) Brain peptides. New Eng. J. Med. *304*, 944 – 951.
14. Jonsson, H. T., Middleditch, B. S. & Desiderio, D. M. (1975) Prostaglandins in human seminal fluid: Two novel compounds. Science *187*, 1093 – 1094.
15. Jonsson, H. T., Middleditch, B. S., Schexnayder, M. A. & Desiderio, D. M. (1976) 11,14,19-trihydroxy-9-ketoprosta-5,13-dienoic acid in human seminal fluid. J. Lip. Res. *17*, 1 – 6.
16. Perry, D. L. and Desiderio, D. M. (1977) Quantification of prostaglandins in human seminal fluid. Prostaglandins *14*, 745 – 750.
17. Akil, H., Young, E., Watson, S. J. & Coy, D. H. (1981) Opiate binding properties of naturally occurring N- and C-terminus modified beta endorphins. Peptides *2*, 289 – 292.
18. Brot, N., Werth, J., Koster, D. & Weissbach, H. (1982) Reduction of N-acetyl methionine sulfoxide: a simple assay for peptide methionine sulfoxide reductase. Anal. Biochem. *122*, 291 – 294.
19. Dayhoff, M. O. (1972) Atlas of Protein Sequence and Structure – 1972, Vol. 5, 124 pp. (plus 418 pp data section), National Biomedical Research Foundation, Washington, D. C.
20. Carr, D. B. (1981) Endorphins at the approach of death., Lancet p. 390.
21. Nauta, W. J. H. & Feirtag, M. (1979) The Brain, p. 47, Freeman, San Francisco.
22. Schwartz, J.-C., Malfroy, B. & de la Baume, S. (1981) Biological inactivation of enkephalins and the role of enkephalin-dipepidyl-carboxypeptidase ("enkephalinase") as neuropeptidase. Life Sci. *29*, 1715 – 1740.
23. Bennet, H. P. J., Browne, C. A. & Solomon. S. (1981) Biosynthesis of phosphorylated forms of corticotropin-related peptides. Proc. Natl. Acad. Sci. USA *78*, 4713 – 4717.
24. Browne, C. A., Bennett, H. P. J. & Solomon, S. (1981) Isolation and characterization of corticotropin- and melantropin-related peptides from the neurointermediary lobe of the rat pituitary by reverse phase liquid chromatography. Biochem. *20*, 4538 – 4546.
25. Eipper, B. A. & Mains, R. E. (1982) Phosphorylation of pro-adrenocorticotropin/endorphin-derived peptides. J. Biol. Chem. *257*, 4907 – 4915.
26. Desiderio, D. M. & Cunningham, M. D. (1981) Triethylamine formate buffer for HPLC-field desorption mass spectrometry of oligopeptides. J. Liq. Chromatogr. *4*, 721 – 733.
27. Desiderio, D. M., Yamada, S., Tanzer, F. S., Horton, J. & Trimble, J. (1981) HPLC and field desorption mass spectrometry measurement of picomole amounts of endogenous neuropeptides in biologic tissue. J. Chromatogr. *217*, 437 – 452.
28. Desiderio, D. M., Kai, M., Tanzer, F. S., Trimble, J. & Wakelyn, C. (1984) Measurement of enkephalin peptides in canine brain regions, teeth, and CSF with HPLC and mass spectrometry. J. Chromatogr. *297*, 245 – 260.
29. Hearn, M. T. W. (1980) in: Advances in Chromatography, Vol. 18 (Giddings, J. C., Grushka, E., Cazes, J., and Brown, P. R., eds.) pp. 59 – 100, Marcel Dekker, New York.
30. Hancock, W. S., Bishop, C. A., Meyer, L. J., Harding, D. R. K. & Hearn, M. T. W. (1978) High pressure liquid chromatography by peptides and proteins. IV. Rapid analysis of peptides by HPLC with hydrophobic ion-pairing of amino groups. J. Chromatogr. *161*, 291 – 298.
31. Hancock, W. S., Bishop, C. A., Prestidge, R. L., Harding, D. R. K. & Hearn, M. T. W. (1978) RP-HPLC of peptides and pro-

teins with ion-pairing agents. Science *200*, 1168–1170.

32. Harding, D. R. K., Bishop, C. A., Tarttelin, M. F. & Hancock, W. S. (1981) Use of perfluoroalkanoic acids as volatile ion pairing reagents in preparative HPLC. Int. J. Pept. Res. *18*, 214–220.
33. Abrahamsson, M. & Groningsson, K. (1980) HPLC of the tetradecapeptide somatostatin. J. Liq. Chromatogr. *3*, 495–511.
34. Morris, H. R., Etienne, A. T., Dell, A. & Albuquerque, R. (1980) A rapid and specific method for the high resolution purification and characterization of neuropeptides. J. Neurochem. *34*, 574–582.
35. Reeve, J., Yamada, T., Chew, P. & Walsh, J. H. (1982) Rapid high yield purification of neuropeptides from canine intestinal muscle. J. Chromatogr. *229*, 57–65.
36. Dunlap, C. E., Gentleman, S. & Lowney, L. I. (1978) Use of trifluoroacetic acid in the separation of opiates and opioid peptides by RP-HPLC. J. Chromatogr. *160*, 191–198.
37. Fridland, G. H. & Desiderio, D. M. (1986) Profiling of neuropeptides using a gradient reversed-phase high-performance liquid chromatography with novel detection methodologies. J. Chromatogr. *379*, 251–268.
38. Felber, J.-P. (1974) Radioimmunoassays. Vol. I of this series, 324–379.
39. Berson, S. A. & Yalow, R. (1958) Radioimmunoassay. Adv. Biol. Med. Phys. *6*, 349–430.
40. Yalow, R. A. (1980) Radioimmunoassay. Ann. Rev. Biophys. Bioeng. *9*, 327–345.
41. Skelly, D. S., Brown, L. P. & Besch, P. K. (1973) Radioimmunoassay. Clin. Chem. *19*, 146–186.
42. Gros, C., Pradelles, P., Rouget, C., Bepoldin, O., Dray, F., Fournie-Zaluski, M. C., Roques, B. P., Pollard, H., Llorens-Cortes, C. & Schwartz, J. C. (1978) Radioimmunoassay of methionine- and leucine-enkephalins in regions of rat brain and comparison with endorphins estimated with a radioreceptor assay. J. Neurochem. *31*, 29–39.
43. Olgart, L., Gazelius, B., Brodin, E. & Nilsson, G. (1977) Release of substance P-like immunoreactivity from the dental pulp. Acta. Physiol. Scand. *101*, 510–512.
44. Hofreiter, B. T., Allen, J. P., Mizera, A. C., Powers, C. D. & Masi, A. M. (1982) HPLC and RIA of rat plasma corticosterone. Steroids *39*, 547–555.
45. Granstom, E. (1978) in: Advances in Prostaglandin and Thromboxane Research, Vol. 5 (Frolick, J., ed.) pp. 119–210, Raven Press, NY.
46. Atweh, S. F. (1982) in: Neurobiology of Opiate Reward Processes (Smith and Lane, eds.) pp. 59–87, Elsevier Biomedical Press, Amsterdam.
47. Simantov, R. & Snyder, S. H. (1976) in: Opiates and Endogenous Opioid Peptides (Kosterlitz, H., ed.) pp. 41–48, Elsevier, Amsterdam.
48. Terenius, L. & Wahlstrom, A. (1975) Search for an endogenous ligand for the opiate receptors. Acta Physiol. Scand. *94*, 74–81.
49. Terenius L. & Wahlstrom, A. (1975) Morphine-like ligand for opiate receptors in human CSF. Life Sci. *16*, 1759–1764.
50. Wahlstrom, A., Johansson, L. & Terenius, L. (1976) in: Opiates and Endogenous Opioid Peptides (Kosterlitz, H., ed.) pp. 49–55, Elsevier, Amsterdam.
51. Terenius, L. (1977) Opioid peptides and opiates differ in receptor selectivity. Psychoneuroendocrinology *2*, 53–58.
52. von Knorring, L. V., Almay, B. G. L., Johansson, F. & Terenius, L., (1978) Pain perception and endorphin levels in cerebrospinal fluid. Pain *5*, 359–365.
53. Nyberg, F. & Terenius, L. (1982) Endorphins in human CSF. Life Sci. *31*, 1737–1740.
54. Nyberg, F., Wahlstrom, A., Sjolund, B. & Terenius, L. (1983) Characterization of electrophoretically separable endorphins in human CSF. Brain Res. *259*, 267–274.
55. Desiderio, D. M. & Takashita, H. (1985) Radioreceptor assay of opioid peptides in selected canine brain regions. Anal. Lett. *18*, 1789–1798.
56. Desiderio, D. M., Tanzer, F. S. & Fridland, G. (1985) Metabolic profiling of opioid peptides in tooth pulp by HPLC and radioreceptor assay. Neuropept. *6*, 463–469.
57. Desiderio, D. M., Cezayirli, R. C., Fridland, G., Robertson, J. T. & Sacks, H. (1985) Metabolic profiling of radioreceptorassayable opioid peptides in a human pituitary ACTH-secreting tumor. Life Sci. *37*, 1823–1828.
58. Takeshita, H., Desiderio, D. M. & Fridland, G. (1986) Metabolic profiling of opioid peptides in canine pituitary and selected brain regions using HPLC with a radioreceptor assay detector. Biomed. Chromatogr. *1*, 126–139.

59. Desiderio, D. M., Onishi, H., Fridland, G., Wood, G. & Pagidipati, D. (1987) HPLC-receptorassay of opioid peptides in the cerebrospinal fluid of lower back pain patients. Biomed. Chromatogr. *2*, 47 – 57.
60. Desiderio, D. M., Fridland, G., Francisco, J., Sacks, H., Robertson, J. T., Cezayirli, R. C., Killmar, J. & Lahren, C. (1987) RP-HPLC-receptorassay profiling of opioid peptides in human pituitary. Clin. Chem. (in press).
61. Heimer, L. (1983) The Human Brain and Spinal Cord, pp. 327 – 329, Springer-Verlag, New York.
62. Atweh, S. F. & Kuhar, M. J. (1983) Distribution and physiological significance of opioid receptors in the brain, Brit. Med. Bull. *39*, 47 – 52.
63. Lord, J. A. H., Waterfield, A. A., Hughes, J. & Kosterlitz, H. W. (1977) Endogenous opioid peptides: multiple agonists and receptors. Nature *267*, 495 – 499.
64. Chang, K. J. & Cuatrecasas, P. (1979) Multiple opiate receptors. J. Biol. Chem. *254*, 2610 – 2618.
65. Chang, K. J., Cooper, B. R., Hazum, E. & Cuatrecasas, P. (1979) Multiple opiate receptors: different regional distribution in the brain and differential binding of opiates and opioid peptides. Molec. Pharm. *16*, 91 – 104.
66. Smith, A. P. & Loh, H. H. (1981) The opiate receptor, in: Hormonal Proteins and Peptides, Vol. 10, pp. 89 – 159, Academic Press, New York.
67. Vaught, J. L., Rothman, R. B. & Westfall, T. C. (1982) Mu and delta receptors: their role in analgesia and in the differential effects of opioid peptides in analgesia. Life Sci. *30*, 1443 – 1455.
68. Tabakoff, B. and Hoffman, P. L. (1983) Alcohol interactions with brain opiate receptors. Life Sci. *32*, 197 – 204.
69. Schulz, R., Wuster, M. & Herz, A. (1982) Endogenous ligands for kappa-opiate receptors. Peptides *3*, 973 – 976.
70. Paterson, S. J., Robson, L. E. & Kosterlitz, H. W. (1983) Classification of opioid receptors. Brit. Med. Bull. *39*, 31 – 36.
71. Dubner, R., Sessle, B. J. & Storey, A. T. (1978) Neural Basis of Oral and Facial Function, Plenum, N. Y.
72. Parris, W. G., Tanzer, F. S., Friedland, G. H., Harris, E. F., Killmar, J. & Desiderio, D. M. (1987) Effects of orthodontic force on methionine enkephalin and substance P concentrations in human pulpal tissue. Amer. J. Orthodont. Dentofac. Orthop. (submitted).
73. Gray, E. G. & Whittaker, V. P. (1962) The isolation of nerve endings from brain: an electron-microscopic study of cell fragments derived by homogenization and centrifugation. J. Anat. *96*, 79 – 88.
74. Walker, J. A. Jr., Tanzer, F. S., Harris, E. F., Wakelyn, C. & Desiderio, D. M. (1987) The enkephalin response in human tooth pulp to orthodontic force. Amer. J. Orthod. Dentofac. Orthop. *92*, 9 – 16.
75. Desiderio, D. M. & Onishi, H. (1986) Canine limbic system substance P receptors. J. Rec. Res. *6*, 141 – 153.
76. Desiderio, D. M., Takeshita, H., Dass, C. & Fridland, G., The analysis of opioid peptides with HPLC, radioreceptor assay, radioimmunoassay, and mass spectrometry. NIDA Res. Monograph, *70*, 304 – 318.
77. Walker, J. A., Jr., (1984) The enkephalin response in tooth pulp to orthdontic force. M. S. Thesis, Univ. Tenn.
78. Desiderio, D. M., Takeshita, H., Onishi, H., Fridland, G. H., Dass, C., Tinsley, P. & Killmar, J. (1985) in: Peptides: Structure and Function (Deber, C. M., Hruby, V. and Kopple, K. D., eds.) pp. 715 – 718, Pierce Chemical Co.
79. Benninghoven, A. (1972) Investigations on the mechanism of secondary ion formation from organic compounds: amino acids. Surf. Sci. *123*, 721 – 727.
80. Benninghoven, A. & Sichtermann, W. (1977) Secondary ion mass spectrometry: a new analytical technique for biologically important compounds. Org. Mass Spectrom. *12*, 595 – 597.
81. Benninghoven, A. & Sichtermann, W. K. (1978) Detection, identification and structural investigation of biologically important compounds by secondary ion mass spectrometry. Anal. Chem. *50*, 1180 – 1184.
82. Surman, D. J. & Vickerman, J. C. (1981) Fast atom bombardment quadrupole mass spectrometry. J. Chem. Soc. Chem. Commun. 324 – 325.
83. Barber, M., Bardoli, R. S., Sedgwick, R. D. & Tyler, A. N. (1981) Fast atom bombardment mass spectrometry of bradykinin and related oligopeptides. Biomed. Mass Spectrom. *8*, 337 – 342.
84. Barber, M., Bordoli, R. S., Sedgwick, R. D. & Tyler, A. N. (1982) Fast atom bombardment mass spectrometry of angiotensin peptides. Biomed. Mass Spectrom. *9*, 208 – 214.

85. Barber, M., Bordoli, R. S., Sedgwick, R. D. & Tyler, A. N. (1981) FAB of solids: a new ion source for MS. J. Chem. Soc. Chem. Commun., p. 325.
86. Barber, M. Bordoli, R. S., Sedgwick, R. D., Tyler, A. N., Garner, G. V., Gordon, D. B., Tetler, L. W. & Hider, R. C. (1982) Fast atom bombardment mass spectrometry of the large oligopeptides melittin, glucagon, and the B chain of bovine insulin. Biomed. Mass Spectrom. *9*, 265–268.
87. Barber, M., Bordoli, R. S., Elliott, G. J., Tyler, A. N., Bill, J. C. & Green, B. N. (1984) FAB-MS: A mass spectral investigation of some of the insulins. Biomed. Mass. Spectrom. *11*, 182–186.
88. Barber, M. Bordoli, R. S., Elliott, G. J. & Horoch, N. J. (1983) FAB-MS of human proinsulin. Biochem. Biophys. Res. Commun. *110*, 753–757.
89. Martin, S. A., Costello, C. E. & Biemann, K. (1982) Optimum experimental conditions for FAB-MS. Anal. Chem. *54*, 2362–2368.
90. Franks, J. (1983) FAB: The fast atomic beam source. Int. J. Mass Spectrom. Ion Phys. *46*, 343–346.
91. Franks, J. & Ganhder, A. M. (1974) A saddle field ion source of spherical configuration for etching and thinning applications. Vacuum *24*, 489–491.
92. Gillian, J. M., Landis, P. W. & Occolowitz, J. (1983) Accurate mass measurement in FAB MS. Anal. Chem. *55*, 1531–1533.
93. Field, F. H. (1982) FAB study of glycerol: mass spectra and radiation chemistry. J. Phys. Chem., *86*, 5115–5123.
94. Burlingame, A. L., Whitney, J. O. & Russell, D. H. (1984) Mass spectrometry. Anal. Chem. *56*, 417 R–467 R.
95. Shackleton, C. H. L. (1985). Mass spectrometry: application to steroid and peptide research. Endocrine Rev. *63*, 441–486.
96. Burlingame, A. L., Baillie, T. A. & Derrick, P. J. (1986) Mass spectrometry. Anal. Chem. *58*, 165 R–211 R.
97. Yergey, J., Heller, D., Hansen, G., Cotter, R. J. & Fenselau, C. (1983) Isotopic distribution in mass spectra of large molecules. Anal. Chem. *55*, 353–356.
98. Dell, A. & Morris, H. R. (1982) FAB-high field magnet MS of 6000 dalton polypeptides. Biochem. Biophys. Res. Commun. *106*, 1456–1462.
99. Barber, M., Bordoli, R. S., Elliott, G. J., Sedgwick, R. D., Tyler, A. N. & Green, B. N. (1982) FAB-MS of bovine insulin and other large peptides. J. Chem. Soc. Chem. Commun. 936–938.
100. Desiderio, D. M. & Katakuse, I. (1984) Fast atom bombardment mass spectrometry of insulin, insulin A-chain, insulin B-chain and glucagon. Biomed. Mass Spectrom. *11*, 55–59.
101. Barber, M., Bordoli, R. S., Garner, G. V., Gordon, D. B., Sedgwick, R. D., Tetler, L. W. & Tyler, A. N. (1981) Fast atom bombardment mass spectra of enkephalins. Biochem. J. *197*, 401–404.
102. Williams, D. H., Bojesen, G., Affret, A. D. & Taylor, L. C. E. (1981) Study of difficult peptides from paracoccus cytochrome c-550 and a dolphin cytochrome C. FEBS Lett. *128*, 37–39.
103. Williams, D. H., Bradley, C. V., Santikarn, S. & Bojesen, G. (1982) FAB-MS. Biochem. J. *201*, 105–117.
104. Dell, A., Etienne, T., Panico, M., Morris, H. R., Vinson, G. P., Whitehouse, B. J., Barber, M., Bordoli, R. S., Sedgwick, R. D. & Tyler, A. N. (1982) The characterization of an adrenal zone glomerulosa-stimulating component of posterior pituitary extract as bis-acetyl alpha-MSH. Neuropeptides *2*, 233–240.
105. Rinehart, K. L., Gaudioso, L. A., Moore, M. L., Pandey, R. C., Cook, J. C., Barber, M., Sedgwick, R. D., Bordoli, R. S., Tyler, A. N. & Green, B. N. (1981) Structures of eleven zervamicin and two emerimicin peptide antibiotics studied by FAB-MS. J. Amer. Chem. Soc. *103*, 6517–6520.
106. Rinehart, K. L. (1982) Fast atom bombardment mass spectrometry. Science *218*, 254–260.
107. Bradley, C. V., Williams, D. H. & Hanley, M. R. (1982) Peptide sequencing using combination of Edman degradation, carboxypeptidase digestion and FAB-MS. Biochem. Biophys. Res. Commun. *104*, 1223–1230.
108. Williams, D. H., Bradley, C., Bojesen, G., Santikarn, S. & Taylor, L. C. E. (1981) FAB MS: a powerful technique for the study of polar molecules. J. Amer. Chem. Soc. *103*, 5700–5704.
109. Levsen, K. & Schwarz, H. (1976) Collision activation mass spectrometry – a new probe for determining the structure of ions in the gas phase. Angew. Chem. Int. Ed. (Engl.) *15*, 509–568.
110. Jennings, K. R. (1978) in: High Performance Mass Spectrometry: Chemical Applications (Gross, M. L., ed.) pp. 3–17, Amer. Chem. Soc., Washington, D. C.

111. Jennings, K. R. & Mason, R. S. (1983) in: Tandem Mass Spectrometry (McLafferty, F. W., ed.) p. 197, Wiley, New York.
112. Dass, C. (1984) The mechanism of ring opening and ring formation reactions of gas-phase radical cations. Ph. D. Thesis, Univ. Nebraska – Lincoln.
113. Hunt, D. F., Shabanowitz, J. & Giordani, A. B. (1980) Collision activated decompositions of negative ions in mixture analysis with a triple quadrupole mass spectrometer. Anal. Chem. *52*, 386–390.
114. Glish, G. L. & Todd, P. J. (1982) Collision region for mass spectrometry/mass spectrometry. Anal. Chem. *54*, 842–843.
115. McLafferty, F. W. (1980) Tandem mass spectrometry (MS/MS): a promising new analytical technique for specific component determination in complex mixtures. Acc. Chem. Res. *13*, 33–39.
116. Amster, I. J., Baldwin, M. A., Cheng, M. T., Proctor, C. H. & McLafferty, F. W. (1983) Tandem mass spectrometry of higher molecular weight compounds. J. Amer. Chem. Soc. *105*, 1654–1655.
117. McLuckey, S. A. & Cooks, R. G. (1983) in: Tandem Mass Spectrometry (McLafferty, F. W., ed.) p. 303, Wiley, NY.
118. McLafferty, F. W., Kornfeld, R. Haddon, W. F., Levsen, K., Sakai, I., Bente III, P. F., Tsai, S.-C. & Schuddemage, H. D. R. (1973) Application of collisional activation spectra to the elucidation of organic ion structures. J. Amer. Chem. Soc. *95*, 3886–3892.
119. Levsen, K. Wipf, H.-K. & McLafferty, F. W. (1974) Mass spectrometric studies of peptides – VI: applications of metastable ion and collisional activation spectra. Org. Mass Spectrom. *8*, 117–128.
120. Levsen, K. & Beckey, H. D. (1974) Collision activation spectra of field ionized molecules. Org. Mass Spectrom. *9*, 570–581.
121. Weber, R. and Levsen, K. (1980) Collision induced dissociation of field desorbed di- and tripeptides. Biomed. Mass Spectrom. *7*, 314–316.
122. Hunt, D. F., Buko, A. M., Ballard, J. M., Shabanowitz, J. & Giordani, A. B. (1981) Sequence analysis of polypeptides by collision activated dissociation on a triple quadrupole mass spectrometer. Biomed. Mass Spectrom. *8*, 397–408.
123. Hunt, D. F., Bone, W. M., Shabanowitz, J., Rhodes, J. & Ballard, J. M. (1981) Sequence analysis of oligopeptides by secondary ion/collision activated dissociation mass spectrometry. Anal. Chem. *53*, 1704–1706.
124. Desiderio, D. & Fridland, G. (1984) A review of combined liquid chromatography and mass spectrometry. J. Liq. Chromatogr. *7*, 317–351.
125. Lacey, M. L. & MacDonald, C. G. (1977) Three dimensional representation of metastable peaks at high mass resolution. Biomed. Mass Spectrom. *12*, 587–594.
126. Lacey, M. J. & MacDonald, C. G. (1983) in: Tandem Mass Spectrometry (McLafferty, F. W., ed.) p. 321, Wiley, NY.
127. Haddon, W. F. (1980) The constant neutral linked magnetic field-electric sector scan. Org. Mass Spectrom. *15*, 539–543.
128. Shushan, B. & Boyd, R. K. (1980) Modulated-field linked-scan studies of metastable ions. Int. J. Mass Spectrom. Ion Phys. *34*, 37–62.
129. Steinauer, R. & Schlunegger, U. P. (1982) Detection of fragment genesis in the MS: amino acid sequence analysis by DADI/MIKE, linked scan and collision activation MS. Biomed. Mass Spectrom. *9*, 153–157.
130. Desiderio, D. M., Sabbatini, J. Z. & Stein, J. L. (1980) FD-MS of hypothalamic oligopeptides. Adv. Mass Spectrom. *8*, 1298–1305.
131. Desiderio, D. M., Stein, J. L., Cunningham, M. D. & Sabbatini, J. Z. (1980) HPLC and FDMS of hypothalamic oligopeptides. J. Chromatogr. *195*, 369–377.
132. Desiderio, D. M., Yamada, S., Sabbatini, J. Z. & Tanzer, F. S. (1981) Quantification of picomole amounts of underivatized leu-enkephalin by means of FD-MS. Biomed. Mass Spectrom. *8*, 10–12.
133. Desiderio, D. M. & Sabbatini, J. Z. (1981) FD-CAD-B/E MS of underivatized oligopeptides. Biomed. Mass Spectrom. *8*, 565–568.
134. Desiderio, D. M. & Katakuse, I. (1985) Fast atom bombardment-collision activation-linked field scanning mass spectrometric analysis of peptides. Mass Spectroscopy *33*, 351–370.
135. Desiderio, D. M. & Yamada, S. (1982) Measurement of endogenous leucine enkephalin in canine thalamus by HPLC and FDMS. J. Chromatogr. *239*, 87–95.
136. Desiderio, D. M. & Kai, M. (1983) Preparation of stable isotope-labeled peptide internal standards for field desorption mass spectrometry quantification of peptides in biologic tissue. Biomed. Mass Spectrom. *10*, 471–479.
137. Desiderio, D. M. & Kai, M. (1983) Field desorption MS of enkephalins in canine

brain with [18]O peptide internal standard. Int. J. Mass Spectrom. Ion Phys. *48*, 261–264.
138. Desiderio, D. M. & Friedland, G. H. (1986) in: Mass spectrometry in Biomedical Research (Gaskell, S., ed.) pp. 43–458, John Wiley and Sons, Chichester, New York, Brisbane, Toronto and Singapore.
139. Desiderio, D. M. & Yamada, S. (1983) Field desorption MS measurement of picomole amounts of leucine enkephalin in canine spinal cord tissue. Biomed. Mass Spectrom. *10*, 358–362.
140. Desiderio, D. M., Katakuse, I. & Kai, M. (1983) Measurement of leucine enkephalin in caudate nucleus tissue with FAB-CAD-B/E MS. Biomed. Mass Spectrom. *10*, 426–429.
141. Desiderio, D. M. (1984) in: Handbook of Use of HPLC for the Separation of Amino Acids, Peptides, and Proteins (Hancock, W. S., ed.) pp. 197–211, CRC Press, Boca Raton, FL.
142. Desiderio, D. M. (1983) in: Adv. Chromatography, Vol. 22 (Giddings, J. C., Grushka, E., Cazes, J., and Brown, P. R., eds.) pp. 1–36, Marcel-Dekker, New York.
143. Tanzer, F. S., Desiderio, D. M. & Yamada, S. (1981) in: Peptides: Synthesis-Structure-Function (Rich, D. H. and Gross, E., eds.) pp. 761–764, Pierce Chem. Co., Rockford, Ill.
144. Yamada, S. & Desiderio, D. M. (1982) Measurement of endogenous leucine enkephalin in caudate nucleus and hypothalamus with HPLC-FDMS. Anal. Biochem. *127*, 213–221.
145. Yamada, S. & Desiderio, D. M. (1982) in: High Performance Liquid Chromatography of Proteins and Peptides (Hearn, M. T. W., Regnier, F. E., and Wehr, C. T., eds.) pp. 211–219, Academic Press, New York.
146. Lawson, A. M., Lim, D. K., Richmond, W., Samson, D. M., Setchell, K. D. R. & Thomas, A. C. S. (1980) in: Current Developments in the Clinical Applications of HPLC, GC and MS (Lawson, A. M., Lim, D. K., and Richmond, W., eds.) pp. 135–153, Academic Press, New York.
147. Millard, B. J. (1978) Quantitative Mass Spectrometry, Heyden, London.
148. Vance, D. E., Krivit, W. & Sweeley, C. C. (1975) Metabolism of neutral glycosphingolipids in plasma of a normal human and a patient with Fabry's disease. J. Biol. Chem. *250*, 8119–8125.
149. Pickett, W. C. & Murphy, R. C. (1981) Enzymatic preparation of carboxyl oxygen-18 labeled prostaglandin $F_{2\alpha}$ and utility for quantitative mass spectrometry. Anal. Biochem. *111*, 115–121.
150. Desiderio, D. M., Laughter, J., Kai, M. & Trimble, J. (1984) A microcomputer data acquisition system for mass spectrometry. J. Comp. Enchanc. Spectros. *2*, 21–31.
151. Desiderio, D. M. & Dass, C. (1986) The measurement of leucine enkephalin at the femtomole level by fast atom bombardment mass spectrometry/mass spectrometry methods. Anal. Lett. *19*, 1963–1971.
152. Tolun, E., Dass, C. & Desiderio, D. M. (1987) Trace level measurement of enkephalin peptides at the amol-fmol level by FAB-MS. Rpd. Commun. Mass Spectrom. *1*, 77–79.
153. Kudo, T., Chang, H.-L., Maeda, S., Uchida, Y., Kakamae, J. & Inoke, R. (1983) Changes of the met-enkephalin-like peptide content induced by noxious stimuli in the rat incisor pulp. Life Sci. *33*, 677–680.
154. Kudo, T., Chang, H.-L., Maeda, S., Uchida, Y., Kakamae J. & Inoke, R. (1983) Changes of the met-enkephalin-like peptide content induced by noxious stimuli in the rat incisor pulp. Life Sci. *33*, 677–680.
155. Delaney, M. F., (1984) The chromatographic uncertainty principle. L. C. Magaz. *2*, 85–86.
156. May, H. E., Tanzer, F. S., Fridland, G. H., Wakelyn, C. & Desiderio, D. M. (1982) HPLC and proteolytic enzyme characterization of peptides in tooth pulp extracts. J. Liq. Chromatogr. *5*, 2135–2154.
157. Dass, C., Friedland, G. H., Tinsley, P. W. Killmar, J. T. & Desiderio, D. M. (1988) Structural characterization and quantitation of β-endorphin in human pituitary by fast atom bombardment mass spectrometry of trypsin generated fragments. Anal. Biochem. (Submitted)
158. Katakuse, I. & Desiderio, D. M. (1983) FAB-CAD-B/E MS of neuropeptide substance P. Anal. Biochem. *129*, 25–29.
159. Dupont, A., Lepine, J., Langelier, P., Merand, Y., Rouleau, D., Vaudry, H., Gros, C. & Barden, N. (1980) Differential distribution of β-endorphin and enkephalins in rat and bovine brain. Regul. Pept. *1*, 43–52.

Chapter 8
Porphyrins and Bile Pigments

A. H. Jackson

1 Introduction

Porphyrins and bile pigments, together with chlorophylls and Vitamin-B_{12}, comprise the tetrapyrrolic 'pigments of life' which play a vital role throughout the natural kingdom. Chlorophylls are intimately involved in photosynthetic processes in plants in which carbon dioxide is reduced to carbohydrates by harvesting the energy of sunlight, and this energy is then utilised by most living organisms in the oxidative processes catalysed by the various haem-enzymes, haemoglobin and myoglobin. Bile pigments are oxidative degradation products from haem, which appear to be essentially waste products in mammalian systems, but in plants the bile pigment phytochrome is a plant growth regulator and in algae they play an important role in photosynthesis together with the chlorophylls. Vitamin-B_{12} is the coenzyme of an important series of enzymes which catalyse a variety of biochemically important rearrangements and methylation reactions.

The major features of the structural chemistry of porphyrins, chlorophylls and the bile pigments bilirubin and biliverdin, were determined largely by classical degradative chemical methods during the first half of this century (1). Visible spectroscopy also played an important part in their detection and analysis during this period. However, during the last thirty years or so the newer methods of mass and NMR spectrometry have begun to play an increasing role in structure determination and quantitative analysis of these tetrapyrrolic pigments. This chapter will be confined largely to a discussion of the uses of mass spectrometry for studies of porphyrins and bile pigments occuring in mammalian systems, but it has proved equally useful in studies of newly discovered chlorophyll and bile pigments in plant systems (2).

Much of the scientific interest in porphyrins and bile pigments in recent years has been concerned with their biosynthesis and metabolism *in vivo*, and clinically, studies of abnormal metabolism have commanded the major attention. In the porphyrias, for example, owing to partial defects in one or more of the enzymes of the haem pathway, a variety of porphyrins may be excreted in urine, faeces or into blood, whilst the causes of jaundice include an overproduction of bilirubin (haemolysis), hepatocellular damage, and decreased excretion of bilirubin due to biliary obstruction. Although a relatively common condition in the newborn, neo-natal jaundice is due to a temporary failure of bilirubin to undergo conjugation with glucuronic acid. Many of the intermediates in these biochemical processes only occur in very small quantities (although the amounts may increase markedly in pathological conditions) and mass spectrometry has been particularly valuable in their analysis and structure determination because it can be applied to sub-microgram quantities of material. In the ensuing discussion it is convenient to discuss the porphyrins and bile pigments separately because of the marked differences in their chemical and mass spectral characteristics, but the final section of this chapter will be concerned with the use of combined gas chromatography/mass spectrometry in the analysis of mono-pyrrolic degradation products obtained from both porphyrin and bile pigments.

2 Porphyrins

Since all the porphyrins of clinical interest are formed directly, or indirectly, in the biosynthesis of haem and haem enzymes the main pathway to haem (3) is outlined in Scheme 1, and the structures of most of the porphyrins discussed in this review are shown in Table 1. Four units of the pyrrolic intermediate porphobilinogen (PBG) couple together to form an open chain tetrapyrrole and the latter then cyclises with rearrangement of the D-ring to form uroporphyrinogen-III, the first macrocyclic intermediate (4). Successive decarboxylations of the four acetic acid side-chains of the macrocycle then occur in a largely 'clockwise' fashion, starting with the acetic acid residue on the D-ring (5). The coproporphyrinogen-III is then converted into protoporphyrinogen-IX by successive oxidative decarboxylations of the propionic acid residues on ring-A, and then on ring-B (6). Finally the macrocycle undergoes dehydrogenation to protoporphyrin-IX (7) and iron is then inserted to form haem.

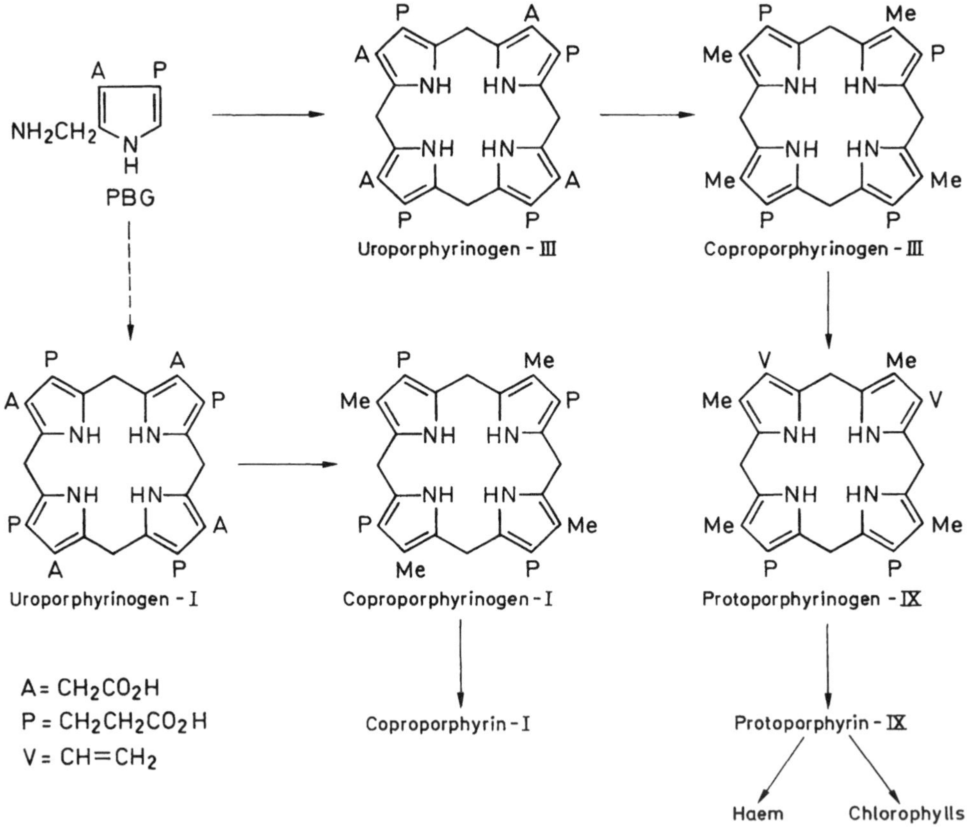

Scheme 1. Biosynthesis of haem (the Fe^{II} complex of protoporphyrin-IX) from porphobilinogen (PBG) via uroporphyrinogen III, and formation of type-I porphyrinogens and porphyrins via uroporphyrinogen-I.

Table 1. Structures of porphyrins and porphyrinogens

	Substituents							
Porphyrin	1	2	3	4	5	6	7	8
Uro III	A	P	A	P	A	P	P	A
Hepta (III series)	A	P	A	P	A	P	P	Me
Hexa (III series)	Me	P	A	P	A	P	P	Me
Penta (III series)	Me	P	Me	P	A	P	P	Me
Copro III	Me	P	Me	P	Me	P	P	Me
S-411	Me	Acrylic	Me	P	Me	P	P	Me
Uro I	A	P	A	P	A	P	A	P
Copro I	Me	P	Me	P	Me	P	Me	P
Isocopro	Me	Et	Me	P	A	P	P	Me
Dehydroisocopro	Me	V	Me	P	A	P	P	Me
Desethylisocopro	Me	H	Me	P	A	P	P	Me
Hydroxyisocopro	Me	$CHOHCH_3$	Me	P	A	P	P	Me
Hardero	Me	V	Me	P	Me	P	P	Me
Isohardero	Me	P	Me	V	Me	P	P	Me
Proto IX	Me	V	Me	V	Me	P	P	Me
Haemato	Me	$CHOHCH_3$	Me	$CHOHCH_3$	Me	P	P	Me
Pempto	Me	H	Me	V	Me	P	P	Me
Isopempto	Me	V	Me	H	Me	P	P	Me

Me = CH_3; Et = CH_2CH_3; V = $CH{=}CH_2$; A = CH_2CO_2H; P = $(CH_2)_2CO_2H$; acrylic = $CH{=}CHCO_2H$.

Scheme 2 summarises the enzymes involved in the various stages of the haem pathway, and indicates the various porphyrinic conditions which may arise as a result of defects of these enzymes (8). The overall process appears to be under feed-back control of haem upon the activity of ALA synthetase, so that a partial blockage in the pathway leads to an increase in the general level of metabolic activity and the excretion of porphyrins related to the (reduced) porphyrinogens which are intermediates in the pathway.

Two of these defects lead to the formation of other porphyrins than those on the main pathway to haem. The first of these is a very rare type of porphyria in which isomeric 'type I' porphyrins are produced; owing to a defect in uroporphyrinogen cosynthetase the open chain bilane cyclises to form uroporphyrinogen-I which then undergoes a series of decarboxylations (catalysed by uroporphyrinogen decarboxylase) to form coproporphyrinogen-I (9) (Scheme 1); the abnormal pathway ceases at this point because it cannot be further metabolised by the later enzymes in the haem pathway and the intermediate type-I porphyrinogens then undergo spontaneous oxidation to form the corresponding porphyrins. The second defect leading to other abnormal porphyrins is porphyria cutanea tarda (10); this is a much more common condition which is sometimes familial, but more often is caused by poisoning (e. g. by various drugs, or by chlorinated hydrocarbons). Uroporphyrinogen decarboxylase is partially inhibited and coproporphyrinogen oxidase converts the intermediate type-III pentacarboxylic porphyrinogen into the vinyl porphyrinogen, *dehydroisocopropor-*

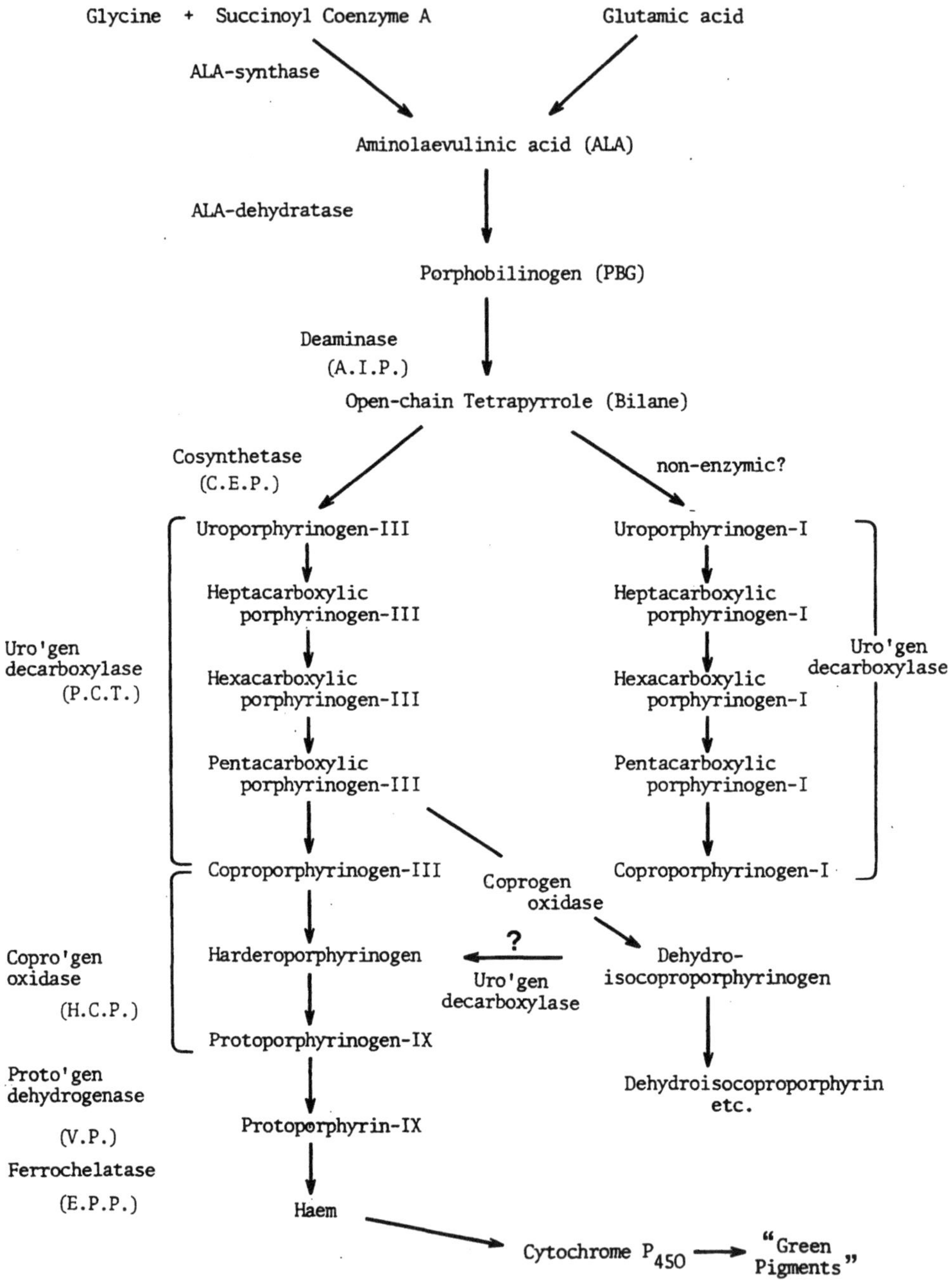

Scheme 2. Defects in haem biosynthesis leading to porphyrias.
The scheme above shows the main pathway of haem biosynthesis from glycine and succinoyl coenzyme-A and also the enzymes responsible for each stage. The pathway appears to be regulated by feedback control by haem on the level of ALA-synthetase, and in normal metabolism very little if any of the intermediates between ALA and haem accumulate. (Very recently, however, evidence has been provided that some 20% of mammalian ALA

may be formed by an alternative pathway from glutamic acid, which was shown several years ago to operate in plants.)

In abnormal metabolism partial defects in these enzymes (either of genetic origin, or due to environmental factors e. g. poisoning by drugs or other toxic substances and heavy metal ions) may lead to different types of *porphyria* and these are indicated in brackets alongside the enzymes affected. These porphyrias are characterised by excretion of relatively large amounts of one, or more, intermediates in the pathway, or of related porphyrins, and the effects of the metabolic lesions are summarised briefly below.

Acute Intermittent Porphyria (AP)
Decreased deaminase activity leading to an increase in ALA-synthetase, and excretion of large amounts of ALA and PBG in the urine.

Congenital Eythropoietic Porphyria (CEP)
A rare form of porphyria in which there is an imbalance of cosynthetase activity in relation to deaminase activity leading to formation of substantial amounts of type-I porphyrins (and acute photosensitivity).

Porphyria Cutanea Tarda (PCT)
Decreased hepatic uroporphyrinogen decarboxylase activity leading to excretion of porphyrins with 8–4 carboxyl groups (including isocoproporphyrin and its congeners in faeces) – some increase in type-I porphyrin excretion is also often observed. Photosensitivity is a common factor in the disorder. This is largely an acquired disorder (e. g. due to polyhalogenated hydrocarbons, acute alcoholism or iron overload) but evidence for a genetic disorder is increasing.

Hereditary Coproporphyria (HCP)
Decreased coproporphyrinogen oxidase activity leads to excretion of large amounts of coproporphyrin-III in faeces (and to a lesser extent in urine).

Variegate Porphyria (VP)
A hepatic form of porphyria in which increased faecal excretion of both proto- and coproporphyrin occurs; in acute attacks excretion of ALA, PBG and porphyrins also occurs, and skin photosensitivity is observed.

Erythropoietic Protoporphyria (EPP or PP)
Deficiency of ferrochelatase activity leading to greatly increased faecal excretion of protoporphyrin; photosensitivity is also commonly observed.

Porphyrias Due to Toxic Substances
Lead and other heavy metal ions may cause decreases in hepatic ALA-dehydratase, coproporphyrinogen oxidase and ferrochelatase activities, whereas iron causes decreases in cosynthetase and uroporphyrinogen decarboxylase activity. Polyhalogenated hydrocarbons and some other toxic organic compounds including excess of alcohol, also affect uroporphyrinogen decarboxylase and these are the main causes of acquired PCT. A wide range of drugs, including barbiturates and other sedatives, have also been shown to induce porphyria by reducing ferrochelatase activity, or by stimulating hepatic cytochrome-P_{450} formation; inactivation of cytochrome-P_{450} is also caused by these drugs (see text).

phyrinogen. The latter undergoes further modifications in the gut, followed by oxidation, resulting in the formation of the corresponding porphyrin and three closely related porphyrins (11) (Scheme 3). It is interesting to note that dehydroisocoproporphyrinogen is metabolised *in vitro* to protoporphyrin-IX by chicken red-cell haemolysates; whether this alternative pathway to haem also occurs *in vivo* remains to be seen (12).

The porphyrins formed in both these abnormal pathways may be excreted in the urine and faeces, and may also find their way into the bloodstream; this may lead to acute photosensitivity and the formation of unpleasant skin lesions etc.

Haem is incorporated unchanged into haemoglobin, myoglobin, and most of the haem enzymes (cytochromes, peroxidases and catalases). However, in cytochrome-C the protein is covalently linked via the sulphur atoms of cysteine residues to the vinyl groups of the haem, and in cytochrome oxidase the haem has been modified by addition of a C_{15}-farnesyl side-chain to the A-ring vinyl group and by oxidation of the methyl group on the D-ring to a formyl group. However the amounts of these enzymes which occur naturally are relatively very small by comparison with haemoglobin and myoglobin.

Small amounts of a number of other porphyrins have also been found in mammalian systems e. g. pemptoporphyrin (13), mesoporphyrin and dihydroprotoporphyrin in faeces, 'S-411' porphyrin (14) (dehydrocoproporphyrin) in meconium and harderoporphyrin (15) (cf. Table 1).

2.1 Sample preparation for mass spectrometry

Mass spectrometry has played a very important role in the structure determination of all the new porphyrins discovered in recent years. However it cannot readily be applied directly to the naturally occurring porphyrin free acids owing to their very low volatility. It is customary, therefore, to convert the porphyrin acids to their rather more volatile methyl esters by esterification with methanol containing 5% concentrated sulphuric acid (v/v), or by heating in methanol containing boron trifluoride etherate, or with diazomethane in ether solution. However, before this can be done with urine samples the porphyrin is first isolated by adsorption on talc, and then the talc is treated directly with methanol/sulphuric acid to esterify any porphyrin present. With faecal and blood specimens it is convenient to freeze-dry the material, and then treat the residue with methanol/sulphuric acid; after leaving to stir overnight insoluble material is filtered off usually with the addition of an inert material such as celite to avoid problems of emulsification by fatty material in the faeces. The methanol/sulphuric acid extracts are then taken up in chloroform, washed with aqueous sodium carbonate solution to neutralise the acid, the chloroform extracts are dried and evaporated to dryness to recover the crude porphyrin esters (16). It is often convenient at this stage to carry out a preliminary chromatographic purification on an alumina column (Brockmann Grade III) to separate the porphyrins from other materials which are also present (especially in faecal material). In most cases mixtures of porphyrins will be present, but preliminary analyses can be carried out by field desorption mass spectrometry (see below), prior to further resolution by thin layer or high performance liquid chromatographic (HPLC) methods followed by mass spectrometric studies.

2.2 Electron impact mass spectrometry (EI-MS)

Because of the high stability of the aromatic macrocyclic nucleus of porphyrins the molecular ion is usually the most intense peak in the mass spectrum of a porphyrin (17, 18). A number of fragment ions are also observed corresponding to cleavages of bonds β- to the aromatic nucleus e. g. loss of the methyl from ethyl side-chains, loss

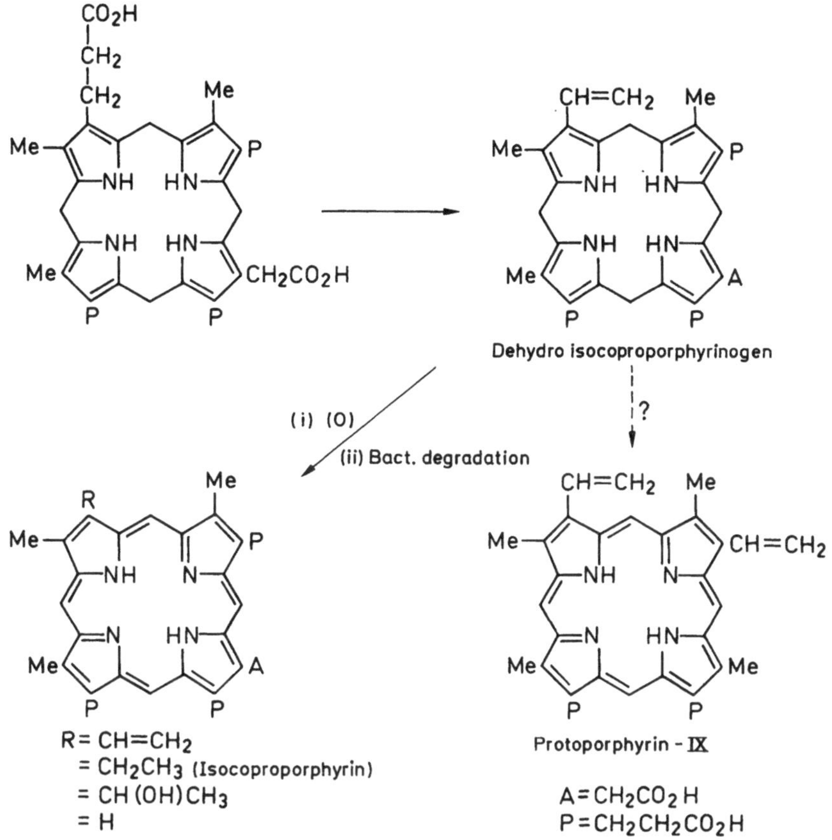

Scheme 3. Formation of isocoproporphyrin and its congeners from the pentacarboxylic porphyrinogen intermediate between uroporphyrinogen-III and coproporphyrinogen-III (cf. Scheme 1) and a possible alternative route to protoporphyrin-IX.

of CO_2Me from CH_2CO_2Me side-chains and loss of CH_2CO_2Me from $CH_2CH_2CO_2Me$ side-chains (Fig. 1). In addition other minor cleavages of the ester groups by loss of OMe or CO_2Me also occur. Thus the spectra give information about the molecular weight of the porphyrin, and the nature of the side-chains, although not in general about the order of substituents around the ring system. A series of lower intensity doubly charged ions are also usually observed in the electron impact mass spectra of most porphyrins, but cleavages of the ring do not occur under normal conditions. Accurate mass determination allows the assignment of formulae to the molecular ion, and to the various fragmentation products, allowing one to confirm preliminary deductions from the low resolution spectra. In some cases metastable ions may also be of value in this respect in helping to define fragmentation pathways.

Metal complexes of porphyrins exhibit very similar features to the parent compounds, and they are usually slightly more volatile. Thus low intensity ions ($\approx 1-2\%$) due to complexes of porphyrins with the first series of transition metals are commonly observed in porphyrin mass spectra (and are presumably due to the ability of por-

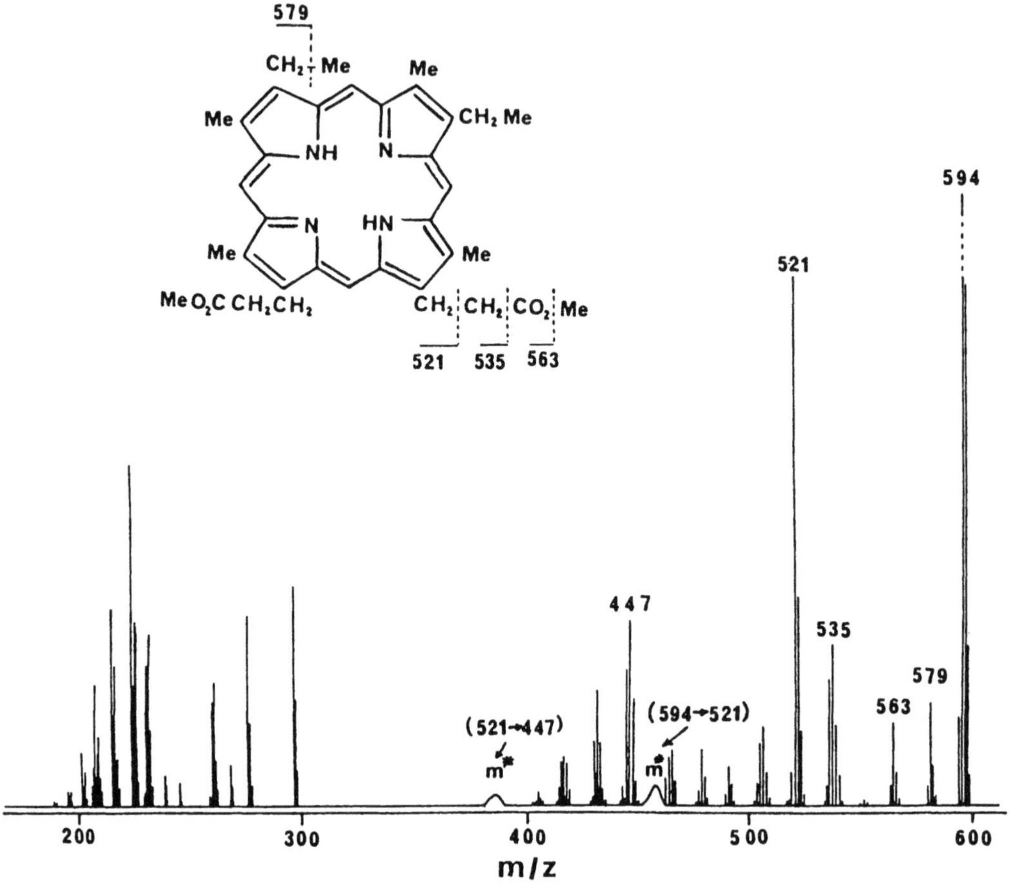

Figure 1. Electron impact mass spectrum of mesoporphyrin-IX dimethyl ester.

phyrins to scavenge metal ions from solvents etc. during work-up etc.). Indeed silicon and germanium complexes of alkyl porphyrins are sufficiently volatile that they can be gas chromatographed (19) but not those of porphyrins with carboxylic ester side-chains (even if converted to their trimethylsilyl esters).

One of the first examples of the use of mass spectrometry in structure determination of porphyrins was pemptoporphyrin (13) (Table 1) which was isolated from faeces; the molecular mass of the dimethylester proved to be 26 daltons less than that of protoporphyrin dimethyl ester and this, together with visible spectral comparisons, led to the conclusion that it was a monovinyl analogue. This was confirmed by the NMR spectrum and subsequently the position of the vinyl group on ring-A was determined by total synthesis (20) of pemptoporphyrin and its isomer, isopemptoporphyrin (with the vinyl group on ring-A). Deuteroporphyrin also occurs in small amounts in faeces as well as pemptoporphyrin and protoporphyrin, and thus pemptoporphyrin was shown to be a partially devinylated protoporphyrin, degradation presumably having occured through the action of bacterial enzymes in the gut.

Another interesting porphyrin of biochemical significance is harderoporphyrin (15), first isolated in small amounts together with coproporphyrin and protoporphyrin from the harderian gland of the rat. Mass spectral determinations of the molecular mass of its dimethyl ester showed that it was a tricarboxylic porphyrin, with three propionate residues and one vinyl group. As with pemptoporphyrin the visible absorption spectrum with a Soret band at λ_{max} 404 nm (intermediate between that of aetioporphyrin λ_{max} 398 nm, and protoporphyrin λ_{max} 408 nm), confirmed the presence of one vinyl group. It was concluded, therefore, that the related porphyrinogen (from which it was presumably derived) was an intermediate between coproporphyrinogen-III and protoporphyrinogen-IX (Scheme 1). The precise structure (Table 1) was deduced by synthesis (15), and subsequent work has confirmed the role of harderoporphyrinogen in mammalian and avian metabolism (21).

A related porphyrin found in meconium is 'S-411' porphyrin (14) (so called because of the position of its Soret band at the edge of the visible region of the spectrum). After isolation and conversion to the tetramethyl ester the mass spectrum showed that the molecular mass was 708 (i. e. 2 daltons less than coproporphyrin tetramethyl ester), and the fragmentation pattern also showed the usual cleavages associated with propionate ester side-chains. The visible spectrum was of the 'rhodo type' showing the presence of a strongly conjugating group in one of the side-chains, and this led to the conclusion that the structure was an analogue of coproporphyrin with an acrylic acid side-chain (probably in the A-ring) instead of a propionic acid side-chain. This was later fully substantiated by total synthesis (22), and the question then arose as to whether the corresponding porphyrinogen was an intermediate between coproporphyrinogen-III and harderoporphyrinogen. Deuterium labelling studies combined with mass spectrometric analysis, however, showed that in biosynthetic experiments a deuterium labelled coproporphyrinogen with CD_2 groups neighbouring the carboxyl groups in the propionate side-chains of the A and B rings of coproporphyrinogen-III afforded protoporphyrin-IX retaining both deuterium atoms in each case (Scheme 4) whereas had the acrylic porphyrinogens been intermediates one of the deuteriums would have been lost in each case (23).

Mass spectrometry also played a key role in the structural studies of the group of four related tetracarboxylic porphyrins occurring as abnormal metabolites in porphyria cutanea tarda (11) (Scheme 3). Thus after chromatographic isolation isocoproporphyrin tetramethyl ester was shown by mass spectrometry to be isomeric with coproporphyrin-III tetramethyl ester as it had the same molecular mass, and similar fragmentation behaviour. Two of the other three porphyrins were shown to have molecular masses 2 and 28 daltons less than isocoproporphyrin, and the fourth, which was more polar than the others, also gave an apparent molecular ion 2 daltons less than isocoproporphyrin. The visible spectrum of the dehydroisocoproporphyrin was consistent with the presence of a vinyl side-chain (cf. pemptoporphyrin and harderoporphyrin discussed above) and this led to the conclusion that the fourth (more polar) porphyrin was a hydroxyethyl derivative which had lost water very readily on mass spectrometry (due to the relatively high source temperatures required for porphyrins). (Haematoporphyrin dimethyl ester likewise undergoes a double dehydration and gives the spectrum of protoporphyrin-IX dimethyl ester). The presence of the labile hydroxyethyl side-chain was confirmed readily, however, by silylation to form a trimethylsilyl ether which gave the expected molecular ion (at M + 90), and fragmentation

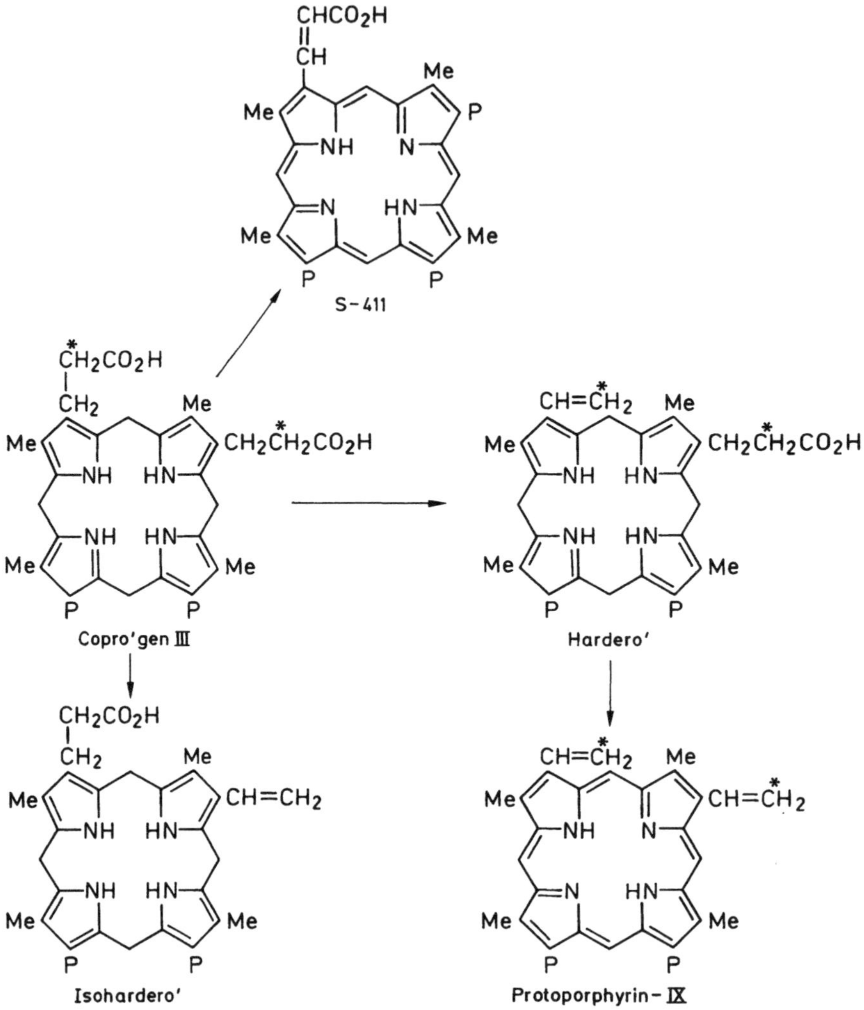

Scheme 4. Biosynthesis of protoporphyrin-IX from coproporphyrinogen-III; isoharderoporphyrinogen is not an intermediate in this process *in vivo*, neither is the porphyrinogen related to the S-411 porphyrin found in meconium (see text for details).

by loss of TMSOH, as well as the other usual modes (11). It was concluded that isocoproporphyrin itself contained an ethyl side-chain (corresponding to the vinyl group of dehydroisocoproporphyrin and to the hydroxyethyl side-chain of the more polar porphyrin) whilst the porphyrin 28 daltons less was the desethyl analoque (Scheme 3). This evidence did not of itself prove the precise structures of isocoproporphyrin and its congeners, but these were shown by a combination of NMR studies together with further mass spectral investigations of mono-pyrrolic degradation products (24) (see below). The structures have since been completely confirmed by total synthesis (25, 26). It is also interesting to note that dehydroisocoproporphyrinogen, isocoproporphyrinogen and desethylisocoproporphyrinogen are converted by red-cell

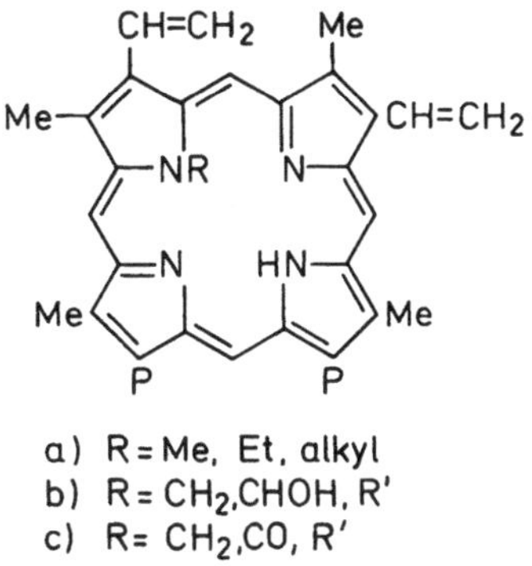

Scheme 5
Structures of the "green pigments" – *N*-substituted porphyrins – derived from liver cytochrome P_{450} by a) alkylation, b) oxidative alkylation with alkenes, and c) oxidative alkylation with alkynes.

haemolysates into protoporphyrin, dihydroprotoporphyrinogen and pemptoporphyrinogen (12), thus raising the possibility that there is an alternative pathway from the usual pentacarboxylic porphyrinogen intermediate (Scheme 1) via dehydroisocoproporphyrinogen (Scheme 3) to protoporphyrin-IX as well as via coproporphyrinogen-III. Interestingly, evidence has recently been obtained that the corresponding acetyl porphyrin (27) ('oxoisocoproporphyrin') is excreted by a patient with porphyria cutanea tarda.

Mass spectrometry has also been useful in confirming the nature of the intermediates between uroporphyrinogens I and III, and coproporphyrinogens I and III respectively (28, 5, 29). The corresponding porphyrins were isolated from porphyric patients and converted to their methyl esters prior to chromatographic separations; mass spectrometric analyses then showed as expected that the intermediate hepta-, hexa- and pentacarboxylic porphyrinogens were derived by successive decarboxylations of the acetic acid side-chains.

One of the most exciting developments of recent years has been the discovery that the so-called 'green pigments' occasionally found in liver, and reports of which go back over twenty years, are nitrogen substituted haems which arise by alkylation of cytochrome P_{450}. Mixtures of isomers are usually found although there may be some preference in the case of particular substituents. After removal of iron by mild acid treatment, the metal free *N*-substituted porphyrins were initially studied by mass spectrometry (field desorption methods – see below), as well as by visible spectroscopy especially in view of the paucity of the natural material. These *N*-substituted porphyrins are of two types a) *N*-methyl and other simple *N*-alkyl derivatives, and b) *N*-hydroxyalkyl or ketoalkyl derivatives of protoporphyrin-IX (30, 31) (Scheme 5). The former series are presumably derived by direct alkylation whereas the latter are derived by an 'oxidative alkylation' process from drugs and other materials with alkene or alkyne residues; for example, *N*-hydroxyethyl protoporphyrins are found in the livers of rats exposed to ethylene.

The EI mass spectra of the dimethyl esters of the *N*-substituted porphyrins are similar to that of protoporphyrin-IX dimethyl-ester itself except that fragmentation of larger *N*-substituents may also be observed. The pharmacology of these *N*-substituted porphyrins has excited great interest because the simple *N*-alkyl derivatives

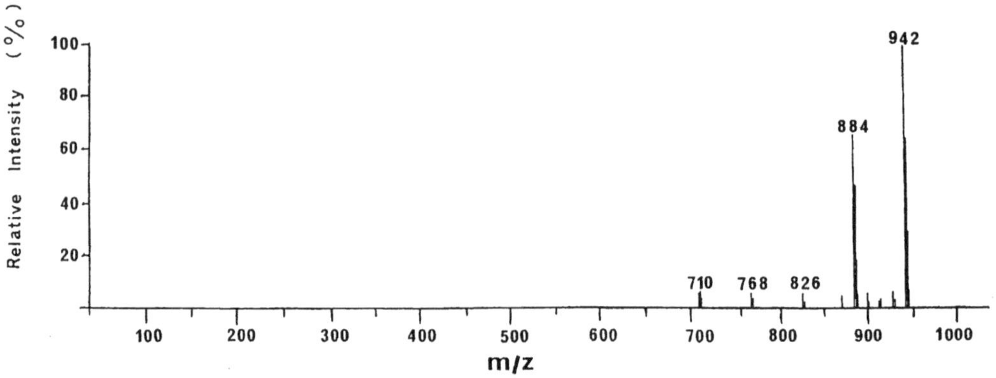

Figure 2. Field desorption mass spectrum of a mixture of porphyrin methyl esters derived from the porphyrins present in the faeces of a porphyric patient.

(especially the *N*-methyl compounds) have proved to be potent inhibitors of ferrochelatase (Scheme 2), and there exists the possibility that small amounts of *N*-methyl protoporphyrin if formed in normal metabolism could exercise a regulatory effect on the overall process of haem metabolism.

2.3 Field desorption mass spectrometry (FD-MS)

This 'soft' method of ionization has proved to be of particular value with relatively involatile compounds such as porphyrins (29, 32). It has the advantage that almost no fragmentation occurs except with very labile side-chains; in FD haematoporphyrin dimethyl ester loses water to afford the molecular ion of protoporphyrin dimethyl ester as in the EI mode.

However, because usually only molecular ions are formed, FD-MS is particularly useful for the preliminary analysis of mixtures of related natural materials. Thus the porphyrins excreted by rats poisoned with hexachlorobenzene (as an inducer of porphyria cutanea tarda) were converted to their methyl esters and analysis of the mixture showed the presence of porphyrins containing from two to eight carboxylic side-chains (29, 32) (Fig. 2). These were subsequently separated chromatographically and the various fractions examined in more detail by FD- and EI-MS and NMR spectroscopy; their structures were finally proved by synthesis (28, 33).

Mass spectra of porphyrins with up to two or three free carboxylic side-chains can also be obtained by FD-MS e. g. haem, itself, gives a molecular ion in this way, but porphyrins with four or more acidic side-chains are not amenable unless converted into their esters.

FD-MS has proved to be a very powerful technique in combination with HPLC especially as the emitter can be dipped into the concentrated eluates, or the solutions can even be syringed directly on the emitter wire. A particularly impressive use of FD-MS was in the determination of the molecular mass of haem-a (35) as its dimethyl ester following esterification and purification by HPLC; in this way it was confirmed that, haem-a, the prosthetic group of cytochrome oxidase contains a labile C_{17} side-chain in the A-ring, and that modification of the side-chain did not occur during isolation.

More recently FD-MS has proved invaluable in the analysis of the *N*-substituted green pigments (31) especially those with the larger and more labile *N*-substituents e. g. evidence has been obtained for an *N*-steroidal protoporphyrin derived from a C-17 ethinyl substituted contraceptive steroid (35) (Scheme 5). The FD mass spectrum of Vitamin-B_{12} shows an intense molecular ion (at *m/z* 1254) with fragments resulting mainly from cleavage of the axial ligands and of an acetamide side-chain (36), whereas extensive thermal degradation occurs on attempts to obtain EI mass spectra. A number of related corrinoids have also been studied, e. g. cobyrinic acid derivatives and again their field desorption spectra proved to more informative than their electron impact spectra (36).

2.4 Fast atom bombardment mass spectrometry (FAB-MS)

This relatively new 'soft' ionization technique is, like field desorption, of particular value for studying involatile and polar materials and has recently been used to obtain mass spectra of porphyrins (37), and of Vitamin-B_{12} and corrins. The molecular ion usually predominates although, by altering the conditions (e. g. changing the matrix or studying negative ion spectra) more fragmentation, FAB induced chemical reactions and exchange of ligands occur (36).

Another very interesting application of FAB-MS of therapeutic interest has been in studies of the tumour-localising components of 'haematoporphyrin derivative' (HPD) which has been used in the photochemotherapy of tumours. This material is a mixture of monomeric derivatives, dimers and trimers of haemotoporphyrin (Table 1) produced by treatment of haematoporphyrin with acetic and sulphuric acids followed by alkali; these have been partially separated by reversed phase HPLC and the mono- and di-meric porphyrins studied by FAB-MS (38—42). The main components obtained before alkaline hydrolysis were mono- and di-acetoxyethyl derivatives as well as mono-hydroxyethyl mono-vinyl derivatives and protoporphyrin; after hydrolysis the main components were haematoporphyrin, monohydroxyethyl monovinyl porphyrins and protoporphyrin (41). In other studies, the main tumour localising component has been isolated by gel filtration and HPLC; FAB mass spectral and NMR studies indicated that it was either "dihaematoporphyrin ether", an ether formed by linkage of hydroxyethyl groups on the two components of the dimer (38, 40), or a "dihaematoporphyrin ester" formed between one of the hydroxyethyl groups and one of the propionic acid groups on each of the two monomeric components (39, 42).

2.5 Other mass spectrometric techniques

As mentioned above the EI-MS of porphyrins usually only gives information about the nature of peripheral substituents but not about the order of substitution. Some new methods have recently been developed which provide at least a partial answer to this problem. Thus if the porphyrins are first reduced to porphyrinogens and then subjected to mass spectrometry, fragments corresponding to one, two and three pyrrolic rings can be observed (43); it cannot be determined in which way a specific di-substituted pyrrole occurs in a sequence of rings, but it is at least possible to confirm that the substitution pattern accords with that expected on biogenetic reasoning for example.

Another approach to this problem has been to oxidise porphyrins with lead dioxide to their *meso*-tetraoxo analogues (44) the mass spectra of which show fragmentations corresponding to mono-, di- and tri-pyrrolic species and which can then be used to determine the sequence of rings. So far this approach has only been applied to alkyl porphyrins.

Chemical ionization mass spectrometry is another promising approach which has recently been applied to the analysis of petroporphyrins (using methane as reagent gas) as it helps to increase sensitivity owing to the lower degree of fragmentation compared with EI (45). The use of hydrogen as reagent gas leads to reduction of the porphyrin *in situ* and fragments corresponding to one, two and three pyrrolic units can be observed (46). Very recently the newer technique of desorption chemical ionization (DCI) has been explored with porphyrins, and this also affords fragments corresponding to cleavages of the macrocyclic ring, which are of potential structural value (47). The use of DCI with ammonia as reagent gas has recently been shown to provide useful sequence information about the pyrrole rings, e. g. from the fragment ions in the dipyrrolic mass region (48). The method has also been extended to the identification of isomeric *N*-alkylated derivatives of protoporphyrin-IX. Secondary cleavages of side-chains occur to a far lesser extent under ammonia DCI, than hydrogen DCI conditions.

A very recent technique of potential value in the quantitative analysis of minute amounts of porphyrins in chemical samples is two-step laser desorption mass spectrometry (LD-MS) (49). An initial laser pulse causes desorption of the substrate molecules by rapid heating of a film of substrate, and then a second, delayed, pulse causes selective multiphoton ionization of the molecules in a time-of-flight mass spectrometer. In this way, protoporphyrin-IX dimethyl ester gives the [M + 1] ion with virtually no fragmentation, and showing linear dependence on sample concentration (over five orders of magnitude!); the detection limit was in the subrange (approx. 4×10^{-17} mol) (49).

LD-MS has also been applied to Vitamin-B_{12} (36), but fragmentation was fairly pronounced, and it will be interesting to see if the two-step laser process provides (as one would expect) a significant improvement. Other techniques such as Californium plasma desorption and secondary ion mass spectrometry have also been applied to Vitamin-B_{12} and other corrinoids, and compared among ten different ionization methods by Scheibl and Schulten (36).

Another potentially useful technique for porphyrins is direct liquid chromatography mass spectrometry (LC-MS). It has already been applied to the analysis of petroporphyrins (50) and preliminary studies in Cardiff have shown that it can be utilised for esters of porphyrins of clinical interest. LC-MS is not as sensitive as GC-MS (nanomole qualities being required) but GC-MS cannot readily be utilised with the relatively less volatile porphyrin esters (in contrast to the alkyl porphyrins obtained from petroleum). Thermospray ionization combined with LC-MS has also been applied to corrins and Vitamin-B_{12} (36).

3 Bile Pigments

The naturally occurring bile pigments are formed by oxidative ring-opening of haem followed by reduction of the initially formed biliverdin, to bilirubin by biliverdin reductase (Scheme 6). The bilirubin subsequently undergoes conjugation by glucuronic acid, and may then undergo further reduction by bacterial enzymes in the gut leading to a series of urobilins and finally to stercobilin (Scheme 6) (51).

As with porphyrins mass spectral studies have mostly been carried out on the methyl esters of the bile pigments, as these are much more volatile than the free acids (52). The methods of isolation from faecal samples are very similar to those used for porphyrins and the bile pigments likewise may also be purified chromatographically before mass spectral studies are carried out. However, biliverdin and especially bilirubin are difficult to extract quantitatively from body fluids because of their tendency to bind to proteins and because of the presence of lipids and special methods have been developed (53).

Unlike porphyrins, however, the bile pigments readily fragment between the pyrrole rings when subjected to EI-MS, as well as undergoing β-cleavages of the side-chain substituents (52) (Fig. 3). Fragmentations between the rings occur most readily at methylene ($-CH_2-$) bridges rather than at the (conjugated) methene ($-CH=$) bridges. Thus the molecular ion is the base peak in the mass spectra of the fully conjugated biliverdin dimethyl ester, although ions are also produced corresponding to cleavage at the central methine carbon, and to monopyrrolic units. In contrast the

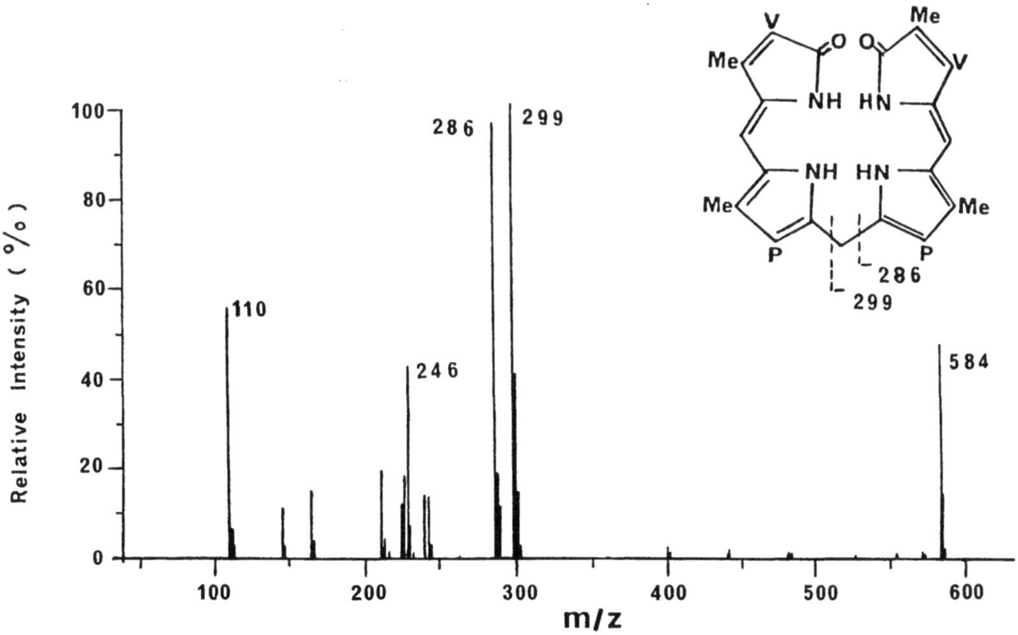

Figure 3. Electron impact mass spectrum of bilirubin.

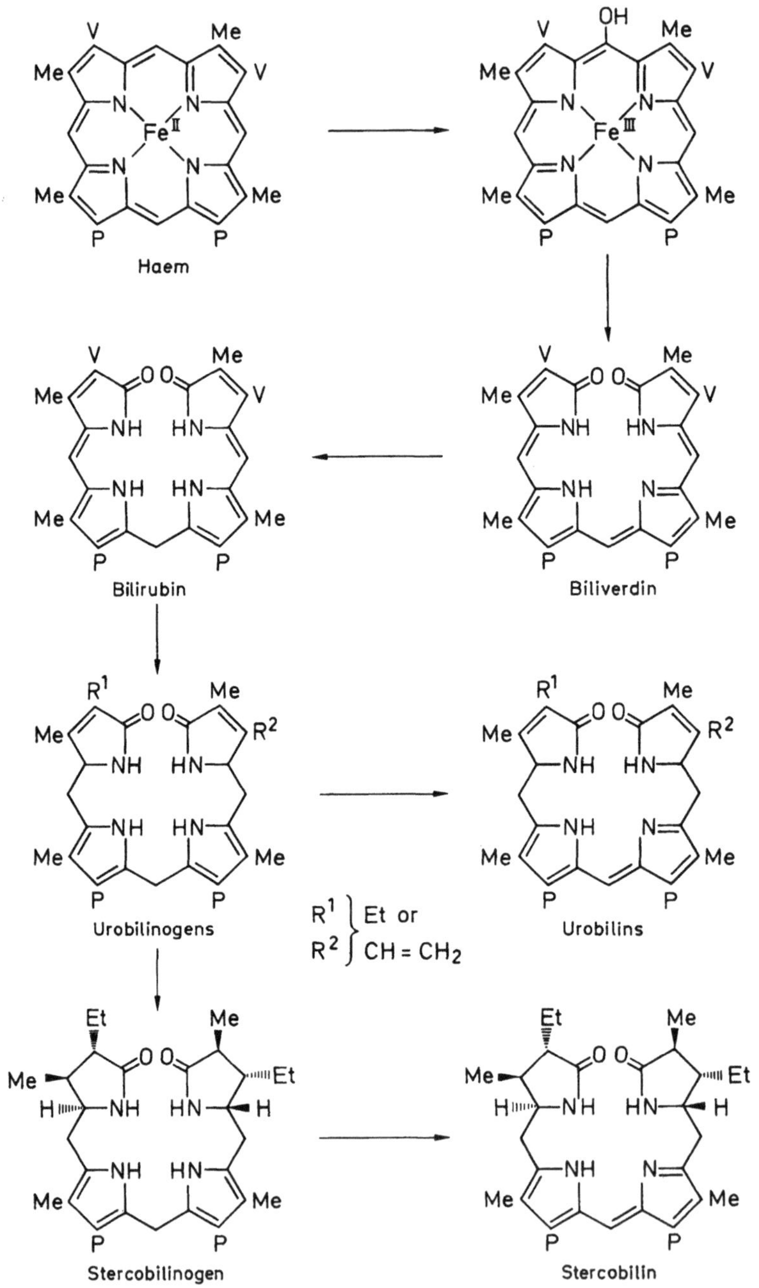

Scheme 6. Formation of biliverdin, bilirubin, urobilins and stercobilin.

molecular ion of bilirubin dimethyl ester is only about 20% of the intensity of the ion corresponding to cleavage at the central methylene group. With both biliverdin and bilirubin esters cleavages of side-chains in the major fragments are also observed, whereas fragmentation of side-chains from the molecular ions are almost non-existent.

Occasionally impurities are formed during the isolation process and chromatography due to formation of adducts from vinyl groups with methanol, or to oxidation (54). Traces of boron complexes have also been observed when borontrifluoride methanol has been used for esterification; the latter are much more volatile than the parent compounds but are readily recognised by peaks at 7 and 8 daltons higher than the expected molecular ion (owing to the boron isotopes ^{10}B and ^{11}B) (54).

Bilirubin free acid gives a good mass spectrum with an intense molecular ion ($\approx$ 30% of the base peak at 299) but biliverdin free acid is too involatile to obtain a spectrum. Indeed most bile pigments free acids are too involatile to obtain EI spectra directly, but the carboxylic acid groups of bilirubin-IXα are strongly *internally* hydrogen bonded (55). However, the FD mass spectra of most bile pigments, whether free acids or esters, give essentially only molecular ions (56); the lack of fragmentation is a disadvantage in some respects, but on the other hand FD spectra have been very useful in giving molecular mass information, where the molecular ions in the EI spectra have been very weak or non-existent.

Some work has also been carried out with the more volatile trimethylsilyl (TMS) esters of bile pigments (57). Thus bilirubin forms a tetra TMS derivative (on treatment with *N,O*-bis(trimethylsilyl)acetamide in pyridine) in which both acidic side-chains are esterified and the terminal lactam residues are also *O*-alkylated

$$(-NH-CO \rightarrow -N = C(OTMS)-).$$

Bilirubin dimethyl ester forms a similar bis OTMS derivative by *O*-alkylation of the lactam residues. These volatile TMS derivatives exhibit very strong molecular ions with relatively weak fragmentations, but their relatively low stability (e. g. to traces of acid) has probably precluded their widespread use.

The β- and δ-isomers of biliverdin and bilirubin (formed by ring-opening of haem at the β- and δ-positions) can be distinguished from the normal α-isomer (and the δ-isomer) by mass spectrometry as their cleavages at the central bridge carbon atom give fragment ions of different molecular masses.

Bilirubin is excreted by man, and many other mammals, in the form of a water soluble conjugate, and this was identified as a diglucuronide (58) using mass spectrometry to help identify the azo pigments formed from the bilirubin and diazotised ethylanthranilate, and glucuronic acid obtained by hydrolysis. The formation of such azopigments is, of course, the basis of the well known Van den Bergh method for the quantitative estimation of bilirubin in serum.

Apart from biliverdin and bilirubin most of the other naturally occurring mammalian bile pigments (Scheme 6) are normally only present in relatively small amounts in reduced form in faeces (51, 59). The end product of bacterial reduction is the colourless stercobilinogen and the main intermediate reduction products are the urobilinogens in which one, or both, vinyl groups on the terminal rings of the original bilirubin are reduced to ethyl groups. These tetrapyrroles, however, readily re-oxidise to the orange yellow pigments, stercobilin and the urobilins, in which the two central rings are conjugated through a methine bridge. Their structural elucidation presented a considerable challenge because of the relative paucity of the natural pigments, as

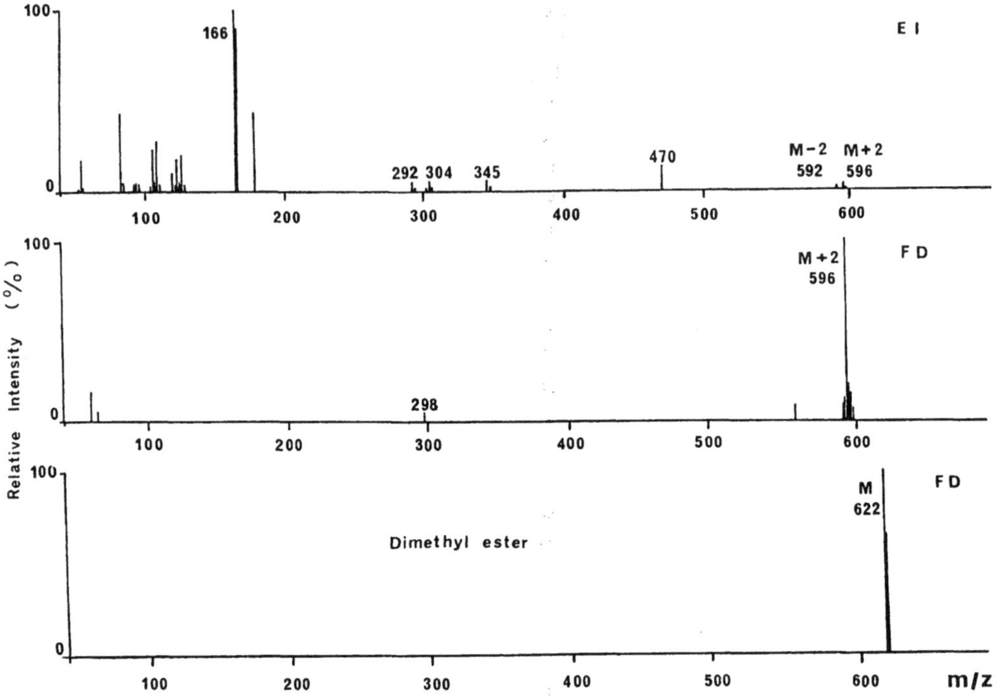

Figure 4. Electron impact and field desorption mass spectra of stercobilin and its dimethyl ester.

well as the ease with which some of them underwent rearrangements on treatment with acid or alkali.

When EI-MS was first applied to the reduced faecal metabolites of bilirubin, however, it was found that not only were the molecular ions very weak, but that disproportion reactions occurred in the mass spectrometer (60). Stercobilin was the classic example of this behaviour, (Fig. 4) affording pseudo molecular ions (1−2%) at *m/z* 592 and 596 (instead of the expected *m/z* 594), the major fragments corresponding to monopyrrolic ions. The dimethyl ester also exhibited a very weak molecular ion (4%), but at *m/z* 622 (corresponding to *m/z* 594 for the free acid) and the tri- and di-pyrrolic fragment ions were much more intense; the presence of a major ion at *m/z* 497 in the spectrum confirmed that both terminal lactam rings were fully reduced. The FD mass spectrum of the dimethyl ester showed only the molecular ion at *m/z* 622 for the dimethyl ester, but interestingly the free acid gave a molecular ion at *m/z* 596 (i. e. M + 2) (61) (Fig. 4).

The urobilin also showed anomalous behaviour on EI-MS, and it was not entirely clear whether similar processes were occurring. For example a specimen of *d*-urobilin studied in Liverpool and Cardiff was originally thought to contain one terminal ring bearing an ethyl group and the other a vinyl group because the EI mass spectrum showed a 'molecular ion' at *m/z* 588 (60). More detailed studies showed that at lower source temperatures a weak ion at *m/z* 590 was observed (and no *m/z* 588) (62). The methyl ester was also examined and this exhibited a molecular ion at *m/z* 618 (corresponding to m/z 590 for the free acid) whereas the FD spectrum exhibited a molecular ion at m/z 619 (i. e. M + 1). Subsequent degradative studies (see below) confirmed that neither of the terminal rings bore vinyl groups, thus substantiating the results with the dimethyl ester.

Because of the relatively small amounts of the natural reduced bile pigments which were available a range of semi-synthetic pigments were prepared by sodium amalgam or catalytic hydrogenation of bilirubin, followed in some cases by acid-catalysed rearrangements. The EI and FD mass spectra of some thirty of these materials were studied, and as expected the molecular ion was the base peak, if not the only peak in most of the spectra (56). The fully conjugated biliverdin type pigments also gave intense molecular ions in their EI spectra, whereas the more reduced pigments such as urobilins, with only two conjugated rings, or the violins with three rings conjugated gave weak molecular ions on EI, and more intense fragments corresponding to the two or three conjugated rings.

4 Degradation Products from Porphyrin and Bile Pigments

The original structure determination of porphyrins relied heavily on two types of degradation: a) hydrogen iodide reduction to monopyrrolic fragments, and b) oxidative degradation to maleimides (1). In the second procedure it was found necessary to reduce peripheral vinyl groups to ethyl groups prior to oxidation, as the rather drastic conditions required led to complete destruction of vinyl bearing pyrrole rings.

These two methods have been refined in recent years with the advent of chromatographic techniques and have been applied to the analysis of very much smaller amounts of material than in the original procedure. A key feature of the early structural work on haem-*a* for example was the oxidation to a maleimide bearing methyl and C_{17} side-chains corresponding to the A-ring of the porphyrin (63).

Later it was found that bile pigments could be oxidised to monopyrrolic fragments under much milder conditions than porphyrins, and the method was applied to the structure determination of plant bile pigments, the oxidation products (maleimide and diformyl pyrroles) being separated and isolated by TLC (64). Gas chromatography is, however, a much more sensitive technique and was applied to the structure determination of chlorobium chlorophyll derivatives using reductive alkylation (65, 66).

More recently this method was extended to the use of GC-MS for the structure determination of isocoproporphyrin (12). Two possible structures were deduced on the basis of mass and NMR spectral studies of the tetramethyl ester, both of which were reasonable possibilities on biosynthetic grounds. They were, however, readily distinguished by degradative methods. Thus oxidation with chromic acid afforded three maleimides which were characterised by GC-MS and comparisons with the retention times and mass spectra of authentic materials obtained by oxidation of known porphyrins (or appropriately substituted pyrroles). This showed that the struc-

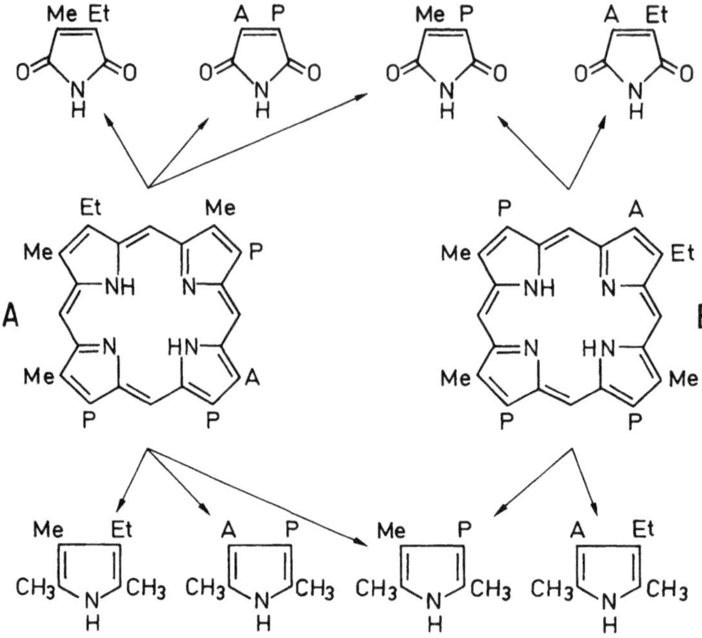

Scheme 7. Degradation of isocoproporphyrin to three maleimides and to three alkyl pyrroles showing that the structure corresponds to formula A rather than formula B.

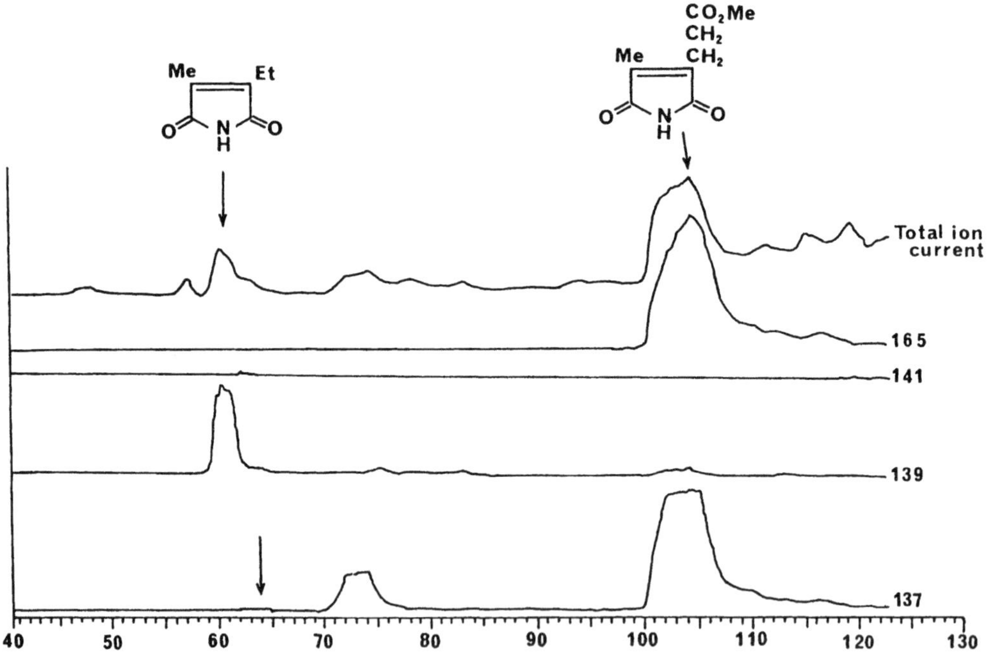

Figure 5. Mass chromatograms of the maleimides obtained by oxidative degradation of a sample of *d*-Urobilin; the absence of a peak in the trace m/z 137, at the position corresponding to methylvinylmaleimide (marked with an arrow) shows that the terminal rings of the urobilin both contain methyl and ethyl groups.

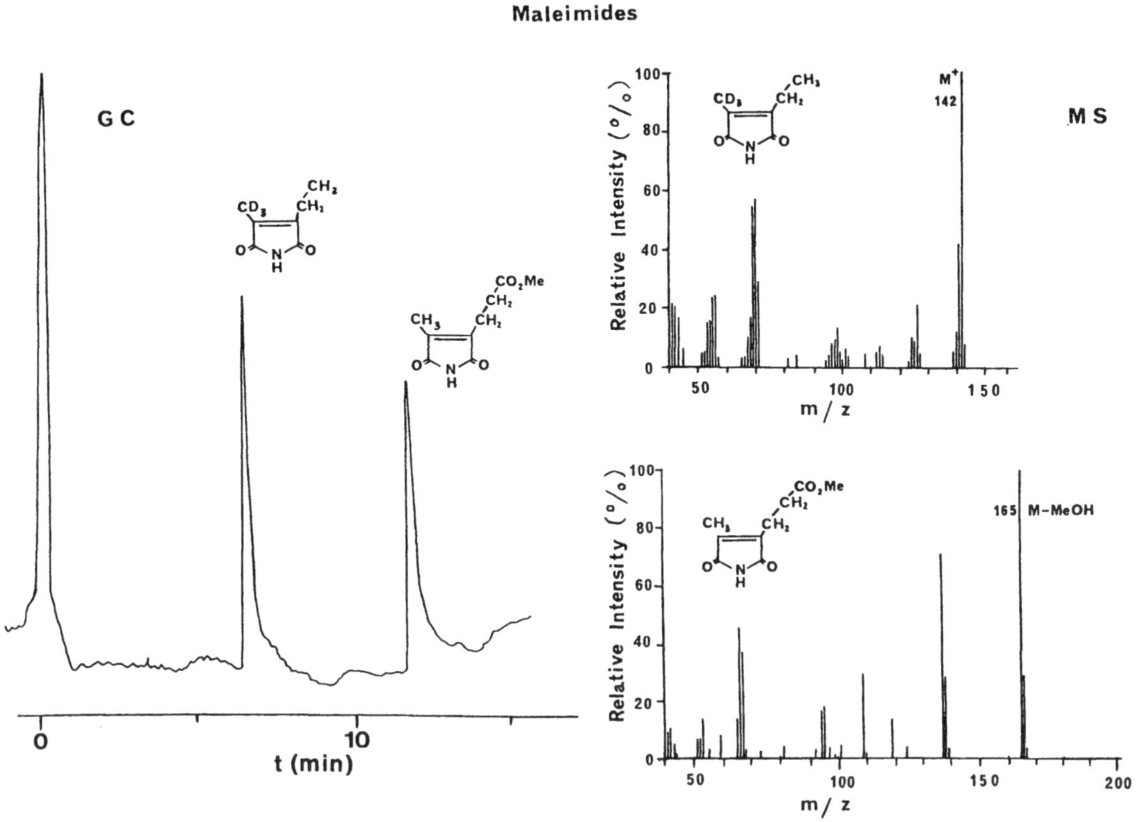

Figure 6. GC-MS of maleimides derived from the tricarboxylic porphyrinogen (harderoporphyrinogen) formed from deuterium labelled coproporphyrinogen (see Scheme 8).

ture of isocoproporphyrin corresponded to Structure A (Scheme 7) as the alternative formulation B would have given rise to only two different maleimides. These conclusions were also confirmed by reductive alkylation of isocoproporphyrin with hydrogen iodide and formaldehyde which afforded a mixture of three pyrroles as expected on the basis of Structure A, rather than the two which would have been derived from Structure B; analysis and characterisation of the alkyl pyrroles was again carried out by GC-MS.

The oxidative degradation procedure has also been applied to studies of bile pigment structure and as milder conditions can be used even vinyl substituents survive (in contrast to porphyrins). Thus as described above, samples of *d*-urobilin showed a molecular ion at *m/z* 591 (M + I) on FD-MS, and in EI mass spectra ions at *m/z* 588 and 590 were observed. The lower molecular mass was attributed to the presence of a vinyl, rather than an ethyl, group in one of the terminal rings (5). However, oxidative degradation to maleimide afforded only methyl ethyl maleimide and haematinic acid (as its methyl ester); careful examination of the GC-MS traces using mass chromatography (and comparisons with methyl vinyl-maleimide obtained from oxi-

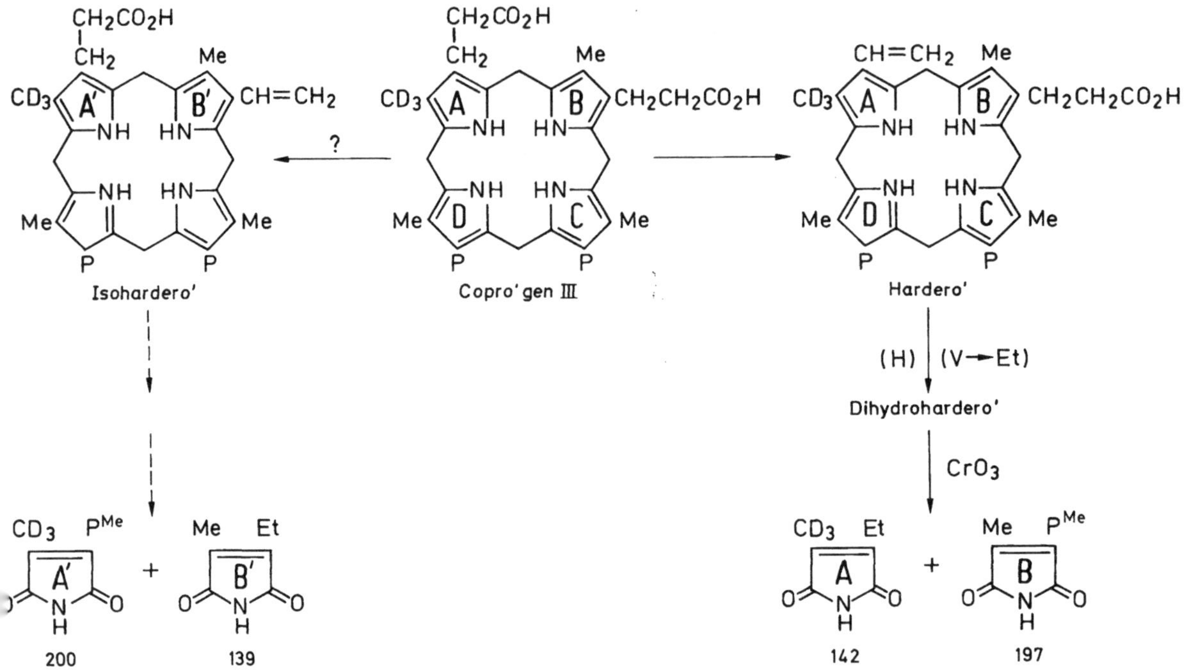

Scheme 8. Oxidative degradation of labelled tricarboxylic porphyrinogen formed from deuterium labelled coproporphyrinogen-III; GC-MS analysis of the maleic imides formed shows that only harderoporphyrinogen is formed as an intermediate (see also Figure 6).

dation of bilirubin) showed that no methyl vinyl maleimide was formed, and thus the *d*-urobilin had molecular mass 590, corresponding to a bile pigment with terminal rings bearing methyl and ethyl groups (Fig. 5).

Further studies of these degradative methods have shown that the chromic acid oxidations of porphyrins can be carried out with as little as 10 μg material, whereas the reductive alkylation method requires at least 100 μg porphyrin. The lower sensitivity in the reductive alkylation is attributed to the instability (ease of oxidation) of the alkyl pyrroles; however, this method has the advantage that the methine bridge carbon atoms are retained, and this proved useful in the detection of *meso*-methyl groups in porphyrins related to the chlorobium chlorophylls (66).

Another recent application of the oxidative degradation method was the confirmation that the pathway between coproporphyrinogen-III and protoporphyrin-IV involved only harderoporphyrinogen (Scheme 4). A ring-A deuterium labelled coproporphyrinogen was synthesised, and on metabolism by chicken red-cell haemolysates afforded protoporphyrin-IX together with a small amount of the intermediate tricarboxylic porphyrinogen; the latter was isolated at the porphyrin trimethyl ester and chromic acid oxidation afforded a trideuteromethylethyl maleic imide and unlabelled haematinic methyl ester which were identified by GC-MS (Fig. 6 and Scheme 8) (67). None of the trideuteromethyl analogue of methyl haematinate was formed, thus showing that isoharderoporphyrinogen was not an intermediate in the biochemical conversion, and this was also cinfirmed by radio labelling and by HPLC (6).

References

1. Fischer, H. & Orth, H. (1937) Die Chemie des Pyrrols, Akademischer Verlag, Leipzig.
2. Jackson, A. H. (1976) Structure, properties and distribution of chlorophylls, in: The Chemistry and Biochemistry of Plant Pigments (Goodwin, T. W., ed.) pp. 1–63, Academic Press, New York.
3. Bogorad, L. (1978) Biosynthesis of porphyrins, in: The Porphyrins, Vol. VI (Dolphin, D. ed.) pp. 125–178, Academic Press, New York.
4. Battersby, A. R., Fookes, C. J. R., Matcham, G. W. J. & McDonald, E. (1980) Biosynthesis of the pigments of life: formation of the macrocycle. Nature, *285*, 17–20.
5. Jackson, A. H., Sancovich, H. A., Ferramola, A. M., Evans, N., Games, D. E., Matlin, S. A., Elder, G. H. & Smith, S. G. (1976) Phil. Trans. Roy. Soc. Lond. B, *273*, 191–206.
6. Jackson, A. H., Elder, G. H. & Smith, S. G. (1978) The metabolism of coproporphyrinogen-III into protoporphyrin-IX, Int. J. Biochem., *9*, 877–882.
7. Couch, P. W., Games, D. E., Jackson, A. H., Jackson, J. R., Belcher, R. V. & Smith, S. G. (1974) Conversion of coproporphyrinogen-III to protoporphyrin-IX, Enzyme, *17*, 82–90.
8. Meyer, U. A. & Schmid, R. (1977) The porphyrins, in: The Metabolic Basis of Inherited Disease, 4th edn. (Stanbury, J. B., Wyngaarden, J. B. & Frederickson, D. S., eds.) pp. 1166–1220, McGraw-Hill, New York.
9. Jackson, A. H., Rao, K. R. N., Supphayen, D. & Smith, S. G. (1977) Intermediates between uroporphyrinogen-I and coproporphyrinogen-I, J. Chem. Soc. Chem. Comm. 696–698.
10. Elder, G. H. (1977) Porphyrin metabolism in porphyria cutanea tarda. Semin. Haematol. *14*, 227–242.
11. Elder, G. H. (1972) Identification of a group of tetracarboxylate porphyrins containing one acetate and three propionate β-substituents, in faeces from patients with symptomatic cutaneous hepatic porphyria and from rats with porphyria due to hexachlorobenzene poisoning. Biochem. J. *126*, 877–891.
12. Jackson, A. H., Lash, T. D., Ryder, D. J. & Smith, S. G. (1980) Synthetic and biosynthetic studies of porphyrins, Part V, Evidence of an alternative pathway in the biosynthesis of haem. Int. J. Biochem. *12*, 775–790.
13. Sano, S., Shingu, T., French, J. M. & Thonger, E. (1965) The chemical structure of pemptoporphyrin. Biochem. J. *97*, 250–256.
14. French, J., Nicholson, D. S. & Rimington, C. (1970) Identification of the acrylate porphyrin S-411 from meconium. Biochem. J. *120*, 393–397.
15. Kennedy, G. Y., Jackson, A. H., Kenner, G. W. & Suckling, C. J. (1970) Isolation, structure and synthesis of a tricarboxylic porphyrin from the harderian glands of the rat. FEBS Lett. *6*, 9–12.
16. Eales, L. (1978) Clinical chemistry of the porphyrins, in: The Porphyrins, Vol. VI (Dolphin, D., ed.) pp. 663–804, Academic Press, New York.
17. Jackson, A. H. Kenner, G. W., Smith, K. M., Aplin, R. T., Budzikiewicz, H. & Djerassi, C. (1965) Pyrroles and related compounds VIII: The mass spectra of porphyrins. Tetrahedron *21*, 2913–2924.
18. Smith, K. M. (1975) Mass Spectrometry of Porphyrins, in: Porphyrins and Metalloporphyrins (Smith, K. M., ed.) pp. 381–398, Elsevier, Amsterdam; and Budzikiewicz, H. (1978) Mass Spectra of Porphyrins and Related Compounds, in: The Porphyrins Vol. III (Dolphin, E., ed.) pp. 395–462, Academic Press, New York.
19. Boylan, D. B. & Calvin, M. (1967) Volatile silicon complexes of etioporphyrin-I. J. Amer. Chem. Soc. *89*, 5472; Games, D. E., Jackson, A. H. & Millington, D. S. (1974) Applications of mass spectrometry to the structure determination of pyrrole pigments, in: Mass Spectrometry in Biochemistry and Medicine (Frigerio, A. & Castagnoli, N., eds.) p. 257–266, Raven Press, New York and Marriott, P. J., Gill, J. P., Evershed, R. P., Hein, C. S. & Eglinton, G. (1984) Computerised gas chromatographic mass spectrometric analysis of complex mixtures of alkyl porphyrins. J. Chromatog. *301*, 107–128.
20. Jackson, A. H., Kenner, G. W. & Wass, J. (1967) Rational syntheses of chlorocruoroporphyrin (Spirographis porphyrin) and pemptoporphyrin. J. Chem. Soc. Chem. Comm. 1027–1029.
21. Cavaleiro, J. A. S., Kenner, G. W. & Smith, K. M. (1974) Porphyrin and related com-

pounds, XXXII. Biosynthesis of protoporphyrin-IX from coproporphyrinogen-III. J. Chem. Soc. Perkin Trans. I, 1188–1194, and Games, D. E., Jackson, A. H., Jackson, J. R., Belcher, R. V. & Smith, S. G. (1976) Biosynthesis of protoporphyrin-IX from coproporphyrin. J. Chem. Soc. Chem. Comm. 187–188.

22. Couch, P. W., Games, D. E. & Jackson, A. H. (1976) Synthetic and biosynthetic studies of porphyrins, Part I. Synthesis of the S-411 porphyrin from meconium. J. Chem. Soc. Perkin Trans. I, 2492–2501.
23. Battersby, A. R., Baldas, J., Collins, J., Grayson, D. A., James, R. J. & McDonald, E. (1972) Mechanism of the biosynthesis of the vinyl groups of protoporphyrin-IX. J. Chem. Soc. Comm. 1265–1266.
24. Stoll, M. S., Elder, G. H., Games, D. E., O'Hanlon, P., Millington, D. S. & Jackson, A. H. (1973) Isocoproporphyrin: Nuclear magnetic resonance and mass spectral methods for the determination of porphyrin structure. Biochem. J. *131*, 429–432.
25. Baptista de Almeida, J. A. P., Kenner, G. W., Rimmer, J. & Smith, K. M. (1976) Pyrroles and related compounds, XXXV. A stepwise general synthesis of unsymmetrically substituted porphyrins. Tetrahedron *32*, 1793–1799.
26. Clezy, P. S. & Hai, T. T. (1976) Synthesis of isocoproporphyrin tetramethyl ester and related compounds. Aust. J. Chem. *29*, 1561–1571.
27. Jackson, A. H., Rao, K. R. N. & Smith, S. G. (1984) unpublished work.
28. Jackson, A. H., Rao, K. R. N., Supphayen, D. & Smith, S. G. (1977) Intermediates between uroporphyrinogen-I and coproporphyrinogen-I. J. Chem. Soc. Chem. Comm. 696–698.
29. Evans, N., Games, D. E., Jackson, A. H. & Matlin, S. A. (1975) Applications of high pressure liquid chromatography and field desorption mass spectrometry in studies of natural porphyrin and chlorophyll derivatives. J. Chromatog. *115*, 325–333.
30. DeMatteis, F., Gibbs, A. H., Jackson, A. H. & Weerasinghe, S. (1980) Conversion of liver haem into N-substituted porphyrins or green pigments: Nature of the substituent at the pyrrole nitrogen atom. FEBS Lett. *119*, 109–112.
31. Ortiz de Montellano, P. R., Mico, B. A., Beilan, H. S. & Kunze, K. L. (1981) Olefins as suicide inhibitors of cytochrome P450. Dev. Biochem. *19*, 151–166.
32. Jackson, A. H. (1979) Applications of mass spectrometry in studies of porphyrins and related tetrapyrroles. Phil. Trans. Roy. Soc. Lond. A *293*, 21–37.
33. Jackson, A. H., Sancovich, H. A. & Ferramola de Sancovich, A. M. (1980) Synthetic and biosynthetic studies of porphyrins, Part III. Structures of the intermediates between uroporphyrinogen-III and coproporphyrinogen-III: Synthesis of fourteen heptacarboxylic, hexacarboxylic and pentacarboxylic porphyrins related to uroporphyrin-III. Bio-organic Chem. *9*, 71–120.
34. Caughey, W. S., Smythe, G. A., O'Keefe, D. H., Maskasky, J. E. & Smith, M. L. (1975) Heme-A of cytochrome-c oxidase II structure and properties: Comparisons with hemes B, C and S and derivatives. J. Biol. Chem. *250*, 7602–7622 and DeFilippi, L. J. & Hultquist, D. E. (1977) High pressure liquid chromatography and field desorption mass spectrometry of heme a, heme a dimethyl ester and acetyl heme a dimethyl ester. Biochim. Biophys. Acta *498*, 395–402.
35. Ortiz de Montellano, P. & Kunze, K. L. (1980) Self-catalysed inactivation of hepatic cytochrome-P450 by ethynyl substrates. J. Biol. Chem. *255*, 5578–5585.
36. Schiebel, H. M. & Schulten, H.-R. (1986) Soft ionization of biomolecules: a comparison of ten ionization methods for corrins and vitamin B_{12}, in: Mass Spectrometry Review, Vol. 5, (Gross, M. L., ed.) pp. 249–311, Wiley, New York and references therein.
37. Meili, J. & Seibl, J. (1984) A new versatile matrix for fast atom bombardment analysis. Org. Mass Spect. *19*, 581–582.
38. Dougherty, T. J., Potter, W. R. & Weishaupt, K. R. (1983) The structure of the active component of hematoporphyrin derivative, in: Porphyrins in Tumour Phototherapy (Adreoni, A. & Cubeddu, R., eds.) pp. 23–35, Plenum, New York.
39. Dougherty, T. J., Potter, W. R. & Weishaupt, K. R. (1984) The structure of the active component of hematoporphyrin derivative. Prog. Clin. Biol. Res. *170*, 301–314.
40. Kessel, D. & Cheng, M. L. (1985) On the preparation and properties of dihaematoporphyrin ether, the tumour-localising component of HPD. Photochem. Photobiol. *41*, 277–282.
41. Meijers, J. C. M., Lim, C. K., Lawson, A. M. & Peters, T. J. (1986) Analysis of tumour-localising haematoporphyrin deriva-

tive by high-performance liquid chromatography and fast atom bombardment mass spectrometry. J. Chromatogr. *352*, 231–239 (see also references contained therein for earlier HPLC studies).
42. Kessel, D. (1986) Proposed structure of the tumour localising fraction of HPD (hematoporphyrin derivative). Photochem. Photobiol. *44*, 193–196.
43. Budzikiewicz, H. & Pesch, R. (1976) Localisation of substituents in porphyrins: The mass spectroscopic fragmentation behaviour of porphyrinogens. Org. Mass Spect. *11*, 821–830.
44. Boylan, D. B. (1970) Mass spectrometry differentiation of isomeric alkyl porphyrins. Fragmentation analysis of meso-tetraoxoporphyrinogens. Org. Mass Spect. *3*, 339–351.
45. Shaw, G. J., Quirke, J. M. E. & Eglinton, G. (1978) Analysis of petroporphyrins by chemical ionisation mass spectrometry. J. Chem. Soc. Perkin Trans. I, 1655–1659.
46. Sundaraman, P., Gallegos, E. J., Baker, E. W., Slayback, J. R. B. & Johnston, M. R. (1982) Hydrogen chemical ionisation tandem mass spectrometry of metalloporphyrins. Anal. Chem. *56*, 2552–2556. and Wolff, G. C., Chicarelli, M. I., Shaw, G. I., Evershed, R. P., Quirke, J. M. E. & Maxwell, J. R. (1984) Structure analysis of naturally occurring alkyl porphyrins by hydrogen chemical ionisation mass spectrometry. Tetrahedron *40*, 3777–3786.
47. Jiang, X. Y., Szente, A. W., Tolf, B. R., Kehres, L. A., Bunnenberg, E. & Djerassi, C. (1984) Mass spectrometry in structural and stereochemical problems 264. Enhanced structure determination of β-pyrrole and N-substituted porphyrins by desorption chemical ionisation mass spectrometry. Tetrahedron Lett. *25*, 4083–4086.
48. Tolf, B. R., Jiang, X. Y., Wegmann-Szente, A., Kehres, L. A., Bunnenberg, E. & Djerassi, C. (1986) Mass spectrometry in structural and stereochemical problems. 266. Enhanced structural determination of substituted porphyrins by ammonia desorption chemical ionization mass spectrometry. J. Amer. Chem. Soc. *108*, 1363–1374.
49. Hahn, J. H., Zenobi, R. & Zare, R. N. (1987). Subfemtomole quantitation of molecular adsorbates by two-step laser mass spectrometry. J. Amer. Chem. Soc. *109*, 2842–2843.
50. Mcfadden, W. H., Bradford, D. C., Eglinton, G., Hajibrahim, S. K & Nicolaides, N. (1979) Application of combined liquid chromatography/mass spectrometry (LC/MS) analysis of petroporphyrins and melbonian gland waxes. J. Chromatog. Sci. *17*, 518–522.
51. Schmid, R. & McDonagh, A. F. (1979) Formation and metabolism of bile pigments *in vivo*, in: The Porphyrins, Vol. VI (Dolphin, D., ed.) pp. 257–292, Academic Press, New York.
52. Jackson, A. H., Kenner, G. W., Budzikiewicz, H., Djerassi, C. & Wilson, J. M. Pyrroles and related compounds, X. Mass spectra of linear di-, tri- and tetra-pyrroles. Tetrahedron *23*, 603–622.
53. McDonagh, A. F. (1979) Bile pigments: bilatrienes and 5,15-biladienes, in: The Porphyrins, Vol. VI (Dolphin, D., ed.) pp. 293–491, Academic Press, New York.
54. Manitto, P. & Manti, D. (1974) The oxidation of bilirubin to biliverdin dimethyl ester. Evidence for acid catalysed addition of methanol to the vinyl groups of biliverdin dimethyl ester isomers. Gazz. Chim. Ital. *104*, 513–521.
55. Bonnett, R., Davies, J. E. & Hursthouse, M. B. (1976) Structure of bilirubin. Nature *262*, 326–328.
56. Games, D. E., Jackson, A. H., Saxton, R. G. & Stoll, M. (1975) Applications of field desorption mass spectrometry, in: Metabolism and Chemistry of Bilirubin and Related Tetrapyrrols (Bakken, A. F. & Fog, J., eds.) pp. 254–257, Pediatric Res. Inst., Rikshospitalet, Oslo, Norway.
57. Salmon, M. & Fenselau, C. (1974) Evaluation of biliverdin derivatives for mass spectral analysis. Biomed. Mass Spectrom. *1*, 212–222, and Compernolle, F., Blanckaert, N. & Heirwegh, K. P. M. (1976) Mass spectral study of derivatives of the four bilirubin-IX isomers. Biomed. Mass Spect. *3*, 155–160.
58. Gordon, E. R., Goresky, C. A., Chang, T. & Perkin, A. S. (1976) Isolation and characterisation of bilirubin diglucuronide as the major bilirubin conjugate in dog and human bile. Biochem. J. *155*, 477–486.
59. Schmid, R. & McDonagh, A. F. (1977) Hyperbilirubinaemia in: The Metabolic Basis of Inherited Disease, 4th bdn. (Stanbury, J. B., Wyngaarden, J. B. & Frederickson, D. S., eds.) pp. 1221–1257, McGraw-Hill, New York.
60. Jackson, A. H., Smith, K. M., Gray, G. H. & Nicholson, D. C. (1966) Molecular species of the urobilins. Nature *209*, 581–583.

61. Cooper, G., Games, D. E., Jackson, A. H., Saxton, R. G. & Stoll, M. S. (1976) Structural studies of bile pigments using mass spectral methods, in: Advances in Mass Spectrometry in Biochemistry and Medicine, Vol. 2 (Frigerio, A., ed.) pp. 257–271, Spectrum, New York.
62. Jackson, A. H., Games, D. E. & Cooper, G. (1976) Intermediates in haem degradation. Biochem. Soc. Trans. *4*, 214–217.
63. Grassl, M., Coy, V., Seyffert, R. & Lynen, F. (1963) Chemical constitution of the cytohemins. Biochem. Z. *338*, 771–795.
64. Rüdiger, W. (1968) Bile pigments, a new degradation technique and its application, in: Porphyrins and related compounds (Goodwin, T. W., ed.) pp. 121–130, Academic Press, London.
65. Gregorovich, B. V., Liang, K. S. V., Clugston, D. M. & McDonald, S. F. (1968) Reductive C-alkylation. Can. J. Chem. *46*, 3291–3300.
66. Chapman, R. A., Roomi, M. W., Morton, T. C., Krajcarski, D. T. & McDonald, S. F. (1971) The analytical reduction of porphyrins to pyrroles. Can. J. Chem., *49*, 3544–3564.
67. Jackson, A. H., Jones, D. M., Philip, G., Lash, T. D., Batlle, A. M. del C. & Smith, S. G. (1980) Synthetic and biosynthetic studies of porphyrins, Part IV. Further studies of the conversion of coproporphyrinogen-III to protoporphyrin-IX: Mass spectrometric investigations of the incubation of specifically deuterated coproporphyrinogen-III with chicken red cell haemolysates. Int. J. Biochim. *12*, 681–688.

Chapter 9
Prostaglandins

R. W. Kelly

1 Introduction

1.1 Problems and successes in the measurement and identification of prostanoids

The metabolism of arachidonic acid continues to provide new challenges to physiologists and biochemists; the last half century has seen the sequential discovery of the stable prostaglandins (D, E, and F), the highly, reactive intermediate endoperoxides, the short lived, potent platelet aggregating agent thromboxane A_2, the antiaggregating PGI_2 (prostacyclin) and most recently the family of leukotrienes and related compounds which form the lipoxygenase pathway. The measurement of all of these compounds has presented methodological problems arising from the closely related structures, their instability and their low levels of production. Other problems have arisen because attempts to measure prostaglandin levels have resulted in the generation of the compounds to be measured.

1.2 Unique problems of measuring prostaglandins in tissue

Prostaglandins (PGs) are local mediators which are not stored but are produced in response to a stimulus. The point at which control of PG synthesis is exercised is mainly at the level of precursor (arachidonic acid) production thus intracellular levels of free arachidonic acid are generally very low although arachidonic acid incorporated into phospholipids is abundant. These low cytosolic levels are maintained by an ATP dependent reincorporation of free arachidonic acid into membrane phospholipids by the action of acyl-coenzyme A ligase and acetyl transferase (1).

The process of tissue collection disturbs the cellular homeostasis by releasing and activating phospholipases in damaged tissue and at the same time interfering with the reincorporation process, thus tissue levels of prostaglandins have to be interpreted with care as has been shown by Green in simultaneous gas chromatography-mass spectrometry (GC-MS) measurements of plasma metabolites and tissue concentrations (2). Moreover when the synthesis of PGs by tissue is investigated using radiolabelled precursors, regard has to be taken of the different stores of arachidonic acid and particularly the availability of free arachidonic acid from endogenous phospholipids. A recent study addressed this problem by using ^{14}C labelled arachidonic acid with autoradiographic quantitation together with ^{12}C measurements of PGs formed from endogenous arachidonic acid using GC-MS (3). This showed that the relative contribution of the exogenous arachidonic acid was very variable with endogenous precursor predominating in some samples and exogenous precursor predominating in others.

PGs and leukotrienes are involved in inflammatory reactions, steroidal anti inflammatory drugs such as corticosteroids (possibly by liberating an inhibitor, lipomodulin) act to inhibit phospholipases whilst non-steroidal antiinflammatory drugs such as aspirin, indomethacin and the fenamates act by inhibiting the cyclooxygenase which converts the arachidonic acid to the key endoperoxide intermediates PGG_2 and PGH_2 (Figs. 1 and 2). Although many of these widely used drugs act on prostaglandin synthesis little is known of the mechanism of action of the stable PGs such as E, F

MECHANISMS OF CELLULAR ACTIVATION PRODUCING FREE ARACHIDONIC ACID

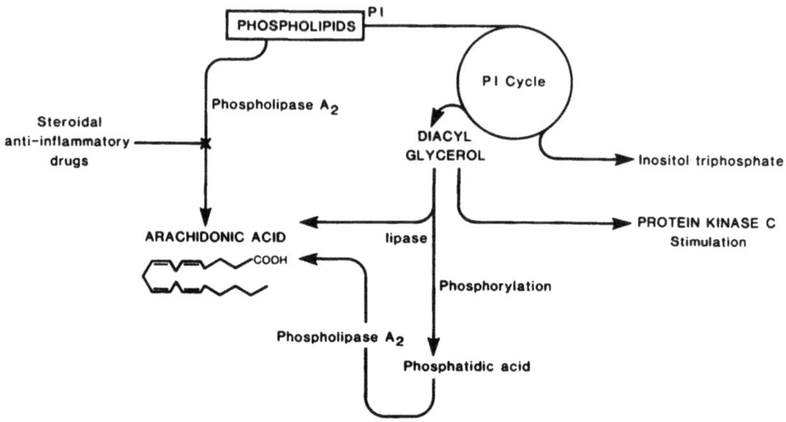

Figure 1. Pathways which lead to the production of free arachidonic acid within the cell. The process of stimulus secretion coupling is thought to produce arachidonic acid via the phosphatidyl inositol (PI) pathway. Arachidonic acid is incorporated in the 2 position of PI after PI synthesis and is present in high concentrations in this minor but important phospholipid. Phospholipase A_2 which is calcium dependent might play a subsequent role in releasing the major amounts of arachidonic acid following increases in cytosolic free calcium levels resulting from cell stimulation and calcium mobilization by inositol triphosphate.

and D. With these compounds being produced locally and catabolized locally much of the research has concentrated on the indirect approach of studying effects when PG synthesis is blocked by the above pharmacological agents and by using radiolabelled precursors in isolated tissue, cells and homogenates. However, many questions such as the levels of circulating PGs and production rates of PGs *in vivo* have required assays of high sensitivity and specificity and GC-MS is an ideal technique to provide some of these answers.

The main contribution of mass spectrometry to this field has undoubtedly been its unparalleled ability to identify unknown lipids when they are only available in microgram and submicrogram amounts. The technique of combined GC-MS using the jet separator devised by Ryhage found one of its very early uses in establishing the structures of PGE, and PGF, (4) and since then GC-MS has been the primary method of identification of all the naturally occurring prostanoids such as endoperoxides (5), thromboxanes (6) prostacyclin (7) and the leukotrienes (8, 9, 10). GC-MS has also made a major contribution to the measurement of prostaglandins since the high sensitivity and specificity of the technique has allowed the measurement of very low circulating levels of PG in plasma (for a review see 11). The identification and quantitation of urinary metabolites of PGs which depends almost entirely on GC-MS has given a valuable estimation of PG production rates.

GC-MS plays a complementary role to other techniques in PG research. Bioassay has made a unique contribution, being the means by which "prostaglandin" was first discovered (12) and being used more recently (as the superfused tissue cascade method)

DIVERGENCE OF LIPOXYGENASE AND CYCLOOXYGENASE PATHWAYS

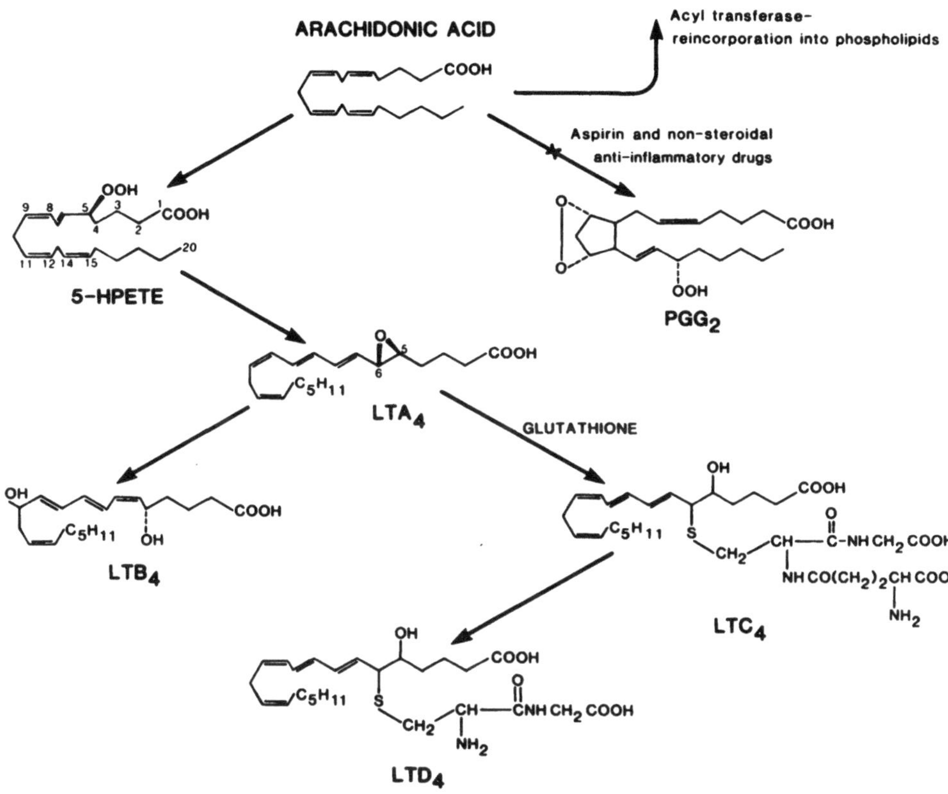

Figure 2. The fate of free arachidonic acid in the resting cell is probably to be reincorporated into the phospholipid pools. Following stimulation of the cell, arachidonic acid may be available for the cyclooxygenase or the lipoxygenase pathway, the latter giving rise to the family of leukotrienes and the former giving the key endoperoxide intermediate PGG which is further transformed to the prostanoids (see Figure 5).

to discover prostacyclin (13). Bioassay has the disadvantage of limited inherent specificity and only being able to handle limited numbers of samples. Radioimmunoassay (RIA) is the most widely used analytical technique for PG research since it combines sensitivity with reasonable specificity and it has the capacity to measure large numbers of samples. RIA has the disadvantage that it suffers from a certain degree of non-specific interference. Misleading results have been obtained from all of the above techniques and the cautionary tales of instability of PGs, interconversion and interference from similar, but at the time unknown, prostanoids apply to all methods of analysis. The greater specificity available with the use of physico-chemical techniques has not prevented its contribution to the appreciable list of artificially high levels of PGs recorded in the literature.

This chapter does not comprise of an exhaustive treatment of the use of MS in PG research since this technique has been used to some degree in virtually all areas of PG research but it will rather concentrate on the areas where the technique has made the most impact.

2 Measurement of PGs by GC-MS Using Stable Isotopes as Internal Standards

2.1 Internal standards

The most widely used internal standards for analysis of the primary PGs are the 3,3,4,4-tetradeutero analogues. Both the 3 and the 4 position are relatively free from proton exchange and these standards have been ideal for the assay of PGE_2, PGF_2 and 6-oxo $PGF_1\alpha$. The 3,3,4,4-tetradeutero compounds are now available commercially (Merck, Sharpe & Dohme). Synthesis of deuterated compounds for other assays have been mainly biochemical. Deuterated PGD_2 was prepared by exposing purified rat peritoneal mast cells to octadeutero arachidonoic acid, this gave predominantly 5,6,8,9,14,15-hexadeutero PGD_2 (14).

The measurement of PG metabolites has required considerable effort directed to the synthesis of the corresponding deuterated compound. Standards for E metabolite have either been prepared by the *in vitro* metabolism of 3,3,4,4-tetradeutero PGE_2 (15) or by the catalytic deuterogenation of 15-keto PGE_2 to give 5,6,13,14-tetradeutero-15-oxo PGE_1 (16). Where urinary metabolites are assayed, the dinor compounds are prepared by microbiological β-oxidation (17). In the case of the PGF_2 metabolite this microbiological step is combined with deuterium exchange into PGE_2 with subsequent borohydride reduction and metabolic degradation (18).

An account (including practical details) has been given by Murphy and Clay (19) of the incorporation of ^{18}O into the carboxylate moiety of arachidonic acid, 5-hydroxyeicosatetraenoate (5-HETE) and PGs using either acid, base or enzyme catalysed exchange. It is reported that catalysis with liver esterase in $H_2{}^{18}O$ can give 92.3% $^{18}O_2$ incorporation into PGF_2 with 0.5% ^{16}O remaining, although incorporation into other PGs was less successful.

The use of deuterated methoxylamine hydrochloride for the preparation of internal standards for the GC-MS analysis of PGs with keto group was proposed by Pace-Asciak and Edwards and this method was applied to the measurement of PGF metabolites (20). However, a recent evaluation of this approach has shown that significant hydrogen exchange can occur (1.5 to 86%) this is a particular problem with acyclic oximes whereas minimal exchange was reported in 5-membered ring oximes as found in the E series (21).

2.2 Sample extraction

Stable isotope-labelled compounds acting as internal standards provide a measure of loss during extraction procedures nevertheless it is important that recovery during extraction should be as high as possible in order to maintain good sensitivity and precision. Many examples can be found of the extraction of PGs using organic solvents but the preferred method is to use short reverse phase (C-18) columns treated with either methanol or acetonitrile. These columns are to some extent reuseable and give very high recoveries with the use of the minimum amount of solvent (22, 23).

Where volumes are large, such as in extraction from urine, then lipophilic resins such as Amberlite XAD-2 are used. In a study on the levels of a PGD_2 metabolite (9-hydroxy-11,15-dioxo-2,3,18,19-tetranorprost-5-ene-1,20-dioic acid) Roberts (24) has used 5−10% of a 24 h urine sample extracted at pH 3.2 and obtained 85−100% recoveries. This recovery was not invariably attained since it is reported that urines with lower recoveries were discarded and a further aliquot extracted.

2.3 Sample purification

Despite the specificity of GC-MS assays, samples for PG analysis have to be purified prior to assay since most extracts of plasma contain large amounts of neutral lipids such as triglycerides and extracts of urine contain many low molecular weight contaminants. The high mass of these contaminants can easily ruin GC capillary columns and therefore some purification step is normally employed. Where PG content of samples is reasonably high the minimum of purification can be used e. g. the measurement of PG metabolites in extracts of tissue (25). However, where the level of PGs is particularly low such as the measurement of 6-oxo $PGF_1\alpha$ in peripheral plasma then more extensive purification was necessary (23). When pregnancy samples were assayed (26) the purification was further extended and involved a C-18 reverse phase column extraction (which by careful choice of solvents can also effect considerable purification), a Sep-Pak (Waters associates) silica column, a thin layer chromatography step and finally a lipidex column to purify the sample from derivatization reagents.

When plasma samples are assayed it is possible to remove neutral lipids prior to assay by extracting with cold petroleum ether (3 × plasma volume) from an acetone − aqueous layer (3 parts acetone : 1 part plasma) as in the measurement of the main PGE_2 metabolites by GC-MS (15). In this assay, extracted plasma was derivatised as the bis-methyl oxime and purified in a silicic acid column followed by separation by HPLC. HPLC purification in this case had the additional advantage that it separated the two PGE_2 metabolites 15-oxo PGE_2 and 13,14-dihydro-15-oxo PGE_2 thus allowing the measurement of both metabolites.

An alternative to extraction and purification prior to GC-MS is the sequestration of the prostaglandin to be assayed using a solid phase antibody. An early report of this technique described the measurement of iloprost (5-((E)-(1S,5S,6R,7R)-7-hydroxy-6-[(E)-(3S,4RS)-3-hydroxy-4-methyl-1-octen-6-inyll-bicyclo[3.3.0]octan-3-yliden) pentanoic acid, a stable postacyclin analogue, using the purified IgG fraction of antiserum coupled to cyanogen bromide activated sepharose 4B (27). This combined extraction and purification procedure was followed by negative ion chemical ionization mass spectrometry (NICI-MS) to give an assay with a sensitivity of 0.69 $fmol \cdot ml^{-1}$ (0.25 $pg \cdot ml^{-1}$) plasma. A recent example of the technique has been reported for the measurement of thromboxane B2 in urine (28) where the partially purified antiserum was coupled to hydroxysuccinimidyl SP500 silica giving a gel with a capacity for 1.62 nmol (570 ng) thromboxane per ml of gel. Subsequent analysis by HRGC-NICI-SIM-MS without further purification gave remarkably clean traces at *m/z* 740 (M-PFB of the *t*-BDMS derivative).

2.4 Derivatization and ionization method

Choice of derivative is usually decided by the choice of ionization method. The majority of GC-MS assays utilize electron impact ionization (EI) and this allows the use of trimethylsilyl ethers or *t*-butyldimethylsilyl (*t*-BDMS) ethers. Although PGs derivatized as TMS ethers show multiple fragmentation pathways (Table 1) and therefore in general contain a smaller proportion of the ionization current in any one ion, combination of TMS ether with derivatives such as butyl boronates can increase the ionization current within a single ion (29) thus increasing sensitivity and specificity. Most assays using TMS ethers are designed for specificity rather than sensitivity and it is the wide choice of available ions which add flexibility to these derivatives and dictate their use.

t-Butyldimethylsilyl ethers (30) undoubtedly have proved extremely useful in EI GC-MS assays of PGs as the $(M-57)$ ion due to the loss of a butyl radical, carries a high proportion of the ion current (Fig. 3). Careful choice of the other derivatives used in the molecule allows the $M-57$ peak to be the base peak in nearly all cases.

Although several sensitive assays of leukotriene B_4 (LTB4) and 5-HETE have been reported using NICI (31, 32) a recent report of the use of bis *t*-BDMS ether/*t*-BDMS ester of LTB_4 shows a useful base peak at *m/z* 629 $(M-57)$ in the fully saturated molecule (31). These authors report a favourable comparison with NICI sensitivity with improved GC characteristics compared with the pentafluorobenzyl ester. Such comparisons are rare and the outcome is perhaps more likely to be decided by the type of instrumentation used. In general magnetic sector instruments can give good sensitivity with the high masses associated with BDMS ethers and exhibit NICI sensitivity which is affected by poorly understood variables (31) whilst quadrupole instruments can cope with higher source pressures and are widely used for NICI but have reduced sensitivity at the high masses needed for EI-MS of *t*-BDMS ethers. The most recent assays make use of the pentafluorobenzyl (PFB) esters with NICI-MS where the loss of the pentafluorobenzyl moiety leaves the negative charge on the lipid carboxylate anion $(M-C_6F_5CH_2)$.

The NICI GC-MS technique appears to be the more sensitive although no direct comparison with EI assays of *t*-BDMS derivatives has been made. The sensitivity of NICI of PFB esters of PGs is two orders of magnitude higher than the measurement of the corresponding methyl esters using the EI mode (14) but in this study the comparison was made with the methyl esters/TMS ethers giving at best, fragment ions with $2-5\%$ of the total ion current. Using the NICI technique sensitivities of $1-8$ pg of PG injected into the GC column have been reported for PGF_2, E_2, D_2, 6-oxo $PGF_1\alpha$ and thromboxane B_2 all as PFB esters, methyl oximes and TMS ethers (14). An

Table 1. Diagnostic ions for identification of prostaglandins derivatised as methyl oximes, methyl esters, trimethyl silyl ethers. Figures in parentheses are typical intensities as % of base peak. (M–15 = loss of CH_3; M–31 = loss of $-OCH_3$ (from methyl oxime); M–90 = loss of $-OSiCH_3$ + H; 116 = loss of $CH_2CHOSi(CH_3)_3$; M–71 = loss of C16–C20; M–159 = loss of C16–C20 together with 19–O–$Si(CH_3)_3$.)

				Loss of C16 to C20				
	M	M–15	M–31	M–71	M–159	M–90	M–(31+90)	M–(90+90)
PGE_2	539 (–)	524 (0.7)	508 (12)			449 (6)		
15-oxo PGE_2	494 (–)	479 (3)	463 (16)					
6-oxo PGE_1	594 (6)		553 (11)			494 (1)	463 (21)	
19-OH PGE_1	629 (2)	614 (2)	598 (6)		470 (6)	539 (2)	508 (6)	
19-OH PGE_2	627 (4)	612 (3)	596 (9)		468 (5)	537 (16)	506 (23)	447 (1)
19-OH $PGF_1(8\alpha)$[a]	674 (7)				515 (47)	584 (38)		494 (24)
19-OH $PGF_1(8\beta)$[a]	674 (2)				515 (24)	584 (42)		494 (13)
$PGF_2\alpha$	584 (–)			513 (8)		494 (10)		404 (8)
6-oxo $PGF_1\alpha$	629 (2)	614 (11)	598 (47)	558 (18)			508 (65)	449 (21)
15-oxo $PGF_2\alpha$	539 (9)	524 (5)	508 (52)			449 (3)	418 (41)	
PG D_2	539 (5)		508 (16)	468 (35)			418 (4)	
Thromboxane B_2	629 (–)						508 (5)	

a) From Ref. 73

	Loss of α side chain (142)			Other useful ions		
	M–(142–31)	M–(142+31+90)				
PGE_2	366 (12)	276 (77)	434 (5)	340 (15)	225 (61)	295 (44)
15-oxo PGE_2	321 (34)		291 (84)			
6-oxo PGE_1			379 (40)	391 (25)		
19-OH PGE_1		366 (30)	297 (64)			
19-OH PGE_2		364 (24)	416 (10)	295 (100)		
19-OH $PGF_1(8\alpha)$[a]			425 (100) [M–(159+90)]	310 (98)	309 (36)	
19-OH $PGF_1(8\beta)$[a]			425 (100)	310 (11)	309 (24)	
$PGF_2\alpha$			423 (20) [M–(71+90)]	333 (15) [M–(71+90+90)]		
6-oxo $PGF_1\alpha$			468 (30) [M–(71+90)]	418 (66) [M–(31+90+90)]	378 (100) [M–(17+90+90)]	
15-oxo $PGF_2\alpha$			392 (57) [M–(116+31)]			
PG D_2			378 (22) [M–(71+90)]			
Thromboxane B_2			301•(51)			

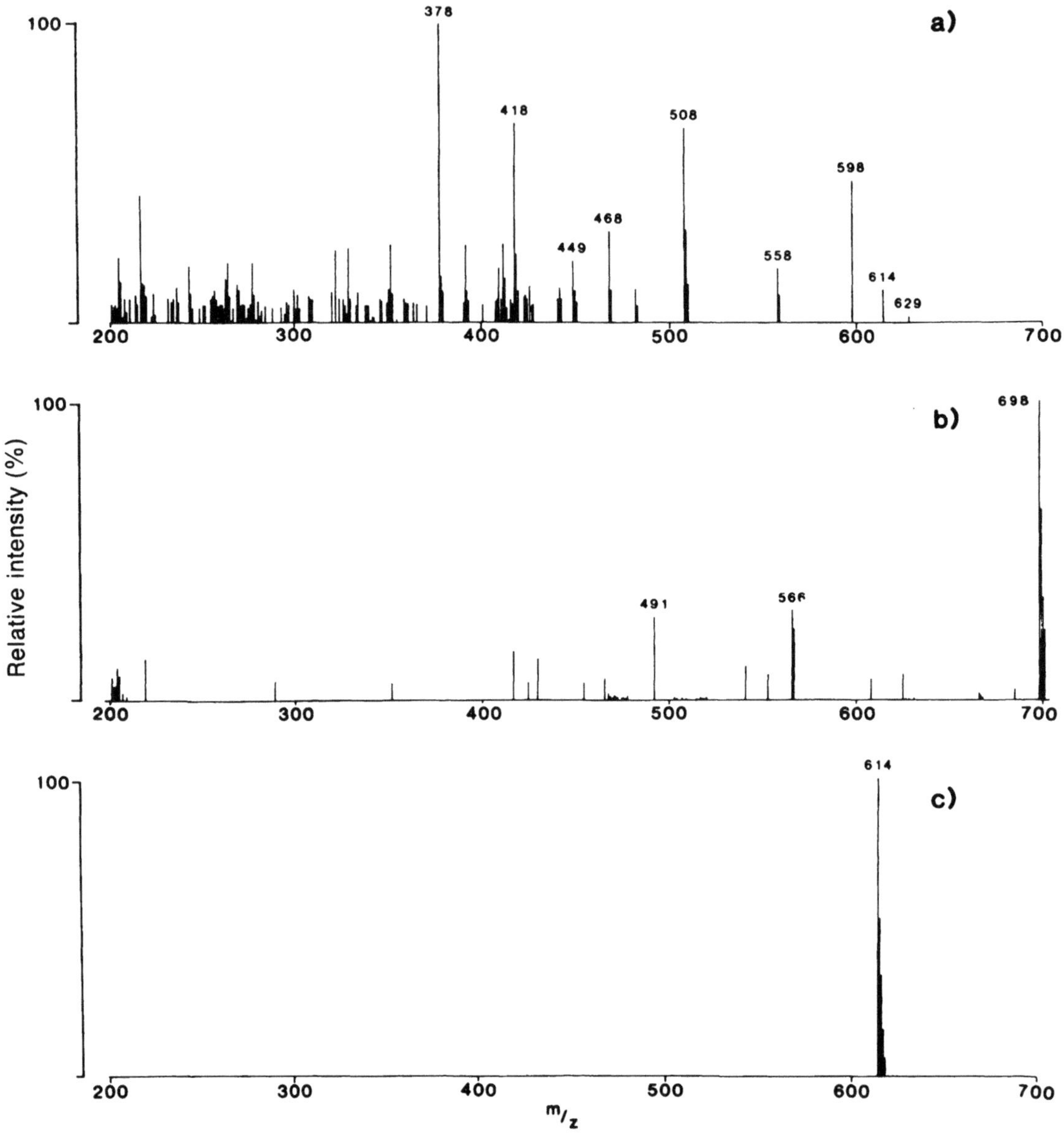

Figure 3. Comparison of mass spectra of 6-oxo $PGF_1\alpha$ as a) EI-MS of methyl ester, methyl oxime, TMS ether – a spectrum ideal for identification. Peaks annotated are: *m/z* 629, M; 598, (M – 31) M – OCH_3, from the methyl oxime; 558 (M – 71) = loss of C16 to C20 inclusive; 508, (M – 90 – 31); 468, (M – 71 – 90); 418, (M – 90 – 90 – 31) and 378, (M – 71 – 90 – 90). b) EI-MS of oxime, methyl ester, *t*-BDMS ether – the exceptionally high mean of the base peak means that this derivative is ideal for selected ion analysis of 6-oxo $F_1\alpha$ in relatively complex mixtures. Molecular ion (755) is not detected. *m/z* 566 represents loss of 132 ($OSi(CH_3)_2C_4H_9$) and *m/z* 491 represents M – (2 × 132). c) NICI-MS of pentafluoro benzyl ester, methyl oxime, TMS ether of 6-oxo F_1 from (44). The base peak *m/z* 614 represents loss of the pentafluorobenzyl group leaving a carboxylate ion.

alternative derivative, the PFB oxime which has been used successfully for electron capture GC of prostaglandins (33) was not useful for NICI GC-MS owing to the charge retention on the PFB fragment (34). Similarly, mixed PFB ester and PFB oxime derivatives resulted in a decreased sensitivity of detection.

2.5 Oximation

β-Ketol systems in PGs and their metabolites provide unstable entities liable to dehydration. Dehydration is accompanied by binding of PGs to proteins and sulphydryl groups (35) and can occur during sample storage and purification. Derivatization protects the β-ketol system and the methyl oximes are widely used for this purpose and to give good GC properties. Both methyl oximation (36) and oximation (30) of PGs can be carried out in aqueous solution in the presence of a high acetate ion concentration and this procedure allows samples to be treated with oximating solution prior to storage. Methyl oximation in aqueous solution has the added advantage that it reduces the polarity of prostanoids, allowing better extraction (36).

The use of the oxime rather than the methyl oxime allows higher mass ions when this derivative is further transformed by *t*-butyldimethylsilylation; this increases specificity and sensitivity of assays using the $(M-57)$ ion (30). Alternatively trimethylsilylation of the oxime has been used to provide high mass ions in the spectra of trimethylsilylated-oximated PGD_2 metabolites (37) where the ion at m/z 511 $(M-89)$ represents the loss of the Me_3SiO group.

3 Prostacyclin and Thromboxane Measurements

3.1 Introduction

Thromboxane A_2 (TXA_2) was discovered with the observation that the unstable endoperoxides (Fig. 2) had a potent aggregating activity when added to platelet suspensions (5). The structure of the unstable TXA_2 was inferred from GC-MS studies of the stable breakdown product TXB_2 and the analogues obtained when TXA_2 decomposed in the presence of methanol and azide to give the corresponding 11-methoxy and 11-azide derivatives (6). Following the discovery of the thromboxanes, measurements with Vane's cascade superfusion bioassay revealed an unexpected tissue response which was traced to a compound produced from endoperoxides by arterial microsomes. Subsequent analysis of this compound by GC-MS revealed that it was similar to a previously reported cyclic ether. Pace-Asciak and coworkers (38, 39) had been reporting compounds with oxy substituents at the 6-position and there is little doubt that these were derivatives of prostacyclin. Unfortunately, the structure assigned to these compounds on the basis of mass spectral evidence had a 7,8-double bond. Thus a report of 6(9)oxy-11,15-dihydroxy-prosta-7,13-dienoic acid production from PGG_2 (39) was agonizingly close to giving the structure of prostacyclin. The significant

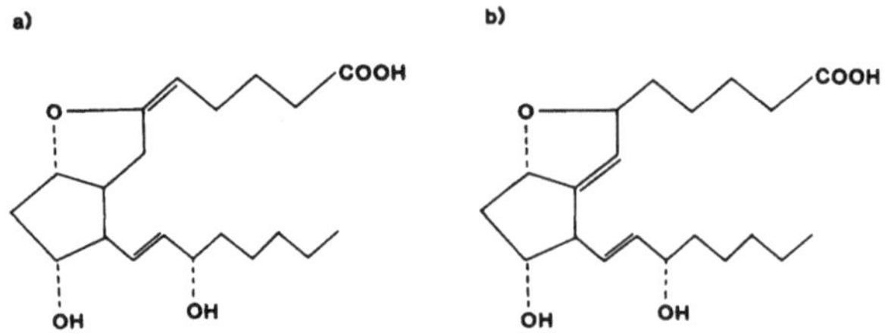

Figure 4. Structure of prostacyclin (a) and compound identified on MS evidence some years previously, and (b) 6(9)oxy-11,15-dihydroxyprosta-7,13-dienoic acid (39).

step forward was the demonstration by Vane (13) of a potent, biologically active compound with a potency in inhibiting platelet aggregation some 30 times that of PGE_1. The compound was unstable at neutral and acidic pH and logical deduction from this property, together with GC-MS studies of deuterium incorporation during hydrolysis in D_2O (40) allowed the assignment shown in Figure 4. With hindsight it is facile to suggest that the fragmentation of the 3,4-bond to give *m/z* 423 as suggested by Pace-Asciak would be more likely with 5,6-unsaturation rather than 7,8-unsaturation.

3.2 Prostacyclin as a circulating hormone

PGs are generally produced locally, act within the same organ and are mostly catabolized locally. Small amounts of PG released into the peripheral circulation are rapidly removed. This has been accepted since the demonstration of highly efficient PG catabolism by the lungs (41) and also by the liver (42). Following the discovery of prostacyclin, experiments measuring its concentration on the venous and arterial sides of perfused lungs showed that PGI_2 was apparently not acted on by the pulmonary PG 15-dehydrogenase. This suggested that PGI_2 was a circulating hormone, that it was synthesised by the lungs and that material released from the lungs could affect aggregation and arterial vasodilatation (43). This hypothesis was finally disproved when the highly sensitive NICI-GC-MS assay developed by Blair et al. (44) showed there to be less than 13.5 $fmol \cdot l^{-1}$ (5 $pg \cdot l^{-1}$) of 6-oxo PGF_1 in peripheral blood. Such a low level was manifestly insufficient to cause any changes in vasodilation or platelet aggregation. These findings have been supported by GC-MS measurements of an urinary PGI_2 metabolite which showed that total daily production of prostacyclin (of around 2 $nmol \cdot 24\ h^{-1}$ [700 $ng \cdot 24\ h^{-1}$] for man and 0.85 $nmol \cdot 24\ h^{-1}$ [300 $ng \cdot 24\ h^{-1}$] for women) (45) was in accord with the above low plasma levels.

One of the problems in any analytical method is that despite highly specific and sensitive assays, pitfalls are still possible and support was given for some time to the hypothesis of the circulating nature of prostacyclin by GC-MS levels showing erroneously high PGI_2 concentrations in blood (46).

Table 2. Platlet aggregatory action of thromboxanes and prostaglandins

Precursor:	Arachidonic acid	Eicosapentaenoic acid
Product:	TXA_2 Proaggregatory	TXA_3 Weakly proaggregatory
	PGI_2 Antiaggregatory	PGI_3 Antiaggregatory

3.3 Factors affecting Prostacyclin and Thromboxane production

The importance of PGI_2 and TXA_2 lies in their relevance to thrombogenic disorders and research has centered on the factors which affect the balance of production of these two compounds. The corollary of the local production and rapid inactivation of PGs is that changes in local production do not have a very marked effect on levels in peripheral blood. Various agents which might raise GC-MS-NICI measurable 6-oxo F_1 have been tested. Mild chemical or mechanical stimulus raises levels of 6-oxo F_1 slightly; in trained athletes the baseline level of $2.7-8.1\ pmol \cdot l^{-1}$ $(1-3\ ng \cdot l^{-1})$ is raised to a mean of $29.7\ pmol \cdot l^{-1}$ $(11\ ng \cdot l^{-1})$ by exercise (47). This rise which is small in absolute terms is not sufficient to account for the improved health attributed to regular exercise or the more objectively assessed protective role of moderate conditioning exercise, as seen in monkeys fed an atherogenic diet (48).

It has been known for some time that Eskimoes fed on a high-fish diet suffer less myocardial infarction than other populations. This improved health has been ascribed to the increased dietary intake of highly unsaturated fatty acids, particularly those with the 20:5 ω 3 structure which are the precursors to the 3 series PGs (having an extra double bond at 17, 18). A study of the effects of a fish oil diet used a control group fed with a vegetable oil and found a significant rheological difference between the two groups (49). The group fed fish oil (containing 18.6% eicosapentaenoic acid) had a significantly reduced whole blood viscosity but no change was seen in plasma viscosity, haemoglobin content, packed cell volume or platelet count. Cholesterol levels both in plasma and in HDL were the same although there was an unexplained fall in plasma triglyceride levels in the fish oil group. These findings support the suggestion that Δ^{17} prostacyclin (PGI_3) has a more effective antiaggregatory action than the corresponding thromboxane A_3 has a proaggregatory effect (Table 2).

Until recently there was no direct evidence for the production of either TXA_3 or PGI_3 in humans but GC-MS evidence for these compounds is now available (50, 51). A group from Munich has demonstrated that platelets from a group taking a fish oil enriched diet are less reactive and produce less thromboxane A_2, moreover the patients exhibit a blunted response to pressor hormones (50). GC-MS analysis of urine extract (derivatised as methyl oxime, methyl ester, TMS ether) showed the presence of Δ^{17}-2,3-dinor-6-oxo $PGF_1\alpha$, the structure being confirmed by hydrogenating this metabolite to the same saturated product as was obtained from the dinor-PGI_2 metabolite. This Δ^{17} compound only appeared in the urine after the commencement of the fish-oil diet. The same group have also produced GC-MS evidence of TXA_3 occurring in human platelets after an eicosapentaenoic acid rich diet (52).

One of the aspirations of pharmacologists has been to produce a thromboxane antagonist which, by inhibiting the proaggregatory prostanoid and redirecting synthesis to antiaggregatory products will provide an alternative therapy for cardiovascular disease. Now that such compounds are apparently available it is disappointing to find that although GC-MS-NICI analysis shows the TXA_2 production to be reduced by more than 99% by such a drug and although production of E_2, D_2 and $F_2\alpha$ is increased by some 20 fold, there is no significance effect on platelet aggregation in response to collagen (53).

3.4 Prostacyclin and Thromboxane in pregnancy

Prostacyclin production in man as determined by urinary metabolites is low; approximately 0.25 pmol $\cdot$ kg^{-1} $\cdot$ min^{-1} (90 pg $\cdot$ kg^{-1} $\cdot$ min^{-1}) of prostacyclin is produced which corresponds to the 1.4–8.5 pmol $\cdot$ l^{-1} (0.5–3 ng $\cdot$ l^{-1}) of plasma found in men and non-pregnant women. However, some of the cardiovascular changes that accompany a normal pregnancy could be explained by increased prostacyclin production. Thus there is an increase in blood volume but this is accompanied by a significant decrease in peripheral vascular resistance giving rise to a decrease in blood pressure which is most pronounced in the third trimester. GC-MS assay of the prostacyclin metabolites, 2,3-dinor-6-oxo $PGF_1\alpha$ (which accounts for 6.8% of infused PGI_2) and 6,15-dioxo-13,14-dihydro-2,3-dinor $PGF_1\alpha$ (accounting for 4.1% of infused material) suggests that prostacyclin production in normotensive women is raised some five times between 25 and 40 weeks of gestation (54) and that patients suffering from pregnancy induced hypertension in the third trimester have PGI metabolite levels significantly lower than corresponding pregnant controls. However the GC-MS measurement of 6-oxo $PGF_1\alpha$ in peripheral plasma showed only a small rise (26). This small (and variable) rise in circulating PGI_2 does not conflict with the increased urinary metabolite levels but suggests that PGI_2 production may be localized within the utero placental unit where catabolism may also occur since it is known that myometrium is active at metabolizing PGs (25). Plasma levels of 6-oxo $PGF_1\alpha$ do rise during labour (particularly during the late 2nd stage) (55) when patterns of PG catabolism may be altered.

In contrast to PGI_2 production, no consistent change in thromboxane metabolite levels was seen in pregnancy although the levels were consistently higher in pregnant women than in normal controls (56).

In early pregnancy the trophoblast has a high capacity for PGI_2 synthesis (57) and it has been suggested that the mechanism by which invading cytotrophoblastic tissue can colonize a maternal blood vessel without being immobilized by platelet aggregates may involve PGI_2. It is known that in preeclampsia (toxaemia of pregnancy) some failure occurs in the trophoblastic invasion of uterine spiral arterioles and the suggestion is that this condition may be a failure to produce prostacyclin. At present there are several studies using bioassay, e. g. (58) which show a decreased PGI_2 production in preeclamptic women at term and these are supported by GC-MS studies showing a reduction in PGI metabolite levels in patients suffering from pregnancy-induced hypertension (59).

4 Prostaglandins and Reproduction

4.1 Prostaglandin production and pathology of the non pregnant uterus

Although endometrial prostaglandin production is under the control of oestradiol and progesterone, and although there is evidence linking PG production at the end of the menstrual cycle with menstruation, the disorders of menstruation are not connected with any obvious imbalance of steroid hormone production. However, since relief from complaints such as dysmenorrhoea and menorrhagia is given by non-steroidal anti inflammatory drugs such as the fenamates, the suggestion is that disorders of PG metabolism play a role in the aetiology of these diseases.

Primary dysmenorrhoea (that which is not secondary to diseases such as endometriosis and infection) is associated with an increase of PGF metabolite (13,14-dihydro-15-oxo $PGF_2\alpha$) in peripheral plasma (60) and with an increase of PGF_2 (43, 44) and E_2 (61) in menstrual fluid. The complexity of menstrual fluid has dictated that the PG measurements above have been carried out with GC-MS. The levels of 6-oxo $PGF_1\alpha$ were also measured in one study (61) and found to be low in comparison with those of $PGF_2\alpha$ and PGE_2. This is in accord with the findings that human endometrium produces predominantly PGE_2 and PGF_2 and little 6-oxo $PGF_1\alpha$ (62, 63). Although elevated PG production is associated with dysmenorrhoea, the actual interrelationship is not clear. GC-MS measurement of PG production has allowed a direct correlation between uterine contractility (as expressed as uterine work) and PG content of the menstrual fluid in a group of dysmenorrhoeic women (64) but it is possible that increased PG production has a direct effect on perceived pain (65). Diagnosis of menorrhagia, which can be a subjective complaint, requires objective measurement of blood loss. This is usually achieved by extracting haematin from menstrual pads by alkali digestion, followed by comparison of the colour with standards obtained in the same way from peripheral blood. When measurements are made in this way the relationship between blood loss and PG production can be assessed. Studies using ^{14}C labelled precursor have shown that blood loss is related to E/F ratios (66) or I_2 production (67) i. e. correlated with production of vasodilator prostaglandins or inversely related to the production of vasoconstrictor PGs. Such findings are apparently difficult to reconcile with the effect of non-steroidal anti inflammatory drugs such as the fenamates which relieve the symptoms of menorrhagia (68). Recent studies have helped to explain this. GC-MS measurements of PGs produced by endometrium were made using the endogenous arachidonic acid as substrate and using exogenous arachidonic acid in additional incubations. The increase in production with exogenous arachidonic acid as a percentage of maximal PG synthesis shows an inverse correlation with menstrual blood loss (30). This suggests that women with menorrhagia have more available free arachidonic acid (possibly released by mild mechanical stimulation such as uterine contractions) and thus produce excessive endoperoxide precursor. Over-production of this intermediate might lead to non enzymic breakdown to PGE or possibly by interaction with myometrial tissue produce the potent vasodilator PGI_2. In either case it can be seen that PG synthesis inhibitors would reduce endoperoxide production (Fig. 2) and restore the balance in favour of the vasoconstricting PGs.

PROSTANOIDS FORMED FROM EDOPEROXIDES G_2 AND H_2

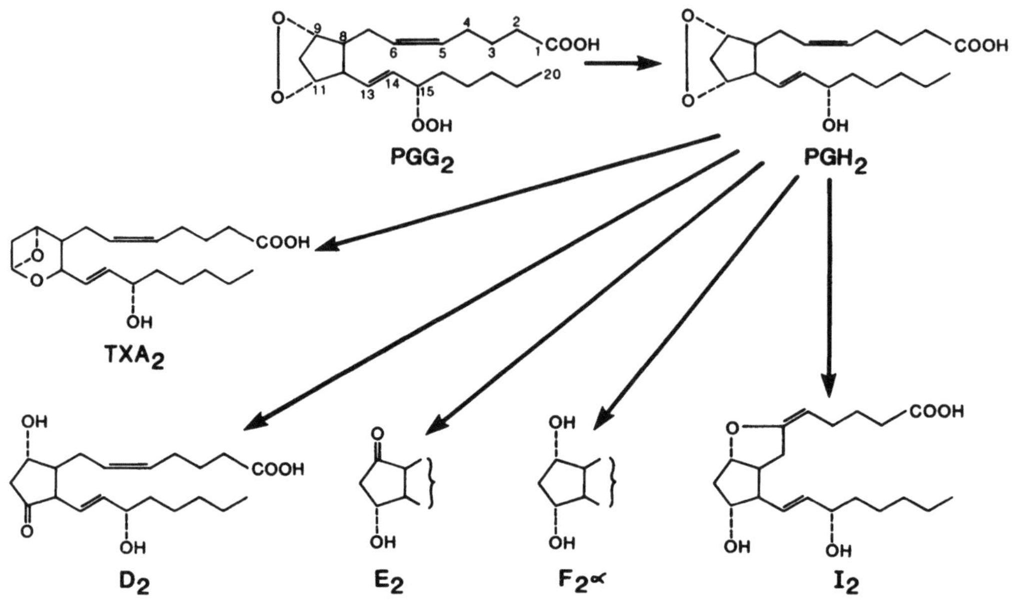

Figure 5. PGG_2 is reduced to PGH_2 which can non enzymically break down to PGE_2 (and a small amount of $F_{2}\alpha$) or be transformed enzymically to PGs D, E, F or I or to thromboxane A_2. PGs A, B and C are dehydration products of PGE and are now thought to be mainly artifacts (101).

4.2 Prostaglandin catabolism within the human uterus

Since prostaglandins are not stored within tissue, the available PG levels will depend on the balance of production and metabolism via the PG 15-dehydrogenase (Figs. 5 and 6). In human endometrium the PG 15-dehydrogenase activity is restricted to the early and mid luteal phases of the cycle (69) and is almost totally absent when PG production is known to be high, that is at menstruation. GC-MS studies of the enzyme of the endometrium and myometrium have allowed K_m values to be determined for these enzymes using both E_2 and F_2 as substrates. PGE is the preferred substrate for the 15-dehydrogenase enzyme but its main metabolite, 13,14-dihydro-15-oxo PGE_2 is unstable, it is readily dehydrated to the A form which then cyclises to form 11-deoxy-13,14-dihydro-15-oxo-11B,16ξ-cyclo prostaglandin E_2 (70). This interconversion, unsuspected for several years, has caused considerable problems and uncertainties in reported levels of this metabolite. However, the technique of aqueous oximation in the presence of a high acetate anion concentration has allowed esimation of the unstable E metabolite in plasma or tissue samples prior to any extraction (16). The internal standard used was prepared by catalytic deuterogenation of 13,14-dihydro-15-keto PGE_2. The addition of the internal standard was followed by oximation in

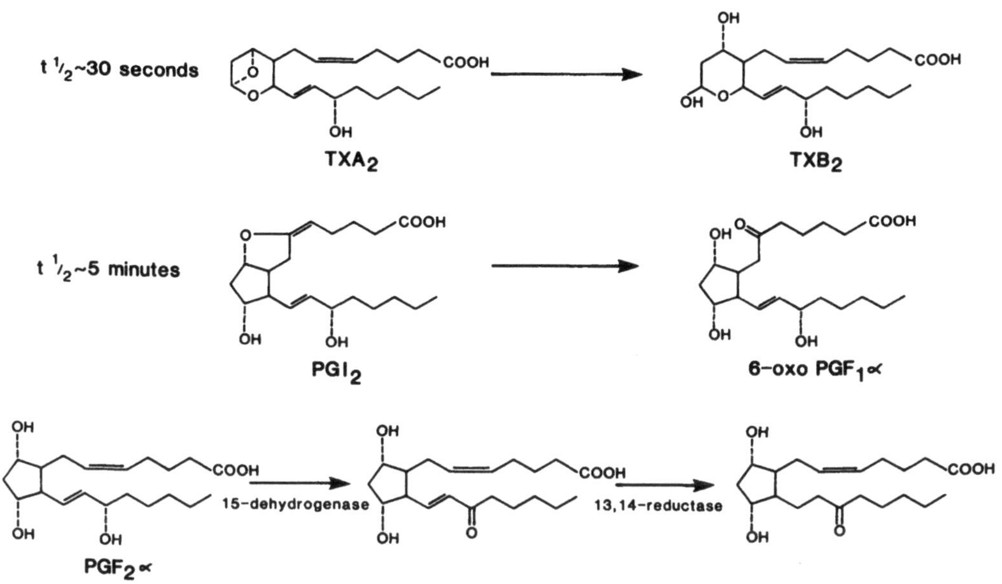

Figure 6. Thromboxane A_2 and PGI_2 (prostacyclin) have short half-lives in aqueous solution and give rise to the stable but biologically inactive TXB_2 and 6-oxo $PGF_{1}\alpha$. All primary prostaglandins are substrates for the main catabolizing enzyme PG 15-dehydrogenase (EC 1.1.1.141). The catabolism of $F_{2}\alpha$ is shown as an example.

aqueous solution (60°C for 30 min at pH 5.1), conventional extraction and derivatization as the methyl ester and *t*-BDMS ethers of hydroxyl and oxime-OH groups. This derivative was monitored using EI-MS at m/z 681 for 13,14-dihydro-15-oxo PGE and 687 for the deuterated internal standard. Measurement of PG 15-dehydrogenase in the uterus demonstrated a high activity of the enzyme in myometrium (25) and that the K_m of the endometrial and myometrial enzymes was similar for both E_2 and $F_{2}\alpha$ as substrates.

4.3 Prostaglandins and male reproduction

Prostaglandins were first discovered because of the potent contractile action of human semen on smooth muscle and for some time the main source of supply of prostaglandins was biosynthesis from sheep accessory glands (seminal vesicles). Indeed the list of PGs produced from ram seminal vesicle incubations and identified by GC-MS is still being extended with the recent identification of 5,6-dihydro PGE_3, a compound with reduced uterotonic activity but similar antiaggregatory properties to PGE_1 (71). Since the original isolation of PGE and PGF from semen many other PGs have been identified in this fluid using GC-MS (36, 72–74). The results of these studies have identified 4 prostaglandins, E_1, E_2, 19-OH E_1 and 19-OH E_2 as the main prostaglandins of human semen with a total of around 1 mg of prostaglandin in the average ejaculate (75). The identification of the components of human semen is a further example of the use of a highly specific technique such as GC-MS giving the wrong biological composition of fresh body fluids because the instability of the components was not fully investigated (76), such is the convincing nature of physico-chemical evidence that the fallacies are reproduced even a decade later (77).

Seminal prostaglandins are of interest for two reasons, first the extraordinarily high levels (e. g. 10^7 times higher levels of E than in peripheral blood) dictate a specific function which might help to elucidate the role of PGs elsewhere; second, the levels of PG in semen may have implications for a man's fertility. Because of the high levels of PGs in semen GC alone can be used for the quantitation of levels and GC-MS has been reserved for the measurement of the more minor components such as the 19-OH PGFs (75, 78). Several studies have demonstrated apparently significant decreases of PG levels in the semen of men with idiopathic infertility (79, 80) but the correlation is not clear because of the wide range and the skewed distribution of PGs in the semen from normal fertile men (75). The relevance of seminal PGs to fertility will have to await the assignment of a definite biological role for these compounds. Although seminal PGs may act on the female tract to aid sperm transport, this hypothesis is unproven and it is possible that they contribute to the immunosuppressive action of seminal plasma (81) or possibly interact with the spermatozoa by raising intracellular cAMP levels and thereby affecting sperm function (82).

5 Identification of Urinary Metabolites of PGS and their Measurement

5.1 Catabolism of the stable prostaglandins

The main pathways of catabolism of PGs are the oxidation of the 15-hydroxyl by 15-dehydrogenase followed by the saturation of the 13,14-double bond (Fig. 6). Both these reactions occur in the lungs or in other tissues of the body. Other stages of metabolism which occur in the liver and perhaps elsewhere are one or two stages of β-oxidation to give the dinor and the tetranor metabolites respectively. ω or ω-1 hydroxylation may occur independently to give the 20-hydroxy or the 19-hydroxy PGs. In most cases these metabolites are found in urine together with unchanged parent PG but the contribution of the kidney to excretion of unchanged PG is thought to be significant in non-labelled measurements.

The metabolism of $F_2\alpha$ was the first to be studied and a GC-MS assay for the measurement of dinor-13,14-dihydro-15-oxo $F_1\alpha$ was reported in 1976 (18). The deuterated internal standard was prepared by equilibrating the 9-ketone of PGE_2 with D_2O followed by borohydride reduction, this was then infused into a monkey and the urine was collected and the deuterated material isolated. The metabolism of D_2 is complicated by many of the metabolites having the F ring structure (83) but a GC-MS assay has been reported for a significant urinary metabolite of D_2, 9-hydroxy-11,15-dioxo-2,3,18,19-tetranorprost-5-ene-1,20-dioic acid (37), which retains the β-ketol system characteristic of D_2. The D metabolite was purified by XAD_2 extraction, silicic acid chromatography and HPLC. The internal standard being a deuterated methyl ester is not added until after extraction and a ^{3}H label is used for estimating loss during this procedure. The derivative used is the methyl ester, TMS ether/TMS oxime which gives a prominent ion at *m/z* 511 (95% of base peak which is at *m/z* 258) due to loss of 89 daltons (-OSi(Me)$_3$ loss from TMS-oxime). Studies using this assay have shown that normal levels of this metabolite range from $0.74-1.3\ nmol \cdot 24\ h^{-1}$ ($240-420\ ng \cdot 24\ h^{-1}$) but patients suffering from systemic mastocytosis, a disorder of mast cell proliferation which results in potentially fatal attacks of flushing and hypotension, have levels of D_2 metabolite in their urine of $25-185\ nmol \cdot 24\ h^{-1}$ ($8-60\ \mu g \cdot 24\ h^{-1}$). Recent studies, in which PGD_2 ($0.4\ nmol \cdot kg^{-1} \cdot min^{-1}$; $128\ ng \cdot kg^{-1} \cdot min^{-1}$) was infused into volunteers, showed little decrease in blood pressure although nasal congestion and vasodilation in the "blush" areas was noticed (84). It was suggested in this study that although PGD_2 levels may be raised in patients with mastocytosis, they are not high enough to account for the symptoms of this disease although they may potentiate the action of other agents such as histamine.

5.2 Metabolism of prostacyclin

The metabolism of prostacyclin itself is probably initiated in the lungs or other organs of the body by the conventional PG 15-dehydrogenase. However, the relative instability of the compound will give a non-enzymic conversion to 6-oxo $PGF_1\alpha$ which is a poor substrate for the 15-dehydrogenase. It is likely therefore that most of the 6,15-dioxo F compounds are formed by non enzymic break down following oxidation at the 15 position.

Early GC-MS studies of PGI_2 metabolism showed that the monkey was not as active at metabolizing I_2 as the rat, with 13% unchanged 6-oxo F_1 being excreted (85). This study also showed about 5% 19-hydroxylation metabolites which have been shown in other studies to be due to hydroxylation by the liver (86). The finding of excretion of unchanged 6-oxo $F_1\alpha$ has been confirmed in man (87) with 23% of eluted label being identified by GC-MS as 6-oxo $PGF_1\alpha$ and 20% being excreted as 2,3-dinor 6-oxo $PGF_1\alpha$. It now seems clear that a second dinor metabolite, 6,15-dioxo-13,14-dihydro-2,3-dinor $PGF_1\alpha$ is also a major metabolite and is useful in reflecting PGI_2 production (88). Internal standards for measurement of the dinor metabolites have been synthesized by the usual transformations available for PG synthesis but with the use of mycobacterium rhodocrous to effect the β-oxidation step.

It is interesting to note a divergent and biologically important pathway observed by Wong et al. (65) where they perfused rabbit liver with prostacyclin in Tyrode's solution and found significant production of 6-oxo PGE_1. This compound is relatively stable and its potent antiaggregatory activity and its demonstration in liver perfusates has stimulated the search for the enzyme responsible for its production (9-hydroxy dehydrogenase) in other situations, for a review see (89). A GC-MS method for the measurement of 6-oxo PGF_1 has been reported (90) and the authors suggest that this shows little conversion of PGI_2 to 6-oxo PGE_1 following I_2 infusion. Unfortunately the sensitivity of this assay (around $81\ fmol \cdot ml^{-1}$ [$30\ pg \cdot ml^{-1}$] plasma) is probably not sufficient to assess the conversion given that we know that circulating I_2 levels are generally well below this level. Measurements of urinary metabolites of prostacyclin have confirmed the relatively low production of PGI_2 in the human and have reinforced the findings that I_2 is not a circulating hormone. However, measurements of urinary 6-oxo F_1 and the corresponding 13,14-dihydro compound show a marked (at least 10 fold) increase in the excretion of these metabolites in the neonate in day 3 of life (91) although the contribution of an immature hepatic catabolizing system to these levels of metabolite is not clear, and the suggestion has been made that such metabolites may reflect renal synthesis (88).

Reports that the levels of 6-oxo $PGF_1\alpha$ were higher in male urine than female urine led to an investigation of prostacyclin metabolite levels (6-oxo PGF and its dinor form) in semen using GC-MS and a synthesised 18,18,19,19-[$^2H_4$]-2,3-dinor-6-oxo $PGF_1\alpha$ internal standard (92). Little of either compound was found in comparison with levels of other PGs in semen (7.56 $pmol \cdot ml^{-1}$ [2.8 $ng \cdot ml^{-1}$] 6-oxo $PGF_1\alpha$ and 5 $pmol \cdot ml^{-1}$ [1.7 $ng \cdot ml^{-1}$] of the dinor compound) suggesting that a contribution from semen was not the reason for the discrepancy between male and female urine levels.

Efforts to identify the antiaggregatory factor present in the human placenta have so far shown that it is neither I_2 nor a recognizable metabolite and appears to be of high polarity (93).

5.3 Thromboxane metabolites in urine

In vitro investigations have shown that thromboxane B_2 is a poor substrate for the PG 15-dehydrogenase but perfusion of guinea pig lungs apparently effects metabolism to 13,14-dihydro-15-keto TXB_2 and this has led to the suggestion that although TXB_2 apparently lacks lipophilicity there is a carrier mediated uptake process operating in the lungs. TXB_2 infused into a man gave at least twenty urinary metabolites (94) and 2.5% of the recovered radioactivity was present as unchanged thromboxane. One interesting additional pathway identified was the dehydration of the 11-hemiacetal hydroxyl to give a δ-lactone ring. Since the hemiacetal ring is in equilibrium with the acyclic form as evidenced by the ready formation of acyclic oximes, it is uncertain whether this oxidation was on the closed ring form or oxidation (which might well have been non enzymic) of the acyclic aldehyde. The relative amounts of the dinor and the tetranor thromboxane derivatives were 23.0 and 5.3% respectively and several GC-MS urinary assays for thromboxane production have subsequently measured the dinor thromboxane B_2 (95, 96). The assay from the Karolinska group (96) uses microbiological β-oxidation to obtain the 19,19,20,20[$^2H_4$]-dinor-TXB_2 internal standard. The metabolite is purified using silicic acid followed by reverse phase HPLC and the derivatised as the methyl ester *t*-BDMS ether. The $(M-57)$ ion of this derivative (m/z 641) was used for detection and the corresponding m/z 645 ion was used for the internal standard. Recent results with this assay have shown that thromboxane production is inhibited by aspirin and indomethacin but not by paracetamol (56) and a preliminary result reported in the same paper suggests that urinary excretion of 2,3-dinor thromboxane B_2 may be elevated, following myocardial infarction.

6 Leukotrienes

The leukotrienes are the most recently discovered group of compounds derived from arachidonic acid (8) and provide the greatest challenge in the methodology of PG quantitation. These compounds were first discovered in polymorphonuclear leukocytes but many other cell types are now known to produce them. Because of the conjugated triene system (Fig. 2) leukotrienes absorb strongly in the UV at 280 nm and therefore HPLC with UV detection is an ideal method of detection and quantitation. However, this technique provides little structural information on the leukotrienes and is not as sensitive as some of the GC-MS methods for the PGs. The structure of slow reacting substance (SRS) from rat-basophil leukemia cells (9) and slow reacting substance of anaphylaxis (SRS-A) from guinea pig lung (10) have been determined by mass spectrometry of samples derivatized as *N*-acetyl TMS ethers and volatized in a direct insertion probe. Both these SRS forms were identified as 5-hydroxy-6-cysteinylglycinyl-7,9,11,14-eicosatetraenoic acid. GC-MS has also aided in the complete stereochemical structural assigment of SRS as 5(S)-hydroxy-6(R)-cysteinylglycinyl-7,9-*trans*-11,14-*cis*-eicosatetraenoic acid (97). This compound is derived from the glutathione lipid leukotriene C_4 by the action of γ-transpeptidase and is further degraded to leukotriene D_4 (Fig. 2).

The sulphur containing leukotrienes are exceptionally polar and do not chromatograph well on GC and therefore there are no good GC-MS methods for their quantitation. One way round the problem is to subject the compounds to Raney nickel desulphurization and then analyse the lipid moiety (98). This is not wholly satisfactory however since the peptide chain variations are not detected. Other approaches such as desorption-chemical ionization of the *N*-acetyl-trimethyl derivative (99) provide qualitative information but are difficult to quantitate.

A sensitive NICI method has been reported (32) for the measurement of the dihydroxy lipid leukotriene B_4 which is formed by the action of leukotriene B synthase on LTA_4. At present $[^2H_4]PGF_2\alpha$ is used as the internal standard for quantitation as the retention times are similar e. g. 9.9 min for LTB_4 and 10.2 min for $[^2H_4]PGF_2\alpha$. This method has allowed quantitation of LTB_4 production from human pulmonary macrophages. The macrophages produce 175.7 ± 45.9 pmol (65 ± 15 ng) TXB_2/mg cell protein but only 21.7 ± 17.9 pmol (7.3 ± 6 ng) LTB_4/mg protein. The use of flurbiprofen which inhibits TXB_2 synthesis by more than 90% gives only a two fold increase in LTB_4 production, this suggests that such nonsteroidal anti inflammatory drugs do not cause a massive switch from the cyclooxygenase pathway to the lipoxygenase pathway which has been suggested.

One of the main routes of catabolism of leukotriene B_4 is by 20-hydroxylation, this was determined in the monkey by the administration of 5,6,8,9,11,12,14,15-$[^3H_8]LTB_4$ and purification of labelled metabolites. Although 25% of the label was recovered in the urine, greater than 80% of the tritium was in water. The major non-volatile metabolite (C value 26.7) was identified as LTB_4 owing to the presence of ions *m/z* 567 (M−181); *m/z* 387 (M−181−90−90) in a NICI spectrum of the PFB TMS derivative (100).

References

1. Lands, W. E. M. & Crawford, C. G. (1976) Enzymes of membrane phospholipid metabolism in animals. In: The Enzymes of Biological membranes, Biosynthesis of Cell Components, Vol. 2 (Martonosi, A., ed.), pp. 3–85, Plenum, N. Y.
2. Green, K. (1979) Determination of prostaglandins in body fluids and tissues. Acta Obstet. Gynec. Scand. Suppl. *87*, 15–20.
3. Dimov, V., Christensen, N. J. & Green, K. (1983) Analysis of prostaglandins formed from endogenous arachidonic acid in homogenates of human reproductive tissues. Biochim. Biophys. Acta. *754*, 38–43.
4. Bergstrom, S., Ryhage, R., Samuelsson, B. & Sjövall, J. (1963) The structure of prostaglandin E_1, F_1 and F_1. J. Biol. Chem. *238*, 3555–3564.
5. Hamberg, M. & Samuelsson, B. (1974) Prostaglandin endoperoxides. Novel transformations of arachidonic acid in human platelets. Proc. Natl. Acad. Sci. U.S.A. *71*, 3400–3404.
6. Hamberg, M., Svensson, J. & Samuelsson, B. (1975). Thromboxanes: A new group of biologically active compounds derived from prostaglandin endoperoxides. Proc. Natl. Acad. Sci. U.S.A. *72*, 2994–2998.
7. Bunting, S., Gryglewski, R., Moncada, S. & Vane, J. R. (1976) Arterial walls generate from prostaglandin endoperoxides a substance (prostaglandin x) which relaxes strips of mesenteric and coeliac arteries and inhibits platelet aggregation. Prostaglandins *12*, 897–913.
8. Borgeat, P. & Samuelsson, B. (1979) Arachidonic acid metabolism in polymorphonuclear leukocytes: unstable intermediate in formation of dihydroxy acids. Proc. Natl. Acad. Sci. U.S.A. *76*, 3213–3217.
9. Morris, H. R., Taylor, G. W., Piper, P. J., Samhoun, M. N. & Tippins, J. R. (1980) Slow reacting substances (SRSs): the structure identification of SRSs from rat basophil leukaemia (RBL-1) cells. Prostaglandins *19*, 185–201.
10. Morris, H. R., Taylor, G. W., Piper, P. J. & Tippins, J. R. (1980) Structure of slow-reacting substance of anaphylaxis from guinea pig lung. Nature *285*, 104–107.
11. Blair, I. A. (1983) Measurement of eicosanoids by gas chromatography and mass spectrometry. Brit. Med. Bulletin. *39*, 223–226.
12. Von Euler, U. S. (1936) On the specific vasodilating and smooth muscle stimulating substances from accessory genital glands in man and certain animals (prostaglandin and vesiglandin). J. Physiol. (Lond), *88*, 213–234.
13. Moncada, S., Gryglewski, R., Bunting, S. & Vane, J. R. (1976) An enzyme isolated from arteries transforms prostaglandin endoperoxides to an unstable substance that inhibits platelet aggregation. Nature (Lond) *263*, 663–665.
14. Barrow, S. E., Heavey, D. K., Ennis, M., Chappell, C., Blair, I. A. & Dollery, C. T. (1984) Measurement of prostaglandin D_2 and identification of metabolites in human plasma during intravenous infusion. Prostaglandins *28*, 743–754.
15. Hubbard, W. C. (1982) Quantitation of 15-keto-13,14-dihydro-PGE_2 in plasma by GCMS. Methods in Enzymol. *86*, 571–578.
16. Kelly, R. W. & Abel, M. H. (1983) The measurement of 13,14-dihydro-15 keto prostaglandin E_2 by combined gas chromatography mass spectrometry. Biomed. Mass Spectrometry *10*, 276–279.
17. Nelson, N. A., Jackson, R. W. & Sebek, O. K. (1978) Synthesis of thromboxane B_2 metabolites. Prostaglandins *16*, 85–92.
18. Brash, A. R., Baillie, T. A., Clare, R. A. & Draffan, G. H. (1976) A new method for the determination of the major metabolite of prostaglandins F_1 and F_2 in human urine by stable isotope dilution and combined gas chromatography mass spectrometry. Biochem. Med. *16*, 77–94.
19. Murphy, R. A. & Clay, K. L. (1982) Preparation of ^{18}O derivatives of eicosanoids for GCMS quantitative analysis. Methods in Enzymol. *86*, 547–551.
20. Pace-Asciak, C. R. & Edwards, N. S. (1982) Preparation of deuterium-labeled urinary catabolites of $PGF_2\alpha$ as standards for GC-MS. Methods in Enzymol. *86*, 553–559.
21. Herold, D. A., Smith, B. J., Ross, R. M., Marquis, F., Ayers, C. R., Wills, M. R. & Savory, J. (1987) Limitation of deuterium labelled methoximes as internal standards, in the mass spectral analysis of prostaglandins. Prostaglandins *33*, 500–602.
22. Powell, W. S. (1982) Rapid extraction of arachidonic acid metabolites from biological samples using octadecyl silica. Methods in Enzymol. *86*, 467–477.

23. Barrow, S. E., Waddell, K. A., Ennis, M., Dollery, C. T. & Blair, I. A. (1982) Analysis of picomolar concentrations of 6-oxo prostaglandin F_1 in biological fluids. J. Chromatog. *239*, 71–80.
24. Roberts, L. J. (1982) Quantification of the PGD_2 urinary metabolite 9-hydroxy-11,15-dioxo-2,3,18,19-tetranorprost-5-ene-1,20-dioic acid by stable isotope dilution mass spectrometric assay. Methods in Enzymol. *86*, 559–570.
25. Abel, M. H. & Kelly, R. W. (1983) Metabolism of prostaglandins by the non-pregnant human uterus. J. Clin. Endocr. Metab. *56*, 678–685.
26. Barrow, S. E., Blair, I. A., Waddell, J. A., Shepherd, G. L., Lewis, P. J. & Dollery, C. T. (1983) Prostacyclin in late pregnancy: analysis of 6-oxo-prostaglandin $F_1\alpha$ in maternal plasma, in: Prostacyclin in Pregnancy (P. J. Lewis, S. Moncada & J. O'Grady, eds.) pp. 79–85, Raven Press. N. Y.
27. Krause, W., Jakobs, U., Schulze, P. E., Nieuweboer, B. & Humpel, M. (1985) Development of antibody mediated extraction followed by CG/MS (antibody GC/MS) and its application to iloprost determination in plasma. Prostaglandins Leukotrienes and Medicine *17*, 167–182.
28. Hubbard, H. L., Eller, T. D., Mais, D. E., Halushka, P. V., Baker, R. H., Blair, I. A., Vrbanac, J. J. & Knapp, D. R. (1987) Extraction of thromboxane B2 from urine using an immobilised antibody column for subsequent analysis by gas chromatography-mass spectrometry. Prostaglandins *33*, 149–160.
29. Kelly, R. W. & Taylor, P. L. (1973) Method for the measurement of prostaglandin F_2 in biological fluids by GCMS. Anal. Chem. *45*, 2079–2082.
30. Kelly, R. W., Lumsden, M. A., Abel, M. H. & Baird, D. T. (1984) The relationship between menstrual blood loss and prostaglandin production in the human: Evidence for increased availability of arachidonic acid in women suffering from menorrhagia. Prostaglandins, Leukotrienes and Medicine *16*, 69–78.
31. Murphy, R. C. (1984) Mass spectrometric quantitation and analysis of leukotrienes and other 5-lipoxygenase metabolites. Prostaglandins *28*, 597–601.
32. MacDermot, J., Kelsey, C. R., Waddell, K. A., Richmond, R., Knight, R. K., Cole, P. J., Dollery, C. T., Landon, D. N. & Blair, I. A. (1984) Synthesis of leukotriene B_4 and prostanoids by human alveolar macrophages: analysis by gas chromatography/mass spectrometry. Prostaglandins *27*, 163–179.
33. Christ-Hazelhoft, E. & Nugteren, D. H. (1981) Prostacyclin is not a circulating hormone. Prostaglandins *22*, 739–781.
34. Waddell, K. A., Blair, I. A. & Wellby, J. (1983) Combined capillary column gas chromatography negative ion chemical ionization mass spectrometry of prostanoids. Biomed. Mass. Spectrom. *10*, 83–88.
35. Granstrom, E., Fitzpatrick, F. A. & Kindahl, M. (1982) Radioimmunologic determination of 15-keto-13,14-dihydro-PGE_2: A method for its stable degradation product, 11-deoxy-15-keto-13,14-dihydro-11B, 16B-cyclo PGE_2. Methods in Enzymol. *86*, 306–320.
36. Taylor, P. L. & Kelly, R. W. (1974) 19-Hydroxylated prostaglandins as the major prostaglandins of human semen. Nature. Lond. *250*, 665–667.
37. Roberts, L. J. (1982) Quantification of the PGD_2 urinary metabolite 9-hydroxy-11,15-dioxo-2,3,18,19-tetranorprost-5-ene-1,20-dioic acid by stable isotope dilution mass spectrometric analysis. Methods in Enzymol. *86*, 559–570.
38. Pace-Asciak, C. & Wolfe, L. S. (1971) A novel prostaglandin derivative formed from arachidonic acid by rat stomach homogenates. Biochemistry *10*, 3657–3664.
39. Pace-Asciak, C., Nashat, M. & Menon, N. K. (1976) Transformation of prostaglandin G_2 into 6(9)-oxy-11,15-dihydroxy-prosta-7,13-dienoic acid by the rat stomach fundus. Biochim. Biophys. Acta. *424*, 323–325.
40. Johnson, R. A., Morton, D. R., Kinner, J. H., Gorman, R. R., McGuire, J. C. & Sun, F. F. (1976) The chemical structure of prostaglandin x (prostacyclin). Prostaglandins *12*, 915–928.
41. Ferreira, S. H. & Vane, J. R. (1967). Prostaglandins: their disappearance from and release into the circulation. Nature (Lond) *216*, 868–873.
42. Dawson, W., Jessup, S. J., McDonald-Gibson, W., Ramwell, P. W. & Shaw, J. E. (1970) Prostaglandin uptake and metabolism by the perfused rat liver. Brit. J. Pharmac. *39*, 585–598.
43. Moncada, S., Korbut, R., Bunting, S. & Vane, J. R. (1978). Prostacyclin is a circulating hormone. Nature (Lond) *273*, 767–768.

44. Blair, I. A., Barrow, S. E., Waddel, K. A., Lewis, P. J. & Dollery, C. T. (1982) Prostacyclin is not a circulating hormone in man. Prostaglandins *23*, 579–589.
45. Falardeau, P., Oates, J. A. & Brash, A. R. (1981) Quantitative analysis of two dinor urinary metabolites of prostaglandin I_2. Anal. Biochem. *115*, 359–367.
46. Hensby, C. N., Fitzgerald, G. A., Friedman, L. A., Lewis, P. J. & Dollery, C. T. (1979) Measurements of 6-oxo PGF_1 in human plasma using gas chromatography-mass spectrometry. Prostaglandins *18*, 731–736.
47. Ritter, J. M., Barrow, S. E., Blair, I. A. & Dollery, C. T. (1983) Release of prostacyclin *in vivo* and its role in man. Lancet, 317–319.
48. Kramsch, D. M., Aspen, A. J., Abramovitz, B. M. Kreimendahl, T. & Hood, W. B. (1981) Reduction of coronary atherosclerosis by moderate conditioning exercise in monkeys on an atherogenic diet. N. Engl. J. Med. *305*, 1483–1489.
49. Woodcock, B. E., Smith, E., Lambert, W. H., Morris Jones, W., Galloway, J. H. Greaves, M. & Preston, F. E. (1984) Beneficial effect of fish oil on blood viscosity in peripheral vascular disease. Brit. Med. J. *288*, 592–594.
50. Knapp, H. R., Roy, L. & Fitzgerald, G. A. (1983) Dietary modification of endogenous arachidonate metabolism in man. Clinical Res. *31*, 841.
51. Fischer, S. & Weber, P. C. (1984) Prostaglandin I_3 is formed *in vivo* in man after dietary eicosapentaenoic acid. Nature *307*, 165–168.
52. Fischer, S. & Weber, P. C. (1983) Thromboxane A3 (TXA3) is formed in human platelets after dietary eicosapentaenoic acid (C20:5 omega 3). Biochem. Biophys. Res. Commun. *116*, 1091–1099.
53. Blair, I. A., Lewis, P. J. Orchard, M. A. & Waddell, K. A. (1983) Prostanoid production by human platelets: redirection by the thromboxane synthetase inhibitor UK 37, 248. Brit. J. Pharmacol. *79*, 356P.
54. Brash, A. R., Goodman, R. P. & Fitzgerald, G. A. (1983) Endogenous prostacyclin production in human pregnancy. In: Prostacyclin and Pregnancy (P. J. Lewis, S. Moncada & J. O'Grady, eds.), pp. 71–77, Raven Press, N. Y.
55. Mitchell, M. D., Keirse, M. J. N. C., Brunt, J. D., Anderson, A. B. M. & Turnbull, A. C. (1979) Concentrations of the prostacyclin metabolite, 6-keto prostaglandin F_1, in amniotic fluid during late pregnancy and labour. Brit. J. Obstet. Gynaecol. *86*, 350–353.
56. Vesterqvist, O. & Green, K. (1984) Urinary excretion of 2,3-dinorthromboxane B_2 in man under normal conditions, following drugs and during some pathological conditions. Prostaglandins *27*, 627–644.
57. Rackoczi, I., Tihanyi, K., Falkay, G., Rozsa, I., Demeter, J. & Gati, I. (1983). Prostacyclin production in trophoblast. In: Prostacyclin in Pregnancy (P. J. Lewis, S. Moncada and J. O'Grady, eds.), pp. 15–23, Raven Press, N. Y.
58. Remuzzi, G., Marchesi, D., Zoja, C., Muratove, D., Mecca, G., Misiani, R., Rossi, E., Barbato, M., Capetta, P., Donati, M. B. & deGaetano, G. (1980) Reduced umbilical and placental vascular prostacyclin in severe preeclampsia. Prostaglandins *20*, 105–110.
59. Goodman, R. P., Killiam, A. P., Brash, A. R. & Branch, R. A. (1982) Prostacyclin production during pregnancy: comparison of production during normal pregnancy and pregnancy complicated by hypertension. Am. J. Obstet. & Gynecol. *142*, 817–822.
60. Lundstrom, V. & Green, K. (1978) Endogenous levels of $PGF_2\alpha$ and its main metabolites in plasma and endometrium of normal and dysmenorrhoeic women. Am. J. Obst. Gynecol. *130*, 640–646.
61. Lumsden, M. A., Kelly, R. W. & Baird, D. T. (1983) Primary dysmenorrhoea: the importance of both prostaglandins E_2 and F_2. Brit. J. Obs. Gynaecol. *90*, 1135–1140.
62. Abel, M. H. & Kelly, R. W. (1979) Differential production of prostaglandin within the human uterus. Prostaglandins *18*, 821–828.
63. Kelly, R. W., Cooper, I., Smith, S. K. & Abel, M. H. (1980) The measurement of the production of 6-oxo prostaglandin F-1α by the uterus. In: The Role of Prostaglandins, Prostacyclin and Thromboxanes Measurement (J. M. Boeynaems & A. G. Herman, eds.), pp. 142–154, Martinus Nijhoff, The Hague.
64. Lumsden, M. A., Kelly, R. W. & Baird, D. T. (1983) Is prostaglandin $F_2\alpha$ involved in the increased myometrial contractility of primary dysmenorrhoea? Prostaglandins *25*, 683–692.
65. Ferreira, S. H., Moncada, S. & Vane, J. R. (1974) Prostaglandins and signs and symptoms of inflammation. In: Prostaglandin Synthetase Inhibitors (Robinson, J. H. &

Vane, J. R., eds.), pp. 175–187, Raven Press, N. Y.

66. Smith, S. K., Abel, M. H., Kelly, R. W. & Baird, D. T. (1981) Prostaglandin synthesis in the endometrium of women with ovular dysfunctional uterine bleeding. Br. J. Obs. Gynaecol. *88*, 434–442.
67. Smith, S. K., Abel, M. H., Kelly, R. W. & Baird, D. T. (1981) A role for prostacyclin (PGI_2) in excessive menstrual bleeding. Lancet i, 522–524.
68. Fraser, I. S., Pearse, C., Shearman, R. P., Elliott, P. M., McIlreen, J. & Markham, R. (1981) Efficacy of mefenamic acid in patients with a complaint of menorrhagia. Obstet. Gynaecol. *68*, 542–551.
69. Casey, M. L., Hemsell, D. L., MacDonald, P. C. & Johnston, J. M. (1980) NAD-dependent 15-hydroxy prostaglandin dehydrogenase activity in human endometrium. Prostaglandins *19*, 115–122.
70. Granstrom, E. & Kindahl, H. (1980) Radioimmunologic determination of 15-keto 13,14-dihydro PGE_2: a method for its stable degradation product, 11-deoxy-13,14-dihydro-15-keto-11,16-cycloprostaglandin E_2. In: Advances in Prostaglandin and Thromboxane Research *6* (Samuelsson, B., Ramwell, P. W. & Paoletti, R., eds.), pp. 181–182, Raven Press. N. Y.
71. Samel, N., Jarving, I., Lohmus, M., Lopp, A., Kobzar, G., Sadovskaya, V., Valimae, T. & Lille, U. (1987) Identification and biological activity of a novel natural prostaglandin, 5,6-dihydro-prostaglandin E3. Prostaglandins *33*, 137–146.
72. Samuelsson, B. (1963) Isolation and identification of prostaglandins from human seminal plasma. J. Biol. Chem. *238*, 3229–3234.
73. Jonsson, H. T., Middleditch, B. S. & Desiderio, D. M. (1975) Prostaglandin in human seminal fluid: two novel compounds. Science (N. Y.) *187*, 1093–1094.
74. Taylor, P. L. (1979) The 8-iso prostaglandins, evidence for eight compounds in human semen. Prostaglandins *175*, 259–267.
75. Templeton, A. A., Cooper, I. & Kelly, R. W. (1978) Prostaglandin concentrations in the semen of fertile men. J. Reprod. Fertil. *52*, 147–150.
76. Hamberg, M. & Samuelsson, B. (1966) Prostaglandins in human seminal plasma. J. Biol. Chem. *241*, 257–263.
77. Mawhinney, M. S. (1983) In: Male Accessory Sex Organs and Androgen Action in Infertility in the Male (L. I. Lipshultz & S. S. Howards, eds.), pp. 135–164, Churchill Livingstone, N. Y.
78. Svanborg, K., Bygdeman, M. & Eneroth, P. (1983) The F and 19-hydroxy F prostaglandins and their 8-isomers in human seminal plasma: Data on chromatography and mass spectrometry. Biomed. Mass Spectrom. *10*, 495–498.
79. Hawkins, D. F. (1968) Relevance of prostaglandins to problems of human subfertility. In: Prostaglandin Symposium of the Worcester Foundation, (Ramwell, P. W. & Shaw, J. E., eds.), pp. 1–12, Interscience N. Y.
80. Svanborg, K., Benvold, E., Bygdeman, M. & Eneroth, P. (1983) Relationship between prostaglandins in human seminal fluid and fertility. In: Advances in Prostaglandin, Thromboxane and Leukotriene Res., Vol. 12 (Samuelsson, B., Paoletti, R. & Ramwell, P., eds.), pp. 455–459, Raven Press N. Y.
81. James, K., Harvey, J., Bradbury, A. W., Hargreaves, T. B., Cullen, R. T. & Donaldson, K. (1983) The effect of seminal plasma on macrophage function – a possible contributory factor in sexually transmitted disease. AIDS Research *1*, 45–57.
82. Aitken, R. J. & Kelly, R. W. (1985) Analysis of the direct effects of prostaglandins in human sperm function. J. Reprod. Fertil. *73*, 139–146.
83. Ellis, C. K., Smigel, M. D., Oates, J. A., Oelz, O. & Sweetman, B. J. (1979) Metabolism of prostaglandin D_2 in the monkey. J. Biol. Chem. *254*, 4512–4163.
84. Heavey, D. J., Lumley, P., Barrow, S. E., Murphy, M. B., Humphrey, P. P. A. & Dollery, C. T. (1984). Effects of intravenous infusions of prostaglandin D_2 in man. Prostaglandins *28*, 755–767.
85. Sun, F. F. & Taylor, B. H. (1981) Metabolism of prostacyclin in Cynomolgus monkey. Prostaglandins *21*, 307–314.
86. Wong, P. Y., Malik, K. U., Desiderio, D. M., Mc.Giff, J. C. & Sun, F. F. (1980) Hepatic metabolism of prostacyclin (PG12) in the rabbit: formation of a potent novel inhibitor of platelet aggregation. Biochem. Biophys. Res. Commun. *93*, 486–494.
87. Rosenkranz, B., Fischer, C., Reimann, I., Weimer, K. E., Beck, G. & Frolich, J. C. (1980) Identification of the major metabolite of prostacyclin and 6-ketoprostaglandin F1 alpha in man. Biochim. Biophys. Acta *619*, 207–213.

88. Falardeau, P., Oates, J. A. & Brash, A. R. (1981) Quantitative analysis of two dinor urinary metabolites of prostaglandin I_2. Anal. Biochem. *115*, 359–367.
89. Moore, P. K. & Griffiths, R. J. (1983) 6-keto prostaglandin E_1, – a review. Prostaglandins *26*, 509–515.
90. Jackson, E. K., Goodman, R. P., Fitzgerald, G. A., Oates, J. A. & Branch, R. A. (1982) Assessment of the extent to which exogenous prootaglandin I_2 is converted to 6-keto-prostaglandin E_1 in human subjects. J. Pharmacol. and Exptl. Therapeutics *221*, 183–187.
91. Fischer, S., Scherer, B. & Weber, P. C. (1982) Prostacyclin metabolites, in urine or adults and neonates, studied by gas chromatography-mass spectrometry and radioimmunoassay. Biochim. Biophys. Acta *710*, 493–501.
92. Fischer, C. & Meese, C. O. (1985) Simultaneous determination of 6-oxo prostaglandin $F_1\alpha$ and 2,3-dinor-6-oxo prostaglandin $F_1\alpha$ in biological fluids by stable isotope dilution and negative ion chemical ionization mass spectrometry. Biomed. Mass Spectrom. *12*, 399–404.
93. Dembele-Duchesne, M. J., Thaler-Dao, H., Chavis, C. & Crastes-de-Paulet, A. (1982) The human placental anti-aggregating factor is neither prostacyclin, nor a prostacyclin metabolite. Prostaglandins *24*, 701–714.
94. Roberts, L. J., Sweetman, B. J. & Oates, J. A. (1981) Metabolism of thromboxane B2 in man. Identification of twenty urinary metabolites. J. Biol. Chem. *256*, 8384–8393.
95. Maas, R. L., Taber, D. F. & Roberts, L. J. (1982) Quantitative assay of urinary 2,3-dinor thromboxane B_2 by GCMS. Methods in Enzymol. *86*, 592–603.
96. Vesterqvist, O., Green, K., Lincoln, F. H. & Sebek, O. K. (1984) Development of a GC-MS method for quantitation of 2,3-dinor-TxB2 and determinations of the daily urinary excretion rates in healthy human. Thromb. Res. *33*, 39–49.
97. Morris, H. R., Taylor, G. W., Rokach, J., Girard, Y., Piper, P. J., Tippins, J. R. & Samhoun, M. W. (1980) Slow reacting substance of anaphylasis, SRS-A: assignment of the stereochemistry. Prostaglandins *20*, 601–607.
98. Hammarstrom, S. (1981) Conversion of 5,8,11-eicosatrienoic acid to leukotrienes C_3 and D_3. J. Biol. Chem. *256*, 2275–2279.
99. Maas, R. L., Lawson, J. A., Brash, A. R. & Oates, J. A. (1983) Analysis of leukotrienes and lipoxygenase products by fourier transform NMR and direct probe mass spectrometry Adv. Prostaglandin Thromboxane Leukotriene Res. *11*, 229–234.
100. Serafin, W. E., Oates, J. A. & Hubbard, W. C. (1984) Metabolism of leukotriene B_4 in the monkey. Identification of the principal non volatile metabolite in the urine. Prostaglandins *27*, 899–911.
101. Middleditch, B. S. (1975) PGA: fact or artifact? Prostaglandins *9*, 409–411.

Chapter 10
Purines and Pyrimidines

K. H. Schram

1 Introduction

Although the value of mass spectrometry in solving problems relating to the structure of nucleic acid components has been recognized for many years, the utility of this sensitive and specific analytical tool in clinical applications pertaining to nucleic acid bases, nucleosides and nucleotides has been realized, with a few exceptions, only in the recent past.

The major limitation of mass spectrometry in the analysis of nucleic acid bases, nucleosides and, most especially, nucleotides and oligonucleotides, is the thermally labile and non-volatile nature of these compound classes. A number of derivatives have been employed to overcome these volatility limits, and the gas chromatographic separation of complex mixtures of nucleic acid bases and nucleosides can be performed on a relatively routine basis. However, the gas phase separation of nucleotides, even with derivatization, is not a generally useful technique. New ionization methods and the very recent introduction of liquid chromatography-mass spectrometry methods applicable to nucleic acid components have overcome the limited volatility of nucleotides and oligonucleotides to a large extent, thus permitting mass spectral analysis of underivatized samples of these classes of compounds. New methods for the preparation of stable isotope labeled analogs of nucleic acid components have been developed which should make internal standards, necessary for quantitation, more readily available. Finally, these new mass spectrometric techniques have been applied to a variety of clinically related problems which were thought to be beyond the capabilities of mass spectrometry only a few years ago.

In selecting material for discussion the main goal is to illustrate a mass spectrometric technique which should be applicable to any biochemical or clinical problem involving the identification or quantitation of a nucleic acid base, nucleoside, nucleotide or oligonucleotide. In some cases, the methods have not yet been applied to a clinical or biochemical problem, but the method is described to illustrate its potential. The work presented is therefore intended to serve as a foundation on which future, expanded research efforts may be built.

Previous studies into the basic mechanisms of fragmentation of nucleic acid components provide the necessary knowledge for the interpretation of the mass spectrum of an unknown nucleic acid component. Excellent reviews covering the fundamental aspects of the mass spectrometry of nucleic acid components have appeared earlier (1 – 3) and are recommended as important references. Methods for the preparation of volatile derivatives necessary for the gas phase separation and analysis of bases and nucleosides have also been reviewed (4 – 8). The biannual reviews of mass spectrometry appearing in Analytical Chemistry (9) and current periodicals, Biomedical and Environmental Mass Spectrometry, Analytical Biochemistry, Mass Spectrometry Reviews, to mention only a few, are important sources for keeping abreast of new developments in the field of mass spectrometry as it applies to the analysis of nucleic acid components.

2 Bases

Interest in purine-type compounds in a biological system derives from the fundamental role of these molecules as components of RNA and DNA (10, 11), their occurrence as metabolic intermediates or end products are purine anabolism or catabolism (12), the established or suspected role of these compounds are regulators of metabolic function (13), their utility as biological probes (14) or their use in the treatment (15) or diagnosis (16) of cancer and other disease states (17).

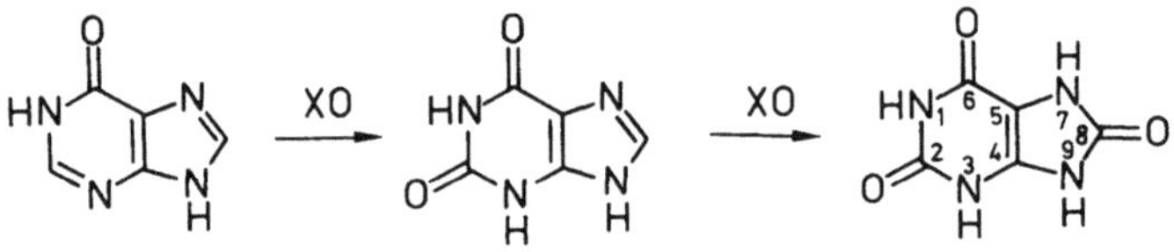

1 Hypoxanthine **2** Xanthine **3** Uric acid

Scheme 1. Metabolic generation of uric acid (**3**) by the action of xanthine oxidase (XO) on xanthine (**2**) and hypoxanthine (**1**).

2.1 Purine bases

The clinical and biochemical importance of purine anabolism and catabolism has prompted the application of mass spectrometry to the analysis and identification of various purine metabolites in human plasma and urine.

The end product of purine metabolism in humans is uric acid (**3**) which will precipitate in joints to cause gout in cases of overproduction. Particular interest has been expressed, therefore, in the quantitation of the precursors of **3**, xanthine (**2**) and hypoxanthine (**1**), which are converted to **3** by the action of xanthine oxidase (XO).

A method for the simultaneous quantitation of **1** and **2** in human serum and urine samples utilized stable isotope dilution and selected ion monitoring (SIM) GC-MS techniques with the 7,9-$^{15}N_2$ labeled analog of **2** as the internal standard (18). Preparation of the butyl derivatives was necessary to eliminate interferences from other endogenous compounds. Quantitation was achieved by monitoring the molecular ions ($M^{+\bullet}$) of the tributyl derivatives of **2** and the internal standard at *m/z* 320 and 322, respectively. No interferences were observed but a correction for the isotope contribution of **2** to the internal standard was necessary. Levels of **1** were determined by monitoring a fragment ion at *m/z* 231 in the mass spectrum of the dibutyl derivative with comparison to a calibration curve generated using the $^{15}N_2$-labeled internal standard. A detection limit of 50 ng $\cdot$ mg^{-1} for **1** and **2** was reported which required 0.5 ml of serum for analysis.

The above studies have been extended to include the quantitation of the xanthine oxidase inhibitor allopurinol (**4**) and its major metabolite oxipurinol (**5**) (19). Following essentially the same experimental procedure as described above, serum and urine levels of **4** and **5** were determined by monitoring the molecular ion of **5** (*m/z* 320) and a

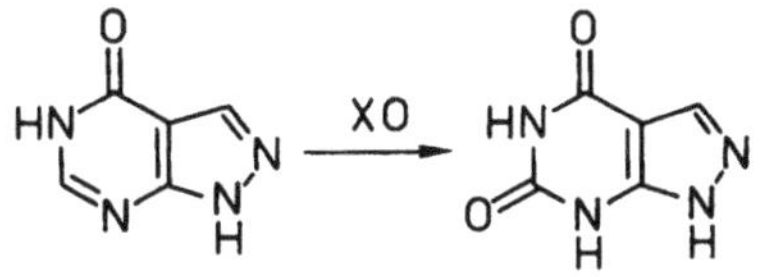

4 Allopurinol 5 Oxipurinol

Scheme 2. Conversion of allopurinol (4) to oxipurinol (5).

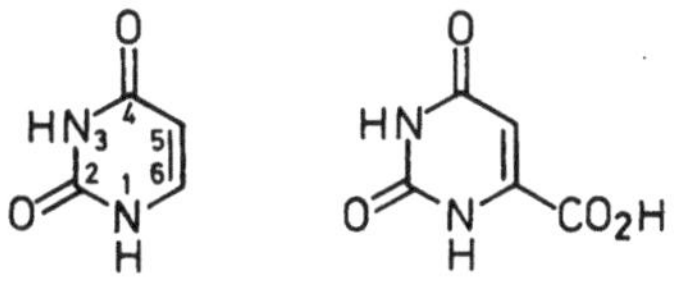

6 Uracil 7 Orotic acid

fragment ion at *m/z* 205 in the spectrum of **4**, both as their butyl derivatives. Although **5** provides a molecular ion at the same nominal and exact mass as **2**, baseline gas chromatographic separation of the two compounds permits their differentiation. Likewise, both **3** and **4** may be distinguished on the basis of their retention times.

During the course of studies to develop an high pressure liquid chromatographic (HPLC) assay for compounds **1**—**5**, a number of unidentified components were observed in the chromatograms of urine extracts obtained from normal and gouty subjects (20). Three of the unidentified components were collected using HPLC and subjected to mass spectral analysis. Electron impact ionization (EI) of the first unknown, introduced by direct probe, gave a suspected molecular ion at *m/z* 113 with major fragment ions at *m/z* 112, 84 and 69 suggesting either creatine or creatinine. The mass spectra of the second and third components, while differing from the first, were essentially identical to each other. Prominent ions at *m/z* 112 and 69 suggested the presence of uracil (**6**) in both samples. Conversion of the samples to their permethyl derivatives (5) followed by GC-MS analysis indicated a probable molecular weight of 198 daltons for component 2 with a major fragment at *m/z* 82. Identity of this second compound was established as the trimethyl analog of orotic acid (**7**) based on the coincidence of the mass spectrum and chromatographic properties with a reference sample. The correspondence in GC and MS data of the remaining unidentified component with those of **7** led to the belief that orotidine, the N 1 riboside monophosphate analog of **7**, was thermally degrading to produce **7**. This supposition was later confirmed using HPLC.

The widespread ingestion of caffeine (**8**) in various beverages and the clinical utility of theophylline (**9**) in the treatment of asthma have stimulated interest in the identification and quantitation of these and related methylxanthines in samples of human and animal origin. Thin-layer chromatographic purification followed by mass spectrometry was used to identify the metabolites of **8** in rat urine (21). A similar approach, using HPLC, was used to separate **8, 9**, theobromine (**10**), and 1,7-dimethylxanthine (**11**) (22). Comparison of the mass spectra of the purified components with known

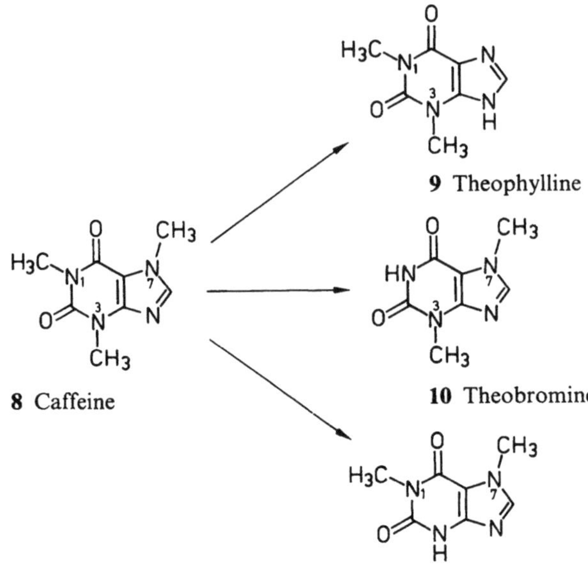

Normal ratio **9/10** = 1.4
when N1 is C^2H_3, ratio 9/10 = 2.6
when N7 is C^2H_3, ratio 9/10 = 0.16

Scheme 3. Metabolism of caffeine (**8**) to theophylline (**9**), theobromine (**10**) and 1,7-dimethylxanthine (**11**). Metabolic switching may occur upon replacement of a CH_3 group with a C^2H_3 function. (Adapted from ref. 24.)

standards and knowledge of the fragmentation pattern of these compounds allowed the positive identification of these caffeine analogs in human plasma.

Chemical ionization mass spectrometry (CI-MS) has been used to quantitate **8** and **9** extracted from plasma, urine, breast milk, saliva and liver homogenates (23). SIM of the MH^+ ions of **8**, ethylated **9** and a trideuteriomethyl analog of **8** provided the desired pharmacokinetic information.

The effect on the metabolic fate of **8** substituted in the *N*1 or *N*7 positions with trideuteriomethyl (C^2H_3) groups provides an excellent example of the process termed metabolic switching (24), which is an alteration in the metabolism of a drug when multiple pathways are available. The substitution of an H atom by a ^2H atom at the site of metabolic transformation decreases the rate of metabolism at that site due to an increased stability of the C-^{2}H versus the C-H bond. Since *N*-demethylation is the major metabolic pathway of caffeine in rats, selective placement of a C^2H_3 group in the *N*1 or *N*7 positions of **8** will alter the ratios of the metabolic products **9** and **10**. The results of this study, summarized in Scheme 3, clearly show an alteration in the ratio of **9**/**10** formed in the metabolism of the [*N*1$-C^2H_3$]-**8** analog relative to **8**, i. e., demethylation of the *N*1 position is suppressed by the C^2H_3 group. An opposite effect is observed in the [*N*7$-C^2H_3$]-labeled analog with the production of **10** now predominating.

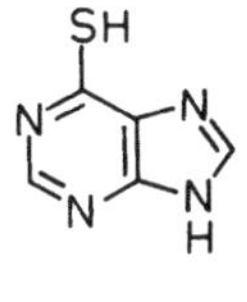

12 6-Mercaptopurine

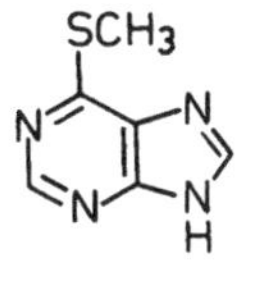

13 6-Methylmercaptopurine

The identification of a new metabolic pathway operating in the metabolism of **9** in newborns, but not adults, was made during the GC-MS analysis of the plasma (50 µl samples) and urine of infants treated with **9** (25). The operation of this new pathway was investigated by administration of a 1 : 1 mixture of [1,3-$^{15}N_2$,2-^{13}C] labeled **9** (the type and location of labels were chosen specifically to avoid the potential of metabolic switching) to a pair of newborn twins. Metabolic products and **9** were extracted from the plasma samples and analyzed by GC-MS. Two major components were present whose mass spectra displayed prominent ions separated by three mass units. The earlier eluting component was identified as **8** based on the appearance of an $M^{+\bullet}$ ion at *m/z* 194 (197) and a mass spectrum identical to that of authentic **8**. The conversion of **9** to **8** was therefore established. The second component was identified as the *N*7-pentyl derivative of **9**, whose possible origin is not discussed. Purification of urine samples by thin layer chromatography (TLC) and GC-MS analysis also revealed the presence of **10** and **11**. The major metabolic pathway established in these studies, i. e., conversion of **9** to **8**, is not observed in adults where 3-methylxanthine, 1-methyl- and 1,3-dimethyluric acid are the major metabolites. No 3-methylxanthine was observed in the newborn samples.

The above pathway in newborns was confirmed in a second study where the levels of **8** and **9** were determined using [3,7$-(C^2H_3)_2$]-**8** and [3$-C^2H_3$]-**9** as the internal standards (26). In this case, the C^2H_3 labeled internal standards were added to the biological sample after sample collection so the potential of metabolic isotope effects was eliminated. By combining an ion pairing extraction step with the derivatization step (ethylation with ethyl iodide) high recoveries of **8** and **9** with detection of 40 and 20 ng · ml^{-1}, respectively, were realized. Ions monitored in the quantitation of **8** and **9** were fragments of *m/z* 194 (200) and 208 (211). Data from these experiments indicated a steady state plasma level of **8** at approximately 1/3 the levels of **9**.

Mass spectrometry has also been used to quantitate levels of 6-mercaptopurine (**12**), an antitumor agent, in the plasma of patients receiving this drug. Although extensive decomposition occurs during the extractive alkylation of the compound, the assay is reproducible and provides the detection limits needed to quantitate the drug in the plasma of children on maintenance therapy with this agent (27). A disadvantage of using the permethyl derivative to quantitate levels of **12** is the formation of the *S,N*9-dimethyl analog. Since the *S*-methyl analog **13** is a metabolite of **12**, quantitation using the permethyl derivative may provide erroneously high levels of **12** caused by interference from endogenous **13**. Use of the pentafluorobenzyl derivative eliminates this interference and permits the quantitation of **12** in plasma samples at the 10 ng · ml^{-1} level (28).

The treatment and cause of immunodeficiency diseases are currently an area of great interest. The 9-alkyl analog of hypoxanthine, NPT 15392 (**14**), is of potential

Table 1. Structures of NPT 15392 (**14**) and its major metabolites (adapted from ref. 29)

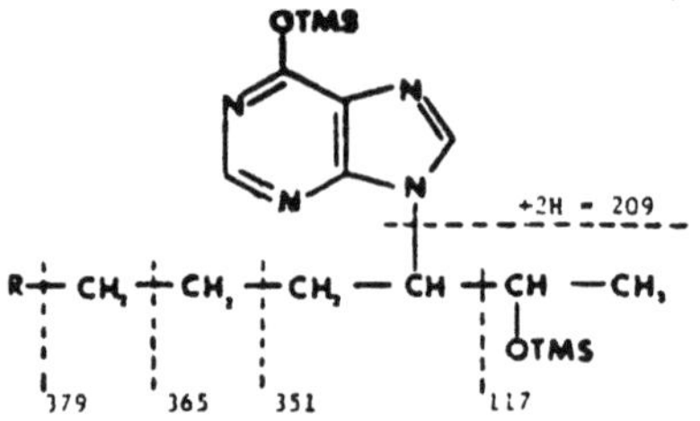

Compound	R
14, Erythro-9-(2-hydroxy-3-nonyl)-hypoxanthine (NPT 15392)	$CH_3CH_2CH_2$-
Metabolite 1	$CH_3CH(OTMS)CH_2$-
Metabolite 2	$TMSOCOCH_2CH_2$-
Metabolite 3	TMSOCO-
Metabolite 4	CH_3COCH_2-
Metabolite 5	Glucuronide of Metabolite 1
Metabolite 6	Clucuronide of NPT 15392
Metabolite 7	$TMSOCOCH_2$-

utility in treating patients with deficiencies in cell-mediated immune function. The metabolism of **14** has been investigated recently in a number of animal species, with metabolites being identified by mass spectrometry (29). Urinary metabolites were identified with the aid of ^{14}C labeled **14** and purified by TLC prior to mass spectral analysis of the trimethylsilyl (TMS) derivatives. A total of seven metabolites were isolated from the various animal species. Each of the metabolites shared a number of common mass spectral fragments which were also in the spectrum of the parent drug. An ion at *m/z* 209 was assigned as the B + 2H ion (30) formed by cleavage of the glycosidic bond with transfer of 2H to the aglycone. The presence of this ion in all spectra indicated that no modifications had occurred in the heterocyclic base. An ion at *m/z* 117 appeared as the most intense peak in the spectra of all samples and was attributed to the $TMSO-CH-CH_3$ portion of the sidechain. This data coupled with high resolution information and the sequential fragmentation of the sidechain led to the structural assignments for the metabolites shown in Table 1.

2.2 Pyrimidine bases

The factors to be considered in the design and synthesis of stable isotopically labeled internal standards for the quantitation of pyrimidine bases in biological samples have been discussed (31). The use of ^{13}C, ^{15}N and ^{18}O is recommended in order to eliminate the biological isotope effects encountered with ^{2}H labels. If the ^{2}H labeled material is not involved in biochemical transformations, the ease of preparation and decreased expense of their synthesis make ^{2}H labeled analogs the label of choice. The isotopic

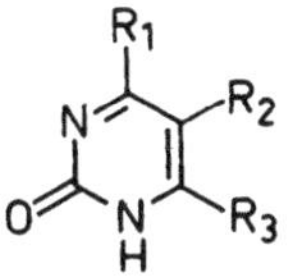

15 Orotic acid-[C4-^{13}C; N1,N3-$^{15}N_2$]; R_1=OH, R_2=H, R_3=CO_2H
16 Uracil-[N1,N3-$^{15}N_2$]; R_1=OH, R_2=R_3=H
17 Thymine-[C^5-$^2H_3$]; R_1=OH, R_2=C^2H_3, R_3=H
18 Uracil-[N1,N3-$^{15}N_2$; C6-^{2}H]; R_1=OH, R_2=H, R_3=^{2}H
19 Thymine; R_1=OH, R_2=CH_3, R_3=H
20 Cytosine; R_1=NH_2, R_2=R_3=H
21 5-Chlorouracil; R_1=OH, R_2=Cl, R_3=H
22 5-Fluorouracil; R_1=OH, R_2=F, R_3=H
23 5-Fluorouracil-[N1,N3-$^{15}N_2$]; R_1=OH, R_2=F, R_3=H

contribution of the atoms present in the analyte, the sensitivity limits needed to perform the analysis and the presence of biological and instrumental noise all combine to determine the number and level of labels needed in the internal standard. The higher the number and level of isotope labels incorporated into the internal standard, the greater the sensitivity and specificity of the analysis. Thus, the use of two or more stable nuclides may, therefore, be advantageous and the synthesis of orotic acid-[$^{15}N_2$,C4$-$^{13}C] (**15**), uracil-[$^{15}N_2$] (**16**) and thymine-[$C^5-{}^2H_3$] (**17**) has been reported and the use of these samples as internal standards in biochemical studies described (31, 32). Because of incomplete labeling, dilution factors were limited to 100/1 with the uracil and thymine standards but a 1000/1 dilution was possible using the orotic acid analog. An additional mass shift could be gained in the uracil compound by exchange of the C6 hydrogen by deuterium to give **18**. The abundance of the M_0 ion of the TMS derivative was seen to drop from 10% in **16** to 2% in **18**. These labeled bases could be converted to their labeled nucleoside analogs by condensation with the appropriate sugar. The above standards have been used in a GC-MS study for the determination of uracil (**6**), thymine (**19**) and their nucleosides and nucleotides in human and rat plasma with a sensitivity of 20$-$40 pmol $\cdot$ ml^{-1}. Levels of **6** and **19** found in human plasma were 0.98 $\pm$ 0.64 and 0.05 $\pm$ 0.06 μmol, respectively. Analysis of the nucleoside and nucleotide analogs of **6** and **19** was performed using the same techniques following quantitative hydrolysis to the respective nucleic acid bases.

The simultaneous detection and quantitation of **6**, **19**, and cytosine (**20**) in biological samples using a combination of gas chromatography-mass fragmentography (GC-MF) and GC-MS for total ion monitoring (GC-MS-TIM) has been described (33). The detection limits for **6**, **19** and **20**, using 5-chlorouracil (**21**) as the internal standard, were 1 ng $\cdot$ ml^{-1} in plasma and 1$-$5 ng $\cdot$ g^{-1} wet tissue for the GC-MF method. Levels of **6** and **19** detectable in plasma and tissues and **20** in plasma using the GC-MS-TIM method were 200 ng $\cdot$ ml^{-1} or g^{-1} wet tissue. Significantly higher levels, 2.5 μg $\cdot$ g^{-1} wet tissue, were required for the detection of **20** in tissue by the second procedure. Construction of two calibration curves, one for each of the two methods used in quantitation, was necessary because the pools of pyrimidine bases in tissues exceeded the upper limit of quantitation (2.5 μg $\cdot$ g^{-1}) in some samples. Quantitation was

Table 2. Pyrimidine base pools in plasma and tissues of nude mice (H − 30). Values are the mean values of five analyses. ND = not detected (adapted from ref. 33)

Tissue	Concentration (per ml of plasma or g of tissue)		
	Uracil (**18**)	*Thymine* (**29**)	*Cytosine* (**20**)
Plasma	2.81 ± 0.78 nmol	ND	ND
Tumour	0.47 ± 0.08 μmol	8.48 ± 1.86 nmol	1.11 ± 0.11 nmol
Liver	0.20 ± 0.05 μmol	0.17 ± 0.09 nmol	0.47 ± 0.11 nmol
Kidney	0.48 ± 0.03 μmol	0.75 ± 0.45 nmol	1.37 ± 0.61 nmol
Lung	0.18 ± 0.03 μmol	1.58 ± 0.11 nmol	0.89 ± 0.20 nmol
Spleen	0.38 ± 0.78 μmol	9.56 ± 2.48 nmol	1.02 ± 0.11 nmol

achieved by monitoring the molecular ions of the TMS derivatives of **6** and **19** and the M − 15 ions in the spectra of **20**-$(TMS)_2$ and **21**-$(TMS)_2$. Concentrations of the pyrimidine base pools in various tissues and plasma of nude mice determined using these methods are summarized in the Table 2.

The quantitation of 5-fluorouracil (**22**), **6**, **19** and **20** in plasma and tissues with procedures similar to those described above, but with sample ionization in the CI mode, has also been reported (34). The gas chromatography-chemical ionization-mass fragmentography (GC-CI-MF) method compared favorably with the GC-EI-MF procedure in terms of precision and sensitivity, except for **20** where 50 $ng \cdot ml^{-1}$ was the lower limit of detection compared to 1 $ng \cdot ml^{-1}$ for the other compounds. The utility of the CI mode is, however, limited to the analysis of plasma samples since tissue components contribute a significant background which was poorly separated from the compounds of interest. The internal standards in this study were **16**, the [1,3 − $^{15}N_2$]analog **23** (for **22**) and **21** (for **19** and **20**). A single peak, MH^+, was observed in the mass spectra of the TMS derivative of each sample, except **20** which displayed an M − 62 ion as the base peak of the spectrum.

2.3 5-Fluoropyrimidine bases and nucleosides

The clinical utility of GC-MS for the quantitation of pyrimidine bases is probably best illustrated in the case of 5-fluorouracil (5-FU) (**22**), analogs of **22**, especially ftorafur (FT) (**24**), and metabolites of 5-FU and FT. Because of the widespread use of these two compounds as antimetabolites in the treatment of various malignancies, a simple, sensitive and specific assay for **22** and **24** in biological samples has considerable importance in designing proper therapeutic regimens for patients receiving these agents and for studying the *in vitro* or *in vivo* biochemical transformations of these drugs.

A method for quantitating **22** in the presence of **24** in the plasma and urine of rats and rabbits utilized GC-MS of the TMS derivatives of **22** and 5-FU-[1,3 − $^{15}N_2$](**23**) (35, 36). Monitoring of the $M^{+\cdot}$ and M − 15 ions provided a sensitivity of 1 $ng \cdot ml^{-1}$ following column chromatographic separation of these components from **24**. The instability of **24** to the derivatization procedure and its subsequent decomposition to

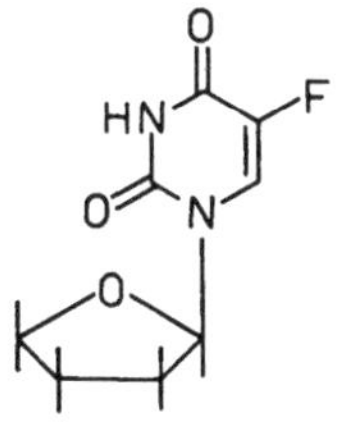

24 Ftorafur

produce small but interfering amounts of **22** necessitated the chromatographic purification step.

A novel extraction procedure for removal of interfering substances in human plasma coupled with the GC-MS analysis of the permethyl derivative provided the required sensitivity for the quantitation of low levels (10 ng · ml^{-1}) of **22** (37). The extraction procedure included heating of the plasma in the presence of buffer, washing with ether and selective concentration of **22** in 16% *n*-propanol-ether using optimized conditions. The internal standard, the dimethyl derivative of **19**, was chosen because of similar mass and the activity of Me_2–**19** as a carrier for Me_2–**22**. The validity of the last property of the internal standard was, however, not substantiated. The permethyl derivative was chosen over the TMS analog because of ease of preparation, stability toward aqueous hydrolysis and a lower increase in mass with the addition of each derivative function.

A five-fold improvement in sensitivity limits with a greatly simplified sample preparation scheme has been reported for the analysis of **22** in plasma and urine samples (38). Sample preparation involved extraction of the 5-FU from a saturated ammonium sulfate solution with 20% isopropanol-ether. After removal of the organic phase and thorough drying, the sample was converted to the dimethyl analog with diazomethane for GC-MS analysis. When C_4H_{10} was used as the reagent gas in the CI mode, the resulting spectra of dimethyl-**22** and the internal standard, dimethyl-**23**, were very simple with most of the ion current being carried by the MH^+ ions at *m/z* 159 and 161 (internal standard). Reproducibility was improved by the addition of **6** to the plasma sample before performing the extraction. The sensitivity reported for this method was 2 ng · ml^{-1}. Higher recovery levels, 90%, may be gained with a more extensive procedure involving an anion-exchange chromatographic purification step (39).

The experimental variables involved in converting **22** to its dimethyl, dibutyl and dihexyl derivatives have been examined (40) and the dibutyl derivative was preferred for analysis of biological samples. Combining the use of the dibutyl derivative with sample purification using anion exchange chromatography, a recovery of 97% is realized and a calibration curve which is linear over the range 5 ng · ml^{-1} to 1 μg · ml^{-1} is obtained. Use of the ^{18}O labeled analog **25** (41) provided results comparable to other work, but incomplete labeling of the internal standard required corrections to be made for unlabeled material present and reduced the sensitivity of the assay (42). An improved method for the preparation of a multiply labeled internal standard containing both ^{18}O and ^{2}H isotopes, i. e., **26**, should overcome the necessity for

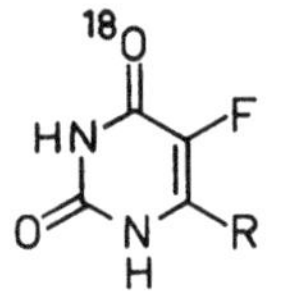

25 5-Fluorouracil-[O^4-^{18}O]; R = H
26 5-Fluorouracil-[O^4-^{18}O; C6-^{2}H]; R = ^{2}H

making corrections for unlabeled material and provide greater sensitivity in future assays (43).

Significantly greater sensitivity for the quantitation of **22** was realized using the pentafluorobenzyl derivative with negative ion chemical ionization (NICI) (44). Detection limits of 50 fg injected on column were realized using this method. Sample preparation consisted of addition of the internal standard, **21**, to the plasma sample, extraction of the 5-halogenated pyrimidine bases with propanol-diethyl ether (22 : 78, v/v) at pH 6 and back extraction into aqueous buffer at pH 10.5. The pentafluorobenzyl derivatives were then prepared by extractive alkylation using pentafluorobenzyl bromide in dichloromethane with tetra-butylammonium as the counter ion. Quantitation of **22** in human plasma was possible down to 390 pg · ml^{-1}, which permitted monitoring of blood levels for 8 hours following administration of a 500 mg · m^{-2} bolus injection.

Extensive *in vivo* metabolism (45) of **22** has stimulated considerable interest in identifying and quantifying the metabolites of 5-FU including 5-fluorouridine (FUR) (**27**) (46), 2′-deoxy-5-fluorouridine (FUdR) (**28**) (47) and 5,6-dihydro-5-fluorouracil (**29**) (48). Other metabolic pathways also exist for **22**, e. g., incorporation into RNA and DNA and conversion to 2′-deoxy-5-fluorouridine-5′-monophosphate (**30**), the active form of the drug, but mass spectrometry has not as yet played a role in the analysis of these metabolic products.

The identification of FUR as a metabolite of both 5-FU and FUdR was based, in part, on capillary GC-MS analysis of urine extracts of rats administered either **22** or **28** (46). Following the specific isolation of **27** from urine by column chromatographic procedures and permethylation, the mass spectrum (CI, CH_4) displayed ions characteristic of permethyl-**27** at *m/z* 319 (MH^+, 100%), 185 (BH · C_3H_5, 7%) and 175 (S, 20%).

Serum levels of **28** in cancer patients have been determined using CI-GC-MS with the 2′-deoxylyxofuranoside of **22** serving as the internal standard (47). Following the high recovery (93%) isolation of **22** and **28** from 1 ml of serum by anion-exchange chromatography, the samples were permethylated, injected onto a capillary column and the MH^+ ion at *m/z* 289 recorded in the single ion monitoring mode. The MH^+ ion was the base peak in the spectra of both **28** and the internal standard when ammonia was used as the reagent gas. Although the isomers could not be differentiated with CI-MS, differences in fragmentation under EI conditions produced major variations in the mass spectra which were characteristic of each compound. Most significant was a change in the base peak in the spectrum of permethyl-**28** from *m/z* 145, attributed to glycosidic bond cleavage with retention of charge by the carbohydrate moiety, to

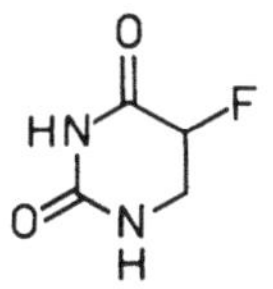

29 5,6-Dihydro-5-fluorouracil

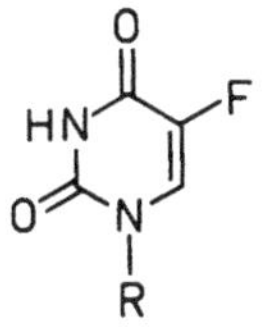

27 5-Fluorouridine (FUR); R = β-D-ribofuranosyl
28 2′-Deoxy-5-fluorouridine (FUdR); R = 2-deoxy-β-D-erythropentofuranosyl
30 5-Fluoro-2′-deoxyuridine-5′-monophosphate (FdUMP); R = 2-deoxy-β-D-erythropentofuranosyl-5′-monophosphate

m/*z* 113 in the spectrum of the internal standard. The authors attributed the intensity of the *m*/*z* 113 ion to the loss of CH_3OH from the sugar ion at *m*/*z* 145. The differences in the EI mass spectra are thus of potential interest in discrimination of epimeric 2′-deoxyfuranosides. The high recoveries observed with the sample extraction method, the base-line separation of the analyte from the internal standard and a linear calibration curve over the concentration range of interest permitted quantitation of **28** down to 1 ng · ml^{-1}.

The identification and quantitation of 5,6-dihydro-5-fluorouracil (**29**) in human plasma samples has also been reported using GC-MS techniques (48). The identity of **29** was established following isolation by HPLC, conversion to the dipentyl derivative and comparison of the EI and CI spectra with a reference sample. Quantitation of the dipentyl derivatives of **22, 29** and the internal standard (5-bromouracil) was performed by SIM of the *m*/*z* 253 (**22**), 203 (**29**) and 261 (internal standard) in the EI mode and the MH^+ ions of **29** (*m*/*z* 273) and internal standard (*m*/*z* 331) in the CI mode. No interferences were present in plasma blanks analyzed by EI, but plasma impurities prevented use of CI for determination of **22**. Detection limits in the EI mode were 10 ng · ml^{-1} and 80 ng · ml^{-1} for **22** and **29**, respectively. CI provided greater sensitivity for determination of **29**.

Mass fragmentography has been used to determine the levels of **22** in plasma and tissue samples following administration of **24** and analogs related to **24** (49–53).

3 Nucleosides

As mentioned earlier, a considerable body of literature exists describing the mass spectrometry of nucleic acid components and most of these references are beyond the scope of this chapter. However, to provide a guide to background material and because of a possible interest in areas peripherally related to the present work Table 3 is included (54–82). The table is not comprehensive.

Table 3. Leading references to work in the mass spectrometry of nucleosides

Topic	Reference
Derivatives	
General references	1–4
Trimethylsilyl	7, 8, 54, 55
Permethyl	5, 6, 56, 57
Trifluoroacetyl	58
Tert-Butyldimethylsilyl	59
Other	60–62
Modified nucleosides in RNA	10, 11, 63–65
Polyoxin antibiotics	66–69
Synthetic or naturally occurring nucleosides of biological interest	63, 70–72
C-Nucleosides	73–75
Cytokinins	76–81
Negative ion EI	82

3.1 Quantitation Studies

With the exception of the 5-fluorouracil nucleosides mentioned in a previous section, few reports have appeared in the literature concerning the qualitative or quantitative use of mass spectrometry in a clinical biochemical sense. The works described below are therefore intended to illustrate the potential of mass spectrometry in solving problems of this type with the expectation that once the feasibility of a method is shown to be practical, applications will follow.

A method for determining the pharmacokinetics of bases and nucleosides possessing antitumor activity using CI-GC-MS mass fragmentographic techniques has been described (83). Samples analyzed in this study included the permethyl derivatives of 5-FU (**22**), FUdR (**28**), 6-MP (**12**), 6-mercaptopurine riboside (6-MPR) (**31**) and arabinosyl cytosine (ara-C) (**32**). Quantitation was achieved by monitoring the MH^+ ion in the methane CI spectrum of each nucleoside and comparing peak height ratios to calibration curves linear over the range 500 pg to 10 ng. Fragmentations in the CI mode were significantly reduced relative to the EI mode and in each case the MH^+ ion was the base peak of the spectrum. No interference from endogenous substances was observed. The utility of the method was further illustrated by following the disappearance of ara-C from serum of rats after IV administration of the drug. Detection limits were approximately $100\ ng \cdot ml^{-1}$ which allowed quantitation of **32** for about 2 hours following administration of the drug. Only an external standard, imipramine, was used in this study.

A significant increase in sensitivity has been achieved in the quantitation of ara-C (**32**)by use of the acetyl-methyl mixed derivative **33** and with cytidine (**34**) or the deuteriomethyl acetyl derivative as the internal standard (84). The acetyl-methyl derivative **33** was chosen because of its relative ease of preparation and because the nucleoside acetates permit partial purification by extraction into chloroform from the aqueous phase, thus simplifying sample preparation. In addition, the GC properties of the acetyl-methyl derivative affords separation of **32** from **34** and other structurally

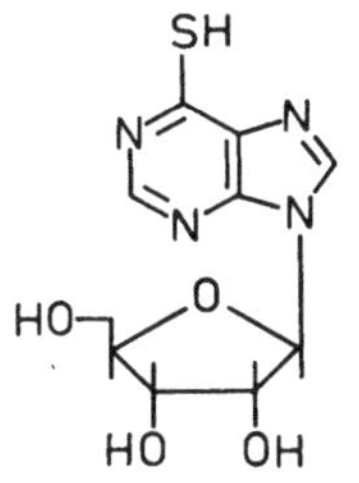

31 6-Mercaptopurine riboside

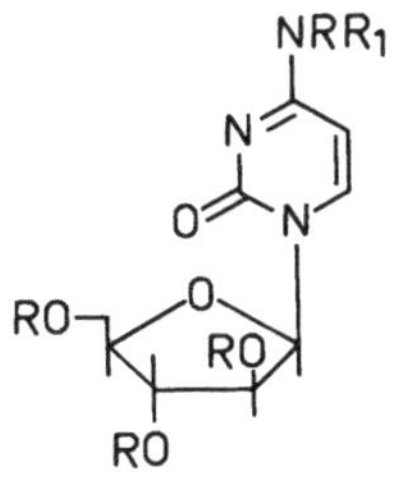

32 Ara-C; $R=R_1=H$
33 Mixed derivative; $R=$Acetyl, $R_1=CH_3$

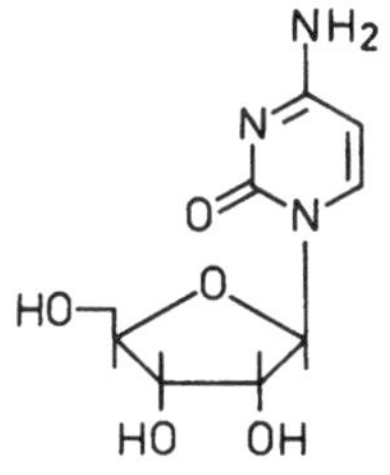

34 Cytidine

related pyrimidine nucleosides. The acetyl-methyl derivative of **34** was used as the internal standard in quantitation studies down to the 50 $ng \cdot ml^{-1}$ level. Below this level the appearance of an endogeous interference, suspected to be cytidine itself, prevented accurate quantitation. At lower levels of **32**, multiple-ion detection with the $[^2H_{12}]$acetyl-methyl derivative as the internal standard enabled **32** to be measured to the $3-10$ $ng \cdot ml^{-1}$ level with detection down to 1 $ng \cdot ml^{-1}$. Using the SIM techniques described, plasma levels of **32** were monitored for about 5 hours following administration of a single intravenous dose (2 $mg \cdot kg^{-1}$) of the drug, a significant increase in sensitivity and/or selectivity relative to other instrumental and biological methods currently available.

The isolation and characterization of a metabolite of the antiviral nucleoside 5-(2-bromo-E-ethenyl)-2′-deoxyuridine (BVDU) has been reported using a combination of HPLC, for isolation, and C_4H_{10} CI-MS, for molecular weight and structural information (85). Relative to the spectrum of BVDU, the metabolite showed an MH^+ ion doublet at *m/z* 217/219 indicating the presence of the bromine atom and a mass difference equivalent to the loss of the deoxyribose function present in the starting drug. The base peak of the metabolite spectrum was at *m/z* 137 indicating expulsion of HBr from the MH^+ ion; this ion was also present in the spectrum of BVDU. Ions characteristic of the deoxysugar moiety were not observed in the metabolite spectrum. Based on this information, the structure of the metabolite was assigned as 5-(2-bromo-E-ethenyl)uracil (BVU) and confirmed by analysis of, and comparison with, a known sample.

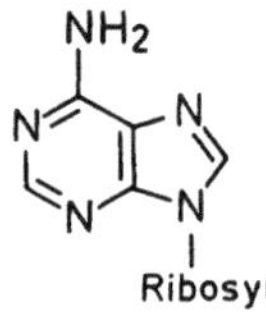

35 Adenosine

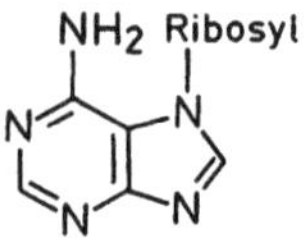

36 7-(β-D-Ribofuranosyl)adenine

The stability of the glycosidic bond of nucleosides is of interest both biochemically and clinically. The transfer of a ribose group to a nucleic acid base is fundamental to the salvage pathway known to operate in cells and is of paramount importance in the activation of a number of nucleic acid base antimetabolites. Most studies of the glycosidic bond strength of nucleosides have utilized classical methods employing UV detection to follow the acid catalyzed hydrolytic reaction. The use of CI-MS to determine the relative glycosyl bond strengths of pairs of 7- and 9-*β*-ribofuranosyl-purines, e. g., adenosine (**35**) and 7-(*β*-D-ribofuranosyl) adenine (**36**), has, however, been reported (86). Essentially, the method compares the $MH^+/(B + 2H)^+$ ion intensity ratios observed in the CI spectrum of five nucleoside pairs examined with the rate coefficients determined from solution experiments. Ammonia was used as the reagent gas in these experiments because of its relatively high proton affinity, which suppressed competing fragmentation pathways (87, 88). Results of the comparison showed the 7-substituted isomers to display a greater tendency to fragment relative to the 9-substituted isomers and, therefore, the relative glycosyl bond strength of the nucleoside pairs was clearly reflected in the CI mass spectrum. Although the CI spectra did not permit the establishment of the site of attachment of the sugar to the base if only a single isomer was available, the suggestion was made that differentiation of isomers does appear feasible on the basis of the data obtained with these experiments. The advantages of the CI-MS procedure versus the more classical approach include the speed and simplicity of data acquisition and the smaller sample size needed for analysis. The major limitation of the mass spectral approach is the requirement for the vapor phase analysis, which excluded a number of available compound pairs from consideration, e. g., guanosine, inosine and 6-mercaptopurine riboside analogs.

3.2 Urine profiling

A considerable interest in the identification and quantitation of nucleic acid bases and nucleosides, endogenous to human urine and blood, has emerged in the last few years (89). The basic premise of these profiling studies is that the presence of a natural or modified base or nucleoside in human urine (plasma) may be of value as a "biological marker" of a disease state and elevated (or decreased) levels of a compound relative to a "normal" value may be of diagnostic value. Most of the profiling studies have utilized HPLC for separation of these complex nucleoside mixtures because of the greater ease of analysis of nucleosides with this chromatographic mode relative to GC. This approach however suffers a number of limitations including a necessary prior knowledge of the compound(s) to be quantitated, the potential for coelution of

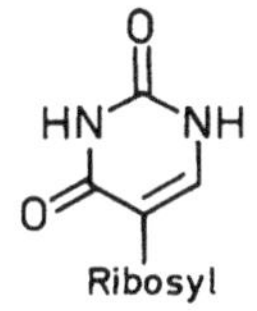

37 Pseudouridine

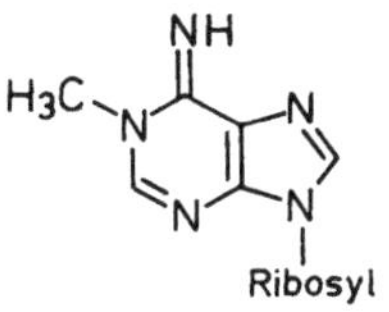

39 N^2-Methylguanosine; $R = H$, $R_1 = CH_3$
40 N^2,N^2-Dimethylguanosine; $R = R_1 = CH_3$

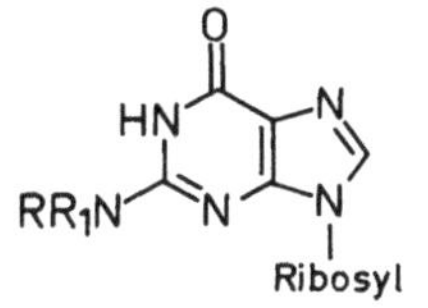

38 1-Methyladenosine

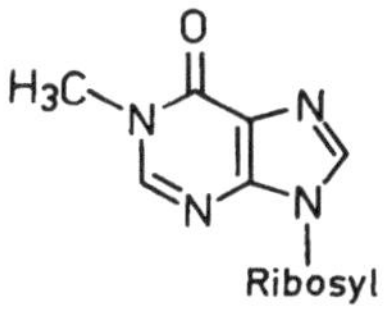

41 1-Methylinosine

interfering artifacts and the inability to determine directly the identity of an unknown component. In an effort to increase the specificity and, where needed, the sensitivity of the analysis, methods for the capillary GC-MS separation, identification and quantitation of nucleosides isolated from human urine are being developed (90). Two approaches are currently being pursued in these studies (91). The first area being investigated involved quantitation of five nucleosides known (16) to be elevated in the urine of patients with various types of cancer and four nucleosides either identified (89, 92–94) or suspected of being present in human urine. The five nucleosides known to be excreted in large amounts in the urine of cancer patients are pseudouridine (**37**), 1-methyladenosine (**38**), N^2-methylguanosine (**39**), N^2,N^2-dimethylguanosine (**40**) and 1-methylinosine (**41**). The four nucleosides identified or suspected as being present in human urine at low levels are the 2′-*O*-methylated analogs of adenosine (**42**), uridine (**43**), cytidine (**44**) and guanosine (**45**). The internal standards which have been prepared (91, 95) for quantitation of these nucleosides are the appropriate C^2H_3 labeled analogs. The use of EI, CI and mass chromatography techniques to locate and identify other modified nucleosides of potential utility as biomarkers in urine samples is the second objective of this work.

In addition to the identification and quantitation of the above mentioned nucleosides, suspected as being produced by degradation of tRNA, a number of other unusual nucleosides have been identified in human biological fluids, primarily urine. Identification of these modified nucleosides and bases in human urine are of interest not only for their possible utility as biological markers but as indicators of heretofore unrecognized biochemical pathways operating in the human biological system.

The structure of a new nucleoside isolated from the urine of chronic myelogenous leukemia patients has been established as 3-carboxamide-1-(β-D-ribofuranosyl)pyridin-4-one (**46**) (96). A molecular mass of 270 daltons for the free nucleoside was established by the preparation of the per-TMS and the $[^2H_9]TMS_2$ derivatives. Major fragment ions suggested the presence of an unmodified ribose attached to a base with a composition of $C_6H_5N_2O_2$ containing five rings plus double bonds. A

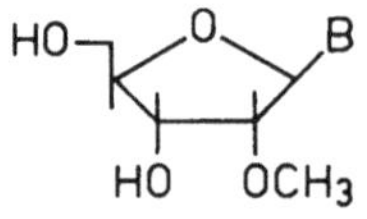

44 2′-*O*-Methylcytidine
45 2′-*O*-Methylguanosine
42 2′-*O*-Methyladenosine
43 2′-*O*-Methyluridine

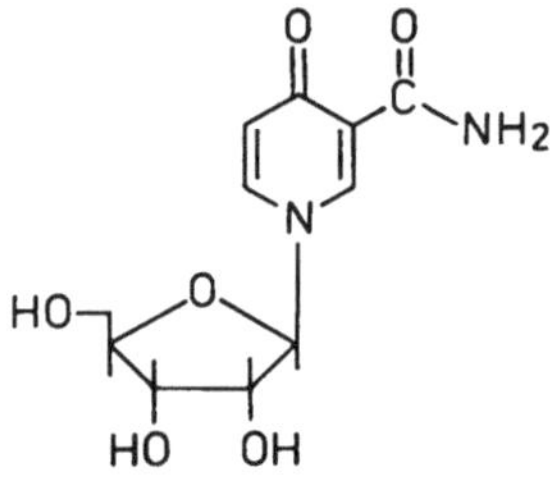

46 1-(β-D-Ribofuranosyl)pyridin-3-carboxamide-4-one

weak ion at *m/z* 530 in the high mass region of the spectrum (TMS derivative) was attributed to the loss of CO from the molecular ion which is characteristic of, but not unique to, a quinone-like structure which was suspected on the basis of UV data. An unusual M−45 ion at *m/z* 513 was established as the elimination of CH_3NO from a sidechain attached to the base. Confirmation of the suspected structure was obtained by preparation of an authentic sample of the nucleoside and a comparison of the spectral properties of the synthetic sample with the compound isolated from the urine sample.

The structure of **46**, the first reported instance of a pyridine nucleoside to be found in human urine (96), suggested a relationship to nicotinamide mononucleotide (NMN) and nicotinamide adenine dinucleotide (NAD). In order to determine if the new nucleoside was derived from these cofactors, rats were administered ^{14}C labeled nicotinamide and NAD and their urines chromatographed in a search for the presence of radiolabeled-**46**. No evidence of the presence of **46** was observed and the origin of **46** is currently being investigated. It was noted that the levels of **46** are lower in the urine of cancer patients ($0.25-0.30\ mg \cdot 24\ h^{-1}$ urine) than in normal human urine ($1.8\ mg \cdot 24\ h^{-1}$ urine).

The identity of one of four isomeric substances isolated from the urine of patients with colon carcinoma and leukemia has been reported to be 6-amino-3-methyl-5-(*N*-formylmethylamino)uracil (**47**) (97). Based on preliminary results using UV, NMR and chromatographic procedures, possible structures were narrowed to **47** and 5-acetamido-6-amino-3-methyluracil (**48**). Mass spectrometry supported, but could not differentiate, these two possibilities. High resolution data indicated a single composition of $C_7H_{10}N_4O_3$. Ions observed in the upper mass range included fragments at *m/z* 180 ($M-H_2O$), 170 (M−CO) and a peak at *m/z* 155. The latter ion corresponds in nominal mass to a retro-Diels-Alder elimination of HNCO from the molecular ion of many uracil-type compounds and would not be observed in the spectrum of **47** since N3 is substituted. Exact mass measurements and metastable ion analysis indicated the *m/z* 155 to result from the sequential expulsion of CO (*m/z* 170) and a CH_3 radical from the $M^{+\bullet}$ ion and therefore this ion is unrelated to the HNCO loss of N3 of unsubstituted uracils. Differentiation of **47** from **48** was made by comparison of the spectral properties of synthetically prepared **47** and **48**. Excretion levels of **47** are much higher in normal subjects ($0.75\ mg \cdot 24\ h^{-1}$ urine) than in cancer patients ($0.2-0.3\ mg \cdot 24\ h^{-1}$ urine), and, thus the origin of **47** does not appear to be related to the degradation of tRNA. Speculations as to be origin of **47** include its possible

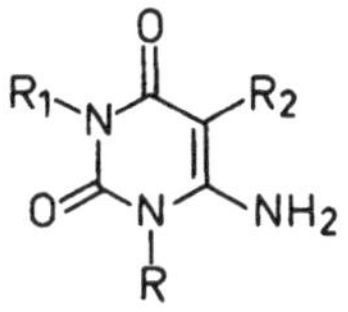

47 R=H, $R_1=CH_3$, $R_2=NCH_3CHO$
48 R=H, $R_1=CH_3$, $R_2=NHCOCH_3$
49 R=H, $R_1=CH_3$, $R_2=NHCHO$
50 $R=CH_3$, $R_1=H$, $R_2=NCH_3CHO$

formation from 1,7-dimethylxanthine (**11**), or the corresponding nucleoside 1,7-dimethylxanthosine, by oxidation to 1,7-dimethyldihydrouric acid which could undergo ring opening to form the new uracil analog **47**. Formation of **11** may occur by demethylation of caffeine **8** or from 1-methyl- or 7-methylguanine derived from tRNA by deamination followed by methylation. The possibility thus exists that **47** may originate from both endogenous and exogenous sources. The origin of **48**, also found in human urine, is unknown at the present time.

The identities of two additional compounds closely related to **47** and **48** have been established as 6-amino-3-methyl-5-(*N*-formylamino)uracil (**49**) and 6-amino-1-methyl-5-(*N*-formylmethylamino)uracil (**50**) (98). Of particular interest in this study is the use of CI, in conjunction with a mass spectrometer modified to permit the direct introduction of solutions into the ion source, to differentiate **49** from the isomeric *N*3 substituted analog **48** on the basis of ion abundance differences. A similar approach was used to distinguish **50** from **47** where the intensity of the MH^+-CO ion (*m*/*z* 171) was noted to be significantly and reproducibly decreased in the case of **50**. The origin of these new uracil analogs in human urine is thought to be from metabolism of methylated xanthines, compounds previously believed to be stable to metabolic processes.

Most recently, this same group has isolated a novel nucleoside from the urine of a chronic myelogenous leukemia (CML) patient and characterized the sample as 7-β-D-ribofuranosylhypoxanthine (99). This is the first report of an *N*7 glycosidically linked nucleoside to be found in human urine and may prove to be of considerable value as a biological marker of cancer since the molecule may not be excreted in the urine of normal patients. Speculations as to the origin of this nucleoside include degradation of a vitamin B_{12} analog present in the sera of CML patients, biosynthesis by the action of a transglycosylase enzyme which transfers a sugar from a pyrimidine nucleoside to a purine base, in this case adenine or hypoxanthine, or the sample may be of bacterial origin.

EI-MS and CI-MS have been used to assist in the structure elucidation of 5′-methylthioadenosine sulfoxide (MTA-sulfoxide) (**51**) in the urine of immunodeficient children (100). Most crucial in establishing the structure of the sample was the presence of an isotope cluster associated with the MH^+ ion at *m*/*z* 314 in the methane CI spectrum indicating the presence of a sulfur atom. Ions at *m*/*z* 296 in the CI spectrum and *m*/*z* 298 in the EI spectrum were assigned as loss of H_2O and a methyl radical, respectively, and established the molecular weight of the compound. Major fragment

Table 4. Nucleic acid base levels in plasma (or serum) and urine of normal individuals. Plasma levels expressed in $mg \cdot l^{-1}$. Urine levels given as either (a) mg excreted in 24 hours or (b) $nmol \cdot \mu mol^{-1}$ creatinine; nq = not quantated. Table compiled from data in references cited below.

Compound	Plasma	Urine	Reference
Adenine		1.4[a]	89
		2.25 ± 2.44[b]	103
—, 1-methyl		0.3 – 0.4[a]	89
—, N^6-methyl		nq	89
Allantoin		11.9[a]	89
Cytosine		trace	89
—, 3-methyl		nq	89
Guanine		0.3 – 1.4[a]	89
—, N^2,N^2-dimethyl		trace	89
—, 1-methyl		0.6[a]	89
—, N^2-methyl		0.3 – 0.6[a]	89
—, 7-methyl		2.7 – 7.8[a]	89
		9.5 ± 6.1[b]	103
—, 8-hydroxy-7-methyl		1.6[a]	89
Hypoxanthine	1.75 ± 0.80	8.5 ± 3.1[a]	18
		6.8[a]	89
	3.2 ± 3.4	16.4 ± 13.7[a]	19
—, 1-methyl		0.4 – 0.9[a]	89
Imidazole, 4-acetic acid		4.5 (mg/l)	89
—, 4-acetic acid-1-methyl		trace	89
—, 5-amino-4-carboxamide		1.05[a]	89
N-(6-Purinyl)aspartic acid		0.5 – 0.9[a]	89
Thymidine	0.05 ± 0.06 (μmol/ml)		32
Uracil	0.98 ± 0.64 (μmol/ml)		32
		7[a]	89
—, 5-acetamino-6-3-methyl		6 – 10[a]	89
—, 6-amino-3-methyl-5-(*N*-formyl-amino)		0.03[a]	98
—, 6-amino-1-methyl-5-(*N*-formyl-methylamino)		0.03[a]	98
—, 6-amino-3-methyl-5-(*N*-formyl-methylamino)		0.75 – 1[a]	97
—, 5-hydroxymethyl		trace	89
Uric acid		390 – 588[a]	89
	0.059 ± 0.005	440 ± 53[a]	18
—, 1,3-dimethyl		trace	89
Xanthine		2.8 – 8.7[a]	89
	0.63 ± 0.10	9.0 ± 2.6[a]	16
	0.9 ± 0.3	25.0 ± 15.0[a]	17
—, 1,3-dimethyl		trace	89
—, 1,7-dimethyl		nq	89
—, 1-methyl		nq	89
—, 7-methyl		nq	89

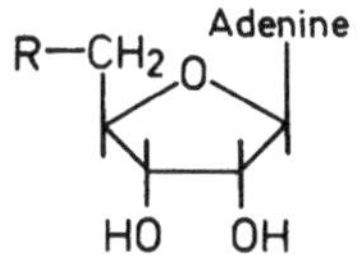

51 MTA-sulfoxide, R = CH_3SO-
52 MTA, R = CH_3S-

ions in the EI spectrum indicated the presence of an unmodified adenine base (B + H and B + 2H ions), a *cis*-hydroxyl containing sugar (borate complex formed) unmodified in the 4′-ring position (B + 30 ion), with the 1′ and 2′ positions of the suspected ribose being unsubstituted (B + 44 ion). This data coupled with the facile loss of a methyl radical from the molecular ion led to the structural assignment of **51**, with unequivocal proof being obtained by comparison of spectral data with a reference sample. The structure **51** suggested an origin from 5′-methylthioadenosine (MTA) (**52**) by oxidation either *in vivo* or during isolation of the sample. In turn, MTA is a product of polyamine synthesis but intracellular levels of MTA are normally low due to rapid metabolism to methionine and adenine by the successive action of the MTA phosphorylase, an unidentified enzyme which converts 3-methylthioribose phosphate to 2-keto-4-methylthiobutyrate and glutamine aminotransferase. The increased levels ($3-5$ nmol · μmol^{-1} creatinine) of **51** in the urine of immunodeficient children therefore may indicate the presence of an heretofore unrecognized metabolic defect in these patients.

Compilations (101 – 117) of "normal" levels of nucleic acid bases and nucleosides in human urine and blood are given in Tables 4 and 5.

The use of liquid chromatography-mass spectrometry (LC-MS) for the analysis of urine samples is described in Section 8.

3.3 New methods of stable isotope labeling

The importance of stable isotopically labeled compounds for structure fragmentation studies and for use as internal standards in quantitative determinations is well established (118 – 123). Although methods have been available for a number of years for the enzymatic and chemical incorporation of various stable nuclides into nucleic acid components (see appropriate sections of refs. 2 and 3), general chemical procedures yielding high levels of isotope incorporation are of relatively recent origin. Many of the references cited in previous sections have utilized stable isotope dilution methods for quantitation and are of importance to investigators studying those specific compounds. The citations listed in Table 3 are of a more general nature in that the methods have been shown to be, or may be expected to be, applicable to the preparation of purine and pyrimidine nucleosides or bases with widely varying structures, assuming stability to the reaction conditions.

A number of methods have been developed for the incorporation of deuterium into the heterocyclic ring of nucleic acid bases and nucleosides. Base catalyzed exchange

Compound	Plasma/Serum	Urine	Reference
Adenosine	0.11 – 0.56	–	101
	–	0.23 ± 0.07[b]	102
		1.04 ± 0.42[b]	103
	–	0.29 ± 0.09[b]	104
	–	trace	89
–, 2′-deoxy	–	0.76 ± 1.13[b]	103
–, N^6-(Δ^2-isopentenyl)	–	0.050 ± 0.019[a]	94
–, 1-methyl	not detected	4.32 ± 1.58[b]	105
	–	1.77 ± 0.58[b]	106
	–	5.91[a]	16
	–	1.8[b]	107
	–	2.02 ± 0.35[b]	102
	–	2.15 ± 0.4[b]	104
–, N^6-methyl	–	2.1 – 5.2[a]	89
–, 5′-methylthio sulfoxide	–	0.34 ± 0.21[b]	100
Cytidine	–	trace	89
–, 2′-deoxy	–	trace	89
–, 2′-*O*-methyl	–	0.2 – 0.5[a]	89
–, 5-methyl	–	0.11 ± 0.03[b]	102
Guanosine	–	trace	89
	0.67 ± 0.27	–	16, 105
	0.88 ± 0.51	–	108
–, N^2,N^2-dimethyl	–	1.36 ± 0.90[b]	105
	–	1.20 ± 0.30[b]	109
	–	0.18 – 2.00[b]	110
	–	2.27[b]	16
	–	0.8[b]	107
	–	105 ± 0.12[b]	102
	–	7.6 ± 5.9[b]	103
	–	3.27 ± 0.40[b]	112
	–	1.25 ± 0.30[b]	104
	–	1.1 – 2.1[a]	89
	–	0.05 ± 0.022[a]	111
–, 1-methyl	–	0.4 – 0.6[a]	89
	–	0.96 ± 0.44[b]	105
	–	1.41[b]	16
	–	0.6[b]	107
	–	0.70 ± 0.04[b]	98
	–	1.83 ± 0.29[b]	112
	–	0.71 ± 0.16[b]	104
	–	0.68[b]	113
–, N^2-methyl	–	0.2 – 0.3[a]	89
	–	1.29 ± 1.08[b]	105
	–	1.20 ± 0.15[b]	109
	–	0.35 ± 0.14[b]	106
	–	2.38[a]	16
	–	0.3[b]	107
–, 2′-*O*-methyl	–	0.17 ± 0.09[a]	94
–, 7-methyl	–	1.46[b]	16

Compound	Plasma/Serum	Urine	Reference
Imidazole, 5-amino-4-carboxamide riboside	–	nq	89
			101
	5.62 ± 2.87	–	108
	–	0.18[a]	89
–, 1-methyl	–	2.1 – 2.9[a]	89
	–	0.051 ± 0.011[a]	114
	–	0.04 ± 0.018[a]	111
	–	1.04 ± 0.42[b]	105
	–	1.15 ± 0.54[b]	109
	–	1.15 – 1.18[b]	106
	–	1.46[b]	16
	–	0 – 2.01[b]	110
	–	1.2[b]	107
	–	0.96 ± 0.08[b]	102
	–	2.38 ± 0.28[b]	112
	–	1.20 ± 0.24[b]	104
Pseudouridine	–	42 – 62[a]	89
	–	0.78 ± 0.206[a]	111
	–	0.07 ± 0.078[a]	114
	2.48 ± 0.13	17.77 ± 6.36[b]	105
	–	22.4 – 26.7[b]	106
	–	22.6 ± 4.32[b]	109
	–	8.57 – 27.64[b]	110
	2.52 ± 0.28	24.13[b]	16
	1.72 ± 0.77	–	115
	–	0.30 ± 0.08 (mmol · 24 h^{-1})	116
	–	23.5[b]	107
	–	24.80 ± 4.8[b]	102
	–	66.7 ± 22.2[b]	103
	–	33.19 ± 3.9[b]	112
	–	20.89 ± 4.32[b]	104
	–	20.33[b]	113
N-9-(*β*-D-Ribofuranosyl)-9H-purin-6-yl)carbamoyl threonine	–	0.3[a]	89
	–	2.2 ± 0.9[a]	94
Uridine	1.9 – 8.4	–	
	2.54 ± 0.63	–	105
	3.17 ± 1.11	–	108
	–	trace	89
–, 2′-*O*-methyl	–	0.1[a]	92
1-(*β*-D-Ribofuranosyl)-pyridin-3-carboxamide-4-one	–	1.8[a]	96

Table 5. Nucleoside levels in human plasma (or serum) and urine. Plasma serum levels expressed in nmol · l^{-1}. Urine levels reported in either (a) mg · 24 h^{-1} or (b) nmol · µmol^{-1} creatinine.

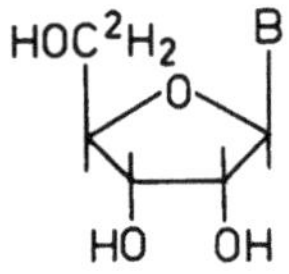

53 Adenosine-[5′,5′-$^2H_2$]; B = adenine
54 Uridine-[5′,5′-$^2H_2$]; B = uracil

of the C5 and C6 positions of pyrimidine nucleosides containing ribose, arabinose and 2′-deoxyribose sugars attached to uracil, thymine, cytosine and 5-fluorouracil bases have been reported (124, 125).

A method of exchanging the aromatic hydrogens with deuterium using platinium oxide catalysis has been applied to a number of nucleosides and nucleotides (126). Excellent yields of adenosine type purine nucleosides may be obtained with high levels of deuterium label in the 1 and 8 positions. Pyrimidines undergo exchange of the C5 and C6 hydrogens and thymidine also exchanges the hydrogens of the 5-methyl group. An extension of this work has revealed that in addition to the simple reaction mentioned above, ring reduction of both purine and pyrimidine nucleosides may be effected by performing the reaction in an atmosphere of deuterium gas (127). In addition, exchange of the non-labile hydrogens of the carbohydrate may also take place.

The specific deuterium labeling of the 5′-methylene group of both purine and pyrimidine nucleosides has been reported. The most direct route to adenosine-[5′,5′-$^2H_2$] (**53**) involves chromic acid oxidation of 2′,3′-*O*-isopropylidene adenosine to the 5′-carboxylate analog which, *in situ*, is converted to the 5′-isopropylester. Reduction with lithium aluminium deuteride to the alcohol and acid-catalyzed removal of the isopropylidene blocking group affords the labeled adenosine **53** in high yield with essentially quantitative incorporation of label (128). Other methods for the preparation of **53** are also available (129–131) and efforts are currently under way to extend this method to other purine type ring systems.

A condensation reaction between 1,2,3,5-tetra-*O*-acetyl-β-D-ribofuranose [5′,5′-$^2H_2$], obtained by hydrolysis of **53**, and uracil has been utilized to prepare uridine-[5′,5′-$^2H_2$] (**54**) in high yield of both product and label (7).

Methods for the incorporation of ^{18}O into nucleic acid bases and nucleosides have also been developed. A simple method of exchanging ^{18}O for ^{16}O in the C4 position of uracil type bases using a catalytic amount of thionyl chloride ($SOCl_2$) has been shown to provide high levels of isotope incorporation (132). Unfortunately, the procedure failed to provide labeled products when applied to nucleosides. Modification of the procedure by substitution of 1 mol · l^{-1} HCl for the $SOCl_2$ however effected an essentially quantitative ^{18}O exchange at the C4 position with additional label being observed in the C2 position (41). These conditions were extended to a variety of 5-substituted pyrimidine nucleosides with the level of ^{18}O incorporation decreasing in the order uridine ≈ 5-methyluridine > 5-fluorouridine > 5-hydroxyuridine > 5-chlorouridine > 5-bromouridine > 5-iodouridine > pseudouridine.

The reaction of O^2,5′-anhydro pyrimidine nucleosides under basic conditions has been used to prepare a number of ^{18}O labeled 2′-deoxy- (133, 134), ribo- (135) and

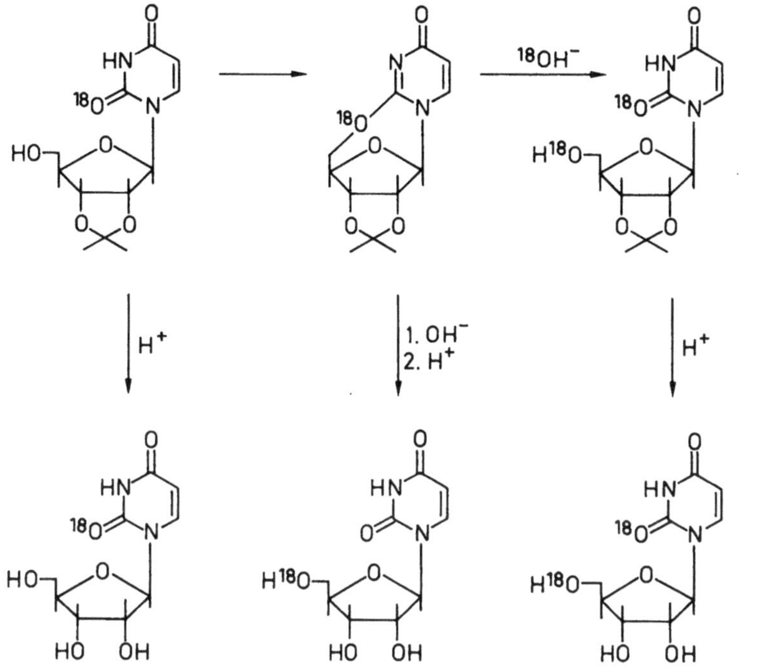

55 Uridine-[O^2-^{18}O] **56** Uridine-[5′-^{18}O] **57** Uridine-[O^2,5′-$^{18}O_2$]

Scheme 4. A synthetic method for the preparation of base or sugar ^{18}0 labeled pyrimidine nucleosides. (Adapted from ref. 135.)

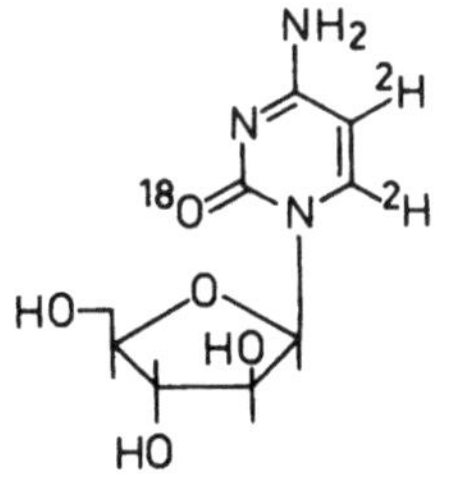

58 Ara-C-[O^2-^{18}O,5,6-$^2H_2$]

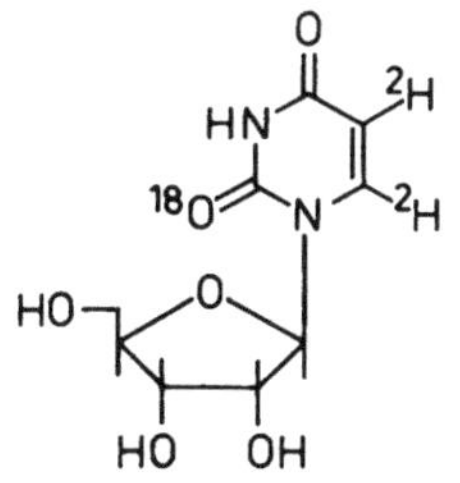

59 Ara-U-[O^2-^{18}O,5,6-$^2H_2$]

arabinosyl (136) nucleosides. The method can be used to prepare nucleosides labeled in the base, the carbohydrate or both portions of the molecule, as illustated with uridine in Scheme 4. The preparation of two monolabeled, **55** and **56**, and one dilabeled, **57**, analogs is possible using this procedure. A combination of deuterium and the ^{18}O ring-opening methods have been employed for the synthesis of multiply labeled samples of ara-C (**58**) and ara-U (**59**) for quantitation studies (91).

Methods have also been described for the ^{15}N labeling of nucleosides (137–142).

3.4 Fast atom bombardment mass spectrometry of nucleosides

The mass spectral analysis of thermally labile or non-volatile compounds of biological interest has undergone a profound revolution with the introduction of fast atom bombardment (FAB) as an ionization technique (143). Many aspects of FAB make this ionization technique the method of choice in the analysis of nucleosides, and the advantages of FAB over other ionization methods have been described (144, 145). The major utility of FAB-MS in the nucleic acid area (146) has been in the analysis of nucleotides and oligonucleotides (see later) with only brief mention having been made to the application of this technique to nucleosides (8).

The positive and negative FAB mass spectra of a series of 30 nucleosides and 2 dinucleotides have been recently examined with and without collisional activation using a tandem mass spectrometer (147). Nucleosides examined included various purine and pyrimidine analogs which served to show the general utility of FAB in the analysis of this class of compounds. A major advantage of using collisionally activated decomposition (CAD) methods with a tandem mass spectrometer is the exclusion of matrix ions from the spectra, which greatly simplifies interpretation of data and permits detection of some fragments not observed in the normal FAB spectra. The results of this work indicate that FAB-MS, especially when used in conjunction with CAD and MS/MS, will have an important role to play in the analysis of nucleosides either in or obtained from biological samples.

A comparison has been made of the FAB mass spectra of the eight major nucleosides found in RNA and DNA, pseudouridine, and the 2',3'-*O*-isopropylidene derivative of adenosine with CI and EI spectra (148). The data obtained in this study are summarized in Table 6.

Two interesting techniques for the characterization of specific features of nucleosides and nucleotides unique to FAB mass spectrometry have been reported. The first is a simple method for determining the number of active hydrogen atoms in a molecule (149, 150) by using [hydroxy-$^2H_3$]glycerol and ^2H_2O as the FAB matrix and following the subsequent hydrogen-deuterium exchange reaction. The method has been shown to be applicable to nucleotides as well as carbohydrates, one peptide and complex antibiotics. The second technique monitors formation of a complex between the *cis*-diol functionality of ribosides with the boronate group of 4-tolueneboronic acid (151). The boronic acid complex, carrying a net negative charge, thus provides a simple method of distinguishing ribosides from deoxyribosides. The MH^+ of adenosine boronate complex appears at *m/z* 366 with no ion appearing at *m/z* 268 for free adenosine while 2'-deoxyadenosine shows a strong MH^+ ion at *m/z* 250, with only a very weak 2'-deoxyadenosine-boronate complex ion (*m/z* 350) being observed. Only

Table 6. A comparison of the major ions and their relative intensities of nucleosides ionized using FAB, EI and CI. (adapted from ref. 148)

		MH^{+} [a,b] *m/z* (%RI)	B + 44[c] *m/z* (%RI)	B + 30[d] *m/z* (%RI)	B + 2H *m/z* (%RI)	S^+ *m/z* (%RI)	Other *m/z* (%RI)
Adenosine	FAB	268 (92)	178 (3)	164 (7)	136 (100)	133 (3)	$[MH + G]^+$: 360 (28)
	CI	268 (59)	178 (3)	164 (3)	136 (100)	133 (2)	
	EI	267 (20	178 (35)	164 (70)	136 (78)	133 (2)	BH^+ : 135 (100)
Guanosine	FAB	284 (62)	194 (5)	180 (3)	152 (100)	133 (2)	$[MH + G]^+$: 386 (12)
	CI	–	–	–	–	–	
	EI	–	–	–	–	–	
Uridine	FAB	245 (58)	155 (6)	141 (1)	113 (100)	133 (11)	$[MH + G]^+$: 337 (13)
	CI	245 (19)	155 (3)	141 (17)	113 (100)	133 (33)	
	EI	244 (9)	155 (6)	141 (20)	113 (100)	133 (57)	BH^+ : 112 (22)
Cytidine	FAB	244 (63)	154 (4)	140 (4)	112 (100)	–	$[MH + G]^+$: 336 (12)
	CI	244 (2)	154 (2)	140 (2)	112 (100)	–	
	EI	243 (1)	154 (6)	140 (43)	112 (100)	133 (2)	BH^+ : 111 (46) [B + 41] : 151 (26)
Deoxyadenosine	FAB	252 (54)	162 (7)	164 (4)	136 (100)	117 (4)	$[MH + G]^+$: 344 (8)
	CI	252 (31)	162 (6)	164 (23)	136 (100)	117 (5)	
	EI	251 (10)	162 (37)	164 (12)	136 (30)	117 (5)	BH^+ : 135 (100)
Deoxyguanosine	FAB	268 (43)	178 (4)	180 (1)	152 (100)	117 (2)	$[MH + G]^+$: 360 (3)
	CI	–	–	–	–	–	
	EI	–	–	–	–	–	
Deoxycytidine	FAB	228 (46)	138 (6)	140 (1)	112 (100)	117 (5)	$[MH + G]^+$: 320 (5)
	CI	–	138 (2)	140 (4)	112 (100)	117 (20)	
	EI	227 (2)	138 (15)	140 (< 1)	112 (63)	117 (15)	BH^+ : 111 (100) 153 (18)
Thymidine	FAB	243 (42)	153 (2)	155 (< 1)	127 (100)	117 (38)	$[MH + G]^+$: 335 (4)
	CI	243 (19)	153 (2)	155 (24)	127 (100)	117 (49)	
	EI	242 (8)	153 (4)	155 (< 1)	127 (27)	117 (100)	BH^+ : 126 (32)
Pseudouridine	FAB	245 (100)	155 (29)	141 (8)	113 (15)	–	$[MH + G]^+$: 337 (17) $[MH - 2H_2O]^+$: 209 (7) $[MH - 3H_2O]^+$: 191 (2)
	CI	245 (64)	155 (100)	141 (19)	113 (28)	133 (2)	$[MH - H_2O]^+$: 227 (5) $[MH - 2H_2O]^+$: 209 (5) $[MH - 3H_2O]^+$: 191 (4)
	EI	244 (< 1)	155 (8)	141 (100)	113 (10)	–	$[M^{+\cdot} - H_2O]$: 226 (4) $[M^{+\cdot} - 2H_2O]$: 208 (6)

[a] Molecular ion species in EI is $M^{+\cdot}$ [b] Methane used as reagent gas in CI mode unless noted otherwise.
[c] In the deoxy series, the ion corresponding to the B + 44 ion is the B + 28 ion.
[d] The relative contribution of $BH \cdot C_2H_5$ and $B \cdot CH_2O$ to the B + 30 ion in the CI mode (methane) are not currently known.

the adenosine-deoxyadenosine example was given in this preliminary account and additional studies are needed to determine the general utility of the method for distinguishing modifications in the sugar function of nucleosides and nucleotides. A similar observation has been made in the FAB mass spectra of nucleosides obtained using only the glycerol matrix (146).

In general, the acquisition of FAB spectra of nucleosides is a relatively straightforward procedure, if sample quantity is not severely limited, and experimental procedures are clearly described in the above referenced papers. However, caution should be used in the interpretation of some spectra because of potential side reactions which may occur. For example, dehalogenation reactions have been observed during the FAB analysis of purine nucleosides substituted in the 2- and 8-positions with chlorine or bromine and pyrimidine nucleosides substituted in the 5-position with iodo, bromo, chloro and fluoro groups (152). The rate of exchange of the halogen follows inversely the carbon-halogen bond strength and proceeds through a radical intermediate. Interaction of the radical site with the matrix, in this case glycerol, affects the hydrogen exchange. Another problem is formation of $[M + 2H]^{+\bullet}$ and $[M + 3H]^{+}$ ions resulting from one- and two-electron reductions during desorption (153). Careful consideration must, therefore, be given to the data before using isotopic ratios to confirm empirical formulas. In addition, if each of the nucleotides in an oligomer undergo reduction, the observed molecular weight of a polynucleotide chain could be significantly overestimated, with errors increasing with the number of residues in the chain.

Although FAB spectra provide important information concerning the molecular weight and mass of the aglycone, the amount of structural information in a normal FAB spectrum is limited relative to the classical EI data. One approach to restoring the structurally informative fragment ions to the FAB spectra has been to perform the analysis on the trimethylsilyl derivatives of the nucleosides (154). Although the derivatization reaction adds an additional step to the analysis, the degree of fragmentation approaches that observed in the EI spectrum with 8 of 10 sugar and 13 of 16 base ion series fragments being observed. A similar, although not as extensive, increase in fragmentation is observed in the FAB spectra of nucleotide-TMS derivatives. Of particular advantage, this approach can be used without the application of heat to the sample and the method should be of considerable value for the analysis of particularly fragile nucleosides of unknown structure.

In spite of the above mentioned problems, the application of FAB mass spectrometry to the structure elucidation of nucleosides of unknown structure has been widely reported. Examples include the structure elucidation of the nucleoside antibiotics ascamycin (155) guanine 7-*N*-oxide (156) and polyoxins (157), for monitoring the production of toyocamycin (158) in a fermentation broth, identification of the adducts of pyrrolizidine alkaloids with nucleosides and nucleotides (159, 160) and in confirmation of the structure of ribosyl-diphthamide (161).

3.5 Other ionization modes for the analysis of nucleosides

In addition to the standard EI, CI, and the recently introduced FAB methods of ionization, a number of other techniques have been developed for the mass spectral analysis of nucleosides. However, these methods do not appear to have significant potential for application to clinical problems, at least at their present stage of development, and are therefore mentioned only for completeness. Field desorption (FD) has been used to provide mass spectra of thirteen nucleoside antibiotics (162). In most cases the molecular ion or quasimolecular ion (MH^+) is the base peak of the spectrum with little other fragmentation being observed. The secondary ion mass spectra (SIMS) of a number of nucleic acid bases, nucleosides and nucleotides have been reported (163, 164). The spectra obtained using SIMS are characterized by intense MH^+ and MAg^+ ions with only a few fragment ions being formed. A number of reports have described the use of desorption chemical ionization (DCI) (165–167) for the production of mass spectra of nucleosides intractable to standard EI or CI methods, and considerations in optimizing the operating conditions for DCI mass spectrometry have been described (168). In addition to the reagent gas, the major variables of concern are the configuration and heating rate of the emitter wire and the temperature of the source; the former conditions determine the desorption efficiency while the latter plays a dominent role in the extent of fragmentation. The desired information from the spectrum thus determines the conditions to be used in the experiment. Optimum conditions for MH^+ formation can be achieved by using a fast heating rate with a low source temperature while increasing the source temperature will increase fragmentation. Under optimum conditions, a 20 ng sample of sucrose, the standard test compound, produced a clear, reproducible mass spectrum. A technique closely related to DCI, in-beam electron ionization, has also been described for the analysis of nucleosides (169–171). In contrast to DCI spectra, however, fragmentation is more extensive, with additional structural information being available. Plasma desorption (PD) has been used to elucidate the structure of adenomycin (172) and a comparison of the pulsed laser and PD spectra of nucleotidic compounds showed no striking differences (173). Finally, electro-hydrodynamic ionization has been applied to the analysis of purine and pyrimidine bases, nucleosides and nucleotides (174).

4 Nucleotides

4.1 Quantitative analyses

Applications of mass spectrometry in studying the biochemical and clinical properties of nucleotides have, until recently, been limited by the volatility of this class of compounds. Although permethylation affords derivatives sufficiently volatile for gas chromatographic separation of some members of this class (175, 176), most investigators studying nucleotide biochemistry have relied on HPLC or radiometric analytical methods. A few exceptions do exist but the methods depend, in general, on the hydrolysis of the nucleotide to a nucleoside or base for quantitation by mass spectrometry. For example, the nucleotide anabolites of 6-mercaptopurine (**12**) have been analyzed in blood samples following separation of the nucleotide from the nucleoside fraction by XAD column chromatography. The nucleotide fraction was then dephosphorylated enzymatically to nucleosides and, following permethylation, analyzed by

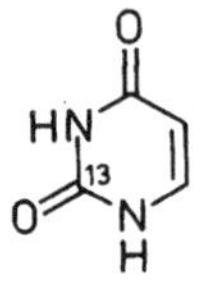

60 Uracil-[C2-^{13}C]

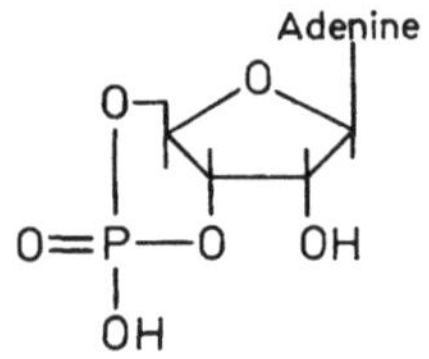

61 cAMP

GC-MS (131). Using this procedure, blood levels of the monophosphate anabolites of **12** and **13** have been measured.

An elegant GC-MS method for measurement of the flux through the *de novo* pyrimidine biosynthetic pathway using ^{13}C labeled bicarbonate and CO_2 has been described and shown to be applicable to both *in vitro* and *in vivo* studies (177). The *in vitro* experiments were conducted by culturing L1210 cells in an appropriate medium containing $NaH^{13}CO_3$ which produced carbamyl phosphate labeled with ^{13}C. Subsequent anabolic steps then yielded uracil nucleotides labeled in the C2 position with ^{13}C, e. g. **60**. Initial attempts to quantitate the flux through the *de novo* pathway by enzymatic hydrolysis of the uracil nucleotides to uridine proved unsuccessful but the enzymatic conversion of uridine to uracil provided the required sensitivity. Levels of ^{13}C incorporation as low as 2% can be measured with confidence by this method with good precision. Parallel experiments using ^{14}C labeling methods indicated the techniques were not statistically different. Modification of the procedure allowed the *in vivo* quantitation of $^{13}CO_2$ incorporated into the uracil nucleotide pool of L1210 tumors in mice. Changes in the flux through the *de novo* pathway were evident in animals pretreated with known inhibitors of the pathway relative to control animals. A number of advantages of the stable isotope dilution method over radioisotope and other instrumental procedures were evident from this work. Relative to the radioisotope methods, the stable isotopes may be used with 100% enrichment without posing a biological hazard, except with deuterium and, thus, may be used in human pharmacological studies after obtaining proper clearance. The use of highly enriched precursors minimizes the problem of dilution which occurs during *in vivo* studies. The use of stable isotopes of N, O and C permits the determination of changes in the *de novo* pathway by following small precursors for which no radioisotopes are available. For example, the above studies are being repeated with ^{15}N labeled glutamine to follow the turnover and size of bicarbonate and glutamine pools. Results may permit differentiation of the respective contributions of salvage versus *de novo* synthesis in the formation of pyrimidines *in vivo* and to quantitate the effects of drugs designed as enzyme inhibitors on the *de novo* pathway.

The direct identification and quantitation of adenosine-3′,5′-cyclic monophosphate (cAMP) (**61**) by mass spectrometry in samples obtained from both animal (178) and plant (179–181) sources has been reported. A specific assay for cAMP based on stable isotope dilution EI-GC-MS methods sensitive to the 3 pmol level has been developed and compared with standard protein-binding and radioimmunoassay (RIA) methods (178). Comparison of the levels of cAMP determined by each of the three methods showed good correlation and established the feasibility of GC-MS in per-

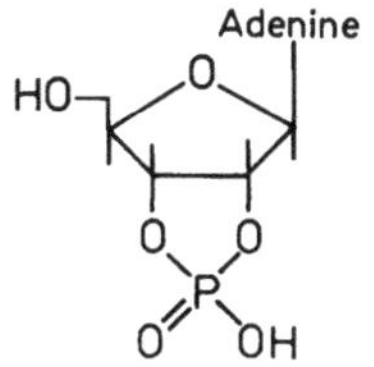

62 2′,3′-cAMP

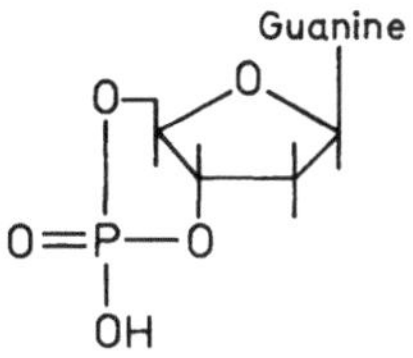

63 2′-Deoxy-cGMP

forming these analyses. The biological methods had an advantage in terms of analyzing multiple samples and provided a greater sensitivity. However, the GC-MS method had the distinct advantage of direct sample identification with the possible extension to other cyclic nucleotides without the necessity of preparing specific antibodies. The importance of the ability to directly characterize a product of unknown structure was illustrated by the appearance of an artifact derived from the biological sample which had a retention time only slightly longer than the cAMP-$(TMS)_3$ analyte but with a different mass spectrum (178). Possible structures suggested by the isolation procedure and mass spectrum for this unknown component were 2′,3′-cAMP (**62**) and 2′-deoxy-3′,5′-cyclic guanosine monophosphate (**63**).

Controversy has existed as to whether or not cAMP is present in the tissues of plants, since the methods used to detect this compound are subject to interference from other endogenous compounds. Unambiguous identification of cAMP in plant tissue was obtained by column chromatographic isolation and mass spectrometric (179–181) identification of cAMP as its TMS derivative (179). Quantitation of cAMP in plant tissue has been achieved using cAMP-[2,8-$^2H_2$,6-^{15}N] as the internal standard (180).

4.2 Fast atom bombardment mass spectrometry of nucleotides

No other ionization technique has had as great an impact on the mass spectral analysis of mono-, di-, tri- and oligonucleotides as fast atom bombardment. The simplicity of sample preparation, at least relative to FD or PD, the relatively low cost of installation of a FAB gun and the ability to retrofit the needed equipment to existing mass spectrometers all combine to make FAB a technique available to almost any laboratory involved in the analysis of this class of compounds. The utility of FAB in providing mass spectra of underivatized nucleotides has been illustrated by the analysis of samples with various structural modifications and experimental conditions (182–186) and this new field has been recently reviewed (146). The results of these studies may be summarized in the following statements: 1. Free nucleotides and various nucleotide salts are amenable to analysis with FAB; 2. Both positive and negative ion detection provide molecular weight and structurally significant fragments. Positive ion spectra are more complex than negative ion spectra, with the methods being complementary; 3. Structural isomers may be differentiated, at least in some cases; 4. Nucleoside di- and triphosphates may also be analyzed using FAB but the diphosphates are best analysed in the negative ion mode and the triphosphates may require the addition of

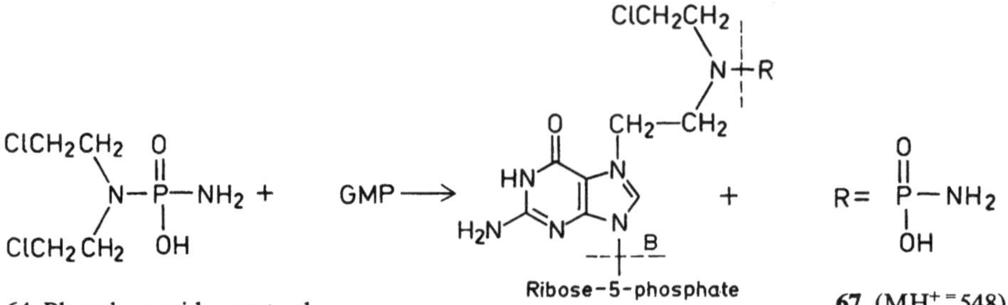

64 Phosphoramide mustard

66 R=H
(MH^+ = 469)
(G=284)
(BH=257)
(BH−HCl=221)

67 (MH^+ = 548)
(M−R=469)
(BH−HCl=300)
(G=284)
(BH−R=257)
(BH−R−HCl=221)

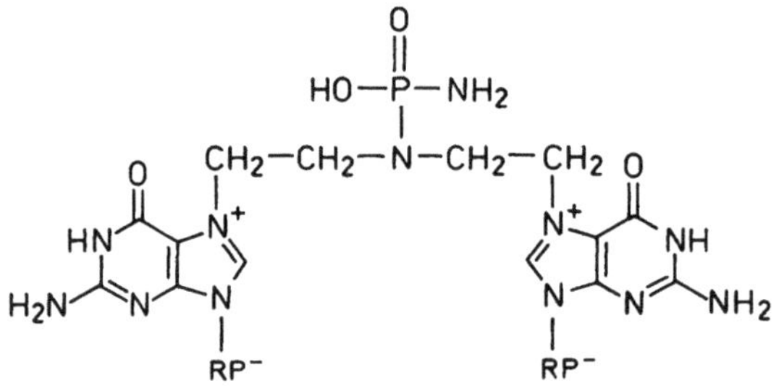

68 $(M+Na)^+$ = 897
$(MH-79)^+$ = 796
(7-Methyl-GMP) = 377
(GMP) = 363

Scheme 5. Reaction of phosphoramide mustard (**64**) with GMP (**65**) to produce **66**–**68**. Major ions of structural importance are indicated.

dilute HCl before usable spectra are produced; 5. Quantitation with FAB requires the use of stable isotope labeled internal standards; 6. Results of FAB analysis are reproducible between laboratories.

Applications of FAB to clinically oriented problems are similar to those described in the section of FAB of nucleosides, i. e., the major power of FAB is evidenced in structure elucidation.

The interaction of alkylating agents with components of RNA and DNA, with resultant formation of covalent bonds, is presumed to be responsible for the cytotoxicity and chemotherapeutic activity of these agents. As an example, the interaction of phosphoramide mustard (**64**), a metabolite of the antitumor agent cyclophosphamide, has been reported to form cross-links between DNA strands but, until recently, no structural information concerning the identity of the products has been available. An examination of the products formed in the reaction of **64** with guanosine 5′-monophosphate (**65**) (GMP) has resulted in the identification of two monomeric species,

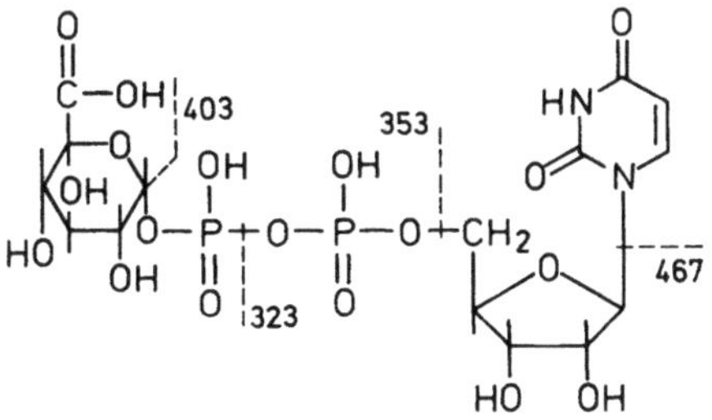

69 Uridine-5′-diphosphoglucuronic acid

N-(2-chloroethyl)-*N*-[2-(5′-phospho-7-guanosinyl)ethyl]phosphorodiamidic acid (**66**) and *N*-(2-chloroethyl)-*N*-[2-(5′-phospho-7-guanosinyl)ethyl] amine (**67**), and a cross-linked dimer, *N,N*-bis-[2-(5′-phospho-7-guanosinyl)-ethyl]phosphorodiamidic acid (**68**) (187). Structures were assigned on the basis of UV data, ^{31}P NMR, and mass spectrometry of the samples following HPLC isolation. Initial attempts to obtain spectra using FD were unsuccessful (188) and FAB was therefore used. The two monomeric species showed similar spectra with MH^+ ions providing molecular weight information, including the chlorine isotope peak, base related fragments (B + H and/or B + H − HCl) and an ion at *m/z* 284 representing a guanosine + H ion formed by elimination of the *N*7 alkyl sidechain and phosphate moiety. The spectrum of **67** also displayed MH − 79, BH − 79 and BH − 79 − HCl fragments indicating cleavage of the N − P bond of the side chain. The FAB spectrum of the dimer produced an ion at *m/z* 897 (M + Na) with fragment ions at MH − 79 (*m/z* 796), GMP + H (*m/z* 363) and 7-methyl-GMP (*m/z* 377).

The utility of FAB-MS in following enzymatic reactions in conjunction with stable isotope labeling has been described during development of a general method for the synthesis of labeled glucuronides (189). Optimum conditions for the acid catalyzed ^{18}O exchange reaction of the enzyme cofactor uridine-5′-diphosphoglucuronic acid (**69**) were determined using FAB to follow the course of stable isotope incorporation. By comparing the negative ion spectra obtained on unlabeled **69** with samples exposed to various exchange conditions, both the site and extent of ^{18}O exchange could be determined by examination of the ions shown in structure **69**. Thus, those ions associated with the presence of the glucuronic acid moiety (*m/z* 579, 467 and 353) show a mass shift indicative of the number and level of ^{18}O atoms exchanged while those ions (*m/z* 403 and 323) lacking the carboxylic acid function did not change. Mass spectral analysis of products obtained under mild acid conditions indicated the preferential incorporation of two ^{18}O atoms into the carboxylate group with 12% unlabeled material, 44% monolabeled and 44% dilabeled product. High concentrations of acid increased the amount and number of labels but produced substantial hydrolysis of the sample. To evaluate the extent of label lost during transglucuronylation, *p*-nitrophenol was incubated with a crude preparation of hepatic uridine diphosphoglucuronyl transferase in the presence of labeled cofactor **69**. Results indicated a less than 10% loss of label occurred in this enzymatic transformation and, thus, the method should provide a means of preparing ^{18}O labeled glucuronides.

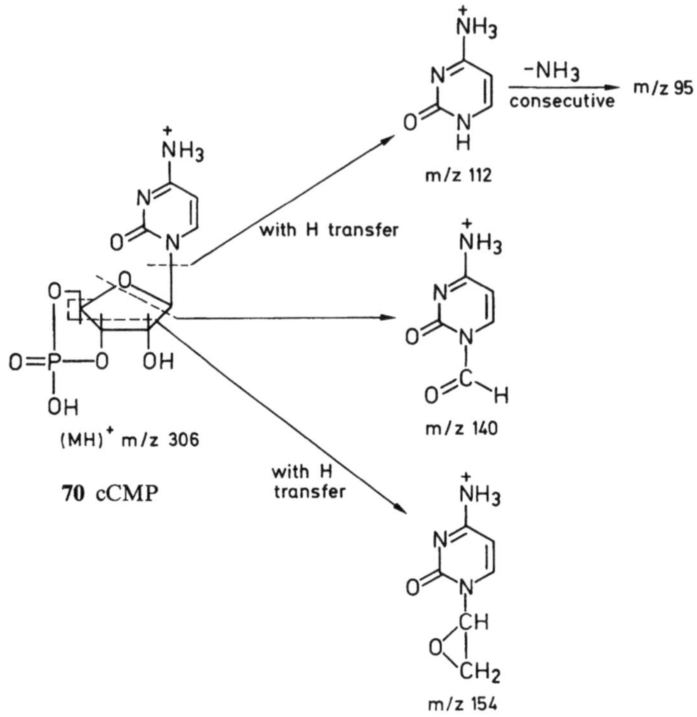

Scheme 6. Major ions in the FAB-CID mass spectrum of cCMP (**70**). (Adapted from ref. 193.)

FAB-MS has also been used to determine the stereochemical course of the yeast phenylalanyl-tRNA synthetase reaction with isotopically labeled substrated (190), for differentiation of isotopically labeled nucleotides in combination with tandem mass spectrometry (191) and to detect impurities in synthetic samples of nucleotide 3′-phosphoramidites (192).

The long-lived ions produced by FAB permit the application of other mass spectrometric techniques including high resolution and CAD of MH^+ ions of nucleotides. The latter technique has been used to differentiate cytidine 3′,5′-cyclic monophosphate (cCMP) (**70**) from the 2′,3′-isomer **71** and guanosine 3′,5′-cyclic monophosphate (cGMP) (**72**) from its 2′,3′-analog **73** isolated from biological sources (193, 194). Under normal operating conditions each of the samples provided the MH^+ ion as the sole ion in the spectrum and the samples could not be distinguished. Using CAD coupled with mass-analysed ion kinetic energy spectra (MIKES) however, significant differences were noted for the isomers which permitted the identification of 3′,5′-cyclic nucleotides from their 2′,3′-analogs. In the case of **70**, major ions appear at *m/z* 154, 140 and 112. The *m/z* 112 ion is the base peak in the spectra of both isomers **70** and **71** and arises from glycosidic bond cleavage. The ions at *m/z* 154 and 140 represent cleavages across the carbohydrate ring to form the commonly observed B + 44 and B + 30 ions. In contrast, the *m/z* 140 ion is weak and the *m/z* 154 ion is absent in the spectrum of **71**. An ion at *m/z* 178, suggested to form by the consecutive loss of the phosphate group

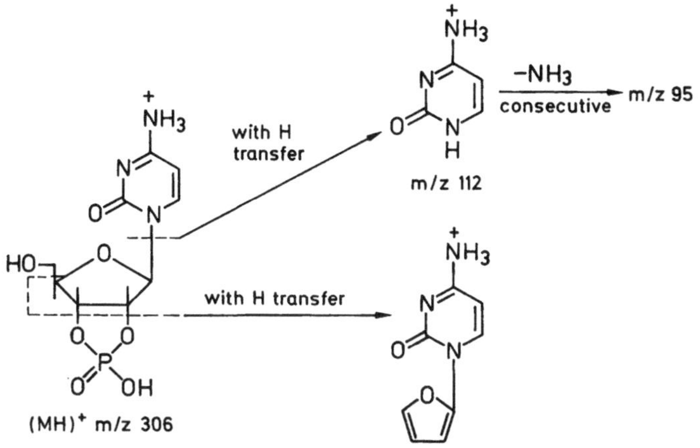

71 2′,3′-cGMP

Scheme 7. Major ions in the FAB-CID mass spectrum of 2′,3′-cCMP (**71**). (Adapted from ref. 193.)

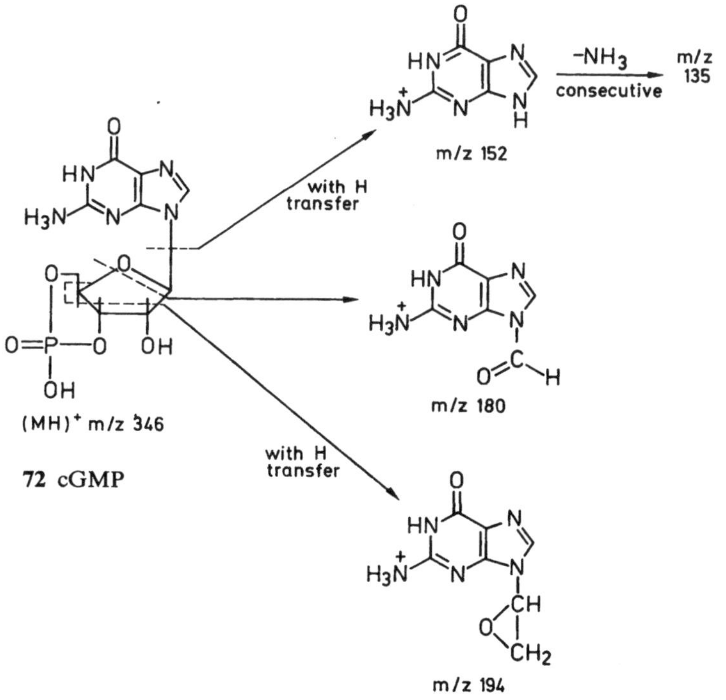

72 cGMP

Scheme 8. Major ions in the FAB-CID mass spectrum of cGMP (**72**). (Adapted from ref. 193.)

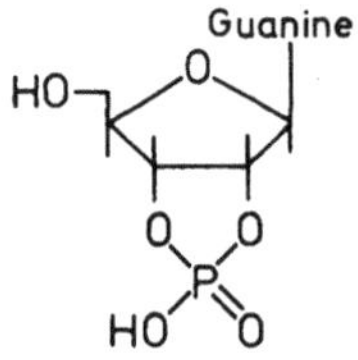

73 2′,3′-cGMP

and methanol, is present in the spectrum of **71** but is absent in **70**. The CAD MIKE spectrum of **72** was similar to cCMP and cAMP in displaying prominent B + 44, B + 30 and B + 2H ions. The 2′,3′ analog **73** was distinctly different in that the B + 30 ion is of significantly lower intensity and the B + 44 ion is virtually absent. Three new ions, *m*/*z* 195, 214, and 218, appear in the spectrum of **73** that are not present in the spectrum of **72**, further serving to distinguish **72** from **73**. The ion at *m*/*z* 195 is possibly formed from MH^+ by glycosidic bond cleavage with transfer of a hydrogen to the base and charge retention by the sugar. The other new ions in the spectrum of **73**, *m*/*z* 218 and 214, are not readily explained. However, the differences in the spectra of the GMP isomers are evident and may be used to differentiate the isomers. Using the differences noted above, the presence of **70** in rat liver and **72** in plant tissues was demonstrated. Attempts to quantitate the levels of **70** and **72** in biological extracts were not successful due to the unavailability of the necessary internal standards.

As was observed in the case of the nucleosides, the nucleotides may undergo modification by interaction with the matrix to form side products which may interfere with interpretation of the spectra. Examples noted in the analysis of nucleoside cyclic phosphates and related compounds include the exchange of a halogen by hydrogen, substitution of an azido group by an amine and elimination of a methoxy function with replacement by hydroxyl (195). Considering the number of side reactions observed in the FAB spectra of both nucleosides and nucleotides, caution is again urged in the interpretation of an unknown sample.

An extensive study of the FAB spectra of dinucleotides obtained using a tandem mass spectrometer concluded that metastable and CAD spectra provide more information than the normal FAB spectrum (196). Following metastable decomposition or collisional activation, the $(M-H)^-$ ions of all dinucleotides studied decompose by preferential elimination of a BH fragment from the 3′- rather than the 5′-terminus, thus permitting differentiation of isomeric dinucleotides. In addition to presenting the major features in the mass spectra of a considerable number of mono- and dinucleotides, the discussion of fragmentation mechanisms makes this an important paper in the field.

4.3 Other ionization methods

Ionization methods, other than FAB, of particular interest in the analysis of nucleotides include plasma desorption (PD), field desorption and DCI. The utility of PD is described in a later section. FD continues to be of considerable value in the characterization of nucleotides and smaller polynucleotide units. The most valuable information provided by FD is the production of intense MH^+ and (M + cation) ions which establish molecular weights, although in some instances fragmentation may be enhanced, e. g., with laser-assisted desorption as was recently reported (197) in a comparative study of FD and FAB of pyrimidine nucleotides and nucleotide triphosphates. The general conclusions of this work were that FD provides good molecular weight information along with structurally informative fragment ions, without the complications of a matrix necessary for FAB analysis. In addition, sample size for FD were reported to be an order of magnitude less than required in FAB analysis of the same samples. In response to the low sensitivity observed using liquid SIMS, another term used to describe FAB, it was pointed out that the spectra described above were acquired using positive ion detection, rather than the preferred negative ion mode. Additional work (198) showed approximately equal sensitivity of liquid SIMS to the FD results mentioned earlier. The question of sensitivity of FAB versus other ionization methods remains to be settled.

DCI in conjunction with tandem mass spectrometry has also been used to characterize and quantitate the dinucleotide thymidyl(3′,5′)thymidine methyl phosphotriester (199). The spectrum obtained using ammonia as the reagent gas with normal scanning of the mass spectrometer displayed ions assigned as being MH^+ (*m/z* 561, 8%), $Tp(Me)H^+$ (*m/z* 337), BH^+ (*m/z* 127, 100%) and ammonium adducts of these ions. Because of the symmetry of the molecule, a distinction between a 3′- and 5′-cleavage to produce the *m/z* 337 ion could not be made. Similar results were obtained in the negative ion mode, but the spectrum was not as intense, although a number of artifact ions were not observed as was the case in the positive ion mode. Tandem mass spectrometry showed considerably more fragmentation with the most intense daughter ion being observed at *m/z* 225, corresponding to protonated thymidine with the loss of water. A major advantage of DCI for the analysis of nucleotides appeared to be a significant gain in sensitivity with a detection limit of around 5 ng. Additional work is warranted in the use of DCI for the identification and quantitation of nucleosides and nucleotides. Recent applications of FD include the identification of intermediates used in the preparation of oligonucleotide chains (200) and for the differentiation of 5′- from 3′- and 2′-monophosphates (201). Attempts to obtain sequence information of dinucleotide phosphates or higher units have not been successful (202). Reviews describing the use of FD in the analysis of nucleotides have recently appeared (203, 204).

5 Macromolecular Modifications

The considerable power of mass spectrometry in elucidating the structure of modified nucleic acid components has been brought to bear on the analysis of products formed, or suspected to form, at the macromolecular level upon exposure of DNA, or RNA, to various carcinogens, UV light, and other agents. Considering the low incidence in the number of modifications occurring in DNA, estimated to occur in one base in 10^5

to 10^6 base pairs (205), and, consequently, the small amount of sample available for analysis, the success, or partial success, of these investigations is noteworthy.

The considerable interest in the structure of these adducts derives from the mutation theory of the origin of cancer (206). Alteration in DNA structure by the formation of covalent bonds between the ultimate carcinogen and base residues in DNA are, thus, thought to be the initiating event in the development of a neoplasm. Knowledge of the structure of the adduct will, therefore, reveal the biochemical pathways operating in transforming a chemical into an ultimate carcinogen. A number of reviews have addressed various aspects of this area (205–209).

5.1 Adducts formed with monomeric units

Investigations into the mechanism of action of the antibiotic neothramycin indicated formation of a 1:1 complex of the antibiotic with guanine base residues in double stranded DNA was the most likely event leading to antitumor activity (210). To test this hypothesis, the antibiotic was reacted with 2′-deoxyguanosine in aqueous dimethyl sulfoxide and the products isolated by TLC. Two peaks at *m*/*z* 511 and 534 in the FD spectrum were assigned to the $M^{+\bullet}$ and $[M + Na]^+$ ions and substantiated the formation of a 1:1 complex. Based on the MS and NMR data, the structure **74** was assigned to the adduct.

FAB-MS has been used to determine the structure of the product formed between cisplatin and deoxyguanosine (211). Attempts to obtain MS data using desorption CI and FD were unsuccessful but both positive and negative ion FAB spectra provided significant fragment ions of value in assigning the structure **75** to this complex. Ion clusters due to the presence of stable isotopes of chlorine and platinum proved especially useful in assigning structures to the fragment ions present in both the positive and negative modes. Major ions in the positive ion mode arose from the sequential loss of Cl and HCl from the MH^+ ion. Additional elimination of NH_3, anhydroribose and guanosine residues, in various combinations, lend additional support for the assigned structure. The identification of adducts formed in the reaction of [Pt(dien)Cl]Cl with model compounds and DNA has also been described (212).

Considerable interest has been evidenced in the identification of base pairs involved in the formation of cross-links of DNA strands exposed to various mutagens since

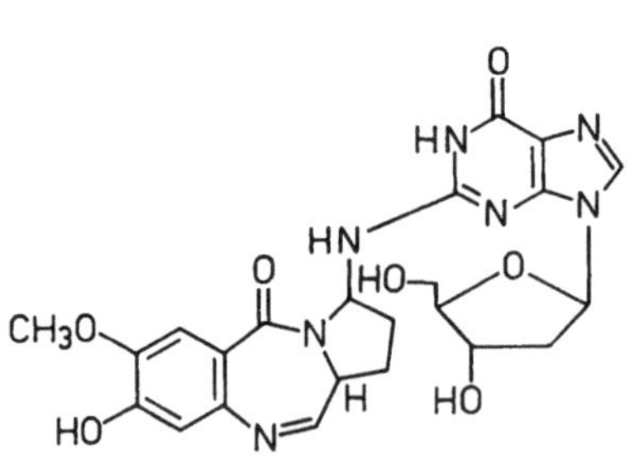

74 Neothramycin-2′-deoxyguanosine adduct

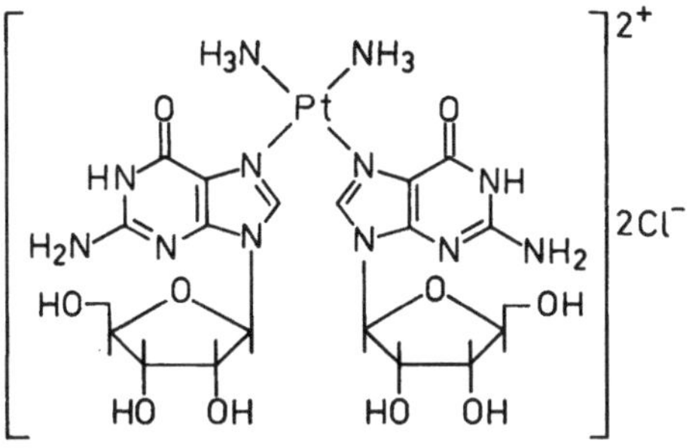

75 *cis*-Diamino-bis-guanosine platinum (II) chloride

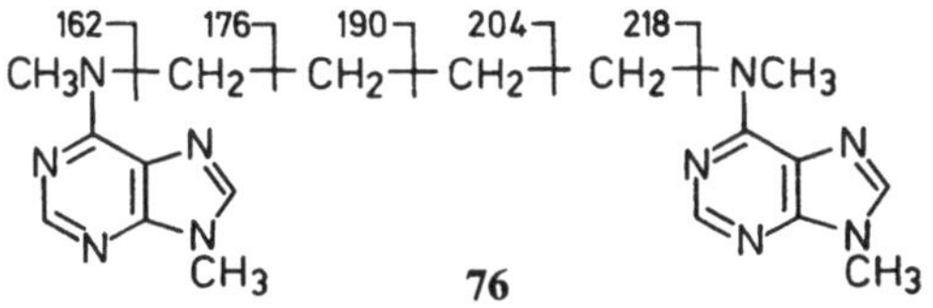

76

Scheme 9. Major fragments in the EI mass spectra of methylene bridged adenine **76**. (Adapted from ref. 213.)

one such event, if unrepaired, can interfere with cell division. The EI mass spectra of a series of adenine and adenosine cross-linked dimers having the general structure **76**, where $n = 1-4$, have therefore been examined in order to establish fragmentation patterns useful in characterizing products which may subsequently be isolated from DNA (213). Major fragment ions in the base series arise from cleavages of the methylene bridge sidechain, illustrated above for the $n = 4$ analog, and the related loss of one or two hydrogen atoms. The predominent ion in this series was the m/z 176 fragment which may be stabilized by the N^6 amino group. Unusual fragments which depend on the presence of both purine rings for their formation included ions at m/z 217 and 163 suggested to form through a McLafferty rearrangement, with retention of charge by both possible fragments. Other ions in the spectra of these samples were formed by loss of a methyl radical from m/z 163 and subsequent degradation of the purine ring. The spectra of the nucleoside analogs were, in general, the same as was observed in the base series with the exception of a mass shift as a result of the sugar moiety. Other cross-linked nucleic acid bases have been characterized using mass spectrometry of DNA preparations treated with nitrous acid (214) or *N,N'*-bis(2-chloroethyl)-*N*-nitrosourea (215).

The selective reaction of the antitumor agent BCNU, 1,3-bis(2-chloroethyl)-1-nitrosourea, with poly(C) and poly(G), but not poly(A) or poly(U), has been reported and the products characterized as their TMS derivatives (216) by mass spectrometry. Two products were isolated from the poly(C) mixture following chemical and enzymatic cleavage to the nucleoside level and were assigned structures **77** and **78**. Mass spectral evidence for the structures included appropriate mass shifts, relative to cytidine, for

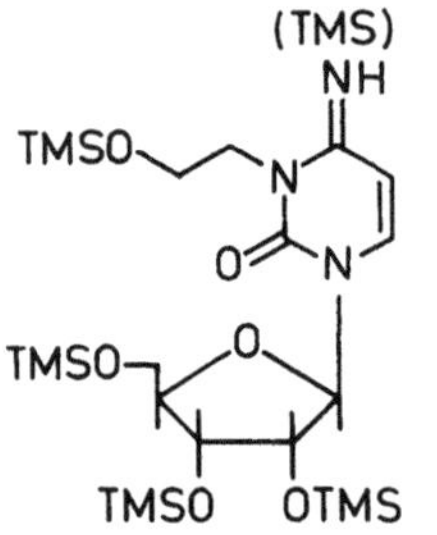

77 3-β-Hydroxyethylcytidine (TMS derivative)

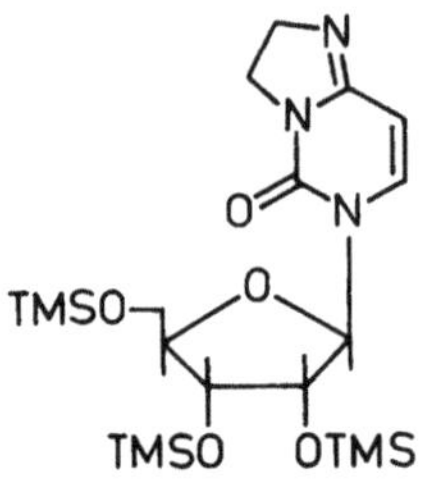

78 3,N^4-Ethenocytidine (TMS derivative)

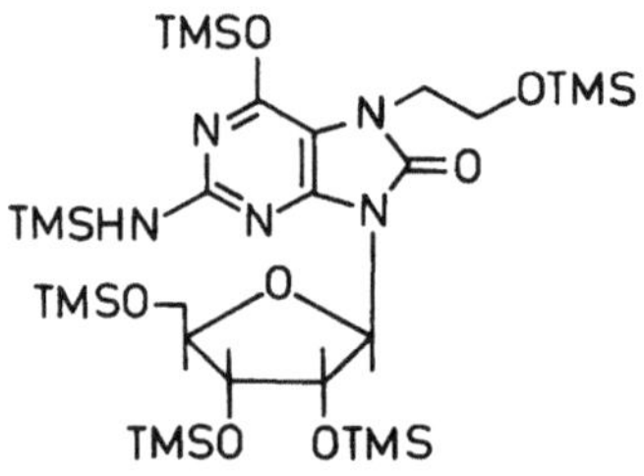

79 7-β-Hydroxyethyl-8-oxoguanosine (TMS derivative)

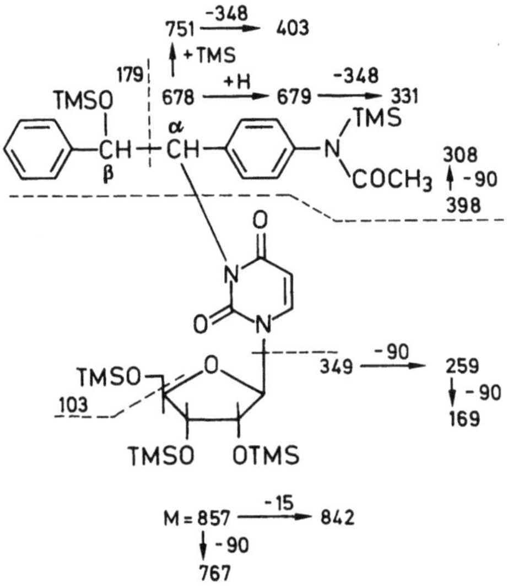

80 1-(4-Acetamidophenyl)-1-(3-uridyl)-2-hydroxy-2-phenylethane

Scheme 10. Structure and major fragment ions in the EI mass spectrum of **80**. (Adapted from ref. 217.)

the presence of one additional (**77**) and one less (**78**) TMS group, the absence of a strong B + 41 ion in the spectra of both compounds indicating substitution at the *N*3 position, the appearance of an ion at M − 43 in the spectrum of **77** for the loss of the sidechain and the absence of an ion at M − 43 in the spectrum of **78**, strongly suggesting a cyclic structure. The appearance of base related ions confirmed that all modifications had occurred in the aglycone portion of the molecules. The single product isolated from the poly(G) reaction was assigned structure **79**. Based on the structures of the products isolated in the above reactions, the suggestion was made that BCNU provides two reactive carbon fragments, possibly chloroethyl carbonium ions, which react with the nucleophilic sites of the cytosine and guanine residues of DNA or RNA.

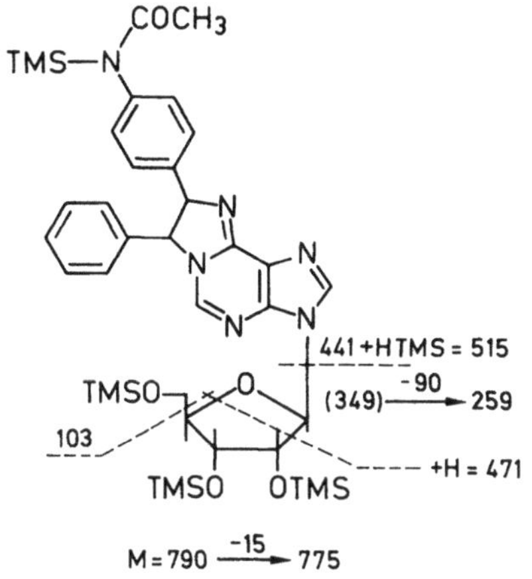

81 3-(β-D-Ribosyl)-7-phenyl-8-(4-acetamidophenyl)-7,8-dihydro-imidazo [2,1-i]purine

Scheme 11. Structure and fragmentation of **81**. (Adapted from ref. 217.)

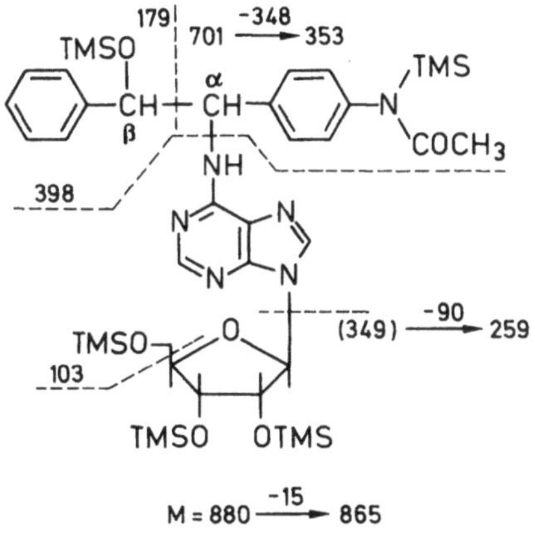

82 1-(4-Acetamidophenyl)-1-(N^6-adenosyl)-2-hydroxy-2-phenylethane

Scheme 12. Structure and fragmentation of **82**. (Adapted from ref. 217.)

The major components formed in the reaction of cytidine, adenosine and guanosine with the carcinogen *N*-acetoxy-4-acetamidostilbene(*N*-AcO-AAS) have been examined using EI-MS (217–219). The major product in the cytidine reaction has been assigned the deaminated structure **80**. Two major products, **81** and **82**, are formed in the condensation of the carcinogen with adenosine. Only one major component was obtained from the guanosine reaction, **83**, but a number of unstable, minor products were also produced. None of the guanosine adducts, however, possessed properties expected if coupling had occurred in the $N7$, $C8$ or N^2 positions, suggesting the $N1$ and O^6 positions as the only sites of modification under these conditions. The product

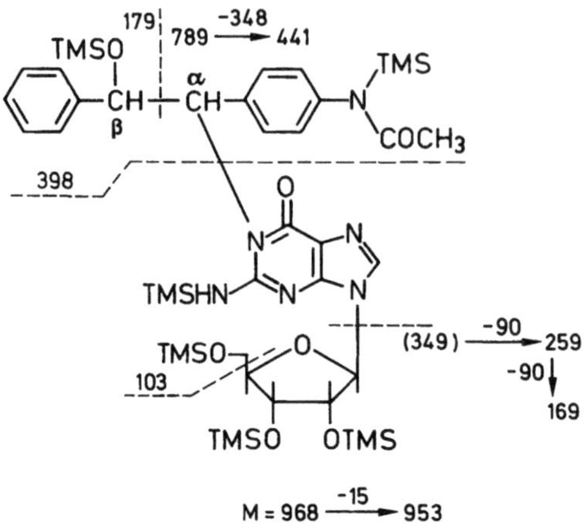

83 1-(4-Acetamidophenyl)-1-(1-guanosyl)-2-hydroxy-2-phenylethane

Scheme 13. Structure and fragmentation of **83**. (Adapted from ref. 217.)

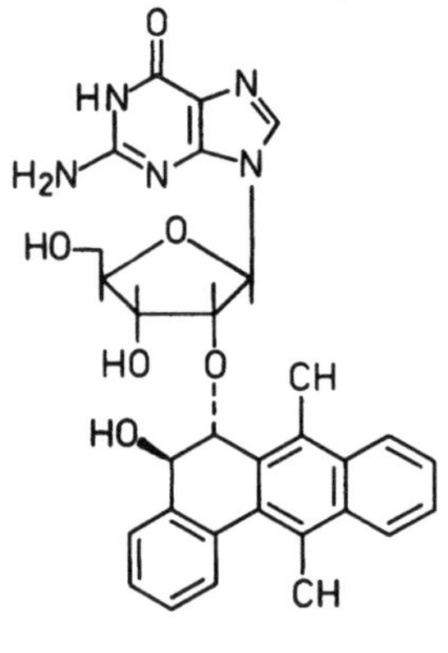

84 7,12-Dimethylbenz[a]anthracene-guanosine adduct

structures strongly suggested an *N*1 to N^6 rearrangement, further indicating that *N*-AcO-AAs behaves more like a classical alkylating agent than other members of this compound class. Inefficient repair of these *N*-alkylated products may therefore be the event leading to mutagenesis.

A number of other carcinogen or suspected carcinogen-nucleic acid base adducts have been examined with mass spectrometry at the nucleoside or polynucleotide stage. Examples include identification of the products formed in the reaction of α-acetoxy-*N*-nitrosopyrrolidene (220), *N*-nitroso-2-hydroxymorpholine (221), *N*-acetyl-2-aminofluorene (222), anthracene-oxides (223, 224) and benzo[a]pyrene analogs (225–228) with either guanine or adenine containing nucleic acid species. In all cases, adduct formation occurs in the aglycone, with the exception of structure **84** (223).

5.2 Direct modification of DNA and RNA

Most work at the macromolecular level has been performed using intact DNA to study the formation of adducts with various toxins, carcinogens or other chemicals and drugs. Studies involving RNA have been sparse with the exception of the structure elucidation of modified nucleosides and a few reports of the incorporation of modified nucleosides or bases into RNA, e. g. benzimidazole nucleoside (229) and the natural hypermodified base queuine (65, 230).

A variety of mass spectral methods have been employed in the structure determination of adducts formed in reactions of DNA with aflatoxin B1 (231), 1,2:5,6-dianhydrogalactitol (232), 5-(3-methy-1-triazeno)imidazole-4-carboxamide (233), *N*-methyl-4-amino-azobenzene (234), *N*-hydroxy-2-napthylamine (235), 15,16-dihydro-11-methylcyclopenta[a]phenanthren-17-one (236), benzo[a]pyrene diol epoxide (237), 4-nitroquinoline 1-oxide (238), styrene oxide (239), methyl methane sulfonate (240), *N*-methyl-*N*-nitorsourea (241), mitomycin (242, 243) and a variety of psoralens (244–246). In each of these cases, the experimental procedure was essentially the same and involved the reaction of DNA with the agent under investigation. The modified DNA is then hydrolyzed and the hydrolysis products, e. g., bases, nucleosides or nucleotides, were separated and isolated using some form of chromatography. Mass spectral analysis was then performed on the purified materials. The base residue most commonly found to be modified by the various species mentioned above was guanine. An interesting exception was the pyrimidine-psoralen photoadducts formed in DNA (244–246).

Interest in the structure of the DNA-psoralen adducts stems from the use of psoralens in the treatment of vitiligo and psoriasis, as tools in determining chromatin structure, for mapping DNA-protein interactions and for "freezing" DNA and RNA secondary and tertiary structures (246). The psoralens have also recently been identified as potential mutagens and carcinogens. The *in vitro* treatment of DNA with four psoralens in the presence of UV irradiation resulted in the isolation of diasteromeric pairs of thymidine-psoralen-thymidine diadducts. The structure **85** is assigned to the product formed by treatment of DNA with 4,5',8-trimethylpsoralen. The highest mass ion in the EI spectrum of the TMS derivatized adduct was present at m/z 776, corresponding to cleavage of the cyclobutane ring to give a psoralen thymidine fragment. Other strong ions in the EI spectrum had m/z values characteristic of the psoralen group. Attempts to obtain unambiguous molecular mass information, $Mr = 1162$ daltons, using FD-MS were unsuccessful, but ions were present corresponding to the monoadduct (psoralen + thymidine), psoralen, thymidine (m/z 242) and thymine (m/z 126). Removal of the carbohydrate groups with a re-analysis using FD established a molecular mass of the aglycone portion of the dimer as 480 daltons. The aglycone sample was also analyzed using high resolution EI to provide elemental compositions substantiating the proposed structure. The stereochemistry and positions of coupling were determined using other instrumental techniques, primarily NMR. Based on the structures of the products isolated and characterized in this study, the sequence of events leading to the photoinduced cross linking of thymine residues with psoralens in DNA requires intercalation of the psoralen followed by the absorption of two photons for production of the thymidine-psoralen-thymidine dimer.

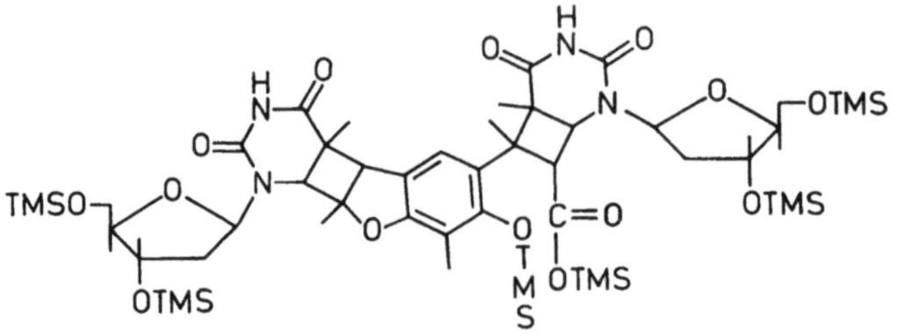

85 Thymidine-psoralen-thymidine photodiadduct

In an excellent series of papers (247–251), Dizdaroglu and co-workers have developed capillary GC-MS techniques for the characterization of ionizing radiation-induced damage to DNA. Following exposure to the radiation, the DNA is hydrolyzed to the nucleosides or bases, the samples converted to either the TMS or *t*-butyldimethylsilyl derivatives, injected onto the capillary column and analyzed by EI. Changes induced to the base, sugar or a combination of both, e. g., 8,5′-cyclo-purine nucleosides, may be identified by the mass spectra. Although the work described in these papers limited discussion to the radiation-induced damage to DNA, the techniques should be applicable to any type of DNA damage, at least if the modified nucleosides are amenable to separation by GC.

6 Quantitative Analysis of Base Residues in DNA and RNA

Characterization of a DNA or RNA molecule requires the identification and quantitation of the base residues contained in the chain and a knowledge of the sequence of the nucleotides in the chain. Until recently, mass spectral methods for the determination of the sequence of a polymeric nucleotide chain had proven to be of little value, even in distinguishing isomeric dinucleotides. A number of approaches, however, have been developed recently for the identification, and in some cases the quantitation, of the bases present in a sample of DNA or RNA.

6.1 Identification and quantitation of bases in DNA or RNA

The classical approach to the identification and quantitation of a specific component known to be present in an oligomer is to purify the large molecular weight entity, hydrolyze the chain to a monomeric unit and separate the monomers using chromatography. An example of this approach is the quantitation of 5-methylcytosine (**86**) in DNA by isotope dilution GC-MS (252). Modifications in the procedure, relative to earlier reports (253–255) include the use of the *t*-butyl-dimethylsilyl derivative and

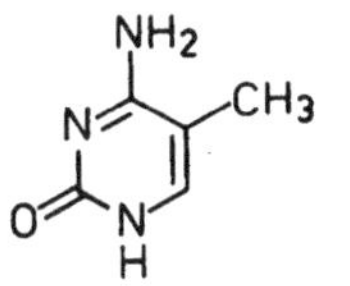

86 5-Methylcytosine (unlabeled)
87 5-Methylcytosine-[C^5-$^2H_3$]

the preparation of 5-($^2H_3$-methyl)cytosine (**87**) for use as the internal standard. Quantitation was achieved by monitoring the *m/z* 299 and 296 ion currents representing the M − 57 ions for the labeled (**87**) and unlabeled (**86**) samples, respectively. The validity of the above method was established by determining the content of **86** in calf thymus, pBR322 and X-174 DNA, and comparing the results to other literature values. The method was shown to be applicable to the analysis of submicrogram samples of DNA in a relatively short time and to be independent of the sequence of the DNA, the latter in contrast to methods based on restriction endonucleases. The quantitation of 1/20 000 bases on a 1 μg sample DNA or to the mmol · g^{-1} level in tens of ng of DNA is estimated to be possible. The method should also be applicable to the quantitation of any base assuming the availability of the stable isotopically labeled internal standards.

A similar approach has been used to identify the site and level of stable isotope incorporated into tRNA of cells grown in the presence of a labeled precursor, ^{13}C-methionine (256). Conditions for the complete hydrolysis of the tRNA samples to the base level with subsequent HPLC separation of the bases permitted mass spectrometry of the purified products.

The extent of replacement of thymine by 5-bromouracil (BU) in DNA upon exposure of cells to 5-bromo-2′-deoxyuridine (BUDR) may be achieved by analyzing as few as 10^5 cells using capillary EI-GC-MS methods (257). DNA is extracted, hydrolyzed enzymatically to the component nucleic acid bases and the bases extracted into an organic solvent. SIM of the M − 15 ions of the TMS derivatized samples of BU (*m/z* 319) and its internal standard, 5-iodouracil (*m/z* 367) and the M + 1 ions of thymine (*m/z* 276) were found to provide optimum results in the quantitation. Because of the small number of cells needed to perform the quantitation, the suggestion is made that this technique may be applicable to clinical studies with patients receiving BUDR as a chemotherapeutic agent. A comparison of the extent of thymine replacement in normal versus tumor cells would be a valuable aid in designing dosage regimens or in evaluating the route of administration of the drug.

6.2 Pyrolysis electron impact mass spectrometry of intact DNA or RNA

An alternative to the classical approach described above is the technique termed pyrolysis-electron impact-mass spectrometry (PYEIMS) which, with variations, has been used for the detection (258 − 260) and quantitative estimation (261 − 262) of bases in DNA and RNA (263). A sample of native DNA (or RNA) is placed on the pyrolysis probe and inserted into the mass spectrometer which, upon thermolytic decomposition of the sample, affords mass spectra permitting the identification of major and minor base residues present in the DNA. Although this approach affords a minimum of sample handling relative to the classical approach, the potential of thermal rearrangements affording spurious ions resulting in a misinterpretation of data cannot be ruled out (264). In addition, the intensity of the ions formed in the degradation of a DNA with one sequence and known base ratio may vary in a second DNA with the same base ratio, but different sequence, i. e., unknown factors cause inconsistent results. Semi-empirical equations have been derived using data obtained on the base ratios observed in the PYEIMS analysis of 22 DNAs of known composition. The utility of PYEIMS in the identification and quantitation of bases of known and unknown structure in DNA or RNA shows promise but is yet to be established.

The analysis of intact DNA samples with DCI in conjunction with pyrolysis of the sample has also been reported (265).

7 Sequencing of Oligonucleotides

No other area in the application of mass spectrometry to the analysis of nucleic acid components illustrates more clearly the impact of technological developments than the recently reported sequencing of oligoribo- and oligodeoxyribonucleotides. The attainment of this long sought goal is primarily a result of the introduction of new ionization methods, although advances in other instrumental components, most especially magnet design, have also been of fundamental value. Two methods have so far shown the capability of providing sequence information from oligonucleotides using mass spectral analysis.

The first report of the sequence determination of a protected oligodeoxyribonucleotide involved ionization using ^{252}Cf-plasma desorption (PD) and a time-of-flight mass analyzer (266). The positive ion mass spectra of four derivatized oligodeoxynucleotides of known sequence were dominated by the $[M + Na]^+$ ion with few fragment ions being formed. In contrast, the negative ion PD spectra provided considerable fragmentation with nested fragments extending from the 3′- or 5′-terminal nucleotide to the other end of the chain. The utility of this technique was illustrated by the successful analysis of a series of protected oligonucleotides of up to 6 residues in length. A number of subsequent reports have expanded the findings of the initial paper but the conclusions remain unaltered (267–269). The largest unit sequenced using PD ionization is a fully protected heptamer with a calculated average mass of 4577.67 daltons. Major fragments observed in the PD mass spectrum of a hexamer are shown below.

Complimentary to the above studies was a report of a detailed analysis of the PD spectra of ten unprotected trideoxyribonucleoside diphosphates (270). The samples examined were designed to answer specific questions concerning the effect of sequence and the identity of the aglycone on the fragmentation of these oligonucleotides. For

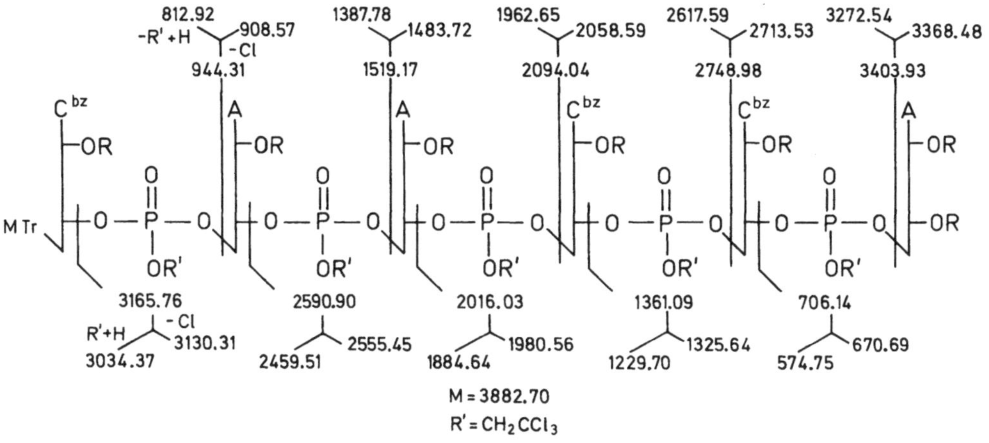

Scheme 14. Major fragments observed in the ^{252}Cf-plasma desorption mass spectrum of a derivatived nucleotide hexamer. (Adapted from ref. 268.)

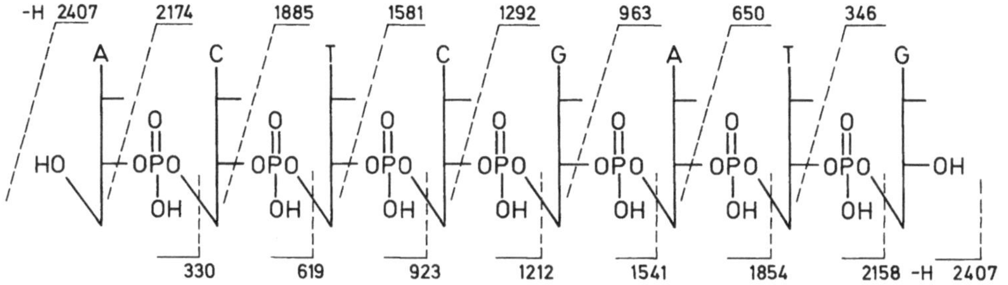

Scheme 15. Major fragments in the negative ion FAB mass spectrum of an underivatized polydeoxynucleotide octamer. (Adapted from ref. 271.)

example, the three trimers GGA, GAG and AGG were prepared to determine the effect of the position of a given base in the oligomer while the trimers GCC, GGT and GGG provided information on the influence of the aglycone. Examination of the mass spectra for ions specific for bases located at the end or internal positions permitted a determination of the base composition and the sequence of the oligomer. The use of a laboratory computer for the analysis of the data facilitated the very rapid determination of the structure of the oligonucleotide.

Fast atom bombardment (FAB) has also shown considerable promise in the sequencing of oligodeoxynucleotides (271 – 274). Results are similar to those obtained using PD, with fragmentation being significantly more informative in the negative ion mode. Degradation occurred from each end of the oligodeoxynucleotide and proceeded sequentially throughout the length of the chain. Fragments containing the 5′-phosphate function may, in general, be distinguished from the 3′-phosphate fragments by a greater intensity, although the validity of this observation has been challenged (275).

The sequencing of oligodeoxynucleotides containing up to 10 residues with a molecular weight of 3024 daltons. (M − H) has been demonstrated (272) as shown below. Based on available literature, a significant advantage of the FAB method, relative to PD, was the ability to perform the analysis on an underivatized sample, although the analysis of blocked samples was also possible (276, 277). Elimination of the need to derivatize has the effect of extending the number of residues contained in a chain which can be accommodated within the working mass range of a given mass spectrometer. Additional advantages of FAB relative to other desorption methods have been mentioned earlier.

Reports of the use of SIMS with fast ions of Cs^+ for the analysis of di- and oligonucleotides have also appeared (278, 279). Molecular weight related ions are observed in both the positive and negative ion modes but the spectra appear weak relative to those produced by FAB or PD. The samples must also be derivatized which limited the number of residues that can be analyzed with a given instrument, although a time-of-flight analyzer was used in these studies. Informative fragment ions were weak, but present, in the positive ion mode and negative ion detection provided little sequence information. The utility of the SIMS method for sequencing oligonucleotides is yet to be demonstrated.

8 Liquid Chromatography-Mass Spectrometry

Although liquid chromatograph-mass spectrometer (LC-MS) interfaces have been available commercially for a number of years and a considerable body of literature exists on the subject (280) few reports of the use of this technique for the analysis of nucleosides or, more importantly nucleotides, from biological samples have appeared until recently, and the full potential of LC-MS, particularly in the analysis of nucleotides and oligonucleotides, is yet to be explored. As has been evidenced by the number of papers describing the separation of nucleosides, nucleotides and oligonucleotides, HPLC is the chromatographic method of choice for the analysis of nucleic acid components.

Three types of interfaces are commonly employed for coupling the liquid chromatograph to the mass spectrometer. The oldest method of performing LC-MS separations depends on deposition of the eluent of the chromatograph onto a moving belt with subsequent heating of the sample to effect volatilization. Initial reports on the separation of a nucleoside mixture using positive or negative ion CI for ionization produced spectra permitting molecular weight assignment (281). The moving belt interface, however, does not produce mass spectra of phosphate containing compounds so cannot be used for the analysis of nucleotides (282, 283), at least at its present stage of development. Results using an improved version of the moving belt were, however, more encouraging (284, 285), as evidenced by the production of MH^+ and B + 2H ions for a sample of 2′-deoxyadenosine-5′-monophosphate. Most exciting, however, was a preliminary report of the combined use of a moving belt with FAB ionization (286); this combination could prove to be extremely important in the analysis of nucleic acid components in a biological or synthetic mixture if sensitivity limits are of an acceptable magnitude.

The applicability of using a direct liquid introduction (DLI) probe for the analysis of nucleosides has been demonstrated (287, 288), although some specific restrictions in the operating conditions exist with the current interface. For example, and in common with the other LC-MS interfaces, the mobile phase must contain only volatile buffers, e. g., ammonium acetate or ammonium formate. Since most mobile phases used in the HPLC analysis of nucleic acid components have historically used non-volatile buffers, some re-education will be necessary for potential LC-MS users and some work, not involving mass spectrometry, has been reported on the reverse phase separation of nucleosides and nucleotides with volatile buffer systems (289, 290). An additional problem with the DLI interface is the formation of cluster ions by the solvent. Since most nucleosides (nucleotides) contain bases in the 110–350 dalton range, the choice of organic component for the mobile phase is somewhat restricted to methanol and acetonitrile. Even with these solvents, the mass range below m/z 130 is contaminated with artifact peaks unless a small desolvation chamber is mounted prior to the ion source. In spite of these limitations, DLI has been successful in the analysis of nucleosides in the CI mode using the organic component in the mobile phase as the reagent gas (287). Mass spectra of the 8 nucleosides and 7 nucleic acid bases examined produced intense MH^+ ions and, with inosine and adenosine, strong M + 2H ions. The amount of sample required for analysis was very high with 25–30 µg of sample being injected onto the column. Lower detection limits have been achieved by modifications to the interface (288) and the analysis of nucleosides in human urine has been reported (291).

The nucleoside fraction of a 20 ml urine sample was isolated, concentrated and injected onto a microbore, reverse phase column with a desolvation chamber located between the exit of the column and the ion source (291). The mobile phase was methanol: water (80 : 20) or 0.01 mol $\cdot$ l^{-1} ammonium formate : methanol (97 : 3) and ionization was performed using EI or CI (NH_3). The mass spectra of reference samples were characterized by a strong MH^+ ion, an S^+ ion of considerable intensity, which indicated the presence of either ribose (m/z 133) or deoxyribose (m/z 117) unit, and ions associated with the aglycone at masses corresponding to B, B + 30, B + 44 (ribosides) and B + 44 – 16 (deoxyribosides). Thus, information concerning the identity of both the base and sugar residues are present in most of the spectra. Analysis of the urinary nucleoside fraction permitted the detection of pseudouridine and a component suspected as being 5,6-dihydrouridine. Other nucleosides known to be in urine, but present in lower quantities than pseudouridine, were not detected. Better chromatographic resolution and lower detection limits are needed before the DLI technique can be applied to a thorough analysis of nucleosides in urine.

The most recent development in interfacing a liquid chromatograph to a mass spectrometer is the thermospray technique (292–294) which, in addition to serving as an interface, also provides a new ionization method. The latter aspect is important since, with the exception of the FAB moving belt system, other standard ionization methods are not applicable to nucleotide analysis because of volatility limitations. The spectra produced by thermospray are reported to be similar to those produced by FAB, and spectra of both nucleosides and nucleotides have appeared in the literature.

Thermospray LC-MS has been used in the analysis of nucleosides obtained from the enzymatic hydrolysis of nucleic acids (295). Sensitivity limits for most nucleosides are approximately 0.1 – 1 ng per component using SIM and 10 – 50 ng for full mass spectra. This sensitivity permits the detection of modified nucleosides in enzymatic hydrolysates of 2.5 μg of tRNA. Other examples of the application of thermospray LC-MS to the analysis of nucleosides and nucleotides include the structure determination of a new fluorescent nucleoside from archaebacterial tRNA (296) and the separation and identification of antitumor drugs or their metabolites, including 5-FU and related compounds (63) and 3-deazauridine (297).

9 Future Potential of Mass Spectrometry for the Analysis of Nucleic Acid Components

The application of mass spectrometry to biochemical and clinical problems involving the analysis of nucleic acid components has only just begun. The recent developments described above have illustrated the potential of mass spectrometry to solving problems of a clinical and biochemical nature. Each of the areas of stable isotope labeling, ionization methods and LC – MS have contributed to the components necessary for the operation of the system.

The most important advances have been the development of FAB and the use of combined techniques, FAB CAD MS/MS and LC – MS, for the analysis of complex biological mixtures containing nucleic acid bases, nucleosides, nucleotides and oligonucleotides. Quantitation using FAB, either alone or with one of the combined techniques, has been illustrated. Sequencing of large oligonucleotides will require enhanced detection of high mass ions and development of methods for increasing the decomposition of large molecules. However, methods currently available may be used to analyse compounds thought impossible only a few years ago. One important aspect of the application of mass spectrometry to clinical and biochemical studies in the future will be recognition of the most relevant areas for investigation and the communication of these by researchers in the biochemical and clinical fields to the laboratory of the mass spectrometrist.

Acknowledgements

Portions of the work described in this chapter were supported by NIH grant CA-24690, CA-42309 and CA-43068 and NSF grant CHE-8105089. My thanks are also extended to Dr. James McCloskey, Debra Slowikowski, Robert Teece and Douglas Strong for reading this chapter and suggesting changes and corrections.

References

1. McCloskey, J. A. (1974) In: Basic Principles in Nucleic Acid Chemistry (Ts'o, P. O. P., ed.) pp. 209–309, Academic Press, New York.
2. Hignite, C. (1980) In: Biochemical Applications of Mass Spectrometry: First Supplementary Volume (Waller, G. R., and Dermer, O. C., eds.) pp. 527–566, Wiley, New York.
3. Hignite, C. (1972) In: Biochemical Applications of Mass Spectrometry (Waller, G. R., ed.) pp. 527–566, Wiley, New York.
4. Schram, K. H. & McCloskey, J. A. (1979) In: GLC and HPLC Analysis of Therapeutic Agents, Part 3 (Tsuji, K., ed.) pp. 1149–1190, Marcel Dekker, Inc., New York.
5. Von Minden, D. L. & McCloskey, J. A. (1973) Mass spectrometry of nucleic acid components: N,O-Permethyl derivatives of nucleosides. J. Am. Chem. Soc. *95*, 7480–7490.
6. Teece, R. G. & Schram, K. H. (1986) Permethylation and gas phase analysis of permethylated nucleosides. In: Nucleic Acid Chemistry, Vol. 3 (Townsend, L. B. & Tipson, S. R., eds.) pp. 311–328, Wiley-Interscience, New York.
7. Pang, H., Schram, K. H., Smith, D. L., Gupta, S. P., Townsend, L. B. & McCloskey, J. A. (1982) Mass spectrometry of nucleic acid constituents, trimethyl derivatives of nucleosides. J. Org. Chem. *47*, 3923–3932.
8. McCloskey, J. A. (1985) Mass spectrometry of nucleic acid constituents and related compounds. In: Mass Spectrometry in the Health and Life Sciences (Burlingame, A. L. & Castagnoli, Jr., N., eds.) pp. 521–546, Elsevier Science Publishers, B. V., Amsterdam.
9. For the latest in this series see Burlingame, A. L., Baillie, T. A. & Derrick, P. J. (1986) Mass spectrometry. Anal. Chem. *58*, 165R–211R.
10. McCloskey, J. A. & Nishimura, S. (1977) Modified nucleosides in transfer RNA. Acc. Chem. Res. *10*, 403–410.
11. Agris, P. F. & Kopper, R. A., eds. (1983) The modified nucleosides of transfer RNA, II, Alan R. Liss, Inc., New York.
12. Jackson, R. C. & Harkrader, R. J. (1981) In: Nucleosides and cancer treatment (Tattersall, M. H. N. and Fox, R. M., eds.) pp. 18–32, Academic Press, Australia, Sydney.
13. Anderson, T. R. (1982) Cyclic cytidine 3′,5′-monophosphate (cCMP) in cell regulation. Mol. and Cell. Endocrin. *28*, 373–385.
14. Suhadolnik, R. J. (1979) Nucleosides as biological probes, Wiley-Interscience, New York.
15. Heidelberger, C. (1982) In: Cancer Medicine, 2nd ed. (Holland, J. F. and Frei, E. III, eds.) pp. 801–824, Lea and Febiger, Philadelphia.
16. Salvatore, F., Colonna, A., Castanzo, F., Tusso, T., Esposito, F. & Cimino, F. (1983) Modified nucleosides in body fluids of tumor bearing patients. Recent Results in Cancer Research *84*, 360–377.
17. Robins, R. K. (1975) Ann. N. Y. Acad. Sci. *255*, 597–610.
18. Chabard, J. L., Lartigue-Mattei, C., Vedrine, F., Petit, J. & Berger, J. A. (1980) Mass fragmentographic determination of xanthine and hypoxanthine in biological fluids. J. Chromatogr. *221*, 9–17.
19. Lartigue-Mattei, C., Chabard, J. L., Bargnoux, H., Petit, J. & Berger, J. A. (1982) Simultaneous determination of allopurinol and oxipurinol in biological fluids by mass fragmentogoraphy. J. Chromatogr. *229*, 211–216.
20. Miyazaki, H., Matsunaga, Y., Yoshida, K., Arakawa, S. & Hashimoto, M. (1983) Simultaneous determination of plasma and urinary uric acid, xanthine, hypoxanthine, allopurinol, oxipurinol, orotic acid, orotidine and creatinine by high-performance liquid chromatography. J. Chromatogr. *274*, 75–85.
21. Khanna, K. L., Rao, G. S. & Cornish, (1972) Metabolism of caffeine-^{3}H in the rat. Toxicol. Appl. Pharmacol. *23*, 70–73.
22. Mihda, K. K., Sved, S., Hossie, R. D. & McGilveray, I. J. (1977) High performance liquid chromatographic and mass spectrometric identification of dimethylxanthine metabolites of caffeine in human plasma. Biomed. Mass Spectrom. *4*, 172–177.
23. Horning, M. G., Nowlin, J., Stafford, M., Lertratanangkoon, K., Sommer, K. R., Hill, R. M. & Stillwell, R. N. (1975) The use of gas chromatographic-mass spectrometric-computer systems in pharmacokinetic studies. J. Chromatogr. *112*, 605–615.
24. Horning, M. G., Haegele, K. D., Sommer, K. R., Nowlin, J., Stafford, M. & Thenot, J.-P. (1975) In: Proceedings of the Second International Conference on Stable Isotopes

(Klein, E. R. & Klein, P. D. eds.) pp. 41 – 54, National Technical Information Service, U.S. Department of Commerce, Springfield, Virginia.
25. Brazier, J. L., Ribon, B., Desage, M. & Salle, B. (1980) Study of theophylline metabolism in premature human newborns using stable isotope labelling. Biomed. Mass Spectrom. *7*, 189 – 192.
26. Floberg, S., Lindstrom, B. & Lonnerholm, G. (1980) Simultaneous determination of theophylline and caffeine after extractive alkylation in small volumes of plasma by gas chromatography-mass spectrometry. J. Chromatogr. *221*, 166 – 169.
27. Rosenfeld, J. F., Taguchi, V. Y., Hillcoat, B. L. & Kawai, M. (1977) Determination of 6-mercaptopurine in plasma by mass spectrometry. Anal. Chem. *49*, 725 – 727.
28. Floberg, S., Hartvig, P., Lindstorm, B., Lonnerholm, G. & Odlind, B. (1981) Extractive alkylation of 6-mercaptopurine and determination in plasma by gas chromatography-mass spectrometry. J. Chromatogr. *225*, 73 – 81.
29. Pfadenhauer, E. H., Bankert, C. S., Jensen, J., Jones, C. E., Jenkins, E. E., and McCloskey, J. A. (1984) Identification of the metabolites of erythro-9-(2-hydroxy-3-nonyl)hypoxanthine from laboratory animals. Drug Metab. Dispos., *12*, 280 – 284.
30. Nomenclature used corresponds to that presented in reference 1.
31. Schwandt, H.-J., Wu, A. T., Finn, C. & Sadee, W. (1978) in: Nucleic Acid Chemistry, Part 2 (Townsend, L. B. & Tipson, R. S., eds.) pp. 901 – 915, Wiley-Interscience, New York.
32. Finn, C., Schwandt, H.-J. & Sadee, W. (1979) Determination of uracil and thymine and their nucleosides and nucleotides in picomole amounts by gas chromatography mass spectrometry selected ion monitoring. Biomed. Mass Spectrom. *6*, 194 – 199.
33. Marunaka, T., Umeno, Y. & Minami, Y. (1980) Analysis of pyrimidine bases in biological materials by gas chromatography-mass spectrometry. J. Chromatogr. *193*, 107 – 112.
34. Marunaka, T. & Umeno, Y. (1980) Determination of 5-fluorouracil and pyrimidine bases in plasma by gas chromatography-chemical ionization-mass fragmentography. J. Chromatogr. *221*, 382 – 386.
35. Wu, A. T., Schwandt, M.-J., Finn, C. & Sadee, W. (1976) Determination of ftorafur and 5-fluorouracil levels in plasma and urine. Res. Commun. Chem. Path. Pharm. *14*, 89 – 102.
36. Hillcoat, B. L., Kawai, M., McCulloch, P. B. Rosenfeld, J. & Williams, J. K. O. (1976) A sensitive assay of 5-fluorouracil in plasma by gas chromatography-mass spectrometry. Br. J. Clin. Pharmacol. *3*, 135 – 143.
37. Min, B. H. & Garland, W. A. (1978) Rapid assay of 5-fluorouracil (5-FU) in plasma by GC-CIMS and stable isotope dilution. Res. Commun. Chem. Path. Pharm. *22*, 145 – 154.
38. Lakings, D. B. & Adamson, R. H. (1978) Quantitative analysis of 5-fluorouracil in human serum by selected ion monitoring gas chromatography-mass spectrometry. J. Chromatogr. *146*, 512 – 517.
39. DeLeenheer, A. P. & Cosyns-Duyck, M. Cl. (1979) Gas-Liquid chromatography of some alkyl derivatives of 5-fluorouracil. J. Chromatogr. *174*, 325 – 339.
40. DeLeenheer, A. P., Cosyns-Duyck, M. Cl. & Cruyl, A. A. M. (1978) in: Quantitative Mass Spectrometry in Life Sciences II, Proceedings of the Second International Symposium on Quantitative Mass Spectrometry in Life Sciences (DeLeenheer, A. P., Roncucci, R. R. & Van Peteghem, C., eds.) pp. 399 – 406, Elsevier, Amsterdam.
41. Puzo, G., Schram, K. H. & McCloskey, J. A. (1977) Incorporation of oxygen-18 into nucleosides and bases. Nucleic Acids Res. *4*, 2075 – 2081.
42. Cosyns-Duyck, M. Cl., Cruyl, A. M., DeLeenheer, A. P., DeSchryver, A., Huys, J. V. & Belpaire, F. M. (1980) Analysis of fluorouracil in human plasma by combined gas liquid chromatography mass spectrometry. Biomed. Mass Spectrom. *7*, 61 – 64.
43. Schram, K. H. and Prince, D. (1982) In: Procedures For the Synthesis of Isotopically Labeled Compounds (Engel, J. F., ed.) pp. 65 – 67, Midwest Research Institute, St. Louis, MO.
44. Kok, R. M., De Jong, A. P. J. M., Van Groeningen, C. J., Peters, G. J. & Lankelma, J. (1985) Highly sensitive determination of 5-fluorouracil in human plasma by capillary gas chromatography and negative ion chemical ionization mass spectrometry. J. Chromatogr. *343*, 59 – 66.
45. Myers, C. E. (1981) The pharmacology of the fluoro-pyrimidines. Pharmacol. Rev. *33*, 1 – 15.
46. Gelijkens, C. F., Sandra, P., Belpaire, F. & DeLeenheer, A. P. (1980) Identification and quantitation of 5-fluorouridine in rat urine

after administration of 5-fluorouracil or 5-fluoro-2'-deoxyuridine. Drug Metab. Dispos. *8*, 363.

47. Gelijkens, C. F., DeLeenheer, A. P. & Sandra, P. (1980) Measurement of 5-fluoro-2'-deoxyuridine serum levels by gas chromatography chemical ionization mass spectrometry. Biomed. Mass Spectrom. *7*, 572–575.
48. Aubert, C., Sommadossi, J. P., Cano, J. P. & Rigault, J. P. (1982) Quantitative analysis of 5-fluorouracil and 5,6-dihydrofluorouracil in plasma by gas chromatography. Biomed. Mass Spectrom. *9*, 336–339.
49. Marunaka, T. & Umeno, Y. (1978) Quantitative determination of 1,3-bis(tetrahydro-2-furanyl)-5-fluoro-2,4-pyrimidinedione and its metabolites by high-pressure liquid chromatography and gas chromatography-mass fragmentography. J. Chromatogr. *157*, 321–330.
50. Marunaka, T., Umeno, Y. & Minami, Y. (1980) Quantitative determination of 1,3-bis(tetrahydro-2-furanyl)-5-fluoro-2,4-pyrimidinedione and its metabolites in visceral tissues by high-performance liquid chromatography and gas chromatography-mass fragmentography. J. Chromatogr. *188*, 270–292.
51. Marunaka, T., Umeno, Y. M., Yoshida, K., Nagamachi, M., Minami, Y. & Fujii, S. (1980) High-pressure liquid chromatographic determination of ftorafur,1-(tetrahydro-2-furanyl)-5-fluorouracil and GLC-mass spectrometric determination of 5-fluorouracil and uracil in biological materials after oral administration of uracil plus ftorafur. J. Pharm. Sci. *69*, 1296–1300.
52. Isomura, H., Higuchi, S. & Kaswamura, S. (1981) Gas chromatographic-mass spectrometric determination of plasma 5-fluorouracil after administration of 1-hexylcarbamoyl-5-fluorouracil to' dogs and humans. J. Chromatogr. *224*, 423–429.
53. Marunaka, T. & Umerno, Y. (1982) Gas chromatographic-mass fragmentographic determination of 5-fluorouracil as a metabolite of new 5-fluorouracil derivatives. Chem. Pharm. Bull. *30*, 1868–1871.
54. Hattox, S. E. & McCloskey, J. A. (1974) Structure-retention relationships in the gas chromatography of nucleosides. Anal. Chem. *46*, 1378–1383.
55. McCloskey. J. A., Stillwell, R. N. & Lawson, A. M. (1968) Use of deuterium-labeled trimethylsilyl derivatives in mass spectrometry. Anal. Chem. *40*, 233–236.
56. Teece, R. G., Slowikowski, D. & Schram, K. H. (1983) Synthesis and utility of deuterium labeled trimethylanilium hydroxide in the mass spectral analysis of nucleosides. Biomed. Mass Spectrom. *10*, 30–33.
57. Panzica, R. P., Townsend, L., B., von Minden, D. L., Wilson, M. S. & McCloskey, J. A. (1973) Formation of 3,5,5-trimethyl derivatives of dihydrouracil nucleosides by reaction with methylsulfinyl carbanion and methyl iodide. Biochim. Biophys. Acta. *331*, 147–153.
58. Koenig, W. A., Smith, L. C., Crain, P. F. & McCloskey, J. A. (1971) Mass spectrometry of trifluoroacetyl derivatives of nucleosides and hydrolysates of deoxyribonucleic acid. Biochem. *190*, 3968–3979.
59. Quilliam, M. A., Ogilvie, K. K., Sadana, K. L. & Westmore, J. B. (1980) Mass spectra of sterically crowded trialkylsilyl derivatives of nucleosides. Org. Mass Spectrom. *15*, 207–219.
60. Krahmer, U. I., Liehr, J. G., Lyman, K. J., Orr, E. A., Stillwell, R. N. & McCloskey, J. A. (1977) Use of a mixed derivative for the gas chromatography and mass spectrometry of cytidine and its analogs. Anal. Biochem. *82*, 217–225.
61. Schram, K. H., Taniguchi, Y. & McCloskey, J. A. (1978) Gas chromatography of nucleotides and nucleosides of cytosine. J. Chromatogr. *155*, 355–361.
62. Cory, H. T., Yamaizumi, K., Smith, D. L., Knowles, D. R., Broom, A. D. & McCloskey, J. A. (1979) Characterization of blocked nucleosides by mass spectrometry. Benzyl derivatives. J. Heterocycl. Chem. *16*, 585–590.
63. McCloskey, J. A. (1986) In: Mass spectrometry in biomedical research (Gaskell, S. J., ed.) pp. 75–95, John Wiley & Sons, Ltd., New York.
64. Gerhardt, K. O., Rice. R. H. & Gehrke, C. W. (1983) in: The Modified Nucleosides of Transfer RNA, II (Agris, P. F. & Kopper, R. A., eds.) pp. 93–128, Alan R. Liss, Inc., New York.
65. Kasai, H., Nakanishi, K., Mac Farlane, R. D., Torgerson, D. F., Ohashi, Z., McCloskey, J. A., Gross, H. J. & Nishimura, S. (1976) The structure of Q* nucleoside isolated from rabbit liver transfer ribonucleic acid. J. Am. Chem. Soc. *98*, 5044–5046.
66. Isono, K., Crain, P. F. & McCloskey, J. A. (1975) Isolation and structure of octosyl

acids, anhydrooctose uronic acid nucleosides. J. Am. Chem. Soc. *97*, 943–945.

67. Uramoto, M., Kobinata, K., Isono, K., Higashijima, T., Miyazawa, T., Jenkins, E. E. & McCloskey, J. A. (1980) Structures of neopolyoxins A, B, and C. Tetrahedron Lett. *21*, 3395–3398.
68. Uramoto, M., Kobinata, K., Isono, K., Jenkins, E. E., McCloskey, J. A., Higashijima, T. & Miyazawa, T. (1980) Neopolyoxins A, B, and C: New inhibitors of fungal cell wall chitin synthetase. Nucleic Acids Res. *558*, s69–s71.
69. Uramoto, M., Kobinata, K., Isono, K., Higashijima, T., Miyazawa, T., Jenkins, E. E. & McCloskey, J. A. (1982) Chemistry of the neopolyoxins, pyrimidine and imidazoline nucleoside peptide antibiotics. Tetrahedron *38*, 1599–1608.
70. Cook, A. F., Bartlett, R. T., Gregson, R. P. & Quinm, R. J. (1980) 1-Methylisoguanosine, a pharmacologically active agent from a marine sponge. J. Org. Chem. *45*, 4020–4025.
71. Fuhrman, F. A., Fuhrman, G. J., Kim, Y. H., Pavelka, L. A. & Mosher, H. S. (1980) Doridosine: A new hypotensive N-methylpurine riboside from the nudibranch Anisodoris nobilis. Science *207*, 193–194.
72. Schram, K. H. & Prince, D. (1983) Mass spectrometry of modified nucleic acid bases and nucleosides. 3-Deazauracil and 3-deazauridine. Biomed. Mass Spectrom. *10*, 34–45.
73. Rice, J. M. & Dudeck, G. O. (1969) Mass spectra of uridine and pseudouridine: Fragmentation patterns characteristic of a carbon-carbon nucleoside bond. Biochem. Biophys. Res. Comm. *35*, 383–388.
74. Townsend, L. B. & Robins, R. K. (1969) The mass spectra of formycin, formycin B and showdomycin, carbon linked nucleoside antibiotics. J. Heterocycl. Chem. *6*, 459–464.
75. Crain, P. F., McCloskey, J. A., Lewis, A. F., Schram, K. H. & Townsend, L. B. (1973) Mass spectra of C-nucleosides II. An unusual fragmentation reaction of the heterocyclic moiety of pyrazomycin and some closely related compounds. J. Heterocycl. Chem. *10*, 843–847.
76. Summons, R. E., Duke, C. C., Eichholzer, J. V., Entsch, B., Letham, D. S., MacLeod, J. K. & Parker, C. W. (1979) Mass spectrometric analysis of cytokinins in plant tissue. Biomed. Mass Spectrom. *6*, 407–412.
77. McCloskey, J. A., Hashizume, T., Basile, B., Sugiyama, T. & Sekiguchi, S. (1979) Determination of cytokinins in bamboo shoots by mass spectrometry using selected ion monitoring. Proc. Japan Acad. *55*, Ser. B., 445–450.
78. Hashizume, T., Sugiyama, T., Imura, M., Cory, H. T., Scott, M. F. & McCloskey, J. A. (1979) Determination of cytokinins by mass spectrometry based on stable isotope dilution. Anal. Biochem. *92*, 111–122.
79. Ludewig, M., Dorffling, K. & Konig, W. A. (1982) Electron-capture gas chromatography and mass spectrometry of trifluoroacetylated cytokinins. J. Chromatogr. *243*, 93–98.
80. Hashizume, T., Suye, S., Soeda, T. & Sugiyama, T. (1982) Isolation and characterization of a new glucopyranosyl derivative of 6-(3-methyl-2-butenylamino) purine from sweet potato tubers. FEBS Lett. *144*, 25–28.
81. Dauphin, B., Teller, G. & Durand, B. (1979) Identification and quantitative analysis of cytokinins from shoot apices of Mercurialis ambigua by gas chromatography-mass spectrometry computer system, Planta *144*, 113–119.
82. Smith, D. L., Schram, K. H. & McCloskey, J. A. (1983) The negative ion mass spectra of selected nucleosides. Biomed. Mass Spectrom. *10*, 269–275.
83. Pantarotto, C., Martini, A., Belvedere, G., Bossi, A., Donelli, M. G. & Frigerio, A. (1974) Application of gas chromatography-chemical ionization mass fragmentography in the evaluation of bases and nucleoside analogues used in cancer chemotherapy. J. Chromatogr. *99*, 519–527.
84. Boutagy, J. & Harvey, D. J. (1978) Determination of cytosine arabinoside in human plasma by gas chromatography with a nitrogen-sensitive detector and by gas chromatography-mass spectrometry. J. Chromatogr. *146*, 283–296.
85. Robinson, Y., Gerry, N., Brownsill, R. D. & Vose, C. W. (1984) Identification of a plasma metabolite of 5-(2-bromo-E-ethenyl)-2′-deoxyuridine. Biomed. Mass. Spectrom. *11*, 199–201.
86. McCloskey, J. A., Futrell, J. H., Elwood, T. A., Schram, K. H., Panzica, R. P. & Townsend, L. B. (1973) Determination of relative glycosyl bond strengths in nucleosides by chemical ionization mass spectrometry. A comparative study of 7- and 9-β-

D-ribofuranosylpurines. J. Am. Chem. Soc. *95*, 5762–5764.
87. Wilson, M. S., Dzidic, L. & McCloskey, J. A. (1971) Chemical ionization mass spectrometry of nucleosides. Biochim. Biophys. Acta *240*, 623–626.
88. Wilson, M. S. & McCloskey, J. A. (1975) Chemical ionization mass spectrometry of nucleosides. Mechanisms of ion formation and estimations of proton affinity. J. Am. Chem. Soc. *97*, 943–945.
89. Chheda, G. B. (1970) In: Handbook of Biochemistry 2nd Edn. (Sober, H. A., ed.) pp. G106–G113, The Chemical Rubber Company, Cleveland, OH.
90. Teece, R. G. & Schram, K. H. (1984) In: Abstr. 32nd Ann. Conf. on Mass Spectrom. and Allied Top., May 27–June 1, 1984, San Antonio, TX, pp. 400.
91. Unpublished results from author's laboratory.
92. De, N. C. & Chheda, G. B. (1979) Isolation and characterization of 2′-O-methyluridine from human urine. J. Carbohydrates Nucleosides Nucleotides *6*, 371–385.
93. Hirschhorn, R., Ratech, H., Rubinstein, A., Papageorgiou, P., Kesarwala, H., Gelfand, E. & Roegner-Maniscalco, V. (1982) Increased excretion of modified adenine nucleosides by children with adenosine deaminase deficiency. Pediatr. Res. *16*, 362–369.
94. Vold, B. S., Keigh, D. E. & Slavik, M. (1982) Urine levels of N-[9-(β-D-ribofuranosyl)purine-6-ylcarbamoyl]-L-threonine, N6-(Δ^2-isopentenyl)adenosine and 2′-O-methylguanosine as determined by radioimmunoassay for normal subjects and cancer patients. Cancer Res. *42*, 5265–5269.
95. Teece, R. G. & Schram, K. H. (1981) In: Abstr. 29th Ann. Conf. on Mass Spectrom. and Allied Top., May 24–29, 1981, Minneapolis, MN, pp. 742–743.
96. Dutta, S. P., Crain, P. F., McCloskey, J. A. & Chheda, G. B. (1979) Isolation and characterization of 1-β-D-ribofuranosylpyridin-4-one-3-carboxamide from human urine. Life Sci. *24*, 1381–1388.
97. De, N. C., Mittelman, A., Jenkins, E. E., Crain, P. F., McCloskey, J. A. & Chheda, G. B. (1980) Isolation of a new modified uracil derivative from human urine. J. Carbohydrates Nucleosides Nucleotides *7*, 113–129.
98. De, N. C., Mittelman, A., Dutta, S. P., Edmonds, C. G., Jenkins, E. E., McCloskey, J. A., Blakley, C. R., Vestal, M. L. & Chheda, G. B. (1981) Isolation and characterization of two new modified uracil derivatives from human urine. J. Carbohydrates Nucleosides Nucleotides *8*, 363–389.
99. Chheda, G. B., Dutta, S. P., Mittelman, A., Montgomery, J. A., Sethi, S. K., McCloskey, J. A. & Patrzyc, H. B. (1985) Isolation and characterization of a novel nucleoside, 7-β-D-ribofuranosylhypoxanthine, from the urine of a chronic myelogenous leukemia patient. Cancer Res. *45*, 5958–5963.
100. Mills, J. S., Mills, G. C. & McAdoo, D. J. (1983) Isolation and identification of 5′-methylthioadenosine sulfoxide from human urine. Nucleosides Nucleotides 2, 465–478.
101. Pfadenhauer, E. H. & Tong, S. (1979) Determination of inosine and adenosine in human plasma using high-performance liquid chromatography and a borate affinity gel. J. Chromatogr. *162*, 585–590.
102. Trewyn, R. W., Glaser, R., Kelley, D. R., Jackson, D. G., Graham, W. P. & Speicher, C. E. (1982) Elevated nucleoside excretion by patients with nasopharyngeal carcinoma. Cancer *49*, 2513–2517.
103. Mills, G. C., Schmalsteig, F. C., Koolkin, R. J. & Goldblum, R. M. (1982) Urinary excretion of purines, purine nucleosides and pseudouridine in immunodeficient children. Biochem. Med. *27*, 37–45.
104. Heldman, D. A., Grever, M. R., Speicher, C. E. & Trewyn, R. W. (1983) Urinary excretion of modified nucleosides in chronic myelogenous leukemia. J. Lab. Clin. Med. *101*, 783–792.
105. Colonna, R., Russo, T., Cimino, F. & Salvatore, F. (1981) Modified nucleosides in biological fluids of cancer patients determined by high performance liquid chromatography. J. Clin. Chem. Clin. Biochem. *19*, 640.
106. Speer, J., Gehrke, C. W., Kuo, K. C., Waalkes, T. P. & Borek, E. (1979) tRNA Breakdown products as markers for cancer. Cancer *44*, 2120–2123.
107. Rasmuson, T., Bjork, G. R., Damber, L., Holm, S. E., Jacobsson, L., Jeppsson, A., Littbrand, B., Stigbrand, T. & Westman, G. (1983) Evaluation of carcinoembryonic antigen, tissue polypeptide antigen, placental alkaline phosphatase and modified nucleosides as biological markers in malignant lymphomas. Recent Results in Cancer Research *86*, 331–343.
108. Hartwick, R. A., Krstulovic, A. M. & Brown, P. R. (1979) Identification and quantitation of nucleosides, bases and other

UV-absorbing materials in serum using reverse-phase high-performance liquid chromatography-II. Evaluation of human sera. J. Chromatogr. *186*, 659–676.
109. Gehrke, C. W., Kuo, K. C., Waalkes, T. P. & Borek, E. (1979) Patterns of urinary excretion of modified nucleosides. Canc. Res. *39*, 1150–1153.
110. Tormey, D. C., Waalkes, T. P. & Gehrke, C. W. (1980) Biological markers in breast carcinoma. Clinical correlations with pseudouridine, N^2,N^2-dimethylguanosine and 1-methylinosine. J. Surg. Oncol. *14*, 267–273.
111. Waalkes, T. P., Gehrke, C. W., Zumwalt, R. W., Chang, S. Y., Lakings, D. B., Tormey, D. C., Ahmann, D. L. & Moertel, C. G. (1975) The urinary excretion of nucleosides of ribonucleic acid by patients with advanced cancer. Cancer *36*, 390–398.
112. Heldman, D. A., Grever, M. R., Miser, J. S. & Trewyn, R. W. (1983) Relationship of urinary excretion of modified nucleosides to disease status in childhood acute lymphoblastic leukemia. J. Natl. Canc. Inst. *71*, 269–273.
113. Heldman, D. A., Grever, M. R. & Trewyn, R. W. (1983) Differential excretion of modified nucleosides in adult acute leukemia. Blood *61*, 291–296.
114. Lehtio, H. & Maenpaa, P. H. (1977) The urinary excretion of the nucleosides pseudouridine and 1-methylinosine during normal menstrual cycle. Clin. Chim. Acta *80*, 181–185.
115. Levine, L., Waalkes, T. P. & Stolbach, L. (1975) Serum levels of N^2,N^2-dimethylguanosine and pseudouridine as determined by radioimmunoassay for patients with malignancy. J. Natl. Cancer Inst. *54*, 341–343.
116. Nielsen, H. R. & Killman, S.-A. (1983) Urinary excretion of β-aminoisobutyrate and pseudouridine in acute and chronic myeloid leukemia. J. Natl. Canc. Inst. *71*, 887–891.
117. Karle, J. M., Anderson, L., Dietrick, D. D. & Cysyk, R. L. (1980) Determination of serum and plasma uridine levels in mice, rats and humans by high pressure liquid chromatography. Anal. Biochem. *109*, 41–46.
118. Baillie, T. A., ed., (1978) Stable Isotopes: Applications in Pharmacology, Toxicology and Clinical Research, University Park Press, Baltimore.
119. Klein, E. R. & Klein, P. D. (1978) A selected bibliography of biomedical and environmental applications of stable isotopes. I. Deuterium 1971–1976. Biomed. Mass Spectrom. *5*, 91–111.
120. Klein, E. R. & Klein, P. D. (1978) A selected bibliography of biomedical and environmental applications of stable isotopes. II. ^{13}C 1971–1976. Biomed. Mass Spectrom. *5*, 321–330.
121. Klein, E. R. & Klein, P. D. (1978) A selected bibliography of biomedical and environmental applications of stable isotopes. III. ^{15}N 1971–1976. Biomed. Mass Spectrom. *5*, 373–379.
122. Klein, E. R. & Klein, P. D. (1978) A selected bibliography of biomedical and environmental applications of stable isotopes. IV. ^{17}O, ^{18}O and ^{34}S, 1971–1976. Biomed. Mass Spectrom. *5*, 425–432.
123. Klein, E. R. & Klein, P. D. (1979) A selected bibliography of biomedical and environmental applications of stable isotopes. V. ^{2}H, ^{13}C, ^{18}O and ^{34}S, 1977–1978. Biomed. Mass Spectrom. *6*, 515–545.
124. Cushley, R. J., Lipsky, S. R. & Fox, J. J. (1968) Reactions of 5-fluorouracil derivatives with sodium deuteroxide. Tetrahedron Lett., 5393–5396.
125. Rabi, J. A. & Fox, J. J. (1973) Nucleosides. LXXIX. Facil base-catalyzed hydrogen isotope labeling at position 6 of pyrimidine nucleosides. J. Am. Chem. Soc. *95*, 1628–1632.
126. Maeda, M. & Kawazoe, Y. (1975) Chemical alteration of nucleic acids and their components. XI. Hydrogen-deuterium exchange of nucleosides and nucleotides by platinum. Tetrahedron Lett., 1643–1646.
127. Kinoshita, T., Schram, K. H. & McCloskey, J. A. (1982) Preparation of deuterium-labeled nucleosides by platinum-catalyzed exchange and reduction. J. Label. Cmpds. Radiopharm. *IX*, 525–534.
128. Schubert, E. M. & Schram, K. H. (1982) In: Procedures for the Synthesis of Isotopically Labeled Compounds (Engel, J. F., ed.) pp. 62–64, Midwest Research Institute, Kansas City, MO.
129. Schmidt, R. R., Schloz, U. & Schwille, D. (1968) Synthese 5′-modifizierter adenosinderivat. Chem. Ber. *101*, 590–594.
130. Puzo, G., Schram, K. H., Liehr, J. G. & McCloskey, J. A. (1978), Mass spectrometry of pyrimidine anhydronucleosides. J. Org. Chem. *43*, 76–769.
131. Jardine, I. & Weidner, M. M. (1980) Synthesis of 5′,5′-dideuterated nucleoside and nucleotide anabolites of 6-mercaptopurine.

J. Label. Cmpds. Radiopharm. *XVII*, 389–399.
132. Wang, S. Y., Hahn, B. S., Fenselau, C. & Zafiriou, O. C. (1972) Enrichment of ^{18}O in the nucleic acid bases. Biochem. Biophys. Res. Commun. *48*, 1630–1635.
133. Schram, K. H., Ratcliff, S. & Neenan, J. (1982) ^{18}O Labeled nucleosides. 1. A general method for the synthesis of specifically labeled pyrimidine deoxyribosides. J. Label. Cmpds. Radiopharm. *XIX*, 399–404.
134. Schubert, E. & Schram, K. H. (in press) In: Nucleic Acid Chemistry, Vol. 4 (Townsend, L. B. & Tipson, R. S., eds.), Wiley-Interscience, New York.
135. Solsten, R. T., McCloskey, J. A. & Schram, K. H. (1982) ^{18}O Labeled nucleosides. 2. A general method for the synthesis of specifically labeled pyrimidine ribonucleosides. Nucleosides Nucleotides *1*, 57–64.
136. Schubert, E. M. & Schram, K. H. (1982) ^{18}O Labeled nucleosides. 3. Preparation and mass spectrometric evaluation of ^{18}O-labeled Ara-C and Ara-U. J. Label. Cmpds. Radiopharm. *XIX*, 929–935.
137. Wilson, M. H. & McCloskey, J. A. (1973) Isotopic studies of the base-catalyzed conversion of 1-methyladenosine to N^6-methyladenosine. J. Org. Chem. *38*, 2247–2249.
138. Leonard, N. J. & Henderson, T. R. (1975) Purine ring rearrangements leading to the development of cytokinin activity. Mechanism of the rearrangement of 3-benzyladenine to N^6-benzyladenine. J. Am. Chem. Soc. *97*, 4990–4999.
139. Scott, L. M. & Horgan, R. (1980) Quantification of cytokinins by selected ion monitoring using ^{15}N labelled internal standards. Biomed. Mass Spectrom. *7*, 446–449.
140. Griffey, R. H. & Poulter, C. D. (1983) Efficient syntheses of [3-^{15}N]-uracil and [3-^{15}N]-thymine. Nucleic Acids Res. *11*, 6497–6504.
141. Lipnick, R. L. & Fissekis, J. D. (1979) Synthesis of nitrogen-15 labelled uracil and its 1-deuteromethyl, 3-deuteromethyl, and 1,3-dideuteromethyl derivatives. J. Label. Cmpds. Radiopharm. *XVII*, 247–254.
142. Gao, X. & Jones, R. A. (1987) Nitrogen-15 labeled deoxynucleosides. Synthesis of 6-^{15}N- and 1-^{15}N deoxyadenosines from deoxyadenosine. J. Am. Chem. Soc. *109*, 1275–1278.
143. Barber, M., Bordoli, R. S., Sedgwick, R. D. & Tyler, A. N. (1981) Fast atom bombardment of solids (F. A. B.): A new ion source for mass spectrometry J. Chem. Soc. Chem. Commun., 325–327.
144. Rinehart, K. L., Jr. (1982) Fast atom bombardment masss spectrometry. Science *218*, 254–260.
145. Barber, M., Bordoli, R. S., Elliott, G. J., Sedgwick, R. D. & Tyler, A. N. (1982) Fast atom bombardment mass spectrometry. Anal. Chem. *54*, 645A–657A.
146. Slowikowski, D. & Schram, K. H. (1985) Fast atom bombardment mass spectrometry of nucleosides, nucleotides and oligonucleotides. Nucleosides Nucleotides *4*, 309–345.
147. Crow, J. W., Tomer, K. B., Gross, M. L., McCloskey, J. A. & Bergstrom, D. E. (1984) Fast atom bombardment combined with tandem mass spectrometry for the determination of nucleosides. Anal. Biochem. *139*, 243–262.
148. Slowikowski, D. & Schram, K. H. (1985) Fast atom bombardment mass spectrometry of nucleosides. Comparison with electron impact and chemical ionization mass spectra. Nucleosides Nucleotides *4*, 347–376.
149. Sethi, S. K., Smith, D. L. & McCloskey, J. A. (1983) Determination of active hydrogen content by fast atom bombardment mass spectrometry following hydrogen-deuterium exchange. Biochem. Biophys Res. Commun. *112*, 126–131.
150. Verma, S., Pomerantz, S. C., Sethi, S. K. & McCloskey, J. A. (1986) Fast atom bombardment mass spectrometry following hydrogen-deuterium exchange. Anal. Chem. *58*, 2898–2902.
151. Rose, M. E., Longstraff, C. & Dean, P. D. G. (1983) Negative ion fast atom bombardment mass spectrometry. In situ reactions of boronic acids with triols and related compounds, sugars and nucleosides. Biomed. Mass Spectrom. *10*, 512–527.
152. Sethi, S. K., Nelson, C. C. & McCloskey, J. A. (1984) Dehalogenation reactions in fast atom bombardment mass spectrometry. Anal. Chem. *56*, 1975–1977.
153. Cerny, R. L. & Gross, M. L. (1985) Abundances of molecular ion species desorbed by fast atom bombardment: observation of $(M + 2H)^+$ and $(M + 3H)^+$. Anal. Chem. *57*, 1160–1163.
154. Schram, K. H. & Slowikowski, D. (1986) Fast atom bombardment of trimethylsilyl derivatives of nucleosides and nucleotides. Biomed. Environ. Mass Spectrom. *13*, 263–264.
155. Isono, K., Uramoto, M., Kusakabe, H., Miyata, N., Koyama, T., Ubukata, M., Sethi,

S. K. & McCloskey, J. A. (1984) Ascamycin and dealanylascamycin, nucleoside antibiotics from Streptomyces sp. J. Antibiot. *37*, 670–672.
156. Nishi, M., Inagaki, J.-I., Nohara, F., Isono, K., Kusakabe, K., Sakurai, T., Sethi, S. K. & McCloskey, J. A. (1985) A new antitumor antibiotic, Guanine 7-N-oxide produced by Streptomyces sp. J. Antibiot. *38*, 1440–1443.
157. Chan, J. A., Sitrin, R. D., Yeung, E. W. K., Simolike, G. C., Roberts, G. D. & Jeffs, P. W. (1986) Facil separation and identification of polyoxins from fermentation sources using high-performance liquid chromatography, ultraviolet spectroscopy and fast atom bombardment mass spectrometry. J. Chromatogr. *358*, 291–295.
158. Tondeur, Y., Shorter, M., Gustafson, M. E. & Pandey, R. C. (1984) Fast atom bombardment mass spectrometry and tandem mass spectrometry in antibiotics: identification of nucleoside antitumor antibiotic toxocamycin in fermentation broth. Biomed. Mass. Spectrom. *11*, 622–628.
159. Wickramanayake, P. P., Arbogast, B. L., Buhler, D. R., Deinzer, M. L. & Burlingame, A. L. (1985) Alkylation of nucleosides and nucleotides by dehydroretronecine; characterization of covalent adducts by liquid secondary ion mass spectrometry. J. Am. Chem. Soc. *107*, 2485–2488.
160. Tomer, K. B., Gross, M. L. & Deinzer, M. L. (1986) Fast atom bombardment and tandem mass spectrometry of covalently modified nucleosides and nucleotides: adducts of pyrrolizidine alkaloid metabolites. Anal. Chem. *58*, 2527–2534.
161. Bodley, J. W., Upham, R., Crow, F. W., Tomer, K. B. & Gross, M. L. (1984) Ribosyl-diphthamide: confirmation of structure by fast atom bombardment mass spectrometry. Arch. Biochem. Biophys. *230*, 590–593.
162. Fukushima, K. & Arai, T. (1978) Field desorption mass spectrometry of nucleoside antibiotics. J. Antibiot. *31*, 377–378.
163. Eicke, A., Sichtermann, W. & Benninghovern, A. (1980) Secondary ion mass spectrometry of nucleic acid components: Pyrimidines, purines, nucleosides and nucleotides. Org. Mass Spectrom. *15*, 289–294.
164. Unger, S. E., Schoen, A. E., Cooks, R. G., Ashworth, D. J., Gomes, J. & Chang, C. (1981) Identification of modified nucleosides by secondary-ion mass spectrometry. J. Org. Chem. *46*, 4765–4769.
165. Hunt, D. F., Shabanowitz, J., Botz, F. K. & Brent, D. A. (1977) Chemical ionization mass spectrometry of salts and thermally labile organics with field desorption emitters as solid probes. Anal. Chem. *49*, 1160–1163.
166. Cotter, R. J. & Fenselau, C. (1979) The effects of heating rate and sample size on the direct exposure/chemical ionization mass spectra of some biological conjugates. Biomed. Mass Spectrom. *6*, 287–293.
167. Esmans, E. L., Freyne, E. J., Vanbroeckhoven, J. H. & Alderweireldt, F. C. (1980) Chemical ionization desorption mass spectrometry as an additional tool for the structure elucidation of nucleosides. Biomed. Mass. Spectrom. *7*, 377–380.
168. Takeda, N., Harada, K.-I., Suzuki, M. & Tatematsu, A. (1985) Optimization of operating conditions for desorption chemical ionization mass spectrometry. Org. Mass. Spectrom. *20*, 236–242.
169. Ohashi, M., Tsujimoto, K. & Yasuda, A. (1976) Detection of molecular ions of thermally unstable compounds by in-beam electron impact. Chem. Lett., (Japan) 439–440.
170. Tsujimoto, K., Nomura, H. & Ohashi, M. (1985) In-beam electron ionization mass spectrometry of nucleosides. Mass Analysis *33*, 255–260.
171. Li, W.-S. & Liu, L. K. (1985) In-beam mass spectral studies of nucleosides. Chemistry (The Chinese Chem. Soc., Taiwan, China) *43*, 120–128.
172. Utake, N., Ogita, T., Niyazaki, Y., Yonehata, H., MacFarlane, R. D. & McNeal, C. J. (1981) ^{252}Cf Plasma desorption mass spectrometry of adenomycin (C19–97 substance). J. Antibiot. *34*, 130–131.
173. Schueler, B. & Krueger, F. R. (1980) Comparative study of pulsed laser and fission fragment induced desorption with nucleotic compounds. Org. Mass. Spectrom. *15*, 295–301.
174. Lai, S.-T. F. & Evans, C. A., Jr. (1979) Electrohydrodynamic ionization mass spectrometry: pyrimidines, purines, nucleosides and nucleotides. Biomed. Mass Spectrom. *6*, 10–14.
175. Baker, K. M. (1976) In: Advances in Mass Spectrometry in Biochemistry and Medicine, Vol. 1 (Frigerio, A. & Castagnoli, N., eds.) pp. 231–237, Spectrum Publications, New York.
176. Pettit, G. R., Einck, J. J. & Brown, P. (1978) Structural biochemistry 15. Mass spectro-

metry of permethylated nucleotides. Biomed. Mass Spectrom. *5*, 153–160.
177. Strong, J. M., Anderson, L. W., Monks, A., Chisena, C. A. & Cysyk, R. L. (1983) A ^{13}C tracer method for quantitating de novo pyrimidine biosynthesis in vitro and in vivo. Anal. Biochem. *132*, 243–253.
178. Johnson, L. P., MacLeod, J. K., Summons, R. E. & Hunt, N. (1980) Design of a stable isotope dilution gas chromatography/mass spectrometric assay for cAMP: Comparison with standard protein-binding and radioimmunoassay methods. Anal. Biochem. *106*, 285–290.
179. Newton, R. P., Gibbs, N., Noyse, C. D., Wiebers, J. L. & Brown, G. (1980) Mass spectrometric identification of adenosine 3′ : 5′-cyclic monophosphate isolated from a higher plant tissue. Phytochem. *19*, 1909–1911.
180. Johnson, L. P., MacLeod, J. K., Parker, C. W. & Letham, D. S. (1981) Quantitation of adenosine 3′ : 5′-cyclic monophosphate in cultured tobacco tissue by mass spectrometry. FEBS Lett. *124*, 119–121.
181. Janistyn, B. (1981) Gas chromatographic, mass- and infrared-spectrometric identification of cyclic adenosine-3′ : 5′-monophosphate (c-AMP) in maize seedlings (Zea mays). Z. Naturforsch. *36*, 193–196.
182. Fenselau, C., Vu, V. T., Cotter, R. J., Hansen, G., Heller, D., Chen, T. & Colvin, O. M. (1982) Characterization of N-7 alkylated guanosine monophosphates by fast atom bombardment mass spectrometry and some observations on the behaviour of zwitterions. Spectros. Int. J. *1*, 132–151.
183. Sindona, G., Ucella, N. & Weclawek, K. (1982) Structure determination of isomeric oligodeoxynucleotide salts by fast-atom-bombardment mass spectrometry. J. Chem. Res. (S), 184–185.
184. Aubagnac, J. L., Devienne, F. M., Combarieu, R., Barscut, J. L., Imbach, J. L. & Lazrek, H. B. (1983) Characterization of underived nucleotides by mass spectrometry using the MBSA-FAB ionization method. Org. Mass Spectrom. *18*, 361–364.
185. Eagles, J., Javanaud, C. & Self, R. (1984) Fast atom bombardment mass spectrometry of nucleosides and nucleotides. Biomed. Mass Spectrom. *11*, 41–46.
186. Fenselau, C. (1984) Fast atom bombardment and middle molecule mass spectrometry. J. Nat. Prod. *47*, 215–225.
187. Vu, V. T., Fenselau, C. C. & Colvin, O. M. (1981) Identification of three alkylated nucleotide adducts from the reaction of guanosine 5′-monophosphate with phosphoramide mustard. J. Am. Chem. Soc. *103*, 7362–7364.
188. Mehta, J. R., Przybylski, M. & Ludlum, D. B. (1980) Alkylation of guanosine and deoxyguanosine by phosphoramide mustard. Cancer Res. *40*, 4183–4186.
189. Fenselau, C., Feng, P. C. C., Chen, P. & Johnson, L. P. (1982) Stable isotope analysis by fast-atom bombardment labeling of UDP-glucuronic acid. Drug Metab. Disposit. *10*, 316–318.
190. Connolly, B. A., Eckstein, F. & Grotjahn, L. (1984) Direct mass spectroscopic method for determination of oxygen isotope position of adenosine 5′-O-(thiotriphosphate). Determination of the stereochemical course of the yeast phenylalanyl-tRNA synthetase reaction. Biochemistry *23*, 2026–2031.
191. Mallis, L. M., Raushel, F. M. & Russell, D. H. (1987) Differentiation of isotopically labeled nucleotides using fast atom bombardment tandem mass spectrometry. Anal. Chem. *59*, 980–984.
192. Toren, P. C., Betsch, D. F., Weith, H. L. & Coull, J. M. (1986) Determination of impurities in nucleoside 3′-phosphoramidites by fast atom bombardment mass spectrometry. Anal. Biochem. *152*, 291–294.
193. Kingston, E. E., Beynon, J. H. & Newton, R. P. (1984) The identification of cyclic nucleotides from living systems using collision-induced dissociation of ions generated by fast atom bombardment mass spectrometry. Biomed. Mass Spectrom. *11*, 367–374.
194. Kingston, E. E., Beynon, J. H., Newton, R. P. & Liehr, J. G. (1985) The differentiation of isomeric biological compounds using collision-induced dissociation of ions generated by fast atom bombardment. Biomed. Mass Spectrom. *12*, 525–534.
195. Schiebel, H.-M., Schulze, P., Stohrer, W.-D., Leibfritz, D., Jastorff, B. & Maurer, K. H. (1985) Fast atom bombardment (FAB) mass spectrometry of nucleoside cyclophosphates and structurally related compounds. FAB-induced reactions before desorption. Biomed. Mass Spectrom. *12*, 170–180.
196. Cerny, R., Gross, M. L. & Grotjahn, L. (1986) Fast atom bombardment combined with tandem mass spectrometry for the study of dinucleotides. Anal. Biochem. *156*, 424–435.
197. Schulten, H.-R. & Schiebel, H. M. (1985) Field desorption and fast atom bombardment mass spectrometry of pyridine nucleo-

tides and nucleoside triphosphates. Fres. Z. Anal. Chem. *321*, 531–537.

198. Ligon, W. V. & Dorn, S. B. (1986) Improved secondary ion mass spectral sensitivity for adenosine triphosphate disodium salt. Fres. Z. Anal. Chem. *325*, 626–627.
199. Isern-Flecha, I., Jiang, X.-Y., Cooks, R. G., Pfleiderer, W., Chae, W.-G. & Chang, C.-J. (1987) Characterization of an alkylated dinucleotide by desorption chemical ionization and tandem mass spectrometry. Biomed. Environ. Mass. Spectrom. *14*, 17–22.
200. Seliger, H., Bach, T. C., Gortz, H. H., Happ. E., Holupirek, M., Seemann-Preising, B., Schiebel, N. M. & Schulten, H. R. (1982) High-performance liquid chromatography in combination with field desorption mass spectrometry: Separation and identification of building blocks for polynucleotide synthesis. J. Chromatogr. *253*, 65–79.
201. Budzikiewicz, H. & Feistner, G. (1978) Mass spectroscopic differentiation between 3'- and 5'-nucleoside monophosphoric acids. Biomed. Mass Spectrom. *5*, 512–513.
202. Budzikiewicz, H. (1978) Mass spectroscopic investigation of nucleic acid degradation products. Pure & Appl. Chem. *50*, 159–169.
203. Schulten, H. R. & Lehmann, W. D. (1980) Field desorption mass spectrometry. Trends Biochem. Sci. *5*, 142–146.
204. Schulten, H. R. (1979) Biochemical, medical and environmental applications of field-ionization and field-desorption mass spectrometry. Int. J. Mass Spectrom. Ion Phys. *32*, 97–283.
205. Straub, K. M. & Burlingame, A. L. (1981) Carcinogen binding to DNA. Biomed. Mass Spectrom. *8*, 431–435.
206. Trosko, J. S. & Chang, C. C. (1979) In: Chemical Carcinogens and DNA, Vol. II (Grover, P. L., ed.) pp. 181–200, CRC Press, Boca Raton, Fl.
207. Hemminki, K. (1983) Nucleic acid adducts of chemical carcinogens and mutagens. Arch. Toxicol. *52*, 249–285.
208. Reif, A. E. (1981) The causes of cancer. Amer. Scientist *69*, 437–447.
209. Ames, B. N. (1983) Dietary carcinogens and anticarcinogens. Science *221*, 1256–1263.
210. Maruyama, I. N. & Tanaka, N. (1981) Structure of a neothramycin-2'-deoxyguanosine adduct. Biochem. Biophys. Res. Commun. *98*, 970–975.
211. Puzo, G., Prome, J. C., Macquet, J. P. & Lewis, I. A. S. (1982) Fast atom bombardment mass spectra of the bis-guanosine adduct with cisplatin. Biomed. Mass Spectrom. *9*, 552–556.
212. Johnson, N. P., Macquet, J. P., Wiebers, J. L. & Monsarrat, B. (1982) Structures of the adducts formed between Pt(dien)Cl Cl and DNA in vitro. Nucleic Acids Res. *10* 5255–5271.
213. Sethi, K. S., Gelijkens, C. F., McCloskey, J. A., Shapiro, R. & Brandes, B. S. (1983) Mass spectrometric study of methylene cross-linked derivatives of adenine and adenosine. Biomed. Mass Spectrom. *10*, 665–670.
214. Shapiro, R., Dubelman, S., Feinberg, A. M., Crain, P. F. & McCloskey, J. A. (1977) Isolation and identification of cross-linked nucleosides from nitrous acid treated deoxyribonucleic acid. J. Am. Chem. Soc. *99*, 302–303.
215. Tong, W. P. & Ludlum, D. B. (1981) Formation of the cross-linked base, diguanylethane in DNA treated with N,N'-bis(2-chloroethyl)-N-nitrosourea. Cancer Res. *41*, 380–382.
216. Ludlum, D. B., Kramer, B. S:, Wang, J. & Fenselau, C. (1975) Reaction of 1,3-bis(2-chloroethyl)-1-nitrosourea with synthetic polynucleotides. Biochem. *14*, 5480–5485.
217. Scribner, N. K., Scribner, J. D., Smith, D. L., Schram, K. H. & McCloskey, J. A. (1979) Reactions of the carcinogen N-acetoxy-4-acetamidostilbene with nucleosides. Chem. Biol. Interactions *26*, 27–46.
218. Scribner, J. D., Smith, D. L. & McCloskey, J. A. (1978) Deamination of 1-methylcytosine by the carcinogen N-acetoxy-4-acetamidostilbene: implications for hydrocarbon carcino-genesis. J. Org. Chem. *43*, 2085–2086.
219. Scribner, J. D., Scribner, N. K., Smith, D. L., Jenkins, E. & McCloskey, J. A. (1982) Acetylation of nucleosides by N-acetoxy-N-arylacetamides: dependence on base, aryl group, and buffer composition. J. Org. Chem. *47*, 3143–3145.
220. Chung, F.-L. & Hecht, S. S. (1983) Formation of cyclic N1,N2-adducts by reaction of deoxyguanosine with α-acetoxy-N-nitrosopyrrolidine, 4-(carbethoxynitrosamino)-butanal or crotonaldehyde. Cancer Res. *43*, 1230–1235.
221. Chung, F.-L. & Hecht, S. S. (1985) Formation of the cyclic 1,N^2-glyoxal-deoxyguanosine adduct upon reaction of N-nitroso-2-hydroxymorpholine with deoxyguanosine. Carcinogenesis *6*, 1671–1673.

222. Mitchum, R. K., Evans, F. E., Freeman, J. P. & Roach, D. (1983) Fast atom bombardment mass spectrometry of nucleoside and nucleotide adducts of chemical carcinogens. J. Mass Spectrom. Ion Phys. *46*, 383–386.
223. Kasai, H., Nakanishi, K., Frenkel, K. & Grunberger, D. (1977) Structures of 7,12-dimethylbenz[a]anthracene 5,6-oxide derivatives linked to the ribose moiety of guanosine. J. Am. Chem. Soc. *99*, 8500–8502.
224. Blobstein, S. H., Weinstein, L. B., Grunberger, D., Weisgras, J. & Harvey, R. G. (1975) Products obtained after in vitro reaction of 7,12-dimethylbenz[a]anthracene 5,6-oxide with nucleic acids. Biochem. *14*, 3451–3458.
225. Koreeda, M., Morre, P. O., Yage, H., Yeh, H. J. C. & Jerina, D. M. (1976) Alkylation of polyguanylic acid at the 1-amino group and phosphate by the potent mutagen (±)-7,8α-dihydroxy-9β,10β-epoxy-7,8,9,10-tetrahydrobenzo[a]pyrene. J. Am. Chem. Soc. *98*, 6720–6722.
226. Jeffrey, A. M., Jennette, K. W., Blobstein, S. H., Weinstein, I. B., Beland, F. A., Harvey, R. G., Kasai, H., Miura, I. & Nakanishi, K. (1976) Benzo[a]pyrene-nucleic acid derivative found in vivo: structure of benzo[a]pyrene tetrahydrodiol epoxide-guanosine adduct. J. Am. Chem. Soc. *98*, 5714–5715.
227. Jeffrey, A. M., Grzeskowiak, K., Weinstein, L. B., Nakanishi, K., Roller, P. & Harvey, R. G. (1979) Benzo[a]pyrene-7,8-dihydrodiol 9,10-oxide adenosine and deoxyadenosine adducts: structure and stereochemistry. Science *206*, 1309–1311.
228. Lhoest, G. (1983) In: Recent Developments in Mass Spectrometry in Biochemistry, Medicine and Environmental Research, Vol. 8 (Frigerio, A., ed.) pp. 191–202, Elsevier Scientific Publishing Company, Amsterdam.
229. Seiler, J. P. (1976) The mutagenicity of benzimidazole and benzimidazole derivatives. V. Gas chromatographic-mass spectrometric analysis of benzimidazole nucleoside in Esherichia coli ribonucleic acid. Anal. Biochem. *75*, 45–52.
230. Katze, J. R., Gunduz, U., Smith, D. L., Cheng, C. S. & McCloskey, J. A. (1984) Evidence that the nucleic acid base queuine is incorporated into RNA by animal cells. Biochemistry *23*, 1171–1176.
231. Essigmann, J. M., Croy, R. G., Nadzan, A. M., Busby, W. F., Jr., Reinhold, V. N., Buchi, G. & Wogan, G. N. (1977) Structural identification of the major DNA adduct formed by aflatoxin B1 in vitro. Proc. Natl. Acad. Sci. USA *74*, 1870–1874.
232. Institoris, E. & Tamas, J. (1980) Alkylation by 1,2:5,6-dianhydrogalactitol of deoxyribonucleic acid and guanosine. Biochem. J. *185*, 659–666.
233. Nagasawa, H. T., Shirota, F. N. & Mizuno, N. S. (1974) The mechanism of alkylation of DNA by 5-(3-methyl-1-triazeno)-imidazole-4-carboxamide (MIC), a metabolite of DIC(NSC-45388). Non-involvement of diazomethane. Chem.-Biol. Interactions *8*, 403–413.
234. Beland, F. A., Tullis, D. L., Kadlubar, F. F., Straub, K. M. & Evans, F. E. (1980) Characterization of DNA adducts of the carcinogen N-methyl-4-aminoazobenzene in vitro and in vivo. Chem.-Biol. Interact. *31*, 1–17.
235. Kadlubar, F. F., Unruh, L. E., Beland, F. A., Straub, K. M. & Evans, F. E. (1980) In vitro reaction of the carcinogen, N-hydroxy-2-naphthylamine, with DNA at the C-8 and N^2 atoms of guanine and the N^6 atom of adenine. Carcinogenesis *1*, 139–150.
236. Wiebers, J. L., Abbott, P. J., Coombs, M. M. & Livingston, D. C. (1981) Mass spectral characterization of the major DNA-carcinogen adduct formed from the metabolically activated carcinogen 15,16-dihydro - 11 - methylcyclopent[a]phenanthren-17-one. Carcinogenesis *2*, 637–643.
237. Straub, K. M., Meehan, T., Burlingham, A. L. & Calvin, M. (1977) Identification of the major adducts formed by reaction of benzo[a]pyrene diol epoxide with DNA in vitro. Proc. Natl. Acad. Sci. *74*, 5285–5289.
238. Galiegue-Zouitina, S., Bailleul, B., Ginot, Y.-M., Perly, B., Vigny, P. & Loucheux-Lefebvre, M.-H. (1986) N^2-Guanyl and N^6-adenyl arylation of chicken erythrocyte DNA by the ultimate carcinogen of 4-nitorquinoline-1-oxide. Cancer Res. *46*, 1858–1863.
239. Savela, K., Hesso, A. & Hemminki, K. (1986) Characterization of reaction products between styrene oxide and deoxynucleosides and DNA. Chem.-Biol. Interact. *60*, 235–246.
240. Chang, C.-J., Ashworth, D. J., Isern-Flecha, I., Jiang, X.-Y. & Cooks, R. G. (1986) Modification of calf thymus DNA by methyl methanesulfonate. Quantitative determination of 7-methyldeoxyguanosine by mass spectrometry. Chem.-Biol. Interact. *57*, 295–300.

241. Ashworth, D. L., Baird, W. M., Chang, C.-J., Ciupek, J. D., Busch, K. L. & Cooks, R. G. (1985) Chemical modification of nucleic acids. Methylation of calf thumus DNA investigated by mass spectrometry and liquid chromatography. Biomed. Mass Spectrom. *12*, 309–318.
242. Tomaz, M., Chowdary, D., Lipman, R., Shimotakahara, S., Veiro, D., Walker, V. & Verdine, G. L. (1986) Reaction of DNA with chemically or enzymatically activated mitomycin C; Isolation and structure of the major covalent adduct. Proc. Natl. Acad. Sci. USA *83*, 6702–6706.
243. Tomaz, M., Lipman, R., Chodary, D., Pawlak, J., Verdine, G. L. & Nakanishi, K. (1987) Isolation and structure of a covalent cross-link adduct between mitomycin C and DNA. Science *235*, 1204–1208.
244. Straub, K., Kanne, D., Hearst, J. E. & Rapoport, H. (1981) Isolation and characterization of pyrimidine-psoralen photoadducts from DNA. J. Am. Chem. Soc. *103*, 2347–2355.
245. Kanne, D., Straub, K., Rapoport, H. & Hearst, J. E. (1982) Psoralen-deoxyribonucleic acid photoreaction. Characterization of the monoaddition products from 8-methoxypsoralen and 4,5′,8-trimethylpsoralen. Biochem. *21*, 861–871.
246. Kanne, D., Strabu, K., Hearst, J. E. & Rapoport, H. (1982) Isolation and characterization of pyrimidine-psoralen-pyrimidine photodiadducts from DNA. J. Am. Chem. Soc. *104*, 6754–6764.
247. Dizdaroglu, M. (1986) Chemical characterization of ionizing radiation-induced damage to DNA. BioTechniques, *4*, 536–546.
248. Dizdaroglu, M. (1986) Characterization of free radical-induced damage to DNA by combined use of enzymatic hydrolysis and gas chromatography-mass spectrometry. J. Chromatogr. *367*, 357–366.
249. Karam, L. R., Simic, M. G. & Dizdaroglu, M. (1986) Free radical-induced cross-linking of polydeoxythymidylic acid in deoxygenated aqueous solution. Int. J. Radiat. Biol. *49*, 67–75.
250. Dizdaroglu, M. (1985) Application of capillary gas chromatography-mass spectrometry to chemical characterization of radiation-induced base damage of DNA: Implications for assessing DNA repair processes. Anal. Biochem. *144*, 593–603.
251. Simic, M. G. & Dizdaroglu, M. (1985) Formation of radiation-induced cross-links between thymine and tyrosine: Possible model for cross-linking DNA and proteins by ionizing radiation. Biochemistry *24*, 233–236.
252. Crain, P. F. & McCloskey, J. A. (1983) Analysis of modified bases in DNA by stable isotope dilution gas chromatography-mass spectrometry: 5-methylcytosine. Anal. Biochem. *132*, 124–131.
253. Deutsch, J., Razin, A. & Sedat, J. (1976) Analysis of 5-methylcytosine in DNA. 1. Mass spectrometry. Anal. Biochem. *72*, 586–592.
254. Singer, J. Schnute, W. C., Shively, J. E., Tod, C. W. & Riggs, A. D. (1979) Sensitive detection of 5-methylcytosine and quantitation of the 5-methylcytosine/cytosine ratio in DNA by gas chromatography-mass spectrometry using multiple specific ion monitoring. Anal. Biochem. *94*, 297–301.
255. Gautier, F., Grotjahn, L. & Scheibel, H. M. (1978) Mass spectrometry of nucleic acids: identification and determination of 5-methyl-cytosine in DNA by low and high resolution mass spectrometry In: Quantitative Mass Spectrometry in Life Sciences, Vol. II (DeLeenheer, A. P., Roncucci, R. R. & Van Peteghem, C., eds.) pp. 493–501, Elsevier, Amsterdam.
256. Agris, P. F., Tompson, J. G., Gehrke, C. W., Kuo, K. C. & Rice, R. H. (1980) High-performance liquid chromatography and mass spectrometry of transfer RNA bases for isotopic abundance. J. Chromatogr. *194*, 205–212.
257. Stetson, P. L., Maybaum, J., Shukla, U. A. & Ensminger, W. E. (1986) Simultaneous determination of thymine and 5-bromouracil in DNA hydrolysates using gas chromatography-mass spectrometry with selected ion monitoring. J. Chromatogr. *375*, 1–9.
258. Wiebers, J. L. (1976) Detection and identification of minor nucleotides in intact deoxyribonucleic acids by mass spectrometry. Nucleic Acids Res. *3*, 2959–2970.
259. Bozorgzadeh, M. H., Beynon, J. H. & Wiebers, J. L. (1980) Structural determination of a hypermodified nucleoside in viral DNAs by pyrolysis-electron impact-mass-analysed ion kinetic energy spectrometry. Adv. Mass Spectrom. *8A*, 971–982.
260. Jankowski, K. (1983) In: Recent Developments in Mass Spectrometry in Biochemistry, Medicine and Environmental Research, Vol. 8 (Frigerio, A., ed.) pp. 203–210, Elsevier, Amsterdam.
261. Jankowski, K. & Soler, F. (1980) Mass spectrometry of DNA: Part 2. Quantitative es-

timation of base composition. Europ. J. Mass. Spectrom. Biochem. Med. Environ. Res. *1*, 45–52.

262. Turkkan, N., Soler, F. & Jankowski (1982) Mass spectra of nucleic acids 4-Base quantitation. Biomed. Mass Spectrom. *9*, 91–93.
263. Puzo, G. & Weibers, J. L. (1981) Identification of modified nucleosides in intact transfer ribonucleic acid by pyrolysis-electron impact-collisional activation mass spectrometry. Nucleic Acid Res. *9*, 4655–4667.
264. Hignite, C. (1980) In: Biochemical Applications of Mass Spectrometry: First Supplementary Volume (Waller, G. R. & Dermer, O. C., eds.) p. 560, Wiley, New York.
265. Virelizier, H., Hagemann, R. & Jankowski, K. (1983) Systematic analysis of desorption chemical ionization (DCI) mass spectra of nucleic acids-7. Biomed. Mass Spectrom. *10*, 559–566.
266. McNeal, C. J., Narang, S. A., MacFarlan, R. D., Hsiung, H. M. & Brousseau, R. (1980) Sequence determination of protected oligodeoxyribonucleotides containing phosphotriester linkages by californium-252 plasma desorption mass spectrometry. Proc. Natl. Acad. Sci. USA *77*, 735–739.
267. McNeal, C. J., Ogilvie, K. K., Theriault, N. Y. & Nemer, M. J. (1982) A new method for sequencing fully protected oligonucleotides using ^{252}Cf-plasma desorption mass spectrometry. 1. Negative ions of dinucleoside monophosphates. J. Am. Chem. Soc. *104*, 972–975.
268. McNeal, C. J., Ogilvie, K. K., Theriault, N. Y. & Nemer, M. J. (1982) A new method for sequencing fully protected oligonucleotides using ^{252}Cf-plasma desorption mass spectromctry. 2. Negative ions of subunits in the stepwise synthesis of a heptaribonucleotide. J. Am. Chem. Soc. *104*, 976–980.
269. McNeal, C. J., Ogilvie, K. K., Theriault, N. Y. & Nemer, M. J. (1982) A new method for the analysis of fully protected oligonucleotides by ^{252}Cf-plasma desorption mass spectrometry. 3. Positive ions. J. Am. Chem. Soc. *104*, 981–984.
270. Viari, A., Ballini, J.-P., Vigny, P., Shire, D. & Dousset, P. (1987) Sequence analysis of unprotected trideoxy-ribonucleoside diphosphates by ^{252}Cf-plasma desorption mass spectrometry. Biomed. Environ. Mass Spectrom. *14*, 83–90.
271. Grotjahn, L., Frank, R. & Blocker, H. (1982) Ultrafast sequencing of oligodeoxyribonucleotides by FAB-mass spectrometry. Nucleic Acids Res. *10*, 4671–4678.
272. Grotjahn, L., Frank, R. & Blocker, H. (1983) Sequencing of oligodeoxyribonucleotides by negative FAB-MS. Int. J. Mass. Spectrom. Ion Phys. *40*, 439–443.
273. Grotjahn, L., Blocker, H. & Frank, R. (1985) Mass spectroscopic sequence analysis of oligonucleotides. Biomed. Mass Spectrom. *12*, 514–524.
274. Grotjahn, L. & Steinert, H. (1985) Mass spectrometry in molecular design. In: Mass Spectrometry in the Health and Life Sciences (Burlinghame, A. L. & Castagnoli, N., eds.) pp. 597–615, Elsevier, Science Publishers B. V., Amsterdam.
275. Panico, M., Sidona, G. & Uccella, N. (1983) Bioorganic applications of mass spectrometry. 3. Fast-atom-bombardment-induced zwitterionic oligonucleotide quasimolecular ions sequenced by MS/MS. J. Am. Chem. Soc. *105*, 5607–5610.
276. Grotjahn, L., Frank, R., Geisterberg-Moutsis, G. & Blocker, H. (1984) Fast identification by FAB-mass spectrometry of building blocks for oligonucleotide synthesis. Tetrahedron Lett. *25*, 5373–5376.
277. Ulrich, J., Guy, A., Molko, D. & Teoule, R. (1984) Fast atom bombardment mass spectra of protected deoxyribonucleotides. Org. Mass. Spectrom. *11*, 585–588.
278. Ens, W., Standing, K. G., Westmore, J. B., Ogilvie, K. K. & Nemer, M. J. (1982) Secondary ion mass spectrometry of protected diribonucleoside monophosphates with a time-of flight mass spectrometer. Anal. Chem. *54*, 960–966.
279. Beavis, R., Ens, W., Nemer, M. J., Ogilvie, K. K., Standing, K. G. & Westmore, J. B. (1983) Secondary ion mass spectrometry of protected oligonucleotides. Int. J. Mass Spectrom. Ion Phys. *40*, 475–478.
280. Edmonds, C. G., McCloskey, J. A. & Edmonds, V. A. (1983) A bibliography of combined liquid chromatography mass spectrometry. Biomed. Mass Spectrom. *10*, 237–252.
281. Games, D. E. & Lewis, E. (1980) Combined liquid chromatography mass spectrometry of glycosides, glucuronides sugars and nucleosides. Biomed. Mass Spectrom. *7*, 433–436.
282. Eckers, C., Games, D. E., Games, M. L., Kuhnz, W., Lewis, E., Weerasinghe, N. C. A. & Westwood, S. A. (1981) In: Recent Developments in Mass Spectrometry in Biochemistry, Medicine and Environmental Re-

search, Vol. 7 (Figerio, A., ed.) pp. 169–182, Elsevier, Amsterdam.

283. Games, D. E., Hirter, P., Kuhnz, W., Lewis, E., Weerasinghe, N. C. A. & Westwood, S. A. (1981) Studies of combined liquid chromatography-mass spectrometry with a moving-belt interface. J. Chromatogr. *203*, 131–138.

284. Games, D. E., McDowall, M. A., Levsen, K., Schafer, K. H., Dobberstein, P. & Gower, J. L. (1984) A comparison of moving belt interfaces for liquid chromatography mass spectrometry. Biomed. Mass Spectrom. *11*, 87–95.

285. Games, D. E. (1983) Combined liquid chromatogoraphy/mass spectrometry (LC/MS), Spectra *9*, 3–8.

286. Stroh J. G., Cook, J. C. & Rinehart, K. J., Jr. (1984) Liquid chromatrography / fast atom bombardment mass spectrometry (LC/FABMS) Abstr. 32nd Ann. Conf. Mass Spec. Allied Top., May 27–June 1, San Antonio, TX, p. 119.

287. Esmans, E. L., Luyten, Y. & Alderweireldt, F. C. (1983) Liquid chromatography mass spectrometry of nucleosides using a commercially available direct liquid introduction probe. Biomed. Mass Spectrom. *10*, 347–351.

288. Alderweireldt, F. C., Esmans, E. L. & Geboes, P. (1985) HPLC and LC-MS of nucleosides. Nucleosides Nucleotides *4*, 135–137.

289. Ip, C. Y., Ha, D., Morris, P. W., Puttemans, M. L. & Venton, D. L. (1985) Separation of nucleosides and nucleotides by reversed-phase high-performance liquid chromatography with volatile buffers allowing sample recovery. Anal. Biochem. *147*, 180–185.

290. Greenhut, J. & Rudolph, F. B. (1985) Preparative separation of nucleotides by high-performance-ion exchange liquid chromatography using a volatile buffer system. J. Chromatogr. *319*, 461–466.

291. Esmans, E. L., Geboes, P., Luyten, Y. & Alderweireldt, F. C. (1985) Direct liquid introduction LC/MS microbore experiments for the analysis of nucleoside material present in human urine. Biomed. Mass Spectrom. *12*, 241–245.

292. Blakley, C. R. & Vestal, M. L. (1983) Thermospray interface for liquid chromatography/mass spectrometry. Anal. Chem. *55*, 750–754.

293. Vestal, M. L. (1983) Studies of ionization mechanism involved in thermospray LC-MS. Int. J. Mass Spectrom. Ion Phys. *46*, 1293–1296.

294. Blakley, C. R., McAdams, M. J. & Vestal, M. L. (1976) Crossed-beam liquid chromatograph-mass spectrometer combination. J. Chromatogr. *158*, 261–276.

295. Edmonds, C. G., Vestal, M. L. and McCloskey, J. A. (1985) Thermospray liquid chromatography-mass spectrometry of nucleosides and of enzymatic hydrolysates of nucleic acids. Nucleic Acids Res. *13*, 8197–8206.

296. McCloskey, J. A., Crain, P. F., Edmonds, C. G., Gupta, R., Hashizume, T., Phillipson, D. W. & Setter, K. O. (1987) Structure determination of a new fluorescent tricyclic nucleoside from archaebacterial tRNA. Nucleic Acids Res. *15*, 683–693.

297. Voyksner, R. D., Williams, F. P. & Hines, J. W. (1985) Analysis of candidate anticancer drugs by thermospray high-performance liquid chromatography mass spectrometry and by high resolution mass spectrometry. J. Chromatogr. *347*, 137–146.

Chapter 11
Steroids

S. J. Gaskell

1 Introduction

The steroids are a functionally and chemically diverse group of compounds. Of the steroids that are of interest in clinical biochemistry, the bile acids and vitamins D are considered elsewhere in this volume. The principal emphasis of this chapter is the analysis of the steroid hormones. The use of mass spectrometry as a quantitative reference technique for the analysis of steroids is described in detail in Chapter 13.

The steroid hormones may be classified according to carbon number. The oestrogens, which are C_{18} steroids, are produced principally in the ovaries and the foeto-placental unit though the testes, adrenal cortices and peripheral tissues represent minor sources. The androgens, C_{19} steroids, derive mainly from the male gonads, to a lesser extent from the ovaries, and from the adrenal cortices of both sexes. The C_{21} steroids

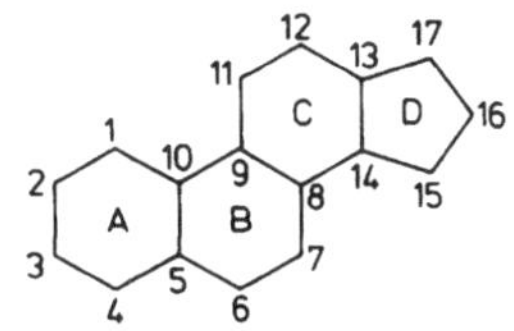

I

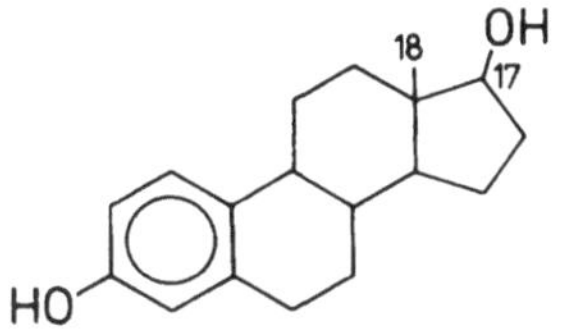

II

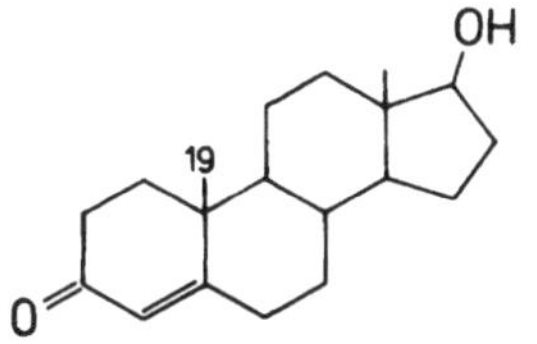

III

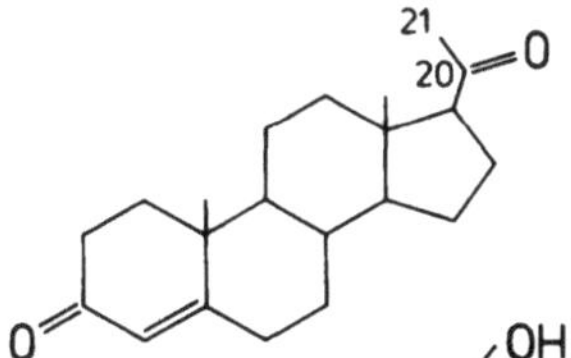

IV

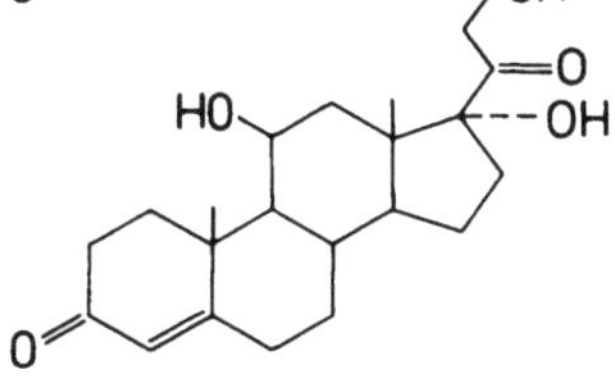

V

Figure 1. Steroid structures (systematic names are given in parentheses). I: gonane nucleus; II: oestradiol-17β (1,3,5-oestratriene-3,17β-diol); III: testosterone (17β-hydroxy-4-androsten-3-one); IV: progesterone (4-pregnene-3,20-dione); V: cortisol (11β,17α,21-trihydroxy-4-pregnene-3,20-dione).

Table 1. Reviews on the biochemical application of mass spectrometry to the analysis of steroids

Topic	References
Isolation procedures	1, 2, 3
Derivatisation	1, 2, 4, 5
Quantitative analyses	4, 6–9
Mass spectrometric methods	1, 2, 4, 6
Steroids in physiological fluids	1, 2, 6, 9, 11–14
Steroids in tissues	2, 6, 12, 16, 17

include the corticosteroids, derived from the adrenal cortex, and progesterone (and its metabolites) which is produced mainly in the ovaries and foeto-placental unit. Structures representative of each group are shown in Figure 1. The oestrogens incorporate an aromatic A-ring with a hydroxy substituent at the 3-position; their consequent acidic nature has important implications for their analysis (Section 2.2). Hydroxyl and keto groups are the major functionalities in steroid hormone structures; conversion of these groups to more stable derivatives is frequently required for gas phase analysis (Section 2.3). Sulphate and glucuronide conjugates of hydroxy steroids are quantitatively important metabolites. Sulphates of the enol forms of keto steroids have also been described. The analytical chemistry of the steroid hormones is further influenced by stereochemistry, both of the ring junctions and of the substituents. The configurations of the B/C and C/D ring junctions are invariably trans but both cis and trans configurations are observed for the A/B junction. The stereochemistry of both the steroid nucleus and the substituents is of fundamental biological importance; thus, for example, 5α-dihydrotestosterone is the principal male sex hormone active at the cellular level but the 5β-analogue has negligible androgenic activity. Similarly, oestradiol-17β is the major female sex hormone while oestradiol-17α is inactive.

The first part of this chapter deals briefly with isolation methods, derivatisation reactions and mass spectrometric techniques of particular importance in the analysis of steroids. Mass spectrometric analyses of steroids in physiological fluids and tissue are then considered. Some discussion is included to indicate the most useful analytical approaches. Reviews which complement the coverage of the present chapter are listed in Table 1.

2 Analytical Procedures

2.1 Extraction techniques

Procedures for the recovery of steroids from biological samples, using solvent and solid-phase extraction, have been recently and comprehensively surveyed by Sjövall and Axelson (1); accordingly, a briefer treatment is given here, with particular emphasis on the most recent procedures.

Table 2. Solid phase extraction techniques

Solid material	Principal applications	References
Polystyrene resins	Extraction of unconjugated and conjugated steroids and lipids	1, 18–20
Substituted dextran gels	Extraction of unconjugated steroids, cholesterol and lipids (extractions from plasma require addition of pentylamine)	1, 25, 26
Substituted porous silica	Extraction of conjugated and unconjugated steroids, but not cholesterol or lipids	1, 22, 23
Diatomaceous earth	Extraction of unconjugated steroids from semen or follicular fluid	27–29
Coupled antisera	Extractions of single or closely-related steroids; applicable to extractions from plasma, urine and saliva	31–34

Solvent extraction remains widely used for the recovery of individual steroids, or groups of steroids of similar polarity, from physiological fluids. Adjustment of the solvent polarity provides a means of influencing the selectivity of extraction. The frequent requirement for large solvent volumes increases the need for high solvent purity and exacerbates problems of solvent removal and toxicity. The risk of emulsion formation constitutes another disadvantage of solvent extraction.

A number of solid phases have been used in extractions of steroids from physiological fluids (Table 2). The neutral polystyrene resin, Amberlite XAD-2, has been widely used for extractions of conjugated and unconjugated steroids from urine since its introduction by Bradlow (18). A modification of the procedure, involving a prior wash of the gel with triethylammonium sulphate, was incorporated following changes in the properties of the commercially-obtained polymer, to ensure good recoveries of steroid sulphates (19). If the extraction is performed at 64°C, to minimise protein binding, then efficient recoveries may also be achieved from plasma after dilution with saline and addition of triethylammonium sulphate (20, 21).

The disadvantages of the procedure (risk of contaminants, slowness and low capacity of the solid phase) have been substantially avoided in the subsequent development of extractions using substituted porous silica. Shackleton and Whitney reported an extensive study of the use of Sep-pak cartridges (containing octadecylsilyl silica, widely used as a solid phase in high pressure liquid chromatography) for the extraction of unconjugated and monoconjugated steroids from urine and urine hydrolysates (22). The procedure comprised rapid (30 ml/min) passage of the sample through the cartridge, water washing and elution with 2 ml methanol. Extraction was therefore simple and rapid, and high recoveries of unconjugated and monoconjugated steroids were obtained. Similar observations were made by Heikkinen et al. on the extraction of oestrogens (including a mixed sulphate/glucuronide conjugate) from urine (23).

Recovery of steroids from blood plasma using this procedure is again influenced by the presence of binding proteins. Heikkinen et al. added triethylammonium sulphate to plasma to counteract steroid-protein interactions; specific binding of oestradiol to sex-hormone binding globulin was suppressed by the addition of the synthetic steroid D-norgestrel (23). Under these conditions, high extraction efficiencies were obtained for a representative set of oestrogens consisting of oestrone, oestradiol, oestrone sulphate, oestriol 16-glucuronide and oestriol 3-sulphate, 16-glucuronide. Axelson and Sahlberg achieved high recoveries of steroids from plasma and milk by performing the extraction at 64°C, following addition of triethylammonium sulphate (in parallel with the strategy adopted for extractions using Amberlite XAD-2) (21). In the procedure reported by Cannell et al. (24), steroid-protein interactions were destroyed by prior acid treatment of plasma at pH 2. Hydrolysis of conjugates was demonstrated not to occur under these conditions but the use of low pH leaves open the question of survival of acid-labile steroids.

Solid phase-coupled antisera are now commonly used in immunoassay to facilitate the separation of free and bound analyte. When used at high concentrations, high efficiencies of extraction from body fluids may be achieved. Dyas et al., for example, used an anti-oestrogen serum to extract ethynyloestradiol from serum prior to immunoassay (30); recoveries of greater than 95% were reported. Glencross et al. reported the use of Sepharose-coupled anti-oestradiol-17β to extract this steroid from bovine milk (31). Gaskell et al. have used the technique to isolate the following steroids from blood plasma or serum, prior to GC-MS analysis: oestradiol-17β (32), cortisol (33) and progesterone (34). Immunoadsorption was also used to isolate dehydroepiandrosterone sulphate from urine and plasma prior to analysis by fast atom bombardment mass spectrometry (Section 2.5.1). Antisera were coupled to cyanogen bromide-activated microcellulose or Sepharose; the choice of solid phase depended on whether a batch or column technique procedure was to be employed. The selectivity of immunoadsorption, based largely on molecular size and configuration, usefully complements the discrimination achieved during GC-MS. Total specificity of extraction, however, is not achieved. Thus, for example, extraction using an anti-oestradiol-17β serum also gives good recoveries of related oestrogens (32), suggesting the application of the technique to the simultaneous analysis of related steroids, provided adequate internal standardisation (ideally through the use of isotopically-labeled analogues) is performed. In general, however, immunoadsorption extraction will be most valuable in analyses of one or two steroids, whereas other forms of solid-phase extraction (for example, octadecylsilyl silica) will be preferred for steroid profile studies.

Recovery of steroids from tissue homogenates requires initial solvent extraction. Assessment of extraction efficiencies is normally not possible; thus, a high recovery of analyte added to a tissue homogenate is a necessary, but insufficient, criterion for good recovery of endogenous analyte. (Rigorous internal standardisation of quantitative analyses is correspondingly difficult). A combination of solvents that penetrates both hydrophilic and hydrophobic structures, such as chloroform/methanol, has been advocated by Sjövall and Axelson (1) and appears to represent the most generally useful approach for the extraction of unconjugated steroids. Preliminary purification of solvent extracts may be performed by evaporation together with Lipidex 1000 and subsequent elution through a Sep-pak cartridge (35). For certain applications, prior digestion of tissue homogenate may provide a fluid sample from which steroids may be more readily extracted using solvent or solid-phase techniques. Batra and Bengtsson (36) extracted progesterone from human and rabbit myometrium following digestion with a mixture of alkali and detergent (0.5 mol $\cdot$ l^{-1} sodium hydroxide containing 5% sodium dodecyl sulphate). Extraction efficiency, using ethyl acetate, was assessed using tissue previously allowed to accumulate [^{3}H]progesterone in vitro and was found to

Table 3. Examples of ion exchange chromatography of steroids using Sephadex gels

Separation	Gel*	References
Unconjugated steroids:		
Oestrogens/neutral steroids	TEAP Lipidex 5000	44
	QAE-Sephadex A-25	45
	DEAE-Sephadex A-25	54
4-ene 3-oximes/saturated	SE-Sephadex LH20	46
3-oximes/others	SP-Sephadex LH20	47
Ethynyl steroids/others	SP-Sephadex LH 20 (Ag^+ form)	48, 49
Conjugated steroids:		
Unconjugated/glucuronides/ monosulphates/disulphates	DEAP-Sephadex LH20	50, 51
	TEAP-Sephadex LH20	52
Unconjugated/oestrogen glucuronides/ neutral steroid sulphates/oestrogen monosulphates/neutral steroid monosulphates/disulphates and mixed conjugates	DEAE-Sephadex A-25	53

* TEAP = triethylammoniohydroxypropyl; QAE = diethyl(2-hydroxypropyl)-aminoethyl; SE = sulphoethyl; SP = sulphohydroxypropyl; DEAP = diethylaminohydroxypropyl; DEAE = diethylaminoethyl.

exceed 93%. Proteolytic enzymes may also be used effectively to solubilise tissue homogenates (37); concurrent reactions mediated by endogenous enzymes cannot, however, be excluded unless appropriate precautions are taken. This approach is, nevertheless, likely to become of increasing importance.

In a detailed study, van Landeghem et al. evaluated procedures for the extraction of steroids from human uterine tissue cytosol and crude nuclear fractions (38). Uterine tissue was obtained from patients infused with [^{3}H]oestradiol-17β or [^{3}H]oestrone before hysterectomy. Diethyl ether extraction of cytosol gave incomplete extraction of unconjugated steroids whereas ethanol/acetone (1 : 1) gave a high recovery of both conjugated and unconjugated steroids. Ethanol/acetone was also effective for the extraction of nuclear fractions. Bonney et al. (39), however, have determined that extraction of oestrone, oestradiol, 5-androstene-3β,17β-diol and dehydroepiandrosterone from human endometrial tissue is achieved with similar efficiency using diethyl ether or ethanol/acetone (1 : 1).

2.2 Fractionation procedures

Liquid-gel chromatography using derivatives of Sephadex has assumed increasing importance in the purification of biological extracts for steroid analysis; the principal advantages of the gels are inertness, low bleed and high sample capacity. Much of the technique development has been by the Stockholm and Helsinki groups; recent reviews are available (3, 40). Hydroxypropyl Sephadex (Sephadex LH-20) swells in non-aqueous solvents; early application (41) was made to the purification of steroid extracts by straight-phase chromatography and its use continues (e. g. ref. 42). Preparation of long-chain hydroxyalkyl Sephadex (commercially available as Lipidex) gave gels which could be used in either straight or reversed-phase modes (43). The particular versatility of the Sephadex gels is apparent from the variety of derivatives that have been prepared for ion exchange chromatography; some derivatives and applications are summarised in Table 3. Despite the mildness of the separation conditions, labile oestrogens require protection by incorporation of ascorbic acid as anti-oxidant; Adlercreutz and co-workers have described in detail procedures based on gel chromatography for the analysis of oestrogens (53–55). Catechol oestrogens may be selectively isolated by chromatography on DEAE-Sephadex in borate form (56).

High pressure liquid chromatography (HPLC) provides wide scope for the isolation of individual steroids prior to mass spectrometric analysis. The topic is the subject of a book (57) and other review articles (58, 59).

2.3 Derivatisation

Reviews which incorporate detailed discussions of the derivatisation of steroids for gas phase analysis have been cited in Table 1. A number of procedures applicable to steroid analysis are included in the comprehensive book on derivatisation in chromatography edited by Blau and King (60). Common derivatives of steroid hydroxy and keto groups are summarised in Tables 4 and 5, respectively. For the identification of steroids, initial analyses using trimethylsilyl (TMS) ether and methyl oxime derivatives are generally advisable in view of the extensive background data which exist concerning diagnostically useful fragmentations. Subsequent analyses using alternative derivatives, such as *tert*-butyldimethylsilyl ethers, may be beneficial, for example to confirm molecular weight. The objective in quantitative analyses is to convert to a derivative which yields a single, or few, abundant ions of high mass, thereby maximising sensitivity during selected ion monitoring. Sterically hindered alkylsilyl ethers and perfluorinated acyl esters are frequently useful in this regard.

The influence of the derivative group on mass spectrometric fragmentation under electron impact is exemplified by the spectrum of the methyl oxime, TMS ether of pregnenolone (Fig. 2). The major fragmentation modes are indicated. Ions of m/z 129 and $[M-129]$ are characteristic of the presence of a Δ^5-3-trimethylsilyloxy group. $[M-31]$ ions indicate a methyl oxime function, while $[M-90]$ ions are associated with TMS ethers. More detailed reviews of the fragmentations of the commoner steroid derivatives have been prepared (1, 61).

Gas phase analysis of steroid glucuronides may be performed after suitable derivatisation, as first demonstrated by Horning and coworkers (81). GC-MS data have been reported for oestrogen glucuronides as per-methyl (82), per-TMS (83), *n*-propyl ester, TMS ether (84) and methyl ester, TMS ether (85) derivatives. Molecular ions formed under electron impact are generally weak. Mass spectrometric differentiation of isomeric conjugates may be achieved, exemplified by the distinction between the aliphatic and aromatic conjugates of oestradiol-17β as per-TMS derivatives (Fig. 3). GC of urinary steroids without prior hydrolysis has been reported by Shackleton (86);

Table 4. Common derivatives of steroid hydroxy groups

Derivative*	Structure	Mass increment	Comment	Principal references**
Isolated hydroxyl:				
TMS	$-OSi(CH_3)_3$	72	Extensive literature on MS and GC properties; MS fragmentation strongly directed by derivative	1, 61, 62
t-BDMS	$-OSi(CH_3)_2C_4H_9$	114	Used mainly for selected ion monitoring analyses; $[M-C_4H_9]^+$ intense under EI; resistant to hydrolysis	2, 63, 64
Other alkylsilyl	$-OSi(CH_3)_2R$ or $-OSiR_3$	–	May provide structural information and intense $M^{+\cdot}$ or $[M-R]^+$	2, 65, 66
Perfluoroacyl:				
pentafluoropropionyl	$-OCOC_2F_5$	146	$M^{+\cdot}$ commonly intense; useful for selected ion monitoring	67
heptafluorobutyryl	$-OCOC_3F_7$	196		67, 68
Methyl ether	$-OCH_3$	14	Phenols readily derivatised	61
Proximal hydroxyls:				
Acetonide	$-O$ \ $C(CH_3)_2$ / $-O$	40		69, 70
Boronate e. g. methyl boronate	$-O$ \ BCH_3 / $-O$	24	Readily prepared under mild conditions; suspectible to hydrolysis	69, 71

* TMS = trimethylsilyl; *t*-BDMS = *tert*-butyldimethylsilyl
** See also references given in the papers cited

methylation, methoximation and trimethylsilylation yielded suitable derivatives of the glucuronides for gas phase analysis; mass spectrometric properties have not, however, been reported. The conditions of trimethylsilylaton also effected displacement of sulphate groups from the major urinary steroid sulphates (principally 3-sulphates of 5-ene steroids) with consequent formation of TMS ethers of the parent steroids. Analogoues studies (87) with perfluoroacylating reagents indicated that the sulphate group was readily displaced from the aromatic sulphates of the oestrogens. For aliphatic steroid 3-sulphates, the presence of Δ^5 unsaturation was required for displacement of the sulphate group to predominate over elimination of H_2SO_4.

Table 5. Common derivatives of steroid keto groups

Derivative*	Structure	Mass increment	Comments	Principal references**
Oxime:			*syn* and *anti* forms frequently observed	
Methyl oxime	$-NOCH_3$	29	extensive reference data available	1, 72–74
Other alkyl oxime	$-NOR$	(R + 14)	Used to achieve mass or retention time shifts	75
TMS oxime	$-NOSi(CH_3)_3$	99		46
t-BDMS oxime	$-NOSi(CH_3)C_4H_9$	141	intense $[M-C_4H_9]^+$ ions under electron impact	64, 76
Enol ether:				
enol-TMS	$-OSi(CH_3)_3$	72	Susceptible to hydrolysis; intense $M^{+\cdot}$ under electron impact	70
enol-t-BDMS	$-OSi(CH_3)_2C_4H_9$	114	Intense $M^{+\cdot}$ under electron impact	78, 79
enol-HFB	$-OCOC_3F_7$	196	Intense $M^{+\cdot}$ under electron impact	67, 80

* TMS = trimethylsilyl: *t*-BDMS = *tert*-butyldimethylsilyl
** See also references given in the papers cited

The site of conjugation in monosulphates of steroid diols may be determined by GC-MS analysis of the products of acetylation of the free hydroxyl group, solvolytic cleavage of the sulphate moiety and trimethylsilylation (88). Mass spectra of the derivatives prepared in this way from 5-pregnene-3,20-diol 3- and 20-sulphates are shown in Figure 4. The two products are clearly distinguished and the spectra again illustrate the major influence of derivative groups on fragmentation of steroids. The origins of the base peaks in the spectra are shown in Figure 5. Losses of 60 and 90 correspond to acetic acid and trimethylsilanol, respectively.

Characterisation of steroid conjugates has been facilitated by the application of secondary ion mass spectrometric techniques, as discussed in Section 2.5.1.

2.4 Mass spectrometry

A full discussion of mass spectrometric procedures is outside the scope of this chapter. This section considers briefly aspects of mass spectrometry particularly pertinent to the analysis of steroids in clinical samples.

2.4.1 Sample introduction and ionization techniques

Where the volatility and stability of the analytes (or simply prepared derivatives) permit, combined GC-MS remains the most effective way of achieving controlled and reproducible sample introduction into the mass spectrometer ion source. The GC of

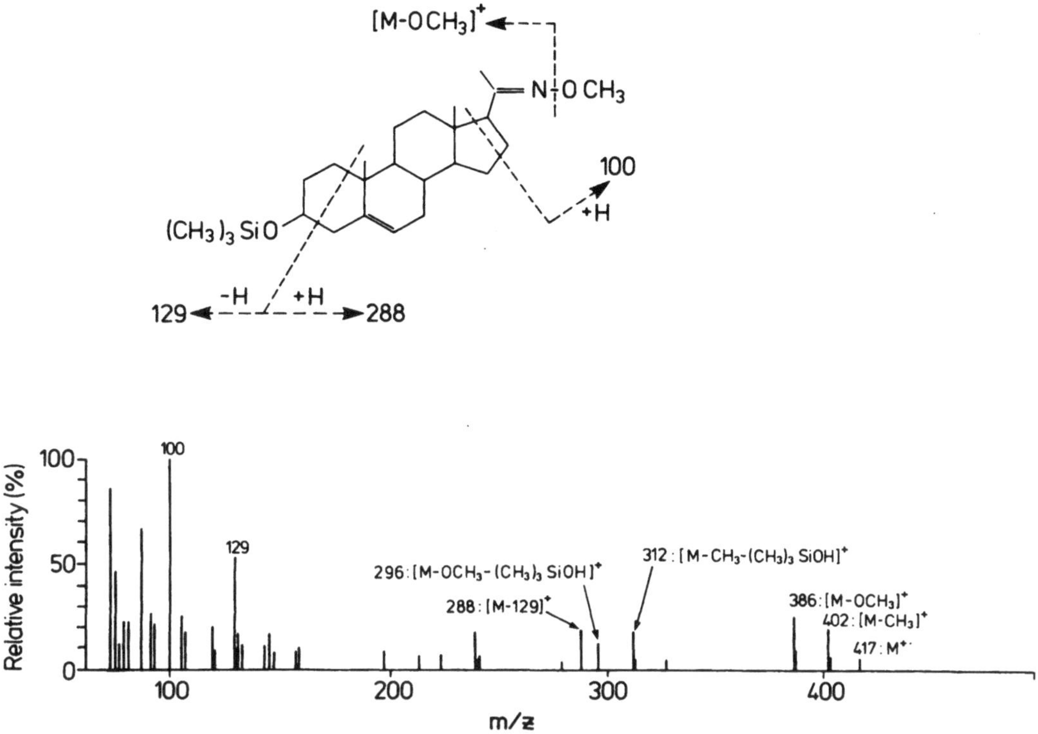

Figure 2. Electron impact mass spectrum of pregnenolone methyl oxime, TMS ether (61). The origins of the major ions are indicated.

steroids was described in 1960 (89) and the first report of combined GC-MS analysis appeared in 1966 (90). The subsequent introduction of open-tubular capillary columns for steroid analysis (91) greatly improved the separation capability. The development of bonded-phase fused silica capillary columns for gas chromatography has been particularly important in facilitating high temperature operation without degradation of column performance or extensive bleed of stationary phase into the ion source. Direct coupling of capillary columns to the source, bypassing the molecular separators which are generally required when packed columns are used, lessens the risk of adsorptive losses or degradation.

Electron impact (EI) is still the most commonly used ionization technique for the analysis of unconjugated steroids; a large body of data now exists for the commoner derivatives and excellent reviews are available (1, 6). Chemical ionization mass spectrometry (CI-MS) of steroids has been reviewed recently by Lin and Smith (92); the number of applications, however, remains relatively few. The principal advantage of the technique is that clear information on molecular mass is generally available. Thus, for example, MH^+ ions constitute the base peak in the methane positive CI spectrum of 20β-trimethylsilyloxy 4-pregnen-3-one (94). The molecular ion is not observed in the EI spectrum, which is dominated by *m/z* 117, formed by cleavage alpha to the TMSO group. Interestingly, in this case the structurally informative fragmentation is

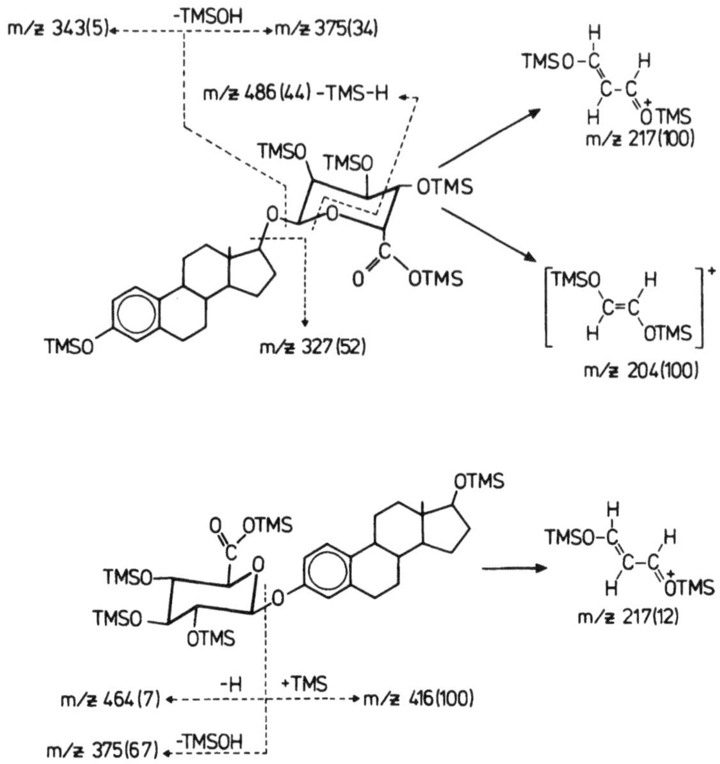

Figure 3. Fragmentation of isomeric oestradiol-17β glucuronides under electron impact (83). The relative abundances of the fragment ions are shown in parentheses. Ions of *m/z* 204 and 217 are derived from the glucuronic acid moiety; their exact mechanisms of formation have not been established.

retained in the CI spectrum. Methane positive CI was used by Gambert and coworkers in the quantification of cholesterol in plasma by GC-MS/ selected ion monitoring (SIM) (95). The ammonia positive CI spectrum of cortisol 3,20-bis-methyl oxime-11,17,21-tris-trimethylsilyl ether is dominated by the protonated molecular ion (Fig. 6); Lindberg et al. reported the quantitative analysis of cortisol in plasma and urine by GC-MS-SIM of MH^+ ions (96). The use of ammonia as reagent may also be beneficial in achieving a degree of selectivity of ionization; the ionization efficiency of 4-cholesten-3-one by NH_4^+, for example, is substantially higher than that of the isomeric 5-cholesten-3-one (97). Negative CI (NICI) of steroids has been reported (98) and reviewed (92). Raju et al. (93) have recently used NICI-MS (with OH^- as reagent ion) for the profiling of androsterone long chain fatty acid esters in human breast cyst fluid. Electron capture (leading to formation of negative ions) is frequently confused with NICI but is mechanistically distinct. Houghton et al. (99) reported analyses of methyl oxime, TMS ether derivatives of synthetic corticosteroids using this technique; abundant diagnostic ions were observed. The improvement of detectability that might accrue during GC-electron capture MS of steroids converted to strongly electron-capturing derivatives has been investigated (100, 101). Analyses of steroid pentafluo-

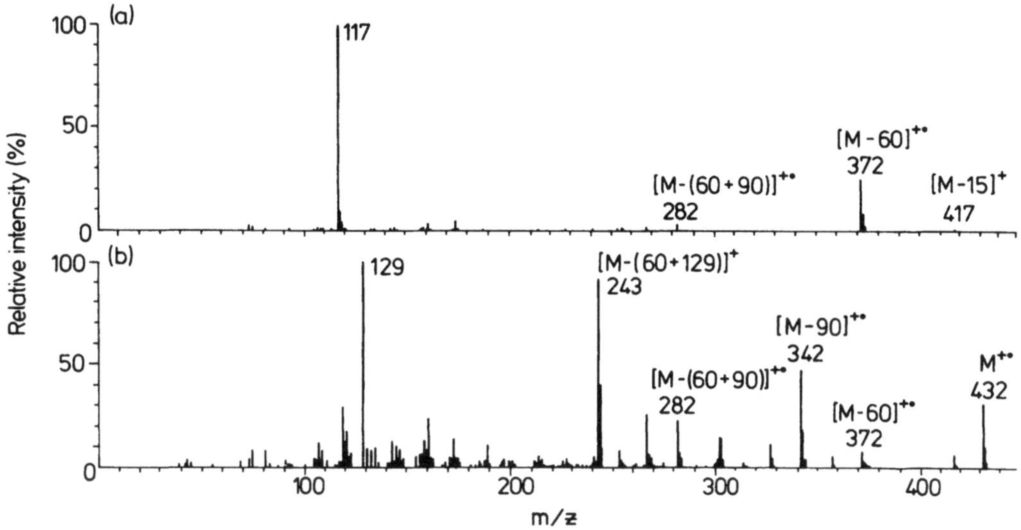

Figure 4. Electron impact mass spectra of 5-pregnene-3β,20α-diol as the 3-acetate, 20-TMS ether (a) and the 3-TMS ether, 20-acetate (b). Losses of 15, 60, and 90 correspond to methyl, acetic acid and trimethylsilanol, respectively (88). The origins of *m/z* 117 and 129 are shown in Figure 5.

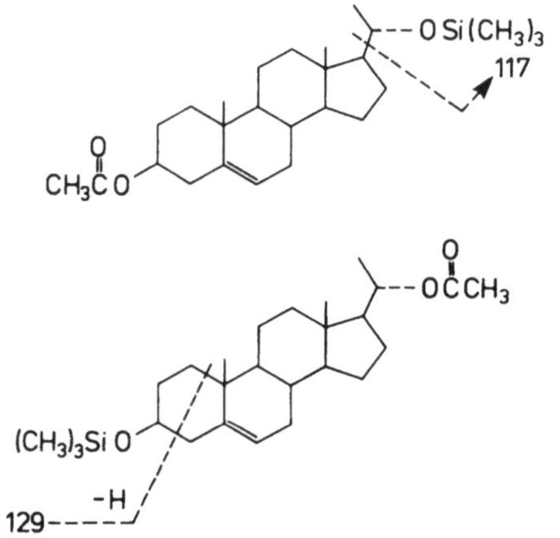

Figure 5. Origins of the base peaks in the spectra of 5-pregnene-3β,20α-diol as the 3-acetate, 20-TMS ether and the 3-TMS ether, 20-acetate (88).

robenzyloximes and heptafluorobutyrates provided satisfactory sensitivities of detection (100) but no improvement over EI-MS was apparent. Conversion of hydroxyl groups to bis-trifluoromethylbenzoyl derivatives (101) has been shown to permit subpicogramme detection during SIM but again this is of the same order as that achieved during EI-MS of suitable derivatives.

CI-MS of underivatised steroid glucuronides has been reported (102, 103) but the newer techniques of secondary ion mass spectrometry (including fast atom bombard-

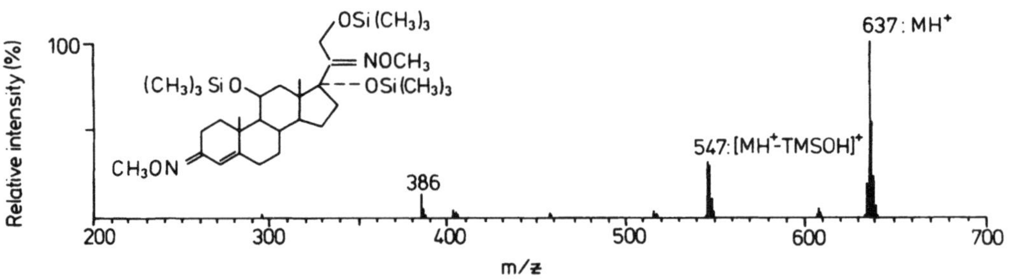

Figure 6. Partial ammonia chemical ionization mass spectrum of cortisol 3,20-bis-methyl oxime, 11,17,21-tris-TMS ether (96).

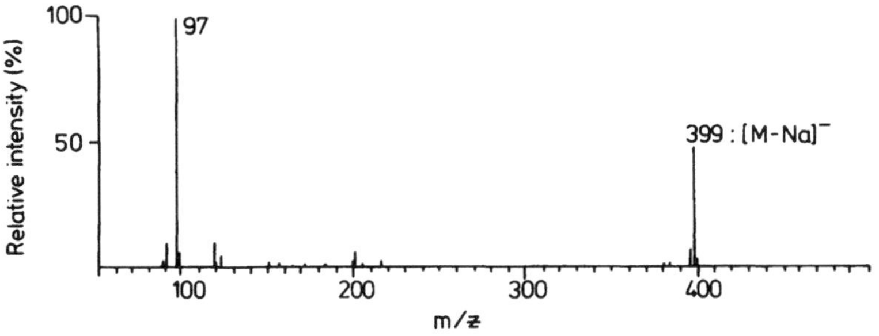

Figure 7. Negative ion secondary ion mass spectrum of 5α-pregnane-3β,20α-diol 20-sulphate, sodium salt (106).

ment; FAB), using a liquid sample matrix, are finding wider applicability in the analysis of intact conjugates. The spectra produced using primary beams consisting of ions (such as Cs^+) and atoms (such as Ar or Xe) have been shown to be similar (104). Positive and negative ion spectra of sulphate and glucuronide conjugates have been reported (105–110); the latter are generally to be preferred since (for mono-conjugates) they are independent of the cationic species. The negative ion spectrum of pregnanediol 20-sulphate is shown in Figure 7. *m/z* 97, attributable to HSO_4^-, is characteristic of sulphate conjugates. Little other structural information is available. A major consequential limitation is the difficulty in distinguishing isomeric structures. Recent studies (111, 112) using collisionally activated decomposition (CAD) and tandem MS have indicated that distinct daughter ion spectra may be obtained from isomeric steroid glucuronides. The distinction is influenced by experimental conditions, however (113), and the method will be of little utility for the analysis of mixtures of isomers.

The application of secondary ion MS (SIMS) to urinary steroid profiling is described in Section 3.1. Combined liquid chromatography (LC)-MS has not been widely applied in steroid analysis but its use is increasing; in most instances, the thermospray (TSP) interface (114) has been employed. Qualitative (111, 115) and quantitative (116) analyses of unconjugated steroids have been reported but the approach is most obviously advantageous for the study of intact conjugates. Analyses of both glucu-

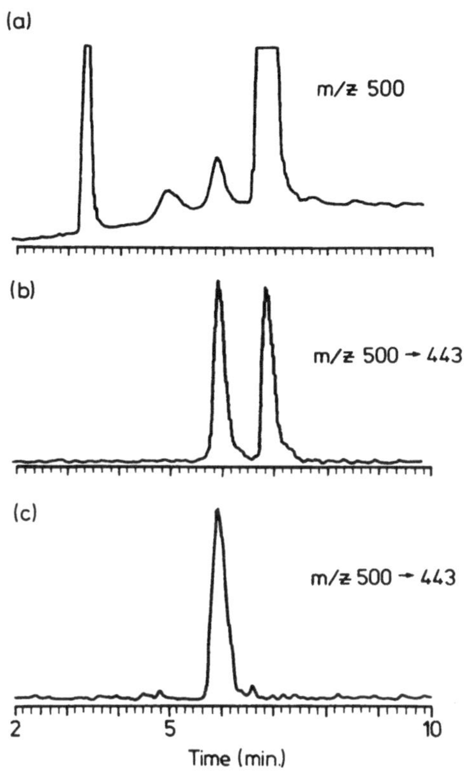

Figure 8. GC-MS analysis for oestradiol-17β in aliquots of an extract of blood plasma, as the *tert*-butyldimethylsilyl derivative (124). (a) Low resolution ($m/\Delta m$ 1000) conventional SIM of m/z 500 ($M^{+\cdot}$). (b) Selected reaction monitoring, using a tandem double-focusing/quadrupole configuration, of the transition m/z 500 $\rightarrow$ 443 ($M^{+\cdot} \rightarrow [M-C_4H_9]^+$). Parent ion resolution 1000. (c) As (b), with a parent ion resolution of 5000.

ronides (111, 117) and sulphates (118) have been described. The recent reports (119, 120) of direct introduction of sample solutions in a FAB source, with the possibility for LC-MS coupling, suggest that LC-FAB-MS should become an important technique.

2.4.2 **Techniques for trace analysis**

During mass spectrometric analyses of biological samples, maximum information is obtained by recording full mass spectra of each component, generally requiring a quantity in excess of 1 ng. Problems of trace analysis are generally characterised by low concentrations of analyte and incomplete separation of extract mixtures prior to mass spectrometric detection. Procedures of selected ion monitoring (SIM) permit specific determination of targeted components with maximum sensitivity. The facile incorporation of internal standards, ideally stable isotope labeled analogues, together with potentially high precision, recommend such techniques as reference methods for the assessment of routine quantitative procedures (Chapter 13).

The selectivity of detection achieved during SIM may be increased where necessary by using enhanced mass spectrometric resolution, thus permitting the selective detection of ions corresponding to a single, or few, elemental compositions rather than all ions of a given nominal mass (121). An alternative approach is to monitor selected ion fragmentations occurring after ejection of a parent ion from the ion source. Metastable peak monitoring, which involves the detection of fragmentations occurring in the first field-free region of a double-focusing mass spectrometer, has been shown to lead to useful improvements in the selectivity of detection of certain derivatives of steroid hormones while maintaining low picogramme sensitivity (122, 123). In selected fragmentation monitoring, simultaneous unit resolution, or better, of parent and daughter ion beams requires the use of a multiple analyser instrument, such as the hybrid double focusing sector/quadrupole combination. Figure 8 illustrates the improvement in selectivity of detection that may be achieved, for the example of the analysis of oestradiol-17β, as the bis-*tert*-butyldimethylsilyl (*t*-BDMS) ether (124).

The advantages of tandem MS have been shown to be decisive in the successful quantification of dehydroepiandrosterone sulphate in serum, using fast atom bombardment (125). A marked improvement in signal/background (by comparison with "conventional" MS) was demonstrated when $[M-H]^- \rightarrow m/z$ 97 (HSO_4^-) fragmentations were monitored. The detection limit was in the low nanogramme range and reliable quantification was achieved using $[^2H_2]$dehydroepiandrosterone sulphate as internal standard. The suppression of ionization, which may result from the presence of salts or other co-extracted material during FAB analysis, was avoided by the use of a selective immunoadsorption extraction (Section 2.1).

3 Analyses of Physiological Fluids and Tissues

3.1 Urine

Steroids in urine are predominantly metabolites of the biologically active hormones and are present mainly as conjugates with sulphuric or glucuronic acids. Gas chromatographic profiling of urinary steroids has long been recognised as a powerful technique in the detection of defects in steroid metabolism (126). Early work on the identification of urinary steroids by GC-MS has been summarised in the reviews by Engel and Orr (13), and Brooks and Gaskell (12).

Early work by Sjövall and coworkers on computerised handling of GC-MS data for steroids facilitated both identification and quantification of components of the steroid profile (127–129). The approach was based on the off-line generation of characteristic mass chromatograms from full scan mass spectral data generated by repetitive scanning during GC-MS. Peak searches of the mass chromatograms could be restricted by specified limits of relative retention time. The approach devised by Sweeley and coworkers (130–132) for computerised evaluation of repetitive scanning

Table. 6. Mass chromatograms generated during analysis of urinary steroid profiles (50)

Characteristic ion	Assignment*	Structural feature
$M^{+\cdot}$	Molecular ion	–
M – 15	$[M-CH_3]^+$	–
M – 90	$[M-TMSOH]^+$	Hydroxy steroid
M – 103	$[M-CH_2OTMS]^+$	18-, 19- or 21-OH
M – 117	$[M-CH_3CHOTMS]^+$	20-OH, C_{21} steroid
M – 129	$[M-CH_2CHCHOTMS]^+$	3β-OH, 5-ene
M – 205	$[M-TMSOCH_2CHOTMS]^+$	20,21-bis-OH, C_{21} steroid

* TMS = trimethylsilyl

GC-MS data also placed explicit emphasis on the importance of GC retention on component identification. Specific retention index windows were searched for the presence of peaks in characteristic mass chromatograms. A more detailed review of computer-based evaluation of GC-MS data for steroid profiles has been made by Sjövall and Axelson (1).

The use of GC-MS in urinary steroid profiling is exemplified by the work of the Stockholm group. Setchell et al. (50) reported the multicomponent analysis of conjugated steroids in male urine. Separation of conjugate classes by gel chromatography (Section 2.2) was followed by conversion to MO, TMS derivatives. Computer processing of the data facilitated the generation of mass chromatograms which enabled the location of individual steroids in the GC profile. Table 6 summarises the principal ion types for which searches were made. Identifications were confirmed by searching additional mass chromatograms and by examination of full mass spectra. Seventy seven C_{19} and C_{21} steroids were fully or partially characterised. Steroids could not be detected in the unconjugated fraction from the normal male; in contrast, abundant corticosteroid metabolites were detected in the same fraction from a patient with Cushing's syndrome. Constituents were quantified using the GC/flame ionization detector profile with two internal standards (5α-androstane-3α,17α-diol and stigmasterol) which bracketed the retention times of the major steroid constituents. Similar studies of the steroid conjugates in female urine have been reported by Axelson et al. (133). Extensive studies of urinary steroid profiles have also been reported by Padieu and coworkers (see, for example, ref. 134).

For the identification of metabolic disorders, it is frequently adequate to perform capillary gas chromatographic analysis, with flame ionization detection, of a total steroid fraction obtained by hydrolysis of conjugates. Methyl oxime, TMS ethers are generally the preferred derivatives. A collection of steroid profiles prepared in this way has been compiled by Shackleton and coworkers (135). Recent developments in extraction techniques (Section 2.1) and direct derivatisation of conjugates (Section 2.3) have markedly improved the ease and speed of such analyses.

Adlercreutz and coworkers have described the profiling of urinary oestrogens (136); the procedure incorporates *in situ* ethoximation, extraction with Sep-pak cartridges, fractionation using ion exchange derivatives of Sephadex, and GC-MS of TMS ethers. Application to a comparison of the urinary oestrogen profiles of vegetarian and omnivorous women has been described (137).

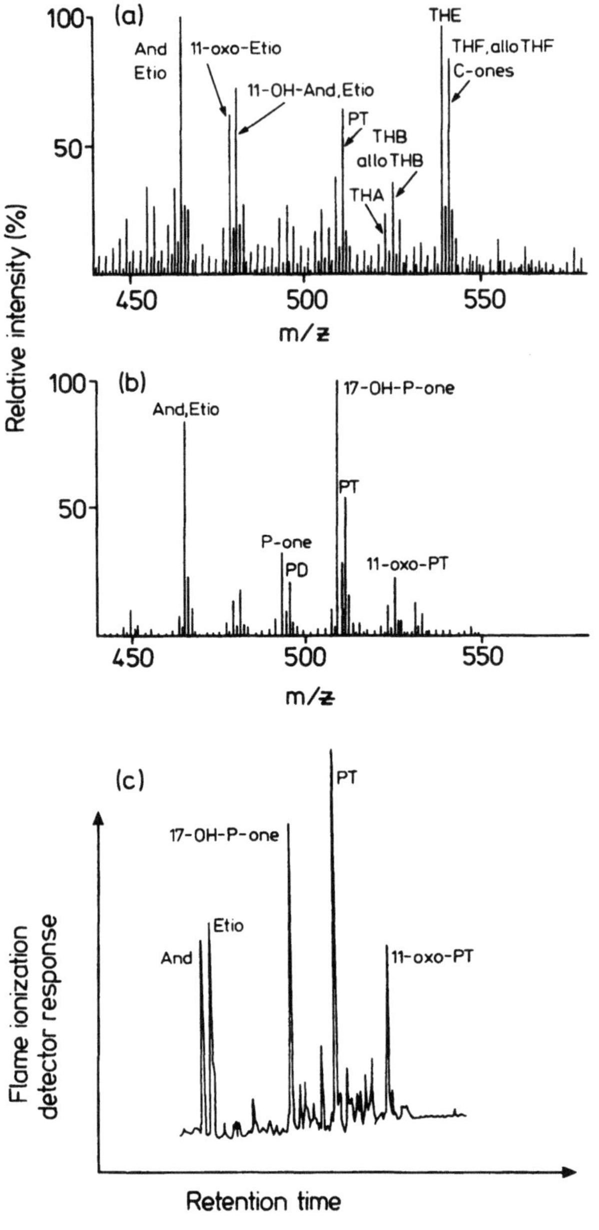

Figure 9. Urinary steroid glucuronide profiles obtained using negative ion secondary ion MS (a, b) or GC of the hydrolysed fraction after methyl oxime, TMS ether formation (c). (a) is a profile from a normal adult; (b) and (c) are profiles from a subject with 21-hydroxylase deficiency (138, 139). GC analysis (c) used a fused silica capillary column (25 m; OV-1), with temperature programming (180° to 290°C, $3°C \cdot min^{-1}$). Retention times were not given in the original paper (138) but analysis times are typically 60 min. And: androsterone; Etio: aetiocholanolone; 11-oxo-Etio: 11-oxo-aetiocholanolone; 11-OH-And, Etio: 11β-hydroxyandrosterone and 11β-hydroxyaetiocholanolone; PT: pregnanetriol; THA: tetrahydro-11-dehydrocorticosterone; THB: tetrahydrocorticosterone; THE: tetrahydrocortisone; THF: tetrahydrocortisol; C-ones: α- and β-cortolone; P-one: pregnanolone; PD: pregnanediol; 17-OH-P-one: 17 hydroxypregnanolone; 11-oxo-PT: 11-oxopregnanetriol.

Direct analysis of underivatised steroid conjugates has been made possible by the introduction of SIMS techniques (Section 2.4.1). Shackleton (138 – 140) has explored the use of SIMS as an alternative to GC for rapid profiling of steroids in urine. Steroids were extracted using octadecylsilyl-substituted silica cartridges (Section 2.1). The total extracts were desalted and analysed directly or separated into conjugate classes (Section 2.2). Figures 9a and 9b compare profiles of glucuronide conjugates determined using SIMS for a normal adult and a subject suffering from 21-hydroxylase deficiency. The equivalence of information obtained by the SIMS method and the (more time-consuming) GC procedure is also illustrated in Figure 9. It is evident that a draw-back of the mass spectrometric method is the failure to distinguish isomeric steroids (Section 2.4.1), such as the glucuronides of androsterone and aetiocholanolone (Fig. 9). In this example, the diagnosis of the disorder is not affected but discrimination between stereoisomers is necessary, for example, in the detection of 5α-reductase deficiency which requires the determination of ratios of 5α- and 5β-epimers.

Moneti et al. (141) have described the quantification of pregnanediol 3α-glucuronide in urine by FAB-MS, using testosterone glucuronide as the internal standard.

GC-MS, possibly in conjunction with selective derivatisations or other microchemical transformations, remains the preferred method for the characterisation of newly recognised urinary constituents. 4-Hydroxy-oestrone, for example was identified in an extract of hydrolysed pregnancy urine (142) on the basis of the following properties: (1.) extraction from benzene/ethyl acetate into borate buffer (consistent with a catechol structure); (2.) GC retention time and mass spectrum of the TMS ether; (3.) GC retention time and mass spectrum of the TMS ether of the product of reduction with sodium borohydride (identified as 4-hydroxy-oestradiol).

When analyses of urinary steroids require the detection of components covering a broad range of concentrations (such as in the analysis of oestrogens (55)), GC-MS with SIM is most suitable. Accurate and precise quantification is best achieved in the same way. Precision is generally maximised by the use of a stable isotope labeled internal standard. Johnson et al. (143), for example, have analysed progesterone in urine using the [7,7-$^2H_2$]analogue as internal standard. GC-MS of the enol-pentafluoropropionate derivative was performed with SIM of molecular ions. A comparison of quantitative data with the results of a radioimmunoassay indicated that the latter procedure was severely positively biased.

3.2 Blood

Determinations of steroid hormones in blood indicate the concentrations potentially available to target organs. In contrast to analyses of urine, where the concentrations represent an average of steroid production and excretion over a period of hours, determinations in blood may provide important data on short-term (such as diurnal) variations in steroid production.

Early studies of steroid sulphates in plasma (144 – 148) employed GC-MS to establish or confirm the identities of components of GC peaks. The careful application of GC-MS and microchemical reactions is illustrated by the identification of 5α-pregnane-3α,20α,21-triol in the monosulphate fraction of human pregnancy plasma (149). GC-MS analysis of the TMS ether suggested a pregnane-3,20,21-triol structure

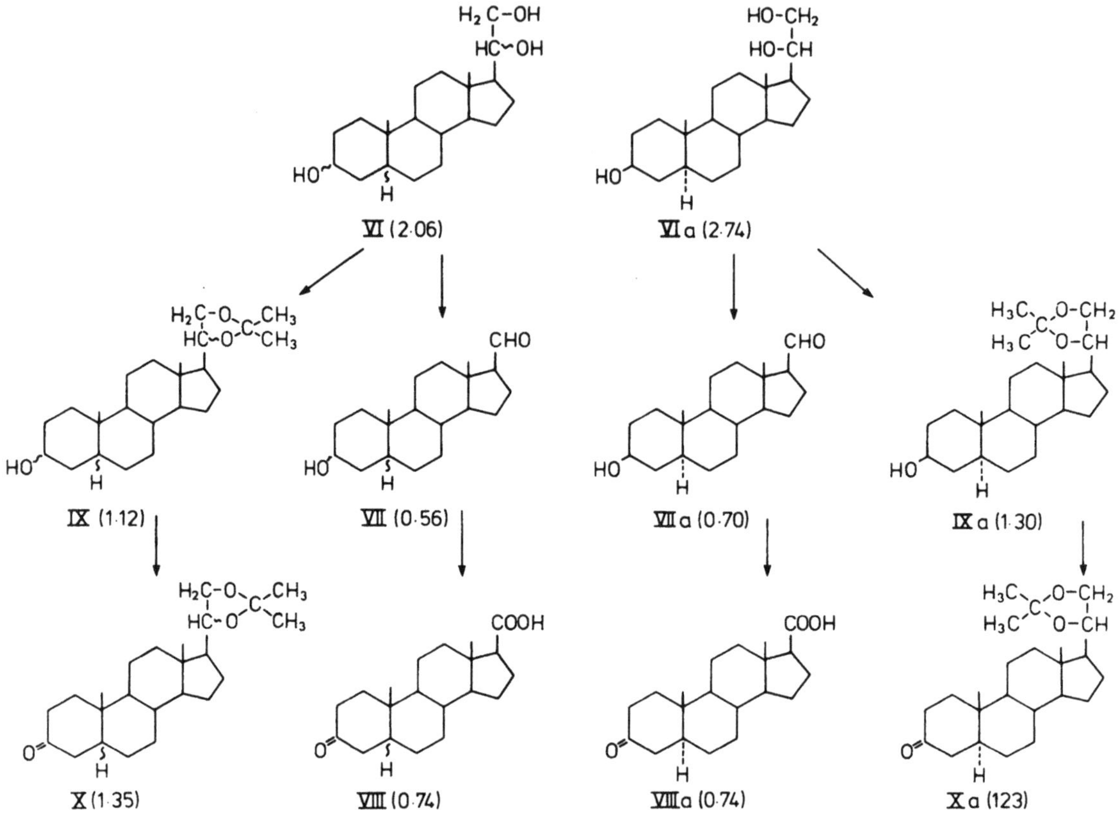

Figure 10. Micro-reactions performed in the characterisation of 5α-pregnane-3α,20α,21-triol (VI) from blood plasma. The equivalent reactions are shown for the authentic standard, 5α-pregnane-3β,20β,21-triol (VIa). Numbers in parentheses refer to the GC retention times of the TMS ethers or methyl esters, relative to 5α-cholestane (149).

but the same derivative of an available standard, 5α-pregnane-3β,20β,21-triol, differed in relative retention time and in the relative abundances of major ions in the mass spectrum. The characterisation of the unknown, following isolation by preparative GC, is summarised in Figure 10. 5α-stereochemistry was established by comparison of GC-MS properties of compounds VIII and VIIIa. The relative retention times of VII and VIIa suggested a 3α-configuration while a comparison of GC-MS data for IX/IXa and X/Xa suggested a 20α-substitution.

Detailed GC-MS profiling of steroids in blood, using computerised procedures similar to those applied to the analysis of urine (Section 3.1), has been reported by Sjövall, Axelson and coworkers (44, 150–152). Purification of extracts using ion exchange derivatives of Sephadex (Section 2.2) permitted analysis of extracts of several ml of plasma using open-tubular capillary columns without overloading. Figure 11 summarises the procedure of Axelson and Sahlberg (152) for the analysis of conjugated steroids. Conjugates of 37 neutral steroids were fully or partially characterised in plasma from a woman in the 34th week of pregnancy. Analysis by GC-MS with

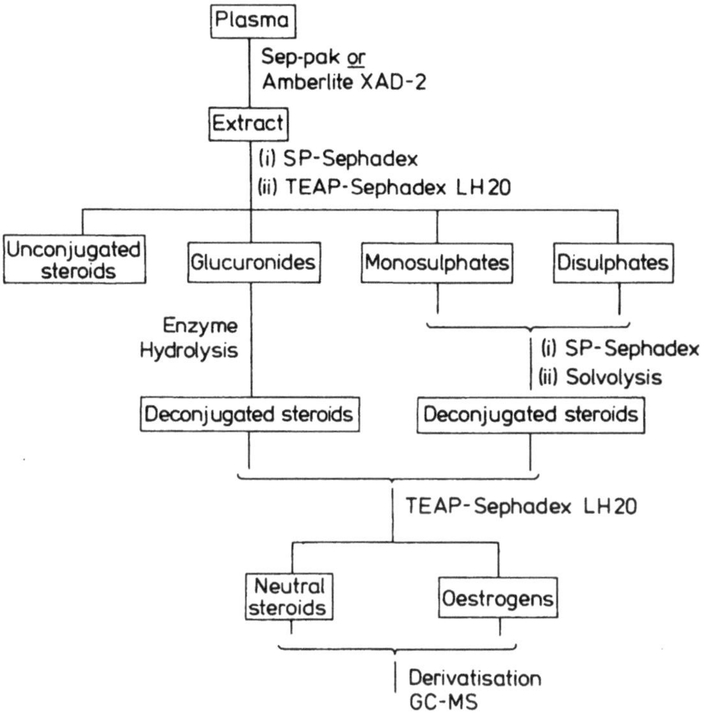

Figure 11. Flow diagram for the analysis of conjugated steroids in blood plasma (152). For abbreviations, see Table 3.

repetitive scanning of the full mass range provided a sensitivity of detection sufficient for the detection of only two oestrogens, oestrone and oestriol. The detection of 6 oestrogens (oestrone, oestradiol, oestriol, 2-methoxy-oestrone, 16α-hydroxy-oestrone, 16-keto-oestradiol), to concentrations of $1-2\ \mu g \cdot l^{-1}$, was achieved during analysis of an unconjugated steroid fraction (44) from pregnancy plasma. Maximum sensitivity during the analysis of targetted steroids, however, is achieved by SIM during GC-MS.

Assays of individual steroids in blood plasma or serum are now very widely performed in routine laboratories, generally using immunoassays, for such purposes as the assessment of adrenal function (by determination of cortisol) or of foetal well-being (by analysis for oestriol). Methods employing isotope dilution and GC-MS-SIM are increasingly recognised as suitable reference methods for determinations of steroids in blood. The topic has been reviewed (4, 7, 11) and is covered in Chapter 13 of this volume.

3.3 Saliva

The analysis of steroids in saliva has attracted considerable recent interest. Non-invasive sampling of a physiologial fluid is attractive in studies which require serial sampling or where the stress associated with venipuncture may invalidate the results. Furthermore, there is evidence to suggest that the saliva concentrations of steroids

Table 7. Analyses of steroids in saliva using mass spectrometry

Steroid	Int. std.*	Isolation procedure	Derivative*	Concentration range	Ref.
Testosterone, dehydroepi-androsterone	*epi*-T	Solvent extn., gel chromatog.	*t*-BDMS MO, *t*-BDMS	200–800 pmol · l^{-1} (male)	154
Testosterone	[$^2H_3$]	Immuno-adsorption, gel chromatog.	*t*-BDMS oxime, *t*-BDMS ether	130–370 pmol · l^{-1} (male) 15–55 pmol · l^{-1} (female)	155
Cortisol	[$^2H_3$]	Solvent extn	MO, TMS	2–60 nmol · l^{-1}	156
Dehydroepi-androsterone sulphate	[$^2H_3$]	Solvent extn., ion exchange chromatog., hydrolysis	*t*-BDMS	0.9–11.0 nmol · l^{-1}	157
Steroid sulphates	–	Solvent extn., ion exchange chromatog., hydrolysis	*t*-BDMS	– – –	158

* *epi*-T = 17-epitestosterone; *t*-BDMS = *tert*-butyldimethylsilyl ether; MO = methyl oxime

and other small molecules (such as a variety of drugs (153)) may correspond to the concentrations of the non-protein bound form in blood plasma.

Since the proportions of non-protein bound steroids in plasma are in the range 1–10%, expected concentrations in saliva are extremely low, placing great demands on the analytical techniques employed. For routine determinations of steroids in saliva, some form of immunoassay is clearly the method of choice. Mass spectrometric methods, however, are useful in establishing concentration ranges to guide the development of routine procedures and in providing reference data to assess the performance of such methods. Table 7 summarises reported analyses of salivary steroids using mass spectrometry.

A direct comparison of radioimmunoassay results for cortisol in parotid fluid (a component of whole saliva) with data obtained by isotope dilution and GC-MS indicated excellent agreement (156). The interpretation of salivary cortisol levels is complicated by evidence for the conversion of cortisol to cortisone in the salivary glands (159). Clarfication of this question is a problem well suited to the unique capabilities of mass spectrometric analysis following administration of stable isotope-labeled steroids; no such studies, however, have yet been performed.

Gaskell et al. reported the identification of testosterone and dehydroepiandrosterone in male saliva using high resolution SIM and metastable peak monitoring of the *t*-BDMS ether and methyl oxime, *t*-BDMS ether derivatives, respectively (154). Approximate quantification, using epitestosterone as the internal standard, indicated saliva concentrations representing 1.5 to 7.5% of blood plasma concentrations in matched samples. More recently, a rigorous isotope dilution mass spectrometric procedure for the determination of testosterone in both female and male saliva has been developed (Table 7). Identification and rigorous quantification of dehydroepian-

Table 8. Identification of oestrogens in seminal plasma (27)

Oestrogen	Derivative	Characteristic ion m/z	Structure
Oestrone	TMS	342	$M^{+\cdot}$
	TMS, TMS oxime	429	$M^{+\cdot}$
		340	$[M-TMSO]^+$
2-Methoxyoestrone	TMS	372	$M^{+\cdot}$
		342	$[M-CH_2O]^+$
	TMS, TMS oxime	359	$M^{+\cdot}$
Oestradiol-17β	bis-TMS	416	$M^{+\cdot}$

drosterone sulphate in saliva has been reported (157). The saliva concentrations represented only 0.1 to 0.2% of the expected serum concentration. Identification and approximate concentrations of a number of other steroid sulphates have been reported (158); the observed levels again represented about 0.1% of the expected serum concentrations. Relative amounts of the various steroid sulphates paralleled the expected serum distributions. Glucuronides, however, were not detected.

The lack of correspondence between saliva concentrations of the steroid conjugates and the expected concentrations of the non-protein bound form in serum, observed in the mass spectrometric assays and others (160), is consistent with the suggestion of Vining and coworkers that the transfer of these constituents from serum takes place only via the "tight" junctions between the acinar cells of the salivary glands (160, 161). These authors suggested that, by contrast, transfer of the lipid-soluble unconjugated steroids is mainly via intra-cellular diffusion. While the clinical value of determinations of saliva concentrations of unconjugated steroids is becoming increasingly widely accepted, it is likely that quantification of conjugated steroids in saliva will be of limited importance, particularly in view of the risk of sample contamination by blood or gingival fluid (160).

3.4 Steroids in other physiological fluids

Application of GC-MS to the analysis of steroids in amniotic fluid and bile has been reviewed (12, 13).

Dehennin and coworkers have studied the steroid constituents of human seminal plasma using GC-MS (27, 162). Steroids were extracted using Extrelut (diatomaceous earth) columns and the total extracts fractionated using columns of Sephadex LH20 (for the neutral steroids) or diethyl(2-hydroxypropyl)-aminoethyl Sephadex (for the oestrogens). For the analysis of total (conjugated and unconjugated) steroids, extraction was preceded by enzymic or acid hydrolysis. The concentrations were too low to permit the recording of full mass spectra so that identifications were based on the detection of characteristic molecular and fragment ions during GC-MS-SIM of appropriate derivatised fractions of the total extracts. Table 8 shows the derivatives and characteristic ions used in the identification of the oestrogens. Quantification was

Table 9. Steroids identified and quantified in human seminal plasma using GC-MS (27, 162)

Steroid	Concentration (mean ± SD; $\mu g \cdot l^{-1}$) unconjugated	total
Oestrone	0.15 ± 0.03	1.8 ± 0.4
2-Methoxyoestrone	1.00 ± 0.20	–
Oestradiol-17β	0.019 ± 0.002	0.24 ± 0.04
Testosterone	0.15 ± 0.02	15.0 ± 1.0
5α-Dihydrotestosterone	0.23 ± 0.03	1.7 ± 0.4
Androsterone	0.32 ± 0.05	
Dehydroepiandrosterone	16.0 ± 2.0	280 ± 20
Pregnenolone	1.1 ± 0.4	8.3 ± 1.3
Progesterone	0.17 ± 0.04	–
17-Hydroxyprogesterone	0.25 ± 0.03	–
Cortisol	7.0 ± 0.2	8.3 ± 0.2

Table 10. Concentrations of oestrogens in human follicular fluid (163)

Oestrogen*	Concentration mean ± SEM	*n*
2-Hydroxyoestrone	141 ± 19 $ng \cdot l^{-1}$	15
4-Hydroxyoestrone	123 ± 12 $ng \cdot l^{-1}$	12
2-Hydroxyoestradiol-17β	360 ± 54 $ng \cdot l^{-1}$	15
4-Hydroxyoestradiol-17β	345 ± 67 $ng \cdot l^{-1}$	12
2-Methoxyoestrone	831 ± 134 $ng \cdot l^{-1}$	15
2-Methoxyoestradiol-17β	1050 ± 90 $ng \cdot l^{-1}$	15
16-Oxo-oestradiol-17β	412 ± 79 $ng \cdot l^{-1}$	15
6α-Hydroxyoestradiol-17β	13.2 ± 0.9 $\mu g \cdot l^{-1}$	11
Oestriol	10.2 ± 2.1 $\mu g \cdot l^{-1}$	15
Oestradiol-17β	1365 ± 267 $\mu g \cdot l^{-1}$	15
Oestrone	211 ± 27 $\mu g \cdot l^{-1}$	15

* Oestradiol-17α and 6β-Hydroxyoestradiol-17β also identified in pooled follicular fluid sample

achieved using deuterium-labeled internal standards. Table 9 lists the observed concentrations of the steroid hormones identified.

A similar analytical strategy was adopted by Dehennin et al. for the identification and quantification of oestrogens in human follicular fluid (163). The isolation procedure incorporated the additional refinement of the separation of catechol oestrogens on a column of DEAE-Sephadex in the borate form (56; Section 2.3). GC-MS analyses involved SIM of ions characteristic of various derivatives, including acetonides (for the catechols) and mixed *t*-BDMS/TMS ethers, formed by transetherification. Table 10 summarises the quantitative data obtained using [^{2}H]-labeled internal standards.

Sahlberg and Axelson (164) have reported a comprehensive study of steroids in human milk; fifty components were identified in unconjugated, glucuronide, mono-

Table 11. Quantitative analyses of steroids in human breast tissue

Steroid	Tissue	Internal standard	Concentration	Ref.
Dehydro-epiandrosterone	Tumours	3α-Hydroxy-5-androsten-17-one	2.5–466 $ng \cdot g^{-1}$ wet weight	172, 173
	Fibro-adenomata		245–1000 $ng \cdot g^{-1}$ wet weight	174
5-Androstene-3,17-diol	Tumours	5α-Androstane-3α,17α-diol	1–40 $ng \cdot g^{-1}$ wet weight	172, 173
	Fibro-adenomata		42–130 $ng \cdot g^{-1}$ wet weight	174
Oestradiol-17β	Tumours:	[16,16,17-$^2H_3$]-Oestradiol-17β		175
	ER+		15.9 ± 7.0 $pg \cdot mg^{-1}$ protein	
	ER−		5.5 ± 5.6 $pg \cdot mg^{-1}$ protein	

sulphate and disulphate fractions. Extraction using a Sep-pak cartridge (Section 2.1) was followed by reverse-phase and ion exchange chromatography steps, using derivatives of Sephadex, to yield fractions corresponding to the unconjugated steroids and the various conjugates. The latter were hydrolysed prior to derivatisation and GC-MS. The steroid pattern was similar to that observed in late pregnancy plasma.

A comprehensive study of steroid sulphates in human breast cyst fluid by capillary GC-MS has identified fifteen major components (165).

3.5 Tissues

The identification and quantification of steroids in tissues are important for the following purposes:

1. Study of in vivo metabolism.
2. Assessment of the hormonal milieu of target tissues.
3. Investigation of steroid mechanism of action by, for example, analysis of subcellular fractions, and comparisons of steroid concentrations with receptor levels.
4. Evaluation of synthetic hormones and anti-hormones.

Immunoassay techniques may provide the most convenient means of quantifying steroids in tissue extracts but a lack of specificity may cast doubt on the validity of the data. The potential selectivity of mass spectrometric methods should provide good characterisations of trace components and reliable quantitative data. The particular value of MS in studies of steroid metabolism, especially using stable isotope labeled precursors, is discussed in Section 4.

Early applications of GC-MS to tissue analysis employed scanning of full spectra for the characterisation of components which were quantified by measurement of GC peaks. Examples include the determination of oestrogens in placenta (166), of C_{19} steroids in testis (167), and of C_{19} and C_{21} steroids in foetal tissues (168).

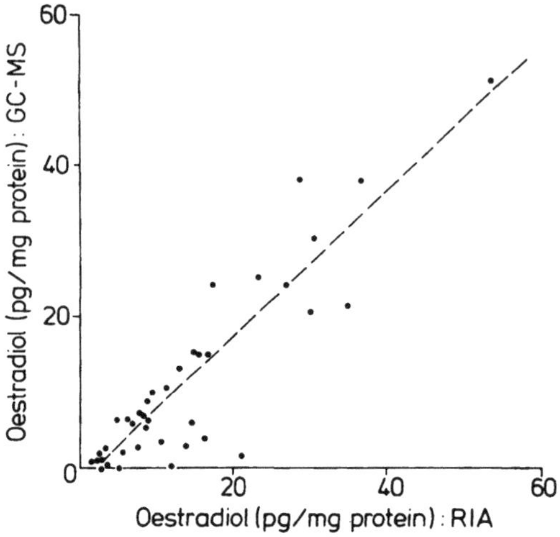

Figure 12
Comparison of concentrations of oestradiol-17β in breast tumour homogenates, determined by GC-MS and radioimmunoassay (RIA). Linear regression gives $y = 0.91\,x - 1.8$ pg/mg protein; $r = 0.88$ (175).

The need for very high sensitivities of detection has prompted the use of SIM techniques for the analysis of steroids in tissues where the expected concentrations are low or where the sample availability is poor. Millington et al. reported the determination of C_{19} steroids in human benign prostatic hyperplasia using SIM at high MS resolution to improve the selectivity of detection (169). Testosterone, 5α-dihydrotestosterone, 5α-androstane-3α,17β-diol and 5α-androstane-3β,17β-diol were identified and quantified using stereoisomeric internal standards. More recently, Moneti et al. (170) have described a more rigorous quantification of the same steroids in prostatic tissue, using [16,16,17-$^2H_3$]-analogues as internal standards.

Millington et al. have also reported the identification (based on detection during high resolution SIM) of the following steroids in human breast tumour tissue (171): dehydroepiandrosterone, testosterone, androsterone, epiandrosterone, oestrone, oestradiol-17β and oestriol. Approximate quantitative date were reported. As with the analysis of physiological fluids, accurate and precise quantification requires the use of an internal standard, ideally a stable isotope labeled analogue. Table 11 summarises quantitative data from several studies of human breast tissue. In their study, Edery et al. (175) concluded that the oestradiol-17β content of whole homogenates of oestrogen receptor positive breast tumour tissue was significantly higher than that of oestrogen receptor negative tissue. This conclusion is consistent with earlier results for tumour tissue cystosols (176), based on radioimmunoassay data. A comparison, by Edery et al. (175) of oestradiol-17β concentrations determined by radioimmunoassay and GC-MS-SIM (Fig. 12) indicated satisfactory agreement at higher concentrations but poor correlation at low concentrations, attributable to the high imprecision of the immunoassay.

A further study of oestradiol-17β in breast tumour cytosols, using isotope dilution and GC-MS, again revealed a tendency to higher concentrations in receptor positive tissue (10); in this study, however, statistical significance ($p = 0.05$) was only achieved when concentrations were expressed relative to the DNA content of the tissue, rather than to cytosolic protein. The same study revealed that in some samples the concentration of 5-androstene-3β,17β-diol (which may compete for binding to the oestrogen receptor) considerably exceeded that of oestradiol-17β.

Determinations of target tissue concentrations of hormonal and anti-hormonal drugs are important in studies of their mechanism of action. Daniel et al. (177) measured levels of the non-steroidal anti-oestrogen, tamoxifen (*trans*-1(4-β-dimethylaminoethoxyphenyl)-1,2-diphenyl-but-1-ene) in breast tissue of treated breast cancer patients. The data indicated a more than one thousand-fold excess of the anti-oestrogen over oestradiol-17β, consistent with effective competition for the oestrogen receptor.

Axelson et al. (178) have developed a procedure, based on anion exchange chromatography and GC-MS, for the determination of oestradiol-17β in the nuclear fraction of uterine tissue. When applied to rat tissue, the GC-MS procedure permitted the assessment of the nuclear uptake and retention of the steroid. A comparison of the GC-MS data with those from a nuclear exchange assay indicated that, at low levels of nuclear oestradiol-17β, there was a significant proportion of unoccupied receptor sites.

4 Metabolism and Production Rate Studies Using Stable Isotopes

As a powerful technique of mixture analysis, GC-MS has a major role in studies of steroid metabolism. Much work has been performed with radioisotopes (see, for example, ref. 179) but this section concentrates on the use of stable isotopes. The advantages of stable isotopes have been discussed by Baillie (180); principal amongst them, from the standpoint of clinical biochemistry, is their greater acceptability for *in vivo* studies in humans.

The preparation of [^{2}H]-labeled steroids has been reviewed (181 – 183). Incorporation of deuterium in a chemically non-labile position is invariably desirable; the siting of the label at a metabolically labile position may, however, provide additional information in studies of steroid metabolism (see below). A method of synthesis of [3,4-^{13}C]steroids has also been described (184).

The identification of metabolites during GC-MS analysis following incubation with an isotope-labeled precursor is facilitated if the isotope pattern is clearly characteristic. Engel and coworkers (13) introduced the "twin ion" technique for steroid metabolism studies in which a characteristic 1 : 1 ratio of labeled to unlabeled species in the precursor is reflected in the isotopic composition of the metabolites. A recent example of this approach is the study of the metabolism of testosterone by isolated human axillary *Corynebacterium spp.* (185).

Baba et al. administered [19,19,19-$^{2}H_3$]testosterone to male subjects and were thus able to distinguish endogenous and exogenous testosterone in the plasma and urine of the subjects (186). Oral administration was found to have little effect on endogenous levels. Quantification of [$^{2}H_3$]androsterone glucuronide and [$^{2}H_3$]aetiocholanolone glucuronide in the urine of the treated subjects indicated that these metabolites accounted for the majority of the administered testosterone (187).

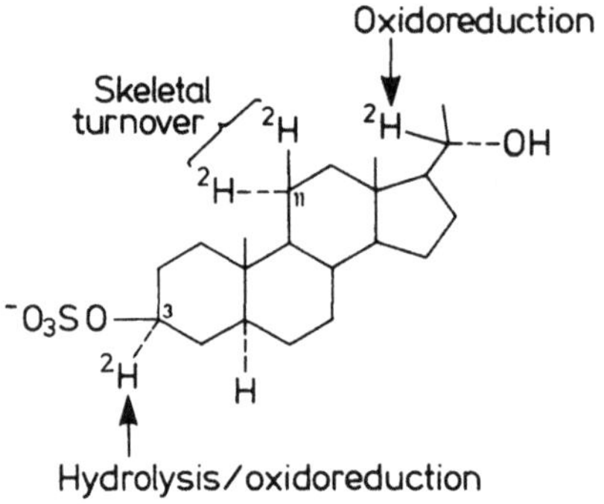

Figure 13
Tracer functions of deuterium labels in [3,11,11,20-$^2H_4$]-5α-pregnane-3,20-diol 3-sulphate (193).

Pinkus et al. (188, 189) pioneered the use of [^{2}H]-labeled steroids in the estimation of urinary production rates. Production rate, *P*, was defined as follows:

$$P = \left(\frac{a}{e} - 1\right)\frac{m}{t}\ \text{mg/day}$$

where

a = labeled species, [^2H_n], in administered steroid, expressed as a percentage of the sum of [^2H_n] and [$^2H_0$] species.

e = labeled form, [^2H_n], of the appropriate steroid from pooled urine, expressed as a percentage of the sum of [^2H_n] and [$^2H_0$] species.

m = mass (mg) of administered [^2H_n] + [$^2H_0$] species.

t = number of days of urine collection.

Urinary production rates for oestrone and oestradiol-17β were determined by Pinkus et al. (188, 189); similar methods were used by Taylor et al. (190) in studies on oestriol. More recently, Johnson and coworkers (191) measured the urinary pregnenolone production rate, following intravenous administration of the [7,7-$^2H_2$] analogue, by GC-MS-SIM analysis of urinary 5-pregnene-3β,20α-diol, a unique metabolite of pregnenolone.

The studies by Baillie et al. (192, 193) provide an elegant example of the wealth of information that may be derived from a single experimental series when the stable isotope labeled analogue is carefully chosen. Analogues were prepared of 3β-hydroxy-5α-pregnan-20-one sulphate and the 3- and 3,20-sulphates of 5α-pregnane-3β,20α-diol, labeled with deuterium in the 3α,11,11- or 3α,11,11,20-positions. The metabolically inert label at C11 provided an indication of skeletal turnover, while the labels at 3 and 20 permitted the assessment of hydrolysis/oxidoreduction and oxidoreduction, respectively (Fig. 13). Following intravenous administration of labeled precursor to pregnant women, plasma steroids were analysed by GC-MS with computerised treatment of data (127–129). Oxidoreduction at C20, 16α-hydroxylation, and sulphoconjugation at C20 were observed but there was no evidence for the sequence involving hydrolysis, oxidoreduction and resulphation at C3. Decay curves were prepared by plotting deuterium excess values versus time for each labeled steroid. Deuterium excess values were calculated as the percentage of labeled species, modified to allow for the isotopic composition of the precursor and the expansion of the endogenous pools caused by the administration of labeled steroid (194). (It is noteworthy that this correction is necessitated by analytical limitations; determination of isotopic compositions by MS with acceptable precision necessitates a significant increase in total steroid levels by administration of labeled steroid. The alternative use of trace quantities of radiolabeled precursor, where this is ethically permissible, in conjunction with scintillation counting, avoids this problem, though in this instance intramolecular location of label is generally not feasible.) Several theoretical models were fitted to the experimental data, the models differing in their assumptions concerning the relative rates of metabolism and of equilibration of administered steroid with the endogenous pools. Using these models, estimates were made of endogenous pool sizes, production rates and half-life times. The half-life times for oxidation and reduction at C20, for example, were approximately 0.5 h and 0.2 h, respectively, compared to half-life times of 2.6 to 3.1 h for turnover of the steroid skeleton.

5 Summary and Assessment of Future Trends

Early applications of mass spectrometry in the clinical biochemistry of steroids were concerned principally with confirmation of the identity of gas chromatographic peaks. With improved understanding of fragmentation patterns and the introduction of computerised data handling, sophisticated techniques of GC-MS profiling of steroids became available and have been applied to both qualitative and quantitative analyses. SIM techniques, in conjunction with isotope dilution, have provided the means for specific and precise quantitative analyses. Such techniques are not widely recognised as appropriate reference methods for the assessment of routine assays for steroids in blood and urine. The distinction between qualitative and quantitative analyses has, however, become increasingly blurred. Thus, the high analytical sensitivites required (due either to low concentrations or poor sample avaliability) in the analysis of steroids in tissues and fluids, such as saliva and follicular fluid, have necessitated the use of SIM techniques for both identification and quantification of steroids. The use of mass spectrometry near its limit of sensitivity (sub-picogramme) is likely to expand, with application to studies of steroid mechanism in target tissues. Such studies will, in common with investigations of metabolism, benefit from the unique capabilities of mass spectrometry in the characterisation and quantification of compounds labeled with stable isotopes. The recent innovations in mass spectrometric techniques, which permit the direct analysis of steroid conjugates, will find increasingly broad application. Increased recognition of the metabolic importance of the conjugates may stimulate studies where improved characterisation and quantification are required.

Acknowledgement

Work at the Tenovus Institute has been generously supported by the Tenovus Organisation. This chapter was prepared, in part, during a leave-of-absence at the Laboratory of Molecular Biophysics, National Institute of Environmental Health Sciences, Research Triangle Park, U.S.A.

References

1. Sjövall, J. & Axelson, M. (1982) Newer approaches to the isolation, identification, and quantitation of steroids in biological materials. Vitamins Hormones *39*, 31 – 129.
2. Sjövall, J. (1985) Analysis of steroids and bile acids by gas chromatography and mass spectrometry. In: Advances in Steroid Analysis '84 (Gorog, S., ed.) Analytical Chemistry Symposia Series, Vol. 23, pp. 359 – 373, Elsevier, Amsterdam.
3. Sjövall, J. (1985) Column techniques in the preparation of biological samples for gas chromatography-mass spectrometry. In: Mass Spectrometry in the Health and Life Sciences (Burlingame, A. L. and Castagnoli, N., eds.) pp. 303 – 317, Elsevier, Amsterdam.
4. Gaskell, S. J. (1983) Analysis of steroids by mass spectrometry. In: Methods of Biochemical Analysis, Vol. 29 (Glick, D., ed.) pp. 385 – 434, Wiley, New York.
5. Brooks, C. J. W. & Middleditch, B. S. (1973) Some aspects of mass spectrometry in steroid analysis. In: Modern Methods of Steroid Analysis (Heftmann, E., ed.) pp. 139 – 198, Academic Press, New York.
6. Shackleton, C. H. L. (1985) Mass spectrometry: application to steroid and peptide research. Endocrine Rev. *6*, 441 – 486.
7. Björkhem, I. (1979) Selective ion monitoring in clinical chemistry. CRC Crit. Rev. Clin. Lab. Sci. *11*, 53 – 105.
8. Siekmann, L. (1979) Determination of steroid hormones by the use of isotope dilution-mass spectrometry: a definitive method in clinical chemistry J. Steroid Biochem. *11*, 117 – 123.
9. Adlercreutz, H. (1977) Quantitative mass spectrometry of endogenous and exogenous steroids in metabolic studies in man. In: Quantitative Mass Spectrometry in Life Sciences (De Leenheer, A. P. & Roncucci, R. R., eds.) pp. 15 – 28, Elsevier, Amsterdam.
10. Gaskell, S. J., Gould, V. J. & Leith, H. M. (1986) High sensitivity determinations of steroids in physiological fluids and tissues. In: Mass Spectrometry in Biomedical Research (Gaskell, S. J., ed.) pp. 347 – 361, Wiley, Chichester (U. K.).
11. Setchell, K. D. R. (1979) Gas chromatography-mass spectrometry. In: Hormones in Blood. 3rd ed. (Gray, C. H. & James, V. H. T., eds.) pp. 63 – 127, Academic, London.
12. Brooks, C. J. W. & Gaskell, S. J. (1980) Hormones. In: Biochemical Applications of Mass Spectrometry, first supplementary volume (Waller, G. & Dermer, O., eds.) pp. 611 – 659, Wiley-Interscience, New York.
13. Engel, L. L. & Orr, J. C. (1972) Hormones. In: Biochemical Applications of Mass Spectrometry (Waller, G., ed.) pp. 537 – 572, Wiley-Interscience, New York.
14. Shackleton, C. H. L. (1987) in: Glass Capillary Chromatography in Clinical Medicine and Pharmacology (Jaeger, H., ed.) pp. 153 – 243, Marcel Dekker, New York.
15. Taylor, N. (1980) in: Current Developments in the Clinical Applications of HPLC, GC and MS (Lawson, A. M., Lim, C. K. & Richmond, W., eds.) pp. 271 – 283, Academic, London.
16. Gaskell, S. J. (1983) Mass spectrometric determination of steroid hormones and related compounds in human endocrine tissues. In: Hormones in Normal and Abnormal Human Tissues. Vol. 3 (Fotherby, K. & Pal, S. B., eds.) pp. 121 – 134, Walter de Gruyter, Berlin.
17. Caprioli, R., Liehr, J. G. & Seifert, W. E. (1980) Applications in clinical medicine. In: Biochemical Applications of Mass Spectrometry, first supplementary volume (Waller, G. & Dermer, O., eds.) pp. 1007 – 1084, Wiley-Interscience, New York.
18. Bradlow, H. L. (1968) Extraction of steroid conjugates with a neutral resin. Steroids *11*, 265 – 272.
19. Bradlow, H. L. (1977) Modified technique for the elution of polar steroid conjugates from Amberlite XAD-2. Steroids *30*, 581 – 582.
20. Axelson, M. & Sjövall, J. (1974) Separation and computerized gas chromatography-mass spectrometry of unconjugated neutral steroids in plasma. J. Steroid Biochem. *5*, 733 – 738.
21. Axelson, M. & Sahlberg, B.-L. (1981) Solid extraction of steroid conjugates from plasma and milk. Anal. Lett. *14*, 771 – 782.
22. Shackleton, C. H. L. & Whitney, J. O. (1980) Use of Sep-pak cartridges for urinary steroid extraction: evaluation of the method for use prior to gas chromatographic analysis. Clin. Chim. Acta *107*, 231 – 243.
23. Heikkinen, R., Fotsis, T. & Adlercreutz, H. (1981) Reversed-phase C_{18} cartridge for extraction of estrogens from urine and plasma. Clin. Chem. *27*, 1186 – 1189.

24. Cannell, G. R., Galligan, J. P., Mortimer, R. H. & Thomas, M. J. (1983) Extraction of steroids from plasma with reversed phase C_{18} cartridges. Clin. Chim. Acta *122*, 419–423.
25. Dyfverman, A. & Sjovall, J. (1978) A novel liquid-gel chromatographic method for extraction of unconjugated steroids from aqueous solutions. Anal. Lett. Ser. *B 11*, 485–499.
26. Dahlberg, E., Snochowski, M. & Gustafsson, J.-A. (1980) Removal of hydrophobic compounds from biological fluids by a simple method. Anal. Biochem. *106*, 380–388.
27. Reiffsteck, A., Dehennin, L. & Scholler, R. (1982) Estrogens in seminal plasma of human and animal species: identification and quantitative estimation by gas chromatography-mass spectrometry associated with stable isotope dilution. J. Steroid Biochem. *17*, 567–572.
28. Reiffsteck, A., Dehennin, L. & Scholler, R. (1982) Identification et dosage de steroides neutres dans le plasma seminal par spectrometrie de masse. Pathol. Biol. *30*, 626–628.
29. Dehennin, L., Blacker, C., Reiffsteck, A. & Scholler, R. (1984) Estrogen 2-, 4-, 6- or 16-hydroxylation by human follicles shown by gas chromatography-mass spectrometry associated with stable isotope dilution. J. Steroid Biochem. *20*, 465–471.
30. Dyas, J., Turkes, A., Read, G. F. & Riad-Fahmy, D. (1981) A radioimmunoassay for ethinyl oestradiol in plasma incorporating an immunosorbent, pre-assay purification procedure. Ann. Clin. Biochem. *18*, 37–41.
31. Glencross, R. G., Abeywardene, S. J., Corney, S. J. & Morris, H. S. (1981) The use of oestradiol-17β antiserum covalently bound to Sepharose to extract oestradiol-17β from biological fluids. J. Chromatog. *223*, 193–197.
32. Gaskell, S. J. & Brownsey, B. G. (1983) Immunoadsorption to improve gas chromatography-high resolution mass spectrometry in the analysis of estradiol-17β in plasma. Clin. Chem. *29*, 677–680.
33. Gaskell, S. J., Collins, C. J., Thorne, G. C. & Groom, G. V. (1983) External quality assessment of assays for cortisol in plasma: the use of target data obtained by gas chromatography-mass spectrometry. Clin. Chem. *29*, 862–867.
34. Gaskell, S. J., Brownsey, B. G. & Groom, G. V. (1984) Analyses for progesterone in serum by gas chromatography-mass spectrometry: target data for external quality assessment of routine assays. Clin. Chem. *30*, 1696–1700.
35. Anderson, S. H. G. & Sjövall, J. (1983) A method combining solvent and gel extraction for isolation and purification of steroids in tissues. Anal. Biochem. *134*, 309–312.
36. Batra, S. & Bengtsson, L. P. (1976) A highly efficient procedure for the extraction of progesterone from uterus and its compatibility with subsequent radioimmunoassay. J. Steroid Biochem. *7*, 599–603.
37. Gould, V. G. & Gaskell, S. J. (1986) Unpublished results.
38. van Landeghem, A. A. J., Poortman, J., Helmond-Agema, A. & Thijssen, J. H. H. (1984) Measurement of endogenous subcellular concentrations of steroids in tissue. J. Steroid Biochem. *20*, 639–644.
39. Bonney, R. C., Sparks, C. A., Cheng, R. W., Reed, M. J. & James, V. H. T. (1984) The measurement of steroid hormones in endometrial tissue – a comparison between two methods of extraction. J. Steroid Biochem. *21*, 479–481.
40. Heikkinen, R., Fotsis, T. & Adlercreutz, H. (1983) Use of ion exchange chromatography in steroid analysis. J. Steroid Biochem. *19*, 175–180.
41. Eneroth, P. & Nystrom, E. (1967) A study on liquid-gel partition of steroids and steroid derivatives on lipophilic Sephadex gels. Biochim. Biophys. Acta *144*, 149–161.
42. Siekmann, L. (1984) Determination of oestradiol-17β in human serum by isotope dilution-mass spectrometry. Definitive methods in clinical chemistry, II. J. Clin. Chem. Clin. Biochem. *22*, 551–557.
43. Ellingboe, J., Nystrom, E. & Sjövall, J. (1968) A versatile lipophilic Sephadex derivative for "reversed-phase" chromatography. Biochim. Biophys. Acta *152*, 803–805.
44. Axelson, M. & Sjövall, J. (1977) Analysis of unconjugated steroids in plasma by liquid-gel chromatography and glass capillary gas chromatography-mass spectrometry. J. Steroid Biochem. *8*, 683–692.
45. Reiffsteck, A., Dehennin, L. & Scholler, R. (1982) Estrogens in seminal plasma of human and animal species: identification and quantitative estimation by gas chromatography-mass spectrometry associated with stable isotope dilution. J. Steroid Biochem. *17*, 567–572.
46. Axelson, M. & Sjövall, J. (1976) Selective liquid chromatographic isolation procedure for gas chromatographic-mass spectromet-

ric analysis of 3-ketosteroids in biological material. J. Chromatog. *126*, 705–716.
47. Axelson, M. & Sjövall, J. (1979) Strong non-polar cation exchangers for the separation of steroids in mixed chromatographic systems. J. Chromatog. *186*, 725–732.
48. Tetsuo, M., Axelson, M. & Sjövall, J. (1980) Selective isolation procedures for GC/MS analysis of ethynyl steroids in biological material. J. Steroid Biochem. *13*, 847–860.
49. Anderson, S. H. G., Axelson, M., Sahlberg, B.-L. & Sjövall, J. (1981) Simplified method for the isolation and analysis of ethynyl steroids in urine. Anal. Lett. *14*, 783–790.
50. Setchell, K. D. R., Almé, B., Axelson, M. & Sjövall, J. (1976) The multicomponent analysis of conjugates of neutral steroids in urine by lipophilic ion exchange chromatography and computerised gas chromatography-mass spectrometry. J. Steroid Biochem. *7*, 615–629.
51. Setchell, K. D. R., Taylor, N. S., Adlercreutz, H., Axelson, M. & Sjövall, J. (1979) The group separation of conjugates of oestrogens using lipophilic ion exchange chromatography. In: Research on Steroids, Vol. 8 (Klopper, A., Lerner, L., van der Molen, H. J. & Sciarra, F., eds.) pp. 131–135, Academic Press, New York.
52. Axelson, M., Sahlberg, B.-L. & Sjövall, J. (1981) Analysis of profiles of conjugated steroids in urine by ion-exchange separation and gas chromatography-mass spectrometry. J. Chromatog. *224*, 355–370.
53. Fotsis, T., Adlercreutz, H., Jarvenpaa, P., Setchell, K. D. R., Axelson, M. & Sjövall, J. (1981) Group separation of steroid conjugates by DEAE-Sephadex anion exchange chromatography. J. Steroid Biochem. *14*, 457–463.
54. Jarvenpaa, P., Fotsis, T. & Adlercreutz, H. (1979) Ion exchange purification of estrogens. J. Steroid Biochem. *11*, 1583–1588.
55. Fotsis, T., Jarvenpaa, P. & Adlercreutz, H. (1980) Purification of urine for quantification of the complete estrogen profile. J. Steroid Biochem. *12*, 503–508.
56. Fotsis, T. & Heikkinen, R. (1983) Selective chromatographic fractionation of catechol estrogens on anion exchangers in borate form. J. Steroid Biochem. *18*, 357–363.
57. Kautsky, M. P., ed. (1981) Steroid Analysis by HPLC. Marcel Dekker, New York.
58. Lim, C. K. (1979) High performance liquid chromatography. In: Hormones in Blood, 3rd Edition, Vol. 3 (Gray, C. H. & James, V. H. T., eds.) pp. 42–62, Academic Press, London, New York.
59. Van der Wal, S. & Huber, J. F. K. (1978) Comparitive study of several phase systems for the separation of estrogen conjugates by high-pressure liquid chromatography. J. Chromatog. *149*, 431–453.
60. Blau, K. & King, G. S., eds. (1977) Handbook of Derivatives for Chromatography, Heyden, London.
61. Brooks, C. J. W. & Middleditch, B. S. (1973) Some aspects of mass spectrometry in steroid analysis. In: Modern Methods of Steroid Analysis (Heftmann, E., ed.) pp. 139–198, Academic Press, New York.
62. Harvey, D. J., Middleditch, B. S. & Brooks, C. J. W. (1982) Mass spectra of the trimethylsilyl ethers of some 3β-hydroxy-Δ^5-C_{21} steroids. Biomed. Mass Spectrom. *9*, 411–418.
63. Phillipou, G., Bigham, D. A. & Seamark, R. F. (1975) Steroid t-butyldimethylsilylethers as derivatives for mass fragmentography. Steroids *26*, 516–524.
64. Kelly, R. W. & Taylor, P. L. (1976) Tert-butyldimethylsilyl ethers as derivatives for qualitative analysis of steroids and prostaglandins by gas phase methods. Anal. Chem. *48*, 465–467.
65. Harvey, D. J. (1978) Comparison of fourteen substituted silyl derivatives for the characterization of alcohols, steroids and cannabinoids by combined gas-liquid chromatography and mass spectrometry. J. Chromatogr. *147*, 291–298.
66. Miyazaki, H., Ishibashi, M. & Yamashita, K. (1979) Dimethylisopropylsilyl ether derivatives in gas chromatography mass spectrometry of hydroxylated steroids. Biomed. Mass Spectrom. *6*, 57–62.
67. Siekmann, L. (1979) Determination of steroid hormones by the use of isotope dilution-mass spectrometry: a definitive method in clinical chemistry. J. Steroid Biochem. *11*, 117–123.
68. Knuppen, R., Haupt, O., Schramm, W. & Hoppen, H.-O. (1979) Selected ion monitoring: a new approach for the specific determination of steroids in the lower picogram range. J. Steroid Biochem. *11*, 153–160.
69. Baillie, T. A., Brooks, C. J. W. & Middleditch, B. S. (1972) Comparison of corticosteroid derivatives by gas chromatography-mass spectrometry. Anal. Chem. *44*, 30–37.
70. Adlercreutz, H., Ervast, H.-S., Tenhunen, A. & Tikkanen, M. J. (1973) Gas chromatographic and mass spectrometric studies on

oestrogens in bile. Part 1. Pregnant women. Acta Endocrinol. *73*, 543–554.

71. Gaskell, S. J. & Brooks, C. J. W. (1978) Studies of aldosterone 20, 21-cyclic boronates by gas liquid chromatography and mass spectrometry. J. Chromatog. *158*, 331–336.
72. Honour, J. W. & Shackleton, C. H. L. (1977) Mass spectrometric analysis for tetrahydroaldosterone. J. Steroid Biochem. *8*, 299–305.
73. Axelson, M. (1977) Deactivation of gas chromatographic systems for quantitative analysis of MO-TMS derivatives of steroids at the picogram level. J. Steroid Biochem. *8*. 693–698.
74. Axelson, M. (1978) Exchange of oxime functions: a useful reaction in GC-MS analysis of steroids. Anal. Biochem. *86*, 133–141.
75. Baillie, T. A., Brooks, C. J. W. & Horning, E. C. (1972) O-butyloximes and O-pentyloximes as derivatives for the study of ketosteroids by gas chromatography. Anal. Lett. *5*, 351–361.
76. Finlay, E. M. H. & Gaskell, S. J. (1981) Determination of testosterone in plasma from men by gas chromatography/mass spectrometry, with high-resolution selected-ion monitoring and metastable peak monitoring. Clin. Chem. *27*, 1165–1170.
77. Chambaz, E. M., Madani, C. & Ros, A. (1972) TMS-enol-TMS: a new type of derivative for the gas phase study of dihydroxyacetone side chain saturated corticosteroid metabolites. J. Steroid Biochem. *3*, 741–747.
78. Blair, I. & Phillipou, G. (1978) Derivatisation of steroid hormones with t-butyldimethylsilyl imidazole. J. Chromatog. Sci. *16*, 201–203.
79. Andersson, S. H. G. & Sjövall, J. (1984) Preparation of enol-tert-butyldimethylsilyl and mixed tert-butyldimethylsilyl-trimethylsilyl ethers for gas chromatography-mass spectrometry of steroids and bile acids. J. Chromatog. *289*, 195–206.
80. Bjorkhem, I., Blomstrand, R. & Lantto, O. (1975) Validation of routine methods for serum progesterone determination using mass fragmentography. Clin. Chim. Acta *65*, 343–350.
81. Jaakomaki, P. I., Yarger, D. A. & Horning, E. C. (1967) Gas-liquid chromatographic separation of human urinary steroid glucuronides. Biochem. Biophys. Acta *137*, 216–219.
82. Thompson, R. M. (1976) Gas chromatographic and mass spectrometric analysis of permethylated estrogen glucuronides. J. Steroid Biochem. *7*, 845–852.
83. Spiegelhalder, B., Rohle, G., Siekmann, L. & Breuer, H. (1976) Mass-spectrometry of steroid glucuronides. J. Steroid Biochem. *7*, 749–756.
84. Miyazaki, H., Ishibashi, M., Itoh, M., Morishita, N., Sudo, M. & Nambara, T. (1976) Analysis of estrogen glucuronides. 1-Characterization of estrogen glucuronides by gas chromatography mass spectrometry. Biomed. Mass Spectrom. *3*, 55–59.
85. Shackleton, C. H. L. (1981) Derivatization of estrogen conjugates for analysis by capillary gas chromatography. Steroids *38*, 485–494.
86. Shackleton, C. H. L. (1981) Urinary steroid metabolite analysis: potential for renaissance. Clin. Chem. *27*, 509–511.
87. Murray, S. & Baillie, T. A. (1979) Direct derivatization of sulphate esters for analysis by gas chromatography mass spectrometry. Biomed. Mass Spectrom *6*, 82–89.
88. Cronholm, T. (1969) Position of the sulfate group in steroid sulfates from human plasma, Steroids *14*, 285–296.
89. VandenHeuvel, W. J. A., Sweeley, C. C. & Horning, E. C. (1960) Separation of steroids by gas chromatography. J. Amer. Chem. Soc. *82*, 3481–3482.
90. Sjövall, J. (1966) Identification of 3β,17β-dihydroxyandrost-5-ene, 3β,20α-dihydroxypregn-5-ene and epiandrosterone in human peripheral blood. Steroids *7*, 447–458.
91. Vollmin, J. A. (1970) Separation and identification of urinary steroids on glass capillary columns in a combination of gas chromatograph and mass spectrometer. Chromatographia *3*, 238–241.
92. Lin, Y. Y. & Smith, L. L. (1984) Chemical ionization mass spectrometry of steroids and other lipids. Mass Spectrom. Rev. *3*, 319–355.
93. Raju, U., Levitz, M., Banerjee, S., Bencsath, F. A. & Field, F. H. (1985) Androsterone long chain fatty acid esters in human breast cyst fluid. J. Clin. Endocrinol. Metab. *60*, 940–946.
94. Smith, A. G., Gaskell, S. J. & Brooks, C. J. W. (1976) Trimethylsilyl group migration during electron impact and chemical ionisation mass spectrometry of trimethylsilyl ethers of 20-hydroxy-5α-pregnan-3-ones and 20-hydroxy-4-pregnen-3-ones. Biomed. Mass Spectrom. *3*, 161–165.

95. Gambert, P., Lallemant, C., Archambault, A. & Padieu, P. (1979) Assessment of serum cholesterol by two methods: gas-liquid chromatography on a capillary column and chemical ionization — mass fragmentography with isotopic dilution of [3,4-^{13}C]cholesterol as internal standard. J. Chromatogr. *162*, 1—6.
96. Lindberg, C. Jonsson, S., Hedner, P. & Gustafsson, A. (1982) Radioimmunoassay and chemical ionization-mass spectrometry compared for plasma cortisol determination. Clin. Chem. *28*, 174—177.
97. Dzidic, I. & McCloskey, J. (1972) Chemical ionization mass spectrometry using ammonia reagent gas. Selective protonation of conjugated ketones. Org. Mass Spectrom. *6*, 939—940.
98. Roy, T. A., Field, F. H., Lin, Y. Y. & Smith, L. L. (1979) Hydroxyl ion negative chemical ionization mass spectra of steroids. Anal Chem. **51**, 272—278.
99. Houghton, E., Teale, P., Dumasia, M. C. & Wellby, J. K. (1982) The use of capillary column gas chromatography and negative ion chemical ionization mass spectrometry to confirm the administration of synthetic corticosteroids to horses. Biomed. Mass Spectrom. *9*, 459—465.
100. Gaskell, S. J. & Brooks, P. W. (1983) High sensitivity analyses of steroids using gas chromatography-negative chemical ionization mass spectrometry. Int. J. Mass Spectrom. Ion Physics *48*, 241—244.
101. Murray, S. & Watson, D. (1986) Bis-trifluoromethylbenzyl derivatives for steroid analysis by gas chromatography electron capture negative ion chemical ionization mass spectrometry. J. Steroid Biochem. *25*, 255—260.
102. Bruins, A. P. (1981) Negative ion desorption chemical ionization mass spectrometry of some undervatized glucuronides. Biomed. Mass Spectrom. *8*, 31—35.
103. Cairns, T. & Sigmund, E. G. (1982) Characterization of glucuronides by chemical ionization mass spectrometry with ammonia as reagent gas. Anal. Chem. *54*, 2456—2461.
104. Aberth, W., Straub, K. M. & Burlingame, A. L. (1982) Secondary ion mass spectrometry with cesium ion primary beam and liquid target matrix for analysis of bioorganic compounds. Anal. Chem. *54*, 2029—2034.
105. Liehr, J. G., Beckner, C. F., Ballatore, A. M. & Caprioli, R. M. (1982) Fast atom bombardment mass spectrometry of estrogen glucuronides and sulphates. Steroids *39*, 399—605.
106. Shackleton, C. H. L. & Straub, K. M. (1982) Direct analysis of steroid conjugates: the use of secondary ion mass spectrometry. Steroids *40*, 35—51.
107. Gaskell, S. J., Brownsey, B. G., Brooks, P. W. & Green, B. N. (1983) Fast atom bombardment mass spectrometry of steroid sulphates: qualitative and quantitative studies. Biomed. Mass Spectrom. *10*, 215—219.
108. Shackleton, C. H. L., Mattox, V. R. & Honour, J. W. (1983) Analysis of intact steroid conjugates by secondary ion mass spectrometry (including FABMS) and by gas chromatography. J. Steroid Biochem. *19*, 209—217.
109. Mattox, V. R., Nelson, A. N., Vrieze, W. D. & Jardine, I. (1983) Synthesis of mono- and diglucosiduronates of metabolites of desoxycorticosterone and corticosterone and analysis by a new mass spectrometric technique. Steroids *42*, 349—364.
110. Shackleton, C. H. L. & Chai, W. (1985) Clinical applications of steroid conjugate analysis by liquid SIMS. In: Mass Spectrometry in the Health and Life Sciences (Burlingame, A. L. and Castagnoli, N., eds.), pp. 491—504, Elsevier, Amsterdam.
111. Shackleton, C. H. L., Gaskell, S. J. & Liberato, D. J. (1987) Chromatographic (GC or LC) and non-chromatographic (FAB) mass spectrometric techniques for clinical steroid analysis. In: Capillary Gas Chromatography — Mass Spectrometry in Medicine and Pharmacology (Jaeger, H., ed.), Hüthig Verlag, Heidelberg.
112. Guenat, C. & Gaskell, S. J. (1986) Application of fast atom bombardment and tandem mass spectrometry to the differentiation of isomeric molecules of biological interest. In: Secondary Ion Mass Spectrometry: SIMS V (Benninghoven, A., Colton, R. J., Simons, D. S. and Werner, H. W., eds.), pp. 524—526, Springer-Verlag, Berlin.
113. Cole, R. B., Guenat, C. R. & Gaskell, S. J. (1987) Effect of experimental conditions on the daughter ion spectra derived from tandem mass spectrometry of steroid glucuronides. Anal. Chem. *59*, 1139—1144.
114. Blakley, C. R. & Vestal, M. L. (1983) Thermospray interface for liquid chromatography/mass spectrometry. Anal. Chem. *55*, 750—754.
115. Watson, D., Taylor, G. W., Laird, S. & Vinson, G. P. (1987) Identification of steroids in rat adrenal glands by liquid chromatog-

raphy-mass spectrometry. Biochem. J. *242*, 109–114.
116. Gaskell, S. J., Rollins, K., Smith, R. W. & Parker, C. E. (1987) Determination of serum cortisol by thermospray liquid chromatography/mass spectrometry: comparison with gas chromatography/mass spectrometry. Biomed. Environ. Mass Spectrom *14*, 717–722.
117. Watson, D., Murray, S. & Taylor, G. W. (1986) Steroid glucuronide conjugates: analysis by thermospray liquid chromatography negative ion mass spectrometry. Biomed. Environ. Mass Spectrom. *13*, 65–69.
118. Watson, D., Murray, S. & Taylor, G. W. (1985) Thermospray liquid chromatography negative ion mass spectrometry of steroid sulphate conjugates. Biomed. Environ. Mass Spectrom. *12*, 610–615.
119. Ito, Y., Takeuchi, T., Ishii, D. & Goto, M. (1985) Direct coupling of micro-high performance liquid chromatography with fast atom bombardment mass spectrometry. J. Chromatogr. *346*, 161–166.
120. Caprioli, R. M., Fan, T. & Cottrell, J. S. (1986) Continuous-flow sample probe for fast atom bombardment mass spectrometry. Anal. Chem. *58*, 2949–2956.
121. Millington, D. S. (1975) Determination of hormonal steroid concentrations in biological extracts by high resolution mass fragmentography. J. Steroid Biochem. *6*, 239–245.
122. Gaskell, S. J. & Millington, D. S. (1978) Selected metastable peak monitoring: a new specific technique in quantitative gas chromatography-mass spectrometry. Biomed. Mass Spectrom. *5*, 557–558.
123. Gaskell, S. J., Finney, R. & Harper, M. E. (1979) The determination of testosterone in hamster prostate by gas chromatography-mass spectrometry with selected metastable peak monitoring. Biomed. Mass Spectrom. *6*, 113–116.
124. Gaskell, S. J., Porter, C. J. & Green, B. N. (1985) Trace analysis by GC/MS using a tandem double focusing/quadrupole instrument. Biomed. Mass Spectrom. *12*, 139–141.
125. Gaskell, S. J. (1988) Evaluation of the quantitative analysis of a steroid sulphate using fast atom bombardment and tandem mass spectrometry. Biomed. Environ. Mass Spectrom. *15*, 99–104.
126. Vollmin, J. A. (1971) High resolution gas chromatography of urinary steroids on glass capillary columns. Clin. Chim. Acta *34*, 207–214.
127. Reimendal, R. & Sjövall, J. (1972) Analysis of steroids by off-line computerized gas chromatography-mass spectrometry. Anal. Chem. *44*, 21–29.
128. Reimendal, R. & Sjövall, J. (1973) Computer evaluation of gas chromatographic-mass spectrometric analyses of steroids from biological materials. Anal. Chem. *45*, 1083–1089.
129. Axelson, M., Cronholm, T., Curstedt, T., Reimendal, R. & Sjövall, J. (1974) Quantitative analysis of unlabelled and polydeuterated compounds by gas chromatography-mass spectrometry. Chromatographia *7*, 502–509.
130. Gates, S. C., Smisko, M. J., Ashendel, C. L., Young, N. D., Holland, J. F. & Sweeley, C. C. (1978) Automated simultaneous qualitative and quantitative analysis of complex mixtures with a gas chromatography-mass spectrometry-computer system. Anal. Chem. *50*, 433–441.
131. Vrbanac, J. J., Braselton, W. E., Holland, J. F. & Sweeley, C. C. (1982) Automated qualitative and quantitative metabolic profiling analysis of urinary steroids by a gas chromatography-mass spectrometry-data system. J. Chromatogr. *239*, 265–276.
132. Vrbanac, J. J., Sweeley, C. C. & Pinkston, J. D. (1983) Automated metabolic profiling analysis of urinary steroids by a gas chromatography mass spectrometry data system. Biomed. Mass Spectrom. *10*, 155–161.
133. Axelson, M., Sahlberg, B.-L. & Sjövall, J. (1981) Analysis of profiles of conjugated steroids in urine by ion-exchange separation and gas chromatography-mass spectrometry. J. Chromatogr. *224*, 355–370.
134. Begue, R.-J., Desgres, J., Gustafsson, J.-A. & Padieu, P. (1976) Analyse modulaire des steroides urinaires pendant la gestation humaine. J. Steroid Biochem. *7*, 211–221.
135. Shackleton, C. H. L., Taylor, N. F. & Honour, J. W. (1980) An Atlas of Gas Chromatographic Profiles of Neutral Urinary Steroids in Health and Disease, Packard-Becker, Delft, The Netherlands.
136. Fotsis, T., Adlercreutz, H. & Bannwart, C. (1985) Determination of the estrogen profile in non-pregnancy urine utilizing ion-exchange chromatography and selected ion monitoring: preliminary results. In: Advances in Steroid Analysis '84 (Gorog, S., ed.) pp. 389–398. Elsevier, Amsterdam.

137. Adlercreutz, H., Fotsis, T., Bannwart, C., Hamalainen, E., Bloigu, S. & Ollus, A. (1986) Urinary estrogen profile determination in young Finnish vegetarian and omnivorous women. J. Steroid Biochem. *24*, 289–296.
138. Shackleton, C. H. L. & Straub, K. M. (1982) Direct analysis of steroid conjugates: the use of secondary ion mass spectrometry. Steroids *40*, 35–51.
139. Shackleton, C. H. L. (1983) Inborn errors of metabolism: detection by a new mass spectrometric method. Clin. Chem. *29*, 246–249.
140. Shackleton, C. H. L., Mattox, V. R. & Honour, J. W. (1983) Analysis of intact steroid conjugates by secondary ion mass spectrometry (including FABMS) and by gas chromatography. J. Steroid Biochem. *19*, 209–217.
141. Moneti, G., Agati, G., Giovannini, M. G., Payzagli, M., Salerno, R., Messeri, G. & Serio, M. (1985) Pregnanediol-3α-glucuronide measured in diluted urine by mass spectrometry with fast atom bombardment/negative-ion ionization. Clin. Chem. *31*, 46–49.
142. Emons, G., Hoppen, H.-O., Ball, P. & Knuppen, R. (1980) 4-hydroxyestrone, isolation and identification in human urine. Steroids *36*, 73–79.
143. Johnson, D. W., Phillipou, G., Ralph, M. M. & Seamark, R. F. (1979) Specific quantitation of urinary progesterone by gas chromatography-mass spectrometry. Clin. Chim. Acta *94*, 207–208.
144. Sjövall, J. & Vihko, R. (1968) Analysis of solvolyzable steroids in human plasma by combined gas chromatography-mass spectrometry. Acta Endocrinol. *57*, 247–260.
145. Sjövall, J., Sjövall, K. & Vihko, R. (1968) Steroid sulphates in human pregnancy plasma. Steroids *11*, 703–715.
146. Janne, O., Vinko, R., Sjövall, J. & Sjövall, R. (1969) Determination of steroid mono- and disulphates in human plasma. Clin. Chim. Acta *23*, 405–412.
147. Hellstrom, K., Sjövall, J. & Vihko, R. (1969) Plasma levels and excretion of solvolyzable steroids. Acta Endocrinol. *60*, 501–511.
148. Laatikainen, T., Peltonen, J. & Nylander, P. (1973) Determination of estriol, estriol sulfate, progesterone and neutral steroid mono- and disulfates in umbilical cord blood plasma. Steroids *21*, 347–359.
149. Sjövall, J. & Sjövall, K. (1968) Identification of 5α-pregnane-3α,20α,21-triol in human pregnancy plasma. Steroids *12*, 359–366.
150. Axelson, M. & Sjövall, J. (1974) Separation and computerized gas chromatography-mass spectrometry of unconjugated neutral steroids in plasma. J. Steroid Biochem. *5*, 773–738.
151. Baillie, T. A., Anderson, R. A., Sjövall, K. & Sjövall, J. (1976) Identification and quantitation of 16α-hydroxy C21 steroid sulphates in plasma from pregnancy women. J. Steroid Biochem. *7*, 203–209.
152. Axelson, M. & Sahlberg, B.-L. (1983) Group separation and gas chromatography-mass spectrometry of conjugated steroids in plasma. J. Steroid Biochem. *18*, 313–321.
153. Dvorchik, B. H. & Vesell, E. S. (1976) Pharmacokinetic interpretation of data gathered during therapeutic drug monitoring. Clin. Chem. *22*, 868–878.
154. Gaskell, S. J., Pike, A. W. & Griffiths, K. (1980) Analysis of testosterone and dehydroepiandrosterone in saliva by gas chromatography-mass spectrometry. Steroids *36*, 219–228.
155. Gould, V. J., Turkes, A. O. & Gaskell, S. J. (1986) Gas chromatography-mass spectrometric analysis of salivary testosterone with reference to diethylstilboestrol treated prostatic cancer patients. J. Steroid Biochem. *24*, 563–567.
156. Gaskell, S. J., Pike, A. W. & Finlay, E. M. H. (1981) New techniques in gas chromatography-mass spectrometry and their use in the validation of routine steroid assays. In: Quality Control in Clinical Endocrinology (Wilson, D. W., Gaskell, S. J., Kemp, K. W., eds.), pp. 67–72, Alpha Omega Publishing, Cardiff, U. K.
157. Finlay, E. M. H. Morton, M. S. & Gaskell, S. J. (1982) Identification and quantification of dehydroepiandrosterone sulphate in saliva. Steroids *39*, 63–71.
158. Gaskell, S. J. & Finlay, E. M. H. (1984) Identification and quantification of steroids in saliva by gas chromatography-mass spectrometry. In: Immunoassays of Steroids in Saliva (Read, G. F., Riad-Fahmy, D., Walker, R. F. & Griffiths, K., eds.), pp. 33–42, Alpha Omega Publishing, Cardiff, U. K.
159. Brooks, F. S. & Brooks, R. V. (1984) Cortisol and cortisone in saliva. In: Immunoassays of Steroids in Saliva (Read, G. F., Riad-Fahmy, D., Walker, R. F. & Griffiths, K., eds.), pp. 322–326, Alpha Omega Publishing, Cardiff, U. K.
160. Vining, R. F., McGinley, R. A. & Symons, R. G. (1983) Hormones in saliva: mode of

entry and consequent implications for clinical interpretation. Clin. Chem. *29*, 1752–1756.

161. Vining, R. F. & McGinley, R. A. (1984) Transport of steroids from blood to saliva. In: Immunoassays of Steroids in Saliva (Read, G. F., Riad-Fahmy, D., Walker, R. F. and Griffiths, K., eds.), pp. 56–63, Alpha Omega Publishing, Cardiff, U. K.
162. Reiffsteck, A., Dehennin, L. & Scholler, R. (1982) Identification et dosage de steroides neutres dans le plasma seminal par spectrometrie de masse. Pathol. Biol. *30*, 626–628.
163. Dehennin, L., Blacker, C., Reiffsteck, A. & Scholler, R. (1984) Estrogen 2-, 4-, 6- or 16-hydroxylation by human follicles shown by gas chromatography-mass spectrometry associated with stable isotope dilution. J. Steroid Biochem. *20*, 465–471.
164. Sahlberg, B.-L. & Axelson, M. (1986) Identification and quantitation of free and conjugated steroids in milk from lactating women. J. Steroid Biochem. *25*, 379–391.
165. Vanluchene, E., Vandekerckhove, D., De Bower, J. & Sandra, P. (1984) Structure and concentrations of fifteen different steroid sulfates in human breast cyst fluids. J. Steroid Biochem. *21*, 367–371.
166. Jakowicki, J., Ervast, H.-S. & Adlercreutz, H. (1973) Gas chromatographic determination and mass spectrometric identification of oestrogens in normal human placental tissue at term. J. Steroid Biochem. *4*, 181–191.
167. Ruokonen, A. & Vihko, R. (1974) Steroid metabolism in testis tissue: concentrations of unconjugated and neutral steroids in boar testis. J. Steroid Biochem. *5*, 33–38.
168. Huhtaniemi, I., Luukkainen, T. & Vihko, R. (1970) Identification and determination of neutral steroid sulphates in human foetal adrenal and liver tissue. Acta Endocrinol. *64*, 273–286.
169. Millington, D. S., Buoy, M. E., Brooks, G., Harper, M. E. & Griffiths, K. (1975) Thin-layer chromatography and high resolution selected ion monitoring for the analysis of C19 steroids in human hyperplastic prostate tissue. Biomed. Mass Spectrom. *2*, 219–224.
170. Moneti, G., Costantini, A., Guama, A., Salerno, R., Pazzagli, M., Natali, A., Goti, A. & Serio, M. (1986) Measurement of testosterone and its 5-alpha-reduced metabolites in human prostatic tissue using isotope dilution mass spectrometry. J. Steroid Biochem. *25*, 765–772.
171. Millington, D. S., Jenner, D. A., Jones, T. & Griffiths, K. (1974) Endogenous steroid concentrations in human breast tumours determined by high-resolution mass fragmentography. Biochem. J. *139*, 473–475.
172. Maynard, P. V., Bird, M., Basu, P. K., Shields, R. & Griffiths, K. (1978) Dehydroepiandrosterone and androstenediol in human primary breast tumours. Europ. J. Cancer *14*, 549–553.
173. Maynard, P. V. & Griffiths, K. (1979) Clinical, pathological and biochemical aspects of the oestrogen receptor in primary human breast cancer. In: Steroid Receptor Assays in Human Breast Tumours (King, R. J. B., ed.), pp. 86–99, Alpha Omega Publishing, Cardiff, U. K.
174. Maynard, P. V., Pike, A. W., Weston, A. & Griffiths, K. (1977) Analysis of dehydroepiandrosterone and androstenediol in human breast tissue using gas chromatography-mass spectrometry. Europ. J. Cancer *13*, 971–975.
175. Edery, M., Goussard, J., Dehennin, L., Scholler, R., Reiffsteck, A. & Drosdowsky, M. A. (1981) Endogenous oestradiol-17β concentrations in breast tumours determined by mass fragmentography and by radioimmunoassay: relationship to receptor content. Europ. J. Cancer *17*, 115–120.
176. Fishman, J., Nisselbaum, J. S., Menedez-Botet, C. J. & Schwartz, M. K. (1977) Estrone and estradiol content in human breast tumors: relationship to estradiol receptors. J. Steroid Biochem. *8*, 893–896.
177. Daniel, C. P., Gaskell, S. J., Bishop, H. & Nicholson, R. I. (1981) Determination of tamoxifen and metabolites in human breast tumours and plasma: biochemical and clinical significance. Europ. J. Cancer Clin. Oncol. *17*, 1183–1189.
178. Axelson, M., Clark, J. H., Eriksson, H. A. & Sjövall, J. (1981) Estrogen binding in target tissues: A GC/MS method for assessing uptake, retention and processing of estrogens in target cell nuclei under in vivo conditions. J. Steroid Biochem. *14*, 1253–1260.
179. Padieu, P. & Maume, B. F. (1977) Evaluation by mass fragmentography of metabolite pathways of endogenous and exogenous compounds in eukaryotic cell cultures. In: Quantitative Mass Spectrometry in Life Sciences (De Leenheer, A. P. & Roncucci, R. R., eds.), pp. 49–82, Elsevier, Amsterdam.
180. Baillie, T. A. (1981) Stable isotopes in pharmacology. Pharmacol. Rev. *33*, 81–132.

181. Tokes, L. & Throop, J. L. (1972) Introduction of deuterium into the steroid system. In: Organic Reactions in Steroid Chemistry, Vol. 1 (Fried, L. & Edwards, J. A., eds.), pp. 145–221, Van Nostrand Reinhold, New York.
182. Johnson, D. W., Phillipou, G. & Seamark, R. F. (1981) Deuterium labeled steroid hormones: synthesis and applications in quantitation and endocrinology. J. Steroid Biochem. *14*, 793–800.
183. Dehennin, L., Reiffsteck, A. & Scholler, R. (1980) Simple methods for the synthesis of twenty different, highly enriched deuterium labelled steroids, suitable as internal standards for isotope dilution mass spectrometry. Biomed. Mass Spectrom. *7*, 493–499.
184. Yuan, S.-S. (1982) Synthesis of 3,4-$^{13}C_2$ steroids. Steroids *39*, 279–289.
185. Nixon, A., Mallett, A. I., Jackman, P. J. H. & Gower, D. B. (1986) Testosterone metabolism by isolated human axillary *Corynebacterium spp.*: a gas-chromatographic mass-spectrometric study. J. Steroid Biochem. *24*, 887–892.
186. Baba, S., Shinohara, Y. & Kasuya, Y. (1980) Differentiation between endogenous and exogenous testosterone in human plasma and urine after oral administration of deuterium-labeled testosterone by mass fragmentography. J. Clin. Endocrinol. Metab. *50*, 889–894.
187. Shinohara, Y., Baba, S. & Kasuya, Y. (1980) Absorption, metabolism, and excretion of oral testosterone in humans by mass fragmentography. J. Clin. Endocrinol. Metab. *51*, 1459–1462.
188. Pinkus, J. L., Charles, D. & Chattoraj, S. D. (1971) Deuterium-labeled steroids for study in humans. I. Estrogen production rates in normal pregnancy. J. Biol. Chem. *246*, 633–636.
189. Pinkus, J. L., Charles, D. & Chattoraj, S. C. (1979) Deuterium-labeled steroids for study in humans. II. Preliminary studies on estrogen production rates in pre- and postmenopausal women. Hormone Res. *10*, 44–56.
190. Taylor, E. S., Hagerman, D. D., Albrecht, B. H. & Penney, L. L. (1974) Estriol production rate measurement studies in pregnant women. Amer. J. Obstet. Gynecol. *120*, 752–758.
191. Johnson, D. W., Phillipou, G., Blair, I. A. & Seamark, R. F. (1979) Measurement of urinary steroid production rates using stable-isotopes and GC-MS. Experientia *35*, 1261–1262.
192. Baillie, T. A., Sjövall, K., Herz, J. E. & Sjövall, J. (1976) Specific deuterium labeling and computerized gas chromatography-mass spectrometry in studies of the metabolism and turnover of steroid sulfates in pregnant women. In: Advances in Mass Spectrometry in Biochemistry and Medicine, Vol. II (Frigerio, A., ed.), pp. 161–170, Spectrum Publications, New York.
193. Baillie, T. A., Curstedt, T., Sjövall, K. & Sjövall, J. (1980) Production rates and metabolism of sulphates of 3β-hydroxy-5α-pregnane derivatives in pregnant women. J. Steroid Biochem. *13*, 1473–1486.
194. Nystedt, L. (1980) On the mathematics of tracer kinetics when pools are not in steady-state. J. Steroid Biochem. *13*, 1487–1488.

Chapter 12
Thyroid Hormones

D. B. Ramsden

1 Introduction

This chapter sets out to illustrate the contribution mass spectrometry has made to our understanding of thyroid hormone physiology. It is divided into five major sections. The first of these is a brief review of the biosynthesis, metabolism and actions of thyroid hormones to set in a general context the succeeding sections for those whose interests are not primarily endocrinological. Later sections attempt to define the nature of the role mass spectrometry has played and can continue to play in thyroidology and the problems of applying the techniques of mass spectrometry to analysis of thyroid hormones, amongst which is the conversion of these highly involatile amino acids to more volatile derivatives. Following a summary of the mass spectral characteristics of the volatile derivatives is a section outlining the work done on the analysis of individual iodothyronines and related compounds. The conclusions contain a brief assessment of the likely impact of newer developments in mass spectrometry.

2 Thyroid Hormone Physiology

2.1 Thyroidal biosynthesis

Thyroid hormones are synthesized by the follicular cells of the thyroid gland as components of a large glycoprotein, thyroglobulin. Initially non-iodinated thyroglobulin is formed within the follicular cell. This is then secreted into the lumen of the follicle. Poorly or non-iodinated thyroglobulin is adsorbed on the apical membrane where tyrosine moieties are converted to 3-mono- and 3,5-diiodotyrosine moieties.

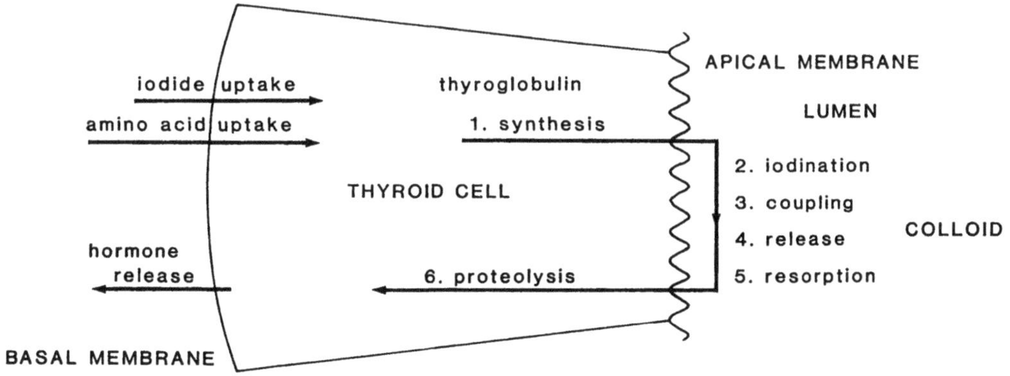

Figure 1. Thyroid cell, showing a) uptake of iodide from and secretion of hormone into extracellular fluid and b) secretion of thyroglobulin into the lumen of the follicle to be stored as a component of the colloid until colloidal resorption takes place.

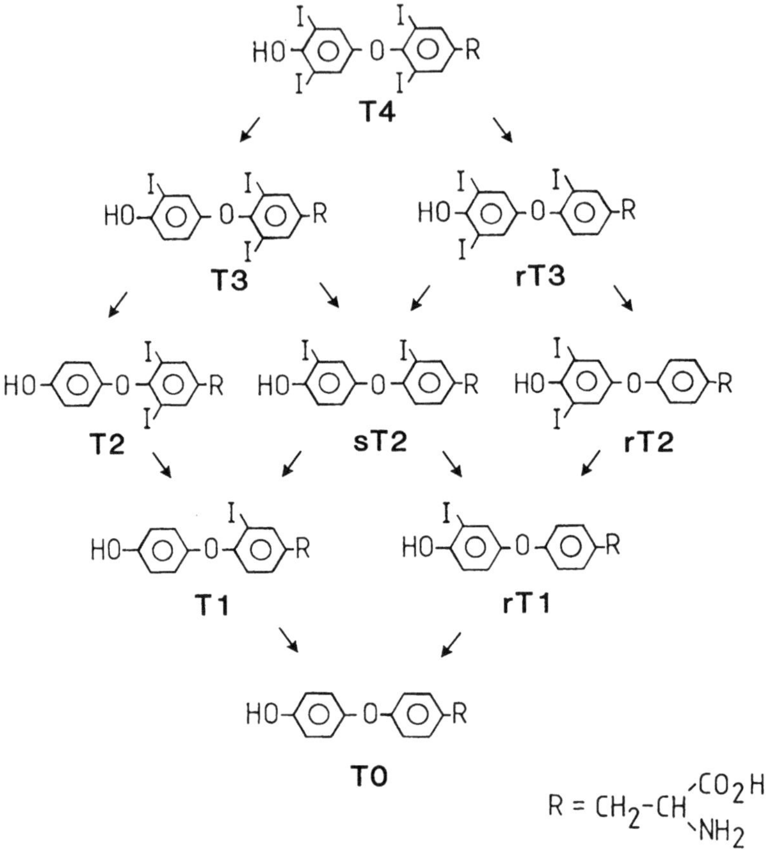

Figure 2. Structures of naturally occurring iodothyronines:
T4 = Thyroxine; T3 = 3,5,3′-Triiodothyronine;
rT3 = 3,3′,5′-Triiodothyronine; sT2 = 3,3′-Diiodothyronine;
T2 = 3,5-Diiodothyronine; rT2 = 3′,5′-Diiodothyronine;
T1 = 3-Monoiodothyronine; rT1 = 3′-Monoiodothyronine;
T0 = Thyronine; s and r are the contractions of "split" and "reverse" respectively.

Iodinated tyrosine residues within thyroglobulin then couple together under oxidising conditions to give the substituted diphenyl ether structure characteristic of iodothyronines. Heavily iodinated thyroglobulin containing thyroxine residues is then released from the apical membrane to be stored in the lumen. Under stimulation by the anterior pituitary hormone thyrotrophin (thyroid stimulating hormone, TSH), thyroglobulin is reabsorbed into the follicular cell, whereupon it is subjected to proteolysis to liberate the thyroid hormones, thyroxine (T4) and 3,5,3′-triiodothyronine (T3) which are secreted into the bloodstream from the basal membrane (1). This process is illustrated in Figure 1 and the structures of the iodothyronines are shown in Figure 2. Under normal conditions iodinated porcine thyroglobulin contains approximately 4 residues of thyroxine, 1 of 3,5,3′-triiodothyronine, 12 of 3,5-diiodotyrosine (DIT) and 11 of 3-

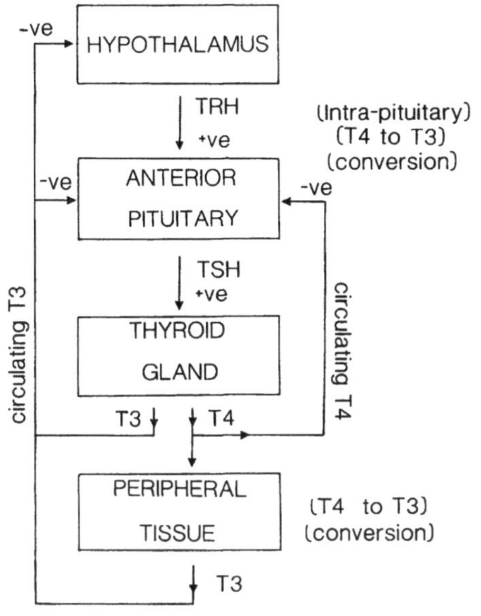

Figure 3. Hypothalamic-pituitary-thyroid feedback axis: + ve denotes stimulatory and −ve inhibitory effects of the various components.

mono-iodotyrosine (MIT) (2). Small amounts of 3,3′,5-triiodothyronine (reverse-T3, rT3) and 3,3′-diiodothyronine (split T2, sT2) are also formed, but thyroidal secretion contributes minimally to the total daily production of these last two compounds.

2.2 Hypothalamic-pituitary thyroid axis

Secretion of thyroid hormones is regulated by a negative feedback system involving inhibition of TSH secretion by thyroid hormones. Trophic stimulation of TSH secretion is achieved by the action of thyrotrophin releasing hormone (TRH) which is synthesized and secreted by the hypothalamus. Most recent evidence suggests that a second negative feedback loop may exist whereby thyroid hormones inhibit the release of TRH (Fig. 3). Many other agents e.g. catecholamines, oestrogens, glucocorticoids, endorphins, neurotensin, somatostatin etc. have also been implicated or proposed as modulators or components of this system, however, at the most, these are of secondary importance to the components outlined above, which form the primary control system for the regulation of thyroid hormone secretion and hence expression (3).

2.3 Serum transport

Upon release into the bloodstream thyroid hormones, which are relatively insoluble in aqueous media, are non-covalently bound by three transport proteins, thyroxine binding globulin (TBG), thyroxine binding prealbumin (TBPA) and albumin (Alb). During the passage around the vascular sytem dissociation of the hormone-protein

Table 1. Serum concentrations of iodothyronines and related compounds in normal euthyroid subjects

Compound	Mean	SD	Units	Reference
Thyroxine	99	28	$nmol \cdot l^{-1}$	+
3,5,3'-Triiodothyronine	1.95	0.51	$nmol \cdot l^{-1}$	+
3,3',5'-Triiodothyronine	570	170	$pmol \cdot l^{-1}$	+
3,3'-Diiodothyronine	144	46	$pmol \cdot l^{-1}$	67
3',5'-Diiodothyronine	130	46	$pmol \cdot l^{-1}$	68
3,5-Diiodothyronine	82	4	$pmol \cdot l^{-1}$	69
3'-Monoiodothyronine	7.3	3.5	$pmol \cdot l^{-1}$	70
3-Monoiodothyronine	70	30	$pmol \cdot l^{-1}$	71
3,5,3',5'-Tetraiodothyroacetic acid	115	26	$pmol \cdot l^{-1}$	32
3,5-Diiodotyrosine	160	130	$pmol \cdot l^{-1}$	56

+ unpublished laboratory reference range, in general agreement with widely accepted values.

complex occurs to allow thyroid hormones to enter target cells. Typical concentrations of the various components are shown in Table 1. From this it can be seen that in serum T4 is the dominant species in terms of concentration, reflecting its higher synthetic rate by the thyroid gland and more avid binding by the transport proteins (4).

2.4 Peripheral metabolism of thyroid hormones

Inside the various target cells thyroid hormones may undergo a number of metabolic reactions. Of these the most important is the conversion of T4 to either T3 or rT3. The importance of these reactions stems from the fact that the majority of the actions of thyroid hormones are mediated by T3, with T4 having little or no intrinsic activity itself but acting solely as a prohormone. The second triiodothyronine, rT3, also has no recognised endocrine activity under physiological conditions. Therefore the conversion of the prohormone T4 to the active T3 or inactive rT3 by peripheral tissue is thought to form the second major control system regulating the expression of thyroid hormone action.

Enzymic monodeiodination of T4 accounts for between 70 and 80% of daily T3 and well over 90% of daily rT3 production. Both T3 and rT3 may then undergo further deiodination to yield ultimately the fully deiodinated compound thyronine (T0). Even within the human and rat, the two most intensively investigated species, the extent to which each of the possible deiodination steps occurs has not been fully elucidated. Figure 4 illustrates the daily production rates of the various iodothyronines where known (5).

Three other metabolic routes of some importance are:

1. conjugation of the phenolic group of iodothyronines with either sulphate or glucuronate (6, 7),
2. deamination followed by oxidative decarboxylation to yield the acetic acid analogues of iodothyronines (8), and
3. oxidation cleavage of the diphenyl ether link of T4 to yield DIT (9).

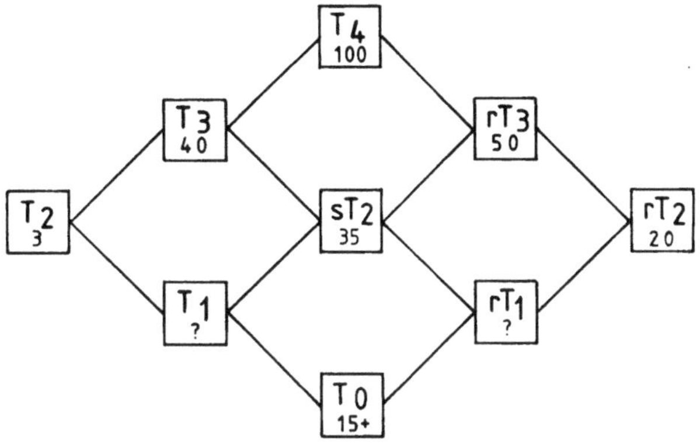

Figure 4. Production rates of thyronines (nmol · 24 $^{-1}$).

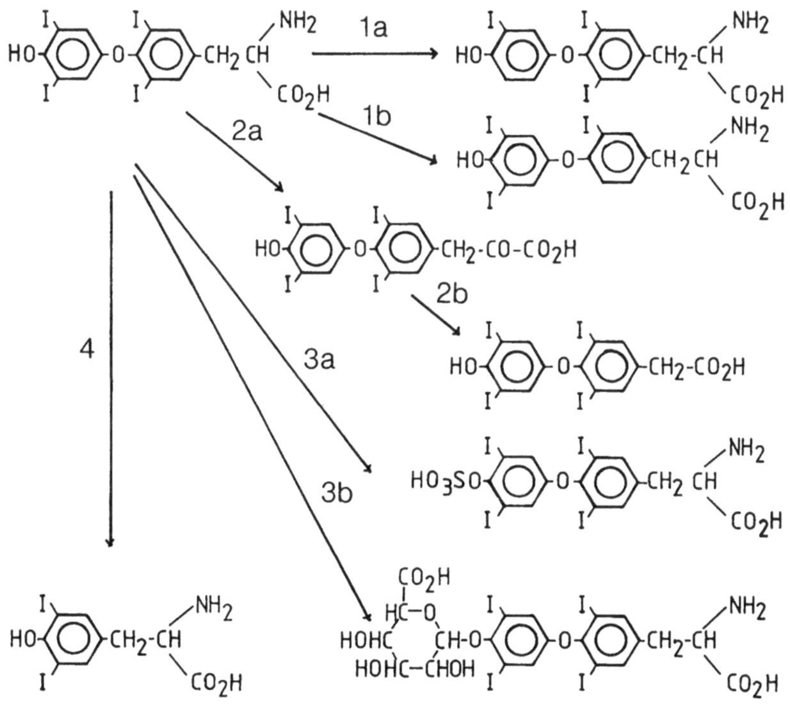

Figure 5. Major metabolic pathways of thyroxine.

These reactions are illustrated in Figure 5. The extent to which deiodination and these other reactions occur varies among different tissues and may vary within a given tissue among different species. In addition to the above a number of minor metabolic pathways have been proposed in individual species e. g. methylation of the phenolic group of iodothyronines in the rat (10). Very little evidence exists concerning the nature and significance of these minor reactions.

2.5 Action of thyroid hormones

The mechanism whereby thyroid hormones exert their actions within the cell is a matter of some debate. However, one major pathway is widely recognised. This involves the interaction of T3 with acidic chromatin protein receptors bound to DNA within the nucleus, which leads to the stimulation of DNA transcription and mRNA translation, culminating in the production of specific enzymes and proteins e. g. α-glycerophosphate dehydrogenase and malic enzyme by the rat liver and Na^+/K^+ ATPase by the rat renal cortex. Modulation of enzymes such as these contribute to changes in basal metabolic rate characteristic of pathological hyper- and hypothyroid states in the human. What contribution the modulation of an individual enzyme makes to the overall changes in basic metabolic rate in thyroid dysfunction is still far from clear (11).

A second general effect of thyroid hormones thought to be mediated via the nucleus of the cell is the regulation of growth and differentiation. This is strikingly demonstrated in the actions of thyroid hormones on amphibian metamorphosis. In mammals the effects of thyroid hormones on cellular differentiation appear most critical in the development of the brain shortly before and for some time after delivery. This has led to the development of large scale screening programmes to detect neonatal hypothyroidism in most developed countries (12).

3 Mass Spectrometry and Thyroidology

The above heading should be regarded in an inquisitorial framework because it can be questioned whether mass spectrometry has any contribution to make to the field of thyroidology. Much is known about thyroid hormone physiology and pathophysiology. Often the data stem from the early phase in the development of modern concepts of medicine, chemistry and physiology before mass spectrometry was available. For example, thyroid dysfunction was the first endocrine abnormality to be described in modern terms (Parry 1825) (13). Thyroxine was the first hormone to be isolated in a pure state (Kendall 1915) (14), and was the first hormone to be synthesized (Harington and Barger 1927) (15). Secondly, adequate assays based on immunological techniques now exist for all the parameters involved in the diagnosis of thyroid dysfunction in humans. Thirdly the prevalence of thyroid disease in humans and the volume of thyroid function assays generated thereby would be too overwhelming for a time-consuming and labour intensive technique such as mass spectrometry to make a contribution to routine thyroid function diagnosis in the clinical chemistry laboratory. Despite the above, the properties of mass spectrometry allow clear areas to be defined in which contributions have been made and potential future contributions can be made to clinical chemistry.

The first role for mass spectrometry stems from the inherently greater precision, accuracy and reproducibility of gas chromatography-mass spectrometry with selected ion monitoring (GC-MS-SIM) and isotope dilution techniques compared to radioimmunoassay and the other immunoassay methods which are the principal techniques

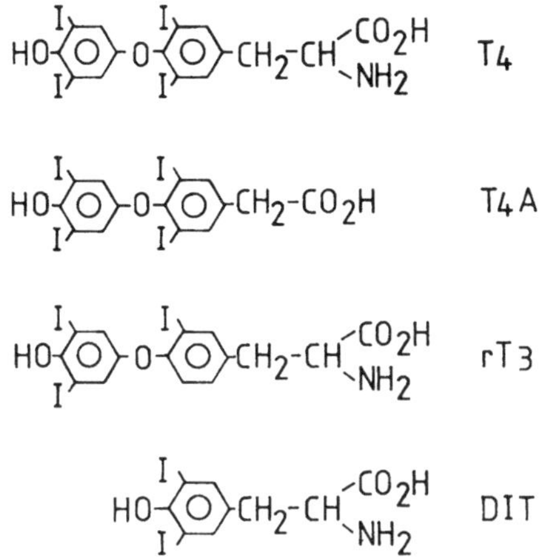

Figure 6. Illustration of common features in the structure of four iodo compounds present in serum.

used in modern biochemical thyroid function testing. Thus mass spectrometric assays for the total T4 and T3 contents of serum can form the basis of reference methods for assessing the performance of radioimmunoassays in routine use (see Chapter 13).

Other applications of mass spectrometric assays lie within research rather than routine spheres of clinical biochemistry. The first of these research areas involves the characterisation of established competitive binding assays for iodothyronines which play little, if any, role in the diagnosis of thyroid dysfunction. The similarities in structure of the various naturally occurring iodothyronines, iodotyrosines and their analogues can pose formidable difficulties for immunological methods of assay. This point is illustrated by reference to assays for four compounds T4, rT3, DIT and 3,5,3′,5′-tetraiodothyroacetic acid (T4A) whose structures are illustrated in Figure 6. As can be seen from this figure all four compounds possess a considerable number of common features. Antibodies or other binding entities directed against the common region of any one of these molecules will tend to cross-react with the other molecules. The degree of cross-reaction will vary with the precise specificity of the binding agent for the molecule as a whole and the influence that the different pK_a's of the four phenolic groups exert on the ligand-binder interaction. Cross-reaction of DIT, T4A, rT3 in an immunological assay for T4 is unimportant because of the very much greater T4 concentration in human serum (1000 × DIT, 1000 × T4A, 100 × rT3). However, the converse is not true because it is obvious that even 1% cross-reactivity of T4 in any of these assays would invalidate the apparent results. It would appear reasonable to suppose that cross-reaction of T4 in rT3 immunoassays, or more precisely the methods by which this cross-reaction is recognised and taken into consideration by different authors, accounts for the relatively wide ranges of rT3 concentration quoted in the literature (16, 17). The same holds true for T4A and DIT. In the latter case, the ranges quoted vary from statements that DIT was totally absent from normal human serum (18) or present in concentrations approaching those of T3 (19) with options between the two extremes. What may, under some conditions, appear to be a minor change in structure in a competitive binding assay, e. g. the difference in the number of iodine atoms between T4 and rT3 rendering cross-reaction a major problem, is in fact a colossal difference in mass which any appropriate mass spectrometer would distinguish, provided the molecules were presented in an acceptable form. Within these terms, mass spectrometry has a role in the more rigorous determination of the concentrations of compounds known to be present in serum and for which assays already exist. Since neither rT3, T4A nor DIT concentrations play any role in the diagnosis of thyroid dysfunction, knowledge of these parameters would appear of value only in terms of understanding thyroid physiology.

An obvious extension of the above is characterisation and assay of metabolites of thyroid hormones whose presence in biological material may be speculated upon with whatever degree of certainty and for which it would be inappropriate to develop an immunological assay as a first step. An example of such an instance would be the assay of thyronine (T0) in human urine (20).

4 GC-MS Analysis of Thyroid Hormones

4.1 Extraction from biological matrix

Even within the thryoid gland iodothyronines and iodotyrosines form only a very minor component of the total organic material. Therefore, before any form of mass spectrometric analysis can be undertaken, some form of extraction must be carried out to enhance the analyte content of the sample. As mentioned before, iodothyronines have relatively low solubility in aqueous media at physiological pH and hence the most common naturally occurring form of these compounds is as a non-covalent complex with a binding protein. The only significant exception to this is in the thyroid gland where iodothyronines are present as an integral part of the primary structure of thyroglobulin. Thus, usually the first step is to submit the sample to conditions which disrupt the non-covalent complex. A range of possibilities exist including:

1. extreme pH,
2. organic solvent extraction and protein precipitation,
3. high temperature protein denaturation,
4. proteolytic digestion of the binding protein.

All four possibilities have been used in the development of non-mass spectrometric assays of various types (21 – 24). Where feasible, dissociation of the complex by low pH offers an efficient and simple system which yields a solution suitable for ion-exchange chromatography as the next step in the extraction regime.

Disruption of the protein-ligand complex with organic solvents such as *n*-butanol/HCl solutions gives rise to the possibility of *n*-butyl ester formation during the removal of the butanol, resulting in reduction of the ultimate efficiency of extraction if the *n*-butyl ester and the parent compound behave differently in later steps. Heat denaturation of the binding protein may be difficult e. g. TBPA withstands boiling for 5 minutes and may give rise to artifactual deiodination (25). If the analyte of interest is in low concentration compared with T4 and can be formed directly from T4 by deiodination, any deiodination of the analyte may be more than compensated for by T4 deiodination yielding a spuriously elevated result. Likewise proteolytic enzymes used for digestion of binding proteins might be contaminated with deiodinase giving rise to similar consequences.

Following disruption of the protein-ligand complex any subsequent extraction regime must have regard to two aims; the removal of as much and as many of the other components as possible whilst retaining the maximum amount of the analyte. This latter point is of crucial importance since unless very large quantities of the biological sample are available, the amount of analyte obtained at the end of an extraction is usually close to the practical working limit of most GC-MS systems, because of the low initial concentrations of iodothyronines.

Having disrupted any protein-analyte complex which may have existed, there are three practical alternatives for subsequent extraction steps:

1. ion-exchange chromatography,
2. adsorption chromatography which may be sub-divided into classical and reverse-phase hydrophobic adsorption, and
3. solvent extraction.

The zwitterion structure of thyronines make them ideally suited for ion-exchange chromatography using highly charged cationic and anionic substituted polystyrene ion-exchange resins.

4.2 Selection of internal standards

Two general possibilities exist in the choice of internal standard for monitoring the extraction and derivatisation steps necessary prior to mass spectrometric assay:

1. Use of a version of the analyte labeled with a stable isotope(s) or
2. Use of an analogue with a similar structure to that of the analyte.

Both approaches have been used successfully but the former is obviously the preferred choice because of the chemical identity of analyte and standard. Lack of commercially available iodothyronines with stable isotopic labels requires these to be synthesized.

4.2.1 Synthesis of iodothyronines containing stable isotopic labels

Regrettably there is no stable isotope of iodine other than naturally occurring ^{127}I. This is unfortunate because incorporation of iodine into the outer phenolic ring of thyronines is a simple procedure (26). Therefore, labeling must involve one or more of the other elements present in thyronines and tyrosines. A cursory glance at the structures will show that uniform labeling with ^{13}C would consistently give the largest mass shifts between analyte and standard. However, since labeled thyronines have to be synthesized either by the investigator or by commercial companies specialising in custom synthesis economic considerations tend to weigh against this objective.

Operating within the constraints of economy and limited isotopic incorporation, at least three viable alternatives exist. The first of these involves the coupling reaction between 3,5-diiodo-4-hydroxyphenylpyruvic acid (DHPPA) and iodotyrosines (27, 29).

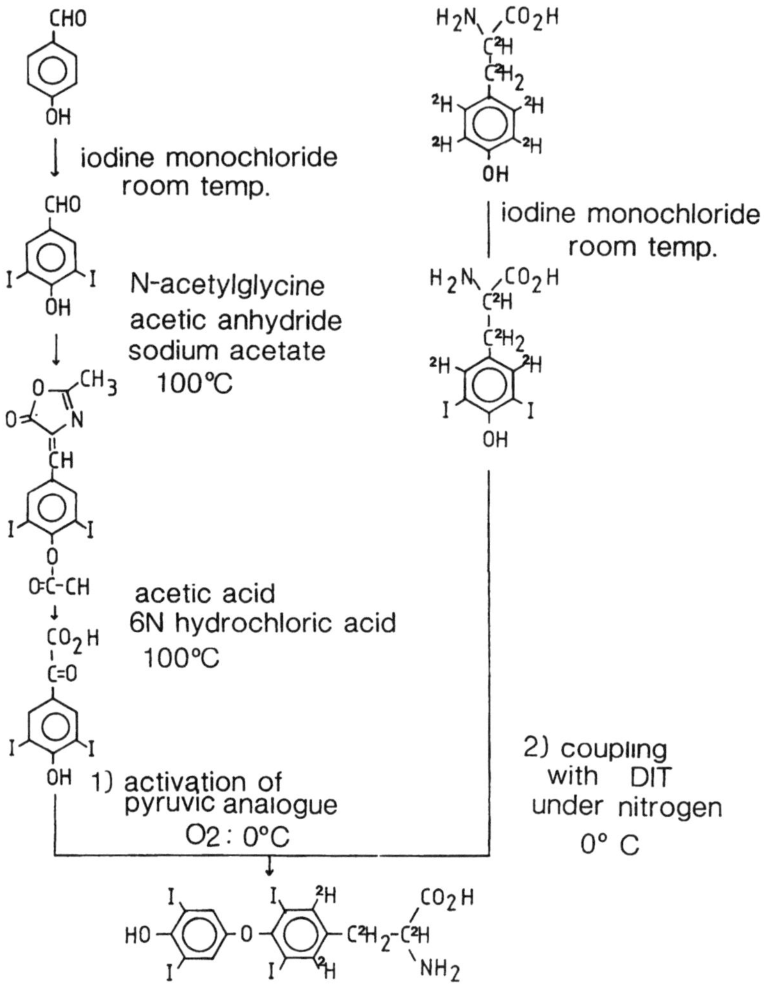

Figure 7. Reaction sequence for the synthesis of ($^2H_5$)-thyroxine from deutero-tyrosine.

Again neither stable labeled DHPPA nor iodotyrosines are commercially available. In fact, DHPPA is not readily available in the natural isotopic form, necessitating its synthesis also. The one compound which is readily available in a stable isotopically labeled form is tyrosine. This may be converted to 3-monoiodotyrosine or DIT and used in a coupling reaction with unlabeled DHPPA. Synthesis of DHPPA from *p*-hydroxybenzaldehyde is also summarised, the complete sequence for the synthesis of T4 being outlined in Figure 7. Frequently small amounts of MIT are formed as well as DIT in the iodination of tyrosine and these give rise to labeled rT3.

Separation of the products of the coupling reaction can be carried out by solvent extraction as described by the original authors or by adsorption chromatography on Sephadex (27). At high pH T4 and rT3 strongly adsorb to Sephadex G 25 whilst most other materials elute rapidly (26). Submission of the products of the coupling reaction

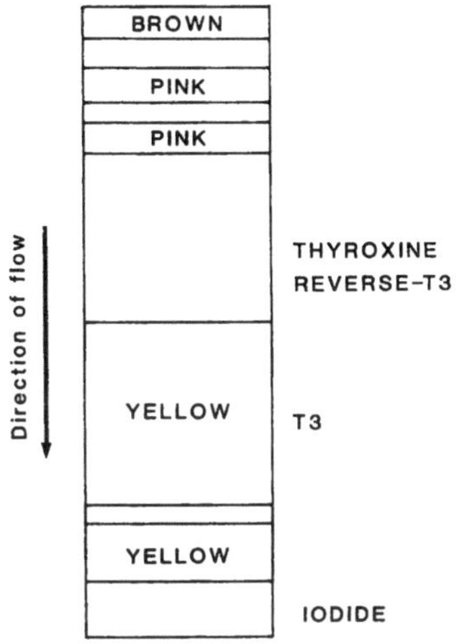

Figure 8
Purification of deutero-thyroxine by adsorption chromatography on Sephadex G25. (Reproduced by permission of John Wiley & Sons Ltd., ref 46.)

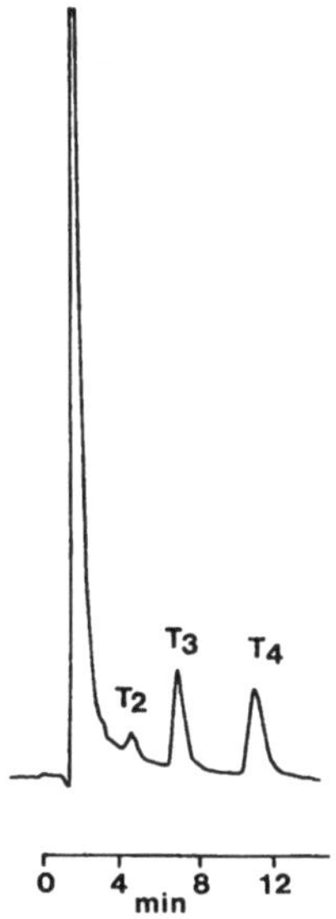

Figure 9
Deiodination of deutero-thyroxine to give deutero-3,5,3′-triiodothyronine and deutero-3,5-diiodothyronine. HPL-chromatogram on Bondapak C18 column; elution with methanol/water (60:40 v/v) at 2 ml · min^{-1}; transmission at wavelength 280 nm.

to this form of chromatography yields T4 (and rT3) with a high degree of purity. Other materials in the reaction mixture elute as a series of intensely coloured bands well separated from the T4 (Fig. 8).

Using 4-hydroxy-3-monoiodophenylpyruvic and 4-hydroxyphenylpyruvic acids as coupling reagents is said to be less efficient than the use of DHPPA (28). Thus alternative methods need to be employed for the synthesis of T3, sT2 and T1. One such method involves first the synthesis of T4 and/or rT3 as outlined above. Selective deiodination of the outer phenolic ring can then be carried out under acid conditions (29). The products of such a deiodination starting with T4 are illustrated in Figure 9. Under similar conditions rT3 would yield sT2 and T1.

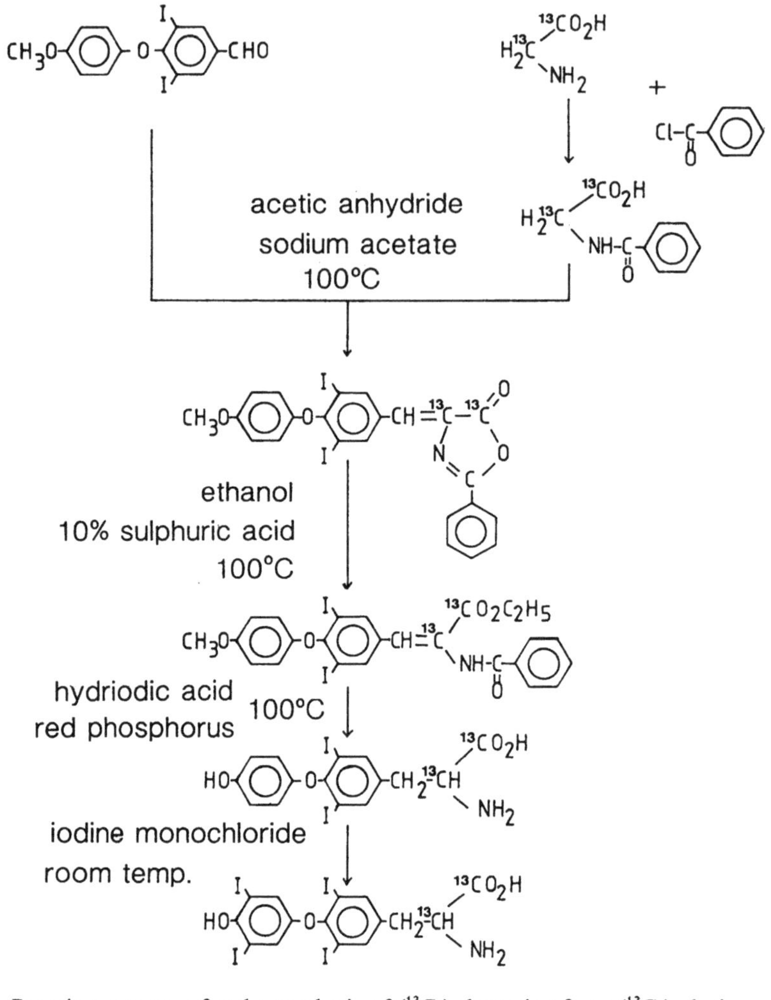

Figure 10. Reaction sequence for the synthesis of ($^{13}C_2$)-thyroxine from ($^{13}C_2$)-glycine.

The second major synthetic route for T4 production restricts isotopic incorporation to the amino acid side-chain of the thyronine structure and utilises isotopically labeled glycine, which can be converted to benzoylglycine, and aldehyde analogues of iodothyronines containing the diphenylether structure. This is the method originally described by Harington and Barger for the synthesis of T4 (15). The reaction sequence is outlined in Figure 10. Once formed the labeled thyroxine may also be selectively deiodinated to yield lesser substituted iodothyronines.

The third synthetic route is limited to rT2, rT1 and T0 and utilises the fact that T4 can be catalytically deiodinated under reducing conditions to yield the fully deiodinated analogue, T0, with deuterium atoms replacing the original iodine substituents. Deutero-T0 may then be reiodinated in the outer phenolic ring to yield rT1 and rT2.

4.2.2 **Chemically related analogues as internal standards**

This approach relies on the fact that relatively minor changes in structure will have minimal effect on the extraction and derivatisation properties of the compounds concerned. For the iodothyronines themselves one method is to use an analogue in which one iodine substituent in one of the aromatic rings is exchanged for a bromine atom. Because bromine itself occurs naturally in isotopic forms, ^{79}Br and ^{81}Br, in almost equal abundance, the relative intensity of any fragment ion used in fragmentographic assays is halved. This disadvantage may be offset by the fact that either mass:charge ratio may be selected to minimise artifactual interference. Moller et al. employed 5′-bromo-3,5,3′-triiodothyronine to estimate T4 levels in human milk (30).

3,5,3′,5′-Tetraiodothyropropionic acid (T4P) was used as the internal standard in the assay of T4A (31, 32) in serum as an increase of one methylene group in the side-chain had minimal influence on extraction and derivatisation properties.

A second important consideration in the use of related analogues is the absence of any of the standard compound in the sample. In the case of T4P, this has been said to be present in serum and other biological fluids. However, it is difficult to envisage the mode of formation of this from T4 *in vivo*, since most deamination reactions go on to yield products with fewer carbon atoms. Further, human serum when investigated yielded no evidence for T4P being present. No naturally occurring brominated thyronines have been reported.

4.3 Formation of volatile derivatives

A range of alternatives are available to convert the amino acid and phenolic groups of iodothyronines to a form suitable for GC and GC-MS studies. These together with details of their preparation are listed in Table 2.

Of the options listed the trimethylsilyl derivatives of iodothyronines are the most easily formed but are one of the least suitable (33–35). In our experience trimethylsilyl derivatives were susceptible to considerable non-specific adsorption and/or degradation on conventional packed columns, with a consequent reduction in sensitivity of the assay. Nevertheless other authors have reported their successful use for both simple gas chromatography (36) and gas chromatography mass spectrometric selected ion monitoring (37). Limiting the use of the silylating reagent to reaction with the phenolic function by prior esterification of the carboxylic acid group of thyrocarboxylic acid analogues gave useful derivatives (38). This reaction could not be translated into use with the thyro amino acids themselves because of the reactivity of the $-NHSi(CH_3)_3$ function. Nevertheless conversion of the carboxylic acids to the methyl esters occurred efficiently at 60 °C using anhydrous methanol saturated with hydrogen chloride and was the first step in a number of the derivatisation schemes outlined below.

Difficulties have been experienced in achieving consistently successful reaction of iodothyronines with either pivaloyl chloride, pivalic anhydride or dichlorotetrafluoacetone yielding the *N,O*-dipivalyl and the oxazolidinone derivatives respectively. These problems were attributed to maintaining the derivatising reagents and sample in an adequately anhydrous state. Although T4 is quite insoluble in aqueous media, it is hygroscopic and water vapour drawn from the atmosphere may be sufficient to inhibit

Table 2. Methods of preparation of volatile derivatives of iodothyronines and related compounds

Derivative	Conditions of formation	Structure of product	References
Trimethylsilyl	Bis(trimethylsilyl)acetamide 5 min, 60 °C	$(CH_3)_3SiO-R-CH$ with branches $NH\,Si(CH_3)_3$ and $CO_2Si\,(CH_3)_3$	Alexander and Scheig (63) Lawson et al. (33) Manius et al. (36)
Trimethylsilyl methyl ester	a) CH_3OH/HCl 20 min, 60 °C b) Bis(trimethylsilyl)acetamide 5 min, 60 °C	$(CH_3)_3SiO-R-CH$ with branches $NH\,Si(CH_3)_3$ and CO_2CH_3	Hoffenberg et al. (42)
Pivalyl methyl ester	a) CH_3OH/HCl 20 min, 60 °C b) pivalic anhydride/triethylamine (4:1 v/v) 60 min, 80 °C or pivalyol chloride/triethylamine (5:1 v/v) 30 min, 60 °C	$(CH_3)_3CO_2-R-CH$ with branches $NH\,CO\,C(CH_3)_3$ and CO_2CH_3	Docter and Hennemann (43)
Permethyl	Methylsulphinyl carbanion (dimethyl-sulphoxide, methyl iodide) RT, 60 min or methylene chloride, diazomethane (1:2 v/v) RT, 30 min	$CH_3O-R-CH$ with branches $N(CH_3)_2$ and CO_2CH_3	Leclercq and Desiderio (64) Zimmer and Grady (65)
	or methanol, diazomethane (1:1 v/v) RT, 15 min	$CH_3-O-R-(CH_2)_n-CO_2CH_3$	Crossley and Ramsden (32)
Oxazolidinone	Dichlorotetrafluoroacetone in acetoni-trile/pyridine (100/1), RT, 30 min	$HO-R-CH-CO$ ring closed through NH and O to $C(CF_2Cl)_2$	Husek (66)

Table 2 (continued)

Derivative	Conditions of formation	Structure of product	References
O-acetyl oxazolidinone	a) Dichlorotetrafluoroacetone as above b) Acetic anhydride	$CH_3CO_2-R-CH-CO$ $\quad\quad\quad\quad\quad | \quad\; |$ $\quad\quad\quad\quad\; NH \;\; O$ $\quad\quad\quad\quad\quad \setminus \; /$ $\quad\quad\quad\quad\quad C(CF_2Cl)_2$	Willetts et al. (20)
O-methyl oxazolidinone	a) Dichlorotetrafluoroacetone as above b) Diazomethane in benzene/diethylether (80 : 20 v/v)	$CH_3O-R-CH-CO$ $\quad\quad\quad\quad | \quad\; |$ $\quad\quad\quad NH \;\; O$ $\quad\quad\quad\quad \setminus \; /$ $\quad\quad\quad\quad C(CF_2Cl)_2$	Willetts (40)
O-Heptafluoro-butyryl oxazolidinone	a) Dichlorotetrafluoroacetone as above b) Heptafluorobutyric anhydride	$C_3F_7CO_2-R-CH-CO$ $\quad\quad\quad\quad\quad | \quad\; |$ $\quad\quad\quad\quad\; NH \;\; O$ $\quad\quad\quad\quad\quad \setminus \; /$ $\quad\quad\quad\quad\quad C(CF_2Cl)_2$	Davis et al. (18)
Trifluoroacetyl methyl ester	a) CH_3OH/HCl 60 min, 60 °C b) Trifluoroacetic anhydride/Acetonitrile (1 : 3 v/v) 30 min, 60 °C	$\quad\quad\quad\quad\quad\quad NHCOCF_3$ $CF_3CO_2-R-CH \langle$ $\quad\quad\quad\quad\quad\quad CO_2CH_3$	Moller et al. (30)
Heptafluorobutyryl methyl ester	a) CH_3OH/HCl 60 min, 60 °C b) Heptafluorobutyric anhydride/Acetonitrile (5 : 4 v/v) 60 min, 60 °C	$\quad\quad\quad\quad\quad\quad NHCOC_3F_7$ $C_3F_7CO_2-R-CH \langle$ $\quad\quad\quad\quad\quad\quad CO_2CH_3$	Petersen and Vouros (41)

these reactions. In addition, to obtain useful chromatographic properties the oxazolidinone derivatives of thyronines require further derivatization to block the phenolic group. In the assay for T0, conversion of the phenolic function to either the acetyl derivatives by reaction with acetic anhydride or the methyl ether by reaction with diazomethane was possible (39). The acetyl ester thus formed was prone to back hydrolysis on stopping the acetylation reaction by addition of phosphate buffer. Methylation of the phenolic group with diazomethane in ethereal solution proceeded too vigorously and over-methylation with disruption of the oxazolidinone ring could result. Diazomethane in benzene reacted too slowly, but controlled reactions could be achieved by a solution of diazomethane in a benzene/ether mixture (80:20 v/v) (40). After initial methylation of the carboxylic acid function, formation of *N,O*-diheptafluorobutyryl (41) and *N,O*-bistrifluoroacetyl (30) derivatives using the respective anhydrides appears relatively trouble-free.

Single step conversion of the thyrocarboxylic acids to dimethyl ether/ester derivatives using diazomethane is also relatively uncomplicated (42). However, reaction of T4 itself with methylsulphinyl carbanion/methyl iodide either did not proceed or resulted in products of undetermined structure which possessed molecular weights in excess of that of tetramethyl T4 (64).

One useful feature of any derivative is that not only should it be suitable for gas chromatography but also that it should be sufficiently stable to allow further purification by liquid chromatography prior to gas chromatography if necessary. The *O*-acetyl and *O*-methyl oxazolidinone derivatives of T0 and the dimethyl derivative of T4A were stable enough to allow simple adsorption chromatography on small silica gel columns (32, 39). The *N,O*-diheptafluorobutyryl derivative of T4 was not so stable and was either destroyed or irreversibly bound on contact with the silica gel. No information is available regarding the other derivatives. Where such an additional post-derivatisation sample preparation step could be carried out, it dramatically improved the results obtained by GC-MS (SIM) analysis.

Methods of preparation of volatile derivatives of iodothyronine and related compounds are summarised in Table 2.

5 Mass Spectrometric Properties of Volatile Derivatives of Iodothyronines and Related Compounds

5.1 General mass spectral features

This section deals with some of the more general mass spectral properties of the volatile derivatives of iodothyronines and related acidic analogues. Five types of derivative are common to both thyro amino acids and carboxylic acids. These are:

1. Trimethylsilyl (33, 36),
2. Trimethylsilyl methyl ester (38, 42),
3. Pivalyl methyl ester (43),
4. Trifluoroacetyl methyl ester (30), and
5. Heptafluorobutyryl methyl ester (41, 44).

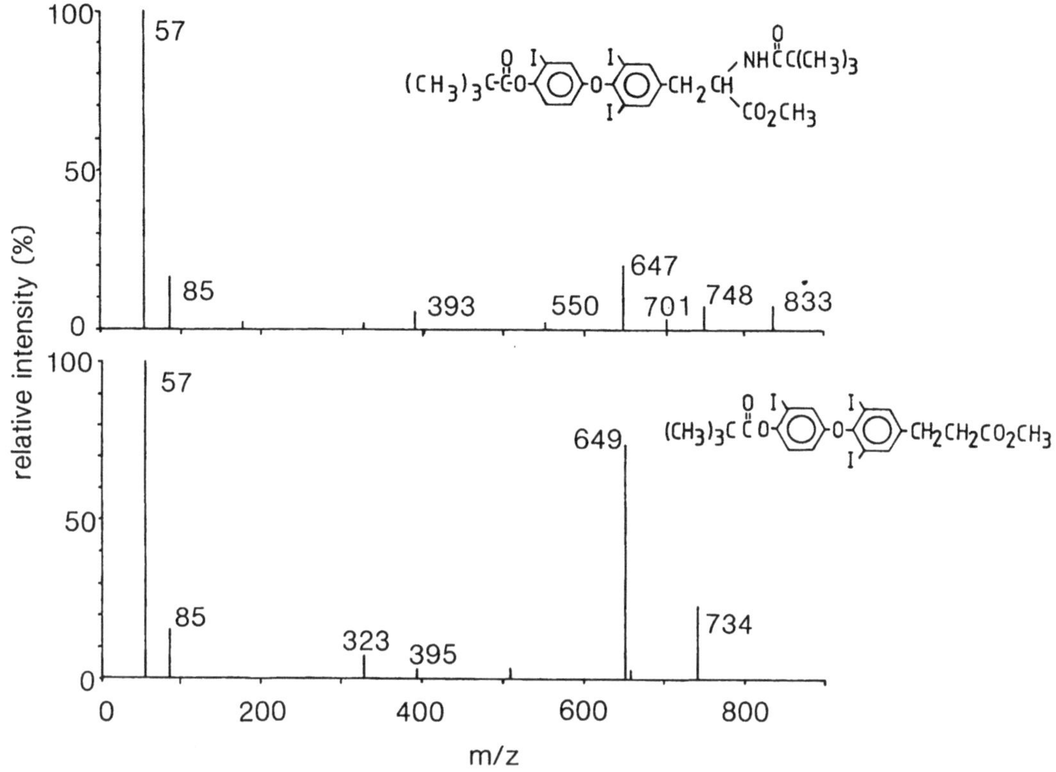

Figure 11. EI mass spectra of *N,O*-dipivalyl-3,5,3′-triiodothyronine methyl ester (upper) and methyl (*O*-pivalyl)-3,5,3′-triiodothyropropionate (lower).

The oxazolidinone series of derivatives are obviously restricted to the thyro amino acids (45), whilst the permethyl derivatives, though theoretically can be formed by both the amino acids and carboxylic acids, in practice were restricted to the acid analogues only, because of the difficulty of synthesis of the amino acid derivatives. For detailed discussion of the spectra of individual compounds the reader should refer to the original papers.

The mass spectral characteristics reflect the properties of the substituted thyronine moiety at the core of these molecules and those of the individual derivatising groups. The extent to which the properties of either dominate the spectra depends on the nature of the derivatising group(s). Varying the degree of iodine substitution appears to have little effect on whether the derivatising group has a dominant effect on the mass spectra within a series of iodothyronines. A second prominent effect is the difference seen in the spectra of the acidic analogues of iodothyronines compared with those of the iodothyronine derivatives themselves. The presence of the amino function in the iodothyronine derivatives gives rise to more facile fragmentation of the structure when subjected to electron impact. Thus the iodothyronines have less intense molecular ions than their acidic counterparts. These two general properties are illustrated in the spectra of *N,O*-dipivalyl T3 methyl ester, *O*-pivalyl T3P methyl ester, *N,O*-diheptafluorobutyryl T4 methyl ester and *O*-heptafluorobutyryl T4A methyl ester (Figs. 11

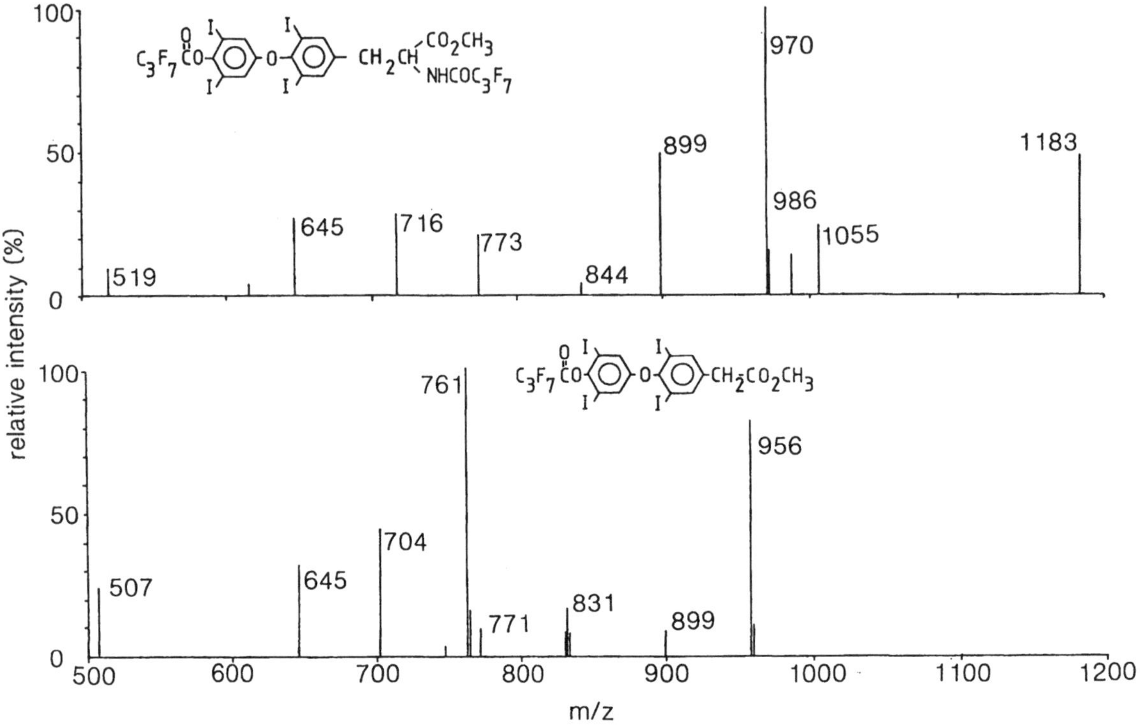

Figure 12. EI mass spectra of *N,O*-diheptafluorobutyryl thyroxine methyl ester (upper) and methyl-(*O*-heptafluorobutyryl)-3,5,3′,5′-iodothyroacetate (lower).

and 12 respectively). The first two of these illustrate the dominance of the derivatising group in retaining charge. This is because of the high stability of the trimethylmethyl carbonium ion $(CH_3)_3C^+$, which is the base peak in both spectra. In contrast the base peaks in the spectra of the heptafluorobutyryl derivatives contain much or all of the substituted thyronine skeleton. The second general point is illustrated by the fact that both the thyro carboxylic acid derivatives have more intense molecular ions than their amino acid counterparts.

Of the other derivatives, the di- and trimethylsilyl derivatives of the thyro acids and amino acids respectively are similar to the pivalyl methyl esters, in that charge tends to be retained on the derivatising group. Thus the base peak of the bis(trimethylsilyl) thyro carboxylic acid derivatives is the ion,

m/z 73, $(CH_3)_3Si^+$

and for the amino acids is the ion,

$$m/z\ 218,\ {}^{+}CH\begin{cases} CO_2Si(CH_3)_3 \\ NHSi(CH_3)_3 \end{cases}$$

The trifluoroacetyl methyl ester derivatives are similar to the heptafluorobutyryl methyl ester derivatives in that the base peaks in the spectra of these compounds are high molecular mass ions.

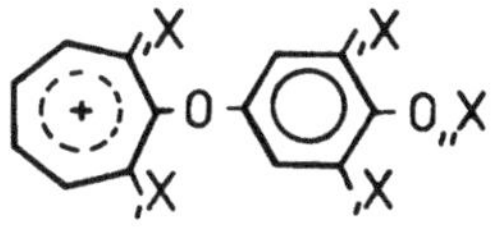

Figure 13
Structure of the substituted tropylium ions which feature in the mass spectra of the volatile derivatives of iodothyronines, where X′ = H or I and X″ = H or derivatising group.

The trimethylsilyl methyl ester derivatives occupy an intermediate position in that the base peak of the *O*-trimethylsilyl thyro carboxylic acid methyl esters are high molecular mass ions (for *O*-trimethylsilyl T4A methyl ester *m*/*z* 834, the molecular ion), whilst the base peak of the spectra of all the *N,O*-bis(trimethylsilyl) thyro amino acid methyl esters is an ion

$$m/z\ 160,\ {}^{+}CH\begin{matrix} \diagup CO_2CH_3 \\ \diagdown NHSi(CH_3)_3 \end{matrix}$$

which is analogous to the base peaks of the tris(trimethylsilyl) derivatives.

The third general feature of these spectra is the presence of substituted tropylium ions of the general formula shown in Figure 13. Ions of this structure are present in the spectra of all eight types of derivatives formed by the amino acids and all six formed by the thyroacetic and thyropropionic acid analogues. They are particularly prominent in the compounds possessing no halogen substitution in the phenyl rings. Thus tropylium ions are the base peaks in the spectra of the oxazolidinone, *O*-methyl oxazolidinone and *O*-acetyl oxazolidinone and *N,O*-diheptafluorobutyryl methyl ester derivatives of T0 and are relatively prominent in the spectra of the tris(trimethylsilyl) derivatives of T0 and tyrosine and the bis(trimethylsilyl) derivative of T0A.

Iodine substitution in the diphenylether structure, particularly into the inner phenyl ring, tends to inhibit the formation of substituted tropylium ions. Where the formation of substituted tropylium ions is inhibited by the presence of iodine, another series of ions can be seen in the spectra of the amino acid derivatives. These are formed by fragmentation of the α-amino acid structure to give ions containing a double bond capable of resonance with the inner phenyl ring. The *N,O*-bis(trifluoroacetyl) methyl esters and the *N,O*-diheptafluorobutyryl esters fragment to give ions containing the substituted methyl acrylate structure $R-CH=CH-CO_2CH_3$, where R = substituted diphenyl ether structure of thyronines. These are formed by loss of the amino function. The tritrimethylsilyl derivatives lose the carboxylic acid function to give ions of the assigned structure

$$R-CH_2-CH=\overset{+}{N}H-(SiCH_3)$$

but may resonate with the structure

$$R-CH=CH-\overset{+}{N}H_2Si(CH_3)_3.$$

In the spectra of more heavily iodinated compounds ions involving the loss of two iodine atoms are observed. The structures of such ions have not been eludicated with any detail in most instances. However, Peterson and Vouros showed using the *N,O*-diheptafluorobutyryl-3′,5′-dibromo-3,5-diiodothyronine methyl ester, that one halogen

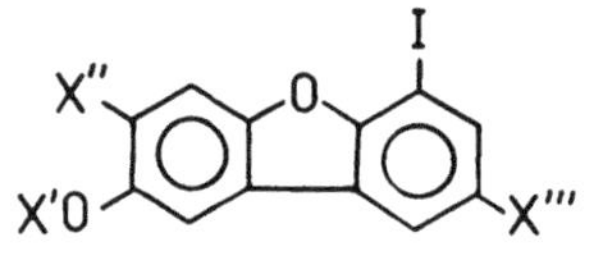

Figure 14
Proposed structure of ion formed by loss of two halogen atoms (ref. 41), where X′ = H or derivatising group, X″ = H or I, and X‴ = Br or I.

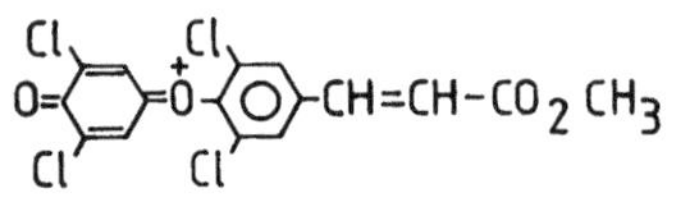

Figure 15:
Quinoid structure of ion forming the base peak of the spectrum of *N,O*-diheptafluorobutyryl-3,5,3′,5′-tetrachlorothyronine methyl ester.

atom was lost from each phenyl ring to form a third cyclic structure in the ion, involving rearrangement of hydrogens (41) (Fig. 14).

The spectra of the trimethylsilyl derivatives of both the thyro amino acids and the thyro carboxylic acids contained a number of doubly positive ions. Usually these were not very intense, however, the ions $[M-15-15]^{2+}$ of bis(trimethylsilyl) T0A and T4P had relative intensities of 10%. The stability of these ions is thought to be due to the degree of spatial separation of the charges that is possible in these structures (33). No comments have been made on the occurrence of doubly positive ions in the spectra of other derivatives. It is not clear whether these do not occur or are merely unreported.

An unusual feature of the spectrum of *N,O*-diheptafluorobutyryl-3,5,3′,5′-tetrachlorothyronine methyl ester is that the base peak has a quinonoid structure (Fig. 15). Ions of similar structure may be seen in the spectra of iodothyronines but are of very much lower relative intensity (41).

5.2 Gas chromatography-mass spectrometry-selected ion monitoring (GC-MS-SIM)

The choice of derivative for use in the assay of iodothyronines by GC-MS-SIM is governed by:
1. the mass spectral properties of the derivative,
2. general chemical and chromatographic properties of the derivative and
3. the nature of original matrix containing the iodothyronine.

When biological samples are being analysed, one must assume that the iodothyronine content is only a minor component of any partially purified extract. Thus chromatographic resolution of the derivatised extract is a prerequisite before any mass spectrometric analysis. The use of high mass spectrometric resolution in combination with direct insertion of the derivatised extract is usually pre-empted by the limitations imposed by sample size.

In order to further minimise artefactual interference advantage can be taken of the relatively high mass of any ions containing more than two iodine atoms. Therefore derivatives such as the *N,O*-heptafluorobutyryl methyl esters are more useful than the trimethylsilyl derivatives, due to the high mass/charge ratios of their base peaks. Compared with the oxazolidinone and trimethylsilyl derivatives both the *N,O*-diheptafluorobutyryl and trifluoroacetyl methyl esters also have superior gas chromatographic properties. Although the *N,O*-dipivalyl methyl esters and the *O*-methyl oxazolidinone derivatives are probably the most stable iodothyronine derivatives, in the author's laboratory these were more difficult to form routinely. The reasons for this were uncertain and, in the case of other authors these difficulties are not apparent from the literature and may therefore not be generally experienced. The above logic ultimately led to the choice of the heptafluorobutyryl derivatives in the development of assays for DIT (44), T0 (45), T4 (46), T3 and rT3 and similar considerations led Moller et al. to the trifluoroacetyl derivatives (30, 47).

The only likely use of either the trimethylsilyl or pivalyl derivatives, which possess low mass/charge ratio base peaks, is in the analysis of mixtures containing several iodothyronines in high concentration and relatively few other compounds e. g. analysis of iodothyronines used in pharmaceutical preparations. The common base peak would avoid alteration of the selected ion during a single chromatographic run.

All the listed derivatives of the thyro carboxylic acids have relatively intense high mass ions in their spectra. Therefore the choice of derivative is governed more by the general chemical and chromatographic properties than by mass spectral ones. Dimethyl derivatives are readily formed by reaction with diazomethane, are stable on exposure to water vapour, have good gas chromatographic properties and have molecular ions which formed the base peaks of their spectra. For these reasons the dimethyl derivatives were used in the identification of T4A and T4F in rat liver extracts and the quantitation of T4A in human serum (32, 38).

6 GC-MS in the Study of Thyroid Hormone Metabolism

6.1 Assays for thyroxine

Before the development of radioimmunoassay methods for the assay of iodothyronines in human serum, gas chromatographic assays for T4 and T3 were attempted with varying degress of success (43). As pointed out earlier, the volume of thyroid function testing precluded the use of such assays in most routine laboratories. Even so these early studies provided much valuable information regarding the conversion of iodothyronines to volatile derivatives. Since the introduction of radioimmunoassay two rigorously worked GC-MS-SIM assays for T4 in human serum have been described (46, 47). The compositions of the two assays are similar. The details of these assays are outlined in Table 3. Essentially both are fragmentographic assays based on the *N,O*-bis(trifluoroacetyl) methyl ester and *N,O*-diheptafluorobutyryl methyl ester respectively. They follow initial qualitative work demonstrating the feasibility of developing GC-MS-SIM assays by Heki and co-workers (34, 35, 37).

The value of these assays is that they provide an accurate alternative method of assaying T4 by which the performance of radioimmunoassays may be assessed. In the first of these papers Moller et al. show the excellent correlation which is possible between a GC-MS-SIM assay and a good radioimmunoassay (47). In the second paper the performance of a radioimmunoassay suspected of giving spurious results

Table 3. Comparison of two GC-MS-SIM assays for thyroxine

Step No.	1st Assay (47)	2nd Assay (46)
1	Addition of internal standard $^2H_2$-T4 to 1 ml of serum	Addition of internal standard $^2H_5$-T4 to 2 ml of serum
2	Protein precipitation with acetone/conc. HCl and T4 extraction with *n*-butanol	Ion-exchange chromatography on Dowex 50. Elution with 5 mol$\cdot^{-1}$ NH_4OH
3	Extraction of *n*-butanol solution with 1. water 2. Na_2CO_3/NaOH	Derivatisation with 1. methanol/HCl 2. heptafluorobutyric anhydride
4	Addition of hexane to butanol solution and extraction with aq. NaOH	
5	Extraction of NaOH sol. with hexane	
6	Acidification of aqueous solution with H_2SO_4 and extraction with ethyl acetate	
7	Derivatisation of ethyl acetate extract with 1. methanol/HCl 2. trifluoroacetic anhydride	

was assessed by the GC-MS-SIM assay. The data presented in the second of these reports suggested that:

1. over the normal T4 range the two assays were responding similarly to changes in analyte concentration but that there was a difference in the absolute concentrations assigned to a given sample,
2. in euthyroid subjects as TBG concentration increased into the supranormal range, the radioimmunoassay gave apparently elevated T4 concentrations over and above the increase in T4 which occurs concomitantly with an increase in TBG. This confirmed suspicions about the assay derived from clinical opinions of the individual patients, and was interpreted as arising from inadequate blocking of serum binding of T4 by 8-anilino-naphthalene-sulphonate, and
3. the increase in T4 in hyperthyroid patients assessed by radioimmunoassay was less than that assessed by GC-MS-SIM. This was the most surprising finding of the three results and the most difficult to interpret. One possible explanation was that the elevated T4 concentrations lay in the region of maximum error on the radioimmunoassay calibration curve and therefore the low level of bound radioactive T4 was subject to a considerable degree of misinterpretation.

The GC-MS-SIM responses were linear over the range 5 to 600 nmol$\cdot l^{-1}$ T4, which covers the vast majority of T4 concentrations encountered in euthyroid and pathological thyroid states, in contrast to the curvilinear nature of radioimmunoassay standard curves (46). The relationship between the GC-MS-SIM and radioimmunoassay results are illustrated in Figure 16.

Moller et al. (30) investigated the T4 content of human breast milk using GC-MS. A number of studies using either radioimmunoassay or competitive protein binding assays have indicated that human milk may contain significant amounts of T4 and that, as a consequence, breast feeding might partially compensate for the hormonal deficiency in congenital neonatal hypothyroidism. This point is clinically important

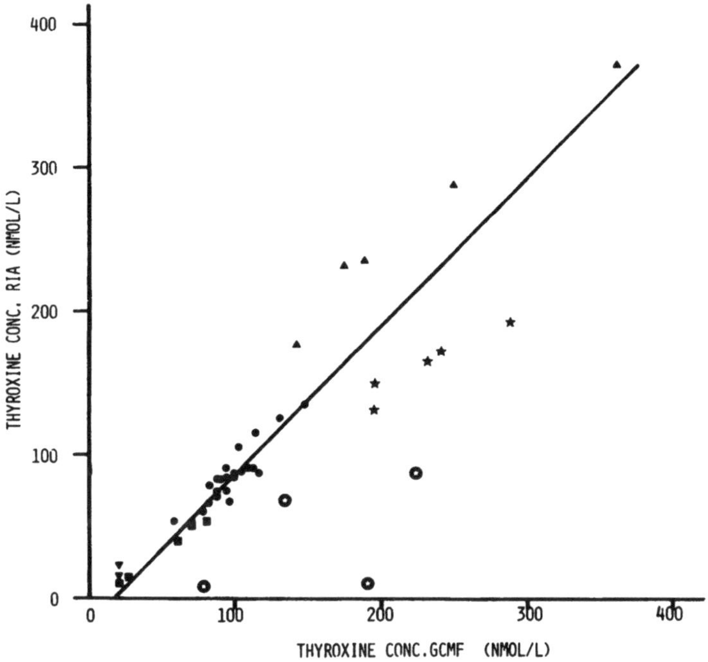

Figure 16. Relationship between serum thyroxine concentration determined by radioimmunoassay and GC-MS-SIM (● normal euthyroid; ■ hypothyroid; ★ hyperthyroid; ▲ congenitally elevated TBG; ▼ congenitally reduced TBG; ☆ serum containing anti-thyroxine autoantibodies). (Reproduced by permission of John Wiley & Son Ltd., ref. 46.)

because of the damaging effects of hypothyroidism on brain development and potentially adverse effects of over-replacement of hormone. Using ion-exchange chromatography to extract T4 from human milk, followed by conversion to the *N,O*-bis-(trifluoroacetyl) methyl ester, T4 levels were found to be less than 14 $nmol \cdot l^{-1}$. The amount of T4 likely to be ingested by the neonatal would be less than 10% of that required for adequate replacement. Therefore, T4 ingestion from breast feeding is unlikely to have a significant masking effect on neonatal hypothyroidism. No attempt was made to determine T3 levels by these means but in view of the relative concentrations of T4 and T3 in serum it is unlikely that the levels of T3 in milk would alter the above conclusions.

6.2 Other iodothyronines

No fully detailed assay for the determination of other iodothyronine concentrations in human serum has yet been described. Nevertheless Heki et al. (37) and results from the author's laboratory indicate the feasibility of such assays. Figure 17a shows the mass fragmentogram of a serum extract with peaks corresponding to T3, rT3 and T4 respectively, whilst Figure 17b shows the mass fragmentogram of the same extract with added rT3. Linear GC-MS-SIM responses have also been shown over the range

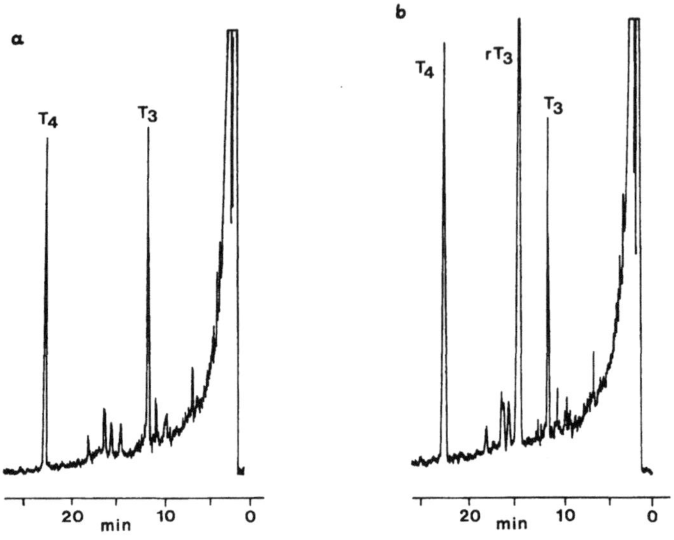

Fig. 17a Fig. 17b

Figure 17. (a) GC-MS-SIM trace (m/z 844) of serum extract derivatised by reaction with methanol/HCl then heptafluorobutyric anhydride.
b) As (a) but original serum contained additional exogenous 3,3′,5′-triiodothyronine.
Gas chromatographic conditions: column OV-1, 20 m × 0.25 mm internal diameter, temperature 280 °C, flow rate 0.5 ml · min^{-1}.

of T3 concentrations likely to be encountered in human serum in euthyroid and pathological states.

One problem inherent in any assay for iodothyronines with fewer than 4 iodine substituents is that of artifactual deiodination. Because of the susceptibility of T4 to undergo deiodination on exposure to UV light, great care must be exercised during extraction and derivatisation procedures to avoid such deiodination. Koya et al. have elegantly identified and quantified the photochemical breakdown products of T4 using GC-MS (48). As pointed out earlier, even deiodination affecting less than 1% of the T4 present in human serum samples could completely invalidate the results of assays for other iodothyronines. Artifactual deiodination was thought to be the cause of the discrepancy between liquid chromatographic assay and radioimmunoassay results for T3 which was evident when radioimmunoassay methods were introduced in the early 1970s. With the lower serum concentrations of the other iodothyronines this problem is magnified. Nevertheless this problem is surmountable. Obviously, appropriate sample handling techniques will minimise deiodination and a number of methods (both GC-MS and non-GC-MS) can be devised to quantitate the extent of deiodination.

6.3 3,5,3′,5′-Tetraiodothyroacetate and 3,5,3′,5′-tetraiodothyroformate

3,5,3′,5′-Tetraiodothyroacetate (T4A) and its 5′-monodeiodination product 3,5,3′-triiodothyroacetate (T3A) are interesting to thyroidologists in so much as they possess the same substituted ring systems as T4 and T3 respectively. Consequently they retain the

Table 4. Details of assay of 3,5,3',5'-tetraiodothyroacetate

Step No.	Procedure	Recovery
1	Addition of internal standard 3,5,3',5'-tetraiodothyropropionate to 2 ml serum	
2	Lyophilisation of serum	
3	Extraction with methanol (4 × 4 ml)	97%
4	Precipitation of protein from methanol solution by addition of ether	96%
5	Evaporation of methanol/ether at 50 °C	
6	Extraction of residue with benzene (2 × 1.5 ml)	88%
7	Adsorption chromatography of benzene extract on silica gel – elution with benzene/glacial acetic acid (9:1 v/v)	97%
8	Derivatisation with CH_2N_2	
9	Adsorption chromatography of derivatised extract on silica gel – elution with benzene	98%

% Recovery denotes approximate recovery of analyte and internal standard in each step.

ability to express hormonal action, unlike most other metabolites of thyroid hormones. In studies of the binding of iodothyronines to nuclear binding protein, which are thought to be the putative receptors for thyroid hormones, T3A has been shown to bind to these proteins with an avidity similar to or greater than that of T3 (49). The reasons why T4A and T3A play no significant role in the overall expression of endocrine activity in the human are:

1. the low concentration of T4A and T3A in human serum and
2. their relatively rapid deiodination compared with T4 and T3 in peripheral tissue, limiting the amount of T3A available to the cell nucleus (50).

Gas chromatographic (51), competitive protein binding (52) and radioimmunoassay methods (53) have been described for T4A in human serum. Both GC and competitive protein binding assays gave somewhat higher results than the radioimmunoassay method suggesting marked artefactual interference, which could not be wholly ruled out of the radioimmunoassay. A GC-MS-SIM assay for T4A has been described (32). This differs in a number of important respects from those for T4. The technical details of the assay are summarised in Table 4. As can be seen an analogue of T4A is used as internal standard, and precipitation of serum proteins and extraction of T4A by organic solvents play a major part in the sample preparation regime. Ion-exchange chromatography was not used because use of a cationic exchange resin (Dowex 50) gave very variable rates of extraction, whereas difficulty was experienced in eluting T4A from synthetic anion-exchange resins. Both pre- and post-derivatisation adsorption chromatography was used as part of the sample preparation. Because of the lengthy and very tedious sample preparation regime, it was also important to demonstrate that analyte and standard behaved similarly. A typical mass fragmentogram is shown in Figure 18. The average injection into the GC column was approximately 30 – 70 pg for a euthyroid serum sample and, assuming a 30% transfer across the jet separator, the peaks generated by the mass spectrometer were in response to 3 – 10 pg of the dimethyl T4A derivative. Successful GC-MS-SIM analysis at these levels

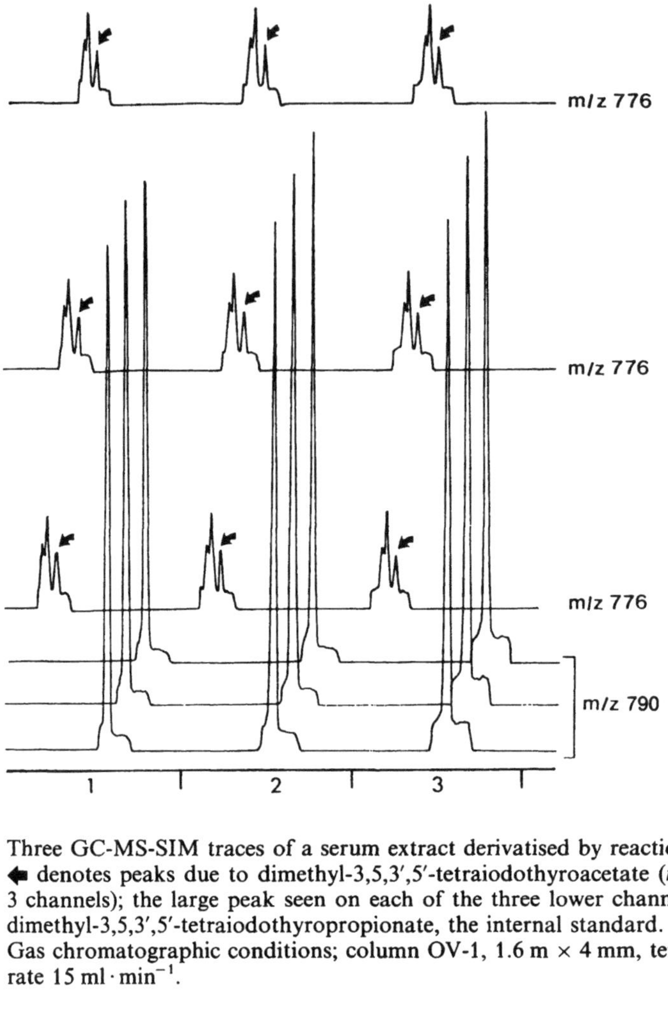

Figure 18. Three GC-MS-SIM traces of a serum extract derivatised by reaction with diazomethane. ⇦ denotes peaks due to dimethyl-3,5,3′,5′-tetraiodothyroacetate (*m/z* 776, monitored on 3 channels); the large peak seen on each of the three lower channels (*m/z* 790) is due to dimethyl-3,5,3′,5′-tetraiodothyropropionate, the internal standard.
Gas chromatographic conditions; column OV-1, 1.6 m × 4 mm, temperature 270 °C, flow rate 15 ml · min^{-1}.

required not only the rigorous sample preparation outlined above, but careful and extensive preparation of the packed GC columns, including repeated siliconisation of the columns themselves prior to packing.

Table 5. Concentration of 3,5,3′,5′-tetraiodothyroacetate in serum

Thyroid Status	No.	Mean (pmol · l^{-1}	SD
Euthyroid	37	116	26
Hyperthyroid	12	204	94
Hypothyroid	5	36	17

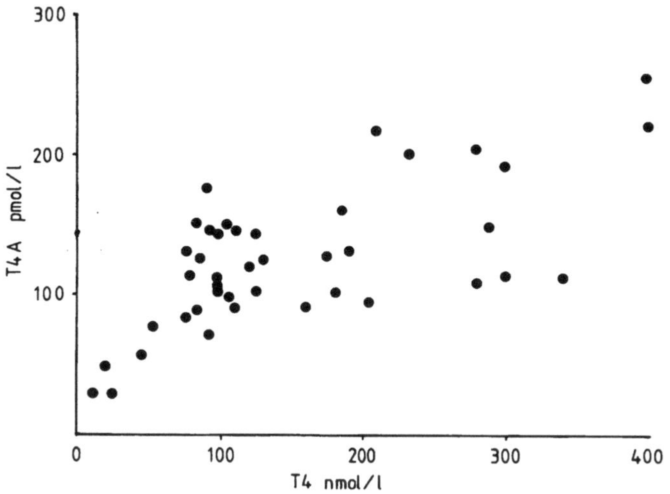

Figure 19. The relationship between serum thyroxine and 3,5,3′,5′-tetraiodothyroacetate concentrations.

The results obtained for T4A levels in normal human serum by GC-MS-SIM were lower than those obtained by either GC or competitive protein binding and somewhat lower than those by radioimmunoassay. Table 5 summarises the levels in normal euthyroid, hypo- and hyperthyroid individuals. Although T4A is strongly bound by TBPA in serum (54), the concentrations of binding protein and ligand are not correlated because of the vast excess of TBPA compared to T4A. Only 1 molecule of TBPA out of 50000 has T4A attached to it in normal serum (32).

T4A is wholly derived from T4 by a two step mechanism involving first transamination to give the pyruvic acid analogue then oxidative decarboxylation to T4A. The reaction is thought to occur in the mitochondria of organs such as the liver and kidneys (8). The relationship between the substrate (T4) and product (T4A) of this reaction sequence over the range of T4 concentrations encountered in hypo-, eu- and hyperthyroidism is illustrated in Figure 19. The fact that this is obviously not linear may indicate that the enzyme(s) responsible for the conversion may become saturated as T4 concentration increases, or that tissue metabolism of T4A changes as the thyroid status alters.

As mentioned above, the liver is a major site of T4 to T4A conversion. The presence of T4A in extracts of rat liver homogenates may be readily demonstrated as evidenced by peak 'A' in Figure 20. Full mass spectra obtained for this material were identical to those obtained from standard T4A preparations.

During the course of the work on T4A in rat liver the possibility of using either T4P or 3,5,3′,5′-tetraiodothyroformic acid (T4F) as internal standards was investigated. As a consequence it was apparent that in these extracts a peak was present which had the same GC retention time as T4F irrespective of whether the *O*-trimethylsilyl ester or dimethyl derivative of T4F was used. The level of this material was very much less than that of T4A in most instances so that full spectra could not be obtained to

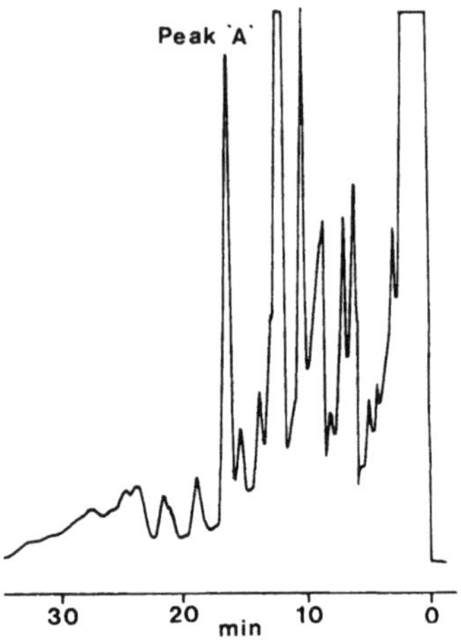

Figure 20
Total ion current trace of permethylated rat liver extract. Peak 'A' had a mass spectrum identical to that of dimethyl-3,5,3',5'-tetraiodothyroacetate, column OV-1, 1.6 m × 4 mm.

confirm the identity of the compound. Nevertheless it is not inconceivable that the material was T4F and a feasible pathway for its formation in the liver can be envisaged. The levels of T4F suggest that it is of very minor significance in the metabolism of T4 (38).

6.4 Thyronine (T0)

The deiodination of T4 proceeds via a series of monodeiodination steps which culminate ultimately in the formation of T0. T0 itself is not bound to any appreciable extent by serum proteins and therefore could reasonably be expected to be filtered through the kidneys and excreted in the urine. To examine this hypothesis assays for T0 in human urine were devised. The initial assay was based on the conversion of T0 to the *O*-acetyl oxazolinone formation (20) and the second on the *N,O*-diheptafluorobutyryl methyl ester (45). The two assay protocols are outlined in Table 6. The internal standard was deuterium-labeled T0 formed by catalytic deiodination of T4. The theoretical deuterium incorporation was 4 deuterium atoms per thyronine molecule but analysis indicated a product with a mass shift of 2 was the dominant species. Because of the natural occurrence of chlorine in two stable isotopic forms ^{35}Cl and ^{37}Cl, a mass difference of 3 had to be used to monitor the analyte and standard when using the *O*-acetyl oxazolidinone derivatives, whereas a mass difference of 2 was used in the assay based on the *N,O*-diheptafluorobutyryl methyl ester derivative.

In healthy adult euthyroid humans the mean T0 excretion was found to be 15 $nmol \cdot 24\,h^{-1}$, with a range of $8-26\ nmol \cdot 24\,h^{-1}$ (20, 39). T0 excretion was significantly higher in hyperthyroid and significantly lower in hypothyroid patients, nevertheless considerable overlap existed between the two pathological ranges and the respective ends of the normal range (45).

The fact that less than 20% of normal daily T4 production was excreted in the urine gave rise to a number of hypotheses regarding the fate of the remaining 80%. These were that:

Table 6. Comparison of GC-MS-SIM assays for thyronine in urine

Step No.	1st Assay (20)	2nd Assay (45)
1	Addition of internal standard ($^2H_4$-T0) to urine (20 ml)	as assay 1
2	Ion-exchange chromatography on Dowex 50 – elution with 5 $mol \cdot l^{-1}$ NH_4OH	as assay 1
3	Ion-exchange chromatography on Dowex 1 – elution with 50% glacial acetic acid	as assay 1
4	Derivatisation with 1. dichlorotetrafluoroacetone 2. acetic anhydride	Hydrophobic chromatography on C_{18}-cartridge – elution with methanol/HCl
5	Adsorption chromatography on silica gel – elution with toluene	Derivatisation with 1. methanol/HCl 2. heptafluorobutyric anhydride

1. deiodination did not proceed to completion,
2. T0 was metabolised further,
3. an alternative metabolic route assumed precedence somewhere along the deiodination pathway,
4. T0 was excreted in faeces rather than in the urine.

These possibilities are not mutually exclusive. The identification of considerable quantities of phenolic conjugates of diiodothyronines in serum suggested that conjugation assumed dominance in preference to deiodination at this point of the deiodination sequence (55). The demonstration that diiodothyronine conjugations were deiodinated preferentially to other iodothyronines by liver deiodinase enzymes may indicate that deiodination is taken to completion and that the T0 conjugate thus formed is excreted into the bile and hence to the faeces.

6.5 3,5-Diiodotyrosine (DIT)

Iodination of tyrosine residues of thyroglobulin is an essential prerequisite in the biosynthesis of T4, however, not all DIT residues of thyroglobulin go on to form T4. Only a minority are thought to do so. During the proteolytic cleavage of thyroglobulin to release T4 and T3, DIT is also liberated. The majority of this DIT is thought to undergo intra-thyroidal deiodination to enable iodine to be re-utilised for hormone production. Some DIT is released into the blood stream however. Recent evidence suggests that thyroidal secretion is the major source of circulating DIT (56, 57). A second metabolic route for the formation of DIT is that of cleavage of the diphenyl ether bond of T4, although this pathway contributes minimally to total daily DIT production. Early attempts to assay the concentration of DIT in human serum gave widely varying results (58 – 60). More reliable information on circulating DIT levels came initially from studies using radioimmunoassay involving chromatographic separation of T4 from DIT to minimise T4 cross-reactivity (19). Recent work has

Table 7. Details of assay of 3,5-diiodotyrosine in urine (44)

Step No.	Experimental Details
1	Addition of internal standard ($^2H_4$-T0) to urine (20 ml)
2	Cation-exchange chromatography on Dowex 50 – elution with 5 mol · l^{-1} NH_4OH
3	Anion-exchange chromatography on Dowex 1 – elution with 1. 0.2 mol · l^{-1} sodium acetate pH 5.6 2. 50% glacial acetic acid
4	Hydrophic chromatography on C_{18}-cartridge – elution with methanol/ammonia (99:1 v/v)
5	Derivatisation with 1. methanol/HCl 2. heptafluorobutyric anhydride

concentrated on the immunological precipitation of DIT to separate it from T4 before proceeding to radioimmunoassay (56, 57). This latter technique gave lower concentrations of DIT than the assay using the chromatographic method. The normal means and ranges for the two systems were: – 2.3; 0 – 7.0 nmol. l^{-1} (19), 161; 23 – 530 pmol. l^{-1} (56), chromatographic and immunoprecipitation assays respectively.

To further eludicate DIT metabolism a GC-MS-SIM assay was developed (44, 61). Urine was used because it was known that DIT is only weakly bound by serum protein therefore a percentage of DIT in serum should be filtered into the urine via the kidneys. In addition, serum DIT concentrations were thought likely to be low and sample size might be a problem. This difficulty was circumvented by urine analysis. The assay protocol was derived from that of the second T0 assay and is outlined in Table 7.

The urinary excretion of DIT by normal healthy euthyroid adults was found to range between 0.21 and 1.61 nmol · $24\,h^{-1}$ with a mean ($\pm$SD) of 0.76 $\pm$ 0.41 nmol · $24\,h^{-1}$. Hypothyroid patients were found to have significantly reduced (0.44 $\pm$ 0.13 nmol · $24\,h^{-1}$; $n = 10$, $2p < 0.01$) and hyperthyroid patients significantly elevated urinary DIT excretion (1.48 $\pm$ 0.72 nmol · 24^{-1}; $n = 10$, $2p < 0.005$) compared with normal subjects (61). Hypothyroid patients receiving thyroxine replacement therapy had levels which were intermediate between those of normal subjects and untreated hypothyroid patients (62). The alterations in the level of DIT excretion in parallel with thyroid status were in accord with the hypothesis that thyroidal synthesis and secretion were the major source of excreted DIT.

Although the thyroid is thought to be the major source of circulating DIT, as mentioned earlier a second means of synthesis exists for its formation, that of cleavage of the diphenylether bridge of T4. This pathway of metabolism is thought to occur in relatively few tissues. In the phagocytosing leucocyte, however, approximately 50% of T4 metabolised proceeds along this pathwway (9). Diphenyl ether cleavage is an oxidative reaction and in the phagocytosing leucocyte is associated with the increased generation of hydrogen peroxide during the respiratory burst. In contrast to ether cleavage, deiodination which overall is the major route of T4 catabolism is a reductive process. In view of the above it was reasoned that in tissue which normally deiodinate T4 but which are also capable of generating peroxide, should deiodination be inhibited T4 may enter oxidative pathways. A tissue which deiodinates considerable amounts of T4 and also possesses peroxisomes is the kidney. In the damaged or diseased kidney deiodination is likely to be inhibited. Therefore, DIT excretion was investigated in patients with chronic renal disease. In a small group of patients with chronic renal disease urinary DIT excretion was significantly elevated (1.43 ± 0.61; $n = 8$, $2p < 0.01$), which was in accord with the hypothesis that T4 may have been diverted into oxidative pathways. However, the study was carried out on a small number of patients with diverse causes of their renal problems, therefore, more detailed and extensive studies are required to confirm the finding before further examination of the hypothesis is undertaken.

7 Future Developments and Conclusions

Several sections of this chapter have outlined specific instances of the part GC-MS studies have played in advancing our understanding of thyroid hormone metabolism. All of these studies have been carried out using electron impact mass spectrometry and samples introduced via gas chromatography. The last few years have seen the introduction of MS instruments which depart quite radically from such conventional GC-MS systems. Developments such as fast atom bombardment sources, coupling of HPLC devices to the mass spectrometer, increased magnet performance and use of higher accelerating voltages offer a number of attractive possibilities for future work. These include avoiding the necessity for volatile derivative formation and increasing the mass scanning range of instruments to increase the number of compounds capable of investigation by mass spectrometry. Unfortunately the usefulness of these new developments in clinical studies is extremely limited. With the exception of negative ion mass spectrometry, most other developments reduce the sensitivity of the mass spectrometer. This is a crippling disadvantage as it is likely that the major areas of interest in thyroid research in the future, and most other areas of endocrinology, will require assay systems with ten or one hundred fold increase in sensitivity compared with the best conventional electron impact GC-MS systems. In the case of thyroid hormones, which are naturally rich in iodine, an element with good electron accepting properties, negative ion mass spectrometry may provide the basis of a system with the required increase in sensitivity. With increased sensitivity, GC-MS or LC-MS systems could provide the means of simultaneously assaying not only all iodo-compounds present in acceptably small serum samples (acceptable from the donors' point of view) but also in samples of thyroid tissue. Small amounts of this material are becoming increasingly available due to growth in popularity of fine needle aspiration of the thyroid for cytological examination. However, developments in other forms of analysis, such as those involving immunological recognition of the analyte in conjunction with enhanced fluorescence or luminescence as means of quantitation, which are more readily automated and less capital intensive, will continue to ensure that the future roles of mass spectrometry are strictly limited to those outlined in an earlier section, i. e. confirmation of the results of more convenient assay systems and the identification and quantitation of some of the more unusual products of thyroid hormone metabolism.

References

1. Tong, W. (1977) In: The thyroid, A Fundamental and Clinical Text (Werner, S. C. & Ingbar, S: H., eds.), pp. 24–40, Harper & Row, New York.
2. Alexander, N. M. & Nishimoto, M. (1979) Rapid analysis for iodotyrosines and iodothyronines in thyroglobulin by reverse phase liquid chromatography. Clin. Chem. *25*, 1757–1760.
3. Morley, J. E. (1981) Neuroendocrine control of thyrotropin secretion. Endocrine Reviews *2*, 396–436.
4. Ramsden, D. B. & Hoffenberg, R. (1981) In: Hormones in Normal and Abnormal Tissues (Fotherby, K. & Pal, S. P., eds.), pp. 257–279, Walter de Gruyter, Berlin.
5. Ramsden, D. B. & Hoffenberg, R. (1982) In: Quantitative Approaches to Metabolism (Cramp, D. G., ed.), pp. 301–328, John Wiley & Sons, Chichester.
6. Taurog, A., Briggs, F. N. & Chaikoff, I. L. (1952) ^{131}I-labeled L-thyroxine. II. Nature of the excretion product in bile. J. Biol. Chem. *194*, 655–668.
7. Roche, J. & Michel R. (1957) Biliary excretion of sulfoconjugated 3,3',5-triiodothyroacetic acid after administration in rats. C. R. Acad. Sci. (Paris) 245, 748–750.
8. Roche, J., Michel, R. & Tata, J. R. (1954) Sur la nature des combinaisons iodées excrétées par le foie et le rein après administration de L-thyroxine et de 1–3:5:3'-triiodothyronine. Biochim. Biophys. Acta *15*, 500–507.
9. Burger, A. G., Engler, D., Buergi, U., Weissel, M., Steiger, G., Ingbar, S. H., Rosin, R E. & Babior, M. (1983) Ether link cleavage is the major pathway of iodothyronine metabolism in the phagocytosing human leucocyte and also occurs in vivo in the rat. J. Clin. Invest. *71*, 935–949.
10. Michel, R. (1961) Données récentes sur la physiologie des hormones thyroidiennes. 18e Congr. Ass. Pédiatres Langue franc. *1*, 1.
11. Ramsden, D. B. & Hoffenberg, R. (1983) In: Clinics in Endocrinology & Metabolism (Clayton, R. N., ed.), pp. 101–115, W. B. Saunders & Co. Ltd., London.
12. Dussault, J. H., Coulombe, P., Laberge, C., Guyda, H. & Khowry, K. (1975) Mass screening of T4 using 40 μl of blood on filter paper. 47,00 measures – 3 with low TBG, seven hypothyroid, hypothyroidism 1:7000 births. J. Pediatr. *86*, 670–674.
13. Parry, C. H. (1825) Diseases of the heart. Elements of Pathology and Therapeutics *2*, 11–128.
14. Kendall, E. C. (1915) The isolation in crystalline form of the compound containing iodine which occurs in the thyroid: its chemical nature and physiologic activity. JAMA *64*, 2042–2043.
15. Harington, C. R. & Barger, G. (1927) XXIII. Chemistry of Thyroxine. III. Constitution and Synthesis of Thyroxine. Biochem. J. *21*, 169–181.
16. Meinhold, H., Wenzel, K. W. & Schürnbrand, P. (1975) Radioimmunoassay of 3,3',5'-triiodo-L-thyronine in human serum and its application in different thyroid states. Z. Klin. Chem. Klin. Biochem. *13*, 571–574.
17. Griffiths, R. S., Black, E. G. & Hoffenberg, R. (1976) Measurement of serum 3,3',5'-(reverse) T3, with comments on its derivation. Clin. Endocrinol. *5*, 679–685.
18. Davis, L. G., Sass, N. L., Manna, B. & Nasynowitz, M. L. (1979) Gas-liquid chromatographic determination of mono- and diiodotyrosines in serum. Clin. Chem. *25*, 218–220.
19. Nelson, J. C., Weiss, R. M., Palmer, F. J., Lewis, J. E. & Wilcox, R. B. (1975) Serum diiodotyrosine. J. Clin. Endocrinol. Metab. *41*, 1118–1124.
20. Willetts, P., Crossley, D. N., Ramsden, D. B. & Hoffenberg, R. (1979) The role of thyronine in thyroid hormone metabolism. J. Clin. Endocrinol. Metab. *49*, 658–660.
21. Premachandra, B. N. (1976) A simple radioimmunoassay of triiodothyronine in unextracted serum. J. Nucl. Med. *17*, 411–416.
22. Chopra, I. H., Solomon, D. H. & Ho, R. S. (1971) A radioimmunoassay of thyroxine. J. Clin. Endocrinol. Metab. *33*, 865–868.
23. Dunn, R. T., & Foster, L. B. (1973) Radioimmunoassay of thyroxine in unextracted serum by a single antibody technique. Clin. Chem. *19*, 1063–1066.
24. Chan, K. H. & Cummins, L. M. (1976) A new extractant for serum thyroxine by enzymatic digestion of thyroxine binding proteins. J. Clin. Endocrinol. Metab. *42*, 189–192.
25. Robbins, J. (1975) In: Thyroid Hormone Metabolism (Harland, W. A. & Orr, J.S., eds.), pp. 1–22. Academic Press, London.
26. Weeke, J. & Orskov, H. (1975) Ultrasensitive radioimmunoassay for direct determination of free triiodothyronine in serum. J. Clin. Lab. Invest. *35*, 237–244.

27. Shiba, T. & Cahnmann, H. J. (1962) Synthesis of specifically iodine 131 and carbon 14-labeled thyroxine. J. Org. Chem. *27*, 1773–1778.
28. Nishinaga, A., Cahnmann, J. H., Koh, H. & Matsuura, T. (1968) Model reactions for the biosynthesis of thyroxine. XII. The nature of thyroxine precursor formed in the synthesis of thyroxine from diiodotyrosine in its keto acid analog. Biochemistry *7*, 388–397.
29. Sorimachi, K. & Cahnmann, H. J. (1977) A simple synthesis of (3,5-^{125}I) diiodothyronine and of (3,5-^{125}I thyroxine of high specific activity. Endocrinology *101*, 1276–1280.
30. Moller, B., Bjorkhem, J., Falk, O., Lantto, O. & Larsson, A. (1983) Identification of thyroxine in human breast milk by gas chromatogoraphy mass spectrometry. J. Clin. Endocrinol. Metab. *56*, 30–34.
31. Crossley, D. N., Ramsden, D. B. & Hoffenberg, R. (1978) In: 2nd Int. Symp. Quantitative Mass Spectometry in Life Sciences (De Leenheer, A. P., Roncucci, R. R. & Van Peteghem, C., eds.), pp. 219–226. Elsevier, Amsterdam.
32. Crossley, D. N. & Ramsden, D. B. (1979) Serum tetraiodothyroacetate (T4A) levels in normal healthy euthyroid individuals determined by gas chromatography-mass fragmentography (GC-MF). Clin. Chim. Acta *94*, 267–272.
33. Lawson, A. M., Ramsden, D. B., Raw, P. J. & Hoffenberg, R. (1975) Mass spectrometric studies of thyroxine and related compounds. Trimethylsilyl derivatives. Biomed. Mass Spec. *1*, 374–380.
34. Heki, N., Noto, M., Hosojima, H., Takahashi, S. & Murata, T. (1976) Analysis of thyroid hormones in serum and urine by mass fragmentography using GC-MS. Folia Endocrinol. Jap. *52*, 149–157.
35. Heki, N. (1978) A simple method for the simultaneous analysis of rT3 and 3,3′-T2 by mass fragmentography using gc-ms. Part I. Folia Endocrinol. Jap. *54*, 1151–1156.
36. Manius, G. J., Fallon, P., Tscherne, R. (1978) Identification and determination of thyroxine using gas liquid chromatography. Anal. Biochem. *87*, 496–505.
37. Heki, N. (1978) A simple method for the simultaneous analysis of rT3 and 3,3′-T2 by mass fragmentography using gc-ms. Part II. Folia Endocrinol. Jap. *54*, 1157–1162.
38. Ramsden, D. B., Lawson, A. M., Raw, P. J. & Hoffenberg, R. (1974) Identification of 3,3′5,5′ tetraiodothyroformic acid within the rat liver. Biochem. J. *143*, 47–50.
39. Willetts, P., Ramsden, D. G., Crossley, D. N. & Hoffenberg, R. (1980) In: Thyroid Research VIII (Stockigt, J. R. & Nagataki, S., eds.), pp. 341–344. Australian Academy of Sciences, Canberra.
40. Willetts, P. (1980) Gas chromatographic-mass fragmentographic measurement of thyronine in human urine. Ph. D. Thesis, University of Birmingham.
41. Petersen, B. A. & Vouros, P. (1977) Analysis of thyroid hormones as their heptafluorobutyryl methyl ester derivatives by gas chromatography-mass spectometry. Anal. Chem. *49*, 1304–1311.
42. Hoffenberg, R., Lawson, A. M., Ramsden, D. B. & Raw, P. J. (1974) In: Mass Spectrometry in Biochemistry and Medicine (Frigerio, A. & Castagnoli, N., eds.), pp. 303–312. Raven Press, New York.
43. Doctor, R. & Hennemann, G. (1971) Estimation of thyroid hormones by gas-liquid chromatography. Clin. Chim. Acta *34*, 297–303.
44. Mohammed, M. N., Farmer, M. & Ramsden, D. B. (1983) 3,5-Diiodotyrosine excretion in human urine in individuals of differing thyroidal status. Biomed. Mass Spec. *10*, 507–510.
45. Ramsden, D. B., Farmer, M., Mohammed, M. N. & Willetts, P. (1984) Mass spectral properties of volatile derivative of thyronine (T0) and use of these in the study of thyronine excretion in eu-, hyper- and hypothyroidism. Biomed. Mass Spec. *11*, 193–198.
46. Ramsden, D. B. & Farmer, M. Development of a gas chromatographic selected ion monitoring assay for thyroxine (T4) in human serum. Biomed. Mass Spec. In press.
47. Moller, B., Falk, O. & Bjorkhem, I. (1983) Isotope dilution-mass spectrometry of thyroxine proposed as a reference method. Clin. Chem. *29*, 2106–2110.
48. Koya, S., Matsuura, K. & Kubota, E. (1982) Studies on the photodegradation of thyroxine with ultraviolet light irradiation: identification of degradated products by gas chromatography-mass spectroscopy. Yakugaku-Zasshi *102*, 923–934.
49. DeGroot, L. J. & Torresani, J. (1975) Triiodothyronine binding to isolated liver cell nuclei. Endocrinology *96*, 357–359.
50. Gosling, B., Schwartz, H. L., Dillman, W., Surks, M. I. & Oppenheimer, J. H. (1976) Comparision of the metabolism and distribution of L-triiodothyronine and triiodothyroacetic acid in the rat; a possible expla-

nation of differential hormonal potency. Endocrinology *98*, 666–675.

51. Ramsden, D. B., Raw, P. J., Carter, P. J. & Hoffenberg, R. (1975) Estimation of tetraiodothyroacetate in human serum. Proc. Roy. Soc. Med. *68*, 69–70.
52. Burger, A., More, A., Sakaloff, C., Vallotton, M. B. & Ingbar, S. H. (1973) A competitive protein binding assay for tetraiodothyroacetic acid. Proc. Vth Annual Meeting of the European Thyroid Assoc. Jerusalem. Abs., p. 38.
53. Burger, A., Nicod, P., Suter, P., Vallotton, M. B., Vagenakis, A. G. & Braverman, L. E. (1976) Reduced thyroid hormone levels in acute illness. Lancet *i*, 653–655.
54. Christensen, L. K. & Litonjua, A. C. (1961) Is thyroxine binding by prealbumin of physiologic importance? J. Clin. Endocrinol. Metab. *21*, 104–106.
55. Rudolph, M., Sakuranda, T., Fang, S. L., Vagenakis, A. G., Braverman, L. E. & Ingbar, S. H. (1978) Appearance of labeled metabolites in the serum of man after the administration of labeled thyroxine, triiodothyronine and reverse triiodothyronine. J. Clin. Endocrinol. Metab. *46*,923–928.
56. Meinhold, V. H., Beckert, A. & Wenzel, K. W. (1980) In: Thyroid Research VIII (Stockigt, J. R. & Nagataki, S., eds.), pp. 345–348. Australian Academy of Science, Canberra.
57. Meinhold, V. H., Beckert, A. & Wenzel, K. W. (1981) Circulating diiodotyrosine: Studies of its serum concentration, source and turnover using radioimmunoassay after immunoextraction. J. Clin. Endocrinol. Metab. *53*, 1171–1178.
58. Pileggi, V. J., Segal, H. A. & Golub, O. J. (1964) Determination of organic iodide compounds in serum III iodotyrosines in normal human serum. J. Clin. Endocrinol. Metab. *24*, 273–280.
59. Werner, S. C. & Block, R. J. (1959) discrepancy between the distribution of iodine in human serum when estimated by iodine-131 and iodine-127. Nature *183*, 406–407.
60. Beale, D. & Whitehead, J. K. (1960) A preliminary investigation of the levels of MIT and DIT in human blood using double isotope dilution techniques. Clin. Chim. Acta *5*, 150–152.
61. Mohammed, M. N. & Ramsden, D. B. (1982) Gas chromatographic-mass fragmentographic (GCMF) assay of 3,5-diiodotyrosine in human urine. Proc. VIIIth Anglo-Dutch Endocrine Meeting, Abs. 30. Supplement to J. Endocrinol. *94*.
62. Mohammed, M. N. (1984) 3,5-Diiodotyrosine in human urine: development and use of a gas chromatographic-mass spectrometric assay. Ph. D. Thesis, Birmingham University.
63. Alexander, N. M. & Scheig, R. (1968) Gas chromatography of the trimethylsilyl derivatives of iodotyrosines and iodothyronines. Anal. Biochem. *22*, 187–194.
64. Leclercq, P. A. & Desiderio, D. M. (1971) A laboratory procedure for the acetylation and permethylation of oligopeptides on the microgram scale. Anal. Lett. *4*, 305–316.
65. Zimmerer, R. O. & Grady, L. T. (1971) GLC of sodium thyroxine. J. Pharm. Sci. *60*, 493–495.
66. Husek, P. (1974) Derivatisation of amino acids with 1,3-dichlorotetrafluoroacetone and its use in gas chromatography. J. Chromatogr. *91*, 475–481.
67. Wu, S. Y., Chopra, I. J., Nakamura, Y., Solomon, D. H. & Bennett, L. R. (1976) A radioimmunoassay for measurement 3,3′-diiodothyronine. J. Clin. Endocrinol. Metab. *43*, 682–685.
68. Chopra, I. J., Geola, F. & Solomon, D. H. (1978) 3′,5′-Diiodothyronine in health and disease: studies by a radioimmunoassay. J. Clin. Endocrinol. Metab. *47*, 1198–1207.
69. Pangora, L., Burman, K. D., Wartofsky, L., Cahnmann, H. J., Smallridge, R. C., O'Brien, J. T., Wright, F. D. & Latham, K. (1980) Radioimmunoassay for 3,5-diiodothyronine and evidence for dependence on conversion from 3,5,3′-triiodothyronine. J. Clin. Endocrinol. Metab. *50*, 1075–1081.
70. Corcoran, J. M. & Eastman, C. J. (1981) Measurement of 3′-monoiodothyronine in human serum. Clin. Endocrinol. (Oxf.) *15*, 11–18.
71. Corcoran, J. M. & Eastman, C. J. (1983) Radioimmunoassay of 3-L-monoiodothyronine: application in normal human physiology and thyroid disease. J. Clin. Endocriniol. Metab. *57*, 66–70.

Chapter 13
Reference Methods

L. Siekmann

1. Introduction

Over the past 15 years an increasing number of methods has been introduced into the everyday analytical routine work of clinical chemical laboratories. Most of these newer methods are superior by virtue of their higher practicability; some of them may even improve the reliability of analytical procedures, e. g. accuracy, specificity, precision and sensitivity. The precision and sensitivity of a new method may be easily checked by a series of measurements and by statistical evaluation. In contrast, it is often difficult to examine accuracy and specificity, which are the most important criteria of reliability. In many papers presenting new methods in clinical chemistry it is assumed that "accuracy" can be demonstrated by measurements before and after addition of a fixed amount of the pure analyte to the biological matrix. However, this procedure describes only the "recovery" of the analyte. Any interference from the biological matrix due to lack of specificity may go undetected in a "recovery" experiment. The analytical result of a particular method may be inaccurate because of a bias originating from interfering substances, even though recovery is demonstrably 100%. Methods comprising a more or less constant bias may still be usefully applied in a single laboratory. However, it is obvious that any progress with respect to the comparability of analytical results from different laboratories and from different methods is hampered as long as methods with a known or unknown bias are accepted. This becomes clear when carrying out external quality control surveys; in some quality control schemes several method-dependent target values are applied to evaluate the participants' results for the same clinical chemical parameter. It is not acceptable to an analytical chemist that for an analyte, which is represented by a well-defined molecular species, different concentration values in the same sample can be officially accepted. In the future, therefore, it will be necessary to examine as far as possible the accuracy of clinical chemical methods and to replace inaccurate analytical procedures with more reliable ones. The only way to ascertain the specificity and accuracy of a clinical chemical method is to compare it with a procedure that has been shown to yield accurate results.

In view of this, clinical chemistry has established its own hierarchy of methods. "Definitive methods" are the most accurate procedures followed by "reference methods", with daily routine methods being ranked last. Definitions for these terms are laid down in an IFCC document (1) as follows:

> "A definitive method is one which after exhaustive investigation is found to have no known source of inaccuracy or ambiguity."

and

> "A reference method is one which after exhaustive investigation has been shown to have negligible inaccuracy in comparison with its imprecision."

These definitions as well as the system of method hierarchy itself have been subject to discussion (2, 3, 4, 5), and several modifications have been proposed (6, 7, 8). In particular, the categorization of definitive methods is still being debated. Since at present there is no way of counting the number of individual molecular particles in a biological matrix according to Dalton's law, it may be questioned whether the true concentration values of substances in biological fluids are attainable. Thus the question arises whether any present method can fulfill the requirements of an absolute or definitive method. Since this discussion has not reached a final conclusion, analytical procedures that claim to be either "definitive" or "reference" methods are described and discussed side by side in this chapter without regard to hierarchy. A number of reference and definitive methods which have been established or are under development were first reviewed in 1979 (9).

2 General Principles of Isotope Dilution-Mass Spectrometry (ID-MS)

Over the past 20 years, mass spectrometry – in particular in combination with gas-liquid chromatography (GC) – has become a powerful tool for the identification of metabolites in biological fluids. A recent review (10) deals with all aspects of mass spectrometry in the field of clinical chemistry. In 1966 Sweeley and co-workers (11) and in 1968 Hammar et al. (12) demonstrated that it is not always necessary to record complete mass spectra for the detection of a compound. They adjusted the mass spectrometer to focus one or a few ions characteristic of the substance and continuously monitored these during GC elution of the biological sample to give one or several simultaneously recorded mass chromatograms. The mass spectrometer used in this mode served as a GC detector of high sensitivity with the outstanding feature that specificity could be selected for the compound under consideration. The conventional technique of recording complete mass spectra was mainly used for the qualitative purpose of identifying substances. In contrast, the selected ion monitoring (SIM) procedure was also applicable to quantitative analysis, since the resulting peaks could be evaluated by peak-area or peak-height measurements.

As an unpredictable portion of the substance under investigation by SIM may be lost in the pre-instrumental part of analysis or even during GC, it is necessary to introduce an internal standard to improve the reliability of quantitative measurements. Substances whose chemical structure are closely related to that of the analyte, i.e. isomers or chemical homologues, were recommended as internal standards. However, from a theoretical and practical point of view an isotopically labeled variant of the substance to be measured is the ideal internal standard. Isotopically labeled analogues may be expected to exhibit the same properties as the non-labeled substances during extraction, purification and gas-liquid chromatography. In contrast, internal standards, even when these show only small differences from the chemical structure of the analytes, are significantly different at least in their chromatographic behaviour. It is worthwhile to note that the principle of isotope dilution-mass spectrometry (ID-MS) was introduced as early as 1940 by Rittenberg and Foster (13), who used ^{2}H- and ^{15}N-labeled compounds for the isotope dilution assay of fatty acids and amino acids by measuring peak height ratios in conventionally scanned spectra. In 1970 the first applications of the isotope dilution principle using the SIM technique for measuring endogenous substances in biolgical fluids were reported (14, 15). These papers were followed by

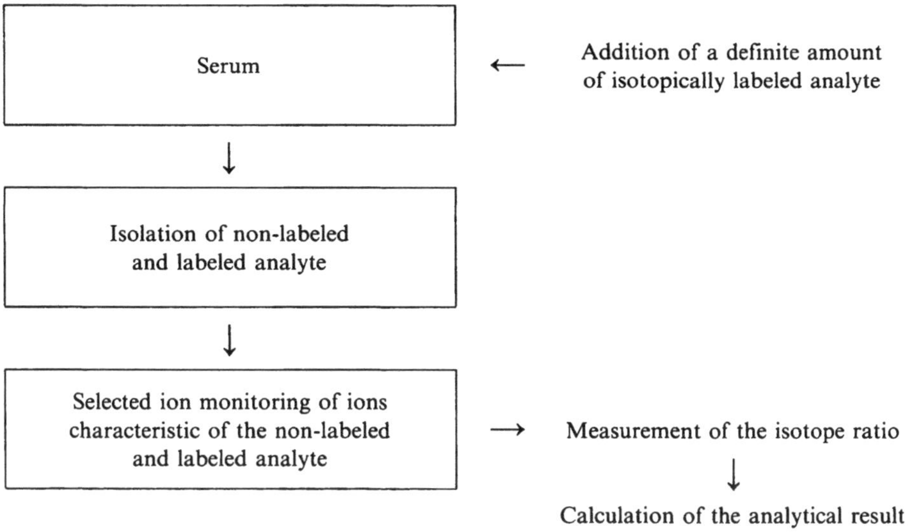

Figure 1. General principle of quantitative analysis by isotope dilution-mass spectrometry.

an increasing number of reports from many fields of the biological sciences. A review of quantitative mass spectrometry in biochemistry and medicine was published in 1978 (16). In recent years ID-MS has won additional recognition as a valuable tool in the development of reference or definitive methods in clinical chemistry. The technical and mathematical principles of isotope dilution-mass spectrometry and its use in the hierarchy of clinical chemical methods were reviewed by De Leenheer et al. (17), and Lawson et al. (18) set up the guidelines of the International Federation of Clinical Chemistry (IFCC) for the use of quantitative mass spectrometry for accuracy control in clinical chemistry.

The general principle of many quantitative procedures employing ID-MS is summarized in Figure 1. At the beginning of analysis a fixed amount of isotopically labeled analyte is equilibrated with the endogenous analyte present in a measured amount of the biological fluid, e. g. serum; the same quantity of the isotopically labeled material is also added to a number of vials containing defined amounts of the pure non-labeled analyte, and these are then used as standards. This initial step is a critical one. Accuracy and precision in measuring the serum sample and the isotopically labeled standard solution as well as accuracy and precision in weighing the primary standard material will determine the reliability of the analytical results. Care should be taken that samples and standards contain the same quantity of labeled substance. Moreover, complete equilibration of labeled and non-labeled analyte in a serum sample should be assured, in particular when a proportion of the substance is bound to protein. Extended equilibration times may be necessary (19).

This first important step is generally followed by a purification procedure which may include extraction and/or column chromatography. For many analytes, derivative formation is advisable or necessary. Most frequently the reaction product is further purified prior to mass spectrometry by using a GC column which is coupled to the mass spectrometer.

Finally the isotope ratio of the non-labeled and labeled analyte is measured by mass spectrometry and the result is calculated from a calibration curve developed from the isotope ratios of the standard mixtures. In the book by Millard (19) many aspects of quantitative mass spectrometry are discussed in detail.

2.1 Selection of isotope label

Internal standard substances labeled with stable as well as with radioactive isotopes have been used for the quantification of endogenous compounds by ID-MS. The radioactive isotopes ^{14}C and ^{3}H (tritium) have been applied mainly in the field of steroid hormone determinations, as radioactive analogues of steroids have been available commercially for many years. Many compounds with stable labeled isotopes suitable for isotope dilution assay have become available only recently and are replacing the radioactive substances in these cases.

Among the non-radioactive isotopes, ^{2}H (deuterium) is the most popular followed by ^{13}C, ^{15}N and ^{18}O. The application of stable isotopes in biomedical research has been reviewed by Haskins (20). There is a risk that an exchange of the isotope label can occur with ^{2}H or ^{3}H labeled compounds during the pre-instrumental part of the analytical procedure, in particular when strong acidic or alkaline conditions prevail in the course of sample preparation. The possibility of exchange obviously depends on the position of the label in the molecule and therefore should be exhaustively tested for each individual substance. ^{13}C- or ^{15}N-labeled compounds do not have this disadvantage as internal standards. A further important consideration is the mass difference between the non-labeled and the isotopically labeled molecular or fragment ion to be recorded because of the possible interference of naturally occurring isotopes. The trimethylsilyl derivative of non-labeled creatinine, for example, in addition to the molelcular ion at m/z 329 (100%) shows $M + 1$, $M + 2$ and $M + 3$ signals of approximately 31.5, 12.0 and 1.0%, respectively. These signals can interfere during recording of the isotopically labeled standard. Though it is true that the degree of interference can be calculated or an experimental calibration curve used, in the case of interfering isotopes from the non-labeled substance, the calibration function becomes non-linear. This results in a limit to the range of isotope ratios from which precise readings of concentration values are then obtainable. The degree of interference becomes smaller the greater the mass difference between the non-labeled and the isotopically labeled substance. A mass difference of at least two daltons is desirable between the analyte and the internal standard. Therefore, at least two deuterium-, two ^{13}C- or ^{15}N-atoms or one tritium-, one ^{14}C- or one ^{18}O-atom need to be present in the labeled internal standard. A difference of three or four daltons is even more favourable although replacing a large number of atoms in a molecule by isotopes may change the chemical properties of the substance. For example, introducing several deuterium atoms into a small molecule usually results in a significantly decreased retention time on GC, indicating a differing chemical behaviour of the two molecular species which is undesirable in the isotope dilution technique.

2.2 Purification and derivative formation of the analyte

In general it is usually necessary to enrich and purify compounds prior to quantitative mass spectrometry. In some instances, e. g. measurement of metal ions, it may be quite sufficient to concentrate the sample to dryness; however, in the majority of cases more efficient methods of enrichment are used. For lipophilic compounds, e. g. cholesterol or steroid hormones, extraction of the analyte from the biological fluid (usually serum) by the use of an organic solvent has proved to be the method of choice. With other substances like creatinine or uric acid, adsorption to an ion exchange material is used for the isolation of the analyte. In particular, when low concentrations of a substance have to be analysed, more extensive procedures for pre-purification must be carried out before mass spectrometry is applied. Thus steroid hormones, after extraction from serum samples by organic solvents, must be further purified e. g. by column chromatography on Sephadex LH-20, by Lipidex 5000 or by an additional step using thin-layer chromatography on silica gel. High-performance liquid chromatography (HPLC) represents a further highly efficient tool for pre-purification of several analytes.

Particularly when mass spectrometry is combined with GC as a final step in detection, chemical derivatisation becomes necessary or advisable. The book by Blau and King (21) has at present one of the most comprehensive collections of methods of derivatisation. For many years formation of trimethylsilyl derivatives has been adopted for protecting analytes containing hydroxy-, fatty acid as well as N-H groups although recently *tert*-butyldimethylsilyl derivatives have been recommended since they are less sensitive to hydrolysis. When keto groups have to be protected, methoximes are frequently used.

Despite the fact that derivative formation is often a cumbersome, time-consuming and labor-intensive step, it is important when reference or definitive methods are being developed as the specificity of the detection procedure may be considerably improved by the use of special derivatives. For example, the molecular mass of steroid hormones is significantly increased by introducing heptafluorobutyric ester groups. When the mass spectrometer is adjusted to the high m/z values of molecular ions of the heptafluorobutyric ester derivatives, few ions from impurities in the biological material occur during SIM (22). The substances which could be possibly interfere with the detection of the analyte may then easily be separated by selecting appropriate GC conditions.

2.3 Instrumentation

The ionization method necessary for substances of biological interest depends largely on the substances' chemical structure. For metal ions, for example, thermal ionization is the method of choice but field desorption and fast atom mass bombardment are also of value. The latter techniques may be applicable when measuring quantitatively substances of high molecular weight and other groups of compounds difficult to evaporate without deterioration. Unforunately these soft ionization techniques are not routinely combined with GC or liquid chromatography (LC). The technical procedures most frequently used for the development of reference methods are conventional electron impact ionization and chemical ionization mass spectrometry (EI and CI respectively) in combination with GC. GC with fused silica capillary columns represents perhaps the most powerful chromatography system currently available for the separation of organic compounds. The effluent of the capillary column is inserted directly into the ion source without the need for a complicated interface. At first glance the on-line combination of LC with mass spectrometry appears especially suitable in view of the fact that derivative formation usually becomes unnecessary. However, the interface (moving belt or thermospray) may lead in some cases to the destruction of sensitive compounds under investigation. It remains to be seen whether LC will replace GC in on-line combinations with mass spectrometry.

Many reference methods described so far were developed by using single focusing magnetic field or quadrupole mass analysers with low resolution. Double focusing high-resolution instruments or tandem mass spectrometers are rarely applied in reference methodology largely on the grounds of cost. In principle, it is possible to introduce relatively crude extracts into the ion source of a high-resolution instrument or a tandem mass spectrometer and use the extremely high specificity of these detectors to avoid pre-instrumental purification. However repeated injection of crude extracts may lead to a reduction of performance of the instrument. It should be noted that with low-resolution instruments, the use of chromatographic pre-purification procedures is not always necessary in order to achieve specificity, but is necessary to prevent an excess of material from the biological matrix from entering the mass spectrometer.

2.4 Calculation procedure

The results of an isotope dilution assay are calculated from the isotope ratios measured by mass spectrometry and from the known amount of isotopically labeled material which has been added to the biological fluid at the beginning of the analytical procedure. The accuracy of the final result depends on the purity and isotope distribution of the labeled substance. Unfortunately, labeled compounds are usually not as well characterized as non-labeled certified reference materials. Therefore it is necessary to calibrate the labeled internal standard substances against primary non-labeled standards which are available, for example, from the National Bureau of Standards, Washington, D. C. In practice this calibration is done by measuring isotope ratios in mixtures of the labeled compound and the certified non-labeled reference material. Standard mixtures of this type are usually included in each series of measurements.

In order to improve the precision of the analytical results, use is recommended of standard mixtures whose isotope ratios are close to the ratios in the samples. In a procedure known as the "bracketing technique" (23), two standard mixtures of labeled and non-labeled substance are used which closely encompass the isotope ratio of the sample to be quantified. The result is then calculated by linear interpolation.

When a calibration curve (amount of analyte versus isotope ratio obtained by ID-MS) is constructed to include the lower range of isotope ratios, the regression line frequently does not pass through the origin of the coordinates. This is usually due to the presence of small amounts of the non-labeled substance. Therefore an isotope ratio greater than zero is obtained even if no non-labeled substance is added to the standard. In particular, radioactively labeled analogues of many compounds contain considerable amounts of non-labeled material (22). In the higher range of isotope ratios the calibration curve may reflect a non-linear relationship. This is explained by the contribution of a natural isotopic form of the non-labeled analyte to the recording of the labeled internal standard. Therefore, when the amount of non-labeled substance becomes infinitely high as compared with the labeled standard in the mixture, a limited isotope ratio will nonetheless be measured reflecting the ratio observed in the non-labeled analyte. This is generally true unless there is a difference of at least four daltons between the two substances. On the basis of these considerations the following equation can be formulated (22, 24):

$$X[\mathrm{ng}] = \frac{I - f_1 I^*}{I^* - f_2 I} f_0 S[\mathrm{ng}] \qquad \text{(Eq. 1)}$$

where

X = amount of analyte to be calculated
I = intensity of the non-labeled analyte in the sample
I^* = intensity of the labeled analyte in the sample
f_1 = ratio I/I^* in the labeled standard
f_2 = ratio I^*/I in the non-labeled standard
S = amount of labeled internal standard in the sample
f_0 = proportional factor

Eq. 1 can be rearranged to give

$$X[\mathrm{ng}] = \frac{R - f_1}{1 - f_2 R} f_0 S[\mathrm{ng}] \qquad \text{(Eq. 2)}$$

where

R = I/I^* (intensity ratio of non-labeled analyte to labeled standard in the sample)

The parameters f_1 and f_2 of the formula are directly measured as isotope ratios in the non-labeled and labeled standard. The third parameter of the formula, f_0S, is obtained after measuring the isotope ratio R in a standard mixture and after insertion of the known amount of non-labeled analyte X as well as f_1 and f_2 into the equation. A number of similar procedures for the calculation of results from ID-MS are reported, based on the same theoretical considerations (25, 26, 27).

Thus the calculation procedure, described here and used in the laboratory of the author for several years (22), can be traced back to theoretical considerations rather than to an empirical curve that may be calculated e. g. by least square regression analysis as recommended by Millard (19).

A disadvantage of the procedure, as far as its use for reference methods is concerned, is that it employs too wide a range of standards. The three standards used for calibration represent the isotope ratios at about 1:1 (standard mixture), at 0 ng of non-labeled (labeled standard) and at an infinite amount of non-labeled analyte (non-labeled standard). For a highly precise determination it would be preferable to employ standards whose isotope ratios are close to that of the sample to be analysed. The calculation procedure described above was accordingly developed to accept three standard mixtures whose isotope ratios bracket the ratio of the sample to be investigated as closely as possible (24). This method takes into consideration both the advantage of a bracketing technique and the fact that there is a non-linear relationship between the isotope ratio and the amount of non-labeled analyte in standards and samples. A similar non-linear bracketing procedure has since been reported (28).

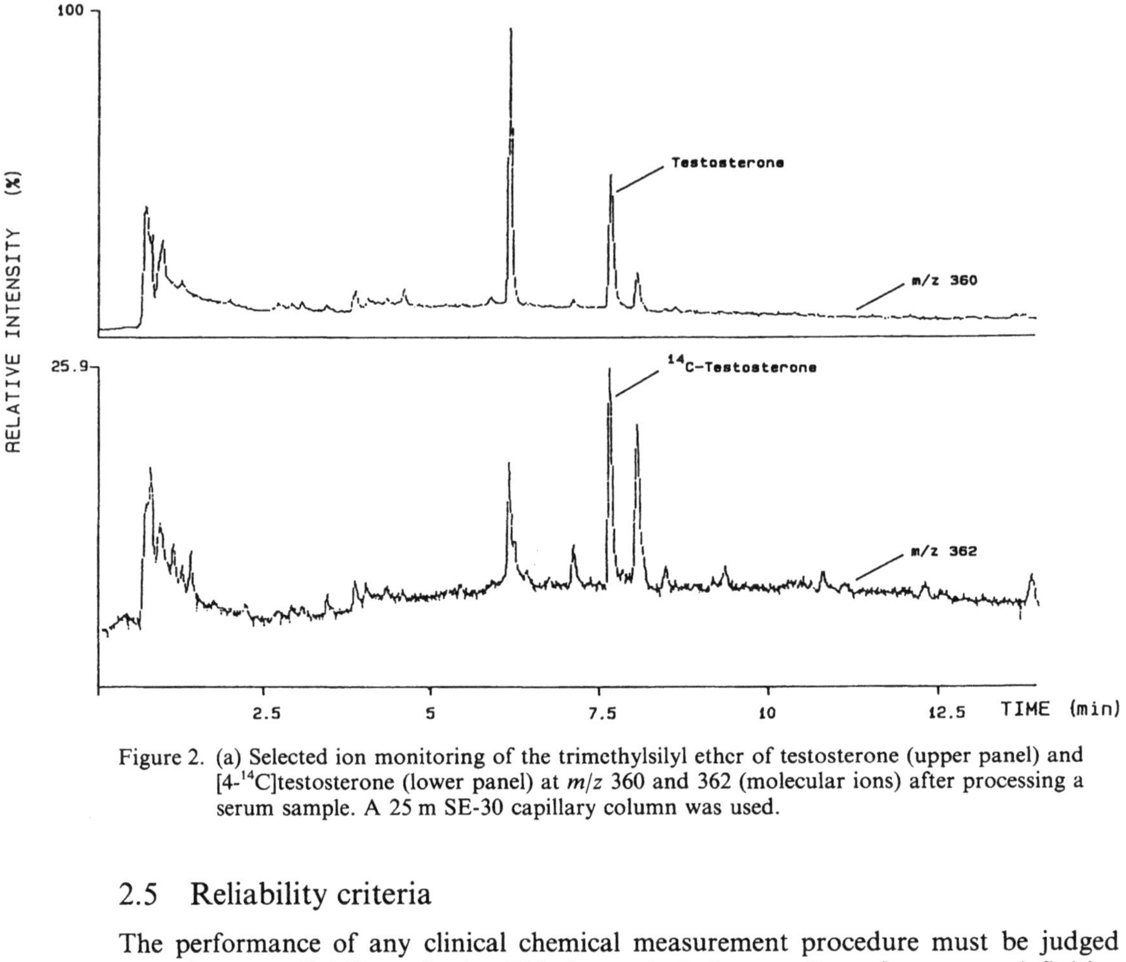

Figure 2. (a) Selected ion monitoring of the trimethylsilyl ether of testosterone (upper panel) and [4-^{14}C]testosterone (lower panel) at *m*/*z* 360 and 362 (molecular ions) after processing a serum sample. A 25 m SE-30 capillary column was used.

2.5 Reliability criteria

The performance of any clinical chemical measurement procedure must be judged according to reliability criteria. This is particularly true for reference or definitive methods.

2.5.1 **Specificity**

With many reference methods described so far, specificity is achieved by GC-MS using the SIM technique. The mass spectrometer is adjusted to detect *m*/*z* -values that are highly characteristic for the substance to be determined. Interference occurs only if compounds from the biological matrix are present which form molecular or fragment ions of the same *m*/*z*-values. As mentioned earlier, the detection of potential interfering compounds can be significantly reduced by using high-resolution mass spectrometry. For many compounds specificity is obtained by combined capillary GC-MS and the use of derivatives which give an increased molecular mass. This principle has been successfully applied to the measurement of steroid hormones (29). It can be demonstrated that SIM of a purified extract from human serum results in a number of peaks, in addition to the steroid to be determined, when the molecular ion of the free steroid

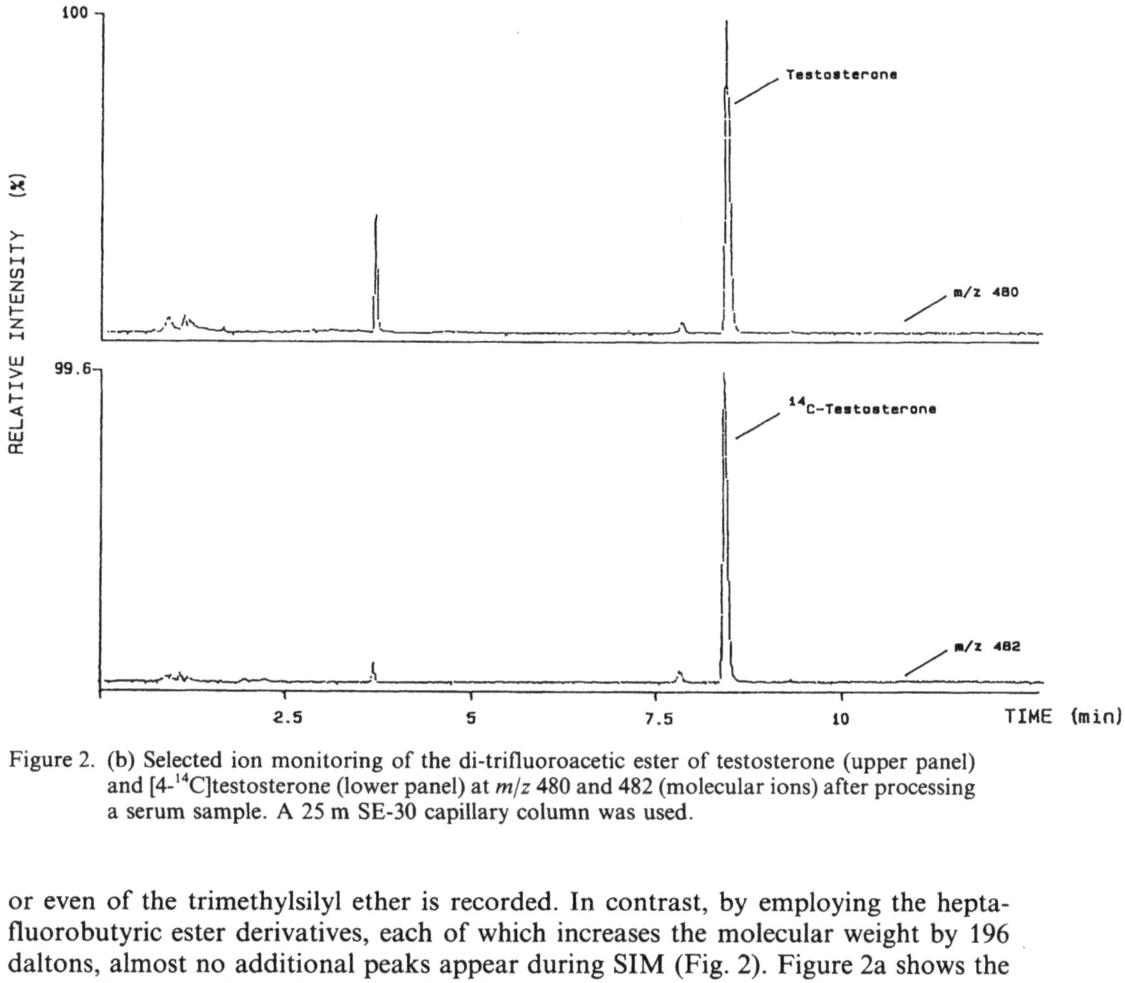

Figure 2. (b) Selected ion monitoring of the di-trifluoroacetic ester of testosterone (upper panel) and [4-^{14}C]testosterone (lower panel) at *m/z* 480 and 482 (molecular ions) after processing a serum sample. A 25 m SE-30 capillary column was used.

or even of the trimethylsilyl ether is recorded. In contrast, by employing the heptafluorobutyric ester derivatives, each of which increases the molecular weight by 196 daltons, almost no additional peaks appear during SIM (Fig. 2). Figure 2a shows the selected ion recording of the trimethylsilyl ether of testosterone at the molecular ion of the non-labeled and the ^{14}C-labeled derivative (*m/z* 360 and *m/z* 362 respectively) after processing a serum sample. The peak obtained after 7 min and 30 s represents the testosterone derivative. Figure 2b records the molecular ions of the trifluoroacetate of the non-labeled and the labeled testosterone at *m/z* 480 and 482, and figure 2c the heptafluorobutyric ester derivatives at *m/z* 680 and 682 after processing the same serum sample.

2.5.2 **Accuracy**

Accuracy is the most important criterion of reliability for the development of reference and definitive methods. A substance is determined accurately when (a) the analytical procedure is specific enough to avoid interference from other compounds and (b) when the losses of the substance during the analytical procedure can be fully measured and calculated. The most reliable control of recovery is to monitor isotopically labeled

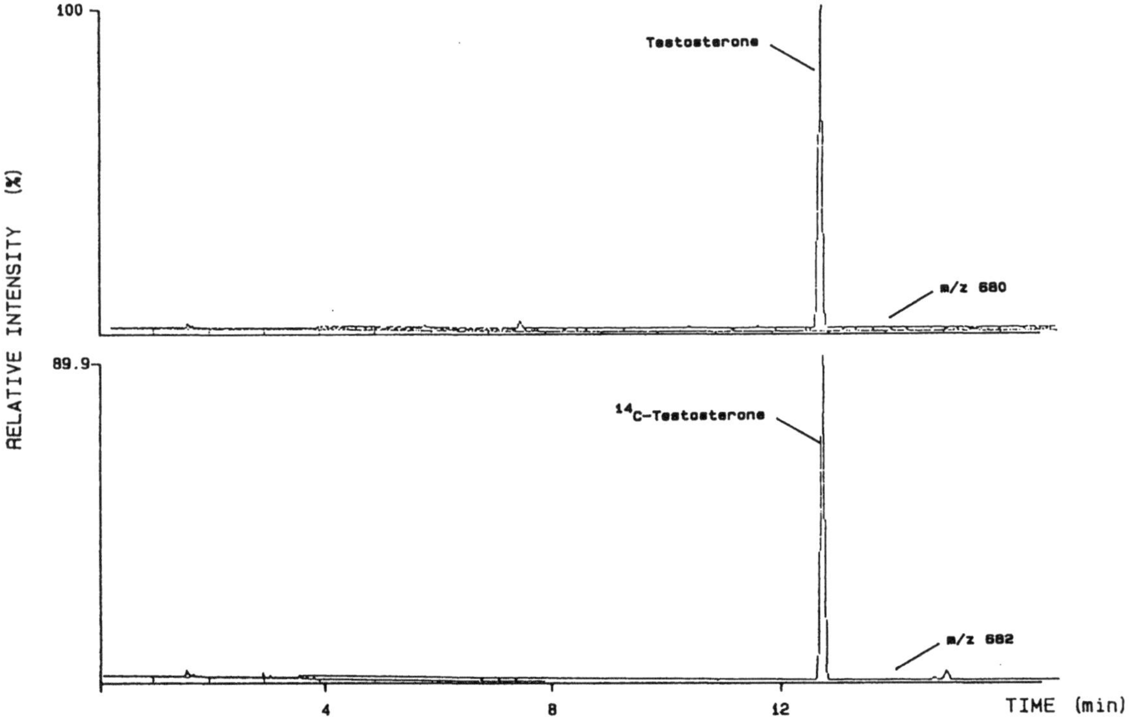

Figure 2. (c) Selected ion monitoring of the di-heptafluorobutyric ester of testosterone (upper panel) and [4-^{14}C]testosterone (lower panel) at *m/z* 680 and 682 (molecular ions) after processing a serum sample. A 25 m SE-30 capillary columns was used.

standard materials using the same detection procedure as the substance to be determined. The isotope dilution technique in combination with mass spectrometry provides, therefore, the most valuable procedure for the highly accurate measurement of many compounds of biological origin.

It should also be noted that the purity of the primary standard as well as the performance of the equipment used for calibration of balances and volumetric glassware are possible sources of systematic error and therefore directly influence the accuracy of the analytical results. Wherever possible, well-characterized certified reference materials, e. g. from the National Bureau of Standards, Washington, should be used as primary standards for calibration. Furthermore, the equipment for volumetric measurements (glass-ware or electronically automated pipettors) should be carefully calibrated by weighing appropriate amounts of water at a definite temperature taking into account the density of water under these conditions and taking care that evaporation does not occur.

2.5.3 **Precision**

Although in reference methodology, precision is a less important parameter than accuracy, it is obvious that a concentration value obtained by means of a highly accurate measurement procedure becomes useless when imprecision makes the results dubious and extends the confidence range unacceptably. The precision necessary for a reference or definitive method may depend on the intended use of the analytical results. For example, sodium in human serum requires a high degree of precision to correspond with its relatively small reference (normal) range in human serum. In contrast, when measuring a hormone such as aldosterone, which occurs in the picogram range in human serum, a much lower precision is acceptable in view of its clinical significance and the lower degree of precision obtained in routine work. This makes it difficult to set up general requirements for precision in reference and definitive methods.

There are several steps in the analytical ID-MS procedure that can influence the precision of the results. There is limited precision involved in sampling the biological fluid and the solution of the isotopically labeled internal standard. The coefficient of variation (CV) for each of these steps depends on the technical equipment and the amount of fluid to be sampled. In order to improve precision some investigators determine the amount of biological fluid by weighing rather than by measuring the volume, a procedure which increases overall precision in the isotope dilution method. However, since it is unsuitable to report concentrations in mass per mass units in clinical chemistry, concentration values have to be converted to mass per volume units. For this the density of the biological fluid has to be carefully measured – again a volumetric measurement. A good compromise is to calibrate the volumetric equipment used for the analytical procedure by weighing appropriate amounts of water at a defined temperature. It may be possible to keep imprecision at less than 0.2% for volumes greater than 0.1 ml by using an electronically automated pipette, and this degree of precision is satisfactory for most applications.

A further source of imprecision stems from the instrument itself. For the measurement of two isotopes the mass spectrometer must be adjusted to switch at least 20 times back and forth between the two ions during the time a peak is eluting from the GC column. With a lower switching frequency, one ion may be recorded on top of the peak and the other on the flank and this would result in imprecision in the peak *height* determinations. Precision in peak *area* measurements is not greatly influenced by the switching frequency, although there is an increased risk of interferences causing inaccurate peak area measurement. Peak height measurements made at the peak maxima are less sensitive to small amounts of interfering substances, as these would usually be resolved sufficiently by the GC column. Depending on the performance of the instrument, the imprecision of isotope ratio measurements using mass spectrometry is about 0.5 to 1%.

2.5.4 **Lower limit of detection**

Amounts as small as 10^{-14} mol may be detectable by mass spectrometry in the SIM mode. Obviously, the lower limit of detection depends on the performance of the mass spectrometer as well as on the substance to be analysed. It should be noted that in some instances the signal-to-noise ratio can be significantly increased by preparing particular derivatives. For example, it may be possible to stabilize the molecular ion of a compound containing hydroxy-groups by the formatiion of a trimethylsilyl ether or perfluoroacyl ester derivative, thereby reducing fragmentation. This usually increases sensitivity when SIM is used to detect the molecular ion. Most substances that are of importance in clinical chemistry occur in μg to $mg \cdot ml^{-1}$ in body fluids, corresponding to a greater than one thousandfold of the lower limit of detection of mass spectrometry. Hence sensitivitiy is not a problem for many groups of clinical chemical analytes although difficulties can arise for substances occurring in lower concentration ranges. This is particularly true when measuring catecholamines as well as many steroid and thyroid hormones in human serum, e.g. oestradiol-17β or aldosterone, where $ng \cdot ml^{-1}$ concentrations have to be quantified. Furthermore some compounds present in low concentrations in body fluids are partially or completely lost during analysis. This is usually due to adsorption of the substance to the surface of glassware, column materials or interfaces used in GC-MS and can be avoided by careful deactivation of active surfaces, e.g. silylation. In some instances the addition of a chemical homologue as carrier substances is a helpful procedure. It may be necessary to use large volumes of the biological fluid when the concentration of an analyte is in the lower range of detection. For example, up to 10 ml serum are used when measuring aldosterone in human serum by ID-MS (30), which results in a large amount of interfering substances in the crude extract. These must be removed by an extensive purification procedure before SIM can be applied successfully. This involves the danger that during this preliminary separation, and before it can be measured, the analyte is lost by adsorption.

The signal-to-noise ratio may be further improved by lowering the resolution of the mass spectrometer. Single focusing, low-resolution mass spectrometers are often used with a resolution of 800 to 1000 (10% valley definition) for recording mass spectra. When the SIM mode is applied, it may be possible to run the instrument at a resolution of 400 to 500 and increase sensitivity.

3 Methods for the Measurement of Electrolytes

3.1 Calcium

Calcium was the first analyte in humen serum for which a definitive method was established. The analytical procedure now considered to be the most accurate method for the measurement of calcium is the ID-MS technique as described by Moore and Machlan (31). The primary function of the method was to provide a set of almost bias-free calcium concentrations in pooled serum samples which were used as target values for an atomic absorption spectrometry procedure.

In the isotope dilution assay, serum samples were spiked with a ^{44}Ca solution and the organic material, from the serum matrix, destroyed by $HClO_4$ and HNO_3. This was followed by ion-exchange chromatography to separate Ca ions from interferences originating from the biological matrix. The resulting $Ca(NO_3)_2$ solution was then deposited on a Re sample filament and subjected to thermal ionization. The isotope ratio measurements were carried out at the National Bureau of Standards, Washington, on mass spectrometers built for this purpose.

Since Ca is one of the most abundant elements in the environment, severe background problems due to contamination may arise in trace analysis of calcium. Large chemical blanks may result unless care is taken in handling the samples and standards. Highly purified reagents and a vertical-flow clean room are necessary prerequisites to a reduction of calcium contamination to an acceptable level. Due to the relatively large difference between the atomic masses of the two isotope species (^{40}Ca and ^{44}Ca), fractionation can occur during their desorption from the sample filament which is largely dependent on the temperature profile during the course of the preheating phase and during analysis. The authors stressed the fact that inter-analysis isotope fractionation can be avoided by using a consistent heating pattern which allows a consistent fractionation of the isotopes.

When this technique was applied to synthetic aqueous calcium solutions, the bias became less than 0.1% of the "true" value. For serum samples the possible error due to imprecision was shown to be less than 0.25% (95% limit of error). This analytical procedure, reported in a detailed protocol, is one of the very few methods currently accepted as definitive. More recently a further isotope dilution method employing a thermal ionization probe in a quadrupole mass spectrometer was described (32) and used for measuring calcium in urine. It is less precise and accurate than the procedure developed by Moore and Machlan (31) and may not prove useful as either a reference or definitive method.

An alternative procedure for the measurement of calcium in human body fluids has been reported by Kownatzki et al. (33), who used volatile β-diketone-complexes and electron impact mass spectrometry. However, this method was designed to serve as an analytical tool for the investigation of special problems in calcium turnover in man rather than for highly accurate measurements. The reliability data reported for this procedure do not justify its use as a definitive or reference method for clinical chemistry.

3.2 Potassium

A definitive method involving ID-MS for the measurement of potassium in human serum was developed at the National Bureau of Standards, Washington (34). ^{41}KCl was employed as an isotopically labeled internal standard. The work-up procedure was similar to the calcium method including destruction of the organic material of the samples by HNO_3 and $HClO_4$ and chromatography on cation-exchange columns. The potassium was analysed by thermal ionization mass spectrometry using a tantalum triple-filament source. Tantalum proved to have a particularly low potassium background and was therefore preferred as a filament material. To measure natural K and and isotopically labeled standard, m/z values 39 and 41 were recorded respectively. Due to the high terrestrial abundance of potassium, care must be taken to reduce blank values to a negligible level compared with serum concentrations. As with the calcium method a possible source of systematic error arises from isotope fractionation, since the lighter ^{39}K isotope is preferentially volatilised and ionized during thermal ionization. Therefore it is necessary to control sample loading and evaporation carefully. Since fractionation was reproduced with a high degree of precision and the analytical procedure applied was identical for standards *and* samples, a systematic error due to isotope discrimination could be excluded. When the method was applied to lyophilised sera (SRM 909, National Bureau of Standards) in a series of vials the imprecision was 0.4% (CV). This was due mainly to a vial-to-vial variation. When the measured potassium concentrations were based on the dry weight of the lyophilised sera in the individual vials, imprecision was reduced to less than 0.2% (CV).

3.3 Phosphate

Following the investigations concerned mainly with the measurement of elements, Pickup illustrated that the technique of ID-MS was also applicable to molecules. A method for the measurement of inorganic phosphate by ID-MS was described (35). Disodium hydrogen orthophosphate enriched in ^{18}O was added to serum samples. The mixtures were subjected to ultrafiltration and the filtrates were transferred to cation exchange columns to replace the cations normally present with NH_4^+. Ammonium phosphate was eluted with distilled water and concentrated to dryness by evaporation under nitrogen. The inorganic phosphate was then converted to tris(trimethylsilyl)phosphate by reaction with *N*-methyl-*N*-trimethylsilyltrifluoroacetic amide (MSTFA) containing 1% trimethylchlorosilane. Isotope ratio measurements were carried out using a combined GC-MS instrument at *m/z* 299 and 307. In spite of the relatively poor precision of the method reported, a substantial discrepancy was observed between the results of routine laboratories participating in a quality control program and the values obtained by isotope dilution mass spectrometry. This discrepancy may have been due to the chemical form of phosphate measured in routine determinations and by isotope dilution mass spectrometry being different. In order to make a correct comparison it would be necessary to determine what fractions of the total phosphate are measured in routine determinations and then to ensure that the added labeled internal standard is equilibrated only with these fractions. The method reported by Pickup may be regarded as a valuable approach to reference methodology but not a definitive method.

4 Methods for the Measurement of Carbohydrates and Lipids

4.1 Glucose

Much effort has been devoted to the highly accurate measurement of glucose in human serum. In an early report, Sweeley et al. (11) described a hepta-deuterated glucose as internal standard and the use of a hexa-trimethylsilyl ether derivative. However, the procedure was not designed as a reference method and the derivative was not entirely suitable for the reliable measurement of glucose. Further methods involving ID-MS have been reported on the metabolism and production of glucose in man (36, 37, 38).

Björkhem and co-workers reported a method for the accurate measurement of glucose using a penta-trimethylsilyl-methoxime derivative (39). 50 μl of human serum was lyophilised overnight together with a deuterated algal sugar mixture containing 300 nmol hepta-deuterated glucose. The crude residue was then reacted to form the methoxime and subsequently the trimethylsilyl derivative. Aliquots of the reaction mixture were subjected to combined GC-MS. A 2.5 m packed SE-30 column was used and SIM was carried out at *m/z* 319 and 323. Although the authors observed the formation of *syn* and *anti* isomers of the methoxime derivatives which were partially separated by the GC column, they contended that this would not influence the results of analyses. Furthermore it was assumed that all sugars present in serum samples could be separated from the glucose derivative under the conditions employed. The relative standard deviation of the method was shown to be 1.9%.

Eldjarn and co-workers (40), using the same derivative and the isotopically labeled standard as Björkhem's group, discussed the various factors limiting the precision of the isotope dilution measurement. In particular, they used a 50 m SE-30 column which was capable of separating the *syn* and *anti* isomers of the penta-trimethylsilyl methoxime derivative of glucose. By optimising many pre-instrumental and instrumental parameters they were able to determine the glucose concentration in a reference serum on several occasions over a period of 5 months with an imprecision of only 0.36% (CV). Unfortunately, no detailed protocol was reported giving the complete measurement procedure.

More recently, a further group of investigators (41) also used the methoxime trimethylsilyl derivative and the deuterium labeled internal standard for the isotope dilution measurement of glucose. The authors demonstrated that a 50 m SP 2100 capillary column was necessary to resolve glucose from other hexoses which may be present in human serum. When aliquots of the same sample were injected repeatedly, the coefficient of variation of peak area ratio measurements was at 0.65%. The imprecision increased to 1.2 – 1.4% when glucose was measured in serum samples on different occasions.

In a very comprehensive investigation a group from the National Bureau of Standards, Washington, presented two different ID-MS methods for the measurement of glucose in human serum (42). Both methods make use of ^{13}C-labeled glucose as internal standard. In the first procedure, glucose was converted into 1,2:5,6-di-*O*-isopropyliden-α-D-glucofuranose by a condensation reaction with acetone. This initial derivative formation was followed by a series of three thin-layer chromatographic steps in order to separate the glucose derivative from other possible reaction products and from other hexoses. The diacetone glucose was then subjected to combined GC-MS. For this procedure a 1.5 m column packed with OV-17 was used and SIM was carried out either with EI using the $[M-15]^+$ ion (*m/z* 245 for the non-labeled and 250 for labeled glucose) or with CI (isobutane) using the $[M+1]^+$ ions at *m/z* 261 and 266, respectively. The second method involved the formation of a cyclic 1,2:3,5-bis(butylboronate)-6-acetate. No additional wet chemistry steps of separation were applied for the isolation of the glucose derivative. A 100 m capillary column coated with SE-30 was used to achieve separation from other monosaccharides possibly present in the serum samples. For the detection of the bis(butylboronate)-6-acetate derivative the electon impact $[M-57]^+$ ions were chosen at *m/z* 297 and *m/z* 302 for the non-labeled and labeled substance, respectively. In addition, CI experiments using isobutane were carried out and the instrument was adjusted to the $[M + 1]^+$ ions at m/z 355 and 361. After considerable developmental work, the authors arrived at excellent precision for both methods ranging from 0.15 to 0.29% (CV). However, when the two methods' results were compared after measuring several serum pools, a bias in the range from 0.48 to 0.93% was observed. In view of the high precision of the procedures, this difference was statistically significant but the authors were unable to offer a satisfactory explanation for this bias. Possibly in one or both methods the isotopic forms were not adequately mixed at the start of the analytical procedure. Nevertheless the two methods, which agree within 1%, both fulfill the criteria of definitive methods. In view of the ultimate purpose of a definitive method, it must nonetheless be asked: Is it justifiable to spend so much effort to increase precision to the degree that differences due to rather small biases become visible and statistically significant but remain insignificant for medical, biological or clinical chemical considerations?

4.2 Cholesterol

Numerous methods for the measurement of total cholesterol (sum of esterified and free), based on the principle of ID-MS, have been reported over the last ten years. The first paper in this series was published by Björkhem's group at the Karolinska Institute (43). The authors synthesized a deuterated cholesterol which contained mainly the tri- and tetra-deuterated compound. This substance was added to a fixed amount of serum. Cholesterol esters were then hydrolyzed by ethanolic potassium hydroxide and cholesterol extracted with hexane. Standards of purified commercial cholesterol were handled in a similar manner. Samples and standards were subjected to GC-MS (4 m SE-30 column) and the molecular ions of the non-labeled and the labeled cholesterol were recorded at m/z 386 and 389 respectively. The coefficient of variation of the method was 1.3%. In Björkhem's early investigation, the photometric Liebermann-Burchard method for the determination of cholesterol in human serum was compared with the isotope dilution procedure. Surprisingly, the two methods were observed to coincide in a limited range of concentrations, although the photometric method is known to exhibit a considerable bias. This may be due to the fact that the Liebermann-Burchard method was calibrated by a serum standard which had been adjusted using the mass spectrometric method. The ID-MS assay described by Björkhem has been applied to evaluate the accuracy of several routine methods for measuring total cholesterol in human serum (44, 45, 46, 47).

A further investigation of the mass spectrometric measurement of cholesterol utilized [4-^{14}C]cholesterol oleate as internal standard (48). Non-labeled and labeled cholesterol esters were hydrolyzed and the resultant free cholesterol was converted to the trimethylsilyl derivative. Selected ion recording was then carried out at m/z 458 and 460 using a packed SE-30 column (1.5 m). Very small amounts of serum (0.5 µl) were used. In order to correct the uncertainty inherent in sampling small volumes, the same syringe equipped with a fixed repeating adapter was used for measuring serum samples and standard solutions. In spite of these precautions a comparatively high degree of imprecision, between 1.3 and 2.8% (CV), was observed, depending on the range of concentration. The procedure was used as a reference method to judge the accuracy of several procedures for cholesterol determination including the Liebermann-Burchard reaction and gas-liquid chromatography.

More recently Gambert and co-workers reported a similar method which was based on the use of [3,4-$^{13}C_2$]cholesterol as internal standard and a 25 m capillary column coated with SE-30 (49). The imprecision measured was 0.8% (CV).

In 1980 Wolthers and co-workers developed a mass spectrometric procedure for the determination of total cholesterol by using hepta-deuterated standard material (50), with the deuterium atoms in the side chain of the molecule. The authors were disappointed by the low reproducibility, 3.1%, of their method and concluded that "at present, mass fragmentographic methods do not fulfill the criteria required for absolute methods". Obviously, they used inadequate mass spectrometric equipment, and no attempts were made to improve the precision of earlier investigations (43, 48, 49), e. g. by employing a revised calibration procedure or by using more precise techniques for sampling sera and standard solutions.

In contrast, Freudenthal's group, using the same analytical principle of spiking the sera with the hepta-deuterated cholesterol and recording the molecular ions of the trimethylsilyl ether derivatives, was able to improve several steps in the analytical procedure (51). In particular they changed the analogue mode of the electron multiplier of the mass spectrometer to a pulse amplifier to enable ion counting. Instrumental sources for the imprecision of the isotope ratio measurement were significantly reduced. The authors noted that theoretical precision as determined by ion statistics was about 0.18%. In the pre-instrumental part of analysis, they used a weighing technique for aliquoting serum samples and standard solutions rather than volumetric measurements. For each serum sample, two standards were prepared containing approximately 97% and 103% of the cholesterol in the sample to be analysed. With these improvements the authors recorded an overall imprecision of 0.5%, which is satisfactory for the application of the procedure as a definitive or reference method.

More recently a detailed protocol for the highly accurate measurement of total cholesterol in human serum was published (23). In this method, serum samples were mixed with hepta-deuterated cholesterol and the cholesterol esters hydrolyzed with potassium hydroxide prior to extraction with dichloromethane. The cholesterol in standards and samples was converted to its trimethylsilyl ether derivative for combined GC-MS on packed OV-101 and OV-17 columns (1.25 m) with both EI and CI. The instrument was focused for m/z 458 and 465, corresponding to the molecular ions of the non-labeled and labeled cholesterol derivative. In principle, the authors applied the same labeled internal standard, the same steps of hydrolysis and derivative formation as described in previous reports and used similar technical equipment. Considerable effort was spent tracing any possible source of error which could affect the accuracy and precision of cholesterol measurements.

Difficulties in the isotope dilution analyses may arise when "total" cholesterol is determined. Cholesterol is present in serum as a mixture of various esters of cholesterol in addition to a small fraction of the non-esterified steroid, whereas isotopically labeled material and non-labeled primary standard cholesterol consist of the pure non-esterified substances. Possible sources of inaccuracy could arise from an incomplete hydrolysis of the various cholesterol esters or from partial destruction of the cholesterol molecule during hydrolysis. The authors were able to demonstrate that under the conditions employed, the recovery of cholesterol after hydrolysis was 100%. Furthermore, no evidence of decomposition of cholesterol during hydrolysis was found. Thus it seems justified to apply the isotope dilution principle using isotopically labeled cholesterol to the measurement of "total" cholesterol. In addition to the standard procedure of recording molecular ions and using EI-MS, they applied SIM to fragment ion peaks and CI-MS to ascertain the accuracy of their method. To optimise the calibration procedure, two standard mixtures were prepared, one of which showed a few % lower and the other a few % higher isotope ratio compared with the ratio measured for a serum sample. The two standards were injected immediately before and after the analysis of the serum sample. Following this bracketing technique and using a weighing procedure for aliquoting the serum samples rather than measuring volumes, a precision of 0.36% was achieved. In view of the excellent precision and the high degree of accuracy, which were confirmed by numerous additional experiments, there can be no doubt that this procedure fulfills the criteria of a definitive method.

Recently, a group of investigators from the Karolinska Institute and the National Bureau of Standards compared the results of their separate isotope dilution methods by applying them to a common set of serum pools (52). In the first trial a small but significant difference between the two methods was observed. A search for the source of this systematic error led to the detection of a small amount of lathosterol in the primary standard used by the Karolinska Institute. In a second trial the Swedish group revised their analytical protocol; instead of tetra-deuterated cholesterol, they used a heptadeuterated substance as internal standard, introduced a trimethylsilyl ether derivative and used the standard reference cholesterol from the National Bureau of Standards. The coefficient of variation, which was about 1.3% in the original method of Björkhem (43), was then calculated to be between 0.5 and 1.5%, an improvement in precision. When the same primary standard cholesterol was used for calibration, the average difference between the values obtained from the two laboratories was reduced to 0.2%.

4.3 Glycerol and triglycerides

An isotope dilution method for the measurement of triglycerides in human serum using SIM was reported by Björkhem in 1976 (53). A mixture of [1,1,2,3,3-$^{2}H_{5}$]glycerol tripalmitate and [1,1,2,3,3-$^{2}H_{5}$]glycerol trioleate in a molar ratio similar to that observed in human triglycerides was added to serum samples. The deuterated and non-labeled triglycerides were extracted with a mixture of chloroform/methanol and the extract was subjected to thin-layer chromatography. Then the glycerol esters were cleaved by acid hydrolysis and the glycerol obtained was converted to the tri-trimethylsilyl ether derivative. SIM was performed on a 3 m packed QF-1 column and the labeled and non-labeled glycerol were recorded at m/z 222 and 218, respectively. The coefficient of variation of the method was 2.4%.

The authors pointed out that serious problems with the analysis of low concentrations of triglycerides may arise from small amounts of glycerol which contaminate organic solvents and reagents. The procedure allowed only glycerol originating from triglycerides to be measured; non-esterified glycerol was not included. In contrast, most of the routine methods are designed to measure "total" glycerol comprising the triglycerides and free glycerol; the phosphate esters are not included since they are not hydrolyzed to glycerol by most of the commonly used lipase enzyme preparations or by ethanolic potassium hydroxide solution.

The isotope dilution procedure as described by Björkhem was used as a reference method to check the accuracy of a fully enzymatic routine method for the measurement of triglycerides (53). In 1978 Eldjarn et al. discussed various factors which influence the accuracy and precision of the isotope dilution measurement of glycerol in human serum (40). They found that the reproducibility of isotope ratio measurements by SIM may be dramatically affected by the choice of isotopes and by ion source conditions. However, a detailed description of the analytical procedure for the measurement of glycerol, and data on the evaluation of accuracy and precision of the method, were not given. More recently a method was published in which [1,3-$^{13}C_2$]glycerol was used as internal standard and tripalmitin from the National Bureau of Standards as certified reference material (54). The triglycerides were hydrolyzed by potassium hydroxide and the resulting hydrolysate contained the fraction originating from the tri-esters as well as the non-esterified glycerol. After formation of the heptafluorobutyric ester derivative, SIM was carried out at m/z 467 and 469. The coefficient of variation was in the range from 0.5 to 1.3% and the procedure was used for the assessment of target values in an external quality control scheme.

5 Methods for the Measurement of Nitrogenous Compounds

5.1 Urea

The quantitative measurement of urea in human serum by ID-MS was reported by Björkhem and co-workers in 1976 (55). A fixed amount of [$^{15}N_2$]urea was added to serum samples, the mixture lyophilised and the labeled and non-labeled urea converted into 5,5-diallyl barbituric acid. The derivative was transferred into organic solvent by ion-pair extraction and then subjected to combined GC-MS. A packed SE-30 column (2.5 m) was used for chromatography, and molecular ions of the labeled and non-labeled derivatives at m/z 238 and 236, respectively, were chosen for SIM. The use of a derivative with a relatively high molecular weight as compared to the free compound was of advantage, since the probability of interference from accompanying substances decreases with the increasing mass of the recorded ion. A drawback of this method was the possible interference from therapeutic use of 5,5-diallyl barbituric acid. Furthermore the yield in derivative formation varied considerably from one series of samples to the next although the authors asserted that this would not affect isotope ratio measurements by mass spectrometry. On the other hand the relatively poor reproducibility of the method (3.6%, CV) may be partially due to the difficulties arising from the formation of the barbituric acid derivative. The procedure was applied to evaluate the accuracy of various routine methods for measuring urea in human serum (44, 45, 46).

Two further reports (56, 57) on ID-MS methods for the determination of [^{15}N]urea in human plasma and in urine have been made. The procedures were developed mainly for measuring ^{15}N-enrichment in human metabolic studies and may therefore be less suitable as reference methods in clinical chemistry.

More recently the National Bureau of Standards' group reported a procedure for measuring serum urea by ID-MS and proposed using it as a definitive method (58). The technique utilized [^{18}O]urea as labeled internal standard and involved conversion to 6-methyl-2,4-bis[(trimethylsilyl)oxy]pyrimidine, capillary GC, and isotope ratio measurement of the $[M-15]^+$ ions. The coefficient of variation was 0.19% for 5 serum pools.

5.2 Creatinine

The concentration of creatinine in human serum is one of the most frequently determined clinical chemical parameters. In common laboratory practice, significant differences are observed when comparing the results of various routine methods applied to the same serum sample. Until recently, no reliable criteria were available to decide which method would yield the more accurate results. To improve this situation, reference and definitive method technology was developed to measure creatinine in serum (59). [$^{15}N_2$]creatinine was synthesized from [$^{15}N_2$]thiourea and a fixed amount of this labeled standard added to a measured volume of serum. The mixture was lyophilised and ethanol was added to the dry residue. After centrifugation, the ethanolic solution was concentrated and submitted to HPLC on a silic acid column. Creatinine was eluted by applying a methanol/ethanol gradient and the appropriate fraction, detected by monitoring the UV-absorbance at 226 nm. This was followed by derivatisation involving formation of the (2-hydoxy-,2-methyl)ethyl derivative and the introduction of two trifluoroacetic groups. This product proved to be the only suitable derivative for GC-MS. Finally, SIM was carried out at m/z 294 and m/z 296, using a 1 m SE-30 column for gas chromatography. A divergence from linearity was observed in the calibration curve due to the contribution of non-labeled creatinine to the signal from the labeled internal standard material. However, in the reported method (59) uncorrected standard curves were used and a coefficient of variation of 3% noted. In the investigation the isotope dilution technique was applied to the evaluation of routine methods based on the Jaffé reaction for measuring creatinine in serum and urine (59). In a more recent paper the comparison was extended to a fully enzymatic method (60).

A different approach to an accurate ID-MS measurement of creatinine in human serum was designed to define creatinine target values in control sera for the quality control scheme of the Deutsche Gesellschaft für Klinische Chemie (61). A labeled creatinine containing two ^{15}N- and one ^{13}C-atoms was synthesized from [$^{15}N_2$,^{13}C]cyanamide and a fixed amount of this product was used as internal standard and added to serum samples. The labeled and non-labeled creatinine were then isolated from the biological material by the use of cation-exchange columns. After addition of the serum samples the columns were washed with distilled water to remove proteins and anionic materials, the creatinine was eluted with an ammonia solution and the solvent evaporated. The dry residue of the eluate was then reacted to form the

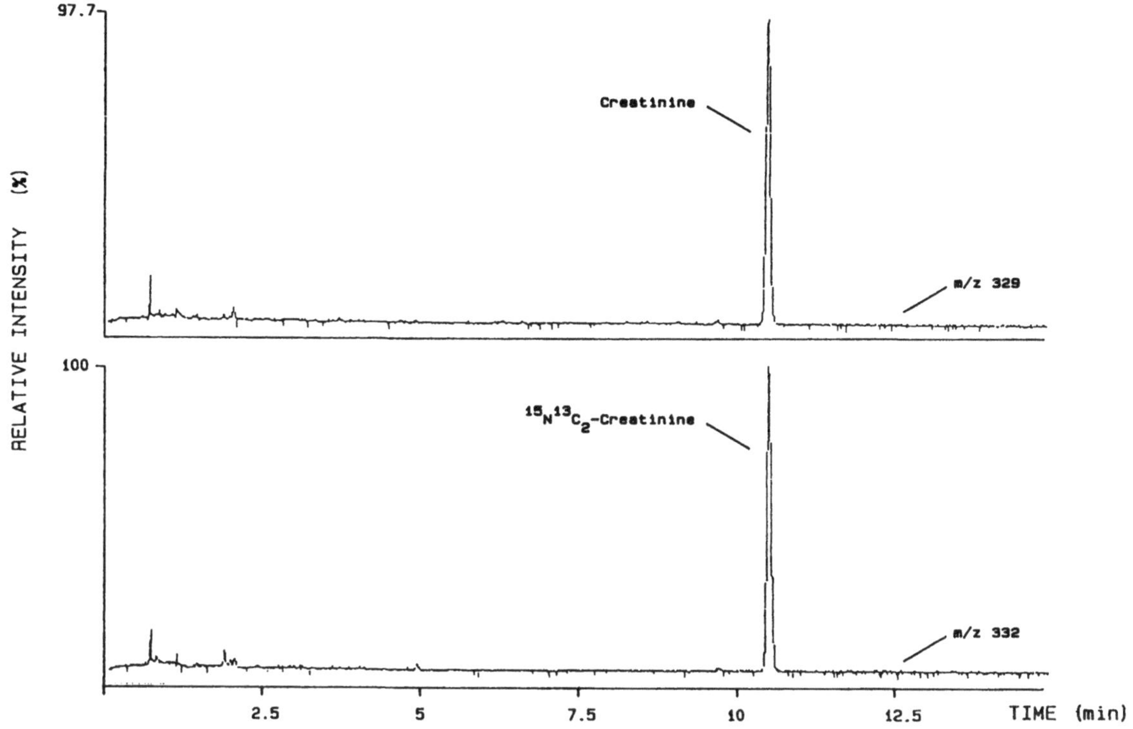

Figure 3. Selected ion monitoring of the trimethylsilyl derivative of creatinine (upper panel) and of [^{13}C,$^{15}N_2$]creatinine (lower panel) after processing a serum sample. A 30 m SE-52 capillary column was used. For experimental details see ref. 57.

trimethylsilyl derivative, previously described by Lawson (63) as suitable for combined GC-MS. The reaction product was separated on a 30 m fused silica capillary column coated with SE-52 coupled to a quadrupole filter mass spectrometer. The molecular ions of the creatinine and labeled creatinine derivatives were recorded at *m/z* 329 and 332, respectively. A certified National Bureau of Standards reference material was used as the primary creatinine standard. In order to increase precision an automated sampler was used whenever possible and all volumetric equipment necessary for measuring sera and standards was carefully calibrated by weighing appropriate amounts of water at a defined temperature. The low contribution of the naturally-occurring isotopes to the signal of the labeled internal standard resulted in an almost linear function between the measured isotope ratio and the amount of creatinine in standards and sera. Furthermore the synthesized labeled creatinine contained less than 1% of the non-labeled compound. Although these mutual isotope contributions were small, they were included in the calculation procedure to enable the use of a non-linear bracketing technique for the calibration. The combination of SIM and high-performance capillary GC columns gave the overall method its high specificity. Figure 3 shows a SIM recording taken after processing a serum sample. For complete series of measurements of different lyophilised control sera on different occasions the imprecision of the method was calculated to be in a range between 0.32 and 1.04% (CV).

Imprecision of around 1% may have been partially caused by an inhomogeneity in the serum pools (vial-to-vial variability).

More recently the group from the National Bureau of Standards proposed, as a definitive method, a procedure which was based on the conversion of creatinine to the ethyl ester of *N*-(4,6-dimethyl-2-pyrimidinyl)-*N*-methyl (62). $^{13}C_2$-labeled creatinine was used as internal standard and capillary GC combined with EI-MS was applied for the isotope ratio measurements. The coefficient of variation was in the range from 0.15 to 0.27%.

5.3 Uric acid

A combination of LC and ID-MS was used for the highly accurate measurement or uric acid in human serum by Lim and co-workers (64). The procedure involved the addition of a known amount of [1,3,9-$^{15}N_3$]uric acid to a serum sample. After equilibration, the labeled and the non-labeled uric acid were purified by chromatography on an anion-exchange column and converted to tetraethyl derivatives by reaction with diethyl sulfate in alkaline solution. Six isomers were formed during derivative formation. These were separated by LC using a reversed-phase column. Two major tetraethyl isomers were isolated for further analysis by GC-MS using a packed OV-1 column, and the molecular ions of the ethyl derivatives of the labeled and non-labeled uric acid at *m/z* 283 and *m/z* 280, respectively. When the method was used to determine uric acid in a tentative reference serum, a high precision (0.28%, CV) was observed on replicate measurements.

A method has been described to avoid the multiple products of derivative formation by introdicing uric acid in its free state into the mass spectrometer by direct inlet (65). Uric acid was isolated by anion exchange chromatography and crystallized from the acidic eluate. The imprecision proved to be 0.4 to 0.5% (CV).

In 1979 Öhman reported an isotope dilution GC-MS method for the determination of serum uric acid for the evaluation of various routine methods (66). [1,3-$^{15}N_2$]uric acid was used as the internal standard and ion-exchange chromatography for isolation of the analyte from serum.

A tetra-trimethylsilyl derivative, previously employed by Walker et al. (67) for measuring the body pool size of uric acid, was formed which was suitable for GC-MS (1 m SE-30 packed column). The molecular ions of the labeled and non-labeled substance were recorded by SIM at *m/z* 458 and 456. The relative standard deviation was reported to be 0.6% as calculated from duplicate analyses of serum samples (within batch imprecision). On the basis of the procedures described by Walker (67) and by Öhman (66) an isotope dilution method was developed (68) which involved the use of a certified reference material (National Bureau of Standards, Washington) as primary uric acid standard and of a 25 m fused silica capillary column (SE-30) for GC-MS. As with the previous methods, a $^{15}N_2$-labeled uric acid preparation was used as internal standard and the tetra-trimethylsilyl derivatives of the labeled and non-labeled compounds were recorded at *m/z* 458 and 456, respectively. The volumetric equipment for sampling standards and samples was calibrated by weighing appropriate amounts of water. The analyses were carried out following a detailed protocol and the results were calculated by the use of a non-linear bracketing technique. The day-to-day imprecision of the method was in the range from 0.6 to 1.2% (CV). The method, which is proposed as a definitive method in clinical chemistry, is used now by the Deutsche Gesellschaft für Klinische Chemie for measuring uric acid in control sera for external quality control.

6 Methods for the Measurement of Hormones

The measurement of hormonal substances is one of the most difficult tasks in the field of clinical chemistry due to their lower concentration in human body fluids than other clinical chemical analytes. Highly sensitive procedures are necessaary for their detection and quantification, particularly in blood. While numerous methods have been described, many have never achieved an adequate level of reliability and, consequently, the reference ranges for many hormones change when new and more reliable methods are introduced. The high sensitivity of radioimmunoassay has made it the method of choice for measuring hormonal substances although in recent years the radioactive label has been partially replaced by enzyme, fluorescent or luminescent labels. Nevertheless the results of several collaborative surveys in recent years indicate that the results of measuring hormone concentrations in the same serum sample vary considerably from one laboratory to the other. This is due in large part to the lack of specificity of various antibodies used in commercial test kits for immunoasssay and to the lack of reliable criteria to determine whether a hormone concentration measured by an immunoassay kit is accurate or not. This has stimulated the development of technology to provide a basis for reference or definitive values which can then represent the best approximation of the "true" value. The only adequate technique currently available for the development of definitive or reference methods in the field of clinical chemistry is ID-MS.

Greatest effort has been devoted to the accurate measurement of steroid hormones and more recently of the thyroid hormone, thyroxine. In the following sections several reference or definitive methods are described in addition to procedures that provide the basis for developing reference method technology.

6.1 Cortisol

The adrenal steroid, cortisol, is one of the most commonly determined hormones in human blood. In 1974 two groups of investigators independently demonstrated that cortisol could be quantified by ID-MS (69, 29). At the Karolinska Institute a method was developed (69) which made use of [4-^{14}C]cortisol as internal standard material. This radioactive steroid, which is commercially available, was added to serum, and the labeled and non-labeled cortisol were extracted by dichloromethane; this extract was reacted first with methoxyamine hydrochloride and secondly with trimethylsilylimidazole to form the di-methoxime-tris(trimethylsilyl) derivative. A 2.5 m SE-30 packed gas chromatography column was used for combined GC-MS and the $[M-31]^+$ fragment ions of the labeled and the non-labeled steroid derivative at m/z 607 and 605 were chosen for SIM. Due to isotopic interference arising from the small difference between the two masses, a non-linear calibration curve resulted. The imprecision of the method as calculated from duplicate analyses was 3.5%.

The other study, which appeared almost simultaneously (29), described the use of [1,2-$^3H_2$]cortisol as internal standard. The methoxime-trimethylsilyl derivative and the conditions for SIM were the same as in the Swedish study. The use of di-tritium labeled cortisol had the advantage of increasing the mass difference (4 daltons) between the fragment ions of sample and standard and produced an almost linear calibration curve. In order to avoid the considerable amounts of radioactivity necessary for each analysis a revised method was established (70) which used ^{14}C-labeled cortisol as an internal standard. After extraction a further a step of purification on Sephadex LH-20 was introduced prior to formation of the methoxime trimethylsilyl derivative, the latter reaction being carried out by addition of trimethylbromosilane to the reagent mixture. To correct the mutual interference of non-labeled and labeled cortisol isotopes a non-linear calibration function was established based on principles described in Section 2.4 (22, 70).

More recently a procedure for the measurement of cortisol in human plasma was reported (71) which made use of the same derivative (di-methoxime-tris(trimethylsilyl)ether) as previous methods. However, the radioactively-labeled internal standard was replaced by [9,12,12-$^2H_3$]cortisol and CI was used instead of EI. The between-run precision of the method (CV 4.6%) was relatively poor. Similarly poor precision was observed from an isotope dilution technique involving high-resolution mass spectrometry (72). The authors investigated different procedures of extraction and purification of the steroid prior to mass spectrometric determination. The imprecision was in the range of 3 to 4% (interassay CV). As discussed in a more recent paper by Gaskell and colleagues (73), the imprecision of ratio measurements is caused mainly by fluctuations of the ion beams and by drift from the peak centres. These effects are proportionately greater in high-resolution mass spectrometry and, in the example quoted, at a resolution of 8000, a coefficient of variation of 2.8% was observed which decreased significantly to 1.6% when the resoltuion was reduced to 1000.

The ID-MS procedures which claim to be definitive methods for the determination of cortisol in human serum (24, 74, 75) each have in common the use a) of a certified reference material (from the National Bureau of Standards) as primary cortisol standard, b) of low-resolution mass spectrometry and c) of the di-methoxime-tris(trimethylsilyl) derivative. The first of these procedures (24) was based on previous results obtained with the SIM technique for the measurement of cortisol (29, 70). ^{14}C-labeled cortisol was used as internal standard and Sephadex LH-20 chromatography employed for purification prior to GC-MS. Since GC columns of high separation power tend to separate the *syn* and *anti* isomers of the methoxime derivative, relatively short capillary (6 m SE-30) or packed columns (1.3 m Dexsil 300) were used. Two pairs of ions at m/z 605 and 607 ($[M-31]^+$) and at 636 and 638 (M^+) were chosen for ion recording. The results calculated independently from each set of ions agreed closely. A non-linear bracketing technique was applied and imprecision was in the range 0.7 to 2.3% (between-assay CV) depending on the serum pool analysed. It was assumed that the imprecision was caused partially by heterogeneity of the lyophilised serum pools used in this investigation.

The second procedure proposed as a definitive method (74) also used ^{14}C-labeled cortisol as internal standard. Serum samples were measured by weighing rather than by volumetric aliquoting and a bracketing technique was applied in the calibration procedure. However, the divergence due to non-linearity of the calibration function was not taken into consideration. The coefficient of variation was reported to be 0.38% for the measurement of a liquid serum pool.

In the third study on the definitive measurement of cortisol in serum (75), [9,12,12-$^{2}H_{3}$]cortisol was utilized as the labeled compound. In addition to a standard procedure which involved extracting labeled and non-labeled cortisol, forming derivatives, and measuring the isotope ratio at *m/z* 608 and 605 by EI-MS, several additional steps of purification including LC or a C_8 mini column pre-purification technique were tested to demonstrate specificity of the standard procedure. In addition, results obtained with negative ionization and with an alternative pair of ions (*m/z* 636 and 639) were compared with those of the standard procedure to establish specificity and accuracy. The imprecision of the method was 0.69% (inter-assay CV).

The ID-MS techniques discussed here have been used for the evaluation of numerous routine methods for measuring cortisol (44, 46, 76), for calibrating radioimmunoassays (77) and for determining target values in external quality control programs (72, 78).

6.2 Aldosterone

There are several difficulties in developing methods for determining aldosterone in human serum. It has a concentration lower than most other hormones and requires extensive purification in order to avoid interference from other compounds of biological origin. Furthermore, the chemical structure of the molecule makes it labile, and only a few methods for the formation of volatile derivatives have been reported. On the basis of earlier investigations (30, 79) a method for the measurement of aldosterone in human serum has been developed which comprised the following steps (70): (1) addition of 1 ng ^{14}C-labeled aldosterone to 8 ml serum; (2) extraction of the steroid with dichloromethane; (3) isolation of aldosterone by column chromatography on Sephadex LH-20; (4) formation of the 18,21-acetal of aldosterone by the use of hydrochloric acid; (5) column chromatography of the derivative on Lipidex 5000; (6) formation of the 3-enol-heptafluorobutyric ester of the 18,21-acetal of aldosterone; (7) SIM at *m/z* 538 and 540 using a 2 m packed SE-30 column. The imprecision of the method was 3.5% (CV) and is presently in use for establishing target values in external quality control for hormone determinations (78).

6.3 Testosterone

One of the earliest studies of ID-MS determination of testosterone in human plasma (80) used ^{14}C-labeled testosterone as internal standard and the labeled and non-labeled steroids were purified by thin-layer chromatography on silica gel. The di-heptafluorobutyric ester derivative of testosterone was prepared for GC-MS and the molecular ions of the labeled and non-labeled testoterone were recorded at *m/z* 682 and 680, respectively.

Chapman and Bailey (81) described a method which involved the use of deuterated testosterone, the formation of a bromomethyldimethylsilyl derivative and SIM at *m/z* 438 and 442. The precision of the method (7.2% CV) was relatively poor.

Björkhem and co-workers (82) utilized ^{14}C-labeled testosterone as internal standard, purification by thin-layer chromatography and SIM of the molecular ions of the di-trimethylsilyl derivative at *m/z* 432 and 434. The authors were able to improve precision to a relative standard deviation of 2.7%.

Baba and co-workers (83) decribed a method which was based on the use of [19, 19,19-$^2H_3$]testosterone as labeled compound and the formation of the di-trifluoroacetic ester derivative. Serum samples were purified by extraction and thin-layer chromatography. SIM was carried out at *m/z* 480 and 483 using a 2 m OV-1 column. The day-to-day imprecision was 3.15% (CV).

Recently, a further investigation of the quantification of testosterone in plasma from men was published (84) using [16,16,17-$^2H_3$]testosterone as labeled standard and the *t*-butyldimethylsilyl ether oxime or methoxime as derivative. High-resolution mass spectrometry was employed for the isotope-ratio measurement. The selectivity of the analytical procedure was further tested by applying various methods of extract fractionation, using Lipidex 5000, Sephadex LH-20 and Sulfoethyl-Sephadex LH-20 chromatography. As an alternative mass spectrometric approach, the testosterone derivatives were detected by metastable peak monitoring. Unfortunately, imprecision proved to be high (3.1 – 8.7%, inter-assay CV), which was mainly attributed to detecting the analyte and the internal standard at different times. The analytical procedure was applied to the evaluation of results obtained from an external quality-assessment scheme.

On the basis of previous investigations (29, 85, 79) an isotope dilution method was elaborated to utilize ^{14}C-labeled testosterone as internal standard, purification of testosterone by Sephadex LH-20 chromatography and formation of the di-heptafluorobutyric ester derivative (70). A 25 m SE-30 capillary column was used for GC, and the molecular ions at *m/z* 680 nd 682 proved to be suitable for SIM of the labeled and non-labeled testosterone derivatives, respectively. The inter-assay coefficient of variation was calculated at 1%. The procedure is presently employed for measuring target values in an external quality control scheme (78).

In a recent report (86), [3,4-$^{13}C_2$] testosterone was used as internal standard and the *t*-butyldimethylsilyl-methoxime was compared with the di-heptafluorobutyric ester derivative with capillary column GC in combination with EI-MS. The precision of this method calculated from duplicate measurements of 35 plasma samples was relatively poor (2.8%).

6.4 Oestradiol-17β

One of the first group of compounds measured by ID-MS was the phenolic steroid hormones, oestradiol-17β, oestrone and oestriol. These were determined in the plasma of pregnant women (15) using 6,7-$^2H_2$-labeled oestrogens as internal standards. For clinical endocrinology, oestradiol-17β has proved to be one of the most important phenolic steroid hormones. In addition to immunoassays, several mass spectrometric methods were reported for measuring oestradiol-17β, some of which may provide a basis for the development of definitive or reference methods.

Wilson and co-workers measured oestradiol-17β after adding a fixed amount of 2,4-di-deuterated oestradiol-17β to the serum samples (87). The ether extract of the steroids was purified by thin-layer chromatography and the bis(trimethylsilyl) ether formed as a suitable derivative for SIM. The coefficient of variation of the method was relatively poor (about 11%).

An ID-MS method for determining oestradiol-17β in human serum was described in 1978 and involved the use of [4-^{14}C]oestradiol-17β as labeled standard (88) and the preparation of di-heptafluorobutyric esters for analysis by combined GC-MS. The results were calculated by a non-linear calibration function, taking into consideration the mutual isotope contribution. The precision was reported to be in the range between 1.5 and 2% (CV).

A further investigation for measuring serum oestradiol-17β was described by Zamecnik and colleagues (89). The authors added a five-hundredfold excess of about 50 ng [2,4,16,16-$^2H_4$]oestradiol-17β to serum samples. The labeled compound simultaneously served as internal standard and carrier substance to protect the small amounts of the phenolic steroid present in the samples. Because the labeled internal standard contained about 0.5% non-labeled oestradiol-17β, in practice about the same amount of non-labeled steroid was added to the samples as was present in the serum. This must have adversely affected precision but no data on the reproducibility for measuring serum samples was given.

Knuppen and co-workers described methods for synthesizing deuterated oestrogens (90) and for measuring low concentrations of oestradiol-17β (91), a procedure based on the use of [2,4,6,6,7,16,16,17-$^2H_8$]oestradiol-17β as internal standard containing only 0.15% of the non-labeled compound. The di-heptafluorobutyric ester was used as volatile derivative for the GC-MS. Small amounts of oestradiol-17β (14 $pg \cdot ml^{-1}$) could be measured by the technique, but due to the low range, an imprecision of about 20% (CV) resulted. For higher concentrations of the phenolic steroid (212 pg/ml), a coefficient of variation of about 7% was calculated. Another group of workers developed methods for the measurement of oestradiol-17β in saliva (92) and plasma (93) using 16,16,17-$^2H_3$-labeled steroid as internal standard. For the analysis of the saliva samples (92), a micro-scale anion-exchange chromatography column was employed for isolating the phenolic steroids. A more recent procedure for measuring oestradiol-17β in plasma has been published (93) which describes an interesting alternative approach using a solid-phase-coupled antiserum for the extraction and fractionation of the steroid. The steroids were converted into the trimethylsilyl ether derivatives and high-resolution SIM was performed at m/z 416.2566 and 419.2755 using a 3 m OV-1 column for GC. The coefficient of variation in a concentration range from about 27 to 300 $pg \cdot ml^{-1}$ was between 4.4 and 10.4%.

The trimethylsilyl derivative was also used by Lawson and co-workers (65). They added [2,4,16,17-$^2H_4$]oestradiol-17β to the serum samples. The steroids were extracted with ethyl ether and then purified on columns of Lipidex 5000 and Sephadex LH-20. The trimethylsilyl ether derivative was analysed by SIM at m/z 416 and 420 using a conventional, packed OV-1 column or a 25 m capillary column coated with OV-1. The coefficient of variation (within batch) was reported to be 2.8%. In 1981 Iwai and colleagues (94) described a method for measuring oestrone, oestradiol-17β and oestriol during pregnancy. They used 2,4- or 9,11-deuterated oestrogens as internal standards and the trifluoroacetic ester derivatives for SIM. The imprecision was at about 3.5% (CV).

Based on previous experience (88, 70), a detailed procedure was elaborated and proposed as a definitive method (95) for the measurement of oestradiol-17β in human serum. This made use of ^{14}C-labeled oestradiol-17β as internal standard. After extraction with dichloromethane, the phenolic steroids were purified by column chromatography on Sephadex LH-20 followed by formation of the di-heptafluorobutyric ester derivative. For SIM a 25 m capillary column coated with SE-30 or with SE-52 was used and the mass spectrometer adjusted to m/z 664 and 666. By introducing two heptafluorobutyric ester groups, the molecular weight of the phenolic steroid was increased from 272 to 664. At this high m/z value there is little if any interference to be expected from accompanying biological material or from the normal GC column bleed. The method is in use for determining target values in collaborative surveys for the measurement of oestradiol-17β (78).

6.5 Oestriol

After an early study dealing with the mass spectrometric measurement of the classical oestrogens in the plasma of pregnant women (15), few additional papers were published reporting methods measuring oestriol in plasma or urine. In 1974 Adlercreutz and coworkers (96) tested several modifications for isolating oestriol and the conjugated steroid from plasma samples. They used deuterated reagents for the silylation of the phenolic steroids *prior* to SIM, a method which in itself does not allow detection of procedural losses during the pre-instrumental work-up. It is probably due to this that the imprecision of the method was relatively high (7.4%, CV).

Björkhem and colleagues reported a procedure for the measurement of total (conjugated and non-conjugated) oestriol in urine during pregnancy (97). [2,4-$^2H_2$]oestriol was used as internal standard. Since part of the deuterium label was lost under acidic conditions, the internal standard could be added only after hydrolysis with sulfuric acid. The non-labeled and the labeled phenolic steroid were extracted from the hydrolysate and converted in to the trimethylsilyl ether derivative. SIM was carried out at m/z 504 and 506, using a 2 m SE-30 column. The imprecision was reported to be 6%. Based on previous experiments (88) a procedure for the measurement of non-conjugated oestriol in human plasma was elaborated (70). It used [4-^{14}C]oestriol as internal standard. After addition of the labeled compound to serum samples the steroids were extracted with dichloromethane and then subjected to column chromatography on Sephadex LH-20. The oestriol fraction was reacted with pentafluoropropionic anhydride and the tri-ester derivative was chromatographed on a 25 m SE-30 capillary column coupled to the mass spectrometer. For SIM the instrument was adjusted to the m/z values 726 and 728, corresponding to the molecular ions of the non-labeled and the isotopically labeled substance, respectively. The inter-assay precision of the method was calculated to be 1.5% (CV).

6.6 Progesterone

Until recently, only very few papers had appeared dealing with ID-MS of progesterone. A detailed procedure for the measurement of serum progesterone was described by Björkhem's group (98). ^{14}C-labeled progesterone served as internal standard and was added to the serum samples. The steroids were extracted with hexane and then purified by thin-layer chromatography. The 3-enol-heptafluorobutyric ester proved to be a valuable derivative for SIM. A 2 m SE-30 column was used and the molecular ions of the derivatives of the isotopomers were recorded at m/z 510 and 512. The coefficient of variation was reported to be 3.7%.

The same isotope label and the heptafluorobutyric ester derivative was also used by Schramm and co-workers (99). For the measurement of progesterone in urine, Johnson et al. (100) described a method which made use of [7,7-$^{2}H_2$]progesterone as isotopically-labeled compound and of pentafluoropropionic ester as derivative. The imprecision of the method was higher than 5% (CV).

On the basis of Björkhem's investigation (98), a similar procedure was reported (70) using ^{14}C-labeled progesterone as standard. Column chromatography on Lipidex 5000 for purification and the heptafluorobutyric ester derivative for SIM were employed. In order to improve specificity, a glass capillary column coated with SE-30 was applied in combination with the mass spectrometer. The imprecision was found to be 2% (CV).

Progesterone was also determined in plasma and serum pools by ID-MS to provide target data for use in a U.K. national external quality assessment scheme (101). Microcellulose-coupled antiserum was used for the extraction of the steroid from the biological matrix. [7,7-$^{2}H_2$]progesterone served as internal standard, and the steroids were converted to the 3-enol-heptafluorobutyrates prior to selected ion recording at a resolution of 5000. The precision was in the range from 1.5 to 5.4% (CV).

Recently LC combined with mass spectrometry (LC-MS) and [3,4-$^{13}C_2$]progesterone as a labeled standard were used to measure progesterone in serum samples (102). The moving belt in the LC-MS interface needed to be deactivated carefully prior to each series of analysis. The precision of this rapid technique was between 2.5 and 6.4%. However, the baseline noise observed in LC-MS was much higher than for GC-MS.

6.7 Thyroxine

Following a preliminary report on the measurement of thyroxine in breast milk (103), the Björkhem group described on ID-MS method for the determination of this thyroid hormone in human serum (104). $^{2}H_2$-labeled thyroxine was used as the internal standard. The thyroid hormones were isolated by an extraction-solvent-distribution method and converted into the *N,O*-bis(trifluoroacetyl)methyl ester of the amino acid. SIM was achieved by the use of a short (0.45 m) GC column packed with an OV-1-coated support. The $[M-184]^+$ fragment ions of the non-labeled and the labeled thyroxine derivative were recorded at m/z 799 and 801, respectively. The interassay-variation was in the range of 1.8 to 3.4% (CV). The procedure has been used as a reference method for the evaluation of a radioimmunometric assay of thyroxine in serum.

In 1984 a further method for the measurement of thyroxine in serum was described by Ramsden and Farmer (105). The authors synthesized a penta-deutero-thyroxine, which was added to the serum samples to be investigated. The thyroid hormones were isolated from the biological matrix by cation-exchange chromatography and the *N,O*-bis(heptafluorobutyric)methyl ester derivative separated on a 6 m OV-1 capillary GC column. A fragment ion of the thyroxine derivative and of the deuterated internal standard containing 4 deuterium atoms were recorded at *m/z* 970 and 974, respectively. The interassay imprecision of the procedure was measured at 2.8% (CV) and a comparison made with results obtained by a radioimmunoassay.

For assessing target data in the external quality control scheme of the Deutsche Gesellschaft für Klinische Chemie a method for the measurement of thyroxine in human serum was developed which makes use of $^{13}C_2$-labeled thyroxine as internal standard (106). The thyroid hormone was isolated from the serum samples by cation exchange resin, followed by the formation of the methyl ester and purification on Sephadex LH-20. Finally, the *N,O*-bis(trifluoroacetyl) derivative was formed and SIM was carried out at *m/z* 870 and 872 using EI-MS in combination with capillary GC. The precision of the method was in the range of 1.4 to 1.9% (CV).

7 Conclusions

In this chapter the potential of ID-MS as a highly accurate method for measuring a number of clinical chemical analytes has been outlined. Many analytical procedures described can be accepted as reference or even definitive methods in clinical chemistry while others provide a suitable basis for further development of definitive or reference technology.

It is true that highly accurate methods such as ID-MS are necessary prerequisites for a reference or definitive method, but in order to set up reference method values or definitive values, it is absolutely essential to maintain a detailed protocol for the certification procedure, including the number of replicates to be analysed on the same or on different occasions plus a thorough estimate of all random and systematic errors.

Reference or definitive methods may be used for the evaluation of individual clinical chemical methods, for the establishment of target data in external quality control surveys as descrived in several investigations (45, 72, 78, 107) and for the assignment of certified values to reference materials which has been discussed recently (108). The National Bureau of Standards, Washington, has issued standard reference material which consists of lyophilised human serum. The concentrations of analytes have been certified in their laboratories using the ID-MS technique and it is expected that in addition to the methods used for calcium, potassium, glucose, cholesterol, urea and creatinine, further procedures will be published dealing with the definitive measurement of chloride, lithium, magnesium and uric acid. Furthermore, the Community Bureau of Reference of the European Communities has initiated a programme for the production of lyophilised human sera as certified reference materials. The cortisol concentration in the first steroid reference material was determined independently by five different laboratories using ID-MS.

References

1. Büttner, J., Borth, R., Boutwell, J. H., Broughton, P. M. G. & Bowyer, R. C. (1979) Approved recommendation on quality control in clinical chemistry. Clin. Chim. Acta *98*, 129F-186F.
2. Cali, J. P. (1976) Rationale for reference methods in clinical chemistry. Pure Appl. Chem. *45*, 63–68.
3. Büttner, J. (1976) The international federation of clinical chemistry (IFCC) and reference methods. Pure Appl. Chem. *45*, 69–73.
4. Mitchell, F. L. (1976) The practical application of reference method technology. Pure Appl. Chem. *45*, 59–62.
5. Neill, D. W. & Doggart, J. R. (1976) When is a reference method a reference method? Clin. Chem. *22*, 285.
6. Brown, S. S., Büttner, J., Mitchell, F. L., Rubin, M. & Cooper, G. R. (1976) Letter to the editor: Response on: "When is a reference method a reference method". Clin. Chem. *22*, 285–286.
7. Cali, J. P., Schaffer, R. & Mitchell, F. L. (1977) Suggested modifiction of the definition for definitive method. Clin. Chem. *23*, 1943–1947.
8. Leijnse, B. (1983) Are definitive (reference) methods and Shewharts principle the best approach towards accuracy? Statistics in Medicine *2*, 131–140.
9. Tietz, N.W. (1979) A model for a comprehensive measurement system in clinical chemistry. Clin. Chem. *25*, 833–839.
10. Hill, R. E. & Whelan, D. T. (1984) Mass spectrometry and clinical chemistry. Clin. Chim. Acta *139*, 231–294.
11. Sweeley, C. C., Elliott, W. H., Fries, I. & Ryhage, R. (1966) Mass spectrometric determination of unresolved components in gas chromatographic effluents. Anal. Chem. *38*, 1549–1553.
12. Hammar, C. G. & Holmstedt, B. (1968) Mass fragmentography. Identification of chlorpromazine and its metabolites in human blood by a new method. Anal. Biochem. *25*, 532–548.
13. Rittenberg, D. & Foster, G. L. (1940) A new procedure for quantitative analysis by isotope dilution, with application to the determination of amino acids and fatty acids. J. Biol. Chem. *133*, 737–744.
14. Samuelsson, B., Hamberg, M. & Sweeley, C. C. (1970) Quantitative gas chromatography of prostaglandin E_1 at the nanogram level. Use of deuterated carrier and multiple-ion analyzer. Anal. Biochem. *38*, 301–304.
15. Siekmann, L., Hoppen, H. O. & Breuer, H. (1970) Zur gas-chromatographisch-massenspektrometrischen Bestimmung von Steroidhormonen in Körperflüssigkeiten unter Verwendung eines Multiple Ion Detectors (Fragmentographie). Z. Anal. Chem. *252*, 294–298.
16. Lehmann, W. D. & Schulten, H. R. (1978) Quantitative Massenspektrometrie in Biochemie und Medizin. Angew. Chem. *90*, 233–250.
17. De Leenheer, A. P., Lefevere, M. F., Lambert, W. E. & Colinet, E. S. (1985) Isotope dilution mass spectrometry in clinical chemistry. Adv. Clin. Chem. *24*, 111–161.
18. Lawson, A. M., Gaskell, S. J. & Hjelm, M. (1985) Methodological aspects on quantitative mass spectrometry used for accuracy control in clinical chemistry. J. Clin. Chem. Clin. Biochem. *23*, 433–441.
19. Millard, B. J. (1978) Quantitative Mass Spectrometry. Heyden, London-Philadelphia-Rheine.
20. Haskins, N. J. (1982) The application of stable isotopes in biomedical research. Biomed. Mass Spectrom. *9*, 269–277.
21. Blau, K. & King, G. S. (1977) Handbook of Derivatives for Chromatography. Heyden, London-Philadelphia-Rheine.
22. Siekmann, L. (1978) In: Quantitative Mass Spectrometry in Life Sciences II (De Leenheer, A. P., Roncucci, R. R. & Van Peteghem, C., eds.), pp. 1–16, Elsevier Scientific Publishing Co., Amsterdam.
23. Cohen, A., Hertz, H. S., Mandel, J., Paule, R. C., Schaffer, R., Sniegoski, L. T., Sun, T., Welch, M. J. & White V, E. (1980) Total serum cholesterol by isotope dilution/mass spectrometry: a candidate definitive method. Clin. Chem. *26*, 854–860.
24. Siekmann, L. & Breuer, H. (1982) Determination of cortisol in human plasma by isotope dilution-mass spectrometry. J. Clin. Chem. Clin. Biochem. *20*, 883–892.
25. Pickup, J. F. & McPherson, K. (1976) Theoretical considerations in stable isotope dilution mass spectrometry for organic analysis. Anal. Chem. *48*, 1885–1890.
26. Lee, W. N. P., Whiting, J. S., Fymat, A. L. & Boettger, H. G. (1983) A least-square approach to the quantitation of stable isotopes

in mass spectrometry. Biomed. Mass Spectrom. *10*, 641–645.

27. Colby, B. N. & McCaman, M. W. (1979) A comparison of calculation procedures for isotope dilution determinations using gas chromatography mass spectrometry. Biomed. Mass Spectrom. *6*, 225–230.
28. Yap, W. T., Schaffer, R., Hertz, H. S., White V, E. & Welch, M. J. (1983) On the difference between using linear and nonlinear models in bracketing procedures in isotope dilution mass spectrometry. Biomed. Mass Spectrom. *10*, 262–264.
29. Siekmann, L. (1974) Mass fragmentography of steroid hormones. J. Steroid Biochem. *5*, 727–732.
30. Siekmann, L., Spiegelhalder, B. & Breuer, H. (1972) Bestimmung von Aldosteron im Plasma des Menschen unter Verwendung der Kombination Gaschromatographie/Massenspektrometrie. Z. Anal. Chem. *261*, 377–381.
31. Moore, L. J. & Machlan, L. A. (1972) High accuracy determination of calcium in blood serum by isotope dilution mass spectrometry. Anal. Chem. *44*, 2291–2296.
32. Yergey, A. L., Vieira, N. E. & Hansen, J. W. (1980) Isotope ratio measurements of urinary calcium with a thermal ionization probe in a quadrupole mass spectrometer. Anal. Chem. *52*, 1811–1814.
33. Kownatzki, R., Peters, F., Reil, H.-H. & Maaß, G (1980) Volatile complexes for the determination of calcium in blood plasma by isotope dilution mass spectrometry. Biomed. Mass Spectrom. *7*, 540–543.
34. Gramlich, J. W., Machlan, L. A., Brletic, K. A. & Kelly, W. R. (1982) Thermal-ionization isotope-dilution mass spectrometry as a definitive method for determination of potassium in serum. Clin. Chem. *28*, 1309–1313.
35. Pickup, J. P. (1976) An accurate method for the determination of inorganic phosphate in serum. Ann. Clin. Biochem. *13*, 306–312.
36. Bier, D. M., Arnold, K. J., Sherman, W. R., Holland, W. H., Holmes, W. F. & Kipnis, D. M. (1977) In-vivo measurement of glucose and alanine metabolism with stable isotopic tracers. Diabetes *26*, 1005–1015.
37. Bier, D. M., Leake, R. D., Haymond, M. W., Arnold, K. J., Gruenke, L. D., Sperling, M. A. & Kipnis, D. M. (1977) Measurement of "true" glucose production rates in infancy and childhood with 6,6-dideuteroglucose. Diabetes 26, 1016–1023.
38. Dan, P., Clemons, P. M., Sperling, M. I., Gelfand, M. J., Chen, I. W., Sperling, M. A. & Norman, E. J. (1983) A method for measurement of glucose enrichment in serum using isotope dilution mass spectrometry and its application for measurement of glucose kinetics. Anal. Lett. *16*, 655–666.
39. Björkhem, I., Blomstrand, R., Falk, O. & Öhman, G. (1976) The use of mass fragmentography in the evaluation of routine methods for glucose determination. Clin. Chim. Acta *72*, 353–362.
40. Eldjarn, L., Helland, P. & Jellum, E. (1978) The influence of GC and MS parameters on accuracy and precision in quantitative mass spectrometry In: Quantitative Mass Spectrometry in Life Sciences II (De Leenheer, A. P., Roncucci, R. R. & Van Peteghem, C., eds.), pp. 151–159, Elsevier Scientific Publishing Co., Amsterdam.
41. Pelletier, O. & Cadieux, S. (1983) Quantitative determination of glucose in serum by isotope dilution mass spectrometry following gas liquid chromatography with fused silica column. Biomed. Mass Spectrom. *10*, 130–135.
42. White V, E., Welch, M. J., Sun, T., Sniegoski, L. T., Schaffer, R., Hertz, H. S. & Cohen, A. (1982) The accurate determination of serum glucose by isotope dilution mass spectrometry — two methods. Biomed. Mass Spectrom. *9*, 395–405.
43. Björkhem, I., Blomstrand, R. & Svensson, L. (1974) Serum cholesterol determination of mass fragmentography. Clin. Chim. Acta *54*, 185–193.
44. Björkhem, I., Blomstrand, R., Lantto, O., Svensson, L. & Öhman, G. (1976) Toward absolute methods in clinical chemistry: application of mass fragmentography to high-accuracy analyses. Clin. Chem. *22*, 1789–1801.
45. Björkhem, I., Blomstrand, R., Eriksson, S., Falk, O., Kallner, A., Svensson, L. & Öhman, G. (1980) Use of isotope dilution-mass spectrometry for accuracy control of different routine methods in clinical chemistry. Scand. J. Clin. Lab. Invest. *40*, 529–534.
46. Björkhem, I., Bergman, A., Falk, O., Kallner, A., Lantto, O., Svensson, L., Akerlöf, E. & Blomstrand, R. (1981) Accuracy of some routine methods used in clinical chemistry as judged by isotope dilution-mass spectrometry. Clin. Chem. *27*, 733–735.
47. Svensson, L., Elg, P., Rasmussen, S., Skrede, S. & Björkhem, I. (1982) A possible

model for accuracy control of determination of serum cholesterol with use of reference methods. Scand. J. Clin. Lab. Invest. *42*, 99–105.
48. Siekmann, L., Hüskes, K. P. & Breuer, H. (1976) Determination of cholesterol in serum using mass fragmentography – a reference method in clinical chemistry. Z. Anal. Chem. *279*, 145–146.
49. Gambert, P., Lallemant, C., Archambault, A., Maume, B. F. & Padieu, P. (1979) Assessment of serum cholesterol by two methods: gas-liquid chromatography on a capillary column and chemical ionization-mass fragmentography with isotopic dilution of [3,4-^{13}C]cholesterol as internal standard. J. Chromatogr. *162*, 1–6.
50. Wolthers, B. G., Hindriks, F. R., Muskiet, F. A. J. & Groen, A. (1980) Mass fragmentographic analysis of total cholesterol in serum using a heptadeuterated internal standard. Clin. Chim. Acta *103*, 305–315.
51. Freudenthal, J., Derks, H. J. G. M., Gramberg, L. G., ten Hove, G. J. & Klaassen, R. (1981) Isotope dilution mass spectrometry of cholesterol in serum. Biomed. Mass Spectrom. *8*, 5–9.
52. Schaffer, R., Sniegoski, L. T., Welch, M. J., White V, E., Cohen, A., Hertz, H. S., Mandel, J., Paule, R. C., Svensson, L., Björkhem I. & Blomstrand, R. (1982) Comparison of two isotope dilution/mass spectrometric methods for determination of total serum cholesterol. Clin. Chem. *28*, 5–8.
53. Björkhem, I., Blomstrand, R. & Svensson, L. (1976) Determination of serum triglycerides by mass fragmentography. Clin. Chim. Acta *71*, 191–198.
54. Siekmann, L., Schönfelder, A. & Siekmann, A. (1986) Isotope dilution-mass spectrometry of total glycerol in human serum – reference method in clinical chemistry. Z. Anal. Chem. *324*, 280–281.
55. Björkhem, I., Blomstrand, R. & Öhman, G. (1976) Determination of serum urea by mass fragmentography. Clin. Chim. Acta *71*, 199–205.
56. Tserng, K.-Y. & Kalhan, S. C. (1982) Gas chromatography/mass spectrometric determination of [^{15}N]urea in plasma and application to urea metabolism study. Anal. Chem. *54*, 489–491.
57. Nissim, I., Yudkoff, M., Yang, W., Terwilliger, T. & Segal, S. (1981) Rapid gas chromatographic-mass spectrometric analysis of [^{15}N]urea: application to human metabolic studies. Clin. Chim. Acta *109*, 295–394.
58. Welch, M. J., Cohen, A., Hertz, H. S., Ruegg, F. C., Schaffer, R., Sniegoski, L. T. & White V, E. (1984) Determination of serum urea by isotope dilution mass spectrometry as a candidate definitive method. Anal. Chem. *56*, 713–719.
59. Björkhem, I., Blomstrand, R. & Öhman, G. (1977) Mass fragmentogoraphy of creatinine proposed as a reference method. Clin. Chem. *23*, 2114–2121.
60. Björkhem, I., Blomstrand, R. & Öhman, G. (1979) Accuracy of a fully enzymatic method for creatinine, as judged by isotope dilution-mass spectrometry. Clin. Chem. *25*, 1666–1667.
61. Siekmann, L. (1985) Determination of creatinine in human serum by isotope dilution-mass spectrometry – definitive methods in clinical chemistry. IV. J. Clin. Chem. Clin. Biochem. *23*, 137–144.
62. Welch, M. J., Cohen, A., Hertz, H. S., Ng, K. J., Schaffer, R., Van der Lijn, P. & White V, E. (1986) Determination of serum creatinine by isotope dilution mass spectrometry as a candidate definitive method. Anal. Chem. *58*, 1681–1685.
63. Lawson, A. M. (1975) Prospects for mass spectrometry in clinical chemistry. Ann. Clin. Biochem. *12*, 51–57.
64. Lim, C. K., Pryde, D. E. & Lawson, A. M. (1978) Specific method for determining uric acid in serum using high-performance liquid chromatography and gas chromatography-mass spectrometry. J. Chromatogr. *149*, 711–720.
65. Lawson, A, M., Kim, C. K., Richmond, W., Samson, D. M., Setchell, K. D. R. & Thomas, A. C. S. (1980) Isotope dilution mass spectrometry as a basis for accuracy in clinical chemistry. In: Current Developments in the Clinical Applications of HPLC, GC and MS (Lawson, A. M., Lim, C. K. & Richmond, W., eds.), pp. 135–153, Academic Press, London, New York, Toronto, Sydney, San Francisco.
66. Öhman, G. (1979) Evaluation of three clinical chemical routine methods for the determination of serum uric acid, by mass fragmentography. Clin. Chim. Acta *95*, 219–226.
67. Walker, R. W., Van den Heuvel, W. J. A., Wolf, F. J., Noll, R. M. & Duggan, D. E. (1977) A stable isotope gas-liquid chromatographic-mass spectrometric assay for determining uric acid body pool size. Anal. Biochem. *77*, 235–242.

68. Siekmann, L. (1985) Determination of uric acid in human serum by isotope dilution-mass spectrometry – definitive methods in clinical chemistry, III. J. Clin. Chem. Clin. Biochem. *23*, 129–135.
69. Björkhem, I., Blomstrand, R., Lantto, O., Löf. A. & Svensson, L. (1974) Plasma cortisol determination by mass fragmentography. Clin. Chim. Acta *56*, 241–248.
70. Siekmann, L. (1979) Determination of steroid hormones by the use of isotope dilution-mass spectrometry: a definitive method in clinical chemistry. J. Steroid Biochem. *11*, 117–123.
71. Lindberg, C., Jönsson, S., Hedner, P. & Gustafsson, A. (1982) Radioimmunoassay and chemical ionisation/mass spectrometry compared for plasma cortisol determination. Clin. Chem. *28*, 174–177.
72. Gaskell, S. J., Collins, C. J., Thorne, G. C. & Groom, G. V. (1983) External quality assessment of assays for cortisol in plasma: Use of target data obtained by gas chromatography/mass spectrometry. Clin. Chem. *29*, 862–867.
73. Thorne, G. C., Gaskell, S. J. & Payne, P. A. (1984) Approaches to the improvement of quantitative precision in selected ion monitoring: High resolution applications. Biomed. Mass Spectrom. *11*, 415–420.
74. Jonckheere, J. A. & De Leenheer, A. P. (1983) Candidate definitive method for the determination of cortisol in human serum. Biomed. Mass Spectrom. *10*, 197–202.
75. Patterson, D. G., Patterson, M. B., Culbreth, P. H., Fast, D. M., Holler, J. S., Sampson, E. J. & Bayse, D. D. (1984) Determination of steroid hormones in a human-serum reference material by isotope dilution-mass spectrometry: A candidate definitive method for cortisol. Clin. Chem. *30*, 619–626.
76. Lantto, O., Björkhem, I., Blomstrand, R. & Kallner, A. (1980) Interlaboratory evaluation of four RIA kits for determination of plasma cortisol, with special reference to accuracy: Influence of matrix in calibration standards. Clin. Chem. *26*, 1899–1902.
77. Lantto, O., Lindback, B., Aakvaag, A., Damkjaer-Nielsen, M., Pomoell, U.-M. & Björkhem, I. (1983) Assay of cortisol with radioimmunoassay method calibrated by isotope dilution-mass spectrometry. A Nordic collaborative study. Scand. J. Clin. Lab. Invest. *43*, 433–437.
78. Röhle, G., Voigt, U., Siekmann, L. & Breuer, H. (1983) Ringversuche für Steroidhormone: Richtigkeit and Präzision der Analysenergebnisse. J. Clin. Chem. Clin. Biochem. *21*, 157–165.
79. Breuer, H. & Siekmann, L. (1975) Mass fragmentography as reference method in clinical steroid assay. J. Steroid Biochem. *6*, 685–688.
80. Siekmann, L., Martin, S. & Breuer, H. (1972) Determination of testosterone in human plasma by gas liquid chromatography-mass spectrometry. Scand. J. Clin. Lab. Invest. *29*, 8.8
81. Chapman, J. R. & Bailey, E. (1974) Determination of plasma testosterone by combined gas chromatography-mass spectrometry. J. Chromatogr. *89*, 215–224.
82. Björkhem, I., Lantto, O. & Svensson, L. (1975) Serum testosterone determination by mass fragmentography. Clin. Chim. Acta *60*, 59–66.
83. Baba, S., Shinohara, Y. & Kasuya, Y. (1979) Determination of plasma testosterone by mass fragmentography using testosterone-19-d_3 as internal standard. J. Chromatogr. *162*, 529–537.
84. Finlay, E. M. H. & Gaskell, S. J. (1981) Determination of testosterone in plasma from men by gas chromatography/mass spectrometry, with high-resolution selected-ion monitoring and metastable peak monitoring. Clin. Chem. *27*, 1165–1170.
85. Siekmann, L., Martin, S., Siekmann, A. & Breuer, H. (1976) Determination of testosterone and 5α-dihydrotestosterone in human plasma by isotopic dilution-mass fragmentography. Acta Endocrinol. *82*, Suppl. 202, 65–67.
86. Sabot, J. F., Deruaz, D., Bernard, P. & Pinatel, H. (1985) Determination of plasma testosterone by mass fragmentography using [3,4-^{13}C] testosterone as an internal standard. J. Chromatogr. *339*, 233–242.
87. Wilson, D. W., John, B. M., Groom, G. V., Pierrepoint, C. G. & Griffiths, K. (1977) Evaluation of an oestradiol radioimmunoassay by high-resolution mass fragmentography. J. Endocr. *74*, 503–504.
88. Siekmann, L., Siekmann, A. & Breuer, H. (1978) Isotopic dilution mass spectrometry of oestriol and oestradiol-17β – an approach to definitive methods. Z. Anal. Chem. *290*, 122–123.
89. Zamecnik, J., Armstrong, D. T. & Green, K. (1978) Serum estradiol-17β as determined

by mass fragmentography and by radioimmunoassay. Clin. Chem. *24*, 627–630.

90. Knuppen, R., Haupt O., Hoppen, H.-O. (1982) Synthesis of [$^2H_8$]estradiol, [$^2H_6$]2-hydroxyestrone and [$^2H_6$]4-hydroxyestrone as internal standards for selected ion monitoring. Steroids *39*, 667–673.
91. Knuppen, R., Haupt, O., Schramm, W. & Hoppen, H.-O. (1979) Selected ion monitoring: a new approach for the specific determination of steroids in the lower picogram range. J. Steroid Biochem. *11*, 153–160.
92. Gaskell, S. J., Finlay, E. M. H. & Pike, A. (1980) Analyses of steroids in saliva using highly selective mass spectrometric techniques. Biomed. Mass Spectrom. *7*, 500–504.
93. Gaskell, S. J. & Brownsey, B. G. (1983) Immunoadsorption to improve gas chromatography/high-resolution mass spectrometry of estradiol-17β in plasma. Clin. Chem. *29*, 677–680.
94. Iwai, M., Kanno, H., Hashino, M., Suzuki, J., Yanaihara, T. & Nakayama, T. (1981) Determination of serum unconjugated estrone, estradiol-17β and estriol during pregnancy by selected ion monitoring. J. Chromatogr. *225*, 275–282.
95. Siekmann, L. (1984) Determination of oestradiol-17β in human serum by isotope dilution-mass spectrometry. Definitive methods in clinical chemistry, II. J. Clin. Chem. Clin. Biochem. *22*, 551–557.
96. Adlercreutz, H., Nylander, P. & Hunneman, D. H., (1974) Studies on the mass fragmentographic determination of plasma estriol. Biomed. Mass Spectrom. *1*, 332–339.
97. Björkhem, I., Blomstrand, R., Svensson, L., Tietz, F. & Carlström, K. (1975) Validation of methods for determination of urinary estriol during pregnancy using mass fragmentography. Clin. Chim. Acta *62*, 385–392.
98. Björkhem, I., Blomstrand, R. & Lantto, O. (1975) Validation of routine methods for serum progesterone determination using mass fragmentography. Clin. Chim. Acta *65*, 343–350.
99. Schramm, W., Louton, T., Schill, W. & Hoppen, H.-O. (1980) An application of gas chromatography mass spectrometry to the quantitation of substances and their labeled variants. Biomed. Mass Spectrom. *7*, 273–276.
100. Johnson, D. W., Phillipou, G., Ralph, M. M. & Seamark, R. F. (1979) Specific quantitation of urinary progesterone by gas chromatography-mass spectrometry. Clin. Chim. Acta *94*, 207–208.
101. Gaskell, S. J., Brownsey, B. G. & Groom, G. V. (1984) Analyses for progesterone in serum by gas chromatography/mass spectrometry: Target data for external quality assessment of routine assays. Clin. Chem. *30*, 1696–1700.
102. Van der Greef, J., Tas, A. C., Rijk, M. A. H., Ten Noever de Brauw, M. C., Höhn, M., Meyerhoff, G. & Rapp, U. (1985) Determination of progesterone by liquid chromatography-mass spectrometry using a moving-belt interface and isotope dilution. J. Chromatogr. *343*, 397–401.
103. Möller, B., Björkhem, I. Falk, O., Lantto, O. & Larsson, A. (1983) Identification of thyroxine in human breast milk by gas chromatography-mass spectrometry. J. Clin. Endocrinol. Metab. *56*, 30–34.
104. Möller, B., Falk, O. & Björkhem, I. (1983) Isotope dilution-mass spectrometry of thyroxine proposed as a reference method. Clin. Chem. *29*, 2106–2110.
105. Ramsden, D. B. & Farmer, M. J. (1984) Development of a gas chromatographic selected ion monitoring assay for thyroxine (T4) in human serum. Biomed. Mass Spectrom. *8*, 421–427.
106. Siekmann, L. (1987) Measurement of thyroxine in human serum by isotope dilution mass spectrometry. Biomed. Mass Spectrom. *14*. 683–688.
107. Gilbert, R. K. (1978) Accuracy of clinical laboratories studied by comparison with definitive mthods. Am. J. Clin. Path. *70*, 450–470.
108. Marchandise, H. & Colinet, E. (1983) Assessment of methods of assigning certified values to reference materials. Z. Anal. Chem. *316*, 669–672.

Chapter 14
Stable Isotopes in Clinical Investigations

D. Halliday and G. C. Ford

1 Introduction

Stable, non-radioactive isotopes are increasingly employed in clinical studies of both a diagnostic and research nature. Stimulation of usage has come from an increased awareness of possible harmful effects in the use of corresponding radionuclides (however irrational this concern may be in specific instances) but in addition, it has come from a realisation of the positive advantages conferred by the use of stable isotopes in their own right. In this context, one might point to the fact that there are no practical radionuclide alternatives for certain elements of clinical interest (especially nitrogen); with certain practical limitations serial studies can be performed in the same patient, and judicial choice of isotope 'cocktails' may be employed in the same patient study without detracting from the analytical precision. Certainly much impetus in their application has stemmed from the increasing diversity of stable-isotope-labeled compounds that are currently available, their relative decrease in cost over the last few years, and not least, the developments and improvements in a range of mass spectrometric and sample preparation techniques required for their analysis.

One can benefit from the classical isotope ratio mass spectrometer (IRMS) for high precision stable isotope ratio analysis (SIRA) which is required for metabolic 'end-product' analysis, and this instrument is presented in some detail as it is not highlighted elsewhere in this volume. Discussion of GC-MS (quadrupole or magnetic) involvement in clinical studies has been limited to two specific areas, namely, qualitative tracer studies and quantitative isotope dilution analysis. The choice of examples in these areas has had to be highly selective in the face of numerous recent publications. Analysis by GC-MS offers the possibility for the simultaneous estimation of isotope enrichment and concentration of labeled and unlabeled compound. In this context, analytical selectivity and specificity is afforded by the choice of derivatising agent, GC column, MS ionization mode, selection of fragment ions to be monitored and dwell time on selected ions. Hybrid combinations of IRMS and GC-MS systems that permit the isotope analysis of samples that IRMS instruments on their own cannot tackle, i. e. samples of low concentration and low enrichment, will be considered where appropriate.

This chapter attempts to present an overview of selected clinical applications of stable isotopes that is biased towards the more recent literature and the authors' personal interests that involve the use of IRMS or GC-MS or both. However, it is hoped that a sufficient breadth of example is provided to stimulate general clinical interest and promote clinician/scientist interaction such that with the aid of stable isotopes and the necessary research panache and ingenuity, as yet unresolved metabolic questions relating to health and disease may be successfully tackled on a quantitative basis.

2 Classical Isotope Ratio Instrumentation

2.1 The isotope ratio mass spectrometer (IRMS)

Dual-inlet, multiple collector, gas isotope ratio mass spectrometers (IRMS) inherently possess low mass range (2 – 100), a resolving power of 100 – 200 (10% valley definition), but extremely high sensitivity (0.2 A ion current $mbar^{-1}$). Design specialisation

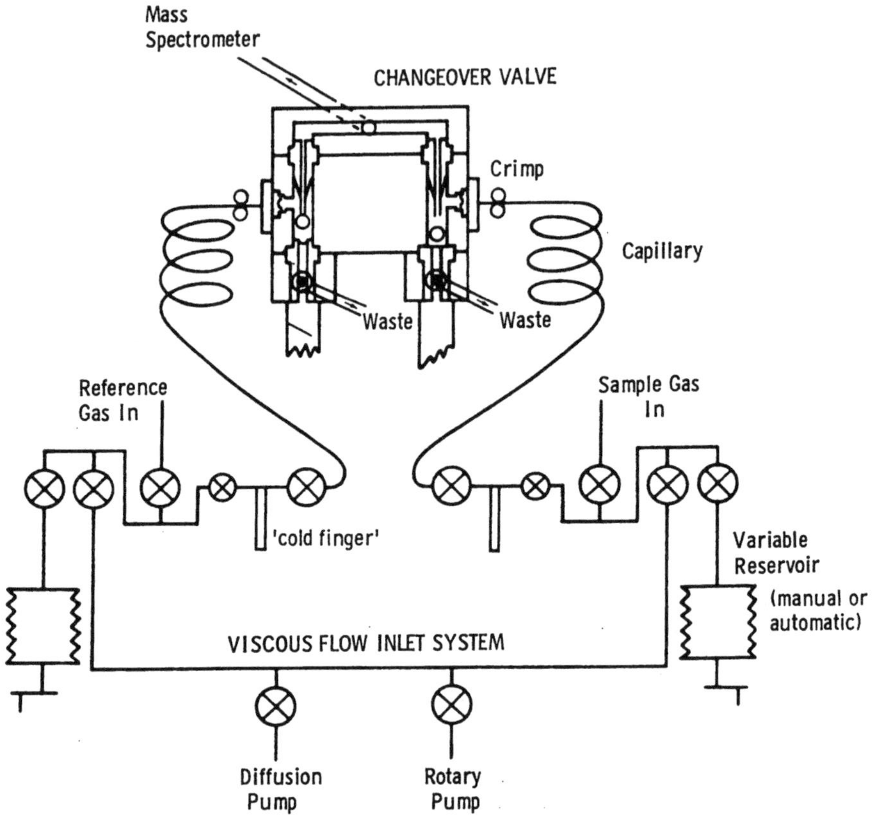

Figure 1. Schematic of isotope ratio mass spectrometer typical standard gas inlet system for precise isotope ratio measurement.

provides the basis for accurate stable isotope ratio analysis (SIRA) of pure, low molecular weight, permanent gases such as H_2, CO_2, N_2 and SO_2, against an initial analyser pressure of $< 10^{-9}$ mbar produced by diffusion or turbomolecular pumps.

Low resolution of these instruments is due primarily to the analyser radius (6 – 30 cm) thus restricting the ion beam path length, the width and position of the resolving slits and the relatively low strength permanent or electro-magnet employed. The high sensitivity stems from '*z*' focussing of the ion beams, the simultaneous collection of 'light' and 'heavy' isotope molecular ion beams in separate Faraday buckets and the rapid alternate isotope ratio comparison of a known reference gas with an unknown sample gas.

Gas sample introduction into the analyser source is by means of a comprehensive dual inlet system. Essentially this consists of all-metal or soft seat inlet valves, variable volume bellows (reservoirs, 2 – 50 ml), metal capillaries and an automatic all-metal changeover valve (Fig. 1). The sample inlet may be connected directly to a range of on-line sample preparation systems (see later) or consist of a multiport manifold system whereby individually prepared samples can be analysed in sequence. Whilst 'routine' sample size for isotope analysis may be considered within the range,

0.05 – 0.5 mg of prepared gas, provision for smaller sample analysis (< 10 μg) is afforded by 'cold finger' trapping of the gas (where practicable) into a small volume prior to capillary entry. The bellows permit the introduction of a wide range of sample size and, as the isotope ratio ultimately obtained is pressure-dependent (1), they provide for the equalization of pressures with respect to sample and reference gas; the capillaries, crimped at their distal end, convert the initially viscous gas flow to molecular flow such that the isotopic content of the gas entering the analyser is identical to that in the reservoirs and provide a low signal/noise ratio for a source operating pressure of some 10^{-7} mbar. Finally the changeover valve serves a dual purpose: it permits the alternate entry of sample and reference gas into the analyser on a strict time basis whilst simultaneously bleeding the gas not being analysed to waste via the pumping system. In this manner the status quo is maintained with respect to the pressure between the reference and sample gases.

The gas tight ion source operates solely in the electron impact mode operating at an electron energy of 70 eV and with a trap current of 300 – 600 μA (dependent on specific analysis). Generated molecular ions are extracted from the source under the influence of an accelerating potential (2 – 10 kV) and the individual ion beams focussed into separate deep Faraday buckets.

The trajectory of ions in the mass spectrometer is dictated by:

$$\frac{m}{z} = \frac{B^2 r^2}{2V}$$

where B is magnetic field in gauss; r is radius of curvature of ion path (cm); and V is the accelerating potential in volts. Thus if B and V are constant, the radius of curvature of the path of an ion will be proportional to the square root of the m/z ratio. It is normal practice to tune the accelerating voltage such that an ion beam is directed into the required Faraday bucket.

Following magnetic field separation of the molecular ion beams, resolving slits situated near the collector system focus the beams into individual deep Faraday buckets. Suppressor electrodes act both to minimize ion drift between the collectors and to reject any secondary electrons formed by ion collisions. Commercially available instruments currently exhibit a range of Faraday collector configurations ranging through single, double, triple and multicollector systems, the latter with provision for external positional adjustment such that up to three isotopes of a single element may be collected simultaneously ($^{28}N_2$ $^{29}N_2$ $^{30}N_2$). Generated currents resulting from ions impinging on the Faraday bucket collector plates are converted into a usable voltage, amplified by a field effect transistor (FET) operational amplifier and thence fed to voltage to frequency converters whose output is linearly proportional to the input voltage. These pulses are passed into separate counters and the required ratio obtained on either a time or count accumulation basis.

Modern 'state of the art' IRMS systems have benefitted by the microcomputer revolution such that current instruments, though retaining manual mode operation, can be run totally unattended under computer control. This control permits keyboard selection of user defined parameters covering all aspects of operational procedures including automatic source tuning, sequence analysis of samples positioned in the manifold, and inlet and analyser bakeout cycles to maintain a low residual gas background.

2.2 Isotope ratio nomenclature

The IRMS provides a measure of the ratio of ion intensities of the ion currents produced by two (or three) isotopic ion beams impinging on the collector plates, whereas isotopic abundances are normally expressed in terms of atom-percent or atom-percent excess of the isotope of interest in clinical or biological studies. Taking the case of nitrogen for example, the measured ratio (R) would be derived from the ion currents corresponding to mass 28 ($^{14}N^{14}N$) and mass 29 ($^{14}N^{15}N$) thus

$$R = \frac{[^{14}N^{15}N]}{[^{14}N^{14}N]} \qquad \text{[Eq. 1]}$$

whereas the atom percent ^{15}N (A) is given by

$$A = \frac{\text{number } ^{15}N \text{ atoms}}{\text{number } ^{14}N \text{ atoms} + \text{number } ^{15}N \text{ atoms}} \times 100 \qquad \text{[Eq. 2]}$$

or

$$A = \frac{[^{14}N^{15}N] + 2\,[^{15}N^{15}N]}{2\,[^{14}N^{14}N] + 2\,[^{14}N^{15}N] + 2[^{15}N^{15}N]} \times 100 \qquad \text{[Eq. 3]}$$

It is unnecessary to measure the ion current corresponding to mass 30 ($^{15}N^{15}N$) to determine atom percent ^{15}N because the term $[^{15}N^{15}N]$ can be eliminated from Eq. 3 as N_2 is in equilibrium with respect to the following

$$^{14}N^{14}N + {}^{15}N^{15}N \rightleftarrows {}^{14}N^{15}N + {}^{14}N^{15}N$$

Theoretical calculation and experimental observation indicate that the equilibrium constant of this reaction at 25°C is 4.0. Therefore

$$\frac{[^{14}N^{15}N]^2}{[^{14}N^{14}N]\,[^{15}N^{15}N]} = 4 \qquad \text{[Eq. 4]}$$

Combination of Eq. 1, 3 and 4 simplifies to the following

$$\text{Atom percent } ^{15}N = \frac{100}{2R + 1}$$

In practical terms this relationship holds for low levels of ^{15}N enrichment, but if relatively high enrichments have to be measured ($< 1\%$), the m/z 28, 29 and 30 ion intensities should be determined. The relative amounts of these ion species follow a binomial distribution $a^2 : ab : b^2$ where a and b are the relative abundances of the two lighter isotopes. These appropriate ratio measurements are facilitated by the use of a triple collector IRMS system. Atom percent excess is obtained by comparison of the unknown sample gas with an international or laboratory reference gas such that:

$$\text{Atom percent excess} = \frac{100}{2R_S + 1} - \frac{100}{2R_R + 1}$$

where the subscripts R and S refer to reference and sample ratios respectively.

^{13}C has a relatively low natural abundance (approximately 1% of total carbon), and exhibits only a small range of variation of isotope abundance in nature as a result of isotope fractionation processes (some 3% variation in the $^{13}C/^{12}C$ ratio). Expressed as atom percent excess, all natural carbon falls within a range of less than 0.1 atom percent excess ^{13}C and the greater majority within the much narrower range of less than 0.02 atom percent excess. Researchers such as geologists, geochemists or hydrologists who are committed to interpreting in a meaningful manner these small variations in natural isotope levels clearly require an expanded scale for convenience. Appropriate amplification of the scale is obtained by use of the 'delta' notation which is based on the mathematical relationship:

$$\delta^{13}C = \left[\frac{(^{13}C/^{12}C) \text{ sample}}{(^{13}C/^{12}C) \text{ reference}}\right] - 1 \times 1000$$

The delta ($\delta\,^{13}C$) value thus measures the variation in parts per thousand of the $^{13}C/^{12}C$ isotope ratio compared with a standard. Internationally the ^{13}C standard adopted is PDB, which is a calcium carbonate of fossil origin (*Belemnitella americana*) from the Peedee formation in South Carolina, having a $^{13}C/^{12}C$ ratio of 0.011237. The same notation is employed for isotope ratios of $^{2}H/^{1}H$, $^{15}N/^{14}N$, $^{18}O/^{16}O$ and $^{34}S/^{32}S$ in naturally occurring samples and is increasingly employed in clinical and biological isotope studies where vast dilution of administered isotope tracer has been experienced, e. g. $^{13}CO_2$ in expired air when monitoring substrate oxidation or ^{15}N-urinary urea estimation when investigating amino acid oxidation.

2.3 Instrumental corrections

As previously indicated, the simultaneous collection of relevant molecular ion beams coupled with the rapid switching between reference and sample gas form the basis of precise SIRA measurements. However, despite recent improvements in electronic and pumping stabilities, certain instrumental corrections relating to peak tailing, changeover valve leakage and 'zero enrichment' bias should be performed routinely for the most accurate SIRA.

Briefly, taking $^{13}CO_2$ as an example, peak tailing refers to the extent of overlap of the mass 44 ion beam with the much less intense mass 45 ion beam, and is a function of both the ion optics and total pressure of the analyser. Effectively, a measured increase in the intensities of both reference and sample mass 45 ion beams will be observed, resulting in a reduction of the apparent relative difference between them. Correction is a simple matter of scanning the mass 45 peak, accurately linking the pre and post peak baseline trace and estimating the mass 44 percentage contribution at the mid-point of the mass 45 peak.

If t represents the absolute mass 44 tail contribution then the apparent delta value (δ_a) is given by

$$\delta_a = \left[\frac{(R_S + t) - (R_R + t)}{R_R + t}\right] \times 1000$$

Division of the true δ by δ_a provides the correction factor T_c; so

$$T_c = 1 + \frac{t}{R_R}$$

Leakage across the seats of the changeover valve, such that small amounts of sample or reference gas leak into the analyser during ratio measurement of reference or sample gas respectively, will introduce errors in measurement of the time ratio. The final machine correction factor relates to the 'zero enrichment' performance of the instrument. If identical gases are placed in both aspects of the inlet system no isotope enrichment or depletion should be observed. If consistent bias is observed then a correction must be applied.

Peak tailing and changeover valve leakage in modern instruments should be $< \times 0.1\%$ (so the observed ratio should be $< \times 1.001$ for each correction). Stable isotope ratio measurements should be corrected in the inverse order of their involvement, i. e. peak tailing, valve leakage and zero enrichment. Currently available instruments, operating within specification, should exhibit such low correction factors that in the context of any clinical study, may be ignored. It is however, sometimes a salutary experience to check these corrections following a major instrument overhaul.

Two specific gas corrections must also be considered: first, during the analysis of deuterium in addition to $^{2}HH^{+}$, H_3^+ ions are formed by the reaction of $H_2^+ + H_2 \rightarrow H_3^+ + H$. The formation of H_3^+ is a second order reaction and thus dependent on the square root of pH_2. Modern instruments possess built-in electronic compensation for this effect – alternatively if a series of ratios obtained at different pressures are graphically plotted, then extrapolation to zero pressure will provide the inherent H_3^+ contribution for a given instrument.

Secondly, when analysing $^{13}CO_2$ (m/z 45/44) the ^{17}O difference between reference and sample gas must be accounted for because the m/z 45 ion beam will contain both $^{13}C^{16}O^{16}O$ and $^{12}C^{16}O^{17}O$. Correcting for the ^{17}O contribution can be achieved by actually monitoring the ^{18}O difference between the sample and reference gases (m/z 46/44) and applying an algebraic correction (2, 3).

3 Sample Preparation of Pure Gases for IRMS

We have felt justified in bringing together a range of methods for the preparation of pure H_2, CO_2, N_2 and SO_2 gases required for isotope analysis by IRMS because the required methodologies tend to be spread throughout the literature. Final choice of method for a particular gas will depend on available equipment and expertise and the necessary precision and accuracy of enrichment measurements required for a particular investigation.

Common to all sample preparations is the theme of isotopic continuity, from starting compound to final gaseous product. Obviously all isotopes of a given element will form a chemical bond with any other appropriate element though the bond strength will vary with the mass of the isotope. In general, light isotopes form weaker bonds than the corresponding heavy isotope which is reflected in their greater readiness to react and this in turn gives rise to the possibility of two types of isotope fractionation occurring. Equilibrium fractionation is related to bond strength whilst kinetic fractionation depends on the greater reactivity of the light isotope. The former fractionation effect is common to all light isotopes whereas kinetic fractionation effectively relates only to carbon and sulphur. As lighter isotopes have an inherent greater reactivity, the product or products formed initially in a chemical or enzymatic reaction will be enriched in that isotope. Clearly, only if a reaction (or sample preparation step) is allowed to proceed or is driven to completion will precursor-product isotope integrity be preserved.

3.1 Hydrogen-2 (deuterium)

Quantitative reduction of water derived from either a body fluid or indirectly from the combustion of human tissue components is required to provide the necessary $^1H/^2H$ gas for precise deuterium estimation. This may be conducted on-line directly to the mass spectrometer sample inlet or perhaps more conveniently using a 'stand alone' preparation line. Historically zinc (4) or platinum-zinc granules (5) have been used to effect the reduction, but subsequently powdered iron (6), magnesium (7) or uranium turnings (8, 9) at 500 – 800 °C have been employed for this reaction. The $^1H/^2H$ gas thus formed requires transfer into an appropriate sample bottle to await isotopic analysis which may be achieved either physically, using an automatic mercury toepler pump or chemically, by reaction of the formed gas with finely cut uranium turnings at about 80 °C (10). At this temperature the hydride UH_3 is formed and when absorption of gas is complete heating to 800°C will break down the hydride such that the $^1H/^2H$ can be bottled (11). Use of the hydride method for gas transfer is undoubtedly safer than a toepler pump though a greater degree of intersample contamination (memory effect) is likely to be experienced. Recently a new single-step approach to the use of zinc as reductant has been described which is more precise than the more conventional methods outlined above and results in an appreciable reduction in sample preparation time (12). This is an extension of an earlier method (13) where the reduction of water vapour is achieved in individual vacuum vessels containing zinc shot, which following evacuation, are heated to 450 °C. Sample tubes can be prepared in batches, the reaction accomplished in a multi-position heating block for 30 minutes to ensure complete reduction, and the sealed sample bottles can then be connected to the mass spectrometer inlet directly.

Introduction of a water sample directly into the mass spectrometer can be by microsyringe (14, 15) or by use of a tapered capillary probe whereby a rapid liquid to water vapour conversion occurs without isotopic fractionation. A further modification has been the development of a hot probe inlet system that exploits the 'Leidenfrost' effect and permits the analysis of impure water samples (16). Severe memory effects and the commitment of the mass spectrometer to one type of analysis constitute the major disadvantages of these latter preparation techniques.

Following initial separation and subsequent purification, compounds of interest may be combusted along classical lines (17) or in closed tube systems (18) to yield water that can be converted to hydrogen gas by one of the above methods. Few workers have attempted to measure H_2O directly in an IRMS primarily due to the formidable 'memory' effects encountered and the ion-molecule reaction forming protonated water molecules (H_3O^+) at a rate proportional to the square of the source pressure (19). Worthy of brief mention in this section, for researchers without access to SIRA analysis, is the estimation of deuterium in aqueous phase using organic mass spectrometry. 1-Propanol exhibits a metastable ion on decomposition such that:

$$C_3H_7OH^+ \rightarrow C_3H_6^+ + H_2O$$
$$m/z\ 60 \qquad m/z\ 42$$

Thus the metastable ions produced from 'natural' and deuterated propanol (obtained by prior isotopic equilibration with the HDO sample of interest) can be used to quantitate the deuterium content of the propanol and indirectly the initial water sample (20). Interestingly, protonated 1-propanol formed in the source does not undergo metastable transition to form the m/z 42 ion which would otherwise interfere with the quantitation. Another approach used successfully in body water estimates has been that of chemical ionization (CI) using methane as reagent gas or alternatively the introduction of the water sample via a gas chromatograph such that the water concentration is sufficiently high to cause complete self-chemical ionization.

3.2 Carbon-13

The ever increasing availability of ^{13}C-labeled substrates has provided a marked impetus in the use of these tracers in clinical studies. Much recent interest centres round the measurement of oxidation rates of specific compounds by assessment of the $^{13}CO_2$ enrichment in expired air.

Available methods for respiratory CO_2 collection include immediate transfer into an evacuated septum-capped tube or greaseless tap sample bottle prior to CO_2 separation (21, 22), the direct cryogenic trapping of exhaled CO_2 in liquid nitrogen (23) and the precipitation of CO_2 as carbonate in barium, sodium or strontium hydroxide (24). Unless the latter base is used the method is suspect in that it results in isotopic fractionation which may not be reproducible, probably due to incomplete trapping. Release of the CO_2 from the carbonate obtained requires the action of 85–100% phosphoric acid under vacuum, with rigorous temperature control for reproducible $^{13}CO_2$ analysis (25). Cryogenic trapping of expired CO_2 and H_2O in liquid nitrogen, pumping away of non-condensible gases, and retention of H_2O using a solid CO_2/alcohol slurry whilst the $^{13}CO_2$ is transferred into the mass spectrometer, is standard procedure though time consuming. When dealing with large sample numbers certain investigators have opted to simply pierce septum-capped tubes in which the expired air was collected with a needle connected directly to the mass spectrometer inlet or the capillaries. In either case the degree of crimp of the capillaries will determine the source pressure observed. A refinement to this approach is to interpose chemical purification steps such that water vapour is removed by a magnesium perchlorate trap and oxygen by passage over copper powder at 300°C. Clearly these clean-up steps will result in a lowering of the abundance sensitivity, a marked reduction in source pressure and a relative increase in the major beam ion currents, effects combining to provide greater precision of isotope measurement.

^{13}C-Glucose oxidation rate can be determined using the above techniques and plasma glucose concentration is normally such that sufficiently small blood samples can be obtained to permit isotope assay by IRMS for estimates of glucose flux. Resin column sample clean-up and glucose isolation provide the basis for combustion techniques employed to generate CO_2. If the total ^{13}C content of glucose is required, conversion to $^{13}CO_2$ can be achieved by wet combustion in vacuo of gluconic acid derived from the parent glucose by enzymatic oxidation (26). Alternatively, if 1-^{13}C glucose is the substrate of choice a method providing specific enzymatic decarboxylation of the 1-C atom has been described (27). Similar decarboxylation, but using ninhydrin, enables the enrichment of 1-^{13}C labeled amino acids to be determined which is particularly useful in monitoring the rate of incorporation of tracer amino acid into body protein (28).

If randomization of label poses no problem to the final experimental calculations then samples requiring analysis of ^{13}C content may be combusted in oxygen to provide the necessary CO_2 which can be cryogenically recovered. Recent trends have been to combine commercial elemental analysers (combustion systems) directly on-line to the mass spectrometer thus providing both quantitative and isotope enrichment data. Another variation on this theme is to achieve initial separation of the compound(s) of interest carrying ^{13}C-label by gas chromatography whose effluent passes directly into the combustion chamber. Removal of unwanted gases arising from the combustion permits pure $^{13}CO_2$ to enter the IRMS or organic mass spectrometer. Several composite systems to cater for this analytical approach have been described (29–33) including the use of an infrared furnace (34) or a static mass spectrometer (35) for measurement of carbon stable isotopes at the nanomole level.

3.3 Nitrogen-15

The preparation of nitrogen gas samples for SIRA analysis generally employs a Kjeldahl digestion stage to convert nitrogenous compounds to ammonium compounds prior to oxidation to molecular nitrogen by the hypobromite reaction. Several variations of the final oxidative stage have been presented that provide for batch handling of samples or semi-automation of the reaction (36–38). Dumas combustion, long favoured by agriculturalists, has found increasing application recently for the oxidation of organic nitrogen in that it lends itself readily to automation. Indeed a series of recent publications describe progressive improvements in sensitivity and precision of ^{15}N gas analysis obtained by interfacing commercial elemental analysers (combustion systems) directly to an IRMS (39–41). One of these combined systems (41) affords complete single sample analysis time of 5 minutes at the 5μg N level with an isotope ratio measurement precision in the range 0.0005–0.001 atom% ^{15}N at natural abundance level – totally adequate for the majority of urinary and plasma nitrogen compounds of clinical interest.

3.4 Oxygen-18

Apart from the increasing use of $H_2{}^{18}O$ and $^2H_2{}^{18}O$ for total body water and energy expenditure measurements respectively (see later) few clinical studies have been reported that have necessitated the measurement of ^{18}O by IRMS. In the former instance the heavy oxygen isotope is collected as $C^{18}O_2$ in expired air, having equilibrated with

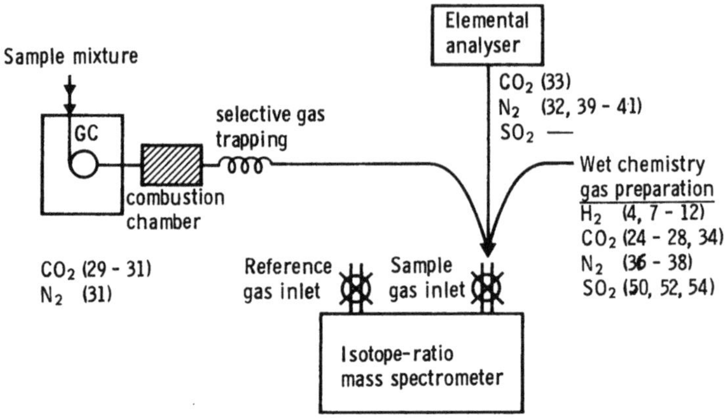

Figure 2. Summary of major options available for the preparation of pure gas samples required for IRMS stable isotope enrichment analysis. Appropriate references are given in parenthesis – for details, see text.

the bicarbonate pool of the body whilst in the latter case a dedicated commercially available IRMS for the simultaneous measurement of ^{2}H and ^{18}O handles the ^{18}O directly as water vapour to monitor the mass 18 ($H_2{}^{16}O$) and ($H_2{}^{18}O$) mass 20 molecular ions (42, 43). Other possibilities exist with modification, for ^{18}O measurements in biological and clinical studies (44, 46). The oxygen isotope content of water is obtained by equilibration with CO_2 (of known isotope composition) followed by analysis of the CO_2 isotope ratio (47). Most precise measurements of this nature require correction for the ^{13}C and ^{17}O differences encountered in both reference and sample gases, and to account for any dissolved cations where relevant (48). IRMS can be employed to monitor ^{18}O enrichment of organic samples (too low to be detected by GC-MS) provided a controlled combustion is undertaken to provide pure CO_2 for isotope analysis.

3.5 Sulphur-34

To date there is scant reference in the literature regarding the use of ^{34}S labeled compounds in clinical or biological studies. Current availability of ^{34}S suggests that it is appropriate at this time to provide an indication of the range of applicable preparation sample techniques drawn from other investigational areas. Analysis of ^{34}S using IRMS is most frequently performed from sulphur dioxide, though the hexafluoride, derived from the fluorination of sulphides, has also been used for mass spectrometric analysis (49, 50).

As with other sample preparation procedures outlined in this section caution must be exercised at each stage to ensure complete conversion to obviate kinetic fractionation. Sulphides can be oxidised directly to SO_2 for isotope analysis (51) and methods are available for the thermal decomposition of sulphates (52, 53). A single combustion train system has been described which is suitable for both sulphide and sulphate conversion to SO_2 using the same procedure; only the oxidation furnace temperature requires regulation (54).

An alternative approach is the initial conversion of either sulphide or sulphate to hydrogen sulphide (55, 56) from which the sulphur is precipitated as silver sulphide which is easily and reproducibly oxidised to provide pure SO_2 for isotope analysis.

Finally, with increasing moves towards automation in SIRA, the possibility exists of coupling a commercially available sulphur analyser on-line to the IRMS thus providing both quantitation and subsequent isotope ratio analysis. A diagrammatic synopsis of these various gas preparations is provided in Figure 2.

4 Selected Ion Monitoring GC/MS – Enrichment Determination of Labeled Metabolites in Body Fluids

Two main areas involving the use of stable isotopes will be considered:
1. as a tracer in a qualitative manner analogous to the use of a radiolabel, e. g. ^{2}H instead of ^{3}H, ^{13}C replacing ^{14}C etc.;
2. as a quantitative tool where an internal standard in the form of a stable isotope labeled analogue of the analyte is used.

4.1 Stable isotopes used as tracers

Stable isotope labeled analogues have been used as tracers to follow the metabolism and excretion of biological components and drugs. The use of ^{14}C or ^{3}H labeling to aid the isolation and identification of metabolites has been accepted for many years in animal studies but even after extensive isolation procedures a considerable mixture of components can remain (for example, plasticizers or greases introduced into the system via TLC plates, plastic sample tubes etc., together with unresolved endogenous material), which may interfere with mass spectrometric analysis. The use of radiolabels in human subjects has been questioned in recent years due to the risk of long term incorporation of the tracer into organs and tissues. As this is felt to be of too great a risk, especially in children and women of child bearing age, stable isotopes have almost completely replaced radioactive compounds for such tracer experiments.

In order to aid the identification of the metabolites by mass spectrometry two main techniques involving stable isotopes have been employed:

4.1.1 **Full scan measurements**

An analogue with a number of stable isotope atoms incorporated into the molecule can be administered producing an easily identified isotope cluster, due to the presence of the stable isotope. This specific technique has been extensively employed in the

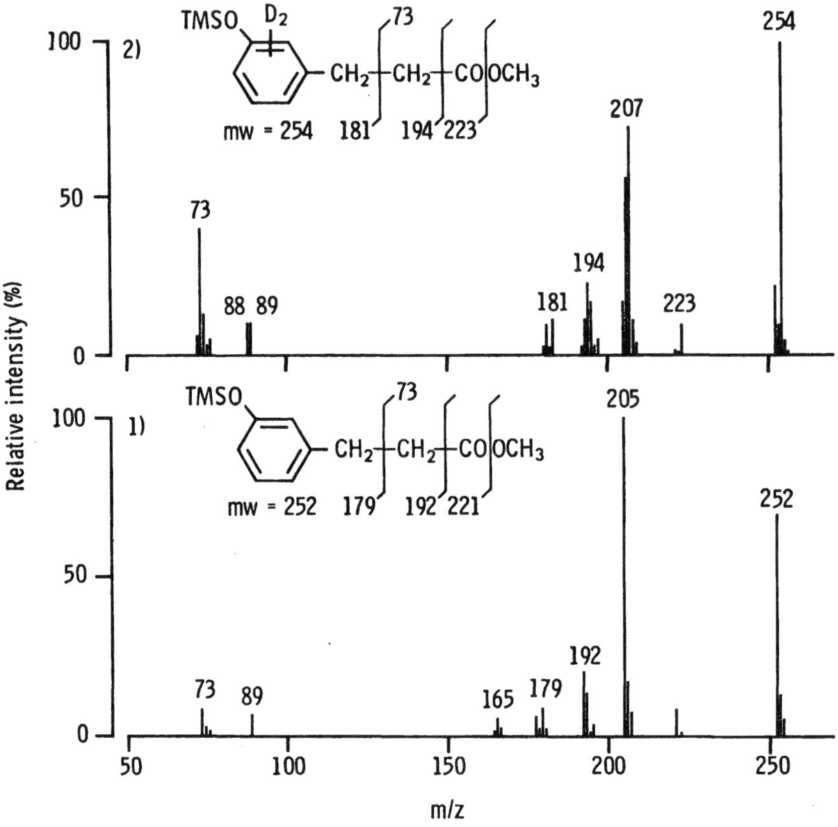

Figure 3. Mass spectrum of deuterated 3-hydroxyphenylpropionic acid obtained after incubation of dideutero-4-hydroxyphenyllactic acid in faeces compared with reference compound. (Reproduced from ref. 59 with permission.)

study of both phenylalanine-tyrosine-dopa metabolism and in *in vivo* enzyme activity in healthy subjects and patients with phenylketonuria (57, 58).

By way of illustration, the intestinal metabolism of tyrosine has been studied by GC-MS using 3,5-dideutero-L-tyrosine as the labeled analogue, both *in vivo* and *in vitro* studies being performed (59). In the latter, incubation of faecal material with a metabolite, dideutero-4-hydroxyphenyllactic acid revealed a number of deuterated aromatic acids. The mass spectrum of the deuterated metabolite 3-hydroxyphenyl propionic acid is shown in Figure 3 compared with the non-deuterated standard.

In a similar experiment, a spectrum of tetradeutero *p*-cresol, obtained after incubation of (3′,5′,2,2-$^{2}H_2$)-4-hydroxyphenylacetic acid is shown in Figure 4 compared again with the non-deuterated standard. The isotopic cluster around the molecular ion (*m*/*z* 112) can be clearly seen.

An alternative method that has been employed to study the metabolism of drugs such as propantheline bromide (I) is to administer an equimolar mixture of the labeled and unlabeled compound (60). Upon analysis of the acquired full scan mass spectra the appearance of characteristic doublets in the spectrum facilitates identification of drug and metabolites.

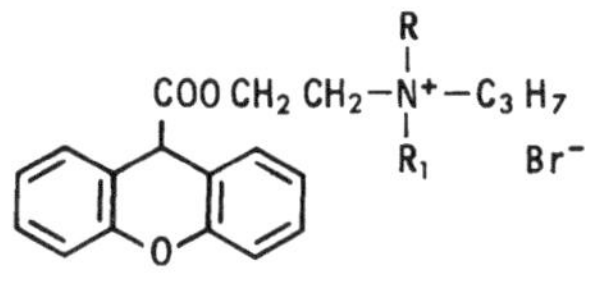

	R	R_1
I	CH_3	C_3H_7
II	CH_3	
d_3 - analogue	C^2H_3	C_3H_7

Figure 5 shows a chromatogram from a methylated urine extract obtained after administering a 1 : 1 mixture of d_0/d_3-propantheline bromide to a single subject. The presence of depropan-2-yl propantheline (II) in the mixture was established by selecting mass chromatograms of specific ions (*m/z* 326, 103, 100) from the total ion chromatogram. A CI spectrum taken from scan 330 (Fig. 6) shows characteristic doublets present for the pseudo molecular ion (*m/z* 326/329) and the major fragment ion at *m/z* 100/103. Ammonia was used as the reactant gas.

This study employed a mixture of stable and radiolabeled analogues to identify minor metabolites. The radiolabel was used to trace the metabolites through the extraction procedures while the stable label acted as a marker for the mass spectrometric identification.

4.1.2 **Selected ion monitoring**

For very small quantities of analyte ($< 1-10$ ng) scanning the mass spectrometer over a wide mass range may not give adequate sensitivity. The use of selected ion monitoring (SIM) by tuning the mass spectrometer to sample only the signal from specific ions can give a large increase in sensitivity and selectivity. For metabolite monitoring this obviously requires some knowledge of the possible nature of the suspected metabolites when choosing ions to monitor.

Several points have to be considered when planning such labeled experiments. To obtain information on as many metabolites as possible the label/labels must be placed in positions largely resistant to biotransformation. Deuterium can be a poor choice of label for in vivo studies due to the large isotope effects associated with it, but has the advantages of being cheap (relative to ^{13}C or ^{15}N) and available in high isotopic purity ($> 99\%$). However, by placing deuterium labels in positions where primary isotope effects may occur, observation of any distortion of the isotope cluster may give useful information about such effects. Information on recycling of parts of a molecule may be of interest, such as in the case of glucose recycling where uniformly labeled ^{13}C-glucose can be used to study exchange of carbon atoms.

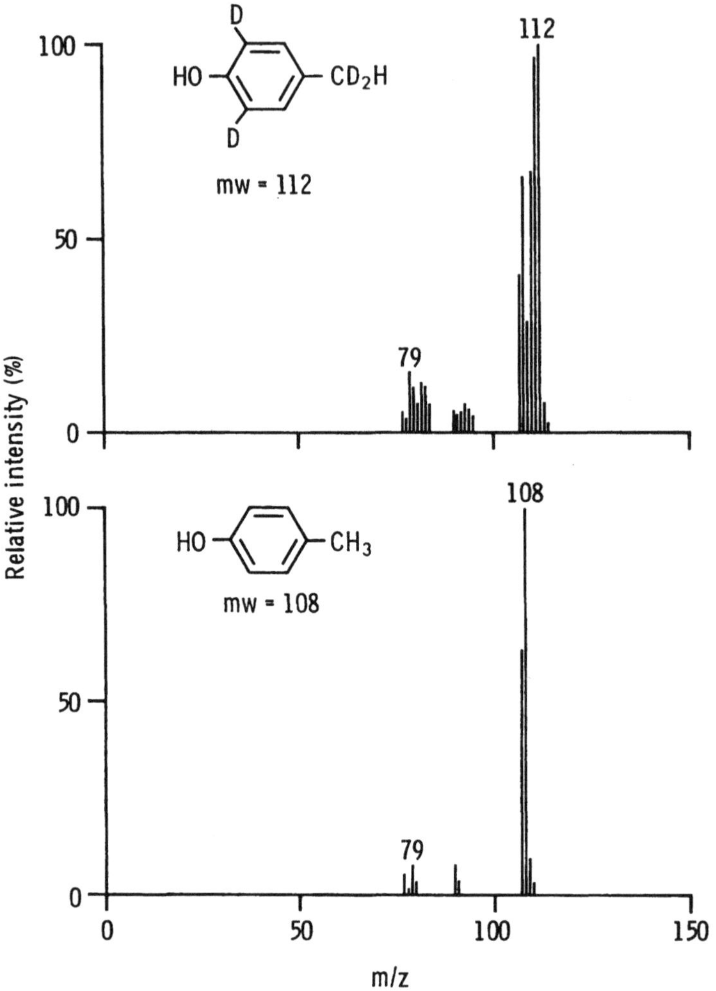

Figure 4. Mass spectrum of tetradeutero *p*-cresol obtained after incubation of (3',5',2,2-$^2H_2$)-4-hydroxyphenylacetic acid compared with standard *p*-cresol. (Reproduced from ref. 59 with permission.)

4.2 Quantitative analysis by isotope-dilution mass spectrometry

The use of a stable isotope labeled compound in SIM by Sweeley et al. in 1966 heralded an approach that has become one of the most widely used techniques in quantitative mass spectrometry (61). The concept involves the dilution of the biological analyte by a stable labeled analogue which acts as an internal standard for correction of losses encountered during the assay procedure.

A number of factors have to be taken into account in the use of such compounds and there has been much debate as to whether in fact an alternative internal standard, such as an isomeric analogue, is preferable.

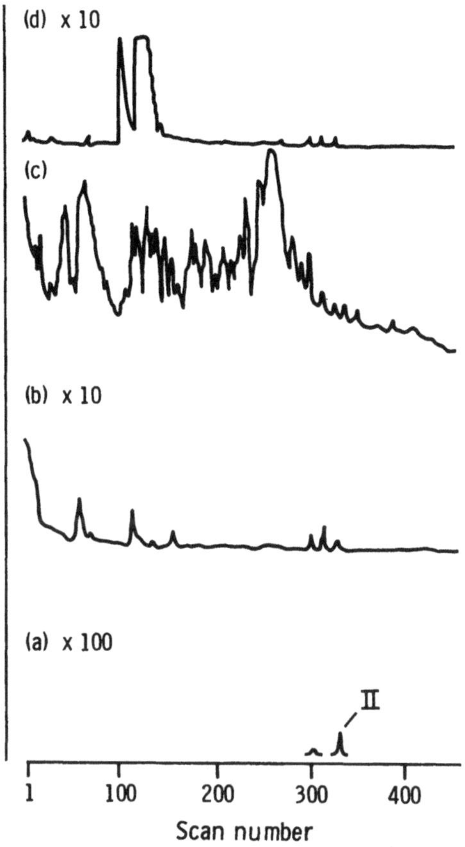

Figure 5. Mass chromatograms at a) m/z 326, b) m/z 103, d) m/z 100 and c) total ion current from methylated 2–4 h urine extract obtained after administration of propantheline bromide. (Reproduced from ref. 60 with permission.)

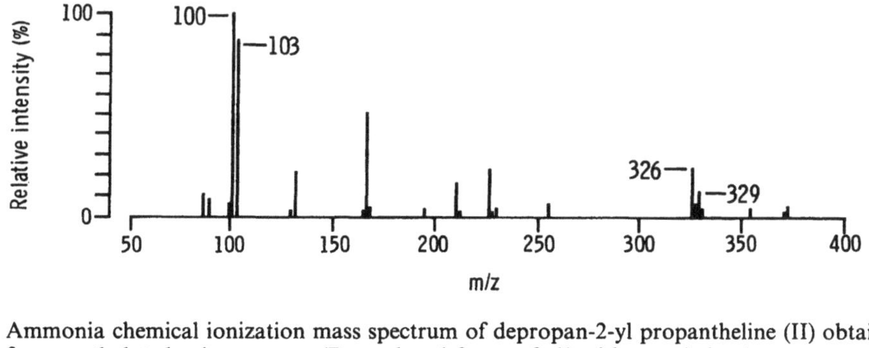

Figure 6. Ammonia chemical ionization mass spectrum of depropan-2-yl propantheline (II) obtained from methylated urine extract. (Reproduced from ref. 60 with permission.)

As the labeled compound is chemically similar to the analyte, it has very similar extraction, chromatographic and ionization characteristics, and addition of the internal standard at the beginning of the extraction procedure acts as a carrier through the method, minimising errors due to procedural losses.

To quantitate the biological analyte a calibration curve is usually prepared in the biological medium by spiking aliquots with varying amounts of the standard analyte and a fixed amount of the labeled internal standard. The mass spectrometer is tuned to monitor a fragment ion from each species (p and $[p + n]$, where n is equal to the number of stable isotopes) and the peak area (or height) ratio determined for each sample. A graph of peak area ratio ($p/[p + n]$) is plotted against concentration of analyte added and a suitable linear or curvilinear equation fitted to the data. Samples containing unknown amounts of the analyte are then analysed after addition of the internal standard, by determination of the peak area ratios and reference to the appropriate curve. Reviews of the use of calibration curves and the interaction of isotopic species are well documented (62–64).

An isotope labeled internal standard should be as isotopically pure as possible, and for this reason deuterium (> 99 atom%) has been the most popular choice. The internal standard should also contain 3 or more isotope atoms in order to reduce any contributions due to naturally occurring heavy isotopes to negligible proportions.

Isotope labeled analogues containing ^{15}N, ^{13}C or ^{18}O are also used, but is is only recently that such labels have been available with enrichments in excess of 95 atom%.

The choice of position of the label in the molecule is very important and a number of aspects need to be considered. If present in labile sites, deuterium exchange can occur during various steps in the analysis procedure. Deuterium atoms can be replaced with hydrogen during the gas chromatographic stage due to exchange from active sites in the injector and GC column. The use of ^{13}C or ^{15}N labels may be more viable alternatives. If derivatization is required prior to analysis, the possibility of losing the isotope label during the reaction must also be considered.

The position of the label influences the method of ionization selected in the mass spectrometric step. For example fragment ions may be of insufficient intensity to be monitored with adequate sensitivity or the label may be lost as part of a neutral species during fragmentation. For high sensitivity and precision a fragment ion of large intensity and in the high mass region of the spectrum is most suitable in order that interference from background peaks such as column bleed or endogenous material does not occur. The selective use of CI to increase the intensity of high mass ions (*vis* pseudo molecular ion) can increase sensitivity dramatically, with a choice of various reactant gases and both positive and NICI being available. Ammonia, being a weak proton donor, is a useful reactant gas and in many compounds the MH^+ ion carries a large proportion of the ion current. The use of this gas has the added bonus in that background components such as column bleed and pump oil vapour are not ionized, leading to lower baseline levels.

When using CI, exchange of the labeled group has to be considered. Ford et al. demonstrated the exchange of trideutero groups in propantheline bromide when using methane as a reactant gas (approx. 7% exchange). This was eliminated by changing to ammonia (65).

Farmer et al. (66) developed a specific and sensitive method for the quantitation of *S*-methylcysteine in haemoglobin using high resolution gas chromatography com-

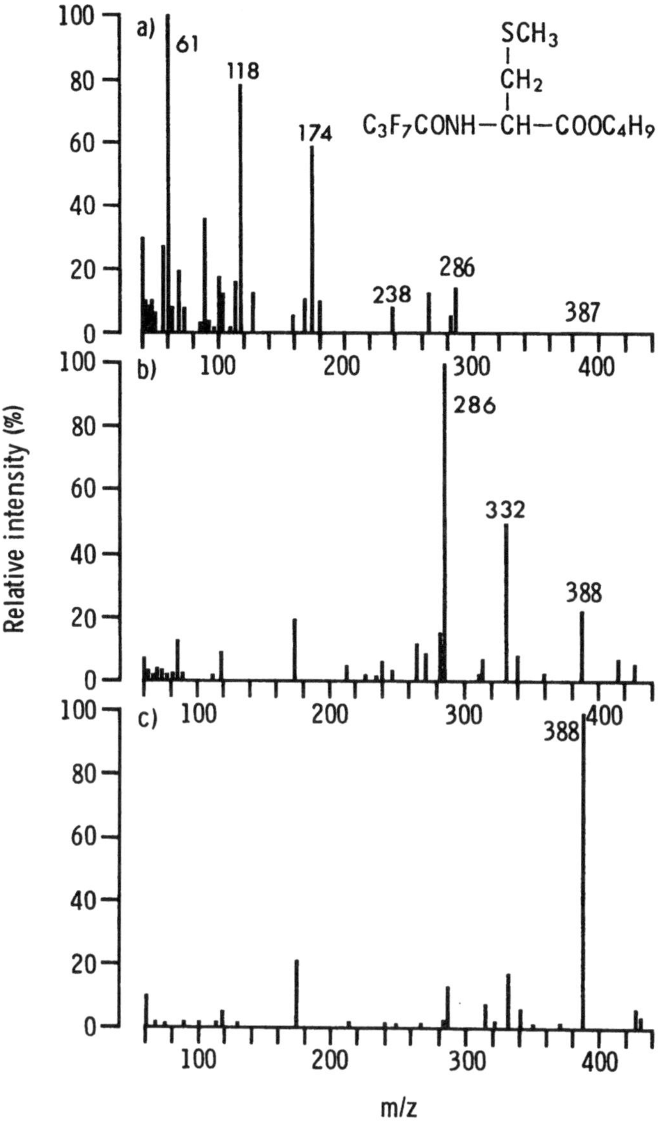

Figure 7. Mass spectrum of *S*-methyl cysteine *N*-HFB, *n*-butyl ester obtained under a) electron impact, b) methane chemical ionization and c) isobutane chemical ionization conditions. (Reproduced from ref. 66 with permission.)

bined with SIM mass spectrometry. The alkylated amino acid was analysed as its *N*-heptafluorobutyryl *n*-butyl ester under isobutane CI conditions using *S*-trideuteromethyl cysteine as internal standard. Linear calibration curves over various concentration ranges were established and sensitivity was in the low pg range, an injection of 10 pg of *S*-methylcysteine giving a signal-to-noise ratio of 20 : 1 when monitoring ions *m*/*z* 388 and 391.

Figure 7 shows a comparison of mass spectra of the *N*-HFB, *n*-butyl ester obtained under a) EI, b) methane CI and c) isobutane CI, and it is observed that there is an enhancement of the molecular ion as ionization conditions become less severe. Under EI conditions, a weak molecular ion (*m*/*z* 387, 0.4%) was observed with major fragmentations leading to *m*/*z* 286 $[M-COOBu]^+$ (13.5%), *m*/*z* 238 $[M-CH_3SHCOOBu]^+$ (7.6%), *m*/*z* 174 $[M-NH_2COC_3F_7]^+$ (57.7%), *m*/*z* 118 $[M-C_4H_8NH_2COC_3F_7]^+$ (76.7%) and *m*/*z* 61 $[CH_3SCH_2]^+$ (100%). Under methane CI conditions the pseudo molecular ion (MH^+) at *m*/*z* 388 is 22.4% of the base peak (*m*/*z* 286 $[M-COOBu]^+$) and the major ion at *m*/*z* 332 results from loss of butene $[MH-CH_2=CHCH_2CH_3]^+$ (49.8%) (Fig. 7).

Using the softer reagent gas, *iso*-butane, fragmentation is reduced still further with the base peak being the MH^+ ion making it very suitable for SIM studies.

The choice of derivative (if necessary) can also be important in terms of both directing fragmentation and choice of ionization technique. The use of electron capturing derivatives such as trifluoroacetate, HFB and pentafluorobenzyl oximes and esters can lead to a dramatic increase in sensitivity by utilising negative CI.

Waddell et al (67) investigated the use of pentafluorobenzyl derivatives in the analysis of prostaglandins in plasma. Figure 8 illustrates mass spectra obtained for 6-oxo $PGF1_a$ after formation of the PFB oxime, methyl ester, the methyl oxime, PFB ester and the PFB oxime, PFB ester derivatives.

Formation of the PFB oxime, methyl ester gives an intense spectrum under methane negative ion conditions, but with extensive fragmentation. A large proportion of the total ion current is carried by two fragment ions, at *m*/*z* 178 $[C_7F_4H_2O]^-$ and *m*/*z* 196 $[C_7F_5HO]^-$ which arise from the PFB moiety. Monitoring the $[M-HF]^-$ fragment gives a limit of detection of 800 pg. Formation of the methyl oxime-PFB ester derivative results in a dramatic redirection of the fragmentation pattern. The base peak is now observed at *m*/*z* 614 $[M-PFB]^-$ with only a small ion (4%) at *m*/*z* 181. The limit of detection monitoring the $[M-PFB]^-$ ion in this case is 2 pg. Finally preparing the PFB oxime-PFB ester produces a negative ion mass spectrum containing four major ions at *m*/*z* 780 $[M-PFB]^-$ (21%), *m*/*z* 196 $(C_7F_5HO)^-$ (100%), *m*/*z* 181 $[C_7F_5H_2]^-$ (12%) and *m*/*z* 178 $[C_7F_4H_2O]^-$ (18%). Efficiency of ionization is similar to the MO-PFB ester but sensitivity is reduced by a factor of three.

Thus although negative ion CI can be a very sensitive and powerful technique, care must be taken in directing the fragmentation pattern to ensure that an intense fragment does in fact contain the site of the label.

Prior to the routine use of capillary columns, a popular procedure was to add the labeled internal standard in large excess (100–1000 fold), not only to act as a carrier through the extraction procedure but also through the GC column and interface, thus minimising losses via adsorption of the column packing and surfaces of the mass spectrometer interface. The evidence for such an effect was of much debate (68, 69), the carrier effect depending on the compound and the site of labeling. As capillary columns, of both glass and fused silica construction are used extensively today, these procedures are of limited use, especially with the high degree of inertness of modern columns and the ability to couple these columns directly into the ion source.

A number of authors have proposed the use of an isomeric analogue as internal standard in place of the labeled analogue (68, 70). A compound is chosen which is chemically similar to the analyte, thus having similar extraction and chromatographic

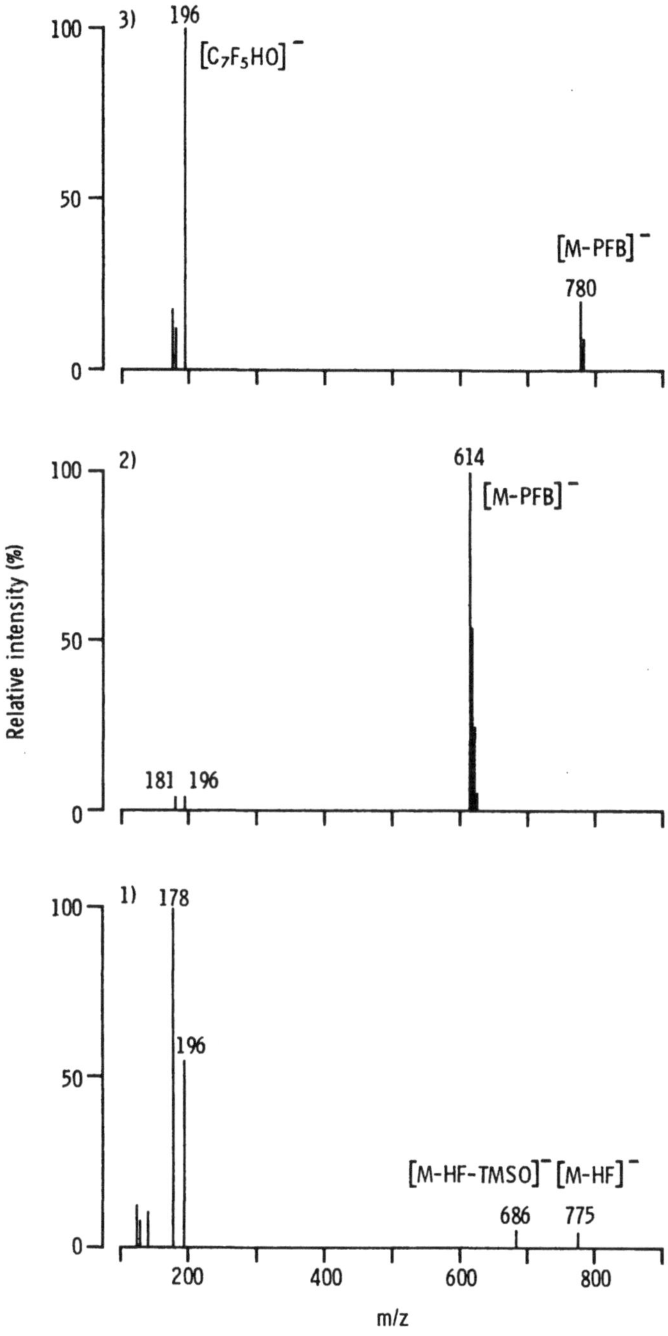

Figure 8. Negative ion mass spectra of 6-oxo $PGF_{1\alpha}$ for 1) PFB oxime, methyl ester, 2) methyl oxime, PFB ester and 3) PFB oxime, PFB ester. All spectra were obtained under methane chemical ionization conditions. (Reproduced from ref. 67 with permission.)

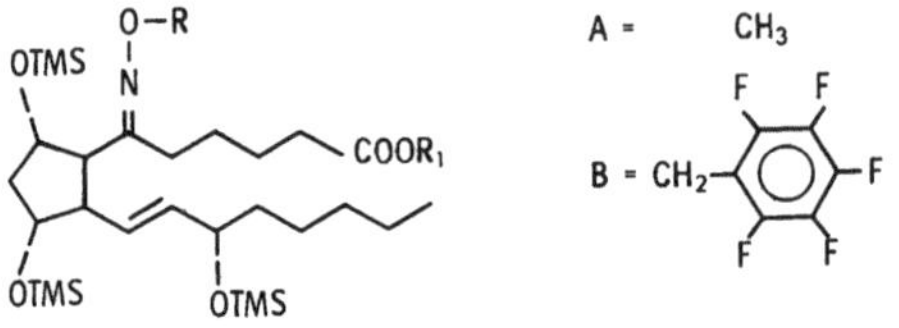

	R	R'
1	B	A
2	A	B
3	B	B

properties, but producing the same fragment ion under ionization. The mass spectrometer is tuned to monitor this single fragment ion with a resultant increase in sensitivity, the two components being separated according to retention time only. The labeled analogue may be added solely as a carrier, if necessary, the only condition being that the labeled compound does not interfere at the selected mass.

A novel approach by Bailey et al. (71) in an extension of the work the group performed on the analysis of *S*-methylcysteine (described above) was to prepare a D-enantiomer of the amino acid as internal standard and perform the separation of the components on a chiral column. In this study the trifluoroacetyl, butyl ester derivative was used in preference to N-HFB, *n*-butyl ester and SIM was performed using *m/z* 288. A three-fold increase in signal-to-noise ratio was observed in comparison with using the trideuteromethyl analogue as internal standard.

Figure 9 illustrates a single ion chromatogram at *m/z* 288 of an extract from guinea pig haemoglobin showing the separation of the D and L forms of *S*-methylcysteine. The background level was found to be 16.1 ng · mg^{-1} haemoglobin.

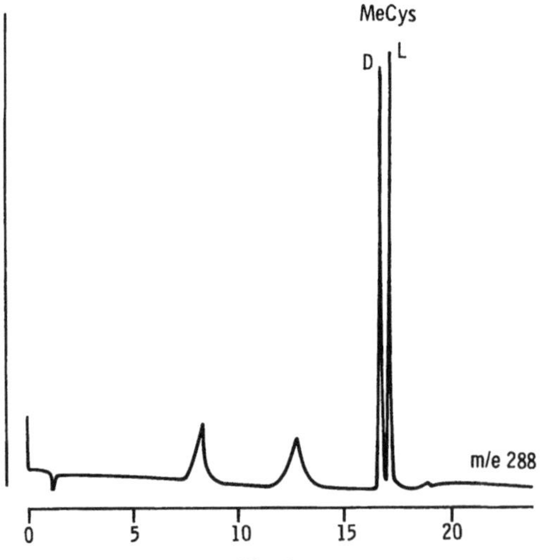

Figure 9. Mass chromatogram of *m/z* 288 showing separation of D and L forms of *S*-methyl cysteine *N*-TFA, *n*-butyl ester using a chiral column coated with Chirasil-Val. (Reproduced from ref. 71 with permission.)

5 Whole Body Stable Isotope Dilution Studies – Variations on a Theme

5.1 Water – ^2H_2O, $H_2{}^{18}O$, $^2H_2{}^{18}O$

Accurate measurement of total body water (TBW) provides indirect information with regard to whole body composition and energy balance. Control mechanisms regulating energy balance in man are little understood and the study of energy balance is complicated by the fact that alterations in body weight may reflect a change in the energy stores, an alteration of TBW, or both. Direct methods for the estimation of body fat or protein in human subjects are lacking and so the degree of hydration assumes importance in energy balance studies. Assuming the TBW of an adult male is some 40 kg and the energy value of adipose tissue is approximately 29 MJ (7000 kcal) per kg, estimation of the former to $\pm$ 200 g (0.5%) will result in an error of only 5.8 MJ (1400 kcal) in measuring energy balance.

Accurate determination of TBW clearly depends both on the uniformity of distribution of chosen label and the precision with which that label can be measured in body fluids. Introduction of ^2H_2O into the body results in the exchange with H_2O molecules for ^2H_2O but in fact the apparent dilution space will be greater due to additional exchange with labile hydrogen atoms of carboxyl, hydroxyl, amino, imino and sulphydryl groups of proteins and other macromolecules. Volume of dilution errors so incurred have been theoretically calculated (72), and experimentally determined to represent a water equivalent of 0.2–0.5% of body weight (73). ^2H_2O has been used to determine total body water in a range of age groups and during the trimesters of pregnancy and post-partum. Pre-dose and equilibrated samples of saliva for deuterium content preclude the necessity of an invasive technique. Similarly if $H_2{}^{18}O$ is employed as label the ^{18}O will rapidly equilibrate with the body bicarbonate pool (carbonic anhydrase activity) and TBW can be estimated from the ^{18}O enrichment of CO_2 in expired air. This approach has been used to compare TBW estimates in the same individuals using ^2H_2O and $H_2{}^{18}O$ (74), and in uncomplicated premature infants and Peruvian infants of 6–36 months (75).

Deuterium dilution methodology has also found expression for the estimation of the volume of milk ingested by solely breastfed infants; clinical information on which nutrient and energy requirements during lactation and infancy are based. Historically, test weighing (babies' weight, before and after feeding), milk expression per unit time and more latterly an ultrasound technique (76) have been used to estimate breast milk intake. Ideally the method of choice should cause minimal interference with the physiology and practice of breast feeding and on these grounds the deuterium dilution technique which follows the disappearance of label in the babies' body fluids (saliva) would appear to fit these criteria. This methodology has been validated against test weighing with good agreement regarding milk intake (77), though not all researchers have found this to be the case (78) – probably due to problems regarding the storage and/or analyses of deuterium content. Reversal of this methodology, maternal deuterium administration and subsequent monitoring the appearance of deuterium in the infant has also been reported (79).

At first sight it is not immediately obvious that a close relationship should exist between body water and respiratory metabolism or indeed that it would be of biological significance. Surprisingly, it is now 30 years since the relationship was theoretically demonstrated and its potential application with regard to the estimation of total CO_2 production (and indirectly energy expenditure) in animals indicated (80). One obvious application in man would be to investigate, under field conditions, anecdotal reports of manual workers in developing countries who apparently maintain body weight despite measured energy intake being reported as less than energy expenditure – thermodynamics would suggest that one of the measurements is incorrect!

Briefly, if an animal (or man) is given a bolus dose of $^2H_2{}^{18}O$, following rapid equilibration with total body water, the ensuing decrease in concentration of the ^{18}O isotope in body water will reflect the sum of both water and respiratory carbon dioxide output; whilst the decrease in concentration of ^2H provides a measure of the water output alone. Determination of the differential disappearance rates of ^2H and ^{18}O permits the calculation of total CO_2 production and hence energy expenditure. Derivation of this relationship is such that:

$$r_{CO_2} = \frac{N}{2}(k_{18} - k_2)$$

where r_{CO_2} is the CO_2 production rate ($mol \cdot d^{-1}$), N is the total body water pool size (mol) and k_2, k_{18} are the $^2H_2{}^{18}O$ elimination rates (d^{-1}). 2 indicates 1 mol O_2 per mol CO_2.

Modification of this original formula accounts for physical isotope fractionation occurring when CO_2 and H_2O vapour leave the body, thus:

$$r_{CO_2} = \frac{N}{2f_3}(k_{18} - k_2) - \frac{f_2 - f_1}{2f_3} r_G$$

where f_1, f_2 are fractionation factors to account for the evaporative water losses of ^2H and ^{18}O respectively, f_3 is the fractionation factor for ^{18}O between H_2O and CO_2 and r_G represents the rate of evaporative water loss.

Validation studies comparing the doubly labeled water method with more conventional metabolic chamber measurements of CO_2 production in small animals, birds, reptiles and insects (81–83) have provided collective methodological agreement within about 5% (range 3–10%), including a recent validation in starlings and sparrows (84); only the scorpion exhibited wide disagreement between the two methods (36 ± 24%) (85).

A decade ago the theoretical rationale of the method was critically dissected and the feasibility of its application to man favourably considered (86). Some five years ago we commented that with the current high level of interest in energy balance and obesity no attempt had been made to validate the $^2H_2{}^{18}O$ method in man (87). Since that time however, there have been four such validation studies performed comparing energy expenditure measurements obtained indirectly from the $^2H_2{}^{18}O$ method with those obtained by direct or indirect calorimetry (88–93). Very encouraging comparative results have been obtained indicating only a 2–5% variation between the methods over periods of 4–12 days. This indirect method of estimating energy expenditure has recently been applied to study diverse groups as breastfed infants (94) and cyclists undergoing heavy sustained exercise (95).

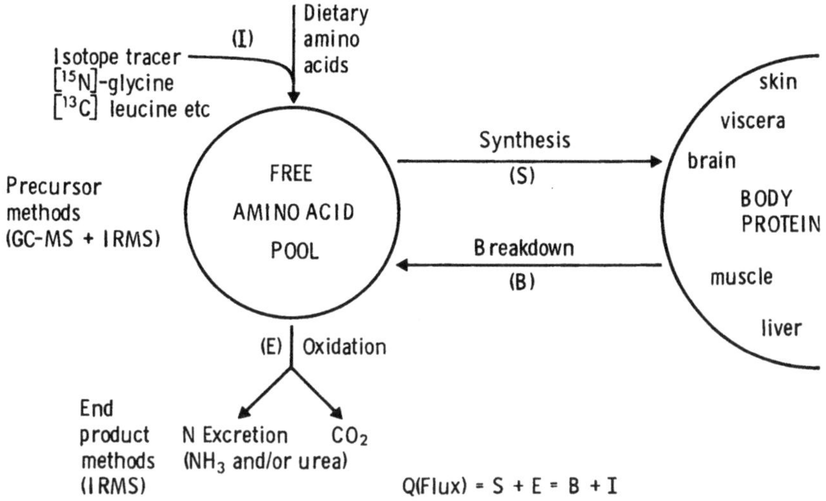

Figure 10. Two-pool model generally employed to study dynamic aspects of whole-body nitrogen protein metabolism involving the use of stable isotopes where dilution is measured by IRMS (end-product methods) or by GC-MS and IRMS (precursor methods).

Certainly the availability of cheaper $^2H_2{}^{18}O$ label and a commercial dual-IRMS for the simultaneous measurement of m/z 3/2, m/z 20/18 ratios have made these validations feasible (42, 43). All authors point out that the above methodological comparisons were performed under ideal laboratory conditions and thus care should be exercised in terms of sample collection, storage and transportation in field studies performed remote from the analytical centre.

5.2 Protein metabolism

The ordered structure of proteins is vital to their specific biological function and once formed why should they be degraded? Tight enzymatic control of a myriad of biochemical reactions within the body can only be acutely maintained if the ability to degrade and resynthesise a given enzyme exists. Clearly in situations of growth and repair protein synthesis must exceed breakdown in individual organs or in the body as a whole. Not surprisingly perhaps stable isotope labeled amino acids have been extensively employed to obtain estimates of whole body protein turnover in normal subjects from neonates to the elderly and in an ever increasing range of pathologic states.

Two major approaches have been followed in an attempt to quantitate whole body protein metabolism: end product methods (using [^{15}N]glycine exclusively) and precursor methods (using a representative amino acid tracer). A simple two pool model relating to a free and fixed amino acid pool is most often employed (Figure 10). Labeled amino acid is administered either orally or intravenously to enter the supposed homogeneous amino acid pool from which it may depart for cellular protein synthesis or be subjected to oxidation, to be eliminated from the body as expired CO_2 or urinary

N. The relatively enormous body protein pool (fixed amino acids), though continuously recycling amino acids through the metabolic pool, is assumed not to cycle during the time course of a study. In the steady state (or as steady a state as can be engineered) the tracer amino acid flux Q is obtained from the equation $Q = S + E = B + I$ (Fig. 10). Measurement of flux is determined by end product methodology, requiring IRMS analysis, exclusively approached using [^{15}N]glycine or the precursor approach in which case [1 – ^{13}C]leucine has been the amino acid of choice in most instances to date and necessarily requires GC-MS techniques for isotope analysis. A discussion concerning the theoretical considerations and assumptions involved in the application of these methods is not possible here but, if required a comprehensive treatise is available (96).

To summarise, modern end product methodology demonstrates that the stochastic approach (constant infusion of [^{15}N]glycine to achieve 'plateau' labeling in [^{15}N]urea) can be used to quantitate whole body protein dynamics though the biggest single drawback to this method of measuring protein flux is the protracted nature of the studies, where in an adult subject a urinary [^{15}N]urea plateau may not be obtained for some 60 h. To overcome this major problem such that acute changes in protein metabolism may be monitored and repeated on a short time basis (2–3 days) the use of a single dose of [^{15}N]glycine has been extensively investigated (97–100). The stochastic technique has provided quantitative data of protein flux in preterm low-birthweight infants (101, 102), neonates (103), normal infants (104), in young adults at two levels of protein intake (105) and in geriatric subjects (106, 107). Alterations in protein flux (synthesis and breakdown) and metabolic consequences have been demonstrated in stress and trauma, including burn injury (108), cancer (109) and the effects of surgery (110–112). In addition, a greater appreciation of the general relationships existing between protein flux, dietary requirements, nitrogen balance, body weight and basal energy expenditure has been achieved from these studies.

Precursor methods for determining whole body protein dynamics are inherently related to individual amino acid metabolism. Thus stochastic precursor methods all aim to achieve a 'plateau' of labeling of infused tracer amino acid in the plasma (amino acid free pool) such that the net entry and exit of the tracee are equal in the steady state. Total tracer and tracer amino acid flux can then be calculated from a knowledge of the 'plateau' labeling (measured by GC-MS) and the rate of infusion of tracer (i. e. isotope dilution). If [1-^{13}C]leucine is the tracer used, then determination of enrichment and production rate of CO_2 in expired air (end product) by IRMS permits flux to be partitioned between synthesis and oxidation. In this manner whole body protein flux has been studied in low-birthweight infants (113), normal subjects, fed and fasted (114, 115), and during exercise (116, 117). In addition, effects of 3-hydroxybutyrate (118) and cortisol (119) on whole body proteolysis have been investigated as have the effect of poor diabetic (insulin) control and obesity on whole body protein metabolism (120).

At present few clinical research centres possess both IRMS and GC-MS analytical facilities required for the above studies – however, with a growing realisation of the quality of metabolic information that can be obtained from such studies the situation is improving rapidly.

5.3 Glucose metabolism

Glucose is the dominant energy source in man and quantitative aspects of glucose metabolism assume a central position in substrate homeostasis in health and disease. The classic glucose tolerance test (oral or intravenous) simply indicates the response of blood glucose concentration to an exogenous load but provides no quantitative information of endogenous glucose production rate or clearance rate.

Resurgence of stable isotope involvement in clinical studies over the last decade has resulted in variously labeled glucose being extensively employed to probe specific aspects of carbohydrate metabolism. Recently the choice of label to differentiate between hepatic glucose production, glucose appearance from recycled C-atoms and gluconeogenesis has been comprehensively detailed (121). For obvious reasons the paediatric population has proved to be the initial focus of attention in terms of glucose kinetic studies. Rates of glucose production and utilisation in this group following a 4 h or overnight fast are some 5–6 mg $\cdot$ kg^{-1} (122, 123) which approximates to a threefold increase in the production rate observed in fasted adults (124, 125). Glucose flux rates immediately post-absorptive have been shown to be linearly correlated with estimated brain weight throughout life, a reflection of the brain being a major glucose-utilising organ. Extensions of these studies have followed glucose production rates in childhood (126), in the last trimester and at term in normal and diabetic pregnancies (127, 128) and in elderly subjects (129, 130). In this latter study glucose recycling and oxidation rates did not differ significantly between young and elderly subjects but the older group required double the amount of exogenous insulin to maintain normoglycaemia when 4 mg $\cdot$ kg^{-1} $\cdot$ min^{-1} 'cold' glucose was infused compared with the younger group. Perturbations in glucose metabolism have been investigated in severely malnourished children (131), in hypoglycaemic low-birthweight infants (132), in infants of diabetic mothers (133) and in children with maple syrup urine disease (134), Duchenne muscular dystrophy (135) and cyanotic congenital heart disease (124). Additional studies indicating the effects on glucose metabolism of burn injury (137, 138), intravenous 'cold' glucose infusion (139) and in response to total parenteral nutrition have been reported (140).

When glucose oxidation measurements are required as an integral part of a study, quantitation of the $^{13}CO_2$ content in expired air and determination of total CO_2 production rate, provide the necessary information. The analytical method (ultimately IRMS analysis) is common, irrespective of the ^{13}C-labeled substrate employed, or if naturally labeled maize glucose is the substrate used. A group of Belgian workers have extensively used this latter substrate, taking advantage of the selective isotope fractionation that occurs at the primary carboxylation stage of plant photosynthetic fixation of atmospheric CO_2 (141, 142). Thus, sugars derived from C_4 plants (maize, sugar cane) contain some 15 parts per thousand more ^{13}C than C_3 plants (sugar beet potato). Using 100 g oral load of maize glucose these workers have investigated the utilisation of glucose in normal obese and diabetic obese subjects (143, 144) and the effect on this utilisation at differing levels of exercise (145, 146). Results from these studies support the concept that in obese subjects, exercise stimulates carbohydrate utilisation, producing an acute improvement of glucose tolerance.

6 General Clinical Applications

6.1 ^{13}C-Breath tests – gastrointestinal malfunction

The lung is demonstrably an extremely efficient excretory organ for the end products of metabolism including metabolic CO_2 originally derived from degradative and/or oxidative pathways. Thus the theoretical basis of any ^{13}C-diagnostic breath test is the use of a specifically labeled substrate that contains a 'target bond', which following enzymatic cleavage will cause the release of a functional group destined to produce $^{13}CO_2$ as a metabolic end product. Oral substrate administration permits staged analysis of components of actual absorption, hepatic transformation, plasma distribution and end organ oxidation. Clearly the diagnostic index of any test will be highest if only one step in the above sequence is rate limiting. The practicalities of performing breath tests and relevant theoretical considerations relating to expired $^{13}CO_2$ collection, sample handling, storage and manual or automatic enrichment measurement are well documented (147–150). Increasing awareness of the possible variations in dietary ^{13}C-intake (variations in natural ^{13}C-abundance) which will be superimposed on a breath test and thereby decrease their signal-to-noise ratio have also been considered in some detail (151, 152).

To date ^{13}C breath tests have been restricted almost exclusively to gastrointestinal malfunction emphasising the fact that they represent, in most instances, a lateral development of existing ^{14}C breath tests rather than original tests per se. Despite the stated high diagnostic value demonstrated in many specific tests cited below, gastroenterologists in general, apparently lack enthusiasm towards implementing such tests into their own clinics. This fact is particularly difficult to comprehend in terms of fat malabsorption studies; the long used classical approach is a 72 h fat balance (fat intake and stool fat content measured) – the alternative, a 6 h [^{13}C]trioctanoin (trioctanoylglycerol) breath test. For the clinical chemist, the choice of 1–1.5 kg stool sample to homogenise, lipid extract and assay, compared with 6–8 gas bottles to be placed on an automated IRMS for $^{13}CO_2$ analysis would present little problem one suspects! This ^{13}C breath test has provided positive discrimination of fat malabsorption from other causes of steatorrhea when in excess of 25% of dietary fat is excreted and discrimination in the majority of patients with 10–20% d^{-1} excretion (153). These authors subsequently presented evidence suggesting that tripalmitin (tripalmiloylglycerol) or triolein (trioleoylglycerol) provide a greater diagnostic discrimination in general fat malabsorption studies than trioctanoin. Sequential use of [^{13}C]triolein, trioctanoin and palmitic acid indicate that fat malabsorption due to pancreatic insufficiency may be differentiated from that due to bile acid deficiency or intestinal mucosal disease (154). ^{13}C stool analysis is a useful addition to the breath test to distinguish between bile acid deconjugation and malabsorption (155). Ileal dysfunction and upper-intestinal bacterial overgrowth can be diagnosed using [^{13}C]glycocholate (156), hepatic microsomal mixed oxidase enzyme activity (drug metabolism) in mild and severe liver disease with [^{13}C]aminopyrine (157, 158), general liver function appraised with related tests (159–161) and the severity of alcoholic cirrhosis employing [^{13}C]galactose (162). One obvious advantage of these tests is that they can be serially repeated in order to monitor the progression of disease or more positively to indicate the efficacy of therapeutic treatment.

Undoubtedly in the near future more use will be made of simpler (and therefore cheaper) ^{13}C labeled molecules such as bicarbonate, acetate (163), proprionate and urea (164) to probe aspects of metabolism using a similar protocol to that outlined above.

6.2 Direct measurement of individual protein synthesis rates

Implicit in the estimation of the fractional protein synthetic rate of an individual protein, or closely related group of proteins is a knowledge or estimate of the precursor pool enrichment from which protein is synthesised; ultimately the appropriate amino acid tRNA enrichment is required but obtaining this measurement is normally at best impractical in terms of tissue requirement and at worst unethical.

Several different approaches have been employed to measure specific body protein fractional synthetic rates in man that have in common the constant infusion of the chosen tracer amino acid. One such approach is the use of [^{15}N]glycine such that hepatic arginine is indirectly labeled (165). As urea is synthesised from the same precursor as hepatic arginine, measurement of urinary [^{15}N]urea indicates the precursor enrichment, whilst the rate of incorporation of ^{15}N into serially obtained plasma albumin is provided by measurement of the arginine bound guanidine-^{15}N enrichment by IRMS, following chemical separation and subsequent nitrogen release. Acute changes in protein synthetic rate cannot be detected by this approach, as urinary urea-^{15}N steady state enrichment is not achieved for 48 – 60 h. However, using a primed continuous infusion of [^{15}N]glycine it has been demonstrated that plasma [^{15}N]glycine exhibits an isotope plateau within 15 min, and urinary [^{15}N]hippurate (an indicator of the hepatic intracellular glycine-free pool) is at plateau within an hour (166). This observation has enabled the direct measurement of plasma apolipoprotein synthesis (VLDL-B fraction) in man employing selected ion monitoring to measure the isotope enrichment in glycine of sequential VLDL-B samples. Direct measurement of total ^{15}N incorporated into specific body proteins (tumour, normal intestinal tissue, liver and fibrinogen) in patients with gastrointestinal malignancies has been compared as a ratio with the ^{15}N enrichment of the tissue-free amino acid pool (167). This approach is theoretically unsound in that the rapid transamination of [^{15}N]glycine to other amino acids that are utilised at different rates for protein synthesis precludes any attempt to monitor the enrichment of the precursor pool enrichment.

We have previously used [α-^{15}N]lysine to measure synthesis rates of plasma albumin and the two major fractions of skeletal muscle protein, namely sarcoplasmic and myofibrillar protein fractions (168). Precursor pool activity was taken as being that of the plasma lysine enrichment at plateau during the 21 – 30 h infusion of label. More recently we have developed a more sophisticated and rapid approach to measure muscle protein synthesis in man using L-[1-^{13}C]leucine, which can be adapted to investigate any tissue obtained by biopsy technique at surgery (28). Briefly, the leucine incorporation into muscle protein is measured by SIRA following a stepwise procedure to isolate free leucine and subsequent decarboxylation to liberate free $^{13}CO_2$, whilst plasma α-ketoisocaproate, leucine's transamination product is taken to most closely approximate muscle intracellular leucine enrichment (169). Interestingly we have demonstrated that in Duchenne and myotonic dystrophy and cancer cachectic patients the muscle wasting observed is a result of decreased protein synthesis compared with normal control subjects as opposed to an increase in protein breakdown (170–173). As skeletal muscle constitutes the largest single protein pool within the body it is reasonable to suppose that muscle protein synthesis will influence estimated whole body protein synthesis markedly despite its relatively slow turnover rate. Further studies are required to define the reproducibility of the method, the possible range of synthetic rates between different skeletal muscles and the effects of exercise and immobilisation on selected muscle groups.

Collagen is an abundant protein in vertebrate animals, located in the extracellular matrix and basement membrane of the majority of tissues. Interest in this protein group stems from the observation that in liver cirrhosis, pulmonary fibrosis, atherosclerosis and rheumatoid arthritis, collagen levels in the affected tissues are elevated. Whether this change results from increased synthesis or decreased protein breakdown can only be answered using tracer technology. Animal studies have indicated that collagen is metabolically inert (174), a conclusion based on radioactive decay from the collagen pool which is known to be subject to error, introduced by the re-utilisation of the labeled precursor amino acid. Recently the possibility of using ^{18}O as a label incorporated into collagen, as a post-translational modification of proline and thus non-reutilisable has been profitably investigated (175). There seems no theoretical reason why [^{13}C] or [^{15}N]proline, as a cheaper alternative, could not be used in man to monitor collagen synthesis employing the loading dose technique where large amounts of 'cold' proline would be administered with the labeled amino acid such that the plasma and tissue-free pool of enriched proline would remain almost constant or follow simple, interpretable kinetics (176).

6.3 Amino acid metabolism

The relatively recent introduction and increasing sophistication of SIM-GC-MS techniques has opened the floodgates to the measurement of plasma substrates labeled with ^{2}H, ^{13}C or ^{15}N and amino acid metabolism, kinetic and quantitative studies being at the forefront (31, 177–179). Emphasis in these studies has been placed on variations in specific amino acid kinetics in the face of dietary nitrogen and energy intake manipulation and in many pathophysiological states. Extensive literature is available on the kinetics of glycine (180), lysine (181), leucine (114, 117, 182), valine (183) and variously labeled alanine (184). Metabolite interconversion can be quantitated from the simultaneous infusion of the two appropriately labeled metabolites. Thus, quantitative data regarding phenylalanine conversion to tyrosine (185–187), leucine nitrogen transfer to form alanine (188) and leucine to keto-acid (α-ketoisocaproate) conversion and reversal have been reported (189).

Amino acid pool size and turnover rate of glycine and alanine have been reported as has the transaminating potential of [^{15}N]glycine (190, 191). In addition to kinetic studies labeled amino acids serve as useful internal standards for appropriate urine and plasma assays (192, 193).

7 Trace Metals – Stable Isotopes

Up to 20 trace elements (minerals, micronutrients), many of which exhibit several stable isotopes, have been identified as being essential to animal growth, reproduction and health. Human requirements have been accurately determined for only a few trace elements whilst the essential nature of others has been inferred from studies in experimental animals fed highly purified diets. Attempts to define human requirements accurately are confounded by the fact that bioavailability is dependent on individual food source, overall dietary mixture, and the metabolic state (e. g. pregnancy, lactation, etc.) at the time of study. Trace elements, although a heterogenous group, generally exist as positively charged cations, cause bond weakening by virtue of electron attraction, and thus possess high reactivity. Few can be unaware of the central role occupied by iron as a vital constituent of the haemoglobin and myoglobin molecules and its involvement in the cytochrome respiratory chain. Many trace elements exist as metalloenzymes, especially copper, zinc and selenium, whilst others act as enzyme activators. The original classification of trace metals as being essential, non-essential or toxic is somewhat misleading as all may be toxic if high intakes are experienced. Elevated blood levels of lead, derived from environmental pollution, result in biochemical abnormalities and neuropsychological behaviour patterns. Of relevance to this presentation is the fact that lead pollution sources may be identified by their isotopic composition as lead-204 indicates the 'original' metal, reflecting elemental formation and is not a radioactive decay product. The remaining three lead isotopes 206, 207, 208 are radiogenic resulting from the decay of uranium or thorium and a given lead source will therefore exhibit a characteristic isotopic composition. Kinetics of lead metabolism in adult subjects have been comprehensively investigated using a 3-compartment model derived from the isotopic analysis of blood, sweat, urine, faeces, hair, nails, bone and the diet (194). Several trace elements deficiencies have been documented; for instance, copper deficiency in infants fed only milk for prolonged periods results in microcytic, normochromic anaemia – a result of a failure to absorb iron in sufficient quantities; magnesium deficiency is manifest by enhanced irritability of the nervous system, peripheral vasodilatation and cardiac arrhythmias whilst chronic zinc deficiency may cause dwarfism and delayed puberty, both of which may be rapidly reversed following zinc supplementation in the diet. Finally, iron, zinc, selenium and vandium deficiencies have been implicated in impairment of the immune response (195).

Trace elements of clinical importance and their stable isotopes are shown in Table 1. Cobalt, fluorine, iodine and manganese, although considered essential dietary components have been omitted as they are all monoisotopic and thus not amenable to investigation by isotope ratio methodology. Table 2 briefly summarises the increasing range of mass spectrometer configurations that are being brought to bear on the analysis of trace elements.

The analytical path chosen will depend on the elements of interest, the intended sample load and the required precision which has been reported to range from some 10% for neutron activation analysis (NAA) (219) to less than 1% when thermal ionization-mass spectrometry (TI-MS) is the analytical method applied to biological samples (220). For particular elements there are additional considerations that influ-

Table 1. Natural abundances of clinically important trace metal isotopes and recent published studies relating to their bioavailability, absorption and metabolism, intrinsic or extrinsic, when administered as labels

Element	Isotopes (natural abundance in %)	References
Calcium	^{40}Ca, 96.94; ^{42}Ca, 0.65; ^{43}Ca, 0.14; ^{44}Ca, 2.08; ^{46}Ca, 0.003; ^{48}Ca, 0.185	196, 197
Chromium	^{50}Cr, 4.31; ^{52}Cr, 83.76; ^{53}Cr, 9.55; ^{54}Cr, 2.38	198
Copper	^{63}Cu, 69.09; ^{65}Cu, 30.91	199, 200
Iron	^{54}Fe, 5.82; ^{56}Fe, 91.66; ^{67}Fe, 2.19; ^{58}Fe, 0.33	200, 201, 202
Lead	^{204}Pb, 1.4; ^{206}Pb, 24.1; ^{207}Pb, 22.1; ^{208}Pb, 52.4	194, 203
Magnesium	^{24}Mg, 78.99; ^{25}Mg, 10.00; ^{26}Mg, 11.01	204, 205, 206
Nickel	^{58}Ni, 68.27; ^{60}Ni, 26.10; ^{61}Ni, 1.13; ^{62}Ni, 3.59; ^{64}Ni, 0.90	207
Selenium	^{74}Se, 0.87; ^{76}Se, 9.02; ^{77}Se, 7.58; ^{78}Se, 23.52; ^{80}Se, 49.82; ^{82}Se, 9.19	208
Vanadium	^{49}V, 0.24; ^{51}V, 99.76	209
Zinc	^{64}Zn, 48.89; ^{66}Zn, 27.81; ^{67}Zn, 4.11; ^{68}Zn, 18.57; ^{70}Zn, 0.62	210

Table 2. Range of mass spectrometric analytical configurations currently employed for isotope abundance (enrichment) quantitation of trace metals in biological and clinical studies

Instrument configuration	Elements	References
Quadrupole-MS		
– GC	Cr, Se	198, 209
– Solid probe	Fe, Zn, Cu, Mg	200, 202, 205
– Inductively coupled plasma source	Cu, Ni, Mg, Zn	208, 210
– Thermal ionization	Ca, Se, Mo	197, 212
Magnetic-MS		
– GC	Mg, Ca, Fe, Ni	213, 214
– FAB	Ca, Fe	196, 215
– Thermal ionization	Zn, Cu, Pb	203, 210, 216
– Solid probe	Cu, Fe, Zn	199, 200, 201
– Field desorption	Th, Li	217
– Spark source	Ni, Fe, Pb, Mg	218

ence the ultimate precision obtainable irrespective of the analytical approach such as the natural isotope abundance, isotope characteristics and the level of enrichment obtained in a particular sample during a clinical study. These respective points are highlighted for copper, having no isotope with low natural abundance, for iron, where the percentage dietary absorption is inherently low, and for zinc, where only low levels of enrichment can be realistically achieved in plasma or urine.

Atomic absorption spectroscopy and neutron activation analysis do not fall within the remit of this chapter and will not be considered further, though one group in particular is extremely active using the latter analytical technique (221).

Of the analytical methods listed in Table 2, GC-MS of volatile metal chelates and TI-MS require expensive dedicated equipment, extensive sample preparation requiring element specific separation in most cases and ultimately the techniques afford only a low sample throughput (10–16 day^{-1}). A very much more attractive analytical approach of great potential is provided by inductively coupled plasma source mass spectrometry. Electrical plasmas generated in an inert gas at atmospheric pressure find wide application in atomic emission and fluorescence spectrometry. When used as the ion source of a quadrupole mass spectrometer, trace metal isotope ratios can be determined with a precision of 1–3% in discrete solution volumes of 50–200 μl containing only 0.2–4 μg of a given metal (207, 211). Additionally, as most elements give rise to only monatomic ions for any single plasma and ion sampling conditions, simultaneous multielement isotope ratio measurements in a complete biological matrix should prove possible (222).

It is of interest to note in passing that lithium distribution between plasma and cerebrospinal fluid (CSF) in cats has been shown to display isotope fractionation between ^{6}Li and ^{7}Li (223). In contrast to lithium, no isotope effect was observed when feeding rats 97.6% ^{54}Fe compared with 'natural' iron (5.8% ^{54}Fe). ^{58}Fe has been used as an extrinsic label to study iron absorption in man (224), whilst iron chelation with 2,4-pentanedione and analysis by quadrupole-MS with a direct inlet probe has shown potential for the use of 3 iron isotopes (^{54}Fe, ^{57}Fe, ^{58}Fe) for biovailability studies as gauged by peak serum abundance labeling following the administration of extrinsically labeled isotopes (202). FAB-MS also shows promise in this particular area (215).

Quadrupole-MS and the use of a special solids probe has been described for the analysis of calcium isotopes in biological fluids (197), as has the use of high resolution mass spectrometry and FAB gun to provide an extremely sensitive analysis for ^{42}Ca/^{40}Ca ratios ($> 5 \times 10^{10}$ ions · μg^{-1} Ca) (196). The authors discuss possibilities of non-calcium isotope interference and suggest that a resolving power of 5000 probably represents a good compromise between selectivity and sensitivity. This analytical system, though perhaps not available to many investigators, provides a rapid sample analysis with little or no initial preparation and with a precision of 0.2–0.5%.

The two radioactive isotopes of copper, ^{67}Cu and ^{64}Cu have $t_{1/2}$ of 61.9 h and 12.9 h respectively; as the former can only be reactor produced neither are suitable for routine studies, but absorption studies have been reported in man using stable isotopes (200).

Thermal ionization mass spectrometry for the analysis of ^{67}Zn has demonstrated the feasibility of using the isotope to quantitate the homeostatic mechanism for this element engineered by alterations in the absorption from and secretion into the gastrointestinal tract (210). Interest in Zn metabolism can be gauged by reports of absorption studies in normal subjects (225), the effects of phytate and α-cellulose on Zn absorption in young men (226), absorption in elderly men (227) and the effects of differential dietary zinc sources in pregnant women fed strictly controlled diets (228).

8 Summary

To take advantage of the several positive attributes of stable isotopes for clinical research, no longer requires justification. The areas of application, current and potential, outlined briefly above and the almost exponential growth in the appearance of publications relating directly to the use of stable isotopes in clinical research and diagnosis have made this a truism. Inevitably we have been selective in chosen examples and have not covered the movement towards intermediary metabolism studies such as the kinetics of lactate (229, 230), fatty acid (231, 232) and ketone body turnover (233, 235), or the more controversial aspects of priming metabolic pools thereby reducing the time of clinical studies to monitor acute metabolic states (236–238).

Synthetic chemists can now produce almost any substrate of clinical interest, labeled to choice (admittedly not cheaply), and if the current armourment of mass spectrometry techniques proves inadequate with time for all analytical requirements, tandem-MS (239) and particle-induced desorption MS (240) etc., offer further potential for circumventing difficult or protracted sample preparation techniques or involatile compound analysis, respectively. Indeed the quality of the isotope data currently derived from clinical studies is such that in many instances the clinical scientist may well have to relinquish strict adherence to oversimplistic biological models in order to achieve a truer picture of biochemical and physiological events occurring within the cell, in tissues and in the whole body.

Acknowledgement

We thank all members of the CRC Nutrition Research Group for their support during the compilation of this chapter, but especially JKH who could not witness its completion.

References

1. Fallick, A. E. & Baxter, M. S. (1977) The pressure effect and peak broadening in gas source stable isotope mass spectrometry. Int. J. Mass Spec. Ion Phys. *25*, 155–165.
2. Craig, H. (1957) Isotopic standards for carbon and oxygen and correction factors for mass-spectrometric analysis of carbon dioxide. Geochim. Cosmochim. Acta *12*, 133–149.
3. Mook, W. G. & Grootes, P. M. (1973) The measuring procedure and corrections for the high-precision mass-spectrometric analysis of isotopic abundance ratios, especially referring to carbon, oxygen and nitrogen. Int. J. Mass Spec. Ion Phys. *12*, 273–298.
4. Graff, J. & Rittenberg, D. (1952) Microdetermination of deuterium in organic compounds. Anal. Chem. *24*, 878–881.
5. Tamiya, N. (1960) Determination of deuterium in water and organic compounds. Anal. Chem. *32*, 724–725.
6. Goldblatt, M. & Jones, W. M. (1964) Determination of the deuterium content of a sample of deuterium oxide by three methods. Anal. Chem. *36*, 431–432.
7. Fraisse, D., Girard, D. & Levy, R. (1973) Dosage microanalytique du deuterium dans les composes organiques. Talanta *20*, 667–680.
8. Bigeleisen, J., Perlman, M. L. & Prosser, H. C. (1952) Conversion of hydrogenic materials to hydrogen for isotopic analysis. Anal. Chem. *24*, 1356–1357.
9. Freidman, I. (1953) Deuterium content of natural water and other substances. Geochim. Cosmochim. Acta *4*, 89–103.
10. Freidman, I. & Hardcastle, K. (1970) A new technique for pumping hydrogen gas. Geochim. Cosmochim. Acta *34*, 125–126.
11. Halliday, D. & Miller, A. G. (1977) Precise measurement of total body water using trace quantities of deuterium oxide. Biomed. Mass Spectrom. *4*, 82–87.
12. Coleman, M. L., Shepard, T. J., Durham, J. J., Rouse, J. E. & Moore, G. R. (1982) Reduction of water with zinc for hydrogen isotope analysis. Anal. Chem. *54*, 995–998.
13. Dubbs, C. A. (1953) Determination of deuterium. Anal. Chem. *25*, 828–829.
14. Hagemann, R. & Lohez, P. (1978) Twin mass spectrometers for simultaneous isotopic analysis of hydrogen and oxygen in water. Adv. Mass Spectrom. *74*, 504–508.
15. Thurston, W. M. (1970) Automatic mass spectrometric analysis of the deuterium to hydrogen ratio in natural water. Rev. Sci. Instr. *41*, 963–966.
16. Thurston, W. M. (1971) Steam film sampling of water for mass spectrometric analysis of

the deuterium content. Rev. Sci. Instr. *42*, 700–703.
17. Schiegl, W. E. & Vogel, J. C. (1969) Deuterium content of organic matter. Earth Plan. Sci. Lett. *7*, 307–313.
18. Trenner, N. R., Arison, B. H. & Walker, R. W. (1956) Sealed-tube combustion technique for determination of deuterium in organic compounds. Anal. Chem. *28*, 530–531.
19. Washburn, H. W., Berry, C. E. & Hall, L. G. (1953) Measurement of deuterium oxide concentration in water samples by the mass spectrometer. Anal. Chem. *25*, 130–138.
20. Bencsath, F. A., Bradlow, H. L. & Field, F. H. (1983) Determination of total body water by deuterium isotope dilution measurements using chemical ionization mass spectrometry. Biomed. Mass Spectrom. *10*, 596–600.
21. Shreeve, W. W., Sharp, J. D., Ott, D. G. & McInteer, B. B. (1976) Test for alcoholic cirrhosis by conversion of [^{14}C]- or [^{13}C]galactose to expired CO_2. Gastroenterol. *71*, 98–101.
22. Ajami, A. M. & Watkins, J. B. (1983) Residual gas contaminants in sterilized serological tubes: a problem re-examined. Clin. Chem. *29*, 725–726.
23. Lacraise, M. (1972) Thesis, University of Liège.
24. Schoeller, D. A., Klein, P. D. & Schneider, J. F. (1975) in: Proceedings of the Second International Conference on Stable Isotopes (Klein, E. R. & Klein, P. D., eds.) pp. 246–251.
25. McCrea, J. M. (1950) The isotopic chemistry of carbonates and the paleotemperature scale. J. Chem. Phys. *18*, 849–857.
26. Kerr, D. S., Stevens, M. C. G. & Picou, D. I. M. (1978) Fasting metabolism in infants: II. The effect of severe undernutrition and infusion of alanine on glucose production estimated with U-^{13}C-glucose. Metabolism *27*, 831–848.
27. Wolfe, R. R., Allsop, J. R. & Burke, J. F. (1979) Glucose metabolism in man: Response to intravenous glucose infusion. Metabolism. *28*, 210–220.
28. Read, W. W., Read, M. A., Rennie, M. J., Griggs, R. C. & Halliday, D. (1984) Preparation of CO_2 from blood and protein-bound amino acid carboxyl groups for quantification and ^{13}C-isotope measurements. Biomed. Mass Spectrom. *11*, 348–352.
29. Unruh, G. E. von, Hanber, D. J., Schoeller, D. A. & Hayes, J. M. (1974) Limits of detection of Carbon-13 labelled drugs and their metabolites in human urine. Biomed. Mass Spectrom. *1*, 345–349.
30. Sano, M., Yotsui, Y., Abe, H. & Sasaki, S. (1976) A new technique for the detection of metabolites labelled by the isotope ^{13}C using mass fragmentography. Biomed. Mass Spectrom. *3*, 1–3.
31. Matthews, D. E. & Hayes, J. M. (1978) Isotope-Ratio-Monitoring gas chromatography-mass spectrometry. Anal. Chem. *50*, 1465–1473.
32. Barrie, A., Bricout, J. & Koziet, J. (1984) Gas chromatography-stable isotope ratio analysis at natural abundance levels. Biomed. Mass Spectrom. *11*, 583–588.
33. Otsuki, A., Ino, Y. & Fujii, T. (1983) Simultaneous measurements and determinations of stable carbon and nitrogen isotope ratios and organic carbon and nitrogen contents in biological samples – coupling of a small quadrupole mass spectrometer and modified carbon-nitrogen elemental analyser. Int. J. Mass Spec. Ion Phys. *48*, 343–346.
34. Mizutani, H. & Wada, E. (1985) Combustion of organic samples by infrared furnace for carbon isotope analysis. Anal. Biochem. *146*, 90–95.
35. Carr, R. H., Wright, I. P., Joines, A. W., Pillinger, C. T. (1986) Measurement of carbon stable isotopes at the nanomole level: a static mass spectrometer and sample preparation technique. J. Phys. E: Sci. Instrum. *19*, 798–808.
36. Ross, P. J. & Martin, A. E. (1970) A rapid procedure for preparing gas samples for nitrogen-15 determination. Analyst *95*, 817–822.
37. Porter, L. K. & O'Deen, W. A. (1977) Apparatus for preparing nitrogen from ammonium chloride for nitrogen-15 determinations. Anal. Chem. *49*, 514–516.
38. Read, W. W. C., Harrison, R. A. & Halliday, D. (1982) A resin-based method for the preparation of molecular nitrogen for ^{15}N analysis from urinary and plasma components. Anal. Biochem. *123*, 249–254.
39. Teuji, O., Masugi, M. & Kosai, Y. (1975) An instrumental method for nitrogen-15 and total nitrogen by combined system of mass spectrometer and automatic nitrogen analyzer. Anal. Biochem. *65*, 19–25.
40. Preston, T. & Owens, J. P. (1983) Interfacing an automatic elemental analyser with an

isotope ratio mass spectrometer: the potential for fully automated total nitrogen and nitrogen-15 analysis. Analyst *108*, 971–977.

41. Barrie, A. & Workman, C. T. (1984) An automated analytical system for nutritional investigations using ^{15}N tracers. Spectros. Int. J. *3*, 439–447.
42. Wong, W. W., Cabrera, M. P. & Klein, P. D. (1984) Evaluation of a dual mass spectrometer system for rapid simultaneous determination of hydrogen-2/hydrogen-1 and oxygen-18/oxygen-16 ratios in aqueous samples. Anal. Chem. *56*, 1852–1858.
43. Barrie, A. & Coward, W. A. (1985) A rapid analytical technique for the determination of energy expenditure by the double labelled water method. Biomed. Mass Spectrom. *12*, 535–541.
44. O'Neil, J. R. & Epstein, S. (1966) A method for oxygen isotope analysis of milligram quantities of water and some of its applications. J. Geophys. Res. *20*, 4955–4961.
45. Kishima, N. & Sakai, H. (1980) Oxygen-18 and deuterium determination on a single water sample of a few milligrams. Anal. Chem. *52*, 356–358.
46. Wong, W. W., Lee, L. S. & Klein, P. D. (1987) Oxygen isotope ratio measurements on carbon dioxide generated by reaction of microliter quantities of biological fluids with guanidine hydrochloride. Anal. Chem. *59*, 690–693.
47. Epstein, S. & Mayeda, T. (1953) Variation of O^{18} content of waters from natural sources. Geochim. Cosmochim. Acta *4*, 213–224.
48. Sofer, Z. & Gat, J. R. (1972) Activities and concentrations of oxygen-18 in concentrated aqueous salt solutions. Analytical and geophysical implications. Earth Planet Sci. Lett. *15*, 232–238.
49. Hulston, J. R. & Thode, H. G. (1965) Variations in the S^{33}, S^{34} and S^{36} contents of meteorites and their relation to chemical and nuclear effects. J. Geophys. Res. *70*, 3475–3484.
50. Puchett, H., Sabels, B. R. & Hoering, T. C. (1971) Preparation of sulphur hexafluoride for isotope geochemical analysis. Geochim. Cosmochim. Acta *35*, 625–628.
51. Robinson, B. W. & Kusakabe, M. (1975) Quantitative preparation of sulfur dioxide for $^{34}S/^{32}S$ analyses, from sulfides by combustion with cuprous oxide. Anal. Chem. *47*, 1179–1181.
52. Bailey, S. A. (1972) Improved method for the preparation of sulphur dioxide from barium sulphate for isotope ratio studies. Anal. Chem. *44*, 1542–1543.
53. Haur, A., Hladikova, J. & Smejkal, V. (1973) Procedure for direct conversion of sulphates into SO_2 for mass spectrometric analysis of sulphur. Isotopenpraxis *9*, 329–331.
54. Coleman, M. L. & Moore, M. P. (1978) Direct reduction of sulfates to sulfur dioxide for isotopic analysis. Anal. Chem. *50*, 1594–1595.
55. Schneider, A. (1970) The sulfur isotope composition of basaltic rocks. Contrib. Mineral Petrol. *25*, 95–124.
56. Thode, H. G., Monster, J. & Dunford, H. B. (1961) Sulphur isotope geochemistry. Geochim. Cosmochim. Acta *25*, 159–174.
57. Curtius, H.-Ch., Völlmin, J. A. & Baerlocher, K. (1972) The use of deuterated phenylalanine for the elucidation of the phenylalanine-tyrosine metabolism. Clin. Chim. Acta *37*, 277–285.
58. Curtius, H.-Ch., Völlmin, J. A. & Baerlocher, K. (1973) Study of Metabolic Pathways *in vivo* using stable isotopes. Anal. Chem. *4*, 1107–1110.
59. Curtius, H.-Ch. & Mettler, M. (1976) Study of the intestinal tyrosine metabolism using stable isotopes and gc/ms. J. Chromatog. *126*, 569–580.
60. Haskins, N. J. (1978) Analysis of drugs and metabolites by GC/MS. Proc. Anal. Div. Chem. Soc. *15*, 253–274.
61. Sweeley, C. C., Elliot, H. H., Fries, I. & Ryhage, R. (1966) Spectrometric determination of unresolved components in gas chromatographic effluents. Anal. Chem. *38*, 1549–1553.
62. Pickup, J. F. & McPherson, K. (1976) Theoretical considerations in stable isotope dilution mass spectrometry for organic analysis. Anal. Chem. *48*, 1885–1890.
63. Campbell, I. M. (1974) Incorporation and dilution values – their calculation in mass spectral assayed stable isotope labeling experiments. Bioorg. Chem. *3*, 386–397.
64. Colby, B. N. & McCaman, M. W. (1979) A comparison of calculation procedures for isotope dilution determinations using gas chromatography mass spectrometry. Biomed. Mass Spectrom. *6*, 225–230.
65. Ford, G. C., Grigson, S. J. & Haskins, N. J. (1976) The replacement of trideuteromethyl by methyl groups when using methane as a reagent gas in chemical ionization mass

spectrometry. Biomed. Mass Spectrom. *3*, 230–231.
66. Farmer, P. B., Bailey, E., Lamb, J. H. & Connors, T. A. (1980) Approach to the quantitation of alkylated amino acids in haemoglobin by gas chromatography mass spectrometry. Biomed. Mass Spectrom. *7*, 41–46.
67. Waddell, K. A., Blair, I. A. & Wellby, J. (1983) Combined capillary column gas chromatography negative ion chemical ionization mass spectrometry of prostanoids. Biomed. Mass Spectrom. *10*, 83–88.
68. Millard, B. J. (1978) Quantitative Mass Spectrometry, pp. 136–148, Heyden Press, London, UK.
69. Haskins, N. J., Ford, G. C., Grigson, S. J. W. & Waddell, K. A. (1978) A carrier effect observed in assays for antidiarrhoeal drug compounds. Biomed. Mass Spectrom. *5*, 423–424.
70. Picart, D., Jacolot, F., Berthou, F. & Floch, H. H. (1978) in: Quantitative mass spectrometry in the life sciences II (de Leenheer, A. P., Roncucci, R. R. & van Peteghem, C., eds.) pp. 105–118. Elsevier. Scientific Publishing Company Amsterdam.
71. Bailey, T., Farmer, P. B. & Lamb, J. H. (1980) The enantomer as internal standard for the quantitation of the alkylated amino acid S-methyl-L-cysteine in haemoglobin by gas chromatography chemical ionization mass spectrometry with single ion detection. J. Chromatog. *200*, 145–152.
72. Culebras, J. M. & Moore, F. O. (1977) Total body water and the exchangeable hydrogen. I. Theoretical calculation of nonaqueous exchangeable hydrogen in man. Am. J. Physiol. *232*, R54–R59.
73. Culebras, J. M., Fitzpatrick, G. F., Brennan, M. F., Boyden, C. M. & Moore, F. O. (1977) Total body water and the exchangeable hydrogen. II. A review of comparative data from animals based on isotope dilution and desiccation, with a report of new data from the rat. Am. J. Physiol. *232*, R60–R65.
74. Schoeller, D. A., van Santen, E., Peterson, D. W., Dietz, W., Jaspan, J. & Klein, P. D. (1980) Total body water measurement in humans with ^{18}O and ^{2}H labeled water. Am. J. Clin. Nutr. *33*, 2686–2693.
75. Trowbridge, F. L., Graham, G. G., Wong, W. W., Mellits, E. D., Rabold, J. D., Lee, L. S., Cabrera, M. P. & Klein, P. D. (1984) Body water measurement in premature and older infants using $H_2{}^{18}O$ isotopic determinations. Pediatr. Res. *18*, 524–527.
76. How, T. V., Ashmore, M. P., Rolfe, P., Lucas, A., Lucas, P. J. & Baum, J. D. (1979) A Doppler ultrasound technique for measuring human milk flow. J. Med. Eng. Technol. *3*, 66–71.
77. Coward, W. A., Sawyer, M. B., Whitehead, R. G., Prentice, A. M. & Evans, J. (1979) New method for measuring milk intakes in breast-fed babies. Lancet *(i)*, 13–14.
78. Butte, N. F., Garza, C., Smith, E. O. & Nichols, B. L. (1983) Evaluation of the deuterium dilution technique against the test-weighing procedure for the determination of breast milk intake. Am. J. Clin. Nutr. *37*, 996–1003.
79. Coward, W. A., Cole, T. J., Sawyer, M. B. & Prentice, A. M. (1982) Breast-milk intake measurement in mixed-fed infants by administration of deuterium oxide to their mothers. Hum. Nutr. Clin. Nutr. *36C*, 141–148.
80. Lifson, N., Gordon, G. B. & McClintock, R. (1955) Measurement of total carbon dioxide production by means of $D_2{}^{18}O$. J. Appl. Physiol. *7*, 704–710.
81. Williams, J. B. & Nagy, K. A. (1984) Validation of the doubly labeled water technique for measuring energy metabolism in Savannah Sparrows. Physiol. Zool. *57*, 325–328.
82. Lifson, N. & McClintock, R. (1966) Theory of use of the turnover rates of body water for measuring energy and material balance. J. Theoret. Biol. *12*, 46–74.
83. Nagy, K. A. (1980) CO_2 production in animals: analysis of potential errors in the doubly labeled water method. Am. J. Physiol. *238*, R466–R473.
84. Williams, J. B. (1985) Validation of the doubly labeled water technique for measuring energy metabolism in starlings and sparrows. Comp. Biochem. Physiol. *80A* (3), 349–353.
85. King, W. W. & Hadley, N. F. (1979) Water flux and metabolic rates of free-roaming scorpions using the doubly labeled water technique. Physiol. Zool. *52*, 176–189.
86. Lifson, N., Little, W. S., Levitt, D. G. & Henderson, R. M. (1975) $D_2{}^{18}O$ method for CO_2 output in small mammals and economic feasibility in man. J. Appl. Physiol. *39*, 657–664.
87. Halliday, D. & Rennie, M. J. (1982) The use of stable isotopes for diagnosis and clinical research. Clin. Sci. *63*, 485–496.
88. Schoeller, D. A., van Santen, E., Petersen, D. W., Dietz, W., Jaspan, J. & Klein, P. D.

(1980) Total body water measurement in humans with ^{18}O and ^{2}H labeled water. Am. J. Clin. Nutr. *33*, 2686–2693.

89. Schoeller, D. A. & van Santen, E. (1982) Measurement of energy expenditure in humans by doubly labeled water method. J. Appl. Physiol. *53*, 955–959.
90. Klein, P. D., James, W. P. T., Wong, W. W., Irving, C. S., Murgatroyd, P. R., Cabrera, M., Dallosso, H. M., Klein, E. R. & Nichols, B. L. (1984) Calorimetric validation of the doubly-labelled water method for determination of energy expenditure in man. Hum. Nutr. Clin. Nutr. *38C*, 95–106.
91. Schoeller, D. A. & Webb, P. (1984) Five day comparison of the doubly labeled water method with respiratory gas exchange. Am. J. Clin. Nutr. *40*, 153–158.
92. Coward, W. A., Prentice, A. M., Murgatroyd, P. R., Davies, H. L., Cole, T. J., Sawyer, M., Goldberg, G. R., Halliday, D. & Macnamara, J. (1984) in: Human Energy Metabolism (van Es, A. J. H., ed.) pp. 126–128, Euro-Nut., Netherlands.
93. Roberts, S. B., Coward, W. A., Schlingenseipen, K.-H., Nohria, V. & Lucas, A. (1986) Comparison of the doubly labeled water ($^{2}H_2{}^{18}O$) method with indirect calorimetry and a nutrient-balance study for simultaneous determination of energy expenditure, water intake, and metabolizable energy intake in preterm infants. Am. J. Clin. Nutr. *44*, 315–322.
94. Lucas, A., Ewing, G., Roberts, S. G. & Coward, W. A. (1987) How much energy does the breast fed infant consume and expend? Clin. Research, *295*, 75–77.
95. Westerterp, K. R., Saris, W. H. M., van Es, M. & ten Hoor, F. (1986) Use of the doubly labeled water technique in humans during heavy sustained exercise. J. Appl. Physiol. *61*, 2162–2167.
96. Waterlow, J. C., Garlick, P. J. & Millward, D. J. (1978) Protein Turnover in Mammalian Tissues and in the Whole Body, chap. 7, North-Holland Publishing Company, Amsterdam.
97. Waterlow, J. C., Golden, M. H. N. & Garlick, P. J. (1978) Protein turnover in man with ^{15}N: comparison of end products and dose regimes. Am. J. Physiol. *235*, E165–E174.
98. Fern, E. B., Garlick, P. J., McNurlan, M. A. & Waterlow, J. C. (1981) The excretion of isotope in urea and ammonia for estimating protein turnover in man with [^{15}N]glycine. Clin. Sci. *61*, 217–228.
99. Fern, E. B., Garlick, P. J., Sheppard, H. G. & Fern, M. (1984) The precision of measuring the rate of whole-body nitrogen flux and protein synthesis in man with a single dose of [^{15}N]-Glycine. Hum. Nutr. Clin. Nutr. *38C*, 63–73.
100. Fern, E. B., Garlick, P. J. & Waterlow, J. C. (1985) Apparent compartmentation of body nitrogen in one human subject: its consequences in measuring the rate of whole-body protein synthesis with ^{15}N. Clin. Sci., *68*, 271–282.
101. Duffy, B., Gunn, T., Collinge, J. & Pencharz, P. (1981) The effect of varying protein quality and energy intake on the nitrogen metabolism of parenterally fed very low birthweight (< 1600 g) infants. Pediatr. Res. *15*, 1040–1044.
102. Plath, Chr., Heine, W., Krienke, L., Richter, I., Wutzke, K. D., Töwe, J. & Krawielitzki, K. (1985) ^{15}N-Tracer-kinetic studies on the nitrogen metabolism of very small preterm infants on a diet of mother's milk. Human Nutr. Clin. Nutr. *39C*, 399–409.
103. Pencharz, P. B., Steffee, W. P., Cochran, W., Scrimshaw, N. S., Rand, W. M. & Young, V. R. (1977) Protein metabolism in human neonate nitrogen-balance studies, estimated obligatory losses of nitrogen and whole-body turnover of nitrogen. Clin. Sci. *52*, 485–498.
104. Picou, D. & Taylor-Roberts, T. (1969) The measurement of total protein synthesis and catabolism and nitrogen turnover in infants in different nutritional states and receiving different amounts of dietary protein. Clin. Sci. *36*, 283–296.
105. Steffee, W. P., Goldsmith, R. S., Pencharz, P. B., Scrimshaw, N. S. & Young, V. R. (1976) Dietary protein intake and dynamic aspects of whole body nitrogen metabolism in adult humans. Metabolism *25*, 281–297.
106. Winterer, J. C., Steffee, W. P., Davy, W., Perera, A., Nany, R., Scrimshaw, N. S. & Young, V. R. (1976) Whole body protein turnover in aging man. Exp. Gerontol. *11*, 79–81.
107. Golden, M. H. N. & Waterlow, J. C. (1977) Total protein synthesis in elderly people: a comparison of results with [^{15}N] glycine and [^{14}C] leucine. Clin. Sci. Mol. Med. *53*, 277–288.
108. Kien, C. L., Rohrbaugh, D. K., Burke, J. F. & Young, V. R. (1978) Whole body protein synthesis in relation to basal energy expenditure in healthy children and in children

recovering from burn injury. Pediatr. Res. *12*, 211–216.
109. Norton, J. A., Stein, T. P. & Brennan, M. F. (1981) Whole body protein synthesis and turnover in normal man and malnourished patients with and without cancer. Ann. Surg. *194*, 123–128.
110. Crane, C. W., Picou, D., Smith, R. & Waterlow, J. C. (1977) Protein turnover in patients before and after elective orthopaedic operations. Brit. J. Surg. *64*, 129–133.
111. Ward, H. C., Johnson, A. W., Halliday, D. & Sim, A. J. W. (1985) Elevated rates of whole body protein metabolism in patients with disseminated malignancy in the immediate post operative period. Brit. J. Surg. *72*, 983–986.
112. Lowry, S. F., Legaspi, A., Jeevanandam, M., Horowitz, G. D., Albert, J. D. & Brennan, M. F. (1986) Body protein kinetics during perioperative intravenous nutritional support. Surgery: Gynecology & Obstetrics *163*, 303–309.
113. De Benoist, B., Abdulrazzak, Y., Brooke, O. F., Halliday, D. & Millward, D. (1984) The measurement of whole body protein turnover in the preterm infant with intragastric infusion of L-[1-^{13}C] leucine and sampling of the urinary leucine pool. Clin. Sci. *66*, 155–164.
114. Motil, K. J., Matthews, D. E., Bier, D. M., Burke, J. F., Munro, H. W. & Young, V. R. (1981) Whole body leucine and lysine metabolism: Response to dietary protein intake in young men. Am. J. Physiol. *240*, E712–E721.
115. Rennie, M. J., Edwards, R. H. T., Halliday, D., Matthews, D. E., Walman, S. L. & Millward, D. J. (1982) Muscle protein synthesis measured by stable isotope techniques in man: the effects of feeding and fasting. Clin. Sci. *63*, 519–523.
116. Rennie, M. J., Edwards, R. H. T., Krywawych, S., Davies, C. T. M., Halliday, D., Waterlow, J. C. & Millward, D. J. (1981) Effect of exercise on protein turnover in man. Clin. Sci. *61*, 627–639.
117. Wolfe, R. R., Goodenough, R. D., Wolfe, M. H., Royale, G. T. & Nadel, E. R. (1982) Isotopic analysis of leucine and urea metabolism in exercising humans. J. App. Physiol. *52*, 458–466.
118. Miles, J. M., Nissen, S. L., Rizza, R. A., Gerich, J. E. & Haymond, M. W. (1983) Failure of infused β-hydroxybutyrate to decrease proteolysis in man. Diabetes *32*, 197–205.
119. Simmons, P. S., Miles, J. M., Gerich, J. E. & Haymond, M. W. (1984) Increased proteolysis, an effect of increases in plasma cortisol within the physiologic range. J. Clin. Invest. *73*, 412–420.
120. Nair, K. S., Garrow, J. S., Ford, C., Mahler, R. F. & Halliday, D. (1983) Effect of poor diabetic control and obesity on whole body protein metabolism in man. Diabetologia *25*, 400–403.
121. Wolfe, R. R. (1984) Tracers in Metabolic Research Radioisotope and Stable Isotope/ Mass Spectrometry Methods. Chapter 9, pp. 113–130 (Alan. R. Liss Inc., New York).
122. Kalhan, S. C., Savin, S. M. & Adam, P. A. J. (1976) Measurement of glucose turnover in the human newborn with glucose-[1-^{13}C]. J. Clin. Endocrinol. Metab. *43*, 704–707.
123. Kalhan, S. C., Bier, D. M., Savin, S. M. & Adam, P. A. J. (1980) Estimation of the glucose turnover and ^{13}C recycling in the human newborn by simultaneous [1-^{13}C]glucose and [6,6-$^{2}H_{2}$]glucose tracers. J. Clin. Endocrinol. Metab. *50*, 456–460.
124. Bier, D. M., Arnold, K. J., Sherman, W. R., Holland, W. H., Holmes, W. F. & Kipnis, D. M. (1977) In vivo measurement of glucose and alanine metabolism with stable isotopic tracers. Diabetes *26*, 1005–1015.
125. Kalhan, S. C., D'Angelo, L. J., Savin, S. M. & Adam, P. A. J. (1979) Glucose production in pregnant women at term gestation. J. Clin. Invest. *63*, 388–394.
126. Bier, D. M., Leake, R. D., Haymond, M. W., Arnold, K. J., Gruenke, L. D., Sperling, M. A. & Kipnis, D. M. (1977) Measurement of "true" glucose production rates in infancy and childhood with 6,6-dideuteroglucose. Diabetes *26*, 1016–1023.
127. Haymond, M. W., Howard, C., Ben-Galim, E. & De Vivo, D. C. (1983) Effects of ketosis on glucose flux in children and adults. Am. J. Physiol. *245*, E373–E378.
128. Kalhan, S. C., Tserng, K.-Y., Gilfillan, C. & Dierker, L. J. (1982) Metabolism of urea and glucose in normal and diabetic pregnancy. Metabolism, *31*, 824–833.
129. Robert, J.-J., Cummins, J. C., Wolfe, R. R., Durkot, M., Matthews, D., Zhao, X. H., Bier, D. M. & Young, V. R. (1982) Quantitative aspects of glucose production and metabolism in healthy elderly subjects. Diabetes *31*, 203–211.
130. Robert, J.-J., Bier, D. M., Schoeller, D., Wolfe, R., Matthews, D. E., Munro, H. W. & Young, V. R. (1984) Effects of intrave-

nous glucose on whole body leucine dynamics, studies with 1-^{13}C-leucine, in healthy young and elderly subjects. J. Gerentol. *39*, 673–681.

131. Kerr, D. S., Stevens, M. C. G. & Robinson, H. M. (1978) Fasting metabolism in infants. I. Effect of severe undernutrition on energy and protein utilization. Metabolism *27*, 411–435.
132. Mehta, A., Halliday, D., Ford, G. C., Wootton, R., Flecknell, P. A. & Stacey, T. E. (1985) Glucose turnover in 3 hypoglycaemic term neonates using 6,6,dideuteroglucose as tracer. Eur. J. Clin. Invest. *15* (2), A25.
133. Kalhan, S. C., Savin, S. M. & Adam, P. A. J. (1977) Attenuated glucose production rate in newborn infants of insulin-dependent diabetic mothers. N. Engl. J. Med. *296*, 375–376.
134. Haymond, M. W., Ben-Galim, E. & Strobel, K. E. (1978) Glucose and alanine metabolism in children with maple syrup urine disease. J. Clin. Invest. *62*, 398–405.
135. Haymond, M. W., Strobel, K. E., De-Vivo, D. C. (1978) Muscle wasting and carbohydrate homeostasis in Duchenne muscular dystrophy. Neurol. *28*, 1221–1231.
136. Haymond, M. W., Strauss, A. W., Arnold, K. J. & Bier, D. M. (1979) Glucose homeostasis in children with severe cyanotic congenital heart disease. J. Ped. *95*, 220–227.
137. Wolfe, R. R., Durkot, M. J., Allsop, J. R. & Burke, J. F. (1979) Glucose metabolism in severely burned patients. Metabolism *28*, 1031–1039.
138. Burke, J. F., Wolfe, R. R., Mullany, C. J., Matthews, D. E. & Bier, D. M. (1979) Glucose requirements following burn injury. Ann. Surg. *190*, 274–283.
139. Wolfe, R. R., Allsop, J. R. & Burke, J. F. (1979) Glucose metabolism in man: Responses to intravenous glucose infusion. Metabolism *3*, 210–220.
140. Wolfe, R. R., O'Donnell, T. F., Stone, M. D., Richmand, D. A. & Burke, J. F. (1980) Investigation of factors determining the optimal glucose infusion rate in total parenteral nutrition. Metabolism *29*, 892–900.
141. Black, C. C. (1973) Photosynthetic carbon fixation in relation to net CO_2 uptake. Ann. Rev. Plan Physiol. *24*, 253–286.
142. De Niro, M. J. & Epstein, S. (1977) Mechanism of carbon isotope fractionation associated with lipid synthesis. Science *197*, 261–263.
143. Lacroix, M., Mosora, F., Pontus, M., Lefebvre, P., Luyckx, A. & Lopez-Habib, G. (1973) Glucose naturally labeled with carbon-13: use for metabolic studies in man. Science *181*, 445–446.
144. Lefebvre, P., Mosora, F., Lacroix, M., Luyckx, A., Lopez-Habib, G. & Duchesne, J. (1975) Naturally labeled ^{13}C-glucose: Metabolic studies in human diabetes and obesity. Diabetes *24*, 185–189.
145. Pirnay, F., Lacroix, M., Mosora, F., Luyckx, A. & Lefebvre, P. (1977) Glucose oxidation during prolonged exercise evaluated with naturally labeled ^{13}C-glucose. J. Appl. Physiol. *43*, 258–261.
146. Ravussin, E., Pakud, P., Thelin-Doerner, A., Arnaud, M. J. & Jéquier, E. (1980) Substrate utilization during prolonged exercise after ingestion of ^{13}C-glucose in obese and control subjects. Int. J. Obesity *4*, 235–242.
147. Schoeller, D. A., Schneider, J. F., Solomons, N. W., Watkins, J. B. & Klein, P. D. (1977) Clinical diagnosis with stable isotope ^{13}C in CO_2 breath tests: methodology and fundamental considerations. J. Lab. Clin. Med. *90*, 412–421.
148. Schoeller, D. A. & Klein, P. D. (1978) A simplified technique for collecting breath CO_2 for isotope ratio mass spectrometry. Biomed. Mass Spectrom. *5*, 29–31.
149. Caspony, W. F. (1978) Breath tests. Clin. Gastroenterol. *7*, 351–374.
150. Schoeller, D. A. & Klein, P. D. (1978) A microprocessor controlled mass spectrometer for the fully automated analysis of breath carbon dioxide. Biomed. Mass Spectrom. *6*, 350–355.
151. Schoeller, D. A., Klein, P. D., Watkins, J. B., Heim, T. & MacLean, W. C. (1980) ^{13}C abundances of nutrients and the effect of variations in ^{13}C isotopic abundances of test meals formulated for $^{13}CO_2$ breath tests. Am. J. Clin. Nutr. *33*, 2375–2385.
152. Schoeller, D. A., Brown, C., Nakamura, K., Nakagawa, A., Mazzeo, R. S., Brooks, G. A. & Budinger, T. F. (1984) Influence of metabolic fuel on the $^{13}C/^{12}C$ ratio of breath CO_2. Biomed. Mass Spectrom. *11*, 557–561.
153. Watkins, J. B., Schoeller, D. A., Klein, P. D., Ott, D. G., Newcomer, A. D. & Hofmann, A. F. (1977) ^{13}C-trioctanoin: A nonradioactive breath test to detect fat malabsorption. J. Lab. Clin. Med. *90*, 442–430.
154. Watkins, J. B., Klein, P. D., Schoeller, D. A., Kirschner, B. S., Park, R. & Perman, J. A. (1982) Diagnosis and differentiation of fat

malabsorption in children using ^{13}C-labeled lipids: trioctanoin, triolein and palmitic acid breath tests. Gastroenterology *82*, 911–917.
155. Fromm, H. & Hofmann, A. F. (1971) Breath test for altered bile acid metabolism. Lancet *(ii)*, 621–625.
156. Solomons, N. W., Schoeller, D. A., Wagonfeld, J. B., Ott, D., Rosenberg, I. H. & Klein, P. D. (1977) Application of a stable isotope (^{13}C)-labeled glycocholate breath test to diagnosis of bacterial overgrowth and ileal dysfunction. J. Lab. Clin. Med. *90*, 431–439.
157. Lauterberg, B. H. & Bircher, J. (1976) Expiratory measurement of maximal aminopyrine demethylation in vivo: effect of phenobarbital, partial hepatectomy, portacaval shunt and bile duct ligation in the rat. J. Pharmacol. Exp. Therap. *196*, 501–509.
158. Macnamara, J. P., Halliday, D., Leung, N., Farrant, P. & Peters, T. J. (1985) ^{13}C-aminopyrine metabolism by control subjects and patients with alcoholic liver disease. Clin. Sci. *68*, Suppl. 11, 75P.
159. Krumbiegel, P., Hirschberg, K., Faust, H., Günther, K. & Schneider, G. (1985) Nuclear medicine liver-function tests for pregnant women and children. 2. A new test method via urine using ^{15}N-methacetin. Eur. J. Nucl. Med. *11*, 58–61.
160. Krumbiegel, P., Günther, K., Faust, H., Möbius, G., Hirschberg, K. & Schneider, G. (1985) Nuclear medicine liver function tests for pregnant women and children. 1. Breath tests with ^{14}C-methacetin and ^{13}C-methacetin. Eur. J. Nucl. Med. *10*, 129–133.
161. Jung, K., Hirschberg, K., Faust, H. & Matkowitz, R. (1985) A liver-function test using ^{15}N-labelled ammonium chloride. Eur. J. Nucl. Med. *11*, 62–64.
162. Shreeve, W. W., Shoop, J. D., Ott, D. G. & McInteer, B. B. (1976) Test for alcoholic cirrhosis by conversion of [^{14}C] or [^{13}C] galactose to expired CO_2. Gastroenterology *71*, 98–101.
163. Lehmann, W. D., Heinrich, H. C., Leonhardt, R., Agarwal, D. P., Goedde, H. W., Kneer, J. & Rating, D. (1986) ^{13}C-Ethanol and ^{13}C-Acetate breath tests in normal and aldehyde dehydrogenase deficient individuals. Alcohol *3*, 227–231.
164. Graham, D. Y., Klein, P. D., Evans, D. J., Evan, D. G., Alpert, L. C., Opekun, A. R. & Boutton, T. W. (1987) Campylobacter pylori detected noninvasively by the ^{13}C-urea breath test. Lancet (*i*), 1174–1177.
165. Gersovitz, M., Munro, H. N., Udall, J. & Young, V. R. (1980) Albumin synthesis in young and elderly subjects using a new stable isotope methodology: Response to level of protein intake. Metabolism *29*, 1075–1086.
166. Cryer, D. R., Yudkoff, M., Matsushima, T., Marsh, J. P., Coates, P. M. & Cortner, J. A. (1985) in: Substrate and Energy Metabolism in Man (Garrow, J. S. & Halliday, D., eds.) p. A3, John Libbey, London.
167. Stein, T. P., Mullen, J. L., Oram-Smith, J. C., Rosato, E. F., Wallace, H. W. & Hargrove, W. C. (1978) Relative rates of tumor, normal gut, liver and fibrinogen protein synthesis in man. Am. J. Physiol. *234*, E648–E652.
168. Halliday, D. & McKeran, R. O. (1975) Measurement of muscle protein synthetic rate from serial muscle biopsies and total body protein turnover in man by continuous intravenous infusion of L[α-^{15}N]lysine. Clin. Sci. Mol. Med. *49*, 581–590.
169. Rennie, M. J. & Halliday, D. (1984) The use of stable isotope tracers as metabolic probes of whole-body and limb metabolism. Proc. Nutr. Soc. *43*, 189–196.
170. Rennie, M. J., Edwards, R. H. T., Emery, P. W., Halliday, D., Lundholm, K. & Millward, D. J. (1983) Hypothesis: Depressed protein synthesis is the dominant characteristic of muscle wasting and cachexia. Clin. Physiol. *3*, 387–398.
171. Rennie, M. J., Edwards, R. H. T., Millward, D. J., Wolman, S. L., Halliday, D. & Matthews, D. E. (1982) Effects of Duchenne muscular dystrophy on muscle protein synthesis. Nature *296*, 165–167.
172. Halliday, D., Ford, G. C., Griggs, R. C., Edwards, R. H. T. & Rennie, M. J. (1985) In vivo estimation of muscle protein synthesis in myotonic dystrophy. Ann. Neurol. *17*, 65–69.
173. Emery, P. W., Halliday, D., Edwards, R. H. T., Souhami, R. L. & Rennie, M. J. (1984) Muscle protein synthesis in cachectic cancer patients. Brit. Med. J. *289*, 584–586.
174. Nissen, R., Cardinale, G. J. & Undenfriend, S. (1978) Increased turnover of arterial collagen in hypertensive rats. Proc. Natl. Acad. Sci. USA *75*, 451–453.
175. Molnar, J. A., Alpert, N. M., Wolfe, M., Burke, J. F. & Young, V. R. (1985) in: Substrate and Energy Metabolism in Man (Gar-

row, J. S. & Halliday, D., eds.) p. A17, John Libbey, London.

176. Garlick, P. J., McNurlan, M. A. & Preedy, V. R. (1980) A rapid and convenient technique for measuring the rate of protein synthesis in tissues by injection of [^{3}H]phenylalanine. Biochem. J. *192*, 719–723.

177. Matthews, D. E., Conway, J. M., Young, V. R. & Bier, D. M. (1981) Glycine metabolism in man. Metabolism *30*, 866–893.

178. Matthews, D. E., Ben-Galim, E. & Bier, D. M. (1979) Determination of stable isotope enrichment in individual plasma amino acids by chemical ionization mass spectrometry. Anal. Chem. *51*, 80–84.

179. Bier, D. M. & Christopherson, H. L. (1979) Rapid micro-method for the determination of ^{15}N enrichment in plasma lysine: application to measurement of whole body protein turnover. Anal. Biochem. *94*, 242–248.

180. Gersovitz, M., Bier, D. M., Matthews, D. E., Udall, J., Munro, H. N. & Young, V. R. (1980) Dynamic aspects of whole body glycine metabolism: Influence of protein intake in young adult and elderly males. Metabolism *29*, 1087–1094.

181. Conway, J. M., Bier, D. M., Motil, K., Burke, J. F. & Young, V. R. (1980) Whole-body lysine flux in young adult men: Effects of reduced total protein and of lysine intake. Am. J. Physiol. *239*, 192–200.

182. Matthews, D. E., Motil, K. J., Rohrbaugh, D. K., Burke, J. F., Young, V. R. & Bier, D. M. (1980) Measurement of leucine metabolism in man from a primed continuous infusion of L-[1-^{13}C]leucine. Am. J. Physiol. *238*, E473–E479.

183. Staten, M. A., Bier, D. M. & Matthews, D. E. (1984) Regulation of valine metabolism in man: a stabile isotope study. Am. J. Clin. Nutr. *40*, 1124–1134.

184. Yang, R. D., Matthews, D. E., Bier, D. M. & Young, V. R. (1984) Alanine kinetics in humans: influence of different isotopic tracers. Am. J. Physiol. *247*, E634–E638.

185. Milstein, S. & Kaufman, S. (1975). Studies on the phenylalanine hydroxylase system in vivo. J. Biol. Chem. *250*, 4782–4785.

186. Matalon, R., Matthews, D. E., Michals, K. & Bier, D. (1982) The use of deuterated phenylalanine for the in vivo assay of phenylalanine hydroxylase activity in children. J. Inher. Metab. Dis. *5*, 17–19.

187. Clarke, T. J. T. & Bier, D. M. (1982) The conversion of phenylalanine to tyrosine in man. Direct measurement by continuous intravenous tracer infusions of L[ring-$^2H_5$]phenylalanine and l-[1-^{13}C]tyrosine in the postabsorptive state. Metabolism *31*, 999–1005.

188. Haymond, M. W. & Miles, J. M. (1982) Branched chain amino acids as a major source of alanine nitrogen in man. Diabetes *31*, 86–89.

189. Matthews, D. E., Schwarz, H. P., Yang, R. D., Motil, K. J., Young, V. R. & Bier, D. M. (1982) Relationship of plasma leucine and α-ketoisocaproate during a L-[1-^{13}C]leucine infusion in man: A method for measuring human intracellular leucine tracer enrichment. Metabolism *31*, 1105–1112.

190. Lapidot, A. & Nissim, I. (1980) Regulation of pool sizes and turnover rate of amino acids in humans: ^{15}N-glycine and ^{15}N-alanine single dose experiments using gas chromatography-mass spectrometry analysis. Metabolism *29*, 230–239.

191. Irving, C. S., Nissim, I. & Lapidot, A. (1978) The determination of amino acid pool sizes and turnover rates by gas chromatographic-mass spectrometric analysis of stable isotope enrichment. Biomed. Mass Spectrom. *5*, 117–122.

192. Kingston, E. E. & Duffield, A. M. (1978) Plasma amino acid quantitation using gas chromatography chemical ionization mass spectrometry and ^{13}C amino acids as internal standards. Biomed. Mass Spectrom. *5*, 621–626.

193. Finalyson, P. J., Christopher, R. K. & Duffield, A. M. (1980) Quantitation of fourteen urinary α-amino acids using isobutane gas chromatography chemical ionization mass spectrometry with ^{13}C amino acids as internal standards. Biomed. Mass Spectrom. *7*, 450–453.

194. Rabinowitz, M. B., Wetherill, G. W. & Kopple, J. D. (1976) Kinetic analysis of lead metabolism in healthy humans. J. Clin. Invest. *58*, 260–270.

195. Rannlind, R. D. & Suskind, R. M. (1983) Iron, zinc & other trace elements' effect on the immune response. J. Ped. Gastro. Nutr. *2*, Suppl. 1, S172–S180.

196. Smith, D. L. (1983) Determination of stable isotopes of calcium in biological fluids by fast atom bombardment mass spectrometry. Anal. Chem. *55*, 2391–2393.

197. Yergey, A. L., Vieira, N. E. & Hansen, J. W. (1980) Isotope ratio measurements of urinary calcium with a thermal ionization

probe in a quadrupole mass spectrometer. Anal. Chem. *52*, 1811–1814.
198. Veillon, C., Wolf, W. R. & Guthrie, B. E. (1979) Determination of chromium in biological materials by stable isotope dilution. Anal. Chem. *51*, 1022–1024.
199. Buckley, W. T., Huckin, S. N. & Budac, J. J. (1982) Mass spectrometric determination of stable isotope tracer for copper in biological materials. Anal. Chem. *54*, 504–510.
200. Johnson, P. E. (1982) A mass spectrometric method for use of stable isotopes as tracers in studies of iron, zinc and copper absorption in human subjects. J. Nutr. *112*, 1414–1424.
201. Schricker, B. R., Gilbert, M. D., Miller, D. D. & Van Campen, D. (1982) Biological effects of substituting enriched ^{54}Fe for natural iron in rats. J. Nutr. *112*, 151–157.
202. Miller, D. D. & Van Campen, D. (1979) A method for the detection and assay of iron stable isotope tracers in blood serum. Am. J. Clin. Nutr. *32*, 2354–2361.
203. Waidmann, E., Hilpert, K., Schladot, J. D. & Stoeppler, M. (1984) Determination of cadmium, lead and thallium in materials of the environmental specimen bank using mass spectrometric isotope dilution analysis (MS-IDA). Fresenius Z. Anal. Chem. *317*, 273–277.
204. Schwartz, R., Spencer, H. & Welsh, J. J. (1984) Magnesium absorption in human subjects from leafy vegetables, intrinsically labeled with stable ^{26}Mg. Am. J. Clin. Nutr. *39*, 571–576.
205. Schwartz, R. & Giesecke, C. C. (1979) Mass spectrometry of a volatile Mg chelate in the measurement of stable ^{26}Mg when used as a tracer. Clin. Chim. Acta *97*, 1–8.
206. Schwartz, R. (1982) ^{26}Mg as a probe in research on the role of magnesium in nutrition and metabolism. Fed. Proc. *41*, 2709–2713.
207. Houk, R. S. & Thompson, J. J. (1983) Trace metal isotope analysis of microliter solution volumes by inductively coupled plasma mass spectrometry. Biomed. Mass Spectrom. *10*, 107–112.
208. Reamer, D. C. & Veillon, C. (1983) A double isotope dilution method for using stable selenium isotopes in metabolic tracer studies: Analysis by gas chromatograph/mass spectrometry (GC/MS). J. Nutr. *113*, 786–792.
209. Failey, M. P., Anderson, D. L., Zoller, W. H. & Gordon, G. E. (1979) Neutron-capture prompt gamma-ray activation analysis for multielement determination in complex samples. Anal. Chem. *51*, 2209–2221.
210. Jackson, M. J., Jones, D. A., Edwards, R. H. T., Swainbank, I. G. & Coleman, M. L. (1984) Zinc homeostasis in man: studies using a new stable isotope-dilution technique. Brit. J. Nutr. *51*, 199–208.
211. Gray, A. L. & Date, A. R. (1983) Inductively coupled plasma source mass spectrometry using continuous flow ion extraction. Analyst *108*, 1033–1050.
212. Heusmann, K. G., Schindlmeier, W., Zeininger, H. & Schmidt, M. (1985) Application of an economical and small thermal ionization mass spectrometer for accurate anion trace analyses. Fresenius Z. Anal. Chem. *320*, 457–462.
213. Hachey, D. L., Blais, J.-C. & Klein, P. D. (1980) High precision isotope ratio analysis of volatile metal chelates. Anal. Chem. *52*, 1131–1135.
214. Kownatzki, R., Peters, F., Reil, G. H. & Maass, G. (1980) Volatile complexes for the determination of calcium in blood plasma by isotope dilution mass spectrometry. Biomed. Mass Spectrom. *7*, 540–543.
215. Eagles, J., Fairweather-Tait, S. J. & Self, R. (1985) Stable isotope ratio mass spectrometry for iron bioavailability studies. Anal. Chem. *57*, 469–471.
216. Turnlund, J. R., Swanson, C. A. & King, J. C. (1983) Copper absorption and retention in pregnant women fed diets based on animal and plant proteins. J. Nutr. *113*, 2346–2352.
217. Schulton, H. R. & Lehmann, W. D. (1978) in: Quantitative Mass Spectrometry in Life Sciences II (de Leenheer, A. P., Roncucci, R. R. & Von Peteghem, C., eds.) pp. 63–82, Elsevier Scientific Publishing Company, Amsterdam.
218. Donohue, D. L., Carter, J. A. & Franklin, J. C. (1977) Separated isotopes as internal standards in spark source mass spectrometry. Anal. Lett. *10*, 371–379.
219. Janghorbani, M., Ting, B. T. G. & Young, V. R. (1981) Use of stable isotopes of selenium in human metabolic studies: development of analytical methodology. Am. J. Clin. Nutr. *34*, 2816–2830.
220. Janghorbani, M., Young, V. R., Gramlich, J. W. & Machlan, L. A. (1981) Comparative measurements of zinc-70 enrichment in human plasma samples with neutron activation and mass spectrometry. Clin. Chim. Acta *114*, 163–171.

221. Janghorbani, M. & Young, V. R. (1982) Advances in the use of stable isotopes of minerals in human studies. Fed. Proc. *41*, 2702–2708.

222. Houk, R. S., Fassel, V. A. & Svec, H. J. (1981) Inductively coupled plasma-mass spectrometry: Sample induction, ionization, ion extraction and analytical results. Dynamic Mass Spectrometry, Vol. 6, Chapter 19, pp. 234–251. Heyen & Sons, London, New York. (Int. Dyn. Mass Spectrom. Conf., Canterbury, UK, 7–10th July 1980.)

223. Stokes, P. E., Okamoto, M., Leiberman, K. W., Alexander, G. & Triana, E. (1982) Stable isotopes of lithium: in vivo differential distribution between plasma and cerebrospinal fluid. Biol. Psych. *17*, 413–421.

224. Fairweather-Tait, S. J., Minski, M. J. & Richardson, D. P. (1983) Iron absorption from a malted cocoa drink fortified with ferric orthophosphate using the stable isotope ^{58}Fe as an extrinsic label. Brit. J. Nutr. *50*, 51–60.

225. Aarmodt, R. L., Rumble, W. F., Johnston, G. S., Markley, E. J. & Henkin, R. I. (1981) Absorption of orally administered ^{65}Zn by normal human subjects. Am. J. Clin. Nutr. *34*, 2648–2652.

226. Turnlund, J. R., King, J. C., Keyes, W. R., Gong, B. & Michel, M. C. (1984) A stable isotope study of zinc absorption in young men: effects of phytate and cellulose. Am. J. Clin. Nutr. *40*, 1071–1077.

227. Turnlund, J. R., Michel, M. C., Keyes, W. R., King, J. C. & Margen, S. (1982) Use of enriched stable isotopes to determine zinc and iron absorption in elderly men. Am. J. Clin. Nutr. *35*, 1033–1040.

228. Swanson, C. A., Turnlund, J. R. & King, J. C. (1983) Effect of dietary zinc sources and pregnancy on zinc utilization in adult women fed controlled diets. J. Nutr. *113*, 2557–2567.

229. Mazzeo, R. S., Brooks, G. A., Budinger, T. F. & Schoeller, D. A. (1982) Pulse injection, ^{13}C tracer studies of lactate metabolism in humans during rest and two levels of exercise. Biomed. Mass Spectrom. *9*, 310–314.

230. Neese, R. A., Gertz, E. W., Wisneski, J. A., Gruenke, L. D. & Craig, J. C. (1983) A stable isotope technique for investigating lactate metabolism in humans. Biomed. Mass Spectrom. *10*, 458–462.

231. Wolfe, R. R., Evans, J. E., Mullany, C. J. & Burke, J. F. (1980) Measurement of plasma free fatty acid turnover and oxidation using [1-^{13}C]palmitic acid. Biomed. Mass Spectrom. *7*, 168–171.

232. Galster, A. D., Clutter, W. E., Cryer, P. E., Collins, J. A. & Bier, D. M. (1981) Epinephrine plasma thresholds for lipolytic effects in man: measurement of fatty acid transport with [1-^{13}C]palmitic acid. J. Clin. Invest. *67*, 1729–1738.

233. Miles, J. M., Schwenk, W. F., McClean, K. L. & Haymond, M. W. (1984) Determination of ketone body turnover in vivo with stable isotopes, utilizing gas chromatogography/mass spectrometry. Anal. Biochem. *141*, 110–115.

234. Bougnères, P. F., Lemmel, C., Ferre, P. & Bier, D. M. (1986) Determination of ketone body kinetics using a D-(–)3-hydroxy-[4,4,4-$^{2}H_3$]butyrate tracer. J. Lipid Res. *27*, 215–220.

235. Bougnères, P. F., Lemmel, C., Ferre, P. & Bier, D. M. (1986) Ketone body transport in the human neonate and infant. J. Clin. Invest. *77*, 42–48.

236. Allsop, J. R., Wolfe, R. R. & Burke, J. F. (1978) Tracer priming the bicarbonate pool. J. App. Physiol. *45*, 137–139.

237. Irving, C. S., Wong, W. E., Shulman, R. J., O'Brian Smith, E. & Klein, P. D. (1983) [^{13}C]bicarbonate kinetics in humans: intra- vs. interindividual variations. Am. J. Physiol. *245*, R190–R202.

238. Irving, C. S., Wong, W. W., Wong, W. M., Boulton, T. W., Shulman, R. J., Lifschitz, C. L., Malphus, E. W., Helge, H. & Klein, P. D. (1984) Rapid determination of wholebody bicarbonate kinetics by use of a digital infusion. Am. J. Physiol. *247*, R709–R716.

239. Yost, R. A. & Fettesolf, D. D. (1983) in: Mass Spectrometry Reviews (Waller, G. R., ed.) pp. 1–46, J. Wiley and Sons, New York.

240. Macfarlane, R. D. (1982) Particle-induced desorption mass spectrometry of large involatile biomolecules: Surface chemistry in the high-energy-short-time domain. Acc. Chem. Res. *15*, 268–275.

Subject Index

CLINICAL BIOCHEMISTRY

PRINCIPLES · METHODS
APPLICATIONS

Series Editors
H. Ch. Curtius
M. Roth

1 Mass Spectrometry
Editor Lawson

2 Data Presentation/Interpretation
Editors Keller, Trendelenburg

3 Human Plasma Lipoproteins
Editors Fruchart, Shepherd

4 Laboratory Organization/Automation
Editor Vonderschmitt

5 Toxicology/Drug Analysis
Editors Brandenberger, Maes

- Stable Isotopes
- Nuclear Magnetic Resonance in Clinical Biochemistry

Further volumes in this series are in preparation.

Walter de Gruyter
Berlin · New York

Bei Fragen zur Produktsicherheit wenden Sie sich bitte an:
If you have any questions regarding product safety,
please contact:

Walter de Gruyter GmbH
Genthiner Straße 13
10785 Berlin
productsafety@degruyterbrill.com